2nd International Conference on Engineering for Sustainable World (ICESW 2018)

IOP Conference Series: Materials Science and Engineering
Volume 413

Ota, Nigeria
9 – 13 July 2018

ISBN: 978-1-5108-7259-2
ISSN: 1757-8981

Printed from e-media with permission by:

Curran Associates, Inc.
57 Morehouse Lane
Red Hook, NY 12571

Some format issues inherent in the e-media version may also appear in this print version.

Printed by Curran Associates, Inc. (2018)

For permission requests, please contact the Institute of Physics
at the address below.

Institute of Physics
Dirac House, Temple Back
Bristol BS1 6BE UK

Phone: 44 1 17 929 7481
Fax: 44 1 17 920 0979

techtracking@iop.org

Additional copies of this publication are available from:

Curran Associates, Inc.
57 Morehouse Lane
Red Hook, NY 12571 USA
Phone: 845-758-0400
Fax: 845-758-2633
Email: curran@proceedings.com
Web: www.proceedings.com

Table of contents

Volume 413

The 2nd International Conference on Engineering for Sustainable World (ICESW 2018)

9–13 July 2018, Mechanical Engineering Department, Covenant University, Ota, Nigeria

Accepted papers received: 9 August 2018
Published online: 10 September 2018

Preface

Papers

IOP Conf. Series: Materials Science and Engineering **413** (2018) 011001 doi:10.1088/1757-899X/413/1/011001

International Conference on Engineering for Sustainable World (ICESW2018)

PREFACE

This is a proceeding of the 2nd International Conference on Engineering for Sustainable World (ICESW).

ICESW brings together international experts, scholars, policy makers, industrialists, etc to discuss contemporary and cutting-edge research in engineering management, innovation, energy development and clean technology. It also provides a golden opportunity to meet experts and develop new collaborations on the fundamentals, applications and products from various fields such as: sustainable energy, sustainable cities, engineering innovations, clean technology/production, sustainable materials and engineering, mechatronics, renewable energy systems etc. The conference strives to foster new collaborative and interdisciplinary relationships among emerging and established engineers, scientists, and policy makers.

ICESW' topics are built on the fundamentals of sustainability, with vigorous attention to all forms of engineering professions and enabling research tools including the following topics.

- Renewable Energy and Sustainability
- Engineering Innovations
- Sustainable Manufacturing Systems and Techniques
- Modelling and Simulations
- Clean Technology
- Local Contents in Manufacturing
- Chemistry and Physics of Material
- Sustainable Materials and Processing
- Engineering Policies and Education for Sustainable Development
- Mechatronics
- Thermal Engineering
- Sustainable Buildings and Cities
- Sustainable Product Design and Development
- Eco-Industrial Parks
- Industrial and Production Engineering

ORGANISING COMMITTEE

Nigeria:

- Oluseyi O. Ajayi (Conference Chair), Covenant University ,Nigeria
- Oluwaseun Kilanko (Co-Chair), Covenant University, Nigeria
- Abiodun A. Abioye (Secretary), Covenant University, Nigeria
- Ojo S.I. Fayomi (Technical Chair), Covenant University, Nigeria
- Imhade P. Okokpuje (Technical Secretary), Covenant University, Nigeria
- Cleophas Loto, Covenant University, Nigeria
- Christian Bolu, Covenant University, Nigeria

IOP Conf. Series: Materials Science and Engineering **413** (2018) 011001 doi:10.1088/1757-899X/413/1/011001

- Festus Oyawale, Covenant University, Nigeria
- Anthony Inegbenebor, Covenant University, Nigeria
- Israel Dunmade, Covenant University, Nigeria
- Mufutau Waheed, Federal University of Agriculture, Nigeria
- David Esezobor, University of Lagos, Nigeria
- Samuel Onawumi, LAUTECH, Ogbomosho, Nigeria
- Richard Fagbenle – University of Ibadan, Nigeria
- Kayode Simonyan, Michael Okpara University of Agriculture, Nigeria
- Moradeyo Odunfa, University of Ibadan, Nigeria
- Sunday Afolalu, Covenant University, Nigeria
- Philip Babalola, Covenant University, Nigeria
- Yinka Ohunakin, Covenant University, Nigeria
- Mfon Udo, Covenant University, Nigeria
- Sociis Okolie, Covenant University, Nigeria
- Salau Enesi, Covenant University, Nigeria
- Olufunmilayo Joseph, Covenant University, Nigeria
- Wahabi Asiru, Federal Institute of Industrial Research, Nigeria
- Samuel Adejuyigbe, Federal University, Oye-Ekiti, Nigeria
- Lasode Muyiwa, University of Ilorin, Nigeria
- John Olorunmaiye, University of Ilorin, Nigeria
- Atanda Pethuel, Obafemi Awolowo University, Ile Ife, Nigeria
- Imasogie Benjamin, Obafemi Awolowo University, Ile Ife, Nigeria
- Knle Oluwasegun, Obafemi Awolowo University, Ile Ife, Nigeria
- Olarewaju Oyewola, University of Ibadan, Nigeria
- Oluwaleke Oluwole, University of Ibadan, Nigeria

International:

- Abimbola P.I. Popoola, Tshwane University of Technology, South Africa
- Olanrewaju Ojo, University of Manitoba, Canada
- Hart Martin, Tomas Bata University, Zlin, Czech Republic
- Michael Packlanather, Cardiff University, Wales, UK
- Giampaolo Campana, University of Bologna, Italy
- Dele Odeshi, University of Saskatchewan, Saakatoon, Canada
- Joshua Yusuf, All Saint's University of Medicine, St. Vincent & Grenadines
- Oladiran Fashina, Auburn University, Alabama, USA
- Michael Ngadi, McGill University, Montreal, Canada
- Klein Ileleji, Purdue University, Indiana, USA
- Luc Hebou, University of Maryland, USA
- Tunde Bello-Ochende, Cape Town University, South Africa
- Muyiwa Adaramola, University of Norway

IOP Conf. Series: Materials Science and Engineering **413** (2018) 011002 doi:10.1088/1757-899X/413/1/011002

Peer review statement

All papers published in this volume of *IOP Conference Series: Materials Science and Engineering* have been peer reviewed through processes administered by the proceedings Editors. Reviews were conducted by expert referees to the professional and scientific standards expected of a proceedings journal published by IOP Publishing.

IOP Conf. Series: Materials Science and Engineering **413** (2018) 012001 doi:10.1088/1757-899X/413/1/012001

Development of a System for Counting of People Using MultiCamera and Sensors

[1]Owodolu A. A., [1]Bolu C., [1]Abioye A. A. and [1]Efemwenkiekie K. U.

[1]Department of Mechanical Engineering, Covenant University, Canaan Land, Ota, Ogun State, Nigeria.

Abstract. There has always been the issue of knowing the number of people present in a place such as symposiums, workshops, religious centres. Manual methods have often been employed. During the use of manual methods, the ushers are required are to stand along the pathways, this can then result to congestion in the pathway or even cause the speaker to lose focus. This paper focuses on to solving the problem of overcrowding and at the same time counting of people available in a place per time. In the paper, image sensing camera were used, the sensing device coupled with the camera were used to capture the image of people in a learning environment. MATLAB/SIMULINK software were used for the experiment. The result showed that the method is 85.4 % accurate with the ground-truth of 254 and TP of 217. Keywords: Multi-Camera; System; Counting; Sensors

1. Introduction

Crowd counting with the computer vision technique in the public has now become focus in research. The basic issue is how to count different people in crowd. Presently, the approaches available for counting crowd are based on monitoring videos which are primarily categorized into three. Firstly, is based on human head or body detection. Secondly, is construction of models for head or body which are used to match similar objects with images detected. Thirdly, is based on classifiers which are used by other studies for human body, face and head detection, with many samples collected with useful features are extracted for training classifier. In addition, there is need for accurate counting of people in a crowded environment. Abdulla et al., (2015) report that at Parade music event which happened at Germany in 2010, 21 people were recorded dead and over 500 recorded injured in that stampede. Ikemura et al (2010) report that image sensing device has the ability to count people by capturing their images. In view of that, this paper focuses on application of image sensing device due to its rapid development, efficiency and possibility to count people accurately, these wide applications make it useful in different sectors. The method of counting and data collection will help the administrators know the number of people in attendance in every meeting for proper planning. Also, this counting system is important due to its diverse applications like access control, surveillance, crowd flow monitoring, management of event and behavior analysis (Luna et al., 2017). Given the needs, these applications are increasingly becoming essential in this era without any doubt especially for detecting and preventing potentially hazardous situations. Crowd counting based systems targeted at counting people indoor and outdoor places automatically. A case study of Faith Tabernacle where we have crowd over 50000 people and Covenant University, Ota is selected. Accurate counting of people and data collection will enable the administrators to have data of people in attendance for proper planning. Owodolu et al., (2018) Likewise, image sensing device captures images. In brief, high density traffic flow of people is time consuming. In this paper, our counting system is based on computer vision with Simulink tool box. Computer vision is an inter- related field which uses computer for videos or digital image processing. Hu et al., (2015) report that from engineering perspective, it automates tasks that the human based visual system does. Automatically computer vision extracts analyses and understands applicable information from a sequence of images or a single image. This image processing comprises of segmentation, background extraction and blobs detection etc. and can be done in MATLAB. The processes are presented and explained in the subsequent sub-headings in this paper and show that, by this processes people can be counted effectively. The sub-headings include: literature review, methodology, analysis and discussion and conclusion.

IOP Conf. Series: Materials Science and Engineering **413** (2018) 012001 doi:10.1088/1757-899X/413/1/012001

2. Literature Review

Abdulla et al., (2015) report that in these decades, the population of people is increasing geometrically worldwide. The growth which indirectly resulted to increased number of people or crowd is due to the urbanization and movement globally. The gathering of crowd can be in close places such as stadium, building halls, airports or open areas such as sport events religious centers, parks and walkways. The reason for gathering has major effect on the crowd large scale behaviours and properties. So, counting is an area of concern in many researches such as science, safety, psychology, public services and computer vision. Abdulla et al., (2015) report that crowd turbulence is one of the reasons for crowd disasters which happens from pressing, pushing, stampede or crowd crushes, and eventually resulting to total loss of emotional control. Such examples of disasters as illustration of this problem are seeing in Colombia Water Festival stampede 2010 where over 380 people died. Likewise, crowd crush of Love Parade music event which happened at Germany in 2010 where 21 people recorded dead and over 500 recorded injured in that stampede. Early work involved locating people by looking for heads in the vertical histograms of the blobs, where the number of peaks was assumed to correspond to the number of heads. Meanwhile there have been different methods to the problem of counting people in crowded environments (Ryan et al., 2014). Also, proposed a system which provides an estimate based on the number of foreground pixels and edge pixels. Fourier transform technique was used to identify motion of the crowd. It is restricted to motion only in the vertical direction in the image. Crowd density estimation is done using elementary techniques such as estimating the area of the image segments which correspond to moving crowds, or using the perimeter of the region occupied by the crowd. Although the system works well for scenes with few people, it is incapable of accurately determining density in case of occlusions and illumination changes.

3. Methodology

Both software and hardware's were used to carry out the experiment. The software includes MATLAB. While the hardware used include: lighting system, Personal Computer, USB cord and High-resolution cameras. Multi-cameras (Three PMW-400K and Six PMW-EX3) have been introduced here for automatic people counting. The images from video camera are processed on compute both software and hardware's were used to carry out the experiment. The software includes MATLAB. While the hardware used include: lighting system, Personal Computer, USB cord and High-resolution cameras. Multi-cameras (Three PMW400K and Six PMW-EX3) have been introduced here for automatic people counting. The images from video camera are processed on computer (RAM 4.00GB; Processor; Intel (R) Celeron CPU 2.16GHz) via USB connection for analysis. The images (data) applied in counting system were collected at Faith Tabernacle and Covenant University Ota.

3.1 Lighting Systems

The most critical and essential aspect in image processing is lighting system used to receive a quality image. This provides homogeneous illumination for quality lighting set-up to make sufficient information from images for vision application.

The methodology for the system contains several steps:

1. Background extraction which uses improved algorithm to extract a current or pure background.
2. Background subtraction which uses background subtraction algorithm, the foreground is separated from the background and forms a foreground mask.
3. The system detects blobs that have more than one in them.
4. Segmentation. These include: image reading, morphological opening to background, background image differencing, balancing of image contrast, image threshold and real image formation. Thus, normalization of the features is important in order to reduce effects of perspective. Moreover, the foreground segment is obtained by the background subtraction

IOP Conf. Series: Materials Science and Engineering **413** (2018) 012001 doi:10.1088/1757-899X/413/1/012001

From Multimedia File block video taken is converted intensity and displayed screen. The link between video and Optical Flow provides detected motion. Noise was decreased in image for best detection by using 2-D Mean. The three running mean check boxes were selected to track mean value. The mean value in a sequence of inputs over a period of time can also be tracked by the mean block.

4. Results and Discussion

Figure 1: **Shows Image Counting with Bounding Boxes and the Blobs formation**

Figure 2: Blobs formation

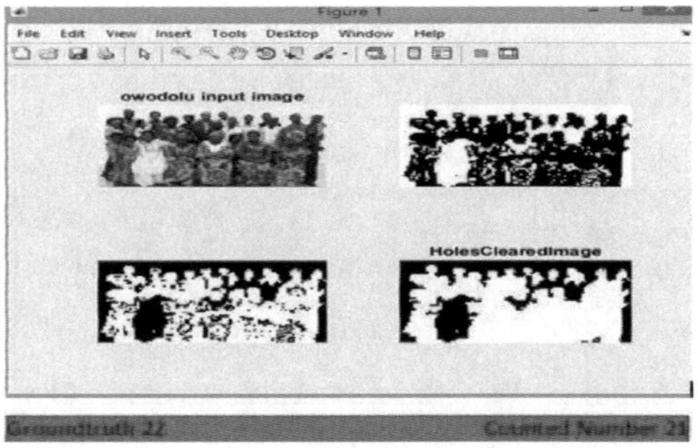

Figure 3: Shows Crowd Counting Images and Results

IOP Conf. Series: Materials Science and Engineering **413** (2018) 012001 doi:10.1088/1757-899X/413/1/012001

Table 1: Shows Crowd Counting Number and the Corresponding Image Size

	Crowd Counting Number Variation			
Image Size	Real Number	True Positive	False Positive	False Negative
302X710	47	35	16	12
130X244	9	9	2	0
149X234	15	10	2	5
152X374	34	33	2	1
164X255	29	19	3	10
126X315	17	17	6	0
137X345	34	31	8	3
367X561	18	18	8	0
509X582	12	12	6	0
86X380	32	26	9	6
198X325	7	7	3	0

Table 1 shows the number of real number of people on ground =254, true positive (TP) =217, false positive (FP) =65, false negative (FN)=37. TP is the number of foreground pixels that correctly detected. FP is the average of false alarms per frame or the number of background pixels that are incorrectly detected as foreground. While FN is the average of false misses which is the number of foreground pixels that are incorrectly detected as background. Pixel with size 302x 710 has highest real number of 47 people, true positive equal 35 people, false positive equal 16 people and false negative of 12 people. Image with size 130x244 has real number of 9 people, true positive of 9 people, false negative equal 0 and false positive equal 2 people which makes it more accurate than the other ones.

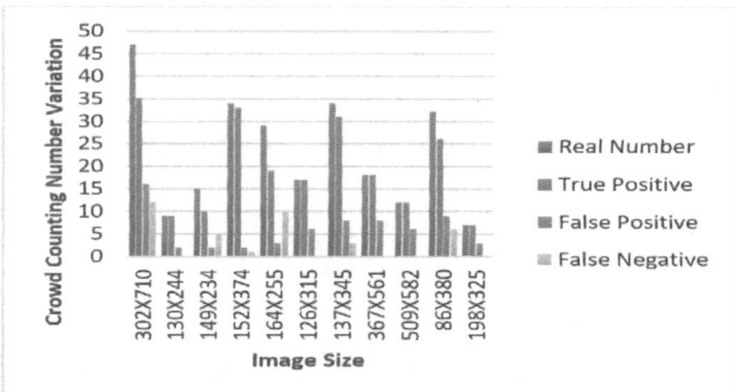

Figure 4: Shows Crowd Counting Number Variation versus Image Size

There is variation in number of people counted. This is as a result of limited images used for system identification This is further displayed in the Figure 5 below

IOP Conf. Series: Materials Science and Engineering **413** (2018) 012001 doi:10.1088/1757-899X/413/1/012001

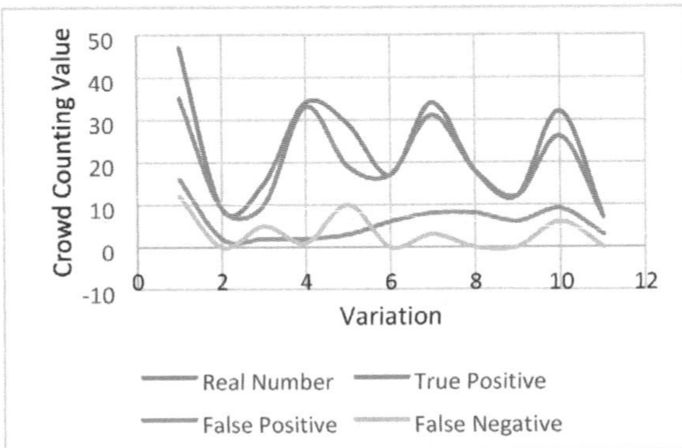

Figure 5: Shows variation in the Crowd Counting

Performance Evaluation

There are various standard measures for the comparison of the candidate binary change mask to the ground truth. These include the following.

1. True positives (TP): This is the number of foreground pixels that correctly detected.
2. False positives (FP): This is the average of false alarms per frame or the number of background pixels that are incorrectly detected as foreground.
3. False negatives (FN): This is the average of false misses which is the number of foreground pixels that are incorrectly detected as background.

Moreover, the statistical measurement used is accuracy to find how well the segmentation process excludes or identifies foreground pixels. Mathematically, this can be expressed as follows:

$$\text{Overall Accuracy} = \frac{\text{Total Processed Counting}}{\text{Total Actual Counting}} \text{ X } 100$$

$$\frac{217}{254} \; x\,100 = 85.4\%$$

5. Conclusion

Computer vision and Simulink library in MATLAB were used as the programming tool boxes or interface on computer that is adaptable for counting implementation processing. The method used was an improvement of an existing image sensing and counting of people in a crowded environment. The image of people was captured at Faith Tabernacle and Covenant University environment with image sensing device for the period. Also, algorithm system was developed and adapted to Viola Jones existing image sensing model for image processing. The non-clarity and high camera resolution resulted to poor counting in some areas which caused computational problem. The overall accuracy of the counting methodology is 85.4%.The true positive is 217 which is greater than the false negative of 37 and false positive of 65 of people. The total ground truth is 254 people. This method can be adopted as alternative to manual counting system.

References

[1] Abdulla, S., Saleh, M., Azmin, S., & Ibrahim, H., Engineering Applications of Artificial Intelligence Recent survey on crowd density estimation and counting for visual surveillance. Engineering Applications of Artificial Intelligence, 2015, 41: 103–114.

[2] Hu, X., Zheng, H., Chen, Y., & Chen, L., Optik Dense crowd counting based on perspective weight model using a fisheye camera. Optik - International Journal for Light and Electron Optics, 2015, 126(1):123–130.

[3] Hu, X., Zheng, H., Wang, W., & Li, X.. Optik A novel approach for crowd video monitoring of subway platforms, 2013,124:5301–5306.

[4] Luna, C. A., Losada-gutierrez, C., Fuentes-jimenez, D., Fernandez-rincon, A., Mazo, M., & Macias-guarasa, J., Robust people detection using depth information from an overhead Timeof-Flight camera. Expert Systems with Applications, 2017, 71: 240–256.

[5] Owodolu.A.A., Christian Bolu, Abioye A.A., Boyo H.O. and Godfrey Onyiagha, A Systematic Human Counting at Guest House using Sensing Device Technique, 2018, 8(3)

[6] Ryan, D., Denman, S., Fookes, C., & Sridharan, S, Scene invariant multi camera crowd counting.Pattern Recognition Letters, 2014, 44: 98–112.

[7] Ryan, D., Denman, S., Sridharan, S., & Fookes, C., An evaluation of crowd counting methods, features and regression models q. Computer Vision and Image Understanding, 2015,130: 1–17.

[8] S. Ikemura, H. Fujiyoshi, Real-time human detection using relational depth similarity features, in: Asian Conference on Computer Vision, 2010:25–38.

IOP Conf. Series: Materials Science and Engineering **413** (2018) 012002 doi:10.1088/1757-899X/413/1/012002

The Role of Research in Attaining Sustainable Development Goals

Fayomi O.S.I.[1, 2], Okokpujie I. P. [1], Udo Mfon[1]

[1]Department of Mechanical Engineering, College of Engineering, Covenant University Ota, Lagos state, Nigeria
[2]Department of Chemical, Metallurgical and Materials Engineering, Faculty of Engineering and the Built Environment, Tshwane University of Technology, Pretoria, South Africa.
Corresponding author: ojosundayfayomi3@gmail.com, ojo.fayomi@covenantuniversity.edu.ng, fayomio@tut.ac.za Tel: (+2348036886783) (+27835828119)

Abstract- A study on the significant roles of research in the realization of the Sustainable Development Goals was carried out to analyze the facts and figures gotten through secondary information's to determine the resultant effect of research in the pursuance of the SDGs. This research explores various opinions of several authors and researchers towards the concept of research in totality. This study gave clear indications that research as a full concept is not a cliché and it is revolving year-in year-out in our contemporary world of interest as the need for human survival on the planet earth gets more prioritized on the agenda of the various sustainability organizations like UNESCO, United Nations, UNEP, IISD. This is the present century's goal and it is all NGOs and other private parastatals are enjoined to be part of this move as it is a collective effort to seeing its realizations.
KEYWORDS: Research, Sustainable, Development, Goals, United Nation.

1. Introduction

The advent of the popularly known SDGs was birthed by the United Nations The concept of development has its significance with respect to its transformed approaches to the social and economic, and it took its root in the previously formed organization called the United Nations in the 1940s and 1950s. Guthrie & Humphrey, [1]. This was further propelled as it continued even during the period of the 1960s and 1970s, although during this period, there were two majorly opposing movements in development concept and they had gained relevant recognition. One was known with broadening the scope and premise of the strategies for development with full considerations for social factors as it includes employment, education, nutrition, health, income distribution, poverty reduction, basic needs, the environment.

In recent years, the United Nations have tried severally to make the word a better place for people to live including the destitute and rich. In pursuance of this, the United Nations had put various polices in place which are all poised towards achieving the set goals. In the documentations of Hwang & Jiwon, [2], Lennon, [3]. and UNDP, [4] it was recorded that owing to the ravaging challenges faced by a lot of people across the globe, ranging from rampant epileptic lifestyle of the people to poverty, gender segregation and dichotomy, ecosystem depletion, and the likes. This lead to the conglomeration of leaders from 189 countries in a meeting that held in the UN headquarters in New York in September 2000. The resolution of the meeting birthed the popularly known MDG (Millennium Development Goals). Rahdari et al., [5]. This was meant to address the development bottlenecks round the various nations of the world. Upon the review of the MDGs made by the United Nation in 2015, several shortcomings were discovered pertaining the

IOP Conf. Series: Materials Science and Engineering **413** (2018) 012002 doi:10.1088/1757-899X/413/1/012002

effectiveness of the MDGs. This review eventually birthed the currently known SDGs (Sustainable Development Goals). Although the millennium goals that was way back established by the SDGs had some level of achievement, but notwithstanding, the Sustainable development goals was initiated to improve on the deficiencies of Millennium goals implemented earlier thus, covering 17 core necessities for living that spans through quality research to environmental sustainability. Several researches have worked on the various contributions made to enhancing the achievement of the SDGs thus far as stated by Ul et al., [6].

However, there cannot be substantial measurable contributions without research and that is why this study is carried out to investigate the significance or the roles of research to aiding the attainment of the Sustainable Development Goals

2. Conceptual Issues: History of Sustainable Development Goals (SDGS), Justification of Problem in Attaining the SDGS

Development according to the Cambridge Dictionary is defined as the process though which an individual or something would grows or transform and gets more advanced. Development in its simplest form is defined to be "good transformation" Boateng et al.,[7] . Development is also defined as a course or process of change and its dynamics, Remenyi, [8]. Development might happen as a result of certain deliberate steps or actions displayed by individual agents or through polices made and orders from some authorities by some authority with the aim attaining substantial improvement, in the utmost favour of surrounding circumstances, Bellù, [9]. Although sustainable development has been largely defined by several scholars, but according to Cerin, [10] and the proposed definition by the Brundtland Commission, Dernbach [11] says Sustainable development is an improvement that addresses the issues of the present without a compromise on meeting the needs of the generations to come the capacity of the future to address their peculiar issues. Stoddart, [12]. According to the National Assembly for Wales [13], It comprises of two ideas:

- The impression of 'needs', definitely the essential needs of the world's destitute, to which prevailing need ought to be given; and
- The likelihood of constraints forced through the state of innovation with social association on the condition's ability to meet the contemporary and forthcoming needs.

Sustainable Development as further classified by Hedenus, et al., [14] and Emas, [15] has three scopes: environmental, financial/economic, and social.

- The natural measurement includes maintaining the earth's creation limit and not surpassing the earth's assimilative limit.
- The monetary measurement includes effectively overseeing limited common assets and capital as long as possible.
- The social measurement includes societal establishments that are essential for human needs, for example, majority rules system, put stock in, law, and global associations.

2.1 History of Sustainable Development Goals (SDGS)

In the year 2000, The UN Millennium Declaration was received, with eight objectives to be achieved by 2015. Every one of the 189 individuals from the UN and 23 global associations conceded to these objectives as reported by Hedenus et al., [14]

The eight Millennium Development Goals are:

- Cut outrageous yearning and destitution significantly
- Achieve widespread important training
- Promote sexual alignment correspondence
- Decline child death
- Improve maternal wellbeing

IOP Conf. Series: Materials Science and Engineering **413** (2018) 012002 doi:10.1088/1757-899X/413/1/012002

- Battle HIV/AIDS, intestinal sickness, and different ailments
- Guarantee ecological supportability
- Build up a universal organization for development

The Millennium Development Goals exhibited a more noteworthy spotlight on advancement contrasted with earlier dialogs; just a single objective (#7) had an immediate association with natural issues. Rather, issues with a greater amount of a natural concentration were tended to through other fora, for example, the worldwide atmosphere arrangements under UN Agenda Convention on Weather Change and significant condition and advancement traditions like the one held in the city of Johannesburg, in South Africa, in the year 2002.

The last came to be called "Rio+10" yet did not get as much consideration as the 1992 meeting and is viewed as less imperative. Be that as it may, concurrences on water what's more, fisheries came out of Johannesburg. Just like the case with different affirmations, the Millennium Declaration and the eight objectives have been viewed as pretty, yet void, words that are not official. In any case, a key distinction from different announcements lies in the work done to recognize quantifiable objectives that are anything but difficult to track and survey. The objectives have had several effects in how worldwide and national aids are taken care of. The outcomes have been blended. Certain objectives, for example, decreasing considerably the quantity of the world's poor, have been accomplished at the worldwide total level, while others, for example, expanding fairness between ladies and men, have not been accomplished and still appear to be far off.

In 2012, the time had wanted the following huge worldwide maintainability tradition, once again in Rio de Janeiro, "Rio+20." The event additionally denoted the start of arrangements on new objectives to supplant those of the Millennium Declaration after 2015. In light of the announcement, that "destitution is the best worldwide test" and Agenda 21 work in progress on new maintainable advancement objectives. In September 2015 the UN embraced the 17 supportable improvement objectives (SDGs) spreading over a wide cluster of branches of knowledge and taking natural issues back to the fore. Not at all like the eight objectives of the Millennium Declaration, have the new objectives incorporated particular focuses for urban areas, maintainable utilization, atmosphere affect, marine assets, and land-based biological systems. Sachs, [16] said from a logical perspective that the MDGs also had a reasonable feat of accomplishment across some of the stipulated scope of the goals during the span of its existence. Some of the efforts made were significant to a measurable extent as the Millennuim Development Goals (MDGs) was well appreciated. However, there were some short comings that made the MDGs not effectively achieved owing to several factor cut across various scopes of influence. Some of the shortcomings were result effects of the basis upon which the goals set initially and some modifications that came alongside the process of execution and pursuance of the goals. Some of the carefully observed shortcomings of the Millenium Development Goals of the United Nation are listed thus;

- Lack of Intermediate objectives and milestone
- Outdated data
- Non-participation of private sectors
- Lack of societal investment/contribution

While the Millennium Declaration objectives were for the most part centered around the poorest countries, this time the reason for existing was to incorporate objectives that draw in and urge all countries to act. As recorded by UNDP, [4], leaders from 193 countries of the world came together in a forum in the year 2015 to face the supposed realities of the future.

Furthermore, what they saw was overwhelming. Starvations. Dry spell. Wars. Diseases. Destitution. Not simply in some faraway place, but rather in their own urban communities and towns and towns. They knew things didn't need to be like this. They knew we had enough nourishment to sustain the world, yet that it wasn't getting shared. They knew there were pharmaceuticals for HIV and different illnesses, yet they cost a considerable measure. They knew that seismic tremors and surges were inescapable, yet that the high passing tolls were most certainly not. They additionally realized that billions of individuals overall shared their expectation for a better future. So, the various representatives from these nations arranged and eventually came to a consensus of what is called the Sustainable Development Goals (SDGs). This arrangement of 17 objectives envisions a future only 15 years off that would be freed of neediness and yearning, and safe from the most exceedingly terrible impacts of environmental change. It's a goal-oriented arrangement. Be that as it may, plentiful confirmation we can succeed. In the previous 15 years, the worldwide group cut outrageous neediness down the middle. Presently we can complete the activity. The United Nations Development Program (UNDP) is one of the main associations attempting to satisfy the SDGs by the year 2030. The realistic achievement of the SDGs will not just be as a dream of the night nor a rocket science, but it will require a very robust contribution from various individuals, groups, organization (both private and non-private), A consolidate effort from almost 170 nations and regions will earnestly enable countries to make the Goals a reality. The requirement of the conscious efforts from the various nations of the earth is to attain and achieve progressive improvement from the previously known MDGs to the presently known SDGs. These goals as stipulated by UNDP, [4] in Hwang & Jiwon, [2] are listed thus;

- End poverty in the entirety
- Termination of hunger, accomplish sustenance security, progressed nourishment, and economical horticulture
- Guarantee sound lives and advance prosperity for all at all ages
- To assure evenhanded and comprehensive quality training what's more, advance deep-rooted learning opportunities for all
- Achieve voluptuous orientation equity and enable ladies.
- Guarantee maintainable and accessibility of water and effective sanitation for all.
- Guarantee sufficient access to reasonable, dependable, maintainable vitality for everyone
- Advance maintained, comprehensive and reasonable economic development, profitable and full. business, also, not too bad work for all
- Construct versatile framework, improved industrialization, and cultivate development
- Decrease imbalance amongst nations and within nations
- Make urban communities and human settlements comprehensive, strong, safe, feasible.
- Guarantee reasonable utilization and creation designs
- Make critical move to combat change in the environmental and effects (observing understandings made by UNFCCC)
- Ration and economically make use of the seas, marine and oceans assets for reasonable developmental progress.
- Ensure, reestablish, advance manageable utilization of earthbound biological systems, economically manage woodlands, battle desertification, turn around arrive corruption, and put an end to biodiversity disaster
- Peace promotion and good social orders for feasible advancement, give room for equity for all, assemble viable, responsible, comprehensive establishments at all levels
- Fortify the approaches for execution and reintroduce the worldwide organization for maintainable advancement

2.2 Justification of Problem

Dhahri and Omri, [17] stated that a causality test that was carried out affirmed the connections among business enterprise and these three dimensions of the SDGs in both small and large. It was basically found that business in these nations emphatically adds to the financial and social measurements of SDGs, but had limitations through negative commitment to the ecological measurement.

Blum et al., [18] discussed the Concept of Sustainable Chemistry and the role it plays in assisting all on-screen characters with identifying doable specialized choices in combination and generation with less unsafe synthetic substances by considering maintainability criteria in zones past the extent of sound administration of synthetic concoctions, as for example asset and vitality productivity, the utilization of inexhaustible feedstock, and the effect on groups of the creation of synthetic concoctions. Along these lines, in every single significant angle towards sustainable development. Bevan et al., [19], represent 0.98 on percentage of development associations in the Australia and 0.80 0n percentage of development yield in the UK. In this manner in the direct of development business, they have high effect to build social obligation Upstill. Nonetheless, in spite of the fact that there are confirmations to help the execution of the Cooperate Social Responsibilities in SME for the development business the level of usage is still very low. Barthorpe said that in the UK setting, there were numerous SME that do not have remorse to insert or formalize CSR into their standard business exercises. Regularly, the SME take part in CSR as occasional accomplices to the bigger development associations in the development store network. The expansive development associations have colossal duty in such manner to support, guide and give assets to the SME to expand their level of CSR Loosemore, [20]. It was expressed that the very known SDGs are specifically significant to social responsibility in business. Suitable inspiration ought to be given to the SME to expand CSR hones.

Elimination of poverty is the first, basic target of sustainable development objectives Ul et al., [6]. Supportable improvement is a vital issue for the provincial territories of Pakistan for the most part connected with expectations for everyday comforts, horticulture and nonagricultural resources, and condition pointers. The measurements of hardship have been distinguished by analyzing the accompanying components: farming and nonagricultural resources, lodging, instruction, vitality sanitation, access to clean drinkable water. The examination reports that 0.44 on percentage of family units in provincial Pakistan are inhabiting the poorest and poor levels, persevering through an absence of unadulterated drinking water, insufficient sanitation offices, poor lodging conditions, dirtied vitality sources, and privation of financial assets. The locale level investigation exhibits that the rustic populaces of Hyderabad. In this manner, to accomplish reasonable advancement and destitution decrease, extra subsidizes from the elected, common, and nearby governments ought to be distributed for social welfare, training, sanitation, water supply, and horticulture improvement in country zones of Pakistan.

3. Summary of Some Method in Carrying Out Research to Attain the SDGS

This study basically focused on the pedigree of the Sustainable Development Goals with a germane route from the critical analogy of the antecedent of the contributions that has been made by several researches though various studies. Amongst many other things, this study further espouses the state on the attainment of the Sustainable Development Goals thus far, why it has not fully succeeded, and those that have played active roles till date.

This paper made use of secondary information's from various articles cut across various fields but targeted towards the Sustainable Development Goals of the United Nations. The benefit of utilizing sets of secondary data or information is that it has been created by groups of master scientists, regularly with vast spending plans what's more, broad assets path past the methods for a solitary understudy, so it removes the requirement for tedious hands on work. Information that has been gathered over an extensive stretch of time will give the chance to complete a longitudinal report (following the improvements after some time), unthinkable to do with information gathered in short tasks Nicholas, [21]. For the scope of the study, the role of research in the attainment of the SDGs is the primary focus. To clearly juxtapose the significance of research in our contemporary world as a tool to positively catalyze the achievement of the clearly defined 2030 agenda of the United Nations.

3.1 The Concept of Research

IOP Conf. Series: Materials Science and Engineering **413** (2018) 012002 doi:10.1088/1757-899X/413/1/012002

Research has been defined by various scholars and authors in series of correspondence, proceedings, publications, research extract and the likes. Amongst others are the few definitions considered for the purpose of this study. Iowa, [22] defined research from a logical point of view a deliberate examination, advancement in research, assessment, poised to create a generalizable fact or information. Research also implies a look for actualities – answers to inquiries and Solutions to Issues. It is a Purposive examination. It is a "Sorted out request" (It looks to discover clarifications to unexplained marvel, to order the dubious suggestions and to revise the misconstrued realities). Exercises which meet this definition constitute explore for motivations behind this strategy. Kothari, [23]

3.2 Types of Research

According to the indications from the study done by Kothari, [23] and the basic types of research are listed thus;

- Descriptive: it incorporates reviews and actuality discovering inquiries of different types. Real reason for engaging examination is depiction of conditions of issues as it is at present.
- Applied: it is a kind of research that goes for locating a solution for prompt issues confronting a general public or industry
- Quantitative: It depends on amount estimation o or on the other hand sum. It is appropriate to marvels that can be communicated as far as amount
- Qualitative: it is worried about subjective marvel, i.e., wonders identifying with or including quality or kind
- Conceptual: is the kind of research that is identified with some abstract idea(s) answers hypothesis. It is for the most part utilized by rationalists and scholars to grow new ideas or to reinterpret existing ones.
- Empirical Research: it depends on understanding or perception alone, often without due reverence for context and hypothesis. It is information-based research, concocting conclusions which are equipped for being checked by perception or investigation.

3.3 Research Approaches

Kothari, [23], highlighted two approaches to research. These approaches are discussed below;

- Qualitative Approach; it is fully focused on the opinion and attitude assessment. The impression and insight of the researcher is a key factor in the kind of research approach.
- Quantitative Approach; it is focused on generating quantifiable data that can be analyzed quantitatively in a seemly rigid and formal way or fashion

3.4 General Significance of Research

Research assumes a key part in forming the world. Through research, we pick up a superior comprehension of the present most squeezing and complex social and logical issues for example, social decent variety, human rights, malady avoidance, and environmental change. Leaders, general society, and even analysts themselves, must believe in how research is directed, what's more, the resulting discoveries. The advantages of the examination undertaking can't be acknowledged unless the information created is both precise and dependable. Opportunity, [25]

Research has its unique significance in solving several operational and planning related problems of business and industry. Kothari, [23]

3.5 Research and the Sustainable Development Goals

The importance of research in various fields of human endeavor cannot be over emphasized as it relates with literarily all that pertains to man and the entire makeup of his existence. At earlier stated, that applied research is one of the types of research that majorly focuses on solving practical real-life problems of different scenarios in the society, organizations, industries, various locality and the likes. Since the sustainable development goals of the United Nation it targeted towards making the world a very conducive and suitable place for humanity to stay in. This means that the seventeen (17) SDGs were born out of a passion to solve the generic problems of humanity. Hence, the role of research becomes extremely germane to positively catalyzing the processes and procedure required to attain the reality of it. The SDGs talks about the level of poverty, the standard of living the people, the ecosystem. All of these are the make ups of the human survival mission.

Research will aid the decision-making process of the policy makers at various cadre of the managerial leadership of the SDGs all around the world where the goals intended for. The essence of research in enhancing growth and development of any system or organization is considerably significant. However, the place of quality and good research is not to be compromised because good research work in any sphere of influence, area of specialization, field of endeavor, for any purpose in the regard of solving relevant and related problems is of utmost importance and significance. Hence, quality research work will birth better result and improved situation.

4. Conclusion

It was discovered during this study that research will forever be a very important ingredient of development and its significance will forever remain relevant for any system that want to grow and merge up with her counterparts across the world. Every improvement is traceable to research because there was a problem and the cause of the problem was consciously and effortful discovered, the required solution was effectively appropriated and the system eventually salvage and something new is being added to knowledge. This is traceable to research because that is all about research.

References

[1] Guthrie, J., & Humphrey, C. . (1999). Public sector financial management developments in Australia and Britain: Trends and contradictions. *Research in Governmental and Nonprofit Accounting*, 9, 283–302. https://doi.org/dx.doi.org/10.1111/j.1468-0408.1993.tb00102.x

[2] Hwang, S., & Jiwon, K. (2016). UN and SDGs : A Handbook for the Youths, 1–74. Retrieved from file:///C:/Users/HP/Downloads/UN and SDGs_A Handbook For Youth.pdf

[3] Lennon, S. (n.d.). The Sustainable Development Goals and Australia – A National and Personal Roadmap to Sustainability.

[4] UNDP. (2016). From the MDGs to Sustainable Development for All. Lessons of 15 Years of Practice. *United Nations Development Programme*, 92.

[5] Rahdari, A., Sepasi, S. & Moradi, M., 2016. Achieving sustainability through Schumpeterian social entrepreneurship: The role of social enterprises. Journal of Cleaner Production, Volume 137, p. 347–360.

[6] Boateng, R., Heeks, R., Molla, A. and Hinson, R. (2008). E-commerce and Socio-Economic Development: Conceptualizing the Link. Internet Research, 18(5), 562–592.

[7] Remenyi, J. O. E. (1941). What is Development ? *Development.*

[8] Bellù, L. (2011). *Development and development paradigms: A (reasoned) review of prevailing visions. Retrieved on.*

[9] Cerin, P. (2006). Bringing economic opportunity into line with environmental influence: A Discussion on the Coase theorem and the Porter and van der Linde hypothesis. Ecological Economics, 209-225.

[10] Dernbach, John C. (2003) "Achieving Sustainable Development: The Centrality and Multiple Facets of Integrated Decisionmaking," Indiana Journal of Global Legal Studies: Vol. 10, Iss. 1,

Article 10.http://www.repository.law.indiana.edu/ijgls/vol10/iss1/10
[11] Stoddart, H. (2011). A Pocket guide to sustainable development governance. Stakeholder Forum.
[12] National Assembly for Wales. (2015). Quick Guide to Sustainable Development: History and Concepts, (March). Retrieved from http://www.assembly.wales/research documents/qg15-003 - sustainable development history and concepts/qg15-003.pdf
[13] Hedenus, F., Persson, M., & Sprei, F. (2016). Sustainable Development: History, Definition & The Role of the Engineer. Retrieved from http://publications.lib.chalmers.se/records/fulltext/230705/local_230705.pdf
[14] Emas, R. (2015). The Concept of Sustainable Development : Definition and Defining Principles. *Brief for GSDR*, 1–3. https://doi.org/10.1016/j.marpol.2014.01.019.
[15] Sachs, J. D. (2012). From millennium development goals to sustainable development goals. *The Lancet*, *379*(9832), 2206–2211. https://doi.org/10.1016/S0140-6736(12)60685-0.
[16] Dhahri, S., & Omri, A. (2018). Entrepreneurship contribution to the three pillars of sustainable development: What does the evidence really say? *World Development*, *106*, 64–77.
[17] Blum, C., Bunke, D., Hungsberg, M., Roelofs, E., Joas, A., Joas, R., … Stolzenberg, H. C. (2017). The concept of sustainable chemistry: Key drivers for the transition towards sustainable development. *Sustainable Chemistry and Pharmacy*, *5*(December 2016), 94–104. https://doi.org/10.1016/j.scp.2017.01.001
[18] Bevan, E. A., & Yung, P. (2015). Implementation of corporate social responsibility in Australian construction SMEs. Engineering, Construction and Architectural Management, 22(3) 295-311. doi: https://doi.org/10.1108/ECAM-05-2014-0071
[19] Loosemore, M. (2016). Social procurement in UK construction projects. International Journal of Project Management, 34(2), 133-144. doi: https://doi.org/10.1016/j.ijproman.2015.10.005
[20] Nicholas, W. (2010). *Research Methods : The Basics. Roultedge.* https://doi.org/doi:10.4324/9780203836071
[21] Iowa, S. U. (2009). Institutional Review Board Research Requiring IRB Review, (February), 2008–2009.
[22] Kothari, C. R. (2004). *Research Methodology: Methods & Techniques. New Age International (P) Ltd.* https://doi.org/10.1017/CBO9781107415324.004
[23] Bodla, D. B. S. (2009). Introduction To Research Methodology, *11ed*(Introduction To Research Methodology), 5. https://doi.org/10.1016/j.worlddev.2018.01.008
[24] Opportunity, C. (2010). Council of Canadian Academies • Report of the Expert Panel on Biodiversity Science November 2010, (November).
[25] Rapportserie, V. (2006). *Good ReseaRch PRactice – What is it?*

IOP Conf. Series: Materials Science and Engineering **413** (2018) 012003 doi:10.1088/1757-899X/413/1/012003

A Review on Automotive Industries and Foundries in Nigeria

[1*]Abioye, A. A., [2]Atanda, P. O., [1]Abioye O. P., [3]Akinlabi S. A., [4]Akinlabi E. T., [1]Bolu, C. A., [1]Afolalu S. A., [1]Ajayi, O. O., [1]Ohijeagbon, I. O.

[1]Department of Mechanical Engineering, Covenant University, Cannanland, Ota. Nigeria.
[2]Department of Materials Science and Engineering, Obafemi Awolowo University, Nigeria.
[3]Department of Mechanical & Industrial Engineering, University of Johannesburg, South Africa.
[4]Department of Mechanical Engineering Science, University of Johannesburg, South Africa
[*]Corresponding author: abiodun.abioye1@covenantuniversity.edu.ng

Abstract. The Federal Government of Nigeria has taken different precautions to support domestic vehicle assembly industries as part of the Nigeria Industrial Revolution Plan (NIRP) to diversify the economy from oil and gas sector to other sectors. The potentials of foundry industries in Africa and most especially in West-Africa Sub-region are under-utilized unlike in the developed countries. Most of the Foundries in Nigeria are operated by Small Scale Industries (SMI) and are not being able to contribute significantly to the automotive industries in the country. This paper is aimed to review several works done by different researchers in the automotive and foundry industries with the view to study the contribution of foundries in the development of automotive industries in Nigeria.
Keywords: Vehicle Assembly, Component Parts, Government, Foundry and Metallurgy

1. Introduction

According to the Global Auto Industry report [1], Africa is considered to be the second fastest growing market for vehicle sales on earth. This is as a result of increasing population, which has been forecasted to grow to 1.25 billion by 2025 and to 2.4 billion in 2050 [2]. Moreover, over the past decade, six out of the ten fastest growing countries were from Africa with GDP of 4.7% growth in the Sub Sahara region as of 2013 [3,4]. In 2014, Africa imported four times more automotive products than it exports, with automotive imports worth $48 billion and exports worth only $11 billion [5]. The major sources of used vehicles in Africa are the United States (US), Europe and Japan [6]. The statistical detail of vehicle in use in Africa is as shown in Figure 1. Despite this enormous potential, Africa has not played a major role in automobile production unlike United State with over 11 million, Japan with over 10 million, Germany and China with almost 6 million each [7]. The total automobile production in the year 2004 from Africa was not up to 600 thousand and was from South Africa and Egypt [7].

Iron and Aluminium are the most commonly cast metal parts used in the automotive due to low cost, ease of machining, ease of casting into complex shapes and desired physical properties [63]. Therefore, iron and aluminium foundries are highly incorporated into the supply chain for a wide range of products automotive industries [8]. There has been reported demand increase in the world for aluminium castings particularly in automotive applications because of their good recycling characteristics, relative strength, and lightweight. The camshaft, crankshaft, gearbox, engine block, automobile suspension, cylinder head and steering systems are some of the casting products from foundries in used automotive industries.

IOP Conf. Series: Materials Science and Engineering **413** (2018) 012003 doi:10.1088/1757-899X/413/1/012003

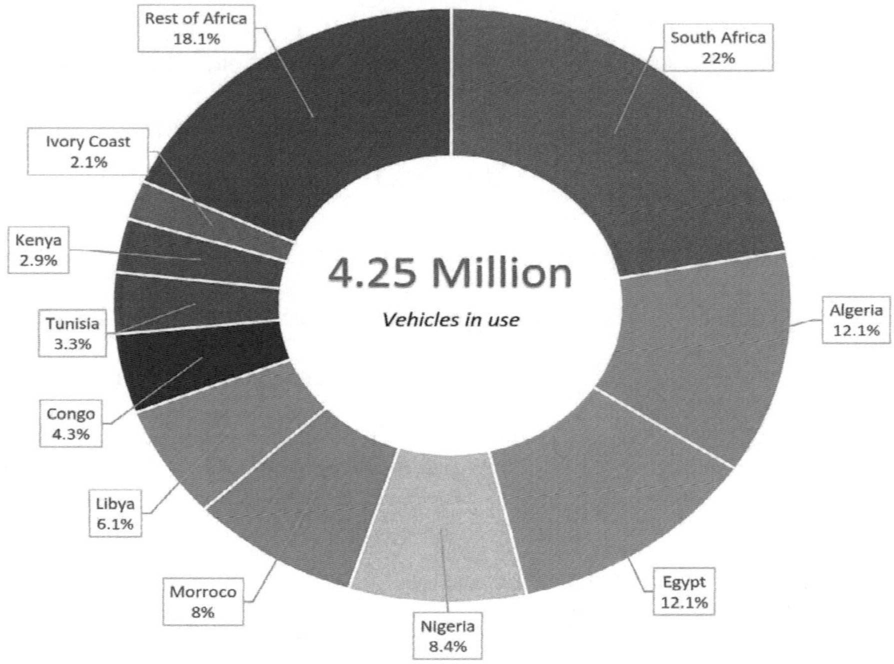

Figure 1: The statistical detail of vehicle in use in Africa [7]

The commonly used foundry automotive components are shown in Table 1.0 [9, 10]. Although the foundry industries in other continents are well developed and has contributed enormously to the development of automotive industries in such countries as the United States, China, India, Mexico, Japan, Germany [11- 14] however, foundries in African countries particularly Nigeria are underdeveloped and do not have enough capacity to contribute significantly to automotive industries in the continent. The purpose of this work is to review the role of foundries in the development of automotive industries in Africa especially in Nigeria.

Table 1: Commonly used Foundry Automotive Components [9, 15, 16, 17]

S/N	Components	Area of Utilization	Material
1	Brake Disc	Front axle	Gray cast Iron
2	Brake drums	Rear axle	Steel Casting
3	Flywheel	Engine	Steel or Gray cast Iron
4	Crankshaft pulley	Engine	Steel or Gray cast Iron
5	Water pump pulley	Engine	Steel or Gray cast Iron
6	Rocker arm bracket	Engine	Ductile cast Iron
7	Admission Valve Seats	Engine	Ductile cast Iron
8	Exhaust valve Seats	Engine	Ductile cast Iron
9	Push rod	Engine	Ductile cast Iron
10	Valve guide	Engine	Ductile cast Iron
11	U-bolt spacer	Engine	Ductile cast Iron
12	Bus Landrail Support	Body	Aluminium alloy
13	Bus door hand parts	Body	Aluminium alloy

14	Passenger door nail bracket	Body	Aluminium alloy
15	Door Hinges	Body	Malleable iron or steel
16	Cylinder sleeves	Engine	Cast Iron
17	Crankshaft timing sprockets	Engine	Cast Steel
18	Camshaft	Engine	Cast Steel
19	Engine Support	Engine	Aluminium Alloy
20	Rear Axle Support	Front axle	Aluminium Alloy
21	Seat slide (Adjuster)	Seat	Aluminium Alloy

2. Foundry and Economy Development

Foundry is one of the main industries prompting the development of world economy. The foundry industry is a main feeder to major manufacturing sectors that drive technological growth and most world economies usually depend on the stability of foundry and steel industries [18, 19]. Foundry is made up of 16-21 process steps depending on the category of castings produced from casting design to pattern making to materials/alloy selection to sand preparation to moulding to melting to casting to fettling and to heat treatment and quality control processes that includes all the production line in Foundry operations [20-22]. The importance of the foundry technology in technological development cannot be underscored and the impact on the economic development of any nation with its high value adding possibilities. Therefore, in the quest for technological development of a country, the foundry industry should be considered a high-value adding possibility [19]. It is described as "the Mother of all Industries" [23-24]. It produces major components for agricultural machines, machine tools, automobile, textile industries, power plants, industrial machinery, oil and gas equipment, cement manufacturing equipment, mining and quarry industry, railway equipment, construction industry and defense equipment [23, 25]. During the Bronze Age, Egypt was believed to have ruled the Western civilized world (nearly 2000 years) due largely to its casting skills that promoted its wealth creation capacity [26]. In the middle age foundry products consisted of baptismal fonts, bells, grave plates, temple portals and war equipment like cannons. The invention of low-melting non-ferrous metal casting technology also led to major breakthrough in the industry.

There have been significant changes in the world map of the greatest casting producers in the last decade. Variations of foundry production in different countries have caused transformation of economic systems in their countries [27]. Automotive industry has been reported as the driving force of the foundry engineering in the world [28]. The automotive industry is the major consumer of the foundry products, which requires quality castings produced by means of practically all available technologies and casting materials. Other industries include; Railways, Power Sector, Tractor Industries, Earth Moving Machineries, Pumps, Compressors, Pipes, Valves, Pipe Fittings, Electrical, Textile, Cement, Agro Machinery, Machine Tools, Engineering Industries, Sanitary Castings and Engineering Exports [19, 26]. According to the census of world casting production, the current estimation of the world's casting production was carried out globally by considering factor encompassing countries of the given continent as shown in Table 2 [27-31].

IOP Conf. Series: Materials Science and Engineering **413** (2018) 012003 doi:10.1088/1757-899X/413/1/012003

Table 2: World's Annual Casting production (Total)- in Kiloton's and Indexes illustrating Changes in Production Volume [27- 31].

Annual Casting Production Volume (total)- in Kilotonnes								
Country	Group	1991	2001	2002	2003	2004	2001:1991	1991:2001
							quotient	% change
USA	I	8839.3	11871.0	11811.7	12069.6	12314.1	1.343	+34.3
Russia		18000.0	6200.0	6200.0	6200.0	6300.0	0.344	-65.6
China		10750.0	14889.0	16261.6	18145.9	22420.4	1.385	+38.5
Japan		7958.7	5841.2	5751.8	6111.4	6386.5	0.735	-26.5
Total (group !)		45548.0 (70.2%)	38801.2 (61.6%)	40025.1 (60.3%)	42526.9 (61.9%)	47421.0 (62.8%)	0.851	-15.0
Germany	II	4336.7	4643.4	4595.4	4722.6	4984.5	1.071	+7.1
France		2351.3	2527.1	3918.2	2484.5	2465.6	1.075	+7.5
Great Britain		1307.7	968.2	886.3	1221.5	1273.0	0.740	-26.0
Italy		1881.5	2393.3	2440.6	2441.0	2442.3	1.224	+22.4
Spain		832.7	1572.3	1628.6	1149.7	1309.2	1.895	+89.5
Total (group II)		10709.9 (16.5%)	12104.3 (19.2%)	13469.1 (20.3%)	12019.3 (17.5)	12473.6 (16.5%)	1.130	13.0
India	III	-	3155.0	3267.0	4038.0	4623.0	-	-
Brazil		1228.9	1760.0	1970.6	2249.4	2829.9	1.432	+43.2
South Korea		1452.5	1683.2	1713.5	1783.8	1857.3	1.159	+15.9
Taiwan		1454.6	1209.6	1441.4	1468.1	1451.8	0.832	-16.8
Mexico		847.0	1680.0	2030.0	1822.9	2185.2	1.983	+98.3
Turkey		553.8	905.8	921.6	955.0	982.0	1.197	+19.7
Total (group III)		5536.8 (8.53%)	10393.6 (16.5%)	11344.1 (17.0%)	12317.2 (17.9%)	13929.2 (18.4%)	-	-
Czech Rep.	IV	1327.0	471.9	441.15	477.1	522.4	0.356	-64.4
Poland		815.8	745.2	660.1	729.4	804.5	0.913	-8.7
Romania		850.5	36.5	36.5	36.5	207.0	0.390	-61.0
Hungary		93.6	121.6	136.8	253.0	164.9	1.299	+29.9
Total (group IV)		3086.9 (4.76%)	1703.7 (2.70%)	1603.65 (2.40%)	1822.0 (2.70%)	1698.8 (2,30%)	0.552	-44.8

Recently, the largest producer of different types of castings of any country in the world is China, with excess of 44.5 million metric tons 2013, which is almost five times the U.S. production and about 40% of the world production. Since 2001, there are has been a consistent increase in Annual Output of Chinese Foundry Products has seen in Figure 2 [13, 32]. China has more than 30,000 foundries as of 2008 with average production per foundry of 883 tons per year compared to 5,943 tons per foundry of the 2,017 foundries in the U.S. Foundries in China are now producing iron castings of 145 mt and steel castings of 520 mt, more than any U.S. foundry taking care of the automotive industries and domestic requirements. Only 4% of total casting productions are for exports, leaving 96% for domestic consumption. China imports about 1% of its castings, mostly of very complex alloys [13, 32, 33]. As of 2011, Foundry capital investment in China were 62 flask type molding machines, 201 flaskless molding machines, 307 no-bake mixers and over 500 electric melt furnaces. These state-of-the-art foundries provided consistent, high quality castings [13]. It has been projected that China would have the largest economy in the world by the year 2030 as a result of the foundry capacities [34].

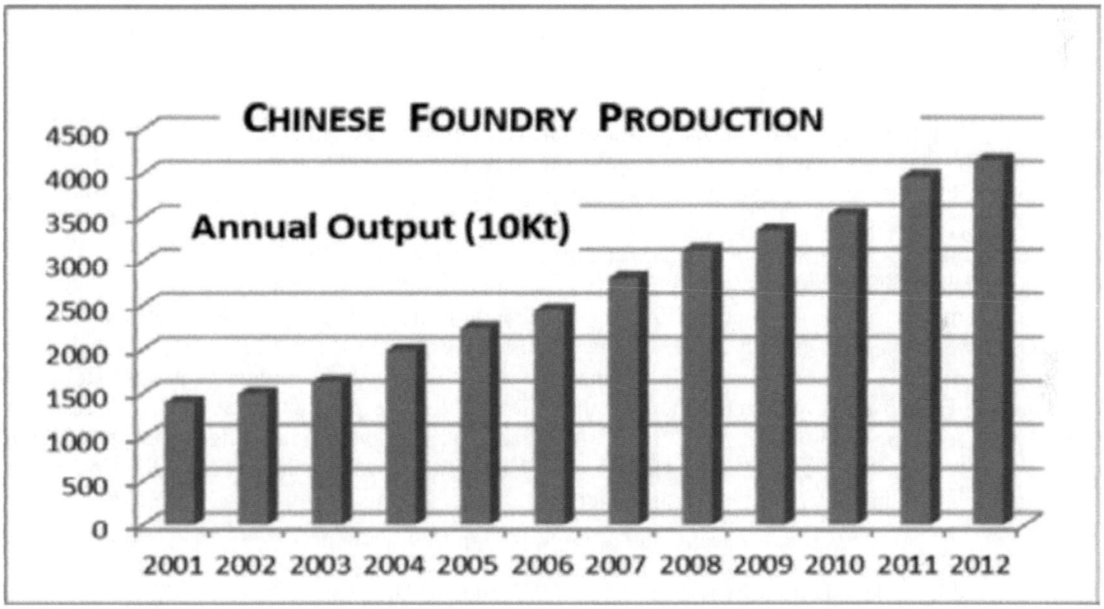

Figure 2: Annual Output of Chinese Foundry Products [13]

However, Indian foundry has been reported in recent research as the second largest in the world [13]. Out of more than 6,000 foundries in India, almost 90% fall under the small and medium scale category and are located in clusters in 19 areas [13, 35, 36]. India tied with the United States as the second largest producer of castings in total weight. Approximately 32% output of Foundry Industry in India drives automotive sector and the rest to downstream Engineering Sector [37]. The consumption of casting in India as % of total production is as shown in Figure 3

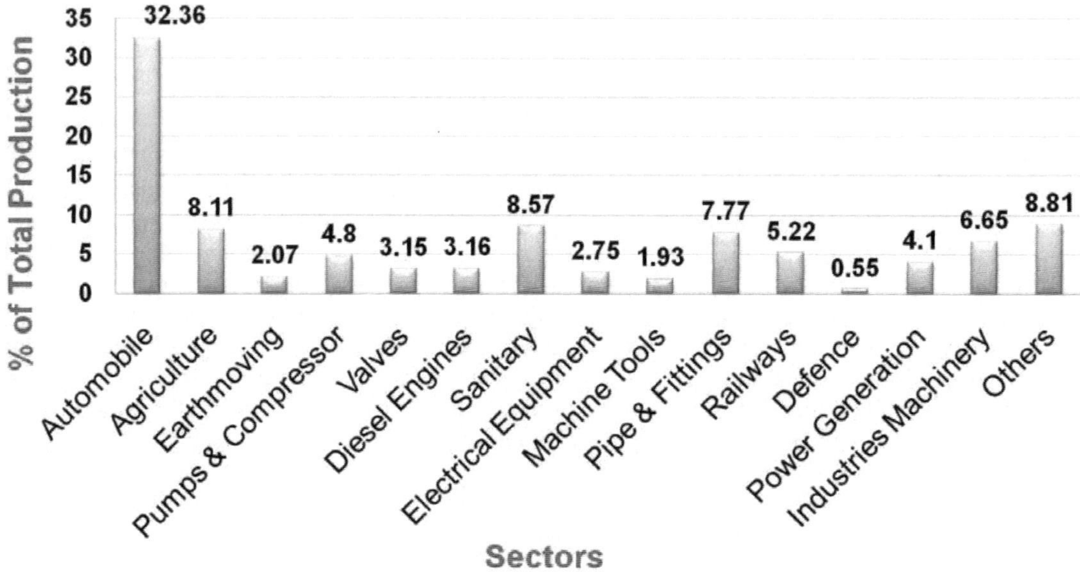

Figure 3: The consumption of casting in India as % of total production [37]

3. Automotive Industry in the World

Auto industry is stated to be synonymous to industrial development in many aspects [38]. This is because of the important role it plays in enhancing research and development, on various technologies associated with the manufacture of automobiles, thereby stimulating technological advancement [38].

Automobile industry in India was rated the seventh largest in the world with an annual production of over 2.6 million units in 2009 and also emerged as Asia's fourth largest exporter of automobiles [37]. According to Global Automotive Executive Survey 2017 by KPMG, India may reach around 33 million orders of sales that are predicted for China in 2030 [39]. The country is expected to lead the world in car volumes with approximately 611 million vehicles on the nation's roads by 2030 [39]. In order to benefit from these predictions, the Foundries in India must increase their % total volume production casting for Automobile industry [37]. According to the Folk group [13], 2014, India has an automobile building mandate to produce vehicles not only for domestic use but also for export. The India Foundrymen Association estimated that in order to meet this projected mandate the country must increase its casting capacity in the next five years [13].

After China and Canada, Mexico is the third largest supplier of castings to the U.S. Engine blocks and automotive castings are the biggest portions of castings from Mexico's foundries [13]. With its foundry capacities, Mexico has been ranked as the principal export market for U.S. auto parts overall and ranked the seventh major vehicle manufacturer in the world in 2015. According to 2016 Top Markets Report Automotive Parts, 2016 [40], Mexico was rated the sixth largest auto part manufacturer in the world with 2,559 auto parts companies with this casting volumes from Mexico continue to increase [13, 40]. Foundries in Canada export about 9.8% of their castings to U.S. Europe has more efficient foundries with very high quality castings and about 350,000 tons of cast products from Europe are exported per year [13]. The automotive and automotive parts producing countries such as United State, Europe, Japan and Korea for consumers around the world are known for the well-equipped and standard foundries [40]. However, as important as the foundry industry appeared in developed countries, the industry's potential is being under-utilised in Africa especially in Nigeria [41].

There is enormous potential for the consumption of steel, aluminium, chrome and PGM in metal castings from foundries in the automotive industry. Aluminium is used to make cast and forged products, such as rims, while stainless steel (that includes chrome) and PGM are used extensively in various components of the exhaust system, particularly in catalytic converters [42, 43, 44].

Algeria, Egypt, Morocco and South Africa with recognized and rapidly emerging automotive industries responsible for close to 80% of overall new vehicle sales in 2015 [5]. These countries have sizeable automotive assembly and manufacturing sectors. However, fewer than 900 000 vehicles were produced on the continent, which accounted for just over 0.9% of global production in 2015 [5, 45]. This may be as result of limited numbers of local foundries in Africa that are capable of supplying such automotive industries with cast products leading to over dependence of such industries on imported castings [64]. Even though domestic vehicle production and assembly may have helped African economies, and could serve as a catalyst for industrialization and economic diversification [5]; the lack of adequate foundry industries in the continent has jeopardized such development.

It has been reported that the number of foundries in South Africa has reduced drastically from 450 in the 1980s to just over 200 in 2003 and 170 in 2015 [43- 47]. The industry is very small compared to its global competitors such as China and India with 30,000 and over 6,000 foundries respectively. The estimated annual production of castings dropped from 506,000 metric tons in 2003 to 375,240 metric tons in 2013 [46, 48] unlike China with an excess of 44.5 million metric tons per year. Approximately 40% to 50% of these casting were used as automotive components, while about 85% of all aluminium castings were for use in the automotive industry [43, 48] Although according to Centre for International Development, 2016 South Africa accounted for three-quarters of Africa's automotive exports with 15% of imports in 2014 [5, 12]; it could be deduced that automotive industries in South Africa depend largely on the imported castings to survive. In Egypt the improvement of the aluminium casting quality has increased the competency of the industrial society in the country [49]. The foundries in Egypt have the potential to produced aluminium automotive cast parts for automotive industry worldwide yet need to improve on quality [50- 53]. The targeted industries in the country for castings produced in the Foundries are automotive industry and spare part industry. The iron foundries in Egypt though present a significant asset to the country's economy; the cast parts are being used in water and drainage, construction, marine application with little or no use in automobile applications [27]. It has been reported that there is a significant prospect for developing casting industries in Egypt but there are a lot of technical challenges such as lack of modern technologies in small Egyptian foundries limiting this development [49].

4. Automobile and Foundry Industries in Nigeria

The automobile is deriving about 75 % of its weight from various metals and alloys. In the Iron, Steel and Aluminium industries, Foundries played a significant part in the success of the automotive industry, in the world. One of the main causes for the slow pace of development of automotive industry in Africa, especially in Nigeria in comparison to the developed countries and countries like Korea, India and Brazil, is that the Foundries in Africa are yet to contribute significantly to the growth of the industry. Research showed that only a few local foundries in Nigeria meet up the International standards of producing auto components castings and forge fittings [54]. Ajaokuta and Delta Steel companies were unable to supply requirement of over 1000 tons of castings and forges required by Peugeot Automobile Nigeria (PAN) as of 1992 [54]. The research survey commissioned by the National Agency for Science and Engineering Infrastructure (NASENI), in 1995, showed that Nigeria has about 160 foundries, out of these 60 are registered commercial and they produce at an average capacity of 55%, accounting for less than 30% of the nation's required castings [23].

According to Okundaye, 1995 [55], the National Demand for cast metal products in Nigeria increased steadily from 72, 000 tonnes to 292, 000 tonnes (1985) to 425,000 in (1990s) and 794, 000 tonnes by the turn of the century, for which the National Committee on Foundry Development estimated 40% of these requirements as automotive components [25, 55]. In Nigeria, 120,000 tons of foundry products is required to meet industrial demand and at present only four foundries have managed to survive out of the sixty registered active ones as of 1995 to cater for this huge demand [19, 56]. This gap (over 70%) in casting products supply has only been met through importation over the years [23, 57]. Oyinlola, 1991, predicted that the actual net demand of castings by the end of the century would be 794, 600 tonnes in Nigeria which is even far less than what China is producing [13, 25, 36, 59]. This may not be possible due few Foundries present in the country. Ekundayo 2015 stated that there will be a strong demand for all major foundry products by the year 2020 due to Government determination to boost farming increasing the demand by the manufacturing and agricultural sectors [25].

Currently installed vehicle assembly capacities in Nigeria stands at more than 70,000 vehicles per year, in total 7 companies have started to assemble vehicles. According to Global Auto Industry Market Report 2015, automotive industries where Vehicles are assembled such as Stallion Motors assemble vehicles for Hyundai, Nissan, Tata and Ashok Leyland at the former Volkswagen of Nigeria (VON) plant in Lagos, which the company took over in 2012 and Nissan have increased their sales from around 1,000 new vehicles in 2012 to around 6,000 in 2014 assembled in the country [1]. Innoson Vehicle Manufacturing (IVM) has an installed capacity of 10,000 vehicles. Peugeot Automobile Nigeria (PAN), currently targeting to assemble around 3,000 vehicles, but having much space for growth at its plant in Kaduna which was originally built in the 1980s. Also assembling passenger cars, Dana Motors, which is part of Nigerian Dana Group revealed current capacities plans to assemble the whole Kia range of passenger cars in 2015. The company assembles Kia Rio, Cerato and Optima at its plant in Lagos. Also, Anammco, Leventis and Sinotrucks Nigeria are assembling trucks in the country [1]. The total vehicles capacity in Nigeria as of 2014 was 78,000 with Truck assembly capacity of 3,000 and car and bus assembly capacity of 75,000 [1]. This just accounted for 8.7% of the vehicles produced on the continent and 0.078 % of global production in 2015 [6, 39, 59]. This production is negligible for a country of 170,000,000 people with the largest economy in Africa; the country GDP of $520bn in 2013 as compared to $350bn of South Africa as the continent's number two [60].

Today almost every fifth African is a Nigerian; by 2050 Nigeria is projected to have 440 million inhabitants, overtaking the US with an estimated population of 400 million [61] making the country a potential for the automotive industry. Without the proper development of foundries, there would be no engineering subsystem and consequently no realistic automotive industrial base; therefore there would be no end to excessive importation dependence to make such an industry economic self-reliance [23, 62]. It was recorded that only a few foundries could supply components to Automotive Plants Nigeria [54] because casting products are not of quality and did not meet up with International Standards of producing auto components. It was also reported that there are no advanced technologies to improve the quality of the casting products produced in the Nigeria Foundries [65]. Only sand system out of other processes in casting has been mechanized to a great extent in few foundries in Nigeria, making the Foundries in the country not geared towards mass production [9]. These may explain the reasons for the choking of the automotive industry in Nigeria and Africa at large.

5. Conclusion

The Nigerian government initiative is aimed at diversification of Nigerian economy from oil industry by developing other sectors including the automotive industry. The Government has taken different precautions such as increasing the import duties on imported vehicles, confiscating the imported used cars to support domestic vehicle assembly industries as part of the Nigeria Industrial Revolution Plan. This may not lead to remarkable breakthroughs in the vehicle assembly industries, if the government did not develop foundry industry and other auxiliary industries which are the sister or complementary industries to the automotive industry in Nigeria. The foundries have not been contributing significantly to the development of automotive industries in Nigeria due to lack of adequate attention to the development of the industry by the government.

References

[1] Global Auto Industry Market Report (2015). The African & Nigerian Automotive Industry [Online] www.xcom-africa.com Accessed 12/11/2017

[2] African Economic Outlook - UNDP, (2013). Special Theme: Structural Transformation and Natural Resources. [Online] www.undp.org/content/dam/.../African%20Economic%20Outlook%202013%20En.p... Accessed on 23/12/2017

[3] World Trade Report, (2014). Trade and development: recent trends and the role of the WTO. [Online] https://www.wto.org/english/res_e/booksp_e/world_trade_report14_e.pdf Accessed on 12/11/2017

[4] Africa's Pulse (2012). An Analysis of issues Shaping Africa's Economic Future. The World Bank. Vol. 6 [Online] www.siteresources.worldbank.org/INTAFRICA/Resources/Africas-Pulse-brochure_Vol6.pd Accessed on 23/12/2017

[5] Deloitte Africa Automotive Insights (2016). Navigating the African Automotive Sector: Ethiopia, Kenya and Nigeria. [Online] https://www2.deloitte.com/.../manufacturing/ZA_Deloitte-Africa-automotive-insights-. Accessed on 23/12/2017

[6] Delliot, (2014). Making an Impact that Matters. [Online] https://www2.deloitte.com/bb/en/pages/about-deloitte/topics/global-report-2014.html Accessed on 12/12/2017

[7] Sales Statistic of New Vehicles 2005-2016 (2016) [Online] www.oica.net/category/sales-statistics/ Accessed on 22/12/2017

[8] Webster, J., McNay, D. A., Lundy, D. (2005). Foundry Products: Competitive Conditions in the U.S. Market. [Online] https://www.usitc.gov/publications/332/pub3771.pdf Accessed on 12/11/2017

[9] Salami, L. A. (1992). Importance of Foundries and Forges to National Development. A paper presented at Nigeria's Industrial Sub-Sector Development and ITAP Subsector Training Workshop: Foundry and Forge Sub-Sector.

[10] Ocheri, C., Mbah A. C., Mbah, C. N. (2017). Focusing Foundry Engineering for Economic Recovery: A Case Study of the Ajaokuta Foundry Shop and Making Shop. International Journal of Swarm Intelligence and Evolutionary Computation. 6(1)

[11] Sturgeon, T. J. and Biesebroeck, J. V. (2010). Effects of the Crisis on the Automotive Industry in Developing Countries. A Global Value Chain Perspective. The World Bank Policy Research Working Paper 5330.

[12] Black, A., Makundi, B. and McLennan T. (2017), Africa's Automotive Industry: Potential and Challenges, Working Paper Series N° 282, African Development Bank, Abidjan, Côte d'Ivoire.

[13] The Folk Group, (2015), State of the Metal Casting Industry by Country, The Folk Group, Doylestown, PA, 18901. 1.215.340.9072. Accessed on 12/11/2017 [Online] www.folkgroup.com

[14] United Nation (2007), Industrial Development for the 21st Century: Sustainable Development Perspectives. Department of Economic and Social Affairs, United Nations, New York. Accessed on 12/11/2017 [Online] www.un.org/esa/sustdev/publications/industrial_development/full_report.pdf

[15] Abioye A. A., Atanda, P. O., Kolawole, O. F., Olorunniwo, O. E., Adetunji, A. R., Abioye O. P. and Akinluwade K. J. (2015). The Thermal Analysis of Fuel Fired Crucible Furnace using Autodesk Inventor Simulation Software. Advances in Research 5(3): 1-7

[16] Abioye, A. A., Atanda, P. O., Abioye O. P., Afolalu S. A. and Dirisu, J. O. (2017) Microstructural Characterization and Some Mechanical Behaviour of Low Manganese Austempered Ferritic Ductile Iron. International Journal of Applied Engineering Research. 12(23), 14435-14441

[17] Atanda, P. O., Olorunniwo, O. E., Abioye, A. A. and Oluwole, O. O. (2014). Modeling heat flow across the fuel-fired crucible furnace using ADINA. International Journal of Scientific & Engineering Research, 6(7).

[18] Panchal, S. (2010). World Foundry Industries: Overview. Barcelona: International Foundry Forum

[19] Alpha, A. A., (2013). Development of Self-Instructional Manual for Sand Casting Technology in the Polytechnics in North Central States of Nigeria. [Online] http://www.unn.edu.ng/publications/files/ALPHA,%20ABDULLAHI20%ALIYU.pdf Accessed: 12 April, 2017

[20] Broom J. R. (1994) Foseco Foundry men's handbook tenth edition. 284-289.

[21] Mudiare E. (1994) A tale of two steel plants, a paper presented at Nigerian Metallurgical Society. 20-35.

[22] Packer KF, Miller WCT (1979) Foundry quality control. (3rd Edn), Mcgraw–hill book company, New York. 200-203.

[23] Agbo, P. O. and Ayeni, A. O. (2017). Determinants of Sales of Foundry Products in Nigeria: A Discriminant Function Analysis. International Journal of Advanced Studies in Economics and Public-Sector Management. 5(1), 2354-4228.

[24] Ezekwe, G. (1995). Foundry is Mother of All Industries. Foundry Chronicle, 1(1), 5.

[25] Okundaye, S.E. (2015). Developing the Foundry Industry for Sustainable Economy in Nigeria. www.seetconf.futminna.edu.ng Accessed: 27 September, 2017

[26] Groover M.P. (2011). Principles of Modern Manufacturing, 4th Edn. John Wiley & Sons Asia.

[27] Dańko, J. and Holtzer, M. (2006). The State of Art and Foresight of World's Casting Production. METABK 45 (4) 333-340.

[28] Gallo, R. (2010). Circumstances of the Automotive Industry Impact on Aluminium Foundries. Proceedings, American Foundry Society, Schaumburg, IL USA 1-17.

[29] Modern Casting Staff Report (2010). The global casting market experienced a marked contraction in 2009, with overall shipments decreasing 14% from 2008. 44th Census of World Casting Production.

[30] Gutowski, T. (2014) Casting http://www.mit.edu/afs/athena/course/2/2.810/oldfiles/ts_temp/Casting.ppt

[31] Census of World Casting Production 1991, 2003, 2004, 2005 Modern Casting No 12, (monthly journal, edited annually with the data concerning the number of casting houses and the world casting production in the year preceding the issue).

[32] Abioye, O. P., Abioye, A. A., Atanda, P. O., Osinkolu, G. A., & Folayan A. J. (2017). Numerical Simulation of Outer Die Angle of Equal Channel Angular, 8(12), 264–273.

[33] Global Steel Trade Monitor (2017). Steel Exports Report: China. Steel Import and Monitoring Analysis Washington, D. C.

[34] The World Bank (2012). China 2030 Building a Modern, Harmonious, and Creative High-Income Society. International Bank for Reconstruction and Development / International Development Association or The World Bank. Conference Edition [Online] https://www.uscc.gov/.../Planning%20for%20InnovationUnderstanding%20China%2... Accessed on 12/12/2017

[35] Ajay, M. (2013). Energy conservation measures in the Foundry Sector. Winrock International India. 65 [Online] Accessed on 12/12/2017. https://www.sidbi.in/files/Energy-Conservation-Measures-in-the-Foundry-Sector.pdf
[36] Pantange, G. and Khond, M. (2013). Some Studies on Energy Consumptions and Identification of Suitable Energy Management Techniques in Indian Foundry Industries. European Scientific Journal 9(24)
[37] Panchal, S. (2010). Indian Foundry Industry Overview. International Foundry Forum, Barcelona
[38] Marcellosabbioni (2015). 49th Census of World Casting Production Modest growth in Worldwide Casting Market. BA Crawford Specialised Publications (Pty) Ltd [Online] Accessed on 12/11/2017
[39] KPMG's 18th Consecutive Report, (2017) Global Automotive Executive Survey 2017. [online] kpmg.com/GAES Accessed on 12/12/2017
[40] Larger Top Markets Report, (2016). 2016 Top Markets Report Automotive Parts Country Case Study. International Trade Administration [Online] www.trade.gov/topmarkets Accessed on 23/12/2017
[41] Agbu, O. (2007). The Iron and Steel Industry and Nigeria Industrialization: Exploring Cooperation with Japan. Institute of Developing Economies, Japan External Trade Organization. 418.
[42] Chatterjee, K. K. (2007) Uses of Metals and Metallic Minerals. New Age International (P) Ltd. 4835/24, Ansari Road, Daryaganj, New Delhi – 110002
[43] Paul R. L. (2013) The state of the local foundry sector. Paper Presented at The Ferrous and Base Metals Development Network Conference 2012, Johannesburg
[44] Paul, L., Johann, M., Shane, G. (2008). Industrial Structure and Skills in the Metals Beneficiation Sector of South Africa. Sector Studies Research Project [Online] www.labour.gov.za/DOL/downloads/.../Metals%20Benification_DoL_Report.pdf Accessed on 23/12/2017
[45] Nizeyimema, E. (2017) Profiling African Auto Industry and Market Perspective on Current State and Growth Potential. African Auto Industry& Market Outlook Q1. Slide Share [Online] https://www.slideshare.net/.../african-auto-industry-market-outlook-q1-2017 Accessed on 23/12/2017
[46] Davies, J. T., (2015). South African Foundry Industry. Paper Presented at 2nd VAMCOSA Foundries Workshop. [Online] www.foundries.org.za/wp.../02/2-John-Davies_Vamcosa_Workshop_20022015.pdf Accessed on 12/12/2017
[47] Zanele, M. and Didier, N. (2014). Improving Operating Methods in a South African Foundry Company - A Case Study. Proceedings of the 2014 International Conference on Industrial Engineering and Operations Management Bali, Indonesia.
[48] Madzivhandila, T. (2015). South African Foundry Industry Present Status asqnd Future Trends. Paper Presented to the V BRICS Foundry Forum [Online] www.globalcastingmagazine.com/wp.../South-Africa-BRICS-Presenation-2015.pdf Accessed on 23/12/2017
[49] El-Mahallawi, I., Khalifa W., A. Shash, A., Aref, N. (2015) Roadmap to Egyptian Casting Industry: Material Technology. Conference Paper
[50] Metalcasting Industry Technology Roadmap, (1998). American Foundrymen's Society, January, 1998.
[51] Dispinar, D., Akhtar, S., Nordmark, A., Di Sabatino, M., Arnberg, L. (2010) Degassing, hydrogen and porosity phenomena in A356, Materials Science and Engineering: A, 527 (16–17), 3719-3725.

[52] Lei Zhao, L., Pan, Y., Liao, H., Wang, Q. (2012) Degassing of aluminium alloys during re-melting", Materials Letters, 66(1) 328-331.

[53] Gruzleski J. E. and Closset, B. M., (1990). The Treatment of Liquid Aluminium Silicon Alloys, American Foundry men's Society, Inc., Des Plaines, IL, 1.

[54] Abubakar, U. A. (1992). Contributions of the Nigerian Foundries and Forges to the Development of the Auto Industry in Nigeria. Paper presented at the National Workshop on Foundry and forge Sub-sector organized by the Policy Analysis Department, Federal Ministry of Industry and Technology.

[55] Okundaye, S. E. (1995). The Role of Small-Scale Foundry Industry in Revitalizing the Depressed Nigerian Economy. Proceedings of the Nigerian Metallurgical Society. 91-98.

[56] Barberopoulos, R. V. (2012, February 14). Nigeria requires 120,000 tonnes of foundry products - Barberopoulos. (N. Egwuonwu, Interviewer)

[57] Inuwa, K. (1995). Foundries Subsystem: A Crucial linkage for the Successful Operations of the Other Industries. The Nigerian Engineers, 33 (4)

[58] Oyinola, A. K. (1991). Non-Destructive Testing Techniques in Foundry Industries. Proceedings of the Nigerian Metallurgical Society, 66-67.

[59] PwC Nigeria Automotive Industry, (2015). Africa's Next Automotive Hub. [Online] www.pwc.com/ng Accessed on 23/12/2017

[60] Onigbinde, O. and Sosami, A. (2017). Examining Effective Budget Communication for Citizens in Nigeria. A study by BudgIT with funding and support from Institute for Development Studies January 2017 [Online] yourbudgit.com/wp-content/uploads/2017/08/BudgIT_MAVC_Research_Report.pdf Accessed on 12/12/2017

[61] The 2017 Revision Key Findings and Advance Tables (2017). UN Department of Economic and Social Affairs, World Population Prospects. [Online] https://esa.un.org/unpd/wpp/Publications/Files/WPP2017_KeyFindings.pdf Accessed on 12/12/2017

[62] Majekodunmi, S. (1997). Foundry: The Mother of all Industries. Guardian Supplement. ,April 3.

[63] Afolalu, S.A., Adejuyigbe, S.B. and Adetunji, O.R. and Olusola, O.I. (2015) Production of cutting tool from recycled steel and palm kernel shell as carbon additives;, International Journal of Innovation and Applied Studies, Vol.12 (1) July, 2015, pp 110-122 http: www.ijoar.org.

[64] Afolalu, S.A., Adejuyigbe, S.B. and Adetunji, O.R. (2015) Impacts of Carburizing Temperature and Holding Time on Wear of High Speed Steel Cutting Tools, International Journal of Scientific and Engineerinsg Research, Vol. 6 (5), pp 905-909, http: www.ijser.org.

[65] Afolalu, S.A., Adejuyigbe, S.B. and Adetunji, O.R. and Olusola, O.I. (2015) Effects of Carburization on Mechanical Properties of recycled steel with palm kernel shell as carbon additives; International Journal of Advance Research, Vol. 3 (5), pp 1-7, http: www.ijoar.org.

Effect of machining variables and coolant application on HSS tool temperature during turning on a CNC lathe

T Ejieji[1], **S M Adedayo***[1], **O W Bello**[1] and **S Abdulkareem**[1]

[1]Department of Mechanical Engineering, University of Ilorin, Ilorin, Nigeria E-mail: adyos1@yahoo.com

Abstract - Cutting variables in the operation of a Computerized Numerical Control (CNC) lathe machine can affect parameters such as the work piece, tool temperature, material surface finish, material properties and vibration. A programmable CNC lathe machine was used in this work to examine the effect of different cutting variables on HSS tool temperature. Carbon steel material of C20 carbon equivalent and ferritic stainless steel was machined at different cutting speeds, depth of cut with an HSS tool. A type K thermocouple was fitted at a distance 5 mm from tool nose by a direct surface contact on tool under an applied pressure. A constant feed rate was maintained throughout the experiments. Temperature profiles were obtained for different depth of cut ranging between 0.5 and 2.5 mm and cutting speed range of 500 to 800 rpm and a constant feed rate of 7 mm /minute. Tool Temperatures were observed to increase with cutting speed and depth of cut. Increase in depth of cut from 0.5 to 1.5 mm resulted into a 64% increase in tool temperature while coolant application reduced tool temperature by as much as 63.2% in stainless steel. Tool life can be significantly enhanced by appropriate machining variables selection.
Keywords: CNC lathe machine; temperature; cutting parameters; type K thermocouple

1. Introduction

Computerized Numerical Control (CNC) lathes are rapidly replacing the older production lathes due to their ease of setting, operation, repeatability and accuracy. Unlike the manually controlled lathes, CNC machines are often totally enclosed thus enhancing safety during application. Substantial amount of heat is generated from the shearing action associated with machining process, the heat is transmitted through the shank of the tool. Friction and shearing exist at the tool – chip interface and below the tool edge respectively with conductive heat transfer through work piece, the chip and the tool. The cutting zone temperature is a key factor which directly affects tool wear, work piece surface integrity and machining precision according to the relative motion between the tool and work piece.

Temperature in the cutting zone depends on contact length between tool and chip, cutting forces and friction between tool and work material. Cutting variables such as cutting velocity, feed and depth of cut affect temperature and which by extension also affect the material strength and hardness [1, 2]. Heat generated is unevenly distributed between chip, cutting tool and the blank work-piece. Mode of distribution depends on configuration, size and thermal conductivity. From 10 – 20% of total heat generated goes into the tool while the remaining heat is dissipated through the chip and blank workpiece [3]. Main techniques used to evaluate the cutting temperature during machining are classified into two, contact and non-contact methods. Contact requires direct conduction through a sensor. These methods are; embedded thermocouple, thermally sensitive paint, Physical vapour deposition (PVD) film method and thermo – chemical powders while non - contact radiation methods are; infrared radiation pyrometers, infrared cameras and optical thermocouple techniques. Other methods are use of thermo-sensitive paint, metallography based on microstructural and microhardness variations, use of thermal cameras [4], of all the various methods, the most widely used for obtaining average chip – tool interface temperature is the thermocouple. Other methods suffer from various disadvantages such as slow response, indirectness and complications in measurement [5].

Choice of any temperature measurement technique depends on the thermal situation, accessibility of sensor to the object location, dynamics of the situation, accuracy needed, instrumentation cost, data collection and analysis [6]. Thermocouples offer the following advantages (a) simplicity, (b) remote signal measurement and low cost [7]. Goyal et al. [8] carried out an exhaustive study of most existing experimental techniques of temperature measurement applicable to metal cutting. Their paper discussed in details seven different methods in practice highlighting the advantages and disadvantages of temperature on cutting tool. Abhang *et al.* [9] predicted temperature at chip / tool interface during lathe turning process. Their results show that increase in cutting speed, feed rate, and depth of cut increases the cutting temperature while increasing the nose radius reduces the cutting temperature [10]. Ghodam *et al.* [11] compared cutting tool temperatures of uncoated and CVD coated tungsten carbide tools. A tool - work thermocouple embedded in the work piece was used to convert the measured emf to the work / tool interfacial temperatures. As feed rate or depth of cut increases temperature of uncoated tool increases at an higher rate than the coated tool. They concluded that coating of the tool increases the tool life and can be used at higher cutting speeds [11]. Knowledge of tool thermal history during machining serves as valuable information for the prediction of tool wear, surface finish, choice of coolant and application rate, control of cutting characteristics and a valuable data base for development of appropriate predictive software on thermal characteristics and cutting variables [12]. An appreciable temperature rise beyond certain standards could be indicative of faulty tool or work-piece setting. This can negatively affect part quality, dimensional accuracy and introduction of residual stresses.

Cutting fluids application during machining are meant to lubricate and cool the tool / work – piece interface region [13]. They are broadly classified as water miscible and straight oils. The straight oils are used primarily for lubrication of tool / work interface. Soluble oils act as both coolant and lubricant for the sliding surfaces during machining. Other researchers that had looked at influence of cutting fluid on temperatures are Shaw *et al.* [14], Smart and Trent [15]. At cutting speeds in excess of 200 m/min, cutting fluids have no significant effect on cutting zone temperature Viera *et al.* [16], studied the cooling ability of different cutting fluids on AISI 1020 steels and deduced that chip – tool interface temperature increased with increasing cutting speed during machining while semi-synthetic cutting fluids was the most effective among the coolants. Researchers had found that high pressure coolant application to chip – tool interface reduces cutting temperature, tool life and surface finishes are improved [17, 22].

In this work the effect of machining variables on tool temperature during dry and wet machining of mild and stainless steel on a CNC lathe was experimentally examined using type K thermocouple sandwiched under pressure between cutting tool and a gasket material. The temperature values on both materials were compared and the results are hereby reported. For every machining variable 5(five) different values were tested and corresponding temperature values taken.

2. Experimental Methods

2.1 Materials and Machine Specification

Temperature history measurement was carried out on Tungsten high speed steel (HSS) tool. The AISI T-1 grade was used with chemical composition shown in Table 1.

Table 1: Chemical Composition of AISI T – 1 HSS Tool in wt % (Source: ASTM/A600 standards)

C	Si	Mn	Cr	V .	W .	Fe
0.65–0.75	0.20-0.40	0.20-0.40	3.75-4.50	0.90-1.30	17.25-18.75	77.05-73.90

Final specification of work piece material was diameter 50 mm against a length of 70 mm. Compositional analysis of the work-pieces was carried out by a specialized laboratory. Atomic Absorption Spectrometer was used for the analysis. Material used for machining was mild steel and stainless steel with the composition given in Table 2.

Table 2: Composition of Machined Mild Steel Material (wt %)

C	Si	S	P	Mn	Ni	Cr	Cu	AL	Ca	Fe
0.10	0.32	0.08	0.01	1.16	0.23	0.03	0.22	0.03	0.03	97.85

The stainless steel used belongs to the Ferritic steel category with less than 0.08%C, 12 – 19% Cr, 0 – 5%Ni, less than 5%Mo and traces of Ti. Temperature was monitored at tool tip at a distance 5 mm from nose tip. A horizontal CNC lathe machine was used. The specifications are: - Make: AJAX Machine; Model number: AJEV 310; Maximum distance between centers is 1500 mm, Tail stock diameter: 95 mm; speed range is 10 – 2500 rpm.

2.2 Temperature measurement

A type K thermocouple (Ni – Cr / Ni - AL) connected to a temperature data logger was used in taking the temperature measurement. The thermocouple hot junction was positioned at a distance 5 mm from tool nose on rake surface. A heat resistant insulating material of the rubberized asbestos cloth type of 4 mm thickness was placed on the thermocouple as an interface with a clamping steel bar. This method is nondestructive as no part of the HSS tool is drilled or welded towards thermocouple probe insertion. The assembly was carefully clamped on the tool post cautiously maintaining the relative distance of thermocouple to tool tip. Figure 1(a) is a schematic of the arrangement of thermocouple and tool while figure 1(b) is the set – up of tool on tool post and work-piece mounted on 3 – Jaw CNC lathe chuck. Pressure resulting from tool clamp caused an effective contact between tool and workpiece.

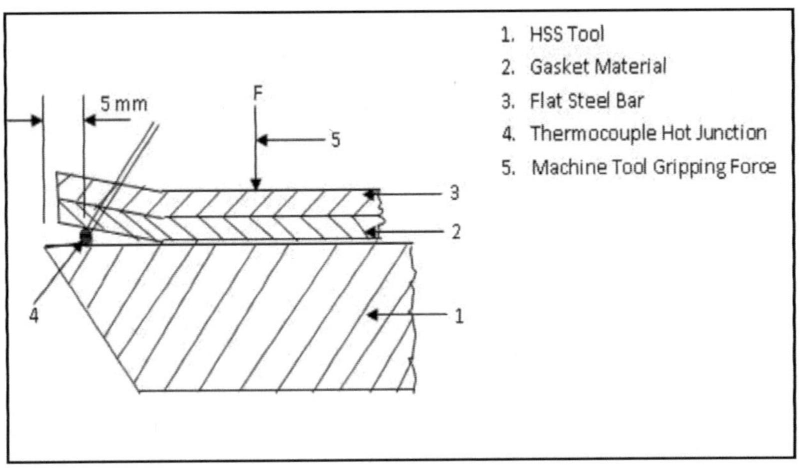

Figure 1(a) Sectional view showing positioning of thermocouple hot junction.

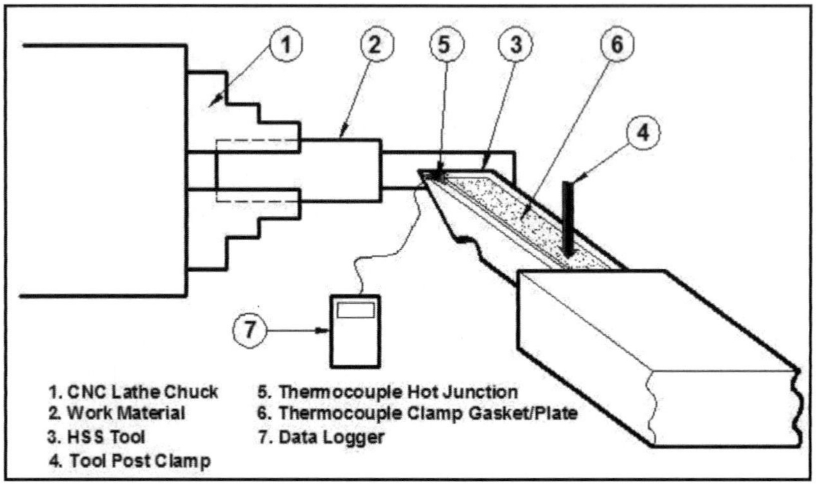

Figure 1(b) Mounting of the Work-piece and tool with thermocouple

Temperature of tool was recorded at short intervals of 2 seconds under the following machining conditions. Table 3 shows the selected machining characteristics for both dry and wet cut.

Table 3: Selected Machining Variables

	Dry	Dry	Wet	Wet
	Cutting Speed (rpm)	Depth of Cut (mm)	Cutting Speed (rpm)	Depth of Cut (mm)
1	500	0.5, 1.0, 1.5, 2.0, 2.5	500	0.5, 1.0, 1.5
2	650	0.5, 1.0, 1.5, 2.0, 2.5	650	0.5, 1.0, 1.5
3	800	0.5, 1.0, 1.5, 2.0, 2.5	800	0.5, 1.0, 1.5

A constant feed rate of 7 mm/min was used while machining duration was limited to 600 seconds. After every experiment, ample time was allowed for the tool to cool to initial state of atmospheric temperature before commencement of the next cut.

Data entry in the data logger were opened and converted into graphical plots of temperature versus time for the various machining variables.

2.3 CNC Lathe Machine Set – Up

Work-piece was mounted on the three jaw chuck and automatically clamped. Machine was programmed using G and M – Codes to set the depth of cut. Reference points were chosen for X and Z axis then subsequently add the depth of cut to the indicated X axis value. Process termination is done with respect to the Z – axis.

3. Results and Discussion

3.1 Tool Thermal History under Varying Depths of Cut

Figure 2 shows temperature variation with time at cutting speed of 500 rpm and depths of cut ranging from 0.5 to 2.5 mm for mild steel. In all cases temperature increased gradually with machining time and also with increasing depth of cut. At a typical machining time of 570 seconds and depths of cut of 1.0 and 2.5 mm, temperature values of 46 and 77°C was observed respectively. Increasing depth of cut by 150% at 500 rpm resulted into a 67.4% increase in temperature.

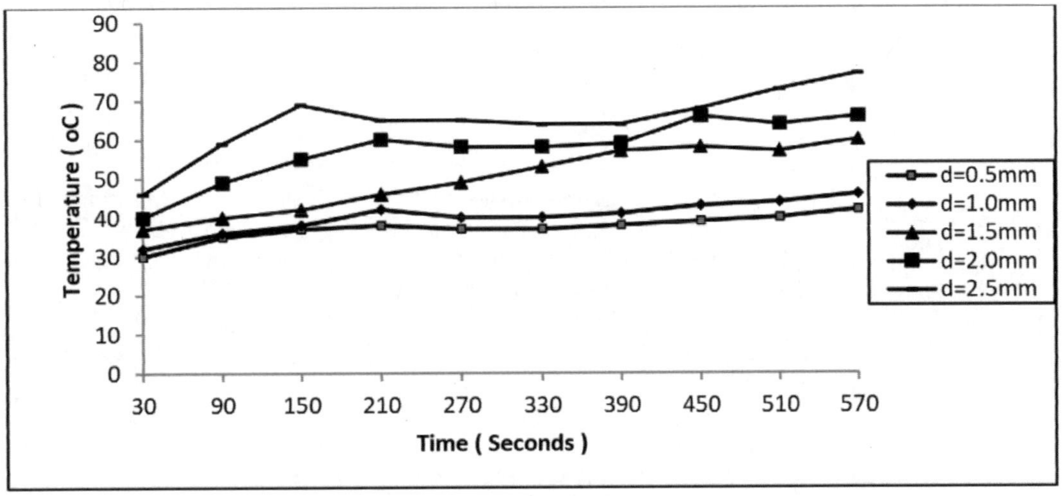

Figure 2: Variation of tool temperature with time at speed 500rpm for mild steel

Figure 3 shows temperature variation with time at cutting speed of 650 rpm and depths of cut ranging between 0.5 and 2.5 mm for mild steel. In all cases temperature increased gradually with machining time and also with increasing depth of cut. At a typical machining time of 570 seconds and depths of cut of 0.5 and 1.5 mm, temperature values of 42°C and 73°C was observed respectively. Increasing depth of cut by 150% at 650 rpm resulted into a 52.5% increase in temperature.

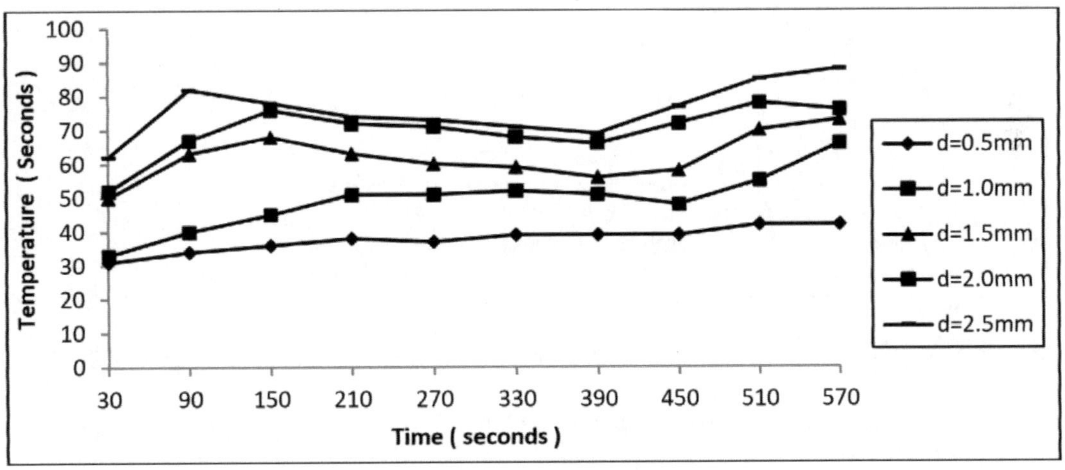

Figure 3: Variation of tool temperature with time at speed 650rpm for dry machining of mild steel

Figure 4 shows temperature variation with time at cutting speed of 800 rpm and depths of cut ranging between 0.5 and 2.5 mm for mild steel. In all cases, temperature increases gradually with machining time and also with increasing depth of cut. At a typical machining time of 570 seconds and depths of cut of 1.0 and 2.5 mm, temperature values of 66 and 115°C was observed respectively. Increasing depth of cut by 150% at 800 rpm resulted into a 74.2% increase in temperature. Akhil *et al.* [2] obtained 61.2 and 90.4°C at depth of cut of 1.0 and 2.0 mm respectively with cutting speed 685 rpm and 12 mm/min feed rate).

Figure 5 shows observed maximum temperature under different depths of cut and machining speeds. Temperature consistently increased with depth of cut and machining speed, attaining the values 75, 86 and 111°C at machining speeds of 500, 650 and 800 rpm respectively. For a typical depth of cut of 2.50 mm and 540 seconds of machining time increase of speed by 60% resulted into a 48% increase in temperature.

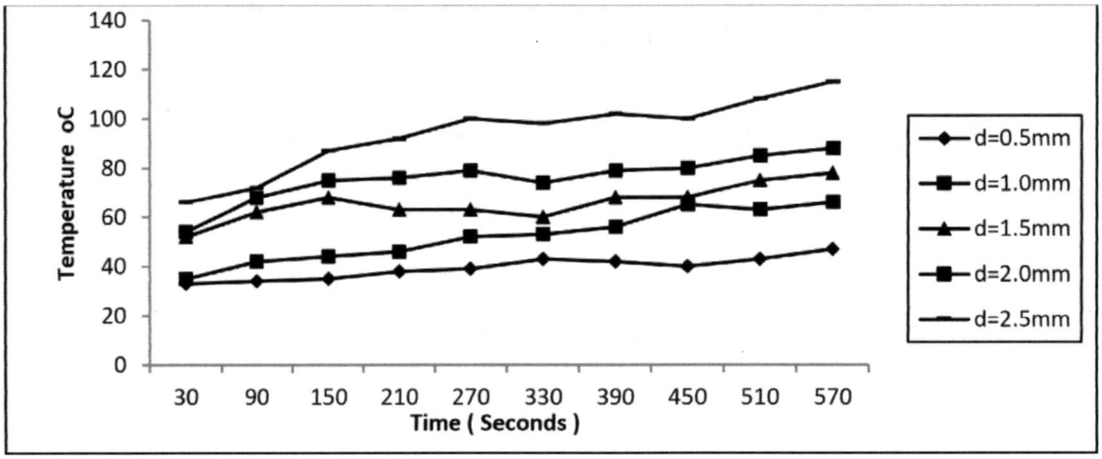

Figure 4: Variation of tool temperature with time at speed 800 rpm in dry machining of mild steel

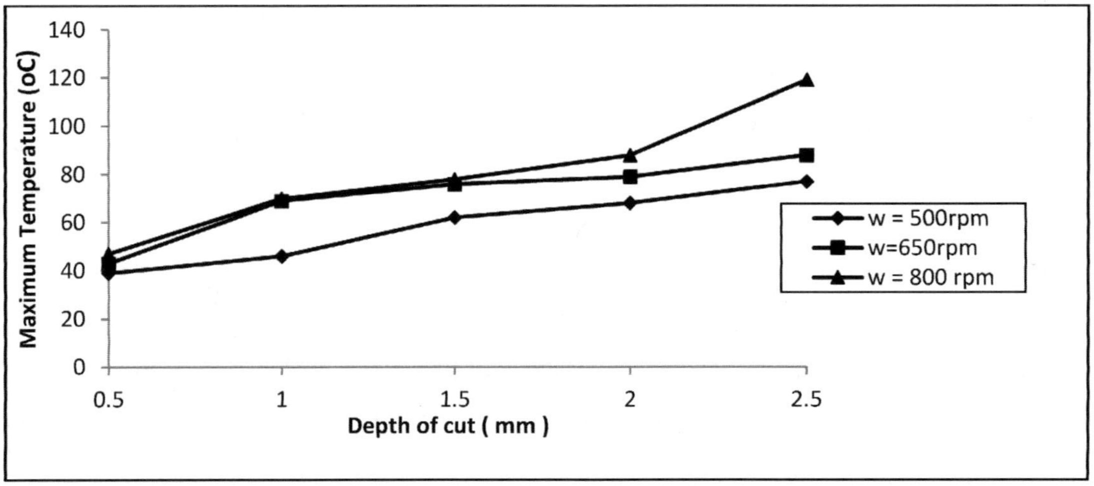

Figure 5: Variation of tool maximum temperature with depth of cut in dry machining of mild steel

3.2 Effect of Material Type on Tool Temperature History

Figure 6 shows temperature variation with time at cutting speed of 650 rpm and depths of cut ranging between 0.5 and 1.5 mm for stainless steel work material. At all depth of cut temperature increases gradually with machining time and also with increasing depth of cut. At a typical machining time of 570 seconds and depths of cut of 0.5 and 1.5 mm, temperature values of 82 and 118°C was observed for stainless steel while mild steel indicated 42 and 73°C respectively. This is explained in terms of the higher tensile strength of stainless steel(680MN/m^2) as compared with mild steel (370 MN/m^2). Higher deformation energy is required to shear the work piece along the shear plane in stainless steel as compared with mild steel thus resulting in higher temperature history. Increasing depth of cut by 150% at 650 rpm resulted into a 43.90% increase in temperature in the machining of stainless steel material.

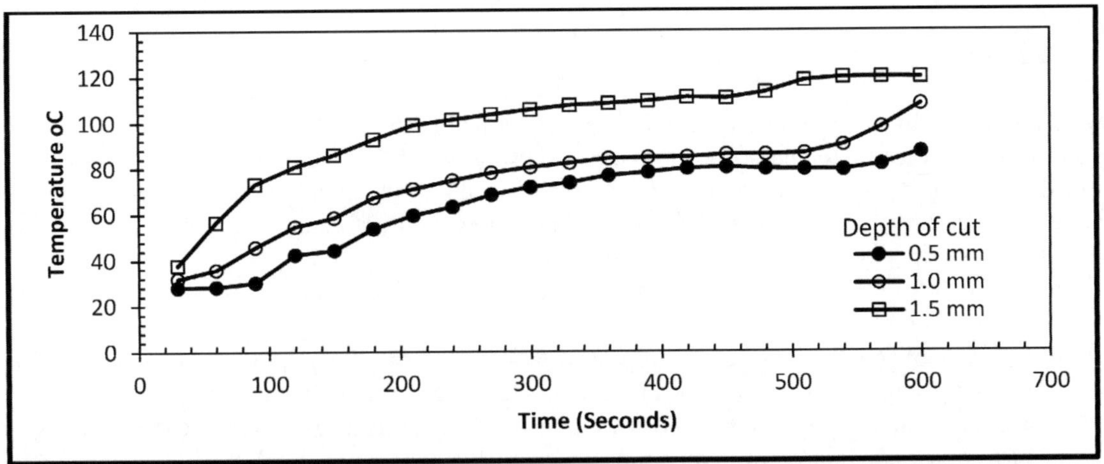

Figure 6: Tool temperature variation with time at 650 rpm using stainless steel materials

3.3 Effect of Coolant Application on Tool Temperature History

Figure 7 shows cutting tool maximum temperatures at various spindle speeds under dry and wet cutting conditions. Peak temperatures of 120°C and 37.4°C were observed under dry and wet machining conditions for stainless steel at 750 rpm. Coolant application reduced peak temperature by as much as 68.83%. Table 4 shows a comparison of HSS tool temperatures in the machining of mild and stainless steel at machining time of 570 seconds

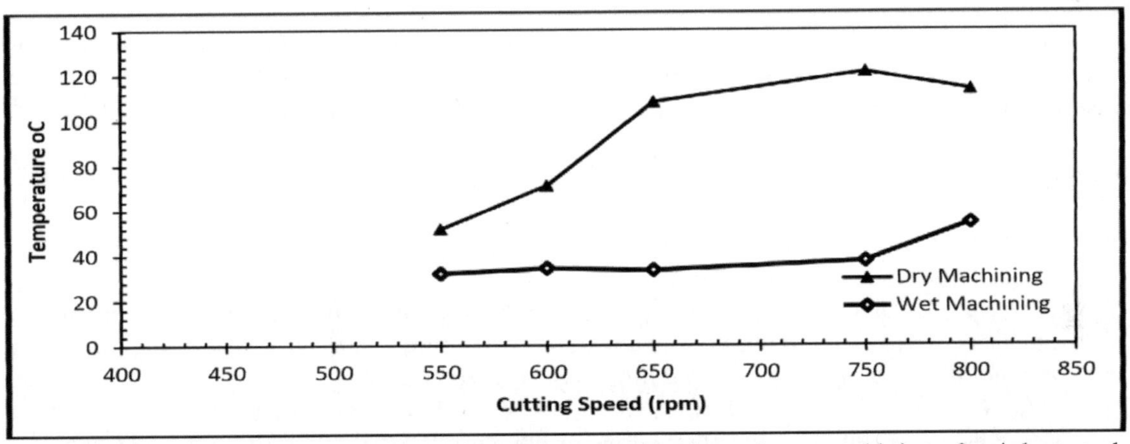

Figure 7: Variation of tool temperature with cutting speed in dry and wet machining of stainless steel

Table 4: Comparison of HSS Tool Temperatures in Machining of Mild and Stainless Steels

Machining Variables	Machining Speed (rpm)					
			500	650	500	650
			Mild Steel		Stainless Steel	
(a) Peak temperature at given depth of cut	0.5 mm	°C	40.0	42.0	39.7	81.6
	1.0 mm	°C	46.0	66.0	51.0	97.9
	1.5 mm	°C	58.0	73.0	54.3	119.7
(b) Percent temperature increase at 0.5 to 1.5 mm depth of cut increase.		%	40	42.5	26.9	31.8
(c) Temperature with coolant application	1.0 mm	°C	-	-	32	36.0
(d)Percent temperature reduction due to cooling		%	-	-	37.3	63.2

4. Conclusions

This investigation has shown the influence of cutting variables and work – material type on HSS tool tip temperature. Tool tip temperature was measured using type K thermocouple with the hot junction pressed on the tool by a modest clamping force with an insulating material interface. Tool temperature generally increased with increasing depth of cut and speed while a change of material affected the thermal history. Specific conclusions from this investigation are as listed:

- HSS tool tip temperature increased by as much as 49.35% by increasing speed by 60%.(500 to 800 rpm)
- Peak temperature attained during machining increased with depth of cut. In mild steel maximum temperature value of 115°C was attained at 2.5 mm depth of cut.
- Temperature gradually increased with time in all cases. Temperature increased from 37.8°C to a maximum of 119.7°C over a time period of 570 seconds in stainless steel machining.
- Work material mechanical strength affected temperature history with stainless steel material of higher mechanical strength than mild steel causing higher tool temperature.
- A temperature reduction from 97.9 to 36°C translating to 63.23% reduction obtains during coolant application in stainless steel machining.
- HSS tool deformation became significant during machining of stainless steel at depths of cut in excess of 1.5 mm. (A stronger cutting tool than HSS is recommended for machining stainless steel beyond 1.5 mm depth of cut.)

Acknowledgement
This work was carried out at the Production laboratory of the Faculty of Engineering and Technology, University of Ilorin, Nigeria. Special thanks to Mr Ibrahim a technologist of the central workshop in assisting with the preparation of the work specimens and setting up of the CNC machine.

References

[1] Isik, Y. (2010). An Experimental investigation on effect of cutting fluids in turning with coated carbides tool. *J. Mech. Eng*, *56*(3), 195-201.

[2] Okokpujie, I., Okonkwo, U., & Okwudibe, C. (2015). Cutting parameters effects on surface roughness during end milling of aluminium 6061 alloy under dry machining operation. *International Journal of Science and Research*, *4*(7), 2030-2036.

[3] Akhil C S, Ananthavishnu M H, Akhil C K, Afeez P M, Akhilesh R and Rahul R (2016) Measurement of cutting temperature during machining, *IOSR J. of Mechanical and Civil Eng.*, **13**(2), Ver. I, 108-22

[4] Kus, A., Isik, Y., Cakir, M. C., Coşkun, S., & Özdemir, K. (2015). Thermocouple and infrared sensor-based measurement of temperature distribution in metal cutting. *Sensors*, *15*(1), 12741291.

[5] Conradie, P. J. T., Oosthuizen, G. A., Treurnicht, N. F., & Al Shaalane, A. (2012). Overview of work piece temperature measurement techniques for machining of Ti6Al4V. *South African Journal of Industrial Engineering*, *23*(2), 116-130.

[6] Kovač, P., Mankova, I., Gostimirovic, M., Sekulic, M., & Savković, B. (2010). A REVIEW OF THE EXPERIMENTAL TECHNIQUES FOR THE MEASUREMENT OF TEMPERATURE GENERATED IN MATERIAL REMOVAL PROCESSES. *Novi Sad, 2010*, *13*(1), 1.

[7] Komanduri, R., & Hou, Z. B. (2001). A review of the experimental techniques for the measurement of heat and temperatures generated in some manufacturing processes and tribology. *Tribology International*, *34*(10), 653-682.

IOP Conf. Series: Materials Science and Engineering **413** (2018) 012004 doi:10.1088/1757-899X/413/1/012004

[8] Goyal, A., Dhiman, S., Kumar, S., & Sharma, R. (2014). A Study of Experimental Temperature Measuring Techniques used in Metal Cutting. *Jordan Journal of Mechanical & Industrial Engineering*, *8*(2).

[9] Abhang, L. B., & Hameedullah, M. (2010). Chip-tool interface temperature prediction model for turning process. *International Journal of Engineering Science and Technology*, *2*(4), 382393.

[10] Nwoke, O. N., Okonkwo, U. C., Okafor, C. E., & Okokpujie, I. P. (2017). Evaluation of Chatter Vibration Frequency in CNC Turning of 4340 Alloy Steel Material. *International Journal of Scientific & Engineering Research*, *8*(2), 487-495.

[11] Chu, T. H., & Wallbank, J. (1998). Determination of the temperature of a machined surface. *Journal of Manufacturing Science and Engineering*, *120*(2), 259-263.

[12] Ghodam, S. D. (2014). Temperature measurement of a cutting tool in turning process by using tool work thermocouple. *Int J Res Eng Technol*, *3*(04).

[13] Okonkwo, U. C., Okokpujie, I. P., Sinebe, J. E., & Ezugwu, C. A. (2015). Comparative analysis of aluminium surface roughness in end-milling under dry and minimum quantity lubrication (MQL) conditions. *Manufacturing Review*, *2*, 30.

[14] Shaw, M. C., & Cookson, J. O. (1984). *Metal cutting principles*(pp. 183-201). Oxford: Clarendon press.

[15] Kurimoto, T., Barrow, G., & Davies, B. J. (1982). The influence of aqueous fluids on the wear characteristics and life of carbide cutting tools. *CIRP Annals-Manufacturing Technology*, *31*(1), 19-23.

[16] Vieira, J. M., Machado, A. R., & Ezugwu, E. O. (2001). Performance of cutting fluids during face milling of steels. *Journal of Materials Processing Technology*, *116*(2-3), 244-251.

[17] Okokpujie, I. P., Ohunakin, O. S., Bolu, C. A., & Okokpujie, K. O. (2018). Experimental dataset for prediction of tool wear during turning of Al-1061 alloy by high speed steel cutting tools. *Data in brief*, *18*, 1196-1203.

[18] Kamruzzaman, M., & Dhar, N. R. (2009). The influence of high pressure coolant on temperature tool wear and surface finish in turning 17CrNiMo6 and 42CrMo4 steels. *J Eng Appl Sci*, *4*(6), 93-103.

[19] Okokpujie, I. P., Ajayi, O. O., Afolalu, S. A., Abioye, A. A., Salawu, E. Y., Udo, M., ... & Ikumapayi, O. M. (2018). Modeling and optimization of surface roughness in end milling of aluminium using least square approximation method and response surface methodology. *International Journal of Mechanical Engineering and Technology (IJMET)*, *9*(1), 587-600.

[20] Okokpujie, I. P., Ikumapayi, O. M., Okonkwo, U. C., Salawu, E. Y., Afolalu, S. A., Dirisu, J. O., ... & Ajayi, O. O. (2017). Experimental and Mathematical Modeling for Prediction of Tool Wear on the Machining of Aluminium 6061 Alloy by High Speed Steel Tools. *Open Engineering*, *7*(1), 461-469.

[21] Okokpujie, I. P., & Okonkwo, U. C. (2015). Effects of cutting parameters on surface roughness during end milling of aluminium under minimum quantity lubrication (MQL). International Journal of Science and Research, 4(5), 2937-2942.

[22] Okokpujie, I. P., Salawu, E. Y., Nwoke, O. N., Okonkwo, U. C., Ohijeagbon, I. O., & Okokpujie, K. (2018). Effects of Process Parameters on Vibration Frequency in Turning Operations of Perspex Material. In *Proceedings of the World Congress on Engineering* Vol. 2. Pp 700-707.

IOP Conf. Series: Materials Science and Engineering **413** (2018) 012005 doi:10.1088/1757-899X/413/1/012005

Passive Slug Attenuation Device: Potential and Operability

Ajibola T Ogunbiyi [1,3] ; Adegboyega B Ehinmowo [1,2] ; Samuel E Sanni [3] ; Hoi Yeung [1]

[1]Oil & Gas Engineering Centre, Cranfield University, Bedfordshire, MK43 0AL, United Kingdom.
[2]Department of Chemical and Petroleum Engineering, University of Lagos, Akoka, Yaba, Lagos, Nigeria.
[3]Department of Chemical Engineering, Covenant University, Canaanland, km 10, Idiroko Road, Ota, Nigeria.
Email: aehinmowo@unilag.edu.ng

Abstract Over the years, there have been some concerted efforts towards developing techniques to mitigate the threat posed by slugs to oil and gas production systems. Passive slug attenuation devices have been known to show promising potential in this regard but can generally be confronted with operability issues. In this study, experimental investigation on the hydrodynamic slug attenuation potential and the operability of a process-intensified passive slug control device- Pseudo Spiral Tube (PST) was carried out. The result showed that the device possesses the capability to partially attenuate slug flow. This was achieved with the help of the swirl flow generated within the device which helps achieve air entrainments in the liquid slug leading to reduction in the effective density of the liquid slug. The results showed that the slug severity was reduced by 24% for the case studied. However, the slug redeveloped few meters downstream the device. Therefore, in order to achieve maximum slug attenuation, the device should be installed immediately upstream the topside separator. It was also observed that the pipeline might be pigged with the device installed. However, during multiphase flows, the pig may get stuck intermittently in the device. Hence, it was conceived that a flexible pig could be more appropriate to overcome this challenge.

Keyword: Passive slug control, pigging, Pseudo spiral tube, Operability, swirl flow; multiphase flow

1 Introduction

1.1 Slug flow in multiphase pipeline-riser systems

In oil and gas industry, liquid and gas are usually transported using a single pipeline. This multiphase transport results to a number of flow assurance challenges and one of such is slug flow. Slug flow belongs to intermittent flow regime characterised by unsteady flow of liquid and gas that manifests in pressure and flow fluctuations. This intermittent behaviour has the potential for pipeline damage, separator trip off and hence, a negative impact on production. Three types of slugging are widely identified: hydrodynamic, terrain/severe, and operation-induced slugging.

Severe/Terrain induced slugging has been known to be of concern to the petroleum industry and several authors have investigated this phenomenon and proffered a number of solutions [1, 2, 3, 4, 5]. Others proposed stability criteria and some further classified severe slugging into classical severe slugging, transitional, oscillation flow etc.[2, 6, 7, 8, 9, 10, 11, 12].

Severe slug flow is known to characteristically manifest in large pressure and flow rate fluctuation leading to poor performance of separators, pipeline fatigue, and sometimes, eventual plant shutdown. It has been reported that as production activities shift to deep and ultra-deep offshore, the effect of severe slugging may become more heightened [13].

Operation-induced slug flow is the type of slugging due to various operational changes such as pigging operations, flow ramp up, system depressurization and restart. The aforementioned operations usually generate a large liquid volume flowing intermittently and received in the topside facility in form of slugs.

Hydrodynamic slug flow is another type of slug flow usually encountered in multiphase pipelines. This kind of slug is generally believed to be of high frequency and is short in length. This phenomenon has been previously studied by some authors and proposed models for its prediction [14-16]. Other researchers including *Issa et al.* [17] have also investigated the characteristics of hydrodynamic slug but only little has been done on its attenuation till date. Hydrodynamic slugging has been reported to possess the tendency to cause problems in pipeline-riser systems [1, 18, 19, 20]. There is therefore the need to investigate possible ways of attenuating it.

The control/attenuation of the first two classes of slug flow has been well researched by many authors. Control systems or devices slug attenuation capability are well documented in Literature. In 2011, *Xing* broadly classified these attenuation techniques into passive and active methods. Passive slug mitigation techniques are ''self-acting'' methods that attenuate slug flow without any external influence [11]. On the other hand, active methods require an external influencer to perform the slug attenuation function. Gas lift, subsea separation, slug catcher, choking (manual and automated) are some of the techniques currently adopted [21-23].

Many of these techniques have been deployed successfully in the industry but there are still few areas that need further attention. Gas lift/injection for example can require huge cost for implementation while the need for an appropriate model for the controller design is typically an issue for automated slug control system. Multiphase slug flow is a complex phenomenon and its modelling can be tasking. Many existing phenomenological models are simplified to be able to meet controller design requirements and usually tuned for specific conditions [24, 25], hence the predictive capability of such models are therefore limited. Also, the controllable measurements which usually include downhole and riser base measurements are not easily obtained. Controllers are usually designed to operate within a limited range (envelope), which means that the back pressure imposed by the choke might not be substantially reduced. Although recently, few studies have been done in some of these areas, there is still the need to further attend to these issues for improved oil production.

Notable progress has been made in the research on passive slug mitigation methods and some success reported from field trials conducted on these devices. For instance, various reports show that passive slug mitigation could be achieved by design modification, riser base mixer [26], slug catcher [27], the use of flow conditioner in the pipeline [22, 28, 29-32], venturi device [34], self-gas lifting method [35, 36], the bubble breaker [37] and the intermittent absorber [12]. The performance of many of these devices however, may be undermined due to operability issues.

1.2 Swirl flow and application

Swirl flow is one of the important drivers of process intensification. It has been widely applied in various industries including process engineering, automobile and environment. Some of the areas of applications include cyclone separators, turbo machinery and pollution control. In separation process, for example, swirl flow helps in drastic size reduction of separators as compared with the scenario in gravity separators [38].

Many authors have investigated various aspects of swirl flow generation, decay and applications. In 2015, *Rocha* and his team, for example, reviewed various swirl generators and revealed the need for the optimization of their geometries [39]. Some of the issues raised include their intrusive nature, maintenance, and potential increase in pressure drop due to flow reversal. Swirl flow decay has been investigated by many authors such as *Najafi et al.*, (2011) [40]. For flow applications, *Yeung and Cao,* in 2007, proposed the device investigated in this work for severe slug flow attenuation, building on the work carried out by *Adedigba* in the same year[41]. In this paper, the hydrodynamic slug attenuation potential of the device and its effect on pigging operation were investigated.

1.3 Pigging

Pigging is one of the cardinal operations in the oil and gas industry. The procedure is usually employed for pipeline inspection and maintenance. This is achieved with the help of pigs, pig launcher, pig receiver and a flow medium which provides the driving force for the pigs. There are various pig designs which could be broadly categorised as gel pigs, utility pigs and inspection pigs [42, 43]. The various types of pigs, and operation have been well documented in the work of *Cordell and Vanzant* written in 2003 [44]. The study investigated pigging operation in a passive slug attenuation device- Pseudo Spiral Tube (PST).

2.0 Experimental Study

The experimental setup for this study is shown in Figure 1. It is one of the horizontal multiphase facilities located in Cranfield University's Oil and Gas Centre. Air flowed in from the central compressors into the metering section via a needle valve. A gas turbine flow meter with a maximum scale of 60 m3/h (±1%) was used for gas flow measurement, while pressure transducers of 0-6 barg range (±0.25%) and thermocouples of 0-100 °C range (±1%) were used for the pressure and temperature measurements respectively, at the gas metering section. Water was supplied from a 4.4 m3 water tank to the experimental rig by means of a centrifugal pump with a maximum capacity of 40 m3/h at a 5 barg discharge pressure. This supply to and from the tank was done via a bypass line. An electromagnetic flow meter with a range of 0-4.524 m3/h (with a ±1% uncertainty) was used to measure the flow rate of water.

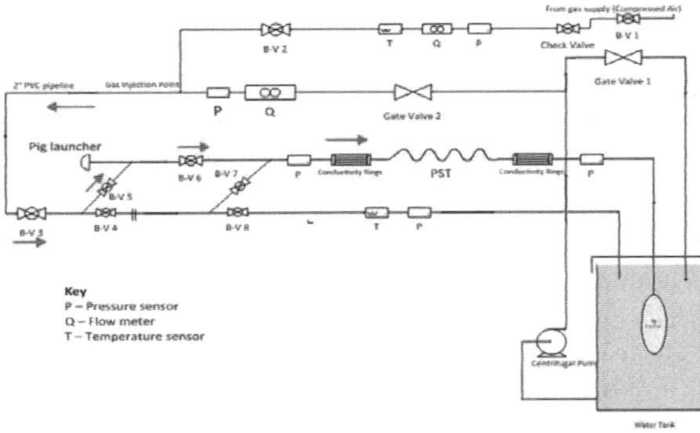

Fig. 1 Schematic of experimental set up

2.1 Experimental procedure on slug attenuation potential of the PST

The passive slug mitigation device (PST) investigated in the study constituted the test section. A number of standard piping elbows joined together in series (end to end) at a particular angle of twist between adjacent elbows, forms the PST. Hence, the geometry of PST is dependent on the elbow's internal angle, the elbow's radius to pipe diameter ratio and the angle of twist between adjacent elbows, as shown in Figures 2a and 2b.

IOP Conf. Series: Materials Science and Engineering **413** (2018) 012005 doi:10.1088/1757-899X/413/1/012005

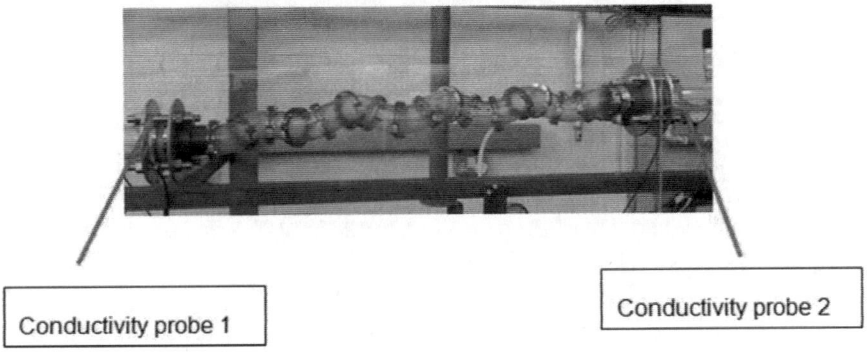

Fig. 2a Test section showing the PST and the conductivity rings

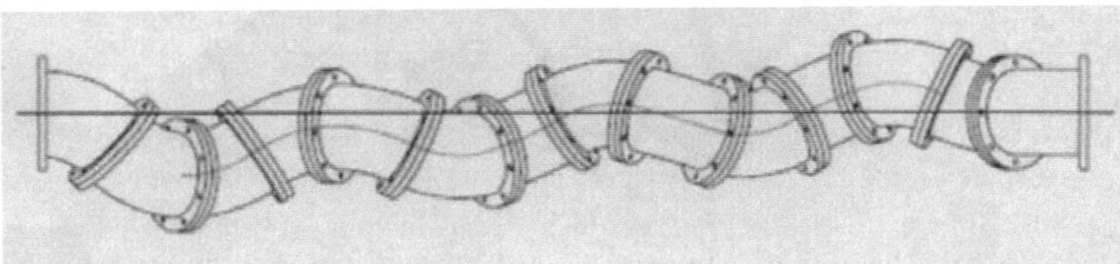

Fig. 2b Schematic diagram of the PST

The design of this device has been documented in a paper by *Yeung and Cao* [41]. The two-phase flow characteristics were observed both at upstream and downstream sections of the test section. Valuable information gathered from the experimental rig include slug attenuation potential of the PST and the generation of flow regime map. The procedure adopted for this work has been reported in the work of *Ogunbiyi* [45].

2.2Experimental procedure on pigging operation through the device

One of the objectives of this study was to investigate the operability of the proposed passive slug attenuation device. This section describes the pigging studies carried out to meet the desired objective. Figure 3 shows a section of the experimental rig and Figure 4 shows the model pigs used in this study.

Fig. 3 Pictorial view of the experimental rig

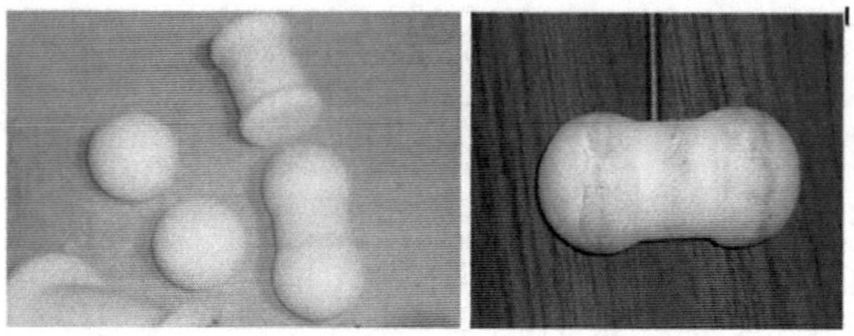

1. A collection of model pigs available 2. Double-sphered pig used in this work

Fig. 4 Model pigs used for the pigging study

The various pigs were used for preliminary investigations of the pigging potential of the passive device. After preliminary studies, the double-sphere shaped pig was observed to perform better than other types and was subsequently used in this study. The model pig had a length of about 8.6 cm with equal width of about 4.1 cm each at both ends.

The pigging operation was carried out under two conditions. The first involved driving the pig using single phase (water) flow while the second was done using two-phase air-water flow. During pigging operation, the system response was monitored using readings from the conductivity rings, visual inspection and display on the high speed camera. The liquid holdup characteristics of the flow at the entrance and exit of the PST were compared in order to understand the phenomenological changes that occurred during each run. The velocity of the pig was computed for each run and compared with the fluid's velocity which is a critical factor for slug formation. Detailed procedure of the pigging operation has been documented in *Ogunbiyi* [45].

3 **Results and discussion**

3.1 Slug flow attenuation potential of Pseudo Spiral Tube (PST)

An experimental matrix covering both slugging and non-slugging regimes was investigated with special interest in the slugging region.

The flow regimes obtained compared well with that of *Mandhane et al.* [46] as shown in Figure 5. However, few conditions existed where slug flow observed appeared on non-slug region on the *Mandhane*'s map, this could be traced to pipeline diameter and geometry difference. Representative slug flow conditions were further investigated for possible attenuation by the PST and the results are presented.

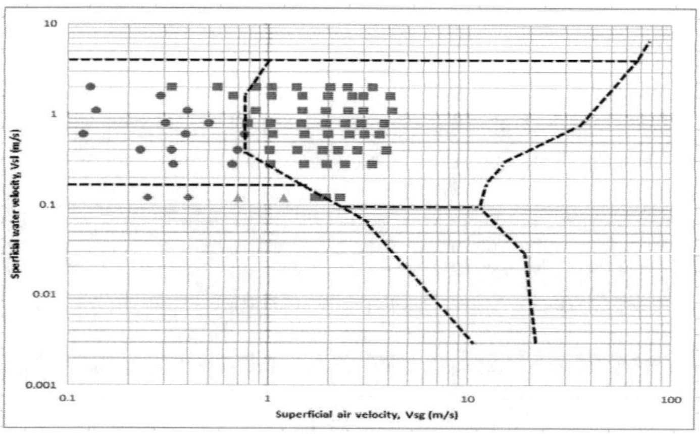

Slug flow Smooth stratified flow

IOP Conf. Series: Materials Science and Engineering **413** (2018) 012005 doi:10.1088/1757-899X/413/1/012005

● Elongated bubble flow ▲ Wavy stratified flow

Fig. 5 Comparison between the flow regime observed and that by Mandhane *et al.* (1974)

Figure 6 shows the liquid holdup response of a representative slug flow condition investigated. This was at a liquid superficial velocity of 0.99 m/s and air superficial velocity of 1.02 m/s.

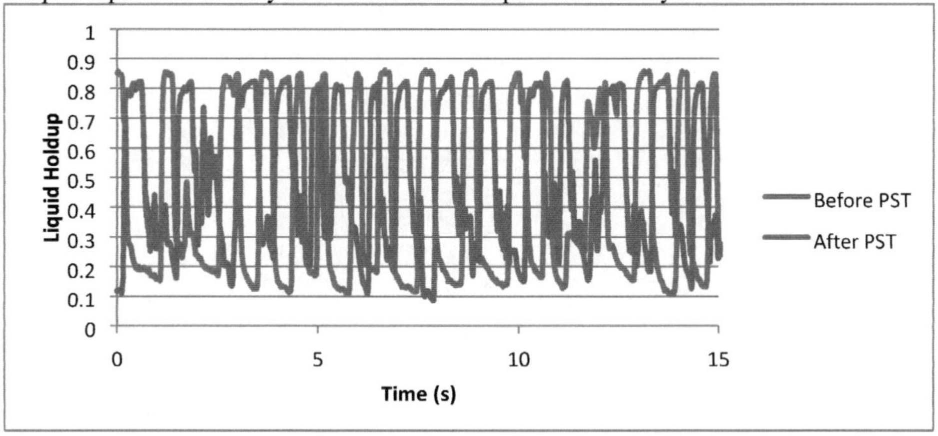

Fig. 6 Slug attenuation impact of PST at 0.99 m/s Vsl and 1.02 m/s Vsg

The liquid holdup time response and visual observation confirmed slug flow at upstream and downstream the device (PST). However, the amplitude of the slug was observed to have decreased at the downstream compared with that at the upstream section of the PST. This shows that the slug characteristics was slightly altered by the PST.

Other representative slug flow conditions investigated showed that slug flow occurred around downstream and upstream sections of the device and similar trend as shown in Figure 6 were observed. This suggests that the PST provides partial slug attenuation and not a total attenuation.

The slug attenuation (SA) [12] index can be defined for the PST as

$$SA = \left\lfloor \frac{(Magnitude\ of\ fluctuation)_{upstream} - (Magnitude\ of\ fluctuation)_{downstream}}{(Magnitude\ of\ fluctuation)_{upstream}} \right\rfloor \quad (1)$$

The magnitude of fluctuation can be estimated using simple arithmetic range or standard deviation of the data. For the case studied for example, the slug attenuation (SA) was 0.24. This implies that the intensity/severity of the slug had been reduced by 24%. This is in agreement with previous studies on passive slug attenuation device even though the PST design in this work (sixteen 45-degrees elbows with an angle of twist of 90o) is not exactly the same as that used in *Xing*'s work of 2011 [5, 11, 12]. This slug attenuation was made possible as a result of the swirl flow generated by the geometry of the PST. The swirl helped achieve air entrainments in the liquid slug body leading to reduction in the effective density of the liquid slug.

3.2 **Pigging of the PST**

The main focus here was on the ability to pig the pipeline when the PST is installed. As discussed in section 1.2, the device has previously been reported to possess potential for severe slug flow attenuation. However, it was not clear if the device would cause operational setbacks when installed on a

IOP Conf. Series: Materials Science and Engineering **413** (2018) 012005 doi:10.1088/1757-899X/413/1/012005

pipelineriser system. This section presents the results on pigging operation of the PST. Single-phase pigging of the PST.

The characterisation of the flow of the pig through the PST was done using the liquid holdup obtained at the upstream and downstream conductivity rings.

3.2.1 Single phase (Water) pigging operation through the PST

Figure 7 shows the liquid holdup plots at the upstream and downstream sections of the device (PST) using only water (single phase pigging operation) at a velocity of 0.56 m/s.

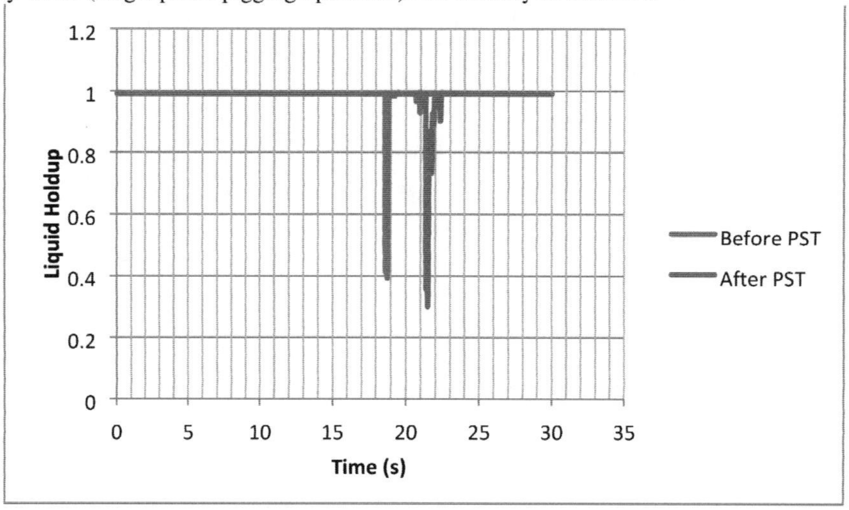

Fig. 7 Time traces of liquid holdup at a liquid superficial velocity of 0.56 m/s

Figure 7 shows that, prior to the arrival of the pig at the entrance of the PST, the pipe was fully filled with water as represented by liquid holdup of 1.0 until at 18.5 s when a sharp drop in the liquid holdup to 0.4 signalled the pig's arrival at the PST. As the pig moved past this point, there was a sharp rise in the liquid holdup back to 1. Similarly, at the second conductivity ring, the liquid hold-up again dropped from 1.0 to 0.3 at 22 s. It was observed however, that the liquid holdup did not drop to zero. This was as a result of the water surrounding the pig body and may also be traced to leakage between the pig and the pipe wall. For this case, the pig was observed to spend 4.06 s in the PST.

The results above show a similar trend with experimental studies for water superficial velocities of 0.80, 1.06, 1.28, 1.47, 1.58, 2.11 and 2.37 m/s and summary of the results are shown in Table 1. The liquid holdup time response taken at upstream and downstream of the device at these liquid superficial velocities were all similar to that shown in Figure 7.

Table 1 Variation of average pig velocity with pig's residence time in PST

Superficial Water velocity, Vsl (m/s)	Time at Ring 1 (s)	Time at Ring 2 (s)	Resident Time in PST (s)	Length between Ring 1 to Ring 2 (m)	Velocity of Pig (m/s)
0.56	17.88	21.94	4.06	1.21	0.30
0.80	15.68	17.82	2.14	1.21	0.57
1.06	12.89	14.25	1.36	1.21	0.85

IOP Conf. Series: Materials Science and Engineering **413** (2018) 012005 doi:10.1088/1757-899X/413/1/012005

1.28	12.07	13.50	1.43	1.21	0.85
1.47	10.94	12.04	1.10	1.21	1.1
1.58	10.64	11.74	1.10	1.21	1.1
2.11	9.03	9.90	0.87	1.21	1.39
2.37	9.69	10.32	0.63	1.21	1.92

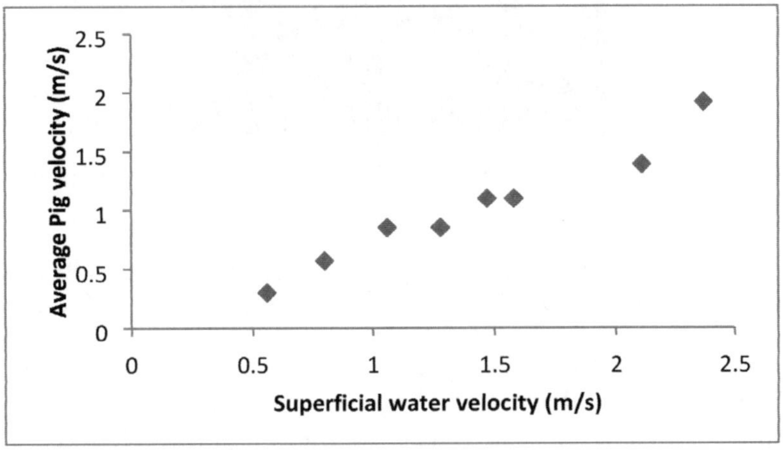

Fig. 8 Variation of average pig velocity with superficial water velocity

Table 1 presents the summary of the results of eight (8) experimental studies on single phase pigging. An interesting observation here was the relationship between the pig's and superficial liquid velocities. Expectedly, the pig was to travel at the liquid superficial velocity but this was not the case; instead, it travelled at a velocity less than this. This discrepancy could be traceable to resistances in terms of drag force caused by the seepage of water through narrow clearances between the pig and pipe walls, and friction between the pig and the internal walls of the pipeline. Figure 8 shows the plot of average pig's velocity against the superficial velocity of the carrier fluid. The pig velocity increased with the liquid superficial velocity in a near linear manner.

3.2.2 Two- phase (Air-Water) pigging through the PST

Figure 9 shows the progress of a pig downstream the device (PST) when (Vsl, Vsg) = (0.80 m/s, 0.27 m/s). Here, air-water, two-phase flow provided the driving force for the pig. Downstream the pig, a bubbly flow (bubble-laden water) regime was observed. At lower velocities, little or no liquid was observed to flow behind the pig while travelling through the second conductivity ring leading to a zero or near zero liquid holdup as shown in Figure 9(c) and (d). It was also observed that sufficient air flow temporarily transports the pig through before the arrival of the new cycle of liquid flow. The intermittency is a known characteristics of slug flow.

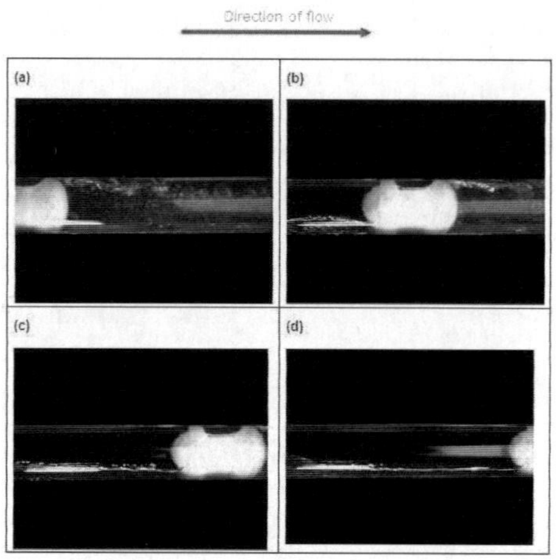

Fig. 9 Movement of a pig downstream the PST at (Vsl, Vsg) = (0.80 m/s, 0.27 m/s)

The upstream and downstream liquid hold up time response are shown in Figure 10. Elongated bubble flow was initially observed before point A where operational induced slug flow ensued. At point B, a liquid holdup of zero was recorded for about 8 s which marked the time when the pig was at the first conductivity ring. A rise in the liquid holdup was observed at point D as liquid re-entered the device (PST). The pig was observed to exit the device at Point C and a period of no liquid flow was observed as shown between C and D. This intermittent flow of liquid and gas flow is primarily a characteristics of slug flow as observed for an operation (pigging) induced slug flow. It was observed that the pig's resident time (time between points B and C) in the PST was much smaller than that spent during the single-phase pigging operation. This can be explained by the difference in magnitude between the mixture velocity and single phase flow velocity. At sufficiently high flow rates, turbulence is expected to enhance the flow of pig. Similar trends/results were observed for other conditions investigated but not reported in this paper. Thus, the interaction between the three-phases - water, air and pig - played an important role in determining the flow behaviour at the entry point of the pig into the PST. Different flow patterns were observed for various flow conditions.

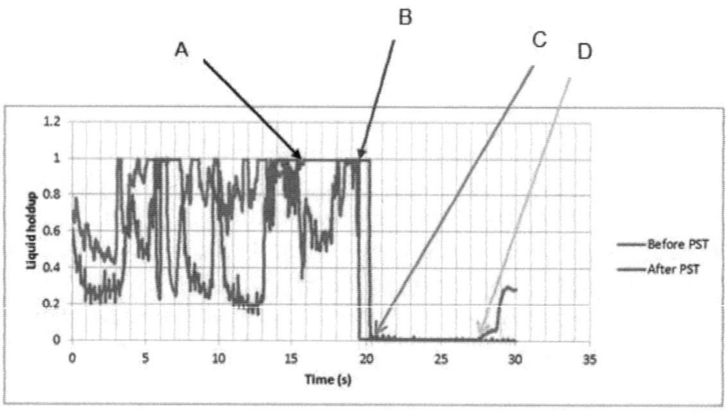

Fig. 10 Variation of liquid holdup with time at (Vsl, Vsg) = (0.37 m/s, 0.39 m/s)

The variation in the average pig velocity with the mixture velocity at a liquid velocity of 0.27 m/s and air superficial velocities of 0.94, 1.32 and 1.93 m/s, was studied and the result is as shown in Figure 11. Unlike in Figure 8, no clear-cut relationship could be established between the average pig velocity and mixture velocities. The average pig velocity increased from 0.98 m/s to 2.09 m/s as the mixture velocity increased from 1.21 to 1.59 m/s, and then dropped from 2.08 m/s to 0.44 m/s as the mixture velocity increased from 1.59 to 2.2 m/s. This abnormal trend was a result of cases of stuck pigs inside the PST at low liquid velocities.

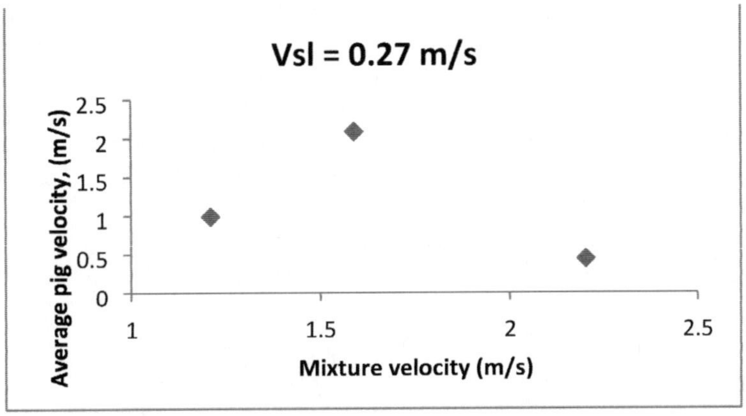

Figure 11: Variation of pig's average velocity with air-water velocity for Vsl 0.27 m/s

However, at a higher liquid velocity of 0.8 m/s over a range of air superficial velocities, there is a clear, linear relationship between the pig's velocity and the air-water mixture velocity as seen in Figure 12. This observed positive correlation between the pig's velocity and the driving fluid velocity at this velocity follows the expected trend.

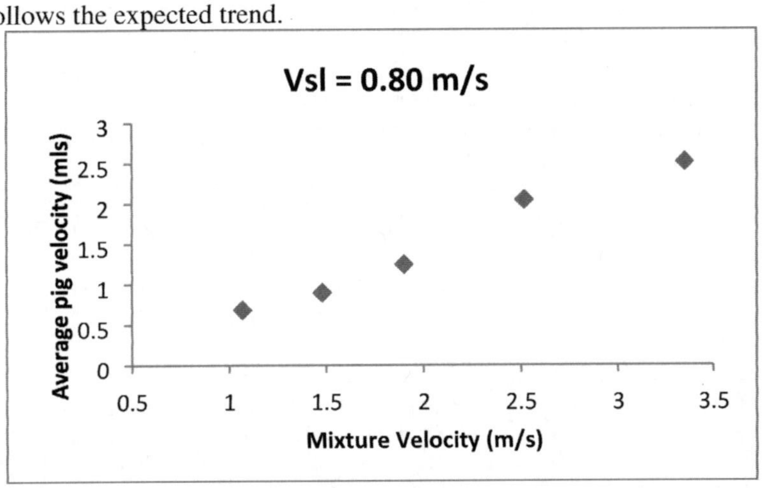

Fig. 12 Variation of pig's average velocity with air-water mixture velocity for Vsl 0.80 m/s

From the results, it can be said that pigging operations show more success at high fluid velocity than at low fluid velocity. This suggests that at these flow conditions (usually very low liquid flow rates), a smooth pigging operation of the device cannot be achieved.

The observed positive correlation between the pig velocity and the driving fluid velocity for the single-phase pigging seems to follow common logic. However, this was not the case for various flow conditions investigated for the two-phase pigging, which was as a result of the occasional and temporal cases of stuck pigs inside the PST, which the oncoming fluid was able to drive out.

4 Conclusion

The hydrodynamic slug attenuation potential of a passive slug attenuation device (PST) has been investigated. The device was observed to partially attenuate slug flow. However, the slugs were observed to redevelop downstream the device. It is therefore necessary to install the device at the immediate upstream end of critical equipment such as first stage separator for effective slug attenuation.

The single-phase pigging operations were successful. However, at low flow rates, the pig was observed to get stuck in the PST. This scenario of stuck pigs was more pronounced during two-phase flow pigging at low liquid rates. Pigging the PST device may be easier for single phase flows relative to multiphase flows. If multiphase pigging must be done, high liquid velocities should be used. In both single-phase and two-phase (air-water) pigging operations, the average pig velocity was observed to be lesser than the liquid or mixture velocity of the driving fluid. The discrepancy was attributed to the resistances due to drag force and frictional forces of the pig on the transporting fluid and wall pipeline.

Foam pig was observed to perform better than other type of pigs investigated in this study. A further work on the efficiency of the use of flexible pigs for smooth pigging operation of the PST device is therefore proposed.

Acknowledgement

The Authors thank Petroleum Technology Development Fund (PTDF) and Niger Delta Development Commission (NDDC), Nigeria, for wholly sponsoring the M.Sc programme of Ajibola Ogunbiyi and supporting the PhD programmes of Adegboyega Ehinmowo respectively.

References

[1] Schmidt, Z., Doty, D. & Dutta-Roy, K. (1985). Severe slugging in offshore pipeline riser-pipe systems. *Old SPE Journal*, 25 (1): 27–38.

[2] Malekzadeh, R., Henkes, R. & Mudde, R.F., (2012). Severe slugging in a long pipeline–riser system: Experiments and predictions. *International Journal of Multiphase Flow*, 46, pp.9–21.

[3] Zheng, G., Brill, J.P. & Taitel, Y. (1994). Slug flow behavior in a hilly terrain pipeline. *International Journal of Multiphase Flow*, 20 (1): 63–79.

[4] Tengesdal, J.O., Thompson, L. & Sarica, C., A (2003). Design Approach for Self-Lifting" Method to Eliminate Severe Slugging in Offshore Productino Systems". In *SPE Annual Technical Comferece and Exhibition*. Denver, Colorado, USA.

[5] Sarica, C., Begen, P., Pereyra, E. and Kouba, G. (2014). Feasibility of surfactants as severe slugging suppression agents. In *9th North American Conference on Multiphase Technology*. Banff, Canada: BHR Group.

[6] Bøe, A. (1981). Severe slugging characteristics; part i: flow regime for severe slugging; part ii: Point model simulation study. *Selected Topics in Two-Phase Flow, NTH, Trondheim, Norway*

[7] Taitel, Y. (1986). Stability of severe slugging. *International Journal of Multiphase Flow*, 12 (2): 203–217.

[8] Pots, B., Bromilow, I. & Konijn, M. (1987). Severe slug flow in offshore flowline/riser systems. *SPE Production Engineering*, 2 (4): 319–324.

[9] Tin, V. (1991). Severe slugging in flexible risers. In *International Conference on multiphase production*. Cannes, France, BHRG, p. 507.

[10] Tin, V. & Sarshar, M.M. (1993). An investigation of severe slugging characteristics in flexible risers. In *Proc. 6th Int. Conf. on Multiphase Production, BHRG, France*. p. 228.

[11] Xing, L. (2011). *Passive slug mitigation by applying wavy pipes*. PhD Thesis. Cranfield University, Bedford, UK.

[12] Ehinmowo, A.B. (2015). *Stabilising slug flow at large valve opening using an intermittent absorber*. PhD Thesis, Cranfield University, UK.

[13] Pickering, P.F., Hewitt, G.F., Watson, M.J. and Hale C.P. (2001). The prediction of flows in production risers-truth & myth. In *Institute of International Research (IIR) Conference*. Aberdeen, UK, p. 1.

[14] Issa, R.I. & Woodburn, P.J. (1998). Numerical prediction of instabilities and slug formation in horizontal two-phase flows. In *3rd International Conference on Multiphase Flow, ICMF98*. Lyon, France.

[15] Issa, R.I. & Kempf, M.H.W. (2003). Simulation of slug flow in horizontal and nearly horizontal pipes with the two-fluid model. *International Journal of Multiphase Flow*, 29 (1): 69–95.

[16] Danielson, T. (2011). A simple model for hydrodynamic slug flow. In *Offshore Technology Conference*. Houston, Texas, USA.

[17] Issa, R.I., Barbeau, S., Hale, C.P., Odozi, U., Hewitt, G.F., Richardson, S.M. and Wong, W.L. (2007). Measurement and prediction of slug characteristics in three-phase flows. In *13th International Conference on Multiphase Production Technology*. Edinburgh, UK: BHR Group.

[18] Brill, J., Zelimir, S., Coberly, W, Herring, J and Moore, D. (1981). Analysis of two-phase tests in largediameter flow lines in Prudhoe Bay field. *Old SPE Journal*, 21 (3): 363–378.

[19] Guzman Vazquez, E. & Fairuzov, Y. (2009). A study of normal slug flow in an offshore production facility with a large diameter flowline. *SPE Production & Operations*, 24 (1): 171–179.

[20] Danielson, T., Bansal K.M., Djoric, B., Larrey D., Johansen, S.T., De Leebeeck, A. and Kjølaas J. (2012). Simulation of Slug Flow in Oil and Gas Pipelines Using a New Transient Simulator. In *Offshore Technology Conference*. Houston, Texas, USA: SPE.

[21] Ogazi, A., Cao, Y., Yeung, H. and Lao, L. (2010). Slug control with large valve openings to maximize oil production. *SPE Journal*, 15 (3): 812–821.

[22] Xing, L., Yeung H., Shen J. and Cao, Y. (2013a). A new flow conditioner for mitigating severe slugging in pipeline/riser system. *International Journal of Multiphase Flow*, 51: 65–72.

[23] Ehinmowo, A.B. & Cao, Y. (2015). Stabilizing slug flow at large valve opening using active feedback control. In *Proceedings of the 21st International Conference on Automation & Computing*. University of Strathclyde, Glasgow, UK.

[24] Storkaas, E. & Skogestad, S. (2003). Cascade control of unstable systems with application to stabilization of slug flow. In *International Symposium on Advanced Control of Chemical Processes AdChem'03*. Hong Kong.

[25] Jahanshahi, E. & Skogestad, S. (2011). Simplified dynamical models for control of severe slugging in multiphase risers. In 18th IFAC *World Congress*, Milano, Italy, pp. 1634–1639.

[26] Yocum, B.T. (1973). Offshore riser slug flow avoidance: mathematical models for design and optimization. In *SPE European Meeting*. London, England.

[27] Miyoshi, M., Doty, D. R. and Schmidt, Z. (1988) 'Slug-catcher design for dynamic slugging in an offshore production facility', *SPE production engineering*. Society of Petroleum Engineers, 3(4), pp. 563–573.

[28] Xing, L., Yeung H., Shen J. and Cao, Y. (2013b). Experimental study on severe slugging mitigation by applying wavy pipes. *Chemical Engineering Research and Design*, 91 (1): 18–28.

[29] Makogan, T. and Brook, G. (2007) 'Device for controlling slugging', *Device for controlling slugging*. USA: WO Patent 2,007,034,142.

[30] Makogon, T., Estanga, D. and Sarica, C. (2011) 'A new passive technique for severe slugging attenuation', in *15th International Conference on Multiphase Production Technology*. Cannes, France: BHR Group.

[31] Adedigba, A.G., Badr, O., Yeung H., Waring, R. and Caro, C. (2006). The use of novel pipe device to reduce severe slugging in pipeline-riser systems. In *5th North American Conference on Multiphase Technology*. Banff, Canada: BHR Group.

[32] Brasjen, B. J. *et al.* (2013) 'Experimental investigation of terrain slugging formation ,evolution and potential for mitigation', in *16th International Conference on Multiphase Production Technology*. Cannes, France: BHR Group.

[33] Adedigba, A. G. (2007) *Two-phase flow of gas-liquid mixtures in horizontal helical pipes*. PhD Thesis. Cranfield University, Bedford, United Kingdom

[34] Almeida, A. R. and Goncalves A.L., M. (1999) 'Venturi for severe slug elimination', in *BHR Group Conference Series Publication*. Bury St. Edmunds; Professional Engineering Publishing; 1999, pp. 149 160.

[35] Sarica, C. and Tengesdal, J. Ø. (2000) 'A new technique to eliminate severe slugging in pipeline/riser systems', in *SPE Annual Technical Conference and Exhibition*. Dallas, Texas, USA.

[36] Tengesdal, J. Ø. (2002) *Investigation of self-lifting concept for severe slugging elimination in deepwater pipeline/riser systems*. Pennsylvania State University.

[37] Schrama, E. and Fernandes, R. (2005) 'The bubble breaker: Breaking up slug flow Into dispersed bubbly flow using a passive mechanical device', in 12th International Conferenc on Multiphase Production Technology. Barcelonna, Spain: BHR Group.

[38] Ehinmowo, A.B, Aliyu, A.M., Alagbe, S.O., Adedipe O. and Olumi O. (2015). Offshore Topside Rotating Packed Bed as Process Intensified Alternative for Natural Gas Sweetening and Dehydration. *American Chemical Science Journal*, 8 (4): 1–12.

[39] Rocha, A.D., Bannwart, A.C. & Ganzarolli, M.M. (2015). Numerical and experimental study of an axially induced swirling pipe flow. *International Journal of Heat and Fluid Flow*, 53, pp.81–90.

IOP Conf. Series: Materials Science and Engineering **413** (2018) 012005 doi:10.1088/1757-899X/413/1/012005

[40] Najafi, A. F., Mousavian, S. M. and Amini, K. (2011) 'Numerical investigations on swirl intensity decay rate for turbulent swirling flow in a fixed pipe', *International Journal of Mechanical Sciences*, 53(10), pp. 801–811. doi: 10.1016/j.ijmecsci.2011.06.011.

[41] Yeung, H. & Cao, Y. (2007). *Effects of Geometry at the Riser Base on Riser Hydrodynamic Behaviour*, Bedford, UK.

[42] Tiratsoo, J.N.H. (1991). *Pipeline pigging and integrity technology* 2nd editio., Beaconsfield.

[43] Cordell, J and Vanzant, H. (2003). *The pipeline pigging handbook* 3rd ed., Houston Texas, USA: Clarion Technical Publishers.

[44] Nwoke, O. N., Okokpujie, I. P., & Ekenyem, S. C. (2017). Investigation of Creep Responses of Selected Engineering Materials. *Journal of Science, Engineering Development, Environmen and Technology (JOSEDET)*, *7*(1), 1-15.

[45] Ogunbiyi, A.T. (2012). *Pigging of a passive slug mitigation device*. MSc Thesis, Cranfield University, UK.

[46] Mandhane, J.M., Gregory, G.A. & Aziz, K., 1974. A flow pattern map for gas—liquid flow in horizontal pipes. *International Journal of Multiphase Flow*, 1 (4): 537–553.

Production and Quality Analysis of Evaporating Dish Using Local Materials

J.A. Adejo and S.A. Lawal*

Department of Mechanical Engineering, School of Infrastructure, Process and Engineering Technology, Federal University of Technology, Minna. Nigeria
Corresponding author email: lawalsunday@futminna.edu.ng

Abstract- Evaporating dish was produced using raw materials such as kaolin, quartz, ball clay and talc that were sourced from Agaie, Kutigi, Kagara and Kadna respectively in Niger state. A control sample was produced using 30% feldspar, 5% quartz, 27% kaolin, 37% plastic clay and 1% talc, while all other samples were formulated by varying the compositions to compare with the control sample. The milled and mixed samples were casted and fired at a temperature of 1250°C. Physical examinations such as linear shrinkage, bulk density, water absorption was carried out on all the samples produced, while mechanical test such as failing load was also carried out on all the samples. The outcome of all the tests shows that the locally manufactured evaporating dish satisfied the qualities required of an evaporating dish for laboratory use.
Keywords: Shrinkage; bulk density; temperature; kaolin; plastic clay.

1. Introduction

The development of a nation depends on the quality of its research output. Nigeria is no doubt an oil rich nation and has solely relied on this resource as her economic backbone to the utter neglect of all other huge naturally endowed mineral resources that abound almost everywhere in the country. The development of ceramic raw materials particularly clay minerals in Nigeria and their applications in manufacturing and production of scientific equipment such as evaporating dish will not only boost the nation's economy, but will enhance her technological development, changes the country's status from consumer to producer nation and create employment for the teeming unemployed youths. The production of evaporating dish which is used for gravimetric analysis (determination of the amount of ion being analysed through the measurement of mass) will aid in improving research and save the country foreign exchange which would have been used in importation. There is constant search for new and relevant sources of raw materials to enhance the production of evaporating dish and other scientific products, reduce the cost of production and improve the quality of the products. The future of production of evaporating dish will depend largely on the national will and determination to succeed. This would be possible if the government facilitate the appropriate channel of establishing man power that can contribute positively in the processing of local raw materials to produce evaporating dish [1].

The exploitation, processing of raw materials and production of evaporating dish can influence the development and capacity building among the local miners and other professionals involved in processing of these raw materials. These include the development of ceramic materials for research purposes as well as for production in industries. The production of quality evaporating dish will enhance quality science education in schools, quality research works and generation of much needed foreign exchange for the country. In Nigeria, most of the raw materials for development of evaporating dish are available in most of the states of the federation. These materials are found in raw state and cannot be used directly for production of evaporating dish.

The lack of milling facilities has hindered research and production of evaporating dish. The present approach to the production of evaporating dish started as experiments to test the locally available raw materials and their suitability for the production of evaporating dish. Therefore, to evaluate raw material composition on the quality of evaporating dish and identify local raw materials for processing and production. Evaporating dish is a small bowl with a spout, usually made of porcelain or borosilicate glass. As its name suggests, it is commonly used to evaporate solvents in a sample. The shape of the evaporating dish makes evaporation easier due to a large surface area with which the solvent is exposed to the atmosphere [2]. Ajakor et al. [3], indigenously produced electrical porcelain from Nigeria mineral. In their study, universal composition of triaxial porcelain (kaolin, 45%; ball clay, 5%; feldspar, 25% and quartz, 25%) were used to produce a test piece. The pulverized and thoroughly mixed composition was compacted and fired at a temperature of 1200°C and the physical and electrical properties were investigated. The results showed that the sample has a breakdown voltage of 26kV/mm, low water absorption and dielectric constant between 9.0 and 10.9 within the frequency range considered. Egbai [4], carried out kaolin quantification in Ukwu Nzu and Ubulu Uku using electrical resistivity method. Ito et, al [5], determined the swelling and shrinkage properties of undisturbed expansive soil. The desire to produce quality evaporating dish for both local consumption and export is what prompted this study.

However, this will help to ignite the Nigeria industrial revolution of the production and good quality evaporating dish using local materials, which is in line with the study carried out by Azeta et al [11] and Salawu et al [12].

2. Materials and Methods

2.1 Materials

The materials and equipment used in this study include digital and mechanical weighing scale, ball mill, electric vibrating sieve, roller pot mill, blunger, spraying gun, moulds, storage tanks/stirrers and the electric kiln. The raw materials for slip preparation like plastic clay, kaolin, feldspar, quartz and talc were beneficiated, milled and used for casting, after which the products were fettled, dried, glazed and fired in the kiln for vitrification to take place.

2.2 Methods

2.2.1 Raw materials preparation

The raw materials were weighed in different proportions using a digital weighing scale of maximum capacity of 30 kilograms. The raw materials were analysed by means of X-Ray Fluorescent Photometer (XRF) method at National Geological Survey Agency, Kaduna, Nigeria.

2.2.2 Beneficiation of raw materials

The raw materials, plastic clay, kaolin, feldspar, quartz, and talc were beneficiated. This involved series of processes whereby the chemical and physical properties and characteristics of raw materials are modified to render the raw material easier to process. The feldspar was milled for four hours in the ball mill and then sieved using an electric vibrating sieve of 300 mesh size. 50kg of feldspar was collected. Also, 50kg of quartz was collected from Kadna in Bosso local government area of Niger state. They were broken into chips with the use of metal hammer and then ball milled. The ball milling lasted for six (6) hours. The milled quartz was then sieved using 300 mesh. The kaolin used in this study was milled in the ball mill and then sieved using the vibrating sieve of 300 mesh and sun dried while the plastic clay was milled using the ball mill and it serves as a binder for the other materials.

IOP Conf. Series: Materials Science and Engineering **413** (2018) 012006 doi:10.1088/1757-899X/413/1/012006

Finally, the talc was beneficiated using the jaw crusher and then milled using the ball mill. The vibrating sieve of 300 mesh was used for sieving and the materials were weighed after beneficiation.

2.2.3 Mould making

Plaster mould casting technique was used to form the mould for the casting. All measurements of the profile have to allow 10% for an element of shrinkage during production. From the profile a model of the actual shape was made, the techniques used in making the plaster mould to replicate a model were in accordance with the procedure outlined by Okpanachi, et al [6]. Mould soap was used to prevent the plaster from sticking to the model. Place the model into the bottom of the container and carefully seal the edges around the model with plastic clay so the plaster cannot run under the model. The liquid was then mixed plaster over the model to obtain at least 1inch plasters thickness all-round the model. Once the plaster was set, it was removed from the model and the mixed liquid plaster into a container and observed carefully as it sets. At the point that the plaster begins to become firm, the model was pressed into the surface and held for a little while as the mix hardens. Once the mould is set, the model was removed carefully so as not to disturb details that remains. Also, before the plaster was prepared, a mould box was built by making a box comprising of four sides held together with external cord on a flat bottom base. The sides and bottom joints of the box were sealed with ordinary plastic clay and model placed upside down on the box base, sealing the edges with clay and cover the outside tightly with soap. Finally, the plaster was then prepared by pouring it over the model carefully until the mould box is filled. The mould was thoroughly dried before use. Spraying gun was then used in applying a coating paint, ink, glaze, through the air onto a surface

2.2.4 Test blends for slip production

The test blends used for slip production is shown in Table 1

Table 1: Test blends of materials

S/N	Feldspar	Quartz	Kaolin	Plastic clay	Talc
1	0	0	41.77	57.23	1
2	5	1	39.23	53.77	1
3	10	2	36.7	50.3	1
4	15	3	34.17	46.83	1
5	20	4	31.64	43.36	1
6	25	5	29.11	39.89	1
7	30	6	26.58	36.42	1
8	35	7	24.05	32.95	1
9	40	8	21.52	29.48	1
10	30	5	27.00	37.00	1

2.2.5 Determination of suitable firing temperature

Evaporating dishes (Fig. 1a) produced were used for testing the firing temperature of the glazes. These products were glazed and fired at the temperature ranges of 1106^{0}C, 1156^{0}C, 1206^{0}C and 1256^{0}C.

IOP Conf. Series: Materials Science and Engineering **413** (2018) 012006 doi:10.1088/1757-899X/413/1/012006

a. Slips casting of evaporating dish

b. Evaporating dish ready to be glazed

c. Spraying gun for application of glaze

Fig. 1: Experimental Set up

2.2.6 Determination of densification parameters

The densification parameters of the evaporating dish were obtained after firing in the firing kiln. The densification parameters of the fired piece were obtained by measuring the linear shrinkage, bulk density and water absorption capacity and failing load was obtained after the ASTM C20 – 00 FORM. The linear shrinkage was determined using the Equ. 1 – 3 [7; 8; 9]. The boiling method was used in water absorption test at 100^0C for 2 hours and was calculated using Equ. 4.

$$\text{Drying shrinkage} = \frac{L_W - L_d}{L_d} X \ 100 \quad (1)$$

$$\text{Fired shrinkage} = \frac{L_W - L_f}{L_d} X \ 100\% \quad (2)$$

$$\text{Total shrinkage} = \frac{L_W - L_f}{L_W} X \ 100\% \quad (3)$$

Where L_W = Wet length, L_F = Fired length and L_D = Dry length.

$$\text{Water Absorption} = \frac{W_s - W_d}{W_d} X \ 100\% \quad (4)$$

Where W_S = Soaked weight after boiling for 100^0C for 2 hours and W_d = Dry weight.

The strength of the porcelain insulators was investigated by determining their failing load according to ASTM, [8]. The insulators were coupled on the tensile machine and allowed to be loaded up to 20KN until failure occurred. The force at which the materials failed were recorded. The failed load was determined using Equ. 5. [10]. The Loss on Ignition (LO1) was calculated using Equ. 6. Finally, the percentage residue was determined using Equ.7;

$$\sigma_T = \frac{p}{A} \quad (5)$$

where $A = b * t$ and $\rho = \sigma_T(b * t)$.

Where σ_T = tensile strength, p = Maximum load, b = width and t = thickness.

$$L.O.I = \left(\frac{m_2 - m_3}{m_2 - m_1}\right) x\ 100 \quad (6)$$

Where m_1 is the mass of the porcelain, m_2 is the mass of crucible and sample weighed while m3 is mass of the crucible and its contents cooled in a desiccator.

$$\%Residue = \frac{Dry\ Weigth\ x\ 100}{A} \quad (7)$$

Where A, is the mass of the sample obtained using; A = Litre. Weight – 1000dm^3 x 1.625.

3. Results and Discussions

3.1 Chemical and percentage compositions of constituent and Evaporating Dish

The percentage composition of the evaporating dish is shown in Table 2 and it was observed that, specimen 10 served as a control specimen and the percentage composition of kaolin as well as plastic clay for other specimens were calculated based on the percentage of composition of the clay (plastic and ball clay) and the percentage composition of others were calculated thereafter. The XRD results are shown in Table 3.

Table 2: Constituent percentage compositions of Evaporating Dish

Specimen	Feldspar (%)	Quartz (%)	Kaolin (%)	Plastic Clay (%)	Talc (%)
Specimen	%	%	%	%	%
1	0	0	41.77	57.23	1
2	5	1	39.23	53.77	1
3	10	2	36.70	50.30	1
4	15	3	34.17	46.83	1
5	20	4	31.64	43.36	1
6	25	5	29.11	39.89	1
7	30	6	26.58	36.42	1
8	35	7	24.05	32.95	1
9	40	8	21.52	29.48	1
10	30	5	27.00	37.00	1

Table 3: Chemical characterization of the materials

Oxide Composition %	Feldspar	Quartz	Kaolin	Plastic Clay	Talc
SiO_2	68.4	94.5	66.7	58.75	52.9
Al_2O_3	17.9	0.06	23.71	27.6	17.38
SO_3	ND	3.07	ND	1.2	ND
K_2O	9.33	0.39	0.31	0.38	0.31
CaO	0.19	0.6	0.8	0.16	2.1
Fe_2O_3	0.73	0.422	0.478	0.637	8.4
MgO	0.16	ND	0.08	ND	3.73

SrO	0.032	ND	0.024	0.014	0.088
PbO	0.017	0.015	ND	0.019	ND
BaO	0.065	0.03	ND	0.08	ND
MnO	0.08	0.025	0.005	0.018	0.37
TiO_2	0.12	0.11	0.133	0.11	0.559
V_2O_5	0.01	ND	ND	0.035	0.02
Cr_2O_3	ND	0.01	ND	0.032	0.772
CuO	0.02	0.029	0.016	0.024	0.077
Cl	ND	ND	ND	0.604	ND
ZnO	ND	0.055	0.002	0.004	0.01
Na_2O	2.13	ND	0.23	ND	0.02
ZrO_2	0.022	0.054	0.02	0.171	ND
Ta_2O_5	ND	ND	ND	ND	0.07
NiO	ND	ND	ND	ND	1.28
L O I	0.73	0.63	7.49	10.16	11.91

ND = NOT DETECTED

From Table 3, it can be observed that the silica content of the clay varies between 58.75% and alumina content of about 27.6% and this therefore shows that is type of clay can be classified as aluminasilicate type of clay [13]. The entire chemical compositions are in qualitative agreement with the requirements of ceramic wares [14]. Quartz with a percentage silica content of 94.5% is highly desirable since it serve as the refractory skeleton and it is usually desirable at a composition of nearly 100%.

3.2 Physical test carried out on the produced specimen

The results of the physical test conducted on each of the specimen is presented in the Table 4, which shows that the values of total shrinkage for all the specimen is within the recommended values for evaporating dish wares [13], though, the shrinkage values for specimen 3, 4 and 9 are a bit high. In the case of specimen 9, it is as a result of the high fluxing i.e. feldspar present in the charged materials as well as high amount of quartz resulting to high liquid flow during firing and consequently. Table 4: Physical properties of the produced Evaporating Dish

Properties	3	4	5	6	7	8	9	10
Total Shrinkage (%)	7.3	6.9	3.52	3.7	3.2	3.07	8.6	2.9
Water Absorption (%)	4.30	3.71	2.68	2.69	2.65	2.63	4.78	2.61
Bulk Density (g/cm3)	1.21	1.23	1.24	1.27	1.30	1.31	1.35	1.28
Failing Load (kN)	0.43	0.67	0.93	0.95	0.95	0.97	0.62	1.10

Also, from Table 4, it can be seen that the water absorption is higher for specimen 3 and specimen 9. In the case of specimen 3, it is as a result of too small feldspar content i.e. 10%, leads to immature evaporating dish formation and high porosity which consequently leads to more water absorption [1516], while in the case of specimen 9, the water absorption was high as a result of too much feldspar i.e. 40%, as it also allows for reaction with impurities to form slag which are burnt off at elevated temperatures during firing, hence creation of more porosity which automatically results in higher water absorption this also result in thermal shock upon cooling. Similarly, as shown in Table 4, the bulk density experienced slight reduction as with increasing amount of kaolin and reduction in feldspar. This phenomenon was exhibited as a result of kaolin's higher tendency for loss of ignition compared to feldspar. From, Table 4, the failing load ranges between 0.43kN to 1.10kN which is in accordance with the standard requirement of an evaporating dish wares [17].

4. Conclusions

Kaolin, ball clay, feldspar, quartz and talc, were sourced locally from Niger state province (Agai, Kutigi, and Kagara and Kaduna state). These materials were then formulated in different compositions and then used in the production of evaporating dish. From this study, it can be concluded that;

i. The control Specimen (Specimen 10) possesses the optimal properties. Thereafter, formulated specimens (specimen 8) with composition 24.05% kaolin, 32.95% plastic clay, 35% feldspar, 7% quartz and 1% talc, was found to possess the highest failing load of about 8.0kN, corresponding to water absorption 2.63%, bulk density 1.31g/cm^3 and total shrinkage of 3.07%.

ii. The raw material used is good for a successful production of high quality evaporating dish.

iii. And among the produced evaporating dish, specimen 4 to 8 possesses the best constituent formulation.

References

[1] Onaji, P. B. and Usman M, (1988), Development of Slip Cast Electrical Porcelain Body: Bomo Plastic Clay, *Nigeria Journal of Engineering* (NJE), 5, (.2), 89-96.

[2] Nisra (2006), [online], Accessed 22nd November 2012, Available from World Wide Web: http://study.com/academy/lesson/evaporating-dish-definition-functions-quiz.html

[3] Ajakor E.M, Anih, L.U, Ogwata C.M (2015), Indigenous production of Electrical porcelain from Nigerian Mineral. *International Journal of Scientific and Research Publications.* 5:6

[4] Egbai, J.C (2013). Kaolin Quantification in Ukwu-nzu and Ubulu-Uku using Electrical Resistivity Method. *IJERS*, 14, (3).

[5] Ito M, Azam S (2009) Engineering characteristics of a glaciolacustrine clay deposit in a semiarid climate. Bull Engineering Geology Environment 68:551–557.

[6] Okpanachi, C. O., Ibrahim, S. I., Okoro, A. C., Dogo, K., Idris, M. K. (2017). The Suitability of Local Quartz Sand in The Production of Bath Crucibles. International Journal of Scientific and Technology Research, Volume 6, Issue 03. Pp. 126-128.

[7] Norton, F.H (1980), Ceramics for the Artist Potter, Addison-Wesley Publishing Company Inc., Massachusetts, USA.

[8] ASTM C356-17, Standard Test Method for Linear Shrinkage of Preformed High-Temperature Thermal Insulation Subjected to Soaking Heat, ASTM International, West Conshohocken, PA, 2017, www.astm.org

[9] Clay Shrinkage Testing (2010), [online], Accessed 13th August 2017, Available from World Wide Web: http://www.potters.org/subject01835.htm

[10] Khurmi, R. S., and J. K. Gupta. 2005. A Textbook of Machine Design. New Delhi, India: Eurasia Publishing House (PVT), LTD.

[11] Azeta, J., Okokpujie, K. O., Okokpujie, I. P., Osemwegie, O., & Chibuzor, A. (2016). A Plan for Igniting Nigeria's Industrial Revolution. *International Journal of Scientific & Engineering Research*, *7*(11), 489.

[12] Salawu, E. Y., Okokpujie, I. P., Afolalu, S. A., Ajayi, O. O., & Azeta, J. (2018). Investigation of production output for improvement. *International Journal of Mechanical and Production Engineering Research and Development*, *8*(1), 915-922.

[13] Chester J.H (1973), Refractories, Production and Properties. The Iron and Steel Institute, London, Pp 4-13, 295-315.
[14] Grimshaw R.W (1971), The Chemistry and Physics of Clay and Allied Ceramic Materials, 4th ed. Revised, New York, Wiley Interscience, P15
[15] Nwoke, O. N., Okokpujie, I. P., & Ekenyem, S. C. (2017). Investigation of Creep Responses of Selected Engineering Materials. *Journal of Science, Engineering Development, Environmen and Technology (JOSEDET)*, *7*(1), 1-15.
[16] Zamek J (2016), Ceramic Industry Magazine available at; https://www.ceramicindustry.com/articles/90093-discontinued-feldspar. Accessed January 2017.
[17] Ryan.W. (1978) "Clay and Glazes for Potter" Pitman, London

IOP Conf. Series: Materials Science and Engineering **413** (2018) 012007 doi:10.1088/1757-899X/413/1/012007

Buckling Behaviour of Three-Dimensional Prestressed Stayed Columns

Peter Hyman[1], Adelaja I. Osofero[1], Srinivas Sriramula[1]

[1]University of Aberdeen, School of Engineering, Fraser Noble Building, King's College, AB24 3UE Aberdeen, UK
E-mail address: aiosofero@abdn.ac.uk

Abstract. Slender columns exhibit instability under compression, which causes buckling. Prestressed stayed system can be used to improve the load carrying capacity of these columns. In this study, three-dimensional prestressed stayed column was investigated through the use of commercially available finite element software- ABAQUS. Although many numerical studies into the buckling behavior of stayed columns exist, these have mainly focused on two-dimensional (2D) systems. In order to fully understand the behavior of these structural systems, three-dimensional (3D) numerical studies are essential. In this work, the buckling behavior of three-dimensional prestressed stayed columns was investigated through linear perturbation method. The critical buckling loads and modes of four-branch and three-branch columns were investigated. The effect of varying the cross-arm length on the buckling load of the four-branch column was also investigated and compared with the theoretical values. It was found that the theoretical buckling loads reported in previous research with 2D systems were conservative and largely underestimated the buckling capacities of prestressed stayed columns. Furthermore, it was observed that transition point from the symmetric to antisymmetric buckling mode occurs at a higher value of cross-arm to column length ratio for the 3D prestressed stayed columns compared to the 2D model.

1. Introduction

Columns are vertical structural members loaded in compression in the axial direction and used extensively in the construction industry. They are constructed from various kinds of materials including steel, wood, stone and concrete. A major problem with steel columns, particularly slender steel columns, is that they are sensitive to buckling under compressive loading i.e. axial loading results in horizontal deflection of the member, due to instability. Leonard Euler first derived the critical buckling load of a slender column, known as the Euler formula (Case et al. 1999) as:

$$P_E = \frac{\pi^2 EI}{L^2}, \quad (1)$$

where P_E is the Euler buckling load, E is the Young's modulus of the column, I is the moment of inertia of the column and L is the column length.

Equation (1) presents the Euler load formula, which is still used today to calculate the buckling load of a slender column. The buckling behavior of columns has necessitated the development of innovative systems in an attempt to improve their performance under loading.

Prestressed stayed columns are a form of slender column which benefit from high strength and low cross-sectional area. The name prestressed comes from the fact that the cable or rod stays are pretensioned, so the whole structure is prestressed.

Due to the tensioned stays, buckling is delayed and the system exhibits enhanced buckling resistance compared to its unstayed counterpart. When the column is loaded axially, the pretension counteracts the compressive force and provides lateral stability, thereby delaying the occurrence of buckling. The stays

IOP Conf. Series: Materials Science and Engineering **413** (2018) 012007 doi:10.1088/1757-899X/413/1/012007

are connected via cross-arms, which are connected directly to the column face (Fig. 1). This type of structure can come in a number of configurations, defined by the number of cross-arms.

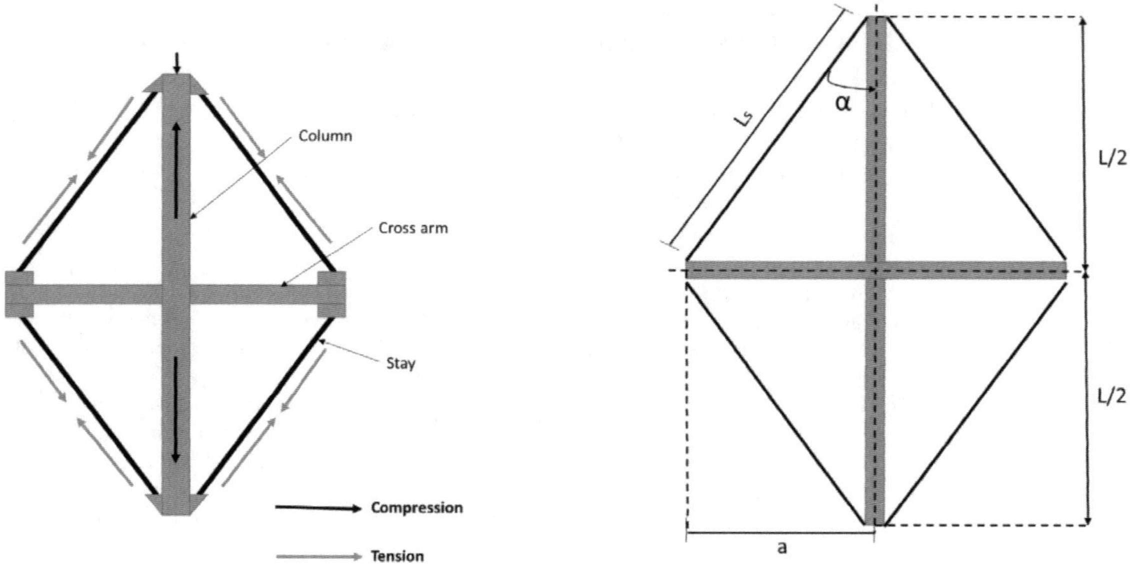

Fig. 1 Typical stayed column with two cross-arms about the mid plane of the column and four stays

A single, two-branch cross-arm stayed column, shown in Fig. 1, is the simplest form of this type of system. In this work, configurations with three or four branches have been investigated, as well as setups with multiple cross-arms along the length of the column (Fig. 2).

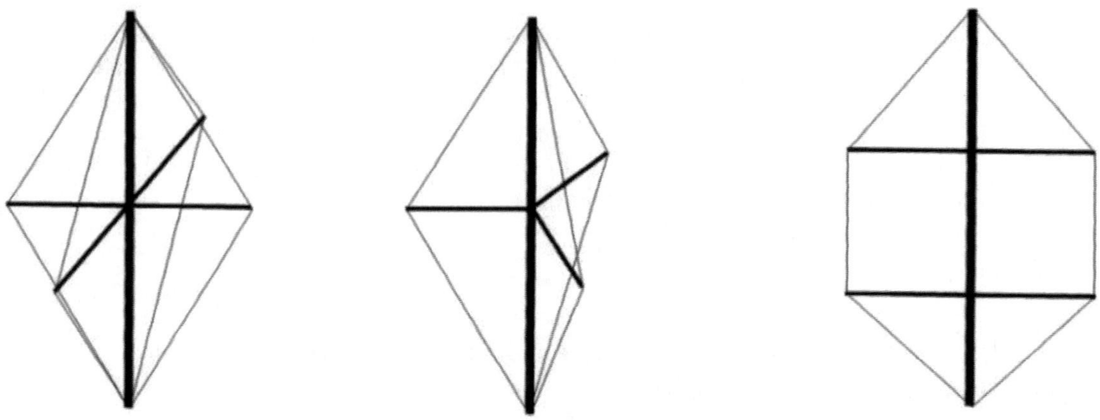

(a) Four-branch cross-arm column, (b) Three-branch cross-arm column, (c) Double cross-arm column.

Fig. 2 Different configurations of prestressed stayed column

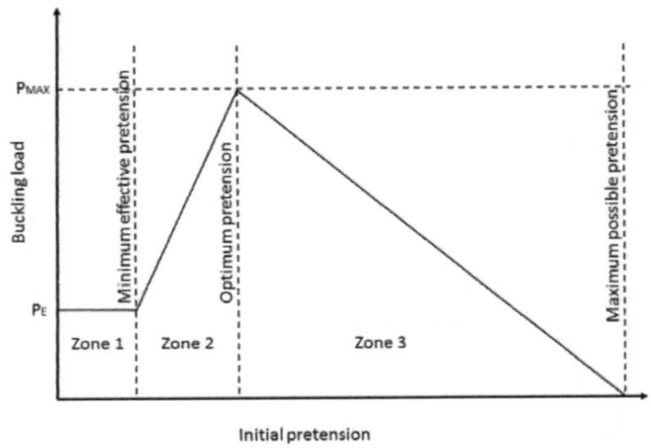

Fig. 3 Zones of buckling behaviour of prestress stayed column (Hafez et al. 1979)

The possible buckling modes of prestressed stayed column under axial compression are shown in Fig. 4. These are the symmetric, antisymmetric and interactive buckling modes. It should be noted that interactive buckling is a post-buckling effect, which occurs as a combination of the other two modes.

(a) Symmetric buckling, (b) Antisymmetric buckling, (c) Interactive buckling

Fig. 4 Buckling modes of the prestressed stayed column

The buckling behavior of prestressed stayed columns has been studied analytically since the 1960's (Chu and Berge, 1963, Ellis 1971). Worthy of note is the work of Hafez et al. (1979), who proposed that the buckling behavior of prestressed stayed columns can be grouped into three zones depending on the level of initial prestress (Fig. 3). The findings of this work have formed the basis for many subsequent studies into the buckling behaviour of stayed columns.

Additional analytical works were carried out by Saito and Wadee (2009), Saito and Wadee (2010), Wadee et al. (2011) and Yu and Wadee (2017). As far as the authors are aware, the work of Temple (1977) was the first attempt to understand prestressed stayed column behavior using numerical simulations. Several subsequent studies investigated the behavior of two-dimensional (2D), single cross-arm stayed columns (Wong and Temple 1982, Chan et al. 2002, Saito and Wadee 2009, Saito and Wadee 2010 and Osofero et al. 2013). These studies focused on the influence of imperfections on the buckling load, the optimal prestress level and parametric studies. However, as far as the authors are aware, the findings of these studies have not been validated for three-dimensional prestressed stayed columns.

Further numerical studies by Yu and Wadee (2017) looked at a triple bay column. Li et al. (2016) also investigated the stability of a three-branch single cross-arm column through numerical modelling using the commercially available finite element software – ABAQUS. It focused solely on the three-branch column, which is covered in the current study and expanded into the four-branch configuration. Although, Pichal and Machacek (2017) utilized ANSYS Finite Element software to examine two-dimensional (2D) and three-dimensional (3D) columns, it employed a sliding connection between the stays and the cross-arms. Comparison with previous work is somewhat impossible due to different connection type. Previous studies have adopted fixed connection between the cross-arm and stays.

A similar structural system has also been studied by Guo et al. (2017) called a pre-tensioned cable stayed buckling-restrained brace. It differs from a 3D stayed column in that it has an inner core running along the length of the column to resist the axial load. Guo et al. (2016) investigated 3D single-cross-arm PCS-BRB using ABAQUS software. It focused on the buckling resistance of these structures. The results highlight that ABAQUS software can be used to effectively study three-dimensional geometries and that these techniques should be applied to prestressed stayed column which is the focus of this study.

2. Numerical modelling

Three-dimensional numerical models were developed for two, three and four-branch columns in ABAQUS with column length, outer diameter and thickness of 2900 mm, 42.4 mm and 3 mm respectively. The cross-arm length was varied between 50 and 600 mm with an outer diameter of 27.9 mm and thickness of 3 mm while the stay area was 60 mm^2. The mechanical properties of the main column, cross-arms and stay are presented in Table 1.

Table 1 Summary of the mechanical properties of column, cross-arm and stay in N/mm^2

Column Young's Modulus	Cross-arm Young's Modulus	Stay Young's Modulus	Column Yield Stress	Cross-arm Yield Stress
209,000	208,000	100,000	342	421

3. Numerical set-up

The 3D columns and cross-arm were modelled using the 3-noded quadratic B32 beam elements while the 2-noded linear T3D2 truss elements were used for the stays. The no-compression option was used in the stay material to prevent any compressive load being carried by the stays and to effectively simulate stay slackening. Single elements were used in the mesh for the stays so that no bending of the stays was allowed, while 10 mm mesh size was adopted for the cross-arms and column. The model had between 338 and 538 elements in total depending on the cross-arm length. A concentric point load was applied to one end of the column, and pinned boundary conditions were used at either end, allowing movement along the column axis at one end.

4. Numerical Results

Using the ABAQUS perturbation analysis, the critical buckling loads of three prestressed stayed column configurations were obtained. Four-branch, three-branch and two-branch stayed columns were investigated with varying cross-arm length. Theoretical buckling loads (N^cmax) for the column geometries were also calculated using Equation (2). The obtained buckling load values from the three-dimensional numerical modeling were compared with the values from a corresponding two-dimensional geometry and the theoretical values. The theoretical values of the critical buckling load, N^cmax, were obtained using Equations (2) – (6) from Osofero et al. (2013), and first derived by Smith et al. (1975).

$$N^{c}_{max} = \frac{4D^2EI}{L^2} \tag{2}$$

Where E is the Young's modulus, I is the second moment of area, L is the column buckling length and the values of term 'D' for the symmetric and antisymmetric modes can be calculated using Equations (3) and (4) respectively.

$$\frac{2K_s}{B}\sin^2\alpha = \frac{D^3}{D - \tan D} \tag{3}$$

$$\frac{B}{\sin^2\alpha}\left(\frac{\cos^2\alpha}{3B_a}+\frac{1}{2K_s}\right)=\frac{D-\tan D}{D^2\tan D} \tag{4}$$

α is the angle between the column and cross-arm. Axial stiffness for the column, cross-arm and stays can be calculated using expressions in Equation (5) if the Young's modulus, cross sectional area and the length of the main column (E, A, L), of the cross arm (E_a, A_a, a) and of the stay (E_s, A_s, L_s) are known.

$$K_c=\frac{EA}{L},\quad K_s=\frac{E_sA_s}{L_s},\quad K_a=\frac{E_aA_a}{a} \tag{5}$$

The bending stiffness of the column and cross-arm B and B_a can also be calculated using Equation (6).

$$B=\frac{8EI}{L^3},\quad B_a=\frac{E_aI_a}{a^3} \tag{6}$$

where, I and I_a are the second moments of area of the column and cross-arm respectively.

Theoretical buckling loads for the symmetric and antisymmetric modes obtained from Equations (2) – (6) and the corresponding optimal prestressing forces for varying cross arm lengths are presented in Table 2. It can be seen that at some value of cross-arm length, the symmetric buckling load becomes greater than the antisymmetric buckling load. This is known as the transition point i.e. when the critical buckling mode transit from the symmetric to the antisymmetric mode.

Table 2 Theoretical buckling loads and optimum prestress forces for column length of 2900 mm and varying cross-arm length

		Symmetric (kN)		Antisymmetric (kN)	
a (mm)	2a/L	N^c_{max}	T_{opt}	N^c_{max}	T_{opt}
50	0.03	29.19	1.95	91.35	6.11
100	0.07	61.17	4.08	108.36	7.23
130	0.09	86.23	5.74	110.69	7.36
150	0.10	102.19	6.78	110.77	7.35
160	0.11	109.12	7.23	110.53	7.33
175	0.12	117.69	7.79	109.99	7.28
180	0.12	120.05	7.94	109.77	7.26
190	0.13	124.07	8.19	109.28	7.21
200	0.14	127.30	8.39	108.73	7.17
250	0.17	136.13	8.89	105.80	6.91
300	0.21	139.66	9.03	102.90	6.65
350	0.24	141.42	9.02	100.34	6.40
400	0.28	142.44	8.95	98.01	6.16
420	0.29	142.73	8.91	97.17	6.07
600	0.41	144.06	8.37	91.42	5.31

Effect of varying cross-arm length on the buckling modes of two-dimensional and three-dimensional models is highlighted in Fig. 5. Theoretical values are also shown for comparison.

It is clear that the critical buckling mode changes at a certain cross-arm length (transition point). However, the transition point is different for the two-dimensional and three-dimensional models. At low cross-arm lengths, the symmetric mode is critical, but at the transition point, the antisymmetric mode becomes the critical buckling mode. Furthermore, the 3D model gives the highest buckling load value, conveying that the theoretical and 2D numerical model may be conservative in their estimates.

As the cross-arm length varies, the buckling mode shape changes. This is due to the change in the rotational stiffness of the cross-arm i.e. at low cross-arm length (before the transition point), the rotational stiffness of the cross-arm is greater than its axial stiffness. Hence, the critical buckling mode is symmetric with the cross arm moving longitudinally (Fig. 4a). However, as the cross-arm length increases, the rotational stiffness of the cross-arm decreases. If the cross-arm length continues to increase, a point is reached, where the rotational stiffness is now lower than the axial stiffness. At this point, which is also known as the transition point, the critical mode becomes antisymmetric, i.e. no longitudinal movement at the cross arm but rotation of the cross-arm with respect to the column surface is experienced (Fig. 4b).

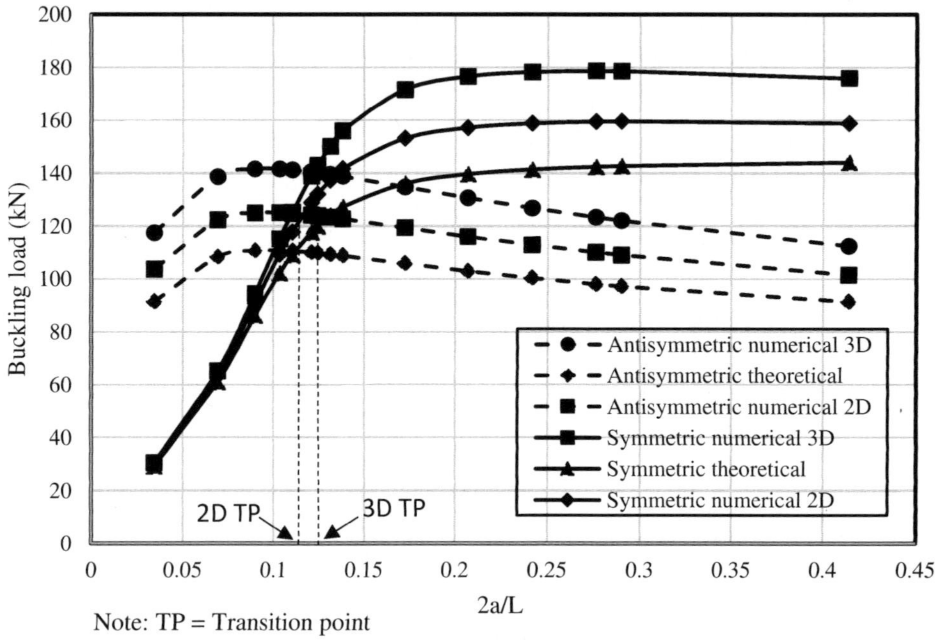

Fig. 5 Comparison of buckling loads from the 2D, 3D and theoretical models

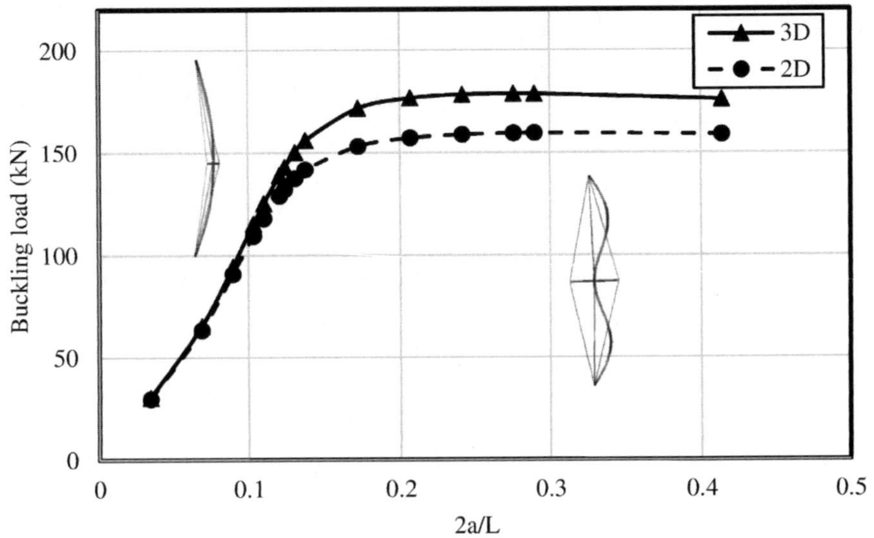

Fig. 6 Symmetric buckling loads for two-dimensional and three-dimensional models at varying 2a/L

Figures 6 and 7 highlight changes in the buckling shapes due to increase in cross-arm length. The transition point is important as the buckling load values for the symmetric and antisymmetric buckling modes are almost equal at this point. This means that interactive buckling may be induced as a combination of both buckling modes can occur.

The buckling shape of the symmetric mode changes dramatically with increase in cross-arm length. This occurs due to changes in the rotational stiffness of the cross-arm. At lower cross-arm length, the symmetric buckling mode has the greatest deflection at mid-height along the column as the rotational stiffness is greater than the axial stiffness. Whereas, as the cross-arm length increases, the mid-height becomes the least deflected point, until it is not deflected at all, as the rotational stiffness is lower than the axial stiffness.

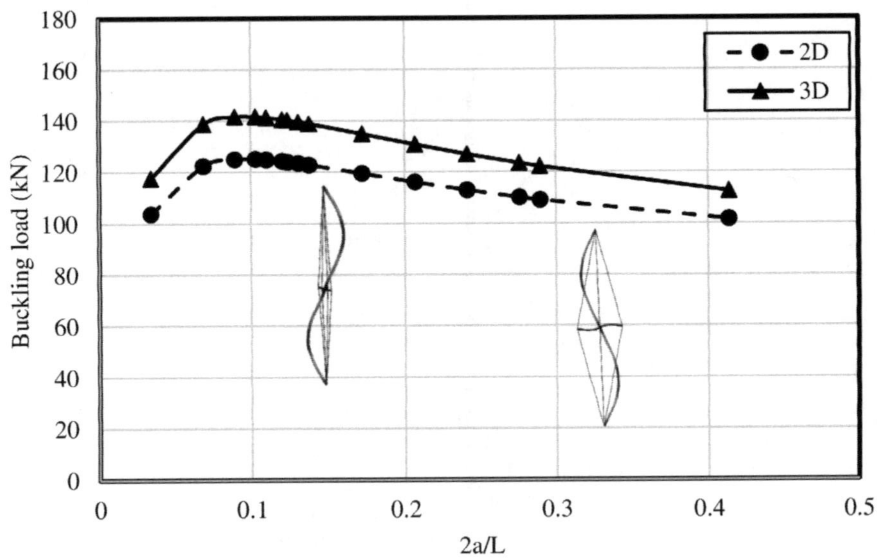

Fig. 7 Antisymmetric buckling loads for two-dimensional and three-dimensional models at varying 2a/L

IOP Conf. Series: Materials Science and Engineering **413** (2018) 012007 doi:10.1088/1757-899X/413/1/012007

Fig. 7 highlights changes in the shape of the antisymmetric buckling mode of a three-dimensional model as the cross-arm length increases. At low cross-arm lengths, the cross-arm deflects at the stay ends. Buckling of the cross-arms is experienced at higher cross-arm length. Representative symmetric and antisymmetric buckling modes for two-dimensional and three-dimensional models are presented in Fig. 8 to highlight the difference in buckling mode shapes.

The comparison presented in Fig. 8 highlights the similarities between the buckling shapes of the 2D and 3D systems as well as the slight differences. However, the main difference between the 2D and 3D buckling shapes is the ability of the 3D column to buckle out of the cross-arms plane. Three-branch columns, with varying angle of orientation, i.e. varying α_i, were also investigated to establish the optimal configuration (Fig. 9).

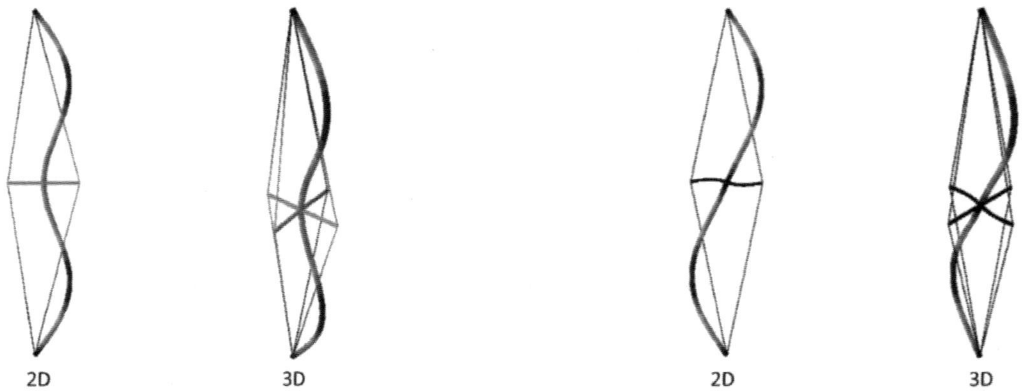

a) Symmetric modes b) Antisymmetric modes

Fig. 8 Comparison of symmetric and antisymmetric buckling mode shapes for 2D and 3D models

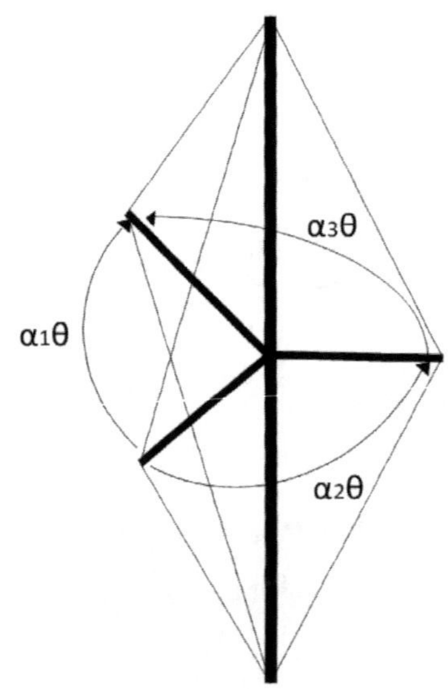

Fig. 9 Orientation of three-branch column

Fig. 10 presents the buckling load for varying angle of orientation for both symmetric and antisymmetric buckling modes.

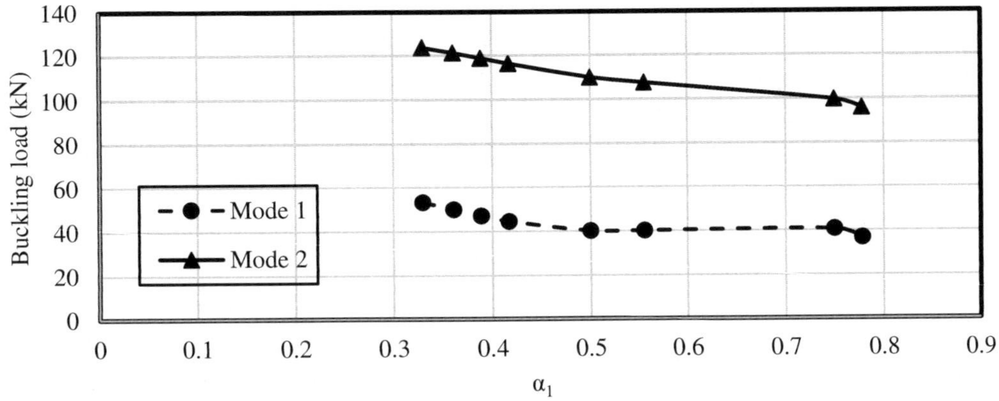

Fig. 10 Buckling load of three-branch columns with varying angle α

As can be seen, the configuration with equal angle of 120° between each arm i.e. $\alpha_1 = \alpha_2 = \alpha_3$ has the greatest buckling load for both the symmetric and antisymmetric buckling modes.

5. Conclusion

Results from the current work shows that buckling load values of prestressed stayed columns obtained from three-dimensional numerical modelling are consistently higher compared to the values obtained from equivalent system when modelled in two dimensions. This shows that both analytical and two-dimensional modelling underestimate the buckling capacity of this structural system which can potentially lead to it being overdesigned. In addition, it further shows that the theoretical buckling values of prestressed stayed columns are somewhat conservative and procedures for obtaining these values should be updated. It is clear that, to fully understand the buckling behaviour of prestressed stayed columns and also maximize its load carrying capacity, three-dimensional numerical modelling instead of the popular two-dimensional modelling must be employed. Although the trends obtained from the present work is encouraging, further investigation into the post buckling behaviour and the sensitivity of the system to change in key parameters, such as prestress value and column imperfection is necessary.

Acknowledgments

Research is supported by the Engineering and Physical Science Research Council (EPSRC) Doctoral Training Partnership Grant.

References

[1] Chan, S., Shu, G. & Lü, Z. 2002, "Stability analysis and parametric study of pre-stressed stayed columns", Engineering Structures, vol. 24, no. 1, pp. 115-124.

[2] Chu, K. & Berge, S.S. 1963, "Analysis and Design of Struts with Tension Ties", Journal of the Structural Division, vol. 89, no. 1, pp. 127-164.

[3] Ellis, J.S. 1971, "R.M.C. design-build-test projects", Journal of Engineering Education, vol. 62, no. 3, pp. 294-296.

[4] Guo, Y., Fu, P., Zhou, P. & Tong, J. 2016, "Elastic buckling and load resistance of a single cross-arm pre-tensioned cable stayed buckling-restrained brace", Engineering Structures, vol. 126, no. Supplement C, pp. 516-530.

[5] Guo, Y., Zhou, P., Bradford, M. A., Pi, Y., Tong, J. & Fu, P. 2017, "Theoretical and numerical studies of elastic buckling and load resistance of double cross-arm pre-tensioned cable stayed buckling-restrained braces", Engineering Structures, vol. 153, no. Supplement C, pp. 674-699.

[6] Hafez, H.H., Temple, M.C. & Ellis, J.S. 1979, "Pretensioning of single-cross-arm stayed columns", J. Struct. Div. ASCE, vol. 105, pp. 359-375.

[7] Case, J., Chilver, L. & Ross, C.T.F. 1999, "18 - Buckling of columns and beams" in Strength of Materials and Structures (Fourth Edition), eds. J. Case, L. Chilver & C.T.F. Ross, Butterworth-Heinemann, London, pp. 424-457.

[8] Li, P., Wadee, M.A., Yu, J., Christie, N.G. & Wu, M. 2016, "Stability of prestressed stayed steel columns with a three branch crossarm system", Journal of Constructional Steel Research, vol. 122, no. Supplement C, pp. 274-291.

[9] Osofero, A.I., Wadee, M.A. & Gardner, L. 2013, "Numerical Studies on the Buckling Resistance of Prestressed Stayed Columns", Advances in Structural Engineering, vol. 16, no. 3, pp. 487-498.

[10] Pichal, R. & Machacek, J. 2017, "Buckling and Post-buckling of Prestressed Stainless Steel Stayed Columns", Procedia Engineering, vol. 172, no. Supplement C, pp. 875-882.

[11] Saito, D. & Wadee, M.A. 2010, "Optimal prestressing and configuration of stayed columns", Proceedings of the Institution of Civil Engineers - Structures and Buildings, vol. 163, no. 5, pp. 343-355.

[12] Saito, D. & Wadee, M.A. 2009, "Numerical studies of interactive buckling in prestressed steel stayed columns", Engineering Structures, vol. 31, no. 2, pp. 432-443.

[13] Smith, R.J., Ellis, J.S. & McCaffrey, G.T. 1975, "Buckling of a single-crossarm stayed column", Journal of the Structural Division, vol. 101, no. 1, pp. 249-268.

[14] Temple, M.C. 1977, "Buckling of stayed columns", ASCE J Struct Div, vol. 103, no. 4, pp. 839-851.

[15] Wadee, M.A., Gardner, L. & Hunt, T.A. 2011, "Mode jumping in prestressed stayed columns: A spring and link model approach", ENOC 2011: 7th European Nonlinear Dynamics Conference, Rome, Italy, European Mechanics Society.

[16] Wong, K.C. & Temple, M.C. 1982, "Stayed column with initial imperfection", Journal of the Structural Division, vol. 108, no. 7, pp. 1623.

[17] Yu, J. & Wadee, M.A. 2017, "Optimal prestressing of triple-bay prestressed stayed columns", Structures, vol. 12, no. Supplement C, pp. 132-144.

IOP Conf. Series: Materials Science and Engineering **413** (2018) 012008 doi:10.1088/1757-899X/413/1/012008

A microcontroller based gas leakage detection and evacuation system

Aderibigbe I. Adekitan*, **Victor O. Matthews, Olayinka Olasunkanmi**

Department of Electrical and Information Engineering, College of Engineering, Covenant University, Ota, Nigeria

*E-mail address: ade_kitan@yahoo.com

Abstract. Liquefied Petroleum Gas (LPG) is a major fuel for cooking in Nigeria. LPG is dispensed into cylinders which are in various sizes and storage capacities. Multiple cases of household fires, injuries and fatalities due to LPG leakage and explosion have been recorded in Nigeria. This undesirable trend persists majorly because households in Nigeria don't have gas leakage detection systems, and at best only smoke detectors are available in selected high class homes. The proactive safety approach is to detect gas leakage and evacuate the gas before it is ignited. In this study, a model gas leakage detector and evacuation system are presented. The implemented, microcontroller-based system, activates a buzzer when a gas leak is detected, it shuts the gas supply solenoid valve to stop the gas flow, and also, it evacuates the gas by switching on evacuator fans. To ensure that the house occupant is adequately notified when there is a gas leakage; the system sends a text via short message service (SMS) to a stored mobile number. The status of the system can be remotely determined by the user, by sending codes to prompt the device via SMS.

Keywords: LPG Gas leakage detection, smart system, fire prevention, safety protection system, remote monitoring, SMS system prompt

1. Introduction

Hydrocarbon is a major source of energy, both for industrial and domestic applications. Liquefied Petroleum Gas (LPG) is a flammable hydrocarbon gas which consists of a mixture of propane and butane hydrocarbons. LPG is a product obtained from the refining of crude oil and gas, and it is a source of energy for domestic cooking, industrial ovens, electric power generators, and it serves as fuel for vehicles. LPG can also be used as a propellant, refrigerant and as a petrochemical feedstock. LPG has no odour and as such, Ethanethiol is added to it as a powerful odorant, for the detection of gas leakage via the sense of smell [1].

In Nigeria and globally, fire accidents and hazards caused by gas leakage have been a major challenge in the industries (e.g. oil and gas industry), and even in homes (in the kitchen). Several cases of fire accidents and explosions due to gas leakage from gas pipes, the burners of gas cookers or the gas cylinders have been reported. The 1984 San Juanico tragedy in Mexico City; one of the most severe LPG disaster in history, was caused by undetected LPG leakage [2].

Although, gas leakage detection is very important, but preventing fire requires adequate control of the detected gas leakage. In order to prevent gas explosion, there should be statutory environmental requirements on the use of LPG so as to prevent accidents, and protect life and property from disaster. The safety of life and property is vital in all our daily activities whether domestic or industrial. Achieving utmost safety requires adequate planning, early fault detection, and the use of appropriate forms of protection

Over the years, there have been several accidents caused by the leakage of LPG in homes and industries (especially oil and gas industries) from gas pipes and gas cylinders, which has led to the loss of several lives and properties through fire outbreaks and explosions, poisoning and suffocation due to lack of oxygen. In 2013, a report on LPG related accidents in Japan was filed by the High-Pressure Gas Safety Institute of Japan (KHK). It showed the number of accidents caused by LPG leakage, and a comparison was made with the previous years. As at 2013, there were 206 recorded accidents which left 3 people dead and 52 people injured [3]. From 2014, up to January 2015, there were over 12 recorded LPG leakage related accidents in Chennai in India, and according to the report, this is an indication that the loss of life and damages caused by LPG accidents is on the increase in India. As at January 2015, in Chennai, two incidents had already been reported which left 3 people dead, and 6 others injured [4].

In Nigeria, although there are no accurate statistics and records on LPG related accidents, there have been cases of fire accidents and explosions in industries, fields, and environments where LPG is processed or used. A good example is an incident that occurred in August 2015, in which a family in Azia community, in Anambra State, took delivery of the bodies of nine of their children, who were roasted to death in their Lagos abode by fire that was caused due to undetected gas leakage [5]. There is no system in place that detects and arrests gas leakage in most kitchens in Nigeria. Even with all the modern building techniques and technologies, installation of gas leakage detection systems in the house, especially in the kitchen, is neglected, and this creates a fire hazard safety gap that can lead to serious disaster.

In this study, a smart, gas leakage detector is designed and implemented to detect gas levels above the normal threshold level, especially in a confined environment. When leakage is detected the system shuts off the gas supply by activating a solenoid valve, and it also evacuates the leaked gas by activating an exhaust fan. The gas leakage detector notifies the residents by activating a loud alarm, and it also sends an SMS text to a pre-configured number, so as to ensure that the residents are notified of any gas leakage even when they are not at home.

2. Smart Systems

Smart systems are systems that combine elements of sensing, actuation, and control for various analyses, and also for making appropriate decisions based on the available information, in a way that promotes versatility and adaptability of the system. Most times, how 'smart' a system is, may be ascribed to autonomous operations which includes; networking capabilities, closed-loop control, and energy efficiency [6]. A smart system should have high degree of reliability, efficiency and sustainability with an intelligent operational management system [6].

Smart systems incorporate many components, some of which are sensors for receiving signal, command and control unit (CCU) which sends instructions and carry out decision based information access and control, and actuators that execute the required task. Smart systems provide undeniable benefits in its use and application [7]. The smart system technology provides a means of addressing highly complex challenges and conditions, for instance, by offering early warning detection capabilities, responding to detected challenges, adapting the response system to manage unexpected challenges, thus enhancing the lifespan of the system. Another benefit is seen in the preventive maintenance of the smart systems. This leads to improved system operation, performance and function.

With the advent of improved technology, there is an increase in the use of smart systems in tackling challenges in the economy, society, and the environment as a whole. Smart systems are employed in various residential, industrial and commercial areas ranging from day to day tasks to very intricate and dangerous missions. Some important areas of application include healthcare, energy sector, manufacturing, transportation, military and defence, logistics, safety and security amongst others. Smart systems in this area are employed to assist in averting and guarding against mishaps, disasters and criminal activities which can negatively affect human lives and properties. Some examples include antitheft sensor systems, hazardous chemical sensing systems, gas sensing systems, radiation sensing systems, and anti-intrusion sensor systems, just to mention a few [8-12].

2.1 Smart Gas Detection

In the study by [13], a smart device was designed to identify and measure methane gas in zones of the flammable gas stockpile site. The device measures the air and water quality, including every parameter that can have deviation as a result of gas leakage in the water or air. The sensors measure the amount of

CH_4 and CO_2 gas in the air while the temperature, pH, and electrical conductivity of the water are monitored. The device is controlled by an Arduino UNO microcontroller that transmits measured data to the database on Raspberry Pi 3.

In [14], a gas detection system employing an MQ-9 sensor was proposed. The study developed an embedded system that triggers a buzzer and a set of LEDs to alert residents in the event of a gas leak. An android based automatic gas detection and indication robot was developed by [15]. When gas leak is detected, the mobile robot instantly interprets the data and sends it to an android mobile phone via wireless Bluetooth communication. An android application for smartphones using the Android OS, which can get data from the robot and also control the robot's motion using Bluetooth, was built. The study by [16], proposed a wireless LPG leakage monitoring system for home safety. The gas sensor sends a signal to the control unit, and after processing the signal, the control unit transmits a signal to the solenoid valve assembly and this triggers an alarm. The status of the gas level can also be determined through a web browser and the GUI.

3. The implemented gas leakage detector and evacuation system

The gas leakage is detected using MQ-2 gas sensor; the sensitive layer of the MQ-2 gas sensor is made of SnO_2. The conductivity of SnO_2 is low at normal operating condition, when the atmosphere is free from gas contamination. When the concentration of target gases such as methane exists in the atmosphere, the conductivity of SnO_2 increases, and this changes the sensor's resistance. The corresponding pulse is fed to the microcontroller. The sensor is powered by a 5V supply. The resistance of the sensor can be determined using equation 1 [17]. The MQ-2 sensor is shown in Figure 1.

$$R_s = \frac{V_c R_l}{V_{rl} - 1} \qquad (1)$$

R_S is the resistance of the sensor
V_C is the voltage supplied to the load cell
V_{RL} is the voltage developed across the load resistance
R_L is the resistance of the sensor

Fig.1 MQ-2 Gas sensor

PIC16F628A; a flash-based, fully static, CMOS, 8-bit microcontroller is deployed as the brain of the whole system to control all the processes. It has an inbuilt ADC. It is connected to the relay, GSM module, buzzer, and the exhaust fans. The SMS based, gas leakage alert message is facilitated through a SIM800L GSM module. The GSM module requires 4.2V DC as the input supply voltage. LM7805 voltage regulator is utilized to regulate the 12V-DC that the power supply delivers to 5V DC. In this design, 4 pins of the SIM800L were used and these are:

- Vcc pin for supply
- Txd pin (transmission pin) connected to the receiving pin of the microcontroller
- Rxd pin (Receiving pin) connected to the transmission pin of the microcontroller
- Gnd pin (Ground)

IOP Conf. Series: Materials Science and Engineering **413** (2018) 012008 doi:10.1088/1757-899X/413/1/012008

The PIC16F628A microcontroller is programed using "MP-Lab" software installed on Windows 8 operating system. "CSS C" compiler is used for building the program. The code developed was dumped to the microcontroller using "PICFLSH" programmer software.

When the gas sensor detects gas leakage, it sends a signal to pin 18 of the microprocessor, the microprocessor processes the information and gives an output to the buzzer, fan 1, fan 2, GSM module and the relay which activates the valve via the following pins; pin 13, pin 10, pin 11, pin 8 and pin 12 respectively. The operation of the gas detector device is explained by the flow chart of Figure 2. The GSM module is connected to the Universal Asynchronous Receiver-Transmitter (UART) pin of the microcontroller. The circuit diagram is shown in Figures 3 and 4.

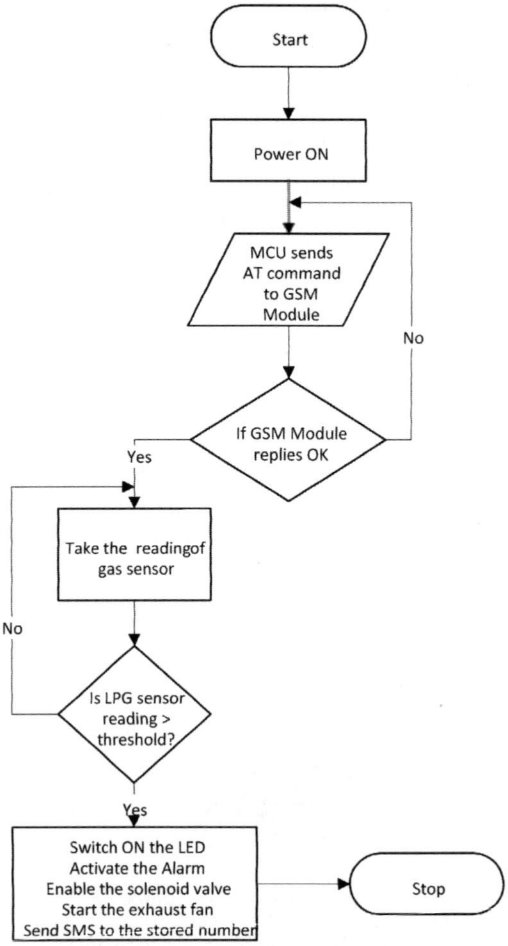

Fig.2 A flow chart of the gas detector operation

The following are the different SMS messages received from the gas detector and evacuation device:

- system initialization successful
- gas detected, evacuation procedure enabled
- gas completely evacuated, environment safe
- alert: system is on and sensing
- alert: gas status is okay, environment is clean

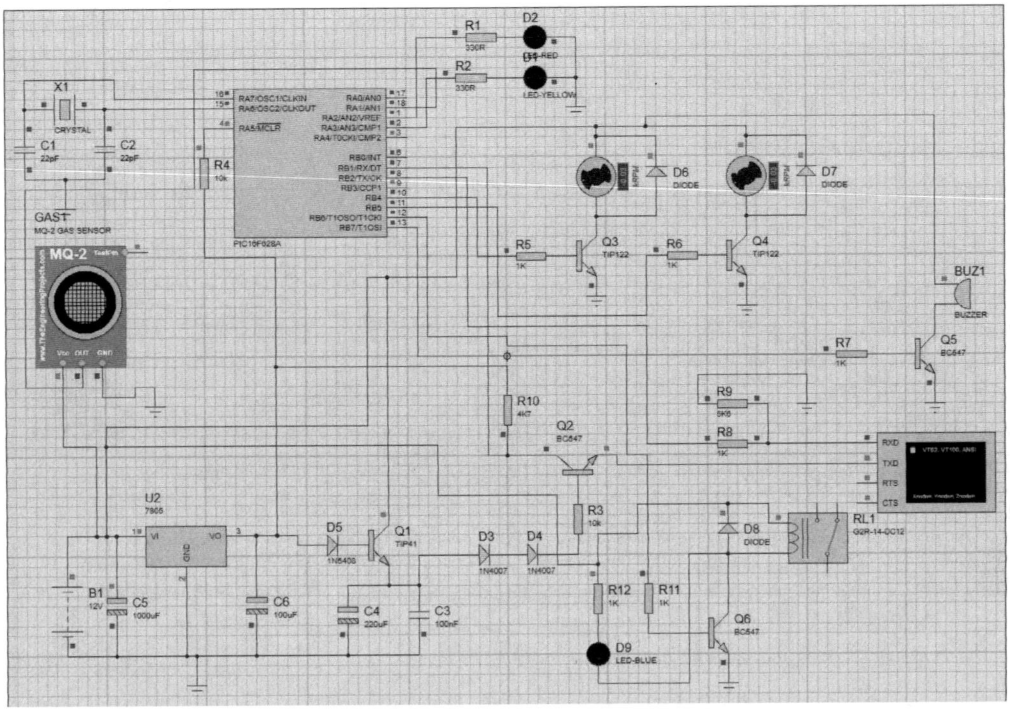

Fig.3 The circuit diagram of the Gas Detection & Evacuation System

4. Benefits of the implemented model

The implemented gas leakage detection and evacuation model device offers a number of advantages over dependence on unreliable detection by human sense of smell. The automated gas detection and evacuation process ensures safety by preventing gas induced house fires. The device performs the following:

- It detects LPG leakage using MQ-2 sensor
- Activates an alarm upon gas leakage detection, and it stops the alarm once the leakage is under control (gas concentration in the atmosphere is within normal range)
- It incorporates an SMS based alert mechanism, and it sends SMS text (alert message) to a specified mobile number
- Shuts off gas supply using a solenoid valve when gas leakage is detected
- It evacuates the leaked gas from the leakage area using an exhaust fan
- It incorporates a feature that allows the user to remotely probe via SMS if the gas detection module is ON, and also determine the current gas leakage status by sending unique codes.

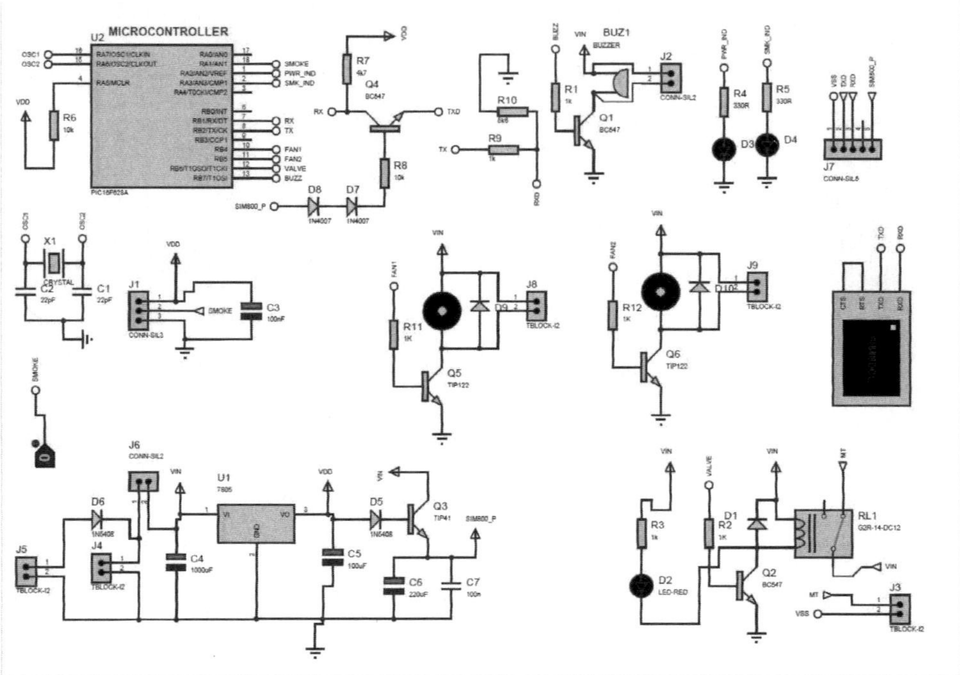

Fig.4 Component arrangement for PC board implementation

5. Conclusion

Safety is the proactive, and the cheapest option to preventing accidents and mishaps. Gas leakage induced fire is a menace that has occurred in different parts of the world. In Nigeria, LPG gas is supplied via cylinders for household use, and cases of gas leakage from the cylinder, the supply hose or the gas burner has been recorded by different users over the years with some resulting in gas explosion, property destruction, injury and fatality. This study presents the design of a model gas detection and evacuation system. After design implementation, the device accurately detected simulated gas leakage, and the SMS and buzzer, user alert system performed exactly as designed. The evacuator fans were automatically activated to suck away the leaked gas from the enclosure so as to prevent a potential ignition and fire. This smart device offers a number of safety benefits that are vital for early gas leakage detection, and response towards preventing LPG leakage related house fires in Nigeria.

Acknowledgments

The Authors appreciate Covenant University Centre for Research, Innovation and Discovery for sponsoring our participation at the conference.

References

[1] B. Sonkar, A. Sood, A. Ranjan, and A. Faisal, "Microcontroller Based LPG Leakage Detector Using GSM Module," International Journal of Electrical and Electronics Research, vol. III, pp. 264-269, 2015.

[2] G. Arturson, "The tragedy of San Juanico—the most severe LPG disaster in history," Burns, vol. 13, pp. 87-102, 1987.

[3] KHK, "Annual Report on Liquefied Petroleum Gas," The High-Pressure Gas Institute of Japan, Tokyo2013.

[4] P. Petlee and R. Deepa, "Fires from LPG leaks on the rise; police cite negligence," in The Hindu, ed: The Hindu, 2015.

[5] S. Eno-Abasi and G. Akutu, "Stemming cooking gas-related accidents/deaths," in The Guardian, ed: The Guardian, 2017.

[6] G. Akhras, "Smart Materials and Smart Systems for The Future," Canadian Military Journal, 2000.

[7] Paul Fanning. (2012, March 15). Smart systems bring benefits to industrial applications. Available: http://www.eurekamagazine.co.uk/design-engineering-features/technology/smart-systems-bring-benefits-to-industrial-applications/40993/

[8] E. Adel and L. Micheal, "Smart cities: safety, security and privacy," Journal of Advanced Research, 2014.

[9] O. Osemwegie, S. John, K. Okokpujie, and I. Shorinwa, "Development of an electronic fare collection system using stationary tap-out devices," in Proceedings - 2016 International Conference on Computational Science and Computational Intelligence, CSCI 2016, 2017, pp. 234-236.

[10] V. O. Matthews, A. O. Ajala, S. I. Popoola, and A. A. Atayero, "Smart vehicular traffic management system using RFID technology," in Lecture Notes in Engineering and Computer Science, 2017, pp. 414-417.

[11] T. O. Takpor, J. Ademola, S. I. Popoola, J. A. Badejo, and A. A. Atayero, "Smart assistive mHealth system for medication adherence in patients with Alzheimer's disease," in Lecture Notes in Engineering and Computer Science, 2017, pp. 1076-1080.

[12] A. U. Adoghe, S. I. Popoola, O. M. Chukwuedo, A. E. Airoboman, and A. A. Atayero, "Smart weather station for rural agriculture using meteorological sensors and solar energy," in Lecture Notes in Engineering and Computer Science, 2017, pp. 323-326.

[13] V. Carmela and I. Ana, "Smart gas detection system," Institute of Electrical and Electronics Engineering, 2017.

[14] A. Falohun, A. Oke, and B. Abolaji, "Dangerous Gas Detection using an Integrated Circuit and MQ-9," International Journal of Computer Applications, 2016.

[15] R. Sushma and R. Manohar, "An Android Based Automatic Gas Detection and Indication Robot," International Journal of Computer Engineering and Applications, 2014.

[16] H. Mujawar, D. Bachuwar, and S. Kasbe, "Design and development of LPG gas leakage detection and controlling system," Solapur University Research Journal, vol. IV, 2015.

[17] S. Rajitha and T. Swapna, "Security alert system using GSM for gas leakage," International Journal of VLSI and Embedded Systems-IJVES, vol. 3, pp. 173-175, 2012.

IOP Conf. Series: Materials Science and Engineering **413** (2018) 012009 doi:10.1088/1757-899X/413/1/012009

A data-based investigation of vehicle maintenance cost components using ANN

Aderibigbe Israel Adekitan[1,*], Adetokun Bukola[2] and Okokpujie Kennedy[1]

[1]Department of Electrical and Information Engineering, College of Engineering, Covenant University, Ota, Nigeria
[2]Department of Electrical Engineering, Pan African University of Basic Sciences, Technology and Innovation, hosted at Jomo Kenyatta University of Agriculture and Technology, Nairobi, Kenya
*E-mail address: ade_kitan@yahoo.com

Abstract. Vehicle maintenance is a critical and vital operational component that determines vehicle performance and service longevity. The severity of vehicle usage has been defined as one of the key factors that determine vehicle maintenance requirements. Nigeria is a tropical country with intense heat, poor road network and quality. These factors severely affect car performance and raises maintenance demands toward ensuring vehicle reliability and optimum performance. In this study, vehicle maintenance cost and fuel consumption data in terms of cost and volume, together with the mileage coverage as the vehicle usage data, for two corporate organizations were analyzed. The data was collected and systematically analyzed. The common faults were categorized and their frequency of occurrence was determined. An artificial Neural Network (ANN) model was developed for predicting future maintenance cost, given a set of anticipated vehicle usage inputs. The model has an overall correlation R-value of 0.76645.

Keywords: ANN model, vehicle maintenance, predictive cost analysis, data pattern recognition, road transportation, statistics

1. Introduction

Supply chain reliability is the back bone of many commercial operations, and it also ensures adequate distribution and availability of various products, supplies and equipment for both domestic and industrial uses. Product and material distribution can be via air, water, rail and road transportation. Transportation is vital for socio-economic interactions. Freight transportation must be cost effective, efficient, safe, fast and reliable to ensure itch-free national distribution of goods and economic development. According to [1], road transportation is the most active mode for haulage transportation. It was estimated that 75% of the total 2200 billion tonne-kilometres of freight transportation in Europe was via road. In the Nigerian transport industry, the road transport sub-sector contributes about 90% to the Gross Domestic Product. As at the end of 2017, the estimated number of vehicles in Nigeria was 11,583,331 [2, 3].

The transportation industry in Nigeria is plagued by many challenges due to poor planning, poor road network and condition, scarce funds, corruption and defective policies, insufficient vehicle fleets, and so forth. Vehicles daily ply poorly constructed, and poorly maintained roads with pot holes and many unpaved road sections which create inherent hazards [4]. This negatively impacts vehicle reliability, resulting in repetitive vehicle breakdowns with an attendant increase in maintenance cost. To ensure vehicle availability for service, corporations need to dedicate a sizable amount of their budget towards vehicle maintenance and overhaul. Most of the vehicles in Nigeria were imported. It is only in recent times, due to new Government policies that both indigenous and foreign car assembling plants are springing up again in the country,

after decades of zero production due to various economic challenges.

Foreign Exchange (FOREX) issue is a recurrent problem in Nigeria due to FOREX scarcity and price fluctuations. Currently, one US dollar is around 365 Naira. This makes vehicle importation very expensive, and as a result of the high cost, most of the cars and trucks imported into the country are fairly-used, and they are rarely brand new. This implies that a good percentage of the service life of the vehicle is used up before arrival in Nigeria, and upon arrival into the country, deterioration sets in due to many factors and realities of the Nigerian society. The only way to ensure vehicle reliability and prolong vehicle service life is through adequate and timely vehicle maintenance [5], but this is often challenged by the need to manage scarce resources.

A visit to the car pack and maintenance warehouse of some organizations will reveal run-down and abandoned vehicles, with the tyres of some of the vehicles already removed, and the chassis supported by stone slabs. A number of times, this occurs as a result of scarce spare part, low maintenance budget and excessive car breakdowns leading to abandonment. Vehicles abandoned in this way end up deteriorating further with time, and this increases the cost of restoration, and reduces the sale value, if the vehicle is sold off in that state. If such cars are parked in open and unfenced locations in front of factories without adequate security, the car parts end up being stolen by scavengers. Cases of stolen head lights, tyres, batteries and so forth, have been reported.

In maintenance engineering and management, a number of maintenance components and parameters can be extracted for objective analysis [6]. This study presents and examines the maintenance data of two corporate organizations for fault frequency trending and for data parameter relationship identification. An ANN model was developed using the fuel consumption, cost of fuel, mileage covered and the maintenance cost data for the sampled period as model parameters.

2. Case study: haulage company truck and corporate pool car maintenance

The effect of the harsh economic conditions and poor infrastructural development is quite severe on haulage company vehicles that need to move bulk load from one point to another [7]. The trucks are the lifeline of the companies and must be operational to deliver value. According to the study in 2009 by AAA Car Care, more than 62% of vehicles are operated on severe service conditions, some of these conditions include:

- Driving in hot weather under frequent stop and start traffic
- Driving on dusty or muddy roads and on roads with gravel spread
- Driving below 50 miles per hour over long distances
- Transporting heavy loads or towing a trailer

Nigeria is a tropical country, with an average temperature of about 28°C, which could rise to about 35°C in the northern region. When haulage trucks travel between local communities, a number of unpaved roads will be encountered. Most of these roads are unpaved, dusty and muddy, and the pot holes in some road sections are filled with stones and pebbles. Poor roads and insufficient road network creates heavy traffic along major roads, and this induces stop and start vehicle movement. Since there is no dedicated lane for trucks and bulk vehicles, heavy loaded trucks caught up in traffic are also forced to stop and start with the slow moving traffic. By these realities, most vehicles in Nigeria are operating under severe conditions, and this requires frequent maintenance and oil change to limit damage, and prolong vehicle service life. Frequent maintenance has an associated cost implication which may result in non-compliance in a bid to reduce expenses. In Europe, road transport accounts for 17.5% of total greenhouse gas emissions [8], and this will be aggravated in developing countries by poorly maintained vehicles with sooty vehicle exhausts.

A. The haulage company truck maintenance analysis

Commercial trucks travel thousands of kilometres with heavy loads every year, and this journey is most times under severe conditions. Prior to embarking on a journey, trucks ought to meet the minimum safety and maintenance requirement to prevent sudden breakdown en route to their destination. This is often not the case; a number of broken down trucks can be seen on major highways and on interstate routes in

Nigeria. Some of these loaded trucks are sometimes at the breakdown spots for weeks, with the driver and driver assistant sleeping in, or around the truck to protect the cargo until the truck is fixed. Often the needed technical assistance, funds and or spare parts is coordinated from the office of the logistic company which is in one state, while the broken down truck is in another. The repair process is sometimes slow and this can lead to cargo theft and product damage for perishable goods.

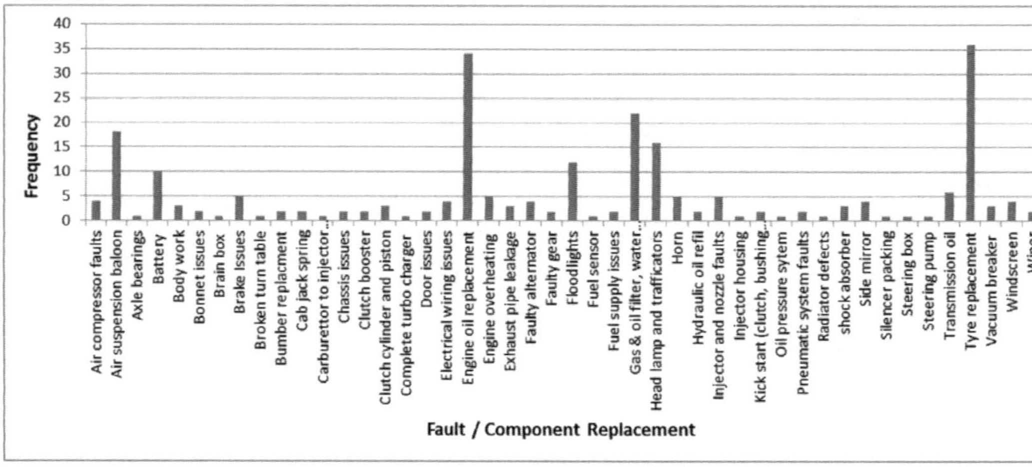

Fig.1 Faulty components and their frequency of occurrence

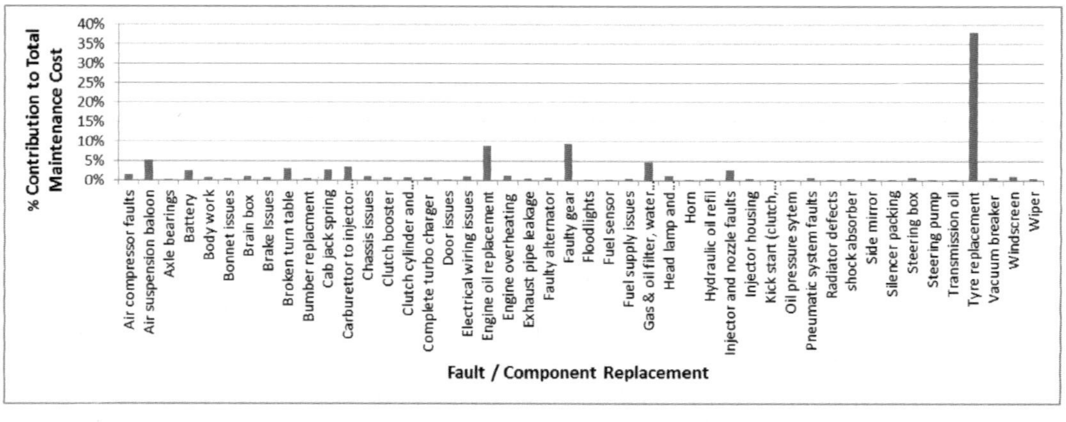

Fig.2 The contribution of each fault and repairs to total maintenance cost

Figure 1 shows the fault log and their frequency of occurrence for 15 haulage trucks, for one operational year. Figure 2 shows the contribution of each fault and repairs to the total maintenance cost, based on the average cost of each, expressed in percentage. The availability for service of each of the truck throughout the year could not be established via records. It is possible that some of the trucks were out of service for extensive periods within the year. The data set shows that the gas filter, oil filter and the water separator were replaced 22 times, while engine oil replacement was done 34 times. For the 15 trucks, if oil and filter replacements were done at least once in 6 months, the filters would have been replace 30 times. This indicates a possible cost saving measure by topping the engine oil without a complete replacement of the filtration system, and this could negatively impact the engine performance of heavy duty trucks. This could also imply that some of the trucks as earlier suggested were out of service for certain periods, and as such there was no need to perform some periodic maintenance.

IOP Conf. Series: Materials Science and Engineering **413** (2018) 012009 doi:10.1088/1757-899X/413/1/012009

Depending on the type and design, the number of trailer truck tyres can be 10 and up to 18 per truck. Substandard tyres are gradually flooding the market, though they are sometimes cheaper but they do not last long, and are usually replaced more frequently. Poor Nigerian roads, heavy loading of the trucks and the spread of low quality tyres are likely factors that may have resulted in the replacement of 36 tyres representing 37.97% of the total maintenance expenses.

B. The corporate pool car maintenance analysis

Pool cars are vital in corporate environment for conveying personnel from one place to another for various business purposes. In some organization, each car is assigned to a specific unit or department, but the maintenance of all the cars is often managed by the maintenance department. This study analyses the maintenance and fuel related data for 25 pool cars of a corporate organization. From the plot in Figure 3 and Figure 4, periodic maintenance (filter and oil replacement) occurred 16 times and this is followed by brake related issues that occurred 14 times. Labour charges for various services contributed significantly to the maintenance cost profile. The highest maintenance cost is 15.6397% of the total maintenance cost, and it was incurred for replacing a single propeller shaft. This emphasises how the nature of the fault and the affected component determines maintenance expenses.

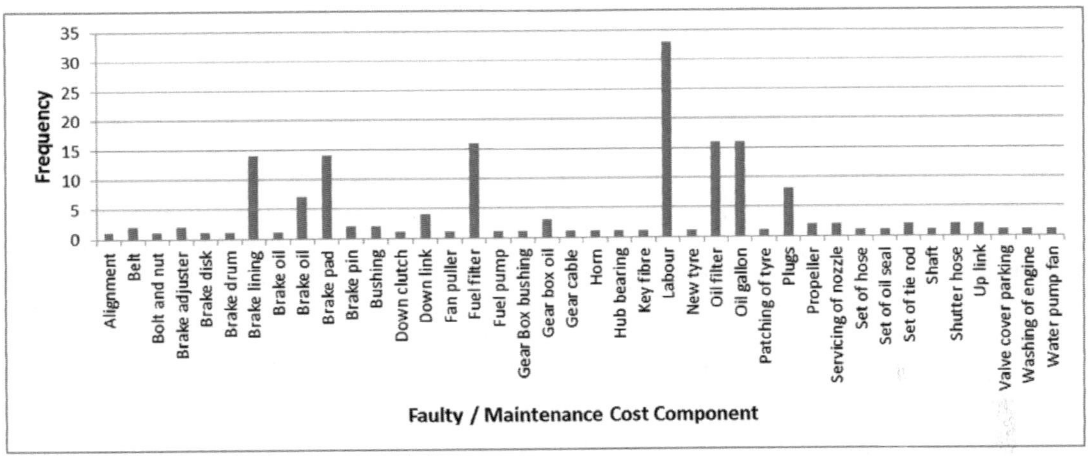

Fig.3 Frequency of occurrence of various car faults

For the 5 months of maintenance cost data for the pool cars, a single tyre was not replaced, this is in contrast to the haulage truck scenario. This could be due to the significant difference in car and truck loading or it could imply that some tyres were recently replaced outside of the 5 months window data. For the cars, the frequency of brake related issues is quite high as compared with that of the trucks. Since the two organizations are in different states, this could be an indication of a more intense stop-and-start traffic in the state of the pool car organization.

Traffic and vehicle statistics can be applied for various studies and analysis. In [9], crash data was collected and analysed for acquiring valuable information about traffic safety. Maintenance planning and management can also be improved by providing valuable statistics about past trends and by showing the relationships between various maintenance parameters. This can help in making informed management decisions. This study carried out prediction analysis for the pool car data using the ANN model in Figure 7. The model was implemented in MATLAB.

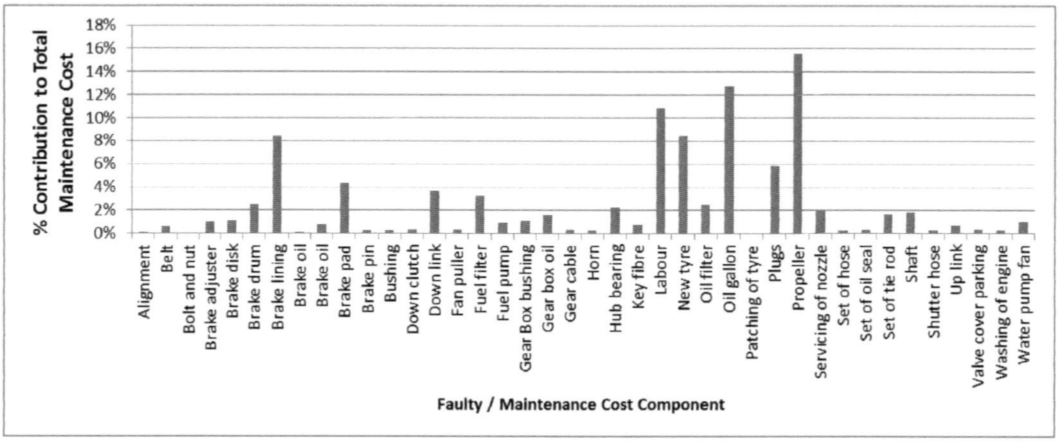

Fig.4 Total car maintenance cost breakdown

3. Maintenance cost prediction using ANN

ANN has found application in various fields of life for data cluster analysis, prediction and so forth. In the study by [10], neural network was applied for vehicle maintenance scheduling. In this study, the goal of the pattern recognition and relationship identification among the parameters is to determine the extent to which the collected vehicle usage parameters can be used to predict the vehicle maintenance cost of the pool cars, and this is achieved using an Artificial Neural Network (ANN) model. The ANN model is deployed using the Nonlinear Input-Output time series tool box in MATLAB. The model is a dynamic neural network, which include tapped delay lines. Through dynamic filtering, the ANN model uses past values of the six vehicle usage parameters, that is, the Fuel Cost (FC) in Nigerian Naira, the Fuel Volume in litres, the car Mileage in km, the Normalized Fuel Cost, the Normalized Fuel Volume, and the Normalized Mileage as model inputs, to predict future Maintenance Cost as the model target.

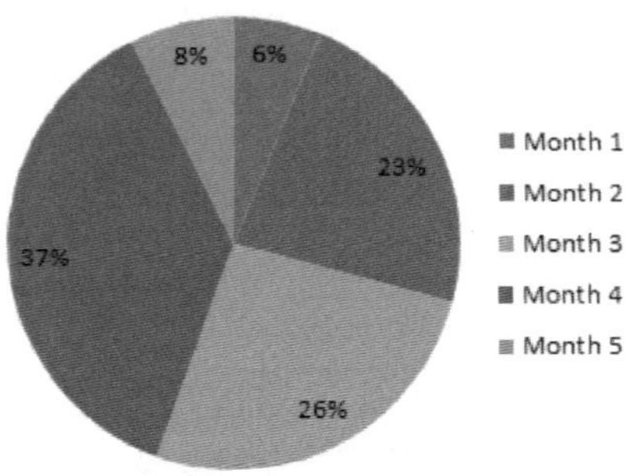

Fig.5 Percentage monthly contribution to the total maintenance cost for 5 consecutive months

The ANN model comprises the inputs, the hidden layer with delays, and the output layer. The hidden layer has 2 delays and 20 neurons with sigmoid activation functions. The analysed data is for a period of 5 months, comprising 111 days of vehicle usage data. The 111 sample dataset is divided into three sets, in the ratio 70:15:15 for training, validation and testing. The ANN network training was performed using the Levenberg-Marquardt back-propagation algorithm, and the performance after the implementation was analysed using Regression Analysis and the Mean Squared Error (MSE).

IOP Conf. Series: Materials Science and Engineering **413** (2018) 012009 doi:10.1088/1757-899X/413/1/012009

The monthly contribution to the total expenditure on fuel and maintenance is shown by the pie charts of Figure 5, while Figure 6 shows the variation in the four key data parameters across the months

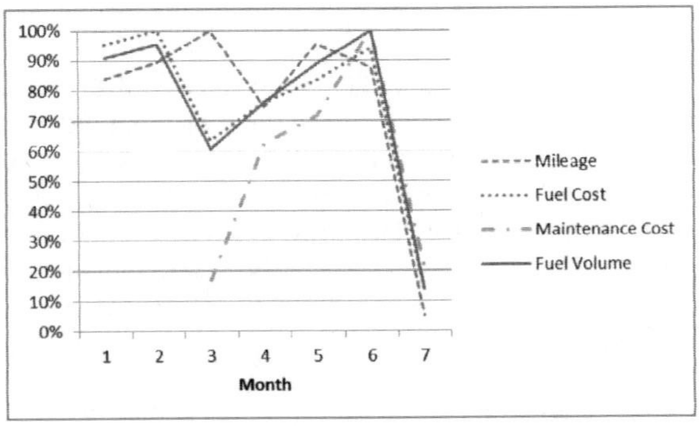

Fig.6 Percentage variation of the 4 study data components across the months

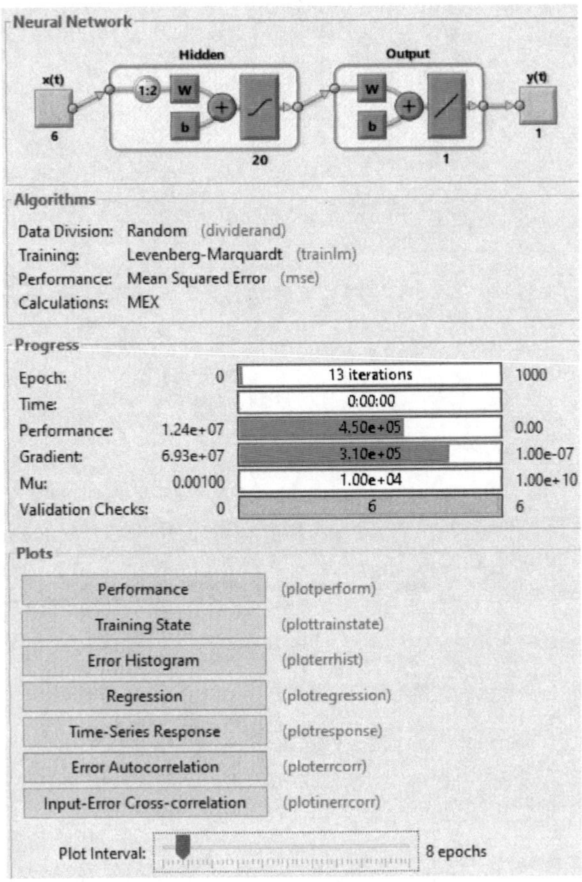

Fig.7 The neural network

4. Results and Discussion

The performance of the model can be analysed using the MSE and the correlation coefficient (R-value) of the regression, which indicates how close the model output prediction is, to the target (maintenance cost). As shown in Figure 8, for the training data set, the R-value is 0.76833, it is 0.76893 for the validation, and 0.80774 for the test. The overall R-value for the three analyses is 0.76645. As shown in Figure 8, the dashed lines in the regression plot represent a perfect line of fit, when the output is exactly equal to the target; this is an ideal situation. The solid lines in the regression plot stands for the real regression line showing the true relationship between the target and the predicted maintenance cost. The R-value can be increased by using more data, say for a period of two years in order to increase the training dataset. The best validation performance of the model as measured using the MSE occurred at epoch (iteration) 7, as shown in Figure 9. The MSE gradually reduced from the peak value until the model execution was terminated after 6 consecutive validation failure checks. Figure 10 shows the variation of the gradient coefficient, the Marquardt adjustment parameter (Mu), the total validation fails and the number of iterations.

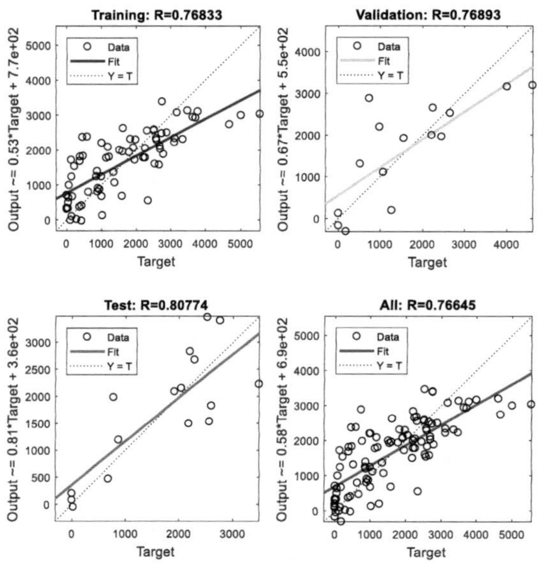

Fig.8 Regression plot

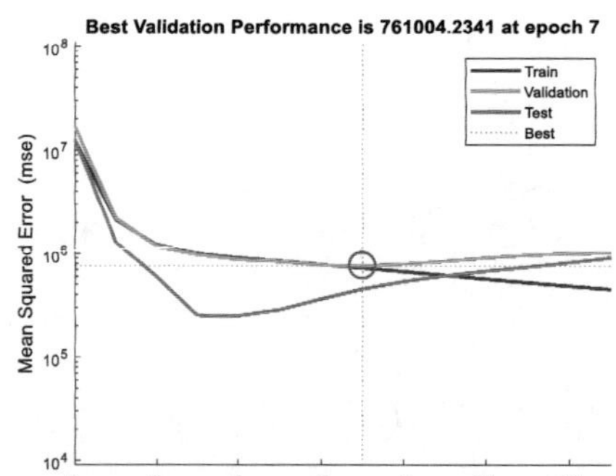

Fig.9 ANN model validation performance plot using mean squared error

IOP Conf. Series: Materials Science and Engineering **413** (2018) 012009 doi:10.1088/1757-899X/413/1/012009

Figure 11 shows the output of the model and the desired target (MC) for the training, validation and model test analysis. The error histogram is shown in Figure 12, and it reveals that the maximum model error for the maintenance cost prediction is 2397 Naira. The error is computed as the difference between the target and the model output. . The magnitude of the error can be reduced by using more training data to improve the accuracy of the model.

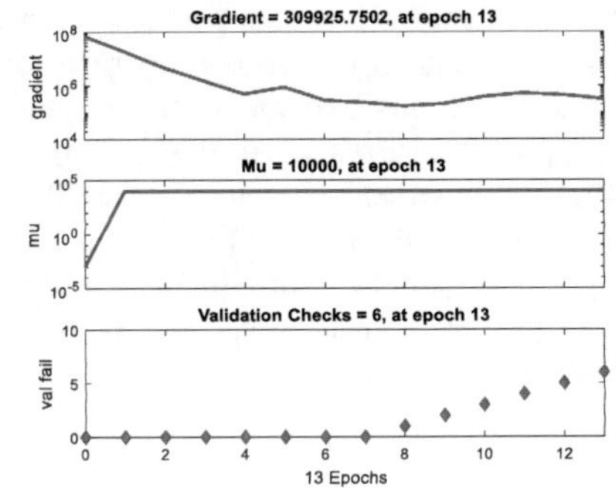

Fig.10 Network Training State

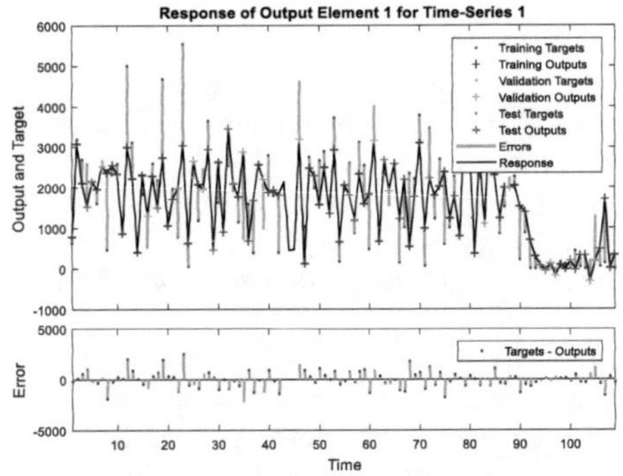

Fig.11 Output element response plot

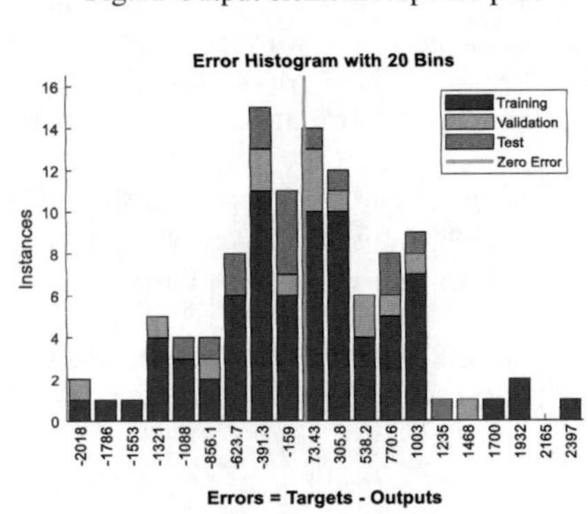

Fig.12 Error Histogram

IOP Conf. Series: Materials Science and Engineering **413** (2018) 012009 doi:10.1088/1757-899X/413/1/012009

Using the trained ANN model, different sets of the vehicle usage parameters were supplied as inputs to the model, and the model predicted the equivalent maintenance cost in Naira. This implies that given an anticipated vehicle mileage, fuel consumption in litres and fuel cost over a period of time, the company can predict how much will be spent on maintaining the pool cars, thereby making this ANN model a useful tool for maintenance budgeting.

5. Conclusion

Maintenance expenses makes up a sizable portion of an organization's budget, and as such it must be cost effectively managed by making informed decisions. Vehicles are vital to corporate business, either as a direct source of income (haulage) or for business support (pool cars). In this study, the maintenance and vehicle fuel consumption operational data of two corporate organizations were collected, studied and analysed. The different types of vehicle faults that make up the maintenance cost components were identified together with their frequency of occurrence. An ANN model was developed using the pool car usage data as input to predict the car maintenance cost. The result shows that there is a significant correlation between the predictor inputs, and the predicted maintenance cost, using a set of fuel volume, fuel cost and car mileage input data. This model can provide useful information for vehicle maintenance cost budget planning. The scope of this research can be extended by collecting other parameters, both qualitative and quantitative to enable the implementation of an improved prediction model.

Acknowledgment

The Authors appreciate Covenant University Centre for Research, Innovation and Discovery for sponsoring our participation at the conference.

References

[1] Eurostat. (2017). Freight Transport Statistics. Available: http://ec.europa.eu/eurostat/statistics-explained/index.php/Freight_transport_statistics
[2] NBS. (2017, Transport Statistics. Available: http://www.nigerianstat.gov.ng/pdfuploads/TRANSPORT.pdf
[3] NBS. (2018, Road Transport Data (Q4 2017). Available: http://www.nigerianstat.gov.ng/pdfuploads/Road_Transport_Data_-_Q4_2017_.pdf
[4] A. Adekitan, "Root Cause Analysis of a Jet Fuel Tanker Accident," International Journal of Applied Engineering Research, vol. 12, pp. 14974-14983, 2017.
[5] E. Baffour-Awuah, "Service Quality in the Motor Vehicle Maintenance and Repair Industry: A Documentary Review."
[6] C. Bolu, Modeling maintenance productivity measurement of engineering production systems: Discrete event simulation approach vol. 13, 2013.
[7] A. O. Adewumi and O. J. Adeleke, "A New Model for Optimizing Waste Disposal Based on Customers' Time Windows and Road Attributes," Applied Mathematical Sciences, vol. 10, pp. 2051-2063, 2016.
[8] EEA. (2011). Most car makers must further improve carbon efficiency by 2015 Available: https://www.eea.europa.eu/highlights/most-carmakers-must-further-improve/
[9] Y. Liu, Z. Li, J. Liu, and H. Patel, "Vehicular crash data used to rank intersections by injury crash frequency and severity," Data in Brief, vol. 8, pp. 930-933, 2016/09/01/ 2016.
[10] S. Kamlu and V. Laxmi, "An effective method for maintenance scheduling of vehicles using neural network," in Lecture Notes in Electrical Engineering vol. 453, ed, 2018, pp. 51-66.

IOP Conf. Series: Materials Science and Engineering **413** (2018) 012010 doi:10.1088/1757-899X/413/1/012010

Mathematical Modelling and Analysis of Human Arm as a Triple Pendulum System using Euler – Lagragian Model

Michael C. Agarana [1, 2], Esther T. Akinlabi [1]

[1]Department of Mechanical Engineering Science, University of Johannesburg, South Africa.
[2]Department of Mathematics, Covenant University, Nigeria.
E- mail address: michael.agarana@covenantuniversity.edu.ng

Abstract. This study attempts to model the human arm as a dynamical triple pendulum system. The equation of motion of te human arm was obtained using Euler-Lagrange equation. The resulted second order differential equation was solved analytically. Simulated results were presented with the aid of a computer software – Maple. It was observed that the angular displacement values of the three segments are directly proportional to their respective angular acceleration, which is in the modelling and analysis of human arm motion as a multiple pendulum system. Generally, the longer the segments of the human arm the longer it takes to swing back and forth, and the fewer back-and-forth swings there are in a second.

1. Introduction

Human parts move when involved in activities such as walking, running dancing, jumping and so on. The proper movement of these parts of the body results in good body balance. Proper functioning of human arms results in a good human activity. Using mechanics and other mathematical concepts for the human body part motion modelling and analysis is constantly expanding and becoming very important in body mechanics via the application of Newtonian mechanics to the human skeletal system [1]. Agarana et al [1] considered the movement of human arm during dance and pointed out the importance of balanced arm movement in a good dance. In dancing, the arm locomotion is one of the most complicated motions of a human body [1-3]. Human body or part of it always strives to maintain balance. So, during any activity the balancing of human arm ensures good and sustained position, at least for a considerable long period of time.

Body mechanics involves the coordinated effort of muscles, bones, the nervous system to maintain balance, posture, and alignment during moving, transferring, and positioning a body. Proper body mechanics allows individual to carry out human activities without excessive use of energy. It also helps in preventing injuries [3, 4]. A balance of the body, during human activities, is maintained via body mechanics. When a vertical line falls from the centre of gravity through the base of support, body balance is achieved, otherwise the body will lose its balance [4]. Balance in this sense means ability to maintain the line of gravity of a body within the base of support with minimal postural sway [4]. In this study, the three human arm parts represent the three segments in the triple pendulum system.

The external forces responsible for the motion of the triple pendulum, can be accounted for by splitting it into a sum of kinetic and potential forces [8, 9]. The Lagrangian [11] was used in this paper to model the equation of motion of the human as a triple pendulum system. The analysis of the dynamics of different segments of the human arm during activities were also carried out. The relevance of this study to body mechanics cannot be overemphasized.

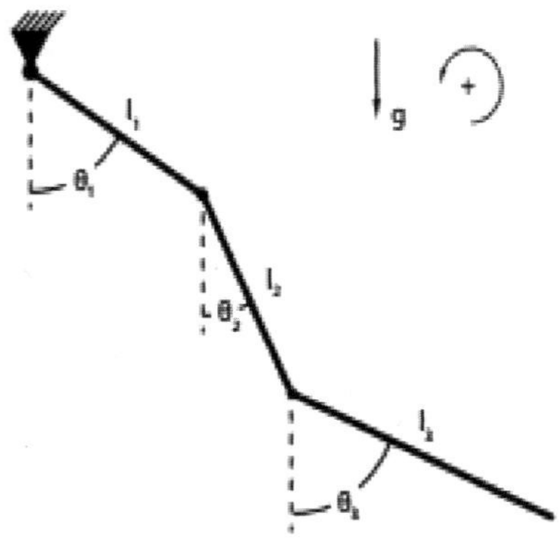

Fig.1. Schematic of Human arm showing the three segments [1]

2. Modelling the Human Arm

The human arm can be simulated as three links, where the upper arm is the first link that is jointed with elbow. The lower part of the arm is connected to the wrist. The third part of the arm is the palm. For simplification sake, the rotation angle from a vertical position of the three human arm parts are denoted as $\theta_1, \theta_2, \theta_3$ respectively. Also the corresponding length of the links are represented by l_1, l_2, l_3 respectively, as shown in schematic of triple pendulum in Agarana's work [1].

2.1. Mathematical Model Formulation

A triple pendulum consists of three pendulum such that one pendulum is attached to another, then to another. Such dynamic system is capable of exhibiting chaotic behaviour. Considering human arm modelled as a triple bob pendulum with masses m_1, m_2, m_3 attached massless wire of lengths l_1, l_2, l_3. The angles the wire make with the vertical, as stated above, are represented as $\theta_1, \theta_2, \theta_3$ respectively. Following the work of Agarana [1], the acceleration due to gravity is g and the positions of the bobs are given respectively as:

(x_1,y_1), (x_2,y_2) and (x_3,y_3)
where

$$x_1 = l_1 sin\theta_1 \quad (1)$$

$$x_2 = l_1 sin\theta_1 + l_2 sin\theta_2 \quad (2)$$

$$x_3 = l_1 sin\theta_1 + l_2 sin\theta_2 + l_3 sin\theta_3 \quad (3)$$

$$y_1 = -l_1 \cos\theta_1 \quad (4)$$

$$y_2 = -l_1 \cos\theta_1 - l_2 \cos\theta_2 \quad (5)$$

$$y_3 = -l_1 \cos\theta_1 - l_2 \cos\theta_2 - l_3 \cos\theta_3 \quad (6)$$

It was assumed in this study that the values of θ_1, θ_2 *and* θ_3 ranges from 0 to 90 degrees. This implies that none of the segments of the human arm during dance should make more than 90 degrees with the vertical.

Going by Agarana [1], the potential energy of the system is given as:

$$V = m_1 g y_1 + m_2 g y_2 + m_3 g y_3 \tag{7}$$

While the kinetic energy of the system is given as:

$$T = \frac{1}{2}\mathrm{m}_1 \mathrm{v}_1^2 + \frac{1}{2}\mathrm{m}_2 \mathrm{v}_2^2 + \frac{1}{2}\mathrm{m}_3 \mathrm{v}_3^2 \tag{8}$$

V and T can be written respectively as:

$$V = -(m_1 + m_2 + m_3) g l_1 \cos\theta_1 - (m_2 + m_3) g l_2 \cos\theta_2 - m_3 g l_3 \cos\theta_3 \tag{9}$$

$$\begin{aligned} T = {} & \frac{1}{2}[(m_1 + m_2 + m_3) l_1^2 \dot{\theta}_1^2 + (m_2 + m_3) l_2^2 \dot{\theta}_2^2 \\ & + m_3 l_3^2 \dot{\theta}_3^2] + m_2 l_1 l_2 \dot{\theta}_1 \dot{\theta}_2 (\cos\theta_1 \cos\theta_2 + \cos\theta_1 \sin\theta_2) \\ & + 2 m_3 l_1 l_3 \dot{\theta}_1 \dot{\theta}_3 (\cos\theta_1 \cos\theta_3) \\ & + m_3 l_2 l_3 \dot{\theta}_2 \dot{\theta}_3 (\cos\theta_2 \cos\theta_3 + \sin\theta_2 \cos\theta_3) \end{aligned} \tag{10}$$

The Lagrangian is given as

$$L = T - V \tag{11}$$

$$\begin{aligned} L = {} & \frac{1}{2}[(m_1 + m_2 + m_3) l_1^2 \dot{\theta}_1^2 + (m_2 + m_3) l_2^2 \dot{\theta}_2^2 + m_3 l_3 \theta_3^2) \\ & + m_2 l_1 l_2 \dot{\theta}_1 \dot{\theta}_2 (\cos\theta_1 \cos\theta_2 + \cos\theta_1 \sin\theta_2) + 2 m_3 l_1 l_3 \dot{\theta}_1 \dot{\theta}_3 (\cos\theta_1 \cos\theta_3) \\ & + m_3 l_2 l_3 \dot{\theta}_2 \dot{\theta}_3 (\cos\theta_2 \cos\theta_3 + \cos\theta_3 \sin\theta_2) \\ & + (m_1 + m_2 + m_3) g l_1 \cos\theta_1 + (m_2 + m_3) g l_2 \cos\theta_2 + m_3 g l_3 \cos\theta_3 \end{aligned} \tag{12}$$

The Euler – Lagrangian equation is given as:

$$\frac{d}{dt}\left(\frac{\partial L}{\partial \dot{\theta}_1}\right) - \frac{\partial L}{\partial \theta_1} = 0$$

Evaluating the Euler-Lagrangian equation for $\dot{\theta}_1, \dot{\theta}_2, \dot{\theta}_3$, for non-stationary values respectively gives:

$$\begin{aligned} & (m_1 + m_2 + m_3)(l_1 \ddot{\theta}_1 - g \sin\theta_1) + m_3 l_2 [\ddot{\theta}_2 \cos\theta_1 \cos\theta_2 \\ & + \ddot{\theta}_2 \cos\theta_1 \sin\theta_2 + \cos\theta_2 \sin\theta_1 + \sin\theta_1 \sin\theta_2 \\ & - \dot{\theta}_2 \cos\theta_1 \sin\theta_2 - \dot{\theta}_2 \cos\theta_2 \sin\theta_1 + \dot{\theta}_2 \cos\theta_1 \cos\theta_2 \\ & - \dot{\theta}_2 \sin\theta_1 \sin\theta_2] + 2 l_2 \dot{\theta}_2 [m_2 \ddot{\theta}_3 \cos\theta_1 \cos\theta_3 \\ & + m_3 \dot{\theta}_1 \dot{\theta}_3 \cos\theta_3 \sin\theta_1 - m_2 \dot{\theta}_3 \cos\theta_1 \sin\theta_3 \\ & - m_2 \dot{\theta}_3 \sin\theta_1 \cos\theta_3] = Q_1 \end{aligned} \tag{13}$$

$$(m_2+m_3)(l_2\ddot{\theta}_2+g\sin\theta_2)+m_3l_1[\ddot{\theta}_1\cos\theta_1\cos\theta_2$$
$$+\ddot{\theta}_1\cos\theta_1\sin\theta_2-\dot{\theta}_1\cos\theta_1\sin\theta_2-\dot{\theta}_1\sin\theta_1\cos\theta_2$$
$$+\dot{\theta}_1\cos\theta_1\cos\theta_2-\dot{\theta}_1\sin\theta_2\sin\theta_1$$
$$+\dot{\theta}_1\dot{\theta}_2\cos\theta_1\sin\theta_2-\dot{\theta}_1\dot{\theta}_2\cos\theta_1\cos\theta_2]$$
$$+m_3l_3[\ddot{\theta}_3\cos\theta_2\cos\theta_3+\ddot{\theta}_3\cos\theta_3\sin\theta_2$$
$$-\dot{\theta}_3\cos\theta_2\sin\theta_3-\dot{\theta}_3\sin\theta_2\cos\theta_3$$
$$+\dot{\theta}_3\cos\theta_2\cos\theta_3-\dot{\theta}_3\sin\theta_3\sin\theta_2$$
$$+\dot{\theta}_2\dot{\theta}_3\sin\theta_2\cos\theta_3-\dot{\theta}_2\dot{\theta}_3\cos\theta_3\cos\theta_2]=Q_2 \quad (14)$$

$$m_3(l_3\ddot{\theta}_3-g\sin\theta_3)+2m_2l_1(\ddot{\theta}_1\cos\theta_1\cos\theta_3$$
$$-\dot{\theta}_1\cos\theta_1\sin\theta_3-\dot{\theta}_1\cos\theta_3\sin\theta_1)$$
$$+m_3l_2(\ddot{\theta}_2\cos\theta_3\cos\theta_2+\ddot{\theta}_2\sin\theta_2\cos\theta_3$$
$$+\dot{\theta}_2\dot{\theta}_3\cos\theta_2\sin\theta_3+\dot{\theta}_2\dot{\theta}_3\sin\theta_2\sin\theta_3$$
$$+\dot{\theta}_2\cos\theta_2\cos\theta_3-\sin\theta_2\sin\theta_3$$
$$-\cos\theta_2\sin\theta_3-\cos\theta_3\sin\theta_2)$$
$$+2m_3l_1\dot{\theta}_1\dot{\theta}_3\cos\theta_1\sin\theta_3=\mathrm{Q}_3 \quad (15)$$

Where, Q_1, Q_2, Q_3 are the (non-conservative) generalized forces.

3. Model Solution

From figure 1 above;

$$Sin\theta_1=\frac{x_1}{l_1}, Sin\theta_2=\frac{x_2}{l_2}, Sin\theta_3=\frac{x_3}{l_3}$$

$$\Rightarrow \theta_1=Sin^{-1}\frac{x_1}{l_1}, \theta_2=Sin^{-1}\frac{x_2}{l_2}, \theta_3=Sin^{-1}\frac{x_3}{l_3}$$

$$\Rightarrow \theta_1=Sin^{-1}\frac{x_1}{\sqrt{y_1^2+x_1^2}}, \theta_2=Sin^{-1}\frac{x_2}{\sqrt{y_2^2+x_2^2}}, \theta_3=Sin^{-1}\frac{x_3}{\sqrt{y_3^2+x_3^2}}$$

Table 1: The angular displacement, lengths of segments and different coordinates

(x_1,y_1)	(1,1)	(1,2)	(2,4)	(3,5)	(5,2)
θ_1	0.707	0.447	0.447	0.514	0.928
l_1	1.414	2.236	4.472	5.831	5.385
(x_2,y_2)	(7,4)	(8,6)	(10,7)	(10,9)	(11,11)
θ_2	0.868	0.800	0.819	0.743	0.707

l_2	8.062	10.000	12.207	13.454	15.556
(x_3,y_3)	(12,12)	(13,12)	(13,14)	(14,15)	(15, 15)
θ_3	0.707	0.735	0.680	0.682	0.707
l_3	16.971	17.692	19.105	20.518	21.213

In order to solve for the masses, m, m, and m, the values of the parameters in table 1 are substituted into equations (31), (32) and (33) to become

$$(m_1+m_2+m_3)(-g\sin\theta_1)+m_3l_2[\cos\theta_2\sin\theta_1+\sin\theta_1\sin\theta_2]=Q_1 \quad (16)$$
$$(m_2+m_3)(g\sin\theta_2)=Q_2 \quad (17)$$
$$m_3(-g\sin\theta_3)+2m_2l_1(-\cos\theta_2\sin\theta_3-\cos\theta_3\sin\theta_2)=Q_3 \quad (18)$$

For the purpose of this research work let Q_1, Q_2, Q_3 be assigned respectively the following values; 3,2,1.
When the three masses are at the position (1,1), (7,4) and (12,12) respectively, the absolute values of the masses can be evaluated from the following equations:

$$-0.121m_1-0.121m_2+15.044m_3=3 \quad (19)$$
$$0.149m_2+0.149m_3=2 \quad (20)$$
$$0.121m_3+0.08m_2=-1 \quad (21)$$

m_1= 6378.68, m_2= 64.01, m_3= 50.59
Similarly, on position (2,4), (10,7) and (13,14), the absolute values of the masses are:

m_1= 26.78, m_2= 11.5, m_3= 26.1

For position (1,2), (8,6) and (13,12), the absolute values of the masses are:

m_1= 4.94, m_2= 22.91, m_3= 37.2

For position (3,5), (10,9) and (14,15), the absolute values of the masses are:

m_1= 8.46, m_2= 6.35, m_3= 7.29

For position (5,2), (11,11) and (15,15), the absolute values of the masses are:

3.1 Periods and Frequency of different segments of the pendulum system

The period of the motion for a pendulum is how long it takes to swing back-and-forth, measured in seconds. The frequency of a pendulum is how many back-and forth swings there are in a second, measured in hertz [12]. Frequency F is the reciprocal of the period T: Looking at the three segments as different simple pendulums joined together, The periods of each segment, using the parameters obtained above, are obtained as follows:

IOP Conf. Series: Materials Science and Engineering **413** (2018) 012010 doi:10.1088/1757-899X/413/1/012010

The period of a simple pendulum is given as [20]:

$$T = 2\pi\sqrt{\frac{l}{g}} \tag{22}$$

Where l is the length and g is the acceleration due to gravity.

The periods for the three segments with different lengths, assuming a fixed initial position of the masses at (3,5), (10,9), (14,15) respectively, are shown in tables 2, 3, and 4 respectively.

Table 2: The values of the periods and frequency of first segment at given lengths.

Length of 1st Segment	Length	T_1 F_1
l_1	5.83	4.82 0.21
$2l_1$	11.66	6.85 0.15
$3l_1$	17.49	8.38 0.12
$4l_1$	23.32	11.39 0.09
$5l_1$	29.15	10.82 0.092
$6l_1$	34.98	11.87 0.084

Table 3: The values of the periods and frequency of second segment at given lengths

Length of 2nd Segment	Value	T_2 F_2
l_2	13.45	7.36 0.14
$2l_2$	26.9	10.41 0.096
$03l_2$	40.35	12.74 0.078
$4l_2$	53.8	14.72 0.068
$5l_2$	67.25	16.46 0.061
$6l_2$	80.7	18.03 0.055

Table 4: The values of the periods and frequency of third segment at given lengths

Length of 3rd Segment	Value	T_3 F_3
l_3	20.52	10.74 0.093
$2l_3$	41.04	15.19 0.066
$3l_3$	61.56	18.60 0.054
$4l_3$	82.08	21.48 0.047
$5l_3$	102.6	24.02 0.042
$6l_3$	123.12	26.31 0.038

4. Results and Discussion

To maintain balance during human activities, the required masses at the end of each segment of the human arm and their positions were calculated analytically. The values revealed that the position of the arm segments at every point in time, the mass at the end of each segment and the length of the segment are all important in the body mechanics analysis. From tables 2, 3, and 4, it can be seen that there is a general positive correlation between the length of the human arm segments and period but a negative correlation with the frequency. It was also observed from the study that the masses required at the end of the segments of the human arm depends on the position of that segment at a point in time, as shown the result from equations 21, 22, and 23. There is also a positive correlation between the angular displacement, angular acceleration and angular acceleration.

5. Conclusion

This paper analytically modelled the dynamics of human arm as a triple pendulum system in motion. The angular displacements were determined by the simulated positions of the three segments of the human arm. Each of the three simple pendulums that form the triple pendulum represents each of the three segments of the human arm, namely; the upper arm, the lower arm and the hand. With Euler - Lagrange equations, the equations of motion of the triple pendulum were obtained. The solution to these equations reveal the dynamics of the segments of human arm. The sensitivity and interrelationship of the parameters were studied. For a good body mechanics to be achieved, especially about the locomotion of human arm, the analytical results of this paper give a clue of how stability and balance of the whole human arm movement during an activity can be supported.

References

[1] Michael C. Agarana, Oluseyi O. Ajayi, Analytical Modelling of Human Arm Locomotion during Dance as a Dynamical Multiple Pendulum System. Lecture notes in Engineering and Computer Science 2017, London, UK.

[2] Loay A. Al-Zu'be et al, Modelling and Simulating Human Arm Movement using a 2 Dimensional 3 Segments Coupled Pendulum System. World Academy of Science, Engineering and Technology International Journal of Medical, Health, Biomedical, Bioengineering and Pharmaceutical Engineering, 2012, 6(11)

[3] Winter, D.A., Biomechanics of human movement with applications to study of human locomotion. Crit Rev Biomed Eng, 1984, 9(4): 287-31

[4] Perry, Potter, Ostendorf, 2014. https://opentextbc.ca/clinicalskills/chapter/3-2-body- mechanics/

[5] Nick Eyre, Jeff Holzgrafe, Modelling of a Damped Triple Pendulum. Franklin W. Olin College of Engineering. ENGR 2340: Dynamics https://www.nickeyre.com/images/triplependulum.pdf

[6] Michael C. Agarana, Olasumbo .O.Agbola, Dynamic Analysis of damped driven pendulum using Laplace transform international journal of Mathematics and computation, 2015. 26(3): 98-107

[7] Michael C. Agarana, Moses .E. Emetere, Solving non-linear damped driven simple pendulum with small amplitude using a semi analytical method. ARPN Journal of Engineering and Applied Sciences, 2016: 11(7)

[8] Bannister Ross, The Double Pendulum. University of Reading, June 2001. Web. 09 Dec. 2012.

[9] Michael C Agarana, S.A Iyase, Analysis of Hermite's Equation Governing the motion of Damped Pendulum with small displacement, International Journal of Physical Sciences, 2015, 10(2): 364 – 37

[10] Agarana Michael C, Bishop Shela .A, Quantitative Analysis of Equilibrium Solution and Stability for Non-linear Differential Equation Governing Pendulum Clock. International Journal of Applied Engineering Research, 2015, 10(24):.44112-44117

[11] Lagrangian mechanics. From Wikipedia, the free encyclopaedia https://en.wikipedia.org/wiki/Lagrangian_mechanic

[12] https://www.school-for-champions.com/science/pendulum_equations.htm#.WpFSKnwo_IU

IOP Conf. Series: Materials Science and Engineering **413** (2018) 012011 doi:10.1088/1757-899X/413/1/012011

Modelling of A36 Steel Plate Dynamic Response to Uniform Partially Distributed Moving Iron Load using Differential Transform Method

Agarana Michael C.[1,2], Akinlabi Esther T.[1]

[1]Department of Mechanical Engineering Science, University of Johannesburg, South Africa.
[2]Department of Mathematics, Covenant University, Nigeria. E-mail address: michael.agarana@covenantuniversity.edu.ng

Abstract. In the present paper the authors focus on dynamic response of A36 steel plate supported by a simple subgrade, under a moving iron load. The fourth order partial differential equations governing the dynamic behavior of the plate were transformed into their algebraic forms using differential transform method. The new set of algebraic equations were solved analytically. Computer software - Maple was deployed to plot the three-dimensional (3D) graphs. The results obtained revealed that the absolute value of the deflection of the steel was very high when the mass of the iron load is highest at constant velocity. Also, the A36 steel plate deflected more under a moving iron load especially with a high velocity and less foundation rigidity.

1. Introduction

A36 steel, usually referred to as ASTM A36 steel, is a common structural steel whose standard was established by the ASTM International [1]. ASTM A36 steel is one of the most widely used low carbon structural steels. Its plate can be used for general structural purposes. For instance, it can be welded in the construction of building and bridges. It is equivalent to EN S275 steel plate [1]. It exhibits good strength and can be galvanized to provide increased corrosion resistance. A36 plate application depends on the thickness and corrosion resistance of the alloy. Some of the products manufactured using A36 structural steel plate are: Buildings, Cabinets, Pipe and tubing [2]. Apart from the fact it can be securely welded, it is easy to machine and fabricate. It can have fabricated into different forms including Plates, Structural Shapes, Bars, Sheets, Girders, Angle iron, T iron. Some properties of A36 steel are represented in table 1[2,3]. During production, sometimes loads made of iron traverses through the ASTM A36 steel plate which causes vibration of the plate. The traversed load is assumed to uniformly partially distribute [4].
Differential transform method (DTM) is a semi analytical method. It involves the transformation of differential equations to their algebraic forms. DTM is relatively simple to apply compared to other methods used in solving such problem, which computationally intensive.
Many researchers [5,6,7,8,9] have worked in this area. Some of the authors include: Y. Kumar [10], who studied the free vibration analysis of isotropic rectangular plates on Winkler foundation using differential transform method. He obtained the Characteristic equations in the form of infinite series and solved them numerically using a computer program developed in C++ to obtain natural frequencies. The results obtained show reliability and fast convergence of the method for rectangular plates.
Agarana et al also studied application of differential transform method to vibration analysis of damped railway bridge on Pasternak foundation under moving train. He applied differential transform method to analysis the vibration of orthotropic damped rectangular railway bridge supported by Pasternak foundation. The results obtained revealed that both the foundation modulus and damping have effects on the deflection of the railway bridges which was modelled as the plate in this work [11,12,13]
None of the authors have considered specifically the dynamic response of A36 steel plate, supported by a subgrade under a moving iron load using DTM, without the effect of damping and considering simple foundation. This is what this paper attempt to address, because of the importance of this type of steel.

IOP Conf. Series: Materials Science and Engineering **413** (2018) 012011 doi:10.1088/1757-899X/413/1/012011

Table 1: Mechanical properties and their values for A36 steel plate.

Mechanical Properties of A36 steel plate	**Values**
Density	7,800 kg/m^3 (0.28 lb/cu in).
Young's modulus	200 GPa (29,000,000 psi)
Poisson's ratio	0.26
shear modulus	75 GPa (10,900,000 psi).
minimum yield strength (thickness of plate <8in)	36,000 psi (250,000 kPa)
ultimate tensile strength (thickness of plate <8in)	58,000–80,000 psi (400–550 MPa)
minimum yield strength (thickness of plate >8in)	32,000 psi (220 MPa)
ultimate tensile strength (thickness of plate >8in)	58,000–80,000 psi (400–550 MPa)

2. Formulation of Problem

Following Agarana et al, the equation governing the vibration of damped simply supported orthotropic plate resting on Pasternak foundation under a moving load can be written as [14,15]:

$$\alpha_1 \frac{\partial^4 w}{\partial x^4} + 2\alpha_2 \frac{\partial^4 w}{\partial x^2 \partial y^2} + \alpha_3 \frac{\partial^4 w}{\partial y^4} + Kw + m\frac{\partial^2 w}{\partial t^2} + 2m\gamma \frac{\partial w}{\partial t} + G_1\left(\frac{\partial^2 w}{\partial x^2} + \frac{\partial^2 w}{\partial y^2}\right) = 0 \quad (1)$$

Where,

w = w(x,y,t) is the deflection of the plate at the point (x, y), m is the mass density per unit area, K, G1 is the foundation stiffness, t is the time in seconds, E is the Young's modulus, v is the velocity, H is the thickness of plate, γ is the viscous damping coefficient, α_1 is the flexural rigidity in the x direction, α_2 is the flexural rigidity in the y direction, and α_3 is the effective torsional rigidity. For this paper, the simplest form of foundation was considered. The effect of damping was neglected also. Therefore equation (1) becomes:

$$\alpha_1 \frac{\partial^4 w}{\partial x^4} + 2\alpha_2 \frac{\partial^4 w}{\partial x^2 \partial y^2} + \alpha_3 \frac{\partial^4 w}{\partial y^4} + Kw + m\frac{\partial^2 w}{\partial t^2} = 0 \quad (2)$$

With initial and boundary conditions [16]:
Following Kumar, the deflection function w, can be written as [17]

$$w = \overline{w}\sin(my\pi), \ -\infty \le \mathrm{m} \le +\infty \quad (3)$$

The last term in equation (2), $m\frac{\partial^2 w}{\partial t^2}$, is a force. It can be written as $\Omega\overline{w}$.
Where Ω is the frequency parameter, given as:

$$\Omega^2 = \frac{12(1-v^2)}{Eh^3} \tag{4}$$

Substituting equation (3) into (2) leads to;

$$\alpha_1 \frac{d^4\overline{w}}{dx^4} - 2\alpha_2\pi^2 m^2 \frac{d^2w}{dx^2} - (\Omega^2\alpha_3\pi^4 m^4 - k)\overline{w} = 0 \tag{5}$$

3. Problem Solution

Equation (5) can be solved using differential transform method (DTM). Before solving the equation, let the following values of the parameters be substituted.

Table 2: Mechanical properties of A36 steel and parameters values

Mechanical properties of A36 steel	Notation	Value
Young modulus	E	200
Shear modulus	G	75
Flexural rigidity	D	0.0021
Density		7800
Poisson ratio	v	0.26
Frequency parameter		21.16
mass	m	5
Foundation stiffness	k	100
Height of the plate	h	0.05
Torsional constant	J	36

Substituting specific A36 steel plate values for flexural rigidity in x and y directions, effective torsional rigidity, mass density per unit area, and the foundation stiffness, into equation (5) after dividing by π^4 gives:

$$\alpha_1 / \pi^4 \frac{d^4\overline{w}}{dx^4} - 2\alpha_2 m^2 / \pi^2 \frac{d^2w}{dx^2} - (\Omega^2\alpha_3 m^4 - k)\overline{w} = 0 \tag{6}$$

$$\alpha_1 / 180^4 \frac{d^4\overline{w}}{dx^4} - 0.0015\alpha_2 \frac{d^2w}{dx^2} - (279841\alpha_3 - 100)\overline{w} = 0 \tag{7}$$

Following the works of Kumar and Agarana;

$$\frac{d^4\overline{w}}{dx^4} = a\frac{d^2w}{dx^2} + b\overline{w} \tag{8}$$

where

$$a = \frac{0.0015\alpha_2}{\alpha_1/180^4}, b = \frac{279841\alpha_3}{\alpha_1/180^4}$$

$$DT[\overline{w^4}] = a\overline{w^2} + b\overline{w} \tag{9}$$

$$\Rightarrow (k+1)(k+2)(k+3)(k+4)Y(k+4) = a(k+1)(k+2)Y(k+2) + bY(k)$$

$$\therefore Y(k+4) = \frac{1}{(k+1)(k+2)(k+3)(k+4)}[a(k+1)(k+2)Y(k+2) + bY(k)] \tag{10}$$

With initial conditions;

Y(0) = 0, Y(1) = -1, Y(2) = 1, Y(3) = -2

For k = 0:

$$Y(4) = \frac{1}{24}[2a] \tag{11}$$

For k= 1:

$$Y(5) = \frac{1}{120}[-12a - b] \tag{12}$$

For k = 2:

$$Y(6) = \frac{1}{360}[a^2 + b] \tag{13}$$

For k = 3:

$$Y(7) = \frac{a}{6}[-12a - b] \tag{14}$$

and so on.

Generally, for k+4 = n, (k = 0,1,2,3…) we have

$$Y(n) = \frac{a}{12}, \frac{-12a-b}{120}, \frac{a^2+b}{360}, a\left(\frac{-12a-b}{6}\right), \ldots \tag{15}$$

It was assumed in this study that the flexural rigidity in both the x and y directions of the plates have the same value of 0.0021. It was also assumed that the plate is a square with length of 4 units. Therefore, the values of a and b can be evaluated as follows:

IOP Conf. Series: Materials Science and Engineering **413** (2018) 012011 doi:10.1088/1757-899X/413/1/012011

$$a = \frac{0.0015\alpha_2}{\alpha_1 / 180^4} = 1575000$$

$$b = \frac{279841\alpha_3}{\alpha_1 / 180^4} \approx 0.$$

Therefore equation (15) becomes:

$$Y(n) = 131250, -157500, 6890625000, -4961250000000, ... \quad (16)$$

Following Edeki's work [9], it leads to:

$$y(x) = \sum_{n=0}^{n=\infty} Y(n)x^n \quad (17)$$

$$\therefore\ y(x) = Y(0) + Y(1)x + Y(2)x^2 + ...$$

$$= 0 + 131250x - 157500x^2 + 6890625000x^3 - 4961250000000x^4 + ... \quad (18)$$

$$\therefore\ w(t) = 0 + 131250t - 157500t^2 + 6890625000t^3 - 4961250000000t^4 + ... \quad (19)$$

For t = 0,1,2,3,... respectively gives;

$$w(0) = 0$$
$$w(1) = -4954359401000$$
$$w(2) = -79324875630000$$
$$w(3) = -401861250000000$$

For velocity, equation (19) is differentiated to give:

$$w^{/}(t) = 0 + 131250 - 315000t + 20671875000t^2 - 19845000000000t^3 + ... \quad (20)$$

For t = 0,1,2,3,..., respectively gives;

$$w^{/}(0) = 131250$$
$$w^{/}(1) = -19824328310000$$
$$w^{/}(2) = -158677313000000$$
$$w^{/}(3) = -535628953900000$$

For acceleration, equation (20) is differentiated to give:

$$w^{//}(t) = 0 + 131250 - 315000 + 41343750000t\ - 59535000000000t^2 + ... \quad (21)$$

For t = 0,1,2,3,..., respectively gives;

$$w''(0) = -183750$$

$$w''(1) = -59493656430000$$

$$w''(2) = -238222687300000$$

$$w''(3) = -535690968900000$$

The above shows the truncated values of the deflection, velocity of the deflection and acceleration of the deflection after the fourth terms.

4. RESULT DISCUSSIONS

Putting the assumptions into consideration, the results obtained show that the absolute values of the deflection, the velocity of the deflection and the acceleration of the deflection increase as time increases, within the range of time considered (t = 0,1,2, 3). Equations (19), (20) and (21) show the huge dynamic response after time zero. The deflection, the velocity of the deflection and the acceleration of the deflection are well pronounced. However, comparing the increase in angular velocity with the increase in angular acceleration at any given time, the increase in angular acceleration is more than the increase in angular velocity. Similarly comparing values of the deflection, velocity of the deflection and acceleration of the deflection, at a time, shows that it is most pronounced in the acceleration, followed by the velocity, then the deflection itself. The reason for the huge deflection and its rate with time can be attributed to the iron load and the type of plate, the A36 steel plate, bearing in mind that damping was neglected in this study.

5. CONCLUSIONS

A36 steel plate's dynamic response to uniform partially distributed moving iron load was modelled and analyzed using differential transform method. The results obtained are put in table form and agree with the ones in literature. Deductions are made from the results. The differential transformation method is an effective method for solving the type of differential equation problem being considered in this paper. It is also easy to handle, because the differential equations are transformed into their algebraic form. The mechanical properties of A36 steel plates have effects on the dynamics of the system.

REFERENCES

[1] https://www.leecosteel.com/a36-steel.html
[2] http://www.onealsteel.com/carbon-steel-plate-a36.html
[3] https://www.leecosteel.com/a36-steel.html
[4] Michael C. Agarana, A.N Ede, Application of Differential Transform Method to Vibration Analysis of Damped Railway Bridge on Pasternak Foundation under Moving Train. Lecture notes in engineering and computer science: proceedings of the world congress on Engineering 2016, June 29 – July 1, London UK.
[5] Z. Odibat, Differential Transform Methods for Solving Volferra Integral Equations with Separable, Kernels, Mathematical and Computer Modeling, 2008, 48: 1144-1146.
[6] M Thongmoon and S. Pusjuso, 2010, The Numerical Solutions of Differential Transform Method and the Laplace Transform Methods for a System of Differential Equations. Nonlinear Analysis: Hybrid Systems, 2010, 4: 425-431

IOP Conf. Series: Materials Science and Engineering **413** (2018) 012011 doi:10.1088/1757-899X/413/1/012011

[7] A Arikoglu, I. Ozkol, Solution of Boundary Value Problems for Integro-Differential Equations by Using Transform Method, Applied Mathematics and Computation, 2005, 168: 1145-1158

[8] H. Abdel - Halim Hassan, Application to differential transformation method for solving systems of differential equations. Applied Mathematical Modeling. 2008 32: 2552-2559.

[9] Sunday O Edeki, et al, A Semi-Analytical Method for Solutions of a Certain Class of Second Order Ordinary Differential Equations, Applied Mathematics, 2014, 5: 2034-20

[10] Y, Kumar, Differential transform method to study free transverse vibration monoclinic rectangular plates restimg on Wimkler foundation, Applied and Computational mechanics, 2013, 7: 145-

[11] Michael.C Agarana, M.E Emetere, Solving Non-Linear Damped Driven Simple Pendulum with Small Amplitude Using A Semi Analytical Method, ARPN Journal of Engineering and Applied Science, 2016, 11(7): 4478-4484

[12] Michael C. Agarana, Olasumbo O. Agboola, Dynamic Analysis of Damped Driven Pendulum using Laplace Transform Method, International journal of Mathematics and Computation, 2015, 26(3): 98-109.

[13] Michael C. Agarana, S. A. Iyase, Analysis of Hermite's Equation Governing the Motion of Damped Pendulum with Small Displacement. International Journal of Physical Sciences, 2015, 10(2): 364-370

[14] L. E. Millet, The Large – angle Pendulum Period. Phys. Teach, 2003, 41: 162-163

[15] Michael C. Agarana, and J. A. Gbadeyan, Finite Difference Dynamic Analysis of Railway Bridges Supported by Pasternak Foundation under Uniform PartiallyDistributed Moving Railway Vehicle. Lecture Notes in Engineering and Computer Science: Proceedings of The World Congress on Engineering and Computer Science, 21-23 October, San Francisco, USA, 2015: 996-100

[16] Michael C. Agarana, J. A. Gbadeyan, O. O. Agboola, T. A. Anake, O. J. Adeleke, Dynamic Response of an inclined Railway bridge supported by Winkler Foundation under a Moving Railway Vehicle. Australian Journal of Basic and Applied Sciences, 2015, 9(11): 355-361.

[17] Y. Kumar, Free Vibration analysis of Isotropic Rectangular plates on Winkler Foundation using differential transform method, International Journal of Applied Mechanics and Engineering, 2013, 18(2): 589-597

IOP Conf. Series: Materials Science and Engineering **413** (2018) 012012 doi:10.1088/1757-899X/413/1/012012

Influence of Cement, Bitumen and Lime on Some Lateritic Soil Samples as Pavement Material

Agashua, Lucia. O; Ogbiye, Adebanji .S

Department of Civil Engineering, Covenant University, Ota, Ogun State, Nigeria
*Corresponding author: ade.ogbiye@covenantuniversity.edu.ng; mobile: +2348033880225

Abstract- This study was carried out in order to determine cement, lime and bitumen prerequisite of some lateritic soil specimens as sub base materials. Soil specimens A, B, C was collected from Dual carriageway road construction project at Estate Iwo road, Osogbo, Osun state and stabilized with 0, 3, 6, 9, and 12% of cement, lime and bitumen. Various laboratory test was conducted on the soil specimen such as California Bearing Ratio (C.B.R), Compaction, Atterberg's limit test and particle size distribution. The investigation revealed that beneficial effects were obtained by the addition of cement, Lime and Bitumen to improve the strength of weak or poor soils. The C.B.R values are 9.88%, 4.36% and 7.27% for sample A, B and C respectively at 0% additive content and at 12% additives, the samples gave C.B.R values of 55%, 50.0% and 50.00% by using cement. All the three samples have maximum C.B.R values of 45.0% with lime as additives. For using bitumen, result gave maximum C.B.R values of 41% for sample A and B and Sample C have 40.0%. Addition of additives improved the samples from a poor subbase material to an excellent subbase material with C.B.R values over 30%. Hence, soil improvement of laterite samples from the studied site can be carried out successfully by using Lime or bitumen judging from the result of the various tests on the different soil – additive mixtures.

1. Introduction

Country developments have stirred knowledge of economic resources management in the nation, hence people are being stimulated to be assiduous at resources earlier reprehend and provide way for such materials to be recycle (Akinje, 2015). Currently, sites such as abandoned sanitary landfills (Garbage dumps), swamps, bays, mashes, hillsides and other poor areas are being used for construction sites, this trend expected to both continue and accelerate when alternative sites are not available or environmental considerations, citizen opposition and zoning regulation severely limit the option available, therefore it becomes more necessary to modify or stabilize the available soil to obtain the needed properties (Amu et al., 2011). In cases such as earthdams, embankments or other fills, where selected materials in sufficient quantities may not be available, selective use of the available materials and understanding of both the function of the earth structure and the mechanics of the earth mass, can produce a satisfactory solution via use of zoned construction (Ali, 2012). Tropical regions lateritic soils fill about 23% land surface, choose for highway materials usage can be economically seen (Akinwumi et al., 2012). Ali (2012) attest that soil stabilization is the technique in improving soil engineering and physical properties so as to acquire predetermined goals. It may be any or a combination of one or more of the following; Mechanical densification with various types of mechanical equipment as rollers, falling weights explosives, static pressure, fabrics, freezing and so on; additives – Gravel to cohesive soils, clay to granular soils, Chemical additives such as Portland cement, Lime fly ash (by product from coal burning) often with lime or Portland cement, asphalt cement, sodium and calcium chlorides paper mill waste and others (Wright and Dixon, 2013; Osinubi and Amadi2010). To stabilize unpaved roads, addition of gravel layers to be partially worked into a clayey road surface is common, in fine grained soils, it involves excavating to a depth and blending the excavated soil with Portland cement, fly ash, lime with sufficient water. Addition of lime, fly ash and sometimes portland cement is made in clayey deposits particularly those subject to large volume changes to effect a ion exchange to reduce the activity of the clay minerals. It could be obtained by addition of aggregate, lime, salt, bitumen, resinous materials, cement and calcium chloride (Akinje, 2015) and (Kadyali and Lai, 2008). Engineers are facing most economical stabilization technique problem at a particular project depends on stabilizer material readily availabe for use and right choice at certain time couple with strength achievement for road design and construction work (Osinubi and Amadi, 2010); Stabilizer like lime, bitumen and cement increases parameters of soil strength, improve capability and reduces settlement of soil at lower cost especially in construction works that needs greater volume of soil improvement (Zhang and Yi, 2011) and (Latifi et al., 2013). Traditional stabilizer such as fly ash, lime, bituminous materials and cement; non traditional stabilizer involves acids enzymes, lignin derivatives, liquid polymers, silicates, resins and ions have lesser water content when soil grain is coarse (Akinwumi et al., 2012; Garber and Hoel,2010). Cement an hydraulic binder; finely grinded inorganic material made from materials which contain the proper proportions of lime, silica, alumina and iron with minor amounts of magnesia and sulphur trioxide, when reacted with water forms a paste that set as well hardens through hydration reactions means and processes, retains strength after hardening and stable under water, whereas bitumen a genetic names applied to various mixtures of hydrocarbons gaseous, liquid, semisolid or solid in nature and is completely soluble in carbon disulphide, most common materials within this family of bitumen are tars, pitches and asphalts. It has a number of properties which make them useful in the construction industry, it has tendency to adhere to a solid surface and this adhesiveness will depend on the surface and the state of the bitumen. Types are: Tar – used to saturate felt paper and to coat Kraft paper to render it waterproof; Pitch-coal tar – used in making pitch (after distillation of coal) and gravel built-up roofs; asphalt- results from the refining of naphtha base crude oils which produce aviation – grade gasoline feel soil, cold – test lubricating oils and asphalt (Wright and dixon, 2013; Latifi et al., 2013).

2. Materials and Methodology

Study Area

The materials used for this research work are lateritic soil specimen, cement, lime, bitumen and water. Lateritic soil specimen used in this research was collected from three location at Okada estate Oke – fia bye pass road, NUJ, Moye petrol station, Dada Estate Osogbo, Osun State, Nigeria and was labelled A – C. Ungoing construction works exposed the soil strata and the specimen was taken at 2m depths using Trial Pit method. The samples were air dried to remove excess water, the additives used are cement, bitumen and lime. The cement and slaked lime used for these tests was Portland cement purchased from the Oja Oba market Osogbo and properly stored in the laboratory to prevent unnecessary hardening while S 125 bitumen used was gotten from Dekit Construction Company, Dada Estate Osogbo, Osun State Nigeria. Potable water that was treated in the laboratory were used throughout the test, Tests on the lateritic soil specimen was carried out in the Department of Civil Engineering Laboratory, Federal University of Technology Akure, Ondo state, Nigeria, collected samples were gently broken down for sieving and soaked for24 hours before carrying out the test. Atterberg samples were mixed to near plastic limit and allowed to equilibrate for 24 hours before testing while compaction and C.B.R were carried out using standard methods by American Association of State and Transportation Officials (AASHTO, 2007) for 0%, 3%, 6%,9% and 12% additive contents for strength comparison.

Sampling and Laboratory Analysis

Particle Size Distribution expresses the size of particle in a soil in terms of percentages by weight of boulders, cobbles, gravel, sand, silt and clay and results given in form of fractions by weight of different size grades. This expressed as percentage of the whole sample and summed to obtain a cumulative percentage, curves are then plotted on a semi-logarithmic paper to give a graphical representation of the particle size distribution, slope of the curve provides an indication of the degree of soiling also determine the percentage by weight of particle within the different size ranges. The weight of soil retained in each sieve is determined and the cumulative percentage by weight passing each sieve is calculated. The Atterberg are water contents which define the various stages of consistency for a given soil. Liquid limit

(W_l) is the water content at which 25 blows of the liquid limit machines closes a standard groove cut in the soil for a distance of 12.7cm. LL were carried out on the samples from different locations, 200gm of the soil was weighed and pulverized by using 425mm sieve with the aid of the spatula. The soil passing through 0.425mm sieve were mixed with water to a stiff consistency until the paste is uniform and left for 24 hours. Plastic limit (Wp) is the water below which the soil no longer behaves as a plastic material. It is range of water contents between W_L and the plastic limit Wp that the soil behaves as a plastic material, this range is termed the plasticity (or plastic) index. Ip is also the water content at which thread of soil when rolled to a diameter of 3mm will just crumble. Compaction the process of increasing the density of a soil by packing the particle closer together with a reduction in the volume of air, there is no significant change in the volume of water in the soil. The curves which are obtaining by plotting moisture content versus dry density for each test are known as compaction curves. As the moisture in the soil is increased, the dry density increases at least up to a point where it reaches a maximum value. From the curve, optimum moisture content which is the moisture content at which the maximum dry density is obtained by determining the highest point on the compaction curve (Apex) and dropping vertical down to the horizontal moisture scale or line. Compaction test were carried out on the natural sample of soils collected from Dada Estate bye pass road after which 3, 6, 9 and 12 percentages of Portland cement, Lime and bitumen were added for soil-additive mixtures. The soil samples were first pulverized, 500gm of each sample was measured and about 4% of water added, thoroughly mixed for the first and divided into three equal parts, immediately compacted in the B.S mould in three layers. Each layer was compacted using 25 uniformly spaced blows of a 5.5Kg rammer dropping at a height of 12 inches (304.8). The compacted specimen and mould were then weighted and a representative sample taken for moisture content and dry density determination. The same process was repeated for all the additive but different is the mixture of the required percentage of Lime, cement and Bitumen. After compaction, the bulk density and water content of the soil were determined and dry density calculated. Dry density was plotted against water content with the formation of a curve. The maximum possible value of dry density is referred to as the "Zero air voids". The OMC determined is used to control the quantity of water to be added during the compaction of soil earthwork in road construction. For California Bearing Ratio (CBR) Test about 6kg of the soil sample passing through N0 4 sieve was mixed with the desired amount of optimum moisture content (OMC) previously obtained from compaction Test and compacted in the C.B.R mould. It involved compacting the soil in three layers given 15 blows with a 2.5kg rammer. The mould and compacted soil were placed under the California bearing ratio machine and tested with penetrations of 25, 50, 15, 100, 125 to 750 inches. The steps were repeated for the bottom surface for each sample. The C.B.R test is essentially a penetration test having the function of measuring the soil resistance to penetration prior to reading to estimate shearing value. This action is an exact duplicate of what vehicles exert on a flexible pavement structure. It can be defined as an indirect way of measuring the shear strength of soil for use as subgrade or base material in flexible pavement construction. The C.B.R is calculated as the ratio of the force required to push the plunge to a certain depth into the soil to that force required to push the same plunge to the depth into a standard sample of compacted crushed stone that is C.B.R % equal (Test load / standard load) x 100.

3. Results and Discussion

Laboratory test were performed on the three samples collected for the purpose of identification, classification and determination of the engineering characteristics of the material used. The samples were treated with 3%, 6%, 9% and 12% of Cement, Lime and Bitumen contents in the laboratory for properties evaluation all in accordance with AASHTO 2007 Pavement manual and Approximate equivalent classification from AASHTO 2007 Pavement manual;

Table 1: Basic and Engineering Properties of the selected laterite soil

IOP Conf. Series: Materials Science and Engineering **413** (2018) 012012 doi:10.1088/1757-899X/413/1/012012

PROPERTIES	**SOIL SAMPLES**		
	A	**B**	**C**
Grain Size Distribution Coarse (%) Fine (%) Bulk density (KN/m^3)	 90.85 09.15 14.64–29.76	 93.45 06.55 12.23–22.24	 93.90 06.10 14.64–22.78
Consistency Limit (%) Liquid Limit Plastic Limit Plasticity Index	 43.00 19.09 23.91	 43.50 25.65 17.85	 21.50 16.33 5.17
Maximum Dry Density (KN/m^3)	18.65	16.62	14.80
Optimum Moisture Content (%)	9.15	9.90	9.15
California Bearing Ratio (%)	9.88	4.36	7.27
Soil Classification	A-2-7	A-2-7	A-2-4
Soil Type	Silty or clayed clayey gravel and sand		

Particle Size Distribution

The results of the particle size analysis are shown in Table 2, graphically represented in Figure 1 for the three samples collected from Okefia- Dada Estate Road, Osogbo, Osun State. The result shows that sample A higher in fine and bulk density than sample B as well higher tha sample C, while sample C greater in Coarse aggregate than Sample B and far greater than sample A.

Table 2 Particle Size Distribution of Control Sample A, B and C

Samples	**Material type**		**Bulk density range (KN/m^3)**
	Fine (%)	**Coarse (%)**	
A	09.15	90.85	14.64 – 29.76
B	06.55	93.45	12.23 – 22.24
C	06.10	93.9	14.64 – 22.78

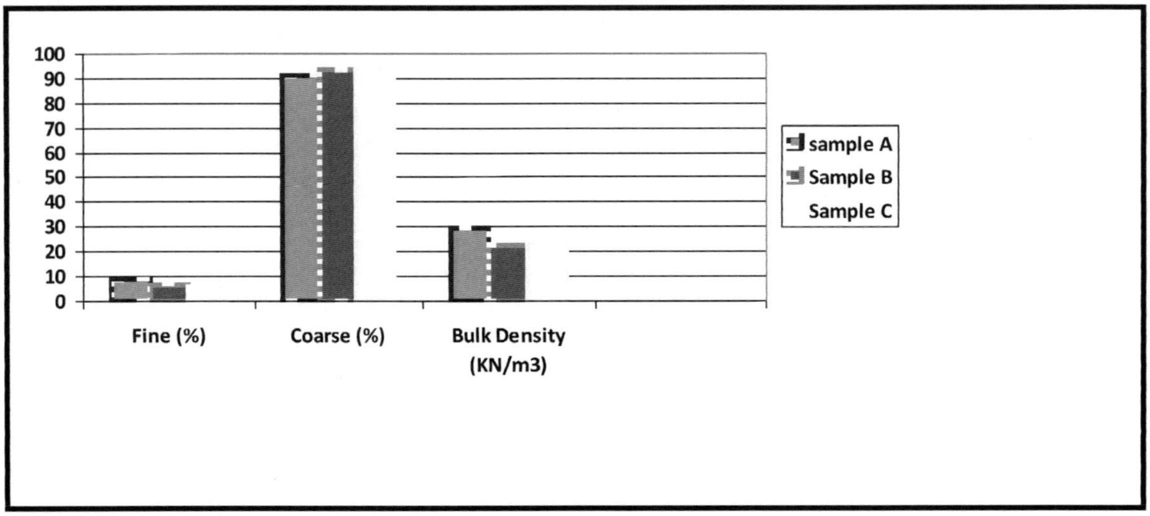

Figure 1: Graph for particle size distribution for samples

Atterberg Limits

The test results are analyse in Table 3 and 4 and graphically illustrated in Figures 2and 3 reveal that the average Lime additive (3%) Liquid limit and plastic limit value is higher than cement additive which is also higher than bitumen additive, while Bitumen additive (3%) average plasticity index greater than lime additive which also greater than cement additive. Also Lime additive (12%) Liquid limit and plastic limit value is higher than bitumen additive which is also higher than cement additive, while Bitumen additive (12%) average plasticity index greater than lime additive which also greater than cement additive.

Table 3 Atterberg Limit test Control, Cement treated, Lime treated and bitumen treated sample at 12%

Samples		**Liquid Limit (LL)%**	**Plastic Limit (PL)%**	**Plasticity index (PI) %**
Control (0%)	A	43.00	19.09	23.91
	B	34.50	25.65	17.85
	C	21.50	16.33	05.17
	Average	**33.00**	**20.36**	**15.64**
Cement treated (12%)	A	35.30	31.00	4.30
	B	34.50	28.50	6.00
	C	33.50	29.50	4.00
	Average	**34.43**	**26.12**	**4.77**
Lime treated (12%)	A	36.42	30.85	5.57
	B	37.24	31.19	6.05
	C	33.50	29.50	4.00
	Average	**35.72**	**30.51**	**5.21**
Bitumen treated (12%)	A	36.49	30.50	5.99
	B	33.00	27.00	6.00
	C	33.40	29.40	4.00
	Average	**34.30**	**29.00**	**5.33**

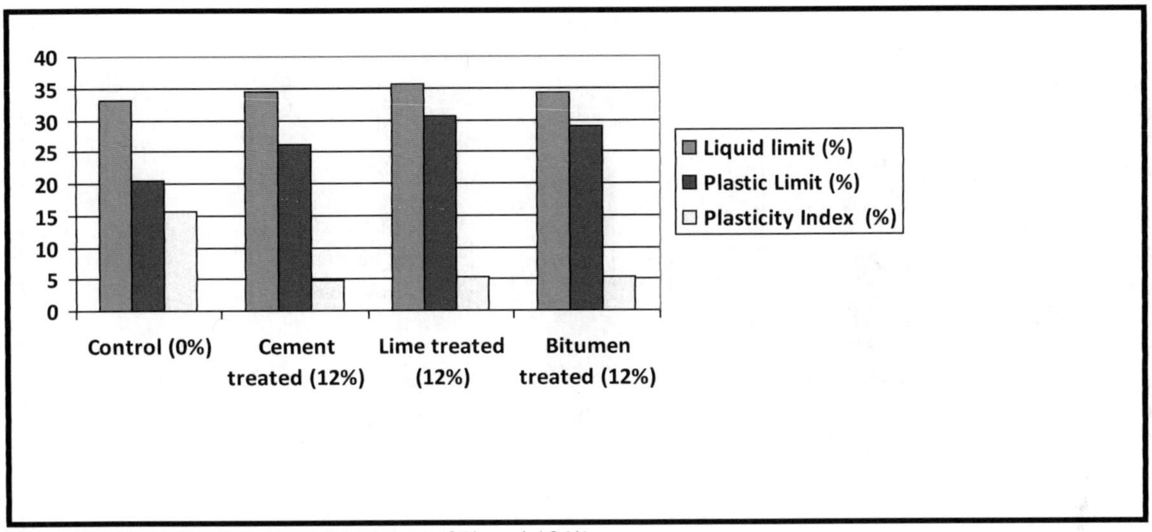

Figure 2: Chart for Atterberg Limit test (0% and 12%)

Table 4 Atterberg Limit test Control, Cement treated, Lime treated and bitumen treated sample at 3%

Samples		**Liquid Limit (LL)%**	**Plastic Limit (PL)%**	**Plasticity index (PI) %**
Control (0%)	A B C	43.00 34.50 21.50	19.09 25.65 16.33	23.91 17.85 05.17
	Average	**33.00**	**20.36**	**15.64**
Cement treated (3%)	A B C	45.19 45.19 42.00	28.81 26.10 24.50	16.38 19.09 17.50
	Average	**44.13**	**26.47**	**17.66**
Lime treated (3%)	A B C	44.40 45.50 42.00	22.49 26.50 24.75	21.91 19.00 17.25
	Average	**43.97**	**24.58**	**19.39**

IOP Conf. Series: Materials Science and Engineering **413** (2018) 012012 doi:10.1088/1757-899X/413/1/012012

Bitumen treated (3%)	A B C	42.00 47.00 40.00	18.09 21.58 24.00	23.91 25.42 16.00
	Average	**43.00**	**21.22**	**21.78**

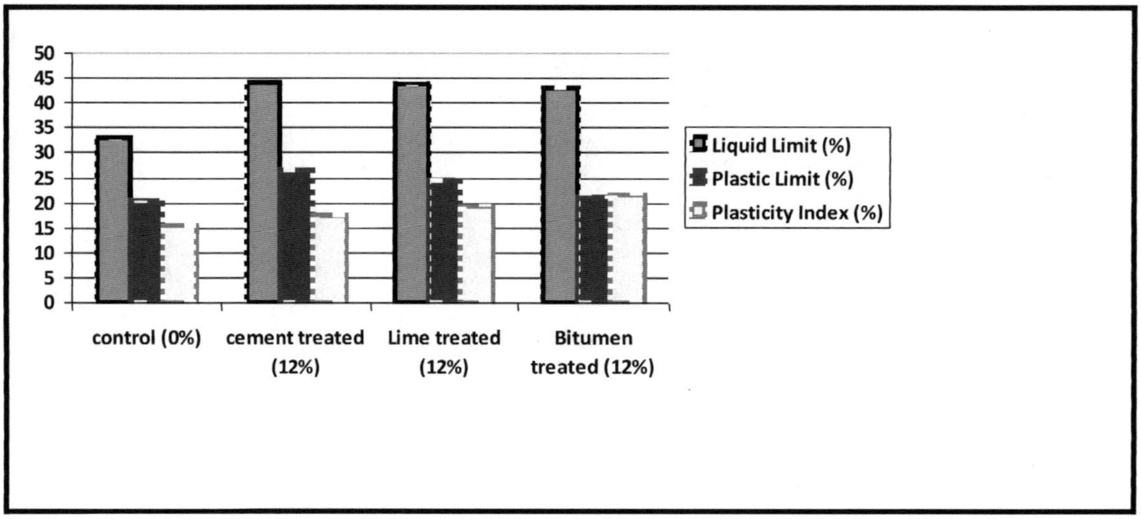

Figure 3: Graph for Atterberg for all samples (0% and 3%)

Compaction Test

The test results are analyzed in Table 5 and 6 and graphically illustrated in Figures 4 and 5 shows three samples show low dry densities which will lead to low strength of the samples if used for construction materials in its natural state. The average Lime additive (3%) Maximum Dry density value is higher than cement additive which is also higher than bitumen additive, while Bitumen additive (3%) average Optimum dry density values greater than lime additive which also greater than cement additive. Also Bitumen additive (12%) Maximum Dry density value is higher than Lime additive which is also higher than cement additive, while cement additive (12%) average Optimum dry density values greater than bitumen additive which also greater than lime additive.

Table 5 **Compaction test Control, Cement treated, Lime treated and bitumen treated sample at 12%**

Samples		**Maximum Dry density (MDD) KN/m^3)**	**Optimum moisture content (OMC) %**
Control (0%)	A B C	18.65 16.62 14.80	09.50 09.90 09.15
	Average	**16.69**	**09.52**

Cement treated (12%)	A B C	13.05 14.70 9.15	30.00 10.00 19.00
	Average	**12.30**	**20.00**
Lime treated (12%)	A B C	17.00 15.30 19.00	17.25 11.00 9.30
	Average	**17.10**	**12.52**
Bitumen treated (12%)	A B C	20.00 13.60 19.00	10.00 22.00 7.00
	Average	**17.53**	**13.00**

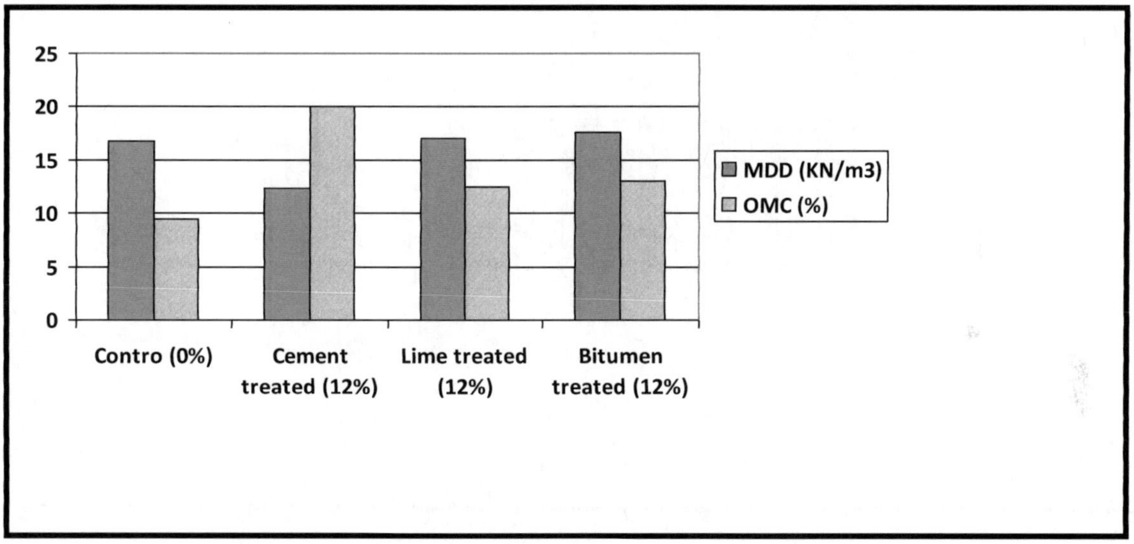

Figure 4: Compaction Chart for samples (0% and 12%)

Table 6 Compaction test Control, Cement treated, Lime treated and bitumen treated sample at 3%

Samples		**Maximum Dry density (MDD) KN/m^3)**	**Optimum moisture content (OMC) %**
Control (0%)	A B C	18.65 16.62 14.80	09.50 09.90 09.15
	Average	**16.69**	**09.52**

Cement treated (3%)	A B C	16.80 19.80 19.60	16.50 11.50 14.80
	Average	**18.73**	**14.27**
Lime treated (3%)	A B C	17.90 18.80 20.20	21.60 13.40 14.00
	Average	**18.97**	**16.33**
Bitumen treated (3%)	A B C	17.25 15.80 19.95	11.40 8.60 10.00
	Average	**17.67**	**10.00**

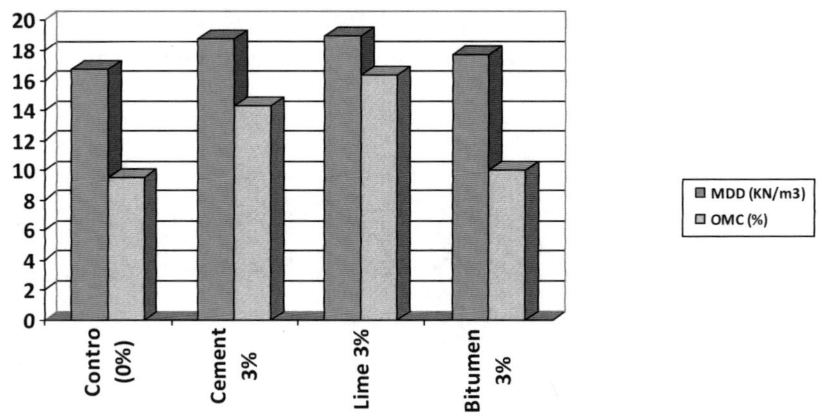

Figure 5: Compaction Chart for samples (0% and 3%)

California Bearing Ratio (C.B.R) Test

The test results are analyse in Table 7 and 8 and graphically illustrated in Figures 6 and7. The three samples categorized as poor sub base materials based on AASHTO 2007 reveal three samples show low dry densities which will lead to low strength of the samples if used for construction materials in its natural state. The average Lime additive (3%) California bearing ratio value is higher than cement additive which is also higher than bitumen additive while average cement additive (12%) California bearing ratio is higher than Lime additive which is also higher than bitumen additive.

Table 7 California bearing ratio Control, Cement treated, Lime treated and bitumen treated sample at 12%

Samples		California Bearing ratio (C.B.R) %
Control (0%)	A B C	9.88 4.36 7.27
	Average	**7.17**
Cement treated (12%)	A B C	55.0 50.0 50.0
	Average	**51.67**
Lime treated (12%)	A B C	45.0 45.0 45.0
	Average	**45.0**
Bitumen treated (12%)	A B C	41.0 40.0 40.0
	Average	**40.33**

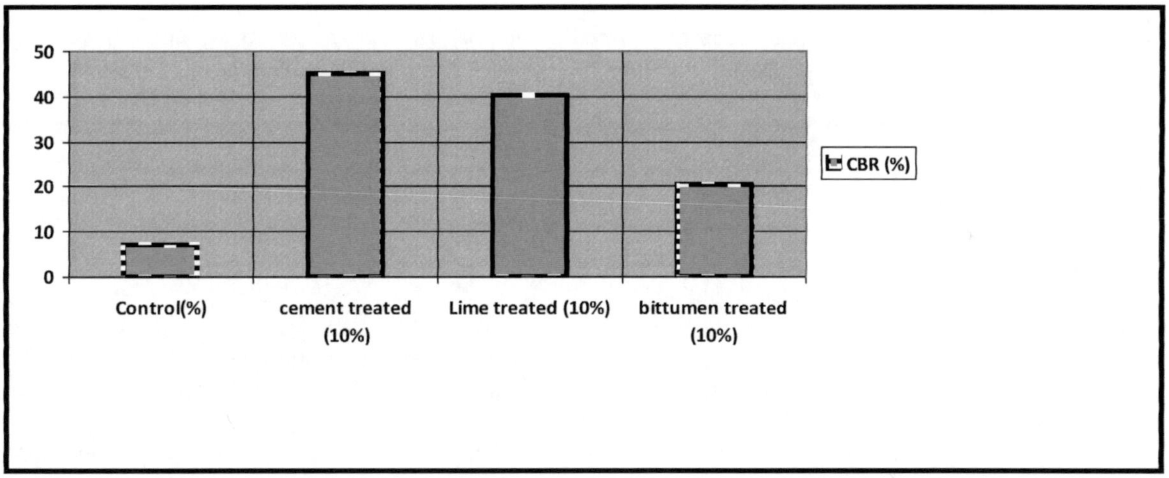

Figure 6: Graph of California bearing ratio for samples (0% and 3%)

Table 8: California bearing ratio Control, Cement treated, Lime treated and bitumen treated sample at 3%

Samples		California Bearing ratio (C.B.R) %
Control (0%)	A B C	9.88 4.36 7.27
	Average	**7.17**
Cement treated (3%)	A B	22.01 20.00

	C	23.00
	Average	**21.67**
Lime treated (3%)	A	24.00
	B	35.00
	C	18.00
	Average	**25.67**
Bitumen treated (3%)	A	18.00
	B	16.00
	C	12.00
	Average	**15.33**

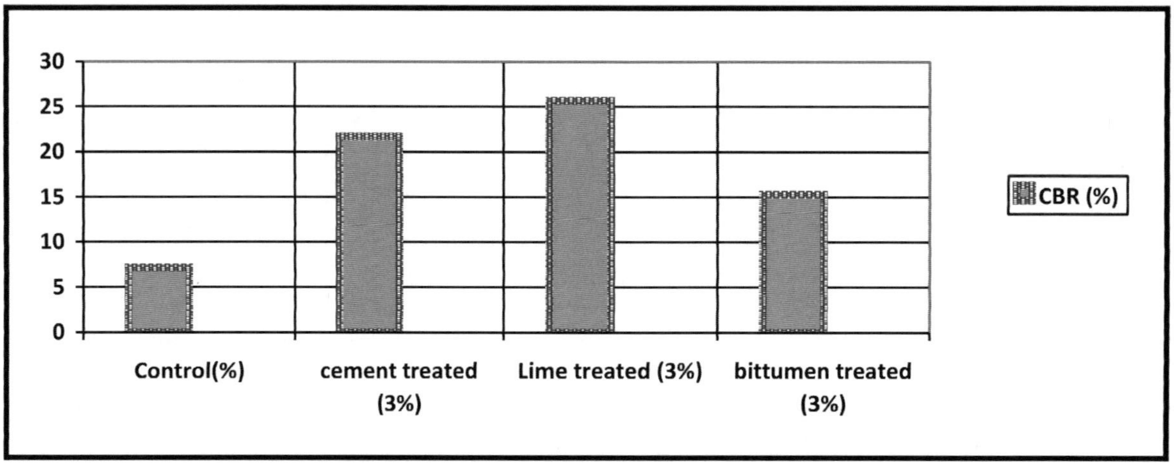

Figure 7: Graph of California bearing ratio for samples (0% and 3%)

4. Conclusion

The study of laterite soil sample from a construction site located at Dada Estate road, Osogbo Osun State shown that beneficial effects are obtained by the addition of cement, lime and bitumen to improve the strength of weak or poor soil. Lateritics soil sample A and B belong to A-2-7 while sample C is categorized as A-2-4 which is silty or clayed gravel sand. The Atterberg Limits of the samples ranges from 29.00% to 47% for Liquid Limit with plasticity index between 3 and 26. Atterberg at 3% and 12% lime additive LL and PL is greater and Bitumen additive higher in PI. For Compaction at 3%; lime additive MDD higher and Bitumen additive higher in OMC while 12%; bitumen additive MDD greater while cement additive higher in OMC. Lime additive at CBR 3% is highest while at 12% is cement additive. Increase in percentage of Lime additive also increase LL and Pl and increase in bitumen content as additive gave rise to PI values. Lower percentage of lime and bitumen additive gave highest values of MDD and OMC respectively; greater percentage of Bitumen and cement additive result to higher values of MDD and OMC respectively. With little amount of Lime additive and moderate quantities for Bitumen there will be great soil strength improvement but cement require large quantities before soil strength can be improve. From comparison with (Olugbenga Oludolapo Amu, Oluwole Fakunle Bamisaye and Iyiola Akanmu Komolafe, 2011) soil improvement of lateritic samples from the studied site at Osogbo, Osun state capital Nigeria can be carried out successfully by using Lime or bitumen additives.

IOP Conf. Series: Materials Science and Engineering **413** (2018) 012012 doi:10.1088/1757-899X/413/1/012012

References

[1] Akinje, I. Comparison Characterization of A-6(10) laterite soil stabilized with powermax cement and hydrated Lime separately. *International journal of Engineering and technology,* 2015, *5* (7): 392 - 401.

[2] Akinwumi, I. Plasticity, Strength and Permeability of Reclaimed Asphalt and Lateritic Soil Blends. *International Journal of Scientific and Enginnering Reseach,* 2014, *5* (6): 631-636.

[3] Akinwumi, I. I; Adeyeri, J . B and Ejohwomu, O . A. Effects of steel slag Addition on the plasticity, strength and permeability of Lateritic soil. *Proceedings of the Second International Conference of Suitanable Design, Engineering and Construction.* Texas, 2012.

[4] Ali, F. Stabilization of Residual Soils Using Liquid Chemical. *The electronic journal of Geotechnical Engineering Science and Technology*, 2012, 1 (2): 102 - 110.

[5] Osinubi K. J and A.A Amadi. Evaluation of Strength of Compacted Lateritic Soil - Bentonite mixture for use as landfill linear and cover. *journal of Engineering Research,* 2010, *3*(13): 1-8.

[6] Kadyali L. R and Lai, N . B. Principles and Practices of Highway Engineering (including expressways and Airport engineering). Romesh Chander Khanna, Delhi, India, 2008.

[7] Latifi, N, Marto A and A Eisazadeh . Structural characteristics of Lateritic soil treated bu SH 85 and TX - 85/ Non traditional stabilizers. *The Electrronic Journal of Geotechnical Engineering,* 2013, 1(3): 1707 - 1718.

[8] Garber N.J and L.A Hoel. *Traffic and Highway Engineering*. Engage learning Stamford. 2010, 4th Ed., U.S.A.

[9] Zhang G and Yi. NP. The experimental study on strength of subbase soil treated with liquid stabilizer. *Advanced Materials Research,* 2011, 19 (4): 985 - 988.

[10] Amu Olugbenga Oludolapo, Oluwole Fakunle Bamisaye and Iyiola Akanmu Komolafe. The Suitability and Lime Stabilization Requirement of some lateritic soil samples as pavement. *International Journal of Pure and Applied Sciences Technology* , 2011, *1*(2): 29 - 46.

[11] Wright P.H and Dixon K.K. *Highway Engineering.* 2013, 7th ed., New york: Wiley and sons.

Performance and Yield Assessment for Renewable Dispersed Generation in Nigeria: Case Study on Grid-Tied Solar PV Systems

[1]Agbetuyi Ayoade Felix, [2]Udeme Eyo Okon, [3]Obiakor Emeka, [4]Kayode Oladipupo. O, [5]Owolabi Bango

[1,2,3,4,5] Department of Electrical & Information Engineering, Covenant University,Ota, Nigeria.
Corresponding Author: ayo.agbetuyi@covenantuniversity.edu.ng

Abstract: The problem of Renewable Dispersed Generation (RDG) and renewable resource harnessing in most cases, is not the inadequacy of resource distribution at a particular location but rather, it is the perpetually deprived exploitation of these resources. The existing dread for the performance of distributed generation systems that utilize renewable sources for power generation has crippled sustainable power generation expansion for several years. Hence, this study performs a practical assessment of the productivity of solar PV systems in Nigeria. In this study, five states with diverse geographical or meteorological data are selected from the cardinal regions of the country. These include Sokoto State (North West), Borno State (North East), Ogun State (South West), Rivers State (South South) and Abuja, the federal capital territory at the center. The global horizontal radiations available from the National Aviation and Space Agency (NASA) for these cities were used in simulating the performance of a 1-MW grid-tied solar PV plant using the PVsyst software. The result shows that the performance parameters with respect to energy production are better in the northern region compared with the southern region with capacity factors decreasing from 20.46% to 16.21% for a 1 MW solar PV plant located at these regions. This was seen to reflect on the corresponding annual energy yield of similar systems sited at these locations. Also, the performance ratios of these systems, given their respective reference yields, were observed to be better in the southern region compared with the northern region of Nigeria. This can be attributed to external factors that can influence system efficiency. These factors tend to favor systems located at the southern region as opposed to those at the northern region. These findings validate this study as a decision tool for predicting the performance of any Renewable Dispersed Generation (RDG) systems utilizing solar energy as the source at other regions of the country.
Keywords: Performance Assessment; Solar PV Systems; global horizontal radiations; Renewable Dispersed Generation **(RDG).**

I. Introduction

The danger posed by traditional power generation to general economic sustainability, development and substantial growth is very obvious. This is reflected on how each country of the world seeks countless alternative energy sources for electric power generation [1]. The basic aim for generation expansion in developed countries has been to advance the economic and commercial performance of the sub- sector which contrasted to the rather critical role played by the macroeconomic status in the developing and transition countries [2]. In Nigeria for example, electric power supply has never kept phase with demand despite the fact that the country is leading among the oil producing countries of Africa. This is supplemented by renewable resources such as water, wind and abundance solar energy that can be used to generate a substantial amount of electricity [2].

Increase in load demand necessitates generation expansion by building new generation power plants and corresponding expansion of the power networks. This is far from real solution from modern economic and environmental perspectives where there is a global struggle to meet the targets set in the Kyoto Protocol in order to reduce greenhouse gas emissions [3]. This has led to an increase of interest in Renewable Dispersed Generation (RDG), which is defined as small-scale electricity generation fueled by renewable energy sources such as wind and solar or by low-emission energy sources like fuel cells and micro-turbines. RDG can also be referred to as embedded generation which means small size generating units connected to the distribution power system [3, 4, 5].

According to Sustainable Energy Regulation Network SERN [6], Nigeria has available capacity of 6,056 MW out of the total installed capacity of 10,396 MW from the 23 grid-connected generating

plants in the country from which the fossil fuel generation accounts for 79% of the total installed capacity and the remaining 21% is derived from the hydro source. There is also a forecasted increment in peak demand from 5,000 MW in 2011 to about 16,000 MW in the year 2020 [1, 6 - 8]. The Energy Information Administration (EIA) estimates state that Nigeria has one of the lowest net electricity generation per capita rates in the world [6]. Nigeria has been estimated to produce about 227,500 tons of fresh animal waste daily, meaning that Nigeria can produce about 6.8 million m3 of biogas, since 1 kg of fresh animal waste is needed to produce 0.03 m3 of biogas [6]. Statistics has also revealed that hydropower generation is the only renewable source of power generation currently utilized in the country from which only about 19% of the potential has been tapped despite the fact that hydropower resource is widely distributed throughout the country.
According to [9, 10] , while studying voltage stabilization for grid-tied RDG systems utilizing wind energy, it was revealed that some places in the northern part of the country have average annual wind speed of about 5.4 m/s at height of 30 m as an opportunity for possible power generation. From the solar energy perspective, Nigeria has been grouped among countries with best solar resource distribution [10], with annual average total solar radiation reaching about 5.5 KWh/m2/day. With 6 hours/day sunshine hours, an annual total of about 1,770 thousand TWh/m2/day can be reached which is very favorable for PV generation in the country [10].
From these findings, one can conclude that, if there is an effective harnessing of available renewable energy resources in Nigeria, a good mix of power generation expansion through renewable dispersed generation can be achieved in a short time. Thus, this study focuses on the performance and yield assessment of grid-tied solar PV plant which, according to the U.S. Department of Energy (DOE), is classified among the renewable dispersed generation.

2. Solar Energy Applications

For a number of good reasons which may include concerns for the environment and the opportunity for demonstration of the willingness to be involved in the green world move, a country may choose to embark on power generation from renewable sources. A Country like Nigeria will count power generation from solar applications as one of its renewable energy source majors [11, 12]. Power generation from solar applications can utilize solar thermal energy systems or solar photovoltaic systems availability. Solar thermal systems are solar energy application systems that can concentrate heat energy from the sunlight and use it at varying temperatures; such as low, medium or high range depending on the applied technology. These applications have found usefulness in areas such as space heating and cooling, crop drying and solar distillation [13]. Solar photovoltaic systems covert solar radiation directly to DC electricity using photovoltaic cells made from semiconductors such as silicon. These systems may be applied as a standalone unit or off-grid unit and can also be applied as grid-connected unit depending on the power need and consumer location [13, 14].

3. The Grid-Tied Solar PV Systems

These are solar PV systems that are connected to the public utility either for metering purpose or as a renewable dispersed generation. These systems have no provision for autonomous supply hence; no energy storage device is required. They may range in size from small decentralized rooftop systems of few kilowatts to very large central grid-tied systems with capacity in the megawatts range [13, 16]. An inverter is used to convert the DC electricity from solar PV array to AC electricity which is fed directly into the utility grid. Since they supply AC electricity into the public grid, their load seems to be unlimited and their performance is usually evaluated on annual interval [15].

4. Proposed System

The proposed system is a 1-MW grid-tied solar PV plant that utilizes polycrystalline silicon modules with inverters in an array configuration to supply power to unlimited load on the utility grid is as shown in figure 1. The location of this proposed system will be at strategic places in the cardinal regions of Nigeria and Abuja (Lat. 9.0oN, Long.7.4oE) at the center. These locations are- Sokoto (Lat. 13.0oN, Long. 15.0oE) in the North West region, Borno State (Lat. 11.9oN, Long. 13.2oE) in the North East region, Ogun State (Lat. 7.1oN, Long. 3.3oE) in the South West region and Rivers State (Lat. 5.1oN, Long. 7.1oE) in the South South region. The site and meteorological data for these locations were imported from the satellite measurements provided by the American National Agency for Space and Aviation NASA for PVSyst software. A summary of annual global horizontal solar radiation on a plane inclined on 15o at these regions are presented in Table 1.0 and the plot of Figure 2. Table 2 shows the basic design parameters for the 1MW grid-tied solar PV system.

Table 1: Annual global solar radiation for the locations
Source: National Agency for Space and Aviation (NASA)

Months	Sokoto (Kwh/m2)	Maiduguri (Kwh/m2)	Ogun State (Kwh/m2)	Rivers State (Kwh/m2)	Abuja FTC (Kwh/m2)
January	171.6	173.9	170.5	171.4	182.3
February	178.9	176.4	159.6	156.5	170.5
March	203.8	207.7	174.8	164.9	194.4
April	201.3	198.6	160.5	152.7	181.8
May	196.7	197.2	157.8	146.3	173.0
June	188.2	179.1	137.1	129.3	151.8
July	179.0	168.3	124.0	119.3	137.6
August	170.8	159.3	117.5	116.9	129.9
September	178.8	167.1	123.3	118.2	141.9
October	190.5	182.6	145.7	132.4	164.6
November	177.9	175.2	153.3	145.2	179.4
December	168.7	165.8	165.8	164.0	181.7
Year	**2206.2**	**2151.3**	**1790.0**	**1717.2**	**1988.8**

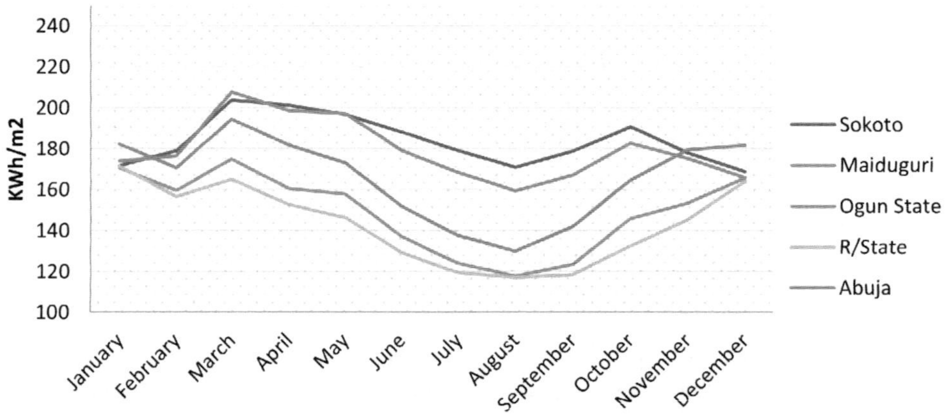

Fig. 1: Comparison of Annual Average Solar Radiation for the selected Cardinal Regions in Nigeria

Table 2: The basic design parameters for the 1MW grid-tied solar PV system.

Module Type	Polycrystalline Silicon (generic)
Space Requirement	6508 m2
Module Efficiency	15.5%
Module Capacity	250Wp

Installed Capacity	1000kWp
Total Number of Modules	4000 (16 series by 250 parallel string)
Inverter Capacity	2 x 500KWac
PV Array Orientation	
Inclination	15o
Operation Temperature	50oC
Configuration	Array Type

5. Result Analysis

The International Energy Agency IEA has laid down measures that can be followed in order to evaluate and compare the performances of solar PV systems. The criteria include the description of the energy conversion chain beginning from the solar radiation input to the final electricity injected into the grid by appropriate and normalized quantities which are based on the IEA Performance Database Standard IEC 61724 for Photovoltaic Power Systems (PVPS) [17, 18].

These performance parameters are illustrated in Figure 4 and include the Reference System Yield (Y_r), Final System Yield (Y_f), Collection Loss (L_c), Array Yield (Y_a), Performance Ratio (P_R) and Capacity Factor for the grid connected system as shown in Figure 3;

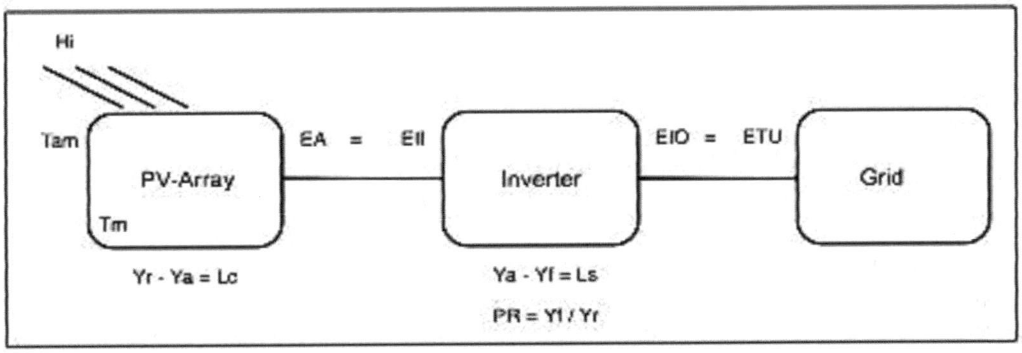

Fig. 2: Illustration of System Performance Parameters

EIO represents the equivalent AC electricity obtained from conversion of the inverter input EII and is fed into the grid as (ETU) as illustrated in figure 4.

a. Total Energy Yield

This is the total energy that the PV array can inject into the utility grid at a particular reference time, usually evaluated at annual bases for grid-connected solar PV systems and it is measured in MWh for a particular time. For the solar PV systems at selected regions, the energy injected into the public grid are- Sokoto: 1792.6 MWh, Borno: 1765.7 MWh, Abuja FCT: 1669.5 MWh, Ogun: 1488.7 MWh and Rivers: 1419.7 MWh.

The monthly comparison of the total energy yield presented in figure 4 showed that total energy yield is better for systems in the northern region than systems located at the southern region.

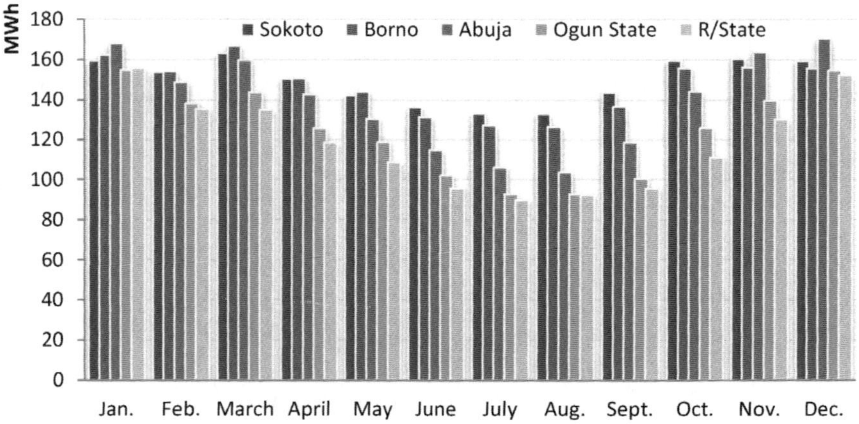

Fig.3: Comparative plot of Total Energy Yield (MWh)

b. Reference System Yield

The Reference System Yield Y_r or the ideal array yield is numerically equal to the incident energy in the array plane, and is expressed in kWh/m²/day. The annual daily average reference yield obtained for the five (5) regions are- Sokoto: 6.27 kWh/m²/day; Borno State: 6.093 kWh/m².day; Ogun State: 5.010 kWh/m².day; Rivers State: 4.768 kWh/m².day and 5.626 kWh/m².day for Abuja FCT.

$$Y_r = \frac{(total\ insolation)}{(reference\ irradiance)} = \frac{KWh/m^2}{1000W/m^2} \qquad 1$$

It is observed from the result that the reference system yield increased as one moves from southern to the northern regions. This is due to the higher clearness index in the northern region [10].

c. Final System Yield

The Final System Yield Y_f is the system daily useful energy or the inverter output, referred to the nominal power expressed in (kWh / KWp / day) [19, 20].

$$Y_f = \frac{Net\ Energ\ Output\ KW\ AC}{Array\ Output\ Power\ KW\ DC} \quad \text{(KWh/kWp/day)} \qquad 2$$

The result for the selected systems are- Sokoto: 4.91 kWh/kWp/day, Borno: 4.84 kWh/kWp/day, Abuja: 4.57 kWh/kWp/day; Ogun: 4.08 kWh/kWp/day and Rivers: 3.89 kWh/kWp/day. It is seen from these results that systems located in the northern region have higher final system yield compared with systems located at the southern region because of the better reference yield in the northern region.

d. Collection Loss

The Collection Loss L_c is the array losses including thermal, wiring, module quality, mismatch and IAM losses, shading, dirt, MPPT, regulation losses, as well as all other inefficiencies. It can be expressed mathematically, according to [19] as;

$$L_c = Y_r - Y_a \quad \text{(kWh/kWp/day)} \qquad 3$$

The collection losses for similar grid-tied 1 MW solar PV system located at selected regions are Sokoto: 1.21 kWh/kWp/day, Borno: 1.11kWh/kWp/day, Abuja FCT: 0.92 kWh/kWp/day, Ogun: 0.81 kWh/kWp/day and Rivers: 0.76 kWh/kWp/day. The result showed that collection loss is better for systems located at the southern region compared with those at the northern regions.

e. The Performance Ratio PR

This is the global system efficiency with respect to the *nominal* installed power and the incident energy expressed mathematically without unit according to [13] as;

$$P_R = \frac{Y_f\ (Final\ Yield)}{Y_r\ (Referecne\ Yield)} \qquad 4$$

The performance ratios obtained for the grid –tied 1-MW solar PV system located at the selected regions are as given below- Sokoto: 0.783, Borno: 0.794, Abuja FCT: 0.813, Ogun: 0.814 and Rivers: 0.816.

From this result, it is observed that the performance ratio improves across the country from the northern region to the southern region for the 1-MW PV plant. This is due to better collection loss that characterized systems located at the southern region.

f. Capacity Factor

The Capacity factor (CF) of a power plant is obtained by dividing the actual yield of a power plant by its potential yield, if it was functional at full nameplate capacity over a given reference time. It can be expressed mathematically as the ratio of the overall energy to the amount of energy the plant would have produced at full capacity in a period of time, usually annually [15]. It can be expressed as a percentage.

$$C.F = \frac{\text{Annual Energy output}}{Nameplate\ Capacity \times 8760\ hours} \quad (5)$$

The capacity factor for the system in Sokoto region is obtained as 20.46%, the system in Borno has 20.16%, the system in Abuja FCT has 19.06%, the system located at Ogun has 17% and similar system in Rivers State has 16.21% capacity factors.

The result for the capacity factors for the 1-MW PV Solar plant at the selected locations showed that systems located at the northern region have better capacity factors compared with similar systems located around the southern region. This is due to the better system reference yield obtained in the northern region of the Country compared with the southern region.

The general performance parameter of the 1 MW grid-connect solar PV plants in the selected cardinal regions of Nigeria is summarized and presented in Table 3.

Table 3: General Performance Parameters for the 1 MW solar PV plants at selected Regions

Performance Parameters	Sokoto System	Borno System	Abuja System	Ogun System	Rivers System	
						HIGH
Total Energy Yield *(MWh)*	**1792.6**	1765.7	1669.5	1488.7	**1419.7**	
Reference Yield *(KWh/m2/day)*	**6.27**	6.093	5.626	5.01	**4.768**	
Final Yield *(KWh/kWp/day)*	**4.91**	4.84	4.57	4.08	**3.89**	**MEDIUM**
Collection Loss *(KWh/kWp/day)*	**1.21**	1.11	0.92	0.81	**0.76**	
Performance Ratio	**0.783**	0.794	0.813	0.814	**0.816**	
Capacity Factor *(%)*	**20.46**	20.16	19.06	17.0	**16.21**	**LOW**

6. Conclusion

Findings from the performance and productivity assessment of the 1-MW grid-connected solar PV systems located at the selected cardinal regions of Nigeria are as presented in Table 3. It can be seen that best values for average daily incident energy in the PV array plane of the 1-MW solar PV plant is obtainable in the country with a comparative daily average values ranging from 6.27 KWh/m2/day in the North to 4.768 KWh/m2/day in the South. This corresponds to the findings reported by [10]. Also, the annual global solar radiation for the different regions presented in Table 1, shows that there is a progressive increment on the average solar energy received on PV array as you measure from the southern region to the northern region of the Country. The high reference yield in the northern region of the Country reflects harmoniously on other performance parameters of the system such as the total energy yield (Table 3). It is interesting to note that the global system efficiency with respect to the nominal installed power referred to as the performance ratio PR, obtained for the 1-MW solar PV system is rather better for the systems in the southern region of the country than the systems located at the northern region. This is due to other factors such as the ambient temperature which affects the functionality of these systems. This, in turn, has a corresponding influence on the thermal loss and component efficiency of these systems [2, 6, 8]. The performance ratio values obtained for the systems are Sokoto: 0.783, Borno: 0.794, Abuja FCT: 0.813, Ogun: 0.814 and Rivers: 0.816. The relevance of this study therefore, can be seen to serve as a decision tool for the predictability of the performance of any renewable dispersed generation, RDG systems utilizing solar energy at other regions of the country.

References

[1] Titus Koledoye Olugbenga, A.-G. J. "The Current and Futrue Challenges of Electricity Market in Nigeria in the Face of Deregulation Process". African Journal of Engineerung Research, 1(2), 33-39, 2013.

[2] Isola, W. A. "An Analysis of Electricity Market structure and its Implications for Energy Sector Reforms and Management in Nigeria". *Global Advanced Research Journal of Management and Business Studies Vol. 1(5), 1*, 141-149. Retrieved from http://garj.org/garjmbs/index.htm, May 2017.

[3] Zeineldin, H., El-Saadany, E., & Salama, M. M. "Impact of DG Interface Control on Islanding Detection and Non Detection Zones." *IEEE Transactions on Power Delivery, Vol. 21*, 1515-1523, 2006.

[4] Kurt D. Jr. "Africa's Renewable Energy Potential:" www.africa.com/africasrenewable- energy potential, 2015.

[5] Zellagui, M., Karimi, M., Mokhlis, H., Benabid, R., & Chaghi, A. "Impact of Renewable Dispersed Generation on Performance of Directional Overcurrent Relay on MV Distribution Power System." *International Conference on Electrical Engineering CEE.* Batna, Algeria: Researchgate, 2014.

[6] National Renewable Energy And Energy Efficiency Policy (Nreeep). "Approved FEC for the Electricity Sector," 2015 also available at www.power.gov.ng/download/NREEE%20POLICY%202015%20FEC%20APPROVED%20COPY.pdf retrieved May 2017.

[7] Onagoruwa, B. " *Nigerian Power Sector Reforms and Privatisation.* Nigeria." Director General, Bureau of Public Enterprises, 2011.

[8] Joseph, I. O. "Issues and Challenges in the Privatized Power Sector in Nigeria". *Journal of Sutainable Development Studies, 6*, 161-174, 2014.

[9] Agbetuyi. A. Felix, Akinbulire Tolulope O, Abdulkareem Ademola and C.O.A Awosope "Wind Power Potential In Nigeria" International Electrical Engineering Journal (IEEJ) 3(1) pp 595-601, 2012.

[10] Olayinka, S. "Estimation of global and diffuse solar radiations for selected cities in Nigeria". *International Journal of Energy and Environmental Emgineering, 2*, 13-33, 2011.

[11] Offiong, A. "Assessing the Economic and Environmental Prospects of Stand-by Solar Powered Systems in Nigeria", J. Applied Sci. and Env. Management, 7 (1): 37–42, 2003.

[12] Akpabio, L.E and Etuk, S.E. "Relationship between Global Solar Radiation and Sunshine Duration for Onne, Nigeria", Turk. J. Phys, 27, pp 161–167, 2003.

[13] Blazev A. "Solar Tecnologies for the 21st Century." Fairmont Press, Inc.ISBN0-88173-697-X, USA, 2013.

[14] Anthony F., Durschner C., and Remers K. "Photovoltaics for Professionals: Solar Electric Systems; Marketing, Design and Installation" Germany, Beuth Verlag, 2010.

[15] The German Solar Energy Society (DGS LV Berlin BRB), Ecofys. "Palnning and Installing Photvoltaic Systems: *A guide for installers, architects and engineers."* USA: James & James Science Publishers Ltd., 2006.

[16] Alsema, E. A. "Energy pay-back time and CO2 emissions of PV systems; *progress in photovoltaic Resource Application,"* 8, pp.17-25, 2000.

[17] Nordmann, T., Clavadetscher, L., Sark, W. G., & Mike Green. "Analysis of Long-Term Performance of PV Systems:" *Different Data Resolution for Different Purposes.* IEA International Energy Agency, 2014.

[18] Quaschning V. "Understanding Renewable Energy Systems," Earthscan Publications Ltd, London, 2005.

[19] PVSyst Version 6.0 Help Documents/ Simulation variable: Standalone System. Accessed May 2017

[20] Hay J. E.. "Calculation of Monthly Mean Solar Radiation for Horizontal and Inclined Surface," Solar Energy, Vol. 23, No. 4, pp. 301-308, 1979.

Comparative Study of the Chemical Compositions of Liquid Fuel from Thermal Cracking of Low and High-Density Polyethylene Plastic Waste.

Abosede A. Ajibola[1], James A. Omoleye[2], Vincent E. Efeovbokhan[3]

Department of Chemical Engineering, College of Engineering,
Covenant University
[1]adesinasibose@gmail.com, [2]james.abiodun@covenantuniversity.edu.ng,
[3]vincent.efeovbokhan@covenantuniversity.edu.ng
[1]08054117177, [2]08033737001, [3]08139599960

Abstract. The increase in the rate of accumulation of plastic waste (PW) has been of great concern to the world especially in the developing countries due to its non-biodegradable nature and improper waste management practices. Hence, efforts towards the conversion of this waste (PW) to resourceful materials have led us to the exploration of pyrolysis (anaerobic thermal cracking) of plastic waste under a controlled condition to produce liquid fuel. A stainless steel batch reactor was used in the cracking of the low and high-density polyethylene (LDPE and HDPE) plastic wastes into liquid fuel components at a temperature of 230°C. The liquid fuel obtained from the pyrolyzed LDPE and HDPE was analyzed using GC-MS. Fifty (50) compounds each was identified for both LDPE and HDPE which revealed the presence of mostly alkenes and aromatics in the hydrocarbon ranges of $C_8 - C_{24}$. This is made up of 36% of gasoline fractions range from C_6 - C_{12}, 32% of diesel fractions range C_{13} - C_{20} , and 14% oil of residual fuel range of $C_{20} - C_{28}$ and 18% of non-hydrocarbons was discovered for the HDPE while 36% of gasoline fractions range of C_6 - C_{12}, 34% of diesel fractions range C_{13} - C_{20}, oil and 12% residual fuel range of $C_{20} - C_{28}$ and 18 % of non-hydrocarbons was discovered for the LDPE . There is little or no difference in the products of pyrolysis of light and heavy polyethylene plastic waste.

Keywords: Polyethylene (PE), Thermal Cracking, Pyrolysis, GC-MS and Clean Technology.

1. Introduction

The use of plastic is of greater importance in our daily lives and its consumption has led to the drastic increase of plastic waste, PW in the twentieth century [1]. There is a higher demand for plastic, which has also increased rapidly because they are considered as solution materials to various sectors like Engineering, medicals, electronics, aerospace, etc. They are essentials for the advancement of technology due to their great physical properties such as its lightweight and flexibility compared to other materials like metals, glass, wood, and concrete [2].

According to UNEP [3], a major components of Municipal Solid Waste (MSW) is the plastic waste and it's fast becoming the third largest MSW in developing countries. In addition, developing countries do not put into consideration the advantages of economic gain involved by utilizing some certain recycling methods but still depend solely on the conventional method of landfilling of MSW disposal. This conventional method has contributed to the major health and environmental hazards such as greenhouse gas emissions, groundwater pollution, and several other human health problems. For this reason, exploration of other methods like recycling has been a main attention for research in which PW is used as raw materials for recovery of valuable products and energy so as to solve the shortage of natural resources in the nearest future [2] [4].

PW recycling methods have been mainly grouped into four major types namely Primary recycling which involves waste scraps being processed into products with similar properties to the original products; Secondary recycling has to do with waste/scrap plastics being processed into materials that have different properties to that of the initial product; Tertiary methods deals with conversion of these wastes scrap in the production of essential fuels and chemicals or as a segregated waste and lastly Quaternary recycling involves the burning of these plastics and retrieving energy contents afterwards[5].
Each of these recycling methods is advantageous and effective for different applications.
The main types of plastic used for daily commodities are thermoplastics which become soft when heat is introduced and later hardens when cooled and the other, thermosets which harden irreversibly when heat is introduced. MSW is of six main component namely, Low-Density Polyethylene (LDPE), High-Density Polyethylene (HDPE), Polyvinyl Chloride (PVC), Polystyrene (PS), Polyethylene Terephthalate PET) and Polypropylene (PP). Polyethylene plastics (both LDPE and HDPE) make up about forty percent of the MSW [2] [6].

Pyrolysis involves the cracking or breakdown of the long chain large Hydrocarbons in an anaerobic atmosphere. Gas, oil, and char are the three major products obtained during pyrolysis process and these products are of great value to production and refinery industries [7]. Compared to other MSW management practices, pyrolysis is more desirable and satisfactory environmentally because it reduces the carbon footprint of products. Pyrolysis process also lessens the amount of carbon monoxide and carbon dioxide emissions by making an effective use of inert atmosphere that is free of oxygen to avoid the formation of dioxins through product reaction with oxygen [8] [9]. This pyrolysis process is a cleaner technology, a drive towards energy security and a measure to combat fossil fuel depletion.
There have been a number of published articles on the pyrolysis of polyethylene plastics (for LDPE and HDPE). For example, Chanashetty and Pau l [10] undertook pyrolysis of LDPE and PP experiment using temperature ranges from 100 OC to 400 OC and obtained a 50-65% conversion to fuel oil. Onwudili et al [6] also investigated the pyrolysis of LDPE, PS and their mixtures between temperatures of 300 OC to 500 OC, LDPE was thermally degraded at 425OC
This work seeks to evaluate the cracking of Low Density Polyethylene (LDPE) Plastic Waste. The success of this work will enhance the effectiveness of plastic waste processing to useful products in Nigeria and Africa in general in an economically affordable manner and hence reduce the environmental hazard posed by their indiscriminate dumping.

2. Materials and Method

Materials

Two different types of polyethylene plastic wastes (PPW), High-Density Polyethylene (HDPE) and Low-Density Polyethylene (LDPE) were used in this experiment. The PPW was obtained from waste bins and garbage from Ota, South-West Nigeria. They were then sorted, cleaned to remove dirt and then shredded into smaller particle sizes.

Experimental Procedures

LDPE and HDPE Plastic wastes were each collected and prepared by cutting into smaller particle sizes of 3 to 4 mm and charged into a semi-batch pyrolysis reactor. From Fig 1, the continuous batch reactor of capacity 300 ml is a lagged cylindrically shaped stainless steel container with a mild steel cover. The reactor was electrically heated with a 2 kW heating element equipped with an automatic temperature controller and connected to it was a coil condenser, liquid and gas collectors. 20g of PPW sample was used in each experiment and fed into the reactor, tightly closed and purged with nitrogen gas at the start of each experiment. The heating element was turned on at the specified temperature to heat the PPW pyrolyzed and cracked it until oil fractions begin to evolve, condensed and collected.

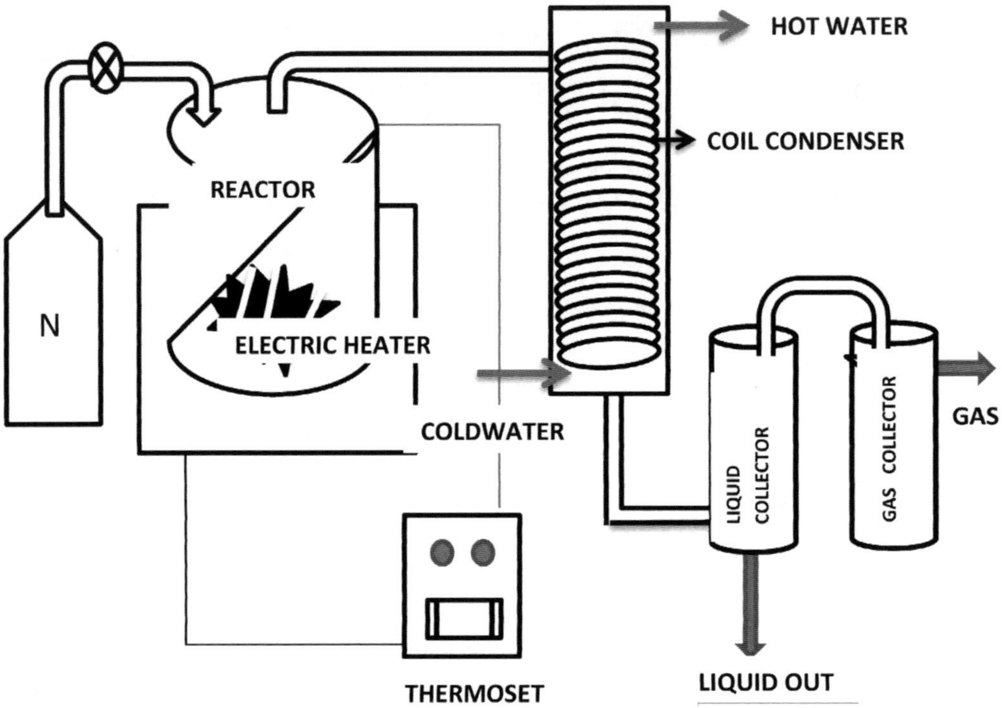

Fig. 1 Schematic Representation of the Pyrolysis Set Up

Analytical Methods

Analysis of the pyrolysed oil was done using Gas Chromatography - Mass Spectrophotometry (GC-MS) to determine the chemical compositions of the hydrocarbon products. The GC-MS and column oven temperature conditions used are stated in the table below

Table 1: GC/MS condition

GC-MS Agilent 433UI HP-5ms Ultra Inert		
Column Oven Temperature	500c	
Injection Mode	Split	
Injection Temperature	1500C	
Injection Volume	1 ml	
Split Ratio	10:1	
Average Velocity	36.445cm/Sec	
Column Pressure	7.6522psi	
Column Flow	1.2211ml/Min	
Carrier Gas	Helium 99.9995% Purity	
Frequency	50Hz	
Electron Multiplier Volts	1024.9	
Colum Oven Temperature Progress		
Rate (0C/min)	Temperature (0C)	Hold Time (min)
-	50	0
8	300	9
Column		

Length Diameter Film Thickness Total runtime	30.0m 250μm 0.25μm 32 mins
Ms Conditions	
Source Temperature Start mass range m/z End mass range m/z	2300C 50 550

3. Identification of Components:

The mass spectra database of National Institute Standard and Technology (NIST) library was used for the interpretation of the GC-MS. The spectrum of component unknown was compared with the spectrum of components known. The names of the materials tested and their molecular weight were confirmed.

Table 2: Chemical Compositions of the Liquid Fuel Products from Pyrolysis of HDPE

Peak	Retention Time	Area %	Compound Name	Molecular Formula	Molecular Weight
1	3.224	1.43	4-Ethyl-5-Methyl-Nonane	C12H26	170.3348
2	3.47	2.6	3-Octyne	C8H14	110.1968
3	4.042	1.29	1,3-Dimethylbenzene	C8H10	106.165
4	4.477	1.46	Trans-2-Nonene	C9H18	126.2392
5	4.695	1.77	Cyclohexane, 2-Propenyl-	C9H16	124.2233
6	5.221	2.72	3,4-Octadiene, 7-Methyl-	C9H16	124.2233
7	6.177	1.54	E-12-Tetradecenal	C14H26O	210.3556
8	6.48	2.17	Methylcycloheptane	C8H16	112.2126
9	6.566	2.43	1-Indanone	C9H14O	138.2069
10	6.726	1.01	1-Cyclohexyl-2-Propene	C9H16	124.2233
11	6.961	1.35	Indene	C9H8	116.1598
12	7.476	1.18	Pinane	C10H18	138.2499
13	8.111	1.04	5-Undecene	C11H22	138.2499
14	8.208	1.22	5-Undecene	C11H22	138.2499
15	8.866	1.34	1-Butynylbenzene	C10H10	130.1864
16	8.693	2.29	1-Methylindene	C10H10	130.1864
17	9.078	1.3	Spiro(4,5)Decane	C10H18	138.2499
18	9.392	3.29	Cyclododecene	C12H22	166.3031
19	9.885	1.31	2-Dodecene	C12H24	168.319
20	9.892	1.6	2-Dodecene	C12H24	168.319
21	11.304	3.22	1-Tridecene	C13H26	182.3455
22	11.698	1.11	Tridecene,	C13H26	182.3455
23	11.836	1.21	2-Chloroethyl Linoleate	C20H35ClO2	342.948
24	12.042	1.1	Bicyclo[3.3.2]Decan-9-One	C10H16O	152.237
25	12.173	1.12	5-Butyl-4-Nonene	C13H26	182.3455
26	12.677	1.25	8-Dodecen-1-Ol, Acetate	C14H26O2	226.355
27	12.814	4.37	1,13-Tetradecadiene	C14H26	194.3562
28	12.963	4.46	2-Tetradecene,	C14H28	196.3721
29	14.393	3.47	1,13-Tetradecadiene	C14H26	194.3562
30	15.887	4.16	E-10-Pentadecenol	C15H30O	226.404
31	17.312	3.62	1,19-Eicosadiene	C20H38	278.5157

IOP Conf. Series: Materials Science and Engineering **413** (2018) 012014 doi:10.1088/1757-899X/413/1/012014

32	17.466	3.73	3-Heptadecene,	C17H34	238.4519
33	18.668	2.54	1-Hexadecyne	C16H30	222.4094
34	19.949	2.35	1,19-Eicosadiene	C20H38	278.5157
35	21.042	4.78	Dibutyl Phthalate	C16H22O4	278.348
36	21.191	2.17	Bicyclo[10.8.0]Eicosane	C20H38	278.524
37	22.353	1.87	1,19-Eicosadiene	C20H38	278.5157
38	23.474	1.51	Bicyclo[10.8.0]Eicosane	C20H38	278.524
39	23.594	1.95	1-Docosene	C22H44	308.5848
40	24.544	1.01	1,19-Eicosadiene	C20H38	278.5157
41	24.653	2.07	9-Tricosene	C23H46	322.6113
42	25.58	0.96	1,21-Docosadiene	C22H42	306.5689
43	25.666	1.65	Cyclotetracosane	C24H48	336.6379
44	26.633	1.24	1-Docosene	C22H44	308.5848
45	27.199	2.18	Di-N-Octyl Phthalate	C24H38O4	390.5561
46	27.571	1.58	9-Tricosene	C23H46	322.6113
47	27.623	1.52	Pentadecane	C15H32	212.4146
48	28.481	1.23	Octacosyl Acetate	C30H60O2	452.7962
49	28.521	1.13	Tetracosane	C24H50	338.6538
50	30.243	1.1	Heptadecane	C24H50	240.4677

Table 3: Chemical compositions of the liquid fuel products from pyrolysis of LDPE

Peak	**Retention Time**	**Area %**	**Compound Name**	**Molecular Formula**	**Molecular Weight**
1	3.264	1.17	1,2-dimethyl cyclohexane	C8H16	112.2126
2	4.506	1.09	Cyclooctanone	C8H14O	126.1962
3	4.643	1.02	4-Nonene	C9H18	126.2392
4	5.004	1.02	2-methylpropyl Cyclohexane	C10H20	140.2658
5	5.244	2.82	7-methyl-3,4-Octadiene,	C9H16	124.2233
6	6.171	1.37	trans-4-Decene	C10H20	140.2658
7	6.491	1.65	methylCycloheptane	C8H16	112.2126
8	6.577	1.94	9-Methylbicyclo[3.3.1]nonane	C10H18	138.254
9	7.024	1.16	9-Methylbicyclo[3.3.1]nonane	C10H18	138.254
10	7.51	1.26	2,5,5-Trimethyl-1,6-heptadiene	C10H18	138.2499
11	8.185	1.22	5-Undecene	C11H22	154.2924
12	8.815	0.99	4,8-dimethyl-1,7-Nonadiene	C11H20	152.277
13	9.072	1.06	2-Pentanylcyclopentane	C10H20	140.2658
14	9.398	3.4	14-methyl-(Z)-8-hexadecen-1-ol	C17H34O	254.4513
15	9.839	1.23	2-Dodecene	C12H24	168.319
16	10. 365	1	Cyclododecane	C12H24	168.319
17	10.754	1.1	1,5-Dimethyldecahydronaphthalene	C12H22	166.3031
18	10.863	1.2	Octacosyl heptafluorobutyrate	C32H57F7O2	606.7826
19	11.132	3.23	1-Octadecyne	C18H34	250.4626
20	11.286	3.68	1-Tridecene	C13H26	182.3455
21	11.59	2.81	2,3,4-Trimethylhexan	C9H20	128.2551
22	11.681	2.2	5-Butyl-4-Nonene	C13H26	182.3455
23	11.819	2.06	1,1-Dimethyl-2-propylcyclohexane	C11H22	154.297
24	12.671	1.79	trans-Pinane	C10H18	138.2499
25	12.797	3.31	14-methyl-(Z)-8-hexadecen-1-ol	C17H34O	254.4513
26	12.963	4.52	2-Tetradecene,	C14H28	196.3721
27	14.376	2.67	1,12-Tridecadiene	C13H24	180.3297

28	14.537	3.58	1-Pentadecene	C15H30	210.3987
29	15.876	3.15	7-Dodecen-1-ol, acetate	C14H26O2	226.355
30	17.306	2.6	trans-2-Dodecen-1-ol,trifluoroacetate	C14H23F3O 2	280.3264
31	17.461	3.91	cis-3-Heptadecene	C17H34	238.4519
32	18.656	1.93	1,19-Eicosadiene	C20H38	278.5157
33	18.811	2.82	1-Octadecene	C18H36	252.4784
34	19.068	1.13	trans-9-Octadecene	C18H36	252.4784
35	19.938	1.88	Pentadecanal	C15H30O	226.3981
36	21.081	2.86	Z-5-Nonadecene	C19H38	266.513
37	21.161	1.03	Octadecane	C18H38	254.4943
38	21.014	4.36	Dibutyl Phthalate	C16H22O4	278.348
39	21.186	1.26	Bicyclo[10.8.0]Eicosane	C20H38	278.524
40	22.3	2.31	Cycloeicosane	C20H40	280.5316
41	22.341	1.16	Bicyclo[10.8.0]Eicosane	C20H38	278.524
42	22.444	2.22	Z-5-Nonadecene	C19H38	266.513
43	23.524	1.01	Heneicosane	C21H44	296.5741
44	23.543	2.23	1-Docosene	C22H44	308.5848
45	23.617	1.475	8-Heptylpentadecane	C22H46	310.6006
46	24.596	1.4	9-Tricosene	C23H46	322.6113
47	25.659	1.01	Pentadecane	C15H32	212.4146
48	25.614	1.06	Cyclotetracosane	C24H48	336.6379
49	25.672	1.56	Tetracosane	C24H50	338.6538
50	27.182	1.83	Di-n-octyl phthalate	C24H50O4	390.5561

4. RESULTS AND DISCUSSION

The results appropriated to the GC-MS analysis has established the identity of the number of compounds or fractions of the liquid fuel obtained from the pyrolysed PPW. Identification of compounds was also done through the mass spectrophotometry attached with the GC. The liquid fuel extracts was dark yellow in colour. Fifty compounds were detected for the pyrolysis of both the LDPE and HDPE. It is interesting to know that liquid fuel for both shows similar compositions. The eight major compounds confirmed for HDPE were 1-Tridecene (3.22%), Cyclododecene (3.29%), 1,19-Eicosadiene (3.62%), 3-Heptadecene (3.73%), E- 10-Pentadecenol (4.16%), 1,13-Tetradecadiene (4.37%), 2-Tetradecene (4.46%), Dibutyl Phthalate (4.78%) with the retention time of 11.304, 9.392, 17.312, 17.466, 15.887, 12.814, 12.963 and 21.042 minutes respectively. The nine major compounds also confirmed for the LDPE were 1-Octadecyne (3.23%), 7-Dodecen-1-ol acetate (3.15%), 14-methyl-(Z)-8-hexadecen-1-ol (3.31% & 3.40%), 1-Pentadecene (3.58%), 1-Tridecene (3.68%), cis-3-Heptadecene (3.91%), Dibutyl Phthalate (4.36%), 2-Tetradecene (4.46%) with the retention time 11.132, 15.876, 12.797, 9.398, 14.537, 11.286, 17.461, 21.014 and 12.963 minutes respectively. This conforms to the findings of Shah et al.[11].

It was observed that at a particular retention time, 12.963, the same compound namely 2-Tetradecene was confirmed for both the HDPE and LDPE. Also Bicyclo[10.8.0] Eicosane was confirmed. The compound obtained here is similar to the compound obtained by Patil, Varma, Gajendra, & Mondal [12].

Characterization of Liquid Products

From the GC-MS analysis results, Fig. 2 and Fig. 3 represent the total ion chromatography graph of the liquid fuel products, which showed the peak areas of all the identified compounds, particularly peak areas of more than 3% for the major compounds detected. Both plastics revealed the presence of hydrocarbons in the C8 – C24 range. By comparing the compounds present in the liquid fuel of the pyrolyzed HDPE and LDPE, it can be deduced from Fig. 4 that the carbon number distribution of HDPE and LDPE contains same fractions but slightly different percentage composition. For HDPE, gasoline

(C6 - C12) fraction had 36%, diesel (C13 – C20) fraction had 32%, C21- C28 fraction 14% and the none hydrocarbon fraction range gave 18% while for LDPE, gasoline (C6 - C12) fraction had 36%, diesel (C13 – C20) fraction had 34%, C21- C28 fraction 12% and the none hydrocarbon fraction range also gave 18% [13].

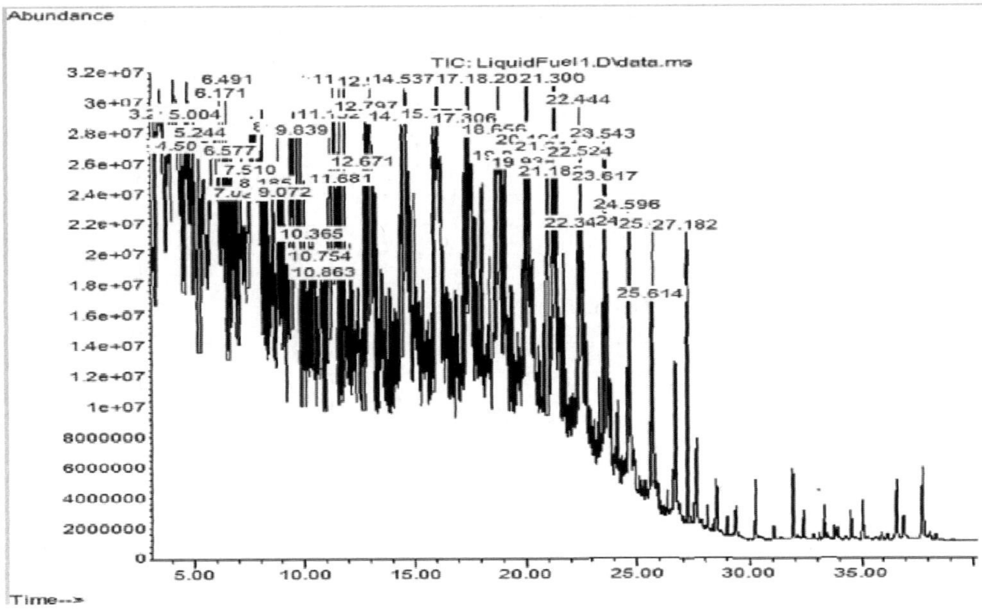

Fig. 2: Total Ion Chromatography of the pyrolyzed LDPE

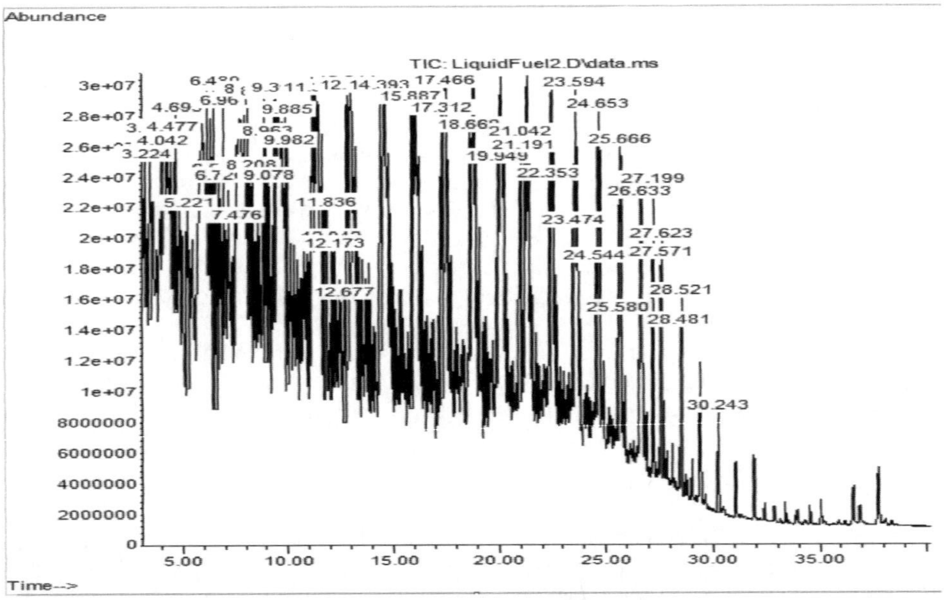

Fig. 3: Total Ion Chromatography of the pyrolyzed HDPE

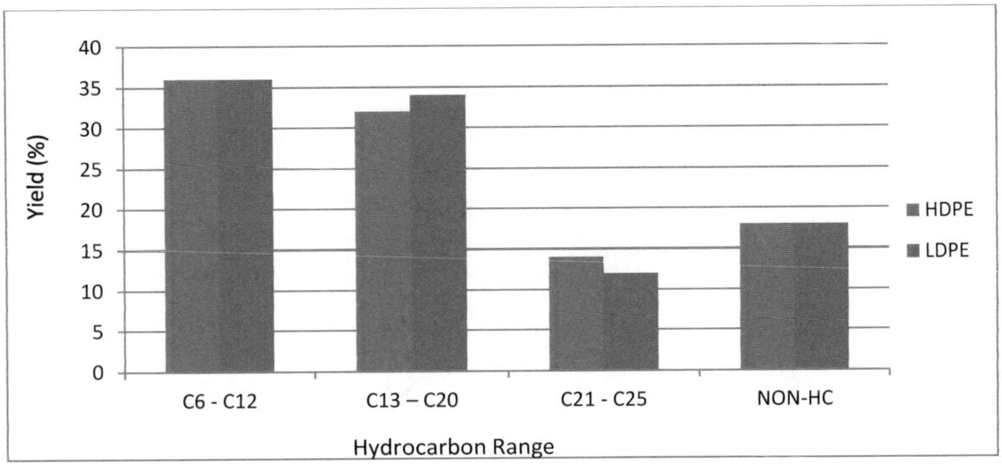

Fig. 4: Comparison of Hydrocarbon Ranges

5. CONCLUSION

In spite of all environmental problems caused by plastic wastes, it is still a valuable raw material for petrochemical and refinery industries. The thermal cracking of LDPE and HDPE operated using a semi-batch reactor has been used to obtain useful hydrocarbon fractions at a temperature of 230oC. GC-MS showed the compositional analysis of the liquid fuel obtained containing mainly aliphatic and aromatic compounds within the carbon range of C8 to C24. The liquid products obtained have similar products to fossil fuels which can be used as alternative fuels for a more sustainable and cleaner environment when necessary blending is done for upgrading.

References

[1] Neves, I. C., Botelho, G., & Machado, A. V. (2007). Catalytic degradation of polyethylene : An evaluation of the effect of dealuminated Y zeolites using thermal analysis, *104*, 5–9. https://doi.org/10.1016/j.matchemphys.2007.02.032

[2] Al-salem, S. M., Antelava, A., Constantinou, A., Manos, G., & Dutta, A. (2017). A review of thermal and catalytic pyrolysis of plastic solid waste. *Journal of Environmental Management*, *197*(1408), 177–198. https://doi.org/10.1016/j.jenvman.2017.03.084

[3] UNEP. (2009). Converting Waste Plastics Into A Resource. Osaka/Shiga. Retrieved from www.unep.fr

[4] Kumar, S., & Singh, R. K. (2011). Recovery of hydrocarbon liquid from waste high density polyethylene by thermal pyrolysis. *Brazilian Journal of Chemical Engineering*, *28*(4), 659–667. https://doi.org/10.1590/S0104-66322011000400011

[5] Al-Salem S.M. (2009). Establishing an integrated databank for plastic manufacturers and converters in Kuwait. *Journal of Waste Management*, 29, 479-484.

[6] Onwudili, J. A., Insura, N., & Williams, P. T. (2009). Composition of products from the pyrolysis of polyethylene and polystyrene in a closed batch reactor: Effects of temperature and residence time. *Journal of Analytical and Applied Pyrolysis*, *86*(2), 293–303. https://doi.org/10.1016/J.JAAP.2009.07.008

[7] Sharuddin, A., Faisal, A., Daud, W. M., & Aroua, M. K. (2016). A review on pyrolysis of plastic wastes. *Energy Conversion and Management*, *115*, 308–326. https://doi.org/10.1016/j.enconman.2016.02.037

[8] Singh R.K. and Ruj B. (2016). Time and temperature depended fuel gas generation from pyrolysis of real-world municipal plastic waste, *Journal of Fuel*, 104, 164-171.

[9] Chen D., Yin L., Wang H. and He P. (2014). Pyrolysis technologies for municipal solid waste: a review. *Journal of Waste Management,* 34, 2466-2486.

[10] Chanashetty, V. B., & Patil, B. (2015). Fuel from Plastic Waste. *International Journal on*

Emerging Technologies, *6*(2), 121–128.
[11] Shah, S. H., Khan, M. Z., Raja, A. I., Mahmood, Q., Bhatti, A. Z., Khan, J., … Wu, D. (2010). Low temperature conversion of plastic waste into light hydrocarbons. *Journal of Hazardous Materials*, *179*(1–3), 15–20. https://doi.org/10.1016/j.jhazmat.2010.01.134
[12] Patil, L., Varma, K., Gajendra, S., & Mondal, P. (2017). Thermocatalytic Degradation of High Density Polyethylene into Liquid Product. *Journal of Polymers and the Environment*, *0*(0), 1–10. https://doi.org/10.1007/s10924-017-1088-0
[13] Marcilla, A., Beltrán, M. I., & Navarro, R. (2009). Thermal and catalytic pyrolysis of polyethylene over HZSM5 and HUSY zeolites in a batch reactor under dynamic conditions. *Applied Catalysis B: Environmental*, *86*(1–2), 78–86. https://doi.org/10.1016/J.APCATB.2008.07.026

IOP Conf. Series: Materials Science and Engineering **413** (2018) 012015 doi:10.1088/1757-899X/413/1/012015

Cost-effectiveness Management of a Compressed-air System Using an Energy Analysis Application

Peter I. Anakhu[1]*, Christian A. Bolu[1], Abiodun A. Abioye[1], Godfrey Onyiagha[2]

[1]Mechanical Engineering Department, Covenant University, Ota, Nigeria

[2]Electrical and Information Engineering Department, Covenant University, Ota, Nigeria

*E-mail address: peter.anakhu@stu.cu.edu.ng

Abstract. This paper developed an application for a foundry enterprise to reduce the delay in fixing identified air leaks and decrease energy consumption in the compressed-air systems. In many instances, air leaks were detected, but the response towards correcting them is usually less expeditious by the various departments responsible. Perhaps, there is a lack of awareness about the opportunities to conserve energy either through prompt repairs of air leaks or by regulating the pressure setting to match demand per time. An energy analysis application was developed using Matlab®. It facilitated increased response to repairs of air leaks and improved supply-demand management of output pressures from compressed-air systems.

Keywords: ***Air leaks, compressed-air, energy cost, waste energy, energy audit***

1. Introduction

Compressed air (CA) is an essential input requirement in process equipment including pneumatic-operated machine, cleaning machine, refrigeration, foundries, and more [1], [2]. Many manufacturing enterprises including sand casting industry require CA to perform a variety of tasks ranging from moulding, cleaning, blowing, and more. It is also used in some burners that generate heat for the preheating of ladle in sand casting. Also, compressed air is the most common energy source for control valve actuators [3]. The compressed air is usually supplied at a guage pressure of between 0.55 to 1.0 MPa [4]. Some specific compressed-air systems require higher pressures in the range of 3.2 to 4.0 MPa [2]. Generally, increased energy cost is associated with CA supplied at high pressures. Perhaps, this explains why the cost of generating CA is relatively high, approximately 10 - 30% of the total electricity consumption in many industries around the world [1], [5]. Obviously, many manufacturers may be quick to identify low-pressure compressed-air systems as a conceivable low-cost alternative that may reduce energy consumption. But lower pressure of compressed air may not guarantee the desired pressures at the various points of end use. This is because the usual practice in many industries is to locate compressed-air systems in adjacent shelters specifically built outside the points of end use and then convey the compressed-air through pipe networks to various points of end use (Fig. 1) [6]. Consequently, pressure drops are not uncommon along the lines due to fittings, valves, filters and air leaks [1]. Several studies have identified methods for promoting energy savings and comprehensive optimisation decisions in compressed-air systems. They include supply-demand management technique, reducing output pressures to match demand per time, reducing leaks, prompt attention to leaks repair, and reducing average inlet temperature by installing intake duct [5], [7]. The comprehensive optimisation decisions involves the use of a combination of exergy and energy efficiencies as well as cost-effective analysis [6].

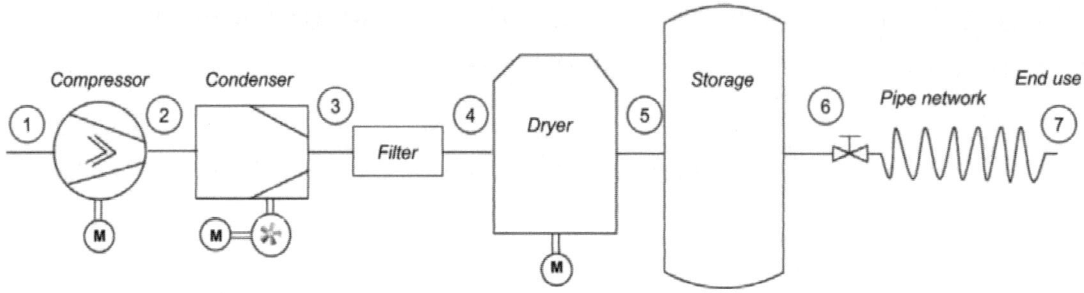

Fig. 1. Typical compressed-air system components, pipe network and points of end use (Source: [6])

Compressed air is more expensive to use than most users would probably want to admit [3]. This is because the general opinion about air is that it is free. Air is actually free of costs but not after it has been compressed. Huge electricity costs are required to filter, clean and raise the pressure of normal air to a status that is useful to operate many devices.

Air leaks have been identified to cause significant pressure drops and energy loss in compressed-air systems [4]. Similar to some accessories such as fittings and filters in the compressed-air lines, air leaks are difficult to avoid or totally eliminate. However, proactive measures are required to fixing identified air leaks. This paper developed an energy analysis application for compressed-air systems to demonstrate the energy savings potential when repairs are promptly carried out on identified air leaks in the system. This is to discourage the usual practice of waiting until the air leaks caused significant pressure losses that interfere with the normal operation of the equipment connected to the compressed-air lines.

It is also believed that old compressed-air systems should be replaced to reduce energy consumption. But having old equipment changed to a new one is always not against the grain to waste energy [8]. Instead, the authors argued in favour of three elements that may be used to promote energy efficiency, viz: To identify and value waste energy. To encourage good practice for resource efficiency. In other words, to adopt supply-demand management technique to match output pressures to needs per time. To understand potential savings and set targets. The developed application may be used to facilitate the accomplishment of some aspects of the listed elements including estimating the energy cost of air leaks and computing the savings in energy cost due to prompt repairs of air leaks.

Furthermore, it is not a usual practice in many industries using compressed-air systems to have the time to study the cost involved in the operation of the utility system [5]. However, with the energy analysis application, energy analysts and maintenance crew may be alert on what better practice to adopt to reduce energy consumption and overall operating cost of a CA system. In this study, an energy analysis application is introduced to facilitate the best improvement decisions about air leakages and management of pressure supply-demand from compressed-air systems.

2. Literature Review

Different legislative and normative measures are usually adopted by various industries to reduce energy cost [9]. Similarly, with the advent of mechatronics measurement systems, many researchers have recently directed attention to developing new devices to retrieve information about air leaks and possible system inefficiencies [2].

Ming Yang [10] developed methodologies that served as a guide on how to analyse compressed-air system energy efficiency. It was found that the energy loss related to air leaks and non-optimised operations of compressed-air systems can be minimised. However, empirical data were largely utilized in the investigation and the conclusions were based on the outcome of the processed data. This approach can be extended to include a predictive analytical model to server as a forewarn to various departments

IOP Conf. Series: Materials Science and Engineering **413** (2018) 012015 doi:10.1088/1757-899X/413/1/012015

about the increased energy cost due to delay in air leaks repair and poor scheduling of output pressures from compress-air systems.
Optimization methods have also been suggested to improve energy efficiency and reduce cost of manufacturing compressed-air systems [2]. According to Jovanovic et al. [2], the method may be focused on different stages including modeling of the overall air supply capacity, capacity validation of all compressors used in a manufacturing line, compressor design optimization, and using simulation to evaluate multiple compressors at a station. The present research work is aimed at developing an application in consideration of stage one in addition to incorporating the analysis of air leaks to demonstrate the energy savings potential involved in promptly correcting air leakages.
Also, simulation of some fittings such as Tees and 90° elbows was carried out to determine the effects of pressure gradient across compressed-air pipe network [11]. The authors reported a reduction in compressor electricity consumption when the pipe junction geometry was redesigned. The redesign involved the replacement of Tees with branch lines of 46° from the main pipe and the substitution of 90° elbows with 150-mm fillet radii. However, replacing the existing pipe fittings with the new junction geometry may be difficult to implement in some orientations that are perpendicular to each other. Nevertheless, the pipe junction redesign may be considered alongside with the supply-demand management model proposed in the current work to maximize energy savings.

The use of mechatronic measurement systems to track, detect and increase the response to fixing air leaks, thereby increasing system efficiency, has recently become a focus of several studies [2]. But in developing countries including Nigeria, high-tech mechatronic systems are only gradually gaining attention and the inclusion of mechatronic devices in compressed-air systems is still being considered with caution. Also, substantial investment may be required to retrofit old compressed-air systems with mechatronic devices. By using energy analysis application to facilitate decision making about air leakages, substantial energy and cost may be saved.
On a general observation, it was stated that conservation measures be adopted in the use of energy in various energy intensive sectors [12]. Therefore, an energy analysis application is being developed for compressed-air systems to facilitate decision making about air leaks, and to substantially reduce energy consumption and cost associated with the use of the systems.

3. Materials and Method

Air leaks in compressed-air lines are considered. In addition, supply-demand management model of output pressure that meets air supply needs per time is incorporated.

3.1 Thermodynamic analysis and energy application model

The higher the discharge pressure from a compressed-air system, the larger the work required for compression. On the contrary, when the discharge pressure, $P_{disch(max)}$, of the system is reduced to $P_{disch(red)}$, the power input requirements of the system is also reduced by a factor defined in Eq. 1 [4].

$$f_{red} = 1 - \frac{\left(P_{disch(red)}/P_{atm}\right)^{(n-1)/n} - 1}{\left(P_{disch(max)}/P_{atm}\right)^{(n-1)/n} - 1} \quad (1)$$

where n is ratio of specific heat, $P_{disch(red)}$ is the reduced discharge pressure, $P_{disch(max)}$ is the maximum discharge pressure possible, P_{atm} is the atmospheric pressure, and f_{red} is the power reduction factor.
A unit mass of air that escapes through leaks on a compressed-air line usually wastes some amount of energy equivalent to the mechanical energy used in compressing it, and is determined in Eq. 2 [6], [13]–[15].

$$w_{comp,in} = \frac{n\,R\,T_{amb}}{n_{comp}\,(n-1)} \left[\left(\frac{P_{disch}}{P_{atm}}\right)^{(n-1)/n} - 1\right] \quad (2)$$

where $w_{comp,in}$ is input energy to compressor, n_{comp} is the compressor efficiency, n is ratio of specific heat, R is the gas constant, T_{amb} is the ambient temperature, P_{atm} is the atmospheric pressure, and P_{disch} is the discharge pressure.

IOP Conf. Series: Materials Science and Engineering **413** (2018) 012015 doi:10.1088/1757-899X/413/1/012015

The unit mass of air usually has a mass flow rate as it escapes through a leak of minimum cross-section. This is determined using Eq. 3 [4].

$$\dot{m}_{air} = A\, C_{disch} \frac{P_{line}}{R\, T_{line}} \left(\frac{2}{n+1}\right)^{1/(n-1)} \sqrt{n\, R \left(\frac{2}{n+1}\right) T_{line}} \qquad (3)$$

where $\dot{m}_{air}$ is the mass flow rate, C_{disch} is the discharge or loss coefficient, A is the cross-sectional area of the leak hole, n is ratio of specific heat, R is the gas constant, P_{line} is the line pressure, and T_{line} is the line temperature.

The waste power when a leak is not repaired is determined as follow:

$$Power_{wasted} = \dot{m}_{air} \text{ x } w_{comp,in} \qquad (4)$$

3.2 Analysis of data

Data were collected from a typical single-stage positive-displacement compressor from a foundry facility. The operating conditions of the compressor and the ambient environmental conditions of the air were recorded. The compressor rating including the mass flow rate and efficiency were obtained from technical book. The operating hours were estimated. The maximum discharge pressure of the compressed-air system is 1000 kPa. The highest-pressure requirement in the connecting points is 600 kPa. A gauge pressure to each of the points was between 640 – 720 kPa. An average 680 kPa was assumed for the line pressure. Table 1 shows the complete data used in the energy analysis.

Table 1: Compressed-air operating conditions data

Description	Unit	Value
Hole diameter	mm	2.50
Atmospheric pressure:	kPa	101.00
Ambient temperature:	C	24.00
Max. discharge pressure	kPa	10000.00
Reduced discharge pressure	kPa	780.00
Discharge temperature	C	43.00
Line pressure	kPa	680.00
Line temperature	C	32.00
Specific heat ratio		1.40
Hours per day	h	10.00
Discharge or loss coeff.		0.65
Motor efficiency	%	86.00
Compressor efficiency	%	72.00
Compressor mass flow rate	kg / s	3.00
Exchange rate	NGN / Dollar	364.00
Unit cost per kWh	$ / kWh	0.0824

3.3 Application development

The application was developed using Matlab®. A typical single-stage positive displacement compression system was considered. The algorithms were based largely on the thermodynamics models for the compression systems. The assumptions made are: 1. The air is an ideal gas. 2. The compression system operates at a steady-state condition. 3. The losses at the fittings, filters and valves are negligible compared to energy lose due to air leaks.

4. Results and discussion

The energy analysis application can be used to facilitate decision making about energy management relating to compressed-air operation. The application revealed that air leaks caused substantial waste energy on compressed-air systems (Table 2). The waste energy increases with the increase in the diameter of hole as shown in Fig 2.

IOP Conf. Series: Materials Science and Engineering **413** (2018) 012015 doi:10.1088/1757-899X/413/1/012015

Table 2: Energy analysis result

Description	Unit	Value
Input energy to compressor	kJ / kg	1.1302 x10^3
Cost of electricity per year	$	8.7190 x 10^5
Cost savings per year	$ / yr	2.1768 x 10^8
Pressure reduced	kPa	9220
Wasted energy	kW	85.4317
Hole cross-sectional area	m^2	4.9087 x 10^6
Mass flow rate	kg / s	0.0756
Wasted power per day	kWh / day	993.3920
Wasted power per year	kWh / yr	3.0994 x 10^5
Cost wasted per day	$	77.4846
Cost wasted per year	$	2.4175 x 10^4
Reduction factor		0.6859

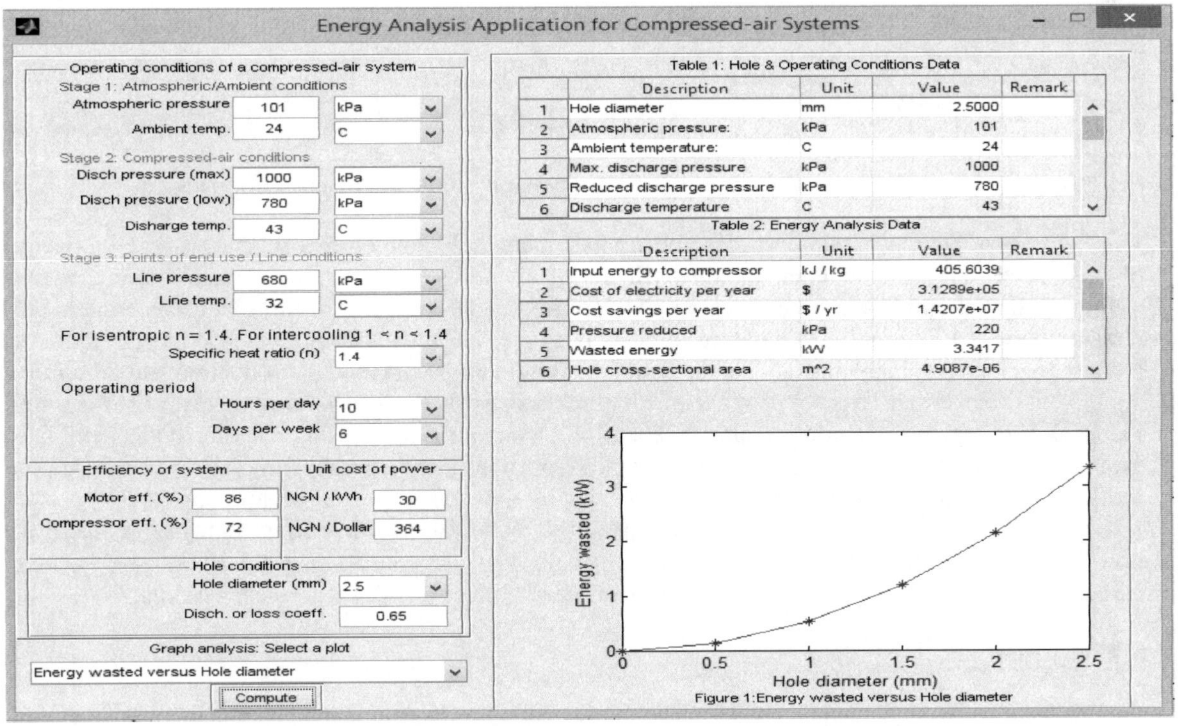

Fig 2: Analysis of waste energy against diameter of air leak

The minimum discharge pressure that can ensure that connecting points are not affected is 780 kPa. This pressure was used as the reduced pressure setting on the compressed-air system. That is, the system has an operating pressure range of 780 – 1000 kPa. For a pressure reduction of 220 kPa, cost of energy saved per annum is $1.42 x 107. The larger the pressure reduction without significantly affecting the operation of the connecting points, the higher the cost savings. As the discharge pressure increases to maximum, cost savings decreases as indicated in Fig 3.

IOP Conf. Series: Materials Science and Engineering **413** (2018) 012015 doi:10.1088/1757-899X/413/1/012015

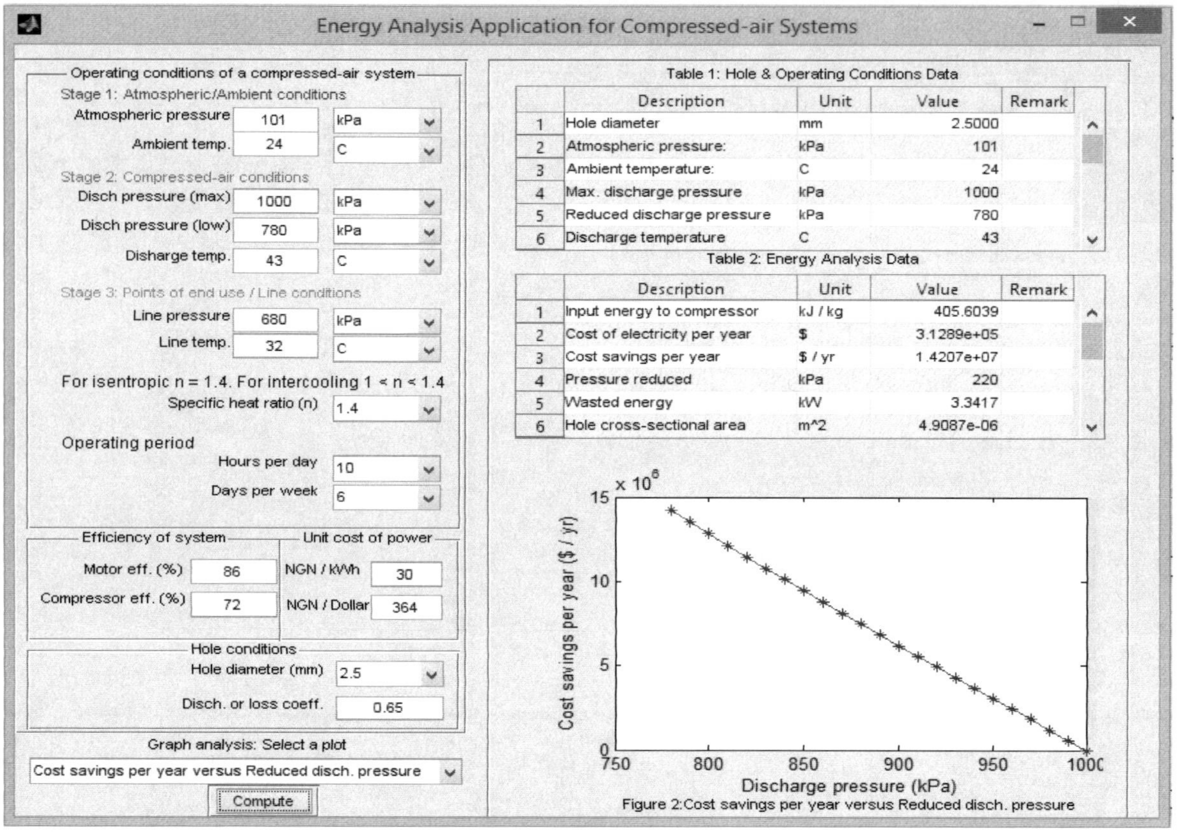

Fig 3: Analysis of the cost savings per year by reducing discharged pressure

This study presents a different cost-effective management technique of energy analysis for the operation of compressed-air system based on the utilization of energy application to demonstrate the energy savings potential associated with quick-fix of identified air leaks and in using reduced pressures at low demand periods Fig 2 and 3. Air is free of costs but only before it is compressed. From the moment of compression, any compressed air has several cost implications including production, distribution and equipment wear costs. Every unit mass of air that escapes through leaks has wasted some amount of mechanical energy used in compressing it. Therefore, when air is allowed to leak due to a general lack of awareness about its negative impact, a considerable portion of the energy or money is wasted. In many industries, this lack of awareness is more rooted in the non-technical departments that are usually responsible for funding maintenance request from the technical crew to correct faulty equipment. Also, both the technical and non-technical teams are usually too occupied to have the time to study the cost associated with air leaks and the cost savings associated with reducing air pressure setting.

5. Conclusion

Supply-demand management of compressed air to match production needs per time is one of the main goals of every compressed-air system to achieve better energy efficiency. Also, prompt repairs of air leakages can have significant energy savings on a compressed-air system. Planning engineers and maintenance crew may have to include air leakages in emergency maintenance records. More importantly, prompt attention is required by various departments involved to reduce energy or money losses. The application was considered for a single-stage compressed-air system. It may be further developed to include multi-stage compressed-air systems. Other elements of the compressed-air system that may be included in the energy analysis model to further increase energy savings are filters, fittings and valves.

Acknowledgments

This research would not have been possible were it not for the collaboration, inspiration, and support from Nigerian Foundries Ltd. We express deep gratitude to the management of the company. Helpful discussion with Moses Ojeirinde at Nigerian Foundries Ltd is gratefully acknowledged.

References

[1] S. Mousavi, S. Kara, and B. Kornfeld, "Energy Efficiency of Compressed Air Systems," *Procedia CIRP*, vol. 15, pp. 313–318, 2014..

[2] V. Jovanovic, B. Stevanov, and D. Se, "Energy ef fi ciency optimization of air supply system in a water bottle manufacturing system," vol. 85, pp. 306–317, 2014.

[3] G. Borden and P. G. Friedmann, *Control Valves*. Instrument Society of America, 1998.

[4] A. C. Yunus and A. B. Michael, *Thermodynamics: An Engineering Approach*. Singapore: McGraw-Hill, 2011.

[5] R. Saidur, N. A. Rahim, and M. Hasanuzzaman, "A review on compressed-air energy use and energy savings," *Renew. Sustain. Energy Rev.*, vol. 14, no. 4, pp. 1135–1153, 2010.

[6] K. Taheri and R. Gadow, "Industrial compressed air system analysis: Exergy and thermoeconomic analysis," *CIRP J. Manuf. Sci. Technol.*, vol. 18, pp. 10–17, 2017.

[7] W. David, "Air filtration and efficiency: Cutting the cost of compressed air," *Filtr. Sep.*, vol. 45, no. 9, pp. 32–34, 2008.

[8] M. Despeisse, A. Davé, L. Litos, S. Roberts, P. Ball, and S. Evans, "13th Global Conference on Sustainable Manufacturing - Decoupling Growth from Resource Use A collection of tools for factory eco-efficiency," *Procedia CIRP*, vol. 40, pp. 542–546, 2016.

[9] M. Hacksteiner, F. Duer, I. Ayatollahi, and F. Bleicher, "Automatic assessment of machine tool energy efficiency and productivity," *Procedia CIRP*, vol. 62, pp. 317–322, 2017.

[10] Ã. Ming Yang, "Air compressor efficiency in a Vietnamese enterprise," *Energy Policy*, vol. 37, pp. 2327–2337, 2009.

[11] M. T. Sambandam, N. A. Madlool, R. Saidur, D. Devaraj, and S. Rajakarunakaran, "Investigation of energy saving potentials in T-junction and elbow in compressed air systems," *Energy Effic.*, vol. 10, no. 5, p. 1099 1113, 2017.

[12] S. O. Oyedepo, "On energy for sustainable development in Nigeria," *Renew. Sustain. Energy Rev.*, vol. 16, no. 5, pp. 2583–2598, Jun. 2012.

[13] T. D. Eastop and A. McConkey, *Applied Thermodynamics for engineering technologists*. Dorling Kindersley (India) Pvt. Ltd, 1993.

[14] C. P. Merle and W. . Craig, *Thermodynamics for engineers*, 2nd ed. Schaum's Outline Series, McGraw-Hill, 2006.

[15] J. Rayner, *Basic engineering thermodynamics*, 5th ed. Addison Longman Limited, China, 1996.

IOP Conf. Series: Materials Science and Engineering **413** (2018) 012016 doi:10.1088/1757-899X/413/1/012016

Rapid Cyclic Heating of Mild Steel and its Effects on Microstructure and Mechanical properties

Aweda*, J.O., Orhadahwe, T.A. and Ohijeagbon, I.O.

Department of Mechanical Engineering, Faculty of Engineering and Technology, University of Ilorin, Ilorin, Nigeria. *E-mails addresses: joaweda@unilorin.edu.ng,

Abstract. The effect of subjecting mild steel to several cycles of rapid heat treatment on its mechanical properties and microstructure was investigated. Mild steel of 0.213 wt % carbon was subjected to transformation heat treatment from austenite to pearlite and quenched in running water. Rapid heating was achieved by preheating an electric muffle furnace to 840 °C before charging the samples into it. Each cyclic heat treatment was for a period of 200 seconds held at 840 ^{0}C and cooled to 700 ^{0}C which was repeated four times. The effects of cycle numbers were evaluated by testing for impact, hardness, ultimate tensile strength and microstructural properties. The results showed that after 4th cycle of rapid heating the samples had impact energy 64.6 J, Brinell hardness number 563 and ultimate tensile strength 1257.78 N/mm^2. The samples after one cycle had ultimate tensile strength 1027.45 N/mm^2, % elongation 10.396% and impact energy 286.174 N.m before failure. Through cyclic heating, grain refinement was achieved by the fast simultaneous nucleation at the grain boundaries and the fast martensite to austenite transformation due to the fast heating rate which prevented austenite grain growth. Mechanical properties of the studied steel sample were improved with the rapid heat treatment cycles given.

Keywords: Cyclic heating, rapid heat treatment, microstructure, quenching.

1. Introduction

Steel is one of the widely used metals and exist as low, medium and high carbon steel. It can be strengthened through additional heat treatment processes Jiang and Yazheng [1] which make it to be adopted for various applications. Low carbon steels (0.002 - 0.25% C) account for a large proportion of the total output of steel due to its relatively low price and adaptable properties similar to that of iron therefore making its application versatile for many engineering applications as observed by Adamczy and Grajcar [2]. Low carbon steels are used for different applications as in automobiles, furniture, refrigerators and tinplate roofing. In between 0.15 and 0.25% C content is used for structural purposes such as beams, channels and angles for construction.

The choice of heat treatment adopted for strengthening steel is very important as faulty heat treatment process may incur high cost of operations and untimely failure of materials while in service which may have effect on human safety (Senthilkumar and Ajiboye [3]. Muszka *et al.*; [4] opined that the specific mechanical properties of metals are the basic requirements for its application. These properties can be obtained from different strengthening mechanisms which act together during deformation process. Metal strengthening procedure depends on its chemical composition, deformation history and resulting microstructure.

There are various methods of strengthening that include solid solution strengthening, dispersion strengthening, grain refinement, dislocation strengthening and textural strengthening as submitted by Lv, *et al.;* [5]. The rate of metal cooling from the austenitic temperature range as in Aweda and Adeyemi [6] has great effect on structure and properties of the material. They observed that fast cooling rates result in phase transformation at lower temperatures thus yielding finer grain structure than slow cooling.

Fast cooling rate transforms steel from its austenitic phase to a new martensitic phase Akinlabi *et al.*; [7]. During quenching, the cooling rate is controlled to avoid soft pearlite or bainite formation. Therefore, the method adopted for quenching depends on the grade of steel, section thickness, distortion allowed and the properties imparted in the opinion of Roney and Loker [8].
Grain refinement can simultaneously enhance strength and toughness of materials which leads to the development of high performance materials, Lv *et al.;* [5]. Calcagnotto *et al.*; [9] achieved grain refinement by subjecting dual-phase steel to large strain warm deformation and subsequent inter-critical annealing that led to improved strength and ductility of the product. In addition to other procedures of steel processing, Jiang and Yazheng [1] reported that cyclic rapid heat treatment can be adopted to improve its properties.

The possibility of refining the austenite grains through shortening the austenitizing cycle by subjecting steel to rapid heat treatment involving heating steel at a fast heating rate to its austenitic temperature and quenching immediately without soaking was attempted by Grange [10]. In his submission, he argued that soaking the steel at the austenizing temperature was not really necessary, adding that the process of very short or rapid austenitizing treatment has the advantages of saving time, cost, minimizing the oxidation and decarburization of steel and minimizing austenite grain growth thereby leading to a finer grain heat treated product. As at the time of his work, he opined that two or more cycles of rapid heat treatment would improve the ductility of mild steel. Improving on Grange [10] submission, Koscielna and Wojciech [11] and Zheng *et al.;* [12], applying cyclic heating process to Ti-48Al-2Cr-2Nb alloy discovered that the average grain diameter reached its minimum value of 149 μm after 5 cycles while further increase in the number of cycles had no effect on the grain size.

With rapid heat treatment, a hard surface with improved fatigue and wear resistance is achieved Boyce *et al.*; [13]. Thermal cycling, as observed by Saha *et al.*; [14] produced fine microstructure with a good combination of strength and ductility. This was in agreement with Smoljan [15] who concluded that repeated heating is required to have improved effects of structure refining than single heating. He further stated that the number of cycles affects the obtained results of heat treatment and suggested five cycles. Above this number, micro-cracks may occur that can affect toughness as well as yield strength, Lin *et al.*; [16].

With insufficient cycle number, structure refining may not be obtained while at too many cycles, defects may occur due to micro-deformation processes which may not be economical. Number of cycles of rapid heat treatment depends on the material and ranged between three to maximum of six cycles depending on the material. This study therefore, examines the effects of cyclic rapid heat treatment behaviour of mild steel on its mechanical and microstructural properties.

2. Experimental procedures

Commercial mild steel was used for the experimental investigation. The analytical chemical composition of the as-received steel is as presented in Table 1 using optical emission spectroscope. The steel samples dimensions 10x10x55 mm were subjected to isothermal heat treatment at 840^0C for 200 seconds and then cooled to room temperature. The as-transformed steel were subjected to different number of rapid heat treatment cycles thus; 1-cycle, 2-cycle, 3-cycle and 4-cycle as shown in fig 1. For each rapid cycle heat treatment, the specimens were inserted into an electric resistance furnace at 840^0C (above Ac3 temperature 754^0C) and holding for 200 s. This was followed by quenching to 700^0C with a cooling time of 60 s. The specimens were thereafter air-cooled to room temperature, which ends 1-cycle heat treatment. Each cycle consisted of inserting the specimen in an electric furnace at 840^0C and holding for 200 seconds followed by forced cooling to 700^0C. This procedure was repeated for the samples subjected to more than 1-cycle rapid heat treatment. Samples with 2-cycle heat treatment were repeated once, 3-cycle repeated twice and 4-cycle heat treatment repeated thrice, see figure 1.

IOP Conf. Series: Materials Science and Engineering **413** (2018) 012016 doi:10.1088/1757-899X/413/1/012016

Table 1: Compositional analysis of tested mild steel (wt%)

Element	Fe	Cr	C	Si	Mn	S	P	Ni	Cu	Nb	Al	B	Mo	Ti
% wt	97.75	.138	.213	.252	.779	.033	.028	.134	.336	.014	.292	.002	.020	.009

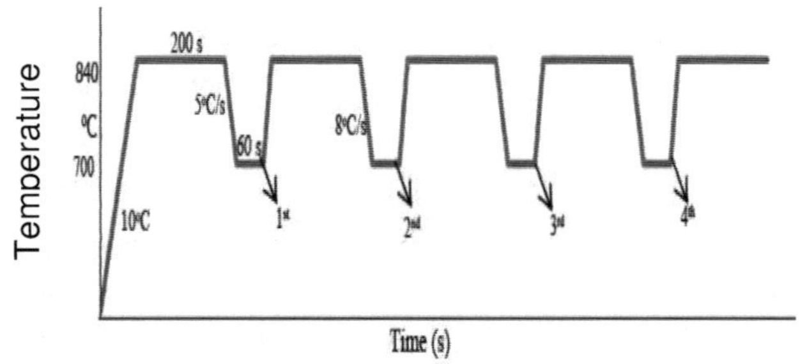

Figure 1 Cyclic heat treatment plots

The hardness values of the specimens were measured on the Brinell hardness testing machine performed in accordance with ASTM E10-15 for metallic materials. A 750 kgf was maintained for 15 seconds on the sample and the indenter was then removed leaving a round indent on the sample. The diameter of the indent was determined optically by measuring two diagonals of the round indent with a portable microscope and the values of Brinell hardness were calculated following equation 1.

-
$$BHN = \frac{2F}{\pi D(D-\sqrt{D^2-d^2})} \quad (1)$$

Where,
BHN = Brinell Hardness Number
F = applied load, kg
D = the diameter of the spherical indenter, mm
d = diameter of the resulting indenter impressions, mm

Impact test was conducted on the standard Charpy V-notch impact tester following ASTM A370, ISO148 and EN 10045-1 procedures. The standard specimens' size of 10 mm by 10 mm by 55 mm and 45° angular grove at the middle with a depth of 2 mm was cut and tested for impact. Following ASTM-E8 procedure for tensile test, a UTM Tensometric S300CT of maximum load of 300 kN at a crosshead speed of 0.03 mms^{-1}. The specimen was loaded gradually until it failed and the value of stress with corresponding strain diagram was plotted electronically for all the tests. From the data obtained the % elongation, yield strength, ultimate tensile strength and Young modulus values were determined. The microstructural observations of the as-transformed and cyclic heat treated specimens were conducted on optical microscope. Silicon carbide papers of different grades placed on the grinding machine were used in the order of 220, 320, 400 and 600, from coarse grade to fine grade. The polishing was performed on a universal polishing machine with a polishing cloth (selvet cloth) and solution of one micron of silicon carbide solution. This was followed by the final polishing stage with selvt cloth swamped with solution of 0.5 μm silicon carbide until a mirror-like surface was attained. The samples, washed and dried were then etched with 4 % NITAL and again washed and dried.

3. Results and discussion
3.1 Hardness values
The effect of Brinell hardness values on cycle numbers is shown on figure 2. The Brinell Hardness Number (BHN) values increase with increase in the number of rapid heating cycles from 271 for as-transformed (control) to 323 after 2-cycles of heat treatment. The increase continued until the value of 563 was reached after four cycles of rapid heating. This increase was probably due to the material's resistance to plastic deformation as the number of grain boundaries increased during quenching. This

trend agrees with Xie *et al*, [17] results where the hardness of mild steel improved through rapid cyclic heating and quenching. This is because subjecting steel to fast heating rate leads to the formation of fine austenite grains which on quenching produces coarse martensite. The coarse martensite grains are refined into finer martensite grains with more grain boundaries upon cyclic rapid heating.

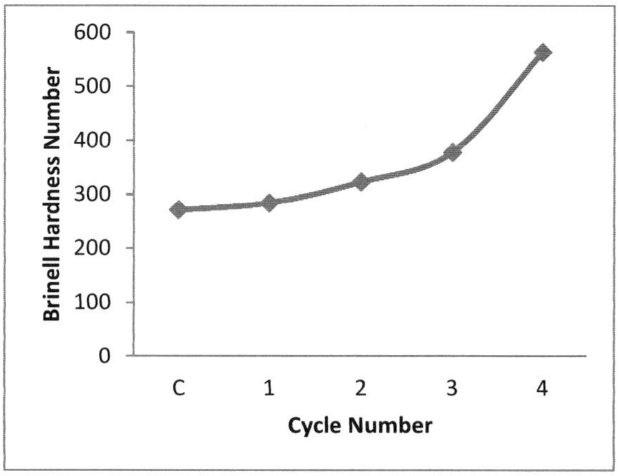

Figure 2 Effect of Brinell hardness on cycle numbers

3.2 Charpy V-Notch Impact Test

Figure 3 is the Charpy V-Notch impact test results with cycle numbers which shows increase in impact energy from 46.8 J for the as-transformed (control) sample to 51.2 after the first cycle heat treatment. There is further increase to 64.6 J with the accomplishment of the 4th cycle of rapid heating. This steady increase indicates that the resilience of the material increases with subsequent increase in cycle number of rapid heat treatment. This is an indication that mild steel possess ability to withstand shock loading with increase in cycle number as its martensitic structure is refined.

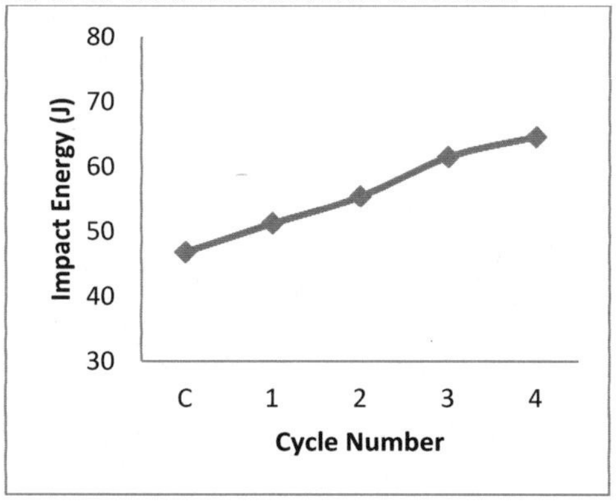

Figure 3 Impact strength of sample with cycle number

The percentage increase in impact strength from the control sample to the first cycle heating is 9.4%, while the increase in impact strength between the first cycle and the second cycle is 8.2%. The percentage increase impact strength between the second and third cycle samples is 11.01% while the difference in the energy between the third and fourth cycle samples was 5.04%. From these results, further increase in cycle numbers of rapid heat treatment may lead to minimal increase in impact strength. After a substantial increase in the 3rd cycle, subsequent increase is marginal with increase in cycle number.

IOP Conf. Series: Materials Science and Engineering **413** (2018) 012016 doi:10.1088/1757-899X/413/1/012016

3.3 Stress-strain curve

The stress-strain curves from the tensile tests of the control (as-transformed) and cycle heat treated samples are as obtained in figure 4. The control sample failed within the elastic limit, suggesting brittle failure but with a high ultimate strength of 1247.78 N/mm^2. This indicates that conventionally, diffusional heating and quenching produces high ultimate strength. The first cycle yielded at lower ultimate strength value of 1027.45 N/mm^2.

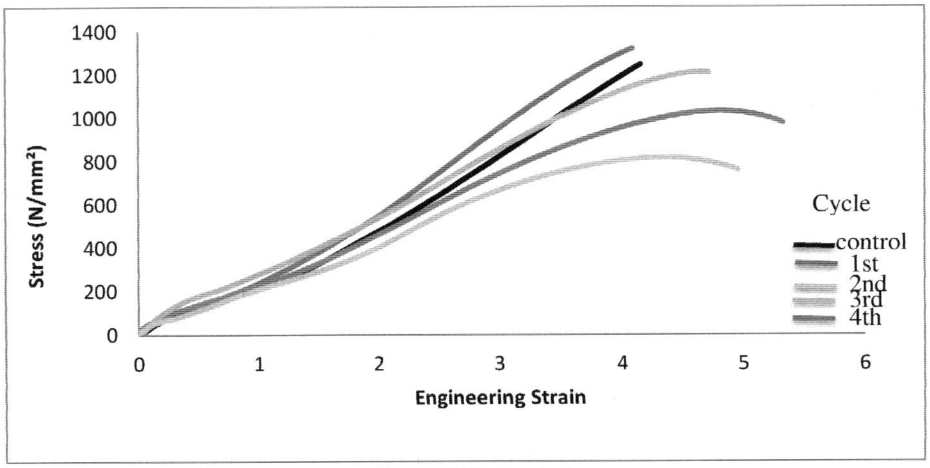

Figure 4 Strain-strain curves

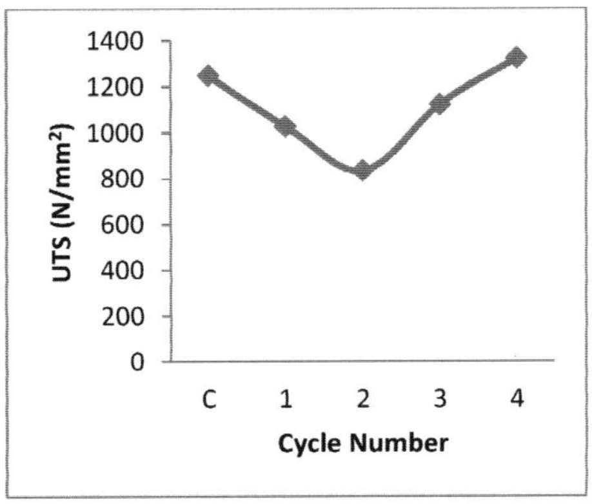

Figure 5 UTS versus cycle numbers

The yield strength initially dropped for the 1st and 2nd cycles only to rise with the 3rd and 4th cycles of heat treatments. The yield strength for the control was 1247.45 N/mm^2 while this value increased to 1321.78 N/mm^2 after the 4th cycle of heat treatment as in figures 4 and 5. After the second cycle, the material experienced plastic deformation with recorded lower ultimate strength value of 833.375 N/mm^2. This shows that the second cycle is more ductile than the first cycle sample. The third cycle also underwent plastic deformation but with a higher ultimate strength than the first and second cycle recording 1122.24 N/mm^2. These results show a reversal of trend as there is a loss in ductility. The fourth cycle sample failed in a brittle manner as it did not undergo plastic deformation. It has the highest ultimate strength of 1321.78 N/mm^2 before failure occurred. Smoljan [14] explained that during rapid cyclic heat treatment, the cycle number should not be too many hence it will lead to loss of ductility in relation with strength. This explains the behaviour resulting from the third and fourth cycle samples.

IOP Conf. Series: Materials Science and Engineering **413** (2018) 012016 doi:10.1088/1757-899X/413/1/012016

3.4 Effect of cyclic heat treatment on yield strength

The control sample with 1247.75 N/mm^2 and 4th cycle sample with a value of 1321.78N/mm^2 becomes the highest yield strength. This shows that the material's resistance to the movement of dislocation was due to their grain boundaries as displaced in figure 5. The yield strength is lower in the 1st cycle been 640 N/mm^2. This occurrence was due to the fact that ductility was beginning to set in as ductile materials have lower yield strength compared to brittle materials. This value reduced further in 2nd cycle to 480N/mm^2 before witnessing a rise in the 3rd cycle to 850 N/mm^2. This is an indication that the 2nd cycle sample is the most ductile while the 3rd cycle experience loss in ductility. The feasible yield point can be attributed to the dislocation pie up on the ferrite grain boundaries and the dislocation activation in the ferrite grains, Lv *et al,;* [5]. The stress-strain curves from the tensile tests of the as-transformed and cycle heat treated samples are as shown in figure 4.

3.5 Effect of cyclic heat treatment on ultimate tensile strength

The values of the ultimate tensile strength for the control and rapid cycle heat treated samples are as presented in figure 5. UTS increases and followed thereafter by a decrease. The initial increase was due to the finer microstructure (finer pearlite and ferrite) originated from non-equilibrium forced cooling in each treatment. These values are however the highest values for the ultimate tensile strength which is a measure of a material's ability to withstand deformation due to tension. The values for the ultimate tensile strength for the 1st, 2nd and 3rd cycles are 1027.45 N/mm^2, 833.375 N/mm^2 and 1122.24 N/mm^2 respectively. The 1st cycle sample has high UTS, while the lowest was recorded at the 2nd cycle. The results show that as the material undergoes more heat cycles, the particles get finer, the materials become more ductile and as a result the ultimate tensile strength decreases in value for 1st and 2nd cycles and thereafter, increasing to 1405 N/mm^2 at the 4th cycle.

3.6 Effect of cyclic heat treatment on % elongation

Figures 6 and 7 show the 2nd cycle has a high ductility as it records high % elongation. The yield point is 8.556 with a percentage elongation of 9.697% and Young modulus value of 15356.3 N/mm^2. With increase in cycle number, the % elongation of the material reduces from 10.4 % for 1st cycle to 8.18% after the 4th cycle while the as-transformed was 8.32%. The 1st cycle sample and 3rd cycle sample also showed ductility with percentage elongations of 10.396% and 8.912% and corresponding Young modulus values of 16398.5 N/mm^2 and 20616 N/mm^2 respectively. The control sample and 4th cycle sample showed the lowest % elongation which also translates to lowest ductility which explains why they have the highest ultimate tensile strength.

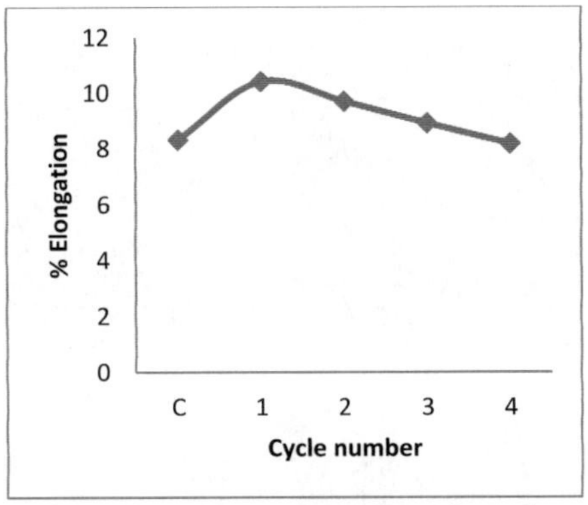

Figure 6 Percentage elongations of samples

IOP Conf. Series: Materials Science and Engineering **413** (2018) 012016 doi:10.1088/1757-899X/413/1/012016

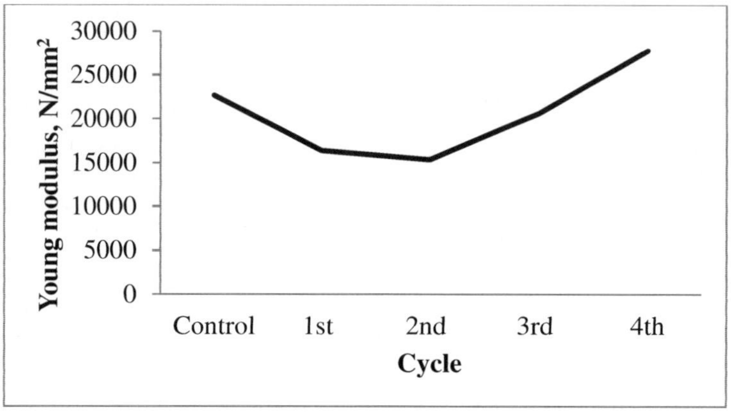

Figure 7 Young modulus values of samples

3.7 Effect of cyclic heat treatment on toughness

The energy absorbed before failure with cycle number on figure 8 showed that the 1st cycle absorbed 296.174 N.m while the control sample had 230.981 N.m before failure. Subsequent cycle heat treatment numbers produced lower energy compared to the 1st cycle. This makes the 1st cycle sample the toughest implying that one cycle of rapid heat treatment is enough to produce toughness in mild steel.

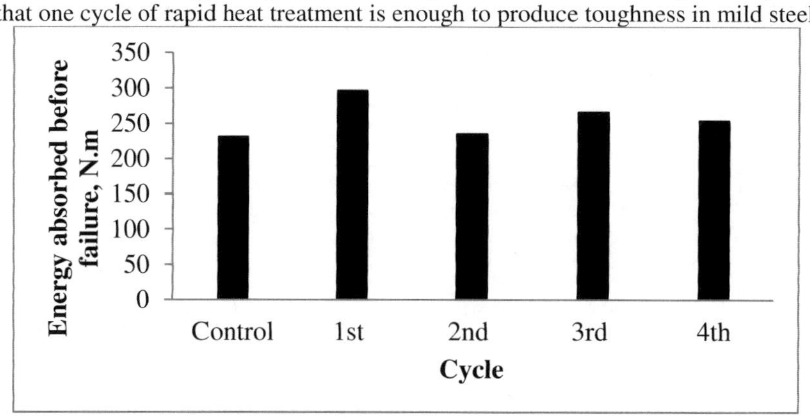

Figure 8 Energy absorbed before failure with cycle number

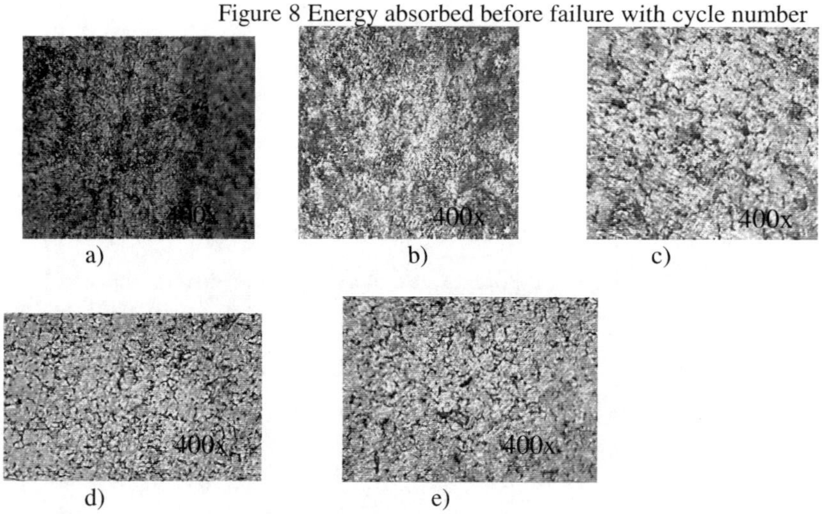

Figure 9 Optical micrographs of amples

a) Control Sample b) 1st Cycle sample c) 2nd Cycle sample d) 3rd Cycle Sample e) 4th Cycle sample

Table 2 Summary of Microstructure Properties

Cycle	Count	Total Area (μm^2)	Grain Size (μm)	Circularity	solidity	Feret (μm)
Control	2499	1153849	1.07	0.801	0.856	0.032
1st	2877	1152582	1.00	0.80	0.853	0.30
2nd	970	249739	0.80	0.769	0.843	0.38
3rd	2633	663536	0.79	0.793	0.854	0.036
4th	2980	859164	0.85	0.799	0.856	0.033

3.8 Effect of cyclic heat treatment on microstructure

As the specimens were held above Ac3 temperature, ferrite-to-austenite transformation occurs and the cementite lamellae begin to dissolve into austenite. The as-transformed indicates isothermally transformed pearlitic characteristics. For the specimen that went through one cycle rapid heat treatment, the cementite is partially dissolved into pearlite while there is further dissolution after 2-cycles of heat treatment. During holding after heating, the cementite dissolves and upon cooling the dissolution is terminated. The microstructure obtained by cyclic rapid heating and cooling of steel as presented in figures 9(a-e) show martensite in solution of ferrite Roney and Loker [8]. The carbon devoid of austenite regions converts into ferrite upon forced cooling as is noted from figures 9 b and c with rapid cyclic heating. With as-transformed figure 9a, cementite phase is present initially within pearlite. With increasing number of cyclic heating figures 9 c and d, the amount of pearlite decreases while the amount of cementite increases. After the 4th cycle figure 9e, the microstructure mainly contains cementite in ferrite with only trace amount of pearlite. In the process of austenization, the ferrite to austenite transformation occurs very fast, whereas the cementite dissolution in austenite is a slow process in accordance with Saha *et al.;* [14]. This is because the fast cooling rate did not allow the carbon particles to diffuse evenly into the matrix of the iron. This results in the jamming of the carbon lattice of the austenitic atomic arrangement. Martensite, an aggregate of iron and cementite has grains that are very coarse hence the mottled contrast seen in figure 9a is due to the high density of dislocation. Rapid heating allows some of the martensite particles to transform directly into austenite without passing through the intermediate phases. The carbon particles however were given more room to diffuse out and realign thereby leading to more evenly distributed microstructure on quenching. The process results in producing finer grains due to nucleation of grains and formation of grain boundaries as observed in figure 9e.

This trend is evident in Table 2 as the control sample possessed the largest grain size of 1.07 μm (microns). There was a reduction in the grain size after the 1st cycle to 1 micron which further reduced to 0.79 microns after the 3rd cycle. This is because more grains are beginning to form due to nucleation as the material is heated rapidly and quenched. The circularity of the grains which indicates the shape of grains, reduced from 0.801 for the control sample to 0.769 after the 2nd cycle before experiencing an increase of 0.793 after the 3rd cycle show grain growth occurring. The solidity of the grains (thickness) also followed similar pattern where there was initial decrease up to the 2nd cycle followed by an increase thereafter. The distance between grain boundaries (Feret) also varied from control cycle through the four cycles. The largest value was recorded in the 2nd cycle of rapid heat treatment with a value of 0.038 μm.

The resultant microstructures obtained and the mechanical properties obtained show that rapid heating leads to a refinement in the grain size and improvement in mechanical properties such as ductility, toughness, ultimate tensile strength, impact and hardness of mild steel. The improvement in mechanical properties of cycle rapid heat treated material was achieved by the microstructural refinement due to the phase interface nucleation as observed by Peng *et al.;* [18]. Rapid heating prevents the growth of austenite grain size which favours nucleation therefore, on quenching, the martensite microstructure is refined. Nucleation rate is accelerated by refining the austenite grain size as the density of nucleation sites increases inversely with the austenite grain size. After 4 cycles of heat treatment, hardness marginally decreases partly due to elimination of pearlite and generation of more cementite spheroids in the microstructure.

4. Conclusion

i. The sample with the 4th cycle of heat treatment possessed high Brinell hardness value and UTS and would find application where high strength is required.
ii. To obtain finer grains and improve toughness and ductility, mild steel should be subjected to a minimum of three and maximum of four cycles of rapid heat treatment. Any further cycle of rapid heat treatment will reverse the trend.
iii. Grain refinement was achieved by the fast simultaneous nucleation at the grain boundaries and the fast martensite to austenite transformation due to the fast heating rate which prevented austenite grain growth.
iv. For the situation requiring tough material in service, one or two cycles of rapid heating is adequate.
v. Young modulus initially decreases due to generation of finer structure and thereafter slightly increases with gradual elimination of pearlite in the microstructure.

References

[1] Jiang, B., and Yazheng, L. Microstructural characterization, strengthening and toughening mechanisms of quenched and tempered steel: Effect of heat treatment parameters. Materials Science and Engineering: A, 2017, 707: 306-314.
[2] Adamczy, J. and Grajcar, A. Heat treatment and mechanical properties of low-carbon steel with dual-phase microstructure. Journal of Achievements in Materials and Manufacturing Engineering, 2007, 22(2): 13-20.
[3] Santhikumar, T. and Ajiboye, T.K. Effect of heat treatment process on the mechanical properties of medium carbon steel. Journal of Mineral & Materials Characterization and Engineering, 2012, 11(2):143-152.
[4] Muszka, K., Majta, J. and Bienas, L. Effect of grain refinement on mechanical properties of micro-alloyed steels. Metallurgy and Foundry Engineering, 2006, 32(2): 87-97.
[5] Lv, Z.Q., Wang, B., Wang, Z.H., Sun, S.H. and Fu, W.T. Effect of cyclic heat treatments on spheroidizing behavior of cementite in high carbon steel. Materials Science and Engineering: A, 2013, 574: 143-148.
[6] Aweda, J.O. and Adeyemi, M.B. Determination of temperature distributions in squeeze cast aluminium using the semi-empirical equations' method. Journal of Materials Processing technology, 2009, 209: 5751-5759.
[7] Akinlabi, E.T and Akinlabi, S.A. Characterising the effects of Heat treatment on 3CR12 and AISI316 Stainless Steels. International Journal of Materials and Metallurgical Engineering, 2014, 8(2): 256-261.
[8] Roney, T.J., and Loker, D.R. Evaluation of quenching methods for the purpose of acoustic data collection. Procedia Manufacturing, 2017, 10: 1118-1128.
[9] Calcagnotto, M., Ponge, D., Adachi, Y. and Raabe, D. Effect of grain refinement on strength and ductility of dual-phase steels. Proceedings of the 2nd international symposium on steel science, 2009, Kyoto, Japan: 195-198.
[10] Grange, R.A. The rapid heat treatment of steel. Metallurgical Transactions, December 1971, 2(1): 65-78.
[11] Koscielna A. and Wojciech S. Effect of cyclic heat treatment parameters on the grain refinement of Ti-48Al-2Cr-2Nb alloy. Materials Characterization, 2009, 60: 1158-1162.
[12] Zheng, L., Xue-ping, R., Zhi-hong, L, Zi-ming L. and Min-Min, G. Effects of two different cyclic heat treatments on microstructure and mechanical properties of Ti-V microalloyed steel. Materials Research, 2015, 18(2): 304-312.
[13] Boyce, B.L., Chen, X., Hutchinson, J.W., and Ritchie, R.O. The residual stress state due to a spherical hard-body impact. Mechanics of Materials, 2001, 33: 441-454.

[14] Saha, A., Mondal, D.K. and Maity, J. Effect of cyclic heat treatment on microstructure and mechanical properties of 0.6wt% carbon steel. Materials Science and Engineering: A, 2010, 527(16-17): 4001-4007.
[15] Smoljan B. An analysis of performance of combined cyclic heat treatment. Journal of Materials Processing Technology, 2004, 1(55): 1704-1707.
[16] Lin H., Lui T. and Chen L. Effect of maximum temperature on the cyclic heating-induced embrittlement of high-silicon ferritic spheroidal-graphite cast iron. Materials Transactions, 2004, 45(2): 569-576.
[17] Xie, G., Zhang, J. and Lou, L. Effect of cyclic recovery heat treatment on surface recrystallization of a directional solidified superalloy. Progress in Natural Science: Materials International, ScienceDirect, 2011, 21: 491-495.
[18] Peng, C., He, Y., Huang, B. and Liaw, P.K. Effects of rapid heating cyclic heat treatment on microstructures and compression mechanical properties of TiAl-based alloy. Transactions of Nonferrous Metals Society of China, 2004, 14(3): 459-463.

IOP Conf. Series: Materials Science and Engineering **413** (2018) 012017 doi:10.1088/1757-899X/413/1/012017

Empirical Models for Mechanical and Electrical Characteristics of Wrought Aluminium Alloy Reinforced with Silicon Carbide Particles

Babalola Philip Olufemi[1,3], Kilanko Oluwaseun[1], Bolu Christian[1], Inegbenebor Anthony[1], Oyawale Festus[1], Adeosun Oluropo[2]

Mechanical Engineering Department, College of Engineering, Covenant University, Ota, Ogun-State, Nigeria 1
Department of Metallurgical and Materials Engineering, University of Lagos, Nigeria 2
E-mail: phillip.babalola@covenantuniversity.edu.ng 3

Abstract. Aluminium matrix composites (AMCs) are advanced engineering materials used for a wide range of applications. AMCs consist of a non-metallic reinforcement incorporated into Aluminium matrix providing advantageous properties over base metal alloys. Wrought aluminium was reinforced with particulate silicon carbide of four different sizes and compositions using stir casting method. The resultant composites were subjected to mechanical and electrical tests. In this paper, empirical models were developed from measured laboratory data of micro-hardness, modulus, and electrical conductivity. Information obtained from the empirical models could be used as guidelines during the conceptual design and optimisation of manufacturing processes and thus, reducing time and costs.
Keywords: aluminium, silicon carbide, stir casting, mechanical properties, electrical properties, empirical models

1. Introduction

Metal Matrix Composite (MMC) is a material consisting of a metallic matrix combined with a ceramic (oxides, carbides) or metallic (lead, tungsten, molybdenum) dispersed phase. Major MMCs are Aluminium Matrix Composites (AMC), Magnesium Matrix Composites, Titanium Matrix Composites and Copper Matrix Composites. However, the widest group of MMC is Aluminium Matrix Composites. Discontinuously reinforced aluminium matrix composites are fast emerging as engineering materials and competing with common metals and alloys. These are gaining significant acceptance because of higher specific strength, specific modulus and good wear resistance compared with unreinforced alloys [1]. AMCs are used in myriad of ways and according to Beffort [2], MMCs of the type 6061/B/50f, and A359/SiC/20p are used as cargo bay struts in space shuttle and brake disks and drums respectively. 2014/Al2O3/10-20p (Al-4.4Cu-0.5Mg-Si-Mn), 6061/ Al2O3/10-20p (Al-1.0Mg-0.6Si-Cu-Cr) and 7005/Al2O3/10p (Al-4.6Zn-1.4Mg-Mn-Cr-Zr-Ti) are used in bicycle frames, drive shafts and cylinder liner. A357/SiC/10-20p (Al-7.0Si-0.5Mg), A359/SiC/10-20p (Al-9.0Si-0.5Mg), A339/SiC/10-20p (Al-12Si-1.0Mg-1.0Ni-2.25Cu), A360/SiC/10-20p (Al-9.5Si-0.5Mg) and A380/SiC/10-20p (Al-8.5Si-3.5Mg) are applicable in brake drums and brake discs. 6061/Al2O3/10p is used in automobile drive shaft. 6092/SiC/17.5p and 2009/SiC/15p-T4 are used in fan exit guide vanes of jet engines. Al/Nextel610/45f is used for electrical conductors.

Production of MMC involves the use of one or more forms of energy. Electrical energy may be obtained from hydropower, nuclear energy, fossil fuel or renewable energy sources. Thermal and other mechanical energy in the manufacturing processes required energy that may also impact on the environment. For example, the furnace used in this work is fired by diesel oil. Other researchers in similar field used gas fired or electric furnaces. In all, there is always carbon imprint directly or indirectly. Furthermore, characterization of the composite; tensile, hardness, impact, fatigue, compression, microstructure tests et cetera also involve energy, time and money. Some or all these resources must be mitigated for sustainable production processes. Proven empirical equations are useful here.

Mathematical modeling of manufacturing process could be used to predict the product quality as well as the forces required. The information from modeling can be helpful in the design of product, tool, machine and process [3]. The mechanical and electrical properties could be modeled from measured properties and used to predict new composite design prior to actual processing and property measurement. For example, Lee and Pyo [4], used micromechanical framework to forecast the effective elastic property and weakened interface evolution of composites with embedded particles. The Eshelby's tensor for an ellipsoidal inclusion with slightly weakened interface was employed to model spherical particles having imperfect interfaces in the composites and is incorporated into the micromechanical framework. Vladimir and Oleksandr [5] worked on equation of state for aluminum silicon carbide using experimental p–V data. Jagadeesh et al. [6], constructed cooling curves for predicting temperature of Al 6061–SiCp composites. The theoretical predictions yielded a lot of information about the cooling rates of the cast composites with enormous time saving in experimentation. Suresha and Sridhara [7] on their part studied wear behaviour of stir cast Al–SiC–Gr hybrid composites considering the percentage of reinforcement (%), applied load (N), sliding distance (m), sliding speed (m/s) and subsequently generated a graphical model. Naher et al. [8], investigated the effect of viscosity during Al–SiC composite development. Focus were on production time (up to 65 min), mixing speeds (50–500 rpm), and particle sizes (13–100 µm) for two different viscosity levels (1 and 300 mPa s) were studied. Computer simulations, ambient temperature analogue fluid simulations, and MMC castings were performed. Volume fraction results of SiC at different locations within the fluids were assessed by each of these methods and compared. In the study, mechanical and electrical properties' models were developed from experimental data for product development by the industry's R and D department. Another investigator [12], used a different matrix (AA6061), different reinforcement (TiC) but the same stir casting method. He only produced model for Ultimate Tensile Strength whereas this present work has for Modulus, Hardness and Electrical Conductivity.

2. Methodology

In this work, stir casting method was used to prepare samples of AMCs using 1170Al reinforced with Silicon Carbide (SiC) particulates of 3 µm, 9 µm, 29 µm, and 45 µm sizes respectively. The chemical composition of Aluminium and Silicon Carbide are presented in Table 1 and Table 2 respectively.

Table 1 The Compositions in Percentage of Aluminium Ingot Obtained from Aluminium Rolling Mills, Ota

Fe	Si	Mn	Cu	Zn	Ti	Mg	Pb	Sn	Al
0.232	0.078	0.000	0.0006	0.0016	0.006	0.0027	0.0012	0.007	99.66

Table 2: The Chemical Composition in Percentage of Silicon Carbide (Logitech)

C	Al	Fe	Si	SiO2	Magnetic Iron	SiC
0.50	0.30	0.20	0.80	0.60	0.04	97.6

IOP Conf. Series: Materials Science and Engineering **413** (2018) 012017 doi:10.1088/1757-899X/413/1/012017

The liquid metallurgy route (stir casting technique) was adopted to prepare the cast composites. The advantages of stir casting method when compared with powder metallurgy and vacuum hot pressing include its simplicity, flexibility, economical, and suitability for mass production (industrial) and production of complex profiled composite components without damaging the reinforcement particles [9]. A batch of 5.0 kg of 1170Al was smelted in a graphite crucible using oil-fired tilting furnace at 750 °C [10]. A K-type thermocouple was used for melt temperature measurement. The molten metal was poured into a preheated mould (450 °C) and the melt agitated with a stirrer to form a fine vortex. The SiC particles was preheated (1100 °C) and added into the vortex with mechanical stirring at 500 rpm as in Abbassipour et al. [11] for about 5mins. Actual mixing occurs when the slurry is at semisolid form. The particle reinforced liquid alloy was allowed to cool. During solidification, the solid-liquid mixture is vigorously stirred to break up the dendritic structure. This procedure was repeated for reinforcement particle sizes of 3, 9, 29 and 45 µm at 2.5, 5.0, 7.5 and10 wt. % of SiC respectively.

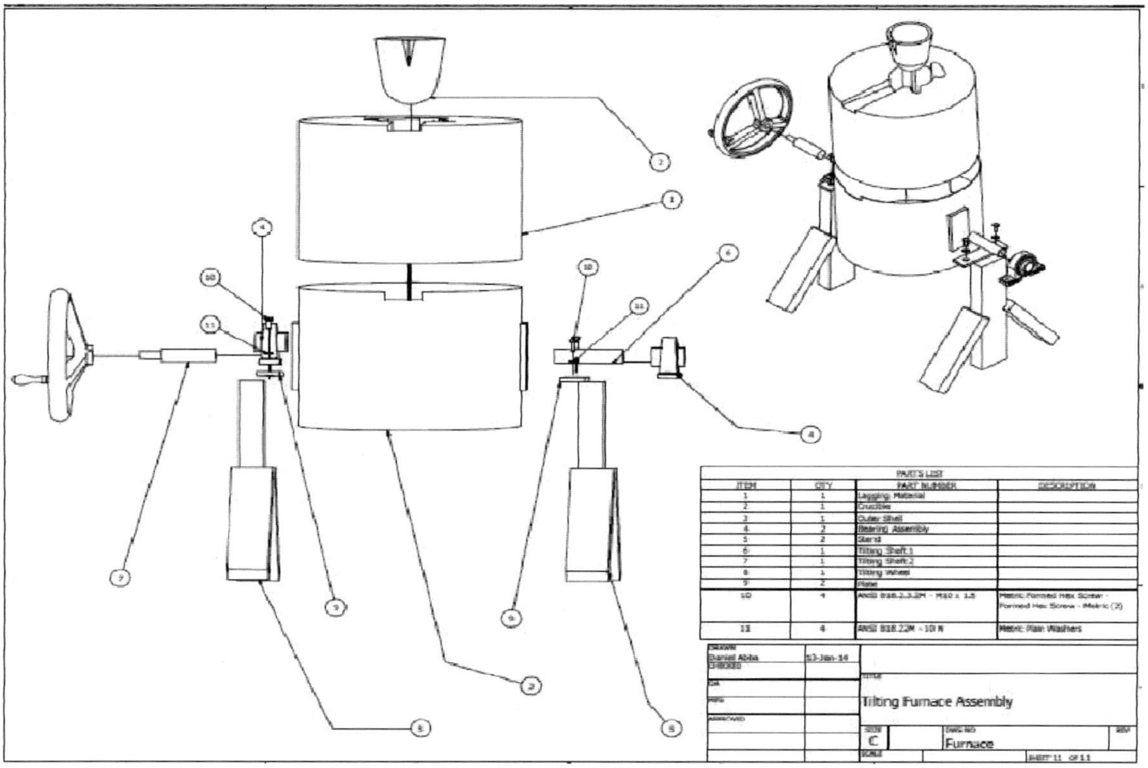

Fig. 1 Tilting Furnace Assembly Drawing [10]

Selected mechanical and electrical tests were conducted on the composites samples. The mechanical tests carried out were tensile and hardness, while the electrical tests were resistivity and conductivity for all the samples. Tensile specimens were machined to dimensions of 5 mm by 10 mm with a gauge length of 25 mm and pulled in Instron Universal Testing Machine (IUTM) with 30 KN load [13]. Micro hardness results were obtained using LECO 700AT tester with a load of 492.3mN and a dwell time of 10 seconds. Before testing, specimen surfaces were polished using emery papers down to 1000 mesh [14]. For electrical conductivity, samples with cross sectional dimensions of 5 mm by 10 mm and a length of 26 mm, were tested using 4-point probe set up machine [14]. The working voltage was 20 mV.

IOP Conf. Series: Materials Science and Engineering **413** (2018) 012017 doi:10.1088/1757-899X/413/1/012017

The results were processed using Excel, Design-Expert, and MATLAB software. From these, mathematical models were generated for simulation purposes. The proportion of the variance in the dependent variable that is predictable from the independent variable is the coefficient of determination, R2. In this work, silicon carbide weight percent (w) and size (μ) were used to predict mechanical and electrical properties of AlSiC. A model equation with an R2 closed to 1 is a good fit.

3. Result and Discussions

Summary of modeling equations obtained for modulus, hardness and electrical conductivity are presented in Tables 3 and 4. In Table 3, particle sizes and weight percentages were separately used to predict modulus, hardness and electrical conductivity. A known SiC weight percent/ particle size when fitted into the models, will predict modulus, hardness and electrical conductivity. The R2 for these models were between 0.7143 to 0.9396 value.

However, in Table 4, particle sizes and weight percentages were simultaneously used to predict modulus, hardness and electrical conductivity. The effect of particle sizes and weight percentages on modulus, hardness and electrical conductivity were shown graphically in Figures 2, 3 and 4 respectively. The models for predicting hardness and electrical conductivity have R2 values of 0.8328 and 0.7513 respectively. The lowest value of R2 was obtained for modulus at 0.4303

Table 3 Model Equations Using either Particle Sizes or Weight Percentages to Predict Composite Properties

S/N	AMC Designation	Fixed Parameter	Modulus (MPa)	Hardness (HV)	Electrical Conductivity (Mς/m)
1	Al/SiC/2.5p/0 -45mm	Particle size	σ=-1.6602m2+70.061m+ 736.63	H = 0.0843m + 20.64, R2 =0.7143	σe=0.0013862m2 - 0.093356m+69.456
2	Al/SiC/5.0p/0 -45mm	Particle size	σ =-0.70001m2+34.179m +642.45	H = 0.239m + 21.19, R2 =0.8249	σe=0.00058141m2 - 0.15024m+68.133
3	Al/SiC/7.5p/0 -45mm	Particle size	σ =-1.1501m2+53.806m+ 809.33	H = 0.2914m + 22.528, R2 =0.849	σe=0.0073248m2 - 0.37581m+66.722
4	Al/SiC/10p/0-45mm	Particle size	σ=-0.84115m2+39.97m+ 545.32	H = 0.3135m + 22.838, R2 =0.8596	σe=0.010671m2 - 0.63267m+59.546
5	Al/SiCp/0-10/3mm	%wt.	σ=-26.348w2+310.57w+ 459.39	H = 0.692w + 19.32, R2 =0.9396	σe = -1.908w + 73.473, R2 =0.7366
6	Al/SiCp/0-10/9mm	%wt.	σ=-20.971w2+230.46w+ 521.04	H = 0.652w + 20.63, R2 =0.8896	σe = -2.0238w + 72.088, R2 =0.9213
7	Al/SiC/0-10/29mm	%wt.	σ=-19.58w2+246.53w+4 86.04	H = 1.576w + 19.37, R2 =0.933	σe = -1.2244w + 71.671, R2 =0.8278
8	Al/SiC/0-10/45mm	%wt.	σ=-11.504w2+148.65w+ 359.56	H = 1.64w + 20.95, R2 =0.8637	σe = -1.7725w + 71.535, R2 =0.8204

IOP Conf. Series: Materials Science and Engineering **413** (2018) 012017 doi:10.1088/1757-899X/413/1/012017

Table 4 Model Equations Using both Particle Sizes and Weight Percentages to Predict Composite Properties

S/N	AMC Designation	Modulus (MPa)	Hardness (HV)	Electrical Conductivity(Mς/m)
1	Al/SiC/wp/mm	*σ =-3.581E+8 +(1.059E+8)m -(3.588E+8)w +(1.059E+8)m w -505.994m2* , R2 =0.4303	H = 19.64284 + 0.56770m + 22369.27468*m w*, R2 =0.8328	*σe=73.41306 – 1.96183m + 23303.67772w*, R2 =0.7513

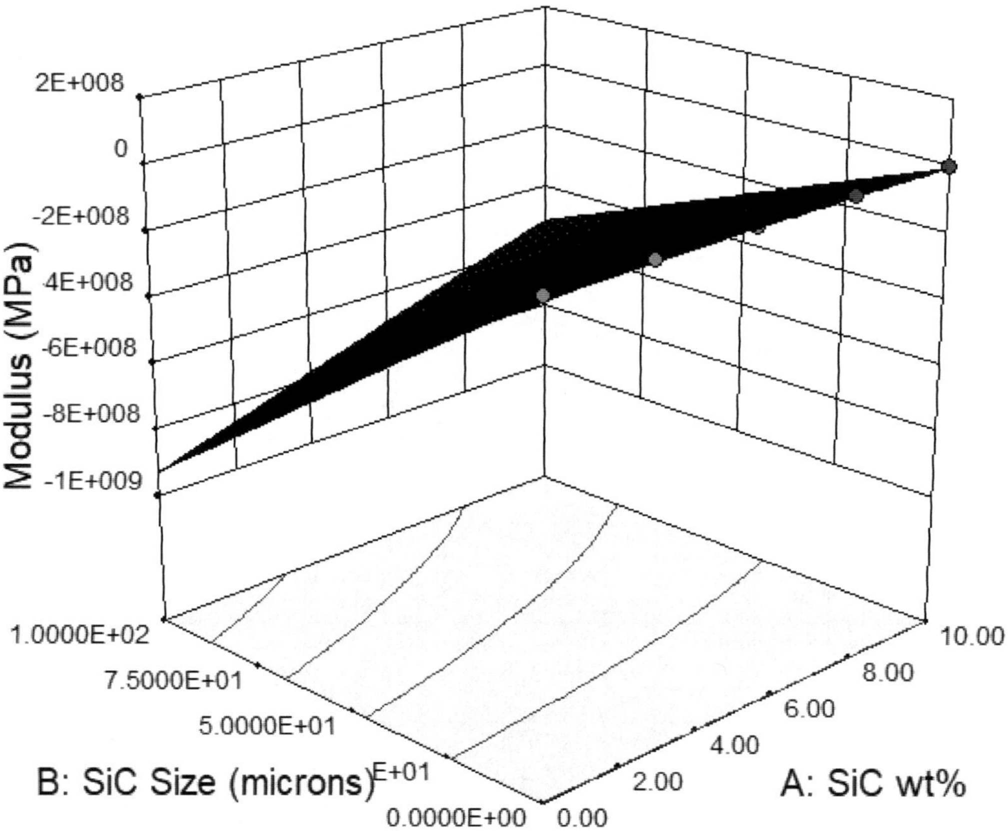

Fig. 2 Effects of silicon carbide particle size (μm) and weight percentages on modulus (MPa) of the AlSiC Composite

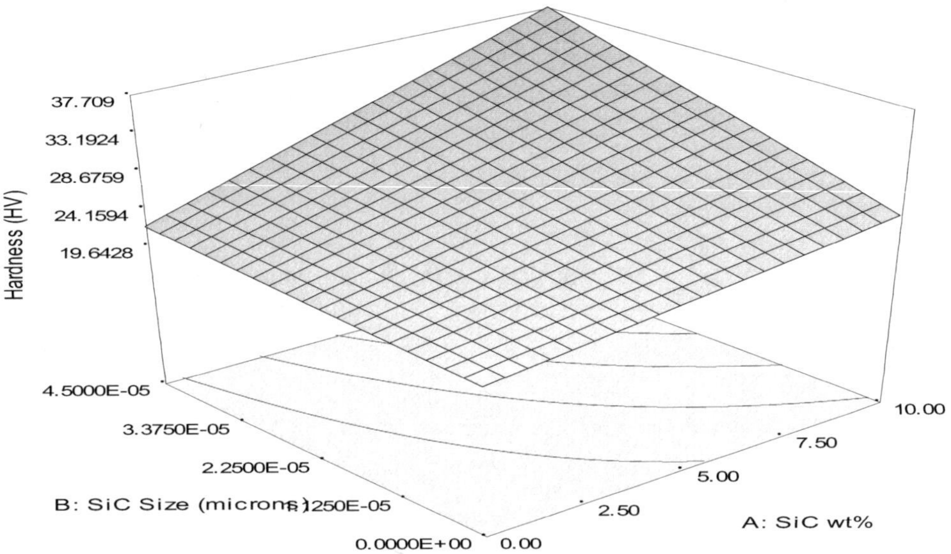

Fig. 3 Effects of silicon carbide particle size (μm) and weight percentages on hardness (HV) of the AlSiC Composite

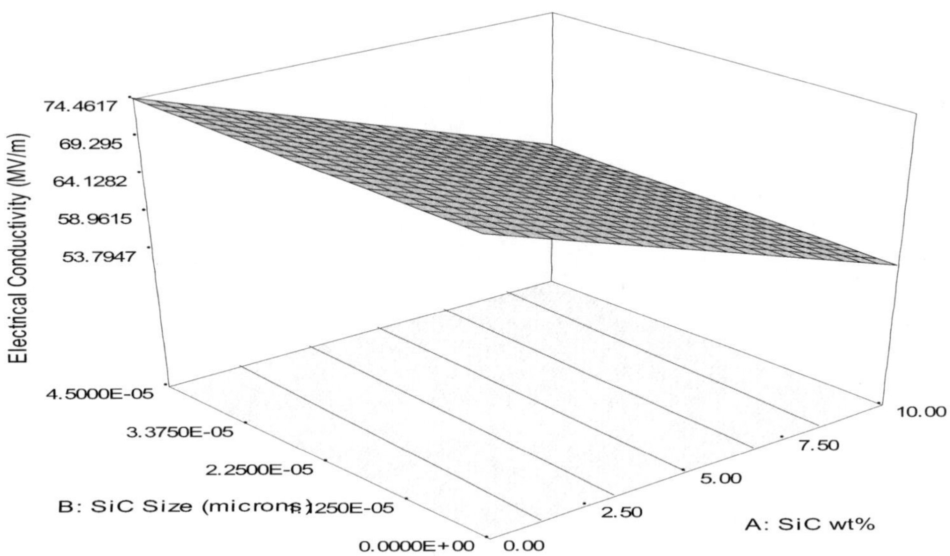

Fig. 4 Effects of silicon carbide particle size (μm) and weight percentages on electrical conductivity of the AlSiC Composite

IOP Conf. Series: Materials Science and Engineering **413** (2018) 012017 doi:10.1088/1757-899X/413/1/012017

4. Conclusions

Unlike the experimental approach which is time consuming, energy sapping with financial implications, the use of Excel, MATLAB and ANN method can generalize the complex relationships and provide approximate solutions. Mechanical properties are related to weight percentage and the size of SiC in the composite. Proven empirical equations obtained in this work are used to make informed decisions during the conceptual design and optimization of manufacturing processes. This makes the production process more sustainable by conserving energy, reducing time and reduction in financial resources.

Acknowledgments

The authors wish to gratefully acknowledge this work was supported by the funding provided by Covenant University Centre for Research, Innovation and Discovery (CUCRID).

References

[1] Mahendra B M, Arulshri K P and Iyandurai N. American Journal of Applied Sciences, 2013,10 (3):219-229.
[2] Beffort O. “Metal Matrix Composites (MMCs) from Space to Earth”. Eiggenossische Materialprufungs-und Forschungsanstalt, EMPA, Abt. Werkstofftechnologie, Feuerwerkerstrasse 39, CH-3602 Thun., http://www.empa.ch/abt126. Accessed October10, (2013).
[3] Dixit U S, Joshi S N, Davim J P. Materials and Design, 2011, 32: 3655–3670.
[4] Lee H K, and Pyo S H. International Journal of Solids and Structures, 2007, 44: 8390–8406.
[5] Vladimir L S, and Oleksandr O K. Solid State Communications, 2005, 135: 87–89.
[6] Jagadeesh S K, Ramesh C S, Mallikarjuna J M and Keshavamurthy R. Journal of Materials Processing Technology, 2010, 210:618–623.
[7] Suresha S, and Sridhara B K. Materials and Design, 2010, 31: 1804–1812.
[8] Naher S, Brabazon D, and Looney L. Composites: Part A, 2007, 38: 719–729.
[9] Kumar B A, and Murugan N. Materials and Design, 2012, 40: 52–58.
[10] Babalola P O, Inegbenebor A O and Bolu C A. Paper presented at International Conference on Clean Technology and Contemporary Engineering Management-ICCEM, Mechanical Engineering, Covenant University, Ota, Nigeria, 2012, pp. 260-271.
[11] Abbassipour A, Niroumand B and Monir V S M. Transactions of Nonferrous Metals Society of China, 2010, 20: 1561−1566.
[12] Jebeen Moses S, Dinaharan I, Joseph Sekhar S. Transactions of Nonferrous Metals Society of China, 2016, 26: 1498−1511.
[13] Inegbenebor A O, Bolu C A, Babalola P O, Inegbenebor A I, Fayomi O S I. Silicon, 2016, pp. 1-5.
[14] Babalola P O, Inegbenebor A O, Bolu C A and Inegbenebor A I, Journal of Materials, 2015, 67(4): 830-833.

IOP Conf. Series: Materials Science and Engineering **413** (2018) 012018 doi:10.1088/1757-899X/413/1/012018

The Development of Fume Extractor for a Welding Booth

Babalola Philip Olufemi[1], Oriyomi Moses Oluwayimika[1]

[1]Department of Mechanical Engineering, Covenant University, Km. 10, Idiroko Road, P.M.B 1023, Canaan land, Ota, Nigeria.
E-mail address: phillip.babalola@covenantuniversity.edu.ng; +2348034843286

Abstract. In our society today, there are several welding activities in the manufacturing establishments, production factories and metal fabrications on our streets daily. During this welding processes, welders are exposed to welding fume that are dangerous to their health. This can be prevented. The aim of the work is to develop a system that remove welding fume from the breathing zone of welder and leave the surrounding safe for other personnel. The system was designed using American Society of Heating, Refrigerating and Air-Conditioning Engineers (ASHRAE) standard among others. The booth volume, hood size, hood location, duct size and fan size were designed, selected and fabricated. The designed static pressure of the system is 164.412 Pa and the design volumetric flow is 6.126 m³/min. The ventilation system was tested by using thermo anemometer, thermometer and a measuring tape. Volumetric flow test on the flexible duct shows that at 150 mm extension, maximum air flow of 4.217 m³/min is achievable with this system while at full extension, 5.376 m³/min is achievable. Results, shows that welding at 0 mm to 100 mm from the hood extract smoke from the source whilst between 200 mm to 400 mm distance from the hood were significant lagged in extracting the smoke but at 500 mm, smoke reaches the welder's breathing zone.
Keywords: welding, fume, airflow, anemometer, extraction

1. Introduction

Most everyday construction processes in our society today engaged at least one welding operation [1]. Welding is an operation in which two or more parts are united by means of heat or pressure or both in such a way that there is continuity in the material (metal) between these parts (Figure 1). A filler metal, whose melting temperature is of the same order as that of the parent material, may or may not be used. There are various types of welding: shielded metal arc welding (SMAW); gas tungsten arc welding (GTAW), also known as tungsten inert gas (TIG); gas metal arc welding (GMAW), commonly termed metal inert gas (MIG); flux-cored arc welding (FCAW); submerged arc welding (SAW); electro-slag welding (ESW); electric resistance welding (ERW), which included spot welding and seam welding; plasma arc welding; aluminothermic welding; electron beam welding (EBW); diffusion welding (DFW); friction welding (FRW); ultrasonic welding (USW); and oxyfuel gas welding (OFW). All these produce welding fume in form of gas or vapor that smell strongly and dangerous to inhale.

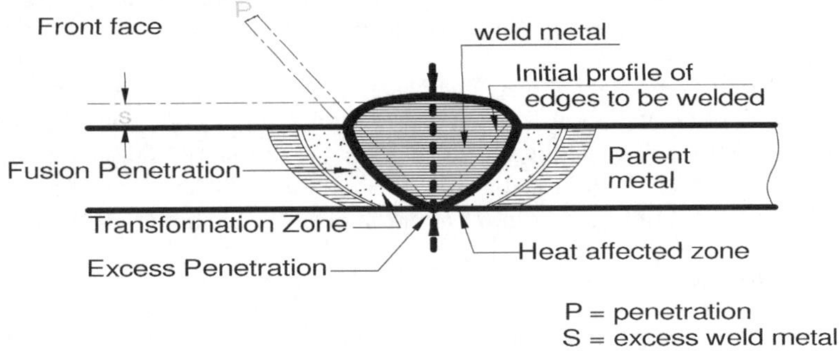

Fig. 1 Butt joint weld [2]

With increase in technological advancements and industrialization during the last few years, there are greater risk of human contact with harmful gases and fumes. Sustainable development must be environment and human friendly with very low emission indexes. Unfortunately, welding operations release harmful gases into the environment due to high energy requirement and hazardous fumes [3].

Welding gases are produced from condensed metal particles and oxides that are volatile due to enormous temperature of the arc [4]. This harmful to the health of individuals around it. For healthier and safe working conditions, this gaseous emission must be reduced to the barest minimum. The working welder is the most vulnerable to this hazard and work on associated risk around the breathing zone is needed [4]. Welding fume impose various health challenges such as cancer, metal fume fever, pneumonia, occupational asthma, irritation of throat and lungs, and temporary reduced lung function.

Amid welding processes, shielded metal arc welding (SMAW) and manual metal arc welding (MMAW) are the most versatile. This is mitigated by enclosing the welding operations, use extractor fan to remove emissions with integral respiratory protective equipment (RPE) by the operator. Furthermore, process parameters such as voltage, current and shielding gas composition may be modified. This is however a compromise on welding production and quality.

Meeker et al. [5] did extensive work on both experimental and field evaluations of a commercially available handy LEV unit. He used an equipment with a small bell-shaped hood and that operated at approximately 103–110 cfm (0.049–0.050 m^3/s). The efficiency of the filter was 99.97. Another investigator, Smargiassi et al. [6] examined the exposure rate of welders working on heavy excavation equipment. The system used GMAW with AIRMIG model 520, and Henlex gun having extractor fan of 70–80 cfm (0.03–0.04 m^3/s) flow rate.

2. Material and Methods

In this work, fume extractor is designed for a welding booth to capture the unwanted gases at the point of production, hence clearing the breathing zone of the worker. The main objective is to keep the welder's breathing area as clean as possible. Welding smoke consists of gases and fine particles, which disperse at relatively low speed from welding point and are best arrested at the source.

2.1 Functional requirement

A welding booth has some functions its meet for welder to properly work and perform optimally as recommended by American Welding Society (AWS). A working space that allows easy manoeuvring within the work or training space. This was met with standard booth sizes of 1,524 mm (5 ft.) x 1,524 mm (5 ft.) x 2,438 mm (8 ft.) according to AWS guidelines for booth width and depth. Inside the booth, is a platform for workpiece that allows operator to be as comfortable as possible while welding. A platform of 1,200 mm x 600 mm x 950 mm (H) is adopted for this purpose. Mechanical ventilation is employed to ventilate the booth. This is discussed later.

The welding booth is made up of main enclosure, work table, hood, duct, air cleaner, air mover, and exhaust stack. Sheet metal for hood, duct and exhaust stack is G-60 coated galvanised steel of lock forming grade as specified by Sheet Metal and Air Conditioning Contractors National Association, Inc. (SMACNA) [7]. This conformed to American Society of Mechanical Engineers (ASTM) A653 and A924 standards. Minimum yield strength for steel sheet and reinforcement is 207 kPa.

2.2 Design approach

For optimum design as recommended by ASHRAE [8] these principles are followed:

- Hood location shall be as close as possible to the source of contamination.

- The hood opening shall be positioned so that it causes the contaminant to deviate the least from its natural path.
- The hood shall be located so that the contaminant is drawn away from the operator's breathing zone.
- Hood size shall be larger than the cross section of flow entering the hood.

2.3 Required volume

Ventilation rate procedure (VRP) is adopted for calculating the required volume. According to Chartered Institution of Building Services Engineers (CIBSE) guide, between 15 to 30 air changes per hour is recommended for welding shops. 30 air changes per hour is selected for the welding booth.

Where:
Booth volume flow rate (Q_b) = Booth volume (V_b) x Air changes per hour
$Q_b = (V_b)$ x Air changes per hour (1)

2.4 Duct sizing

Constant velocity method is used for sizing duct that convey particles as recommended by ASHRAE [9]. From American Conference of Governmental Industrial Hygienists (ACGIH) [10], it recommended 10 to 13 m/s minimum transport velocity for welding fume. For this work, 13 m/s was selected.

For duct; using continuity equation,
$Q = A \times V$ (2)
$A = Q/V$ (3)
Q = Fume flow rate
A = Area of the duct
V = Fume velocity
For round duct

$$A = \pi \frac{d^2}{4} \quad (4)$$

$$d = \sqrt{\frac{4A}{\pi}} \quad (5)$$

2.5 Hood

Side draft hoods was used to draw contaminant away from the operator's breathing zone. ASHRAE Handbook - HVAC Applications, recommended that face area of hood must have at least twice the area of the duct [9]. Hence, the minimum area of hood (A_h) = 2 x Area of Duct (A_d)

$$A_h = 2 \times \pi \times \frac{d^2}{4} \quad (6)$$

2.6 Fan sizing

Welding booth developed is shown in Figure 2 and the exhaust system is as shown in Figure 3. The airflow rate was calculated using continuity equation.
$Q = A \times V$ (2)
Q = Fume flow rate
A = Area of the duct
V = Fume velocity
Given:
V = 13 m/s

IOP Conf. Series: Materials Science and Engineering **413** (2018) 012018 doi:10.1088/1757-899X/413/1/012018

For round duct of d=0.1m,

$$A = \pi r^2 = \pi \frac{d^2}{4} \tag{4}$$

$A = 0.007854\ m^2$

Therefore,

Q = 0.007854 x 13 = 0.102102 m^3/s = **6.126 m^3/min**

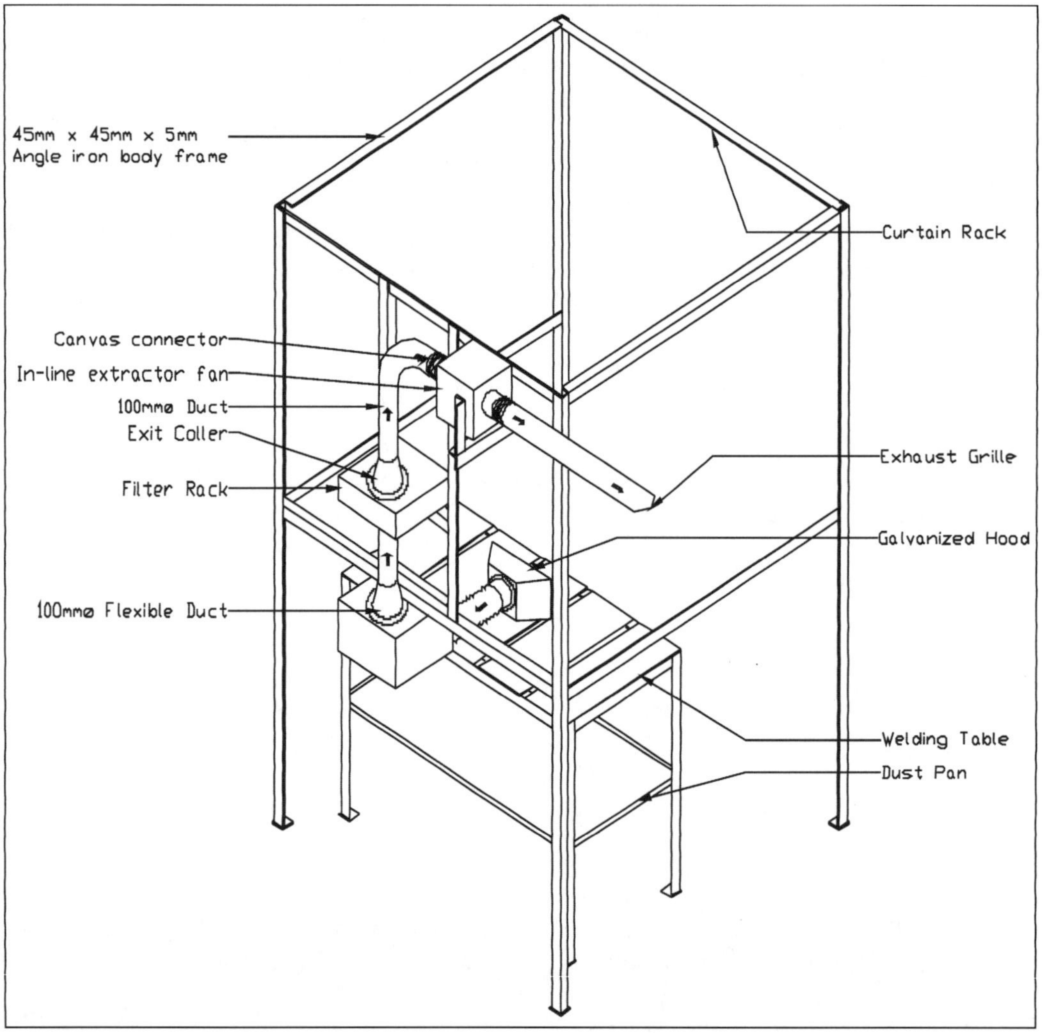

Fig. 2 Exposed welding booth isometric

IOP Conf. Series: Materials Science and Engineering **413** (2018) 012018 doi:10.1088/1757-899X/413/1/012018

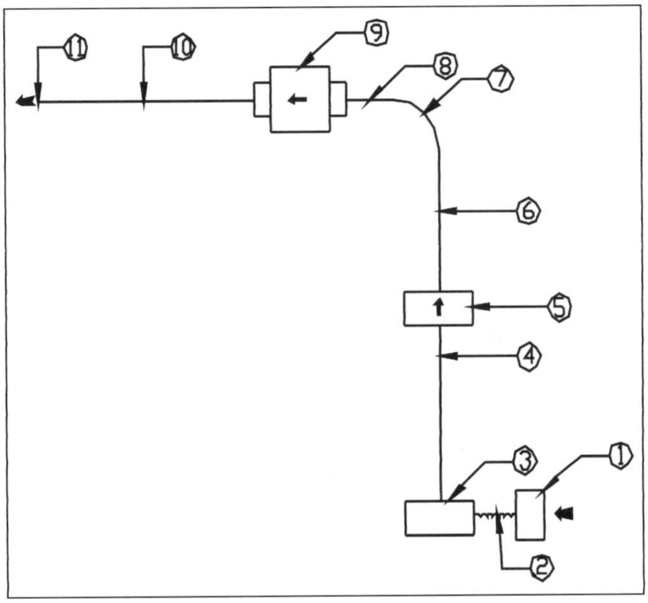

Parts List		
S/No	**Description**	**Material**
1	Hood	Stainless Steel
2	Flexible Duct	PVC/Aluminium
3	Plenum Box	Galvanized Iron
4	Duct	Galvanized Iron
5	Filter	G.I. / Aluminium
6	Duct	Galvanized Iron
7	Elbow	Galvanized Iron
8	Duct	Galvanized Iron
9	Extractor Fan	Galvanized Iron
10	Duct	PVC/ Aluminium
11	Exhaust Grille	Stainless Steel

Fig. 3 Welding booth extractor system schematic

Duct friction factor and pressure loss were calculated by using Darcy and Colebrook equations.
Darcy equations:

$$\Delta p_f = \frac{1000fL}{D_h} \times \frac{\rho V^2}{2} \quad (7)$$

where,
Δp_f= friction losses in terms of total pressure, Pa
f = friction factor, dimensionless
L = duct length, m
D_h= hydraulic diameter, mm
V = velocity, m/s
ρ = density, kg/m³

Colebrook equations:

$$\frac{1}{\sqrt{f}} = -2\log\left(\frac{\varepsilon}{3.7D_h} + \frac{2.51}{\mathrm{Re}\sqrt{f}}\right) \quad (8)$$

where,
ε = material absolute roughness factor, mm
Re = Reynolds number.

$$P_t = \sum_{i\varepsilon F_{up}} \Delta P_{t_i} + \sum_{i\varepsilon F_{dn}} \Delta P_{t_i} \quad \text{for i} = 1, 2,\text{- - - -}, n_{up} + n_{dn} \quad (9)$$

Where,
F_{up} and F_{dn} = sets of duct sections upstream and downstream of fan
P_t = fan total pressure, Pa
ε = symbol that ties duct sections into system paths from exhaust/return air terminals to supply terminals
Also, total pressure is the sum of static pressure and velocity pressure:

$$P_t = P_s + \frac{\rho V^2}{2} \quad (10)$$

Or

$$P_t = P_s + P_v \tag{11}$$

where,

P_t = total pressure, Pa

P_s = static pressure, Pa

P_v = Velocity pressure, Pa

ρ = density of air or gas in duct, kg/m³

V = fluid mean velocity, m/s

For air at standard conditions (ρ = 1.204 kg/m³),

Therefore,

$$P_v = 0.602V^2 \tag{12}$$

Fan static pressure,

$$P_s = P_t - P_v \tag{13}$$

P_s = **164.412Pa**

3. Airflow and Air Velocity Test

The effect of compressing and extending the flexible duct on the air flowrate and air velocity were investigated. Also, the efficacy of fume extraction from various distances away from the hood entrance were examined. The air flowrate and air velocity were measured using Extech (AN 100 Model) Thermo Anemometer together along with digital multi-meter (DT9205A), thermometer and measuring tape.

3.1 Volumetric flow through flexible duct

The volumetric flow test was conducted by mounting the ball bearing vane wheel of the anemometer on the hood, and then move the hood from 150 mm to 200 mm, 250 mm, 300 mm, 400 mm and 500 mm when the in-line extraction fan is in operation. The minimum and maximum values of airflow and air velocity were measured at each of the locations (Figure 4).

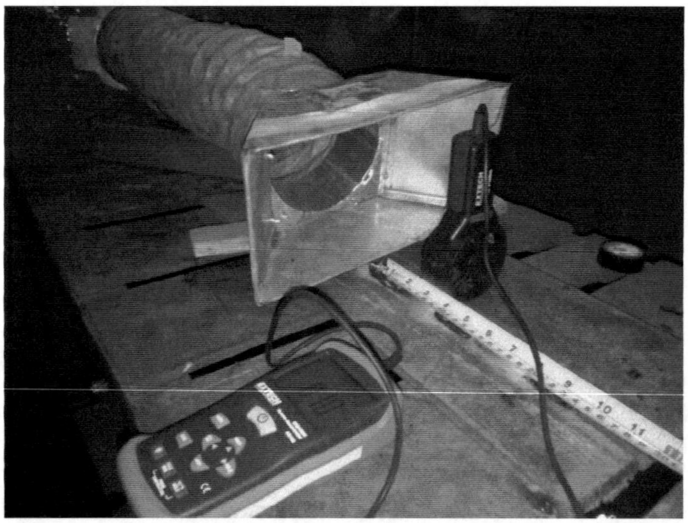

Fig. 4 Test bed showing flexible duct, hood, vane wheel, anemometer and measuring tape

3.2 Welding point proximity to hood test

The efficacy of the extraction is put to test by measuring air flow rate and air velocity from distance to hood entrance during welding operations. Two scenarios were evaluated. One, with the flexible duct un-stretched and the other with the flexible duct fully stretched out. In each case, the thermo anemometer ball bearing vane is placed at varied distances between 20 mm to 120 mm as shown in Figure 5.

Fig. 5 Exhaust hood, anemometer vane wheel, and measuring tape during welding operation

4. Results and Discussion

4.1 Volumetric flow through flexible duct

This test shows the significant impact of flexible duct used in relation to the extension of the duct. The total length achievable was 500 mm and minimum contraction was 150 mm. The result of the test is showed in Figure 6. It is observed that, the degree of extension of flexible duct have impact on the flow of the fluid due to friction introduced by the duct surface. At 150 mm extension, maximum air flow achievable was 4.217 m^3/min and at full extension, the flow increased to 5.376 m^3/min. When the duct is not fully extended, the zig-zag surface increased the surface roughness which led to increased in friction and reduction in airflow and velocity. With the full extension of the duct, it approximates a smooth circular pipe with reduced surface roughness and subsequent increase in airflow and velocity as observed.

4.2 Welding point proximity to hood test

The efficacy of the extraction is put to test by measuring air flow rate and air velocity from distance to hood entrance during welding operations. Two scenarios were evaluated. One, with the flexible duct un-stretched and the other with the flexible duct fully stretched out. In each case, the thermo anemometer ball bearing vane is placed at varied distances between 20 mm to 120 mm as shown in Figure 5.

The results of extraction airflow rate and air velocity from various distances to hood entrance during welding operations are shown in Figure 7 and Figure 8. While Figure 7 results were obtained when the duct was 30% extended, Figure 8 results were obtained when the duct was 100% extended. It was observed that, at distance 20 mm away from the hood, airflow rate and air velocity were of the same value irrespective of the extension rate of the flexible duct. Also, at 120 mm away from the hood, the Thermo Anemometer were unable to read the airflow rate and velocity of air.

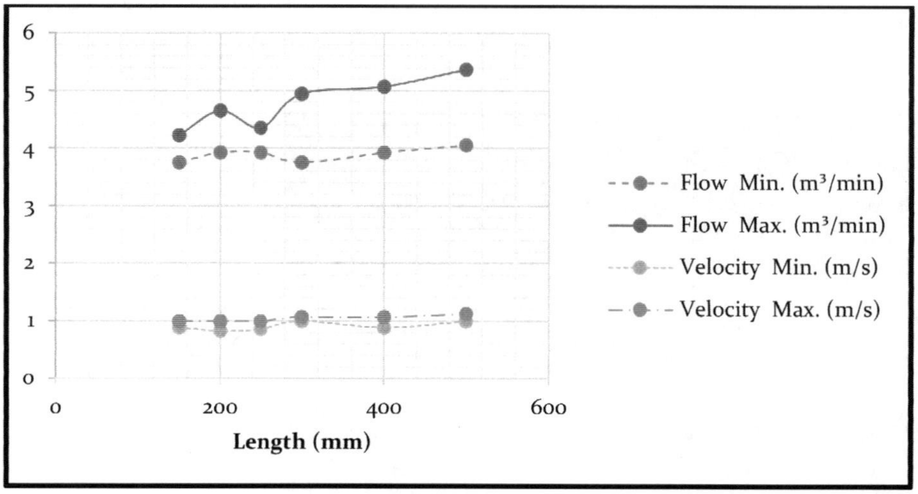

Fig. 6 Effects of flexible duct extension on airflow and air velocity

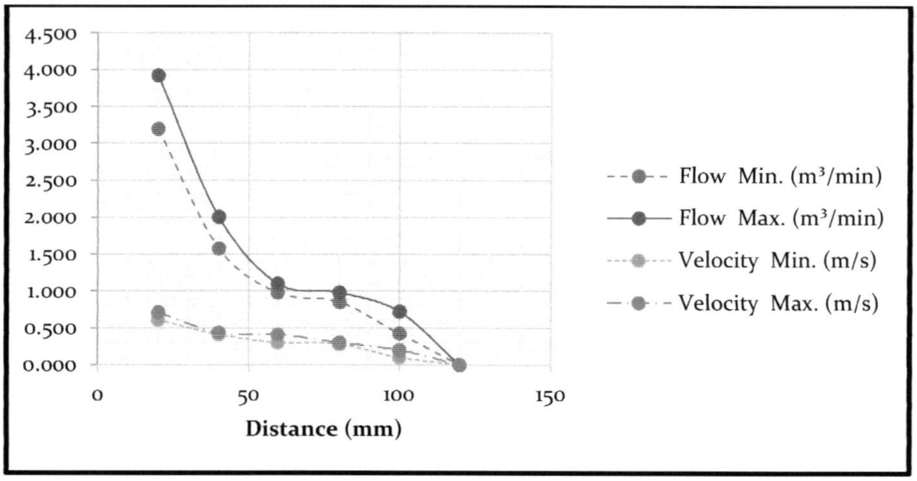

Fig. 7 Airflow and proximity to hood at 30% flexible duct extension

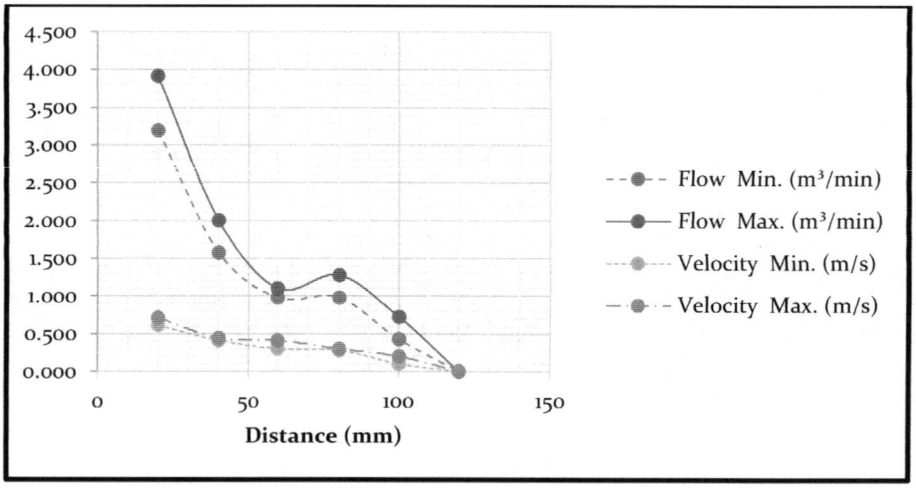

Fig. 8 Airflow and proximity to hood at 100% flexible duct extended

5. Conclusion

The ventilation system design and developed for the welding booth were tested and it showed that the use of flexible duct needed to be avoided where it is possible. This is due to the impart of the surface deformation as result of contraction and extension of flexible duct that increases the frictional force on the air flow thereby reducing the air quantity.

From experiment conducted on proximity of welding point, it was found that extracting hood work well at range between 20 mm to 100 mm irrespective of degree of extension of the flexible duct. Welding operation should be restricted to this range for best fume extraction, as this keeps welding smoke away from breathing zone of welder as from the adjoining welding area.

It was observed that confined welding area restricted gases and fumes from escaping to the environment due to the negative pressure provided by the ventilation system.

Acknowledgements

The authors acknowledge Covenant University Ota, Ogun State, Nigeria, for the sponsorship and provision of research facilities for this project.

References

[1] Han-Qing Wang, Chun-Hua Huang, Di Liu, Fu-Yun Zhao, Hai-Bo Sun, Feng-Feng Wang, Can Li, Guang-Xiao Kou, Ming-Qiang Ye. Fume transports in a high rise industrial welding hall with displacement ventilation system and individual ventilation units. Building and Environment, 2012, 52: 119-128.

[2] Sindo Kou. Welding Metallurgy, Second Edition. John Wiley & Sons, Inc., Publication, New York; 2003, 1: 1-33.

[3] Małgorzata Szewczyńska, Emilia Pągowska, Krystyna Pyrzyńska. Emissions of fluorides from welding processes. Journal of Environmental Sciences, 2015, 37: 179 – 183.

[4] Renaud Persoons, Damien Arnoux, Théodora Monssu, Olivier Culié, Gaëlle Roche, Béatrice Duffaud, Denis Chalaye, Anne Maitre. Determinants of occupational exposure to metals by gas metal arc welding and risk management measures: A biomonitoring study. Toxicology Letters, 2014, 231: 135–141.

[5] Meeker JD, Susi P, Flynn MR. Manganese and welding fume exposures and control in construction. J Occup Env Hyg, 2007, 4: 943–51.

[6] Smargiassi A, Baldwin M, Savard S. Assessment of exposure to manganese in welding operations during the assembly of heavy excavation machinery accessories. Appl Occup Env Hyg, 2000, 15: 746–50.

[7] SMACNA. 2005. HVAC duct construction standards—Metal and flexible, 3rd ed. ANSI. Sheet Metal and Air Conditioning Contractors' National Association, Chantilly, VA; 3.1

[8] ASHRAE, 2009 American Society of Heating, Refrigerating and Air-Conditioning Engineers, Inc Handbook – Fundamentals; F21

[9] ASHRAE, 2007 American Society of Heating, Refrigerating and Air-Conditioning Engineers, Inc Handbook - HVAC Applications; A30

[10] ACGIH. (2010) Industrial ventilation - a manual of recommended practice. 27th edn. Cincinnati, OH: ACGIH.

IOP Conf. Series: Materials Science and Engineering **413** (2018) 012019 doi:10.1088/1757-899X/413/1/012019

A Review of Working Fluids for Organic Rankine Cycle (ORC) Applications

Babatunde, A. Fakeye[1] and [1]*Sunday, O. Oyedepo[1]

[1]Mechanical Engineering Department, Covenant University, Ota
E-mail address: Sunday.oyedepo@covenantuniversity.edu.ng

Abstract. The organic Rankine cycle (ORC) systems are commercially employed for small scale and large scale thermal conversion to electricity of a large variety of abundant heat sources such as exhaust waste heat and some renewable energy sources where conventional steam Rankine and open-gas cycle turbine cycles cannot provide any viable, sustainable, techno-economic solution for power generation. The ORC operates the conventional steam Rankine cycle in subcritical level but can as well operate trans critical cycle while employing heavy molecular organic fluids in place of steam which makes the ORC suitable for medium and low grade heat conversion into electricity. No singular ideal fluid exists for any application and hence over 600 pure and zeotropic mixtures have been investigated by various researches for their best suitability for different applications and operating conditions based on performance characteristics such as efficiency, cost and environmental impacts. In this study a review of working fluids selection for different applications has been conducted. This study helps in identifying the possible most suitable organic fluids for various ORC applications depending on the operating conditions.
Keywords: risk assessment; failure mode

1. Introduction

Organic Rankine Cycles (ORCs) are the most prevalent low-grade waste heat recovery cycles largely because of their simplicity and readily available components [1]. ORC characteristically performs better than the conventional steam Rankine cycle at the low operating temperatures due to the five factors explained by [2] as highlighted below:

- favourable thermodynamic cycle modification/architecture
- practical enthalpy drop and volume flow rate in the turbine
- favourable operating condition of the turbine
- possibility of lower maximum operating pressure and consequently, reduced costs of associated components
- the possibility of selecting positive gauge condensing pressure, hence, avoiding air infiltration.

Designing the best system configuration alongside selecting the best appropriate choice of working fluid for the viable operating condition is therefore the most essential requirement for any ORC system design for any application [3, 4]. An essential advantage of ORC systems is however, the utilization of organic working fluids that can simply employ a single stage expander hence offering advantages of simple design, low cost capital cost and low maintenance [3].

Because of relatively low specific enthalpy drop of organic vapours in turbine compared to water vapour, efficient ORC systems as small as 25 kW utilizing single stage turbo-expanders with reasonable tip speed have been possible. Whereas, water vapour process in most cases require three or four stage turbine with practical minimum size limited to 2 MW [5]. On the whole, the Organic Rankine Cycle units are most appropriate and most widely used for medium and low grade heat sources with wider flexibilities of reconfiguring and modifying the primary cycle to adapt it to the peculiarities of the heat sources [6].

However, the optimal cycle performance and system architecture primarily depends on the selection of the best appropriate working fluid [7]. The use of optimally mixed zeotropic fluids in place of customary pure fluids is a remarkable development to further improve the cycle performance but rather increases the complexity of the system. Pure organic fluids have fixed boiling temperatures and this gives rise to an incompatibility between the temperature profile of the working fluid and that of the heat source [3]. Two pure organic fluids with sufficient difference in their boiling points when optimally mixed produces a temperature glide at phase change which provides a better temperature compatibilities in aimed at either the condenser or the evaporator [8]. Such zeotropic mixtures have advantages of safety, environmental compatibility, and volumetric flow rate [9]but exhibit reduced heat transfer characteristics compared to pure fluids [10, 11].

2 ORC Working Fluids and Applications

The core task in ORC systems design primarily involves the selection of the best appropriate working fluid for the operation condition in either subcritical or supercritical level. This is supreme to obtain optimal performance indexes, simple cycle structure, small components sizes, and in general, safe operation of the system [12, 13]. Suitable working fluids permit proper volume flow rates which is essential for optimum sizing of the turbine for any power level [14], and also determines the proper type and size of the pump, heat exchangers and regeneration in advanced cycles. Xi et al,. [15] projected based on the experimental data and environmental factors that Hydrocarbons (HCs) are generally the best choice as working fluids for ORCs because of high energy efficiency, low cost and good environmental compatibility, especially the high molecular weight alkanes, which are always stable and their flammability and explosive characteristics suppressible by adding fire retardant.
The choice of organic working fluids are primarily influenced by the heat source temperature, the cooling medium or ambient temperature, and also the ORC performance assessment criterion employed for the optimization [16, 17]. Chintala et al. [18] summarized the essential considerations for selecting an appropriate organic working fluid. These include:

- Ability to dissolve in lubricant oils
- Safety and Ecological factors
- Ease of handling
- Abundant availability, and
- Economic considerations.

However, there are conflicting stances imposed by some of the desired features of the organic fluid on the efficiency of the ORC, safe operation, system performance and component sizing [19]. For example, high molecular weights and compressibility of working fluids enhance turbine efficiency [17] and reduces the number of stages for axial turbines, but however, high molecular weights fluids with high critical pressures require high rate of heat transfers and consequently, require bigger heat exchangers [20]. Hence, there is need for trade-offs in optimizing the ORC systems configuration and architecture for the best overall performance of the system.
Guo et al., [9], investigated the effects of different pure fluids and their zeotropic mixtures considering the vital operation parameters such as expander intake temperature, degree of superheat, mass flow rate, volumetric flow, exergy destruction, as well as the effect of recuperation on the performance characteristics of an ORC utilizing the exhaust heat of a 240 MW boiler of a pulverized coal-fired power plant. They explained that there is no optimal working fluid satisfying all the indicators and recommended from their findings that:

- Working fluids with high critical temperature showed better thermal efficiencies for the same operating condition,
- Insufficiently large temperature difference between the critical temperature of the constituent working fluids will result into minor temperature glide which will be unable to coherently match neither heat source nor sink.
- Mixture with the matching the heat sink indicates best efficiency and should be considered first when selecting working fluid, while mixture matching the heat source will yield the least total exergy destruction and reduced heat exchange area in the evaporator.

Zeotropic mixtures have severally been established to provide improved first efficiency and second law efficiencies as well as significant increase in work output as a result of better matching with the source and sink temperatures. The cycle performance however, depends on the proportion of the pure fluids and can be optimally extended to cover a wide range of selection requirements [21]. The thermo-physical properties of zeotropic mixtures hence depends primarily on the composition and the varying concentrations of the constituents, however, Abadi & Kim, [10] identified four major concerns about the use of azeotropic mixtures as follows:

- the conventional means of determining the thermodynamics properties of zeotropes are not sufficiently accurate,
- heat transfer coefficients of the designated optimized zeotropes from studies are unreliable if at all available,
- heat transfer coefficients of zeotropic mixtures typically get degraded and as a result of relatively lower heat transfer coefficients require bigger evaporator and condenser sizes compared to those of pure fluids, and
- composition shift (or separation) and fractionating of the zeotropic mixture during heat transfer compel a limit on the allowable temperature glide which in a way, reduces the deliverable power and first law efficiency of the system.

Additionally, composition shift hinders azeotropic mixtures from having stable thermodynamic properties in both the evaporator and condenser. Also, steeper attention must be paid to make the system leak-proof, especially the evaporator and condenser, in order to forestall undesirable composition shift [22]. Comparative study by Liu et al., [23] who investigated the influence reheat on ORC utilizing pure and zeotropic mixtures showed that zeotropic mixtures can commendably improve the net power output and also enhance the enthalpy drop between the turbine intake and exit of a reheat ORC system when the mixture's constituents and component ratio are both appropriate. However, uncertainty about the mixture stability under different working conditions, coupled with all of the above mentioned concerns limit the use and study of zeotropic mixtures [9]. Nonetheless, zeotropic mixtures have demonstrated wide prospective applications in both subcritical and supercritical ORCs [24].

2.1 Classification and Application of ORC Working Fluids

Recent literatures generally classify fluids as wet, isentropic, or dry as shown in Figure 1 below. Dry fluids are high molecular mass organic fluids exhibiting positive slope on the T-s diagram. Wet fluids are low molecular mass organic fluids exhibiting negative slopes on the T-s diagram. Isentropic fluids are generally of medium molecular mass, exhibiting infinite or nearly vertical on the T-s diagram [24 – 26]. Hence, wet fluids normally require superheating [27]. Liu et al., [23] attributed the presence of hydrogen bond in the molecules of some organic working fluids like water, ammonia, and ethanol, as a probable cause of some fluids being wet as a result of larger vaporizing enthalpies which is considered inappropriate for ORC systems. The measure of wetness or dryness of a fluid is measured by the inverse of the slope, defined as $\xi = ds/dT$. Therefore, the value of $\xi > 0$ implies a dry fluid, $\xi \approx 0$ implies an isentropic fluid and $\xi < 0$ implies a wet fluid [28].

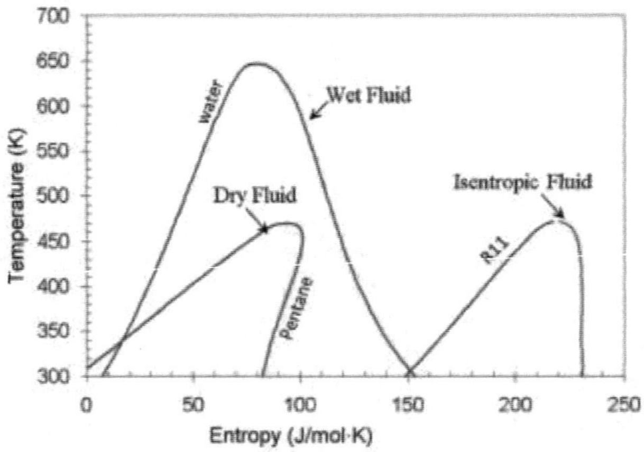

Figure 1: Dry, wet and isentropic working fluids

Isentropic and dry fluids have been widely proposed for subcritical ORC basically to avoid formation of liquid droplet. Nevertheless, if the saturated vapour curve of the dry fluid sharply deflects inwards, the vapor will wastefully exit from the turbine with significant degree of superheat thereby unnecessarily adding to the condenser cooling load. The exiting superheated vapour can however be better utilized for regeneration between the feed pump and the boiler/evaporator to improve the efficiency of the system.

For supercritical Rankine cycles utilizing wet or dry fluids, the turbine intake temperature must be sufficiently high to avoid the double phase region as shown in Figure 2 by propyne, a dry fluid, pentane, a wet fluid [13]. If the temperatures are sufficiently high, the expansion process does not go through the double phase state represented by the dashed lines in figures 2a & b. On the contrary, wet fluids are less affected by superheating after the expansion.

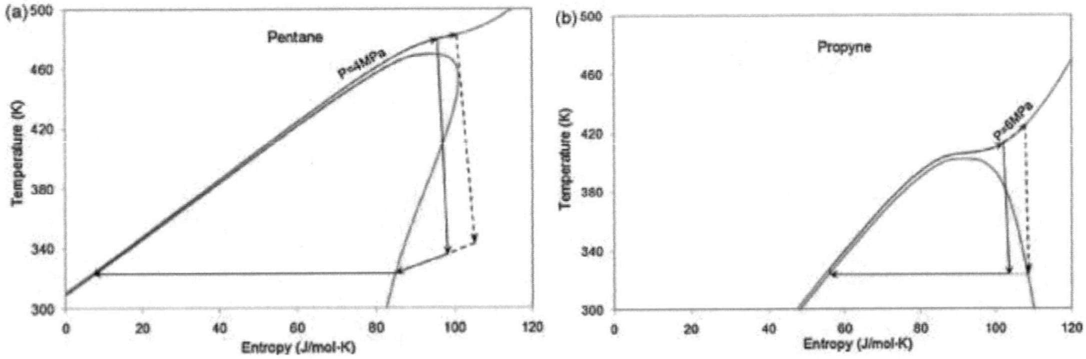

Fig. 2: T–s diagram of working fluids at supercritical levels for a dry and a wet fluid. (a) Pentane. (b) Propyne [13]

Patrick et al.. [29] in Chen [13] was mentioned to discover that wet fluids subcool and nucleate in rapid succession to form a double phase fluid, which characteristically delimits the performance and life of the turbine. On the contrary however, dry fluids passing through the two-phase region as referenced, was established by Shu et al. [30] to have inconsequential effect on the performance and life of the turbine. Bahrami et al., [31] projected the suitable thermo-physical characteristics for good organic working fluids and using all of them as the criteria for the selection of the best appropriate fluid in their work, they concluded that the procedure was a rather complicated one. Chen et al. [13] in his review referring to the proposition of Galloni et al. [32] that working fluids with high latent heat, low specific heat and high density are expected to deliver higher turbine work output with smaller equipment setup for fixed operating conditions. Chen et al. [13] however stated that Richard et al. [33] differed that low latent heat fluids were otherwise more desirable; reasoning that saturated vapor condition at the turbine intake presented best operating condition. Chen et al. [13] hence showed a theoretical inquiry by combining Clausius–Clapayron relation, $\frac{dP}{dT} = \frac{L}{T\Delta V}$, ideal gas law, $V_{gas} = \frac{RT}{P}$, and the widely accepted equation of enthalpy drop,

$$\Delta h_{isentropic} = C_p T_{in}\left[1 - \left(\frac{p_{dis}}{P_{in}}\right)^{\gamma-1/\gamma}\right] \quad (1)$$

to arrive at:

$$\Delta h_{isentropic} = C_p\left[1 - e^{\frac{L}{C_p}\left(\frac{1}{T_1}-\frac{1}{T_2}\right)}\right] \quad (2)$$

He showed that substituting the ideal gas equation into Clausius–Clapayron equation and then integrating yields:

$$ln\frac{P_2}{P_1} = \frac{L}{R}\left(\frac{1}{T_1} - \frac{1}{T_2}\right) \quad (3)$$

This in fact depicts that for T_1 and T_2 fixed by the temperature of the heat source and ambient temperature respectively, the pressure ratio is a function of only the latent heat. Chen [13] however supported that from the equation of enthalpy drop, working fluids with higher latent heat gave higher unit work output but on the contrary, from the said equation, higher heat capacity rather the high latent heat fluids appear to correspond to higher unit work output. Borsukiewicz-gozdur & Nowak, [34] in their studies on the contrary recommended that to maintain the low latent heat criterion for the working fluid, a working fluid with critical point mildly above the heat source temperature should selected for a subcritical cycle in order to maintain the smallest possible latent heat of vapourization.

According to their studies, the critical temperature of working fluids has a vital effect on the effectiveness of operation of ORC cycle and the power output in particular. They maintained that low latent heat of evaporation is however the most desirable feature which is sustained by the fact that the latent heat of evaporation reduces when the critical point is approached. How closely from a practical point of view can the critical point be approached for evaporation to occur of is therefore an advantage. This was afterwards corroborated by Zhang, [26]. Bao & Zhao, [35] held that when using waste heat as the heat source, working fluids having lower latent heats are preferred because the reduced vaporization heat of the organic fluids enables bulk of the heat exchange in the evaporator to result in sensible heat as shown in Figure 3.

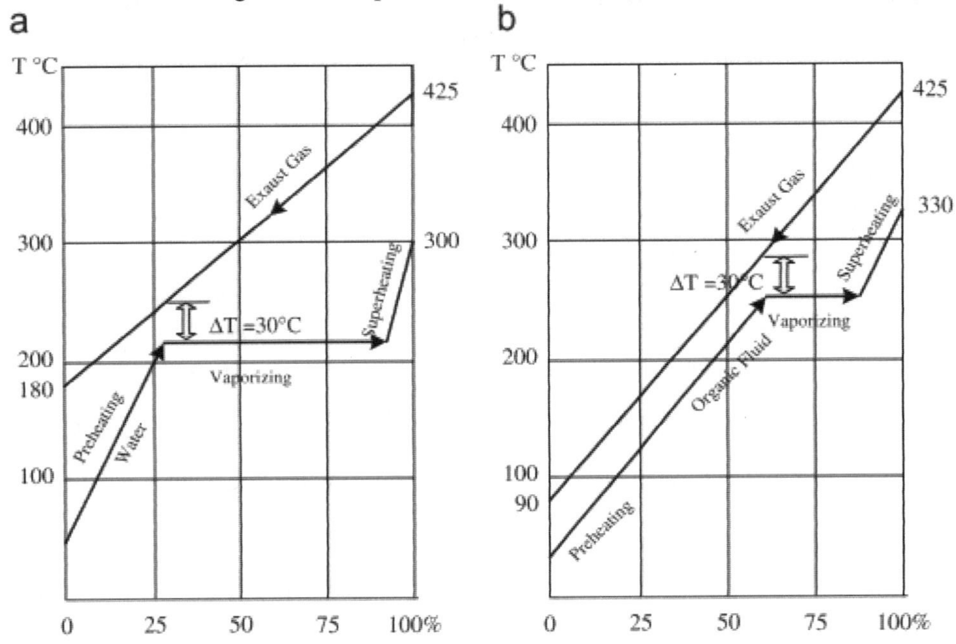

Fig. 3: Influence of latent heat of vaporization on the irreversibility during heat transfer [35].

Zhang [26] studied the thermodynamic effect of the inflection point of saturated vapour curve for 38 different types of dry and isentropic fluids. He defined the inflection point by a model of near-critical point triangle and investigated performance analysis. The results were in agreement with Borsukiewicz-gozdur & Nowak, [34] by concluding that heptane, cyclohexane, octane, nonane, decane, and dodecane in that they had low latent heats of evaporation at the inflection point to be the suitable working fluids. Furthermore, He et al., [36], proposed organic fluids showing low latent heat but high sensible heat.

In the same manner, Liu et al. [23] in He et al., [36] stated that organic fluids found wide applications in waste heat recovery systems because they meet the requirements of low latent heat and high sensible heat. However, the ratio of vapourization latent heat to the sensible heat, described as the vapourization enthalpy ratio, is a close ratio to the thermodynamic index, Jacob Number, which was analytically derived by Mikielewicz & Mikielewicz, [37] by considering twenty pure organic fluids for subcritical and supercritical ORCs. Singh & Pedersen [38] in Bao & Zhao [35] recommended a derivative of the Jacob Number but encompassing condensing and evaporating temperatures, called Figure of merit given as:

$$\text{FOM} = Ja^{0.1}\left(\frac{T_{cond}}{T_{evap}}\right)^{0.8} \text{ and Ja} = C_P\frac{dT}{H_v}, \quad (4)$$

where $C_P dT$ is the vapourization sensible heat and H_v is the vapourization latent heat.

Whereas, Mikielewicz, [39] conducted an appraisal on critical performance indicators of ORC. They introduced an evaluation indicator- the Jacob number, recommended by Mikielewicz and Mikielewicz [37] to set up a modified thermodynamic method and model for ORC and to demonstrate interaction between ORC and Carnot cycle as well as the thermodynamic cycle triangle in order to develop a more convenient and simplified alternative for the complex traditional numerical method. The Jacob number, Ja, represents the ratio of sensible heat capacity to the latent heat capacity of the organic working fluid. The preeminence of this parameter consists in the blend of two key factors influencing cycle performance which are the evaporator and condenser temperatures, as well as organic fluid properties such as specific heat C_P and latent heat h_{fg}, simultaneously.

$$\text{Jacob number is given by Ja} = \frac{\text{Cp}(Te-Tc)}{hfg}. \quad (5)$$

IOP Conf. Series: Materials Science and Engineering **413** (2018) 012019 doi:10.1088/1757-899X/413/1/012019

The work of Wang et al. [40] however generated more explorable correlations between Ja, the thermal efficiency and exergetic efficiency of the considered cycles and the network output and also with possibility of further modifications to include more parameters. The thermophysical properties of the selected working fluid were in agreement with the work of Borsukiewicz-gozdur & Nowak, [34].

In another study, Amicabile et al., [41] employed thermodynamic analysis to show that the enthalpy of vapourization is inversely proportional to the molecular weight of the fluid and thereby suggested that low molecular weight fluids should be avoided because of their high enthalpy of vapourization. Due to the large amount of working fluid and the diversity of criteria for selecting appropriate working fluids for ORC applications, there is no one single optimal fluid for a given temperature level and a given application. Quoilin et al [42] identified two approaches for fluid selection: the screening method and the operating map method.

Screening method is though the popular approach in the scientific literature but it is impaired by the use of objective function void of influences of fluid properties on the practical design of the cycle. The thermodynamic model developed is limited to optimizing working fluid performances in terms of thermal efficiency, output power or exergy efficiency inherently capable of proffering unrealistic working fluids. The operating map approach proposed by Quoilin et al. [42] integrated the interaction between expander, heat exchangers and the working fluid into a selection process that leads to the selection of appropriate types and sizes of the basic components of the ORC. The approach maps out acceptable conditions and limits on component size as shown in Figures 4 to 7, therefore does not yield impractical working fluids. However, the operating map method is only a pre-selection tool to narrow down the possible choices because the operating maps of some working fluids overlap with some others. The fluid selection process therefore has to be further optimized by a more accurate procedure.

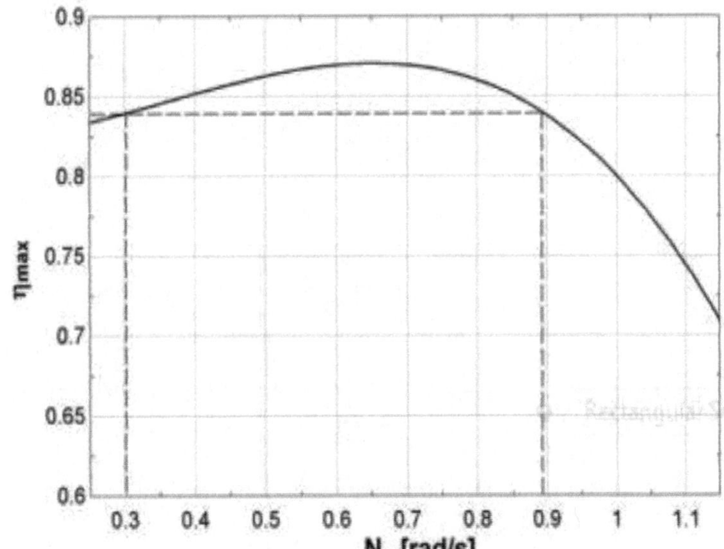

Figure 4: Characteristic maximum efficiency curve as a function of its specific speed for a radial turbine [42]

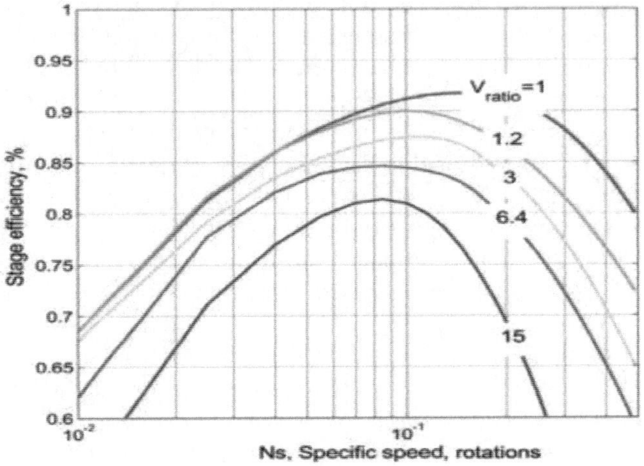

Figure 5: Parametric curves proposed for turbine stage efficiency prediction of axial turbines [2]

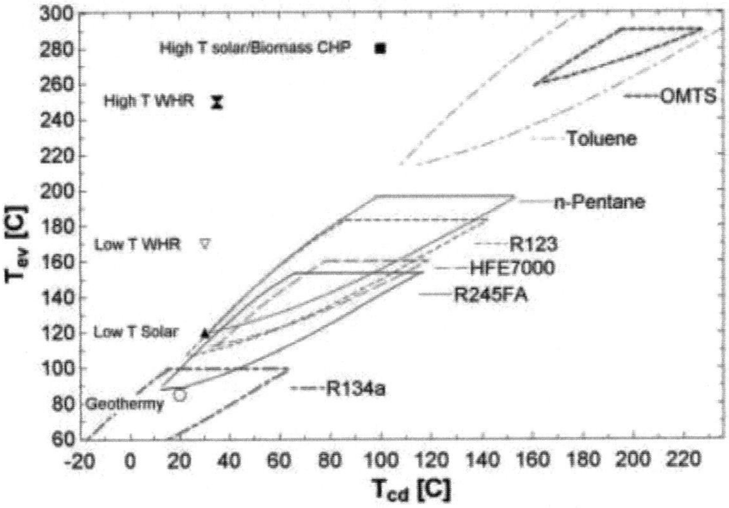

Figure 6: Screw expander operating map [42]

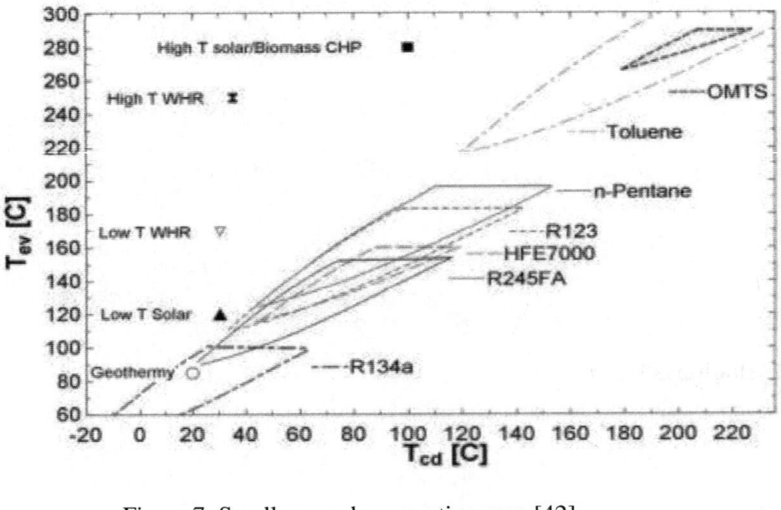

Figure 7: Scroll expander operating map [42]

The top left-hand corner in Figure 7 above shows the region of excessively large expansion ratio while the down right-hand corner is the region of too high volume flow rate.

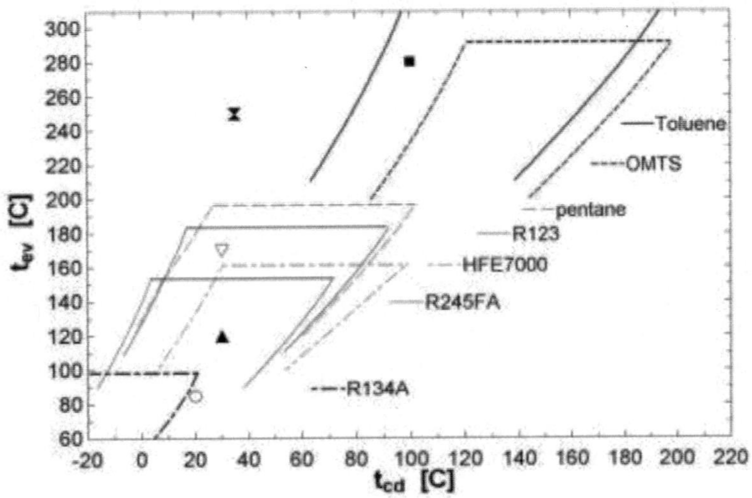

Figure 8: Radial inflow turbine operating map [42]

Of all the fluids investigated, only R123 and R245fa showed capacity to achieve desirable cycle performances. R245fa operates at superatmospheric pressure at condenser temperature of 40°C eliminating the possibility of infiltration of non-condensable gases into the condenser [31] and it is the most utilized working fluid in ORC experiments and the most considered for waste heat recovery applications [43].

3 Effects of Critical Temperature on Working Fluids Pre-selection

The critical temperature is a function of the strength of the intermolecular interactions that binds the molecules of a substance together as a liquid but it sets a limit on the evaporation temperature in subcritical cycles and also determines the temperature glide in zeotropic mixtures. According to [44], the comparatively high critical temperature, high thermal stability and low vapour pressure characterize the suitability of organic fluids for ORC applications. High critical temperature working fluids have been found to be essential to maintain ORC systems within the subcritical range and to avoid disintegration of the organic working fluids at the elevated temperatures of waste exhaust gases [45]. Low critical temperature fluids perform better for supercritical cycles [46], however, the chemical stability of organic working fluids operating on supercritical cycle also depends on their critical temperatures because of the tendency to degenerate with high degree of superheat. High critical temperature permits high turbine inlet temperature fluids to expand to lower pressures at the turbine exit, hence larger enthalpy difference to deliver more power [47] but are however, subatmospheric at ambient temperature [42] which exposes the condenser to the possibility of infiltration of non-condensable gases.

Saadatfar et al., [48] from similar investigation of a number of working fluids based on the correlation between critical temperature of working fluids and the cycle efficiency likewise reported that higher critical temperature produces higher efficiency but lower condensing pressure and hence proposed classification of working fluids as a function of their critical temperature by means of acentric factor and molecular complexity parameters. High critical temperature organic fluids, generally with low condensing pressures, exhibit volume ratio (ratio of the specific volume at the turbine outlet to the specific volume at the inlet during an isentropic expansion) as high as 200 to 300. The specific volume range is undesirable for the cycle architecture because the high volume flow rates require bigger size turbine resulting in low rotational speeds well below 300 rpm and thus require multi-pole generator [2].

Also, working fluid with higher critical temperature exhibits lesser optimal evaporating pinch point and greater optimal condensing pinch point [49]. The optimal evaporating temperatures of different working fluids was observed to display an increasing relationship with increasing critical temperature, but on the other hand, the optimal condensing temperatures had no significant effect. This accordingly was an indication that condensing temperature does not respond to cost-effective performance and hence, according to Li et al [49], this further corroborates the choice of parameters optimization at a given condensing temperature in many studies. Henrik & Per [50] hence clearly stated that higher critical temperature fluids provide superior performance and are less susceptible to changes in condenser temperature and pressure.

Henrik Ö & Per [50] also demonstrated that there is a direct positive correlation between increase in expansion ratio and increased critical temperature of fluids. The fluids with higher critical temperatures showed greater expansion ratio potentials and hence, greater work output potential. They also demonstrated that:

(i) fluids with higher critical points offers higher expander enthalpy drop, and consequently greater work output, and
(ii) for a fluid expanding between two fixed pressures, the volume expansion ratio decreases with increasing maximum cycle temperature but gives rise to increase in enthalpy drop.

Similarly, Wang et al. [40], based their investigation of the performance of ORC systems for WHR solely on the critical temperature criterion, proposed that the critical temperature can be a singular factor for fluid selection. The authors proposed based on their finding that thermal efficiencies have dispersed distribution against every other physical parameter other except for the critical temperature alone. In the investigation, five organic fluids were employed; R245fa, R123, R601, isohexane and hexane, with critical temperatures from low to high.

Their results showed correlation between thermal efficiency increase and increase in critical temperature while R601 provided the maximum efficiency because of the closeness between the critical temperature and the heat source temperature. It was in a similar trend discovered by Agromayor & Nord [51] that transcritical-recuperated cycle architecture employing either dry or isentropic organic fluids exhibited optimal performance when the critical temperature of the working fluid was slightly lower than the temperature of the heat source. Conversely, no common relationship was established for the optimal cycle architecture for heat source temperatures higher than the critical temperatures of the organic fluids. Bianchi et al., [52] also reported from studies that specific pumping work reduces as the critical temperature of organic working fluids increases, especially when the critical temperature is greater than 150°C, but no direct relationship with liquid specific heat.

4. Guidelines for Screening of Working Fluids

Since the efficiency of the Organic Rankine Cycle for a fixed working condition and the life of the turbine as well significantly depends on the selected dry working fluid [53], it therefore requires an optimization process for the probable working fluids in order to determine the most suitable working fluid in concurrently with the expander selection and the ORC architecture. [42] hence identified thirteen basic guidelines for screening of working fluids highlighted below:

- Thermodynamic performance
- Positive or isentropic saturation vapor curve
- High vapor density
- Low viscosity
- High conductivity
- Acceptable evaporating pressure
- Positive condensing gauge pressure
- High stability temperature
- Melting point
- High safety level

5. Conclusion

In this study, comprehensive review of working fluids selection for ORC applications has been carried out. Pure and azeotropic fluids have been discussed. The thermo-physical properties, stability, environmental effects, safety and compatibility, and availability and cost which are essential factors to be considered when making a choice of the most appropriate organic working fluids and appropriate trade-off are discussed comprehensively.

Acknowledgements
The authors acknowledged the support received from Covenant University Management through CUCRID in carrying out this research work.

References

[1] Ziviani, D., Dickes, R., Quoilin, S., & Lemort, V. Organic Rankine cycle modelling and the ORCmKit library : analysis of R1234ze (Z) as drop-in replacement of R245fa for low-grade waste heat recovery. *The 29th International Conference on Efficiency, Cost, Optimization, Simulation and Environmental Impact of Energy Systems*, 2016, 1–13

[2] Astolfi, M., Romano, M. C., Bombarda, P., & Macchi, E. Binary ORC (Organic Rankine Cycles) power plants for the exploitation of medium-low temperature geothermal sources - Part B: Techno-economic optimization. *Energy*, 2014, *66*, 435–446. https://doi.org/10.1016/j.energy.2013.11.057

[3] Chen, H., Yogi Goswami, D., Rahman, M. M., & Stefanakos, E. K. Energetic and exergetic analysis of CO2- and R32-based transcritical Rankine cycles for low-grade heat conversion. *Applied Energy*, 2011, *88*(8), 2802–2808. https://doi.org/10.1016/j.apenergy.2011.01.029

[4] Habeeb, A. Low-Grade Waste Heat Recovery-To- Power Generation Low-Grade Waste Heat Recovery-To-, (October), 2014.

[5] Larjola, J., Uusitalo, A., & Turunen-saaresti, T. Background and Summary of Commercial ORC Development and Exploitation, 2011.

[6] Mocarsk, S., & Borsukiewicz-Gozdur, A. Selected aspects of operation of supercritical (transcritical) organic Rankine cycle. *Archives of Thermodynamics*, 2015, *36*(2), 85–103. https://doi.org/10.1515/aoter-2015-0017

[7] Simone L, Constantine N. M, Ioannis V & Rodolfo T. A thermodynamic feasibility study of an Organic Rankine Cycle (ORC) for heavy-duty diesel engine waste heat recovery in off-highway applications, Int J Energy Environ Eng 2017, 8:81–98

[8] Heberle, F., Preißinger, M., & Brüggemann, D. Zeotropic mixtures as working fluids in Organic Rankine Cycles for low-enthalpy geothermal resources. *Renewable Energy*, 2012, *37*(1), 364–370. https://doi.org/10.1016/j.renene.2011.06.044

[9] Guo, C., Du, X., Yang, L., & Yang, Y. Organic Rankine cycle for power recovery of exhaust flue gas. *Applied Thermal Engineering*, 2015, *75*, 135–144. https://doi.org/10.1016/j.applthermaleng.2014.09.080

[10] Abadi, B.G., & Kim, K. C. Investigation of organic Rankine cycles with zeotropic mixtures as a working fluid: Advantages and issues. *Renewable and Sustainable Energy Reviews*, 2017, *73*(February), 1000– 1013. https://doi.org/10.1016/j.rser.2017.02.020

[11] Weith, T., Heberle, F., Preißinger, M., & Brüggemann, D. Performance of siloxane mixtures in a high-temperature organic rankine cycle considering the heat transfer characteristics during evaporation. *Energies*, 2014, *7*(9), 5548–5565. https://doi.org/10.3390/en7095548

[12] Bândean, D. C., Smolen, S., & Cieslinski, J. T. Working fluid selection for Organic Rankine Cycle applied to heat recovery systems. *World Renewable Energy Congress 2011*, 772–779.

[13] Chen, H., Goswami, D. Y., & Stefanakos, E. K. A review of thermodynamic cycles and working fluids for the conversion of low-grade heat. *Renewable and Sustainable Energy Reviews*, 2010, *14*(9), 3059–3067. https://doi.org/10.1016/j.rser.2010.07.006

[14] Angelino, G. A review of italian activities in the field of organic rankine cycle.pdf. VDI BERICHTE, 1984.

[15] Xi, X., Zhou, Y., Guo, C., Yang, L., & Du, X. Characteristics of Organic Rankine Cycles with Zeotropic Mixture for Heat Recovery of Exhaust Gas of Boiler. *Energy Procedia*, 2015, *75*, 1093–1101. https://doi.org/10.1016/j.egypro.2015.07.496

[16] Shengjun, Z., Huaixin, W., & Tao, G. Performance comparison and parametric optimization of subcritical Organic Rankine Cycle (ORC) and transcritical power cycle system for low-temperature geothermal power generation. *Applied Energy*, 2011, *88*(8), 2740–2754. https://doi.org/10.1016/j.apenergy.2011.02.034

[17] Stijepovic, M. Z., Linke, P., Papadopoulos, A. I., & Grujic, A. S. On the role of working fluid properties in Organic Rankine Cycle performance. *Applied Thermal Engineering*, 2012, *36*(1), 406–413. https://doi.org/10.1016/j.applthermaleng.2011.10.057

[18] Chintala, V., Kumar, S., & Pandey, J. K. A technical review on waste heat recovery from compression ignition engines using organic Rankine cycle. *Renewable and Sustainable Energy Reviews*, 2017, *81*, 493–509. https://doi.org/10.1016/j.rser.2017.08.016

[19] Hettiarachchi, M. H. D., Golubovic, M., Worek, W. M., & Ikegami, Y. Optimum design criteria for an Organic Rankine cycle using low-temperature geothermal heat sources. *Energy*, 2007, *32*(9), 1698–1706. https://doi.org/10.1016/j.energy.2007.01.005

[20] Saadatfar, B., Fakhrai, R., & Fransson, T. The Journal of MacroTrends in Energy and Sustainability Waste heat recovery Organic Rankine cycles in sustainable energy conversion: A state-of-the-art review. *Jmes*, 2013, *1*(1), 161–188.

[21] Andreasen, J. G., Pierobon, L., Larsen, U., Haglind, F., & Author, C. Multi-Objective Optimization of Organic Rankine Cycle Power Plants Using Pure and Mixed Working Fluids. *Proceedings of the 3rd International Seminar on ORC Power Systems*, 2015, 1–11. https://doi.org/10.3390/en9050322

[22] Liu, C., Gao, T., Xu, J., Zhu, J., & Xu, X. Analysis of Pure Fluid and Zeotropic Mixtures Uesd in Low-Temperrature Reheating Organic Rankine Cycles. *Proceedings of the 3rd International Seminar on ORC Power Systems*, 2015, 1–10.

[23] Nouman, J. Comparative studies and analyses of working fluids for Organic Rankine Cycles -ORC. 2012, Retrieved from https://www.diva-portal.org/smash/get/diva2:555314/FULLTEXT01.pdf

[24] Habibzadeh, A., & Rashidi, M. M. Thermodynamic analysis of different working fluids used in organic rankine cycle for recovering waste heat from GT-MHR. *Journal of Engineering Science and Technology*, *11*(1), 2016, 121–135

[25] Zhang, X. Thermodynamic Study of Inflection Point of Saturated Vapor Curve for Dry and Isentropic Working Fluids. *Proceedings of the 3rd International Seminar on ORC Power Systems*, 2015, 1–10.

[26] Kandathil, A. K. A Guide to working fluid selection for Organic Rankine Cycle ORC generators. *Genixx, HEATCATCHER*. Retrieved from http://www.heatcatcher.com/guide-working-fluid-selection-organic-rankine-cycle-orc-generators/, 2016.

[27] Liu, B. T., Chien, K. H., & Wang, C. C. Effect of working fluids on organic Rankine cycle for waste heat recovery. *Energy*, 2004, *29*(8), 1207–1217. https://doi.org/10.1016/j.energy.2004.01.004

[28] Dhar, H., Kumar, S., & Kumar, R. A review on organic waste to energy systems in India. *Bioresource Technology*, (August), 2017, https://doi.org/10.1016/j.biortech.2017.08.159

[29] Patrick L, Athanasios I. P & Panos S. Systematic Methods for Working Fluid Selection and the Design, Integration and Control of Organic Rankine Cycles—A Review, *Energies* **2015**, *8*, 4755 - 4801

[30] Shu, G, Liu, L, Tian, H, Wei, H & Yu, G. Parametric and working fluid analysis of a dual-loop organic Rankine cycle (DORC) used in engine waste heat recovery, Applied Energy 113: 2014: 1188 – 1198

[31] Bahrami, M., Hamidi, A. A., & Porkhial, S. Investigation of the effect of organic working fluids on thermodynamic performance of combined cycle Stirling-ORC. *International Journal of Energy and Environmental Engineering*, 2013, *4*(12), 1–9. https://doi.org/10.1186/2251-6832-4-12

[32] Galloni, E, Fontana, G & Staccone, S. Design and experimental analysis of a mini ORC (organic Rankine cycle) power plant based on R245fa working fluid, Energy 90: 2015: 768 – 775

[33] Richard L, Adam H & David R. A knowledge-based system for low-grade waste heat recovery in the process industries, Applied Thermal Engineering 94: 2016: 590–599

[34] Borsukiewicz-gozdur, A., & Nowak, W. Desirable Thermophysical Properties of Working Fluids in Organic Rankine Cycle. *European Geothermal Congress 2007*, (June), 1–5.

[35] Bao, J., & Zhao, L. A review of working fluid and expander selections for organic Rankine cycle. *Renewable and Sustainable Energy Reviews*, 2013, *24*, 325–342. https://doi.org/10.1016/j.rser.2013.03.040

[36] He, M., Zhang, X., Zeng, K., & Gao, K. A combined thermodynamic cycle used for waste heat recovery of internal combustion engine. *Energy*, 2011, *36*(12), 6821–6829. https://doi.org/10.1016/j.energy.2011.10.014

[37] Mikielewicz, D., & Mikielewicz, J. A thermodynamic criterion for selection of working fluid for subcritical and supercritical domestic micro CHP. *Applied Thermal Engineering*, 2010, *30*(16), 2357–2362. https://doi.org/10.1016/j.applthermaleng.2010.05.035

[38] Singh, D.V & Pedersen, E. A review of waste heat recovery technologies for maritime applications, Energy Conversion and Management 111, 2016: 315–328

[39] Mikielewicz, D. M.. A thermodynamic criterion for selection of working fluid for subcritical and supercritical domestic micro CHP. *Applied Thermal Engineering, Elsevier*, *105*(10.1016/j.applthermaleng.2010.05.035), 2356. https://doi.org/10.1016/j.egypro.2017.03.330

[40] Wang, H., Xu, J., Yang, X., Miao, Z., & Yu, C. (2015). Organic Rankine cycle saves energy and reduces gas emissions for cement production. *Energy*, *86*, 59–73. https://doi.org/10.1016/j.energy.2015.03.112

IOP Conf. Series: Materials Science and Engineering **413** (2018) 012019 doi:10.1088/1757-899X/413/1/012019

[42] Amicabile, S., Lee, J. I., & Kum, D. A comprehensive design methodology of organic Rankine cycles for the waste heat recovery of automotive heavy-duty diesel engines. *Applied Thermal Engineering*, 2015, *87*, 574–585. https://doi.org/10.1016/j.applthermaleng.2015.04.034

[43] Quoilin, S., Declaye, S., Legros, A., & Guillaume, L. Working fluid selection and operating maps for Organic Rankine Cycle expansion machines. *International Compressor Engineering Conference*, 2012, 1–10.

[44] Quoilin, S., Broek, M. Van Den, Declaye, S., Dewallef, P., & Lemort, V. Techno-economic survey of organic rankine cycle (ORC) systems. *Renewable and Sustainable Energy Reviews*, 2013, *22*, 168–186. https://doi.org/10.1016/j.rser.2013.01.028

[44] Ziviani, D., Beyene, A., & Venturini, M. Advances and challenges in ORC systems modeling for low grade thermal energy recovery. *Applied Energy*,2014, *121*, 79–95. https://doi.org/10.1016/j.apenergy.2014.01.074

[45] Song, J., & Gu, C. W. Analysis of ORC (Organic Rankine Cycle) systems with pure hydrocarbons and mixtures of hydrocarbon and retardant for engine waste heat recovery. *Applied Thermal Engineering*, 2015, *89*, 693–702. https://doi.org/10.1016/j.applthermaleng.2015.06.055

[46] Jumel, S., & Feidt, M. Working fluid selection and performance comparison of subcritical and supercritical organic Rankine cycle (ORC) for low- temperature waste heat recovery, *2011*(Iea 2011), 2012, 559–569.

[47] Rusev, T. M. (2014). Comparative Study of Different Organic Rankine Cycle Models : Simulations and Thermo-Economic Analysis for a Gas Engine Waste Heat Recovery Application, 1–77.

[48] Saadatfar, B., Fakhrai, R., & Fransson, T. The Journal of MacroTrends in Energy and Sustainability Waste heat recovery Organic Rankine cycles in sustainable energy conversion: A state-of-the-art review. *Jmes*, 2013, *1*(1), 161–188.

[49] Li, Y. R., Du, M. T., Wu, C. M., Wu, S. Y., Liu, C., & Xu, J. L. Economical evaluation and optimization of subcritical organic Rankine cycle based on temperature matching analysis. *Energy*, 2014, *68*, 238–247. https://doi.org/10.1016/j.energy.2014.02.038

[50] Henrik Ö & Per L, Comparison and analysis of performance using Low Temperature Power Cycles, Applied Thermal Engineering 52 (2013) 160 – 169

[51] Agromayor, R., & Nord, L. O. Fluid selection and thermodynamic optimization of organic Rankine cycles for waste heat recovery applications. *Energy Procedia*, 2017, *129*, 527–534. https://doi.org/10.1016/j.egypro.2017.09.180

[52] Bianchi, G., Fatigati, F., Murgia, S., Cipollone, R., & Contaldi, G. Modeling and Experimental Activities on a Small-scale Sliding Vane Pump for ORC-based Waste heat Recovery Applications. *Energy Procedia*, 2016, *101*(September), 1240–1247. https://doi.org/10.1016/j.egypro.2016.11.139

[53] Algieri, A., & Morrone, P. Energy analysis of organic rankine cycles for biomass applications. *Thermal Science*, *19*(1), 2015, 193–205. https://doi.org/10.2298/TSCI120706030A

IOP Conf. Series: Materials Science and Engineering **413** (2018) 012020 doi:10.1088/1757-899X/413/1/012020

OPTIMIZING THE MOVEMENT OF PEOPLE AND GOODS IN A LOCAL COMMUNITY USING TRANSPORTATION MODEL

Bamigboye Oreofeoluwa[1,1], Olufunke Oladipupo[2,1], Agarana Michael[3,1], Olabode Oluwasanmi[4,1]

Department of Computer and Information Science, Covenant University
E-mail address: bamigboye.oreofeoluwa@yahoo.com

Abstract- Transportation problem is one of the real regions of utilization of linear programming, in this paper, we would take a gander at how to upgrade the separation and time taken to make a trip starting with one point then onto the next in a Local community, with Otta Local government area as our concentration, utilizing the transportation display. Having the capacity to accomplish this accomplishment would improve the development of individuals and merchandise inside the organization by sparing time and separation it takes to move starting with one point then onto the next. Google delineate used to figure the separation and time it takes to move from one intersection to the next and a manual test was completed to check the outcomes from google. The ideal arrangement demonstrated that time and separation can be lessened if certain intersections were utilized and if new street is to be made. Three methods would be utilized to locate the underlying achievable arrangement, they are: Northwest corner method, least cost method, Vogel method, Modi method and Best candidate method would be utilized to locate the ideal arrangement.

Keyword: Transportation Model, Optimal Solution, Best candidate method

1. Introduction

Transportation problem is an uncommon sort of direct programming issue. Transportation demonstrating is an iterative strategy for taking care of issues that include limiting the cost of delivery items from a progression of sources to a progression of goals (Sami Fethi,2015). In a transportation problem, we have certain birthplaces, which may speak to manufacturing plants where we deliver things also, supply a required measure of the things to a particular number of objectives.

The linear programming model has limitations for supply at each source and request at every goal. All limitations are equities in an adjusted transportation display where supply breaks even with request. Limitations contain imbalances in unequal models where supply does not equivalent request. Transportation models assume a vital part in coordination and inventory network administration for lessening cost and enhancing administration (Kamrul Hasan). Some past investigations have formulated arrangement system for the transportation issue with exact free market activity parameters.

Transportation problems are being faced in urban areas due to increase in human activities and most world class universities are like small cities because of their size and compositions. Also, today's campuses have become as complex as urban areas, because of their unique characteristics and inter-related influences, most main universities' campuses are planned as small cities/communities (Coulson. et al., 2011).

(F.L.Hitchcock,2013) firstly presented transportation problem in his paper "The Distribution of a Product from Several sources to numerous Localities" and after that it's presenting by T. C. Koopmans in his historic paper "Optimum Utilization of the Transportation System". These researchers' contributions helped in the development of transportation methods which involve a number of shipping sources and a number of destinations.

Transportation problem is specified by a set of *m* supply points from which a good/service is shipped. Supply point *i* can supply at most *si* units and a set of *n* demand points to which the good/service is shipped. Demand point *j* must receive at least *dj* units. Each unit produced at supply point *i* and shipped to demand point *j* incurs a variable cost of *cij*.

The target of this transportation demonstrate is to decide the sum to be delivered from each source to every goal to keep up the free market activity prerequisites at the least transportation cost.

2. Methodology

We are considering six different origins and six different destinations, they are:

	Origin		Destination
O1	Joju	D1	Oja
O2	Sango	D2	Bells
O3	Ojuore	D3	Davol
O4	Tollgate	D4	Babio
O5	Sango market	D5	Obasanjo

The Best Candidates Method (BCM):

BCM process incorporates three stages; these means are appeared as takes after:

Step1: Prepare the BCM network, if the framework uneven, at that point the lattice will be adjusted without utilizing the additional line or segment candidates in arrangement technique.

Step2: Select the best candidates, which are for limiting problems to the base cost, and expanding benefit to the most extreme cost. In this way, this progression should be possible by choosing the best two candidates in each column. On the off chance that the candidate rehashed in excess of two times, at that point the candidate ought to be chosen once more. And in addition, the sections must be checked to such an extent that on the off chance that it isn't have candidates with the goal that the candidates will be chosen for them. Be that as it may, if the candidate is rehashed in excess of one time, the choose it once more.

Step3: Find the mixes by deciding one candidate for each line and segment, this ought to be finished by beginning from the line that have the minimum candidates, and afterward erase that line and section. On the off chance that there are circumstances that have no candidate for a few lines or segments, at that point specifically choose the best accessible candidate.

Rehash Step 3 by deciding the following candidate in the line that began from. Figure and think about the summation of candidates for every blend. This is to decide the best mix that gives the ideal arrangement.

IOP Conf. Series: Materials Science and Engineering **413** (2018) 012020 doi:10.1088/1757-899X/413/1/012020

THE FOLLOWING METHODS WERE USED

- North - west corner method
- Least cost method
- Vogel's approximation
- MODI method(Modified distribution)

Table 1: shows the distance of the various destinations from Joju (O_1) and the time it takes from Joju to the destinations.

Table 1: Distance and time of various destinations from joju

Destination	Time (Min)	Distance (km)
Oja	6	1.4
Bells	10	2.1
Davol	9	1.6
Babio	4	1.0
Obasanjo	7	2.0

Table 2 shows the distance of the various destinations from Sango (O_2) and the time it takes from Sango to the destinations.

Table 2: Distance and time of various destination from sango

Destination	Time (Min)	Distance (km)
Oja	4	1.9
Bells	8	2.0
Davol	4	1.8
Babio	5	2.8
Obasanjo	4	1.6

Table 3 shows the distance of the various destinations from Ojuore (O_3) and the time it takes from Ojuore to the destinations.

Table 3: Distance and time of various destination from ojuore

Destination	Time (Min)	Distance (km)
Oja	5	2.5
Bells	6	2.0
Davol	6	1.9
Babio	8	2.7
Obasanjo	6	2.1

Table 4 shows the distance of the various destinations from Toll gate (O_4) and the time it takes from Toll gate to the destinations.

Table 4: Distance and time of various destination from tollgate

Destination	Time (Min)	Distance (km)

Oja	7	1.8
Bells	9	1.4
Davol	8	1.3
Babio	6	0.8
Obasanjo	4	2.0

Table 5 shows the distance of the various destinations from Sango market (O_5) and the time it takes from Sango market to the destinations.

Table 5: Distance and time of various destination from sango market

Destination	Time (Min)	Distance (km)
Oja	5	1.9
Bells	7	2.4
Davol	6	2.8
Babio	2	2.6
Obasanjo	8	3.0

Table 6: Table for Distance

	D_1	D_2	D_3	D_4	D_5	Available time
O_1	1.4	2.1	1.6	1.0	2.0	36
O_2	1.9	2.0	1.8	2.8	1.6	25
O_3	2.5	2.0	1.9	2.7	2.1	31
O_4	1.8	1.4	1.3	0.8	2.0	34
O_5	1.9	3.4	2.8	2.6	3.0	28
Required time	27	40	33	25	29	

Table 7: Table for Time

	D_1	D_2	D_3	D_4	D_5	Available time
O_1	6	10	9	4	7	36
O_2	4	8	4	5	4	25
O_3	5	6	6	8	6	31
O_4	7	9	8	6	4	34
O_5	5	7	6	2	8	28
Available time	27	40	33	25	29	

3. Result and discussion

Table 8: For Distance using North West Corner Method

	D_1	D_2	D_3	D_4	D_5	Available time

IOP Conf. Series: Materials Science and Engineering **413** (2018) 012020 doi:10.1088/1757-899X/413/1/012020

O1	1.4^{27}	2.1^{9}	1.6	1.0	2.0	36
O2	1.9	2.0^{25}	1.8	2.8	1.6	25
O3	2.5	2.0^{6}	1.9^{25}	2.7	2.1	31
O4	1.8	1.4	1.3^{8}	0.8^{25}	2.0^{1}	34
O5	1.9	2.4	2.8	2.6	3.0^{28}	28
Required time	27	40	33	25	29	

Total = 27*1.4+9*2.1+25*2.0+6*2.0+25*1.9+8*1.3+25*0.8+1*2.0+28*3.0

37.8+18.9+50+12+47.5+10.4+20+2+84=282.6

Step 1: Start with north-west corner (Uppermost left-hand corner) of the transportation table allocate maximum possible amount of it, so that either the capacity of row is exhausted or the destination requirement of the first column is satisfied.

Step II

a) If the supply for the first row is exhausted, then move down vertically to the cell in the second row.

b) If the demand for the first column is satisfied, then move right horizontally to the second cell in the first row.

c) If both supply in the row and demand in the column is satisfied, then, move diagonally.

Step III: Continuing this way till an allocation is made in the south-east corner cell of the transportation table.

Table 9: For Distance using Least Cost Method

	D1	**D2**	**D3**	**D4**	**D5**	**Available time**
O1	1.4^{11}	2.1	1.6	1.0^{25}	2.0	36
O2	1.9	2.0	1.8	2.8	1.6^{25}	25
O3	2.5	2.0	1.9^{31}	2.7	2.1	31
O4	1.8	1.4^{32}	1.3^{2}	0.8	2.0	34
O5	1.9^{16}	2.4^{8}	2.8	2.6	3.0^{4}	28
Required time	27	40	33	25	29	

Total = 11*1.4+25*1.0+25*1.6+31*1.9+32*1.4+2*1.3+16*1.9+8*2.4+4*3.0

15.4+25+40+58.9+44.8+2.6+30.4+19.2+12=248.3

Step I: Select the cell with the lowest cost and allocate maximum possible amount of it; if there is a tie make an arbitrary selection.

Step II: Delete the row or column or both which are satisfied by the allocation.

IOP Conf. Series: Materials Science and Engineering **413** (2018) 012020 doi:10.1088/1757-899X/413/1/012020

Step III: Again, search for the next lowest cost empty cell for which demand and supply are not exhausted and make the allocations. Repeat step I and II until all supply and demand conditions are satisfied.

Table 10: For Distance using Vogel's Approximation Method

	D_1	**D_2**	**D_3**	**D_4**	**D_5**	**Available time**
O_1	1.4	2.1^{36}	1.6	1.0	2.0	36
O_2	1.9	2.0	1.8	2.8^{25}	1.6	25
O_3	2.5^{27}	2.0^{4}	1.9	2.7	2.1	31
O_4	1.8	1.4	1.3^{5}	0.8	2.0^{29}	34
O_5	1.9	2.4	2.8^{28}	2.6	3.0	28
Required time	27	40	33	25	29	

Total = 35*2.1+25*2.8+27*2.5+4*2.0+5*1.3+29*2.0+28*2.8

73.5+70+67.5+8+6.5+58+78.4=361.9

Step I: Calculate the penalty cost i.e. the difference between the lowest and the second lowest cost values for each row and each column.

Step II: Enter the penalty cost of each row to the right of corresponding row and the penalty cost of each column under the corresponding column.

Step III: Select the row or column with the largest penalty and allocate maximum possible amount to the cell with the minimum cost in the selected row or column. If there are more than one largest penalty rows or columns we select any of them arbitrarily.

Step IV: Cross out the particular row or column on which the requirement is satisfied.

Step V: Recompute the penalty cost for each row and column for the reduced transportation table and go to the previous steps.

Step VI: Repeat step I to V until all the requirement are satisfied.

Table 11: Using Modi Method to find the Optimal Solution

	D_1	**D_2**	**D_3**	**D_4**	**D_5**	**Available time**
O_1	1.4^{11-}	2.1	1.6	1.0^{25+}	2.0	36
O_2	1.9	2.0	1.8	2.8	1.6^{25}	25
O_3	2.5	2.0	1.9^{31}	2.7	2.1	31
O_4	1.8	1.4^{32}	1.3^{2}	0.8	2.0	34
O_5	1.9^{16+}	2.4^{8}	2.8	2.6^{-}	3.0^{4}	28
Required time	27	40	33	25	29	

	D_1	**D_2**	**D_3**	**D_4**	**D_5**	**Available time**
O_1	1.4^{10}	2.1	1.6	1.0^{26}	2.0	36

IOP Conf. Series: Materials Science and Engineering **413** (2018) 012020 doi:10.1088/1757-899X/413/1/012020

O_2	1.9	2.0	1.8	2.8	1.6^{25}	25
O_3	2.5	2.0	1.9^{31}	2.7	2.1	31
O_4	1.8	1.4^{32}	1.3^{2}	0.8	2.0	34
O_5	1.9^{17}	2.4^{8}	2.8	2.6^{-}	3.0^{4}	28
Required time	27	40	33	25	29	

Total = 10*1.4+17*1.9+26*1.0+25*1.6+31*1.9+32*1.4+2*1.3+8*2.4-1*2.6+4*3.0

14+32.3+26+40+58.9+44.8+2.6+19.2-2.6+12=247.2

Step I: To compute the values for each row and column, set Di + Oj = Cij but only for those squares that are currently used or occupied. For example, if the square at the intersection of row 2 and column 1 is occupied, we set D2 + O1 = C21.

Step II: After all equations have been written, set O1 = 0.

Step III: Solve the system of equations for all D and O values.

Step IV: Compute the improvement index for each unused square by the formula improvement index (Iij) = Cij _ Di _ Oj.

Step V: Select the largest negative index and proceed to solve the problem as you did using the stepping-stone method.

Table 12: For Time using North West Corner Method

	D_1	**D_2**	**D_3**	**D_4**	**D_5**	**Available time**
O_1	6^{27}	10^{9}	9	4	7	36
O_2	4	8^{25}	4	5	4	25
O_3	5	6^{6}	6^{25}	8	6	31
O_4	7	9	8^{8}	6^{25}	4^{1}	34
O_5	5	7	6	2	8^{28}	28
Available time	27	40	33	25	29	

Total = 27*6 + 9*10 +25*8+6*6+25*6+8*8+25*6+1*4+28*8

162+90+200+36+150+64+150+4+224= 1080

Table 13: For Time using Least Cost Method

	D_1	**D_2**	**D_3**	**D_4**	**D_5**	**Available time**
O_1	6^{11}	10	9	4^{25}	7	36
O_2	4	8	4^{25}	5	4	25

O_3	5^{16}	6^{7}	6^{8}	8	6	31
O_4	7	9^{5}	8	6	4^{29}	34
O_5	5	7^{28}	6	2	8	28
Available time	27	40	33	25	29	

Total = 11*6+25*4+25*4+16*5+7*6+8*6+5*9+29*4+28*7

66+100+100+80+42+48+45+116+196=793

Table 14: For Time using Vogel's Approximation Method

	D_1	**D_2**	**D_3**	**D_4**	**D_5**	**Available time**
O_1	6	10^{36}	9	4	7	36
O_2	4	8^{4}	4	5^{21}	4	25
O_3	5	6	6^{27}	8^{4}	6	31
O_4	7^{27}	9	8^{6}	6	4^{1}	34
O_5	5	7	6	2	8^{28}	28
Available time	27	40	33	25	29	

Total = 36*10+4*8+21*5+27*6+4*8+27*7+6*8+1*4+28*8

360+32+105+162+32+189+48+4+224= 1156

Table 15: Using Modi Method to find the Optimal Solution

	D_1	**D_2**	**D_3**	**D_4**	**D_5**	**Available time**
O_1	6^{11}	10	9	4^{25}	7	36
O_2	4	8	4^{25}	5	4	25
O_3	5^{16-}	6^{7+}	6^{8}	8	6	31
O_4	7	9^{5}	8	6	4^{29}	34
O_5	5^{+}	7^{28-}	6	2	8	28
Available time	27	40	33	25	29	
	D_1	**D_2**	**D_3**	**D_4**	**D_5**	**Available time**
O_1	6^{11}	10	9	4^{25}	7	36
O_2	4	8	4^{25}	5	4	25
O_3	5^{15}	6^{8}	6^{8}	8	6	31
O_4	7	9^{5}	8	6	4^{29}	34
O_5	5^{1}	7^{27}	6	2	8	28
Available time	27	40	33	25	29	

Total = 11*6+25*4+25*4+15*5+8*6+8*6+5*9+29*4+1*5+27*7

66+100+100+75+48+48+45+116+5+189=792

Table 16: Using **BCM** to find Optimal solution.

The matric is balanced, row equals column.

	D_1	**D_2**	**D_3**	**D_4**	**D_5**	**Available time**
O_1	6	10	9	4	7	36
O_2	4	8	4	5	4	25
O_3	5	6	6	8	6	31
O_4	7	9	8	6	4	34
O_5	5	7	6	2	8	28
Available time	27	40	33	25	29	

Table 17: Elect the best candidate

	D_1	**D_2**	**D_3**	**D_4**	**D_5**	**Available time**
O_1	6	10	9	4	7	36
O_2	4	8	4	5	4	25
O_3	5	6	6	8	6	31
O_4	7	9	8	6	4	34
O_5	5	7	6	2	8	28
Available time	27	40		25	29	

Table 18: Find the combination

	D_1	**D_2**	**D_3**	**D_4**	**D_5**	**Available time**
O_1	6	10	9	4	7	36
O_2	4	8	4	5	4	25
O_3	5	6	6	8	6	31
O_4	7	9	8	6	4	34
O_5	5	7	6	2	8	28
Available time	27	40	33	25	29	

Table 19: Find the combination

	D_1	**D_2**	**D_3**	**D_4**	**D_5**	**Available time**
O_1	6	10	9	4	7	36
O_2	4	8	4	5	4	25
O_3	5	6	6	8	6	31

O_4	7	9	8	6	4	34
O_5	5	7	6	2	8	28
Available time	27	40	33	25	29	

Combination 1: A3,B1,C2,D5,E4= 4+4+6+4+6= 14

Combination 2: A1,B3,C2,D5,E4= 6+4+6+4+2= 12

4. Discussion

The transportation problem from origin to various destinations is computed by estimation but the data collected were calculated and the results were obtained. More accurate results were equally obtained using most transportation problem solution method that gave the optimal solution for time and distance.

Based on these results, it is recommended that the local government chairman should allocate people to always carryout survey periodically to find out if there are variations in the distributions of passengers in their routes. Data obtained in surveys should be analysed with this method to obtain current optimal solution for the current transportation situation and compare it with previous one. If any alteration is made to the number of routes, then the system will need to be modified to meet the specified routes.

The BCM (Best Candidate Method) based on election the best candidates and the alternative in each row and cover all columns with at least one candidate, then we can obtain the combinations that must be have no any intersect means, one candidate for each row and column.

5. Conclusion

This study on enhancing the movement of people and goods in a Local Community was carried out mainly to identify the transportation routes and then structure the system to obtain a model that will enable the management and individuals to operate at minimal cost and reduced distance travelled so that transportation of people would be at a reduced cost and time to reach the destination would also be reduced and furthermore if new roads are to be constructed some of this junctions would be considered to reduce cost and time. As a result of this the community can operate at a minimum cost.

We have established the uniqueness and existence of optimal solution of the transportation problem for Otta Local community. This has been brought out through developed transportation problem into linear programming problem and applying the discuss methods in paper which yields an initial feasible solution and an optimal solution and have stated the optimality conditions of the problem. And we found out that the BCM method gives the optimal solution.

References

[1] Hlayel Abdallah Ahmad Journal of Computer Science 8 (5): 711-715, 2012 ISSN 1549-3636 © 2012 Science Publications

[2] Taghrid, I., Mohamed G., Iman S., 2009. Solving Transportation Problem Using Object-Oriented Model. IJCSNS International Journal of Computer Science and Network Security, VOL.9 No.2

[3] Papamanthou, C., Paparrizos, K., Samaras N. "Computational experience with exterior point algorithms for the transportation problem", Applied Mathematics and Computation, Vol. 158, (2004), pp459-475

[4] Zangibadi, M and Maleki, H. R. "Fuzzy goal programming for multi-objective transportation problem", Applied Mathematics and Computation, Vol. 24, (2007), pp 449-460.

[5] Amir, S., H.Z. Aashtiani and K.A. Mohammadian, 2009. A Shor-term Management strategy for Improving transit network efficiency. Am. J. Applied Sci., 6: 241-246. DOI: 10.3844/ajassp.2009.241.246.
[6] The MODI and VAM Methods of Solving Transportation Problems
[7] C. Oliveira and C.H. Antunes, Multiple objective linear programming models with interval coefficients – an illustrated overview, European Journal of Operational Research, 181(2007), 1434–1463.
[8] Ishibuchi, H., Tanaka, H., Multiobjective programming in optimization of the interval objective function. European Journal of Operational Research, 48 (1990), 219–225.

Development of an Automatic Mini-Conveyor System for Product Monitoring

J. E. Bassey*, K. C. Bala

Department of Mechanical Engineering, School of Infrastructure, Process and Engineering Technology, Federal University of Technology, Minna. Nigeria
Corresponding author email:_pajoebanet.jb@gmail.com

Abstract- This thesis presents the development of an automatic mini-conveyor system for product monitoring. It is aimed at solving relative industrial and manufacturing problems such as product pilferage, inefficient supervision, monitoring and workmanship. It is also to improve production management system, using ICT based platform to enhance monitoring control and data acquisition (MCDA) system in manufacturing process. ASSEMBLY programming language was used on a programmable logic controller (Atmel 89C52) microcontroller for control and monitoring protocol. The method includes; simulation, hardware development; to include; sensing, counting and signalling wireless system containing sensors and a wireless radio frequency transmitter and receiver. Design calculation analysis, fabrication and operational tests were carried out on the developed mini-conveyor. The operational and test result shows that, the time the product travels along the conveyor through a distance of 0.8m between entrance and exit point sensors was between 15.7sec to 16.3sec. At a distance of 0.096metres between sensors, the response to display time on LCD is approximately 1.0 second. The developed mini-conveyor system is able to remotely monitor a product on a communication distance of 22 metres in which the required information is displayed on computer screen on a telnet hyper-terminal platform, through a baud rate of 1200. Therefore, the developed machine has open opportunities for product monitoring control and data acquisition in material handling systems in the industries for more accountability.
Keywords: Mini-conveyor, Automatic, LCD, Wireless system, Sensors, Industries

1. Introduction

Material handling is a field involving the transportation, storage, and control of goods and products throughout the processes of manufacturing, distribution, consumption and disposal of all related materials [1]. The focus of the material handling industry is on the methods, mechanical equipment, systems and related controls used to achieve necessary functions. The use of conveyors in material handling has been in practice since the early 20th century and is known to be the back bone of material handling [2]. Conveyors is commonly known as a piece of equipment that moves material from one place to another and are specifically for quick and efficient transportation of wide variety of materials of all shapes and sizes [3,4]. In a production system where, large quantities of products are produced with high production rates, the products are to be transported with material handling system that is sophisticated yet safe and reliable. According to [5], "material handling looks at the problems of moving, transporting, storing materials and product; improvement of handling methods". Automated materials handling can be referred to as the management of material handling system by use of automated machineries and electronic equipments to reduce or eliminates the need for humans to do certain or all the activities in a material handling process manually. It also helps to increase the efficiency and speed by which materials or products are shipped, stored, and handled. This can significantly cut down on costs, human error or injury, and lost man hours [6, 7]. Wireless network may be referring to any type of computer network

IOP Conf. Series: Materials Science and Engineering **413** (2018) 012021 doi:10.1088/1757-899X/413/1/012021

that is connected without cables of any kind. Nevertheless, Wireless telecommunications networks are adopted and administered using a transmission system which is called radio waves [8].

This work explores the adoption of ASSEMBLY programming language instead of C++, JAVA, and FORTRAN on a Programmable Logic Controllers (PLC); Atmel 89C52 (AT89C52) microcontroller for control and monitoring protocol for the mini-conveyor. The system is developed on the back drop that in the future many production systems will need to adopt ICT in its operations with a choice to pick and adopt monitoring, control and data acquisition systems (MCADS) in the mist of diverse ICT base wireless production monitoring system that is effective, efficient and at low cost of implementation. A conveyor machine is developed on which products can be monitored during production run. There are several literatures concerning belt design with more focus on structural modification, maintenance efficiency, safety and some other research focuses on the automation area of maintenance, inspection and monitoring using different wireless sensor network (WSN) system and programmable logic controllers (PLC) structure and devices [13, 14]. There are many WSN system, PLC algorithm structural unit design and devices that are yet to be fully tested within a semi or full automated conveyor system as an alternative to the already implemented ones [20, 21]. The emphasis in recent time is on the application of ICT in manufacturing and industrial process to help improve production process and increase productivity.

2 Materials and Method

2.1 Materials

The materials used for this work includes:

Mild steel square pipes, angle iron, flat sheet, sprockets, chain and steel rollers. Other materials are PVC flexible (leather), electrical sensors, infra-ray (IR) transmitters, photodiodes, and quarter wave whip antennas.

2.2 Method

In this project, consideration was made concerning the program language adopted and the basic essential electronic and electrical devices for the development of the wireless systems used. The conveyor is developed using essential mathematical engineering models, equations, design data from belt conveyor manuals and required material properties.

The development of this project involves the following methods;

1. Electrical-electronic circuit design for the automatic counting system.
2. Simulation of electrical-electronic circuit based on the assembly language as to evaluate it performance in real life and making necessary adjustment where necessary.
3. Hardware development
4. Design analysis
5. Fabrication
6. Testing

2.3 Design analysis

The design analysis of the major components of the mini belt conveyor to achieve a functional structure as shown in Figure 1, were considered and hereby presented.

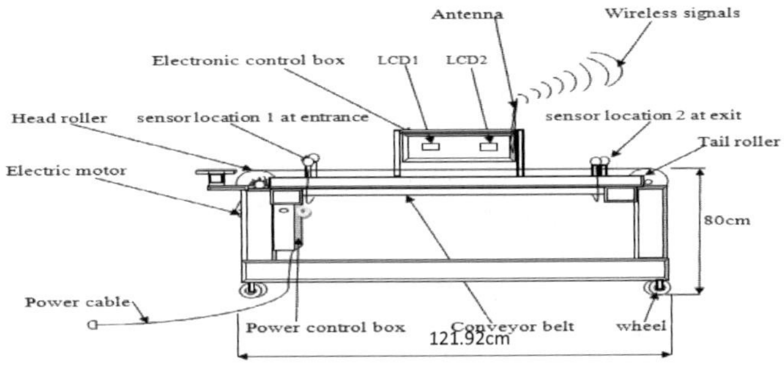

Figure 1: Machine structure

2.3.1 Conveyor Belt Design

The following are design parameters needed for conveyor machine development.

Belt capacity, C is obtained from equation (1).

$$C = 3.6 \times A_l \times V_b \times \rho_b \qquad 1$$

Where A_l = Load cross section perpendicular to belt;
V_b = Belt speed;
ρ_b = Density of conveyed material.

The belt width (W_b) as given by [14] is:

$$W_b = \frac{T_1}{\sigma_b} \qquad 2$$

Where T_1 = belt carrying side tension
σ_b = (allowable design stress of belt material)

The belt thickness (t_b) is as given from equation (3 and 4) by [13] as:

$$P_A = A_b\left(6 - \frac{\rho_b \, x \, V_b^2}{10^6}\right)\left(1 - \frac{1}{e^{\mu\theta}}\right)V_b \qquad 3$$

Where P_A= Absorbed power required by the system
A_b = belt area
6 = Design stress for the belt material
ρ_b= belt material density
V_b= Belt speed;
θ = angle of contact
μ = 0.30 (coefficient of friction for bare roller under dry condition) Fenner Dunlop conveyor manual, 2009.

$$A_b = W_b \, x \, t_b \qquad 4$$

Where t_b = Belt thickness.

2.3.2 Belt Friction and Tension

The frictional and tension factors on the conveyor belt are obtained from effective tension (T_e) from equation (4) as given by [13] are as follows:

$$T_e = F_{te} + F_l + T_{LS} \qquad 5$$

Where F_e = total empty friction;
F_l = load friction.
Similarly, the return side friction (F_r) [14] is given as:
$F_r = (F_e \, x \, M_Q \, x \, L_{cd} \, x \, 0.4 \, x \, g \, x \, 10^{-3})$ kN 6
Where: F_e = 0.020 (for horizontal conveyor equipment friction factor)
M_Q = Mass of moving parts per metre length
L_{cd} = Centre to centre distance of rollers
g = acceleration due to gravity

Therefore, total empty friction (F_{te}) is:
$F_{te} = (F_e \, x \, (L_{cd} + t_f) \, x \, M_Q \, x \, (g \, x \, 10^{-3}))$ kN 7
Where F_e = 0.020 (for horizontal conveyor equipment friction factor)
t_f = 45 (Fenner Dunlop conveyor manual, 2009)

Also the load friction (F_l) is given as [14]:
$F_l = (F_t \, (L_{cd} + t_f) \, x \, \frac{C}{3.6V_b} \, x \, (g \, x \, 10^{-3}))$ kN 8
Where F_t = 0.025 for horizontal and elevating conveyor
C = Transport or capacity of the belt conveyor

Load slope tension (T_{LS});
The slope tension is the product of the belt weight and the vertical lift and has its maximum value at the highest point (height) of the conveyor.

T_{LS} can be calculated thus as [14]:
$T_{LS} = \frac{CH}{3.6 \, x \, V_b} (g \, X \, 10^{-3})$kN 9

The effective tension from equation (4) is:
$$T_e = F_{te} + F_l + T_{LS} \quad 10$$
(The values of F_{te}, F_l and are substituted to get the right relation for T_e)

Carrying side tension, T_1
$T_1 = T_e [\frac{\xi}{e^{\mu\alpha} - 1} + 1]$ 11
Where:
ξ = 1.66 (Drive coefficient), [15]
μ = 0.30 (coefficient of friction for bare roller under dry condition) [15].
α = 180 = (angle of belt wrap or contact)
Return/slack side belt tension, T_2
$T_2 = T_1 - T_e$ [13, 14] 12

2.3.3 Absorbed Power Required by the System

Power (P_A) of the system is calculated from effective tension thus [15]:
$$P_A = T_e \times V_b \, (kW) \quad 13$$
Where V_b = Velocity of conveyor belt

Power requirement of the driving motor power (P_m) is given by [16] as:
$$P_m = \frac{P_A}{\eta} \quad 14$$

2.3.4 Design of Chain Length

The length (L) of chain for an open chain drive connecting two sprockets is obtained from the equation (11) [17] thus;

$$L = \frac{P}{2}(T_1 + T_2) + 2X + \frac{[\frac{P}{2}Cosec(\frac{180}{T_1}) - \frac{P}{2}Cosec(\frac{180}{T_2})]^2}{X} \quad 15$$

Where P = Pitch

T_1 = Number of teeth on driving sprocket

T_2 = Number of teeth on driven sprocket

X = Centre to centre between sprocket

3 Experimental Analysis on sensors

3.1 Analysis of Effect of Distance between Sensors on power Density between Sensors

Sensors distance analysis is based on power transmitted and/or power density across sensors and magnitude of the incident current generated at the photodiode. The table 1 below shows the data obtained for the effect of distance between sensors on power density between sensors.

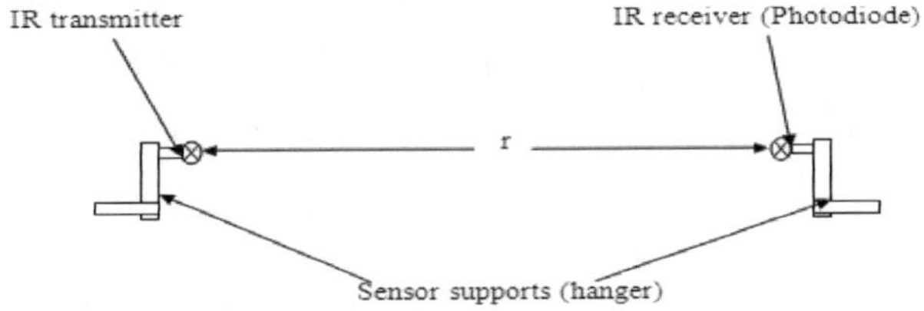

Figure 2: Distance between sensors

P_t is the power transmitted by the IR sensor, which comes with the manufacturer specification. The data for P_d is gotten from the formula;

$P_d = \frac{P_t}{4\pi r^2}$ [18] for charge carrying points 16

3.2 Effect of Distance between Sensors on Response Display time on LCD

Test was carried out to know the shortest time to display on LCD 2 as the product between the IR and photodiode sensor at the exit sensors point with reference to distance between the sensors, before signalling the data to the computer. Data obtained were tabulated and analyzed.

4 Results and Discussion

4.1 Simulation of Designed Circuit

Simulation was done on the ISIS professional animation platform, using a virtual circuit designed containing AT89C52 PLC microcontroller, on proteus 7.0 shown in figure 3 [19,20]. The components in the designed circuit layout are shown in figure 4.

IOP Conf. Series: Materials Science and Engineering **413** (2018) 012021 doi:10.1088/1757-899X/413/1/012021

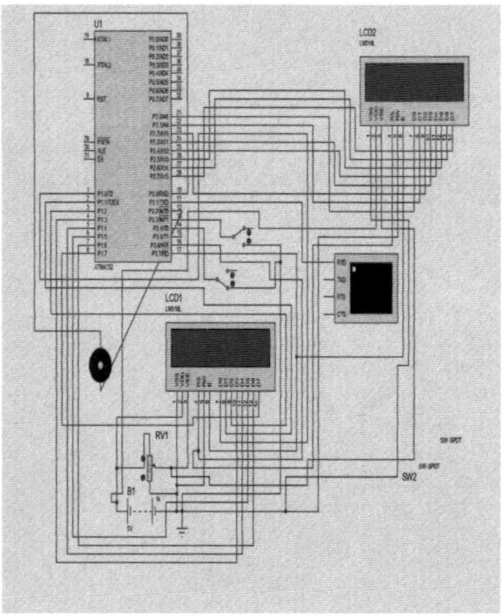

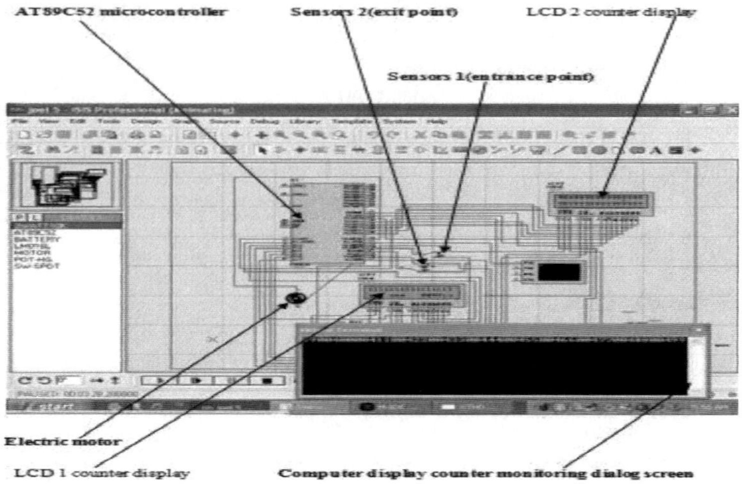

Figure 3: Circuit design on ISIS professional (Proteus 7.0 software)

Figure 4: Simulation layout (Products through first sensors (entry) on conveyor-IR system counting)

4.2 Design Results Used in the Development of the Mini-conveyor

The table1 shows essential parameters obtained from design calculations and used for the development of the mini-conveyor.

Table 1: Essential parameters used in the development of the conveyor

S/N	ESSENTIAL DESIGN PARAMETERS	VALUES
1	Conveyor length (C_L)	1.2192
2	Belt length (L_b)	2.41m
3	Center to centre distance	
	Between	1.1252m
4	pulleys/rollers(Lcd)	200mm
5	Belt width (W_b)	2.0mm
6	Belt thickness (t_b)	230mm
7	Pulley/roller Width (R_W)	50.8mm
8	Pulley/rollerdiameter (D_R)	452mm
9	chain length (L)	1.64kg
10	Belt mass (M_R)	180degree
11	Angle of contact or wrap (	3.3kg
12	α)	10Watts
13	Mass of roller (M_R) x 2	0.005m/s
14	Motor power (P_m)	0.05m/s
15	Motor speed (N_1)	379.4N
16	Belt speed (V_b)	194.9N

17	Carrying side tension (T_1)	0.8m
18	Return side tension (T_2)	
	Conveyor height (H)	14
19	Driven sprocket number of teeth (N_2)	24
	Driving sprocket number of teeth (N_1)	

4.3 System Performance Results

Plate I and II shows the conveyor on off and on mode. Plate III shows product between sensors at the entrance point, at this point the product block the photodiode from the light emitted by the IR sensor, this blockage is seen as an interruption by the photodiode sensing system and it is transformed into counting sequence by the micro-controller and is reflected as a number on the LCD 1(yellow coloured).

Plate IV shows the object or product in between the sensors at the exit point. At this point also, the product blocks the photodiode from the light emitted by the IR sensor, this blockage is seen as an interruption by the photodiode sensing system and it is transformed into counting sequence by the micro-controller and is reflected as a number on the LCD 2 (blue coloured). At this point the second microcontroller compares the count between the two sensors, and sends the count data or count imbalance error data (inform of a code number 5) through the data transmitting wireless signalling unit to the MAX 232 interfacing wireless data receiver modem.

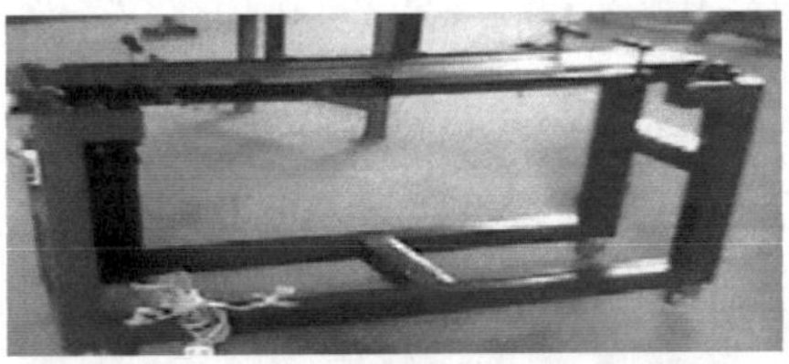

Plate I: conveyor after fabrication and finishing

Plate II: conveyor with counting electronic un with power on and Product on takeoff tray on conveyor

Plate III: Product between sensors at the entrance Point

Plate IV: Product between sensors at the exit point

The modem is connected to computer by means of a DB9 (COM 1 port) to USB conversion cable. The links that leads to the display of the monitoring dialog box on the telnet hyper terminal connectivity, on which the data received by the modem is displayed on the computer screen with window XP operating system, is as follows: Plate V shows the links and path taken to open the telnet hyper terminal

IOP Conf. Series: Materials Science and Engineering **413** (2018) 012021 doi:10.1088/1757-899X/413/1/012021

connectivity dialog box on the computer. Plate VI, shows the expanded view of the extracted the computer screen in plate V link for the display of the monitoring hyper terminal dialog box.

Taskbar → Start button → program → accessory → communication → hyper terminal

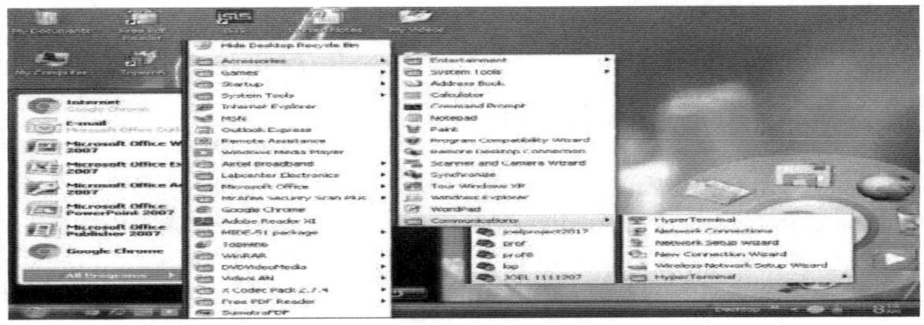

Plate V: MAX 232 data receiving wireless modem connected to a laptop the monitoring

Plate VI: The links that leads to the display dialog box on the telnet hyper terminal connectivity

Plate VII shows the hyper terminal monitoring dialog box initiated and Plate VIII shows the modem connected to computer system with count data being displayed in monitoring dialog box. The pattern of data display on the product monitoring display dialog box is show below in figure 5. The 00000 count is displayed when the modem is turned on and it is not receiving data or information from the signalling system on the conveyor. Immediately the modem connects wirelessly to the conveyor system, connection code 15 is displayed continuously on the monitoring dialog box, but immediately product or object passes between the sensors, the actual numbering begins, such as; 0000100020000300004000050006 etc, extracted from the dialog box on the screen in plate IX, and is shown in Figure 5. The computer receives this information on a telnet wireless protocol, on baud rate of 1200. The baud rate 1200 is the frequency at which the computer receives information and displays it.

Plate VII: The monitoring display dialog box initialized

Plate VIII: MAX 232 data receiving wireless modem turned on and laptop receiving data

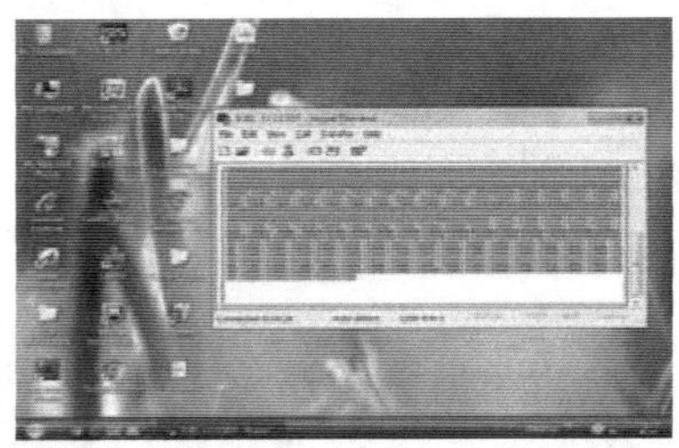

Plate IX: Product monitoring display dialog box screen extracted from computer on plate VIII

Figure 5 shows the data display before counting begins, when the modem is connected to the conveyor counting and signaling unit 0001500015and when the counting data begins, showing 00001 00001 00002 00002 00003 00003 etc.

0000000000000000000000000000000000000000000000000000000000000000000000000000000000000000
00

000001500015000150001500015000150001500015000

**0000100001000010000100001000010000100001000010000200002000020000200002000020000200002
0000200002000020000200002000020000200002000020000200002000030000300003000030000300004
0000400004000040000400004000050000500005000050000500005000050000500005000050000500005
0000600006000060000600007000070000700007000070000700007000080000800008000080000800008
0000900009000090000 9**

Figure 5: Pre-count and count data display array

4.4 Experiential Results and Discussions

4.4.1 Effect of Distance between Sensors on Power Density

Table 2, shows the data obtained by studying the effect of distance between IR and photodiode sensors on power density generated between the sensors.

Table 2: The effect of distance between sensors on power density

Power transmitted P_t (mW)	distance between Sensors, r (m)	r^2	$\frac{1}{r^2}$	power density (P_d)
8	0.10	0.0100	100.0	63.66
8	0.15	0.0225	400.0	28.29
8	0.20	0.0400	25.0	15.92
8	0.25	0.0625	16.0	10.19
8	0.30	0.0900	11.1	7.07
8	0.35	0.1225	8.2	5.20
8	0.45	0.2025	4.9	3.14

The graph obtained reveals that as the distance between sensors increases the power density deceases. This result gives a guide as to how far apart the sensors must be positioned in order to obtain optimum sensitivity when placed in their line of sight. Figure 6 reveals that power density has a great effect on magnitude of the incident current generated.

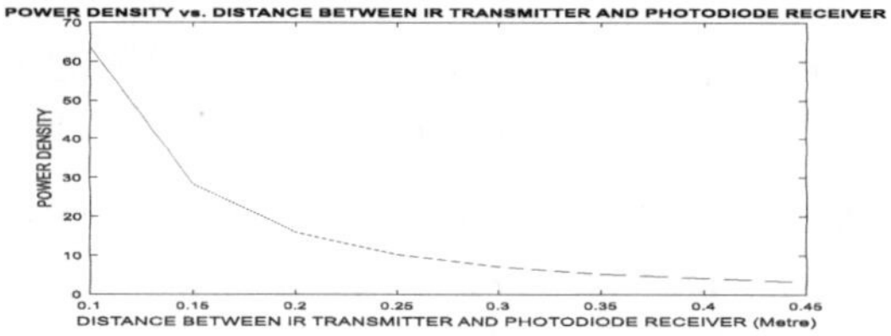

Figure 6: A graph of Effect of distance between sensors on power density

4.4.2 Effect of distance between sensors on Response display time on LCD

The Table 3 shows the values of the response display time on LCD 2 obtained by varying the distance between the IR and photodiode sensor. The graph was plotted sing MATLAB and is shown in Figure 7. This result reveals that at the of 0.096m the time it takes the system to sense the object and display the count on the LCD is approximately 1.0second. Also determined is the time it takes the product to move from sensor at entrance to sensor at exit point and was measured to be between 15.7sec to 16.3sec depending on the power in the motor from the power source. The distance between the entrance sensors and the exit sensor is 0.8m; it therefore means that the product is moving at the speed of 0.05m/s. This is the speed at which the conveyor belt is moving.

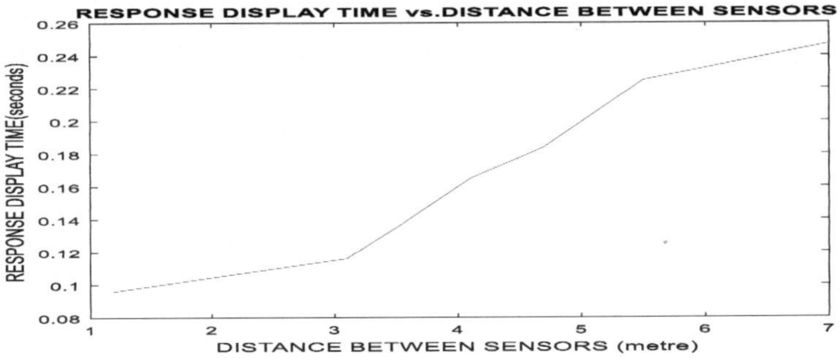

Figure 7: A graph of response display time vs. distance between sensors

4.4.3 Test Distance

During the test, the modem placed on the computer and connected to it was moved away from the conveyor machine and far away from the building in which the conveyor machine is stationed. The maximum distance at which there is no loss of connection signal was measured using tape. The distance measured is 22metres. This test reveals that, the wireless system developed will be operational at a maximum distance of at least 22metres between the factory location and the management or administration departments were the monitoring and collection of production data is carried out.

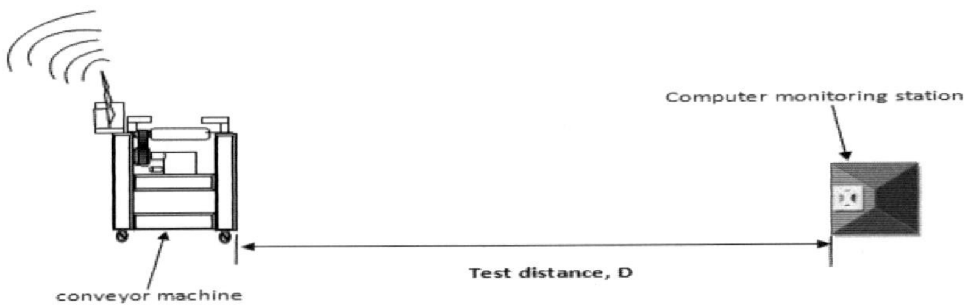

Figure 8: Distance between conveyor machine and the computer monitoring system

It was pointed out by [21], that SCADA was developed on an internet based protocols to enable monitoring and optimization of the process through the web system. However, the main objective of establishing an internet based system is to enhance rather than to replace computer based process control systems. Also, though SCADA and other systems shown in Table 4, are been used as solutions in controlling and managing processes in a wide range of industrial application. This X-system developed is a typical MCDA system and is more advantageous, because of the adoptability of an RFID module architectural structure with counter system in its wireless protocol design platform and is for a typical manufacturing based sector for monitoring, data acquisition and feedback, to reduced error in human intervention, theft reduction and elimination.

4.5 Comparison of MCDA System and other Data Acquisition Systems

The features of X-System developed with a typical MCDAS and other data acquisition systems are presented on Table 3.

Table 3: Comparison of X-System developed and a typical MCDA system and other data acquisition system

System Name	Embedded RFID and WSN	Web-based system	Low cost	Automatic monitoring compared standard time, target	M2M wireless communication	communication range	flexible design, adding modification of software and hardware
SCADA	x	√	x	x	x	-	√
ABACUS	x	√	x	x	x	-	x
BARCO	x	x	x	x	x	15m	x
MCDAS	√	√	√	√	√	50m to 200m	√
X-SYSTEM	√	√	√	√	√	22m	√

Table 4, shows a simplified comparison of few data acquisition systems renowned with the MCDA system developed. SCADA is widely known to provide monitoring and data acquisition system; it is mostly adopted in the gas and piping industry. BARCO, is known for supplying system monitoring to the industries especially for carpet manufacturing company, it uses the Bluetooth technology and has limited accessibility communication distance of about 15metres. The MCDA system is more flexible, robust and gives room to both software and hardware modification, lower cost, longer range of distance of communication and operates on a web base internet protocol using WSN system and embedded RFID than other known monitoring system. The communication distance tested for the wireless monitoring system developed X-system is 22metres. It is greater than the communication distance recorded for BARCO system. Although, the distance is less than that recorded for a typical MCDA system, it is still a typical MCDA system due to similarity in feature adaptation production line output monitoring system application in the industries.

5. Conclusion

This wireless-based monitoring control, data acquisition system, MCDAS, enable immediate update of the detail of production line activities as the production runs are being initiated from one production shift to another. For every production run, information is gathered, therefore, with this system there is no downtime in data reporting and transmission and this data can be printed at the end of every production shift. And also has data log, where previous data can still be assessed. It was also found from experimental analysis that the time it takes the product to move from the entrance sensor to the exit sensor covering a distance of 0.8m is between 15.7second to 16.3second. At a distance of 0.096metres between IR and photodiode sensors the response time to display on LCD is approximately 1.0 second. Therefore, showing that the belt round the rollers is running at a speed of 0.05m/s, which is the designed speed.

Therefore, with rapid advance in wireless information technologies, many solutions are emerging and its application in industries have been researched and developed. It can be concluded that this system which merges four technology that is; mini-conveyor, sensing and counting and signalling wireless network system into one solution is highly compatible and suitable to meet the requirement of any production line layout output monitoring system application in the manufacturing industries.

References

[1] Kay M. G., (2012). Material handling equipment. North Carolina State University. USA.

[2] Conveyor Equipment Manufacturer Association (CEMA). Belt Conveyor Calculations. Retrieved from http://www.kmg.agh.edu.pl/sites/default/files/pdf_BeltConveyorsCalculationsCEMA5_eng.pdf.
[3] Austin Trusty. (2014). Choosing the right conveyor. Retrieved from https://www.bastiansolutions.com/blog/index.php/2014/02/13/choosing-the-right-conveyors-for-your-system/#.V-AHOFsrJdg.com. On the 27th September 2016, by 2.00am
[4] Sanket, P and Thakur, A. G. (2015). A review Paper on Redesign of Gravity Roller Conveyor System for Weight Reduction through Optimization. *International Journal of Scientific and Engineering Research.* Vol 6(2). ISSN 2229-5518. http://www.ijser.org
[5] Abu, J. O. (2011). Manufacturing Engineering and Production Management. Markurdi, Benue State, Nigeria. Aboki Publishers. 223-228.
[6] Wise Greek (2010). What is Automated Materials Handling? http://www.wisegeek.com/what-is-automated-materialshandling.html. Accessed: 14th April 2010.
[7] Salawu, E. Y., Okokpujie, I. P., Afolalu, S. A., Ajayi, O. O., & Azeta, J. (2018). Investigation of production output for improvement. *International Journal of Mechanical and Production Engineering Research and Development*, *8*(1), 915-922.
[8] Madhav and Anagha (2012). Wireless Sensor Network: A Promising Approach for Distributed Sensing Tasks. *Excel Journal of Engineering Technology and Management Science.*
[9] Swartz, R.A., lyunch, J. P., Sweetman, R., Rolfes, R. and Zerbts, S. (2008). Structural Monitoring of Wind Turbine Using Wireless Sensor Networks. *Environment*, 6(3), 1-8.
[10] Jiang, X., Dawson-Haggert, S., Dutta, P. and Culler, D. (2009). Design And Implementation of a High-fidelity ac Metering Network. International Conference on Information Processing in Sensor Networks, IPSN (09), *IEEE Computer Society,* Washington, DC, USA, 253-264.
[11] Dijwaard, Kleumen, Havinga, Kheiboer and Bijl. (2011). Industry: Using Dynamic WSNs In Smart Logistics For Fruits And Pharmacy. *In Sense System,* 218-231.
[12] Kanmani, M., Nivedha, J. and Sundar, G. (2014). Belt Conveyor Monitoring and Fault Detecting Using PLC And SCADA. International Journal of Advanced Research in Electrical Electronics and Instrumentation Engineering, *(IJAREEIE),* 3(4), 234-248. Retrieved on October, 2016 from www.ijareeie.com
[13] Kaushal, P. S., Priyanka S. (2014). Design of an Automated sorting of object rejection and counting machine. *International Journal of Emerging Trends in Engineering and Development*, 3(4), 252-264. Retrieved from http://www.rspublication.com/ijeted/ijeted_index.htm, October 2016.
[14] Okokpujie, K. O., Odusami, M., Okokpujie, I. P., & Abayomi-Alli, O. (2017). A Model for Automatic Control of Home Appliances using DTMF Technique. *International Journal of Scientific & Engineering Research*, *8*(1), 266-272.
[15] Deshmukh, P. M., Trikal, S. P., (2015). Design, Analysis and Failure of Actual Charging Belt Conveyor System Used in the Industry to Set the Optimum Results. *International Conference on Recent Trends in Engineering Science and Management*, 2377- 2382. Retrieved from http://www.conferenceworld.in. October, 2016
[16] Fenner Dunlop conveyor hand book (2009). Retrieved from http://www2.hcmuaf.edu.vn/data/dangnh/file/5_Fenner%20Dunlop_%202009_%20Conveyor%20 Handbook.pdf.
[17] Ananth, K. N., Rakesh, V. and Visweswarao P. K. (2013). Design and Selecting the Proper Conveyor-Belt. *International Journal of Advanced Engineering Technology (IJAET). Vol. 5(2). 43-49.*
[18] Okokpujie, I. P., Okokpujie, K. O., Ajayi, O. O., Azeta, J., & Obinna, N. N. (2017). Design, Construction and Evaluation of a Cylinder Lawn Mower. *Journal of Engineering and Applied Sciences*, 1254-1260.
[19] Khurmi, R.S., Gupta, J.K. (2012). Theory of Machines. Ram Nagar, New Delhi. S. Chand and Company Publishing Ltd.

[20] Adediran, Y. A. (1997). Telecommunication Principles and System. Minna, Nigeria. Finom Associates Publisher.
[21] Okokpujie, I. P., Okokpujie, K. O., Nwoke, O. N., & Azeta, J. (2018). Development of A 0.5 Kw Horizontal Axis Wind Turbine. Journal of Engineering and Applied Sciences, 13(8), 2202-2208.
[22] John J. (2016) Proteus PCB Design and Simulation Software: Introduction. http//www.circuitstoday.com/proteus-softwares introduction. Retrieved 16th February, 2017.

IOP Conf. Series: Materials Science and Engineering **413** (2018) 012022 doi:10.1088/1757-899X/413/1/012022

Nigerian Rain Gauge Station Optimization and National Development: The importance of Head Count

Ngene Ben U[1]* and Obianigwe Njideka[1]

[1]Department of Civil Engineering, Covenant University, Ota
*Corresponding author email: ben.ngene@cu.edu.ng

Abstract. Effective planning for water resources infrastructures requires accurate and up to date data of the water source. The aim of the research is the optimization of the number of rain gauges required for planning for the volume of water in a watershed which is an absolute necessity for the economic development of the nation. To achieve this, the statistical tolerable error method of non-weighted approach was used to determine the optimum number of rain gauges required for an area that efficiently estimates the rainfall of the watershed. The result obtained indicates that Nigeria require a minimum of 1057 rain gauges to cover its land area of 923300 Km^2 and meet the World Meteorological Organisation (WMO) requirement for the Tropical, Mediterranean and Temperate zone. However, Nigeria has 87 existing and operating rain gauge stations presently. This number is grossly inadequate and need to be increased substantially by the opening up of extra 970 stations to meet the minimum requirement of WMO and hence enhance national development through proper water resources planning.
Keywords: rain gauge, watershed, rainfall, network density, statistical accuracy and optimum

1. Introduction

The number and spread of rain gauge network are important parameters of measuring the development of a nation. This is because the effective use of a water resource of a country is dependent on accurate estimation of the amount of rainfall and use of such data in nature planning. Areas of interest in planning of water resources include: Agricultural practices, irrigation, forestry, fisheries, tourism and disaster management especially during flooding, construction activities, flight operations and navigation.

Critical factor to the use of precipitation data according to Mutreja, 1990 as stated by [2] is to have a properly designed network of rain gauges in the watershed to collect the data. This translates to the planning for adequate number required in a given area that will help capture the necessary information for reliable analysis. Network density is therefore an absolute necessity in water resources management as it provides estimates of the expectation of rainfall that will be needed for all the functions mentioned above. Wekena, in his work in 2006 had pointed out that expansion of irrigation development for examples without the knowledge of the water sources will have the negative effect of reduction of agricultural yield and water stress. World Meteorological Organisation report WMO, in 1965 has set the standard requirement of minimum number of gauges stations for various regions of the world. The standard is to ensure accurate measurement of rainfall in a catchment that will enable proper planning and development of nation's water resources. Where there is a shortfall as it is currently the case in Nigeria, effective development and management of the country's water resources that is on scale commensurate with expectation will be impaired.

Figure 1 shows the current location of rain gauge stations in Nigeria according to records obtained from the Nigerian Meteorological Agency.

Fig. 1 Nigeria rain gauge stations (source: Ngene, 2009)

The general lack of proper record keeping, storage and retrieval system in the country is the greatest challenge facing a study of this nature. This work therefore assesses the adequacy of the existing rain gauge network in Nigeria taking cognizance of the WMO requirement and make suggestion as the adequacy or otherwise of the current number.

As noted by [6]; [7], every aspect of Nigeria development as indeed all nations of the world is closely tied to adequate supply of water. The consequences of low agricultural output, limited water supply, inadequate water reserve for power generation and currently the challenge of Herdsmen farmers' conflict in the country can attest to the need for planning for what is available. Proper management of available water resources in Nigeria lead the Federal Government in 1976 to establish the River Basin Development Authorities by decree no 25, 87 of 1979 and Act 7 of 1981. Table 1, shows the 12 River Basins in Nigeria with the head office location. The grouping of states was to ensure that areas with similar geographical and hydrological attributes will harness what is common to them.

Table 1: River Basin Development Authorities in Nigeria and their Headquarters Locations.

S/N	NAME OF RIVER BASIN	HEADQUARTER LOCATION	STATES COVERED
1	Anambra / Imo River	Aba-Owerri Road Agballa-Owerri	Anambra, Imo, Enugu Abia, Ebonyi
2	Benin-Owena Basin	Benin - Warri Road, Obayaton Benin City	Edo, Part of Delta North, Ondo, Ekiti
3	Chad Basin	Maiduguri-Marte Road, Maiduguri	Borno, Yobe, and Part of Adamawa
4	Cross River Basin	Murtala Muhammed Way, Calabar	Cross River, Akwa Ibom
5	Hadejia-Jamaáre	Maiduguri Road, Hotoro, Kano	Kano, Jigawa, Parts of Bauchi
6	Lower Benue River	Oturkpo Road, Makurdi	Benue, Plateau, Nasarawa and Part of Kogi
7	Lower Niger River	GRA Ilorin	Kwara, Kogi
8	Niger Delta Basin	Isaac Boro Park, PH	Rivers, Bayelsa, Parts of Delta
9	Ogun-Osun River	Abeokuta	Lagos, Oyo, Ogun, Osun
10	Sokoto – Rima	Guzan Sokoto Road, Sokoto	Sokoto, Kebbi, Zamfara, Katsina
11	Upper Benue	Yola-Fufore Road, Yola	Adamawa, Taraba, Gombe Parts of Bauchi State
12	Upper Niger	Bosso-Kontagora Road, Minna	Niger, Kaduna and FCT

Source: Ngene, 2009

The stated objective development of the nation water resources infrastructure is still not being effectively realized four decades after for reason that is outside the scope of this work.

Combining the efforts of the River Basins, Nigerian Meteorological Agency, State Agricultural Development Agencies, Schools (as it used to be years back) and other Research Institutes should give Nigeria the required head counts in terms of rain gauge station numbers to meet the WMO standards. No coordination amongst these organisations, have ensured that the Nigeria dream of having an efficient network of rain gauges remains a dream. According to the WMO 1965 report in the study by Ngene *et al* in 2015b, Nigeria which is in the Tropical, Mediterranean and Temperate zone require to cover 600-900 Km^2/gauge. With an existing and operating rain gauge number of 87, covering an area of 923300 Km^2, this translates to 10600 Km^2/gauge. To minimally capture Nigeria rainfall volumes to meet WMO standard for proper planning, the country will need the injection of minimum of 970 gauge stations according to [5].

Planning with the small number that is currently available to the country today accounts for flood devastations in the country, roads and rail lines being washed away, failures of waterway programmes other water resources infrastructure and generally errors in economic analysis and development as witnesses in the nation. The drought of the nineteen seventies that swept most of the Sahel region of Africa, including Nigeria, has left the region with general water resources scarcity [8]. The consequence of this is the low Agricultural out-put, limited water supply and inadequate water reserve for power generation that the region is faced with [5].

2. Methodology

Precipitation is the moisture that falls from the atmosphere to the earth's surface in form of rain, snow, hail, sleet, mist etc. which is produced from the water vapour present in the air [8]. Rainfall constitutes the major form of precipitation in Nigeria in the absence of snow and other forms. The installation and observation of rain gauges in the country is controlled by the Federal Ministry of Aviation through the Nigeria Meteorological Agency (NIMET). The River Basin Authorities and Agricultural Research based organisations collaborate with NIMET in terms of data acquisition, record keeping and dissemination. NIMET has been collecting, archiving, interpreting weather and climate information for many years. To do this, it uses three forms of rainfall measuring instruments such as recording rain gauges, non-recording rain gauges and weather radar where available.

2.1 Manual Measurement

These are mainly with non-recording type of rain gauge. Any receptacle with vertical sides may serve as a rain gauge, but in order to permit more accurate observations certain refinements are necessary [9]. The standard rain gauges consists essentially of a funnel of special design supported upon an outer vessel, partly sunk into the ground within which is placed a receiver to collect the rain entering the funnel. On the rain gauges water collected by the funnel runs into a bottle standing in a copper can. The capacity of the bottle is sufficient for most occasions, but if the bottle overflows, the can retains the surplus for measurement

2.2 Automated Measurement

Automated rain gauges can be in form of recording gauges, weather Radar satellite and sensors.

(i) **Recording or Automatic Rain Gauges.** Three types of these are: (1) Tipping buckets rain gauge (2) weighing bucket rain gauges and (3) siphon rain gauges. It is often important to know, not only the total amount of precipitation, but also its intensity, or rate of fall. Recording gauges give a continuous record of precipitation in the form of a pen trace on a clock driven chart.

(ii) Weather Radar: The radar is based on the principle of echo sounding. High frequency electromagnetic waves are sent out which travel at the speed of light. An extremely small portion of this energy is reflected by objects in the sky and is detected by the radar. By calibration of the echo intensity with rainfall, it is possible to measure the intensity of rainfall [1]. Weather radar can be used to find out rainfall in hill, headwater reaches of streams where rain gauges cannot be installed due to in accessibility of the sites.

(iii) Automatic Weather stations (AWS): Recent advances in technology have led to the establishment of automatic weather stations such as the Australian Bureau of meteorology, automatic weather stations. The AWS are designed to serve the dual purposes of providing real time data for forecasting, warning and information services, as well as high quality data for the climate data base [10].

(iv) Sensors: Sensors such as the 1-wire analogue-to-digital converter (ADCs) can be used to measure a wide range of environmental properties or parameters by converting the signals locally and transmitting the digital data over a common communication link. Wind speed, rainfall may be measured by an ADI computer. Each sensor has a unique ID address by which the master keeps track of it and the environmental parameter it is measuring. Sensor-specific or calibration information may be stored in the ADI's memory or eventually, by downloading from the web [11]. The information contained in the ADI may also be sent to the Automatic weather station database.

(v) Weather Satellite: Information on weather, rainfall can also be collected via satellite orbiting the earth. A satellite is a platform for remote sensing instruments. The instruments, which may be measuring surface or atmospheric temperature or water vapour amounts in various atmospheric layers or simply making a digitized image of cloud and the surface actually senses the radiation emitted or reflected by the object under investigation. The great advantage of satellite measurement is that it provides access to remote areas with no alternative data sources and that they give continuous global coverage [12].

In order to establish the adequacy and efficiency of the number of rainfall measuring instruments, a network design of the number and location of the gauges is required. The design is accomplished through either (a) weighted approach or (b) non-weighted approach.

(a) Weighted Approach

In the method, according by Shih, in 1982, the accuracy in mean estimation and rain gauge density is achieved by the use of tolerable error method. The mean rainfall over an area is established using one of (i) Random sample (ii) Stratified sample without optimum allocation of gauges to stratum and (iii) Stratified sample method with optimum allocation of gauges to the stratum. Weighting is done using either special stratum weighted ratio, used for relocation of rain gauges to data or the weight of each stratum, that is, ratio of stratum area to the total catchment area.

(b) Non-weighted Approach

The method relates number of gauges to either the variance of the rainfall, coefficient of variation, correlation coefficient between stations or error of estimation. Mutreja in his work in 1990, used the monthly total rainfall record in a network of gauges that were fairly uniformly distributed, the coefficient of variation of the mean (C_v) for each month is given as $C_v(C') > 5\%$ to determine the adequacy of the network where C' is less than 10, but

$$N = \left(C'/_{10}\right)^2 n \quad (1)$$

is used instead if C' is more than 10

Where n represent the number of stations for which records are available.

A second non-weighted approach by [15] reported a simple method that relates the number of gauges and the coefficient of variation with the error to be tolerated as

$$N = \left(C_v/_e\right)^2 \quad (2)$$

3. Results and Discussion

From table 2, optimum number of gauges at statistical accuracy for Anambra/Imo River Basin, it is shown that depending on the statistical accuracy desired, the optimum number of rain gauges for an area can be determined. In the table, Anambra/Imo River Basin is used as a case study, where it shown that the number of existing gauges is 14, but the existing and operating station is 8. The area covered by the River Basin represents 3% of the Nigeria land mass or 26175 Km^2. Therefore the existing and operating eight rain gauge stations will translate to 3272 Km^2/gauge which is far lower than the WMO required minimum of 600-900 Km^2/gauge.

For optimum design of network for Anambra/Imo River Basin, a tolerable error with statistical accuracy required is 10% to give 42 gauge stations. The 42 gauge stations gives the region a required minimum area per gauge of 623, however, the economic challenge of providing this optimum number will need to

IOP Conf. Series: Materials Science and Engineering **413** (2018) 012022 doi:10.1088/1757-899X/413/1/012022

be given serious consideration. It is noted that the deficiency between what is existing and what is required is high, therefore to enhance the economic development of the nation, the difference need to be provided since the value is actually the minimum requirement.

Table 2: Optimum number of gauges at statistical accuracy for Anambra/Imo River Basin

	Number of Existing and Operating Raingauge Station	Statistical Accuracy	Zone I No of Gauges Required	Zone II No of Gauges Required	Zone(I + II) Optimum Design
Confidence Level	8	80%	7	6	13
		90%	9	8	17
		95%	11	10	21
Tolerable Error	8	10%	26	16	42
		15%	12	7	19

Source: Ngene, 2009

The impact of using the current Nigeria number of existing and operating rain gauges at 87 is against a projected optimum minimum value of 1057 gauges will be a risk of failure error above 14%. The increase in number to the desired minimum of 1057 will reduce the error risk to 4%. Figure 2 below shows error risk against number of gauge station and from the figure and table 2, it is seen that the lower the tolerable error the higher the number of gauges to be deployed to an area.

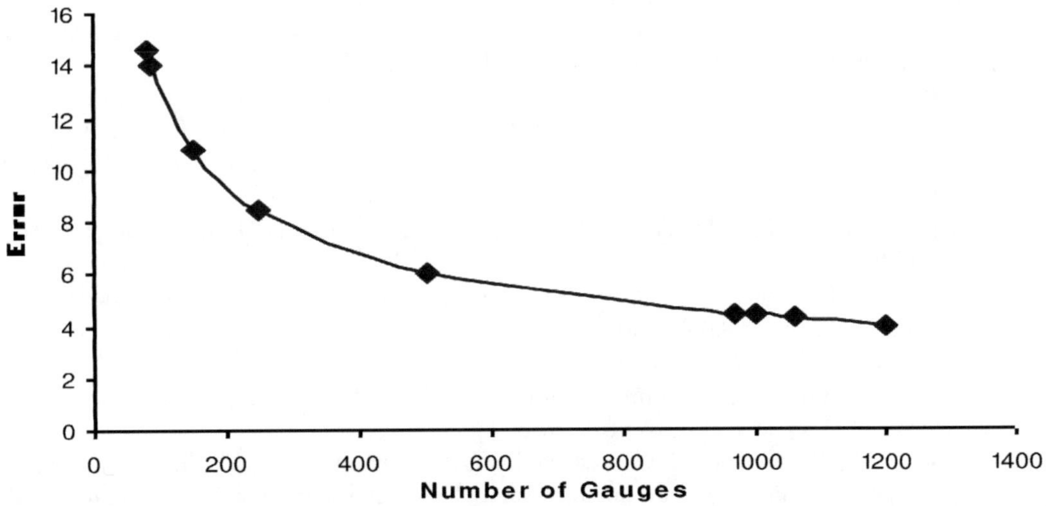

Fig 2: Number of Gauge against Network Error

The economic and financial issues involved must be given adequate consideration in a developing economy such as in Nigeria. There is no doubt that the Nigeria Rain Gauge Stations need dramatic number increase to make way for higher efficiency and reliability in qualitative data collection and use for rainfall measurement. The increase must however be weighed in with available funding options.

A typical example of this is the design of a culvert at 4% risk of error of failure and design life of 15 years will give an estimated return period of flood over the culvert at 368 years. However if the risk is increased to 14% over the same design life the return period of flood will be reduced to 100 years. This

IOP Conf. Series: Materials Science and Engineering **413** (2018) 012022 doi:10.1088/1757-899X/413/1/012022

gives a 4 times probability of occurrence hence more likelihood of failure within the design life of the structure.

4. Conclusion

Nigeria obviously need number (head count of rain gauge stations) increase in its rain gauge station for a reliable design of structures that require the use of rainfall information in the country. The increase is realizable through improved budgetary provision and utilization of funds meant for this purpose through the Nigerian Meteorological Agency, 12 River Basin Development Authorities, Agricultural and Research based agencies of government that need the data and private organizations such as Universities, Schools and Colleges and individual farmers.

The colonial Schools of the old had such plants as part of the School system for effective teaching and learning, the need for it today is even more compelling because of the importance of accuracy in planning for national development.

Reference

[1] Mutreja, K. N. Applied Hydrology. Tata Mcgraw-Hill pub. Co ltd. India, 1990.

[2] Ngene, B. U., Agunwamba, J. C., Nwachukwu, B. A and Okoro, B. C. Comparing Network Design Approaches in Areal Rainfall Estimate of Nigeria River Basins. ARPN Journal of Engrg. and Applied Science, Vol. 10, No. 16, Sept. 2015 (b), pp 7033-7042, ISSN 1819-6608.

[3] Wekena T.O. Operational analysis of the Cascaded Wadecha Belbela Reserve system. M.Eng. Thesis, Arbaminch University, Ethiopia, 2006. (Sourced from the Net).

[4] World Meteorological Organisation. Guide to Hydrometeorological Practice. WMO, Geneva, 1965.

[5] Ngene B. U. Optimization of Rain Gauge Stations in Nigeria. (Unpublished Ph.D. Thesis). Federal University of Technology, Owerri, 2009.

[6] Revenga, C. and A. Cassar. Freshwater trends and projections: Focus on Africa. A Report Prepared for World Wide Fund for Nature (WWF) International, 2002. pp: 4.

[7] Sonuga, J.O. Planning for water supply without adequate data. Proceedings of the International Drinking Water and Sanitation Decade (1981 - 1990) Workshop, Kano Nigeria, 1981.

[8] Offordile, M. E. An Approach to Groundwater Study and Development in Nigeria. Mecon Services Ltd., 1992, P. 69-70.

[9] Linsley, R. K.., Kohler, M. A. and Paulhus, J. L. H. Applied Hydrology. McGraw-Hill New York, 1949.

[10] Bureau of Meteorology – Automatic Weather Station. Automatic Weather Station for Agricultural and Other Application, 1995.

[11] Awtrey, D. "I-Wire Addressable Digital Instrument" Sensors: May 2001

[12] Buba, L. F. Climate Data and Information Technology: The challenges. Journal of the Nigeria Meteorological Society, 2002, Vol. 3, No 2, pg. 53 – 59.

[13] Ngene, B. U., Agunwamba, J. C., Nwachukwu, B. A and Okoro, B. C. The Challenges to Nigeria Raingauge Network Improvement. Research Journal of Environmental and Earth Sciences, Vol. 7, No. 4, Nov 2015(g), pp 68-74, ISSN 2041- 0484; e-ISSN 2041- 0492.

[14] Shih S. F. Rainfall Variation Analysis and Optimization of Gauge Network. Water Res. Research, 1982, 18(4): 1269-1277.

[15] Raghunath H. M. Precipitation. Hydrology Principles, Analysis and Design. Wiley Eastern Limited, 1986, pp. 16-35

IOP Conf. Series: Materials Science and Engineering **413** (2018) 012023 doi:10.1088/1757-899X/413/1/012023

Soil Stabilization for Road Construction: Comparative Analysis of a Three-Prong Approach

Obianigwe Njideka[1*] and Ngene Ben U[1]

[1]Department of Civil Engineering, Covenant University, Ota
*Corresponding author email: ben.ngene@cu.edu.ng

Abstract. The paper aim to assess and compare the stabilizing effect of three materials, cement, sodium chloride and brick dust on clay soil found at locations during road construction. To achieve this, the powder forms of samples of each of the three materials are mixed to clay soils in various percentages of 2%, 6%, 10% and 14%. The moisture content, specific gravity, sieve analysis, atterberg limit tests were carried out to classify the soil using the AASHTO classification system. Based on the results obtained the soil sample corresponds to group A-2-7 (Clayey Sand) and poorly graded sand with clay (SP-SC) soil type for use as drainage and subgrade material. According to the classification, this is acceptable for stabilization of soil. Specific gravity, sieve analysis, atterberg limit, compaction, California bearing ratio (CBR), swell index were carried out on the soil with and without the addition of (cement, sodium chloride, brick dust). The results showed improvement in the maximum dry density value on addition of cement, sodium chloride with gradual increase from 2% to 14%, the brick dust content showed gradual increase up to 6%and after which it started to decrease at 10% and 14%. The three materials are therefore recommended as stabilizing agents for use in road construction where clay soils play a significant part in early deterioration of this important fulcrum of Nigeria national development.
Keywords: soil stabilization, silty clay soil, cement, sodium chloride and brick dust

1. Introduction

Soil may be defined as the natural loose surface material of the earth that supports plant growth when dug or ploughed. Clay is also a natural material composed mainly of fine grain materials that consist of tiny particles that have plastic and adhesive properties. Clay possesses small voids and pores which make it possible to retain water and water tend to cause the expansion and shrinkage, which could lead to settlement.

When exposed to water in high quantity, clay tends to soften and liquefy and this property of clay makes construction difficult due to its low strength and stiffness. This has posed serious challenges in geotechnical engineering because weak soil may cause damage to the foundation of buildings and cause cracks to buildings and along road pavement.

Soil stabilization may be defined as the alteration or preservation of one or more soil properties to improve the engineering characteristics and performance of the soil. Soil stabilization generally refers to the procedure in which a special soil, cementing material, or other chemical materials are added to a natural soil to improve one or more of its properties. Stabilization therefore looks at the various methods employed for modifying the properties of a soil to improve its engineering performance. One may achieve soil stabilization through mechanically mixing the natural soil and stabilizing material together so as to achieve a homogeneous mix or by adding stabilizing material to

undisturbed soil deposit and obtaining interaction by letting it permeate through soil voids. When soil is stabilized by use of additives it improves the properties of less-desirable road soils. When used these stabilizing agents can improve and maintain soil moisture content, increase soil particle cohesion and serve as cementing and water proofing agents.

Since foundation of any structure including road is a critical aspect of the structure in terms of load transfer to the earth, it is usually a difficult problem in civil engineering works when the sub-grade is found to be clay soil. Soils having high clay content have the tendency to swell when their moisture content is allowed to increase. The most common methods of soil stabilization of clay soils in pavement work are cement and lime stabilization. Lime or calcium carbonate is oldest traditional chemical stabilizer used for soil stabilization.

The work presented in this paper is an addition to the application of chemical stabilization technique, by comparing the use of sodium chloride (Nacl), cement, brick dust at different percentages to the clay soil gotten from in a borrow pit in a University environment in Ota, Ogun state. The concentrations by weight of each of the stabilizers used are 2%, 6%, 10%, and 14%. The sodium chloride (Nacl), cement and the brick dust used were all open market purchased from Abeokuta.

To underscore the challenge, it is necessary to understand that Nigerian roads are characterized with potholes, longitudinal cracks and lots of other pavement defects known to road construction. In Nigeria you can hardly travel a mile without coming across potholes which have contributed to greatly to the high rate of accidents occurrence and a dip on the nation's economic development. All roads designed and built have a stipulated design life but roads most times fail long before the expected date, some fail immediately after construction others after heavy flooding, some stay up to half of its life cycle while others last to its full life expectancy with proper maintenance. A major cause of these failures is the increase in water content of granular subgrade materials. Since soil stabilization is generally any method used to improving the ability of the soil to carry load, California Bearing Ratio (CBR) values of soil and hence render it able to resist the loads and adverse effects of construction activities, therefore, this study is to conduct tests to ascertain the best additive amongst cement, sodium chloride or brick dust that improves the geotechnical properties of clay soil.

The use of waste products or by products of industrial production is now being used as it helps to reduce carbon foot print. This justifies the use of brick dusts as alternative material to cement and salt in soil stabilization when it is understood that it contribute to saving of time used for total excavation and replacement as was done in the past in case of weak soils. However, before decisions are made to stabilize soil for Engineering purpose, it is necessary to understand the nature of the soil by carrying out prescribed tests on the soil sample.

(a) SOIL

Soil can be looked at as consisting of a wide range of relatively smaller particles derived from a parent rock through mechanical weathering process that includes air and/or abrasion, freeze-thaw cycles, temperature changes, plant and animal activity such as burrowing and by chemical weathering process that include oxidation and carbonation[1]. Soil can also be defined as the solid material on the Earth's surface that results from the interaction of weathering and biological activity on the parent material or underlying hard rock. Different people define soil to suit their purpose. To a geologist it represents the products of past surface processes; to a pedologist it represents currently occurring physical and chemical processes; to an engineer it is a material that can be built on (foundations to buildings, bridges), building of (tunnels, culverts, basements), building of (roads, runways, embarkments, dams) and supporting (retaining walls, quays) [2]. It was also noted by Sabat in 2012 that for an Engineer, there are a number of techniques available to improve the engineering properties of soil to make it suitable for construction purposes. These include stabilization using dust/powder like waste materials with and without a binder like lime, cement etc. Others include Quarry Dust [3]; marble dust [4], [5], [6], [7]; baryte powder [8]; pyroclastic dust [9] and brick powder [10] are some of the prominent dust/powder like waste materials which have been successfully utilized for stabilization of expansive soil.

(b) CLAY SOIL

Clay has the smallest particle size of any soil type, with individual particles being so small that they can only be viewed by an electron microscope. This allows a large quantity of clay particles to exist in a relatively small space, without the gaps that would normally be present between larger soil particles. Because of the small particle size of clay soils, the structure of clay-heavy soil tends to be very dense.

(c) CLAY MINERALOGY

Clay refers to naturally occurring material composed mainly of fine- grained minerals, which is generally plastic in nature at appropriate water contents and will harden when fired or dried. The minerals found in clay are generally silicates less than 2 microns (one millionth of a meter) in size, about the same size as a virus. Clay is very abundant at the earth surface therefore they tend to form rocks known as shales and are a major component in nearly all sedimentary rocks. The small size of the particles of clay are their unique crystal structures giving clay minerals special properties, including cation exchange capabilities, plastic behavior when wet, catalytic abilities, swelling behavior and low permeability.

Clay minerals are hydrous silicates, largely of aluminum, magnesium, and iron that on heating lose adsorbed and constitutional water and yield refractory material at high temperatures. Plasticity is a characteristic of clay minerals and is largely due to an affinity of the clay surface for water, resulting from a net negative charge on the surface of a clay particle that causes it to adsorb water and other fluids. Clay minerals can therefore profoundly affect a soil's engineering behavior, even when present in small quantities. As the clay content of a soil sample increases, the influence the clay fraction will have on the behaviour of the sample also increases. The strong influence of clay minerals on soil behaviour can be illustrated by the addition of bentonite to a granular soil. Bentonite is a clay mineral composed largely of the clay mineral sodium montmorillonite. In 2012, Sabat conducted series of test and concluded that the addition of brick dust decrease liquid limit, plastic limit, plasticity index, optimum moisture content, maximum dry density and angle of internal friction of clay soil.

In 2012, Mukesh and Patel studied materials and soils derived from the Feuerletten (Keuper) and Amaltheenton (Jura) formations along the new Nuernberg Ingolstadt railway line (Germany) and mentioned earlier work by Ismail in 2004 on this. The work included petrological, mineralogical studies and scanning electron microscope materials related to road construction using lime (10%), cement (10%), and lime/cement (2.5%/7.5%). It was determined that consistency limits, with proper compaction, that by increasing the optimum moisture content (%) of the treated soil, the density (g/cm^3) decreased. The cohesion and the friction angle of the improved materials increased for all the treated mixtures. In case of the lime materials, uniaxial strength increased strongly using lime and cement together. For the Amaltheenton formation, uniaxial strength increased strongly with cement alone. The author also noticed that the loss of weight during freezing and thawing test was low and depended on the material type.

The work by Ampera and Aydogmust in 2005 treated Chemnitz clayey soil (Association of State Highway and Transportation Officials (AASHTO) using lime at (2, 4, and 6%) stabilization and cement at (3, 6, and 9%). The authors conducted compaction-unconfined compressive strength and direct shear test on untreated and treated specimens. They concluded that the strength of cement treated soil was generally greater than the lime treated soil.

Also the work by Tamadher and Mohamed in 2014, aimed to investigate the effect of adding chloride salts including (NaCl, $MgCl_2$, and $CaCl_2$) on the engineering properties of silty clay soil. Three amount percentage of salt (2%, 4%, and 8%) were added to the soil to study the effect of salts on the compaction characteristics, atterberg limits, and unconfined compressive strength. The results showed that the increase in the percentage of each of the chloride salts increased the maximum dry density and decreased the optimum water content. The liquid limit, plastic limit, and plasticity index decreased with increasing salt content. Also the results showed that the unconfined compressive strength increased when the salt content increased too.

IOP Conf. Series: Materials Science and Engineering **413** (2018) 012023 doi:10.1088/1757-899X/413/1/012023

The researchers listed above investigated the use of cement, salt and brick dust at between 2 – 8% in soil stabilization, the current effort therefore extended the argument up to 14% to observe the effect of such increase on the engineering properties of soil.

(d) Cement

Cement has been described as a finely ground inorganic material [14], which when mixed with water, form a paste which sets and hardens by means of hydration reaction and processes and retains its strength and stability even under water. Ordinary Portland cement is the commonest type in use. The raw material from which it is made is lime, silica, alumina and iron oxide. These constituents are crushed and blended in correct proportions and burnt in a rotary kiln. The clinker is cooled, mixed with gypsum and ground to a fine powder to give cement. The main chemical compounds in cement are calcium silicates and aluminates.

Cement is highly used as a stabilizing material for soils, particularly for the construction of highways and earth dams. It can be used to stabilize sandy and clayey soils. As in the case of lime, the cement has an effect to decrease the liquid limit and to increase the plasticity index and workability of clayey soils

(e) Salt

Soil water salinity can affect soil physical properties by causing fine particles to bind together into aggregates. This process is known as flocculation and is beneficial in terms of modifying the required engineering properties of the soil. Taking into consideration the abundance of salt and economic viability, the use of sodium chloride is preferable. Sodium has the opposite effect of salinity on soils. The forces that bind clay particles together are disrupted when too many large sodium ions come between them. When this separation occurs, the clay particles expand, causing swelling and soil dispersion. In a study by Işık and Berrin in 2009, it is stated, that the performance of the gypsum act as an additive for the treatment of the clay soils by means of swell potential and strength.

The effect of adding chloride salts including ($NaCl$, $MgCl_2$, and $CaCl_2$) on the engineering properties of silty clay soil has been studied. The results showed that the increase in the percentage of each of the chloride salts increased the maximum dry density and decreased the optimum water content. The liquid limit, plastic limit, and plasticity index is known to decreased with increasing salt content. Also the results showed that the unconfined compressive strength increased when the salt content increased too.

(f) Brick dust

Brick dust is known to increase the amorphous silica and alumina present in soil especially as burnt clay dust. According to [16], clay bricks when fired at a temperature below 950°C if ground into fine powder results in a pozzolanic reaction when combined with lime because of the presence of silica and alumina. It is however necessary to note that raw clay on its own has no pozzolanic value. A general rule is that clay, though the composition is source specific, should contain 20-30% alumina, 50-60% silica and the remainder consisting of magnesia carbonate , calcium carbonate and iron oxide.

Bhavsar, Joshi, Shrof and Patel in 2014, studied the use of industrial waste such as brick dust in soil stabilization, the advantages presented include; proper disposal of such waste, saving biodiversity, increasing soil properties such as strength, reduce permeability etc., preservation of natural soil and design and construction of economic structures. The work concludes a positive result with the use of brick dust on black cotton soil where increasing content of brick dust reduces the swelling index and hence stabilizing the soil.

IOP Conf. Series: Materials Science and Engineering **413** (2018) 012023 doi:10.1088/1757-899X/413/1/012023

Brick dust is locally available material which is formed from the crushing of industrial clay bricks into dust particles. This admixture is available and sold in the open market in Nigeria, which include Ota, though various companies in Lagos produce them as building bricks. The work by Kumar, Kumar and Prakash in 2016, also reported decrease in swelling index of clay soils when mixed with lime and brick dust due to decrease in plasticity of the soil as a result of the reduction in clay content of soil because of the replacement of the clay with brick dust.

2. Methodology

Clay soil samples, cement, sodium chloride, brick dust and water were used for this study. The clay soil samples were gotten from a borrow pit in Ota. The soil was collected at a depth not less than 1m from the ground. The sample was kept safe and packed in an air-tight sack bag to retain its natural moisture. The natural moisture contents of the samples were immediately determined on getting to the laboratory and the sample was kept to dry in the ovens in the soil laboratory of a Construction Company in Lagos state.

Marks were placed on them to indicate soil descriptions, sampling depths and dates of sampling. The samples were spread on different matting to facilitate air drying. All the clods and lumps in the samples were broken down and reduced to fine particles before performing various tests on them. The tests carried out were done under BS1377 [19] specification. The cement was gotten from the concrete laboratory of the Company, the sodium chloride was purchased from the local market, the brick dust was gotten from Ota Ogun state and crushed at the Company Laboratory, and portable water available in the laboratory was used for tests performed during this study.

The following tests and analyses were carried out in the laboratory according to [19] specification to:

(a) determine the index properties of the soil sample before stabilization

(i) natural moisture content

(ii) sieve analysis

(iii) atterberg Limits

(iv) specific gravity

(b) determine the strength properties of the clay soil sample before stabilization

(i) compaction

The analysis on the strength properties and index properties of the soil samples as in (a) and (b) above were carried out after soil stabilization with various percentages of cement, sodium chloride and brick dust. Microsoft Excel statistical package was used to analyze the data collected by comparing the results of different percentages used in the three approaches.

3. Results and Discussion

(i) Natural Soil Moisture Content

The summary of the result of the geotechnical properties of the natural soil is presented in table 1. The natural moisture content of the soil was found to be 10.3 as presented in the table on the geotechnical properties of natural soil.

IOP Conf. Series: Materials Science and Engineering **413** (2018) 012023 doi:10.1088/1757-899X/413/1/012023

Table 1: Geotechnical properties of natural soil

Natural moisture content (%)	10.3
Specific gravity (g/cm^3)	2.62
Liquid limit (%)	43.8
Plastic limit (%)	23.5
Plasticity index (%)	20.3
Maximum dry unit weight (kN/m^3)	16.7
Optimum moisture content (%)	17.1
Soaked California bearing ratio (%)	8

(ii) SIEVE ANALYSIS

In the particle size analysis of the sample, more than 6.18% of the soil sample passed the No. 200 sieve as shown in table 2. Using AASHTO classification, the soil belongs to the sub group A-2-7 (Clayey Sand) and poorly graded sand with clay (SP-SC).

Table 2: Particle size distribution of natural soil.

Sieve	Weight of sample	Percentage retained	Cumulative % retained	Percent finer
7- 2.36	6.0	1.20	1.20	98.80
14- 1.18	3.0	0.60	1.80	98.20
25- 0.60	43.3	8.66	10.46	89.64
36- 0.425	211.4	42.28	52.74	47.26
52- 0.30	143.1	28.62	81.36	18.64
100- 0.15	56.5	11.30	92.60	7.34
200- 0.075	5.8	1.16	93.80	6.18
-200 (T)	30.9	6.18	100	-
Total	500.0		100	

SPECIFIC GRAVITY

Variation of the specific gravity test for the soil sample with changes in cement, sodium chloride and brick dust content are presented in Figure 1.

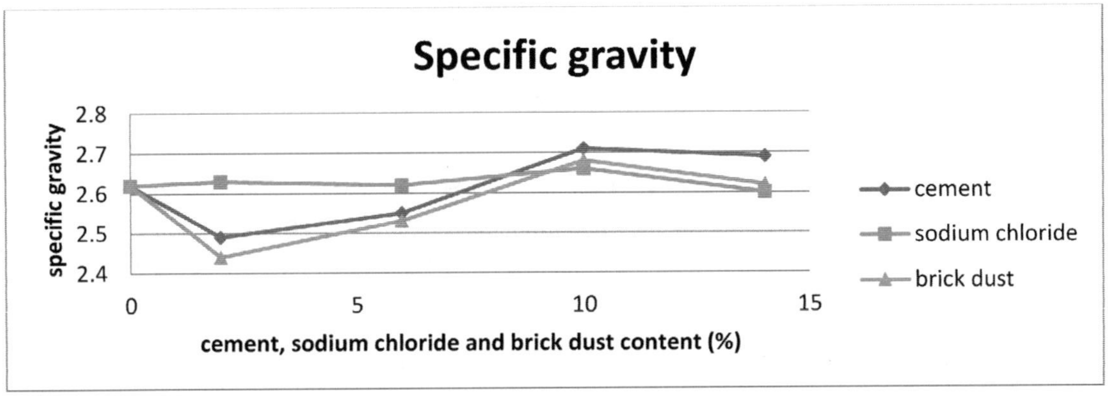

Figure 1: Variation of specific gravity with cement (%), sodium chloride (%), brick dust (%)

Higher cement content expectedly increased the specific gravity and at 14% it reduced it, also a higher sodium chloride reduces the specific gravity. Higher brick dust content increased the specific gravity and at 14% it decreases. The decrease in specific gravity at 14% is also observed for sodium chloride and brick dust.

ATTERBERG LIMITS

The test was performed to determine the liquid limit, plastic limits and plasticity index of the clay soil sample. This is done to characterize its condition by water content.

With progressive increment in percentage of cement content in the clay soil, liquid limit, plastic limit, and plasticity index all reduce but stabilizes at 14% according to Kumar *et al* in 2016.

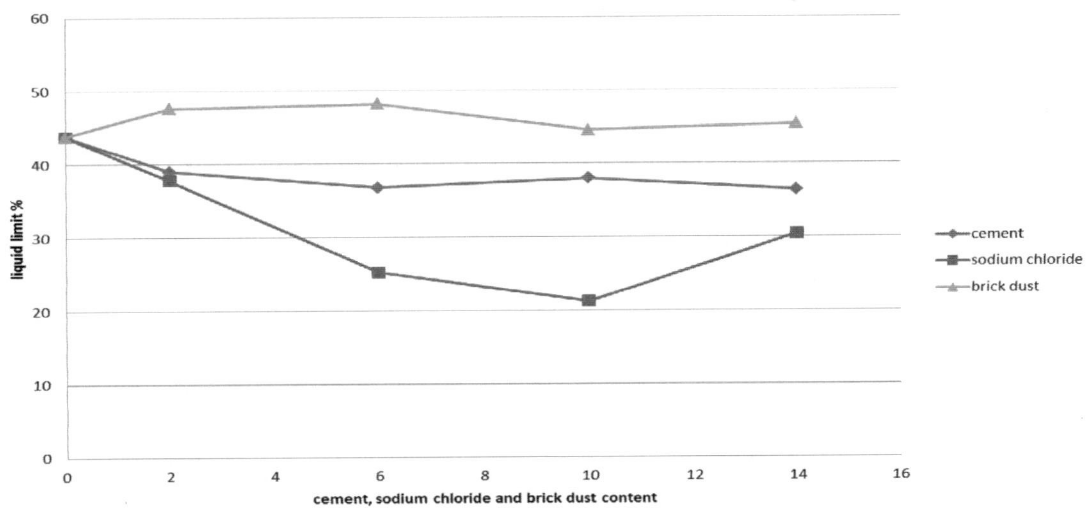

Figure 2: Graph of percentage increase of cement, sodium chloride and brick dust content for liquid limit

According to figure 2 - 4 liquid limits reduces with increase in cement and sodium chloride this is as a result the ability of cement to absorb water and the salinity in the sodium chloride while there was a decrease with increase in brick dust content this could be as a result of removal of the carbonate in fine grained soil.

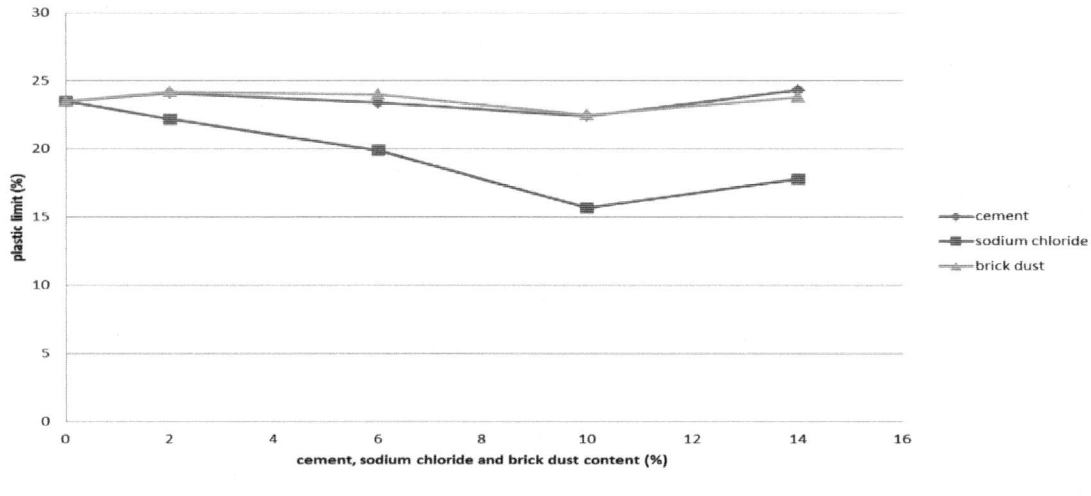

Figure 3: Graph of percentage increase of cement. Sodium chloride and brick dust content for plastic limit

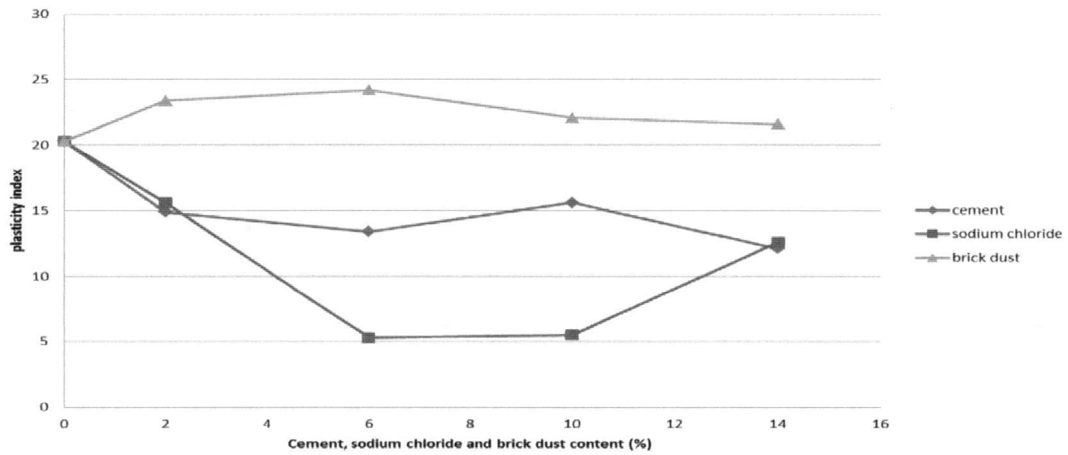

Figure 4: Graph of percentage increase of cement, sodium chloride and brick dust content for plasticity index

COMPACTION

The compaction curve of the soil sample for each 0%, 2%, 6%, 10%, and 14% cement, sodium chloride and brick dust contents are shown in figure 5 - 9. The change in OMC and maximum dry unit weight with brick dust, sodium chloride and cement content can be seen in Figures 6 to 10.

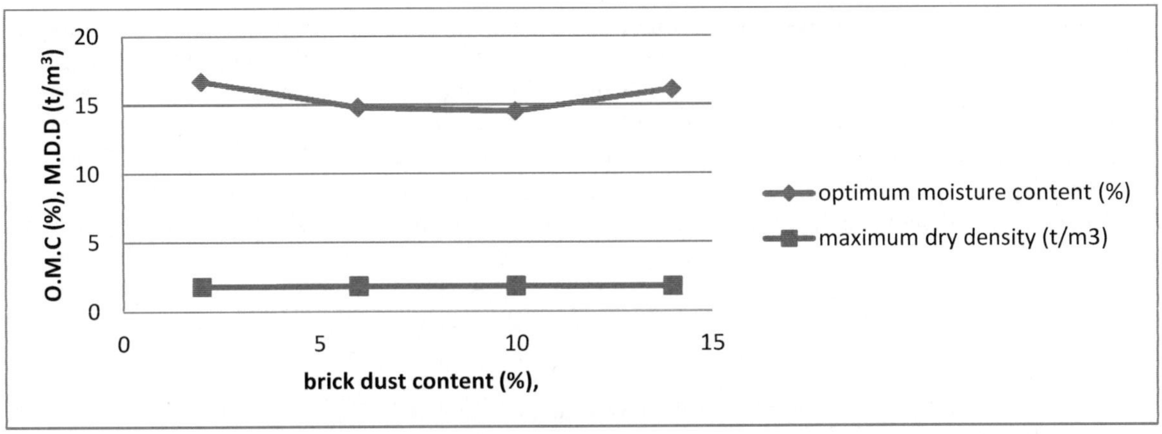

Figure 5: Variations of brick dust content based on compaction characteristics

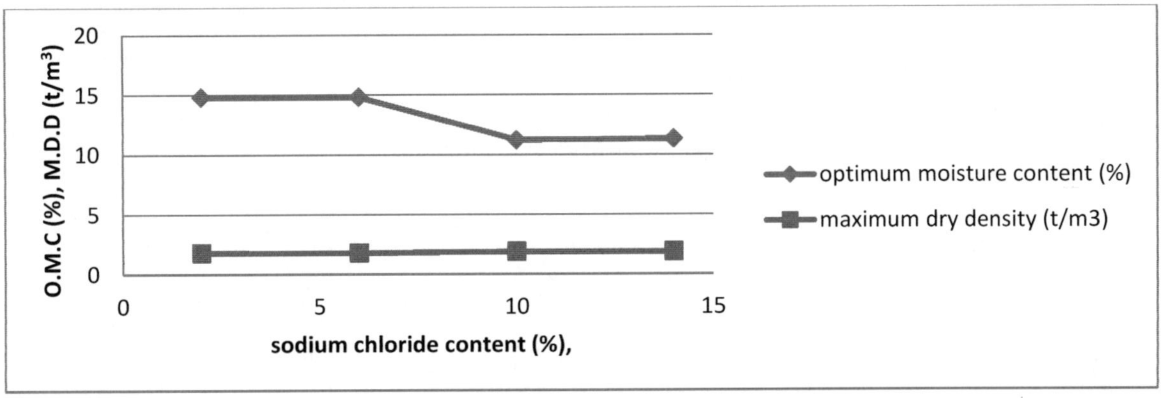

Figure 6: Variation of sodium chloride content based on compaction characteristics

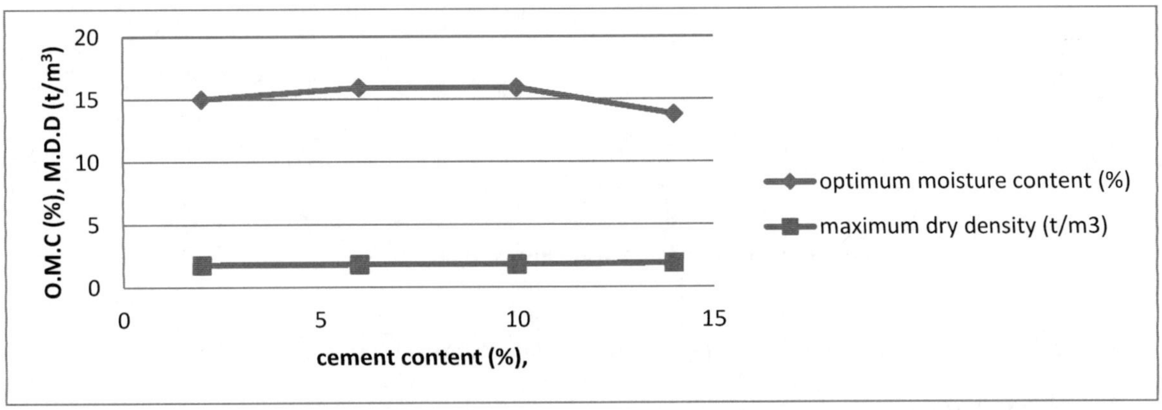

Figure 7: Variation of cement content based on compaction characteristics

The maximum dry unit weight of soil–cement mixture increased with cement contents. The maximum dry unit weight of soil-sodium chloride mixtures expectedly increased with higher salt content although was stable between 10% and 14%. The maximum dry unit weight of soil-brick dust mixtures, expectedly increased with higher brick dust content from 0%-6%, then a slight decrease from 10% to 14%.

IOP Conf. Series: Materials Science and Engineering **413** (2018) 012023 doi:10.1088/1757-899X/413/1/012023

From the figures 5 – 7, while the optimum moisture content for brick dust content is concave upward that of cement content is convex. While optimum moisture content of soil-cement mixture figure 9 decreased from 0% to 2%, then increased as the amount of cement in the mixture increased from 2% to 10% there was slight stability between 6% and 10% before a decrease for the 14% cement content. Again maximum dry density of soil-sodium chloride mixtures figure 10 decreased from 0% to 2%, showed slight stability between 2% and 6%, then a decreased at 10% and a slight increase 14%. Optimum moisture content of soil-brick dust content decrease from 0% to 10%, then increased at 14%.

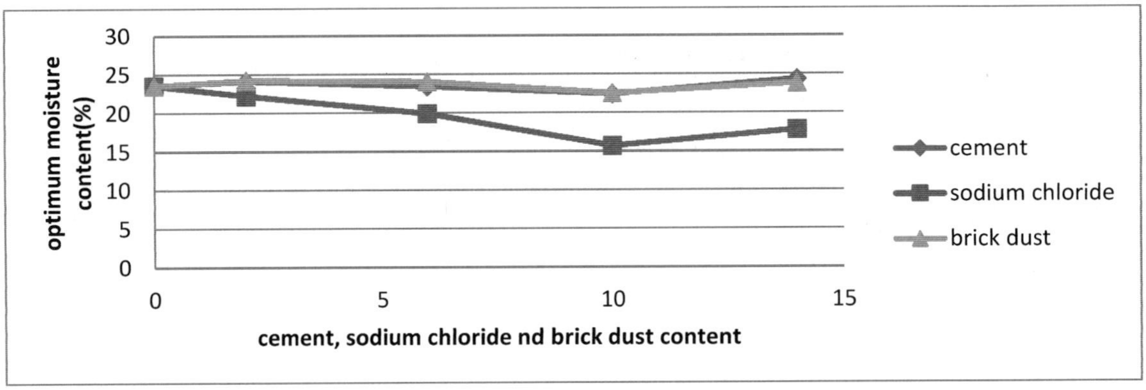

Figure 8: Variations of optimum moisture content with cement, sodium chloride and brick dust content.

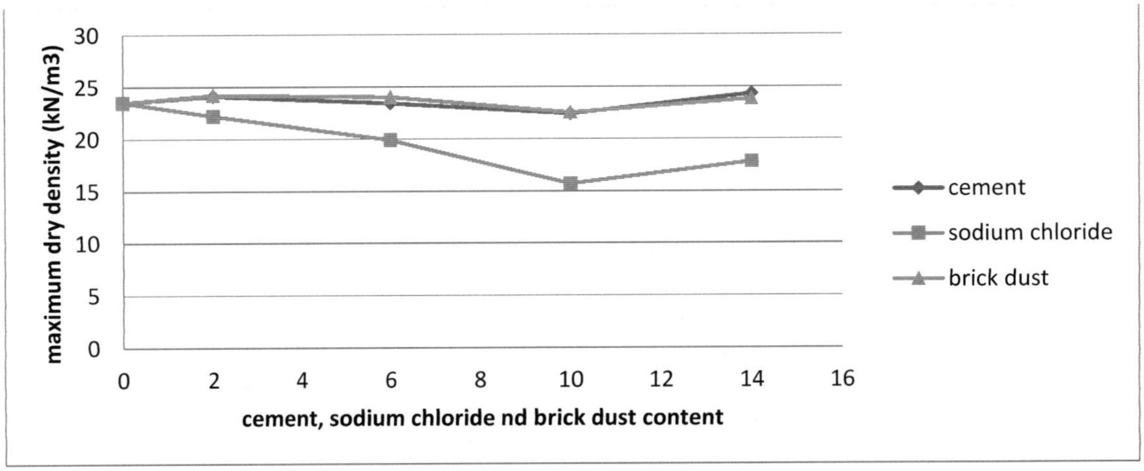

Figure 9: Variation of maximum dry density with cement, sodium chloride and brick dust content

CALIFORNIA BEARING RATIO

The change in CBR value with cement, sodium chloride and brick dust content can easily be seen in figure 10. The figure indicates that the CBR value for soil-cement mixture progressively increased from 8% for the 0% cement stabilization to 29% for the 2% cement addition, then increased to 112% for the 6% cement addition, a slight decrease to 99% for 10% cement addition, later an increase to 123% at 14% cement addition. Figure 8 also indicates that the CBR value for soil- sodium chloride mixture decreased from 8% for the 0% stabilization to 6% for the 2% sodium chloride addition, it then remain stable at 6% for the 6% sodium chloride addition, a slight decrease to 4% for 10% sodium chloride addition, later an increase to 8% at 14% sodium chloride addition.

The CBR value for soil-brick dust mixture increased from 8% for the 0% stabilizer to 9% for the 2% brick dust addition, then increased to 25% for the 6% brick dust addition, a decrease to 5% for 10% cement addition, later an increase to 28% at 14% brick dust addition. The improvement in CBR value may be attributed to change of soil structure from dispersed to flocculate.

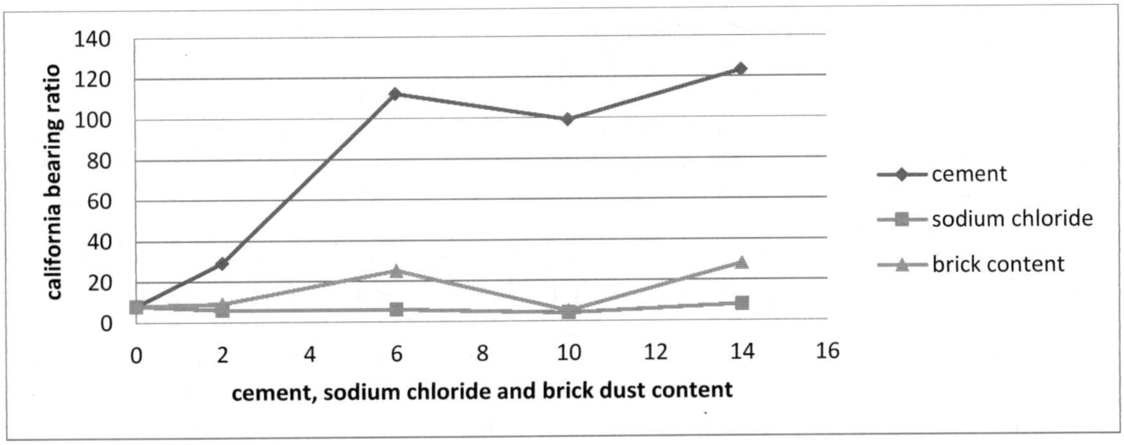

Figure 10: Variation of percentage increase in cement, brick dust and sodium chloride content for California bearing ratio.

4. Conclusion

The clay soil used in this study was classified as A-2-7 by AASTHO system or CL by USC system. The natural soil is silt clay with approximately 6.18% passing the BS No. 200 sieve. Test results generally indicate that the addition of cement and sodium chloride individually reduced the plasticity of clay soil and thereby improved its workability and reduced its moisture holding capacity and swell potential, the addition of brick dust increased the plasticity of clay soil which did not improve the workability.

Therefore, 14% optimal stabilization of A-2-7 soil with cement and sodium chloride will effectively reduce the plasticity of the natural soil to meet the requirement for use as subgrade, subbase and base course materials. Sodium chloride reduced the CBR value of the natural soil to meet the requirement for use only as subgrade material, according to Nigeria General Specification (1997). According to these results cement is the best and suitable stabilizer but because of the cost sodium chloride is more economical. More work is recommended to be done on silty clay stabilization using brick dust and brick dust lime combination at greater percentages than the 14% used in this study.

Reference

[1] Samtani, N. and Nowatzki, E. Soils and Foundation Reference Manual, Volumes I and II, Report No. FHWA NHI-06-088. (2006).

[2] Davison, L and Springman, S (2000) Compaction http://fbe.uwe.ac.uk/public/geocal/soilmech/compaction/compaction.htm#PACTCON (21 Oct 2003).

[3] Sabat, Akshaya Kumar. Stabilization of Expansive Soil Using Waste Ceramic Dust. Electronic Journal of Geotechnical Engineering. (2012). Vol 17. Bund.z, pp 3915-3926

[4] Sabat, A.K. and Nanda, R.P. Effect of marble dust on strength and durability of rice husk ash stabilised expansive soil. International Journal of Civil and Structural Engineering. (2011). Vol.1 (4), pp 939-948.

[5] Baser, O. Stabilization of expansive soils using waste marble dust. Master of Science thesis submitted to Civil Engineering Department, Middle East, Technical University. (2009).

[6] Palaniappan, K. A. and Stalin, V. K. Utility effect of solid wastes in problematic soils. International Journal of Engineering Research and Industrial Applications. (2009). 2(1), pp 313-321.
[7] Swami, B.L. Feasibility study of marble dust in highway sector. Highway Research Bulletin. (2002). Number 67, December, pp 27-36.
[8] Srinivasulu, G. and Rao, A.V.N. Efficacy of baryte powder as a soil stabilizer. Journal of the Institution of Engineers (I). (1995). Vol.76, Nov. pp 129-131.
[9] Ene, E. and Okagbue, C. Some basic geotechnical properties of expansive soil modified using pyroclastic dust. Engineering Geology. (2009). Vol.107 (1-2), pp 61-65.
[10] Abd El-Aziz, M. and Abo-Hashema, M.A. Measured effects on engineering properties of clayey subgrade using lime-homra stabiliser. International Journal of Pavement Engineering, (2012). DOI: 10.1080/10298436.2012.655739.
[11] Mukesh A. Patel and H. S. Patel. A Review on Effects of Stabilizing Agents for Stabilization of Weak Soils. Civil and Environmental Research. (2012). ISSN 2222-1719 (Paper) ISSN 2222-2863 (Online) Vol 2, No.6. pp 1-7
[12] Ampera B. and Aydogmus T. Recent experiences with cement and lime-stabilization of local typical poor cohesive soil. Geotechnik-Kolloquium Freiberg, March 11, 2005, Heft 2005-2, pp. 121- 144.
[13] Tamadher Abood and Mohamed A. S. Mohamed. A Laboratory Evaluation of Stabilization of Salty Clay Soil by Using Chloride Compounds. International Journal of Civil and Structural Engineering Research. (2014). ISSN 2348-7607 (Online) Vol. 2, Issue 2, pp: 47-52.
[14] Oyenuga, V.O. Lecture notes on simple tests to determine/monitor the quality of reinforced concrete materials. (2002).
[15] Işık Yilmaz and Berrin Civelekoglu. Gypsum: An additive for stabilization of swelling clay soils Applied Clay Science**,** Volume 44, Issues 1–2, April 2009, Pages 166–172.
[16] Rogers, Sara B. Evaluation and Testing of Brick Dust as a Pozzolanic Additive to Lime Mortars for Architectural Conservation (Master's Thesis). (2011). University of Pennsylvania, Philadelphia, PA, USA.
[17] Bhavsar, Sachin N., Joshi, Hiral B., Shrof, Priyanka K. and Patel Ankit J. Effect of Burnt Brick Dust on Engineering Properties on Expansive Soil. Int. Journal of Research in Engineering and Technology. (2014). ISSN 2319-1163. Vol.3, No. 4, Pp 433-44.
[18] Kumar, Ajay, Kumar, Ashok and Prakash, Ved Stabilization of Expansive Soil with Lime and Brick Dust. Int. Journal of All Research Education and Scientific Methods. (2016). ISSN: 2455-6211, vol. 4, No. 4, Pp 36-42.
[19] BS 1377 (1990) Method of Test for soil for civil engineering purpose, British Standard Institute, London.

IOP Conf. Series: Materials Science and Engineering **413** (2018) 012024 doi:10.1088/1757-899X/413/1/012024

Evaluation of Mechanical Properties of Locally Recycled Aluminium Alloy Scrap

Benjamin U. Oreko[1], Olusegun D. Samuel[1], Stanley Idi[2], Ojo S. I. Fayomi[3,4]

[1] Federal University of Petroleum Resources Effurun, Delta State, Nigeria
[2] Petroleum Training Institute Effurun, Delta State, Nigeria
[3]Department of Chemical, Metallurgical and Materials Engineering, Tshwane University of Technology, P.M.B. X680, Pretoria, South Africa.
[4]Department of Mechanical Engineering, Covenant University, PMB 1023 Ota Nigeria
E-mail address: oreko.benjamin@fupre.edu.ng

Abstract. Recycling of aluminium alloy scraps reduce the amount of waste directed into landfills. It is a vital part of the aluminium smelting industries. However, some local aluminium smelting craftsmen/artisans in Nigeria cast their aluminium scraps without considering the mechanical properties of the casting obtained from aluminium scraps. This study investigated some of the mechanical properties of cast aluminium alloy namely; tensile strength, impact, and hardness, and the microstructure of the locally recycled aluminium alloy in Nigeria. A permanent mould casting process was adopted in the smelting of the scraped aluminium and their properties were examined. The metallographic specimen of the sample was grinded and polished in a Dia-Pro suspension solution (3μm) on an MD-DAC and DP-DAC 300mm polishing cloth while the surface was exposed to a Keller enchant of 15mil of hydrochloric acid, 10mil of water and 10mil of hydrofluoric acid before viewing under the microscope. From the result, the ultimate tensile strength, impact, and Brinell hardness were 54.87MN/m^2, 6.1J and 64.33BHN, respectively. The metallographic result from the smelted sample revealed a coarse and rough surface microstructure.
Keywords: Aluminium alloy scrap; mechanical properties; locally recycled aluminium

1. Introduction

Recycled aluminium scrap is becoming a more significant component of aluminium production because of its economic, resource savings, environmental benefits and sustainability [1-2]. Advances in aluminium recycling have a remarkable low production cost as compared with primary aluminium production [3-4]. The method and techniques for processing and producing recycled aluminium scrap and dross has been studied [5-11]. The traditional method of re-melting and subsequent refinement of waste/scrap aluminium is still relevant among aluminium smelting shops/cottage industries and local smelting craftsmen/artisans in Nigeria. However, little attention has been given to the mechanical properties of recycled aluminium casting by locally aluminium smelting craftsmen/artisans in Nigeria. Hence, in this study attempt is made to demonstrate the process and pattern of casting scraped aluminium by local craftsmen/artisans and to evaluate the mechanical properties of this recycled aluminium.

2. Materials and Methods

The study was carried out at the Petroleum Training Institute Effurun, Delta State, Nigeria. The raw material (aluminium scraps) was sourced around the dump site close the city where the Institute was situated. The following materials and equipment were used in this study; Aluminium scraps, locally made furnace, specimen mounting machine, abrasive cutter, vernier calliper, dial callipers, steel rule, industrial lathe machine, universal testing machine, grinding and polishing machine, macro hardness testing machine, hack-saw, impact testing machine, metallurgical microscope, and an extensometer. The chemical composition of the recycled aluminium was analyzed at the Quality Analytical Laboratory Service Limited, Warri, Delta State, Nigeria. Table 1 is the chemical composition of recycled aluminium of 1000 series [12].

IOP Conf. Series: Materials Science and Engineering **413** (2018) 012024 doi:10.1088/1757-899X/413/1/012024

Table 1 Chemical composition of recycled aluminium scrap (weight % in mg/kg)

S/N	Parameters	weight %
1	Al	99.13
2	Pb	0.32
3	Cu	0.44
4	Zn	0.11

2.1 Experimental Procedures

i. Casting of Sample: A local furnace was fabricated. A permanent mould casting process was adopted in the smelting of the scraped aluminium (as common used by local aluminium smelt craftsmen/artisans in Nigeria). The scraps were melted at a temperature of 680°C and cooled gradually at an ambient temperature. The casting obtained from aluminium alloy scrap is of dimension 400 mm x 30 mm x 30mm as shown in figure 1.

Fig. 1 Sample of Cast Aluminium Alloy Scrap

ii. Tensile Testing: The castings (3 specimens) obtained from the mould were machined to the required standard. A 20-ton digital universal testing machine with model SM1000-TQ007329-02 was used to determine the tensile test experiment as shown in figure 2. Figure 3 is the fractured sample of the specimen. From the experimental data, the following parameters were obtained; percentage elongation, the percentage reduction in area, ultimate tensile strength (UTS) of the recycled aluminium scraps.

Fig. 2 Universal Testing Machine

Fig. 3 Fractured Specimen

iii. Charpy Impact Test: A charpy impact testing machine with model number: MAT21/IT3U, serial number:010340, was used for the experiment as shown in figure 4. The impact test specimens were prepared according to ASTM E23. The dimensions of the test samples were 55 × 10 × 10 mm and it was v-notched 2mm deep with the aid of a triangular file, at an angle of 45^0 with 0.25mm radius along the base. The impact strength of the specimen was the energy absorbed per unit cross-sectional area at the notch (J). Figure 5 is the fractured specimen after testing.

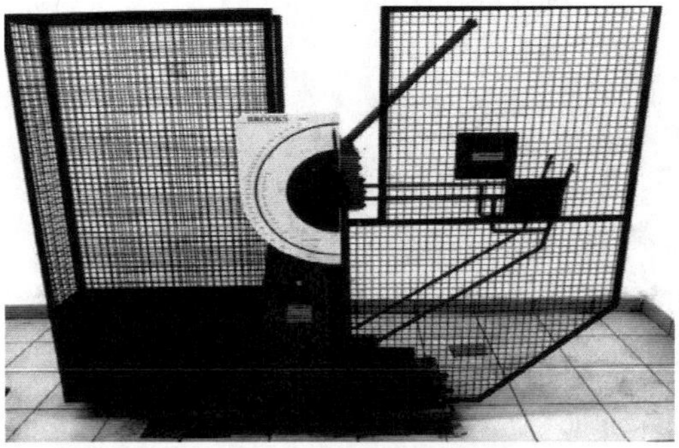

Fig. 4 Charpy Impact Testing Machine

Fig. 5 Fractured Specimens

iv. Brinell Hardness Test: First, the specimen was grinded and polished to attain a smooth surface of 600 microns. The hardness test was carried out using the digital universal testing machine with an applied force of 20 kN for 15 seconds each on three (3) different positions of the sample as shown in figure 6.

Fig. 6 Sample of Brinell Hardness Test on Specimen

v. Metallurgical Examination: The specimen was mounted in an emery mounting powder to form a holding height of 15 mm and a diameter of 30 mm. Specimen attained a mirror-like face, after grinding and polishing in a Dia-Pro suspension solution (3μm) on an MD-DAC and DP-DAC 300 mm polishing cloth in the Struss machine. Thereafter, the sample was exposed to Keller enchant of 15 mil of hydrochloric acid, 10 mil of water and 10 mil of hydrofluoric acid and was quickly viewed under the microscope (with identification number:195-040389) using an Amscope software to analyze the microstructure. Figures 7 to 10 are the grinding and polishing machine, polished specimen, optical microscope, and the microstructural test specimen, respectively.

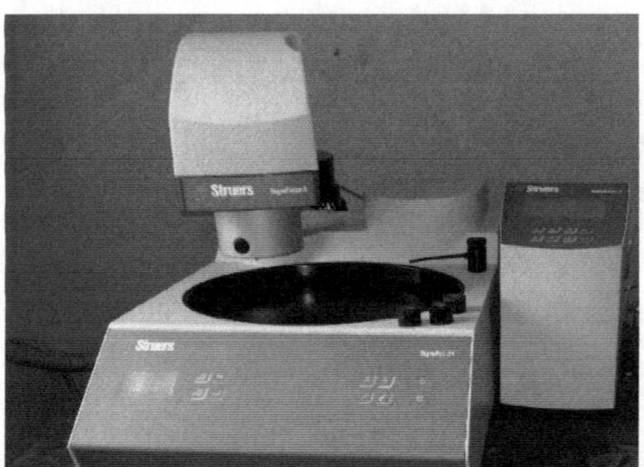

Fig. 7 Grinding and Polishing Machine

Fig. 8 Sample of Polished Specimen

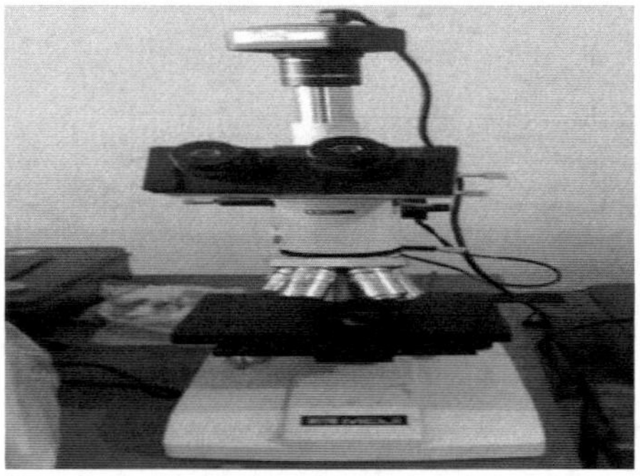

Fig. 9 Optical Microscope

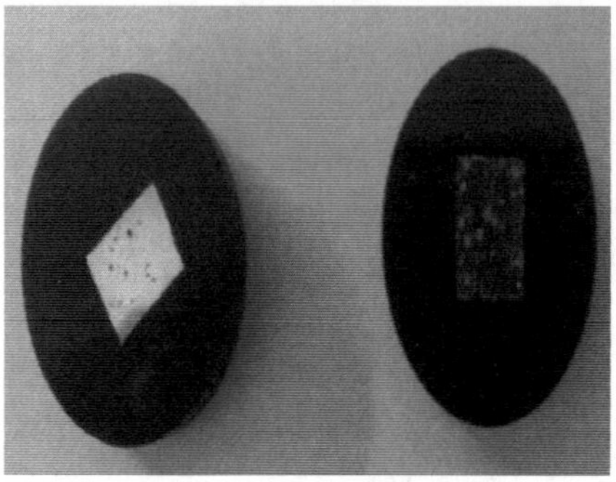

Fig. 10 Microstructural Test Specimen

3. Results and Discussion

i. Tensile Testing Result

Table 2 shows the value obtained from the universal testing machine. The ultimate tensile strength, percentage elongation and reduction in area were 54.86 MPa, 4.63 %, and 23.3 %, respectively. The ultimate tensile strength of the recycled aluminium scrap recorded a lower value as compared with the aluminium 1000 series ultimate tensile strength range of 45.0- 205 MPa [12].

Table 2: Result for the Mechanical Property of Locally Cast Aluminium alloy

No.	Original length (mm)	Final length (mm)	% Elongation	Original Diameter (mm)	Final Diameter (mm)	% Reduction in Area	UTS (MPa)
1	172	180	4.6	9.75	8.54	23	55.86
2	172	181	4.7	9.75	8.42	25	53.98
3	172	180	4.6	9.75	8.61	22	54.77
Average			4.63			23.3	54.87

ii. Brinell Hardness Result

Table 3 shows the result from the universal testing machine for BHN hardness. It was observed that a load of 20 kN produced an indentation diameter of 6 mm from the indenter of 10 mm diameter which gives the Brinell hardness number of 64.33BHN.

Table 3 BHN Test Results of Locally Cast Aluminium

No.	Load (kN)	Dial Diameter (mm)	Indentation Diameter	Brinell Hardness (BHN)
1	20.0	10	6.00	65.0
2	20.1	10	6.05	64.0
3	20.1	10	6.10	64.0
Average	20.07	10	6.05	64.33

iii. Impact Energy

Table 4 shows the maximum energy absorbed by the cast aluminium alloy. At the ambient temperature, the impact resistance of the notched samples shows an average value of 1.6 J.

Table 4: Impact Energy Tests Results of Locally Cast Aluminium

No.	Impact Energy (J)	Average Impact Energy (J)
1	1.6	1.6
2	1.6	
3	1.6	

iv. Result and Analysis for the Metallographic Testing
From the figure 11, the metallographic result indicates coarse structure with a visible grain boundary, pore holes as a result of casting defect and some rough surface finishing in the specimen.

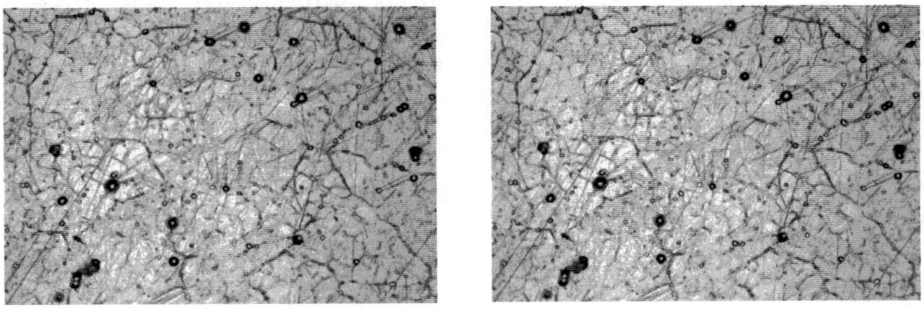

Fig. 11: Microstructure of Locally Cast Alloy

4. Conclusion

In this work, the tensile, hardness, impact strength and the microstructure of recycled aluminium alloy scrap were experimentally studied. It was observed that the ultimate tensile strength of the recycled aluminium alloy scrap has a low value when compared with the conventional value of the same aluminium 1000 series. It is recommended that further work been done on the method of cooling, casting and processing conditions of locally recycled aluminium scrap on the mechanical properties of the material.

References

[1] Jirang Cui & Hans J. Roven. Recycling of Automotive Aluminium [J]. Transactions of Nonferrous Metals Society of China, Elsevier, 2010, 20,(11):2057-2063 doi.org/10.1016/S1003-6326(09)60417-9

[2] Amini Mashhadi, H., Moloodi, A., Golestanipour, M. & Karimi, E.Z.V. Recycling Of Aluminium Alloy Turning Scrap via Cold Pressing and Melting with Salt Flux [J]. Journal of Materials Processing Technology,2009, 209:3138–3142, DOI: 10.1016/j.jmatprotec.2008.07.020

[3] Logozar, K., Radonjic, G., Bastic, M. Incorporation of Reverse Logistics Model into In-plant Recycling Process: A Case of Aluminium Industry [J]. J. Resour. Conserv. Recy. 2006, 49: 49–67.

[4] Zhou, B., Yang, Y., Reuter, M.A., Boin, U.M.J. Modelling of Aluminium Scrap Melting in a Rotary Furnace[J]. J. Miner. Eng., 2006, 19: 299–308.

[5] Takehito Hiraki, Satoru Yamauchi, Masayasu Iida, Hiroshi Uesugi, and Tomohiro Akiyama. Process for Recycling Waste Aluminium with Generation of High-Pressure Hydrogen [J]. Environmental Science & Technology, 2007, 41 (12): 4454–4457. DOI: 10.1021/es062883l

[6] Arunabh Meshram Kamalesh KumarSingh. Recovery of Valuable Products from Hazardous Aluminium Dross: A Review [J]. Resources, Conservation and Recycling, Science Direct, 2018, 130: 95-108. doi.org/10.1016/j.resconrec.2017.11.026

[7] Mbaryana, G.N. & Shkolnikova, E.I. Hydrogen Generation by Oxidation of Coarse Aluminium in Low Content Alkali Aqueous Solution Under Intensive Mixing [J]. International Journal of Hydrogen Energy, 2016, 41(39):17216-17224. doi.org/10.1016/j.ijhydene.2016.08.005

[8] Ansari, M. S., Puttewar, S. P., Agnihotri, A. Studies on Process for Conversion of Waste Aluminium Dross into Value Added Products [J]. Russian Journal of Non-Ferrous Metals, Springer, 2016,

[9] Takehito Hiraki and Tetsuya Nagasaka. An Easier Upgrading Process of Aluminium Dross Residue by Screening Technique [J]. Journal of Material Cycles and Waste Management, springer, 2015, 17(3): 566–573.

[10] Takehito Hiraki, Atsushi Nosaka, Noriyuki Okinaka, Tomohiro Akiyama. Synthesis of Zeolite-X from Waste Metals [J]. Isij International, 2009, 49(10):1644-1648. doi.org/10.2355 /isijinternational. 49.1644

[11] Huihu Wanga Zhixiong Xiea. Investigation on Hydrogen Production Using Multicomponent Aluminium Alloys at Mild Conditions and its Mechanism [J]. International Journal of Hydrogen Energy, Science Direct, 2013, 38,(3): 1236-1243. doi.org/10.1016/j.ijhydene.2012.11.034

[12] MatWEB.com/datasheettext.aspx. Material Property Data, Overview of Material for 1000 series Aluminium. Assessed 02/07/2018

IOP Conf. Series: Materials Science and Engineering **413** (2018) 012025 doi:10.1088/1757-899X/413/1/012025

Application of Accelerometer in the Design and Adaptation of Active Suspension in Automobile

Chujor O. Obarijima[1], Sociis T.A. Okolie[2], Babaremu Kunle O. [3]

Industrial Engineering Unit, Department of Mechanical Engineering, Covenant University, Km. 10 Idiroko Road, Canaan Land, Ota, Ogun State, Nigeria.

Email addresses: [1]okolie.sociis@covenantuniversity.edu.ng
[1]chujorobj@yahoo.com, [3]kunle.babaremu@covenantuniversity.edu.ng

Abstract. Upon the pedigree of technological advancement till our contemporary world, there has been an immense increment and improvement in the automotive industry like the development of vehicle suspension systems to suit the need of their customers whether it is soft (for comfort) or stiff (for good road holding capability for sports vehicle) or even a vehicle suspension system that can switch between these two modes. This research work focused on the effect and application of accelerometer in the design of an active suspension system using a microcontroller. It presents the control method implemented to keep the model vehicle leveled over pitching motions of the vehicle and the design of its component parts as well as the computer program to perform this action. The result showed that over different inclined surfaces the actuator was able to change the wheel –travel of both the rear and front wheels to keep the vehicle chassis leveled. These suggest possible adaptation of the design and improvement efforts which has capacity in addressing challenges of vehicles plying different road surfaces.
Keywords: Suspension; Active suspension; Accelerometer; Microcontroller.

1 Introduction

Vehicle suspension systems are being designed constantly to make travelling in motor-vehicles safer and more comfortable, from the inception of transport vehicles, animal-drawn coaches had suspension systems designed into the coaches, and this was because the paths on which these vehicles travel on are always uneven. Prior to the design and installation of suspension systems on wheeled carts and coaches, the axle of the road wheels was fixed to the body of the cart, this meant that for every vertical distance travelled by the wheel over bumps or down depressions on the road surface, this causes the body of the vehicle to rotate in different directions relative to the wheel travel of the vehicle.

The earliest suspension system was the Leaf Spring Axle suspension, with the level of technology available at that period, beam spring made of wrought-iron were arranged together in layers until the required stiffness was obtained, their beams are usually in curved shape and single spring beams were also used [2]. Suspension systems are of two types namely the Dependent and Independent Suspension systems, in the dependent suspension system, the motion of one wheel affects the other wheel as they are both connected to a rigid member, whereas in the Independent Suspension system, each wheel can move on its own accord without transmitting any force to the other road wheel.

1.1 Suspension Classification

1.1.1 Passive Suspension

A conventional suspension system (passive suspension system) is made of three elemental components according to [7] which are the:

- Elastic component (most commonly a coil springs and leaf springs then there is the more recent air springs).
- Damping component to reduce the oscillation (vibration).

Mechanical Linkages that join the body of the vehicle to the road wheels.

The passive suspension is limited in its design as different road conditions cause wheels to move vertically with different forces and speeds and the stiffness of the spring and damping factor of the damper cannot cause the wheel to move with a force that is equal to the force caused by the bump.

1.1.2 Active Suspension

Active suspension in automobiles is any suspension system in which the wheel-travel is controlled electronically independent of the response to the road surface condition. This system comprises of electronic control unit which receives signals about the road condition via sensors, the electronic control unit sends signals to the actuator to alter the wheel-travel on each road wheel. They are classified according to two main features [9]:

1.1.2.1 Energy input:

Passive, Active and Semi-active suspensions are the three types of suspension systems under this class. Suspension system that does not possess any energy input on any of its suspension components is Passive suspension. However, when energy is added to the system it is an Active suspension and if the system is electronically modified but energy is not added, it is a Semi-active or Adaptive suspension. Active suspensions are those that can change the ride height of the vehicle while Semi-active suspensions are those that cannot change the ride height.

1.1.2.2 Bandwidth:

It is specific reaction-time of the electronically controlled elements. For an active suspension, it has every elemental component of the conventional and has a suspension electronic control unit (ECU) and sensors that detects physical variables. These two elements work together to change or alter the reaction of the road wheels to the road surface conditions. The suspension ECU controls the actuator and other components. The ECAS (Electronically Controlled Active Suspension) vehicle has capabilities to change the acceleration of the sprung mass continuously [1].

1.1.3 Active Stabilizer System

Active stabilizer system is an active suspension system that involves using an electronic system to control the deflection of the stabilizer bar. The passive stabilizer bar is used to reduce vehicle roll by transmitting the part of the vertical motion of one wheel on a side of the vehicle to the other (from right to left and vice versa). The portion of the motion transmitted is proportional to the stiffness of the stabilizer bar. When one wheel experiences force acting on it, the force acts on one end of the bar to produce a torque which is transmitted to the other end of the bar thereby causing that wheel to move in the same direction as the other. Although, this system is not favorable in all situations, for instance when driving off-road, this stabilizer bar system causes one or more wheels to lose contact with the ground surface. The stabilizer bar stiffness cannot be changed to fit all road conditions; the Active Stabilizer System was designed to overcome this challenge by changing the stiffness of the stabilizer automatically or on demand by the driver [10].

1.1.4 Predictive Suspension

The predictive suspension system is the future active suspension system, it uses stereo cameras to scan the surface of the road ahead of the vehicle, this image is fed into an onboard computer that processes the image to determine the nature of the surface, the onboard computer controls the suspension actuators to raise or lower the individual wheels thereby keeping the chassis of the vehicle near horizontal level [4-5].
Active suspension systems have the capability to reduce the motion of the body of the vehicle which is the sprung mass of the suspension system thereby increasing comfort, braking, traction and maneuverability. For a passive suspension system to provide comfort, parameters like the damping factor, the wheel travel and the spring stiffness have to be changed, the wheel travel has to be increased, the damping coefficient and the spring stiffness constant has to be reduced. For the active suspension system to vary these

parameters, there are quite a number of control methods, methods such as Non-linear control, Linear Quadratic Gaussian (LQG) control, Proportional Integral Derivative control (PID) and adaptive control, however in the past Fuzzy Logic Control (FLC) which was developed by Zadeh [1].

1.2 Skyhook Theory

The Skyhook theory is an idealized active suspension system that allows a vehicle to maintain its leveled position as if it were suspended by an imaginary hook from the sky, regardless of the road conditions (Skyhook Theory, n.d.). This theory is not practical because the vehicle's wheel travel for an active suspension system is controlled by actuators. For a vehicle to maintain its leveled position it must have zero vertical acceleration, the acceleration is measured by an accelerometer which is placed on the vehicle.

2. Materials and Method

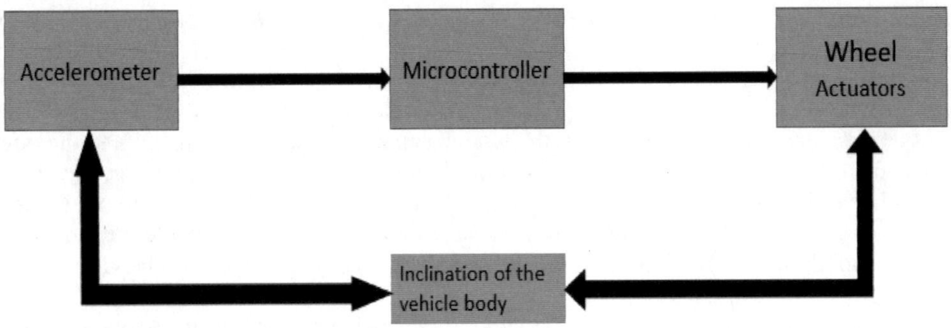

Figure 1: Flow Chart of Active Suspension Control System

In order for a vehicle suspension to be 'active' there must be exist within the system sensors which detect the orientation of the vehicle and the physical environment on which the vehicle drives upon. In this study however, the sensor used is an accelerometer, there are other sensors that may be used like an Inertia Measuring Unit (IMU), Gyroscope etc. for instance, in the gyroscope sensor there exist an error that occurs as a result of compounding null bias stability as the period of integration increases which causes a seeming rotation even when the device is has zero change in velocity [3].

2.1 Major Components Used

- Accelerometer: An accelerometer is a Micro-Electro-Mechanical Systems, (or MEMS) device that measures the acceleration of forces on an object. An accelerometer in this way, measures the forces applied by restrictions on a reference mass to hold its position (The Editors of Encyclopedia Britannica, 2017). It is a capacitive device, the distance between the adjacent plates causes a change in the voltage across the plates, and this voltage is proportional to the acceleration of the forces acting on the body in the x, y, and z axes. The accelerometer used in this project was the ADXL 335 accelerometer which is a triple axis accelerometer [6].
- Microcontroller: Microcontrollers are programmable electronic devices made up of a Central Processing Unit (CPU), Input-Output ports for receiving and sending out electronic signals, the memory for storing the source code of programs written, timer, Analogue-to-Digital converters and Digital-to-analogue converters. [7] The microcontroller used in this project is the Arduino Mega 2650 board.
- Wheel actuators: these are electromechanical devices with a control system that receives and converts the signals from the microcontroller, in order to change the wheel-travel of the vehicle thereby lifting or lowering the chassis of the vehicle. The wheel actuators used in this project

consisted of a servo motor which used a rack and pinion system to convert its rotary motion to linear motion to raise or lower the vehicle.

2.2 Implementation

An accelerometer measures the acceleration of forces as earlier discussed about a reference mass, the values of these accelerations in the x, y and z-axes can be computed to derive the angle of inclination of each axis of the sensor. From the Figure 1, the accelerometer detects the orientation of the body of the vehicle then sends the value of its measurements to the microcontroller. Note that for this implementation the x-axis is parallel to the longitudinal axis of the vehicle.
Formula

$$Angle\ of\ pitch\ of\ the\ x-axis\ be\ \emptyset$$

$$\emptyset = \tan^{-1}\frac{-A_x}{A_z}, \text{ Equation 1}$$

where Ax is the magnitude of the acceleration measured along the x-axis and Az is the magnitude of the acceleration measured along the z-axis of the accelerometer which is equivalent to the acceleration due to gravity.
The microcontroller receives the signals from the accelerometer every millisecond and these values are processed to give the acceleration along each axis and the angle of inclination of the required axis.
As the angles of pitch are determined per time, the wheel actuators which are connected to the output ports of the microcontroller receive electronic signals to either increase or decrease the wheel travel of both the front and rear wheels. The right and left wheels of either front or rear move together and not independently, this is because only the pitching (i.e. nose up or nose down) motion was considered during the active suspension designed during for the tests carried out, however, it is possible to design for a combined motion of the vehicle as the road surfaces are uneven.
With the angle of inclination determined it is possible to determine the height of the bump and determine to what length the wheel-travel of the front and rear wheels must be increased or decreased. The height of the bump can be determined using trigonometry in the Equation 2 below:

$$h = w \sin \emptyset \quad \text{Equation 2}$$

Where h is the height of the bump, w is the wheel base (centre-to-centre distance between the front and rear axles) and $\emptyset$ is the angle of inclination.

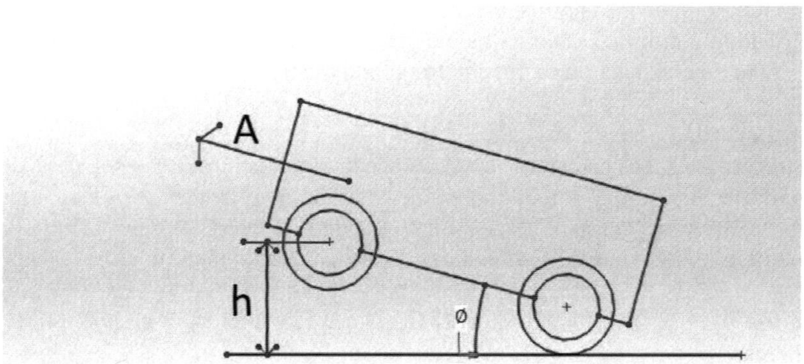

Figure 2: Inclination of Vehicle Chassis in Pitch-up Position (arrow A pointing to the front of the vehicle).

Once the height of the bump is determined, the wheel-travel can be changed to counter change in inclination, this change in wheel-travel is shared by the front and rear wheels by an equal amount and in the opposite direction. To counter the rotation caused by a bump of height **h** as shown in **Figure 2**, the front wheel-travel reduces by half the height of the bump thereby lowering the front of the model vehicle and the

rear wheel-travel is increased by half the height **h** thereby raising the rear of the vehicle. The reverse is the case for a ditch of depth **h.** These combined change in wheel-travels caused the chassis of the model vehicle to counter the inclination motion caused by the bump.
The readings from the accelerometer had a lot of noise, in order to reduce the noise and get a smoother curve, the Arduino Sketch was written to find the average of twenty consecutive inputs from the accelerometer, and the average was used to determine the angle of inclination.

3. Results and Discussion

For a vehicle to maintain its leveled position, it must maintain zero vertical acceleration (from Skyhook Theory). The value of this acceleration is measured by an accelerometer. Although the "vertical" acceleration was not directly measured in this project, the acceleration measured is that of the x-axis which is parallel to the longitudinal axis of the vehicle and computed with the acceleration due to gravity (along the Z-axis) measured earlier on a flat horizontal surface in order to determine the angle of inclination of the body of the vehicle.
The active suspension system of this project was designed to correct pitching motion caused by bumps or ditches or slopes on the road surface by causing a forward or backward pitch of the vehicle chassis by concurrently increasing and decreasing the wheel travel.

Table 3.1: Angle of Inclination, Acceleration and Height of Bump

Angle of Inclination of Chassis	Acceleration (g)	Height of bump(cm)
10o clockwise	0.20	+2.90
0o	0.06	0
10o anticlockwise	0.40	-2.90

From the table above, the values of the acceleration varies with the orientation of the x-axis being measured. With the angle of the inclination derived from the formula stated in the previous section 2, the wheel travel required to compensate the inclination is also determined.

Table 3.2: Height of Bump, Servo position, Wheel-Travel Changes

Height of bump (cm)	Servo position (degrees)	Change in front wheel travel (cm)	Change in rear wheel travel (cm)
+2.90	105.00	-1.44	+1.44
0(flat surface)	90.00	0	0
-2.90	65.00	+1.44	-1.44

4. Conclusion

In this project, an active suspension was designed to keep the model vehicle leveled during pitching motions (while it moved on an inclined surface). The program for the microcontroller to control the active suspension was developed and uploaded into the same. Tests carried out showed that the suspension was active as there the wheel-travel changed. In this project carried out attention was not given to the other suspension elements which are the springs and dampers.
Although the active suspension did not totally eliminate the pitching motions, it reduced the magnitude of the movement of the chassis of the model vehicle. This was due to:

- the response time of the control system,
- speed of the wheel actuators,

IOP Conf. Series: Materials Science and Engineering **413** (2018) 012025 doi:10.1088/1757-899X/413/1/012025

- and the stability of the control system as there was a lot of noise.

Accelerometer application in electronically controlled active suspension system is best used for constant velocity vehicles, because the acceleration of the vehicle will be sensed by the accelerometer and interpreted by the microcontroller program as a change in angle of tilt of the vehicle body, in order to eliminate this error other sensors such as gyroscopes or inertia measuring unit must be integrated into the design.

References

[1] Aly, A., & Salem, F. (2013). Vehicle suspension systems control: a review. *International Journal of Control, Automation and Systems*, *2*(2), 46–54. Retrieved from http://researchpub.org/journal/jac/number/vol2-no2/vol2-no2-6.pdf

[2] Dixon, J. (2009). *Suspension Analysis and Computational Geometry*. Wiltshire: John Wiley. Retrieved from https://books.google.com/books?id=L_dqjIjOGBcC&pgis=1

[3] Fisher, C. (2010). Using an accelerometer for inclination sensing. Retrieved June 21, 2018, from http://healthcare.analog.com/static/imported-files/application_notes/AN-1057.pdf

[4] Jens Meiners. (2010). Mercedes-Benz News: Magic Body Control System Driven – Car and Driver. Retrieved March 25, 2018, from https://www.caranddriver.com/news/mercedes-benz-news-magic-body-control-system-driven.

[5] Okokpujie, K. O., Odusami, M., Okokpujie, I. P., & Abayomi-Alli, O. (2017). A Model for Automatic Control of Home Appliances using DTMF Technique. *International Journal of Scientific & Engineering Research*, *8*(1), 266-272.

[6] Lis, T. (2010). Application note, (April), 1–18. https://doi.org/10.1002/em.20587.Genotoxicity

[7] Loop, S. (n.d.). Introduction to Microcontroller 8051. Retrieved June 21, 2018, from https://openlabpro.com/guide/basics-of-microcontroller/

[8] Savaresi, S. M., Poussot-Vassal, C., Spelta, C., Sename, O., & Dugard, L. (2010a). *Semi-Active Suspension Control Design for Vehicles*. *Semi-Active Suspension Control Design for Vehicles*.

[9] Savaresi, S. M., Poussot-Vassal, C., Spelta, C., Sename, O., & Dugard, L. (2010b). *SEMI ACTIVE SUSPENSION CONTROL DESIGN FOR VEHICLES*. Elsevier.

[10] Toyota Jidōsha Kōgyō Kabushiki Kaisha. (1991). *Toyota technical review*. Toyota Motor Corp.Retrievedfrom https://web.archive.org/web/20120229032944/http://sciencelinks.jp/j-east/article/200609/000020060906A0261266.php

IOP Conf. Series: Materials Science and Engineering **413** (2018) 012026 doi:10.1088/1757-899X/413/1/012026

Enhancement of Transformative Learning in Large Classes of Leaners in Engineering Education

Michael Olawale Daramola

School of Chemical and Metallurgical Engineering, Faculty of Engineering and the Built Environment, University of the Witwatersrand, Wits2050. Johannesburg, South Africa
E-mail address: Michael.Daramola@wits.ac.za

Abstract. Engineering education in the universities and colleges involves the combination of cognitive learning theory and experiential learning theory to ensure transformative learning. These theories form part of the conceptual frameworks that describe how information is absorbed, processed, and retained during engineering education and training. While the application of cognitive learning theory in engineering education focuses on understanding the learners with regards to the complexities of human memory, experiential learning theory aims at developing in the learners the societal technical observation and community commitment skills necessary to shape and transform them into competent young engineers. In addition, transformative learning theory seeks to explain and enhance how the learners revise and reinterpret information acquired during engineering education. Therefore, the combination of the cognitive learning theory and the experiential learning makes the application of transformative learning effective in engineering education. In addition, transformative learning provides an individual the opportunity to acquire experience, question some assumptions during learning, and reflect on them at the later stage. However, the degree of the enhanced transformative learning in engineering education is forestalled by the size of class we have in our universities today. Therefore, it is essential to know the effects of large class of learners on transformative learning in engineering education and how these effects could be minimized to promote transformative learning. In this article, effects of large classes on effective transformative learning in engineering education are enumerated, and suggested solutions to alleviating these effects are discussed.

1. Introduction

Engineering education in the universities and colleges involves the combination of cognitive learning theory and experiential learning theory of learning to ensure transformative learning. These theories form part of the conceptual frameworks that describe how information is absorbed, processed, and retained during engineering education and training. While the application of cognitive learning theory in engineering education focuses on understanding the learners with regards to the complexities of human memory, experiential learning theory aims at developing in the learners the societal technical observation and community commitment skills necessary to shape and transform them into competent young engineers. Furthermore, transformative learning theory seeks to explain and enhance how the learners revise and reinterpret information acquired during engineering education and training [1]. Figure 1 presents an illustration of the dependence of transformative learning on cognitive and experiential theories of learning in engineering education because knowledge is continuously amassed via both personal and environmental experiences [2].

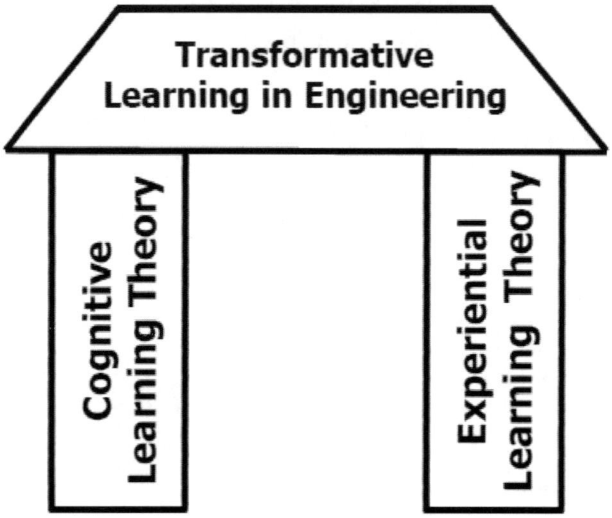

Figure 1. Cognitive learning theory (CLT) and Experiential Learning Theory (ELT), the two strong pillars of the transformative learning in engineering education and training

It is very interesting to know how the combination of the cognitive learning theory and the experiential learning makes the application of transformative learning effective in engineering education. In addition, transformative learning provides an individual the opportunity to acquire experience, question some assumptions during the learning and reflect on them at the later stage. Therefore, learning in engineering education is expected to transform the mind and the transformation is expected to be noticed by the learners and the society because transformative learning should shape people and re-orientate their minds. Therefore, the curriculum of engineering education and training in the higher institutions is cognitively designed with experiential learning as its core, and tailored towards developing critical and creative thinking skills that are essential in engineering practice.

Furthermore, variety of assessments administered in engineering education are purposed to provide the engineers-in-the-making the ability to evaluate evidence, appraise situations and circumstances to draw logical conclusions with respect to the analysis. Consequently, these assessments, that encourage transition from assessment of learning to assessment for learning, should transform the learners. However, the degree of the enhanced transformative learning in engineering education is forestalled by the size of the classes we have in our universities today. Therefore, it is high time we investigated the impact of large classes on transformative learning in engineering education and training in our higher institutions. Because knowing and understanding the effects of the size of classes on the degree of transformative learning in engineering education could pave the way to proffer lasting solutions to the problems. Several questions came to my mind in the process of investigating this problem. The key questions that should be addressed are: What are the effects of large class on transformative learning in engineering education? How can these effects be eradicated or minimized to promote transformative learning? How can the ways of minimizing these effects be integrated into engineering education curriculum to promote the acquisition of required skills in engineering profession?

IOP Conf. Series: Materials Science and Engineering **413** (2018) 012026 doi:10.1088/1757-899X/413/1/012026

2. Effects of large classes on transformative learning in engineering education

2.1. Untimely feedback

In engineering education, assessment and feedback on assessment are targeted at competence development. Therefore, it is imperative to understand the effect of feedback on student's conscious competence and unconscious competence towards enhancing transformative learning. Moving the unconscious competence to conscious competence in engineering education could enhance the acquisition of critical thinking skills expected of an engineer. However, this could be achieved effectively during one-on-one feedback session which could be time-consuming in a large class.

In agreement with Gibbs and Simpson [3], timely provision of feedback to leaners could provide enough time for them to receive further assistance, thereby enhancing the conception of knowledge and acquisition of skills. This is very true in engineering education and training where experiential learning is crucial to developing critical and creative thinking skills required as a professional engineer. But how will a teacher who must be involved in active research and other administrative issues provide timely feedback to students? In fact, this problem is aggravated if the class is a large class. How do we expect the teacher to strike the balance between teaching, research and administration knowing that his/her career development and promotion depends heavily on the research output? How do we effectively transform the minds of the learners via an enabling environment that promotes and supports transformative learning in our universities without having to reduce the yearly in-takes into the universities?

2.2. Unsustainable assessment

Improving standard of education depends strongly on improving student learning efficiency [3]. Furthermore; assessment should be a medium to encourage experiential learning that could lead to transformative learning. For example, training of young embryonic engineers through engineering education requires a form of assessment that will enable them to have refection-in-action and reflection-on-action during the training [4]. To acquire transformative learning that will with impact positively on the society depends strongly on the assessment methods employed during the teaching and learning. A learner will benefit a lot from assessment that is formative in nature because the knowledge and skills acquired through this assessment method are retained for a long time. However, the most reliable, rigorous and cheat-proof assessments system, that considers assessment as a yardstick for standard measurement, results into lifeless learning and unsustainable assessment [3]. Therefore, assessment that will support sustainable and transformative learning should not be too rigorous and not accompanied by lifeless learning since assessment system exerts great influence on students [5].

Thus, the type of assessment required in engineering education and training is the one that provides the engineers-in-the-making the ability to evaluate evidence, appraise situations and circumstances to draw logical conclusions with respect to the analysis. But the key question is: could the current assessment methods in engineering education in our universities promote transformative learning in large classes? As a means of providing conditions that will support transformative learning in higher education, Boud and Soler suggested coursework assignment assessment system rather than the summative assessment [4]. This is well applauded because the performance of students could be enhanced via coursework assignments. In addition, the coursework assessment could promote transformative learning if conducted in formative way. Furthermore, the coursework assignment could promote sustainable learning if detailed personalized feedback is frequently provided to the learners. However, the effectiveness of the coursework assignment, as a formative assessment method, might be hampered by the size of the classes. Administration of the coursework assignment in a large class is time consuming and rigorous. Therefore, the size of classes in our universities in the recent times has contributed to the dramatic decline in the use of formative assessment method in universities. Furthermore, large class constrains the available resource resulting in the decline of quantity and quality of feedback.

My recent experiences in the use of formative assessment in large classes to enhance transformative and experiential learning in chemical engineering education and training has made me to query the effectiveness of employing formative assessment in large classes. After the experiences, my mind has been clouded with one-million questions, a few of which are: are there ways by which transformative learning via formative assessment could be achieved in large classes in engineering education? What are the constraints mitigating effective transformative learning via formative assessment in a large class? How can the different forms of assessment and feedback be integrated into engineering education curriculum to promote acquisition of required skills in engineering profession via experiential learning? Furthermore, if providing quality and timely feedback to students could optimize the projected performance of a formative assessment in learning, then how can a quality and timely feedback be provided in a large class?

3. Enhancing transformative learning in a large class in engineering education

3.1. Providing timely feedback in a large class

In the field of engineering, fundamental knowledge is transferred to the learners by the teachers to help the learners cultivate or build critical thinking skills through the understanding and application of various engineering concepts. Gibbs and Simpson suggested that timely provision of feedback to leaners could provide enough time for them to receive further assistance, thereby enhancing the conception of knowledge and acquisition of skills [3]. How could quality and timely feedback be provided to students in a large class?

Gibbs and Simpson suggested that audiotaping during feedback could assist in providing timely and quality feedback to students because students could re-play the recorded feedback repeatedly until they fully understand the feedback [3]. This could be a good option in providing timely and quality feedback in a large class in engineering education. However, the use of audiotaping application could be hampered by insufficient technical devices and in some cases non-availability of required technologies to embark on it. In addition, the use of videotaping in giving feedback might require an extra assistance from another person who will videotape the discussion. In addition, this method could be effective when providing feedback to the whole class but not for one-on-one feedback. To alleviate this effect, Race suggests the use of shorter assignments and the number of assignments should be considerably reduced [6].

3.2. Developing a sustainable assessment in a large class

As a way of developing sustainable assessment in large classes in engineering education, Beck et al. suggest that course tutorials and lectures at the universities should be targeted at developing independence, intellectual maturity and creativity [7]. To ensure this is achieved at the universities, periodic monitoring of performance and progress of students using rubrics and reflective sessions is essential. Self-assessment by students proposed by McDonald [8] could be an excellent technique by which sustainable assessment could be promoted in a large class in higher education, especially in engineering education. Self-assessment could enable the teachers and the leaners to identify the needs and ways to meet them. Sustainable assessment in large classes in engineering education also could be realizable using “Flipped classrooms”. Furthermore, any assessment that will support sustainable and transformative learning in large classes should not be too rigorous and not be accompanied by lifeless learning [5] .

3.3. The use of a flipped classroom

Recently the use of flipped classroom was proposed and canvassed for as an excellent option to promote transformative learning in large classes in higher institutions. The flipped classroom is defined as a pedagogical approach whereby the traditional teaching approach and homework elements of a course are reversed. In this approach, students are made to view short videos or listen to audiotapes of lectures at home before the class session, and the normal class time is devoted to class exercises, discussions and case studies [9,10]. Compared to the traditional teacher-centered teaching and learning approach where the teacher is the primary source of information, the flipped classroom learning approach uses the learner-centered mechanism. Detailed comparison between the traditional teaching and learning model and the flipped classroom teaching and learning model is presented in

IOP Conf. Series: Materials Science and Engineering **413** (2018) 012026 doi:10.1088/1757-899X/413/1/012026

Table 1. In the learner-centered approach, the normal class time is dedicated to exploring topics in greater depth through exercises and case studies, thereby creating rich learning opportunities for the learners. In addition, this approach encourages active participation of the students and provides an avenue for the learners for personal evaluation.

Table 1. Comparison between traditional classroom and flipped classroom teaching and learning models [9,10]

Traditional classroom teaching and learning model	Flipped classroom teaching and learning model
Teacher prepares material to be delivered in class	Teacher records and shares lectures outside the class
Students listen to lectures and other guided instruction in class and take notes	Students watch/listen to lectures before coming to class
Homework is assigned to demonstrate understanding	Class time is devoted to applied learning activities and higher-order thinking tasks
Little time for students to receive supports from teacher and peers	Students receive support from teacher and peers as needed
Students may miss classes due to legitimate reasons	Missing classes due to legitimate reasons is no longer a problem
It does not help busy students to plan their learning properly	It helps busy students to plan their learning very well
It does not enhance student-teacher interaction	It enhances student-teacher interaction
No flexibility for students to choose when and where they learn	It creates flexibility for students to choose when and where they learn
Educators are not flexible in their expectations of student timelines for learning and in their assessments of students learning	Educators are flexible in their expectations of student timelines for learning and in their assessments of student learning

Despite the advantages of a flipped classroom teaching and learning approach over the traditional teaching and learning approach, it cannot replace the teacher. In addition, it does not encourage students to work alone. It is worthy to mention that it is not an online course but just a way to change what is done in class to promote transformative teaching and learning in large classes. As a matter of fact, the effective implementation of the flipped classroom approach could be thwarted by the affordability of required technology by the learners. Despite the aforementioned drawback, the flipped could be an excellent option to enhance teaching and learning in large classes in engineering education and training. In addition, the implementation of the model could pave the way for the development of other approaches that could enhance transformative learning in large classes in our higher institutions.

4. Conclusion

As much as one might be passionate about making teaching and learning a transformative medium through which minds of students are transformed, achieving transformative learning in large classes has been a mirage. As much as one could always purpose to make teaching and learning in large classes a medium through which the learners could acquire experiential learning that will shape their lives, but one is often discouraged by the outcome after using the traditional teaching and learning model. However, the embrace of flipped classroom model integrated with sustainable assessment and timely feedback could be instrumental to achieving the transformative learning in large classes in engineering education and training in the higher institutions.

Acknowledgments

Research is supported by the National Research Foundation of South Africa (NRF) through the KIC grant No. KIC180404317478.

References

[1] Taylor, E.W. Transformative learning theory [J]. New Directions for Adult and Continuing Education, Jossey-Bass., 2008, 5–15.
[2] Merriam, S. B., Caffarella, R. S., Baumgartner, L. M. Learning in adulthood: a comprehensive guide [B]. San Francisco: John Wiley & Sons, Inc, 2007.

IOP Conf. Series: Materials Science and Engineering **413** (2018) 012026 doi:10.1088/1757-899X/413/1/012026

[3] Gibbs, G., Simpson,C. Conditions under which Assessment Supports Students' Learning [J]. Learning and Teaching in Higher Education, 2004, 1:3-31.
[4] Boud, D., Soler, R. Sustainable Assessment revisited [J]. Assessment & Evaluation in Higher Education, 2015, 1-14.
[5] Miller, C.M.I., Parlett, M. Up to the mark: a study of the examination game [B], Guildford: Society for Research into Higher Education, 1974.
[6] Race, P. Designing assessment and feedback to enhance learning, The lecturer's Toolkit: A practical guide to assessment learning and teaching" [B]. 3rd Edition, Routledge: London, 2006, 79-84.
[7] Beck, R.J., Skinner, W.F., Schwabrow, L.A. A study of sustainable assessment theory in higher education tutorials [J]. Assessment & Evaluation in Higher education, 2013, 38 (3): 326-34.
[8] McDonald, B. Self-assessment for understanding [J]. Journal of Education, 188(1): 25-40.
[9] Bishop, J.L., Verleger, M.A. The flipped classroom: A survey of the research [C]. In proceedings of the 120th ASEE Annual Conference & Exposition, 23-26 June, Atlanta, USA.Paper ID 6219, 2013.
[10] Kurtz, G., Tsimerman, A., Steiner-Lavi, O. The flipped classroom approach: The answer to future learning [J]. European J. Open, Distance & E-learning, 17(2): 171-181.

IOP Conf. Series: Materials Science and Engineering **413** (2018) 012027 doi:10.1088/1757-899X/413/1/012027

Production of CNT yarns for use as filaments in incandescent bulb: effect of carbon source and state of catalyst on production of CNT

Veronica Bodiba, Ezinne Igbokwe, Ndanganeni Mahangani, Oluseyi Aberefa, Michael Olawale Daramola*, Sunny E. Iyuke

School of Chemical and Metallurgical Engineering, Faculty of Engineering and the Built Environment, University of the Witwatersrand, Wits 2050, Johannesburg, South Africa.

*Corresponding author e-mail address: Michael.Daramola@wits.ac.za

Abstract. Presence of highly aligned carbon nanotubes (CNTs) bundles is a prerequisite to producing good quality CNTs that will retain their original good electrical and mechanical properties when spun into yarn. In this study, effects of carbon source and state of catalyst on the CNT yield and quality during the production of CNTs were investigated. CNTs were produced in a Swirled Continuous CVD reactor using acetylene and methane as carbon sources and ferrocene as a catalyst to investigate the effect of carbon source and the production was carried out with the state of the catalyst altered to investigate effect of state of catalyst on the yield and quality of the CNTs. The quality and purity of the produced CNTs were checked with Raman and Scanning electron microscopy (SEM) provides information on the morphology of the produced CNTs. SEM images of the produced CNTs using acetylene gas reveal the presence of more nanoballs than those produced using methane under the similar conditions. However, amount of CNTs produced using methane is very low compared to the amount obtained when acetylene was used. The CNT yield from methane was enhanced when ferrocene (catalyst) was dissolved in cyclopentane. Results from SEM and Raman Spectroscopy indicates CNT bundles with longer lengths and reduced number of nanoballs when methane was used together with ferrocene dissolved in cyclopentane as compared to using ferrocene in a solid state.

1. Introduction

To maximize the excellent mechanical, electrical and thermal properties possessed by CNTs in the nanoscale in fabrication of components, devices and other applications has increased efforts and interest in production macroscopic assemblies of CNTs. Research aimed at optimizing the production of well aligned, good quality and quantity CNTs to meet the requirements for spinning into yarns are currently on top gear. The catalytic chemical vapor deposition (CCVD) is currently the most viable process for the synthesis of carbon nanotubes. The influence of numerous growth parameters on the resulting nanotubes characteristics, such as diameter, length, number of graphene layers, defect density etc., has been studied [1,2].

Generally, nano-sized transition metal particles, e.g., nickel, iron, cobalt, molybdenum and copper, have been successfully used in CCVD either in oxide or metallic forms or as mixtures [1,2]. The solubility of carbon in catalyst particles determines the ability of metal to catalytically decompose gaseous carbon-containing molecules. As Moisala et al. highlighted in their review, the solubility of carbon in iron and nickel significantly increases for metal particles with a diameter of less than approximately 10 nm [1]. Seidel and co-workers investigated the difference of the carbon nanotubes growth behaviour when Fe, Co, and Ni catalysts were used [3].

The authors observed that the order of the lowest growth temperatures agrees with the order of the bulk melting points of the three transition metals (Ni, 1,450 °C; Co, 1,490 °C; Fe, 1,540 °C). For binary compounds, e.g., cobalt–molybdenum [4], iron–molybdenum [5] and iron–cobalt [6,7], the yield of carbon nanotubes has been observed to increase significantly. However, the precise catalyst composition has a great influence on the final product. $Fe_{1-x}Co_x$ alloys are among the most efficient catalysts for the synthesis of multi-walled carbon nanotubes [6,7]. The resulting yield, as well as the nanotubes characteristics, strongly depends on the parameter x.

Most used carbon precursors in CVD method are methane, carbon monoxide, acetylene, ethylene, benzene and toluene [8]. It has been discovered that the use of unsaturated hydrocarbons as carbon precursors results in high yield deposition rate as compared to the use of saturated hydrocarbons [9]. Saturated hydrocarbons are favoured for the growth of single walled carbon nanotubes (SWCNTs) because they synthesize graphitized filaments with fewer walls than when using unsaturated hydrocarbons. Furthermore, excessive carbon precursor concentration or hydrocarbon self-pyrolysis are some of the causes of catalyst poisoning during high temperature processing. This can be avoided by limiting carbon precursor feed rate to the catalyst particles by controlling partial pressure of carbon precursor and choosing the carbon precursor according to the decomposition rate [10].

Obtaining high quality nanoyarns that will exhibit the same electrical and mechanical properties as the original carbon nanotubes (CNTs) bundles from which they are made has been a great challenge, possibly due to inconsistence in the results reported in literature on the production of high quality CNTs, intermediates for nanoyarn production. Against this background, effects of carbon source and state of catalyst on the CNT yield and quality during the production of CNTs in a Swirled Continuous CVD reactor were investigated in this study.

2.Experimental

CNTs were produced in a swirled continuous CVD reactor using acetylene and methane as carbon sources and ferrocene as a catalyst to investigate the effect of carbon source on the yield and quality of produced CNTs. During the production, 10 g of ferrocene ($F_e(CO)_5$) catalyst was vaporizer at 300°C in the vapourizer attached to the reactor and the vapourized catalyst was carried to the reaction zone, where acetylene or methane and H_2 gas were introduced, by argon gas. The flow rate of the carbon source, the argon gas and the H_2 gas into the reactor were 150 ml/min, 150 ml/min and 50 ml/min, respectively. The reaction zone (inside the CVD reactor) was maintained at reaction temperature of 850°C and the reaction was run for 12 minutes. The temperature at the reaction zone was measured with a PT-100 thermocouple and controlled with PID controller built with the reactor. To gain insight into the effect of temperature on the yield and quality of produced CNTs, the reaction temperature was varied from 850°C to 950°C at a step change of 50°C.

To investigate the effect of changing the state of catalyst from solid state to liquid state on the quality and yield of CNTs produced in the reactor, 10 g of ferrocene powder was dissolved in 10 ml of cyclopentane. The liquid solution was vaporizer at 300°C in the vapourizer attached to the reactor and the vapourized catalyst was carried to the reaction zone, where acetylene or methane and H_2 gas were introduced, by argon gas. The flow rate of the carbon source, the argon gas and the H_2 gas into the reactor were 150 ml/min, 150 ml/min and 50 ml/min, respectively. The reaction zone (inside the CVD reactor) was maintained at reaction temperature of 850°C and the reaction was run for 12 minutes. Figure 1 depicts the experimental set-up used in the investigation. The yield of the produced CNTs was calculated by measuring the amount of CNT collected after the production. The quality of the produced CNTs was checked using scanning electron microscope (SEM) equipped with energy-dispersive X-ray spectroscopy (EDS) for morphology and elemental composition; and Raman spectroscopy to check the degree of defect and graphitic formation. The results obtained from the characterization were compared to obtain logical conclusions.

Figure 1: An experimental set-up for CNT production.

3. Results and discussion

Figure 2 depicts the CNT yield obtained when acetylene and methane were used as the carbon sources. The use of acetylene displayed a higher yield compared to methane at the various temperatures studied. For both the maximum CNT yield was recorded at 900°C, where the CNT yield with acetylene at this temperature is about 14.5 times higher than that of methane. Acetylene gives high yield because it is easier to break carbon chains in unsaturated hydrocarbons than in saturated hydrocarbons [11]. The SEM images of CNTs produced are presented in Figure 3. From the images, it is evident that CNTs produced with acetylene contain more of nanoballs than the nanotubes and the length of the nanotubes is short, making production of good quality yarns from these nanotubes a mirage. However, better CNT bundles with fewer nanoballs were obtained when methane was used as a carbon source. The CNTs from methane contain a substantial number of nanotubes with reasonable length and excellent spinnable array required to produce yarns.

EDX could be employed to determine the purity and elemental composition of the produced CNTs from different carbon sources. Elemental compositions of the produced CNTs obtained from EDX are shown in Figure 4. The major component of the products obtained from the CVD using acetylene and methane as carbon sources are carbonaceous in nature (see Figure 4). Acetylene products contain more carbon but the accompanying SEM images indicate the quality of the product. Methane has more impurities from the iron based catalyst but this can be easily purified. The ratio of the intensities of the D and G bands is a suitable identification of the quality of bulk samples. Therefore, from the Raman spectra in Fig 5, it is evident that the use of methane produces CNTs with a fewer structural defects as compared to the CNTs from acetylene. There is also an indication that the methane source produced mostly single-walled carbon nanotubes (SWCNT) and double-walled carbon nanotubes (DWCNT). Both SWCNT and DWCNT show higher differences in intensities of D and G bands [12]. The presence of radial breathing mode (RBM) seen in Fig 5b is a strong indication of the presence of SWCNT in CNTs produced from methane carbon source [13].

IOP Conf. Series: Materials Science and Engineering **413** (2018) 012027 doi:10.1088/1757-899X/413/1/012027

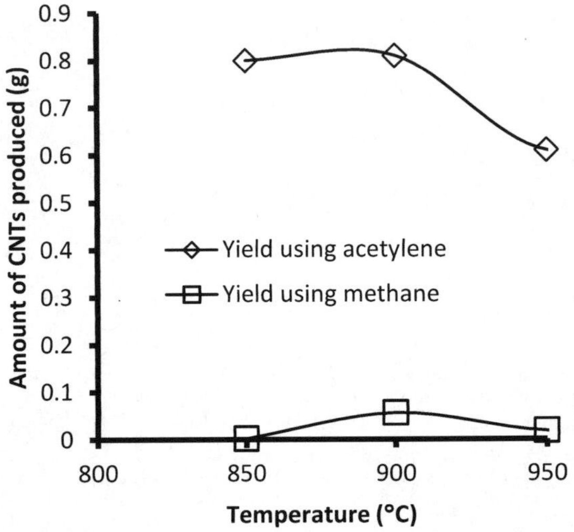

Figure 2: CNT production yield using various carbon sources-(methane and acetylene).

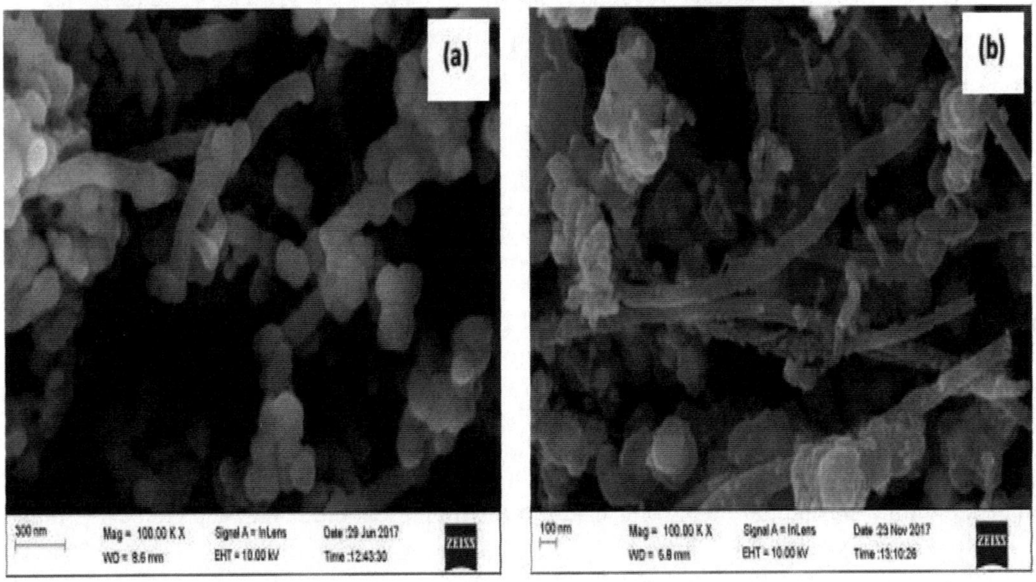

Figure 3: SEM images of CNT produced from (a) Acetylene (b) Methane

IOP Conf. Series: Materials Science and Engineering **413** (2018) 012027 doi:10.1088/1757-899X/413/1/012027

(a)

Element	Weight%	Atomic%
C K	89.08	94.37
O K	5.55	4.41
Fe L	5.37	1.22
Totals	100.00	

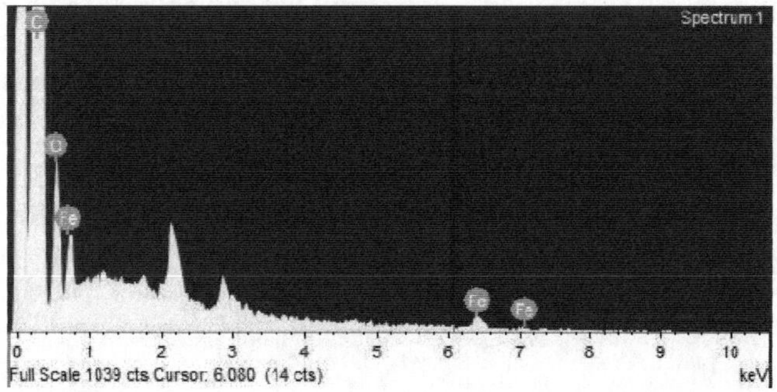

(b)

Element	Weight%	Atomic%
C K	78.43	90.31
O K	7.02	6.07
Si K	0.09	0.05
Fe K	14.46	3.58
Totals	100.00	

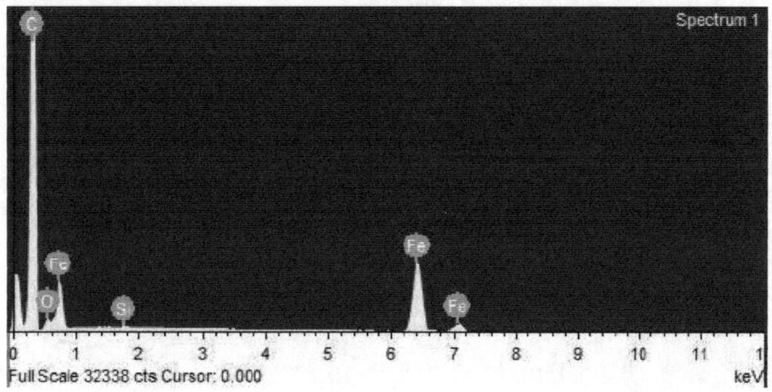

Figure 4: Energy Dispersive X-ray spectroscopy (EDX) of CNT produced from (a) Acetylene and (b) Methane

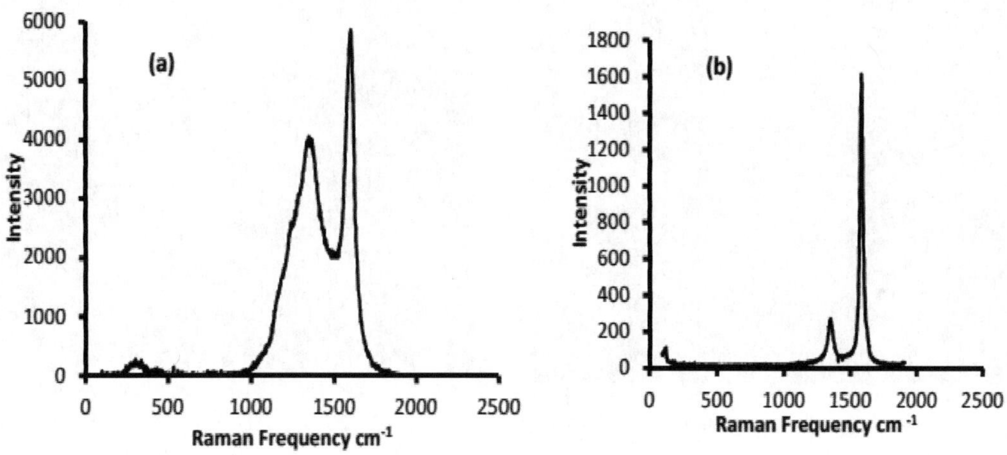

Figure 5: Raman spectroscopy of a) Acetylene and b) Methane

Fig 6 shows a significant improvement in the yield of CNT produced from ferrocene dissolved in cyclopentane. The maximum yield occurred at 900°C for both the ferrocene in solid state and the ferrocene in liquid state. When ferrocene was dissolved in cyclopentane, the CNT yield was 6.8 times higher than the CNT yield when ferrocene was in solid state. The higher yield could be attributed to the participation of cyclopentane in the reaction since it is a hydrocarbon, thereby enhancing the carbon concentration in the reaction [14]. Comparing the SEM images of the produced CNT bundles when the catalyst was in solid state (Fig 7a) with the SEM image when it was dissolved in cyclopentane (Fig 7b) reveals that there is presence of CNTs together with nanoballs in Fig 7a while Fig 7b shows a better formation of CNTs characterized with long array of spinnable multi-walled and single-walled CNTs with a few number of nanoballs. These types of CNTs formed when ferrocene was dissolved in cyclopentane will result in the formation of high quality CNT yarns would possess the same quality as the original CNTs.

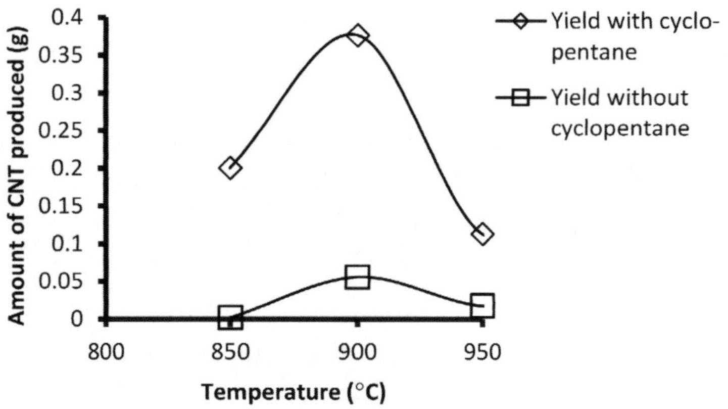

Figure 6: CNT yield with varying state of catalyst using methane as carbon source.

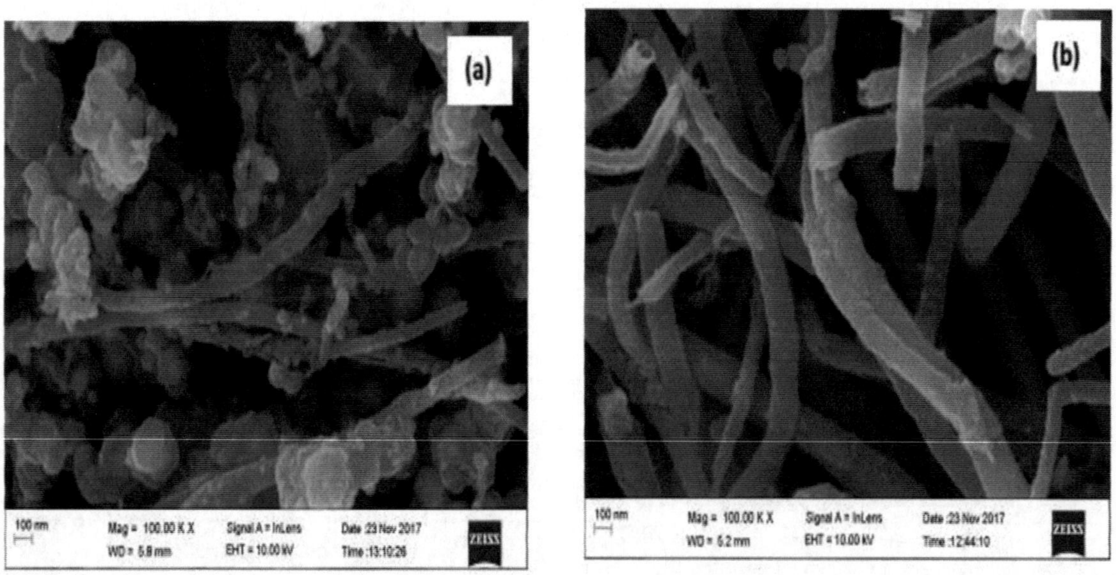

Figure 7: SEM images for CNT bundles produced at various conditions: (a) Methane used as a carbon source with ferrocene in a solid state and (b) Methane used as a carbon source with ferrocene dissolved in cyclopentane.

4.Conclusion

In this study, effect of carbon source and the state of catalyst on the yield and quality of CNTs has been investigated. Results of the investigation reveal that using acetylene as a carbon source yielded higher amount of CNTs compared to using methane gas. However, the quality of CNTs produced when methane gas is used is far better than when acetylene gas is used. In addition, pre-dissolution of the ferrocene catalyst in cyclopentane prior to vapourization enhances the quality and yield of CNTs produced as compared to non-dissolution of the catalyst prior to vapourization.

Acknowledgments

Research is supported by the National Research Foundation of South Africa grant No. NRF-CPRR106053.

References

[1] Moisala, A., Nasibulin, A.G., Kauppinen, E.I. The role of metal nanoparticles in the catalytic production of single-walled carbon nanotubes - a review [J]. J. Phys.: Condens. Matter, 2003, 14: S3011-S3035.

[2] Dupuis, A.C. The catalyst in the CCVD of carbon nanotubes - a review [J]. Prog. Mater. Sci., 2005, 50:929-961

[3] Seidel, R., Duesberg, G.S., Unger, E., Graham, A.P., Liebau, M., Kreupl, F. Chemical vapor deposition growth of single-walled carbon nanotubes at 600 °C and a simple growth model. J. Phys. Chem. B, 2004,108:1888-1893.

[4] Alvarez, W.E., Kitiyanan, B., Borgna, A., Resasco, D.E. Synergism of Co and Mo in the catalytic production of single-wall carbon nanotubes by decomposition of CO[J]. Carbon, 2001,39:547-558.

[5] Harutyunyan, A.R., Pradhan, B.K., Kim, U.J., Chen, G., Eklund, P.E. CVD synthesis of single wall carbon nanotubes under "soft" conditions [J]. Nano Lett., 2002, 2:525-530.

[6] Flahaut, E., Govindaraj, A., Peigney, A., Laurent, Ch., Rousset, A., Rao, C.N.R. Synthesis of single-walled carbon nanotubes using binary (Fe, Co, Ni) alloy nanoparticles prepared in situ by the reduction of oxide solid solutions [J]. Chem. Phys. Lett., 1999, 300: 236-242.

[7] Magrez, A., Seo, J.W., Mikó, Cs., Hernadi, K., Forró, L. Growth of carbon nanotubes with alkaline earth carbonate as support [J]. J. Phys. Chem. B, 2005,109: 10087-10091.

[8] Behabtu, N., Green, M. J., Pasquali, M. Carbon nanotube-based neat fibers [J]. Nano Today, 2008, 3(5–6):24–34.

[9] Awasthi, K., Srivastava, A., & Srivastava, O. N. Synthesis of carbon nanotubes [J]. Journal of Nanoscience and Nanotechnology, 2005, 5(10):1616-1636.

[10] Eatemadi, A., Daraee, H., Karimkhanloo, H., Kouhi, M., Zarghami, N., Akbarzadeh, A., Abasi, M., Younes Hanifehpour Y., Joo, S. W. Carbon nanotubes: properties, synthesis, purification, and medical applications [J]. Nanoscale research letters, 2014, 9(1): 393.

[11] Baker R.T.K. and Harris, P.S. Chemistry and Physics of Carbon [B]. J.P.L. Walker and P.A. Thrower (eds.), Deeker, New York : Basel , 1978.

[12] Costa, S., Borowiak-Palen, E., Kruszyńska, M., Bachmatiuk, A., Kaleńczuk, R. J. Characterization of carbon nanotubes by Raman spectroscopy [J]. Materials Science-Poland,2008, 26 (2): 433-441

[13] Dresselhaus, M.S., Dresselhaus, G., Saitoc, R. and Joriod A. Raman spectroscopy of carbon nanotubes [J]. Physics Reports, 2005, 409:47–99.

[14] Weissker, U., Silke, H., Albrecht, L. & Bernd, B. Carbon nanotubes filled with ferromagnetic materials. Materials, 2010, 3:4383-4427.

IOP Conf. Series: Materials Science and Engineering **413** (2018) 012028 doi:10.1088/1757-899X/413/1/012028

The Study of the Impact of Nano Carbon Additives on Astm A53 Mild Steel During Machining

Sunday A. Afolalu[1], Orenuga Oluwaseyifunmi[1], Samson O. Ongbali[1], Abiodun A. Abioye [1], Imhade P. Okokpujie[1], Oloyede Olamilekan R[2]

[1]Mechanical Engineering Department, Covenant University, Ota. Ogun-State, Nigeria.
[2]Mechanical and Mechatronics Engineering Department, Afe Babalola University, Ado. Ekiti-State

Abstract. In the world today, most of the treated steels were done using chemicals which in turn degrades rapidly. This study developed treated metal from ASTM A53 mild steel using pulverized palm kernel shell as carburizer and egg shell as energizer due to their high carbon and calcium carbonate content. Samples of hundred (100) pieces of ASTM A53 mild steel with dimensional size of 100 x 30 mm each were used. The elemental composition of the sample was checked using UV-VIZ spectrometric machine. Four sets of steel boxes were fabricated in the dimensions of 200 x 50 x 200 mm with two faces detachable and small extension for carrying the boxes at high temperature. Prepared samples buried in pulverized carburizer and energizer at a mixed percentage ratio of 80 to 20 were put together in a muffle furnace (2,500°C max) for case hardening at different holding temperature and time of (HT- 60, 90, 120 and 180 minutes) and (CT- 950, 1000, 1050, 1100°C). This procedure was repeated severally with varying holding time and temperatures. The initial experiment started from 950°C and rose up to1100°C. The hardness for surface, interface and core were taken using hardness tester. Surface hardness of 139.4 HV at holding temperature 980°C and time 180 minutes gave best increase of 43% over the control of 79.7 HV while hardness at interface and core were 146 HV and 147 HV with percentages increase of 9 and 15.3 was achieved. The findings in this study also confirmed that waste materials with some element of carbon can be used to replace foreign chemicals intended for the same purpose.
Keywords: carburization; ASTM A53 Steel; Carbon additives; palm kernel shell; eggshell. Hardness

1. Introduction

The term steel is typically interpreted as meaning an iron-based amalgam containing carbon and steels of under 0.25% carbon (regularly alluded to as gentle steel) have to some degree higher quality close to the upper carbon level. Low carbon steel also called mild steel has lower carbon content than other steels and so naturally it is easier for it for cold-forming processes to be applied to it due to their soft and ductile nature. Low carbon steel becomes a reasonable choice because of its low cost and ease of handling especially when strength is not the most important factor in consideration [1-2]. Medium carbon steels have great machining qualities, and one of the more famous evaluations utilized as a part of machined steel item is AISI 1045. AISI 1045 can likewise be solidified by heating the material too roughly 820-850C (1508 - 1562 F) and held until the point that the material achieves a uniform temperature. It ought to be drenched for one hour for every 25 mm area of material and in this manner cooled in still air [3]. High carbon steels are those with carbon substance in the vicinity of 0.60% and 1.4% of the general weight. The composites in this specific class constitute the most grounded and hardest inside the three gatherings, yet they are additionally the minimum pliable. The essential room temperature structure of iron and steel is called "Ferrite" or alpha iron, and has a body focused cubic (BCC) precious stone structure [4-5]. Ferrite can break up to .0025% carbon (by weight—builds dependably utilize weight not particle %); any additional carbon introduce accelerates out as iron carbide, which turns out to be a piece of the microstructure and expands quality. Over 912C (1674F) Iron changes to a face focused cubic (FCC) structure called Austenite or gamma press, which can break up considerably more carbon—up to 2%. The change from Austenite to Ferrite gives the premise to steel heat treatment [6]. Heat treatment is a procedure that includes a blend of time-controlled heating and cooling operations of metal without changing the item shape that will deliver wants mechanical properties and to watch the microstructure after heat treatment [7-8]. Carburizing similarly implied as case hardening is a heat treatment process that makes a surface which is impenetrable to wear, while keeping up toughness and nature of the inside. This treatment is associated with low carbon steel parts in the wake of machining, and furthermore high composite steel bearing, gears, and distinctive portions [9-10]. Most carburizing is done by heating parts in either a pit warmer, or settled atmosphere radiator, and displaying carburizing gases at temperature. Gas carburizing considers exact control of both the system temperature and carburizing condition (carbon potential) [11]. The primary reason for the carburizing procedure is to get a hard and wear protection surface on machine parts by improvement of the surface layer with carbon to a fixation from 0.75 to 1.2 % and ensuing extinguishing. Steel which has been carburized and extinguished (case – solidified) has a higher weariness confine [12]. Low carbon steels, containing from 0.1 to 0.18 % carbon, might be subjected to carburizing. Steel with a higher carbon content (0.2 to 0.3 %c) might be utilized for vast segments Steel with expanded carbon content (0.25 to 0.35 %) has a more grounded center this empowers the profundity of the carburized case to be decreased along these lines streamlining resulting heat treatment [13-15]. It is notable that composites having high hardness tend to be weak and helpless to fast crack under load, and in this way an article or part made altogether from such an amalgam would not perform acceptably under affect or different sorts of quickly connected anxieties. This is especially valid in steels where high carbon substances are utilized keeping in mind the end goal to acquire high hardness. To overcome this inborn issue, the procedure of case solidifying has been created for steels [16]. In one process for case solidifying steel, the surface of an 'article or parts produced using a low-carbon steel is enhanced in carbon by warming the article or part in con thoughtfulness with a carburizing medium. Amid this carburizing treatment, carbon is caused to diffuse into the low-carbon steel to deliver an advanced layer, ordinarily between 0.005 inch and 0.100-inch profound relying on a definitive use for the article or part [17-18]. The carbon-enhanced layer, known as the case, more often than not contains around 0.6% to about 1.3% carbon, while the low-carbon divide remaining, known as the center, more often than not contains just around 0.05% to around 0.27% carbon. The real case solidifying is affected by cooling' the carburized article or part from a temperature at which the carburized case is totally or generously austenitic (austenitizing temperature) to a temperature at which the case changes to martensite [19-22]. The accomplishment of this austenitizing or solidifying temperature might be affected in a few ways. The article or part can be authorized at the austenitizing temperature or at a higher temperature and cooled specifically to cause change to martensite, or the article or part can be heated to an austenitizing temperature from room temperature after past carburizing or reining heat medications [23-25]. The impact of case solidifying is, hence,' to create a hard, wear safe surface (case) which is likewise unavoidably fragile, furthermore, to consolidate this with a center which has great flexibility despite the fact that of lower hardness [26-28]. The significant part of eggshells is calcium carbonate, $CaCO3(s)$ and it can be achieved volumetrically by utilizing a trademark response of carbonate mixes, in particular their response with acids. Calcium carbonate (limestone) is exceptionally insoluble in unadulterated water yet promptly responds in corrosive as indicated by the response underneath. Charcoal has turned out to be extremely attractive throughout the years and it is acknowledged by and large. Charcoal and also kindling have shaped a noteworthy piece of the sustainable power hotspot for creating territories of the world. Resellers evaluated that the market for charcoal can create up to $1.8 billion dollars regarding vitality making its gross utilization higher than that of power [28-29]. Animal charcoal has additionally been utilized as carburization operator and it is gotten by scorching bones, meat and blood of creatures. The theme of testing mechanical materials has turned out to be such an imperative piece of designing, ending up similarly as vital as applying building strategies to enhance mechanical properties of metals and different materials utilized as a part of training [30], As of late, more respect is given to the translation of the outcomes given by mechanical test regard to how they enhance execution in benefit and furthermore giving sensible understanding into the capacity of a material to play out the require work [31].

2. Materials and Method

The materials used in this research work were palm kernel shell and eggshell as shown in figure 1 which was first dried to remove the moisture content thereafter grinded, milled and sieved to pulverized fine smooth particles to increase the surface area of the agents and to increase the efficiency of the release of the carbon. Egg shell contains a good measure of calcium carbonate and it was used as an energizer. Palm Kernel Shell was recuperated as remaining waste in the extraction of the bit from the nut, after palm oil had been extracted from the monocarp of the oil palm organic product. Samples of hundred (100) pieces of ASTM A53 mild steel

IOP Conf. Series: Materials Science and Engineering **413** (2018) 012028 doi:10.1088/1757-899X/413/1/012028

with dimensional size of 100 x 30 mm each were used. The elemental composition of the sample steel material was checked using UV-VIZ spectrometric machine as shown in Table 3. Four sets of steel boxes were fabricated in the dimensions of 200 x 50 x 200 mm with two faces detachable and small extension for carrying the boxes at high temperature. Mixture of pulverized palm kernel and egg shell at a percentage ratio of 80 to 20 was engaged in the treatment. Muffle furnace of 2,500oC max was used for the carburization. An adequate precaution safety rule was followed in order to achieve high hardness with a high level of consistency and precision, while safeguarding the surrounding of the chamber condition. Prepared samples buried in pulverized carburizer and energizer at a mixed percentage ratio of 80 to 20 were put together in a muffle furnace for carburization process at different holding temperature and time of (HT- 60, 90, 120 and 180 minutes) and (CT- 950, 1000, 1050, 1100OC). This procedure was repeated severally with varying holding time and temperatures. The initial experiment begins from 950oC and goes up to 1100oC. It was ensured that the temperature range of the carburization process was below the melting point of the grade of steel chosen however over the austenizing temperature of the steel grade involved. The temperature blends were explained in the experimental design as shown in Table 1 and 2.

2.1 Experimental Design

Table 1- Case-hardening soaking time with temperatures

Holding Time / Number of minutes			
60 minutes/950oC	60 minutes/1000oC	60 minutes/1050oC	60 minutes/1100oC
90 minutes/950oC	90 minutes/1000oC	90 minutes/1050oC	90 minutes/1100oC
120 minutes/950oC	120 minutes/1000oC	120 minutes/1050oC	120 minutes/1100oC
180 minutes/950oC	180 minutes/1000oC	180 minutes/1050oC	180 minutes/1100oC

Table 2: Sample labels for the experiment

Temperature (°C)	Time (Mins)	Hardness Test	Compositional Analysis
950	60	1B	1C
950	90	2B	2C
950	120	3B	3C
950	180	4B	4C
1000	60	5B	5C
1000	90	6B	6C
1050	60	7B	7C
1050	90	8B	8C
1100	60	9B	9C
1100	90	10B	10C
CONTROL	CONTNROL	CONTROLB	CONTROLC

Table 3: Compositional analysis

ELEMENTS	COMPOSITION (%)
C	0.25
Si	0.852
Mn	0.95
P	0.05
S	0.045

IOP Conf. Series: Materials Science and Engineering **413** (2018) 012028 doi:10.1088/1757-899X/413/1/012028

Cr	0.382
Ni	0.210
Mo	0.206
Al	0.022

2.2 Core, Surface and Interface Hardness Test

The hardness for core, interface and core was performed by utilizing an instrument called hardness tester, which applies distinctive burdens and appropriately arranged. The load was connected through an indenter relying upon the scale utilized. The hardness esteem was finished by estimating the extent of the mark left by the indenter in the wake of having impressed the load for a controlled time for the Vickers. The results for the hardness were shown in Tables 4, 5 and 6

3. Experimental Results and Discussion.

Table 4: Temperature and Time variation effects on surface hardness

Temperature	60 minutes	90 minutes	120 minutes	180 minutes	control
950	117.1	124.5	131.9	139.3	79.7
1000	139.15	139	119.24	99.47	79.7
1050	127.7	116.4	104.16	91.93	79.7
1100	108.1	118.5	105.56	92.63	79.7

Table 5: Temperature and Time variation effects on interface hardness

Temperature	60 minutes	90 minutes	120 minutes	180 minutes	Control
950	125.4	127.2	129	130.8	132.5
1000	127.35	123.9	119.24	99.5	132.5
1050	120.95	118.0	114.16	100.9	132.5
1100	127.4	146.0	125.6	102.6	132.5

Table 6: Temperature and time variation effects on core hardness

Temperature	60 minutes	90 minutes	120 minutes	180 minutes	control
950	128.1	123.76	119.43	115.1	124.5
1000	120.3	125.5	125.2	124.9	124.5
1050	122.2	118.9	120.77	122.64	124.5
1100	147.7	133.5	130.5	127.5	124.5

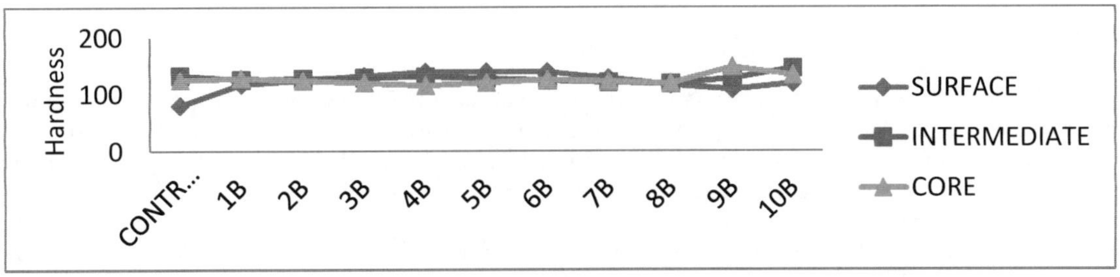

Figure 1: Hardness variation behaviour on case-hardened samples.

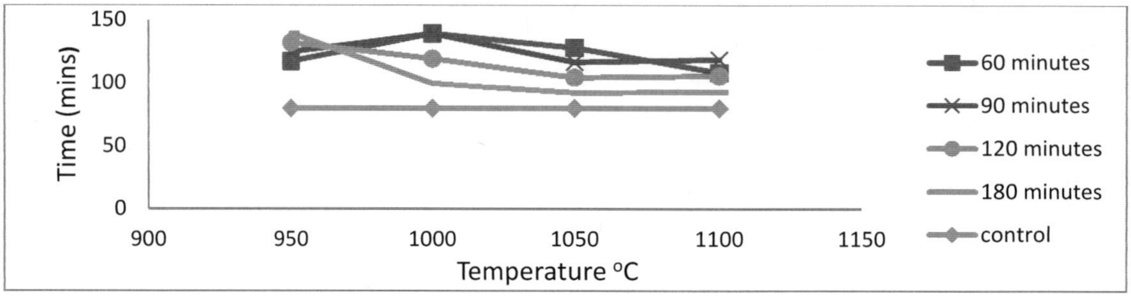

Figure 2: Effects of time and temperature variance on surface hardness.

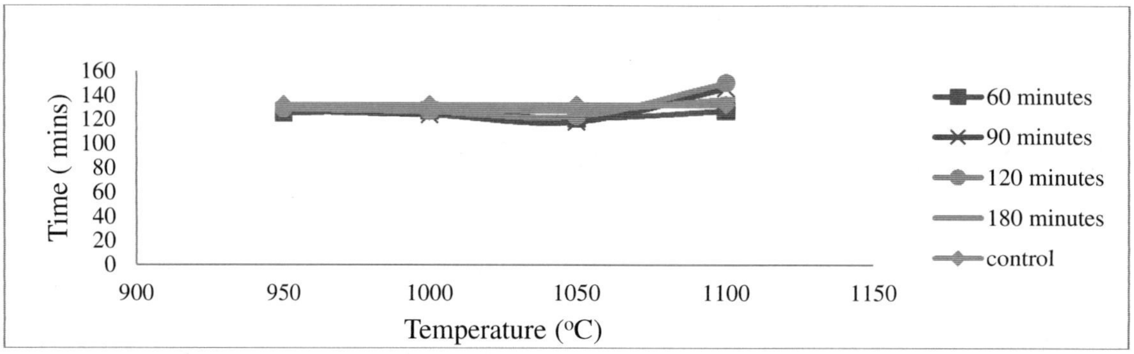

Figure 3: Effects of time and temperature variance on interface hardness

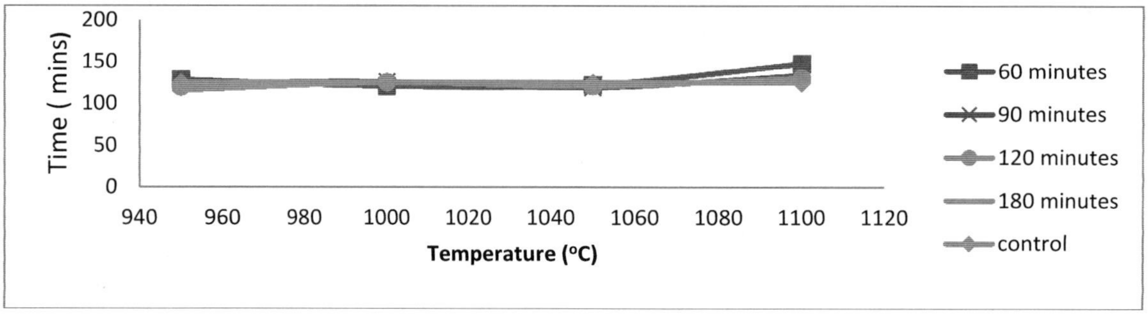

Figure 4: Effects of time and temperature variance on core hardness.

3.1 Discussion

The behaviour of the results shown in Figure 1 provides us with the knowledge that the more holding time applied to the case hardening at different temperature, the more carbon will be infused into the surface of the steel. Considering the hardness variation behaviour on case-hardened samples in Figure 2 with surface hardness of 139.4 HV at holding temperature 980oC and time 180 minutes gave best increase of 43% over the control of 79.7 HV. Checking through Figures 3 and 4, hardness at interface and core were 146 HV and 147 HV with percentages increase of 9 and 15.3 achieved. The best hardness test result shown above was due to the application of eggshell as an energizer which improved the mechanical properties of the carburized steel. The first of which was the surface hardness that increased generally at every sample test because of the high penetration of infused carbon into the surface of the material and formed a deeper case in agreement with Afolalu et al., (2015). Also, from the graphs it can be seen that as the holding time increases, there was an increase in hardness but some cases, increase occurs at a diminishing rate which was in relation to what was postulated by Aramide (2009).

4. Conclusion

This project focused on ASTM A53 mild steel and with the positive results obtained, similar method can be engaged on other grades of steel and recycling of wastes with element of carbon should be encouraged. The findings in this study also confirmed that locally source materials with the required chemical composition can be used to replace foreign chemicals intended for the same purpose.

Acknowledgements

The authors acknowledge the financial support offered by Covenant University in actualization of this research work for publication.

Reference

[1] Alaneme, K. K., Olanrewaju, S. O., & Bodunrin, M. O. (2011). Development and performance evaluation of a salt bath furnace. *International Journal of HNYJMechanical and Materials Engineering*, *6*(1), 67–74.

[2] Aramide, F. O., Ibitoye, S. A., Oladele, I. O., & Borode, J. O. (2009). Effects of Carburization Time and Temperature on the Mechanical Properties of Carburized Mild Steel, Using Activated Carbon as Carburizer. *Materials Research*, *12*(4), 483–487. https://doi.org/10.1590/S1516-14392009000400018

[3] Awopetu, O. O., & Ayodeji, S. P. (2008). Effect of Type of Workpiece Material on Chip Formation Process, *11*(3), 181–186.

[4] Afolalu, S. A., Adejuyigbe, S. B., Adetunji, O. R., & Olusola, O. I. (2015). Production of Cutting Tools from Recycled Steel with Palm Kernel Shell as Carbon Additives. *International Journal of Innovation and Applied Studies*, *12*(1), 110.

[5] Afolalu, S. A., Adejuyigbe, S. B., & Adetunji, O. R. (2015) Impacts of Carburizing Temperature and Holding Time on Wear of High Speed Steel Cutting Tools. *International Journal of Scientific and Engineering Research.* 6 (5), 905-909. 847–854. https://doi.org/10.1051/jphyscol:19824139

[6] Afolalu, S. A., Adejuyigbe, S. B., Adetunji, O. R., & Olusola, O. I. (2015). Production of Cutting Tools from Recycled Steel with Palm Kernel Shell as Carbon Additives. *International Journal of Innovation and Applied Studies*, *12*(1), 110.

[7] Orisanmi, B. O., Afolalu, S. A., Adetunji, O. R., Salawu, E. Y., & Okokpujie, I. P. (2017). Cost of Corrosion of Metallic Products in Federal University of Agriculture, Abeokuta. *International Journal of Applied Engineering Research*, *12*(24), 14141-14147.

[8] Abioye, A. A., Atanda, P. O., Abioye, O. P., Afolalu, S. A., & Dirisu, J. O. (2017). Microstructural Characterization and Some Mechanical Behaviour of Low Manganese Austempered Ferritic Ductile Iron. *International Journal of Applied Engineering Research*, *12*(23), 14435-14441.

[9] Adetunji, O. R., Musa, A. A., & Afolalu, S. A. (2015). Computational Modelling of Chromium Steel in High Temperature Applications. *International Journal of Innovation and Applied Studies*, *12*(4), 1015.

[10] Oyinbo, S. T., Ikumapayi, O. M., Ajiboye, J. S., & Afolalu, S. A. (2015). Numerical Simulation of Axisymmetric and Asymmetric Extrusion Process Using Finite Element Method. *International Journal of Scientific & Engineering Research*, *6*(6), 1246-1259

[11] Adetunji, O. R., Ude, O. O., Kuye, S. I., Dare, E. O., Alamu, K. O., & Afolalu, S. A. (2016). Potentiodynamic Polarization of Brass, Stainless and Coated Mild Steel in 1M Sodium Chloride

Solution. In *International Journal of Engineering Research in Africa* (Vol. 23, pp. 1-6). Trans Tech Publications.

[12] Adetunji O.R., Adegbola A.O., & Afolalu, S.A. (2015). Comparative Study of Case-Hardening and Water-Quenching of Mild Steel Rod on Its Mechanical Properties. *International Journal of Advance Research,* 3(6), 1-9.

[13] Dirisu, J.O., Asere, A.A., Oyekunle, J.A., Adewole, B.Z., Ajayi, O.O., Afolalu, S.A., Joseph, O.O., & Abioye, A.A. (2017). Comparison of the Elemental Structure and Emission Characteristics of Selected PVC and Non PVC Ceiling Materials Available in Nigerian Markets. *International Journal of Applied Engineering Research.* 12(23), 13755-13758.

[14] A.A. Abioye, O. P. Abioye, O. O. Ajayi, S. A. Afolalu, M. A. Fajobi and P. O. Atanda, Mechanical and Microstructural Characterization of Ductile Iron Produced from Fuel- Fired Rotary Furnace, International Journal of Mechanical Engineering and Technology 9(1), 2018. pp. 694–704

[15] Afolalu, S. A., Abioye, A. A., Dirisu, J. O., Okokpujie, I. P., Ajayi, O. O., & Adetunji, O. R. (2018, April). Investigation of wear land and rate of locally made HSS cutting tool. In AIP *Conference Proceedings* (Vol. 1957, No. 1, p. 050002). AIP Publishing

[16] Afolalu, S. A., Abioye, O. P., Salawu, E. Y., Okokpujie, I. P., Abioye, A. A., Omotosho, O. A., & Ajayi, O. O. (2018, April). Impact of heat treatment on HSS cutting tool (ASTM A600) and its behaviour during machining of mild steel (ASTM A36). In *AIP Conference Proceedings* (Vol. 1957, No. 1, p. 050003). AIP Publishing.

[17] Afolalu, S. A., Okokpujie, I. P., Salawu, E. Y., Abioye, A. A., Abioye, O. P., & Ikumapayi, O. M. (2018, April). Study of the performances of nano-case treatment cutting tools on carbon steel work material during turning operation. In *AIP Conference Proceedings* (Vol. 1957, No. 1, p. 050001). AIP Publishing.

[18] Udo, M. O., Esezobor, D. E., Apeh, F. I., & Afolalu, A. S. (2018). Factors Affecting Ballability of Mixture Iron Ore Concentrates and Iron Oxide Bearing Wastes in Metallurgical Processing. *Journal of Ecological Engineering Vol, 19*, 3.

[19] Afolalu, S. A., Abioye, A. A., Udo, M. O., Adetunji, O. R., Ikumapayi, O. M., & Adejuyigbe, S. B. (2018). Data showing the effects of temperature and time variances on nano-additives treatment of mild steel during machining. Data in Brief 19(2018)456–461.

[20] Okokpujie, I. P., Ikumapayi, O. M., Okonkwo, U. C., Salawu, E. Y., Afolalu, S. A., Dirisu, J. O., ... & Ajayi, O. O. (2017). Experimental and Mathematical Modeling for Prediction of Tool Wear on the Machining of Aluminium 6061 Alloy by High Speed Steel Tools. *Open Engineering*, *7*(1), 461-469.

[21] Babatunde, O. G., Fayose, R. S., Daniyan, A. A., Ezenwafor, T. C., Oniya, E. O., Ajayi, N. O., … Olusunle, S. O. O. (2017). Development of a Novel Oven-Furnace. *Journal of Minerals and Materials Characterization and Engineering*, *5*(2), 62–73. https://doi.org/10.4236/jmmce.2017.52006

[22] Fono-Tamo, R. ., & Koya, O. . (2013). Characterisation of Pulverised Palm Kernel Shell for Sustainable Waste. *International Journal of Scientific & Engineering Research*, *4*(4), 6–10. https://doi.org/2229-5518

[23] I. E., A., R. Kamal, A., & A. Ogunjirin, O. (2013). Effects of Heat Treatment on the Properties of Mild Steel Using Different Quenchants. *Frontiers in Science*, *2*(6), 153–158. https://doi.org/10.5923/j.fs.20120206.04

[24] K. A, O., R., A., & Adewale, M. D. (2012). Effective Modeling and Simulation of Engineering Problems with COMSOL Multiphysics. *International Journal of Science and Technology*, *2*(10), 742–748.

[25] Kingsley, O., Silas, U., Samson, A., & Oluwafemi, A. (2012). Development of Low Heat Treatment Furnace, *2*(7), 188–194.

[26] MF, H. (2016). Analysis of Mechanical Behavior and Microstructural Characteristics Change of ASTM A-36 Steel Applying Various Heat Treatment. *Journal of Material Science & Engineering*, *5*(2). https://doi.org/10.4172/2169-0022.1000227

[27] Momoh, I. M., Akinribide, O. J., Olowonubi, J. A., & Ayanleke, J. O. (2015). Effect of walnut shell powder on the mechanical properties of case-hardened steel, *4*(2), 209–218.

[28] Ogundele, A. T., Eludoyin, O. S., & Oladapo, O. S. (2011). Assessment of impacts of charcoal

IOP Conf. Series: Materials Science and Engineering **413** (2018) 012028 doi:10.1088/1757-899X/413/1/012028

production on soil properties in the derived savanna , Oyo state , Nigeria. *Journal of Soil Science and Environmental Management*, *2*(May), 142–146.

[29] Ohize, E. J. (2009). Effects of Local Carbonaceous Materials on the Mechanical Properties of Mild Steel, *13*(2), 107–113.

[30] Paul, I. A., Bem, N. G., Justine, N. I., & Joy, O. N. (2013). Investigation of Egg Shell Waste as an Enhancer in the Carburization of Mild Steel, *1*(2), 29–33. https://doi.org/10.12691/ajmse-1-2-3

[31] Pereira, B. L. C., Oliveira, A. C., Carvalho, A. M. M. L., Carneiro, A. de C. O., Santos, L. C., & Vital, B. R. (2012). Quality of Wood and Charcoal from *Eucalyptus* Clones for Ironmaster Use. *International Journal of Forestry Research*, *2012*, 1–8. https://doi.org/10.1155/2012/523025

[32] Roell, Z. (n.d.). Vickers Hardness Test. *Indentec Hardness Testing Machines Limited*, 8–9. https://doi.org/http://dx.doi.org/10.1016/S1553-4650(13)01241-71

IOP Conf. Series: Materials Science and Engineering **413** (2018) 012029 doi:10.1088/1757-899X/413/1/012029

200 Watts Portable Solar Generator for a Typical Barbing Salon

Edeghe, M. B[1*], Babalola P. O[1]

[1]Department of Mechanical Engineering, Covenant University, Ota, Nigeria.
E-mail address: marvelous.edeghe@stu.cu.edu.ng, phillip.babalola@covenantuniversity.edu.ng

Abstract. The need for power alternatives due to the instability of the on-grid supply becomes very essential. A portable solar generator is proffered to solve that problem as it would replace the environmentally unfriendly petrol and diesel generator being used by many in Nigeria today. The main aim of this work is to design and construct a portable solar generator capable of satisfying the load requirements for a typical barbing salon. Two barbing salons, one at Daniel Hall and the other at Chapel in Covenant University were surveyed for this project to determine their load requirements. The survey indicated a barest demand per hour of 215Wh and a peak demand of 525Wh. The peak demand was chosen as the target. The system has features such as metal frame box and wheel to enhance mobility and appropriately sized components such as the inverter, charge controller and battery. Furthermore, the system can be appropriately sized for other load categories not limited to homes as well as industrial applications. On completion, the final product will weight 45Kg that supplies both 230V 1000W AC and a regulated 12V DC power for a maximum time of 3 hours. With the average body weight indicated as 60.7 Kg this device can be conveniently moved by an average adult.
Keywords: solar panel, inverter, battery

1. Introduction

Nations in the developing world are facing the problem of limited access to energy and application, especially in the rural areas and communities. Hence, there is an urgent need to solve this problem. These problems act as a barrier, preventing serious economic and social development. In Nigeria today, the epileptic power supply makes reliance on the electric grid very unadvisable. To solve this problem, petrol or diesel generators are commonly used. These however, consume fuel in large quantities and pollute the environment. Solar energy as a means of solving this energy challenge has proved to be the most promising [1]. Harnessing solar energy in Nigeria is especially advantageous as it is situated on the sunniest place on the planet, theoretically it's estimated that the concentrated solar power (CSP) and photo-voltaic (PV) energy available in Africa is about 470 and 660 petawatt hours (PWh) [2]. Solar technology has come a long way since it was first conceived with the conversion efficiency of recent state of the art PV modules within the range of 15-20%, and a payback period of 2-4 years and improved panel life time of about twenty-five (25) years

2. System Components and Selection

In carrying out the design of a 200 Watts portable solar generator, the following processes are required: load determination for the intended typical barbing salon; component sizing; construction and assembly, and final economic analysis and justification of the project.

2.1 Battery

The batteries enable the PV array to operate at its optimum power output by keeping the electric load nearly constant. For the purpose of this project, a 12V 75AH Deep Cycle Gel Battery (Fig. 1) was selected [2,3].

IOP Conf. Series: Materials Science and Engineering **413** (2018) 012029 doi:10.1088/1757-899X/413/1/012029

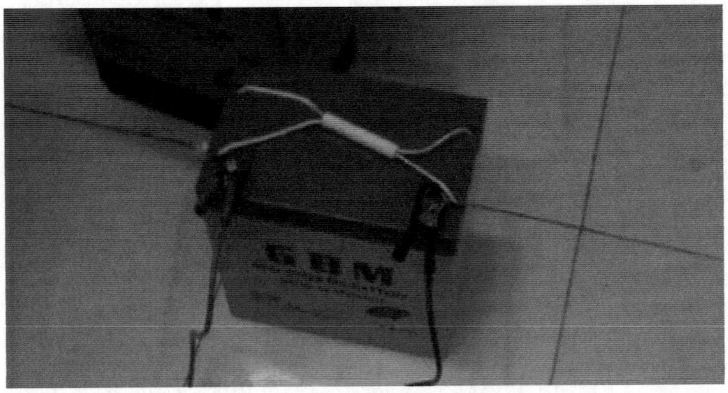

Figure 1: 12V 75AH Battery

2.2. Inverter

This device is installed in the PV system to convert direct current to alternating current. Several solar power systems generate direct current (DC) which is stored in batteries while nearly all lighting, appliances, motors etc. require alternate current (AC) power to operate. As a result, an inverter is required to switch from battery-stored DC to standard power [3]. A 12V, 1KVA inverter (Fig. 2) was selected for this project.

Figure 2: Intelligent Power Inverter 12V,1KVA

2.3 Charge Controller

A charge controller helps to prevent the solar panels from forcing too much electricity into the battery and overcharging it. Charge controllers (Fig. 3) are connected between the panels and the battery. It works by constantly checking the voltage of the battery. If the voltage gets to high the controllers stops electricity from entering the battery [4,5].

Figure 3: Charge Controller

IOP Conf. Series: Materials Science and Engineering **413** (2018) 012029 doi:10.1088/1757-899X/413/1/012029

2.4 Solar panel

A 200 Watts Monocrystalline solar panel (Fig. 4) was selected for the purposes of this project.

Figure 4: 200 Watts Monocrytalline Solar Panel

2.5 Component sizing and load determination for typical barbing salon

The load used was selected from the table below. This table indicates the typical load values for different domestic appliances.

Table 1: Typical load table for a household

Electrical Appliance Typical Energy Consumption Table

Appliance	Consumption (Watts)	Appliance	Consumption (Watts)
Toaster	800-1500	Television - 25" color	150
Popcorn Popper	250	Television - 19" color	70
Blender	300	Television - 12" black and white	20
Electric cooker with oven	1000-2500	VCR	40
Microwave	600-1500	CD player	35
Waffle Iron	1200	Stereo	20
Hot Plate	1200	Clock radio	1
Frying Pan	1200	AM/FM auto cassette player	8
Dishwasher	1200-1500	Satellite dish	30
Sink waste disposal	450	CB radio	5
Washing machine - Automatic	500	Electric clock	3
Washing machine - Manual	300	Radiotelephone - Receiving mode	5
Vacuum cleaner - High Power	1600-2000	Radiotelephone - Transmitting mode	40-150
Vacuum cleaner - Upright	200-700	Lights:100 watt incandescent	100
Vacuum cleaner - Hand	100	Lights:25 watt compact fluorescent	28
Sewing machine	100	Lights:50 watt DC incandescent	50
Iron	1000	Lights:40 watt DC halogen	40
Clothes dryer - Electric	4000	Lights:20 watt DC compact flourescent	22
Clothes dryer - Gas heated	300-400	Lights: Compact fluorescent 40 watt Incandescent equivalent	11
Heater - Electric water heater	4000	Lights: Compact fluorescent 60 watt Incandescent equivalent	16
Heater - Engine block	150-1000	Lights: Compact fluorescent 75 watt Incandescent equivalent	20
Heater - Portable	1500	Lights: Compact fluorescent 100 watt Incandescent equivalent	30
Heater - Waterbed	400	Electric mower	1500
Heater - Stock tank	100	Hedge trimmer	450
Furnace blower	300-1000	Weed eater	500

From the table the load required for barbing salon was selected. The items selected include the following; a television set, standing fan, Clippers, energy saving bulbs and a radio [6,7].

IOP Conf. Series: Materials Science and Engineering **413** (2018) 012029 doi:10.1088/1757-899X/413/1/012029

Table 2: Load calculation to determine panel size and number

S/N	ITEM	QUANTITY	SIZE (Watts)	HOURS IN OPERATION (HRS)	TOTAL (Watts-hour)
1	Television	1	60W	3	180Wh
2	Fan	1	50W	3	150Wh
3	Clippers	2	30W	1	30Wh
4	Bulbs	3	45W	3	135Wh
5	*Radio*	*1*	*30W*	*3*	*90Wh*
TOTAL			215W		525Wh

2.5.1. Calculation of Solar Panel Size and Number

The panel that would have to supply 525Wh to satisfy the watt-hour per day for each appliance.

525Wh × 1.3 = 682.5 Wh

The panel must provide at least 682.5 Watt-hour per day

Panel Generator factor supplied by textbook for sunny areas is 3.43

= 198.97 Wh

The peak solar panel available to me is 200W

= 0.99

=1

Hence, the number of solar panel required is one.

2.5.2. Load Sizing for the Battery

The size of the battery required is obtained by determining the Watt-hours per day used by all appliances.

682.5Wh × 5 = 3,412.5Wh

Step iv:

= 284.375Ah

However, the actual battery size is a fourth of the indicated value as the value required. Hence a battery of 12V, 75AH is used.

2.5.3. The Charge Controller

In selecting the charge controller, it should have enough ampere capacity to pass the current that the panel can provide [8]. The minimum ampere capacity of a discharge controller should be equal to the sum of the ampere from all appliances without motors times by 1.5.

Table 3: Charge controller load values

S/N	ITEM	QUANTITY	POWER (Watts)	CURRENT (A)
1	Television	1	60W	5
2	Fan	1	50W	4.17
3	Clippers	2	30W	2.5
4	Bulbs	3	45W	3.75
5	Radio	1	30W	2.5
TOTAL			215W	17.92

Hence the charge controller required;

17.92 x 1.5 = 26.88A at 12V

3. Construction and Assembly

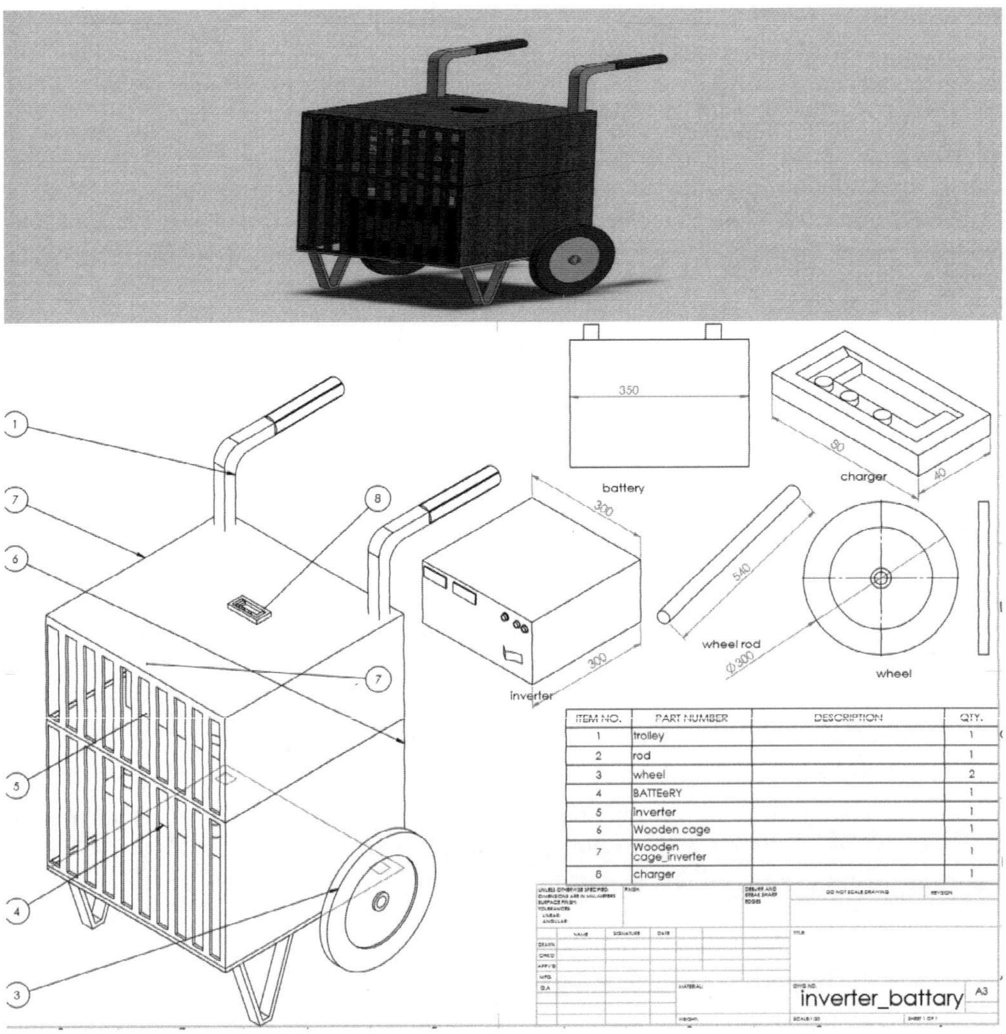

Figure 5: Assembly drawing of portable solar generato

IOP Conf. Series: Materials Science and Engineering **413** (2018) 012029 doi:10.1088/1757-899X/413/1/012029

4. Results and Discussion

During preliminary inspection of the barbing salon, it was discovered that the barber spent an average of 11 minutes per haircut. However, the time spent is determined by the complexity of the haircut, the rapport the barber has with the customer etc., below is a table indicating actual time spent per haircut on each customer.

Table 4: Time per haircut

S/N	TIME (Minutes)
1	04:58
2	18:11
3	27:11
4	15:11
5	21:11
6	09:11
7	10:10
8	15:00
9	06:00
10	09:00
11	14:00
12	12:00
13	05:00
14	06:15
15	05:11
TOTAL	178

IOP Conf. Series: Materials Science and Engineering **413** (2018) 012029 doi:10.1088/1757-899X/413/1/012029

Table 5: Typical values of time against solar panel voltage

S/N	TIME	VOLTAGE
1	11:50	20.00
2	12:00	19.00
3	12:10	19.70
4	12:20	20.40
5	12:31	19.70
6	12:40	19.20
7	12:50	19.40
8	1:00	19.50
9	1:10	19.79
10	1:20	19.20
11	1:30	19.10

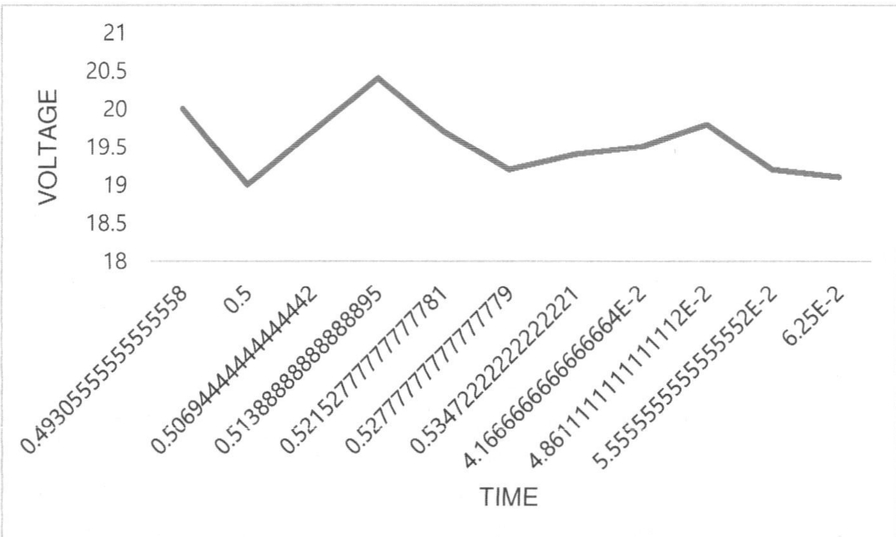

Figure 6: Graph of solar panel voltage against time at noon (typical)

Hence, from the graph we can determine the voltage being supplied by the solar panel at any time of the day. Where Y is voltage and X is time. The polynomial model for the system is denoted below;

$$y = 0.0008x^6-0.03x^5+0.4416x^4-3.1793x^3+11.507x^2-19.09x+30.362 \qquad (1)$$

At full load placed on the battery, the Table 6 and Figure 7 showed the results.

Table 6: Voltage against time at full load on battery

S/N	TIME	VOLTAGE
1	11:46	12.76
2	11:57	11.64
3	12:05	11.50
4	12:10	11.35
5	12:15	11.20
6	12:20	11.00
7	12:25	10.63
8	12:30	10.4

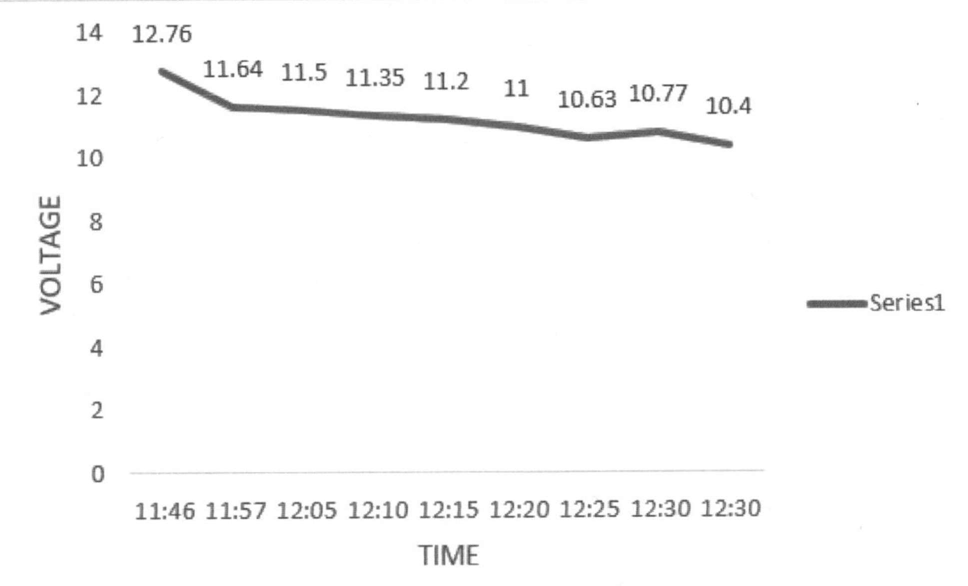

Figure 7: Graph of voltage against time at full load on battery

Microsoft Excel software was used to plot the graph and also to arrive at a model for the system. Where Y is voltage and X is time. It is denoted by a linear equation below;

$$y = -0.2357x + 12.428 \qquad (2)$$

IOP Conf. Series: Materials Science and Engineering **413** (2018) 012029 doi:10.1088/1757-899X/413/1/012029

5. Conclusion

In this project a portable solar generator was presented, the typical load demand for a barbing salon was considered and the system was designed to address this load. After the experiments was completed, a model indicating voltage variation with time of day was arrived at when on battery and when under direct sun light collected for the 200W solar panel installed. The product components included a charge controller to monitor the charging process of the rechargeable 12V battery, and an inverter that can safety switch between a 12V DC supply to 230V 1KVA AC supply. The system weight is 45Kg.

References

[1] Kumar, V., Shiravastava, R.L and Untawade, S.P. (2015). Solar Energy: Review of Potential Green & Clean Energy for Coastal and Offshore Applications. *International Conference on water resources, coastal and ocean Engineering* (ICWRCOE).

[2] Kabir, E., Kumar, P., Kumar S., Adelodun, A.A., Kim, K. (2018). Solar energy: Potential and future prospects, *Renewable and Sustainable Energy Reviews Elsevier.*

[3] Li, Q., Liu, Y., Guo, S., Zhou, H. (2017). Solar energy storage in the rechargeable batteries, *Nano Today*, 46-60 https://doi.org/10.1016/.nantod.2017.08.007.

[4] Soh, L. and Tiew, C. (2014). Building of a Portable Solar AC & DC Power Supply. *Fifth International Conference on Intelligent Systems, Modelling and Simulation*. pp. 445-451.

[5] Majid, Z.A., Hazali, N., Hanafiah, M.A., Abdullah A.A., Ismail, A.F, Ruslan, M.H., Sopian, K. and Azmi, M.S. (2011). Design and Performance of 20 Watts Portable Solar Generator. *1st International Conference on Mechanical Engineering Research (ICMER2011)*. pp. 1-6.

[6] Fraas, L., Minkin, L., Avery, J., Huang, H. X., Fraas, J., Uppal, P., JX Crystals Inc. U.S Army Research Laboratory. (2014) Portable Concentrating Solar Power Supplies. *IEEE Photovoltaic Specialists.*

[7] Riyanto, I, Octaviano, A. Suparmoko, Obara, K (2016). Portable Photovoltaic Powerplant with Solar Tracker for Disaster Affected Area Emergency Power Supply. *Indonesia Japan Joint Scientific Symposium.*

[8] Walpole, S.C., Edwards, P., Cleland J., Stevens, G., Roberts, I. (2012). The weight of nation: an estimation of adult human biomass. *BMC Public Health.*

Experimental study and Analysis of Variance of Material Removal Rate in High Speed Turning of AISI 304L Alloy Steel

O. Ogundimu[1], S. A. Lawal[1*] and I. P. Okokpujie[2]

[1]Department of Mechanical Engineering, School of Infrastructure, Process Engineering and Technology, Federal University of Technology, Minna- Nigeria
[2]Department of Mechanical Engineering, Covenant University, P.M.B 1023, Ota, Ogun State, Nigeria
*Corresponding author's email: lawalsunday@futminna.edu.ng

Abstract- Material removal rate is one of the challenges encounters in the manufacturing process, due to the transformation of a raw solid material into desired shapes. This investigation involved studying the effects of the three variables (cutting speed, feed rate and depth of cut and their interaction) for material removal rate during turning of AISI 304L stainless steel with cemented carbide inserts cutting tool. Response surface methodology was employed in the experimental design and total number of 20 experiments runs were conducted using central composite design. The level of importance and performance characteristics of the machining parameters on MRR was determined by using analysis of variance (ANOVA). It was observed that the cutting speed had the most significant effects on the MRR (66.27%) followed by the depth of cut (21.55%), while feed rate (8.25%) had the least effect on MRR. The result also showed that the cutting speed and the depth of cut has good interactions during the turning operation.
Keywords: Turning operation; Material removal rate; Response surface methodology (RSM); Analysis of variance

1 Introduction

The reason for metal cutting process usually known as machining is to manufacture a preferred shape, dimension and finish of an element by eliminating the excess materials in the form of chips from raw chunk of work piece [1-3]. Today industry requires the production of less expensive and good quality products in short time [4-6]. The issue of Surface finishing, tool wear rate and material removal rate (MRR) are the major response parameters which attention is given to by most industries. Material removal rate is vital parameter that greatly influences production rate and cost [7-13]. Therefore, it is necessary to assess the MRR before the cutting process is started. MRR is a useful factor in the manufacturing, which will contribute to the minimization of cost of manufacturing and reduces the idle time to attain a preferred surface quality.

Ashish et al [1] studied surface roughness, tool wear and material removal rate in CNC turning of EN24 alloy steel and found out that the higher the machining rate the better surface finish are desirable for better performance of any machining process. In turning procedure, it is expected to have high metal removal rate in order to obtain optimum quality at reduced time and cost, the process of trying to remove this unwanted part of the work piece, it can lead to the damage of the cutting tool and the work piece as a result of vibration. Nwoke, et al., [14]; Okonkwo et al., [15] carried out studies on chatter vibration frequency on AISI 4340 alloy during turning operation. The result shows that there is need to reduce vibration in turning operation in order to have high material removal rate and good surface finishing.

Turning operation is one of the leading, highly proficient predictable machining procedures. Normally the work piece is rotating on a spindle and the cutting tool is fed into it radially, axially or both ways concurrently to obtain the required shape or sides. Lawal, et al., [16] work on vegetable and mineral oil-in-water in turning operation of AISI 4340 alloy steel and found out that vegetable oil when applied in machining process reduces surface roughness, tool wear, vibration and increases the rate of metal removal. Authors agreed that vegetable oil will bring significant improvement on MRR and will also help during the investigation of the creep response on different materials [17-20].

Tejas *et al.* [21] work on the investigation and prediction of milling procedure on vertical milling machine by using RSM. The machining factors were examined to establish their importance on mild steel by carrying out experiments using response surface methodology to obtain material removal rate (MRR) and surface roughness (SR), while investigating the effects of spindle speed, depth of cut and feed on mild steel. The experimental results were evaluated using ANOVA and the importance of parameters upon performance measures was achieved. The results revealed that depth of cut and feed rate has superior effect on MRR

Jaganjeet and Sanjeev [22] carried out an investigation on the effects of WEDM process factors for optima cutting quality for the machined components. In this work the effects of numerous process variables of WEDM, such as pulse on time, pulse off time, Servo voltage, peak current, Wire feed, and Wire tension, were study to know their influence on material removal rate of H13 Steel. And the instrument use for the surface roughness measurement is MarSurf PS1. Taguchi L18 Orthogonal Array was employed and the results of the testing were evaluated using MINITAB software. Material removal rate is very significant in machining process, because if not properly study it will affect the surface produce during the operation which can lead to high surface roughness, high wear rate of the cutting tools and the production output of the industry will be affected [23-30].

Moreover, the accomplishments of most researchers in related to the current study are recognized. Development in assessment, experimental examination of dry, wet and MQL in turning operation on material removal rate were reviewed. However, there is need to study MRR using ANOVA to determine the extent the cutting variables will have on the MRR and the interaction between the three variables has with the response parameters.

2 Material and Method

The work piece material used in this research study was AISI 304L stainless steel. It is a commonly used steel grade which has a minimum of 18% chromium and 8% nickel. Cemented carbide cutting tool was applied in the researcher, and Lactuca LT 3000 cutting fluid was also used.

This experiment was conducted using DOE via response surface methodology (RSM). Minitab 16 software was used to analyze the data obtained from the experiment. The selection of the factors values depended on the response investigating and the manufactural. RSM was used in the experimental plan, applying central composite design (CCD). By taking all the factorial angle points, some of the central duplicates and all the axial points, central composite design required between 20 to 23 experimental runs depending on the number of the central duplicates considered, the experimental plan entail of three parameters at three levels such as cutting speed, feed rate and depth of cut as shown in Table 1. The experimental analysis was carry out to assess the effect of cutting speed (N), feed rate (f) and depth of cut on the material removal rate (MRR), and a total number of twenty (20) experiment were run according to central composite design as shown in Table 2.

Table 1: Process variables and Levels

Variables	Unit	Matrix levels		
		-1	0	**+1**
Cutting speed	(rpm)	600	1200	1800

Feed rate	(mm/rev)	0.2	0.25	0.3
Depth of cut	(mm)	0.4	0.6	0.8

Table 2: Design matrix of variables

Runs	Cutting speed (rpm)	Feed rate (mm/rev)	Depth of cut (mm)
1.	1200	0.25	0.6
2.	1200	0.165	0.6
3.	1200	0.25	0.6
4.	600	0.3	0.4
5.	190.9	0.25	0.6
6.	1200	0.25	0.6
7.	600	0.2	0.4
8.	1200	0.25	0.6
9.	1800	0.3	0.4
10.	600	0.2	0.8
11.	600	0.3	0.8
12.	1200	0.25	0.6
13.	1200	0.25	0.936
14.	2209	0.25	0.6
15.	1200	0.334	0.6
16.	1200	0.25	0.263
17.	1800	0.3	0.8
18.	1800	0.2	0.8
19.	1800	0.2	0.4
20.	1200	0.25	0.6

2.1 Experimental procedure for turning operation

The experimental analysis was carried out in CNC lathe machine and the specification are model SPT-32; Mas –Czech Republic having maximum cutting speed of 4000 rpm, rapid oblique of 15,000 mm/min, 28 KW motor drive, diameter of 490 mm, length of 1500 mm and cemented carbide cutting tool (CNMG 120408-PM) were employed in the turning operation. The experiment was carried out under flood cutting condition with Lactuca LT 3000 conventional cutting fluid.
The dimensions of work piece used were 40 x 750 mm. To achieve the material removal rate, the work piece diameter is measured before and after each experimental run. The results are then substituted into equation (1) and the material removal rate gotten for each experimental operation

$$MRR = \frac{\pi}{4} \times \left(D_i^2 - D_f^2\right) \times F \times N \quad (1)$$

Where D_i = Initial diameter, mm, D_f = final diameter, mm, f = Feed rate, mm/rev, N=Spindle speed, rpm [3].

3 Result and Discussion

The design matrix and the result from experiment with the three cutting variables such as cutting speed, feed rate and depth of cut are shown in Table 3 respectively.

Table 3: Experimental results for the material removal rate

Runs	Cutting speed (rpm)	Feed rate (mm/rev)	Depth of cut (mm)	D1 (mm)	D2 (mm)	MRR (mm^3/min)
1	600	0.2	0.4	40	39.2	5971.54
2	1800	0.2	0.4	39.2	38.4	17552.71
3	600	0.3	0.4	38.4	37.6	8595.39
4	1800	0.3	0.4	37.6	36.8	25243.33
5	600	0.2	0.8	36.8	35.2	10857.34
6	1800	0.2	0.8	35.2	33.6	31124.39
7	600	0.3	0.8	33.6	32	14838.37
8	1800	0.3	0.6	32	30.4	42343.64
9	190.92	0.25	0.6	40	38.8	3544.78
10	2209	0.25	0.6	38.8	37.6	39764.93
11	1200	0.16	0.6	37.6	36.4	13390.72
12	1200	0.33	0.6	36.4	35.2	26722.64
13	1200	0.25	0.263	35.2	34.68	8561.94
14	1200	0.25	0.936	34.68	32.80	29891.25
15	1200	0.25	0.6	32.80	31.6	18208.67
16	1200	0.25	0.6	31.6	30.4	17530.08
17	1200	0.25	0.6	40	38.8	22280.18
18	1200	0.25	0.6	38.8	37.6	21601.59
19	1200	0.25	0.6	37.6	36.4	209.23
20	1200	0.25	0.6	36.4	35.2	20211.42

3.1 Analysis of Variance for MRR

A response material removal system was designed and analyzed using Minitab 16 software, Table 4 and 5 shows the Analysis of Variance (ANOVA) result used to analyzed the effect of cutting variables on the material removal rate and their interactions, lack-of-Fit and residual error on the MRR. Table 5 can also be used to predicted the MRR when the need arises.

Table 4: The material removal rate ANOVA for AISI 304L Alloy Steel

Source	DF	Seq SS	Adj SS	Adj MS	F
Regression	9	1189155910	1189155910	132128434	1.04
Linear	3	146553375	146553375	48851125	0.38
Cutting speed	1	11491378	11491378	11491378	0.09
Feed rate	1	99779069	99779069	99779069	0.79
Depth of cut	1	35282928	35282928	35282928	0.28
Square	3	551593797	551593797	183864599	1.45
Cutting speed*Cutting	1	223385	2855184	2855184	0.02
Feed rate*Feed rate	1	98453745	60220465	60220465	0.47
Depth of cut*Depth of cut	1	452916667	452916667	452916667	3.57
Interaction	3	491008737	491008737	163669579	1.29
Cutting speed*Feed rate	1	344910273	344910273	344910273	2.72
Cutting speed*Depth of cut	1	35048834	35048834	35048834	0.28
Feed rate*Depth of cut	1	111049630	111049630	111049630	0.88
Residual Error	10	1268939042	1268939042	126893904	
Lack-of-Fit	5	698108363	698108363	139621673	1.22
Pure Error	5	570830678	570830678	114166136	
Total		19	2458094951		

Table 5: The Predictable regression coefficients for MRR using data in uncoded units

Term	Coef
Constant	14696.9
Cutting speed	69.6793
Feed rate	-423798
Depth of cut	87932.5
Cutting speed*Cutting speed	0.00123641
Feed rate*Feed rate	817675
Depth of cut*Depth of cut	140151
Cutting speed*Feed rate	218.870
Cutting speed*Depth of cut	17.4426
Feed rate*Depth of cut	372575

Figure 1-4 showed the interaction and the effects of the three variables on the material removal rate, it can be seen that the most interactive parameters were cutting speed and depth of cut. From Table 4, the cutting speed was the most significant parameters follow by the depth of cut, this also means that at the increase of the cutting speed, the MRR increase and at increase of the feed rate the MRR decreases, why the depth of cut increases the MRR increases. This result is in line with the observation made by Jaganjeet and Sanjeev [22]. Therefore, with a good combination and interactions between the cutting speed and the depth of cut will lead to a more preferred material removal rate.

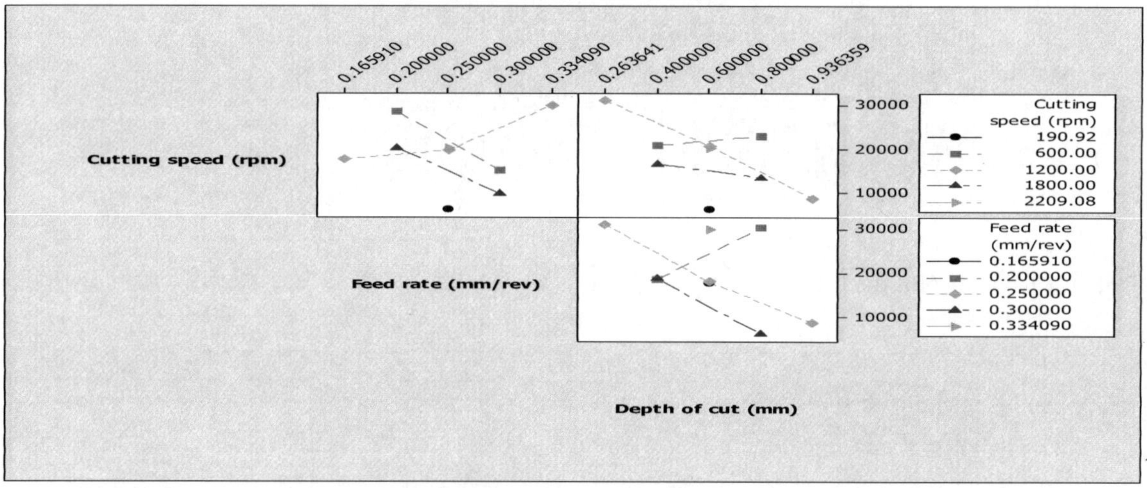

Figure 1: Interaction plot for MRR and the three cutting variables

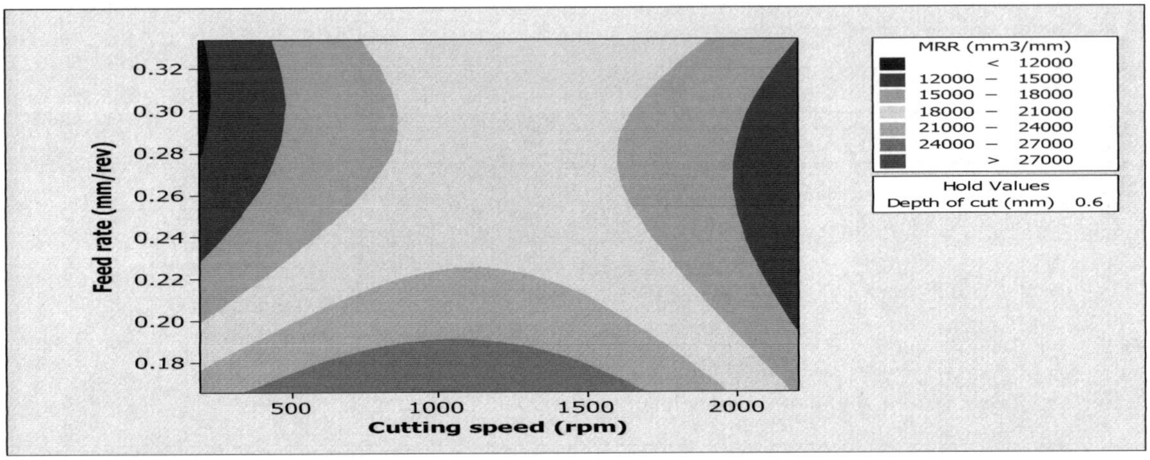

Figure 2: Contour plot for MRR vs. cutting speed and feed rate while depth of cut is constant at 0.6mm

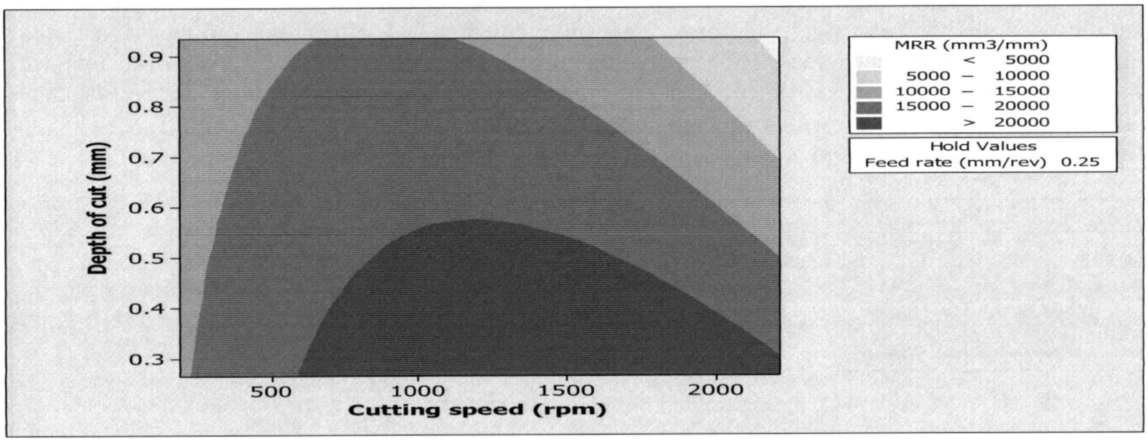

Figure 3: Contour plot for MRR vs. cutting speed and depth of cut while feed rate is constant at 0.25 mm/rev.

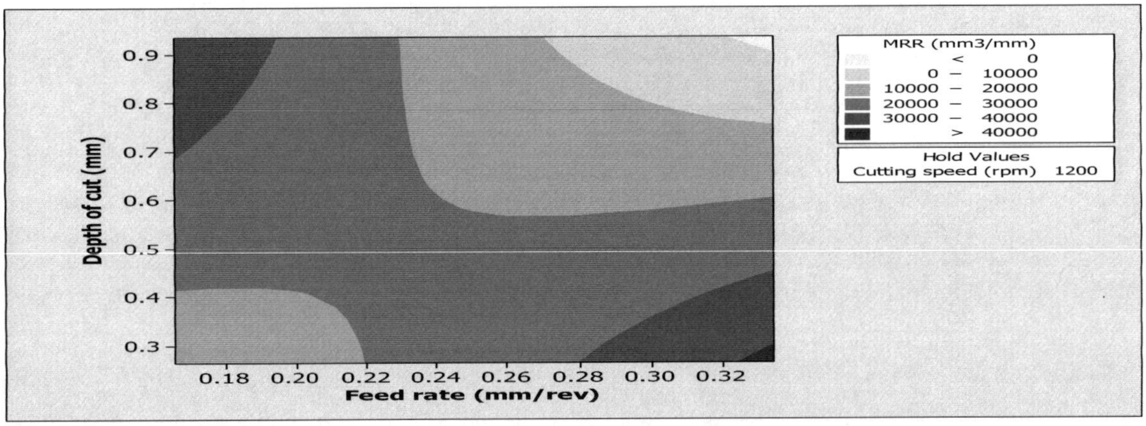

Figure 4: Contour plot for MRR vs. feed rate and depth of cut while cutting speed is constant at 1200 rmp.

IOP Conf. Series: Materials Science and Engineering **413** (2018) 012030 doi:10.1088/1757-899X/413/1/012030

4. Conclusions

The material removal rate could be effectively predicted by using predictable regression coefficients for the uncoded factors such as: cutting speed, depth of cut feed rate as the independent variables. In analyzing the individual parameters, it was found that cutting speed has the most significant effect on MRR with 66.27%, followed by depth of cut with 21.55% and feed rate has the least significant effect on MRR with 8.25%. In terms of interactions the cutting speed and the depth of cut has good level of interactions with each other than the combination of feed rate in other to achieve good process for the material removal rate.

References

[1]. Ashish, B., Bhardwaj, J., Maninder, S., and Sandeep, K. (2013). Optimization of Different Performance Parameters i.e. Surface Roughness, Tool Wear Rate & Material Removal Rate with the Selection of Various Process Parameters such as Speed Rate, Feed Rate, Specimen Wear, Dept. of cut in CNC Turning of EN24 alloy Steel – An Empirical Approach. *The International Journal of Engineering and Science IJES*. 2(1), 103-113.

[2]. Ezugwu, C. A., Okonkwo, U. C., Sinebe, J. E., & Okokpujie, I. P. (2016). Stability Analysis of Model Regenerative Chatter of Milling Process Using First Order Least Square Full Discretization Method. *International Journal of Mechanics and Applications*, *6*(3), 49-62.

[3]. Dhabale, R., and Vijay Kumar, (2015). Optimization of material removal rate of Almg1sicu in turning operation using genetic Algorithm, WSEAS Transactions on Applied and theoretical mechanics, 10(1), 95-101.

[4]. Okokpujie, I., Okonkwo, U., & Okwudibe, C. (2015). Cutting parameters effects on surface roughness during end milling of aluminium 6061 alloy under dry machining operation. *International Journal of Science and Research*, *4*(7), 2030-2036.

[5]. Ekinovic, S., and Ekinovic, E. (2000). High Speed Machining, Masinstvo, 4 (1), 15 – 50

[6]. Okokpujie, I. P., Ikumapayi, O. M., Okonkwo, U. C., Salawu, E. Y., Afolalu, S. A., Dirisu, J. O., ... & Ajayi, O. O. (2017). Experimental and Mathematical Modeling for Prediction of Tool Wear on the Machining of Aluminium 6061 Alloy by High Speed Steel Tools. *Open Engineering*, *7*(1), 461-469.

[7]. Prajapati, N.K., Patel, S. M. (2013). Optimization of Process Parameters for Surface Roughness and Material Removal Rate for Stainless steel 316 on CNC Turning Machine. *International Journal of Research in Modern Engineering and Engineering Technology*, 1(3), ISSN: 2320-6586.

[8]. Okokpujie, I. P., Ajayi, O. O., Afolalu, S. A., Abioye, A. A., Salawu, E. Y., Udo, M., ... & Ikumapayi, O. M. (2018). Modeling and Optimization of Surface Roughness In End Milling of Aluminium Using Least Square Approximation Method and Response Surface Methodology. *International Journal of Mechanical Engineering and Technology (IJMET)*, *9*(1), 587-600.

[9]. Okokpujie, I. P., Ohunakin, O. S., Bolu, C. A., & Okokpujie, K. O. (2018). Experimental data-set for prediction of tool wear during turning of Al-1061 alloy by high speed steel cutting tools. *Data in brief*, *18*, 1196-1203.

[10]. Lawal, S. A., Choudhury, I. A., & Nukman, Y. (2015). Experimental evaluation and optimization of flank wear during turning of AISI 4340 steel with coated carbide inserts using different cutting fluids. *Journal of The Institution of Engineers (India): Series C*, *96*(1), 21-28.

[11]. Lawal, S. A., & Ndaliman, M. B. (2014). Surface Roughness Characteristics in Finish Electro-Discharge Machining Process.

[12]. Okonkwo, U. C., Okokpujie, I. P., Sinebe, J. E., & Ezugwu, C. A. (2015). Comparative analysis of aluminium surface roughness in end-milling under dry and minimum quantity lubrication (MQL) conditions. *Manufacturing Review*, *2*, 30.

IOP Conf. Series: Materials Science and Engineering **413** (2018) 012030 doi:10.1088/1757-899X/413/1/012030

[13]. Afolalu, S. A., Salawu, E. Y., Okokpujie, I. P., Abioye, A. A., Abioye, O. P., Udo, M., ... & Ikumapayi, O. M. (2017). Experimental Analysis of the Wear Properties of Carburized HSS (ASTM A600) Cutting Tool. *International Journal of Applied Engineering Research*, *12*(19), 8995-9003.

[14]. Nwoke, O. N., Okonkwo, U. C., Okafor, C. E., & Okokpujie, I. P. (2017). EVALUATION OF CHATTER VIBRATION FREQUENCY IN CNC TURNING OF 4340 ALLOY STEEL MATERIAL. *International Journal of Scientific & Engineering Research*, *8*(2), 487-495.

[15]. Okonkwo Ugochukwu, C., Nwoke Obinna, N., & Okokpujie Imhade, P. (2018). Comparative Analysis of Chatter Vibration Frequency in CNC Turning of AISI 4340 Alloy Steel with Different Boundary Conditions. *Journal of Covenant Engineering Technology (CJET) Vol*, *1*(1).

[16]. Lawal, S. A., Choudhury, I. A., & Nukman, Y. (2014). Evaluation of vegetable and mineral oil-in-water emulsion cutting fluids in turning AISI 4340 steel with coated carbide tools. *Journal of cleaner production*, *66*, 610-618.

[17]. Lawal, S. A., Ajani, E. A., Namessan, N. A., Ugheoke, B. I., & Egbe, E. A. P. (2017). Experimental evaluation of coated carbide insert on alloy of steel materials during high speed turning process. *Arid Zone Journal of Engineering, Technology and Environment*, *13*(3), 325-335.

[18]. Lawal, S. A., Choudhury, I. A., & Nukman, Y. (2012). Application of vegetable oil-based metalworking fluids in machining ferrous metals—a review. *International Journal of Machine Tools and Manufacture*, *52*(1), 1-12

[19]. Nwoke, O. N., Okokpujie, I. P., & Ekenyem, S. C. (2017). INVESTIGATION OF CREEP RESPONSES OF SELECTED ENGINEERING MATERIALS. *Journal of Science, Engineering Development, Environment and Technology (JOSEDET)*, *7*(1), 1-15.

[20]. Orisanmi, B. O., Afolalu, S. A., Adetunji, O. R., Salawu, E. Y., Okokpujie, I. P., Abioye, A. A., ... & Abioye, O. P. (2017). Cost of Corrosion of Metallic Products in Federal University of Agriculture, Abeokuta. *International Journal of Applied Engineering Research*, *12*(24), 14141-14147.

[21]. Tejas, C. P., Lalit, S. P. and Bhavesh, C. P. (2014). Analysis and prediction of milling process on vertical milling centre (VMC) by using response surface methodology (RSM). *International Journal for Innovative Research in Science and Technology, 1(7): 205-212.*

[22]. Jaganjeet Singh and Sanjeev Sharma (2013). Effects of Process Parameters on Material Removal Rate and Surface Roughness in WEDM of H13 Tool Steel. International Journal of Current Engineering and Technology, Vol.3, No.5 pp 1852-1857.

[23]. Okokpujie, I. P., & Okonkwo, U. C. (2015). Effects of cutting parameters on surface roughness during end milling of aluminium under minimum quantity lubrication (MQL). *International Journal of Science and Research*, *4*(5), 2937-2942.

[24]. Azeta, J., Okokpujie, K. O., Okokpujie, I. P., Osemwegie, O., & Chibuzor, A. (2016). A Plan for Igniting Nigeria's Industrial Revolution. *International Journal of Scientific & Engineering Research*, *7*(11), 489.

[25]. Salawu, E. Y., Okokpujie, I. P., Afolalu, S. A., Ajayi, O. O., & Azeta, J. (2018). INVESTIGATION OF PRODUCTION OUTPUT FOR IMPROVEMENT. *International Journal of Mechanical and Production Engineering Research and Development*, *8*(1), 915-922.

[26]. Imhade P. Okokpujie, E.Y. Salawu, O. N. Nwoke, U. C. Okonkwo, I. O. Ohijeagbon, and Kennedy Okokpujie, "Effects of Process Parameters on Vibration Frequency in Turning Operations of Perspex Material," Lecture Notes in Engineering and Computer Science: Proceedings of The World Congress on Engineering 2018, 4-6 July, 2018, London, U.K., pp700-707.

[27]. Afolalu, S. A., Abioye, A. A., Dirisu, J. O., Okokpujie, I. P., Ajayi, O. O., & Adetunji, O. R. (2018, April). Investigation of wear land and rate of locally made HSS cutting tool. In *AIP Conference Proceedings* (Vol. 1957, No. 1, p. 050002). AIP Publishing.

IOP Conf. Series: Materials Science and Engineering **413** (2018) 012030 doi:10.1088/1757-899X/413/1/012030

[28]. Salawu, E. Y., Okokpujie, I. P., Ajayi, O. O., & Agarana, M. C. (2018). Analytical Technique for the Determination of Hoop Stress and Radial Stress on the Tooth Spur Gear under Vertical Loading in a Food Packaging Machine.

[29]. Afolalu, S. A., Abioye, O. P., Salawu, E. Y., Okokpujie, I. P., Abioye, A. A., Omotosho, O. A., & Ajayi, O. O. (2018, April). Impact of heat treatment on HSS cutting tool (ASTM A600) and its behaviour during machining of mild steel (ASTM A36). In *AIP Conference Proceedings* (Vol. 1957, No. 1, p. 050003). AIP Publishing.

[30]. Afolalu, S. A., Okokpujie, I. P., Salawu, E. Y., Abioye, A. A., Abioye, O. P., & Ikumapayi, O. M. (2018, April). Study of the performances of nano-case treatment cutting tools on carbon steel work material during turning operation. In *AIP Conference Proceedings* (Vol. 1957, No. 1, p. 050001). AIP Publishing.

IOP Conf. Series: Materials Science and Engineering **413** (2018) 012031 doi:10.1088/1757-899X/413/1/012031

Comparative study of biogas production in composite of poultry droppings and lemon grass using pressure computed from strain gage rosette

Adekoya Olaoluwa[1], Akan Williams[1], Ndubuisi Amanda[2], Adebayo Timileyin[3], Henry Boyo[2]

[1]Department of Chemistry, Covenant University, Ota, Ogun State, Nigeria
[2]Department of Physics, Covenant University, Ota, Ogun State, Nigeria
[3]Department of Physics, University of Lagos, Lagos State, Nigeria
E-mail address: olaoluwaadekoya2@gmail.com

Abstract. The advantages of recording micro change in pressure every second inside a biogas production digester using strain gage measurement was well demonstrated in this work. The production of biogas from a composite of Lemon grass and Poultry droppings was carried out by measuring the total gas pressure exerted on the surface of sealed near-cylindrical plastic container used as a digester. The suitability of using a combination of lemon grass and poultry droppings for biogas generation and the determination of an optimal mixing ratio to maximize gas production was studied. The aerobically pre-fermented substrates were mixed in five different ratios labelled 'Digester A-E' respectively to form composite and the formed slurry was digested for 30 days in the five (5) different digesters lagged with fiberglass wool. Pressure values were determined using the principles of 'thin walled cylindrical pressure vessel' computed from the strain gage rosette readings. The results after 30 days show that Digester B with lemon grass had the highest gas production with a pressure of 41.8 kP. Digester C which contained 100% chicken droppings had a pressure of about 11.5 kP, but when mixed with lemon grass inside digester E, in a mixing ratio of 70% to 30% respectively, the pressure of the gas produced increased to 26.5 kP. This is an increase of about 15 kP which is a little over double the amount of gas produced with only chicken droppings. Thus, the result of mixing 30% of lemon grass and 70% of chicken droppings doubled the gas production potential of chicken droppings. Digester A (with 50% lemon grass and 50% chicken droppings) and Digester D (with 70% lemon grass and 30% chicken droppings) had pressures of 17.4 kP and 32.7 kP respectively.
Keywords: biogas production; pressure computed

1. Introduction

Most of organic wastes, such as crop straws, livestock manure, municipal solid wastes and the by-products of food industry, can be used as biogas materials [9]. Study has shown that the methane content of biogas from animal manure substrates could be improved by co-digestion with energy plants [10]. The digestion of organic materials by micro-organisms produces biogas and production of the biogas can be very small in laboratory scale experiments [1].

The level of biogas measured may be used to predict the state of degradation in anaerobic digestion. Unfortunately, several researchers working on biogas in laboratories in most developing countries lack access to sophisticated gas measuring equipment. Accurate measurement of the produced gas pressure could be very difficult task. Most calibrated volume measuring devices may not be sensitive enough to detect micro changes in biogas volume. Direct measurement of the pressure activity inside the digester tells more about biogas formation rather than volume displacement measurement [6].

Pressure measurements and controls are important in digesters because they can influence key processes such as liquid-vapour equilibrium, fluid dynamics and the rate of chemical reaction. A well designed pressure measuring method can be used for comparison of biodegradability caused by microbial activities in biogas production using composite materials [3]. The mechanical Bourdon tube pressure gauge is simple with excellent sensitivity for higher pressure applications but its slow response to change in pressure and hysteresis makes it unsuitable for low pressure application.

The study was intended to develop the method for measuring biogas produced in a plastic digester, by using strain gage rosette to detect changes in biogas production pressure. Micro changes in biogas production per second, per minute and per hour, on a daily basis could be more informative than volumetric measurement. The study helps to overcome volumetric measurement insensitivity problem at low volume or pressure. A set of three (3) strain gauges (rosette) were attached to the surface of five (5) digesters, to measure strains in different directions, providing precise evaluation of the surface stresses, from which pressure was deduced. The study also aims at measuring the production of biogas using different mixing ratio of chicken droppings and *Cymbopogon citratus* (lemon grass) as feedstock in a plastic bio digester, since co-digestion of both digestible materials may produce different gas quantity in the five (5) digesters.

2. Materials

The biomass used were lemon grass and chicken droppings. The lemon grass was obtained from Covenant University, while the chicken droppings were collected from a poultry farm around Lagos. They were mixed homogenously. Other materials used for the study were the following:

- Five of 20 litre size water dispenser plastic bottle as anaerobic digester.
- Five set of Strain gauge sensor BF350 (3 pieces) and temperature sensor LM 35 (Figures 1, 3).
- Five set of Bridge amplifiers HX711 (3 modules), as shown (in Figures 2, 3).
- Five Arduino microcontroller slave boards for data capturing and transmission enabled with WiFi (Figure 2).
- One Arduino microcontroller master board for data reception and logging to PC, enabled through WiFi.

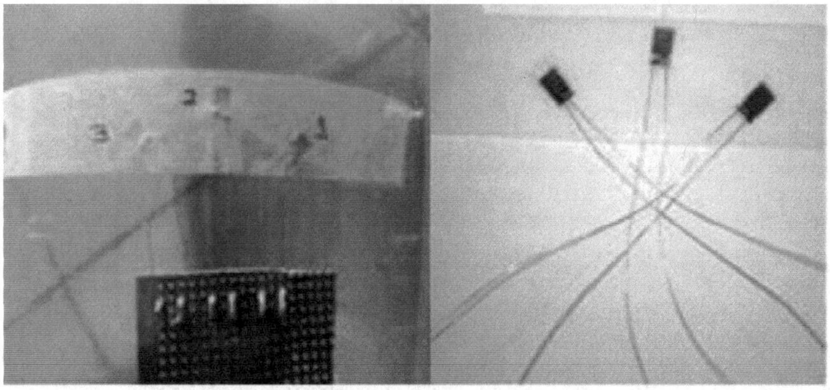

Figure 1: A set of strain gage sensors assembled at an angle of 45 degrees apart (LHS) and glued to the plastic digester (RHS)

.

IOP Conf. Series: Materials Science and Engineering **413** (2018) 012031 doi:10.1088/1757-899X/413/1/012031

Figure 2: Strain gage bridge amplifiers on board with ADC, microcontroller and Wifi customized for data logging

Figure 3: Digester plastic before loading it with feedstocks. Attached to the lower side of the plastics are the strain gages which are connected to the data acquisition board placed in front of the digester

3. Method

The design capacity of the anaerobic digesters was chosen due to the required quantity of slurry and the number of days the material would stay in the digesters (retention time). The main structure consisted of a plastic digestion tank, tough enough to bear the weight and pressure of the slurry, and insulated so as to eliminate the effect of ambient temperature perturbations on the activities of the anaerobes. The tank was air tight, wrapped with fibre glass wool as thermal insulator and placed well above the ground level inside a room.

Two (2) feedstocks (chicken droppings and lemon grass) were prepared into five (5) samples with three (3) mixing ratios, and were loaded into the respective digesters as shown in Figure 3 for a period of thirty (30) days during the study. The digesters were clearly labelled viz:

Digester 1 (A): 50% chicken droppings and 50% lemon grass.
Digester 2 (B): 100% lemon grass.
Digester 3 (C): 100% chicken droppings.
Digester 4 (D): 30% chicken droppings and 70% lemon grass.
Digester 5 (E): 70% chicken droppings and 30% lemon grass.

Chicken droppings and lemon grass (*Cymbopogon citratus*) were the biomass resources used as feed materials for the study. Lemon grass was crushed and milled. A hammer mill was used for the crushing process before samples were transported to the laboratory for further treatment. Partially decayed butchery house discarded waste was used to seed the digester for all the feedstocks digested.
About 2.4 kg dry weight of the respective pre-fermented wastes was mixed with water to form slurry in the ratio 1:1 by weight and then separately poured into each tank through a funnel. The slurry was permitted to fill half of the digester space, leaving a clear height of half the digester as space for the biogas production. The research was carried out for 30 days in batch fermentation and the following was carried out in the course of the experiment:

- Minute-by-minute measurement of the pressure exerted by the biogas generated till the termination of experiment.
- The daily temperature of the digesters was taken along with the ambient.
- Chemical analysis was carried out on each sample of feedstock before and after digestion.

3.1 Chemical Analysis of Feedstocks
The chemical composition of the substrates both before and after digestion was analyzed using standard methods (APHA, 2012). Parameters analyzed included pH, total solids, moisture content, total nitrogen, ash content, organic carbon, biochemical oxygen demand and chemical oxygen demand (Figure 5).

3.2 Analog to digital count (ADC_count) to Bridge output voltage to Strain value
The bridge output voltage 'e' was scaled from the ADC_count using the fact that, a fixed Bridge excitation voltage 'Vin' (= 3.9V) will result to an ADC count of 224 after it has been amplified with a Voltage gain (=60).
The Gage factor 'Ks' of the strain gage is Ks=2.0, and Strain 'ε', can be determined from the expression

$$\varepsilon = \frac{4}{K_s} * \frac{e}{V_{in}} \quad (1)$$

3.3 Measured Strain to Principal Strain Conversion
The signal from each strain gage is given as ε1, ε2, and ε3 respectively [2][4].
The maximum principal strain is given as,

$$\varepsilon_{max} = \frac{1}{2}\left\{\varepsilon_1 + \varepsilon_3 + \sqrt{2|(\varepsilon_1 - \varepsilon_2)^2 + (\varepsilon_2 - \varepsilon_3)^2|}\right\} \quad (2)$$

Minimum principal strain is,

$$\varepsilon_{min} = \frac{1}{2}\left\{\varepsilon_1 + \varepsilon_3 - \sqrt{2|(\varepsilon_1 - \varepsilon_2)^2 + (\varepsilon_2 - \varepsilon_3)^2|}\right\} \quad (3)$$

Direction of principal strain from the ε1-axis is given as,

$$\theta = \frac{1}{2}\tan^{-1}\left[\frac{2\varepsilon_2 - \varepsilon_1 - \varepsilon_3}{\varepsilon_1 - \varepsilon_3}\right] \quad (4)$$

3.4 Principal Strain to Principal Stress Conversion
Let 'v' be the Poisson's ratio (=0.3) and 'E' the Young's modulus (=30*106 psi),
Maximum Principal Stress is[7][8],

$$\sigma_{max} = \frac{E}{2(1-v^2)}\left\{(1+v)(\varepsilon_1+\varepsilon_3) + (1-v) * \sqrt{2|(\varepsilon_1-\varepsilon_2)^2 + (\varepsilon_2-\varepsilon_3)^2|}\right\} \quad (5)$$

Minimum Principal Stress is,

$$\sigma_{min} = \frac{E}{2(1-v^2)}\left\{(1+v)(\varepsilon_1+\varepsilon_3) - (1-v) * \sqrt{2|(\varepsilon_1-\varepsilon_2)^2 + (\varepsilon_2-\varepsilon_3)^2|}\right\} \quad (6)$$

3.5 Stress to Pressure Conversion

For the thin-walled cylindrical digester of thickness 't' (=9.466E-4 meter) and radius 'R' (=12.32E-3 meter), the forces exerted by biogas in the hoop or circumferential direction will result to a pressure 'p', given as

$$p = \frac{t}{R} * \sigma_{max} \quad (7)$$

And the forces exerted by biogas in the longitudinal direction will result to a pressure of

$$p = \frac{2t}{R} * \sigma_{min} \quad (8)$$

The average of both pressures in equation 7 and equation 8 was computed. Although if the bottle is isotropic and have uniform thickness all over, both pressure values will be equal.

4. Results and Discussion

The results after 30 days (Figure 4) shows that Digester B with lemon grass had the highest gas production with a pressure of 41.8 kP. Digester C which contained 100% chicken droppings showed slightly high biogas production with a pressure of about 11.5 kP, but when mixed with lemon grass inside digester E, the biogas pressure increased to 26.5 kP. This is an increase of about 15 kP, which is slightly more than double the amount of gas produced with only chicken droppings. Thus, the result of mixing 30% of lemon grass and 70% of chicken droppings doubled the gas production potential of chicken droppings.

Digester A (with 50% lemon grass and 50% chicken droppings) and Digester D (with 70% lemon grass and 30% chicken droppings) have pressures of 17.4 kP and 32.7 kP respectively. D produced more gas than A because it has higher percentage of lemon grass.

There was a progressive increase in the production of gas within the first fifteen (15) days in all the digesters. Subsequently gas production declined in Digester C (which contained 100% chicken droppings) about the 18th day and in Digester A (with 50% lemon grass and 50% chicken droppings) about the 22nd day indicating low anaerobic activities in both digesters.

Chemical analysis of the digester feedstock before and after digestion is shown in Figure 5 and provided as a reference to describe the chemical state of the feedstock used in this study.

In Figure 6, the temperature fluctuation between maximum and minimum in all five digesters remained within five (5) to ten (10) degrees Celsius and is mesophilic. Higher temperatures above the ambient is an indication of microbial activities in the digester. It is noted that, although Digester B (100% lemon grass) and Digester C (100% chicken droppings) had the same temperature ranges (Figure 6), Digester B produced gas the most (Figure 7) while Digester C produced gas the least.

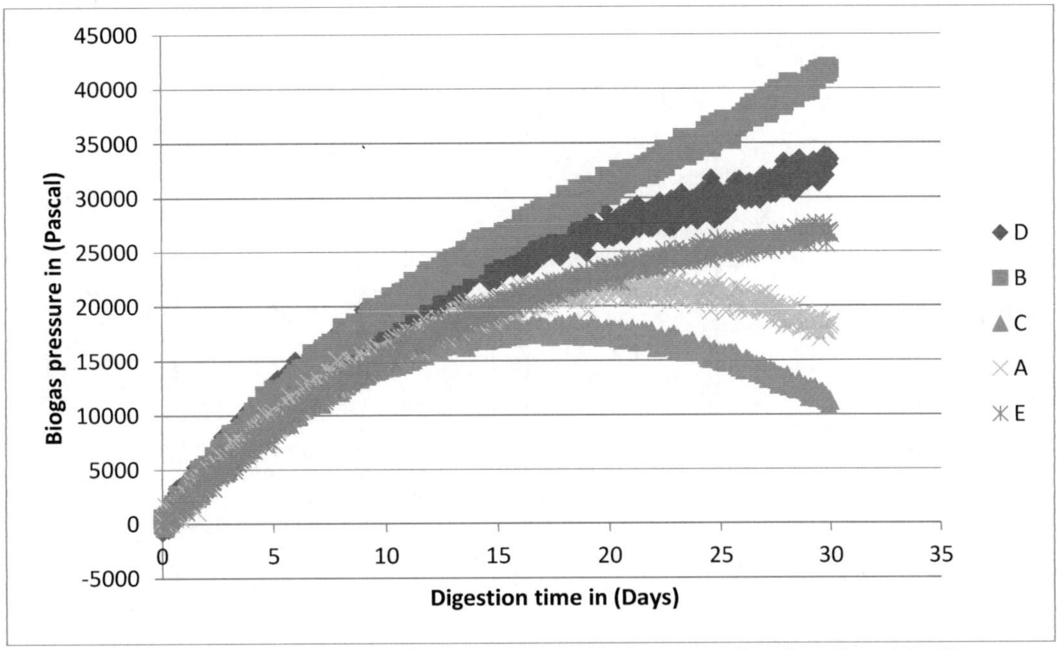

Figure 4: Shows the daily average gas pressure in the five bio-digesters (A-E) computed from strain gage measurements

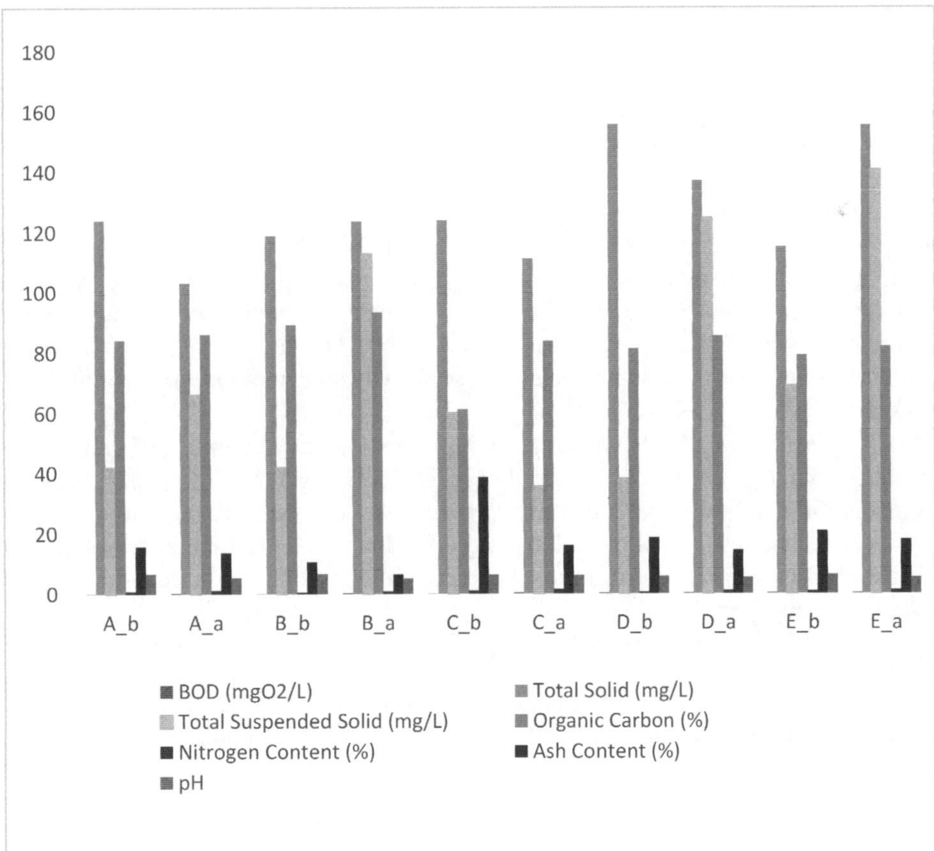

Figure 5: Chemical analysis of the digester feedstock before and after digestion

IOP Conf. Series: Materials Science and Engineering **413** (2018) 012031 doi:10.1088/1757-899X/413/1/012031

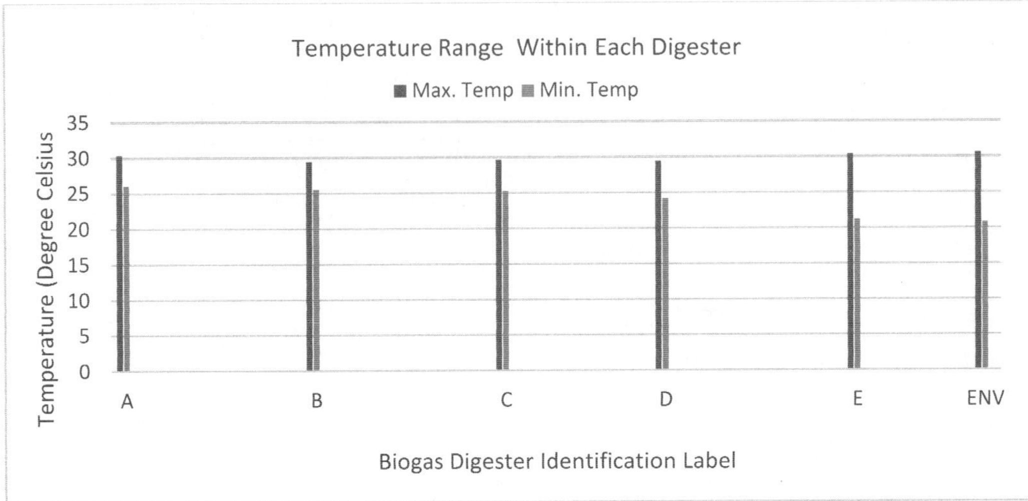

Figure 6: Average temperatures range of the five digesters (A-E) variation and that of its ambient environment (ENV), showing the effect of insulating the digester with Fibre glass wool

5. Conclusion

The following contributions have been added to the existing knowledge about the anaerobic digestion of biomass for energy generation:

- Although, the downward displacement of water for gas collection has been in use before now, one major challenge faced over the years is its inefficiency for gas volume measurement at very low pressure. The vapor pressure contributed by evaporating surface water may also increase the measurement error at low pressure [5]. Also harvesting from the displacement trough usually end up in wastage of gas and leakages. This research has effectively surmounted this challenge by using a specially designed set of strain gage to measure the strain on the plastic (digester) and strain used to calculate the pressure of biogas in the digester. The strain gage has an advantage of recording micro change in pressure every second. The novel usage of this sensor for biogas production at low pressure improves the sensitivity and measurement accuracies to change in pressure of gas.
- This project has equally demonstrated the possibility and suitability of using a combination of lemon grass and poultry droppings for biogas generation in other to enhance the biogas production potential of chicken waste and also provides an optimal mixing ratio that maximizes gas production. The results will go a long way to reduce the challenge of raw material selection for biogas energy in most home or family size biogas system that can be constructed and installed into every home in Nigeria.

Acknowledgments

Research is supported by the Chinese National Foundation under Grants No.51179147 and the ministry of transport construction projects of science and technology (2015318J34090).

References

[1] "A simple device to measure biogas production in laboratory scale digesters," Biotechnology Letters, Vol. 8, pp. 719–720, 1986.

[2] Beer, F. P. and Johnston Jr., E. R. (1981) Mechanics of Materials. McGraw-Hill, New York, pp. 1-

70, pp. 325-330.

[3] F. Tagliapietra, M. Cattani, L. Bailoni, S. Schiavon, "In vitro rumen fermentation: Effect of headspace pressure on the gas production kinetics of corn meal and meadow hay," Animal Feed Science and Technology,Vol. 158, p.p197–201, 2010.

[4] Figliola, R. S. and Beasley, D. E. (2000) Theory and Design for Mechanical Measurements. 3rd edition, Wiley, New York, pp. 432-438, pp. 292-293.

[5] M. Walker, Y. Zhang, S. Heaven, C. Banks, "Potential errors in the quantitative evaluation of biogas production in anaerobic digestion processes," Bioresource Technology, Vol. 100, p.p 6339-6346, 2009.

[6] M. K. Theodorou, B. A. Williams, M. S. Dhanoa, A. B. McAllan, J. France, "A simple gas production method using a pressure transducer to determine the fermentation kinetics of ruminant feeds," Animal Feed Science and Technology, Vol. 48,p.p 185-197, 1994.

[7] Perry, C.C., "Data-Reduction Algorithms for Strain Gage Rosette Measurements", Experimental Techniques. May, 1989, pp. 13-18.

[8] Users Manual for DAQ 6023E/6024E/6025E, National Instruments, Inc., Austin, Texas.

[9] Achinas S., Achinas V. and Euverink G. J. W. (2017). A technological Overview of biogas production from biowaste. *Engineering* **3**(3):299-307

[10] Owamah, H. I., Alfa, M.I. and Dahunsi, S. O. (2014). Opimization of biogas from chicken droppings with *Cymbopogon citratus*. *Renewable Energy* 68:366-371.

Investigating the Viability of Hydrogen Production from Sodium Borohydride Using Acetic Acid and Extracts from Spoilt Lemon, Lime and Orange

[1]Idama, Enyomeji Ademu., [2]Efeovbokhan, Vincent Enontiemonria., [3]Alagbe, Edith Egbimhanlu.

[1,2,3]Department of Chemical Engineering, Covenant University Ota, Ogun State, Nigeria
Email address: Enyomeji.idama@stu.cu.edu.ng Telephone +234(0)8182425523

Abstract. With the fast depletion of fossil fuel sources for energy production and coupled with their negative impacts on the environment, the continued search for alternative sources that are renewable, sustainable and environmentally friendly has become imperative. This research examined the performance of sodium borohydride for the production of hydrogen gas using 5 ml each of organic acid (acetic acid), spoilt extracts of citrus fruits (lime, lemon and orange) without the use of neither catalyst nor heat source. A three neck flat bottom flask was used in which sodium borohydride was reacted with extracts from the three spoilt fruits and acetic acid with concentrations of 1, 5, 7, 12 and 17.5 M. Volume of hydrogen gas produced were recorded and the results revealed that using 7 M solution of acetic acid and 1.0 gram sodium borohydride generated the highest volume of 2460 ml of hydrogen gas in 63.72 min while the undiluted extracts of spoilt orange, lemon and lime gave 100 ml in 0.68 min, 90 ml in 1.67 min and 60 ml in 0.5 min respectively. For the diluted fruit extracts, lime, orange and lemon at a dilution factor of 2 each, generated hydrogen gas of 80ml in 0.5 min, 70ml in 1.3 min and 70ml in 0.62 min respectively. All the reactions took place at an ambient temperature of 27°C. Being able to conveniently retrieve hydrogen from its combined state when needed would advance the use of hydrogen as a source of energy.
Keywords: sodium borohydride, acetic acid, hydrogen, spoilt citrus extract.

1. Introduction

Given the importance of energy to human existence it has become imperative to find sustainable alternative sources of energy given the ever growing population of the human race and its increased demand for energy [1]. Human existence on earth is possible due to man's ability to access energy and transform it from one form to another to meet his needs; most of the energy accessed by man has been transformed into either electrical energy, heat energy or mechanical energy. Electrical energy is the energy used in powering technology such as mobile phones, air conditioners, electric cars, light bulbs, computers, electric cookers, washing machines and various other types of machinery that require electricity for operation, heat energy is used for cooking our foods, keeping warm in winter, melting and molding of metals, drying substances, sterilization and numerous other activities which require heat, some chemical reactions require heat energy to be initiated and completed [2]. At present crude oil, coal and natural gas are the predominant sources of energy in our everyday lives from electricity generation, transportation (air, land & sea) and domestic cooking unfortunately none of these sources of energy is sustainable nor renewable thus prompting the need for the search for sustainable and renewable sources of energy. The consequence of the combustion of petroleum, coal and natural gas for energy production; contributes to the release of greenhouse gases in the atmosphere, which are not just detrimental to our environment but also to both human and animal life [3]. An alternative source of energy which proves extremely attractive is hydrogen firstly because it is the most abundant chemical substance on the earth, constituting about 75% of all baryonic mass and secondly when combusted releases steam (2H2 + O2 $\rightarrow$ 2H2O) hence is environmentally friendly, sustainable and renewable. Thirdly hydrogen has the highest energy per mass of any fuel [4]. Hydrogen has an energy density of 142 MJ/Kg which is three times greater than that of gasoline fuel [5]. Storage of hydrogen in portable forms such that it is easily accessible when needed has been one of the concerns of researchers in the field of energy which include stationary power, portable power and transportation, hydrogen can be stored in three forms, as a gas, liquid or a solid [6]. As a gas hydrogen is stored in high pressure tanks of 691 atmospheres tank pressure, for comparison the environment in which humans live is about one atmosphere depending on altitude and temperature, as a liquid hydrogen is stored at very low temperatures because the boiling point of hydrogen at atmospheric pressure (1 atm) is; -252.8°C and as a solid hydrogen can be adsorbed on the surfaces of solids or absorbed within the solid [7]. To achieve the most cost-effective and safe hydrogen storage method for on-board and off-board hydrogen production and usage, storage as a solid appears most attractive. Most of the electrical energy generated from hydrogen is generated through hydrogen fuel cells; compared to batteries as long as the is a steady source of hydrogen to the fuel cell it continues to generate electricity without ceasing [8] unlike batteries that stop once the chemical reaction producing the electricity stops. Metal hydrides have been found to be a good source for hydrogen absorption hence could be used for hydrogen storage, in addition to being a good store for hydrogen it should be able to easily release the stored hydrogen when needed safely an added advantage would be the characteristic of the metal hydride to be easily replenished back with hydrogen after desorption of the hydrogen absorbed in it [9]. On an examination between hydrogen and the s-block elements of the periodic table (s-block elements are group 1 and group 2 elements of the periodic table alkali metals in group one of the periodic table being strongly electropositive readily accept hydrogen atoms to form metal hydrides which react vigorously with cold water to liberate hydrogen gas, compared to alkaline earth metals in group two of the periodic table the group one metal hydrides react with cold water to liberate one mole of hydrogen gas while the group two metal hydrides react with cold water to liberate two moles of hydrogen gas which makes group two metal hydrides a better choice for hydrogen absorption/desorption than group one metal hydrides[10]. Even though hydrogen has the highest energy per mass of any fuel its density at ambient temperature is considerable low [11] hence resulting in a low energy per unit volume output. Among the group two metal hydrides magnesium hydride appears to be the most favorable considering that it has 7.66 weight percent of hydrogen [12] which gradually decreases down the group; calcium hydride (CaH2) having 4.75 weight percent hydrogen and strontium hydride (SrH2) having 2.25 weight percent hydrogen. Beryllium hydride being the first metal hydride from group two preceding magnesium hydride would have been a much preferred choice being that it has 18.28 weight percent hydrogen, but being that it is extremely toxic makes it environmentally unfriendly due to its toxicity the use of beryllium requires adequate dust control equipment because if inhaled results in berylliosis a life-threatening ailment. Secondly all the elements of groups one and two in the periodic table form hydrides by direct combination with hydrogen except beryllium making its cyclic ability expensive. Thirdly beryllium is a relatively rare element in the universe [10]. Other metals of interest are the metal borohydrides which have a high weight percentage of hydrogen when compared with various hydrogen storage and production compounds with sodium borohydride having 10.6 wt% hydrogen content, lithium borohydride having 18.3 wt% hydrogen content and potassium borohydride having 7.4 wt% hydrogen content[13] the lightweight of the borohydrides is an added advantage; particular interest is taken in sodium borohydride because of its stability in dry air enabling it to be safely handled some draw-backs being the sluggish kinetics of the reaction and high thermodynamics associated with the reactions[14]. Attempts to improve on the slow kinetics and high temperature a composite mixture of sodium borohydride and zinc fluoride with heat not exceeding 100°C was used during the dehydrogenation process [15]. Other metal hydrides which have been investigated for the production and storage of hydrogen are NaBH4, LiBH4, KBH4, LiH, NaH and NaAlH4 one of the metal hydrides which was of interest was sodium borohydride (NaBH4) due to certain properties the compound exhibited such as being non-toxic, non-flammable and low cost when compared with the other metal hydrides making it economical, the research focused on the hydrolysis of sodium borohydride (NaBH4) using bimetallic Co-Ni based complex catalysts[16]. Being able to conveniently retrieve hydrogen from its combined state when needed has limited the effective use of hydrogen has a fuel. Reacting sodium borohydride with acids and various spoilt fruit extracts and observing their performance for fast kinetics at ambient temperature without the use of any external heat source; taking into consideration acids that are

IOP Conf. Series: Materials Science and Engineering **413** (2018) 012032 doi:10.1088/1757-899X/413/1/012032

cheap and easily accessible without the use of any catalyst would improve the use of hydrogen has a source of energy

1.1 Objectives of the Study

Various concentrations of acetic acid will be reacted with sodium borohydride. Extracts of spoilt lime, lemon and orange will be diluted by various amounts and reacted with sodium borohydride. Volume of hydrogen produced for each concentration / dilution is recorded. The concentration with the best performance in hydrogen production with time is noted

2.0 Materials and Method

2.1 Materials

Sodium borohydride (98% purity; sigma-aldrich), Silica gel; (IND), Glacial acetic acid (100% purity; EMSURE), Citric acid (99.5% purity; BDH), Distilled water (water distiller; W4000, SN M17035), Spoilt lemon, lime and oranges sourced from local market; Sango-Ota, Ogun State, Nigeria.

2.2 Equipment / Apparatus

Analytical Balance (RadwagWagiElektroniczne AS 310/C/2, SN 351204/12), Fume Cupboard (Frontier Junior EFB-4A1, SN 2012-68013), Hot plate magnetic stirrer (Stuart UC152), Thermometer (GHZeal), Stoppers, Stopwatch, Teflon tube, Retort stand, Spatula, Test tube rack. Separating funnel 500 ml, 3 – Neck flat bottom flask, Beakers, Measuring cylinder, Laboratory funnel, Pipette, Burette, & Petri Dish (J-Sil Borosilicate glass)

2.3 Experimental Set-up

The 3-neck flat bottom flask is fitted with an exit for generated hydrogen gas, a burette through which the spoilt fruit extract and acetic acid is charged into the reactor and a thermometer. The reactor is placed on a magnetic stirrer which is switched-on as soon as either acetic acid or spoilt fruit extract is charged into the reactor.

2.4 Procedures

2.4.1 Reaction of Acetic Acid with Sodium Borohydride

NaBH4 (s) + 3CH3COOH (aq) → NaBH (CH3COO)3 (aq) + 3H2 (g)
This gives stoichiometric amounts of
1 mole of NaBH4 (s) is 37.83 g/mole and 3 moles of CH3COOH (aq) is 180.15 g/mole.
For ease of management the amounts were scaled down as shown in table 1

Table 1: Scaled-down amounts of sodium borohydride and acetic acid reactants used

S/N	Reactants	Molar amounts	Stoichiometric amounts	Scaled-down amounts	Scaled-down amounts
1	NaBH4 (s)	37.83 g/mole	37.83 g/mole	$\frac{37.83\ g/mol}{37}$	1.022 g/mole
2	3CH3COOH (aq)	60.05 g/mole	180.15 g/mole	$\frac{180.15\ g/mol}{37}$	4.869 g/mole

2.4.2 Procedure for the Production of Hydrogen from Sodium Borohydride using Acetic acid

0.04 grams of sodium borohydride was added into the reactor after which 5 ml of 1 M solution of acetic acid was charged into the reactor, volume of hydrogen produced was measured and time interval during which it was produced. Another 0.04 g of sodium borohydride was added into the same reaction mixture in the reactor and volume of hydrogen produced measured with time. 0.04 g of sodium borohydride was continuously added to the reaction mixture in the reactor until 1.022 g/mole the stoichiometric amount was

IOP Conf. Series: Materials Science and Engineering **413** (2018) 012032 doi:10.1088/1757-899X/413/1/012032

added and total volume of hydrogen produced measured with time. The same procedure was repeated for different concentrations of acetic acid; the concentrations chosen were 17.5 M, 12 M, 7 M, 5 M and 1 M concentrations all with a volume of 5 ml.

2.4.3 Obtaining Extracts from Spoilt Citrus Fruits

Spoilt oranges, lime and lemon were sourced from the local market and their juices extracted and filtered using a sieve to ensure no particle is left in the fruit extracts. The spoilt fruit extracts were diluted with water as shown in table 2, reacted with 0.04 g of sodium borohydride and volume of hydrogen produced during each dilution noted and recorded.

2.4.4 Procedure for the Production of Hydrogen from Sodium Borohydride using Spoilt Citrus Extract

0.04 g of sodium borohydride was charged into the reactor after which it was corked airtight, the burette was filled with 5 ml of undiluted spoilt fruit extract. The undiluted spoilt fruit extract was charged into the reactor after which the magnetic stirrer was switched-on; volume of hydrogen produced was noted and recorded. The same procedure was repeated with diluted spoilt fruit extract, the spoilt fruit extract was diluted as shown in table 2 and volume of hydrogen produced during each dilution recorded.

Table 2: Dilution factor with corresponding volume of water used for spoilt citrus fruit extracts

S/N	Volume of undiluted spoilt fruit extract (ml)	Volume of water added (ml)	Dilution Factor
1	5	0	1
2	5	2.5	1.5
3	5	5	2
4	5	7.5	2.5
5	5	10	3
6	5	12.5	3.5
7	5	15	4
8	5	17.5	4.5
9	5	20	5
10	5	22.5	5.5
11	5	25	6

3.0 Results and Discussion

3.1 Results

Table 3: Variation of hydrogen gas generation from the reaction of 1M of 5 ml acetic acid with sodium borohydride at an ambient temperature of 27°C

Number of runs	Cumulative NaBH4(g) used	Cumulative Volume Hydrogen produced (ml)	Time (min)	Cumulative Time (min)
1	0.04	105	0.07	0.07
2	0.08	205	0.1	0.17
3	0.12	290	0.33	0.5
4	0.16	400	0.62	1.12
5	0.20	480	2.75	3.87
6	0.24	560	3.75	7.62
7	0.28	615	4.33	11.95
8	0.32	645	6.73	18.68
TOTAL	**0.32**	**645**		**18.68**

IOP Conf. Series: Materials Science and Engineering **413** (2018) 012032 doi:10.1088/1757-899X/413/1/012032

Table 4: Variation of hydrogen gas generation from the reaction of 5M of 5 ml acetic acid with sodium borohydride at an ambient temperature of 27°C

Number of runs	Cumulative NaBH4(g) used	Cumulative Volume Hydrogen produced (ml)	Time (min)	Cumulative Time (min)
1	0.04	100	0.2	0.2
2	0.08	210	0.083	0.28
3	0.12	290	0.22	0.50
4	0.16	400	0.28	0.78
5	0.20	510	0.33	1.11
6	0.24	600	0.25	1.36
7	0.28	655	0.2	1.56
8	0.32	740	0.083	1.65
9	0.36	830	0.2	1.85
10	0.40	900	0.28	2.13
11	0.44	1000	0.18	2.31
12	0.48	1080	0.15	2.46
13	0.52	1180	0.22	2.68
14	0.56	1255	0.2	2.88
15	0.60	1360	0.25	3.13
16	0.64	1430	0.47	3.60
17	0.68	1540	0.55	4.15
18	0.72	1640	0.55	4.70
19	0.76	1720	0.78	5.48
20	0.80	1800	0.97	6.45
21	0.84	1905	1.17	7.62
22	0.88	2005	1.67	9.29
23	0.92	2105	1.95	11.24
24	0.96	2205	2.4	13.64
25	1.00	2300	2.62	16.26
TOTAL	**1.00**	**2300**		**16.26**

Table 5: Variation of hydrogen gas generation from the reaction of 7M of 5 ml acetic acid with sodium borohydride at an ambient temperature of 27°C

Number of runs	Cumulative NaBH4(g) used	Cumulative Volume Hydrogen produced (ml)	Time (min)	Cumulative Time (min)
1	0.04	100	0.28	0.28
2	0.08	210	0.40	0.68
3	0.12	280	0.35	1.03
4	0.16	390	0.63	1.66
5	0.20	480	0.33	1.99
6	0.24	580	0.53	2.52
7	0.28	655	0.58	3.1
8	0.32	765	0.67	3.77
9	0.36	835	0.63	4.4
10	0.40	925	0.78	5.18
11	0.44	1035	0.83	6.01
12	0.48	1125	0.78	6.79
13	0.52	1215	1.83	8.62

IOP Conf. Series: Materials Science and Engineering **413** (2018) 012032 doi:10.1088/1757-899X/413/1/012032

14	0.56	1325	2.15	10.77
15	0.60	1415	1.22	11.99
16	0.64	1525	1.83	13.82
17	0.68	1625	3.1	16.92
18	0.72	1735	4.53	21.45
19	0.76	1835	3.73	25.18
20	0.80	1935	4.5	29.68
21	0.84	2035	4.67	34.35
22	0.88	2145	5.83	40.18
23	0.92	2255	7.7	47.88
24	0.96	2350	7.47	55.35
25	1.00	2460	8.37	63.72
TOTAL	**1.00**	**2460**		**63.72**

Table 6: Variation of hydrogen gas generation from the reaction of 12M of 5 ml acetic acid with sodium borohydride at an ambient temperature of 27°C

Number of runs	Cumulative NaBH4(g) Used	Cumulative Volume Hydrogen produced (ml)	Time (min)	Cumulative Time (min)
1	0.04	110	0.40	0.40
2	0.08	205	0.43	0.83
3	0.12	280	0.42	1.25
4	0.16	380	0.53	1.78
5	0.20	450	0.33	2.11
6	0.24	530	0.40	2.51
7	0.28	605	0.38	2.89
8	0.32	705	0.42	3.31
9	0.36	795	0.33	3.64
10	0.40	895	0.47	4.11
11	0.44	995	0.48	4.59
12	0.48	1095	0.53	5.12
13	0.52	1195	0.70	5.82
14	0.56	1295	1.58	7.40
15	0.60	1395	2.13	9.53
16	0.64	1495	3.00	12.53
17	0.68	1595	3.67	16.20
TOTAL	**0.68**	**1595**		**16.20**

Table 7: Variation of hydrogen gas generation from the reaction of 17.5M of 5ml acetic acid with sodium borohydride at an ambient temperature of 27°C

Number of runs	Cumulative NaBH4(g) used	Cumulative Volume Hydrogen produced (ml)	Time (min)	Cumulative Time (min)
1	0.04	110	1.48	1.48
2	0.08	205	1.67	3.15
3	0.12	295	2.47	5.62
4	0.16	395	3.48	9.10
5	0.20	505	4.50	13.60
6	0.24	595	16.67	30.27
7	0.28	705	12.87	43.14

8	0.32	795	19.27	62.41
9	0.36	895	20.85	83.26
10	0.40	995	27.38	110.64
TOTAL	**0.40**	**995**		**110.64**

Table 8: Results for Reaction between Sodium Borohydride and Extract of Spoilt Lime at an ambient temperature of 27°C

Number of Runs	NaBH4 (grams)	Volume of undiluted extract (ml)	Volume of dilution water (ml)	Time per run (min)	Cumulative Time (min)	Volume of Hydrogen per run (ml)	Cumulative Volume of Hydrogen Produced (ml)
1	0.04	5	0	0.5	0.5	60	60
2	0.04	5	2.5	0.77	1.27	70	130
3	0.04	5	5	0.5	1.77	80	210
4	0.04	5	7.5	0.8	2.57	60	270
5	0.04	5	10	0.83	3.4	60	330
6	0.04	5	12.5	0.83	4.23	70	400
7	0.04	5	15	0.83	5.06	70	470
8	0.04	5	17.5	1.08	6.14	60	530
9	0.04	5	20	0.67	6.81	30	560
10	0.04	5	22.5	0.5	7.31	50	610
11	0.04	5	25	0.77	8.08	50	660
TOTAL					**8.08**		**660**

Table 9: Results for Reaction between Sodium Borohydride and Extract of Spoilt Lemon at an ambient temperature of 27°C

Number of Runs	NaBH4 (s) (grams)	Volume of undiluted extract (ml)	Volume of dilution water (ml)	Time per run (min)	Cumulative Time (min)	Volume of Hydrogen per run (ml)	Cumulative Volume of Hydrogen Produced (ml)
1	0.04	5	0	1.67	1.67	90	90
2	0.04	5	2.5	0.5	2.17	50	140
3	0.04	5	5	0.62	2.79	70	210
4	0.04	5	7.5	0.57	3.36	50	260
5	0.04	5	10	0.58	3.94	60	320
6	0.04	5	12.5	0.52	4.46	50	370
7	0.04	5	15	0.5	4.96	50	420
8	0.04	5	17.5	1.82	6.78	55	475
9	0.04	5	20	1.67	8.45	55	530
10	0.04	5	22.5	1.5	9.95	50	580
11	0.04	5	25	2.3	12.25	50	630
TOTAL					**12.25**		**630**

Table 10: Results for Reaction between Sodium Borohydride and Extract of Spoilt Oranges at an ambient temperature of 27°C

Number of Runs	NaBH4 (s) (grams)	Volume of undiluted extract (ml)	Volume of dilution water (ml)	Time per run (min)	Cumulative Time (min)	Volume per run (ml)	Cumulative Volume of Hydrogen Produced (ml)
1	0.04	5	0	0.68	0.68	100	100

2	0.04	5	2.5	0.75	1.43	60	160
3	0.04	5	5	1.3	2.73	70	230
4	0.04	5	7.5	0.63	3.36	60	290
5	0.04	5	10	1.33	4.69	60	350
6	0.04	5	12.5	0.95	5.64	60	410
7	0.04	5	15	1.12	6.76	60	470
8	0.04	5	17.5	1.33	8.09	60	530
9	0.04	5	20	1.42	9.51	60	590
10	0.04	5	22.5	1.52	11.03	60	650
11	0.04	5	25	1.97	13.00	60	710
TOTAL					**13.00**		**710**

3.2 Discussion of Results

3.2.1 Dilution Factors
The spoilt fruit extracts were diluted with water using various dilution factors as shown in table 2 to investigate its hydrogen gas production when reacted with sodium borohydride, does it increase in as dilution increases or does it decrease; is there a particular dilution that is more productive when compared to the other dilution factors. It was observed from table 8, 9, & 10 that dilution factors of 1 & 2 produced the largest volumes of hydrogen gas when compared with other dilution volumes signifying the dilutions which yield optimum performance for hydrogen gas production from the extracts of spoilt lime, lemon & orange. A dilution factor of 2 represents an equal volume of solute mixed with an equal volume of diluent

3.2.2 Acetic acid solutions (Tables 3 to 7)
Tables: 3, 4, 5, 6 & 7 are results for acetic acid solutions of 1 M, 5 M, 7 M, 12 M & 17.5 M respectively reacted with sodium borohydride with the sole objective of investigating its viability for hydrogen production and if viable which of the solutions has the optimum performance for hydrogen production. 1 M solution of acetic acid shown in table 3 produced 645 ml of hydrogen gas on a cumulative addition of 0.32 g of sodium borohydride, on further addition of sodium borohydride molecules no hydrogen gas was produced. 5 M solution yielded a cumulative volume of 2300 ml of hydrogen gas from a cumulative amount of 1.00 g of sodium borohydride, 7 M solution yielded a cumulative volume of 2460 ml of hydrogen gas from a cumulative amount of 1.00 g of sodium borohydride, 12 M solution yielded a cumulative volume of 1595 ml of hydrogen gas from a cumulative amount of 0.68 g of sodium borohydride while 17.5 M solution produced a cumulative volume of 995 ml of hydrogen gas from a cumulative amount of 0.40 g of sodium borohydride shown in tables 4, 5, 6 & 7 respectively. It was observed from tables 3, 4, 5, 6 & 7 that table 5 containing the data for the 7 M solution of acetic acid produced the largest volume of hydrogen gas 2460 ml, this could be attributed to the dilution as compared to table 7 which has 17.5 M solution acetic acid and table 3 which has 1 M solution of acetic acid. 7 M solution is the solution which has the largest effective concentration of hydrogen ions as compared to the actual concentration in solution [17] [18].

3.2.3 Temperature
This research was aimed at producing hydrogen without the use of any neither catalyst nor external heat source making it cost-effective and readily available for on-board usage. Ambient temperature of 27°C was the laboratory temperature at the time of carrying-out the investigation.

3.2.4 The effect of acetic acid concentration on the volume of hydrogen generated
Acetic acid was reacted with sodium borohydride in various concentrations of 17.5 M, 12 M, 7 M, 5 M and 1 M shown in Figure 1

IOP Conf. Series: Materials Science and Engineering **413** (2018) 012032 doi:10.1088/1757-899X/413/1/012032

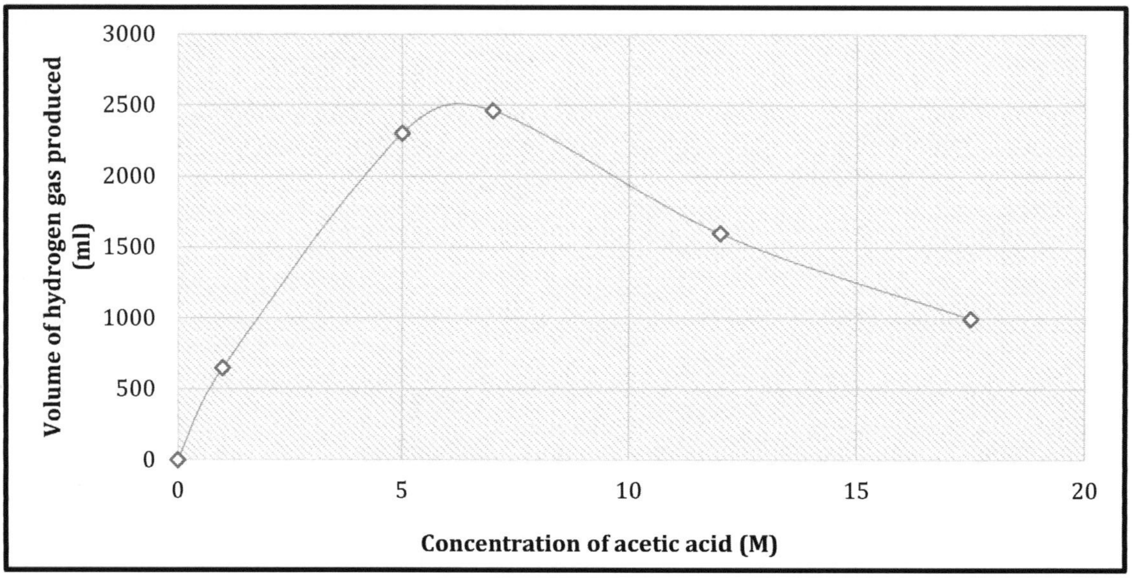

Figure 1: Variation of hydrogen gas produced with the concentration of acetic acid

From figure 1 it was observed that the volume of hydrogen generated increased steadily with the concentration of acetic acid. The graph increased until it got to a maximum at 2500 ml of hydrogen generated at an optimum concentration of 6.3 M and began to decline, hydrogen generation beyond the optimum acetic acid concentration was counterproductive because they all yielded lower volume of hydrogen gas. This may be adduced to the reduced activity of the acid at high concentration; in acidic solutions with high concentrations activity is much lower than the actual concentration because the effective concentration of hydrogen ions is lower than the actual concentration in the solution, as some of the hydrogen ions present in the solution are unable to participate in the reaction [17][18].

3.2.5 The effect of spoilt citrus extracts on hydrogen gas generation from sodium borohydride

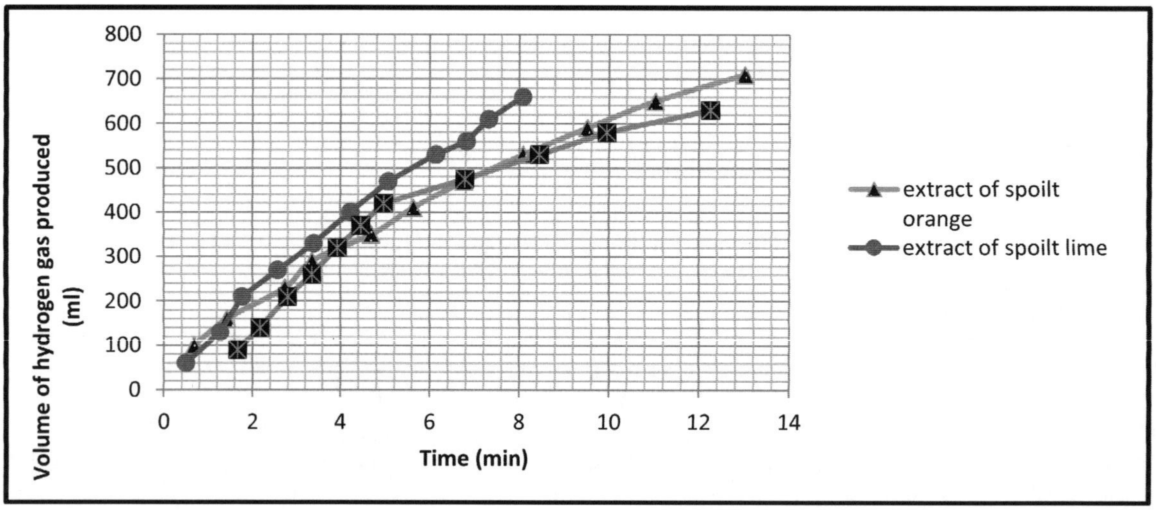

Figure 2: Volume of hydrogen gas produced against time for spoilt lime, lemon & orange extract.

From Figure 2 it was observed that as the spoilt fruit extracts were diluted and reacted with sodium borohydride it continues to generate hydrogen gas. 660 ml in 8.08 min, 630 ml in 12.25 min and 710 ml in 13.00 min for spoilt fruit extracts of lime, lemon and orange respectively. Each point on the graphs represents a new dilution as shown in table 2, it was observed from tables 8, 9 and 10 that dilution factors of 1 and 2 yields better performance for hydrogen gas production compared to the other dilution factors. Cumulatively the spoilt orange extract yields largest volume of hydrogen gas, as the dilution of the spoilt fruit extract increases hydrogen production gradually decreases; this could be attributed to the fact that the active ingredients in the spoilt fruit extracts are gradually being diluted as well as reacting with the sodium borohydride molecules.

3.2.6 Confirmatory Test

To ensure it was hydrogen gas being produced the flame test was carried-out on the gas. Burning flames could be used to identify whether a gas is flammable or oxidizing, hydrogen gas is flammable over a wide range of concentrations in air. For an oxidizing gas such as oxygen, the flame on the splint is allowed to burn for five seconds then extinguished by shaking or blowing out. While the ember is still glowing hot it is introduced into the suspected oxidizing gas, if it where oxygen for example; the glowing ember re-ignites to produce a sustained flame[19]. The gas from the reactor was tubed to a burner; a match stick was lit and placed over the burner after which it was opened releasing the trapped gas. The gas ignited with a pop sound and was sustained for a few seconds before extinguishing; confirming the presence of hydrogen gas.

4. Conclusion

A summary of the findings of this research enables these conclusions to be drawn. Sodium borohydride was successfully reacted with acetic acid and hydrogen gas was produced. A solution of 7 M of acetic acid produced the best performance for hydrogen gas production when reacted with 1.00 gram of sodium borohydride yielding 2460 ml of hydrogen gas in 63.72 min. The undiluted extracts of spoilt orange, lemon and lime gave 100 ml in 0.68 min, 90 ml in 1.67 min and 60 ml in 0.5 min respectively. For the diluted fruit extracts, lime, orange and lemon at a dilution factor of 2 each, generated hydrogen gas of 80 ml in 0.5 min, 70 ml in 1.3 min and 70 ml in 0.62 min respectively. Reactions between sodium borohydride, acetic acid, and spoilt fruit extracts of lime, lemon and orange were safely carried out at an ambient temperature of 27°C. The reaction of these selected organic acids (acetic and citric acids) proceeded without the use of catalyst.

References

[1] Rao, A. *Sustainable energy conversion for electricity and coproducts: Principles, technologies and equipment*. Hoboken, NJ: John Wiley, 2015, pp. 9-22.

[2] Dincer, I., & Rosen, M.A. *Exergy: Energy, environment and sustainable development,* 1st edition. Oxford, England: Elsevier. 2007, pp.1-59.

[3] Freris, L., & Infield, D. *Renewable energy: In power systems* 2nd edition. West Sussex, England: Wiley. 2009, pp. 1-4.

[4] Tester, W.J., Drake, E.M., Driscoll, M.J., Golay, M.W., & Peters, W.A. (2012). *Sustainable Energy: Choosing among options,* 2nd edition. Massachusetts, MA: Massachusetts Institute of Technology Press Books. 2012, pp.782-796

[5] Huang, M., Ouyang, L., Wang, H., Liu, J., & Zhu, M. Hydrogen generation by hydrolysis of MgH2 and enhanced kinetics performance of ammonium chloride introducing. *International Journal of Hydrogen Energy,* 2015, *40* (18), 6145-6150.

[6] Sorensen, B. *Renewable energy: Conversion, transmission and storage*. Oxford, England: Elsevier. 2007, pp. 283-287.

[7] Yerramilli, A., & Tuluri, F. *Energy resources utilization and technologies*, Hyderabad, India: CRC Press. 2012, pp.345- 377.

[8] Belz, S. A synergetic use of hydrogen and fuel cells in human spaceflight power systems. *Acta Astronautica,* 2016, *121* (2016), 323-331.

[9] Rusman, N.A.A., & Dahari, M. A review on the current progress of metal hydrides material for solid-state hydrogen storage applications. *International Journal of Hydrogen Energy, 2016, 41* (28), 12108-12126.

[10] Housecroft, C.E., & Constable, E.C. *Chemistry* 4th edition. Essex, England: Pearson Education Limited. 2010, pp. 691-738.

[11] Sorensen, B. *Renewable energy: Conversion, transmission and storage.* Oxford, England: Elsevier. 2007, pp. 283 – 287.

[12] Perejon, A., Sanchez-Jimenez, P.E., Criado, J.M., & Perez-Maqueda, L.A. Magnesium hydride for energy storage applications: The kinetics of dehydrogenation under different working conditions. *Journal of Alloys and Compounds*, 2016, *681* (2016), 571-579.

[13] Arzac, G.M., & Fernandez, A. Hydrogen production through sodium borohydride ethanolysis. *International Journal of Hydrogen Energy,* 2015, *40* (15), 5326 – 5332.

[14] Li, H-W., Akiba, E., & Orimo, S-I. Comparative study on the reversibility of pure metal borohydrides. *Journal of Alloys and Compounds,* 2013, *580* (1), S292 – S295.

[15] Zhang, Z.G., Wang, H., & Zhu, M. Hydrogen release from sodium borohydrides at low temperature by the addition of zinc fluoride. *International Journal of Hydrogen Energy,* 2011, *36* (14), 8203 – 8208.

[16] Sahin, Ö.,Kilinc, D., & Saka, C. Bimetallic Co – Ni based complex catalyst for hydrogen production by catalytic hydrolysis of sodium borohydride with an alternative approach. *Journal of the Energy Institute,* 2016, *89* (4), 617 – 626.

[17] Deutsch, W.J. *Groundwater geochemistry*: Fundamentals and applications to contamination Boca Raton, FL: CRC Press. 1997, pp. 19.

[18] Aqion. *Activity & ionic strength.* 2012 [Online] Retrieved from http://www.aqion.de/site/69

[19] Chemiscience. *Test for oxygen, hydrogen and carbondioxide.* 2013 [Online] Retrieved from https://chemiscience.wordpress.com/2013/05/01/test-for-oxygen-hydrogen-and-carbon-dioxide/

IOP Conf. Series: Materials Science and Engineering **413** (2018) 012033 doi:10.1088/1757-899X/413/1/012033

Impact of Flood Danger in Built-Up Areas in Nigeria and Floor Management Systems for Espousal

Igibah, Ehizemhen. C; Sadiq, Abubakar. A.

Department of Civil Engineering, university of Abuja, F.C.T Abuja, Nigeria.
*Corresponding author: igibahchrist1@gmail.com, lightjesuniyi@yahoo.com +234 7052275257

Abstract- Tendency of flood danger happening often and size in built up areas in Nigeria, caused man to take steps of providing flood mitigation as remedy against flooding. Diverse flood danger regulation standard was implemented in various region of the universe yet flood occurrence perseveres. This paper appraises key sources of floor hazard in cities along with different floor risk regulation standard in order to develop acceptable floor danger mitigation standard for built up areas. Pertinent information for analyzing the component of flood risk and tendency in flood danger mitigation standard were obtained through corresponding empirical learning and theories. The outcome shows the incapability of non-structural as well as structural standard to totally terminate flood happenings. Flood hazard in cities can sustainably minimize by consolidating present measures together with partial logical transformation of land operation to its natural position.

1. Introduction

Flood danger is one among destructive natural hazards that leads to environmental and resources deteriorating, waste of lives and harm to goods (Forton, 2011). Above 3,000 floods tragic happened in a range of twenty years i.e 1990 – 2010 which liable for 200,000 lives loss and three billion mankind were vagrant across the globe (Smith, 2013). Nearly 200 million human beings averagely in over ninety nations are revealed to ruinous flood incident yearly and it assumed to increase subsequently due to climate change and stable demographic growth in addition to urbanization (UNESCO, 2008). World natural hazards key concern is the making of systems, scheme and strategies of the universal government at diverse levels to mitigate the flood events. Persistence occurrence of flood happenings apart from the undertaken standard shows the incapacity of flood regulation measures to sufficiently limit floods. A justifiable floor danger management needs flood danger evaluation to locate forces and factors that generate possible flood hazard. This article reviewed diverse empirical and theoretical literatures as a system for gathering fact concerning sources of floor danger in cities with the associate flood mitigation standard to locate a sound floor danger management standard for built up areas. Some literatures ascribe floor danger to change in climate (karley, 2009). Affirm that source of flooding in Ghana are enormous rainfall that produce prodigious run – off that causes floods. (crisis, 2009) attested that the flood occurrences frequently expanding cannot be separated from climate change. Different researchers insist that the flood happenings growth was not only assignable to the intense climate changes (Hooijer, 2004). In principle flood hazard is normally predict as the probability of risk (climate change) and the exposure of the elements at danger. Simple grasp of the three features that produce hazard given the vital details for factoring most floods correlated. Majority flood correlated aspects is the overall administration of flood danger simultaneously bestow substantially to the growth and wellbeing of mankind (WMO, 2009). Since floor danger means probability function of flood risk and the likely damage, most flood minimizing standards focus at lessen the probability of flooding and diminish the possible damage (Hooijer 2004). Hence options available to handle flood dangers involve structural (embankments, dyke and dams). For example (Sultana, 2007) avow that structural standards like embankments can proffer safety against innumerable kinds of flooding. Another view on flood mitigation highlights the unification of water resource and land-use administration, risk management and transform the floor danger Management (FDM) paradigm from protective to pro-active, from impromptu to integrated flood management. Centers on handle and living with floods, balancing floods for sound development and near the decision – making procedure divergently by learning to handle danger and stay with the floods (WMO, 2009). Whether retort to flood danger capture structural or mixture of both structural and non-structural standards, public participation of communities at danger in decision formation is necessary for flood hazard management sustainability in built up areas. Beyond all the standards above the review uncover flood event and persevere frequency in cities areas and it became perceptible that absolute avert from flood danger is inexorable. The article also reveals that a slowly orderly land use transformation to its original state could sustainably lessen flood danger and flood frequency (Scilling, 2014).

2. Literature Review

2.1 Theoretical System for Flood Danger Ineluctable

There is no hesitation for an expansion of flood danger in diverse built up area and metropolis across the globe. Particularly river floods that account for nearly complete loss of lives and one – third of all economic diminution from natural risks globally (UNESCO, 2008). Between 1990 – 2010, range of twenty years floods are liable for 200,000 loss of lives and attacked 3 billion mankind by making them vagrant (Smith, 2013). Constant rising of flood events worldwide and the increasing flood danger make environmental scholar contemplate on what could be the potential standards against it. Innumerable diverse views exist explaining the key causes of flood danger and these are analyzed below and shown in Fig. 1.

Fig 1: Benue flood August 2017 displacing over 250,000 people.

2.1.1 The Climate Factors

Several findings were discovered from floor danger research works carried out in numerous built up areas concerning factors accountable for flood danger, catastrophic problems may place at the boundary of natural and social environment. Reality flood danger has been viewed theoretically as outcome of the climate change, the socio cultural environment and periodically jointly mixture of built environment along with climate. Certain environmental scientist insists that climate change together with universe warming is directly or indirectly expanding the quantity of ice melting as well as rain thus increasing run off vastness and later flooding. For example, flood catastrophic in Zimbabwe are correlated to two various circumstances: localized ponderous seasonal rainfall trigger river overfull and the cyclone induced floods prompt to constant and seasonal flood (Ghimbi, 2007). As claimed by (Karley, 2009) the enormous rainfall instigate tidal waves, run- off together with dam – burst and widen flooding was outcome of cyclone Eline 2000 across Southern Africa like Malawi, Botswana, South Africa and so on (Vaz, 2000). (Crisis, 2009) Affirmed that expanding flood occurrences frequency cannot be disparate from change in climate.

Nevertheless, floor risk along with its danger could not be reckon with as only factor answerable to flood hazard, consequently other factors are required to give requisite cognizance and illumination of the procedure.

2.1.2 The Socio – Cultural Factors

The second views assert that cultural events have meaningful impact on the operation of the physical natural environment with measures that only answerable to environment. Of all land utilization reforms that influence hydrology of a region substantially, urbanization is most energetic changes producer in peak flow features, water quality, hydrological utilities and total run off (Leopold, 1968). Expansion of mankind population in the built-up areas, encroachment and alteration of river structure flood plains are responsible for increasing destruction and danger sources from floods (al S. e., 2010). Runoff volume control mainly by infiltration features and affiliated to vegetation, land, soil type together with slope, then directly related to percentages of regions covers by roads, roofs and other water resistance surface at period of hydrograph expanding during storms (Leopold, 1968). Rising of impermeable surface like the consequence of expanding flood peaks throughout storm period and diminishing low flows betwixt storms (Leopold, 1968). For instance, abolition of vegetated land engulfs by substitute it with concrete surfaces in cities areas of rising water proof surfaces, trigger overland flow expansion and lessen infiltration, elude the natural storage along with weakened of the surface causing flooding and rapid run-off. Consequently, urbanization volume rising and surface runoff frequency through amendment of natural drainage structure and changing of run off to streams, concluding result is massive volume of runoff, releasing shorter space of time and possibly causing serious increase in floor peaks (Smith, 2013). For example, empirical studies reveal that climate is not momentous factors in the scrutinized floor destruction growth in Africa based on (al B. e., 2010) study on flood mortality in Africa. The procedure includes but not restrained to physical environment substandard planning, faulty waste management, insufficient drains for cities also flood plain regions profession (Hooijer, 2004). Scrutinize the effect of urbanization for thirty years on two sub-catchments of the Thames, revealing obvious flood occurrence expansion with growth escorted by diminished in storage. (Daniel, 2012) as well attest that the common notion of heavy rainfall as been the principal source of cities flood is disprove, but absence of infrastructure in built up area play a key role for flood hazard in Gombe region, Nigeria. This view theorized established on empirical research that the rising flood danger is not only have root in climate change but the expanding build-up environment influence how environment works as principal cause for flooding. Hence advance in flood studies integrate two factors for flood occurrence in the world.

2.1.3 Collaborate effect of climate and socio-cultural factors

(Pielke, 2000) contend that, there is a fragile relationship betwixt hydrological factor and the destructive floods, because the destructive floods happen from a collaborate impact of physical and societal procedures, and floods emerge from integration of meteorological and hydrological extremes (WMO and GWP 2008). The flood danger is expanding as a result of continued encroachment of mankind and properties as well as climate change in regions at risk of flooding outcome is increasing possible destruction. (Hooijer, 2004) and (Criss, 2015) Finalize that flood volume increases due to in- channel structures along with climatic change outcome. In line with the above views, flood danger can be grasping as a function of hazard, subjection and vulnerability (Hooijer, 2004). Climatic change could be seen as risk together with socio-cultural environment could be appraised as subjection to hazard and the helplessness people in the danger zone. However, in order to effectively comprehend the idea of flood dangers in the city, it is important to study various types of danger components surrounding flood administrative for existing of man in his environment setting. Thus, flood danger hazard can be signalized by climatic change leading to expectation and potency high river flows that result to influx in an area. Unsafe and exposure introduce dubiety of mankind and his values in flood waters domain together with population and valuables located in dangerous region. Hence, flood danger can be seen as a decussate integration of vulnerability, danger together with exposure and if any one among these three components rises or reduces, then danger expands or diminishes consequently. As such apprehension flood danger notion could be an effective standard for hazard minimizing.

2.1.4 SOME STATISTICS OF FLOOD OCCURENCES IN NIGERIA

S/N	Flood Location	Impact on the community	Period
1	North-east region Borno state Rann	Flood cut off roads and made helicopter landings difficult. Aid convoys by road to Rann, in Borno sate which hosts 43,000 emigrant populaces, has become enormously tough. About 800 metric tons of food, nourishment, health, WASH and shelter supplies are awaiting delivery. Both logistics segment and WFP are exploring the possibilities of bringing in relief from Cameroon.	OCHA, 11 Aug 2017
2	North -east region Borno state LGAs like Gwoza, Ngala, kaga balge, Dikwa, Damboa, and Bama	Flood has greatly lessened access to places that are in urgent severe need of sector and emergency shelter interventions attributable to privation of response capacities, security, access, land obtainability and logistics required to set up apposite sites.	IOM, UNHCR, CCCM Cluster, Shelter Cluster, 21 Aug 2017
3	North central 21 LGAs in Benue state, Nigeria	Heavy rains on 4th September, 2017, have led to showy floods, discharges and river spilling over which affected 250,000 people in Benue State, Nigeria. Hundreds of houses besides crop fields destroyed. Both the State Government and the National Emergency Management Agency (NEMA) set up two camps in Makurdi, for populaces that have lost their homes. Also, the Nigerian Red Cross supplied food and non-consumable items to the displaced people while UNICEF increasing access to drinking water in the two camps as well carried out a rapid assessment which discovered these floods as the most severe since 2012.	OCHA, 6 Sep 2017
4	Lagos State	There was a snarl-up on major roads, instigating people to annul or postpone appointments they might have had. Thousands of stranded travelers had to pay increased fees to the limited bus drivers who were agreeable to risk travelling on the roads.	On 2 July 2012
5	Ibadan, Oyo State	Flooding caused some inhabitants at Eleyele, Challenge and Oke-Ayo to abscond from their residences so as to save their lives. Some Christians were prevented from attending churches in the morning because few bridges caved in.	Mid-July 2012
6	Jos, Plateau State	No less than 39 people were exterminated due to flooding in the North- central Nigerian Plateau state. Heavy rainfall instigated the Lamingo dam to overflow which sweeping across a number of localities in Jos, and roughly 200 households were submerged. Besides, about 35 people were missing, when Manasie Phampe, the state head of the Red Cross, publicized that relief efforts were	Late July 2012

		continuing. The floods left 3,000 populaces destitute that are taking shelter in government buildings in Jos.	
7	Delta and Bayelsa State	About 120,000 people rendered homeless, according to state authorities and the Nigeria Red Cross. Several camps set up were also flooded forcing populaces to flee. In Yenagoa, 3,000 people were taken shelter at the Ovom state Sports Complex. In Delta State, 20 health clinics, 5 hospitals, numerous schools, churches and government buildings were destroyed by flood. Schools were either closed or engaged by internally emigrant people.	Early october, 2012
8	Kogi State	Approximately 623,900 persons were being displaced and 152,575 hectares of farmland damaged. It is the worst flooding besides a national disaster according to a NEMA coordinator this prompt Nigerian President Goodluck Jonathan to released 17.6 billion naira (US$111 million) to many states and agencies for rehabilitation, mutilation response and flood relief.	October, 2012

2.1.5 Viability models for flood danger management (FDM)

Hold the divergence betwixt the three components that generates hazard, vulnerability, danger together with exposure – confer the valuable information for factoring in almost all flood associated aspects in the entire management of flood dangers and instantaneously provide considerably progress and welfare of mankind (WMO, 2009). The models accessible for flood danger mitigation comprises of orthodox structural standards (embankments, dams, reservoirs, dyke) and unsegregated non-structural (planning of land, flood caution systems, expulsion, insurance and preparedness) alternatives; government, individual and agency. The first element is the risk and the structural model aim at especially the flood danger zone. The ancient structural flood hazard lessen policies is normally a fundamental on river education, embankment construction as well as reservoirs retention, targeted at minimizing flood risk, which means flooding expectation. For example, (Sultana, 2007) ascribed to structural policies such as (dam, dyke and slope) could afford protective measures against various kinds of flooding. Cities water management in nearly all advanced technological countries of America, China and Europe is been specified by wide margins of safety involves large infrastructure and technical amenities (Pielke, 2000). Recently flood regulation policies include flood-proofing standards like homesteads foundation, plinths upgrading, creating domestics environment for animals (livestock) in flood shelters (Sultana, 2007), the policy has been on protection and regulation rather than on administration.

Secondly, other models examined flood danger in a diverse angle by integrating exposure and unprotected features of the flood danger. Since, it is identified that structural flood regulation cannot provide solution to flood danger and risk challenges, because flood regulating standards have been usually organized in separation from other development. Thus, they are reactive instead of proactive, by concentrating on structural policies, and in search of solutions from mono-disciplines (Shresha, 2012). Also, to contend that flood regulation policies must be mixed with non-structural methods, like planning of land usage to generate extensive flood management (Smith, 2013). Subsequently the paradigm moves from post-disaster rejoinder and relief centric system to pre-disaster proactive readiness and mitigation centric system center on catastrophe as direct perturb and a general understanding of the notion of vulnerability as necessity for developing a major concept. Nevertheless, order and sustainability need transpose from the natural structural flood protection to more extensive Flood danger Management (FDM) system that encompass of retrieval, prevention, guarding, readiness and response. The most effective and sustainable minimizing flood dangers can be obtained through decreasing possible damage (vulnerability) in the region prone to flooding through land use adaptation together with spatial planning. For instance, in recent time, a

IOP Conf. Series: Materials Science and Engineering **413** (2018) 012033 doi:10.1088/1757-899X/413/1/012033

noticeable paradigm move in this aspect is flood plains design and hovering residents' practices became more effective to cope with a highly unforeseeable surrounding in Netherlands (Wostl, 2005). More so 20,000 loss of lives in Indian East Coast in 1977 originated from principal storm, but after this calamity, an advance warning system was introduced when the same location was strike with selfsame flood enormity, in 1996 and 2005, the casualties' number was 1000 (UNESCO, 2008). The principal aim of cities FDM is to reduce loss of lives and destruction of economic, by using natural resources for the welfare and benefit of the mankind.
The third prospective insisted that flood dangers tend to be grasp better by local community's participation, on account of their nearness to the waterways acts as consistent to the risks they are uncovered to. Thus, readiness to be involved in the flood administrative planning is important (al c. e.). Perhaps flood danger takes structural, integrated, or non-structural standards; there still need of public participation techniques in making common resolution about flood danger sustainability. The objective is to provide mitigate flood danger measures which involves how to lessen vulnerability and resilience expansion and ability of local people that has been affected by the flood. Sound FDM system incorporate water resource, environmental and hazard administration, replacing flood mitigation regulation paradigm from defensive to pro-active, from extempore to integrated flood administration and concentrate on how to control and live with floods, stabilize floods for credible progress, and tackle the decision-making procedure in various ways by training on danger management and living with the floods. Because cities flooding is ineluctable as much as built up areas keep expanding together with the rising flood hazard across the globe.

3. Result and Discussion

Established on the prevailing literature three panoramas predict the elements accountable for flood danger in built up areas, climate change reckoned as one of the elements that generates flood risk. Secondly socio-cultural happenings like built up areas expansion seriously lessen infiltration potential and thereby producing more runoff and the next flood occurrence. In addition, propinquity of cities structures subject to flood zone ravage flood hazard in built up environment. Lastly combination of socio – cultural and climatic effects as the key point for cities flood occurrence. Thereby flood danger normally incorporate three components; unprotected, hazard and vulnerability. Climate is the main initiator of flood hazard and profession of mankind at floodplains and their socio-economic state expanding unprotected and vulnerability of flood danger. To effectively mitigate flood danger, apprehension of the three components is needful. Structural flood standards focus on hazard, while non-structural standards target unprotected and vulnerability components. And modern FDM prescribe a universal method of structural and non-structural standards with local people participation of elements at risk for effective and sound FDM in urban settings. A serious evaluation of the available flood danger standards, the reviews unveils that a slow systematic changing of cities land use to its original state could maintain flood danger reduction and flood persistence in city environment (Scilling, 2004).

4. Conclusion

Nigeria has a showery season as well suffers from periodic flash overflows, these flash overflows are sometimes disastrous, specifically in the countryside areas or congested purlieus, where there is lack of drainage. Also, tendency of swift mankind, city development, the growing flood danger in built up areas, most environmental scholars ascertain that complete flood security in nearly all cases could wholly unavoidable, but FDM can provide mitigate flood danger measures which involves how to lessen vulnerability, reduce loss of lives and destruction of economic, by using natural resources for the welfare and benefit of the mankind specially communities participation of local people that are nearness to the waterways so as to aware of consistent to the risks they are uncovered to because through local communities participation, they can grasp flood dangers better. Conclusively, flood dangers could not be completely ignored but can only be minimized to an expected stage and FDM would not attempt to eradicate flood dangers wholly in city areas but will only strive to create strategies to mitigate non-structural standards with community participation.

IOP Conf. Series: Materials Science and Engineering **413** (2018) 012033 doi:10.1088/1757-899X/413/1/012033

References

[1] Baldassare G, Momtanari A, Lins H, Kontsoyiannis D, Brandimarte L and Bloscl G. Flood Fatalities in Africa: From Diagnosis to Mitigation, Geophysical Research Letters, 37. 2010.
[2] Criss R. Increased flooding of large and small watersheds of the central USA and the consequences for flood frequency predictions, In Criss RE, Kusky TM, Eds, Finding the Balance between Floods, Flood Protection, and River Navigation, Saint Louis University, Center for Environmental Sciences, 2009, 1 (2):16-21.
[3] Criss R. Statistics of Flood Populations in a Changing. World. J.Earth Sci., 2015 6(4):1-9.
[4] Karley K. Flooding and Physical Planning in Urban Areas in West Africa: Situational Analysis of Accra, Ghana, Theoretical and Empirical Research in Urban Management, 2009, 4(13):25-41.
[5] Samarasinghe M, Nandala K, Weliwitiya P, Fowzi M, Hazarika M and Samarakoon L. Application of Remota Sensing and GIS for Flood Risk Analysis: A Case of Kalu-Ganga River, 2010, 2(3): 1-6.
[6] Scilling K, Gassman P, Kling C, Cambell T, Wolter C and Anold G. The Potential for Agricultural Land Use Change to Reduce Flood Risk in a Large Watershed. Hydrological Processes, 2014, 28(8): 3314-3332.
[7] Sultana P. Can England Learn Lessons from Bangladesh in Introducing Partcipatory Floodplain Management? Water Resources Management, 2007, 22 (3): 357-376.
[8] Sri Lanka. International Archives of the Photogarmetry, Remote Sensing and Spatial Information Science Vol.38, Part 8 Kyoto.
[9] Smith K. Environmental Hazards: Assessing Risk and Reducing Disaster, 2013, 6th Ed., New York, Rouledge.
[10] Shresha B, Chapag S and Thapa R. Flash flood risk management: A Training of the Trainers Manual, International Centre for Integrated Mountain Development Kathmandu, Nepal, 2014.
[11] Pielke R. Flood Impacts on Society, Damaging Floods as a Framework for Assessment, in: Floods, edited by: Parker DJ, Routledge Hazards and Disasters Series, 2000, 1 (2): 133–155.
[12] Vaz C. Coping with Floods; The Experience of Mozambique. Retrieved December, 15, 2016 from www.thewaterpage.com/floods.
[13] WMO. Integrated flood management: APFM Concept Paper, Geneva, Switzerland: World Meteorological Organization, Associated Programme on Flood Management, 2009.
[14] Wisner B, Blaikie P, Cannon T and Davies I. At Risk: Natural hazards: people's Vulnerability and Disasters, London, Routledge, 2004.
[15] Woslt P. Information, Public Empowerment and the Management of Urban Watersheds, Environmental Modeling and Software, 2005, 20 (1): 457-467.
[16] Correla F, Forham M, Saraiva M and Bernarrdo F (1998). Flood Hazard Assessment and Management: Interface with the Public, Water Resour. Manage.12:209-227.
[17] Forkuo K. Flood Hazard Mapping Using Aster Image with GIS, J. Geomant. Geosci., 2011, 1 (4): 932-950.
[18] Hooijer A, Klijn F, Pedroli B and Van O. Towards Sustainable Flood Risk Management in the Rhine and Meuse River Basins: Synopsis of the Findings of IRMA- SPONG. River Res. Applicat. 2004, 20 (6): 343-357.

Mechanical Characterization of Welded API X70 Steel Exposed to Air and Seawater: A review

T. M. Ike*, O. Adedipe, M. S. Abolarin, S. A. Lawal

Mechanical Engineering Department, School of Infrastructure, Process Engineering and Technology, Federal University of Technology, Minna.
*Corresponding Author's e-mail: madiniike@yahoo.com

Abstract- A review of the mechanical characterization carried out on welded API X70 steel, and other related steels are presented in this paper. The major problems facing pipeline network are failures demonstrated in form of leaks, ageing, ruptures and explosions. However, corrosion-induced failure also occurs in pipelines that are installed in harsh environments such as seawater. The research works under review considered the welded joints of the API X70 pipelines to be a more critical region than the base metal. The mechanical properties of the welds examined include tensile strength, hardness, and Charpy impact absorbed energy. The experimental methods used to characterize the work that are reviewed include scanning electron microscope (SEM), X-ray diffraction and electron backscatter diffraction (EBSD) and ThermoARL optical emission spectroscopic analysis. Welding technique such as submerged arc welding (SAW) and Manual metal arc (MMA) and parameters used in producing welds are also discussed briefly. Discussion of seawater environment is also presented. It is important to mention that most of the works reported in the literature are those that considered weldments that are exposed to ambient air while just a few considered welds exposed to the corrosive environment such as seawater. Therefore, this review concludes with the need for more research to determine more empirical data on the effect of seawater on mechanical properties of welded API X70.
Keywords: Pipeline; Failure; Corrosion; Weld; HAZ

1. Introduction

Pipelines used for the transport of natural resources such as oil and gas are installed in various service environments some of which present enormous challenges as the pipelines experience severe environmental degradation and eventual failure resulting to heavy losses [1]. The pipeline industry has shown an increasing demand for increasingly larger pipe diameters, higher operating pressures and diversity of products transported. This leads to the use of higher strength steel grades to avoid large wall thicknesses [2]. High strength X70 grade steel (indicating minimum 70.3ksi/483MPa yield strength) was introduced to the pipeline industry, and the development of the grade X70 started about 30 years ago along with the introduction of thermo-mechanical (TM) rolling practices [3]. Current pipeline materials in the world market are often regulated according to the American Petroleum Institute's (API) standard 5L. The main design considerations outlined in API-5L are based on alloy chemistry and tensile strength. The chemical compositions of the X-grade steels are fairly simple, with maximum limits on C, Mn, S, P and other elements such as niobium and vanadium [4]. The variations in strength (e.g., between X70 and X80) do not result primarily from variations in alloy composition, but from variations in the processing route of the steel [4]. TM processing allows the yield strength of the pipe steels to be tailored through combinations of grain refinement, precipitation hardening (micro-alloying) and phase formation [5]. The risk associated with the pipeline regarding the safety of people, damage to the environment and loss of income has been a major concern to pipeline integrity managers. Previous research works have established that in most cases, failure of pipeline starts at welds [6], [7], [8]. It is therefore important to determine the quality of the weld, and as such, the mechanical properties of the weld need to be investigated for better understanding. Various mechanical related tests which include Charpy test, hardness test, tensile test, compressive test, microstructural analysis, and spectroscopic analysis could be used to characterize the mechanical properties of weldments. In this paper, review literature on the investigation of the mechanical properties of welded API X70 pipeline steel in air and seawater environments is presented.

2. Failure in pipelines

Researchers have conducted some studies on the causes of pipeline failures in the oil and gas industry [9], [7]. Discussions on the failure of the pipeline are quite broad, the fact due to several sources of failures that are associated with pipelines. Achebe et al., [7] stated that the causes of pipeline failure include outside force damage 27%, mechanically induced 40%, operator error 6%, others 25%, and lastly control problems 2%. But the discussion here is primarily restricted to mechanically induced failure arising from the metallurgical/mechanical properties of the pipeline material and most importantly welded joints. Strength, hardness, toughness, plasticity, brittleness, ductility and malleability are mechanical properties used as measurements of how metals behave under a load [8]. These properties are described in terms of the types of force or stress that the metal must withstand and how these are resisted. Common types of external stresses experienced by pipelines as illustrated in figure 1(a-f) are compression, tension, shear, impact, or a combination of these stresses, such as fatigue [9].

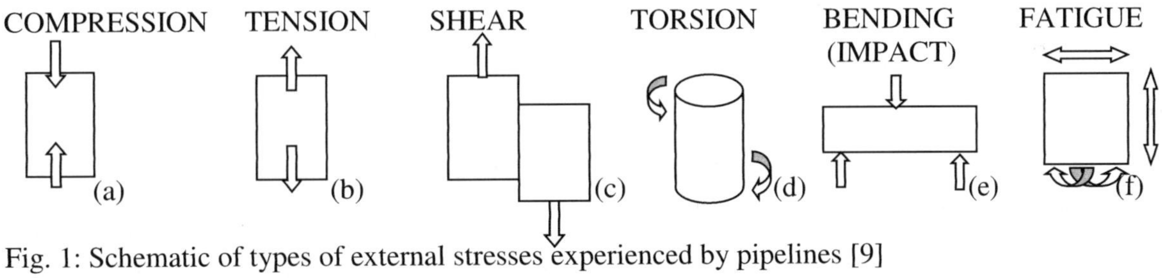

Fig. 1: Schematic of types of external stresses experienced by pipelines [9]

3. Chemical composition and mechanical properties of X70 pipeline steel

Idokoh [5] studied the microstructure and mechanical properties of welds in steel pipeline and discovered that the hardness was highest in the weld bead, a bit lower in the base metal and lowest in the heat affected zone. In this study, a higher density of voids was observed at the top of the weld bead than in the middle. It was also demonstrated that large grains and dislocation density in heat affected zone is related to the high susceptibility of the steel weld to fail under tensile load. Aliu [10] investigated the mechanical and microstructural properties of welded joint of Nigerian National Petroleum Corporation pipelines and observed that the Pipeline was made of High Strength Low Alloy (HSLA-A105) steel. At the end of the investigations, it was concluded that inadequate mechanical tests and lack of microstructural examination of the welded joints before installations were possible sources of pipeline failures. Omale et al., [11] studied the evolution of microstructure, texture and mechanical properties of API X70 steel pipeline after different thermomechanical treatments using a combination of X-ray diffraction and electron backscatter diffraction (EBSD). The investigations revealed that different microstructure consisting of polygonal ferrite, bainite, coarse and fine acicular ferrite grains were obtained with centre line segregation traversing through and parallel to the rolling direction. EBSD investigations confirmed that both dynamic recovery and partial recrystallization occurred during hot rolling requiring further annealing for a more homogenous grain structure. It was also observed that after accelerated cooling, the fast cooling rate and low temperature interruption allows the formation of more bainite, which in turn increased the tensile strength of the steel. The tensile and impact properties of X70 steels pipeline exposed to wet hydrogen sulphide (H_2S) environments were investigated by Pengyan et al. [12], through corrosion and mechanical testing. The results suggested that the Charpy absorbed energy, tensile strength and reduction in area of X70 steel decreased after H_2S corrosion. When hydrogen was released, Charpy absorbed energy and reduction in the area exhibited a certain recovery. Fractography of the tensile and impact specimens showed mixed ductile–brittle failure after corrosion and hydrogen release. Fractured regions with brittle features diminished after the release of hydrogen. Hydrogen was closely associated with mechanical property degradation and fracture features.

Leonardo et al. [13] investigated the microstructure and mechanical properties of two API steels for iron ore pipelines. A chemical analysis of X70 steel was performed using a Thermo ARL optical emission spectrometer, and the microstructure was observed by a JEOL scanning electron microscope (SEM). The chemical composition of the X70 steel used for the experiment is presented in Table 1.

Table 1: Chemical composition of API-5L X70 steel (wt%)

C	Si	S	P	Mn	Ni
0.109	0.239	0.004	0.023	1.536	0.011
Cr	Cu	Al	Ti	Nb	V
0.024	0.011	0.026	0.016	0.045	0.045

The SEM micrograph of X70 steel revealed the presence of polygonal ferrite/pearlite banding (pancake type), an occurrence in hot-rolled, low alloy steels and a ferritic grain size. The presence of a small volumetric fraction of bainite / degenerated pearlite and the absence of acicular ferrite in X70 steel was also noted. The results of the tensile, impact and hardness tests showed yield strength of 586MPa, the ultimate tensile strength of 640MPa, a total elongation of 38%, Rockwell hardness A, HR_A 65, and Charpy impact absorbed the energy of 184J.

3.1 Effect of welding techniques and weld parameters in pipeline steels

The welding technique used in producing welds play very vital roles in controlling the occurrence of welding imperfections. Several welding methods exist but there is need to review the methods that are frequently used in producing steel pipeline. Loureiro [14] suggested that submerged arc welding (SAW) is used extensively in industries to join metals for the manufacture of pipes of different diameters and lengths. In this work, response surface methodology (RSM) technique of design of experiment (DOE) has been applied for the selection of the optimum input variables.

Dearden [15] suggested that in submerged arc welding process parameters play a significant role in determining the quality of a weld. So for such applications, optimum welding process parameters must be selected providing desired weld properties. In the work; beads were taken on 15mm thick plate by varying heat input and wire feed rate. Prediction equations were developed for penetration, reinforcement height and width using the multi-regression method and artificial neural networks; a comparative study of both techniques has also been performed.
Loureiro [14] suggested that in submerged arc welding (SAW), selection of appropriate values of process variables is essential to control heat-affected zone (HAZ) dimensions coupled with the required bead size and quality ensuring a predictable and reproducible weld bead. In the investigation, mathematical models were developed to study the effects of process variables and heat input on various metallurgical aspects; width of the HAZ, weld interface, grain growth and grain refinement regions of the HAZ. From the study, it was concluded that heat input and wire feed rate have positive effects, but welding speed has a negative impact on all HAZ characteristics. Width of grain growth and grain refinement zones increased but weld interface decreased with an increase in arc voltage.
Ito and Bessyo [16] analysed the influence of welding heat input on the mechanical properties of weld joints using heat input from 2.5 to 4.0 kJ/mm for submerged arc welded of duplex steel. The study revealed that the heat input from 2.5 up to 4.0 kJ/mm has no adverse influence on the mechanical properties of the joints in submerged arc welding of duplex steels.
Hosseini et al [17] used the digital image processing techniques to assess the heat-affected zone of submerged arc welding of structural steel plates through the analysis of the grain structure. It was concluded from the study that HAZ in submerged arc welding was of relatively less size due to 10% less grain growth.

3.2 Seawater Environment

Seawater, or salt water, is water from a sea or ocean. On average, seawater in the world's oceans has a salinity of about 3.5% (35 g/L), this means that every kilogram (roughly one litre by volume) of seawater has approximately 35 grams of dissolved salts - predominantly sodium (Na^+) and chloride (Cl^-) ions, average density at the surface is 1.025 kg/L [18]. The presence of dissolved salts mainly NaCl makes seawater one of the most corrosive and most abundant naturally occurring electrolytes. The highly corrosive nature of the seawater is reflected by the fact that most of the common structural metals and alloys are attacked by these dissolved salts and aerated water under varying levels of turbulent flow conditions [19, 25]. Seawater is denser than both freshwater and distilled water (density 1.0 kg/L at 4 °C (39 °F) because the dissolved salts increase the mass by a more significant proportion than the volume. The freezing point of seawater decreases as salt concentration increases. At typical salinity, it freezes at about −2 °C (28 °F) [20]. The coldest seawater ever recorded (in a liquid state) was in 2010, in a stream under an Antarctic glacier, and measured −2.6 °C (27.3 °F) [20]. Joyde et al. [18] stated that Seawater pH is typically limited to a range between 7.5 and 8.4. However, there is no universally accepted reference pH-scale for seawater, and the difference between measurements based on different reference scales may be up to 0.14 units [21].
According to Saleh and Malik [20], the seawater environments can be divided into five zones namely: subsoil, continuously submerged, tidal, splash zone above high tidal and atmospheric zone. The same authors observed in their studies that the corrosion behaviour of metals and alloys differ from one zone to another. In splash zone, the stainless steels usually have

IOP Conf. Series: Materials Science and Engineering **413** (2018) 012034 doi:10.1088/1757-899X/413/1/012034

satisfactory performance while the carbon and low alloy steels do not. Anderson and Rose [22] had found that the austenitic grades performed much better than martensitic and ferritic grades. The Ni, Cu and P alloyed steels were found to be much more resistant than carbon steel in splash zone [23]. Also, it was found that Mn, P and Al had a measurable influence on corrosion rates of low carbon steels under tidal exposure. After five years exposure test it was found that the rate of attack in splash zone was much higher than the atmosphere and deep submerged zones [24]. The approved composition for standard seawater prepared with 41.953 grams of sea salt with enough distilled water to yield on liter artificial seawater is as presented in Table 2.

Table 2: ASTM D114 standard composition for substitute seawater

Sodium Chloride	NaCl	58.49%
Magnesium Chloride	$MgCl_2$-$6H_2O$	26.46%
Sodium Sulfate	Na_2SO_4	9.7%
Calcium Chloride	$CaCl_2$	2.765%
Potassium Chloride	KCl	1.645%
Sodium Bicarbonate	$NaHCO_3$	0.477%
Potassium Bromide	KBr	0.238%
Boric Acid	H_3BO_3	0.071%
Strontium Chloride	$SrCL_2$-$6H_2O$	0.095%
Sodium Fluoride	NaF	0.007%

Note: (1) Density of Seawater equals 1.025 at 15°C; (2) pH adjusted to 8.2 using 0.1N solution of sodium hydroxide or hydrochloric acid (ASTM D114- 2013)

The American Society for Testing and Materials has put up a standard for seawater conditions for many uses including accelerated corrosion studies, oceanographic research, ocean instrument calibration and chemical testing. ASTM D114 is the updated standard "sea salt" which is used in the preparation of substitute/artificial ocean water.

4. Conclusions

The following conclusions are drawn from this work:

1. Welds and corrosion are critical points of consideration in terms of design and initial fabrication and any subsequent repair during the life of a pipeline;
2. Welded joints are more susceptible to failure than the parent material due to the effect of welding on the mechanical properties of the weld and heat affected zone;
3. Installations of pipelines in corrosive environment require an understanding of the corrosion performance of such new weldments, which may be used in the marine environments. Generally, if the corrosion mitigation measures are correctly applied and proper inspection and maintenance adhered to, then welds should not pose any increased risk of corrosion failure, but owners of the steel pipeline must understand the potential issues associated with welds and likelihood of corrosion in the event of coating failure at welds;
4. In optimising maintenance strategies, it is essential therefore to determine with more empirical data the rate at which corrosion degrades the mechanical properties of welded joints of API X70 pipeline in corrosive environment (sea water).

IOP Conf. Series: Materials Science and Engineering **413** (2018) 012034 doi:10.1088/1757-899X/413/1/012034

References

[1] Hopkins P. (2007), "Pipeline stream, past, present, and future' from the 5th Asian Pacific IIW international congress". Sydney, Australia, Key note paper.

[2] Fabian Grimpe, Heike Meuser, Florian Gerdemann, Elke Muthmann (2010), "2nd International conference on Super-High Strength Steel". Peschiera Del Garda, Italy.

[3] Hillenbrand H. G., Niederhoff K. A., Amoris E., Perdrix, C., Streisselberg A and Zeislmair U (1997), "Development of Linepipe in Grades up to X-100". EPRG/PRC Biennial Joint Technical Meeting on Linepipe Research, Washington, D. C.

[4] Omale J. I. (2016), "Microstructure and Mechanical properties of welds in pipeline steel", 18-21, 28-31, 100-116. M.sc Thesis, University of Saskatchewan.

[5] Cheng, Y. F. and Xue, H.B., (2011), "Characterization of inclusion of X80 pipeline steel and its correlation with hydrogen-induced cracking". Corrosion Science, 53(4), 1201-1208.

[6] Nanninga N., Slifa A., levy Y., White C. (2010), "A Review of fatigue crack growth for pipeline steel exposed to Hydrogen". International journal of Hydrogen Energy. 114 – 121

[7] Achebe C. H., Nneke U. C., and Anisiji O. E., (2012), "Analysis of Oil Pipeline Failures in the Oil and Gas Industries in the Niger Delta Area of Nigeria". Proceedings from the International Conference of Engineers and Computer Scientists, Hongkong. Vol. II.

[8] Baudin, T., Boumerzoug Z., Derfouf C., (2010), "Effect of welding on microstructure and mechanical properties of industrial low carbon steel". Material Science and Engineering 2(07), 502.

[9] Muneer Baig, Ehab El-Danaf, Mahmoud Soliman, Abdulhakim Almajid, Waleed Alshalfan, and Khaled Alhajeri (2015), "Mechanical properties, microstructure and toughness assessment of an X70 pipeline steel". Materials Testing: 57, (10) 897-903.

[10] Aliu S., (2012), "Investigation of mechanical and microstructural properties of welded joint of Nigerian National Petroleum Corporation pipelines". M. Eng, Thesis, Federal University of Technology Minna, Nigeria. 9-15.

[11] Omale J. I., Oheri E. G., Tiamiyu A. A., Eskandari, M., Szpunar J.A., (2017), "Microstructure, texture evolution and mechanical properties of X70 pipeline steel after different thermo mechanical treatments". Material Science and Engineering; 703, 497 – 485.

[12] Pengyan Wang, Zheng Lu, Shuqi Zheng, Yameng Qi, Jian Wang, Xanjun Zheng (2015), "Tensile and impact properties of X70 pipeline steel exposed to wet H2S environments". International journal of Hydrogen Energy. 1152 – 1154.

[13] Leonardo Barbosa G, Luiz Claudio C, Rodrigo Vicante B, Luiz Henrique Soares (2014), "Microstructure and mechanical properties of two API steels for Iron ore pipelines". Journal of Universidade Federal de ouro Preto, MG, Brazil, Materials Research. 17, 1516 – 1639.

[14] Loureiro, A. J., (2002)." Effect of heat input on plastic deformation of undermatched welds". Journal of materials processing Technology. 128(1), 240-249.

[15] Dearden O. H., (1940), "A guide to the selection and welding of low alloy structural steel' Transactions of the Institute of Welding, 3, 203-214.

[16] Ito Y and Bessyo K., (1968), "Weldability formula of high strength steels related to heat affected zone cracking". Published by the International Institute of Welding, Doc IX-576-68.

[17] Hosseini, M., Fattahi, M., Nabhani, N., Arabian, N., and Rahimi, E (2010), "Effect of Ti containing inclusions on the nucleation of acicular ferrite and mechanical properties of multipass weld metals". Micron, 45, 107-114.

[18] Jodye I. S., Julian L. R., and Daniel B. W., (2003). The analysis of seawater: A Laboratory Centered Learning Project in General Chemistry. Vol. 1.

[19] Basim O. H., (2010), "Effect of salt content on the corrosion rate of steel pipe in turbulent flowing solution", Acta Metallurgical Slovaca Conference, 1, 66 – 73.

[20] Saleh A. A. and Malik A. V., (2005). "Effect of Seawater level on corrosion behavior of different Alloys". A paper presented at International Desalination Association (IDA) World Congress Conference held at Singapore.

[21] Stumm W. M., (1981): Aquatic chemistry, An introduction emphasizing chemical equilibria in Natural Waters. 414-416.

[22] Anderson D. B. and Ross R. W., (1976), "Protection of steel pitting in marine splash and sprayzone – Metallic Sheathing Concept", 46 – 59.

[23] Larablee C. P., (1958), "Corrosion Resistant experimental steels for marine application", International Conference on Pipelines and Trenches Technology, 23, 501 -504.

[24] O. M. Ikumapayi, I. P. Okokpujie, S. A. Afolalu, O. O. Ajayi, E. T. Akilabi, O. P Bodunde. "Effects of Quenchants on Impact Strength of Single-Vee Butt Welded Joint of Mild Steel" IOP Conf. Series: Materials Science and Engineering 391 (2018) 012007 doi:10.1088/1757-899X/391/1/012007

[25] Humbles A. A. (2009), "The cathodic protection of steel in seawater", corrosion" 2nd International conference on Super-High Strength Steels", Peschiera Del Garda, Italy.

IOP Conf. Series: Materials Science and Engineering **413** (2018) 012035 doi:10.1088/1757-899X/413/1/012035

Evaluation of Friction Stir Welds by X-ray Digital Radiographic Non- Destructive Approach

Itai Mumvenge[1, a], Stephen A. Akinlabi[2, b], Peter M. Mashinini[2, c] Olawale S. Fatoba[1d], Joshua Okeniyi[1e,3f], and Esther T. Akinlabi[1g]

[1]Department of Mechanical Engineering Sciences, University of Johannesburg, South Africa, [2] Department of Mechanical and Industrial Engineering Technology, University of Johannesburg, South Africa, [3]Department of Mechanical Engineering, Covenant University, Ota, Nigeria.
Email: [a]itaimumvenge@gmail.com, [b]stephenakinlabi@gmail.com, [c]mmashinini@uj.ac.za, [d]drfatobasameni@gmail.com, [e]jokeniyi@uj.ac.za, [f]joshua.okeniyi@covenantuniversity.edu.ng, [g]etakinlabi@uj.ac.za

Abstract. This paper focuses on the microstructure evolution and the relationship to the microhardness profiling of the friction stir welded butt joint on similar AA6061-T6 Aluminium alloys. The only parameters used and varied for this study were rotational speed and feed rate. The geometry of the tool was kept constant and the material used was tool steel, W302. Friction stir welds were evaluated by both visual inspection and the X-ray digital radiography method. Evaluation allowed for assessment of the weld integrity by examining for the presence and absence of weld defects. The results indicated that the welds do not have any root defects. The X-rays showed complete penetration as evidenced by the evaluation of microstructural joint interface.

1. Introduction

Friction stir welding is a novel solid-state welding technique that was invented at The Welding Institute (TWI) of UK in 1991 by Wayne Thomas [1], and was initially applied to aluminium alloys [2]. The welds are made by plunging a rotating non-consumable Friction Stir Welding (FSW) tool pin on the abutting edges of a work-piece where the rotational speed causes friction to generate localized heat energy to achieve plastic deformation of the work-piece material. As the FSW tool traverses, the frictional heat also traverses along the joint-line [3].

Like any new technology, material flow complexities and weld processes and application can result in weld flaws. In the case of FSW, inappropriate processing conditions can cause formation of various visible or potential defects around the weld nugget zone (WNZ), the thermo-mechanically affected zone (TMAZ), or sometimes at the WNZ/TMAZ interface [4]. This results in FSW weld-defect of different types, which includes voids, defective tightness, surface groove, excessive flash, 'kissing-bond' [5] and crack-like root-flaws, which are quite different from conventional fusion welding flaws [6].

Defects are inherent to most joint during welding processes [7]. The formation of weld defects within the friction stir welds is mainly caused by inappropriate FSW process parameters. Some of these process parameters that may contribute to the formation of weld defects are mostly tool rotational speed [8], travel speed, tilt angle, shoulder target depth [9], applied load and pin configuration [10].

With recent developments in technology, weld integrity can be evaluated without physical destruction of the weld by non-destructive approach. Non-destructive testing (NDT) relies on the use of physics principles such as acoustic waves and radiation, to ascertain the physical appearance of materials, detecting material flaws and critical defects [11]. The important characteristics of non-destructive testing is the ability to evaluate material and weld flaws without altering the serviceability or effectiveness of the said components, on the other word, it does not damage the material. NDT has vast benefits, some of which include cost effectiveness, great reliability and safer applicability for a wide range of processes.

Commercial use of NDT is widely spread in industrial fields, especially on railway, petrochemical pressure tanks, power generation, aviation, automotive and mining infrastructure inspections [12]. On the contrary, the weld bending test and the micro-optical observation by bending weld samples will both make the materials ineffectual, and they only give understanding into the weld features at the damaged point where the rupture or the material appearance information is acquired, the better option is thus non-destructive testing [13].

Radiographic testing (RT) is one typical non-destructive testing technique used to analyze hidden flaws by using short wavelength electromagnetic radiation to penetrate materials. A radioactive source uses photons such as Ir-192, Co-60 and Cs-137. Neutrons can also be used as a replacement for photons to penetrate the material. X-ray radiography can also be used to analyze weld defects [14]. Digital image processing is carried out on a local or neighbor-hood operation point using several pixels in a limited area of the input image to obtain a point in the processed or final image" [15].

Gaurav *et al.*, [15], performed a digital radiography test for Gas Tungsten Arc Welding (GTAW) and FSW weldments of AA7075-T651 aluminum alloy and they observed the defect of porosity, cluster porosity and small tunnel defect. Edwards et al., [16], used X-ray radiography and validated with metallographic images to examine friction stir welded 6mm thick Ti-6Al-4V butt joints processed and embedded with a tracer material. A variety of rotational and traverse speeds were implemented. The conclusion was that X-ray radiography scans could depict the characterization of the material flow behavior [16].

Patil *et al.*, [17], applied X-ray radiography testing technique to analyze weld defects in similar and dissimilar friction stir welded aluminum alloy butt weld joints. The material used was AA7075 T651 in butt weld joint configuration for similar welds. Dissimilar weld joints were processed using aluminum alloys AA77075 T651 and AA6061 T6. Different tool rotational speeds (800, 900, 1000 rpm) and feed rates (30, 35, 40mm/min) were employed while all the other process parameters were kept constant. The visual inspection was only able to identify flash formation on all the samples. The X-ray radiographic testing techniques revealed a lack of penetration in all weld samples such as cracks, voids and wormhole defects. It was concluded that the X-ray radiography was successful in evaluating the weld defects of the friction stir weld sample for similar and dissimilar materials. It was also observed that increasing the transverse speed, increases the occurrence of weld defects [17].

Some of the weld defects can be difficult to detect using visual inspection techniques or certain types of non-destructive evaluation techniques. It is therefore imperative that the correct selection of inspection method and its parameters are well defined to effectively evaluate the microstructure, defects and the integrity of the friction stir weld. This research work focuses and reports on the non-destructive testing to identify the root defect in friction stir welded joints as applied to AA6061-T6 Aluminium alloys.

2. Experimental Set-Up

The friction stir weld process was conducted using a computer controlled 2-Axis FSW machine at the Indian Institute of Science (IISC), Bangalore, India. The parent material used in this study was an AA6061-T6 aluminum alloy with dimensions measuring 200 x 150 x 4.09 mm3 and having a friction stir welded butt joint length of 120mm. The predominantly concave shoulder with a pin attachment provides confinement for the heated volume of the material.

A spectrometer was used to analyze the chemical composition of the parent material, which was found to be in accordance with the AA6061-T6 standard specifications [21]. The chemical composition of the welded material is shown in Table 1.

Table 1 Chemical Composition of Aluminium AA6061-T6

Element	Si	Fe	Cu	Mn	Mg	Cr	Ni	Zn	Ti	Al	Ag
Wt %	0.68	0.49	0.21	0.08	0.84	0.06	0.01	0.07	0.07	97.40	Balance

A concave tool shoulder with a threaded pin length of 3.8mm was produced from tool steel W302 and used to produce the welds. A weld matrix with criteria category of low, medium and high settings was used to define the combinations of rotational and feed rates. Table 2 shows the optimum process parameters matrix developed for the friction stir welded samples. The pin movement and geometry stir the material, facilitated the downward mixing auger effect [18]. The FSW schematic is illustrated in Figure 1. The fine microstructure generated helped achieve mechanical properties of sufficiently high strength, fatigue and fracture resistant welds [19].

Table 2. Process Parameter Matrix of the FSW process

Weld Designation	Rotational Speed (rpm)	Feed Rate (mm/min)	Code
BW1	700	60	Ll
BW2	700	80	Lm
BW3	700	100	Lh
BW4	900	60	Ml
BW5	900	80	Mm
BW6	900	100	Mh
BW7	1100	60	Hl
BW8	1100	80	Hm
BW9	1100	100	Hh

L = low rotation speed, l = low feed rate
M= Medium rotational speed, m = medium feed rate
H= High rotational speed, h = high feed rate

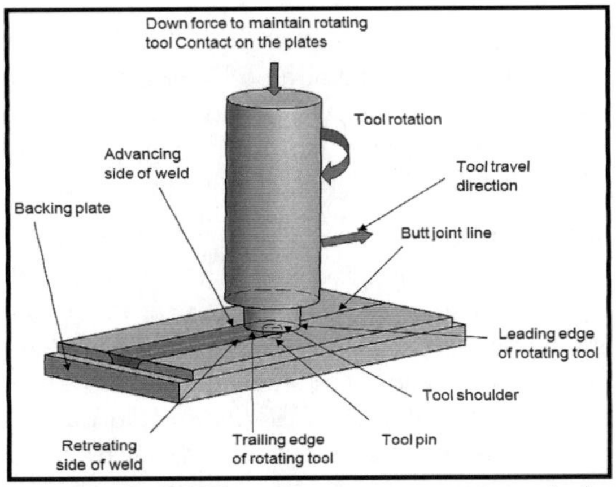

Fig. 1 Schematic of a FSW tool movements on a workpiece **[20]**

Samples were cut and sectioned from the weld coupons. These samples were etched using Weck's reagent. Microstructure observations were conducted using the Olympus BX51M and macrographs for the welds were taken using the Olympus SZX16.

Vickers microhardness were measured using a TH713 Vickers microhardness tester, by Beijing Cap High Technology Co. Ltd in accordance to ASTM 384-16 [22]. Tensile testing was also conducted. The tensile samples specification dimensions were in accordance to the ASTM B557-14 standard [23] while the tensile testing procedure was conducted in accordance to the ASTM E8M-13 standard [24].

The visual inspections on the welds were conducted and the X-ray digital radiography was performed on the test samples to inspect weld defects using the equipment Vidico fox-Raysor, 80kV, 0.6A and Balteau X-ray source. The radiography procedure was conducted according to the ASTM E155-15 standard specification [25]. The foX-Rayzor FPD was used in combination with a Vidisco supplied 270 kV pulsed x-ray source or almost any X-ray source / isotope. The recommended X-ray sources must be up to 160 kV, otherwise shielding is required for higher voltages. Table 3 shows the exposure parameters that were used to produce the X-ray radiographs scan images.

Table 3 X-ray Radiography Exposure Parameters for FSW joints

Process Parameter	Value
Electrical Voltage Range	700
Tube Current Range	700
Exposure Time	700
X-ray tube beam angle	900
Focal spot size	900
X-ray inherent filtration mm	900
Source to Object Distance (SOD)	1100

3. Results and Discussion

3.1. Weld Defects Analysis

Table 4 shows the physical appearances and X-ray scans of the friction stir processed weld joints with the aim of showing defects caused by the generation of high temperature combined with severe local shearing action due to the tool movement. The curling wavelike flashes are a result of the movement of plasticized material displaced by the pin with the tool shoulder not confining the material. The tool rotational speed of 700 rpm and 100 mm/min appear to have processed a weld with the least flash sizes.

It was demonstrated from the results that weld flash is necessitated by material flow behavior which is predominantly influenced by the FSW process parameters such as the tool rotational speed. It is noticeable, that rotational speed appears to be the predominant influencing process variable since it is associated with the translational velocity. Murr, *et al.*, [7], reported that an increase in strain rate can influence the re-crystallization process and these strain rates are associated with very high rotational speeds, which influence the generation of excessively large weld flash [7]. Also, low heat inputs are predominantly associated with higher welding speeds which causes the welded joint to be cooled at a faster rate. The curled flash could thus be attributed to incorrect combinations of the rotational speed and the feed rate which in instances yields low axial forces. Such low axial forces, contribute to the formation of non-symmetrical semi-circular features and in instances, large flash at the top surface of the weld due to poor plasticization and consolidation of the material under the influence of the tool shoulder [9]. The non-symmetrical semi-circular features were not evident with the visual inspection on all the welds. This implies that the welds were homogenous and of sound quality. This inspection needed to be verified by an optical microscopy and other analytical techniques.

Table 4 contains the X-ray radiograph scan images and the respective comments on each examined sample. All the friction stir welds were observed to be defect free from voids and porosity, which could be attributed to good material flow ability due to optimal heat energy. A keyhole was observed in the workpiece at the pin's point of retraction which is a known phenomenon in FSW. Research literature shows that an exit hole can be repaired by eliminated by active passing filling friction stir repair [21, 22].

Table 4 Physical Appearance of the FSW samples

Rotational Speed (rpm)	60mm/min		80mm/min		100mm/min		Comments
	photo	X-ray Image	photo	X-ray Image	photo	X-ray Image	
700			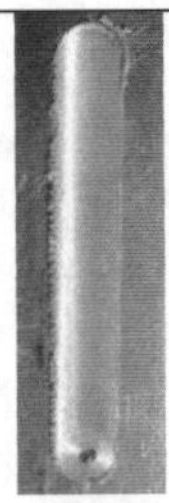		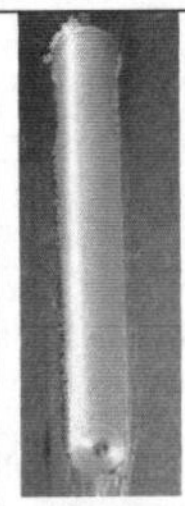		No evidence of root defects was observed on all the welds produced using the 700-rpm rotation speed and the respective feed rates. Keyhole observed.
900	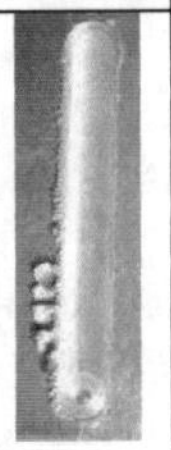						No evidence of root defects was observed on all the welds produced using the 900-rpm rotation speed with the 60 and 80 mm/min feed rates. An insignificant root defect can be seen for the weld produced with the feed rate of 100 mm/min. Keyhole observed.
1100	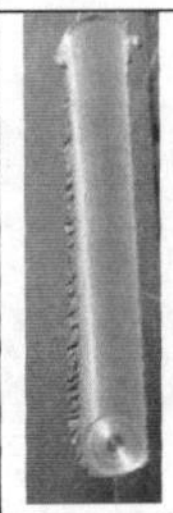						No evidence of root defects was observed on all the welds produced using the 1100 rpm rotation speed and the respective feed rate speeds. Keyhole observed.

3.2. *Macrostructure and Microstructure Analysis*

Figure 2 shows the typical macrograph of the friction stir welded sample produced at 700 rpm and feed rate of 80 mm/min. All the macrographs showed no internal defect present in any of the weld samples done for this research work. The morphological features noticed using the optical microscope consisted of all the four distinct FSW microstructural zones namely: TMAZ, HAZ and SZ or NZ, and the Parent material.

An optical microscopy examination was used to identify the FSW microstructural zones of all the welds produced. Figure 3 shows the typical microstructures on the different zones obtained in the weld's stabilized region. From observation, the evolution of the morphology from the base metal is quite remarkable.

IOP Conf. Series: Materials Science and Engineering **413** (2018) 012035 doi:10.1088/1757-899X/413/1/012035

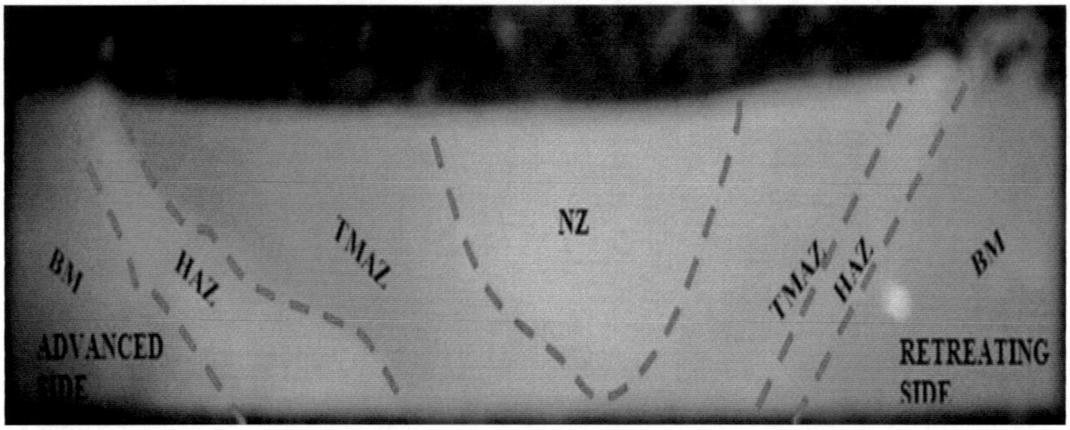

Fig. 2 Typical Macrograph of FSW butt joint produced at weld speeds 700 rpm and 80 mm/min

Optical microscopy and Scanning Electron Microscopy (SEM) examinations were used to study the FSW microstructure zones for all the welds produced. Figure 3 (a) showed a base metal microstructure with a honeycomb-like shape, also revealed in the SEM micrograph presented in Figure 3(b). The presence of the second phase precipitates can be seen in the microstructure analyzed using SEM. These precipitates are mostly elements of Mg2Si [12]. Figure 4 presents the microstructural evolution of the weld sample produced at 700rpm and 80mm/min. The weld zones' diffuse contrast can be observed indicating the different material thermo-flow and stresses [23] not analyzed in this study.

Optical microscopy and Scanning Electron Microscopy (SEM) examinations were used to study the FSW microstructure zones for all the welds produced. Figure 3 (a) shows the base metal microstructure with a honeycomb-like shape as revealed in the SEM micrograph presented in Figure 3(b). The presence of the second phase precipitates can be seen in the microstructure analyzed using SEM. These precipitates are mostly elements of Mg2Si [12]. Figure 4 presents the microstructural evolution of the weld sample produced at 700rpm and 80mm/min. The weld zones' diffuse contrast can be observed indicating the different material thermo-flow and stresses [23] not analyzed in this study.

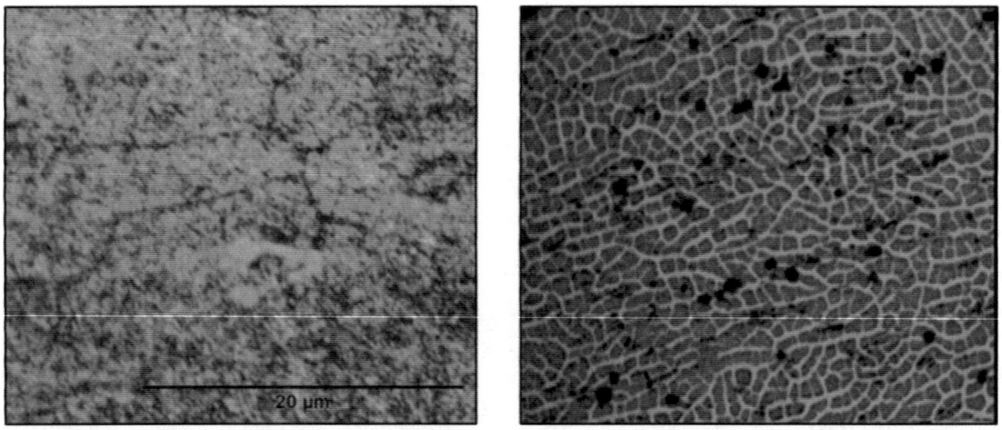

Fig. 3 AA6061-T6 Base Metal Microstructure (a) OM micrograph (b) SEM

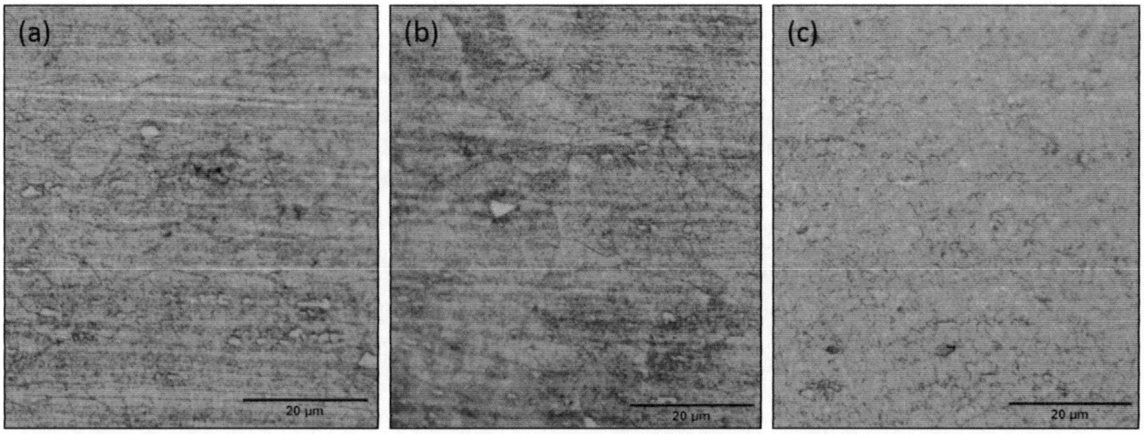

Fig. 4 Typical Microstructure micrographs of the (a) BM (b) HAZ (c) TMAZ (d) WN produced at weld speeds 700 rpm and 80 mm/min

SEM was used to analyze the weld zone microstructure and the presence of intermetallic particles. The resultant micrographs are shown in Figure 5. All the micrographs have white and grey spots which are evident within the grains. These spots represent intermetallic particle constituents. Weck's reagent can be seen to have attacked some of the intermetallic particles and the grains during the etching process. These attacks should not be confused with the weld defects such as voids, whose absence has already been concluded using non-destructive X-ray testing.

The HAZ (Figure 5(a)), which is common to all welding processes subjected to thermal cycle with no deformation, has slight changes from the base metal. The TMAZ (Figure 5b) grain structure was characterized with mixed rotated, elongated and equiaxed grains which reflects grains of partially recrystallized morphology. The WN (Figure 5(c)), has much smaller fine equiaxed grains due to dynamic recrystallization. Although the micrographs for the retreating side are not presented with this paper, a sharp transition was seen between the TMAZ and the SZ on the advancing side while a much more diffuse interface was observed at the RS. This contrast is in relation to the different material flow conditions on the two sides.

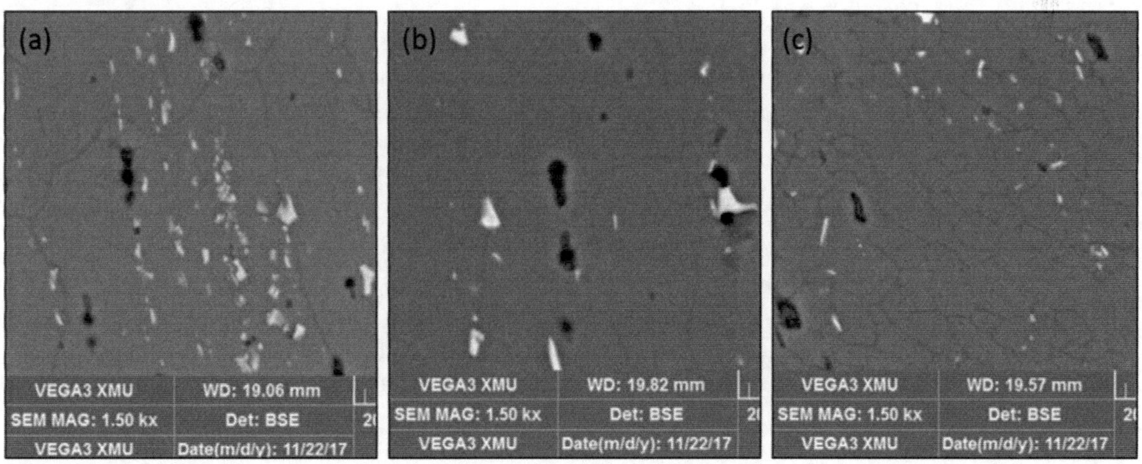

Fig. 5 Typical SEM images of the (a) HAZ (b) TMAZ (c) WN produced at weld speeds 700 rpm and 80 mm/min

As presented in Table 5 below, in the HAZ, exhibiting similar trend to the NZ and TMAZ, the average grain sizes decreased with every increase in the welding speed though it was minimal. For instance, at rotational speed of 700 rpm, the average grain size in the HAZ decreased from 129 μm to 127 μm and finally 126 μm in correspondence with the increase of welding speed from 60 mm/min, 80 mm/min and 100 mm/min respectively. These results further confirm the findings by Fadaeifard *et al.*, [23], in which it was report the grain size decreased with the increase in weld speeds. The increase in deformation rate caused by the rotating tool plays a major role in reducing grain size. However, it is important to note that both frictional heat and deformation rate reduce with the increase in traverse speed [28].

Table 5 Average grain size in various regions

Weld Designation	Rotational Speed (rpm)	Feed Rate (mm/min)	NZ *μm*	TMAZ *μm*	HAZ *μm*	BM *μm*
BW1	700	60	4.4	85.4	129	125
BW2	700	80	3.7	83.1	127	125
BW3	700	100	3.43	79.2	126	125
BW4	900	60	7.4	92.7	122	125
BW5	900	80	6.8	95.1	121	125
BW6	900	100	5.43	80.2	118	125
BW7	1100	60	14	95.3	133	125
BW8	1100	80	10.1	95.1	128	125
BW9	1100	100	8.7	80.2	125	125

In most of these weld samples, the HAZ grain size is slightly larger than in the base metal (BM). This grain growth is highly influenced by heat input in the weldment. Unlike the rotational speed, the weld traverse speed shows a reverse effect on the heat generated during FSW [24], as such the heat energy reduction is relative to the increase in traverse rate. Previous studies have shown that the stirred zone experiences lower temperatures. A short duration is therefore available for grain growth accompanied with recrystallization effects, which cause smaller grain size in the WN.

Akinlabi *et al.*, [25], reported that there is a derived condition that changes the grain size necessitated by a competition between frictional heat and deformation rate at each welding condition such that the grain size is a net effect of deformation rate and frictional effect [31. In this study, it was deduced that the lower welding speeds were effective in increasing the deformation rate and frictional heat resulting in reduced grain size. However, the reduction of grain sizes with increase in welding speeds suggests that frictional heat is dominant than the deformation rate. The increase in weld speeds also necessitated the reduction in peak temperatures and faster cooling rates which assisted in effecting the reduced grain sizes. It is therefore imperative to conclude that the welding parameters such as rotational and traverse speeds play a crucial role in reduced grain sizes.

In Figure 3 (b), small rod/needle-like shaped β" precipitates were observed in the base metal microstructure. The dissolution of these precipitates in the weld samples during FSW can be attributed to the feed rate and the tool rotational speed, which are some of the key parameters, as they respectively influence the heating time and the peak temperature. As such dynamic recrystallization occurs due to intense plastic flow at high temperatures, more so, in the weld nugget to give smaller grain size as recorded in Table 5 and demonstrated in the SEM results in Figure 5(c). The precipitates, which are known to have intense interactions with dislocations and grain boundaries, may start to have effect on dislocation density (driving force) and the recrystallization kinetics. This leads to a microstructural evolution that results in dissolution of precipitates. The smaller grain size in the weld nugget can also be attributed to grain recrystallization because of the pinning effect of these particles. The β'' precipitates are obviously dissolved in the nugget which gave rise to some supersaturated solution in the stirred zone. These heterogeneously distributed precipitates observed in weld nugget gives rise precipitates hardening. The next section will discuss the hardness distribution across the weld traverse which will be seen to be high in the weld nugget when compared to the TMAZ and HAZ. Sauvage *et al.*, [26] contributed that "two mechanisms exist that lead to the formation of heterogeneously distributed precipitates in the weld nugget". The first consideration is the heterogeneous precipitation due to the great increase of the temperature causing the β'' precipitates dissolve. This was explained to be experienced when the FSW pin is moved away, which drops the temperature and increases the driving force for precipitation. The other scenario is during interaction between the coarsening precipitates and dislocations. "In hardened aluminium alloys, the dramatic increase of the temperature in the weld nugget has effect on both the solubility of Mg and Si and the atomic mobility. The solubility of Mg and Si increases in the Al matrix, and this leads to a high driving force for the decomposition of β'' needle" [26]. This effect plays a crucial role in grain sizes and grain boundaries as well as the associated mechanical properties.

3.3. Mechanical Properties Analysis

3.3.1. Microhardness Analysis.

Figure 6 shows the results of the microhardness test carried out in the stabilized region of the weld. The microhardness profiling was conducted across the weld traverse at a depth of 1.5 mm from the weld top surface. The unaffected parent metal average Vickers microhardness values is ~99 HV.

The plot shows a hardness decrease towards the weld retreating side for lower rotational speeds. The most affected hardness values are approximately 6.6 to 11 mm from the weld center on both sides of the welds. The rotational speed of 1100-rpm_60 mm/min recorded the least hardness value of ~ 49 Hv in the HAZ/TMAZ interface. This 48 % degradation in hardness is mostly characterized in the TMAZ. This is attributed to a combination of high stresses and large strains resulting in the deformation of the grain structure, where re-crystallization did not take place, caused a coarse grain structure [27].

Figure 6 demonstrated the degradation in hardness values at the interface of the regions attributed to softening of the material. Employing the rotational speed of 700-rpm achieved the highest peak values with distinct hardness variations in the weld nugget, more-so, for the lower feed rates of 60 and 80 mm/min. This is attributed to higher than normal fast cooling rates and varying thermal cycles resulting in poor material flow. Increasing the feed rate to 100mm/min improved the hardness values uniformity, however, with a decrease noticeable in the weld nugget towards the retreating side. The decrease is possibly due to unimproved fast cooling rates. Further increasing the rotational speed to 900-rpm, results in more uniform hardness values in the weld nugget attributed to optimum heat energy and stable thermal cycle effecting good material flow and stabilized cooling rates. However, the hardness value seems to decrease and this can probably be attributed mainly to the coarsening and dissolution of strengthening precipitates induced by the thermal cycle at the lower feed rates of 60 and 80 mm/min.

IOP Conf. Series: Materials Science and Engineering **413** (2018) 012035 doi:10.1088/1757-899X/413/1/012035

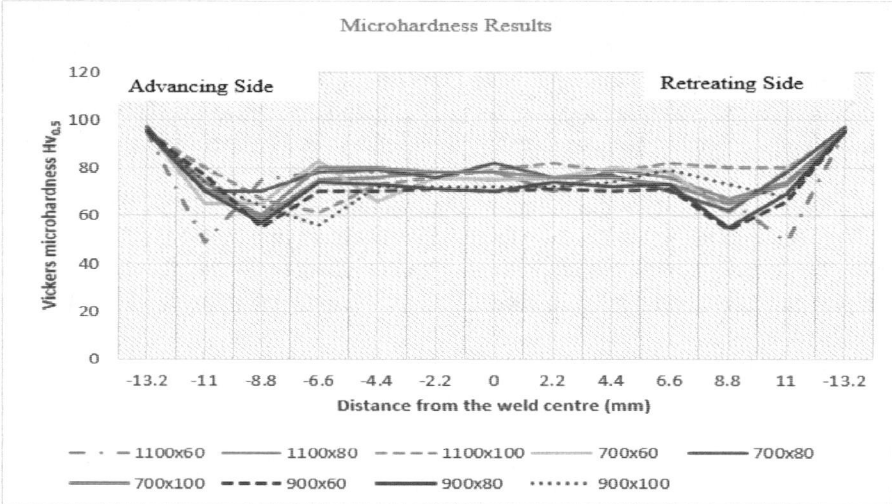

Fig. 6 Microharness values across weld traverse

Increasing the feed rates to 100mm/min, increased the hardness values which was a result of more plasticized material generated due to sufficient heat energy and permissible cooling rates. However, increasing the rotational speed to 1100 rpm at the lower feed rates improved the hardness in the weld nugget, however, with a decrease of hardness in the HAZ/TMAZ interface due to softening of the material. Further maintaining the rotational speed at 1100 rpm and increasing the feed rate to 100mm/min, shows a steep increase in the hardness values toward the retreating side, ranging from 67Hv in the HAZ/TMAZ advanced interface to 80Hv in the retreating HAZ side. This could be an effect of excessive plasticized material.

3.3.2. Mechanical Strength Analysis

The stress-strain behavior was studied and compared against each weld matrix as shown in Figure 7.

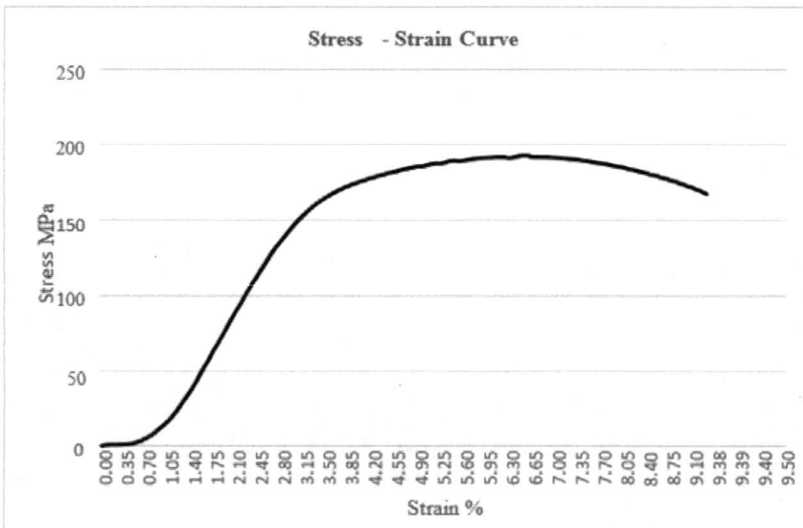

Fig. 7 Typical Stress-strain curve obtained from tensile test results of the weld sample for the process parameter 1100 rpm and 100 mm/min.

The weld sample processed using the combination of 900 rpm and 100mm/min gave the highest extension value of 4.225mm, and had the second highest ultimate tensile strength value of 191MPa (Table 6) and achieving the highest elongation. The parent metal's ductility, yield strength and ultimate tensile strength were 15%, ~271 and ~300 MPa respectively. Chen et al., reported the AA6061-T6 base metal elongation, yield strength and ultimate tensile strength of 12.1%, 242 MPa and 272 MPa respectively.

Table 6 demonstrates that all the weld samples exhibited reduction in the ultimate tensile strength and yield strength by between 34.8± 7% compared to the parent metal. The parent metal of AA6061-T6 is a hardened alloy which relies on precipitates. The thermal gradients and mechanical workings of the pin generated during FSW processing causes dissolution of the precipitates. This decomposition of precipitates and softened material gives rise to the reduction of the mechanical properties such as UTS and yield strength when compared with the base. The reduction of tensile strength inadvertently affects ductility.

As tabulated in Table 6, the FSW sample processed with the tool rotational and weld speeds of 1100rpm and 100mm/min respectively, recorded the maximum tensile strength, yield strength and weld efficiency of 192.51 MPa, 149.51 and 64.2% respectively. The variation of the ultimate tensile strength, yield strength and efficiency was within close range of between 181.73-192.51 MPa, 116.79-149.51 MPa, 60.5 % - 64.2%, respectively, for all the welded samples. It is also distinctively clear that for all the different tool rotation speeds, the ultimate tensile strength, yield strength and weld efficiency improved with increase in the weld speeds. It can be observed that all weld achieved their maximum UTS (192.51MPa, 191.17MPa, 189.94MPa) at the highest combinations of rotational and weld speeds. This could be attributed to the precipitates size and distribution improvement relative to optimum heat generation, which results in increase in mechanical properties relative to the increase in weld speeds. It can be noticed that a decrease in ultimate tensile strength is also uniform with the decrease in the yield strength and is relative to the decrease of the rotational and weld speeds.

Table 6 Mechanical Properties - Tensile Strength Values

Sample	Rotational Speed (rpm)	Welding Speed (rpm)	UTS (MPa)	Yield Strength (MPa)	η comparison to BM
BW1	700	60	183.63	114.26	61.2
BW2	700	80	186.57	147.88	62.2
BW3	700	100	189.45	127.24	63.3
BW4	900	60	181.73	116.79	60.5
BW5	900	80	189.45	145.37	63.2
BW6	900	100	191.17	122.46	63.7
BW7	1100	60	184.49	113.76	61.5
BW8	1100	80	188.65	134.08	62.9
BW9	1100	100	192.51	149.51	64.2

4. Conclusion

- X-ray radiography imaging was successfully used to demonstrate that FSW achieved full weld penetration.

- The X-ray images also enabled conclusion that no root defects were observed on all the welds, also confirming full weld penetration, as evidenced by the microstructural analysis and mechanical properties evaluations.
- Selection of suitable process speeds helped to achieve optimum generation of heat energy sufficient to plasticize the material. The correct process parameters thus contributed to improved material flow, a condition important for improving uniformity of microhardness across the weld traverse.
- Higher rotational speeds and higher feed rates resulted in increased hardness values, with the highest values recorded in the weld nugget and the least values recorded in the SZ/TMAZ interfaces. This was evidenced by a W-shaped hardness distribution data across the welds traverse in some of the welds.
- The base metal showed higher mechanical properties when compared to all the weld regions. The degradation in tensile strength and ductility notable in the interfaces was commonly due to the precipitates decomposition caused by thermal variations from high heat inputs and mechanical workings.

Acknowledgements

The authors would like to acknowledge the University of Johannesburg, South Africa, for allowing us to use their facility to conduct the experimental investigation and acknowledgements extended to Professor Kailas Satish of Indian Institute of Science, Bangalore, India for producing the welds.

Reference

[1] W. Thomas, "Friction Stir Welding". UK Patent 5,460,317, 6 December 1991.

[2] W. Arbergast and P. Hartlet, "Friction stir welding technology development," in *Lockheed Martin Michoud Space system*, Georgia, US, 1998.

[3] E. Akinlabi, R. Mahamood and S. Akinlabi, "Processing Parameters Influence on Wear Resistance Behaviour," *Advances in Material Science and Engineering,* vol. 2014, no. Article ID 724590, pp. 1-2, 2014.

[4] M. Patil, "MTech Dissertation on study of effect of process parameters in friction stir welding of 5083 Al alloy on power input and microhardness contours in weld cross - sections," Indian Institute of Technology, Kharagpur, 2012.

[5] N. Khan, A. Siddiquee and S. Shihab, "Investigations on tunneling and kissing bond defects in FSW joints for dissimilar aluminium," *Alloys and Compunds,* vol. 648, pp. 360-367, November 2015.

[6] P. Kah, R. Rajan, K. Martkainen and R. Suoranta, "Investigation of weld defects in friction stir welding and fusion welding of aluminium alloys," *International Journal of Mechanical and Materials Science,* vol. 4, no. ISSN:1823-0334, December 2016.

[7] H. Chena, K. Yanb, L. Tao, S. Chena, C. Jiangb and Y. Zhaob, "The investigation of typical welding defects for 5456 aluminium alloy friction stir welds," *Elsivier Material Science and Engineering,* vol. 433, no. 1-2, p. October 2006, 64-69.

[8] M. Kathrotiya and K. Bhavsar, "Various Parameter Effects of Friction Stir welds," *International Research Journal of Engineering and Technology,* vol. 04, no. 02, February 2017.

[9] C. Devanathan and S. Babu, "Effect of plunge depth on Friction Stir Welding of Al 6063," in *2nd International Conference*, INCAMA, 2013.

[10] S. Bayazida and H. Farhangia, "Effect of Pin profile of depth of defects of Friction Stir welded 7075 Aluminium Alloy," *Elsevier Material Science,* vol. 11, pp. 12-26, 2015.

[11] C. Brook and C. Hanstead, "Impact of Non Destructive Testing," in *Proceeding on the 28th Annual British Conference of NDT*, Bristol, UK, September 1989.

[12] K. Oyindamola, "Master's Dissertation on Effects of Multi-Pass Friction Stir Processing on Aluminium," University of Johannesburg, Johannesburg, SA, 2016.

[13] C. Meola, A. Squillace, F. Memola, C. Minutolo and R. Morace, "Analysis of stainless steel welded joints: a comparison between destructive and non-destructive techniques," *Elsevier Material Processing Technology,* Vols. 155-156, pp. 1893-1899, 2004.

[14] T. Savaranana, B. Lahirial, V. Pillaib, J. Philip, B. Rao and T. Jayakumar, "Non-Destructive Evaluation of Friction Stir Welded Joints by X-Ray Radiography and Infrared Thermography," in *International Conference on Structural Integrity*, 2014.

[15] N. Gaurav, K. Manu and M. Sivashanmugam, "Digital Radiography Testing for GTAW and FSW Weldments on AA7075-T651 Aluminium Alloy," *Research india publications International Journal of Advanced Mechanical Engineering,* vol. 04, no. 7, pp. 747-754, 2014.

[16] P. Edwards and M. Ramulub, "Material Flow during friction stir welding of Ti-6Al-4V," *Elsevier Science Inc Journal of Materials Processing Technology,* vol. 218, pp. 107-115, April 2015.

[17] C. Patil, H. Patil and P. H, "Investigation on Weld Defects in Similar and Dissimilar Friction Stir Welded joints of Aluminium Alloys of AA7075 and AA6061 by X-ray Radiography," *American Journal of Material Engineering and Technology,* vol. 4, no. 1, pp. 11-15, 2016.

[18] B. Singh, "A handbook on Friction Stir Welding," in *978-3-659-10762-7*, UK, LAP Lambert Academic Publishing, June 2012, pp. 44-54.

[19] B. Kumar, V. Umapathi and J. Kumar, "A comparative study of mechanical properties and microstructures of FSW joints in Conventional welding joints," *Internation Journal of Engineering Trends and Technology,* vol. 35, 2016.

[20] P. Mashinini, "MTech Dissertation in Process window for Friction Stir Welding of 3 mm Titanium (Ti-6Al-4V)," Nelson mandela Metropolitan University, South Africa, May 2010.

[21] Z. Georgeou, "ME dissertation in the Analysis of Material Flow around retractable pin in FSW," Port Elizabeth Technikon, Port Elizabeth, South Africa, 2003.

[22] M. Lin, J. Shude, M. Xiangchen, Z. Yaming and G. Shuangsheng, "New technique for eliminating keyhole by active-passive filling friction stir repairing," *Researchgate Materials & design 97,* February 2016.

[23] F. Fadaeifard, K. Matori, S. Aziz, L. Zolkarnain and M. B. A. Rahim, "Effect of the Welding Speed on the Macrostructure Microstructure and Mechanical Properties of AA6061-T6 Friction Stir Butt Welds," *MDPI Journal of Metals,* vol. 7, no. 48, pp. 2-16, 2017.

[24] A. Sedmak, R. Kumar, S. Chattopadhyaya, S. Hloch, S. Tadic´, A. Djurdjevic´, I. Cˇ ekovic´ and E. Doncˇeva, "Heat input effect of friction stir welding on aluminum alloy AA 6061-T6 welded joint," *Thermal Sciences,* vol. 20, p. 637–641, 2016.

[25] E. Akinlabi, "Doctoral Thesis on Characterisation of dissimilar Friction Stir Welds Between 5754 Aluminium Alloy and C11000 Copper," Nelson Mandela Metropolitan University, Port Elizabeth, South Africa, December 2010.

[26] X. Sauvage, A. Dédé, A. C. Muñoz and B. Huneau, "Precipitate stability and recrystallisation in the weld nuggets of friction stir welded Al-Mg-Si and Al-Mg-Sc alloys.," *Elsvier Science Inc Material Science Engineering,* vol. 491, pp. 364-371, 2008.

[27] Z. Chen, S. Li and L. Hihara, "Microstructure, mechanical properties and corrosion of friction stir welded 6061 Aluminum Alloy," in *University of Hawaii*, Manoa, Honolulu, HI 96822, USA.

[28] P. Kah, B. Mvola, J. Martikainen and R. Suoranta, "Real Time non destructive testing of welding," *Trans Tech Publications Advanced Material Research,* vol. 933, no. 01, pp. 109-116, 2013.

[29] *Standard: ASTM B308/B308M Specification for Aluminium Alloy 6061-T6 Standard Structural profiles,* October 2010.

[30] *Standard Test Methods for Microidentation hardness of material ASTM384-16,* West Conshohocken, PA: ASTM International, 2012.

[31] *Standard Test Methods for tension Testing Wrought and Cast, ASTM B557-14,* 2014: ASTM International, West Conshokocken, PA.

[32] *Standard Test Method for tension testing of Mettalic Material,* West Conshohocken, PA: ASTM Test Methods , 2013.

[33] *Standard Reference Radiographs for Inspection of Aluminium and Magnesium Castings, ASTM E155-15,* West Conshohocken, PA: ASTM International, 2015.

[34] Z. Chen, S. Li and S. Hihara, "Microstructure, mechanical properties and corrosion of friction stir welded 6061 Aluminum Alloy," Hawaii Corrosion Laboratory Department of Mechanical Engineering University of Hawaii, Manoa, Honolulu, HI 96822, USA.

[35] "Wikipedia," Friction Stir Welding, [Online]. Available: https:/en.wikipedia.org/wiki/Friction_stir_welding.

[36] C. Brook and C. Hanstead, "Impact of Non DestructiveTesting," in *Proceeding on the 28th Annual British Conference of NDT,* Bristol, UK, September 1989.

[37] P. Kah, B. Mvola, J. Martikainen and R. Suoranta, "Real Time non destructive testing of welding," *Trans Tech Publications Advanced Material Research,* vol. 933, no. 01, pp. 109-116, 2013.

[38] C. Bird, "Ultrasonic phased array inspection technology for the evaluation of friction stir welds," *The British Institute of non-Destructive testing and Condition Monitoring,* vol. 46, no. 1, pp. 31-36, 1 January 2004.

[39] C. Meola, A. Squillace, F. Memola, C. Minutolo and R. Morace, "Analysis of stainless steel welded joints: a comparison between destructive and non- destructive rechniques," *Elsevier Material Processing Technology,* vol. 155, no. 156, pp. 1893-1899, 2004.

[40] "The Welding Institute," [Online]. Available: http:/www.twi-global.com/technical-technology/job-konwledge/non-destructive-examination-nde-part-1-liquid-penentrant-and-magnetic-particle-inspection-122.

[41] C. Bird, "Ultrasonic phased array inspection technology for the evaluation of friction stir welds," *The British Institute of Non-Destructive Testing and Condition Monitoring,* vol. 46, no. 1, pp. 31-36, 1 January 2004.

IOP Conf. Series: Materials Science and Engineering **413** (2018) 012036 doi:10.1088/1757-899X/413/1/012036

Probabilistic Seismic Hazard Analysis of Nigeria: The Extent of Future Devastating Earthquake.

John Oluwafemi [1], Olatokunbo Ofuyatan [1], Solomon Oyebisi [1], Tolulope Alayande [1], John Abolarin [2]

Department of Civil Engineering, Covenant University, Ota, Ogun-State, Nigeria.
E-mail address: john.oluwafemi@covenantuniversity.edu.ng; spiritjohn20@gmail.com

Abstract. The several past seismic occurrences in Nigeria has recently led to warnings from research agencies and the forecast of large earthquake from researchers in Nigeria. Nevertheless, the major forecast from researchers has appeared to be open ended. To this end, this paper aimed at the probabilistic seismic hazard analysis of Nigeria and the limit to probable future earthquake magnitudes in Nigeria. The Gutenberg-Richter recurrence law was majorly employed for the purpose of this research. The Findings of this research established that Nigeria is at the risk of experiencing earthquake magnitudes as high 6.0 in the year 2020; 6.5 between the year 2021 and 2022; 7.0 between the year 2025 and 2026 and 7.1 in the year 2028 with a 36.79% probability. The probability that an earthquake of magnitude 7.1 will be experienced from 2019 to 2028 also ranges from 9% to 36.79%. The findings of this work inform on the sizes of probable future earthquake magnitude and it is recommended that the government of Nigeria pays rapt attention to earthquakes in Nigeria.
Keywords: earthquakes; magnitudes; recurrence interval; probability; intensity

1. Introduction

Earthquake is a global phenomenon experienced in most regions of the world. It is classified as one of the most devastating natural disasters that pose threat and has the capability to impact negatively on both human lives and the built environment. While other natural disasters such as drought, flood, famine, hurricane, tsunami and more can be devastating at the occurrence, an earthquake can cause much more damage in a swipe, depending on its magnitude. Hence, the continuous occurrence of earthquake events across the globe has continued to be a concern to humanity and some researchers have self-tasked themselves to work tirelessly to device means to mitigate the aftermaths of earthquake occurrence in the nations of the earth [1].

When an earthquake occurs, the trail of the adverse resulting impacts is not limited to the moment of its occurrence. The impact of an earthquake occurred in a nation can lead to a major setback in her development and her economic status and this can even linger for years after the event had occurred based on the magnitude of the earthquake [2] [3]. According to [4], Earthquakes occur due to the sudden release of built-up energy within the rocks. Other activities carried out on land surface can also stimulate earthquake. Some of these activities include drilling of boreholes and erection of heavy buildings.

Earthquake is experienced by most nations of the earth depending on their level of seismicity, while some nations are known for high seismicity, some are known for none and some are known for low seismicity. Nigeria among other nations of the earth has reflected dispersed seismic events within the period of 1933 to recent times. Many of the events that occurred in time past were not captured due to non-availability of seismic recording instruments.

Nigeria in the past was believed to be certified free from earthquake hazards but the instrumental and historical records of Nigeria as presented in Table 1 has shown contrary to this belief. While the past earthquakes are believed to have occurred along the faults in Nigeria, future earthquake occurrences in Nigeria are also expected to occur along the seismic faults.

IOP Conf. Series: Materials Science and Engineering **413** (2018) 012036 doi:10.1088/1757-899X/413/1/012036

Table 1: Historical/Instrumental Earthquakes in Nigeria (Source: [5])

S/N	Year-Month-Day	Origin Time	Felt Areas	Intensity/Magnitude	Probable Epicenter	Coordinates	
1	1933	-	Warri	-	-	$05^{o} 45^{1} 23^{11}$E	$05^{o} 31^{1} 42^{11}$N
2	1939-06-22	19:19:26	Lagos, Ibadan, Ile-Ife	6.5 (Ml)	Akwapin fault in Ghana	$03^{o} 23^{1} 00^{11}$E	$06^{o} 30^{1} 11^{11}$N
3	1948-07-28	-	Ibadan	-	Close to Ibadan	-	-
4	1961-07-02	15:42	Ohafia	-	Close to ijebu-Ode	-	-
5	1963-12-21	18:30	Ijebu-Ode	V	Close to Ijebu-Ode	-	-
6	1981-04-23	12:00	Kundunu	III	At Kundunu Village	-	-
7	1982-10-16	-	Jalingo, Gembu	III	Close to Cameroun Volcanic Line	-	-
8	1984-07-28	12:10	Ijebu-Ode, Ibadan, Shagamu, Abeokuta	VI	Close to Ijebu Ode	-	-
9	1984-07-12		Ijebu Remo	IV	Close to Ijebu Ode	$03^{o} 23^{1} 00^{11}$E	$07^{o} 11^{1} 45^{11}$N
10	1984-08-02	10:20	Ijebu-Ode, Ibadan, Shagamu, Abeokuta	V	Close to Ijebu Ode	-	-
11	1984-12-08	-	Yola	III	Close to Cameroun Volcanic Line	-	-
12	1985-06-18	21:00	Kombani Yaya	IV	Kombani Yaya	-	-
13	1986-07-15	10:45	Obi	III	Close to Obi town	$08^{o} 46^{1}$E	$08^{o} 22^{1}$N
14	1987-01-27	-	Gembu	V	Close to Cameroun Volcanic Line	$11^{o} 15^{1}$E	$06^{o} 42^{1}$N
15	1987-03-19	-	Akko	IV	Close to Akko	$10^{o} 57^{1}$E	$10^{o} 17^{1}$N
16	1987-05-24	-	Kurba	III	Close to Kurba village	$10^{o} 12^{1}$E	$11^{o} 29^{1}$N
17	1988-05-14	12:17	Lagos	V	Close to Lagos	-	-
18	1990-06-27	-	Ibadan	3.7(ML)	Close to Ijebu-Ode	$03^{o} 58^{1}$E	$17^{o} 22^{1}$N
19	1990-04-05	-	Jerre	V	Close to Jerre Village	-	-
20	1994-11-07	05:07:51	Ojebu-Ode	4.2(ML)	Dan Gulbi	-	-

21	1997	-	Okitipupa	IV	Close to Okitipupa Ridge	-	-
22	2000-08-15	-	Jushi-Kwari	III	Close to JushiKwari village	07° 42'E	14° 03'N
23	2000-03-13	-	Benin	IV	Benin City (55km from Benin)	-	-
24	2000-03-07	15:53:54	Ibadan, Akure, Abeokuta, Ijebu-Ode, Oyo	4.7 (ML)	Close to Okitipupa	-	-
25	2000-05-07	11:00	Akure	IV	Close to Okitipupa Ridge	-	-
26	2001-05-19	-	Lagos	IV	Close to Lagos city	-	-
27	2002-08-08	-	Lagos	IV	Lagos city	-	-
28	2005-03	-	Yola	III	Close to Cameroun Volcanic Line	-	-
29	2006-03-25	11:20	Lupma	III	Close to Ifewara	-	-
30	2009-09-11	-	Abomey-Calavi	II	Close to Benin	-	-
31	2011-11-05	-	Abeokuta	4.4	Close to Abeokuta	-	-
32	2016-07-10	-	Saki	IV	Oyo State	-	-
33	2016-08-10	-	Igbogene	III	Bayelsa	-	-
34	2016-09-11	-	Kwoi	III	Kaduna State	-	-
35	2016-09-12	-	SambangDagi	III	Kaduna	-	-

There are also indications from the development of recent times that devastating earthquakes are likely to occur in Nigeria and in West Africa in the nearest future and it is, therefore, important that these indications be taken into cognizance [6][7]. [8] explained that the reason why Nigeria hardly experiences earthquake hazard like other high earthquake-prone nations is likely to be as a result of the location of her geological framework which is believed to be situated in her continent's mobile belt between the Congo and West Africa Craton where notable damage was reported to have occurred in the past leading to bit of impact on the adjacent craton.

The prevention of earthquake occurrence in any country is impossible, different approaches that help to assess their likelihood of occurrence go a long way in helping authorities to make decisions and to implement certain rules that will help minimize or mitigate the aftermath of the hazards [9]. Despite the fact that researchers have warned that Nigeria is at the risk of experiencing devastating earthquakes in the nearest future, there is still no knowledge of the extent of the probable future earthquake magnitude. Hence, probabilistic seismic hazard analysis has been carried out for Nigeria in this research with the employment of the Gutenberg-Richter recurrence law.

1.1 Earthquake Forecast

In the study of earthquakes, it is possible to pinpoint locations where future earthquakes have the likelihood of occurrence by examining tectonic plates' movement in a location. Nevertheless, this approach can still turn out to be inaccurate. Hence, scientists are yet to successfully discover an approach to predict future earthquakes, considering time and location with accuracy and precision [10]. According [11], Earthquake forecast is a statement of probability that an earthquake of a particular magnitude will occur at a particular time and at a particular location. Earthquake forecast is distinctively different from earthquake prediction as many confuse the two to be the same. Unlike in forecasting, prediction comes as a statement of certainty.

Several researchers have related different approaches to assess future earthquakes in times past. As follow, some of these approaches include the study of radon gas emission, seismic waves study, and future earthquake assessment using animal unusual behavior prior to major earthquake event:

i. [12] pointed out that radon emission as earthquake precursor is a good facilitator of knowledge towards earthquake prediction. Studies have shown that there is always very high radon concentration at the site of earthquake occurrence. Hence, its concentration level in groundwater and soil is a good indicator.

ii. The earliest unusual behavior of the animal in response to earthquake occurrence was that of 373BC in Greece in which animals migrated from their different homes to safe locations few days to a major earthquake [13]. [14] described the responses of animals to the earthquake as an inbuilt escape instinct or survivor instinct due to the ability of the animals to sense seismic signals.

iii. Seismic waves carry earthquake energy which can be analyzed for some seismic parameters that will make earthquake assessment feasible. The seismogram is used for capturing these waves and the output is presented as seismograph which is analyzed for the extraction of necessary parameters such as earthquake magnitude, frequency and wavelength [12].

1.2 Gutenberg-Richter Recurrence Law

The Gutenberg-Richter recurrence law was first proposed by [15]. This relationship relates the cumulative number of the earthquake to the magnitude of the earthquake in a region. The Gutenberg-Richter recurrence law is presented in the equation 1.

$$\log_{10} N = a - bM \tag{1}$$

According to [16], global earthquake occurrences including the ones with great depth follow the trend of the G-R recurrence law. The "N" in the equation 1 represents the total number of earthquakes, a and b are referred to as seismicity parameters while "M" is the magnitude of earthquakes. The establishment of the G-R recurrence law's linearity has remained unchanged for more than fifty years [17].

2. Material and Methods

The data used for the purpose of this assessment was the instrumental and the historical data presented in Table 1. The equation 2 as related by [18] was employed for the conversion of every event reported using intensity scale to the Richter magnitude scale.

$$M_L = 1 + 0.667 I_o \tag{2}$$

The Gutenberg-Richter recurrence law presented in the equation 1 was used to generate the Gutenberg-Richter plot of Fig. 1 for Nigeria by plotting the cumulative number of earthquakes in Nigeria against the corresponding magnitudes. The seismicity parameter "a" was derived from the intercept of the plot in Fig. 1 and the b-value was derived by finding the gradient of the plot.

The logarithmic values of the vertical axis of the G-R plot gave the yearly occurrence for each earthquake magnitude while the recurrence interval for each magnitude was obtained by using the equation 3.

$$\text{Recurrence Interval} = \frac{1}{\text{Number of earthquake occurrence per year}} \tag{3}$$

The equation 4-5 related by [19] for the bounded Gutenberg-Richter relationship was used to establish the probability distribution function and the cumulative distribution function for the yearly occurrence of earthquake magnitudes of interest.

$$\beta = 2.303b \tag{4}$$

$$PDF = f_i(m) = P[M = m] = \frac{\beta e^{[-\beta(m - m_{min})]}}{[1 - e[-\beta(m_{max,i} - m_{min})]} \tag{5}$$

$$CDF = f_i(m) = P[M < m] = \frac{1 - e^{[-\beta(m - m_{min})]}}{[1 - e[-\beta(m_{max,i} - m_{min})]} \tag{6}$$

The M_{max} was set as 7.1 as obtained from the G-R plot of the Fig. 1 while the M_{min} was set at the magnitude of 3.0. Other distribution models such as Burr, t Location-Scale, Rayleigh, and Gamma distribution models were used to test for the distribution of earthquake events in Nigeria and the PDF and CDF for the best fitting distribution models was compared to that of the bounded Gutenberg-Richter distribution model.

The poison probability distribution model of equation 7 was employed to establish the probability that certain selected earthquake magnitude will occur in the recurrence year and also that earthquake magnitude of 7.1 will occur in Nigeria from 2019-2028.

$$P[N = n] = \frac{(\lambda t)^n e^{-\lambda t}}{n!} \tag{7}$$

This was done for each consecutive year starting from the time this research was carried out to the recurrence interval year for an earthquake of magnitude 7.1.Matlab 2017a version software was very instrumental in generating all the plots presented in the paper.

3. Results and Discussion

The results obtained from this assessment is well outlined out as follows:

3.1 The Gutenberg-Richter Relationship Plot for Nigeria

The Gutenberg-Richter plot generated for Nigeria using her instrumental and historical data is presented in the Figure 1. From the critical analyses of the G-R plot, it shows that earthquake magnitude that is as high 7.1 is likely going to be experienced in Nigeria in the future. The G-R approach also gave 7.1 magnitudes as the limit to the possible earthquake magnitude in Nigeria. As a basis for comparison, [20] forecasted that earthquake magnitude greater than 5.0 will be experienced in the nearest future. While the findings of [20] is open-ended, the results of this research gave a guide to the extent of earthquake magnitude to be expected in Nigeria. The b-value for Nigeria from the plot is 0.99.

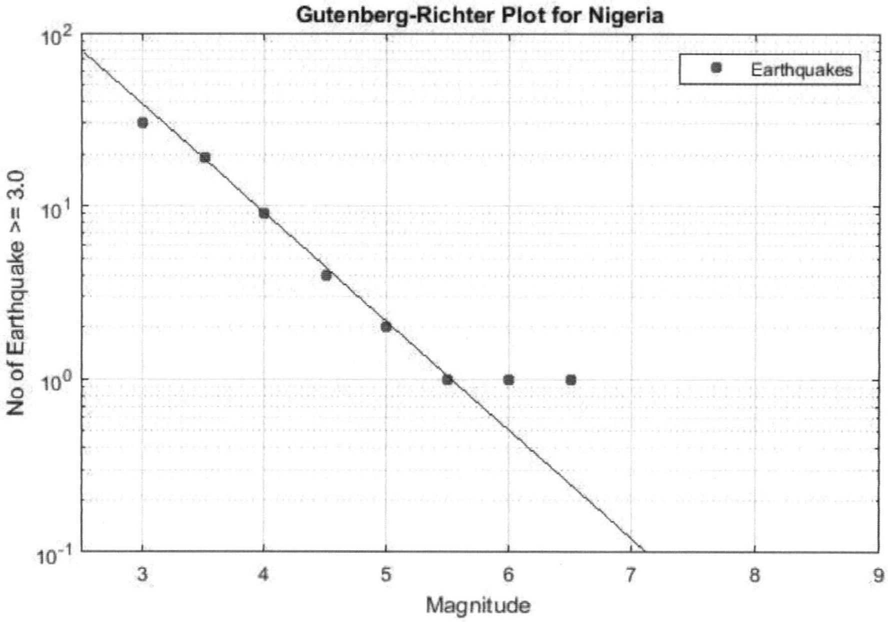

Fig. 1 Gutenberg-Richter relationship Plot

According to [21], when the b-value ranges between 0.72 ± 0.07 to 1.20 ± 0.015, the seismotectonic setting of such region is that of an intraplate. Hence, Nigeria's seismotectonic setting is of intraplate. This suggests that the seismic events that ever occurred in Nigeria lie without the margin plate's region. The value of "a" is found to be 16.8. The value of "a" is high and due to infrequent earthquake occurrences in Nigeria.

$$\log_{10} \lambda m = 16.8 - 0.99M \quad (8)$$

The equation 8 is therefore proposed for the determination of the mean annual rate of exceedance for earthquake magnitude of interest. The equation 8 is only valid for Nigeria provided that it obeys equation 9.

$$0 \leq M \leq 7.1 \quad (9)$$

The "a" and b-value solely depends on the number of events being considered. Therefore, these values are subject to slight changes if more seismic events are considered in the future.

3.2 Yearly Earthquake Occurrence and Earthquake Recurrence Interval for Nigeria

The yearly occurrence of earthquake magnitude and the recurrence intervals for earthquake magnitudes are shown in the Figure 2 and the Figure 3 respectively. Earthquakes with smaller magnitudes have a higher number of occurrences yearly than earthquakes with high magnitudes. This is similar to nature of earthquakes globally. From Figure 2, earthquake magnitudes of 3.0, 3.5, 4.0, 4.5, 5.0, 5.5, 6.0, 6.5, 7.0, and 7.1 are likely to occur 38 times, 19 times, 9 times, 4.2 times, 2.3 times, 1 time, 0.5 times, 0.27 times, 0.13 times, and 0.1 times respectively in a year.

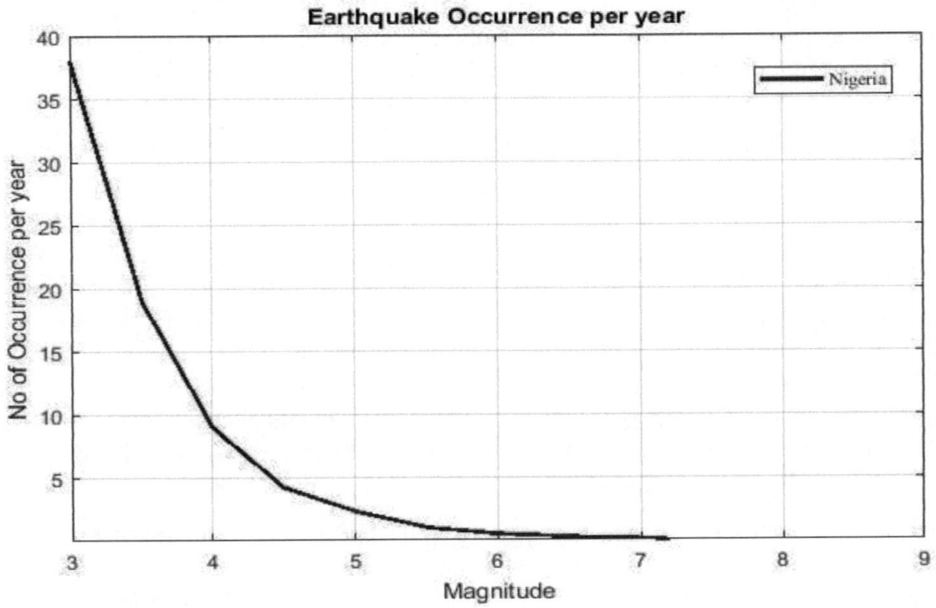

Fig. 2 Earthquake Yearly Occurrence

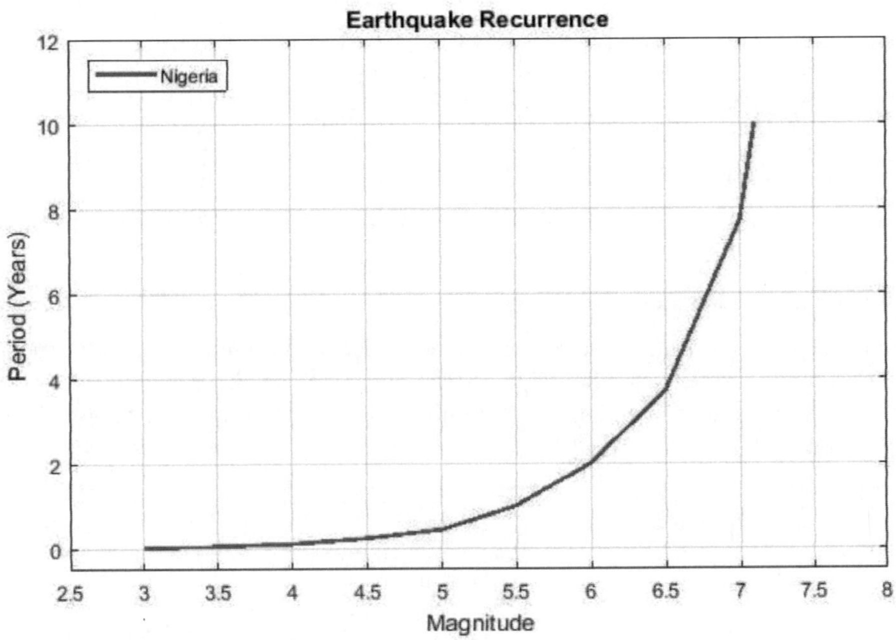

Fig. 3 Earthquake Recurrence interval

Likely also, the recurrence interval of earthquakes as shown in the Figure 3 reveals that earthquake magnitudes of 3.0, 3.5, 4.0, 4.5, 5.0, 5.5, 6.0, 6.5, 7.0 and 7.1 have the recurrence intervals of 0.026 year, 0.053 year, 0.111 year, 0.2938 year, 0.435 year, 1 year, 2 years, 3.704 years, 7.692 years and 10 years respectively in Nigeria. Therefore, assessing the results from the year 2018 shows that there is the likelihood of occurrence for earthquake magnitudes as high as 6.0 in the year 2020; 6.5 between the year 2021 and 2022; 7.0 between the year 2025 and 2026 and 7.1 in the year 2028. The probability that these earthquakes will occur in the recurrence year is 36.79%. [20] also stated 6.0% to 91.1% probability for earthquake greater 5.0 magnitude to occur between the year 2017 and 2028.

3.3 Annual Probability of Occurrence Using Bounded G-R Law

The probability distribution function and the cumulative distribution function using the engagement of the bounded G-R recurrence law is shown for the discontinuous earthquake events in Table 2 and Table 3 respectively. Earthquake magnitudes with small values exhibited a higher probability of occurrence and vice versa.

Table 2 Probability Distribution Function of Earthquake in Nigeria

Earthquake Magnitude	3.5	4.0	4.5	5.0	5.5	6.0	6.5	7.0	7.1
PDF=X=x	0.7293	0.2332	0.0746	0.0239	0.0076	0.00244	0.0008	0.0003	0.0002
CDF= X<x	0.6802	0.8978	0.9674	0.9896	0.9967	0.9990	0.9997	0.9999	1.0000

The least earthquake magnitude under consideration being 3.5 in Table 2 has an annual 72.94% probability of occurrence while magnitude 7.1 has an annual 0.02% probability of occurrence. Thus, the chances of yearly occurrence for an earthquake of magnitude 7.1 is low.

The cumulative distribution function on the other hand in Table 2 appears to have high probabilities for values below magnitudes of interest. This is because the probability is open to arrange of several events below the certain magnitude of interest. Hence, CDF for bounded range as presented in Table 3 is further generated.

Table 3 Cumulative Distribution function of earthquake within bounded range

Earthquake Magnitude	4.0<M<4.5	4.5<M<5.0	5.0<M<5.5	5.5<M<6.0	6.0<M<6.5	6.5<M<7.1
CDF	0.0696	0.0222	0.0071	0.0023	0.0007	0.0003

The CDF for magnitude 4.0 < M < 4.5 is 6.96% annually and CDF for 6.5< M< 7.1 is as low as 0.03% annually.

3.4 Probability Distributions for Earthquakes in Nigeria

The Plots for the PDF and CDF using the Burr, t Location-Scale, Rayleigh, and Gamma distribution models are presented in the Figure 4 and Figure 5. While Rayleigh distribution appears to have no correlation with the seismic data in Nigeria, the Burr and t Location distributions tend to fit better by observation. This is the same both for the PDF and the CDF.

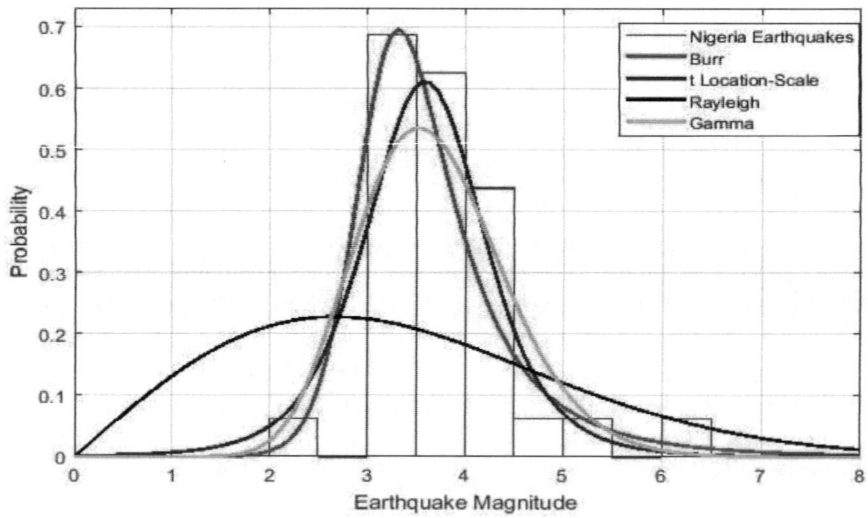

IOP Conf. Series: Materials Science and Engineering **413** (2018) 012036 doi:10.1088/1757-899X/413/1/012036

Fig. 4 Probability Distribution Models for Earthquake

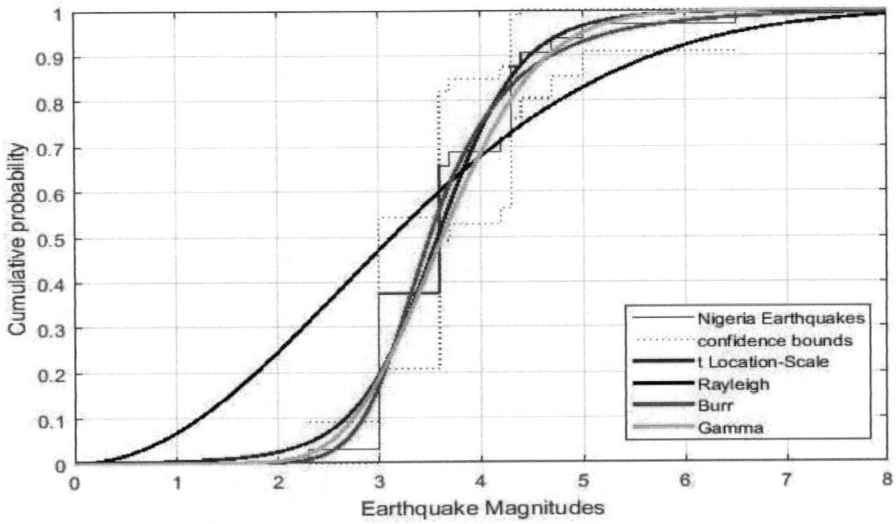

Fig. 5 Cumulative Distribution Models for Earthquake

Hence the corresponding PDF and CDF for the Burr and t Location-Scale distribution models are presented in Table 4 and Table 5 respectively.

Table 4 PDF for Burr &t Location Distribution model for earthquakes in Nigeria

Earthquake Magnitude	4.0	4.5	5.0	5.5	6.0	6.5	7.0
Burr	0.349938	0.166391	0.082108	0.042954	0.023723	0.013731	0.008275
t Location-Scale	0.472999	0.207019	0.071628	0.024539	0.00907	0.003686	0.00164

Table 5 CDF for Burr &t Location Distribution model for earthquakes in Nigeria

Earthquake Magnitude	4.0<M<4.5	4.5<M<5.0	5.0<M<5.5	5.5<M<6.0	6.0<M<6.5	6.5<M<7.0
Burr	0.123435	0.059386	0.030072	0.016129	0.009104	0.00537
t Location-scale	0.16647	0.064278	0.021889	0.007713	0.002967	0.001254

The results of the PDF and the CDF using Burr and t Location-scale distribution model are relatedly close and slightly higher to the PDF and CDF given by the bounded GR recurrence law.

3.5 Probability of 7.1 Magnitude from 2019 to 2028 in Nigeria

The Figure 7 presents the plot showing the probability that earthquake magnitude of 7.1 will occur in Nigeria from 2019 to 2028.

IOP Conf. Series: Materials Science and Engineering **413** (2018) 012036 doi:10.1088/1757-899X/413/1/012036

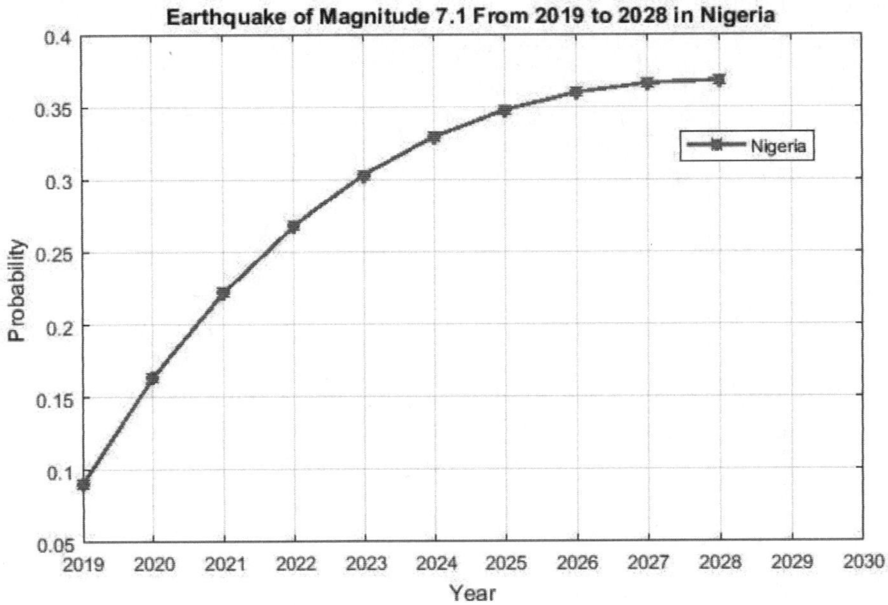

Fig. 7 The probability of 7.1 earthquake magnitude occurrence from 2019 to 2028

The Plot of Fig. 7 shows a yearly and gradual leap in the probability of magnitude 7.1 from 9% in 2019 to 36.79% in 2028.

4. Conclusion

The findings of this assessment established that Nigeria is at the risk of experiencing devastating earthquakes in the future. These Probable earthquake magnitudes are as high as 6.0 in the year 2020; 6.5 between the year 2021 and 2022; 7.0 between the year 2025 and 2026 and 7.1 in the year 2028. The probability that these events will take place in the forecasted year is 36.79%. The probability that an earthquake of magnitude 7.1 will also occur from 2019 to 2028 is between 9% and 36.79%. The b-value for Nigeria is 0.99 and this suggests that the seismotectonic setting of Nigeria is that of an Intraplate. The equation 8-9 is also proposed for the assessment of the annual mean rate of exceedance for earthquakes in Nigeria. It is therefore recommended that the Nigerian authority begins to enforce the law as regards earthquake considerations in structural designs in places such as the South-West region that has displayed most seismic events in times past. It is also recommended that activities such as heavy rock blasting should be relocated to non-residential places to reduce earthquake stimulations in residential areas.

References

[1] DAR M A, DAR A, QURESHI A, RAJU A. A Study on Earthquake Resistant Construction Techniques [J]. American Journal of Engineering Research, 2013, 20(12): 258-264.

[2] AKPAN A E, ILORI A O, ESSIEN N U. Geophysical investigation of Obot Ekpo Landslide site, Cross River State, Nigeria [J]. Journal of African Earth Sciences, 2015, 109: 154-167.

[3] OFUYATAN O, ADEOLA A, SULYMON N, EDE A, OYEBISI S, ALAYANDE T, OGUNDIPE J. Pseudo-Dynamic Earthquake Response Model of Wood-Frame with Plastered Typha (Minima) Bale Masonry-Infill [J]. International Journal of Civil Engineering and Technology, 2018, 9(2): 27-35.

[4] SHANNON D, AMY D, CATHERINE P, KIRSCH T D. The Human Impact of Earthquakes: a Historical Review of Events 1980-2009 and Systematic Literature Review [J]. PLOS Currents Disasters, 2013.

[5] ABOLARIN J, ADEDEJI A. Investigating Earthquake Magnitude by Seismic Signals and Wavelet Transform in Optimal Design [J]. Websjournal of Science and Engineering Application, 2016, 5: 305-322.

[6] UMAR A K, TAHIR A Y, OFONIME U A, DUNCAN D, SATURDAY U E. Towards an integrated seismic hazard monitoring in Nigeria using geophysical and geodetic techniques [J]. International Journal of the Physical Sciences, 2011: 6385-6393.

[7] AWOYERA P, OGUNDEJI P, ADERONMU P. Simulated Combined Earthquake and Dead Load Lateral Resistance Building Systems using Nigeria Seismic Data [J]. J. Mater. Environ. Sci., 2016, 7(3): 781-789.

[8] TSALHA M S, LAR U, YAKUBU T A, KADIRI U A, DUNCAN D. The Review of the Historical and Recent Seismic Activity in Nigeria [J]. Journal of Applied Geology and Geophysics, 2015: 48-56.

[9] GAUDIO C D, RICCI P, VERDERAME G M, MANFREDI G. Urban-scale seismic fragility assessment of RC buildings subjected to L'Aquila earthquake [J]. Soil Dynamics and Earthquake Engineering, 2017: 49-63.

[10] National Earthquake Prediction Evaluation Council. Evaluation of Earthquake Predictions, 2016: 1-6.

[11] CONSOLE R, PANTOSTI D, ADDEZIO G. Probabilistic approach to earthquake prediction [J]. Annals of Geophysics, 2002, 45(6): 723-731.

[12] SAJJAD M, FAISAL A. Computational seismic algorithmic comparison for earthquake prediction [J]. International Journal of Geology Issue, 2011, 3(5): 53-59.

[13] QUAMMEN D. Animals and earthquakes: This World [J]. San Francisco Chronicle, 1985: 15-16.

[14] KIRSCHVINK J. Earthquake Prediction by Animals: Evolution and Sensory Perception [J]. Bulletin of the Seismological Society of America, 2000, 90(12): 312-323.

[15] GUTENBERG B, RICHTER C. Magnitude and Energy of Earthquake [J]. Annali di Geofisica, 1956: 1-15.

[16] ZHAN Z. Gutenberg–Richter law for deep earthquakes revisited: Adual-mechanism hypothesis [J]. Earth and Planetary Science Letters, 2017: 1-7.

[17] CRAMPIN S, GAO Y. The Physics Underlying Gutenberg-Richter [J]. Journal of Earth Science, 2015, 26(1): 134-139.

[18] MURTY C. What are Magnitude and Intensity? [J]. Indian Institute of Technology Kanpur, 2004: 5-6.

[19] ROSHAN A, BASU P. Application of PSHA in low seismic region: A case study on NPP site in peninsular India [J]. Nuclear Engineering and Design, 2010: 3443-3454.

[20] ADEPELUMI A A, ONIBIYO O, ISOGUN M A. Short-term probabilistic prediction of earthquake occurrence in Southwestern Nigeria [J]. Envirotropica, 2010:1-11.

[21] EMAD A. Variation of b – value in the earthquake frequency – magnitude distribution with depth in the intraplate regions [J]. International Journal of Basic & Applied Sciences, 2011, 11(06): 29-37.

IOP Conf. Series: Materials Science and Engineering **413** (2018) 012037 doi:10.1088/1757-899X/413/1/012037

Power Sector Reform in Nigeria: Challenges and Solutions

Kayode Oladipo[1], Agbetuyi A. Felix[1], Owolabi Bango[1], Obiakor Chukwuemeka[1], Fagbuaro Olawale[2]

[1]Department of Electrical and Information Engineering, Covenant University, Ota, Nigeria.
[2]Department of Electrical Engineering, Federal Polytechnic, Ilaro, Ogun State, Nigeria.
E-mail address: oluwafemi.kayode@accessbankplc.com

Abstract. Deregulation of electricity sector in Nigeria has brought to the fore, the need to explore power generation options for the improvement of power capacity, reliability, and availability. One of such options is the paradigm shift from centrally generated electricity to generation at the distribution level. Embedded generation is an emerging concept in the electricity sector, which represents good alternates for electricity supply instead of the traditional centralized power generation concept, though relatively new as a viable option in Nigeria. Embedded generation (EG) refer to any electric power production technology that is integrated within distribution systems, close to the point of use. EG saves cost due to on-site production which avoids transmission and distribution costs and hence, a saving of about 30% of the cost of delivered electricity. Also, on-site production of energy reduces loss of generated power due to transmission and distribution. It also improves the level of power quality and reliability aside from serving niche applications for remote sites. Distributed or embedded generation can provide standby generation (emergency power), peak shaving capability and cogeneration. It improves the efficiency of the power grid by having multiple micro-sources added to the system. Embedded generators are usually connected to the medium or low voltage grid. This paper therefore reviews power sector reform in Nigeria and its challenges. It presents the way forward for epileptic power supply by embracing embedded generation as a potential option to improving power supply in Nigeria.
Keywords: embedded generation; electricity; distribution; reliability; power reform

1. Introduction

Deregulation of electricity sector in Nigeria has brought to the fore, the need to explore power generation options for the improvement of power capacity, reliability, and availability. One of such options is the paradigm shift from centrally generated electricity to generation at the distribution level. Embedded generation is an emerging concept in the electricity sector, which represents good alternates for electricity supply instead of the traditional centralized power generation concept, though relatively new as a viable option, embedded generation is not a completely new concept globally. Embedded generation (EG) refer to any electric power production technology that is integrated within distribution systems, close to the point of use, and these generators are connected to the medium or low voltage grid.
In Nigeria, electricity is received from the transmission grid at 330/132kV and stepped down to lower voltages of 33kV, 11kV, 0.415kV and 240V, as it is distributed to end users. Domestic consumers receive their supply at 240Volts while industrial consumers (heavy users) can take their supply from either the 11kV or 33kV lines. Consequently, 330kVand 132kV are referred to as high voltage while the 33kV and 11kV and below range is referred to as medium/low voltage. Figure 1 and Figure 2 show the layout of a typical Nigeria transmission line.

IOP Conf. Series: Materials Science and Engineering **413** (2018) 012037 doi:10.1088/1757-899X/413/1/012037

Fig. 1. A typical primary transmission [1]

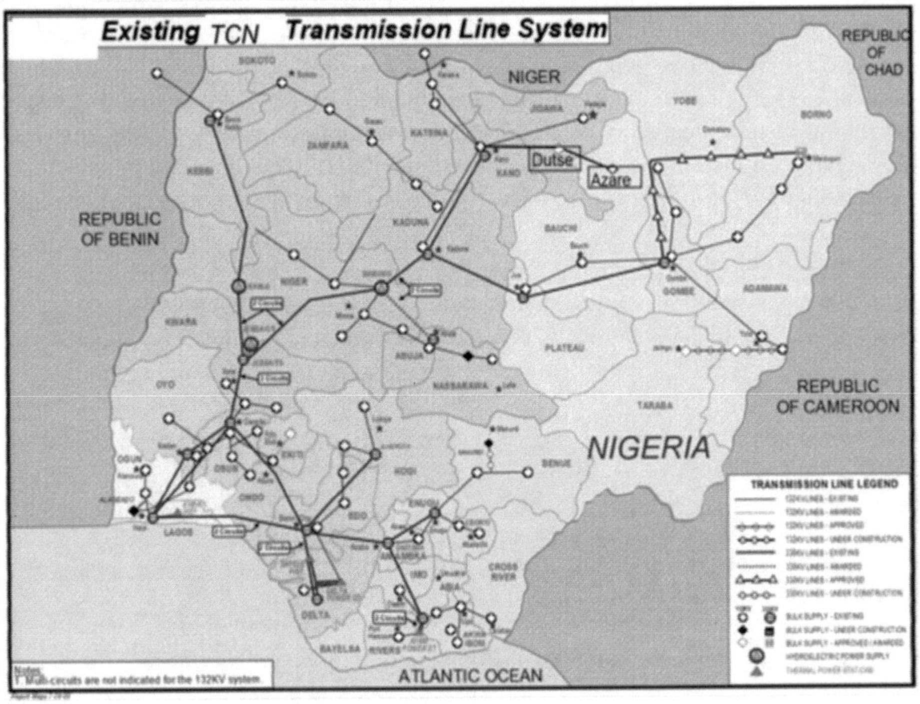

Fig. 2. Nigeria's transmission line system [2]

In recent times, numerous challenges facing central power generation is forcing a rethink of other viable options of getting electricity supply to the end-users. One such mechanism is to generate and power and supply proximal to the distribution network. This mode of power generation at the distribution level is called embedded or distributed power generation. The term distributed generation is often used to depict a small-scale electricity generation. However, there is currently no consensus on how embedded generation (EG) should be exactly defined as shown by the survey conducted by CIRED [3] and the assertions by [4]. The working group of CIGRE [5] devoted to distributed generation defines EG as all generation units with a maximum capacity of 50MW to 100MW, that are usually connected to the distribution network and that are neither centrally planned, nor dispatched. The International Energy Agency (IEA) on the other hand, considers EG as units producing power on a customer's site or within local distribution utilities, and supplying power directly to the local distribution network. In this case, no reference is made to the generation capacity level.

IOP Conf. Series: Materials Science and Engineering **413** (2018) 012037 doi:10.1088/1757-899X/413/1/012037

1.1. Power Generation in Nigeria

Power generation in Nigeria dates back to 1886 when two (2) generating sets were installed to serve the then Colony of Lagos. By an Act of Parliament in 1951, the Electricity Corporation of Nigeria (ECN) was established, and in 1962, the Niger Dams Authority (NDA) was also established for the development of hydroelectric power. A merger of the two (2) organizations in 1972 resulted in the formation of the National Electric Power Authority (NEPA) which was saddled with the responsibility of generating, transmitting and distributing electricity for the whole country. In 2005, as a result of the power sector reform process, NEPA was unbundled and renamed Power Holding Company of Nigeria (PHCN). The Electric Power Sector Reform (EPSR) Act was signed into law in March 2005, enabling private companies to participate in electricity generation, transmission, and distribution. The government unbundled PHCN into eleven electricity distribution companies (Discos), six generating companies (GenCos), and a transmission company (TCN). The Act also created the Nigerian Electricity Regulatory Commission (NERC) as an independent regulator for the sector.

At present, the Federal Government has fully divested its interest in the seven GenCos while 60% of its shares in the eleven (11) DisCos have been sold to the private operators. The Transmission Company still remains under government ownership. The generation companies created following the unbundling of PHCN are as shown in Table 1. The generation sub-sector presently includes 23 grid-connected generating plants in operation with a total installed capacity of 10,396 MW (available capacity of 6,056 MW) with thermal based generation having an installed capacity of 8,457.6 MW (available capacity of 4,996 MW) and hydropower having 1,938.4 MW of total installed capacity with an available capacity of 1,060 MW.

Table 1. Privatization status of seven power stations in Nigeria [5]

GenCo	Installed Capacity (MW)	Type	Privatization Status
Afam Power Plc	776MW	Gas	100% Sold
Sapele Power Plc	414MW	Gas	51% Sold
Egbin Power Plc	1,020MW	Gas	100% Sold
Ughelli Power Plc	900MW	Gas	100% Sold
Kainji Power Plant	760MW	Hydro	Long Term Concession
Jebba Power Plant	578MW	Hydro	Long Term Concession
Shiroro Power Plc	600MW	Hydro	Long Term Concession

IOP Conf. Series: Materials Science and Engineering **413** (2018) 012037 doi:10.1088/1757-899X/413/1/012037

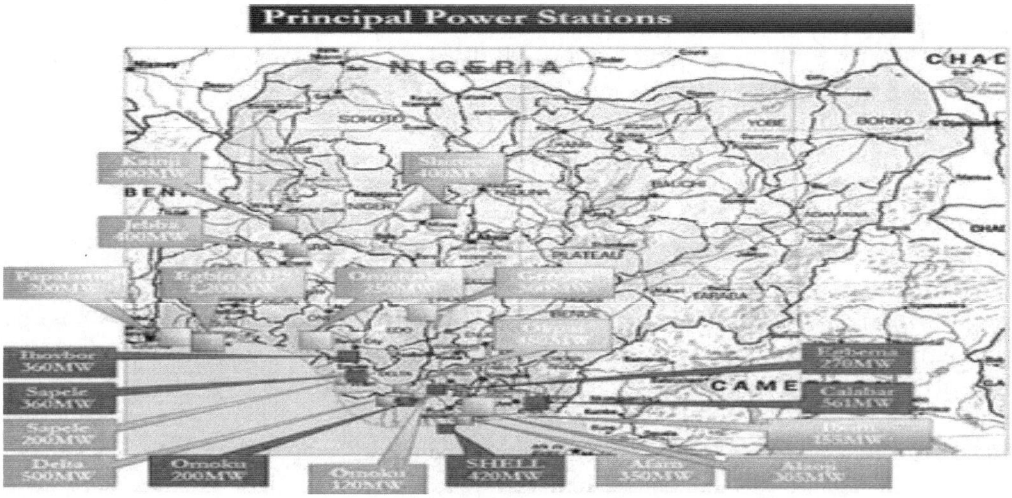

Fig. 3. Nigeria's principal power station system [6]

Figure 3 is the location of principal power stations in Nigeria and the installed capacity. However, in its effort to increase the level of power generation, the Federal Government in 2004, incorporated the Niger Delta Power Holding Company (NDPHC) as a public sector funded emergency intervention scheme. The company has a mandate to manage the National Integrated Power Projects (NIPP) which essentially involves the construction of identified critical infrastructure in the generation, transmission, distribution and natural gas supply sub-sectors of the electric power value chain [7]. In the generation sub-sector, NDPHC is expected to add ten (10) new gas fired power stations to the grid some of which have already been completed and commissioned, while others are at different stages of construction in different parts of the country. In total, the NIPP power stations will add about 4,774MW of power to the national grid network after completion [8].

The NIPP stations are listed below:

- Alaoji (1,074MW) in Abia state,
- Benin (Ihovbor) (451MW) in Edo state,
- Calabar (563MW) in Cross River state,
- Egbema (338 MW) in Imo state,
- Gbarain (225 MW) in Bayelsa State
- Geregu (434 MW) in Kogi State
- Olorunsogo (335MW)in Ogun state
- Omotosho (451 MW) in Ondo state,
- Omoku (225MW) in Rivers state.
- Sapele (450MW) (Ogorode)

Some of the NIPP power stations have already been privatized while plans are under way to sell the rest to interested investors in order to increase private sector participation in the sector thus strengthening the reform programme of the government.

In furtherance of the reform programme, the Nigerian Electricity Regulatory Commission (NERC) has in the past licensed several private Independent Power Producers (IPPs) some of which are at various stages of project development. The Commission has also enacted the Bulk Procurement Guidelines that will ensure the efficient and orderly procurement of large capacity generation in the future. This will enable the Commission to effectively predict the amount of power that can be added to the grid every year. The Commission has in addition developed a Regulation on embedded generation which allows for power generation plants (including renewable energy) to be directly connected to and evacuated through a distribution network. It provides a window for investors, communities, state and local governments to generate and sell or utilize power without going through the transmission grid. This also offers a means for DisCos to increase the amount of power available to sell while eliminating the transmission cost component of the tariff. By implementing reforms, Nigeria targets 40,000MW generating capacity by 2020

IOP Conf. Series: Materials Science and Engineering **413** (2018) 012037 doi:10.1088/1757-899X/413/1/012037

and will need to spend approximately $10bn per annum on the power sector for the next 10 years to achieve this [9] . In the post-privatized power sector, the Nigerian Bulk Electricity Trading PLC (NBET) purchases power generated by the GenCos and IPPs at agreed prices stated in Power Purchase Agreements (PPA) and resells to the DisCos who deliver the power to the end consumer

1.1.1. List of power stations in Nigeria

Twenty three (23) power plants are connected to the national grid. Below is a list of the power generation stations in Nigeria [14]:

1.1.1.1. Kanji power station reservoir.

Kanji power station is hydro-electric powered. It has an installed generating capacity of 760MW. It was completed by 1968 and is located at Niger state. This plant is presently generating 450MW, the plant cannot produce optimally due to old age and lack of proper Maintenance.

1.1.1.2. Jebba hydro power station reservoir

Jebba power station is hydro-electric powered. It has an installed generating capacity of 576.8MW and was commissioned in 1985. It is located in Niger state. This plant is presently generating 450MW, some of the turbine are bad due to lack of proper maintenance and well-trained personnel that can handle the machine.

1.1.1.3. Shiroro power station reservoir

Shiroro power station is hydro-electric powered. It has an installed generating capacity of 600 MW. It was completed in the year 1990 and is in full operation. This reservoir is situated at Niger state. This plant is presently generating 450MW, due to fund and proper maintenance.

1.1.1.4. Egbin thermal power station

Egbin power station operates on gas-fired steam turbine. It has an installed generating capacity of 1320 MW. The power station was completed between 1985 and 1986. It is located in Ikorodu of Lagos State. This plant is presently generating 1100MW, due to lack of proper Maintenance and well-trained personnel that could handle the machine.

1.1.1.5. Sapele power station

Sapele power station operates on gas-fired steam turbine and simple cycle gas turbine. It has an installed generating capacity of 1020 MW. Sapele power station was completed by 1981. It is situated in Ogorode – Sapele in Delta State. This plant is presently generating 90MW, due to faulty turbine and inappropriate funding of the power station.

1.1.1.6. Transcorp-Ughelli power station

Transcorp–Ughelli power station operates a simple cycle gas turbine. It has an installed generating capacity of 972 MW. Ughelli Phase 1 was built in 1966, Phase 2 in 1975, Phase 3 in 1978 and Phase 4 in 1990. The station is located in Ughelli of Delta state.This plant is presently generating 320MW.

1.1.1.7. Afam I-IV & V power station

Afam I-IV and V are thermal power plants operating on simple cycle gas turbine. Both plants were completed between 1982 and 2002. The power station has an installed capacity of 977MW. It is located in Afam in River state. This plant is presently generating 60MW, due to in adequate funding of the power station for sustainability.

1.1.1.8. Geregui power station

Geregu I Power station operates on open cycle gas turbine. It has an installed generating capacity of 414 MW. Geregu I Power station was completed by the year 2007. It is situated near Ajaokuta in Kogi state. This plant is presently generating 276MW due to inadequate funding.

1.1.1.9. Omotosho I power station

Omotosho I power station operates a simple cycle gas turbine with an installed generating capacity of 335 MW. It was completed by 2006. It is situated at Omotosho in Ondo state. This plant is presently generating 76MW, due to inadequate funding and shortage in gas supply.

1.1.1.10. Papalanto (olorunsogo) power station

Olorunsogo power station operates a simple cycle gas turbine. It has a total installed generating capacity of 335 MW. It was completed in the year 2007. This power station is located at Papalanto in Ogun state. This plant is presently generating 76MW, due to inadequate supply of gas and funding.

1.1.1.11. Kwale okpai power station (ipp)

IOP Conf. Series: Materials Science and Engineering **413** (2018) 012037 doi:10.1088/1757-899X/413/1/012037

Okpai power station operates on combined cycle gas turbine. It has an installed generating capacity of 480 MW. It was completed in 2005 as Independent Power Project. Okpai power station is owned by Agip Oil Company in Delta State.This plant is presently generating 361MW, due to inadequate funding.

1.1.1.12. Afam vi power station (IPP)

The combined cycle gas technology used in Afam VI Power plant contributes significantly to the reduction of greenhouse gas emissions. It has an installed generating capacity of 642MW. The plant was completed by 2010 and is situated in Rivers state. It is owned by the Shell Petroleum Development Company Joint Ventures. This plant is presently generating 450MW due to faulty turbine and non-availability of spare parts.

1.1.1.13. Ibom power station (ipp)

Ibom power station operates on simple cycle gas turbine. It has an installed generating capacity of 190MW. It was incorporated in 2001 and completed in 2009. Ibom power station has a proposed phase (II) with installed generating capacity of 500MW. This power plant is located at Ikot Abasi in Akwa Ibom state. This plant is presently generating 0MW, due to inadequate fund for completion.

1.1.1.14. AES Barge (IPP)

AES barge IPP operates on simple cycle gas turbine. It has an installed capacity of 270MW. The power plant was completed by 2001 and is located at Egbin in Lagos state. This plant is presently generating 224MW due to inadequate gas.

1.1.1.15. Omoku power station (IPP)

Omoku power station operates on simple cycle gas turbine. It has an installed generating capacity of 150 MW. It is currently under operation and was completed in 2005 and is situated in Rivers State. This plant is presently generating 60MW, due to defective turbine and inadequate gas supply.

1.1.1.16. Trans-Amadi power station (IPP)

Trans-Amadi power station operates a gas-fired system. It has an installed generating capacity of 136MW. It is situated in Rivers state. This plant is presently generating 60MW, due to defective turbine and inadequate gas supply.

1.1.1.17. Rivers IPP

Rivers IPP situated in Rivers state operates a gas-fired system. It has an installed generating capacity of 180MW. This plant is presently generating 0MW, due to lack of fund.

1.1.1.18. Aba power station (IPP)

Aba power station operates on simple cycle gas turbine. It has an installed generating capacity of 140MW. The power station was completed by 2012 and is situated at Aba in Abia state. This plant is presently generating 140MW, due to lack of fund and in adequate gas supply.

1.1.1.19. Geregu ii power station (NIPP)

Geregu II Power Station (NIPP) operates a simple cycle gas turbine. It has an installed generating capacity of 414MW. It was completed in the year 2012 and is situated in Kogi state. This plant is presently generating 276MW, due to inadequate funding and proper Maintenance of the turbine.

1.1.1.20. Sapele power plant (NIPP)

Sapele power plant is operated in simple cycle gas turbine. It has an installed generating capacity of 450MW. It was brought to completion by 2012. It is situated at Sapele in Delta state.

1.1.1.21. Alaoji power station (NIPP)

Alaoji power station (NIPP) operates on combined cycle gas turbine. It has an installed generating capacity of 1074 MW. The power station was completed by 2015. It is situated at Abia state. This plant is presently generating 0 MW.

1.1.1.22. Olorunsogo II power station (NIPP)

Olorunsogo II Power Station (NIPP) is situated at Olorunsogo in Ogun state. The power plant is operated on combined cycle gas turbine with an installed generating capacity of 675 MW. This plant is presently generating 0 MW.

1.1.1.23. Omotosho II power station (NIPP)

Omotosho power station operates on simple cycle gas turbine. It has installed generating capacity of 500MW. It was completed in the year 2012 and is currently under partial operation. It is situated in Omotosho Ondo state. The plant station is presently generating 76 MW.

1.1.1.24. Omoku II power station (NIPP)

Omoku II power station operates a simple cycle gas turbine. It has installed generating capacity of 225 MW. It is presently connected to the national grid. It is situated in Rivers state. The plant is presently generating 60 MW.

1.1.1.25. Ihovbor power station (NIPP)

Ihovbor power station operates on simple cycle gas turbine. It has a total installed generating capacity of 450MW. It was completed between 2012 and 2013. It is situated at Benin. This power station is presently generating 0 MW.

1.1.1.26. Egbema power station (NIPP)

Egbema power station operates on simple cycle gas turbine. It has an installed capacity of 338 MW. The power station was completed by 2013. It is situated at Imo state. This power station is not generating any megawatt to the grid.

1.1.1.27. Calabar power station (nipp)

Calabar power station operates on simple cycle gas turbine. It has an installed capacity of 561MW. It is located in Cross River state. This plant is presently generating 0 MW.

1.1.1.28. Gbarain power station (nipp)

Gbarain Power Plant has an installed capacity of 225MW. It is situated in Bayelsa State. Power Stations in Nigeria under Construction/Proposed. This plant is presently generating 0 MW.

1.1.1.29. Mambilla power station

Mambilla Power Station is a reservoir having an installed generating capacity of 3050MW. It is to be completed by the year 2018. The reservoir is located at Taraba state. It has not been completed due to lack of fund due to increase in spare parts.

1.1.1.30. Itobe power station

Itobe power plant located at Itobe Kogi State operates on circulating fluidized bed technology. It is coal fired power plant. It has an installed generating capacity of 1200 MW. It is yet to be completed by 2018.

1.1.1.31. Azura thermal power station (IPP)

Azura Thermal power station operates on simple gas turbine. It has an installed generating capacity of 450 MW. It is under development and is situated in Benin City.

1.1.1.32. Kano power station

Kano power station is hydro-powered. It has an installed capacity of 100 MW. The proposed completion date for the plant is the first quarter of 2017. It has not been completed due to present economy of the country.

1.1.1.33. Qua Iboe power plant

Quo Iboe power plant is a proposed power plant to be built in Akwa Ibom state. It is to have an installed generating capacity of 540MW. It will be one of the lowest cost thermal plants in Nigeria.

1.1.1.34. Zungeru power plant

Zungeru power plant is hydro-powered. It is to have an installed generating capacity of 700 MW and is to be completed by 2020. The power station is owned by the Ministry of Mines, Industry and Energy. It is located along the Kaduna River in Niger state.

1.1.1.35. Kaduna power plant

Kaduna power station is a dual fired power (gas/diesel) plant having an installed generating capacity of 215 MW. Its proposed completion is June 2017.

1.1.1.36. Gurara power plant

Gurara power plant is hydro powered. It has an installed generating capacity of 40 MW. It is located in Kaduna state.

1.1.1.37. Dadin kowa hydro power plant

Dadin Kowa power plant is hydro-powered with an installed capacity of 29 MW. Its proposed completion date is November 2019.

1.1.1.38. Okpai power plant phase 2

IOP Conf. Series: Materials Science and Engineering **413** (2018) 012037 doi:10.1088/1757-899X/413/1/012037

The Okpai Power Plant is to be completed by 2019. It has an installed generating capacity of 480MW.

1.1.2. Challenges Facing the Nigerian Power Sector

The poor performance of Nigeria's hitherto state – controlled power sector, resulting in unstable electricity supply and frequent blackouts, has been seen by ordinary Nigerians as evidence of the ineffectiveness of their governments. However, the situation has not improved much since the privatization of much power sector in recent years, even with continued government subsidies for some users.

Some other challenges are as detailed below [15]:

- Lack of proper funding faced by dwindling income due mainly to the collapse of global oil prices.
- The administration has the challenge of convincing frustrated electricity consumers that they must accept substantial increases in energy tariff if Nigeria is to achieve constant, stable and nationwide supply.
- This sector is facing myriad of structural problems that continue to hamper growth in the power sector. This include shortage of gas supply for thermal plants, high level of unpaid electricity bills and the country's outdated band poorly maintained transmission network, which the government still owns but put under private management.
- Also the existing transmission network cannot handle much more load than current peak electricity production
- Also many of the new power operators have struggled to make progress, especially as they have had to contend with aging facilities requiring substantial amounts of investments to upgrade and expand.

Fig. 4. Nigeria's power station [16]

In Nigeria, the operation of the power generation sector is driven by the privatized generation companies (GenCos), Independent Power Producers (IPP), and the generation stations under National Integrated Power Project (NIPP). Currently, the combined installed generating capacity is 12,500MW with about 74% contribution from the gas-powered plants [17]. The installed capacity of Hydro power station in Nigeria is 1,900 MW but the available capacity is 1,350 MW and this is due to aging and lack of proper maintenance of the turbine. The installed power station capacity in Nigeria is as shown in Table 2.

Table 2. Installed power station capacity in Nigeria, type and the available capacity [18]

IOP Conf. Series: Materials Science and Engineering **413** (2018) 012037 doi:10.1088/1757-899X/413/1/012037

Name	Site Location	Type	Installed Capacity (MW)	Available Capacity
AES Nigeria Barge Limited		Thermal	270	224
Afam Power PLC	Afam Rivers State	Thermal	987.2	60
Agbara Shoreline Power Limited	Agbara Ogun	Thermal	100	
Alaoji Generation Company Limited (NIPP)	Alaoji Abia State	Thermal	1074	
Anita Energy Limited	Agbara Lagos State	Thermal	90	
Azura Power West Africa Limited	Ihovbor Benin, Edo State	Thermal	450	
Benin Generation Company Limited	Ihonvbor Edo State	Thermal	450	
Calabar Generation Company Limited	Calabar Cross State	Thermal	561	
Century Power Generation Limited	Okija Anambra State	Thermal	495	
Enersys Nigeria Limited	Ado Ekiti State	Thermal	10	
Delta Electric Power Limited	Oghareki Etiope West LGA	Thermal	116	
DIL Power PLC	Obajana Kogi State	Thermal	135	
Egbema Generation Company Limited	Egbema Imo State	Thermal	338	
Egbin Power PLC	Egbin Lagos State	Thermal	1320	1100
Energy Company of Nigeria (NEGRIS)	Ikorodu, Lagos State	Thermal	140	
Enersys Nigeria Limited	Ado Ekiti Ekiti State	Thermal	10	
Ethiope Energy Limited	Ogorode, Sapele Delta State	Thermal	2800	300
Farm Electric Supply Limited	Ota-Ogun State	Thermal	150	
First Independent Power Company Limited	Omokun Rivers State	Thermal	150	60
First Independent Power Company Limited	Trans Amadi, Rivers State	Thermal	136	
First Independent Power Company Limited	Elewe, Rivers State	Thermal	95	
Fortune Electric Power Company Limited	Odukpani, Cross Rivers State	Thermal	500	
Gbarain Generation Company Limited	Gbarain, Bayelsa State	Thermal	225	
Geometric Power Limited	Aba, Abia State	Thermal	140	140
Geregu Power PLC (BPE)	Geregu Kogi State	Thermal	414	276
Hudson Power Ltd.	Warawa, Ogun State	Thermal	150	
Ibafo Power Station Ltd.	Ibafo, Ogun State	Thermal	200	
Ibom Power Ltd	Ikot Abasi Akwa Ibom State	Thermal	190	
ICS Power Ltd	Alaoji, Abia State	Thermal	624	
Isolo Power Generation Ltd.	Isolo, Lagos State	Thermal	20	
JBS Wind Power Ltd.	Maranban Pushit, Mangu, Plateau State	Wind	100	
Kainji Hydro Electric Plc (Kainji Station)	Kainji, Niger State	Hydro	760	450
Kainji Hydro Electric Plc (Kainji Station)	Jebba, Niger State	Hydro	540	450
Knox J&L Energy Solutions Ltd	Ajaokuta, Kogi State	Thermal	1000	
Lotus & Bresson Nigeria Ltd	Magboro Ogun State	Thermal	60	
MBH Ltd.	Ikorodu Lagos State	Thermal	300	
Minaj Holdings Ltd.	Agu-Amorji Nike Enugu East LGA, Enugu State	Thermal	115	
Nigeria Agip Oil Ltd.	Okpai Delta State	Thermal	480	361
Nigerian Electricity Supply Corporation (Nigeria) Ltd. (NESCO)	Bukuru, Plateau	Thermal	30	
Notore Power Ltd	Onne, Rivers State	Thermal	50	
Ogorode Generation Company Ltd (NIPP)	Ogorode Delta State	Thermal	450	
Olorunsogo Generation Company Ltd (NIPP)	Olorunsogo, Ogun State	Thermal	750	
Olorunsogo Power Plc (BPE)	Olorunsogo, Ogun State	Thermal	335	76
Omoku Generation Company Ltd.	Omoku, Rivers State	Thermal	250	60
Omotosho Generation Company Ltd	Omotosho II Ondo State	Thermal	500	76
Omotosho Power Plc (BPE)	Omotosho Ogun State	Thermal	335	35
Paras Energy & Natural Resources Development Ltd	Ogijo, Ogun State	Thermal	96	
Sapele Power Plc	Sapele, Delta State	Thermal	1020	90
Shell Petroleum Development Company Ltd	Afam VI	Thermal	642	450
Shiroro Hydro Electric Plc	Shiroro, Niger State	Hydro	600	450
Supertek Electric Ltd	Ajaokuta, Kogi State	Thermal	500	
Supertek Nigeria Ltd	Akwete, Abia State	Thermal	1000	
Ughelli Power Plc.	Ughelli, Delta State	Thermal	942	320
Western Technologies & Energy Services Ltd	Sagamu, Ogun State	Thermal	1000	
Zuma Energy Nigeria Ltd. (Gas Plant)	Ohaji Egbema, Owerri, Imo	Thermal	400	
Zuma Energy Ltd. (Coal Plant)	Itobe, Kogi State	Thermal	1200	
TOTAL			**25,255.2**	**4978**

1.2.2. The Overview of Nigeria Embedded Generation

Despite the huge investments by the government over the last years, it is an open secret that the nation is unable to boast of reasonable power supply to its citizens. The Nigerian population is above one hundred and eighty-three million and about 55% of the populations have no access to grid-connected electricity Access to electricity in the rural areas is about 35% and about 55% in the urban areas [19].

It has been estimated that developing economies would need about 1,000MW per million people to meet their electricity demand. Invariably, Nigeria would require more than 160,000MW to achieve the desired electricity generation capacity. Nigeria projects that by the year 2020, the country's generation capacity would be in excess of 40GW (40,000MW), and the energy mix will constitute 69% thermal generation; 17% hydro; 10% coal; and about 4% of renewable [20].

Faced with the myriad of problems, stakeholders in the nation's electricity industry have now adopted the embedded generation concept as a urgent strategy to immediately procure electricity for customers, while waiting for national grid power improve significantly. Presently, the installed power generation capacity in Nigeria is 12,522MW; out of which 10,592MW is gas fired; and 1,930MW is from hydro. It is worth noting that out of the total installed capacity, the maximum peak generation by power plants as at February 13, 2018 was 4,083.34MW [21]. The integration of distributed sources into existing networks is becoming a viable option in modern power system networks considering its numerous advantages such as: cost saving due to on-site production which avoids transmission and distribution costs and hence, a saving of about 30% of the cost of delivered electricity. Also, on-site production of energy reduces loss of generated power due to transmission and distribution [22].

Further, embedded generation improves the level of power quality and reliability aside from serving niche applications such as remote sites. Distributed or embedded generation can also provide standby generation (emergency power), peak shaving capability and cogeneration. It also improves the efficiency of the power grid by having multiple micro-sources added to the system [23]. The Nigeria Electricity Regulatory commission (NERC)'s embedded generation regulation allows an independent power producer to integrate power with the network of the local distribution company without going through the trouble of connecting to the transmission network. This allows distribution companies to procure small power and dedicate it to ring – fenced customers 24/7 electricity supply [24].

Power investment expert said the Federal, states and private sector power plant investors have immense benefits to gain from investing in embedded generation. He also noted that embedded generation has the prospects to achieve national aspirations within a shorter time; reduce technical losses because of proximity to the network, and deepen the electricity market – capacity, standards, contracts and more bankable deals [25]. With embedded generation, Discos would have access to more power supply more cash flows while more customers would be willing to pay. According to him, "It presents opportunity for discos to improve distribution network – either by themselves or by EGs – and performance. The recent Regulations on Embedded Generation and independent Electricity Distribution Networks issued by the Nigeria Electricity Regulation Commission (NERC) (Pursuant to its power to make regulations under the Electric power sector reform Act, 2005) confirm the commitment of the Federal Government to tackling the inefficiencies in Nigeria power sector [26].

The Regulations were issued as part of the effort to improve generation and realize the Federal Government's target of achieving a generation capacity of 40,000MW by the year 2020. Another compelling reason was the need to provide electricity for about 55% of Nigeria population without access to electricity as many communities have no access to the national grid. Furthermore, many of the area connected to the grid experience very frequent power outages [27]. For power developers, IPPs can sell excess power to discos. It guarantees cost reflective tariff. There is immense potential market for power supply to housing estates, industrial estates/clusters, state governments and telecom installations. There is also option to also supply power to eligible customers. No distribution lines are ordinarily required when connected to a disco. No transmission costs would be incurred and no distribution license required according to [28].

The chairman of Nigeria Electricity Regulatory Commission (NERC) recently issued license for embedded power generation to Bauchi state government. Also, Eko Electricity Company of Nigeria has concluded plans to generate over 1000MW of electricity through embedded power generation model 2019 [29]. Embedded, dispersed or distributed generation (DG) are synonyms for an upcoming probable paradigm shift in electric power generation, including smaller-scale generation, combined heat and power (CHP), energy storage, load management and energy efficiency [30]. Recently, the Lagos State government is working towards passing into law Lagos State Electric Power Sector Reform Bill.

2.2.3. Benefits of Embedded Generation

To the end-user, Embedded Generation offers numerous benefits when properly sited, sized, installed, and operated. Among these benefits are:

- It can significantly improve reliability of energy supply, providing a very critical solution to businesses and industries with continuous processes and in situations where health and safety is impacted.
- The wide variety of EG technologies offers the opportunity of selecting the right energy solution at the right location. EG technologies can provide a stand-alone power option for areas where transmission and distribution infrastructure does not exist or is too expensive to build.
- It may offer efficiency gains for on-site applications by avoiding line losses, and using both electricity and the heat produced in power generation for processes or heating and air conditioning.
- Its flexibility of operation because of small modular units enables savings on electricity rates by self-generating during high-cost peak power periods and adopting relatively low cost interruptible power rates;
- Benefits for environmental quality may come from embedded generation's role in promoting renewable energy sources, less-polluting forms of fossil energy, and high efficiency technologies. EG allows power to be delivered in environmentally sensitive and pristine areas by having characteristically high efficiency and near-zero pollutant emissions;
- EG limits capital exposure and risk because of the size, siting flexibility, and rapid installation time afforded by the small, modularly constructed, environmentally friendly, and fuel flexible systems;
- Unnecessary capital expenditure can be prevented by closely matching capacity increases to growth in demand;
- EG avoids major investments in transmission and distribution system upgrades by siting new generation near the customer;
- It offers a relatively low-cost entry point into a new and competitive market;
- Opens markets in remote areas without transmission and distribution systems, and areas without power because of environmental concerns.
- EG responds to increasing energy demands and pollutant emission concerns while providing low-cost, reliable energy essential to maintaining competitiveness in the world market; and
- Establishes new industry worth billions of dollars in sales and hundreds of thousands of jobs and enhances productivity through improved reliability and quality of power delivered, valued at billions of dollars per year.

3. Suggested Solutions to erratic Power Supply in Nigeria

There is a need to create an investor friendly environment for embedded generation solutions, which will allow for government support, tax incentives and reliefs to encourage investments in the power sector. Examples of efforts in this regard include the National Renewable Energy and Energy Efficiency Policy for the Electricity Sector newly approved by the Federal Government; and the Renewable Energy Master Plan developed for Nigeria [29].

It is important for Nigeria to intensify efforts in seeking funding support from the World Bank, Development Finance Institutions and NGOs for the purpose of electrifying Nigeria. The Bangladesh off-grid success story was possible through the support of these institutions.

Nigeria should invest in renewable energy resources. Nigeria is reported to have potentials for harvesting strong wind energy throughout the year from coastal areas like Lagos State through Ondo, Delta, Rivers, Bayelsa to Akwa-Ibom States [31]. Though the initial cost of installing a turbine generator is huge but thereafter the benefits are huge. This will go a long way to improve power supply in Nigeria.

The embedded generation concept should be decentralized among the federal, state and local governments. There should be an increase in energy tariff if Nigeria is to achieve constant, stable and nationwide supply. The customers are willing to pay as long as the power is stable and reliable.

4. Conclusion

Proper funding of the power generation station in Nigeria would improve power generation. The gas lines conveying gas to gas station should be remotely secured against vandalization. The huge outstanding

IOP Conf. Series: Materials Science and Engineering **413** (2018) 012037 doi:10.1088/1757-899X/413/1/012037

electricity bills owned by individuals and government agencies must be paid. Also, the existing transmission network must be adequately maintained and revamped to accommodate much more load than current peak electricity production. Investment in renewable energy resources will help in power stability in Nigeria. Finally, integration of embedded or distributed generation into the existing injection substation will guarantee stable power supply in Nigeria. This has worked in other countries of the world and so it can also work in Nigeria.

References

[1] Newsletter> Overview of the NERC's Regulations on Embedded generation and Independent Electricity Network Accessed 27th January, 2018
[2] Newsletter> Overview of the NERC's Regulations on Embedded generation and Independent Electricity Network Accessed February ,2018
[3] CIGRE, International Council on Large Electricity Systems, http://www.cigre.org Accessed March 2017
[4] G. Pepermans, J. Driesen, D. Haeseldonckx, R. Belmans, W. D'haeseleer, "Distributed generation: definition, benefits and issues", Energy Policy Vol. 33, p.p. 787–798, 2005.
[5] CIGRE, International Council on Large Electricity Systems, http://www.cigre.org
[6] M. Hager et al., "The impact of distributed energy resources on distribution system protection". STRI AB, Ludvika, Sweden. http://www.stri.se
[7] Report from Nigeria Electricity Regulatory Commission. Accessed July,2017
[8] A sambo and E commission, "Electricity Generation and present challenges in the Nigeria power sector", April 2017 and 2010. Accessed March, 2016.
[9] Nigeria Power Baseline Report -nesistats.org. Accessed May ,2016
[10] IKEDC Request for Expression of Interest for Embedded Power Generation Project, 2015
[11] A sambo and E commission, "Electricity Generation and present challenges in the Nigeria power sector", April 2017 and 2010. Accessed March,2016
[12] A sambo and E commission, "Electricity Generation and present challenges in the Nigeria power sector", April 2017 and 2010. Accessed May,2018
[13] Nigeria Electricity Regulatory Commission (NERC) Accessed November 1,2017
[14] Nigeria Electricity Regulatory Commission (NERC. Accessed May 1,2017
[15] IKEDC Request for Expression of Interest for Embedded Power Generation Project, 2015.
[16] Report of CIRED Working Group No 4 on Dispersed Generation, CIRED 1999, Nice. Available at: http://www.cired.be Accessed January,2016
[17] www.worldometers.info), Accessed March ,2018
[18] Nigeria Power Baseline Report -nesistats.org). Accessed May ,2016
[19] IKEDC Request for Expression of Interest for Embedded Power Generation Project, 2015.
[20] Solutions for the Connection and Operation of Distributed generation, EA Technology Ltd, July 2003. Available at www.distributed-generation.gov.uk Accessed August 2017
[21] IKEDC Request for Expression of Interest for Embedded Power Generation Project, 2015 Accessed May 2017
[22] Report from Nigeria Electricity Regulatory Commission. Accessed June,2017
[23] IKEDC Request for Expression of Interest for Embedded Power Generation Project, 2015 Accessed May 2017
[24] I. A Ikem, I. Ibeh O. E Nyong, S. A Takim, M Engineering, and C. River," Integration of Renewable Energy Sources to the Nigeria National Grid – way out of power crisis. Accessed March 13,2017
[25] Report from Nigeria Electricity Regulatory commission. Accessed July,2017
[26] Report from Nigeria Electricity Regulatory Commission. Accessed February 2016
[27] A Sambo and E commission, "Electricity Generation and present challenges in the Nigeria power sector", April 2017 and 2010. Access May 2018
[28] Report from Nigeria Electricity Regulatory Commission. Accessed May 2018
[29] Report from Nigeria Electricity Regulatory Commission. [NERC] Accessed December 2018

IOP Conf. Series: Materials Science and Engineering **413** (2018) 012037 doi:10.1088/1757-899X/413/1/012037

[30] Nigeria Electricity Regulatory Commission (NERC) Accessed November 1,2017
[31] IKEDC Request for Expression of Interest for Embedded Power Generation Project, 2015 Accessed May 2017
[32] Report from Nigeria Electricity Regulatory Commission. Accessed July,2017
[33] Agbetuyi A.Felix, Akinbulire T.O, Abdulkareem A and Awosope COA " Wind Energy Potential in Nigeria" International Electrical Engineering Journal(IEEJ)Vol.3(2012) No1,pp595-601, ISSN 2078-2365.

Effect of Pre-Shelling Treatment on Physical and Mechanical Properties of Cashew Nut

Kilanko Oluwaseun[1], Ojolo Sunday Joshua[2], Inegbenebor Anthony Omieraokholen[1], Ilori Titus Adeyinka[3], Leramo Richard Oluwafemi[1], Babalola Philip Olufemi[1], Onwordi Patrick Nwanne[1]

[1]Mechanical Engineering Department, Covenant University, PMB 1023 Ota, Ogun State, Nigeria
[2]Mechanical Engineering Department, University of Lagos, PMB 56, Akoka, Yaba, Lagos State, Nigeria
[3]Department of Agricultural Engineering, Federal College of Agriculture, Moor Plantation, Apata, PMB 5029, Ibadan, Oyo State, Nigeria
Corresponding author: oluwaseun.kilanko@covenantuniversity.edu.ng

Abstract. This study was conducted to investigate the effect of roasting in hot oil on physical and mechanical properties of large (26-35mm), medium (23-25mm) and small (18-22mm) cashew nuts at moisture content of 7.00% w.b. The physical properties considered were length, width, thickness, geometric mean diameter, sphericity, true density, bulk density, porosity, mass, and coefficient of friction. The mechanical properties considered were, rupture force, deformation at rupture point, energy used for rupture and young modulus determined from compression test carried out in three major direction of loadings. It was observed that all the dimensional characteristics, mass and densities of cashew nuts before roasting were significantly different ($P < 0.05$) from after roasting. The pre-treatment has significant effect on coefficient of friction of cashew nut which has higher values on mild-steel and wood than other materials. The compression test results showed that the rupture force decreases with decrease in nut size for the three loading directions, however the transverse loading requires higher force than longitudinal and lateral loading for the three nut sizes. For large and medium size nuts, the highest rupture energy was needed in transverse loading direction, followed by lateral loading direction. The compression along the longitudinal direction has the least deformation for the three nut sizes. For the three nut sizes, the highest Young Modulus was exhibited in longitudinal loading direction followed by transverse loading direction.
Keywords: cashew nut, nut size, physical properties, coefficient of friction, rupture force, hot-oil roasting

1. Introduction

Cashew nut ((*Anacardium Occidentale Linn.*)) is one of the topmost edible tree nuts of the world. It consists of an outer shell (Epicarp), honey combed structure (Mesocarp), inner shell (Endocarp), Testa and Kernel. The Epicarp is greenish to pinkish brown depending on the degree of dryness. The mesocarp contains a natural resin, known commercially as a Cashew Nut Shell Liquid (CNSL). The CNSL is viscous and blister the human skin. The endocarp is hard but brittle. It protects the kernel from the Natural Resin. The epicarp, mesocarp and endocarp are known as pericarp [1- 2]. The art of processing cashew nuts is to shell them and extract the kernels 'whole' without breaking or damaging them. The methods employed by cashew nut farmers and manufacturers to shell the nuts have mainly included simple shelling devices, which usually are not efficient in their performances, and few cases of highly complex and expensive shelling machinery, usually through imports from Europe [3]. For the conditions in Nigeria which is among the largest producers of cashew nut in the world, there is a need of local shelling machinery with efficient performance and economical cost. Determining the physical and mechanical properties of cashew is of paramount importance to the development of the local shelling machinery.

Shelling of cashew nut to remove the kernel inside it is one of the major processes of the nut. Developing a device or machine to do this requires a comprehensive knowledge of the physical and mechanical properties of the cashew nuts and their mechanical behavior under pressure. Data of physical and mechanical properties of cashew nut such as length, width, thickness, geometric mean diameter, sphericity, true density, bulk density, porosity, mass, coefficient of friction, angle of repose, compressive loads (in lateral, longitudinal and transverse directions), energy at yield, deformation and young modulus are needed for designing shelling machines, harvesting, handling and storage systems for cashew nuts. Some of these data are also needed as input to mathematical model developed for prediction of force required to break the cashew nut.

Physical properties of cashew nuts have unique characteristics which make the nut to behave differently from mechanical or engineering materials. For many biological materials, their awkward or irregular shapes make difficult the analysis of their behavior. For the design of process equipment, cashew nut physical properties are needed. In controlling machines and processes, sensors are used. In developing such sensors, the information of the physical properties of the nut plays an important role. It also helps in storage, handling and quality difference detection during harvesting. Physical properties have also helped in designing many types of separation, grading and cleaning equipment [4]. In the design of storage structures, processing equipment and machines, the knowledge of physical properties make up an essential and important engineering data [5-6].

The behaviors of cashew nut in compression between two parallel plates is revealed by its mechanical properties. For textural analysis and good understanding of the fracture mechanics of the cashew nut, the investigation of mechanical properties is highly needed. The storage, processing techniques and shelling of the cashew nut all require its mechanical properties. In the design of machines for agricultural products, one of the essential considerations is to make sure that the stresses in machine elements do not go beyond the agricultural material's strength. To investigate damage caused by handling and harvesting of the vegetables, fruits, seeds, grains and cashew nut, the force deformation relationship of biological materials is being employed. The results or data of the behavior of an agricultural material from its test helps a great deal in the design of engineering equipment for its handling [6].

2. Materials and methods

Raw cashew nuts were procured from farm gate at Ilorin area of Kwara State, Nigeria at the average moisture content (MC) of 7.00% wet basis. Foreign materials such as stones, spoilt and immature nuts, leaves were separated from the nuts procured for use. The cashew nuts were graded into three different nut sizes using their axial dimensions: large (26-35mm), medium (23-25mm), and small (18-22mm) following the method developed by Balasubramanian [7]. 300 cashew nuts were selected from each nut size at random and the physical and mechanical properties of the nuts were determined before and after roasting in hot palm-kernel oil. The roasting time used was 90s [8] and the temperature used was 190°C-210°C [9-10].

The equipment used were: Carrera Precision Vernier Digital Caliper which measure to an accuracy of 0.01mm; oven (masterchef); Electronic Balance which measure to an accuracy of 0.0001g; measuring cylinder; thermometer; incline plane; stopwatch; a wire mesh basket designed to contain 25 nuts; universal testing machine (Testometric-AX model) which has a capacity from 0 – 25 kN and oil-bath of palm-kernel oil.

The cashew nuts were treated by frying in a bath of hot palm-kernel oil. The raw cashew nut was kept and used as the control. The cashew nuts were prepared and roasted as described below.
300 cashew nuts were selected from each nut size (i.e. large, medium and small). From these nuts, 25 nuts were selected, labeled and put in wire mesh basket which was divided into 25 compartment (see Figure 1). This was done in order to track each nut before and after roasting. The wire mesh basket was dipped into bath of hot oil at the temperature of 190°C-210°C for 90 seconds. This was done several times to complete the roasting of 300 nuts for each of the large, medium and small nut sizes. After roasting, the nuts were allowed to cool naturally for 18 hours [9, 11-12].

Fig. 1 Wire mesh basket

The following properties of the nuts were measured for the same nut before and after roasting in hot oil using standard methods [13] – length, width, thickness, geometric mean diameter, sphericity, true density, bulk density, porosity, mass and coefficient of friction on the following materials: glass, mild steel, galvanized steel, stainless steel, and ply wood. The mechanical properties were determined for only the roasted nuts.

(i) For each of the nuts, the principal axes were measured before and after roasting. The length-a of the nut was measured longitudinally. The width-b and thickness-c were measured perpendicular to the major axis of the nut. The geometric mean diameter (*GMD*) and sphericity (S_p) were determined by the following equations as documented by Mohsenin [13]:

$$GMD = (abc)^{1/3}; \quad ---(1)$$

$$S_p = \frac{(abc)^{1/3}}{a} \times 100\% \quad --- (2)$$

The dimensions were determined for all the nuts (300) of each nut size.

(ii) To determine the true density, the water displacement method was used. The mass of the nut was weighed and immersed into the measuring cylinder with known volume of water. The true density ρ_t of the cashew nut was determined by dividing the mass of the nut by the volume of water displaced when the nut was immersed in water (equation 3). For each of the nut size, true density of 60 nuts were determined.

$$\rho_t = \frac{mass}{volume} \quad --- (3)$$

(iii) The mass of a cylinder was measure. Its volume was determined by filling it to the brim with water of known volume. The mass of the cylinder was determined again after it was filled with cashew nut in order to determine the mass of nut that fill the cylinder. The mass of the nut that fill the cylinder was divided by the volume of water that filled the cylinder to give the bulk density (ρ_b) [13]. This was repeated 60 times for each of the nut size.

$$\rho_b = \frac{weight\ of\ nuts\ in\ the\ container}{volume\ of\ the\ container} \quad --- (4)$$

(iv) With known values of true density (ρ_t) and bulk density (ρ_b), the porosity (p) of the cashew nut was determined using the equation by Mohsenin [13].

$$p = \frac{\rho_t - \rho_b}{\rho_t} \times 100\% \quad --- (5)$$

(v) The angle for the coefficient of friction for cashew nut was determined using the inclined plane method. This was done by placing each of the surfaces on the inclined plane, putting a cashew nut on it and then beginning to lift the inclined plane slowly and carefully until the nut just started sliding. The angle of inclination was measured by the calibrated protractor attached to the side of the inclined plane. Taking the tangent of the angle of inclination at which the cashew nut begins to slide on a surface gives the Coefficient of static friction according to the method described by Ogunsina & Bamgboye [9] and Dutta *et al.* [14]. For the ply wood the angle of inclination was determined in two directions. One was in the direction of the grain of the wood and the other was in the direction across the grain.

$$Coefficient\ of\ friction = \tan\theta \quad --- (6)$$

For each of the nut size, 60 nuts were used to determine the coefficient of friction for each material.

(vi) The moisture content of the raw nuts was determined using ASAE standard 2003. The nuts were placed in moisture dishes and weighed before they were put in oven at 103 C for 6 hours. At the end of oven drying, the final weight of the moisture dishes and the nuts was weighed after reaching room temperature. The moisture content (wet basis) was determined by dividing the loss in weight due to oven drying by the initial weight [15].

$$Mc = \frac{M_l}{M_i} \times 100 - - - - (7)$$

Where Mc = moisture content; M_l = moisture loss; M_i = Initial weight of the sample.

(vii) Compression tests were carried out on the roasted cashew nuts between two parallel plates at a constant rate of 50 mm min^{-1} to determine the force at yield, energy at yield, deformation and Young's modulus at which the cashew nut would crack open without damaging the kernel within. The roasted cashew nuts were compressed in three different orientations (Figure 2): transverse (orientation 1), longitudinal (orientation 2), and lateral or natural (orientation 3). For each nut size, 60 nuts were selected and compressed in each orientation.

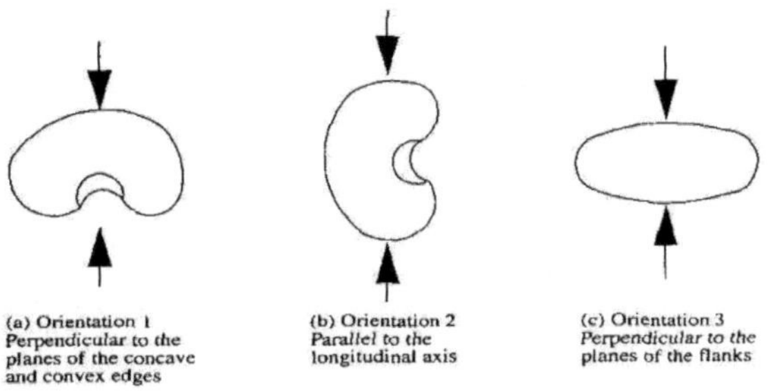

Fig. 2 Orientations of nut in relation to the direction of loading
(Source: Oloso and Clarke, (1993))

3. Results and discussion

Table 1 shows the physical properties of the cashew nuts before and after roasting in bath of hot oil. It should be noted that all the nuts were marked with labels, and their properties were measured before and after roasting. Table 2 shows the t-statistics of the paired sample t-test used to compare the properties before and after roasting. The table shows that the pre-treatment affects the properties of the cashew nut significantly. The negative t-value indicates an increase in the mean value of the properties of the nut when the nut was roasted. The positive t-value indicates a decrease. For all the large, medium and small nut sizes considered, it was observed that there was increase in the dimensions when the nuts were roasted. The increase in the dimensions of the nuts is attributed to the expansion of the shell as it came in contact with the hot oil. The dry shell ruptures and releases the cashew nut shell liquid (CNSL) in the oil bearing cells with high pressure, thus expanding the shell and reducing the weight because of the loss of CNSL. The increase in dimensions of the nut is contrary to the observations made by Balasubramanian [7] who carried out a similar work on raw cashew nut grown in India and Ogunsina & Bamgboye [9] who carried out similar work on cashew nuts grown in Nigeria. In their works, there was a decrease in the dimensions of the nuts which was due to the loss of moisture content and shrinkage during roasting. The difference in the observations was as a result of difference in the moisture content of the raw cashew nut that was roasted. In this work, the dry cashew nut was roasted (at 7.00%w.b.) while the moisture content of the cashew nut roasted in their work was greater than 12.0%w.b. The removal of the moisture present in the pericarp of the cashew nut during roasting led to the shrinkage of the nut. Table 2 shows that the difference in the means of length, width, thickness, geometric mean diameter and sphericity of raw and roasted nuts of all nut sizes are all significant ($P < 0.05$). The table shows that the true and bulk densities are influenced by the treatment. There was significant difference ($P < 0.05$) in the true and bulk densities of all nut sizes. The loss of CNSL of the cashew nut into the oil bath during roasting obviously account for the reduction in weight which in turn led to reduction in density of the nuts. The difference in the means of porosity of medium size nut is significant ($P < 0.05$), however there was no significant difference in the means of porosity of large and small nut sizes. The pre-treatment affects the coefficients of friction significant ($P < 0.05$), however for large size nuts on mild steel and stainless steel and medium size nuts on glass and galvanized steel, there was no significant difference in the means of the coefficients of friction. Tables 1 and 2 show that there was an increase in the coefficient of friction of medium and small size nuts. It was observed that the coefficient of friction on mild-steel and wood are higher than other materials for all the sizes of the nuts. The effect of the pre-treatment on the properties of the cashew nut can be further seen on Figure 3.

Table 1 Physical properties of cashew nuts before and after roasting

Properties		**Large**		**Medium**		**Small**	
	N	**Raw**	**Roasted**	**Raw**	**Roasted**	**Raw**	**Roasted**
Length (mm)	300	32.73(1.9)	33.56(1.9)	30.82(1.4)	31.66(1.5)	25.60(1.7)	27.32(1.3)
Width (mm)	300	24.97(2.0)	26.10(1.9)	23.46(1.2)	24.39(1.2)	20.44(1.3)	21.58(1.0)
Thickness (mm)	300	17.18(1.6)	18.74(1.6)	16.26(1.3)	17.68(1.2)	14.53(1.3)	16.14(1.1)
GMD (mm)	300	24.10(1.5)	25.39(1.4)	22.71(0.9)	23.88(0.9)	19.64(1.1)	21.18(0.9)
Sphericity (%)	300	73.71(3.6)	75.73(3.4)	73.80(3.6)	75.52(3.2)	76.85(3.5)	77.60(2.7)
True density (g/cm^3)	60	0.997(0.08)	0.727(0.09)	1.04(0.07)	0.733(0.08)	1.06(0.1)	0.685(0.09)
Bulk density (g/cm^3)	60	0.63(0.02)	0.467(0.03)	0.634(0.02)	0.462(0.02)	0.623(0.02)	0.400(0.03)
Porosity (%)	60	36.13(6.90)	34.77(9.99)	38.95(5.07)	36.19(7.7)	40.89(7.11)	40.62(9.5)
Mass of the nut (g)	60	7.81(0.94)	6.75(0.92)	5.80(0.59)	5.06(0.61)	3.82(0.67)	3.38(0.58)
COF							
Glass	60	0.38(0.09)	0.27(0.09)	0.26(0.07)	0.27(0.1)	0.29(0.07)	0.36(0.12)
Mild Steel	60	0.54(0.09)	0.52(0.09)	0.47(0.06)	0.54(0.07)	0.51(0.08)	0.59(0.08)
Galvanised Steel	60	0.46(0.08)	0.42(0.07)	0.44(0.05)	0.42(0.06)	0.44(0.06)	0.48(0.06)
Stainless Steel	60	0.45(0.08)	0.44(0.08)	0.43(0.06)	0.48(0.07)	0.43(0.07)	0.57(0.1)
Wood Along	60	0.46(0.09)	0.57(0.09)	0.44(0.07)	0.60(0.09)	0.46(0.07)	0.59(0.08)
Wood Across	60	0.47(0.07)	0.58(0.08)	0.44(0.07)	0.59(0.11)	0.46(0.06)	0.62(0.09)

GMD - Geometric mean diameter, COF – Coefficient of friction

Table 2 Effect of pre-shelling treatment on Physical properties of cashew nuts

t-statistics				
Properties	**df**	**Large**	**Medium**	**Small**
Length (mm)	299	-12.648*	-20.673*	-35.078*
Width (mm)	299	-16.492*	-22.082*	-21.200*
Thickness (mm)	299	-29.360*	-30.073*	-31.660*
Geometric mean diameter (mm)	299	-30.161*	-40.595*	-42.930*
Sphericity (%)	299	-16.936*	-16.544*	-6.550*
True density (kg/m^3)	59	23.470*	24.177*	32.364*
Bulk density (kg/m^3)	59	46.017*	55.824*	59.149*
Porosity (%)	59	1.387	2.366*	0.334
Mass of the nut (g)	59	14.052*	9.411*	14.554*
Coefficient of Friction				
Glass	59	6.641*	-0.128	-4.608*
Mild Steel	59	1.344	-5.624*	-5.922*
Galvanised Steel	59	2.558*	1.752	-3.966*
Stainless Steel	59	1.135	-4.506*	-9.681*

IOP Conf. Series: Materials Science and Engineering **413** (2018) 012038 doi:10.1088/1757-899X/413/1/012038

Wood Along	59	-6.464*	-11.193*	-9.141*
Wood Across	59	-9.430*	-10.684*	-11.041*

*t-Statistics are significant at 5% probability level

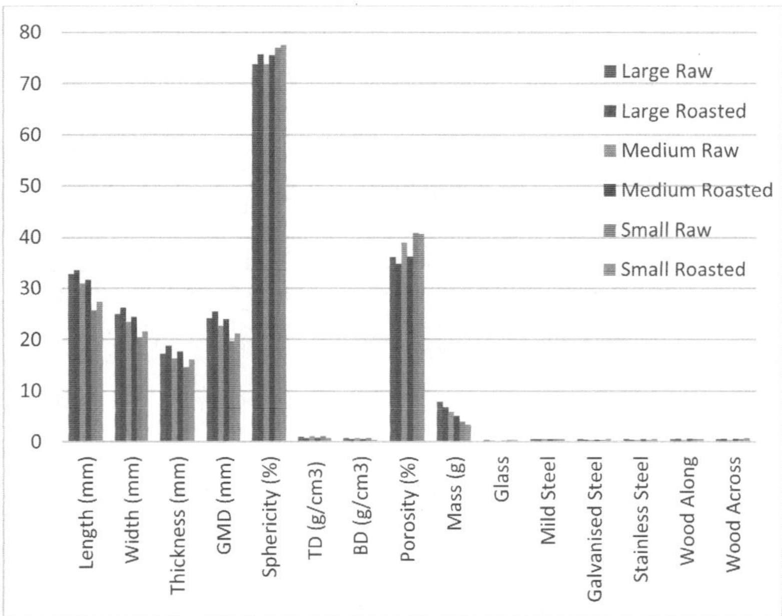

Fig. 3 Effect of pre-treatment on the physical properties of cashew nut

The compression test was carried out to determine the mechanical properties of cashew nut which include rupture force, energy absorbed during the rupture, deformation at rupture point and young modulus of the nut on three major direction of loadings (transverse, longitudinal and lateral). The results for the compression test is presented in Table 3. The rupture force decreases with decrease in nut size for the three loading directions, however the transverse loading requires higher force than longitudinal and lateral loading for the three nut sizes. This indicates that the larger nuts may require higher force to fracture than the smaller nuts, as it was also observed in sunflower seeds [16] and Jatropha curcas seed [17].

For the large and medium size nuts, the highest rupture energy was needed in transverse loading direction, followed by lateral loading direction. However, for the three nut sizes, longitudinal loading requires the least rupture energy. Similar result was reported by Ojolo and Ogunsina [18], however, it was not reported for different loading directions.

The compression along the longitudinal direction has the least deformation for the three nut sizes. This indicates that the nuts need lower compression in this direction to rupture. For the three nut sizes, the highest Young Modulus was exhibited in longitudinal loading direction, followed by transverse loading direction.

Table 3 Mechanical properties of the roasted cashew nuts

LARGE					
Compressive load	**N**	**Force (N)**	**Energy at Yield (N.m)**	**Deformation (mm)**	**Young modulus (N/mm²)**
Lateral	60	345.12(119.4)	1.27(0.58)	7.17(1.86)	1724.44(980.17)
Longitudinal	60	362.44(139.60)	0.86(0.75)	4.72(3.39)	3654.28(1367.89)
Transverse	60	527.14(247.18)	2.07(1.17)	7.78(1.95)	2125.44(714.38)
MEDIUM					
Compressive load	**N**	**Force (N)**	**Energy at Yield (N.m)**	**Deformation (mm)**	**Young modulus (N/mm²)**
Lateral	60	278.03(105.90)	1.09(0.49)	7.81(1.68)	1421.86(327.06)
Longitudinal	60	276.48(142.14)	0.69(0.39)	5.05(1.40)	3247.61(1016.19)
Transverse	60	379.31(216.03)	1.43(0.87)	7.56(1.75)	1905.56(442.11)
SMALL					
Compressive load	**N**	**Force (N)**	**Energy at Yield (N.m)**	**Deformation (mm)**	**Young modulus (N/mm²)**
Lateral	60	273.69(91.87)	0.98(0.42)	7.03(1.62)	1385.08(384.23)
Longitudinal	60	175.18(120.79)	0.40(0.29)	4.74(0.92)	2842.90(1546.29)
Transverse	60	285.42(161.94)	0.88(0.55)	6.33(1.85)	1878.27(541.90)

4. Conclusions

1. The Physical properties of cashew nut was affected by pre-treatment by roasting in hot oil.
2. The dimensions of dry cashew nut increased with roasting and decreased in weight.
3. The rupture force increases with increase in nut size and this was found to be highest among the three directions of loadings when the nuts were loaded transversely. This finding will be useful in the development of shelling equipment for cashew nuts.

Acknowledgements
The author acknowledges the management and staff of Olam Cashew Nut Processing Factories in Kilometre 3, Afon Road Ogbondoroku, Ilorin West, Kwara State, and Olam Cashew Nut Processing Factories, Iseyin, Oyo state, Nigeria for granting access to their facilities during this investigation. Mr. Ufot Eddidiong and Mr Ekwere Akan (students of Covenant University) are highly acknowledged for their great contributions in the determination of the properties of cashew nut. I also appreciate the management of Covenant University for the conference support they give to this paper.

References

[1] Russel D.C. (1969). Cashew nut processing. Agricultural Services Bulletin 6, Agricultural Services Division, F.A.O., Rome, 1969. p 21-50
[2] Bambang S. (2000). Roasting experiment of cashew nut in traditional industry. Journal of Agricultural Technology, December 2000. Vol. 1, Issue 3, p 64-69.
[3] Thivavarnvongs T., Okamoto T. and Kitani O.(1995). Development of compact sized cashew nut shelling machinery (part 1); Synthesis of effective manual and semi-automatic shelling methods. *J. Japanese Soc. Agric. Machinery*. Volume 57, Issue 2, p 57-76.
[4] Teye, E., and Abano, E. E. (2012) Physical properties of two varieties of sweet potato

IOP Conf. Series: Materials Science and Engineering **413** (2018) 012038 doi:10.1088/1757-899X/413/1/012038

grown in the coastal savannah zone of Ghana. International Journal of Science and nature 3 (1) 105-109.

[5] Esrif I. and Halil, Ü (2007) Moisture-dependent physical properties of white speckled red kidney bean grains. Journal of food Engineering. 82, 209-216

[6] Bart-Plange A., Mohammed-Kamil A. P., Addo A. and Teye E. (2012b). Some physical and mechanical properties of cashew nut and kernel grown in ghana. *International Journal of Science and Nature*, Volume 3, Issue 2, p 406-415.

[7] Balasubramanian D. (2001). Physical properties of raw cashew nut. *J. Agric. Eng. Res.*, Volume 78, Issue 3, p 291-297.

[8] Oloso A.O. and Clarke B. (1993). Some aspects of strength properties of cashew nut. *Journal of Agricultural Engineering Research, UK*. Volume 55, Issue 1, p 27-43.

[9] Ogunsina, B.S., Bamgboye, A.I., 2007. Effect of pre-shelling treatment on the physical properties of cashew nut (Anacardium occidentale). International Agrophysics 21 (4), 385–389.

[10] Araujo, M. C., and A. C. O. Ferraz. 2008. Physical and mechanical characteristics of the endocarp and kernel cashew CCP 76 'before and after heat treatment. *Journal of Agricultural Engineering*, 28(3): 565-578.

[11] Andrighetti L., Bassi G.F., Capella P., De Logu A.M., Deolalikar A.B., Haeusler G., Franca F.M.C., Rivoira G., Vannini L. and Deserti R. N. (1994). *The World Cashew Economy, (2nd ed.)*. Linchiostroblu, Italy. p11-16, 33-48, 79-80.

[12] Balasubramania D. (2006). Improving whole kernel recovery of cashew nuts processing specific to Nigerian Nuts. *Journal of Agricultural Mechanization in Asia, Africa and Latin America*. Volume 37, Issue 1, p 58-64.

[13] Mohsenin N.N. (1986). Physical Properties of Plant and Animal Materials. Gordon and Breach Publ. Inc., New York.

[14] Dutta S.K., Nema V.K., and Bhardway R.K. (1988). Physical properties of grain. *J. Agric. Eng. Res.*, Volume 39, Issue 9, p 259-268.

[15] ASAE Standards, 2003. Moisture measurement - unground grain and seeds, St. Joseph, Michigan, ASAE S, 352.2 (593).

[16] Subramanian, R., Sastry, M.C.S., Venkateshmurthy, K., 1990. Impact dehulling of sunflower seeds: effect of operating conditions and seed characteristics. J. Food Eng. 12, 83–94.

[17] Karaj S and Müller J (2010). Determination of physical, mechanical and chemical properties of seeds and kernels of Jatropha curcas L. J. Industrial Crops and Products. 32, 129–138

[18] Ojolo S.J. and Ogunsina B.S. (2007). Development of a Cashew Nut Cracking Device. Agricultural Engineeering International: *The CIGR Ejournal*. Manuscript PM 06 030. Vol. IX. June, 2007.

IOP Conf. Series: Materials Science and Engineering **413** (2018) 012039 doi:10.1088/1757-899X/413/1/012039

A Review of Onshore and Offshore Wind Energy Potential in Nigeria

Adedipe O.[1,2], Abolarin M.S.[1,2], Mamman R. O[1]

Department of Mechanical Engineering, School of Infrastructure, Process and Engineering Technology, Federal University of Technology, Minna Nigeria.
E-mail address: talk2rab@yahoo.com

Abstract. As a result of continued growth in demand for energy in Nigeria and the need to augment fossil fuel sources with renewable energy, this paper review research trends on Offshore and Onshore wind energy potentials in Nigeria, for possible power generation. Researches on development and design of small scale-wind turbine as well as their application for small-scale electrification were also discussed. This paper further examined the suitability of different geographical zones in Nigeria for offshore and onshore power generation, using results of prior researches and also stressed that Wind Energy Technology (WET) is widely underutilized in Nigeria. Several factors militating against this development were briefly highlighted and some suggestions were raised for future work within the realm of wind energy technology.
Keywords: Offshore wind; Onshore wind; Wind speed; Windfarm.

1. Introduction

In order to fulfill the requirement of Sustainable Development Goal (SDG) 7, which encourages the United Nation member states to focus on access to ecofriendly, sustainable and readily available renewable energy source [1], it becomes necessary to assess the potential in the renewable energy sources available at our disposal such as Wind, Waves, Solar, Tides and Geothermal. Wind as a source of renewable energy is gaining prominence around the world since it can be can be harness in small or commercial to meet the present-day energy demand. Wind Energy Technology (WET) serves as a suitable supplement and alternative to the rising cost of power generation from fossil fuel source and as well contribute towards global legislation against Greenhouse Gas (GHG) emission.

Wind as a source of power generation provides clean, ecofriendly, nontoxic and readily available renewable energy source. For effective generation of energy from wind, certain conditions must be met such as; availability of substantial amount of wind within a site, appropriate statistical analysis of the wind to study the nature of wind speed distribution within a site, multi-criteria assessment of the site to ensure the viability of the location for installation of wind energy conversion systems (WECs), cost implication of installing wind energy conversion systems and other logistic analysis. Furthermore, wind is stochastic in nature, hence it is necessary to develop models in accessing long term variation of wind within a particular site before concluding on whether the site poses the requisite wind resources for power generation.

Another important parameter in wind energy studies is the wind power density. Wind power density is the amount of energy available at a particular site for conversion to useful wind energy by a wind turbine and it is measured in watts per meter square, and it is a direct function of the mean value of the wind speed available in that particular site. It is necessary to ascertain the wind power density of particular site prior to installation of wind turbines.

In order to convert the available wind resources in a region into useful energy, an important device; Wind turbine is required. A wind turbine converts the available wind speed in a particular location into wind energy which is then fed into the electricity supply systems. Wind Energy Conversion Systems (WECS), operates at its maximum capacity if its design is based on the wind characteristics of a particular site [2].

1.1 Wind Energy Research Trends in Nigeria

Wind as a renewable energy source is underutilized in West Africa most especially in Nigeria. According to a recent study by African Development Bank (AFDB) [3], North African countries like Morocco, Egypt and Tunisia remain the leaders in wind energy market in Africa, while South and Eastern African countries are projected to reduce the gap soon. Central Africa and West African countries are still lacking in terms of wind farm implementation. Taking Nigeria as a case study, the notion to seek for a sustainable alternative to the intermittent energy situation of Nigeria has prompted the government as well as independent researchers to evaluate the nation's potentials for power generation using wind energy.

Reports on researches shows that some regions within Nigeria have potential for power generation. Oyedepo *et al.* [4], carried out an analysis of wind speed data and as well, evaluate the wind power density in three south-eastern locations in Nigeria (Enugu, Owerri and Onitsha) at a hub height of 10 m above sea level with an annual mean wind speed of 5.42, 3.36 and 3.59 m/s respectively using wind speed data that spanned between 24 and 27 years. From this research, a wind speed value of 5.42 m/s recorded from Enugu is sufficient for power generation according to Batelle-Pacific Northwest Laboratory (PNL) wind power classification scheme [5]. A statistical assessment of wind energy potential in Ibadan, Nigeria was carried out by Fadare [6], and the analysis was based on Weibull distribution model using 10 years (1995-2004) daily wind speed data. It was inferred from this research that the site analyzed, possesses an annual power density of about 15.484 W/m^2 which according to Akpinar and Akpinar [7], is only suitable for lower power consumption application such as battery charging and water pumping.

Ogbonnaya *et al.* [8] Accessed wind energy potential in Nigeria, by utilizing 4 years' wind data collected from seven different states in (Enugu, Jos, Ikeja, Abuja, Warri, Sokoto and Calabar). The annual wind speed at 10 m above the ground varied from 2.3 to 3.4 m/s for sites along the coastal areas and 3.0-3.9 m/s for high land areas and semi-arid regions. Ojosu and Salawu [9], carried out research on the identification of potential wind energy sites in Nigeria. An atmospheric model simulation was developed and applied in studying the flow of wind, so as to identify potential wind-based electrification in different locations and as well determine the viability of the prevailing wind for small and commercial scale application. Igboekwe *et al.* [10] on the other hand, researched on the stochastic modelling of hourly average wind data collected from Umudike, South-East Nigeria. Ten years of hourly average wind speed data were utilized in this research to develop a periodic autoregressive integrated moving average model. The model was then applied in simulating the collected wind data and the result showed that the simulated wind behavior was reproducible and matched well with the characteristics of the experimental values.

More researches are emerging in the analysis of wind speed for different locations in Nigeria and each of these in the limit of their uncertainties have proven that great potential exist for wind power generation in Nigeria.

It can also be seen that most of the sites analyzed are basically onshore sites, and they possess wind power for small scale power generation. In order to meet the commercial demand for wind power, there is need to research on offshore wind resources as studies have shown that offshore wind energy is twice as that onshore wind energy.

1.2 Onshore and Offshore Wind Characteristics in Nigeria

Unlike the typical usage of "Offshore", within the marine industry and wind farm sector, offshore refers to inshore water zones such as seas, lakes, sheltered coastal areas etc. hence offshore wind refers to wind resources within these locations. Contrary, onshore wind refers to wind resources on land.

Offshore wind power or offshore wind energy is the use of turbines installed offshore, usually on the Continental shelf, to harvest Wind energy for electricity generation.

As a result of the high wind speed available in offshore sites compared to onshore, offshore wind power generation is higher [11]. Offshore wind turbines offer certain advantages compared to onshore wind turbines which include: (a) When a strong wind blows, it produces electricity around 3-5 MW per hour, (b) Little variation in wind speed, (c) An offshore wind turbine is more durable compared to onshore wind turbine. It persists within a period of 25-30 years and generates about 50% more energy than an onshore turbine. The most substantial drawbacks for potential offshore wind farm constructions are costs and public acceptance.

According to results from researches and wind data from Nigerian meteorological stations, it has been established that wind speeds are generally weak in the southern part of Nigeria, except for offshore areas from Lagos through Ondo, Delta, Rivers, Bayelsa to Akwa Ibom state which were reported to have prospects for harvesting strong wind energy throughout the year. States like Jos, Katsina and Maiduguri, possesses high wind speed amidst condition such as topography and roughness of surfaces. Further research revealed that wind speed of about 8.07 m/s can be harnessed from the northern part of Nigeria [12].

According to latest report from NIMET, based on the result of 40 years (1968-2007) available average of wind data from the whole 44 stations in Nigeria (Figure 1), it was observed that high onshore wind speeds are paramount in the northern part of Nigeria.

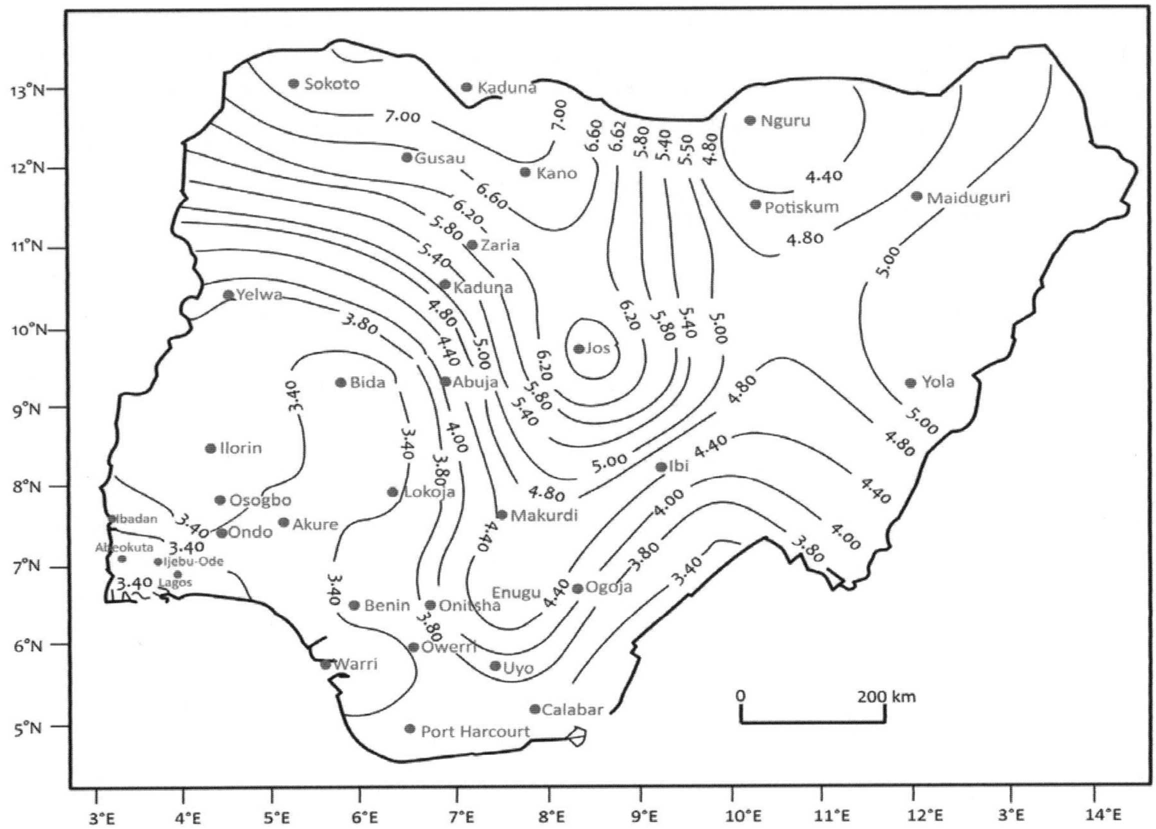

Figure 1:Isovents in m/s determined from 40 year's measurements at 10 m height, obtained from Nigeria meteorological department, Oshodi, Lagos State, Nigeria [13].

From Figure 1, wind speed within the north varies between the range of 4.0-7.5 m/s at 10 m height. Similarly, the southern part of Nigeria has moderate wind speed within the range of 3.0-3.5 m/s. Among the 44 stations operated by Nigerian Metrological Agency, there is no single offshore station despite the reported potential of offshore regions. Offshore wind data can be measured using anemometer installed in a suspended buoy system in water, as adopted in other countries around the world. Few results obtained from oil producing companies in Nigeria, with suspended buoy facilities shows that there is huge potential for offshore wind installation.

2 Research Trend on Wind Turbine Design in Nigeria and Power in Wind

Appropriate selections of design parameters for wind turbine is an important parameter in wind energy technology. Parameters such as the rated power, the cut-in speed and the cut-off wind speed must be selected based on the wind characteristics of the site and this will ensure maximum functionality of the turbine. The efficiency of a wind turbine in a particular site can be determined by the power output ($P_{ave.}$) and capacity factor (C_f) [14].

Small scale wind turbine has been designed and utilized for different purposes in Nigeria, such as water pumping, battery charging, small scale enterprise electrification and experiments. The only large scale

IOP Conf. Series: Materials Science and Engineering **413** (2018) 012039 doi:10.1088/1757-899X/413/1/012039

onshore wind turbine in Nigeria is located in Rimi Local Government Area of Katsina state, with an estimated output capacity of 10 MW [15].

Owing to the low wind speed characteristics in some locations in Nigeria, several works have been done in terms of wind turbine design modification, so as to help harness the available wind resource within the locations. Osadolor *et al.* [16], Carried out a research on the practical analysis of a small scale wind turbine in Auchi, Edo state. This research was focused on the design and construction of small wind turbine to generate 250 Watts of power at a rated speed of 6.5m/s. The result of the performance evaluation shows that the turbine is efficient for power generation.

This research is limited to the fact that the wind turbine produced cannot be used for high wind speed regions except the design parameters are altered to suit the respective locations. Similarly, further work was done on the development of low cost wind turbine generators for developing countries [17, 18]. This work was focused on developing suitable low cost generators for small wind turbine applications and the result showed that the developed generators performed better than the automotive alternator system. Thou, it was observed that the generators were not performing to their maximum capacity, hence proper design optimization was recommended. Ogunjuyigbe *et al.* [19], Worked on the dynamic performance of a wind-driven, self-excited reluctance generator to obtained the desired output voltage under a stochastic speed variation and varying load. The result obtained showed that despite the variation in wind speed and load, a suitable output voltage within preset limit was obtained and this further showed that this technology provides an effective and less expensive solution for output voltage control of wind-driven, self-excited reluctance generator in rural areas. Okechukwu *et al.* [20], this study was focused on the performance evaluation of the horizontal axis wind farm turbine rotor aerodynamic system, using blade element momentum theory required for power generation at the University of Port Harcourt. The result revealed that in order to meet the total power requirement of 21 MW in the university for an estimated period of 20 years, rotor blades of each of the wind farm turbines which are designed for 1.5 MW, wind speed of 17m/s, and for airfoil shape of NACA 2412, are desired.

Thou a lot of work has been done in small scale wind turbine modeling to serve domestic purpose, but the pressing need to supplement the nation's energy demand calls for large scale wind turbine installations. Onshore and Offshore wind turbines will provide large scale energy harvest if appropriate design parameters corresponding to the wind characteristics in a particular location is carefully selected. Ideally, the aerodynamic power in wind is given as [21]

$$\boldsymbol{P_{ae}} = \frac{1}{2}\boldsymbol{\rho\pi R^2 V_{eq}^3 C_p\,(\theta_{Pitch,}\,\lambda)} \qquad (1)$$

where P_{ae} is the aerodynamic power extracted from the airflow,

$\boldsymbol{\rho}$ is the density of air which is usually 1.225 Kg/m^3,

$\boldsymbol{C_p}$ is the power coefficient representing the fraction of power in the wind captured by a turbine blade depending on the pitch angle $\boldsymbol{\theta_{Pitch,}}$ measured in degree and on the tip speed radius $\boldsymbol{\lambda}$.

The tip speed radius is given by the expression [22];

$$\boldsymbol{\lambda} = \boldsymbol{\omega_{rot.}R/V_{eq}} \qquad (2)$$

where $\boldsymbol{\omega_{rot.}R}$ is the blade tip speed, $\boldsymbol{R}$ is the rotor radius and $\boldsymbol{\omega_{rot.}}$ is the rotor speed,

$\boldsymbol{V_{eq}}$ is the equivalent wind speed in m/s,

IOP Conf. Series: Materials Science and Engineering **413** (2018) 012039 doi:10.1088/1757-899X/413/1/012039

$C_p = 0.59$ which suggests that 59% of wind power is the maximum power that a wind turbine can exploit.

Equation (1) indicate that a wind turbine maximum power is a cubic function of the average wind speed.

The aerodynamic torque of the wind turbine is given as [23];

$$T_{ae} = \frac{1}{2}\rho\pi R^3 V_{eq}^2 C_p\,(\theta_{Pitch}, \lambda) \qquad (3).$$

In order to obtain high aerodynamic power output, high wind speed is required so as to drive the turbine. Apart from the high wind speed resources available in the north, offshore wind will effectively drive the turbine for better output efficiency.

Considering the stochastic nature of wind i.e. occurs randomly, wind behavior in a particular site is better described using statistical models. Some among such models include but not limited to; Weibull distribution, Rayleigh distribution, Log Normal distribution, Exponential distribution, Gaussian and Inverse Gaussian distribution. Among the aforementioned probability distribution models, the most widely used Probability Density Functions (PDFs) correspond to the Weibull and Rayleigh distributions because they represent, with high precision, wind speed data from various geographical locations around the World [23].

Weibull distribution is characterized by two parameters known as the shape (k) and scale (c) parameters which gives a good prediction of the energy output of a particular location. In other words, Rayleigh distribution is a modified form of Weibull distribution with a constant shape parameter of 2 and varying scale parameter. One of the limitations of Weibull distribution is its low response to low wind speed values.

3 Challenges Hindering the Prospect of Wind Energy Technology in Nigeria

Some among the several challenges militating against the prospect of wind energy technology in Nigeria includes inadequate research and technology in the field of wind energy, poor government motivation on the idea of wind energy technology, problem of land ownership and acquisition in Nigeria to mention a few, hence government at all level should work together to address issue of wind energy technology, sensitization of local community on the advantages of this initiative, developing a very good business model to attract willing investors and private sectors and as well incorporate wind energy technology into the nations independent power project plan. Synergizing the aforementioned will go a long way in promoting wind power generation technology in Nigeria.

4 Conclusion

This study reviews previous work done on wind energy technology in Nigeria and as well emphasizes on the possibility of harnessing both onshore and offshore wind in Nigeria, for power generation. In addition to the wind resources available on land, more and efficient wind resources are available in offshore sites especially in the south-south, south-east and south-west part of Nigeria. Thou a lot of researches have been done, ranging from analysis of several locations for wind power generations, logistic and cost analysis of wind energy conversion systems to development and design of small scale wind conversion systems for power generations, but wind as a renewable energy source, is not utilized to its full potential in Nigeria. The Nigerian Metrological Agency should have buoy systems on offshores water bodies for wind speed measurement, multi-criteria evaluation of prospective cite should be done to fully ascertain that the location is really suitable for windfarm installation, appropriate cost and logistic analysis should be carried out and finally, Nigeria should include windfarm technologies in her renewable energy policies and masterplan.

References

[1] United Nations, 2015. Transforming our world: The 2030 agenda for sustainable development. https://sustainabledevelopment.un.org/content/documents/21252030%20Agenda%20for%20Sustainable%20Development%20web.pdf

[2] Akpinar, EK., Akpinar, S. (2005). An assessment on seasonal analysis of wind energy characteristics and wind turbine characteristics. Energy Conversion and Management 46, 1848–67.

[3] Mukasa, D., E., Mutambatsere, Y., Arvanitis, T. (2013). Development of Wind Energy in Africa", Working Paper Series No. 170 African Development Bank, Tunis, Tunisia.

[4] Oyedepo, S. O., Adaramola, M., Paul, S. (2012). Analysis of wind speed data and wind energy potential in three selected locations in south-east Nigeria. *International journal of energy and environmental engineering*, 3:7. http://www.journal ijeee.com/content/3/1/7

[5] Ilinca, A., McCarthy E., Chaumel J., Retiveau L. (2003). Wind potential assessment of Quebec Province", *Renewable Energy* 28, 1881–1897.

[6] Fadare, D.A. (2008). A Statistical analysis of wind energy potential in Ibadan, Nigeria, based on Weibull distribution function. *The pacific journal of science and technology*, 9(1), 110–119.

[7] Akpinar, E.K., Akpinar, S. (2004). Statistical Analysis of Wind Energy Potential on the Basis of the Weibull and Rayleigh Distributions for Agin-Elazig, Turkey. *Journal of Power & Energy*. 218:557-565.

[8] Ogbonnaya, I. O., Chikuni, E., Govender P. (2007). Prospect of wind energy in Nigeria, http://active.cput.ac.za/energy/web/due/papers/2007/023O_Okoro.pdf.

[9] Ojosu, J.O., Salawu, R.I. (1990). An evaluation of wind energy potential as a power generation source in Nigeria. *Solar and Wind Technology*. 7: 663-673. June

[10] Igbokwe, M.U., Omekara C.O. (2002). Stochastic Simulation of Hourly Average Wind Speed in Umudike, South-East Nigeria. *Global Journal of Mathematical Sciences,* 1(1&2):59 66.

[11] Madsen & Krogsgaard (2017). Offshore Wind Power 2010 Archived June 30, 2011, at the Wayback Machine. *BTM Consult.*

[12] Ahmad, A. (2016). Wind Energy, Available at https://www.slideshare.net/avaise/wind-power-69583552, Accessed: November 2017.

[13] Nigeria Meteorological Agency (NIMET) Meteorological data, Nigeria Meteorological Agency, Oshodi, 2009.

[14] Celik, AN. (2004). A statistical analysis of wind power density based on the Weibull and Rayleigh models at the southern region of Turkey. *Renewable energy journal*, 29, 593–604.

[15] Burton, T.S., Jenkins, N., Bossanyi, E. (2001). Wind Energy Handbook. Wiley, New Jersey, USA.

[16] Odia, O. O., Ososomi, S. A., & Okokpujie, P. I. (2016). Practical Analysis of a Small Wind Turbine for Domestic Use on Latitude 7.0670 N, Longitude 6.2670 E. *Journal of Research in Mechanical Engineering*, *2*(11), 8-10.

[17] Okokpujie, I. P., Okokpujie, K. O., Nwoke, O. N., & Azeta, J. Development of a 0.5 KW Horizontal Axis Wind Turbine. *Journal of Engineering and Applied Sciences*, *13*(8), 2202-2208.

[18] Anih, S. O. (2013). Low cost small wind turbine generators for developing countries. Available at http://doi.org/10.4233/uuid:8ef37eca-6931-47f5-b2c4-794739e6edc6

[19] Ogunjuyigbe, A.S.O., Ayodele, T.R., Adetokun, B.B., Jimoh, A.A. (2016). Dynamic Performance of Wind-Driven Self-Excited Reluctance Generator under Varying Wind Speed and Load. *IEEE International Conference on Renewable Energy Research and Applications*, 506-51.

[20] Okechukwu, I.C., Larry, A. O, Bode, O. O. (2013). Wind Resource Assessment for Wind Energy Utilization in Port Harcourt, River State, Nigeria, Based on Weibull Probability Distribution Function. *International Journal of Renewable Energy Research*, 3:1.

[21] Muller, H., Poller, M., Basteck, A., Tilscher, M., Pfister, J. (2006). Grid Compatibility of variable speed wind turbines with directly coupled synchronous Generator and Hydro-dynamically controlled gearbox. *6th Int'l Workshop on Large-scale Integration of wind power and transmission Networks for offshore wind farms*, 308:26-28.

[22] Knudsen, H., Nielsen, J. N. (2005). Introduction to the Modelling of Wind Turbines. Wind Power in Power Systems, John Wiley & Sons Ltd.

[23] Acosta, J. L., Djokic, S. Z. (2010). Assessment of renewable wind resources in UK urban areas. *15th IEEE Mediterranean Electrotechnical Conference*, Malta.

Effect of Sulphate and Acid on Self-Compacting Concrete Containing Corn Cob Ash

Masi Tumba[1], Olatokunbo Ofuyatan[1], Osefunaimhen Uwadiale[1], John Oluwafemi[1], Solomon Oyebisi[1]

Department of Civil Engineering, Covenant University, Ota, Ogun-State, Nigeria E-mail address: masitumbalassa@gmail.com

Abstract. Self-compacting concrete requires high powder content which implies increased cost and increased CO_2 emission into the environment; hence, the need for the inculcation of a supplementary cementitious material (SCM) to the concrete mix such as corn cob ash (CCA) which is obtained by drying, grinding and burning the corn cobs. The chemical composition of the corn cob ash showed that it was pozzolanic as it conformed to the requirements stated in ASTM C618. Replacement of cement with CCA was carried out at levels of 0% (control), 5%, 10%, 15% and 20% and workability test was conducted at each replacement levels. The curing of the cubes was conducted for 7, 14 and 28 days in 3 mediums; water, hydrochloric acid (HCl) and magnesium sulphate ($MgSo_4$). The cubes were weighed and crushed from which it was observed that 5% replacement showed better compressive strength, proving to be the optimum result. Results from 15% and above were unable to attain close enough values to the required strength but showed better resistance when submerged in the sulphate and acid solution. It was also observed that increasing CCA content, reduced the strength deterioration factor (SDF) of the concrete, indicating improved durability and the possible use of CCA for partial replacement of cement in SCC. However, replacement levels from 15% and above require longer curing period to attain better strengths as increase in CCA levels decreases the compressive strength of the concrete.
Keywords: Self Compacting Concrete, Corn Cob Ash, Hardened Concrete, Fresh Concrete, Strength

1. Introduction

Self-compacting concrete was developed as a result of the reduction of skilled labor in japan. It was developed in the 1988 to solve the problem of the decline in available work force in the construction industry [1]. The characteristics of self-compacting concrete are affected by the aggregate content, the ratio of water to binder content, and the introduction of superplasticizers [2]. The use of SCC is however limited due to the high cost of this material. This high cost is as a result of the incorporation of increased powder content. The high cost of this powder content (cement) can be diminished by replacing cement with mineral admixtures [3].

Admixtures generally, control setting and hardening properties of concrete, improve the workability and provide extra cementing properties [4]. The introduction of pozzolanic waste ash in the production of concrete improves the properties of the concrete in its fresh and hardened state [5] and also helps to reduce the negative implications of waste in the environment and improve the productivity of such products. Corn cob is an agricultural waste product that is derived from maze or corn [6]. It is the core of the corn to which the kennels are. Corn cob ash (CCA) is obtained from drying the cob of the corn. Grinding it and burning the ash obtained is used as cement replacement [7].

Concrete is susceptible to attacks due to the environment. Hence it is of great importance to determine the effect of SCC to chloride, sulfate and other chemicals on the durability properties of concrete. These substances come in contact with the concrete through waste products mostly from industrial effluents. Concrete strength can be affected and drastically reduced when it comes into contact with such Chloride, sulphate and other chemicals; hence the need to determine its resistant capacity to such harmful substances.

2. Materials and Methods

2.1 Cement

IOP Conf. Series: Materials Science and Engineering **413** (2018) 012040 doi:10.1088/1757-899X/413/1/012040

The cement used for this research work is Dangote brand of cement.

2.2 Coarse Aggregates

The coarse aggregate used was granite having a particle size of 20mm.

2.3 Fine Aggregates

Sharp sand was used. It was air dried for 72 hours in other to reduce the moisture present in it. The sand was also clean and sharp, free from clay, loam, dirt or organic matters and conforms to the requirement of BS EN 12620 (2008).

2.4 Superplasticizer

The super plasticizer used is CONPLAST SP430. It is a brown viscous fluid which spreads out evenly when added in water.

2.5 Water

Tap water from the department of civil engineering's laboratory, Covenant University, Ota, Ogun state was utilized for mixing and curing of the concrete

2.6 Corn Cob Ash

The corn cob used for this experiment was gotten from a farm in Oyo state, Nigeria. The Corn Cob Ash was obtained from the burning of the corn cobs. The cobs were dried and broken down to smaller pieces and burned using open air burning. The ash resulting from the combustion was passed through a number 200 sieves and used in the experiment. The Corn Cob ash was examined using an x-ray fluorescence (XRF) test to determine the oxide composition of the ash.

2.7 Mix proportion

The concrete was mixed with 0% CCA, 5% CCA, 10% CCA, 15% CCA, and 20% CCA replacement levels. The Table 1 presents the mix proportion for the SCC. The superplasticizer percent was increased from 1% to 1.5% to improve the workability of the SCC + CCA concrete as a result of the high absorbency of CCA.

Table 1 Mix proportion for self-compacting concrete

MIX/MATERIALS	CEMENT (Kg)	CORN COB ASH (Kg)	FINE AGGREGATES (Kg)	COARSE AGGREGATES (Kg)	WATER-CEMENT RATIO (%)	SUPERPLASTICIZER (%)
SCC	484.38	0.00	843.75	796.88	0.4	1.00
SCC + CCA 5%	460.16	24.22	843.75	796.88	0.4	1.5
SCC + CCA 10%	435.94	48.44	843.75	796.88	0.4	1.5
SCC + CCA 15%	411.72	72.66	843.75	796.88	0.4	1.5
SCC + CCA 20%	387.50	96.88	843.75	796.88	0.4	1.5

L- box test, V-funnel test and slump flow test was conducted on the freshly mixed concrete to determine the passing ability, filling and flow abilities of the freshly mixed concrete in accordance to the [8] specifications.

The fresh concrete was poured into a mold of 150mm × 150mm × 150 mm and allowed to harden for 24 hours. After the cubes had become hardened, they were taken out of the mold and placed in 3 different solutions; water, 5% HCl acid solution and 3% $MgSO_4$. The cubes were immersed for the duration of 7 days, 14 days and 28 days. The cubes were taken out after the durations in which they were immersed in the solutions, they were weighed and compressive strength test was conducted.

3. Results

3.1 Chemical composition of the corn cob ash

The determination of the chemical composition of the CCA by using an X-ray fluorescence (XRF) machine is presented in the Table 2.

Table 2 Chemical composition of corn cob ash

Oxide	Percentage Composition (%)
Silicon dioxide (SiO_2)	59.54
Aluminium Oxide (Al_2O_2)	6.53
Ferrric Oxide (Fe_2O_3)	4.93
Calcium Oxide (CaO)	5.96
Magnesium Oxide (MgO)	2.32
Potassium Oxide (K_2O)	6.67
Sodium Oxide (Na_2O)	0.43
Sulphur trioxide (SO_3)	1.04
Loss on Ignition (LOI)	9.37

3.1 Fresh properties results

The results for the different test carried out on the fresh SCC is as related in the Table 3.

Table 3. Workability test results for SCC

Mix	Slump (mm)	Blocking ratio (H2/H1)	V-funnel (sec)
SCC (Control)	612.50	0.89	4.41
SCC + CCA 5%	588.50	0.87	4.97
SCC + CCA 10%	552.50	0.78	5.64
SCC + CCA 15%	537.00	0.7	6.13
SCC + CCA 20%	519.50	0.63	7.76

According to the EFNARC standard for SCC slump flow test, all mixes fall within the range of acceptable values with the exception of SCC + CCA 15% and 20%. It can be seen that increased percentage of CCA content, led to a reduction in workability of the SCC fresh concrete.

The results show decreasing blocking ratio with increasing CCA replacement levels. According to the EFNARC recommended standard, only 5% replacement satisfied the limits ($H_2/H_1 \geq 0.8$). This indicates that increasing percent replacement levels reduced the passing ability of the fresh concrete and thus indicating decreased workability. However, according to [9], blocking ratios falling within the range of 0.6-1.0 is acceptable to attain filling abilities that are satisfactory. Hence, it can be said that all replacement levels showed satisfactory results based on the above stated conditions. The results above show increasing flow time with increasing percent replacement levels of CCA. All replacement levels can be seen to fall

IOP Conf. Series: Materials Science and Engineering **413** (2018) 012040 doi:10.1088/1757-899X/413/1/012040

within the recommended EFNARC standards and all fall into the category VF1 (≤ 8sec). Hence, they all satisfy the v-funnel requirements.

3.2 Compressive strength

The findings from the compressive strength test performed on the hardened concrete is presented in Figure 1.

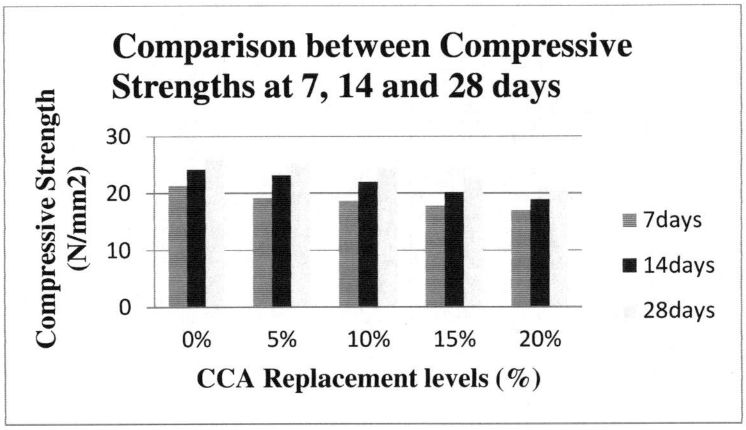

Fig. 1 Comparison between compressive strengths at 7, 14 and 28 days

Replacements levels up to 10% showed strengths very close to the targeted class at 28 days (M 25 and M30). However, all replacement levels attained good strength levels. In contrast to rice husk ash (RHA), increase levels of CCA replacement led to decreased compressive strengths but increased with increasing curing duration. Whereas [3] recorded increased strengths with increasing levels of RHA. Optimum levels were established at 5% for replacement with CCA. Previous literatures inculcated corn cob ash in ordinary concrete. [10] recorded optimum compressive strength at 5% replacement for 28 days attaining a strength of 21.44N/mm^2. [11] also recorded optimum results at 7.5% replacement. It was observed that increasing curing days increased the compressive strength of the corn cob ash concrete.

3.3 Cubes immersed in HCl solution

The Figure 2 and the Figure. 3 present the loss in weight of SCC and the comparison between SDF for the selected days of curing respectively.

Csw – Compressive strength in water

Csa – Compressive strength in acid

$$\text{SDF}(\%) = \frac{\text{Csw} - \text{Csa}}{\text{Csw}} \times 100$$

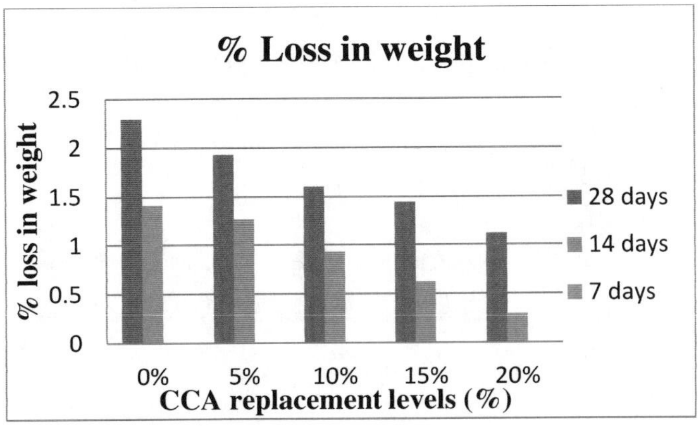

Fig. 2 loss of weight of cubes immersed in HCl solution

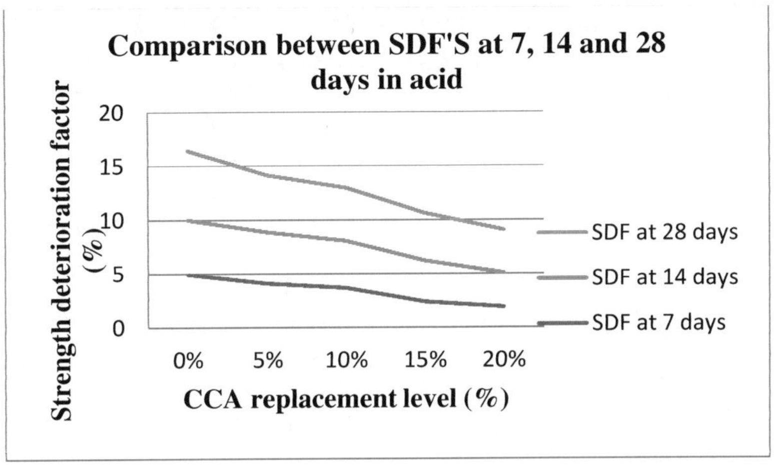

Fig. 3 Comparison between SDF at 7, 14 and 28 days in HCl solution

Admixtures improve the durability properties of the concrete. From Figure 2 and the Figure 3, the increasing replacement levels of CCA showed reduced effects of the acid on the strength and weight properties of the concrete cubes. The details from the graph and table show a decreasing difference in weight and compressive strength of the cubes cured in acid when compared to that cured in water. Hence, improved durability was observed.

3.4 Cubes immersed in $MgSO_4$

The observations from the curing of the cubes in sulphate solution is presented in Figure 4. And the Figure 5.

Csw – Compressive strength in water

Css – Compressive strength in sulphate

$$\text{SDF (\%)} = \frac{\text{Csw} - \text{Css}}{\text{Csw}} \times 100$$

IOP Conf. Series: Materials Science and Engineering **413** (2018) 012040 doi:10.1088/1757-899X/413/1/012040

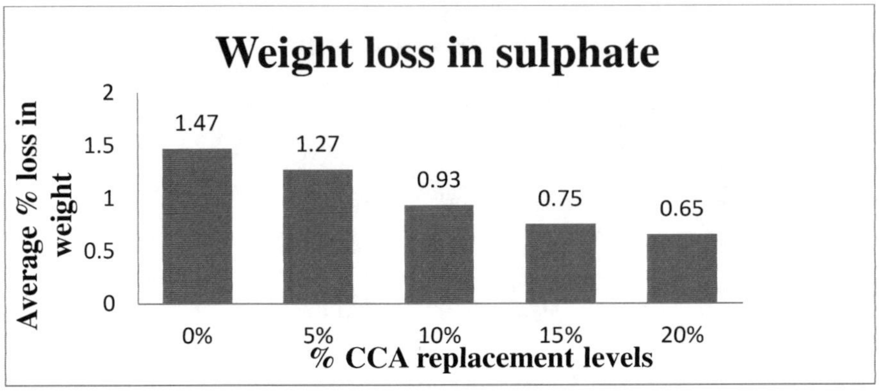

Fig. 4 Weight loss in $MgSO_4$

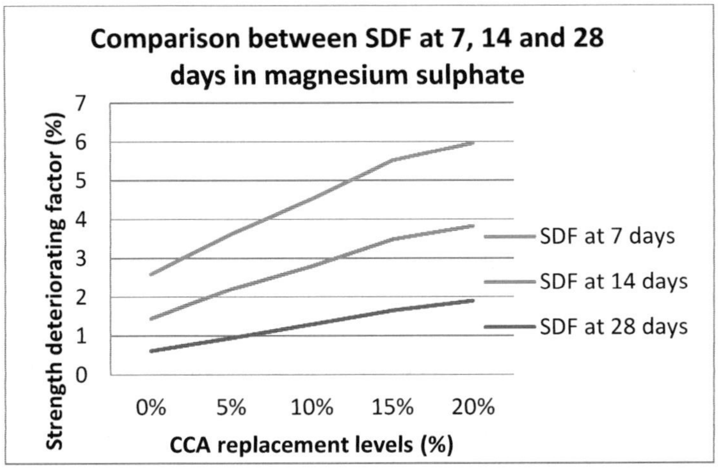

Fig. 5 Comparison between SDF's at 7, 14 and 28 days of cubes immersed in $MgSO_4$

The weight loss of the cubes immersed in sulphate solution increased as curing duration increased and decreased with increasing percentage replacement levels. The weight loss at 7 and 14 days was infinitesimal until the 28th day before a significant difference could be spotted. This could be as a result of low concentration of the sulphate solution or because expansion is not usually associated with magnesium sulphate attack ($MgSO_4$). $MgSO_4$ manifests itself in loss of strength of concrete and is more prominent in pozzolanic mixes [11]. Hence as observed in Figure 5, increase in CCA replacement levels, leads to an increase in the SDF and increase in curing days also increases the SDF. The results when compare with [3] are similar as it was also recorded in their work that there was an increase in the strength deterioration factor with increased CCA replacement levels.

4. Conclusion

Based on the findings from this research, CCA at 5% replacement satisfies the requirements of EFNARC for fresh concrete and hardened SCC. The 5% CCA replacement has its results within the required limits for the slump test, blocking ratio and the v-funnel test while its resulting compressive strength at 28days of curing is also satisfactory. The chemical analysis performed on the CCA also established that CCA is a pozzolanic material in line with the specification of ASTM C618 under class N natural pozzolans. The effect of acid and sulphate on SCC reduces with increase in the replacement level of CCA as regards weight loss and SDF.

Reference

[1] DINESH A, SHARINI S, JASMINE J P, JINEY J, SHAGUFTA J. Experimental Study on self-compacting concrete [J]. International Journal of Science and Research, 2017:42-50.

[2] HAJIME O, MASAHIRO O. Self-Compacting Concrete [J]. Journal of Advanced Concrete Technology, 2003, 1(1): 5-15.

[3] KANNAN V, GANESAN C. Chloride and chemical resistance of self-compacting concrete containing rice husk ash metakaolin[J]. Construction and Building Materials, 2013: 225-234.

[4] JAMAL H. aboutcivil.com, 18 March 2017. [Online]. Available: https://www.aboutcivil.org/concrete-technology-admixtures.html.

[5] TAIWO A. Properties of Corn Cob Ash Concrete, zaria, 2012.

[6] RAHEEM A A, OYEBISI O S, AKINTAYO O S, OYENIRAN I M. Effects of Admixtures on the Properties of Corn Cob Ash Cement Concrete [J]. Leonardo Electronic Journal of Practices and Technologies, 2010: 13-20.

[7] OFUYATAN T, OLUTOGE F, OLOWOFOYEKU A. Durability Properties of Palm Oil Fuel Ash Self Compacting Concrete [J]. Engineering, Technology & Applied Science Research, 2015, 5(1): 753-756.

[8] EFNARC, Specification and guidelines for self-compacting concrete, European Federation, 2002.

[9] FELEKOGLU B, TURKEL S, BARADAN B. Effect of Water/ Cement Ratio on the Fresh and Hardened Properties of Self-Compacting Concrete [J]. Build Environment, 2007: 1795-1802.

[10] OWOLABI T, OLADIPO I, POPOOLA A. Effect Of Corncob Ash As Partial Substitute For Cement In Concrete [J]. New York Science Journal, 2015, 8(11): 1-4.

[11] JOHN K, ASH A, PAUL H, JOSEPH K. Suitability of Corncob Ash as a Supplementary Cementitious Material [J]. International Journal of Materials Science and Engineering, 2016:215-228.

IOP Conf. Series: Materials Science and Engineering **413** (2018) 012041 doi:10.1088/1757-899X/413/1/012041

Multi-Criteria Evaluation of the Appropriate Offshore Wind Farm Location in Nigeria

E. A. Mayaki*, O. Adedipe and S A. Lawal

School of Infrastructure, Process Engineering and Technology, Federal University of Technology Minna, Nigeria
Corresponding author's email: mayakionimisi@gmail.com

Abstract - This paper presents a multi-criteria evaluation of the appropriate offshore wind farm location in Nigeria. The proposed site selection criterion takes into consideration the technical aspect of the Environment consideration. The four attributes considered in this work are proximity to power demand, Distance from Airport, Average wind speed and Interference with existing shipping route. These attributes were analyzed for three alternative locations (Victoria Island, Koko and Abbonema in Lagos, Warri and Port-Harcourt respectively) using Analytical Hierarchy Process to assign weight through pair-wise comparison and fuzzy TOPSIS Multi-criteria decision-making tool to make an appropriate selection. The results from this work showed that Victoria Island (Lagos) has the highest figure cumulative value of 894.23, Koko (Warri) has 740.04 and Abonnema (Port-Harcourt) with a value of 756.96.

1. Introduction

Energy provides an essential ingredient for almost all human activities: such as cooking and space/water heating, lighting, healthcare, food production and storage, education, mineral extraction, industrial production and transportation [1]. Nigeria's energy is supplied from different hydro-power and thermal power stations. The extent of water available at these different hydro-power stations varies due to change in seasonality and this causes irregular supply at low water levels periods. The thermal power stations are also affected by lack of adequate supplies of natural gas [2]. The solution for the country is energy diversification. Harnessing offshore wind power was talked about as early as the 1930s, not until 1990, when the first modern offshore wind turbine was constructed 250 m offshore of Northern Sweden [3]. According to Okokpujie et al [4] and Odia et al. [5] the authors stated that wind turbine will serve as one of the major way to resolve the issue of energy supply in Nigeria. Multiple-criteria decision making (MCDM) or Multiple-criteria decision analysis (MCDA) is a sub-discipline of operations research that explicitly evaluates multiple conflicting criteria in decision making (both in daily life and in settings such as business, government and medicine) [6]. Conflicting criteria are typical in evaluating options: cost or price is usually one of the main criteria, and some measure of quality is typically another criterion, easily in conflict with the cost. There have been important advances in this field since the start of the modern multiple-criteria decision-making discipline in the early 1960s [7]. The MCDM problem can be represented in the criterion space or the decision space. Alternatively, if different criteria are combined by a weighted linear function, it is also possible to represent the problem in the weight space. The analytic hierarchy process (AHP) is a structured technique for organizing and analyzing complex decisions, based on mathematics and psychology. It was developed by Thomas L. Saaty in the 1970s and has been extensively studied and refined since then. It has particular application in group decision making [8] and is used around the world in a wide variety of decision situations, in fields such as government, business, industry, healthcare, ship building and education, [9]. Once the hierarchy is built, the decision makers systematically evaluate its various elements by comparing them to each other two at a time, with respect to their impact on an element above them in the hierarchy. In making the comparisons, the decision makers can use concrete data about the elements, but they typically use their judgments about the elements' relative meaning and importance. It is the essence of the AHP that human judgments, and not just the underlying information, can be used in performing the evaluations [9]. The

AHP converts these evaluations to numerical values that can be processed and compared over the entire range of the problem. A numerical weight or priority is derived for each element of the hierarchy, allowing diverse and often incommensurable elements to be compared to one another in a rational and consistent way [10].

2. Data Collection

This study is limited to technical attribute required for the location of a wind farm. The considered data are Power Demand, Distance from Airport, Average wind speed and Shipping route. These data were collected for the three states i.e. Lagos, Warri and Port-Harcourt, which are the alternatives for which their attributes were analysed. The required Average Wind Speed was collected from the Nigerian Meteorological Agency (NIMET), while the quantitative attributes were sourced for and qualitative attributes were obtained using Questionnaires.

Table 1 Ten years' average wind speed for Lagos Warri and Port-Harcourt

Alternatives	Years										Average
	2002	2003	2004	2005	2006	2007	2008	2009	2010	2011	
Lagos	7.21	6.49	7.27	7.32	7.21	7.6	5.88	4.83	4.7	4	6.251
Warri	7.38	7.46	7.35	7.32	7.32	7.17	7.48	7.07	7.2	7.19	7.294
Port harcourt	7.3	7.37	6.89	7.13	7.38	7.06	7.54	7.3	7.56	7.94	7.347

Table 2 Attributes Value for the three Alternatives

Attributes	Alternatives		
	Lagos	Warri	Port-Harcourt
Power Demand	4.5	3.875	4.375
Distance from Airport	533	438.5	448.23
Average Wind Speed	6.25	7.29	7.35
Shipping Route	1.75	3.875	4.375

3. Multi-criteria Evaluation Process

The pair-wise element matrix is the first step to the application of Analytical Hierarchy Process: The matrix if formed by comparing one criterion against another. Two criteria with same level of importance will have a score of 1, a criterion that is more important than that which it's been compared with will have its actual value on the Effective Criteria and Pair Wise Comparison while a criterion less important than that which it is been compared to will be score the inverse of its actual value on the Effective Criteria and Pair Wise Comparison as shown in Table 2.

$$\begin{bmatrix} C_{11} & C_{12} & C_{13} & C_{14} \\ C_{21} & C_{22} & C_{23} & C_{24} \\ C_{31} & C_{32} & C_{33} & C_{34} \\ C_{41} & C_{42} & C_{43} & C_{44} \end{bmatrix}$$

Having completed the comparison matrix next is to normalize the matrix: The sum of the pair-wise criteria matrix column was calculated. Each entry in the column is then divided by the column sum to yield its normalized score, the sum of each column of the normalized matrix. These procedures are done with the equations (1) and (2).

$$C_{ij} = \sum_{i=1}^{n} c_{ij} \tag{1}$$

$$X_{ij} = \frac{C_{ij}}{\sum_{i=1}^{n} C_{ij}} \begin{bmatrix} x_{11} & x_{12} & . & x_{1m} \\ x_{21} & . & . & . \\ . & . & . & . \\ x_{n1} & . & . & x_{nm} \end{bmatrix} \quad (2)$$

Consistency Analysis

Consistency analysis involves the calculation of the Consistency Ratio (CR), Consistency Index (CI): while the Random Index has already been generated by Satty, 2008 as

Table 3 Random inconsistency indices for n = 10

N	1	2	3	4	5	6	7	8	9	10
RI	0.00	0.00	0.58	0.9	1.12	1.24	1.32	1.41	1.46	1.49

Notes: n = order of Matrix

To calculate the CI and CR the Matrix Multiplication Function was calculated using the EXCEL (=MMULT()) function.

$$\begin{bmatrix} C_{11} & C_{12} & C_{13} & C_{14} \\ C_{21} & C_{22} & C_{23} & C_{24} \\ C_{31} & C_{32} & C_{33} & C_{34} \\ C_{41} & C_{42} & C_{43} & C_{44} \end{bmatrix} \times \begin{bmatrix} W_{11} \\ W_{21} \\ W_{31} \\ W_{41} \end{bmatrix} = \begin{bmatrix} C_{V11} \\ C_{V21} \\ C_{V31} \\ C_{V41} \end{bmatrix} \quad (3)$$

$$Average(\mu) = \frac{\sum_{i=1}^{n} C_{ij}}{n} \quad (4)$$

$$\sigma = \sqrt{\frac{1}{n}\sum_{i=1}^{n}(C_{ij}-\mu)^2} \quad (5)$$

Below are the results for the Multi- criteria analysis of four (4) criteria for a 3 alternatives. Rating scale is as shown in Table 1.

Table 4 Effective Criteria and Pair Wise Comparison

Factor	Factor weighting Score											Factor
	More Important than					Equal	Less Important than					
C1	1	2	3	4	5	1	1	2	3	4	5	C2
C2	1	2	3	4	5	1	1	2	3	4	5	C3
C3	1	2	3	4	5	1	1	2	3	4	5	C4
C4	1	2	3	4	5	1	1	2	3	4	5	C1

IOP Conf. Series: Materials Science and Engineering **413** (2018) 012041 doi:10.1088/1757-899X/413/1/012041

Table 5 Pair Wise Comparison Matrix

Factor	C1	C2	C3	C4
C1	1	3	2	2
C2	0.3333	1	0.5	0.5
C3	0.5	2	1	1
C4	0.5	2	1	1

Table 6 Column Total For Pair Wise Comparison Matrix

Factor	**C1**	**C2**	**C3**	**C4**
C1	1	3	2	2
C2	0.3333	1	0.5	0.5
C3	0.5	2	1	1
C4	0.5	2	1	1
Total	2.3333	8	4.5	4.5

Table 7 Row Total For Normalized Pair Wise Comparison Matrix

Factor	C1	C2	C3	C4	Total	Average
C1	0.4285776	0.375	0.4444444	0.4444444	1.6924664	0.4231166
C2	0.1428449	0.125	0.1111111	0.1111111	0.4900671	0.1225168
C3	0.2142888	0.25	0.2222222	0.2222222	0.9087332	0.2271833
C4	0.2142888	0.25	0.2222222	0.2222222	0.9087332	0.2271833
Total	1	1	1	1		

Table 8 Consistency Analysis For Normalized Pair Wise Comparison Matrix

Factor	C1	C2	C3	C4	Total	Average	Consistency Measure
C1	0.4285776	0.375	0.4444444	0.4444444	1.6924664	0.4231166	5.014431
C2	0.1428449	0.125	0.1111111	0.1111111	0.4900671	0.1225168	5.030388
C3	0.2142888	0.25	0.2222222	0.2222222	0.9087332	0.2271833	4.0978367
C4	0.2142888	0.25	0.2222222	0.2222222	0.9087332	0.2271833	4.0978367
Total	1	1	1	1		CI	0.186708
						RI	0.9
						CR (CI/RI)	0.207453

Table 9 Criteria Weight from AHP analysis

Citeria	C1	C2	C3	C4
Weight	0.48405	1.659624	0.933539	0.933539

Table 10 Decision Matrix

		Lagos	Warri	Port-Harcourt	Criteria Weight
C1	Power Demand	4.5	3.875	4.375	0.48405
C2	Distance from Airport	533	438.5	448.23	1.659624
C3	Average Wind Speed	6.25	7.29	7.35	0.933539
C4	Shipping Route	1.75	3.875	4.375	0.933539

Table 11 Result of Decision Matrix analysis

Attributes	Alternatives		
	Lagos	Warri	Port-Harcourt

Power Demand	2.178225	1.875694	2.11771875
Distance from Airport	884.5796	727.7451	743.8932655
Average Wind Speed	5.834619	6.805499	6.86151165
Shipping Route	1.633693	3.617464	4.084233125
Total	894.226	740.044	756.95673

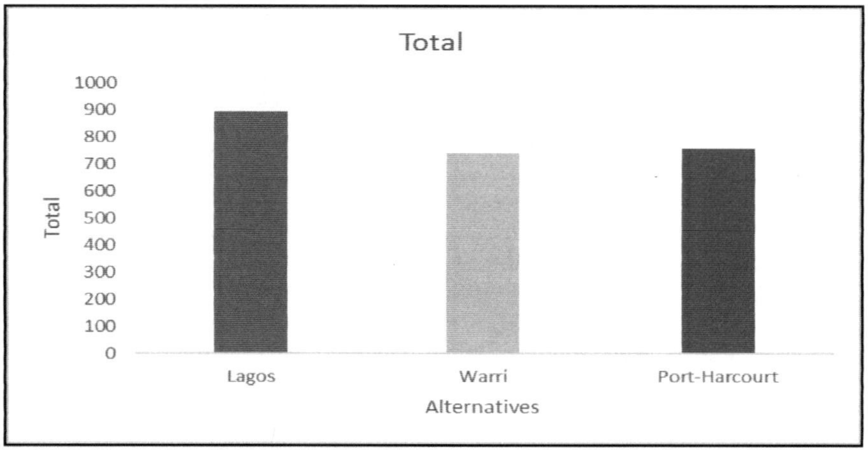

Figure 1 Bar chart of the cumulative result for the three alternatives

4. Conclusions

Nigeria has a great potential for offshore wind farm projects and this should be considered to solve the issue epileptic power supply in the country and the Analytic Hierarchy Process and fuzzy TOPSIS are applicable for the multi-criteria selection of the appropriate offshore wind farm location in Nigeria. The collected wind farm siting attributes revealed that the three alternatives showed a good wind speed characteristic and this was shown by the 10 years' average wind speeds collected for three locations. The application of Analytic Hierarchy Process gave a normalized column total of 1, consistency index and consistency ratio within the expected range as stated by Satty, [10] and finally from the multi-criteria analysis, Victoria Island was found to show the highest consideration value.

References

[1] Gelbspan, R. (2004), Towards a global energy transition. Foreign Policy in Focus: FPIF – PetroPolitics Special Report. www.petropolitics.org

[2] Iwayemi, A., Nigeria's dual energy problems: Policy issues and challenges, International Association for energy economics, Fourth Quarter, 2008, available online [www.iaee.org/en/publications/newsletterdl.aspx? id=53, accessed 21 December 2009], 17–21

[3] Bilgili, M., A. Yasar & E. Simsek (2011) Offshore wind power development in Europe and its comparison with onshore counterpart. Renewable and Sustainable Energy Reviews, 15, 905-915.

[4] Okokpujie, I. P., Okokpujie, K. O., Nwoke, O. N., & Azeta, J. DEVELOPMENT OF A 0.5 KW HORIZONTAL AXIS WIND TURBINE. *Journal of Engineering and Applied Sciences*, *13*(8), 2202-2208.

[5] Odia, O. O., Ososomi, S. A., & Okokpujie, P. I. (2016). Practical Analysis of A Small Wind Turbine for Domestic Use on Latitude 7.0670 N, Longitude 6.2670 E. *Journal of Research in Mechanical Engineering*, *2*(11), 8-10.

[6] Weistroffer, H. R., Smith, C. H., and Narula, S. C., "Multiple criteria decision support software", Ch 24 in: Figueira, J., Greco, S., and Ehrgott, M., eds, Multiple Criteria Decision Analysis: State of the Art Surveys Series, Springer: New York, 2005

[7] *Mardani, Abbas; Jusoh, Ahmad; Zavadskas, Edmundas Kazimieras (2015-05-15).* "Fuzzy multiple criteria decision-making techniques and applications – Two decades review from 1994 to 2014". *Expert Systems with Applications. 42 (8): 4126–4148.* doi:10.1016/j.eswa.2015.01.003

[8] Kao, C. (2010). Weight determination for consistently ranking alternatives in multiple criteria decision analysis. Applied Mathematical Modeling, 34, 1779–1787

[9] Pohekar SD, Ramachandran M (2004) Application of multi-criteria decision-making to sustainable energy planning–a review. Renew Sustain Energy Rev 8:365–381

[10] Saaty, T.L. (2008) 'Decision making with the analytic hierarchy process', *Int. J. Services Sciences*, Vol. 1, No. 1, pp.83–98.

IOP Conf. Series: Materials Science and Engineering **413** (2018) 012042 doi:10.1088/1757-899X/413/1/012042

Investigation of Balling Characteristics of Mixture of Iron Oxide Bearing Wastes and Iron Ore Concentrates

Mfon Udo[1, *], David Esezobor[2], Adeniran Afolalu[1], Harrison Onovo[2], Samson Ongbali[1], Imhade P. Okokpujie[1]

[1] Department of Mechanical Engineering, Covenant University Ota, Ogun State, Nigeria

[2] Department of Metallurgical and Materials Engineering, University of Lagos, Lagos, Nigeria
[*]mfon.udo@covenantuniversity.edu.ng, esezobordave@yahoo.com, adeniran.afolalu@covenantuniversity.edu.ng, harrinov@yahoo.com, samson.ongbali@covenantuniversity.edu.ng, imhade.okokpujie@covenantuniversity.edu.ng

Abstract - Iron oxide bearing wastes in form of dust and sludges are hard to handle because of their micron size particles and moisture content in case of sludge. More often they are stockpiled in large quantities that can occupy large area of real and agricultural estates and cause pollution. Balling or palettization, an agglomeration process was used to process the wastes in order to address the problem of micron size particles and to make them fit for recycling back into metallic iron production route like blast furnace. Balling or green pelletization is the process of forming nearly spherical shaped granules by tumbling moistened particulates with or without binders in balling drums or discs disc. For a pellet to be effective either for being transported or for being recycled in blast furnace to produce metallic iron without disintegrating to dust its balling characteristics should measure up to required standard. Most outstanding of those balling characteristics include Drop Number, Green or Wet Compression Strength, Dry Compression Strength, Abrasion and Tumbler Indices. In this work iron oxide bearing wastes was mixed with iron ore concentrates in various proportions. These mixes were taken through balling or wet pelletization process using Radicon Balling Disc. The balls formed balls were taken through Drop number tests adopting the Free Fall method, where balls are made to fall freely from a height of 50 mm on steel surface, Green compression and Dry compression tests using a 5 kN Universal Testing Machine (INSTRON Corp., model 1011 UK) System while Abrasion and Tumbler indices tests were conducted using Tumbler Index cylinder or drum and adopting ASTM method. It was found that Drop number as high as 7.8 times, Green compression strength and Dry compression strength up to 11.7 N/pellet and 25.99 N/pellet respectively were attained by some of the pellets. The Tumbler and Abrasion indices recorded were up to above 95% and 5% respectively. These values are higher than the minimum recommended.
KEYWORDS Iron oxide bearing wastes, Balling, Balling characteristics

1. Introduction

In the steel industry, iron oxide bearing wastes in form of dusts and sludge are produced at every segment of processing stage of sinter, molten iron and steel production. Several technologies are essentially employed to reuse them in the steel industry. The determining factors of the choice of technology include the size distribution (97 – 100 % minus 20 microns), moisture content (25 – 40%) and chemistry (0.03 – 7.8%, zinc or lead) of the wastes (Esezobor, et al, 2008).

Agglomeration and metallization remain the widely used methods for re – utilization of these wastes (Abioye, et al, 2018) . However, introduction of these materials to sinter and pellet charge through conventional methods are restricted due to the poor quality of the prepared charge and the relatively depreciation of the efficiency of sintering machine or pelletizer as well as the quality of sinter or balls (Blast Jurgen, 2007, Esezobor, et al, 2008). These materials are normally taken through balling process to address these deficiencies.

Balling is the process of forming nearly spherical shaped grannules by tumbling moistened particulates with or without binders in balling drums or discs. Balling is very important in the iron and steel industry. For instance, the fine grained iron ore dust that are produced in the beneficiation process of low grade iron ore are granulated into balls by balling process before being reduced to metallic iron in the blast furnace (Satry, K.V.S, 1984, Afolalu, et al, 2017). Similarly, powdery iron ore or iron oxide materials are usually processed into balls for easy transportation from one place to another. Balls have the advantage over lump iron ores in that they are characterized by narrow grain spectrum, constant quality and good gas permeation during reduction (Werner, et al, 1979, Indian Bureau of Mines, 2011).

The ability of the moistened particulate materials to form balls of good compressive strength is known as the ballability of the material (Udo, et al, 2018, Abioye, et al, 2017). Materials of high ballability will possess high balling characteristics such as drop number, wet and dry compressive strengths, tumbler and abrasive indices. Ballability of material is influenced by factors such as grain size of the material, moisture content and the binder content in the materials' mix, wettability of the material as well as the processing parameters of the granulator (Udo, et al, 2018).

This work investigates the balling characteristics of iron oxide bearing wastes in order to enhance large scale production of iron oxide pellets from these wastes.

2. Experimental Method

2.1 Materials and Methods

For this study iron ore concentrates were obtained from National Iron Ore Mining Company at Itakpe, Kogi State of Nigeria, electric arc furnace (EAF) dust from Universal Steel (casting and rolling) Company, Ogba, Lagos State, Nigeria and pellet dust from Delta Steel Plant in Aladja, Warri, Delta State, Nigeria. Bentonite and lime were purchased from the local market.These materials were pulverized using a ball mill and classified on Dorr – Oliver Sieve Analysis apparatus with ten sizes ranging from 63 microns to 1000 microns. Chemical analyses of the materials were conducted on an Ex 310 Jordan Valley X–ray fluorescent.

The pulverized materials were divided into 30 different compositional mixes as shown in Table1 and rigorously mixed to ensure uniformity of the mixes. Each of the mixes was granulated into balls using Radicon balling disc of diameter 40cm, depth 10cm, inclined at 45° to the horizontal and rotating at a speed of 30 revolutions per minute. The resident time for balling of each mix with varied moisture content was 5 minutes. The ball diameter was controlled at an average of 10 mm.

IOP Conf. Series: Materials Science and Engineering **413** (2018) 012042 doi:10.1088/1757-899X/413/1/012042

Table 1 Weight Percentage of Various Balling Mixes

Moisture Content (wt. - %)	Concentrate (wt. - %)					
	A	B	C	D	E	F
6	89.5	80	70	-	-	-
8	87.5	78	68	-	-	-
9	86.5	77	67	47	27	87
10	85.5	76	66	46	26	-
11	84.5	75	65	45	25	85
12	-	-	-	44	24	-
13	-	-	-	-	-	83
14	-	-	-	42	22	-
15	-	-	-	-	-	81
16	-	-	-	-	-	80

A – control Experiment with 4 % lime and 0.5 % bentonite

B – mix with 10 % dust and 4 % lime

C – mix with 20 % dust and 4 % lime

D – mix with 40 % dust and 4 % lime

E – mix with 60 % dust and 4 % lime

F – mix with 4 % lime and dust

2.2 Experiment

The resulting balls from each granulator were air – dried for 12 hours for drop number, green strength and dry strength tests as well as abrasion and tumbler indices tests

Drop Number Test

Drop number test indicates the drop strength of the balls. It gives indications of the ability of the green pellets to withstand the drops they will encounter from the point of discharge after balling to the firing unit. 20 balls from each mix were tested to obtain the value of drop number. The balls were dropped repeatedly from a height of 50cm onto a steel surface. The number of times each ball dropped before shattering is known as the drop number and was recorded.

Compressive Strength Test

Green strength test evaluates the ability of green balls or pellets to retain their shape during handling. Dry compressive strength test evaluates the ability of balls to retain their shape and withstand thermal stresses from the point of induration in the furnace to blast furnace without disintegrating into dust. A minimum of 20 green and 20 dry balls were used for these tests with their average values taken as the compressive strength (wet or dry respectively).
For green compressive strength test the balls were air dried for 12 hours before used for the test. While the green balls were indurated in electric resistance furnace to a temperature of 300°C for 1 hour and then allowed to cool in the furnace before being used to carry out the dry compressive strength test. Each of the balls was allowed to dry under atmospheric condition for 12 hours and then subjected to compressive test on a 5 kN Universal Testing Machine (INSTRON Corp., model 1011 UK) System. The stress attained at the point of shattering of the ball was recorded as the ball's green or dry compressive strength in Newton/pellet.

Tumbler and Abrasion Tests

Tumbler and abrasion indices tests were carried out simultaneously. For these tests, 500g of balls of diameter between 6mm and 10mm were selected from oven dried pellet into a well-sealed Tumbler Index cylinder or drum. Then the drum was rotated at 20 rpm for 5 minutes. The product of the drum was passed through a sieve of 500µm. The quantity of the shattered pieces that passed through the sieve as well as the quantity that did not pass through the sieve was measured. The ratio of the weight of quantity that did not pass through the sieve to the total quantity of the material gave the tumbler index while the ratio of the weight of the quantity of the material that passed through the sieve to the total quantity of the material gave the abrasion index (Topiary, et al, 2007).

3. Results and Discussion

The results of the sieve and chemical analyses of the iron bearing materials and the binders are presented in Tables 2 – 4.

The result in Table 2 indicates that over 80 % and 85 % of pulverized iron ore concentrates and iron oxide bearing wastes respectively are within the sizes ranging between 125 to 63 microns – a suitable sieve range for balling (Esezobor, et al, 2015) . While Table 3 shows that iron oxide bearing wastes consist of useful compounds such as iron oxides (FeO, Fe_2O_3), carbon, and slagging oxides in quantities close to what is obtained in iron ore concentrates. This indicates that these wastes are suitable as charge as well as additives to sinter and pellet burden by chemical and granulametrical constituents (Esezobor, et al, 2008).

It can be clearly seen from the results of drop number test in Figure 1 and that of green or wet compression strength test in Figure 2 that the drop number and green compression strength of the balls increase as the moisture content as well as that of iron oxide bearing wastes in the mixture increases at constant mechanical force of the balling disc.

Table2: Sieve Analysis of Iron Concentrate, Iron Oxide Bearing Wastes

S/ N	**Materials**	**Sieve Size, µm**									
		+1000	**+710 -1000**	**+500 -710**	**+355 -500**	**+250 -355**	**+180 -250**	**+125 -180**	**+90 -125**	**+63 -90**	**-63**

IOP Conf. Series: Materials Science and Engineering **413** (2018) 012042 doi:10.1088/1757-899X/413/1/012042

1.	Itakpe Iron Ore Concentrate	14.23	4.44	10.8	20.99	36.63	6.46	3.23	1.82	0.71	0.20
2	Ground Concentate	-	-	-	0.31	0.61	4.18	14.29	45.71	26.43	8.47
3	Ground IROBEWAS	-	0.10	0.20	0.61	2.04	4.18	8.36	39.65	31.81	13.05

Table 3: Chemical Analysis of the Iron Ore Concentrate and IROBEWAS

S/N	Compound	Composition, wt %														
		Fe_T	Fe_20_3	L.O.I	Al_20_3	$Si0_2$	MgO	CaO	S	TiO_2	Na_2O	K_2O	P_2O_5	ZnO	MnO	C
1*	Itakpe Iron ore concentrate	63.0	88.9	0.26	0.6	8.6	0.02	0.15	0.02	0.3	0.11	0.01	0.06	ND	ND	-
2	Universal Steel EAF Dust	35.73	51.09	7.08	0.02	3.2	0.25	1.28	0.39	0.3	0.39	0.94	0.10	29.50	2.17	2.6
3	Delta Steel Pellet Dust	54.34	77.64	6.43	1.30	7.03	0.25	4.17	-	0.27	0.27	0.03	ND	ND	0.14	2.5
4	IROBEWAS	47.3	74.5		0.66	5.1	0.24	2.41	0.21	0.3	0.34	0.49	0.10	13.2	1.16	2.6

*FeO = 1.0, SO_3 = 0.01

Table 4: Chemical Analysis of Bentonite and Lime

S/No	Compound	Composition, wt %								
		Fe_2O_3	Al_2O_3	CaO	SiO_2	Na_2O	K_2O	MnO	Moist	Imp
1	Bentonite	0.32	18.85	0.24	62.24	0.91	10.65	2.86	1.67	-
2	Lime	ND	ND	52.06	ND	ND	ND	ND	20.24	15.32

ND – Not identifiable; Moist – Moisture Content; Imp - Impurity

IOP Conf. Series: Materials Science and Engineering **413** (2018) 012042 doi:10.1088/1757-899X/413/1/012042

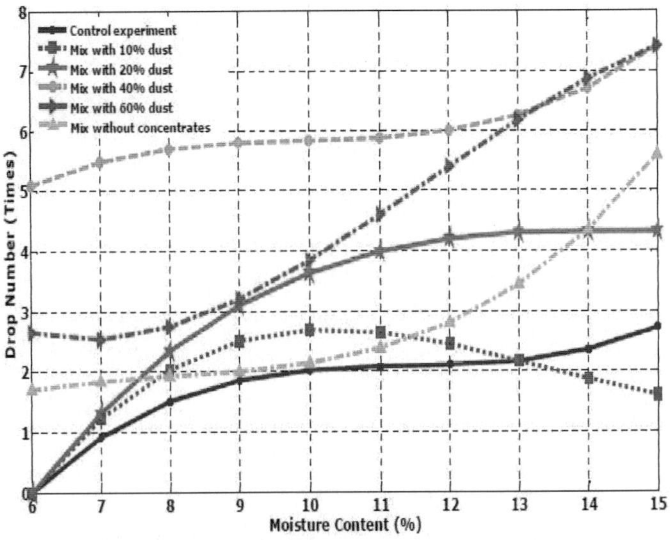

Figure 1: Relationship Between Drop Number of Balls and the Moisture Content for various mixes

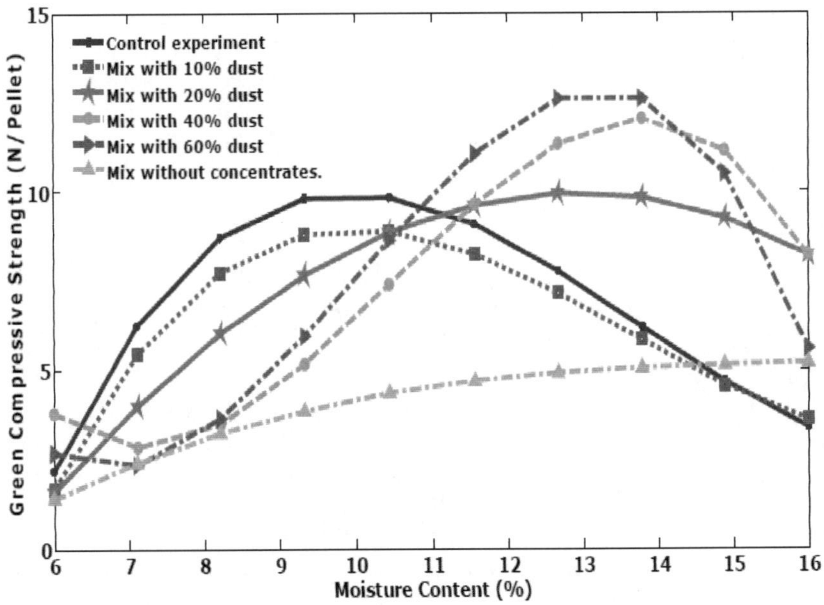

Figure 2: Variation of Green compressive strength with Moisture content

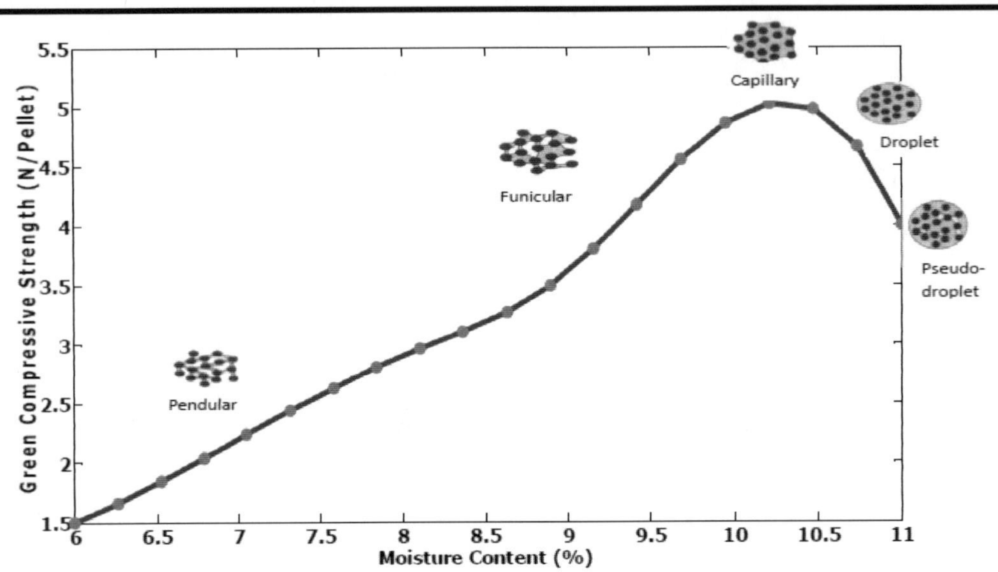

Figure 3. Effects of Moisture Content on the Compressive Strength of Balls

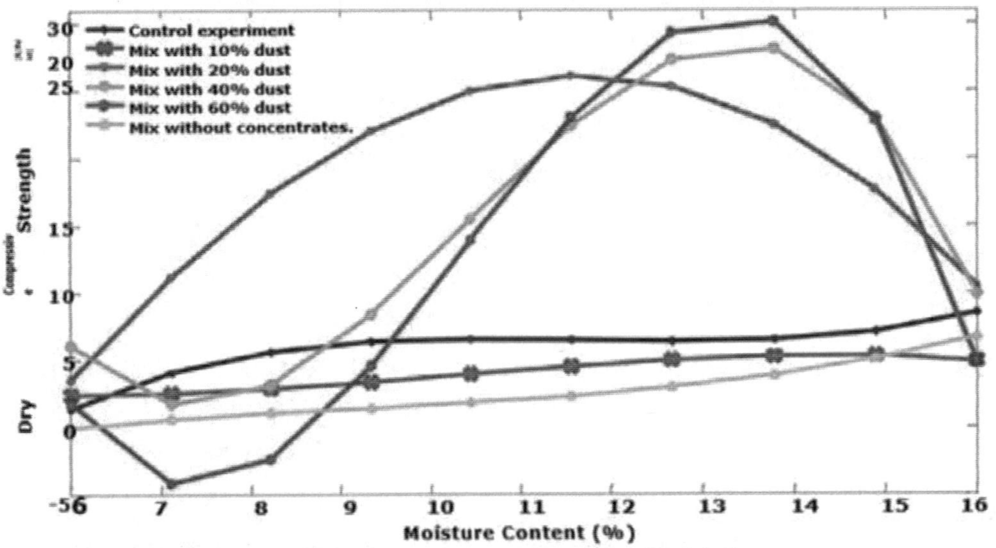

Figure 4: Variation of Dry Compressive Strength with Moisture

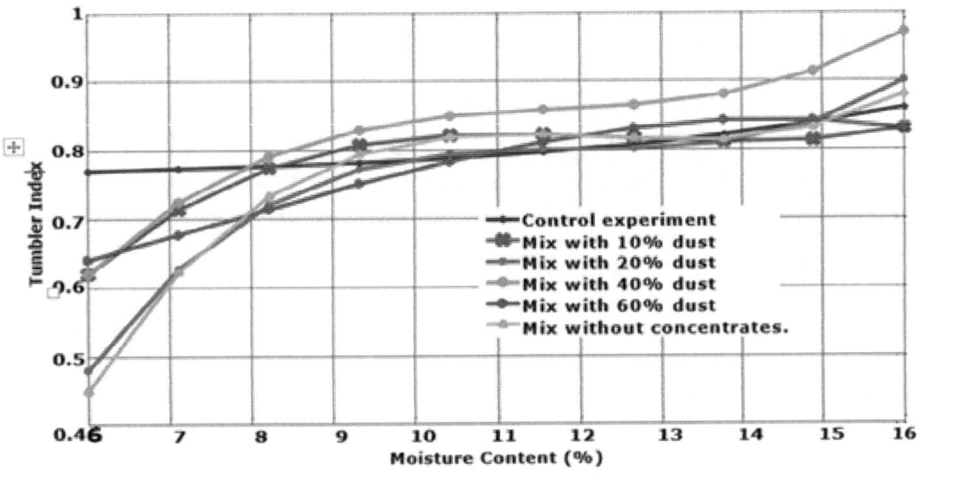

Figure 4: Variation of Tumbler Index with Moisture Content

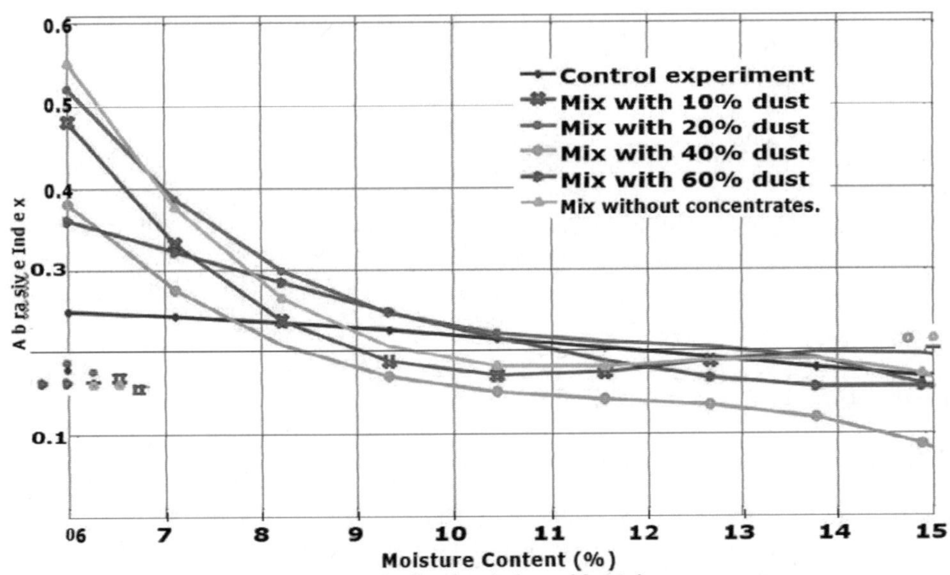

Figure 5: Variation of Abrasion index with Moisture content

The rise in the values of green compression strength and the drop number with increase in moisture content can be attributed to the binding forces (capillary forces) which depend on the rate at which the particulate materials in the balls are saturated with water. The reverse trend occurs when the optimum level of the water saturation is exceeded. This phenomenon can be explained by the capillary theory put forth by Newitt, et al,

1958 (Udo, et al, 2018). The theory postulates five levels of saturation of ball with water which include pendular, furnicular, capillary, droplet and pseudo - droplet levels or stages.

At pendular level , balls have the least liquid saturation: particulate materials are held together by liquid bridges and the capillary force at this stage is at its lowest value. At funicular stage, the liquid saturation is higher and the capillary force is stronger than at pendular stage. At capillary stage all the capillaries in the balls' pore structure are saturated with water and concave surface is found in the opening of the capillaries due to capillary force tending to draw in the water on the balls periphery. The capillary force and therefore, binding force is highest at this stage. There is a rise in water saturations at droplet and pseudo – droplet stages that give rise to low binding force (see Figure 3). This results in the collapse of the balls under much weight.
The green compressive strength and drop number increase as the percentage of IROBEWAS in the mix increases due to the presence of slaked lime in iron oxide bearing wastes (Figures 1and 2). The reaction of slaked lime with water results in the formation of calcium hydroxide [$Ca(OH)_2$)] which is a good binder for

iron oxide materials (Orlov, et al, 1993; Ahmed, et al, 2005). Therefore, the presence of iron oxide bearing wastes in the mix enhances the balling processes. This results in higher wet or green strength values. It should also be noted that the more the quantity of iron oxide bearing wastes in the mix the more the quantity of water that would be required to give the optimum value of green strength for that amount of the wastes. This is because of the hydrophobic nature of iron oxide bearing wastes. When balls' particulate materials are well bonded by capillary forces from water and plasticity forces from the binder, the balls will have the strength to withstand the compressive stresses and the drops they will encounter without disintegrating back into dust.

The highest green strength and drop number values in this work are 11.7 N/pellet and 7.8 times respectively recorded in the mix with 60 % iron oxide bearing wastes at 14 % moisture contents.

The dry compressive strength of the balls increases as the percentages of moisture and iron oxide bearing wastes increase in the mix (Figure 4). This can be attributed to the hardening of binding agents, precipitation of dissolved salts, friction and interlocking between particles, and electrostatic, magnetic and Vander Waals interaction (Satry, 1984). The highest dry strength value of 25.99 N/pellet is obtained from mix with 14 % moisture and 60 % iron oxide bearing wastes contents.

Figures 5 and 6 show the results of tumbler index and abrasion index tests respectively. They show that balls made with iron oxide bearing wastes can have good values of these indices. High values of tumbler index and low values of abrasion index of balls indicate a good binding quality in the ball structure that will disallow easy disintegration of the balls into dust. A high tumbler index of over 95% and corresponding low abrasive index of below 5% were attained by balls made from the mixture of iron oxide bearing wastes and iron ore concentrates.

4. Conclusion

The study revealed

- Iron oxide bearing wastes exhibit good balling properties:
  - Drop Number up to 7.8 times (standard minimum = 3 times)
  - Wet Compressive Strength up to 11.7 N/pellet (standard minimum value = 9 N/pellet)
  - Dry Compressive Strength up to 25.99 N/pellet (standard minimum value = 15 N/pellet)
  - Tumbler Indices up to 95% (standard minimum value = 85%)
  - Abrasive Index up to 4% (standard minimum value = 15%)
- Major balling properties of iron oxide bearing wastes that is Drop Number, Wet and Dry Compressive Strengths, Tumbler and Abrasive Indices are affected by moisture content of the mix
- Pulverized iron oxide bearing wastes is similar in composition to pellet charge. It may therefore be used as suitable additives to the charge.
- The grain sizes of iron oxide bearing wastes are poly-dimensional with narrower intervals of dispersion. The pulverized wastes consist of particles less than 125 μm (about 84.5 % with average diameters of 90 μm. The materials are in sizes identical to those of the pulverized concentrates for balling (Table2).

❖ Iron oxide bearing wastes exhibit good binding property due to the presence of slaked lime in form of Ca $(OH)_2$

References

[1] A. Abioye, O. P. Abioye, O. O. Ajayi, S. A. Afolalu, M. A. Fajobi and P. O. Atanda, Mechanical and Microstructural Characterization of Ductile Iron Produced from Fuel- Fired Rotary Furnace, International Journal of Mechanical Engineering and Technology 9(1), 2018. pp. 694–704

[2] Abioye, A. A., Atanda, P. O., Abioye, O. P., Afolalu, S. A., & Dirisu, J. O. (2017). Microstructural Characterization and Some Mechanical Behaviour of Low Manganese Austempered Ferritic Ductile Iron. *International Journal of Applied Engineering Research*, *12*(23), 14435-14441.

[3] Afolalu, S. A., Salawu, E. Y., Okokpujie, I. P., Abioye, A. A., Abioye, O. P., Udo, M., & Ikumapayi, O. M. (2017). Experimental Analysis of the Wear Properties of Carburized HSS (ASTM A600) Cutting Tool. *International Journal of Applied Engineering Research*, *12*(19), 8995-9003.

[4] Ahmed Y.M.Z., Mohammed, F.M., 2005, *la Metallugigia, Italiana*, pp. 31-33

[5] Blast Jurgen, 2007, *Metallurgical Plant and Technology (MPT) International*, vol. **3,** pp. 32-38.

[6] Esezobor D. E., Balogun S. A., Adeosun S. O., 2008. Development of Waste-free Technology in Steel Industry, 2008 Global Symposium on Recycling, Waste Treatment, and Clean Technology (REWAS 2008) October 12 – 15, 2008 Hilton Cancum Golf and Beach Resort, Cancum, Mexico, pp. 267 – 276.

[7] Esezobor, D. E., Apeh, F. I., Udo, M. O., Fabiyi, M., & Apeh, E. S. (2015). Evaluation of Cost Effectiveness of Onibode Fire-Clay for Production of High Quality Refractory Bricks. *Journal of Minerals and Materials Characterization and Engineering*, *3*(05), 399.

[8] Indian Bureau of Mines, 2011, Iron & Steel Vision 2020.Government of India, Ministry of Mines. Indian Bureau of Mines, August, Nagpur, India

[9] Newitt, D.M., Conway – Jones, J.M., 1958,. *Transaction of Institution of Chemical Engineers*, vol. **36**, pp. 422 – 442.

[10] Orlov, A. V., Rostovsky, D. E., Esezobor, D. E., 1993, *Journal of the Republic Environment Control and Rational Utilization of Natural Resources,* vol. **1719**, series 4, pp.88 – 91.

[11] E. O. Obidiegu, M. A. Bodude and M. O. Udo, 2014, International Journal of Multidisciplinary and Current Research. .Influence of Periwinkle Shell addition on Mechanical Properties of Gray Cast Iron vol. 2, pp 1116 – 1118.

[12] Rashid Ar, 2001, Centrifugal Granulating Process for Preparing Drug-layered Pellets Based on Microcrystalline Cellulose Beads. PhD Dissertation to Department of Pharmacy, University of Finland, pp. 11 – 14

[13] Satry, K.V.S., 1984, *Metallurgists*, vol. **28**, series 5, pp. 151 – 154

[14] Topiary, R. H., Topiary, V.R., 2007, Instruction to Modern Iron making, Khan Publishers, 2 –B, Nath Market, NaiSarak, Delhi - 110006 [8] ASTM, 2010, Standard Test Method for Determination of Abrasion Resistance of Iron Ore Pellets and Sinter by the Tumbler Test. ASTM International, West Conshohocken, PA, USA

[15] Werner Kaas, Heinz Maas, Harry Serbent, Horst Steinnofel, 1979, *Metallurgical Plant and Technology,* vol. **4**, pp. 11 – 14.

[16] Udo, M. O., Esezobor, D. E., Apeh, F. I., & Afolalu, A. S. (2018). Factors Affecting Ballability of Mixture Iron Ore Concentrates and Iron Oxide Bearing Wastes in Metallurgical Processing. *Journal of Ecological Engineering Vol*, *19*, 3.

IOP Conf. Series: Materials Science and Engineering **413** (2018) 012043 doi:10.1088/1757-899X/413/1/012043

Synthesizing a Metal Matrix Composite (Al- Al_2O_3) from Aluminium (AA1237) Reinforced with Alumina (Al_2O_3) Particulates

Nwanji, E. C[1*], Babalola, P. O[1]

Department of Mechanical Engineering, Covenant University, Ota, Nigeria E-mail address: Emmanuel.nwanji@stu.cu.edu.ng, phillip.babalola@covenantuniversity.edu.ng

Abstract. Developing new materials with sophisticated material properties has been one of man's greatest achievement in recent times. Composite has been helping the aerospace, automobile and marine industries where weight shedding is an added advantage. New materials like carbon fibres and carbon nanotubes are now being used extensively in the manufacture of certain vehicles parts and components. This research project dealt with the synthesis of a metal matrix composite using aluminium (AA1237) as the base matrix and reinforcing it with alumina particulate (5% and 10%); a ceramic material of both 80nm and 500nm sizes. The composites were developed using the stir casting method and the casted samples were prepared and subjected to various material testing. For hardness test, the specimen C1 (reinforced with 10% of alumina 80nm) has the highest value with 14.9HB, for electrical test, it was seen that an increase in the amount of reinforcement further reduces the electrical properties such as conductivity. Other tests that could not be carried were predicted using the Rule of Mixture concept.
Keywords: Materials, composite, rule of mixture, properties

1. Introduction

Composite is made up of two major constituents, the matrix and the reinforcement. The main purpose of the matrix is to transfer and distribute the load to the reinforcement. Reinforcement takes different forms of fibres, particles, whiskers, and sheets. The composite load-carrying capacity depends majorly on the bonding type between the matrix and reinforcement and the fabrication route [1]. For example, in reinforced concrete, the matrix is the concrete (sand, cement, stone) mixture while the reinforcement is the steel bars. The reinforcement gives the matrix additional strength since concrete is weak in tensile but strong for compressive stress. Composite are classified based on the matrix material and they include metal matrix composite (MMC), ceramic matrix composite (CMC), polymer metal matrix (PMC) and carbon/graphite metal matrix (CGMC). In this work, metal matrix composite was synthesized using aluminium matrix (AA1237) and alumina particulates using stir casting method. Other processing routes are squeeze casting method, vapour deposition, powder metallurgy, vacuum hot pressing, and co-spray deposition process. Stir casting method is preferred to other methods because it is cost effective and processing parameters could be readily varied and monitored [2,3,5]. Below is a flowchart that describes the steps taken in the research.

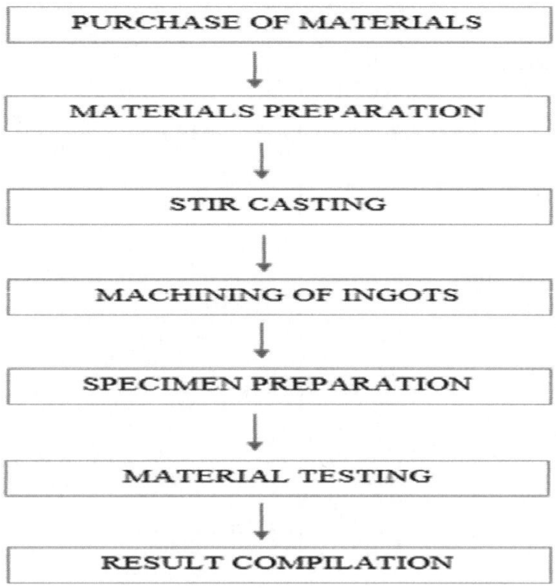

Figure 1: Show research flow chart timeline.

2. Material Preparation

Materials needed were sourced both locally and international, the aluminium ingot of 8.4kg was purchased from the local market here in Nigeria while the alumina particulate of two different grain sizes (80nm and 500nm) 100g each were purchased from US Research Materials, an American company that specialises in the processing, treatment and sales of chemical elements and compounds in the nanoscale. The weight percentage of reinforcement were at 5%, and 10% respectively. Using the volume fraction formula (Equation 1), the mass of alumina particulate required were calculated.

$$\frac{R_m}{M_m + R_m} = \%R_m \tag{1}$$

Where,
Rm = Mass of Reinforcement
Mm = Mass of Matrix

After calculating the required mass of reinforcement needed to achieve 5% and 10% mass fraction for both grain size (80nm and 500nm), the required mass was measured using a digital chemical balance, the result is populated below in the material table;

IOP Conf. Series: Materials Science and Engineering **413** (2018) 012043 doi:10.1088/1757-899X/413/1/012043

Table 1: Materials Preparation

S/N	SPECIMEN	SUB SPECIMEN	ALUMINIUM %	ALUMINA %
1.	**A**	-	100	-
	MASS	660g	660g	-
2.	**B (500nm)**	**B1**	95	5
		MASS	450g	23.69g
		B2	90	10
		MASS	430g	47.78g
3.	**C (80nm)**	**C1**	95	5
		MASS	510g	26.84g
		C2	90	10
		MASS	450g	50g

3. Stir Casting

Stir casting is one of the novel methods to produce metal matrix composites with more uniform distribution of matrix and reinforcement constituents. This approach involves mechanical mixing of the reinforcement particulates/particles into a molten metal bath [4]. The stir casting was done using an oil-fired furnace with graphite crucible to melt aluminium to 750^{0}C [5] with subsequent mechanical stirring. Many factors affect the final properties of the composite like how the variation in the geometry of the stirrer and the feeding mechanism affects the homogeneity of the material.

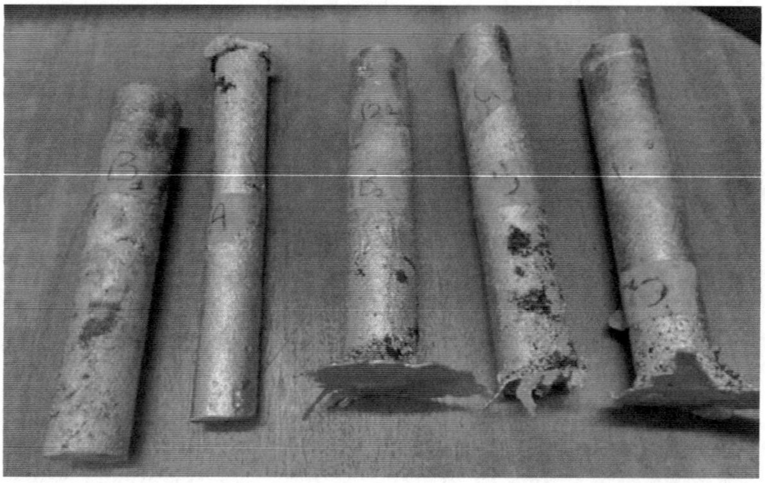

Figure 2: Showing the casted samples (A, B1, B2, C1 and C2)

4. Testing

Previous researcher [6] have shown that there was significant improvement in material properties after the addition of such reinforcement. The tensile strength of Aluminium metal matrix composite was improved by the addition of the Al_2O_3 particles, the tensile strength and porosity of 6% vol of Al- Al_2O_3 composite decreased with increasing reinforcement. Percentage elongation of the composite decreased with increase in Al_2O_3 content which confirms that alumina increases brittleness. Increasing of hardness with increasing weight percentage of Al_2O_3 particles is mainly due to grain reinforcement and particle strengthen effects.

4.1 Hardness Test

The Brinell hardness test was done using a TQ SM1000 Universal Testing Machine with a steel ball of 10mm diameter indenter, the small circular samples were placed in a cavity aligned to the axis as the steel ball indenter, a lever is pushed dropping the steel ball on the sample using some form of hydraulic power, count for 15secs and then take the reading of the applied force on the digital readout display. Using a granule (magnifying glass with a measuring scale), the diameter of the indentation on the samples were measured. The Brinell hardness value was calculated using the Brinell scale formula.

$$HBS = 0.102.\frac{2.F}{\pi.D\ (D-\sqrt{D^2-d^2}} \qquad (2)$$

Where,
HBS = H(Hardness), B(Brinell), S(Steel)
F = Applied force
D = Indenter diameter
d = Indentation diameter

Table 2: The Brinell Hardness Result

SAMPLE	APPLIED FORCE (N)	IDENTER DIAMETER(D) (mm)	IDENTATION DIAMETER(d) (mm)	HB (ROM PREDICTIED)	HB (ACTUAL)
A	6628	10	7.5	-	12.7
B1	6432	10	7.3	17.3	13.6
B2	6631	10	7.0	20.6	14.6
C1	6424	10	7.0	17.3	14.6
C2	6570	10	7.0	20.6	14.9

4.2 Electrical Test

For the electrical test, this was carried out in the Physics department of Covenant University, the aim was to determine the resistance offered by each of the five samples, then use the value to calculate conductance, conductivity and resistivity.

SAMPLE	VOLTAGE (V)	CURRENT (A)	RESISTANCE($\frac{V}{I}$) (Ω)	CONDUCTANCE($\frac{1}{R}$) (S)	CONDUCTIVITY ($\frac{1}{\rho}$) (S/m)	RESISTIVITY($\frac{RA}{L}$) (Ω*m) X 10^{-3}
A	0.38	2.17	0.175	5.714	94.876	10.54
B1	0.10	2.05	0.049	8	75.1	6.24
B2	0.62	2.17	0.03	3.5	40.16	10
C1	0.34	2.18	0.05	6.41	81.96	9.5
C2	0.10	2.13	0.047	5.13	78.1	4.03

Table 3: Electrical Result

4.3 Elemental Analysis

The spectrometer analysis was done at Aluminium Rolling Mills, Ota Ogun State. The elemental analysis was done using a Metavision 108N+ spark spectrometer for all samples. This was done to determine the exact sub-series of the 1000 series which contain the aluminium we used, from the analysis it was since that the aluminium was AA1237.

5. Results and Discussion

Rule of mixture is a predictive tool used to speculate some specific material properties of discontinuous and continuous composite using the volume fraction of both the matrix and reinforcement. The only hindrance to the rule of mixture concept is that it does not take in consideration the grain size of reinforcement used like the Hall – Petch model relation. The rule of mixture formula can be used to speculate properties like elastic modulus, ultimate tensile strength, mass density, thermal conductivity and electrical conductivity is given as:

$$E_c = ME_m + (1 - M)E_f \qquad (3)$$

Where,
Ec = Designated material property of composite
Em = Designated material property of the matrix
Ef = Designated material property of the fibres
M = Mass fraction of the matrix in percentage = $\frac{M_m}{M_m + M_f}$

Using the rule of mixture concept, we were able to predict some important material properties.

5.1 Predicted Properties

Table 4: Predicted Brinell Hardness

S/N	VOLUME FRACTION (MATRIX) %	VOLUME FRACTION (REINFORCEMENT) %	MATERIAL PROPERTY (MATRIX) Mpa	MATERIAL PROPERTY (REINFORCEMENT) Mpa	COMPOSITE MATERIAL PROPERTY Mpa
1.	0.90	0.10	20	80	26
2.	0.95	0.05	20	80	23

Table 5: Predicted Young's Modulus

S/N	VOLUME FRACTION (MATRIX) %	VOLUME FRACTION (REINFORCEMENT) %	MATERIAL PROPERTY (MATRIX) Mpa	MATERIAL PROPERTY (REINFORCEMENT) Mpa	COMPOSITE MATERIAL PROPERTY Mpa
1.	0.90	0.10	68.5	375	99.15
2.	0.95	0.05	68.5	375	83.83

Table 6: Predicted Density

S/N	VOLUME FRACTION (MATRIX) %	VOLUME FRACTION (REINFORCEMENT) %	MATERIAL PROPERTY (MATRIX) kg/m^3	MATERIAL PROPERTY (REINFORCEMENT) kg/m^3	COMPOSITE MATERIAL PROPERTY kg/m^3
1.	0.90	0.10	2700	3960	2826
2.	0.95	0.05	2700	3960	2763

5.2 Hardness Result

From the graph below, it is seen that the sample C2 has the highest hardness value with a value of 14.9 MPa. It is also seen that there is a positive trend to the graph which suggests that an increase in the volume fraction of reinforcement leads to an increase in hardness value of samples and C2 having the highest value means that nanocomposite have better hardness value than micro composites with the same volume fraction of reinforcement.

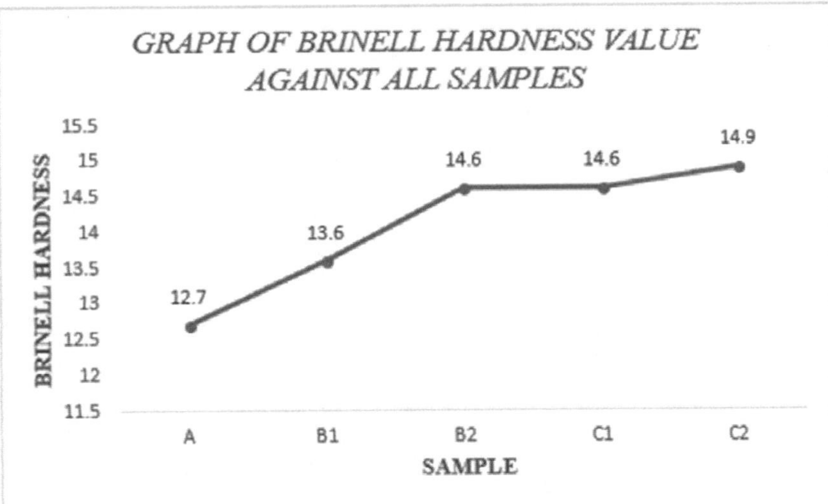

Figure 3: A graph of Brinell Hardness against all samples

IOP Conf. Series: Materials Science and Engineering **413** (2018) 012043 doi:10.1088/1757-899X/413/1/012043

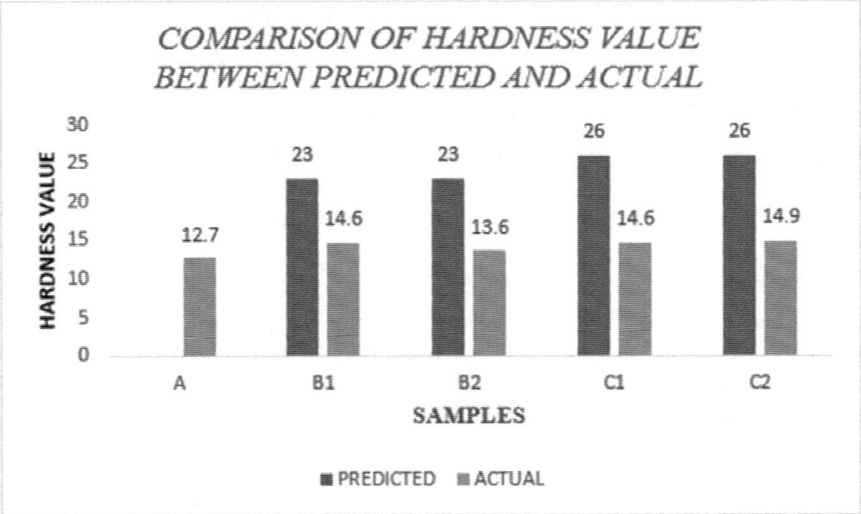

Figure 4: A comparative graph between ROM and actual Brinell Hardness value.

From graph, it is seen that the predicted value for hardness exceeds the actual for both the 5% and 10% reinforcement of both sizes, this is not questionable or strange because the rule of mixture does not take into consideration factors that can affect the final properties of the composite such as grain size, type of matrix and reinforcement used, processing route, physical and chemical properties of both matrix and reinforcement.

5.3 Electrical Result

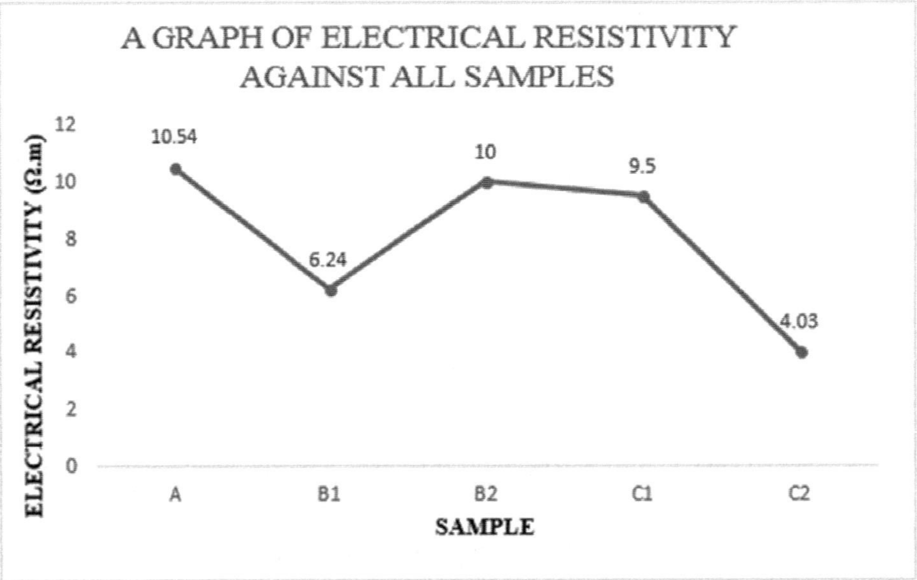

Figure 5: A graph of Electrical Resistivity against all samples.

From the graph above, it is seen that the sample with the highest value for electrical resistivity is sample A (pure aluminium AA1237) and the lowest value was sample C2 (reinforced with 10% of 80nm). Ideally, electrical resistivity is meant to be either sample C2 or B2 because alumina is a ceramic material and also an electrical insulator and an increase in volume fraction used should increase the electrical resistance of the composite.

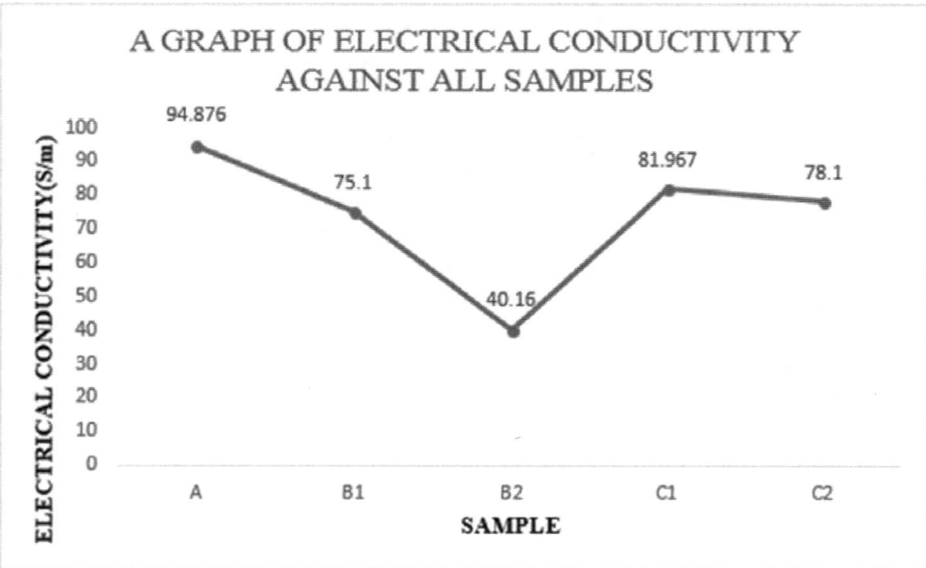

Figure 6: A graph of Electrical Conductivity against all samples

6 Conclusion and Recommendations

6.1 Conclusion

For this research work, a metal matrix composite was developed form aluminium (AA1237) reinforced with alumina particulate (80nm and 500nm) sizes, using various material testing method and from the experiments carried out in line with experimental results, it can be seen that specific material properties like hardness, conductivity and resistivity was improved and was also better than the monolithic material. Specific material properties like hardness was seen to improve by a large margin, suggest that it can be used for an application requiring high hardness. Other material test that could not be done was estimated using rule of mixture concept.

6.2 Recommendations

- All casting should be carried out the same day to avoid variance in various properties of composites.
- When quenching the casted sample, it is advisable to allow the samples to cool down in air.
- More test such as wear test, thermal conductivity test and mechanical test should be carried out for further verification of improvement in material properties.
- More focus on developing bio-composites using biological materials like fly ash, rice husk, and animal bones as reinforcement. These are not expensive, easily accessed and available in large quantities.

IOP Conf. Series: Materials Science and Engineering **413** (2018) 012043 doi:10.1088/1757-899X/413/1/012043

References

[1] Bhanot, V. K., & Singh, D. (2013). Research work on Composite Epoxy Matrix & Ep Polyester reinforced material. *International Journal of Engineering Research & Technology*.
[2] Kumar, B.A. and Murugan, N., *Mater. Des.*, **40**, 52 (2012);
[3] https://doi.org/10.1016/j.matdes.2012.03.038.
[4] Adams, M.A.R., Kevin F. and Robert, A.H., "Aluminum silicon carbide
[5] (AlSiC) for advanced microelectronic packages", in *Ceramics Process Systems*, Corp. Chartley, MA 02712, 1998.
[6] Ramesh, B. T., Koppad, V., & Hemanth, R. T. (2017). Fabrication of Stir casting setup for metal matrix composite. *International Journal for Scientific Research and Development, 5*(06), 944 - 949.
[7] Babalola, P. O., Bolu, C. A., Inegbenebor, A. O., & Odunfa, K. M. (2014, April). Development of Aluminium Matrix Composite: A Review. *Online International Journal of Engineering and Technology Research, 2*, 1 - 11.
[8] Hemalatha, K., Venkatachalapathy, S. K., & Alagumurthy, N. (2013, Nov 6). Processing and Synthesis of Metal Matrix Al 6063/ Al2O3 Metal Matrix Composite by Stir Casting Process. *International Journal of Engineering Research and Application, 3*(6), 1390 - 1394.

IOP Conf. Series: Materials Science and Engineering **413** (2018) 012044 doi:10.1088/1757-899X/413/1/012044

Weight Optimization of Square Hollow Steel Trusses Using Genetic Algorithm

*** A.N. Ede[1], O.O. Oshokoya[1], J.O. Oluwafemi[1], S.O. Oyebisi[1], O.M. Olofinnade[1], A. I. Akpabot**

[1]Civil Engineering Department, College of Engineering, Covenant University, Ota, Ogun State, Nigeria,

Abstract. Conceptual design in structural engineering entails a large amount of trial and errors or extensive expertise to obtain the most economical and functional design solutions for large engineering projects. In this paper a modern optimization technique called Genetic algorithm, adopting its concept from genetic evolution is used to optimize the shape, size and topology of a plane truss structure with the aim of minimizing the total weight of the truss. A genetic algorithm developed in MATLAB was implemented in this paper to optimize the weight of plane truss structures. The objective function of the optimization problem is subjected to constraints such as stress limits, buckling constraints, tension and compression capacity according to British steel design code BS 5950. The plane trusses which were subject to point loads were tested in the genetic algorithm, the resulting optimized truss structures were then subject to real life loading to determine their feasibility to withstand real life loading. The optimized trusses presented by the algorithm were modelled in a structural analysis and design software called SAP 2000, where they were subjected to dead and live loads. After design the weight saving discovered between the original trusses and the optimized version was between 37 - 47%. The results show that the genetic algorithm implemented in this study is useful in optimizing the weight of a plane truss structure.
Keywords: Steel, Structural optimization, Genetic Algorithm

1. Introduction

Sustainability of the environment involves the optimum usage of resources to create infrastructures or projects that meet present day needs and will still be useful to meet future needs [1] An important topic in the sustainability of civil engineering structures is the topic of structural optimization. Structural optimization is a concept that is introduced during the conceptual design stage of engineering structures. During the conceptual design stage, a lot of trial and error or intuition by experts is required to obtain a good structural solution. However, with structural optimization the designer can define the objectives of the design and the constraints to help obtain good and optimal structural solutions [2].
Structural optimization can be categorized into shape optimization, topology optimization and size optimization. Shape Optimization treats the geometry of the truss as the design objective and the design variable considered is the node coordinates. In topology optimization the connectivity of the members is the objective while the number of nodes is the variables. For size optimization the members size is the objective while the design variable is area of available cross-sectional profiles (discrete) or a specified range of member area (continuous) [3].
Researches in structural optimization can be dated back to the 1900's when Michell presented theoretical optimum shapes for statically determinate trusses. Since then, many other researches have continued in that path, it became more intense with the increase of computational capabilities [4]. This study was therefore carried out to optimize the weight of plane steel trusses by considering topology, size and shape using genetic algorithm technique. When the optimization problem is without constraints it is called a non-constrained optimization but when the optimization process is subjected to one or more constraints, it is referred to as a constrained optimization [5].

2. Truss Optimization using Genetic algorithm

A truss system consists of straight bars joined together at ends to produce rigid framework which will aid easy load distribution and transfer to the corresponding support in form of purely axial force [6]. Optimization techniques are categorized into classical and modern methods of optimization. Classical

Methods are analytical and make use of differential calculus for locating optimum solutions e.g. Linear programming, nonlinear programming etc. While modern methods adopt their operation from nature. It is a probabilistic based approach e.g. genetic algorithm, particle swamp optimization, ant colony optimization etc. [7].

The genetic algorithm is a method for solving both constrained and unconstrained optimization problems that is based on natural selection, the process that drives biological evolution. Some of the concepts adopted form this field are chromosome, genes, reproduction, mutation, crossover, populations, genotype etc. The technique was first consolidated by John Holland in 1975 [8].

Deb and Gulati [9] used binary encoded genetic algorithms to optimize an 11-member and six node 2D truss structure that was constrained by stress limits and buckling capacity according to Eurocode 3. The optimized structure gave an overall weight of 4899.15 kg, a much lighter weight than previous researches. Osman et al. [10] also developed a genetic algorithm using a different optimization process to find the optimum weight of plane and space trusses. The proposed genetic algorithm was used to reduce computation time significantly by reducing the required design variables for the optimization.

2. Development of genetic algorithm in MATLAB

The genetic algorithm and direct search toolbox embedded in the MATLAB software was used as a tool in developing the genetic algorithm. MATLAB is a powerful language and it has the relevant built-in functions to run the algorithm [13]. The finite element method was used for the analysis of the truss structures produced during operation of the algorithm. It is used to determine the fitness values of the various truss solutions. Finite element method is a robust and effective analysis technique for computer solutions of various engineering problems. It is used to solve complex engineering problems that analytical methods cannot accurately solve [11].

For the optimization of the size, topology and shape of the plane truss structures, design constraints were considered to ensure feasibility of the truss structures. The constraints considered were; material constraints in which the sections used were limited to the ones available commercially, the truss structure must have all basic nodes (support nodes and load nodes), the truss structure must be kinematically stable and the displacement limit of the node is between Lx/250 and Lx/240, as stated in BS 5950. After the genetic algorithm produces results of an optimized truss structure, the configuration of the optimized truss and original truss were analyzed in an analysis and design software called SAP 2000. Figure 1 shows a cross-section of the square hollow steel section used in this paper.

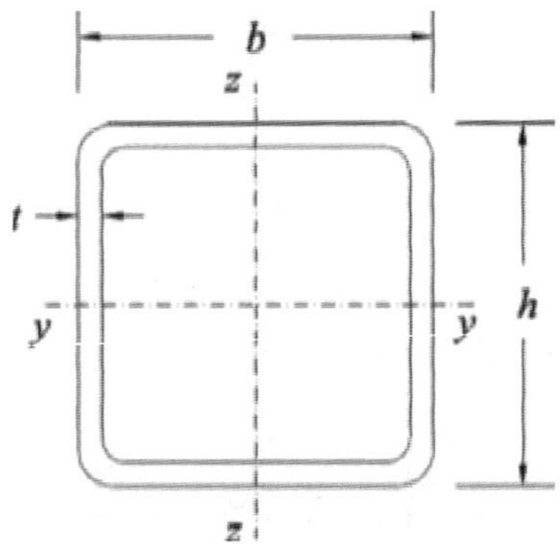

Fig 1: Cross section of a square hollow steel section [12]

3. Results and Discussion

The genetic algorithm developed in MATLAB was tested on two truss problems. The trusses are, a cantilever truss structure and a pitched truss. Figures 2a and 2b show a representation of the trusses.

IOP Conf. Series: Materials Science and Engineering **413** (2018) 012044 doi:10.1088/1757-899X/413/1/012044

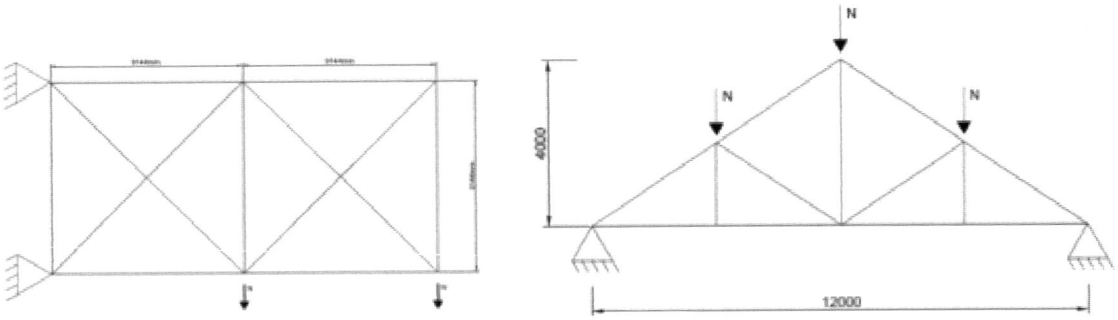

(a) Cantilevered truss with 6 nodes 2-point 2-point and 2 support nodes (b) Pitched truss structure With 6 Nodes

Fig 2: Plane truss test problems

Table 1 Result of optimized cantilever truss in the algorithm

Member	Cross-sectional profile	Weight	Start coordinates (x1, y1)	End coordinates (X2, y2)	% allowable stress	Stress (N/mm2)
2	300x300x6.3	528.5232	0, 0	9.144, 0	97.2612	-180.5268
3	150x150x6.3	363.3771	0, 9.144	9.144, 0	90.283	248.2783
4	160x160x5	425.7522	0, 9.144	16.6, 3.1	87.4321	240.4383
5	200x200x5	245.4731	9.144, 0	16.6, 3.1	84.6533	-127.1448
8	120x120x4	61.7711	16.6, 3.1	18.288, 0	99.3574	273.2328
10	160x160x5	220.9704	18.288, 0	9.144, 0	92.5886	-78.2301
	Total weight	1849.9258kg				
	Displacement	Horizontally 11.267 mm Vertically 71.3408 mm				

Figures 3,4 and 5 show the original and optimized cantilever truss structure with the details being presented in tables 1, 2 and 3. The optimized cantilever structure was thereafter modelled in SAP 2000 software and subjected to dead and live loads. The result of analysis and design showed that all truss members successfully passed stress checks in accordance to BS 5950 requirements. The total weight saving of the optimized cantilever truss structure compared to the original truss is 47%. For the cantilever truss the weight of the original truss when subjected to dead and live loads gave a value of 5970.22 kg and the optimized truss gave a weight value of 3146.2Kg. From the original truss the number of members were 10 but when the algorithm optimized the truss the number of members reduced to 6 showing that 4 members in the original truss were non-critical.

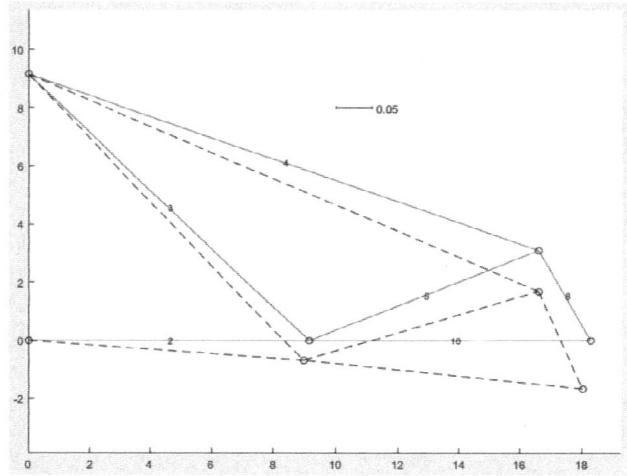

Fig 3: Optimized cantilever truss in the algorithm

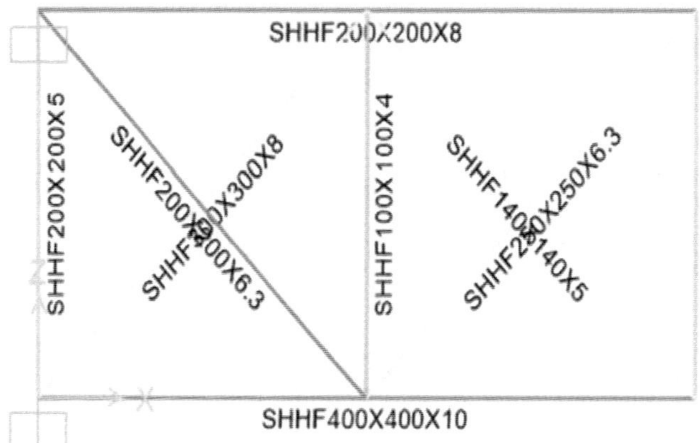

Fig 4: SAP 2000 model and section sizes of the original cantilever truss

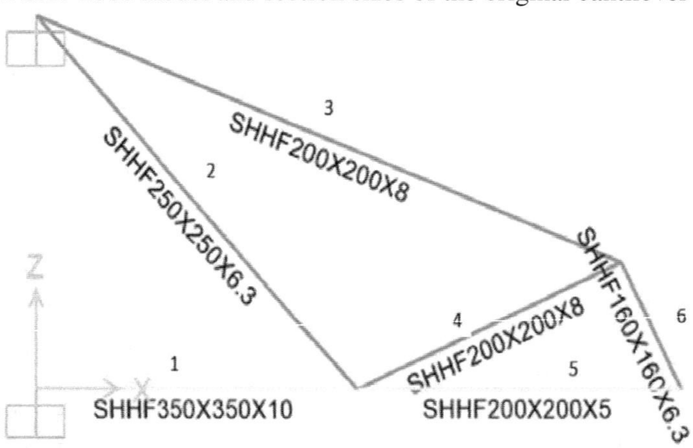

Fig 5: SAP 2000 model and section sizes of the optimized cantilever truss

Table 2 Sap 2000 design result for the original truss

Member	Cross-sectional profile	Length (m)	Weight (Kg)	Stress (N/mm2)
1	400X400X10	18.288	2231.136	-147.416
2	150X150X5	9.144	206.6544	114.576
3	200X200X8	18.288	872.3376	256.576

4	200X200X5	9.144	277.976	0.492
5	100X100X4	9.144	108.8136	-122.206
6	300X300X8	12.93157	941.418296	160.037
7	250X250X6.3	12.93157	619.4222	-85.804
8	200X200X6.3	12.93157	491.3997	263.567
9	140X140X5	12.93157	271.5657	244.275

	Total weight	5970.723496Kg
	Displacement	Vertical = 53.916mm Horizontal = 12.21mm

Table 3 Sap 2000 design result for the optimized cantilever truss

Member	Cross-sectional profile	Length (m)	Weight (Kg)	Stress (N/mm2)
1	350X350X10	9.144	969.264	-122.945
2	250X250X6.3	12.93157	619.44	152.684
4	200X200X8	17.66607	842.6715	125.359
5	200X200X8	8.07477	381.6	-81.612
6	160X160X6.3	3.52978	106.2463	135.073
7	200X200X5	9.144	227.9776	-63.862

	Total weight	3147.1994Kg
	Displacement	Vertical = 38.82mm Horizontal = 7.02mm

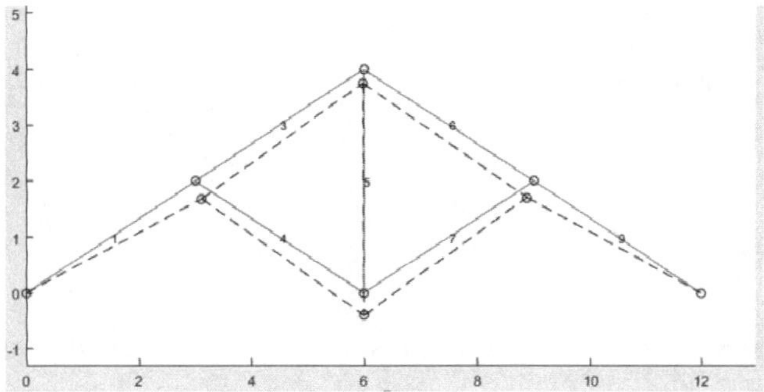

Fig 6: Optimized pitched truss in the algorithm

Table 4 Result of optimized pitched truss in the algorithm

Member	Cross sectional profile	weight	Start coordinates (x1, y1)	End coordinates (X2, y2)	% of allowable stress	Stress (N/mm2)
1	120x120x4.9	63.0971	0,0	3,2	78.6274	- 159.8234
3	100x100x5	53.0016	3,2	6,4	85.2432	- 142.3244
4	80x80x3.2	27.5104	3,2	6,0	73.992	-91.6939
5	50x50x4	22.5600	6,4	6,0	98.8142	271.7391

6	120x120x4	51.9199	6,4	9,2	71.1524	-145.6985
7	80x80x3.2	27.5104	6,0	9,2	73.992	-91.6939
9	160x160x5	86.8938	9,2	12,0	48.8696	-116.3081

	Total weight	336.6586kg
	Displacement	Horizontally 4.3712 mm, 27.32% of the limit Vertically 13.9069 mm, 28.9728% of the limit

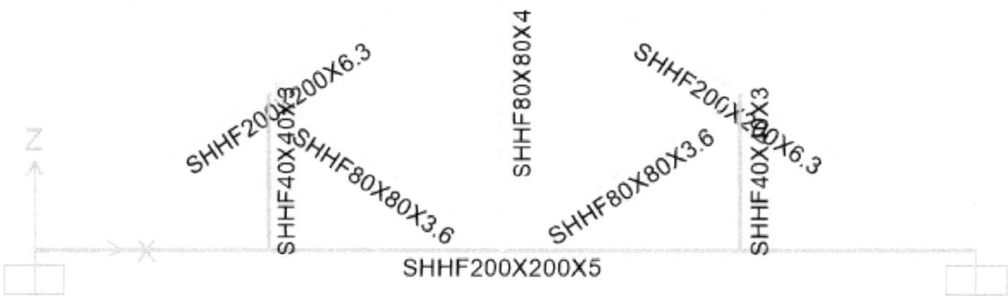

Fig 7: SAP 2000 model and section sizes of the original pitched truss

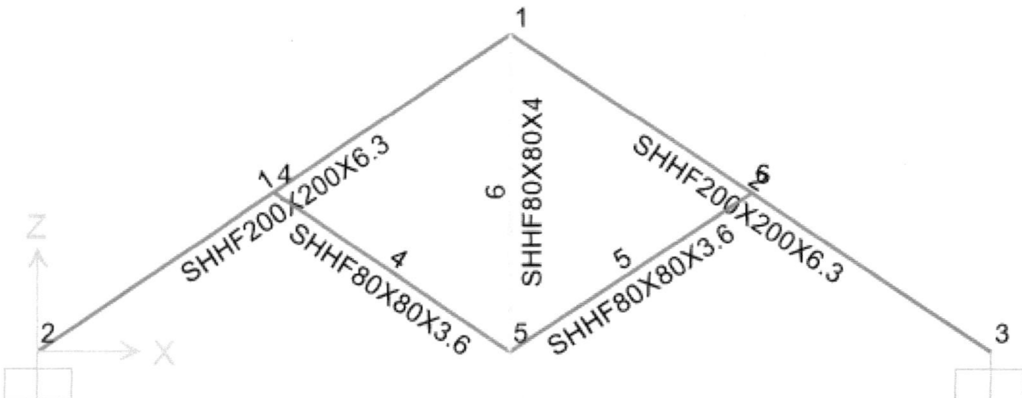

Fig 8: SAP 2000 model and section sizes of the original pitched truss

Figure 6 shows the optimized pitched truss after the optimization process in the algorithm. The original truss structure had 10 members but after the optimization the truss structure had 5 members. The optimized pitched truss with the original truss structure were modelled in SAP 2000 as shown in figures 7 and 8. The reduction in weight of the pitched roof truss is about 37-47 %. As presented in tables 4, 5 and 6, the weight of the original truss structure after being subject to dead and live load was 1097.9 kg, but after optimization the truss weight was reduced to 688.08 kg.

Table 5 Sap 2000 design result for the original pitched truss

Member	Cross-sectional profile	Length (m)	Weight (Kg)	Stress (N/mm2)
1	200X200X6.3	7.2111	270.218	-116.383

2	200X200X6.3	7.2111	274.0218	-116.383
3	200X200X5	12	406.8	48.484.
4	80X80X3.6	3.60555	51.19881	-68.769
5	80X80X3.6	3.60555	51.19881	-68.769
6	80X80X4	4	37.64	150.498
7	40X40X3	2	6.82	-22.084
8	40X40X3	2	6.82	-22.084
	Total weight	1097.89742Kg		
	Displacement	Vertical – 6.708mm Horizontal – 2.16mm		

Table 6 Sap 2000 design result for the optimized pitched truss

Member	Cross sectional profile	length	Weight (Kg)	Stress (N/mm^2)
1	200X200X6.3	7.2111	274.0218	-120.391
2	200X200X6.3	7.2111	274.0218	-120.391
4	80X80X3.6	3.60555	51.19881	-74.952
5	80X80X3.6	3.60555	51.19881	-74.952
6	80X80X4	4	37.64	153.105
	Total Weight	688.08122Kg		
	Max displacement	Vertical – 6.85mm Horizontal – 2.25mm		

4. Conclusion

A genetic algorithm was developed in MATLAB to optimize the weight of plane truss structures and the optimized trusses were modelled in a structural analysis and design software called SAP 2000 where they were analyzed and designed under real life loads. The trusses that were optimized showed significant reduction in the total weight by 30 - 40% showing that the algorithm is a useful tool for engineers in obtaining structures of optimal weight during design. All structures optimized were within acceptable stress and displacement limits. For further research works the genetic algorithm used in this paper can be implemented in the design of other structural elements using other available steel profiles provided in the design codes. Also, further researches are suggested for the optimization of 3D trusses such as tower cranes using genetic algorithm technique.

Acknowledgement
The authors wish to show their appreciation to the chancellor and the management of Covenant University for the Platform created to ensure the successful completion of this research work.

References

[1] Anthony, N. E., Gideon, O. B., Oluwarotimi, M. O., David, O. O., Gideon, A. A., & Ben, U. N. (2016). mpact of Reliable Built Structures in Driving the Sustainable Development Goals: A look at Nigerian Building Structures. *3rd International Conference on African Development Issues (CU-ICADI 2016)* (pp. 350-353). Ota: Covenant University Press.

[2] Mark, S., Eric, L., Chung-Soo, D., & David, S. (2009). Optimization Tools for the design of Structures. SEAOC 2009 Coventio Proceedings, 1-17.

IOP Conf. Series: Materials Science and Engineering **413** (2018) 012044 doi:10.1088/1757-899X/413/1/012044

[3] Kaveh, A., & Kalatjari, V. (2004). Size/geometry optimization of trusses by the force methodand genetic algorithm. ZAMM – Zeitschrift für Angewandte Mathematik und Mechanik., 84(5), 347-357.
[4] Rory, C. (2013). Algorithm Selection in Structural Optimization (1st ed.). Massachusetts: Massachusetts Institute of Technology.
[5] Margaret, E. K. (2012). A Two-Phase Genetic Algorithm for Simultaneous Dimension, Topology, and Shape Optimization (1st ed.). Ohio: The Ohio State University.
[6] Anthony, N. E., Olatunbosu, A., & Oluwarotimi, M. O. (2015). Modelling, Analysis and Design of a MultiStorey Storey Helipad-Car Park: a Proposal for Canaan. International Journal of Innovative Science and Modern Engineering (IJISME), Volume - 3 Issue-4, pp43-47.
[7] Singiresu, S. R. (2009). Engineering Optimization (1st ed.). New Jersey: John Wiley & Sons,Inc.
[8] Reeves, C. R., & Rowe, J. E. (2002). Genetic Algorithm: Principles and Perspectives- A Guideto GA Theory. New York.
[9] Deb, K., & Gulati, S. (2000). Design of Truss-Structures for Minimum Weight using Genetic Algorithms. Finite Elements in Analysis and Design, 447-465.
[10] Osman, S., Atef, E., Tharwat, S., & Osman, H. (2014). Optimization of Plane and Space TrussesUsing Genetic Algorithms. International Journal of Engineering and Innovative Technology (IJEIT), 3(7), 66-73.
[11] Austrell, P. E., Dahlblom, O., Lindemann, J., & Olsson, A. (2004). CALFEM- A Finite Element Toolbox. Lund: Structural Mechanics.
[12] BSI (2000). Structural use of steelwork in building - Part 1: Code of practice for design - Rolledand welded sections. BSI.
[13] MathWorks. (2017). Genetic Algorithm and Direct Search Toolbox 2.4.2. Retrieved November 17, 2017, from www.mathworks.com

Developmental Design of an Orthopaedic Recovery System

I. O. Ohijeagbon[1*], O. C. Omoragbon[2], J. O. Aweda[3], A. S. Adekunle[4], J. F. Opadiji[5]

[1, 2, 3, 4]Department of Mechanical Engineering, University of Ilorin, Ilorin, Nigeria
[5]Department of Electrical Engineering, University of Ilorin, Ilorin, Nigeria

E-mail addresses: [1, *]idehaiohi@yahoo.com, [2]cynthy_8@yahoo.com, [3]jacobaweda@gmail.com, [4]adekunlebayo@yahoo.com, [5]jopajidi@yahoo.com

Abstract- A developmental design of an orthopaedic recovery system has been conducted. Anthropological data of mass and length distribution of the body segments were used to estimate the components and total length of the main supporting frame and maximum body load that may be accommodated by the system. Vital geometrical, operational load and power parameters were also designed for the required hydraulic subsystems. Results show that the total length of the main supporting frame of the system is 2.237 m and the maximum body load that can be accommodate on the system is 371 kg. The load and power requirements were highest at the hip joint, with the following values: maximum supporting load 2,313.16 N, internal pressure of hydraulic system 379.89 kN/m^2, buckling load of connecting rod 9252.64 N, critical buckling load of hydraulic system 13370.06 N, required power 86.56 W, expected electrical power input 149.93 W and spring stiffness 6.18 kN/m respectively. When constructed, the developed design is expected to facilitate treatment and recovery of orthopaedic patients. Physiotherapeutic services for body joints related problems would be greatly aided to offer better quality services. Required human involvement and effort on the side of the service providers shall be greatly reduced.

Keywords: Recovery system, anthropological data, operational load, hydraulic system, power

1. Introduction

Patients having neurologic injuries such as stroke, traumatic brain and spinal cord injuries form the largest inpatient populations in a typical comprehensive rehabilitation hospital. These patients typically receive intensive therapeutic care, mostly 2-3 times per week depending on patient recovery progress or other logistics constraints [1]. The effects of musculoskeletal ailment is not central only to the victim, rather, musculoskeletal conditions are a major burden on family members, workplace, health systems, and social care systems [2-3]. Aside from the direct cost associated with the treatment and recuperation of musculoskeletal conditions, the indirect costs are usually predominant. Musculoskeletal disorders, such as low back pain have been found to be a major cause of temporary work disability in many demographics, and consequently contribute to permanent disability pensions [4].

The disciplines primarily concerned with the treatment of muscle-joint-skeleton malfunction and motion impairment are orthopaedics and physiotherapy. Orthopaedics is the diagnosis, care, and treatment of patients with disorders of the bones, joints, muscles, ligaments, tendons, nerves, and skin which make up the musculoskeletal system [5-6]. While physiotherapy has to do with provision of services to individuals and populations to develop, maintain and restore maximum movement and functional ability where movement and function are threatened by ageing, injury, pain, diseases, disorders, conditions or environmental factors [7]. Several factors could be responsible for movement impairment in humans, such as, arthritis, neck pain, rheumatoid arthritis, fractures, low back pain, and osteoarthritis, to mention a few [5-6, 8-10].

Known methods for physiotherapeutic treatment of orthopaedic related problems over the years include: Maitland-the assessment and treatment of the patient's signs and symptoms to develop an effective treatment plan [11]; Cyriax (Orthopedic Medicine)-based on logic and reasoning, that all pain has a source, all treatment must reach the source, and all treatment must benefit the lesion [11-13]; McKenzie-based on "cause and effect", a systematic progression of mechanical forces applied (the cause) utilizes pain response (the effect) to monitor changes in motion, named as the three mechanical syndromes: postural, dysfunction and derangement [14]. Others are acupuncture, muscle energy technique, Mulligan, natural apophyseal glides (NAGS), myofascial therapy, exercise therapy, and massage [15-27]. Tremendous progress has been made on orthopaedic implants for patients

IOP Conf. Series: Materials Science and Engineering **413** (2018) 012045 doi:10.1088/1757-899X/413/1/012045

needing surgery or partial replacements of their musculoskeletal system [28-29], and also with orthopaedic rehabilitation devices, actuated with pneumatic muscles to help the recovery of the significant numbers of patients with lower limb bearing joint disabilities [30]. The field of biomechanics applications to human movement is found to be useful for improvement of performance and the reduction or treatment of injury. It is observed that human joints, namely, wrist, elbow, shoulder, ankle, knee, neck and girdle, exhibits motion similar to machine elements, such as, rolling, sliding, combined roll-sliding, spin, compression, and traction respectively [31-34]. Hence, the design and development of a holistic orthopaedic recovery system would be a perfect synergy with the joints in a human body during recovery.

Constant effort is being made in the field of rehabilitation robotics to facilitate the use of robotic devices to assist therapeutic intervention and recovery of patients with movement or posture problems [35-37]. Such devices as robotic machines operate on the principle of electro-mechanics or mechatronics, which combines mechanical, electrical and electronics engineering for electromechanical energy conversion processes [38-39]. Although laudable, robotic machines may require very high skill-set and expertise to operate them, and such level of professionalism may not always be easy to come by especially amongst remotely located health-care facilities or low-income earners.

The motivation for this work is identifying the need to enhance access to physiotherapy and restore movement and musculoskeletal functions, so that patients experiencing orthopaedic related problems can recover effectively. The study focuses on the development of a simple physiotherapeutic mechanical system for orthopaedic patients which can be electrically operated. It involves a design to produce a feasible machine; with a multifaceted system device for enhanced therapeutic applications at different joints in the human body as opposed to a localized application device or system.

2. Preliminary Design Analysis

2.1 Description of the orthopaedic recovery system

For the design under consideration, the orthopaedic recovery system consist of six major supports; namely, head, back, thigh, leg, foot and arm supports. The adjustments of the supports to desired positions shall be based on electromechanical principles, in which the various joints and supports will be controlled automatically when exercising or moving the body joints at a suitable range of movement. An electrical interface is expected to direct and regulate the mechanical system to aid the movement of various components for each intersecting joint member to experience the required range of motion depending on the particular need and rate of recovery of the patient.

2.2 Dimensional considerations for the orthopaedic recovery system

2.2.1 Sectional dimensions

The body dimensions used for design were estimated from the anthropometric, mass distribution and skeletal joint centres data obtained by stereo photographic techniques [40]. The design dimensions as shown in Figure 1 for the five major sections of the orthopaedic recovery system are listed as:

1. Length of head support : AB = 0.329 m
2. Length of back support: BC = 0.587 m
3. Length of thigh support: CD = 0.434 m
4. Length of leg support: DE = 0.600 m
5. Length of foot support: 0.287 m
6. Length of arm support (from shoulder to tip of middle finger): 0.841m (upper arm-0.352 m, fore arm-0.288 m, hand length-0.201 m)

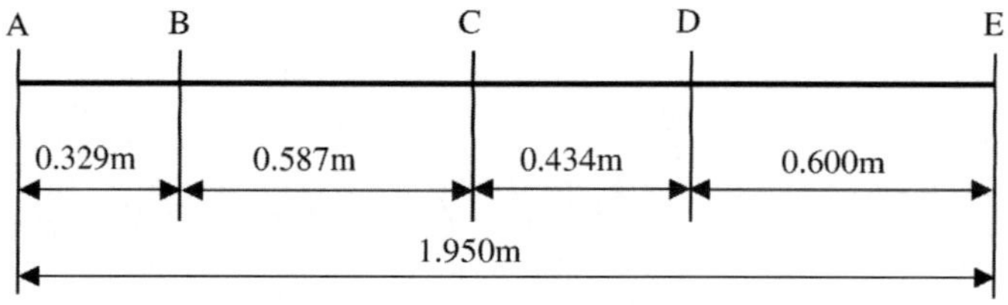

Figure 1: Diagram showing the total length of the system

2.2.2 Mass of body segments

The load that each segment of the system would carry was estimated as the design load. The design load is the upper estimate of forces that a load bearing member is expected to carry [41]. An average adult body mass is range between 63.3-97.7 kg [40, 42]. A design load of 120 kg was used for analysis. Various masses of body segments given in Table 1 were used to estimate mass distribution for each segment of the body by interpolation.

Table 1: Mass distribution of the body segments

Body segment	*Total body mass			Estimated design load	
	Small (63.30 kg)	Medium (85.10 kg)	Large (97.70 kg)	Mass (120.00 kg)	Weight (1,177.20 N)
Head	4.00	4.20	4.40	5.40	52.97
Neck	0.90	1.10	1.20	1.47	14.42
Thorax	18.60	24.90	30.50	37.46	367.48
Abdomen	1.90	2.40	2.90	3.56	34.92
Pelvis	8.60	11.80	14.60	17.93	175.89
Thigh	7.70	9.80	11.80	14.49	142.15
Calf	3.10	3.80	4.50	5.53	54.23
Foot	0.80	1.00	1.10	1.35	13.24
Upper arm	1.50	2.00	2.40	2.95	28.94
Fore arm	1.10	1.40	1.60	1.97	19.33
Hand palm	0.50	0.50	0.60	0.74	7.26

*Source: Anthropology Research Project [40]

3. Detailed Design Analysis of Critical Parts

3.1 Design of the structural frame

3.1.1 Determination of bending moments and shear forces of the frames

In order to conduct the structural and strength analysis of the system so as to obtain the total geometrical cross-sectional area and complete physical dimensions of members of the system, the schematic and free body diagrams of the system was set-up as shown in Figures 2 and 3. The parts numbered (1) to (6) in Figure 2 represent the major body supports of members of the system.

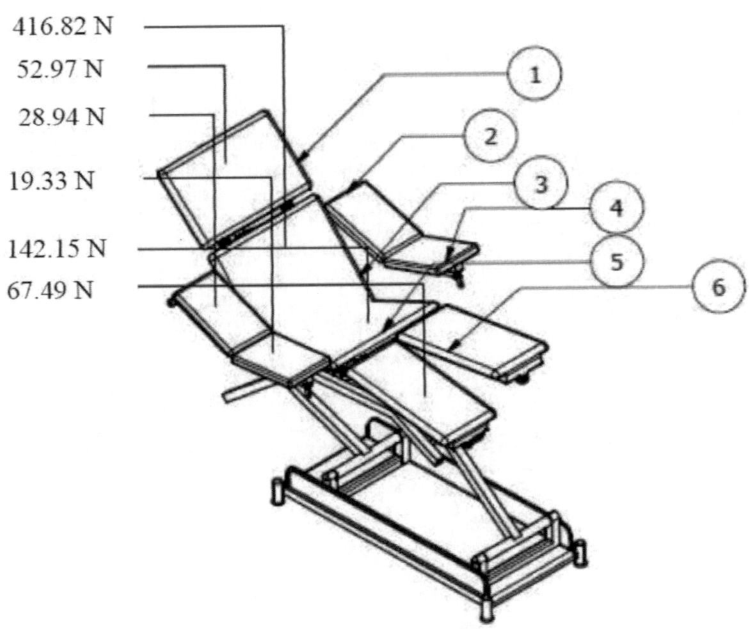

Figure 2: Schematic diagram of major body supports of the system

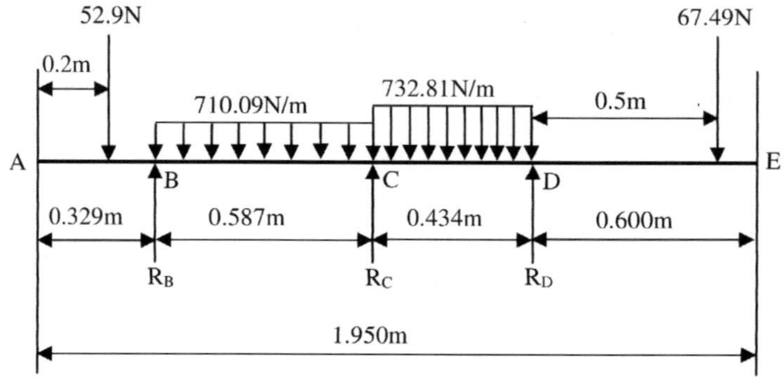

Figure 3: Free body diagram of the system

Applying the equations of static equilibrium and Clapeyron's theorem [41] of three moment equation for a two-span continuous beam of three consecutive supporting points (BCD) to the free body diagram of Figure 3, the bending moment at point C was obtained by equation (1). The method allows the bending moments at B and D to be obtained at first by taking moments about points B and D to obtain $M_A = -M_B$ and $M_E = -M_D$ thereafter the bending moment at point C.

$$M_B BC + 2M_C(BC + CD) + M_D CD = \frac{\omega_{BC}(BC)^3}{4} + \frac{\omega_{CD}(CD)^3}{4} \tag{1}$$

Where,
$\omega_{BC} = 710.09N/m$, $\omega_{CD} = 732.81N/m$

The moments at the supports B, C and D where determined as

$M_B = 6.833Nm$, $M_C = 15.783Nm$, $M_D = 33.745Nm$

The reactions at the supports B, C and D where determined as

$R_B = 288.271N$, $R_C = 376.834N$, $R_D = 190.145N$

The values of the bending moment and shear forces determined above were used to obtain the bending moment and shear force diagrams of the orthopaedic recovery system as shown in Figures 4 and 54 respectively. Furthermore, they were also used to estimate the cross-sectional dimensions of members of the system.

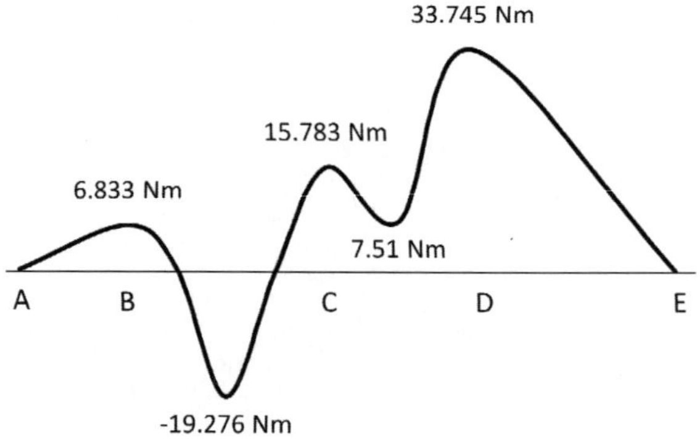

Figure 4: Bending moment diagram of the system

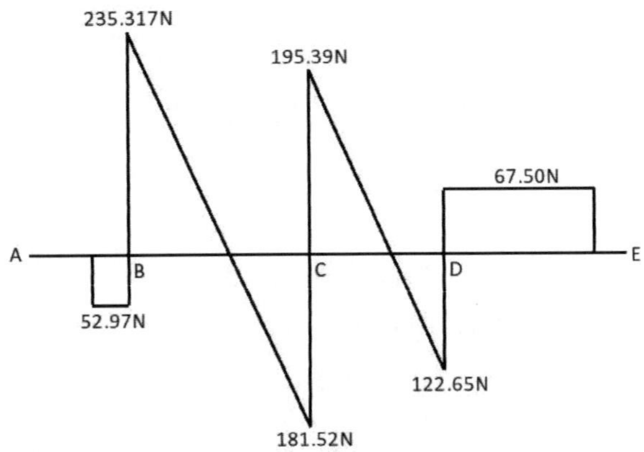

Figure 5: Shear force diagram of the system

3.1.2 Cross-sectional analysis of the frames

If each section of the frame structure is assumed to be a simply supported beam, then the maximum defection of a section can be expressed as:

$$\delta_{\max} = \frac{5\omega L^4}{384EI} \tag{2}$$

where, $\delta_{\max}$ is the maximum deflection, L is the length of the section considered, E and I are the Young's modulus of elasticity and second moment of area respectively. The allowable deflection of structural members is given by:

$$\delta_{allowable} = \frac{L}{240} \tag{3}$$

Using a structural steel pipe with the following specification conforming to IS: 4923 – 1997; minimum yield strength, 210 MPa, and minimum ultimate tensile strength (UTS), 330 MPA [43]. And the Young's modulus of elasticity for steel is taken as 200 GPa. Hence, by equating equations (2) and (3), the second moment of area can be expressed as:

$$I = \frac{240}{L} \times \frac{5\omega L^4}{384EI} = \frac{240 \times 5\omega L^3}{384E} \tag{4}$$

Therefore for the specified length CD of 0.434 m, the second moment of area was determined as $I = 9.36 \times 10^{-10} m^2$.

For a solid cross-section of uniform dimension, the second moment of area can be expressed as

$$I = \frac{bh^3}{12} = \frac{b^4}{12} \tag{5}$$

where b and h are the width and height of the cross-section respectively. Consequently, the width and height of the cross-section was determined as: $b = h = 0.0103m = 10.3mm$. Hence a minimum solid cross-sectional dimension of $b = h = 15mm$ was selected.

3.2 Design of the hydraulic system

3.2.1 Analysis of sectional maximum operating loads

The orthopaedic recovery system was designed to operate on the principle of hydraulic transmission on the basis of the maximum operating loads. Values of the working loads (F_w) for each hydraulic transmission accessory required for various parts of the system were used together with a factor of safety (n) of 4 to determine the maximum operating loads. The maximum operating loads (F_{max}) for each required hydraulic transmission accessory were obtained as presented in Table 2 by equation (6):

$$F_{max} = n \times F_w \tag{6}$$

Table 2: Maximum load distribution for hydraulic transmission accessories

Body segment	Estimated working load, F_w (N)	Maximum operating load, F_{max} (N)
Head	52.97	211.88
Neck	14.42	57.68
Thorax	367.48	1,469.92
Abdomen	34.92	139.68
Pelvis	175.89	703.56
Thigh	142.15	568.60
Calf	54.23	216.92
Foot	13.24	52.96
Upper arm	28.94	115.76
Fore arm	19.33	77.32
Hand palm	7.26	29.04

3.2.2 Design features

The design of the hydraulic system is on the basis of a single acting spring retracting cylinder as exemplified by the cross-sectional representation shown in Figure 6. A single-acting cylinder is pressurized at one end only, with the opposite end vented to the atmosphere through a breather filter or air cylinder or vented to a reservoir (hydraulic cylinder) [44]. The retraction or return stroke of the cylinder is accomplished by an incorporated compression spring, while the ends of the cylinder and piston-rod unit are provided with bearing connections to the main body of the system [44-45]. When the cylinder is pressurized (suction of hydraulic fluid), the hydraulic system experiences an extension due to the displacement of the piston-rod unit from the resting position at one end to the pressurized state at the other end, in which case the free-length of the compression spring is compressed and shortens in length. On reaching the compression limit, the check valve is actuated and opened for the exit of hydraulic fluid and release of pressure. Consequently, the piston-rod unit is then retracted to the rest position due to the extension of the compression spring back to its original length.

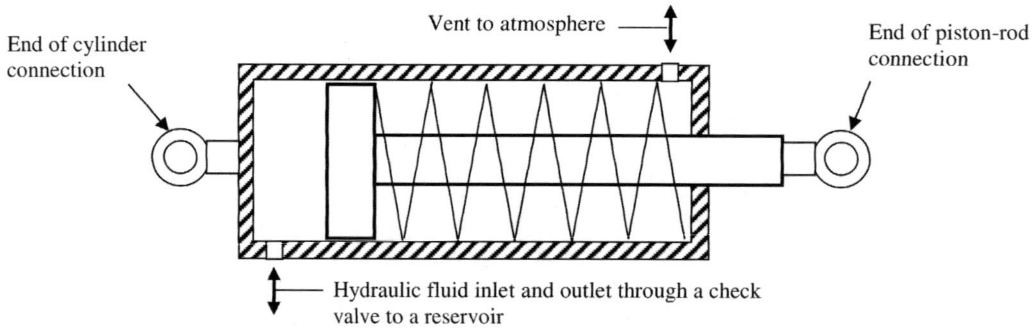

Figure 6: Single acting spring retracting cylinder

3.2.3 Buckling load analysis

The buckling load analysis was conducted in line with Standard for Certification 2.9 [45]. The moments of inertia for the cylinder and the piston-rod can be evaluated as follows:

IOP Conf. Series: Materials Science and Engineering **413** (2018) 012045 doi:10.1088/1757-899X/413/1/012045

$$I_C = \frac{\pi(D_o^4 - D_i^4)}{64} mm^4 \tag{7}$$

$$I_R = \frac{\pi d^4}{64} mm^4 \tag{8}$$

where,
I_C = moment of inertia for the cylinder, mm^4
I_R = moment of inertia for the piston rod, mm^4
D_o = outer diameter of the cylinder, mm
D_i = inner diameter of the cylinder, mm
d = diameter of the piston rod, mm

Since the piston-rod is clamped at one end to the main frame of the system and free to move in the cylinder, therefore the buckling load of the piston-rod can be determined as a clamped-free condition given by [46]:

$$P_B = \frac{\pi^2 EI}{4L^2} = \frac{\pi^2 EI_R}{4L_R^2} \tag{9}$$

where,
E = Young's Modulus of Elasticity = 200 GPa for steel
L_R, length of piston-rod

Assumptions:

1. L, length between mountings in fully extracted position (L_C + $L_{r,v}$) equals 1.5 times of supporting member of the system, L_m:
   $$L = L_C + L_{r,v} = 1.5 \times L_m \text{ (where, } L_{r,v} = L_C\text{)} \tag{10}$$
   Hence,
   $$L_C = 0.5 \times L = 0.5 \times 1.5 \times L_m = 0.75 \times L_m \tag{11}$$
2. L_S, compressed solid length of retracting spring is 50% of L_C, length of piston-rod or:
   $$L_S = 0.5 \times L_C \tag{12}$$
3. D_P, L_P, diameter/thickness of piston is 20% of L_C, length of cylinder or:
   $$D_P = L_P = 0.2 \times L_C \tag{13}$$
   (internal diameter of cylinder, $D_i = D_P$, tolerance = 1 mm)
4. $L_{S,f}$, free length of spring:
   $$L_{S,f} = L_C - L_P \tag{14}$$
5. L_R, length of piston-rod is given as:
   $$L_R = 1.5 \times L_C + L_P \tag{15}$$

The acceptable criterion for the buckling load is expressed as [45]:

$$\frac{P_B}{P_a} \geq 4 \tag{16}$$

where,
P_a = actual maximum load = F_{max} (see Table 2)
P_B = buckling load, N

IOP Conf. Series: Materials Science and Engineering **413** (2018) 012045 doi:10.1088/1757-899X/413/1/012045

The diameter of the piston-rod, d was therefore determined from equations (9) and (16), by first stating $P_B = 4 \times P_a$, and consequently determining I_R and d respectively.

The actual maximum load can be expressed in terms of hydraulic system pressure, p and the cross-sectional area of the piston as follows:

$$P_a = \frac{p \times \pi D^2}{4} \tag{17}$$

where, D = diameter of the piston, mm

For control purposes, the Euler breaking or critical load, P_{eu} for a bar with the same dimensions as the piston-rod and with a length corresponding to the fully extracted cylinder is expressed as [45]:

$$P_{eu} = \frac{\pi^2 EI_R}{2 \times L^2} \tag{18}$$

where, $P_{eu} > P_B$

The wall thickness of the cylinder can be determined from the allowable longitudinal stress in the cylinder, σ_a and internal fluid pressure, p as follows [47]:

$$t = \frac{p \times D_i}{4 \times \sigma_a} + 2\,mm \tag{19}$$

$$\sigma_a = 4 \times p \tag{20}$$

The outer diameter of the cylinder is expressed as:

$$D_o = D_i + 2t \tag{21}$$

3.2.4 Spring stiffness analysis

The analysis of the spring stiffness of the hydraulic systems and consequently spring selection was carried out based on Hook's law, expressed by equation (22):

$$k = \frac{F}{x} \tag{22}$$

where,
k = the spring stiffness
$F = F_{max}$ = compression load exerted on the springs = the load that must be overcome by the springs during retraction of each component members of the system
= maximum deflection experienced by the springs during operation

From previous analysis above, the maximum deflection that can be experienced by the springs during operation is 50% of L_R, length of piston-rod:

$$x = 0.5L_R \tag{23}$$

3.2.5 Pumping system analysis

The pumping system is designed on the basis of a volumetric flow rate $V_C / 10\sec$. That is, a central pumping system with the aid of checking valves would be incorporated in the recovery system to deliver an internal fluid pressure sufficient for a positive displacement of any of the piston within 10 seconds of application of pump power. The fluid power P_o can be related to the fluid pressure, p and volumetric flow rate, $\frac{V_C}{10}$ by:

$$P_o = p \times \frac{V_C}{10} \tag{24}$$

Where, the maximum swept volume of each hydraulic cylinder is determined from the inner cross-sectional area of the cylinders and the maximum deflection experienced by the springs during operation. Hence, required power for each of the hydraulic system can be expressed as:

$$P_o = p \times \frac{\pi D_i^2 x}{4 \times 10} = \frac{p \pi D_i^2 x}{40} \tag{25}$$

The expected electrical power input of the pump at a unit power factor of for the hydraulic systems can therefore be determined as:

$$P_{el} = \sqrt{3} \times P_o \tag{26}$$

4. Results and Discussion

4.1 Cross-sectional parameters of the frame

The total length of the orthopaedic recovery system in its completely stretched state was determined from the sectional dimensions shown in Figure 1. Hence, the total length of the system used for the bending moment and shear force analysis was 1.950 m. Furthermore, the total length was also used to estimate the cross-sectional dimension of the frame structure.

The cross-sectional dimensions of the frames could either be circular, square or rectangular, and may be solid or hollow. Considering each section of the fame structure that that bears different parts of the body load to comprise of solid square members, and also the most adverse condition of the uniformly distributed load and bending moment exerted on the frame structure with values of $\omega = 732.81 N/m$ and $M = 33.745 Nm$ respectively; the cross-sectional dimensions of the frame structure were determined.

The minimum cross-sectional area of the frame was obtained as 15 x 15 mm (225 mm^2). The solid cross section may however be replaced with a hollow square cross-section to enhance stability during fabrication and for better functional purposes. According to the specification IS: 4923 – 1997 [43], a 32 x 32 mm hollow cross-section of 4 mm thickness, and 3.19 kg/m is hereby recommended for the design of the framework of the system. The recommended cross-section results in a cross-sectional area of 240 mm^2, which is greater than the minimum of 225 mm^2.

The information on sectional dimensions and mass of body segments earlier determined was used for the analysis of bending moment and shear force analysis of the frames. From Figures 2 and 3, the part AB supports the load of the head, 52.97 N acting at a distance 0.20 m from A; the part BC supports the combined load of the neck, thorax and abdomen (416.82 N) and acts as a uniform distributed load of 710.09 N/m over the span BC; the part CD supports the combined load of the pelvis and thigh (318.04 N) and acts as a uniform distributed load of 732.81 N/m over the span CD; and the part DE supports the combined load of the calf and foot, 67.49 N acting at a distance 0.50 m from D. The above listed spans and loads were used to obtain other geometrical design parameters of the orthopaedic recovery system as presented in Table 3.

4.2 Geometrical parameters of the system

With the aid of equations (7) to (23), the detailed geometrical parameters of the orthopaedic recovery system were determined and presented in Table 3. The length of the supporting frames indicates highest values of 0.600, 0.587 and 0.434 m for the knee, hip and pelvis/thigh joints respectively, while the lowest supporting frames are those of the wrist, ankle and elbow joints with dimensions 0.201, 0.287 and 0.288 m respectively. The lengths of the supporting frames were found to directly correlate with other geometrical parameters of the system. The knee joint resulted in the highest value of length between mountings in fully extracted position (0.90 m), length of cylinder (0.45 m), length of piston-rod (0.765 m), compressed solid length of retracting spring (0.225 m) and free length of spring (0.36 m). The same goes for other geometrical parameters, such as inner diameter of piston, diameter and thickness of piston (90.00 mm), and outer diameter of piston (101.25 m). The thickness of cylinder was however the same for the knee and hip joints, while the hip joint resulted in the highest value of minimum diameter of piston-rod with a dimension of 22 mm, followed by the pelvis/thigh and knee joints which both has a value of 13.00 mm. The maximum deflection of spring occurred at the knee and hip joints at 0.38 and 0.37 mm respectively.

4.3 Operational load and power of the system

The operating load and power requirements for the orthopaedic recovery system was conducted using equations (9), (16) to (18), (20), (24) and (25) accordingly as presented in Table 4. The load and power characteristic shows that hip joint possesses the highest values. The maximum supporting load and buckling load of connecting rod were obtained as 2,313.16 and 9252.64 N respectively, while the internal pressure of hydraulic system and expected electrical power input were obtained as 379.89 kN/m^2 and 149.93 W respectively. The maximum value of spring stiffness for retraction of the hydraulic systems was also at the hip joint with a value of 6.18 kN/m. This was followed by the pelvis/thigh joint, having a value of 2.06 kN/m, while the lowest value of designed spring stiffness occurred at the wrist with a value of 0.23 kN/m accordingly.

These trends were due to high concentration of body load on the supporting frame between the neck and the hip. Consequently, the critical operating load and power requirements for the orthopaedic recovery system are those associated with the hip joint. Table 4 also reveal that values of the critical buckling load were higher than those of the buckling load of the connecting rod of the hydraulic system, hence, validating determined values of the diameter of the connecting rod.

The developmental design conducted in this study affords the option of using a centralised hydraulic pump which is capable of feeding different parts of the recovery system via a distribution network and checking vales as occasion demands. Alternatively, each part can be operated independent of others, since their individual operating parameters have been determined. A decentralised operational system may be preferred when energy is to be conserved. However, a multifaceted network system would be advantageous in instances whereby a particular patient requires physiotherapy at different body joints at the same time.

4.4 Developed orthopaedic recovery system

A 3-D model of the orthopaedic recovery system are presented in Figures 7 and 8, which respectively highlights the partial back/side isometric view, and the partial side/front isometric view of the orthopaedic recovery system. The Figures clearly presents a pictorial representation of the assembled system with required hydraulic systems installed in their respective positions. The Figures also shows installed upholstery material on the system to ensure comfort to users of the system. The maximum body weight the designed system can accommodate should not exceed 371 kg. This maximum value of body weight can be determined from the total sum of the maximum supporting loads of different body parts analysed and used in the design analysis as indicated in Table 4.

Table 3: Geometrical design parameters of the orthopaedic recovery system

Body joints	Length of supporting frame L_m (m)	Length between mountings in fully extracted position L (m)	Length of cylinder L_C (m)	Diameter, thickness of piston D_P, L_P (mm)	Length of piston-rod L_R (m)	Compressed solid length of retracting spring L_S (m)	Free length of spring $L_{S,f}$ (m)	Inner diameter of cylinder, D_i (mm)	Thickness of cylinder t (mm)	Outer diameter of cylinder D_o (mm)	Minimum diameter of piston-rod d (mm)	Maximum deflection of spring (m)
Neck	0.329	0.494	0.247	49.35	0.419	0.123	0.197	49.35	5	55.52	9	0.21
Hip	0.587	0.881	0.440	88.05	0.748	0.220	0.352	88.05	8	99.06	22	0.37
Pelvis/thigh	0.434	0.651	0.326	65.10	0.553	0.163	0.260	65.10	6	73.24	13	0.28
Knee	0.600	0.900	0.450	90.00	0.765	0.225	0.360	90.00	8	101.25	13	0.38
Ankle	0.287	0.431	0.215	43.05	0.366	0.108	0.172	43.05	5	48.43	6	0.18
Shoulder	0.352	0.528	0.264	52.80	0.449	0.132	0.211	52.80	5	59.40	8	0.22
Elbow	0.288	0.432	0.216	43.20	0.367	0.108	0.173	43.20	5	48.60	6	0.18
Wrist	0.201	0.302	0.151	30.15	0.256	0.075	0.121	30.15	4	33.92	4	0.13

Table 4: Operational design load and power parameters for the orthopaedic recovery system

Body joints	Maximum supporting load F_{max} (N)	Internal pressure of hydraulic system p (kN/m^2)	Buckling load of connecting rod P_B (N)	Critical buckling load of hydraulic system P_{eu} (N)	Required power P_o (W)	Expected electrical power input P_{el} (W)	Spring stiffness k (kN/m)
Neck	211.80	110.73	847.20	1224.20	4.44	7.69	1.01
Hip	2,313.16	379.89	9252.64	13370.06	86.56	149.93	6.18
Pelvis/thigh	568.60	170.83	2274.40	3286.51	15.73	27.25	2.06
Knee	269.88	42.42	1079.52	1559.91	10.32	17.88	0.71
Ankle	52.96	36.38	211.84	306.11	0.97	1.68	0.29
Shoulder	115.76	52.87	463.04	669.09	2.60	4.50	0.52
Elbow	77.32	52.75	309.28	446.91	1.42	2.46	0.42
Wrist	29.04	40.68	116.16	167.85	0.37	0.64	0.23

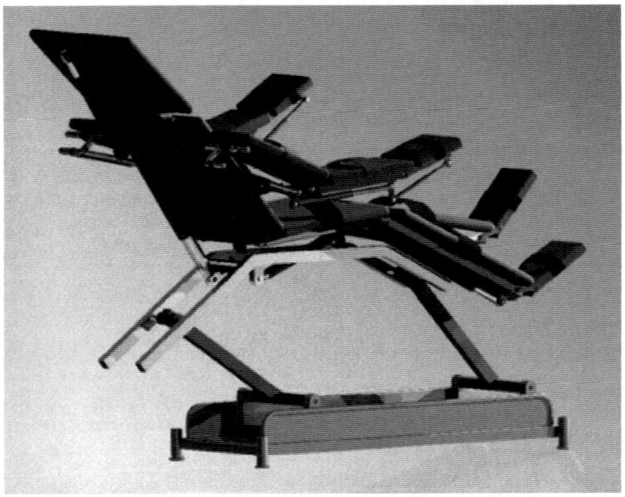

Figure 7: Partial back/side isometric view of the orthopaedic recovery system

Figure 8: Partial side/front isometric view of the orthopaedic recovery system

5. Conclusion

A feasible design of an orthopaedic recovery system has been developed in this study. The design of critical parts such as the main frame and hydraulic systems for operating the motion at various joint would facilitate the eventual construction of the system. The total length of the main supporting frame of the system is 2.237 m, while the length of the entire hand supporting frame is 0.841 m. The maximum body load the designed orthopaedic recovery system can accommodate is 371 kg, which far exceed the recommended body size of 97.70 kg for large persons [40]. Results obtained indicate that the hip joint has the highest design specification for load and power requirements. And these were reported in Table 4 as maximum supporting load 2,313.16 N, internal pressure of hydraulic system 379.89 kN/m^2, buckling load of connecting rod 9252.64 N, critical buckling load of hydraulic system 13370.06 N, required power 86.56 W, expected electrical power input 149.93 W and spring stiffness 6.18 kN/m respectively.

Furthermore, a 3-D representation of the orthopaedic recovery system showing installed upholstery and required hydraulic systems in their respective positions are also included in this study. It is to be noteworthy to mention that the developmental design conducted in this study would go a long way to facilitate treatment and recovery of orthopaedic patients. The system would also aid practitioners in the medical profession in dispensing better quality service at reduced human intervention and labour. Also, the system would contribute immensely to the socioeconomic status of any community and the world at large by ensuring that more and more people are able to become productive once again after experiencing the trauma of musculoskeletal function disorder or body joints related problems.

The study was limited to the design of critical geometric, load and power requirements. Hence, a substantive design of the hydraulic network functionality and detailed electrical application requirements are necessary to enhance the production of the system.

References

[1] David R, Lum P, Winters J. Emerging Technologies for Improving Access to Movement Therapy following Neurologic injury. Journal IEEE Press, 2002, 2: 1-2.

[2] Gauthy R. Musculoskeletal disorders: An ill-understood "pandemic". European Trade Union Institute, Belgium. 2007.

[3] Dembe A E. The Social Consequences of Occupational Injuries and Illnesses American Journal of Industrial Medicine, 2001, 40: 403-417.

[4] Woolf A D, Pfleger B. Burden of major musculoskeletal conditions, Bulletin of the World Health Organization. 2013, 81 (9): 646-656

[5] Health Status. Orthopaedic Disorders [online]. [Accessed 5th December 2013]. www.healthstatus.com/health_blog-shoulderjoint-pain/Orthopaedic-disorders.

[6] American Academy of Orthopaedic surgeons. Career in orthopaedics [online]. [Accessed 16th December 2013]. www.orthoinfo.aaos.org/topic.cpm.

[7] Physical Therapy [online]. [Accessed 17th February 2014]. http://en.wikipedia.org/wiki/physical_therapy.

[8] John Hopkins Medicine [online]. [Accessed 16th December 2013]. http://www.hopkinsmedicine.org.

[9] IASP. Global year against musculoskeletal pain, Rheumatoid Arthritis. Journal IASP Press. 2009, 1: 1.

[10] Bone Fracture [online]. [Accessed 7th January 2013]. http://en.wikipedia.org/wiki/Fracture.

[11] Maitland Australian Physiotherapy seminar [Online]. [Accessed 17thFebruary 2014]. http://www.ozpt.com/about.php

[12] Southorn N. Musculoskeletal physiotherapy. 2nd Edition. 2010, 52-58.

[13] Mashad University of Medical Sciences. Cyriax Method in physiotherapy, [online]. [Accessed 17th February 2014]. www.researchgate.netpost/ i_want_to_know_ more about Cyriax in physiotherapy.

[14] Mckenzie Institute. The McKenzie Method [online]. [Accessed 17th February 2014]. www.mckenziemdt.org/approach.

[15] Hodges I, Maskill C. Effectiveness of acupuncture for the treatment and rehabilitation of accident related Musculoskeletal disorders, Nzhta Report. 2002, 5: 55-59.

[16] Ernst E. Acupuncture – A critical Analysis. Journal of Internal Medicine. Blackwell Publishing Ltd. 2006, 3: 125-137.

[17] Massage college. Muscle Energy Technique [online]. [Accessed 17th February 2014]. www.massageclinic.com.au/service/MuscleEnergytechnique.

[18] Webster G. The physiology and application of Muscle Energy Techniques. Bridge of Allan co. 2002, 2: 19.
[19] John G. Muscle Energy Techniques, Lotus publishing. 2011, 1: 26-27.
[20] Mulligan J. The Mulligan Concept; The next step in the evolution of manual therapy. NZ Journal. 1999, 5: 1-2.
[21] Fernandez-de-las-penus C, Monica S, Josue F, Juan C M. Manual Therapies in Myofascia Trigger point treatment. Journal of body and movement therapies. 2003, 9: 28.
[22] John W. The effectiveness of myofascial Techniques in the treatment of myofascial Pain. 2011, 2: 45.
[23] NCBI Resources. Exercise Therapy [online]. [Accessed 19th February 2014]. www.ncbi.nlm.gov/mesh/68005081
[24] Medscape Therapeutic Exercise. Exercise Therapy [online]. [Accessed 19th February 2014]. www.medicine.medscape.com/article.
[25] Holey E, Cook E. Therapeutic Massage. WB Saunders, London. 1997.
[26] Massage [online]. [Accessed 19th 2014]. http://en.wikipedia.org/wiki/massage.
[27] Scott D. The Extraordinary benefits of Daily Massage. 5th Edition. 1999, 3-9.
[28] Standard of Care: Total Knee Arthroplasty. The Brigham and Women's Hospital, Inc., Department of Rehabilitation Services. 2008.
[29] McLaughlin R E. Total Joint Replacement and Other Orthopaedic Implants, Journal of Rehabilitation Research and Development Rehabilitation R & D Progress Reports. 1985, 42-63.
[30] Petre I, Deaconescu A, Rogozea L. Orthopaedic Rehabilitation Device Actuated with Pneumatic Muscles, International Journal of Advanced Robotic Systems. 2014, 1-8.
[31] Duane K. Fundamentals of Biomechanics, Springer Science and Business Media, LLC, 2nd Edition. 2007, 3-9, 23.
[32] Thompson C W, Floyd R T. Structural kinesiology. Journal IEEE Press. 1994, 1: 1.
[33] Cheng H. Introduction to Manual Therapy. Hung Kuang University Department of Physical Therapy, FA Davis Co., 4th ed. 2013, 10-14.
[34] Luttgens K, Hamilton N. Kinesiology: Scientific Basis of Human Motion, 9th Edition. 1997.
[35] Spasticity [online]. [Accessed 5th December 2013]. http://en.wikipedia.org/wiki/Spasticity.
[36] Hillman M. Rehabilitation Robotics from Past to Present: A historical perspective. Advances in Rehabilitation Robotics. Berlin: Springer-Verlag. 2004, 8: 25-44.
[37] Krebs H I. Rehabilitation robotics: Performance-based progressive robotassisted therapy. Automatic Robots. 2003, 2: 15, 7-20.
[38] Paulo D J. Mechatronics, John Wiley and Sons. 2011.
[39] Quang H. Principles of Electromechanical Energy Conversion, 2nd Edition, 2007.
[40] Joe H. Anthropometry and Mass Distribution for Human Analogues. 1988, 1: 6-38.
[41] Design load [online]. [Accessed 28th February 2014]. http://en.wikipedia.org/wiki/Design_load.
[42] Average weight [online]. [Aceessed 28th February 2014]. http://en.wikipedia.org/wiki/Averageweight.
[43] Pipes Tubular World. Bhushan Power and Steel Ltd [online]. http://www.bhushanpowersteel.com/assets/Pipe_Catalog.pdf . [Aceessed 20th June 2018]
[44] Milwaukee Cylinder. Design Engineering Guide. www.milwaukeecylinder.com. 2012.
[45] Standard for Certification 2.9. Hydraulic Cylinders, Det Norske Veritas, Norway. 2002.
[46] Patnaik S N, Hopkins D A. Strength of Materials. Elsevier, USA. 2004.
[47] Huda Z, Ajani M H. Evaluation of Longitudinal and Hoop Stresses and a Critical Study of Factor of Safety (FoS) in Design of a Glass-Fiber Pressure Vessel, International Journal of Civil, Environmental, Structural, Construction and Architectural Engineering. 2015, 9 (1): 39-42.

Suitability of Pulverized Cow Bones as a Paving Tile Constituent

J. O. Aweda[1], P. O. Omoniyi[2], I. O. Ohijeagbon[3,*]

Department of Mechanical Engineering, University of Ilorin, Ilorin, Nigeria

E-mail addresses: [1]jacobaweda@gmail.com, [2]lakepete05@gmail.com, [3, *]idehaiohi@yahoo.com

***Corresponding author**

Abstract- Economic advantage of waste has been clamoured severally in order to improve the environment and reduce pollution. In Nigeria, the common materials used for the production of paving tiles are granite dust and cement, in some cases silica sand is added. This study is aimed at investigating the feasibility of using available raw materials such as laterite, silica sand and pulverized cow bones to develop paving tiles. Tiles of 200 mm × 100 mm × 60 mm were produced and the effects of various materials added were investigated and analysed. The production method employed was casting and curing was done for twenty-eight (28) days. Physical and mechanical analyses of the paving tiles showed that the tiles with mixing ratio of 4:1:3 (laterite: pulverized cow bones: silica sand) with 20% cement stabilization was the optimum mix ratio, producing the maximum compressive strength and modulus of rupture of 5.05 MPa and 1.83 MPa respectively.
Keywords: Cow-bones, Compressive-strength, Laterite, Paving-tiles, Silica-sand

1. Introduction

The need to make judicious use of waste materials in the surroundings for wealth creation has been clamoured in order to reduce cost and sustain the environment. Animal bones disposal has been a major challenge in abattoirs across Nigeria especially in urban cities with huge population. Lagos, a city in south western Nigeria, accounts for 6,000 slaughtered cattle per day [1], in a country producing 13.8 million cattle yearly [2]. The time taken for bones to decay, which could be millions of years [3] add to the defacing of the environment when not properly disposed off. Recently, researchers have developed several ways to economically use animal bones to create wealth, some of which are in stonewares and animal feeds, where they are charred and pulverized. The process of charring, further constitute environmental pollution due to the exhaust caused by burning. Bones consists mainly calcium oxide (32.1%) and phosphorus pentoxide (28.3%) [4], giving it its load bearing capacity, which may be explored as building construction constituents and products.

Kim *et al.* [5] analyzed the average breaking strength of dry bones of cattle, which is the required weight to break a cattle bone such as femur, scapular etc. as 9.9 kg. Also, the potential of bones as pozzollanic substance was also established by [6].
Paving tiles are of different shapes which could be (rectangular, square, polygon), and are of different weights and dimensions [7]. They are also known to be of good load bearing capacity and have proved useful in marshy regions due to its good water absorption and permeability [8].
Conventionally, paving tiles are usually made of granite, silica sand and stone dust with cement as binder [9]. With increase in the rate of urbanization in Nigeria, there is huge demand for paving tiles. Granite one of the core material for production of paving tiles is decreasing, coupled with the high cost of crushing and transportation, therefore, the need to have other local alternative materials to compensate for this problem will be of good economic advantage.

Laterite is found almost everywhere in the tropics [10], so also is silica sand is found at river beds, which is gritty in texture, appropriate for mixing concrete and could be used as soil amendment, making tighter soils looser and could be used as stabilizers [11]. Several locally sourced raw materials which are alternatives to the conventional granite-cement mixture has been used in production of paving tiles. Raheem *et al.* [12] used laterite with cement as binder in the production of paving tiles. Works have been done on the use of agricultural wastes such as palm kernel shell as coarse aggregate [13], and corncobs addition as a substitute in paving tiles [14]. All have proved feasible but provides a gap that the agricultural wastes are seasonal i.e. not available all year round. The conventional materials had proved expensive and fast getting exhausted [7]. Humans are known to consume animal all year round, the cost and availability of laterite, silica sand, also makes the consideration of cow bones as a paving tile constituent sustainable.

This research aims at using available raw materials such as laterite, silica sand and pulverized cow bones, to produce paving tiles and study both its physical and mechanical properties, and its objectives were to analyse effects of various materials on the physical and mechanical properties of paving tiles. Furthermore, the optimum mixing ratio was also determined.

2. Materials and Method

Cow bones used in this study consist of femurs, scapulars and rib of cow, and were procured from Ipata abattoir in Ilorin Kwara State Nigeria. They were washed and sundried for 4 weeks, in order to remove the organic matter in the marrows of the bones. Thereafter they were crushed and pulverized using laboratory ball mill. Laterite, was sourced in the vicinity of University of Ilorin Teaching Hospital and, the physical properties were analysed and determined as colour, specific gravity, moisture content as received, sieve analyses, plastic limit, liquid limit and plasticity index. The laboratory tests carried on laterite were done in accordance with [15].

Silica sand, was also collected at a river bed along University of Ilorin Teaching Hospital road and its physical properties determined. Ordinary Portland cement was used as binder with varying percentage addition. The mixing proportions of cement, laterite, silica sand and pulverized cow bones are shown in Table 1 with a water cement ratio of 0.7 by volume.

Table 1 Aggregate mix for experimental tiles

Pulverized Cow Bones, % wt	Laterite, % wt	Silica Sand, % wt	Cement, % wt
30	40	10	20
20	25	35	20
10	20	50	20
5	20	55	20
30	35	20	15
20	35	30	15
10	35	40	15
5	30	50	15
30	30	30	10
20	45	25	10
10	50	30	10
5	40	45	10

2.1 Specimen Preparation

Quantity of materials required to cast a single tile, were weighed, mixed thoroughly for each mix ratio before the addition of water. Each tile was mixed separately to avoid rapid setting of the mixture. The mould of 200 mm × 100 mm × 60 mm was filled to its maximum volume, rammed with 2.5 kg rammer for 28 blows; the surface was levelled and trimmed. This was done in accordance to Agunwa [15]. It was also ensured that the moulds were filled in three layers. to the filling up of the mould, it was smeared with used automobile engine oil, in order to prevent the sticking of the cast to the walls of the mould when demoulding. The paving tiles were demoulded after 24 hours under atmospheric condition and cured in water pond for twenty one days (21) then sundried for another seven (7) days making up an entire curing process of twenty eight (28) days [15].

2.2 Experimental Tests

2.2.1 Water absorption

The mass of each tile specimen was weighed and re-weighed before and after submerging in water for 24 hours. The specimens were taken out of water and the surfaces carefully wiped to remove excess water [16]. The percentage water absorption A was calculated using equation (1):

$$A = \frac{M_S - M_D}{M_D} \times 100\% \quad (1)$$

Where,
M_S = mass of the saturated tile and
M_D = mass of the dried tile

2.2.2 Bulk density

Tile specimens were dried in oven at 110°C for 24 hours [16], and the tile test specimen's dry mass, M_D was determined. The bulk density, B was then calculated using equation (2):

$$B = \frac{M_D}{V} \quad (2)$$

Where;
V = volume of the tile
M_D = dry mass of the tile.

2.2.3 Determination of percentage drying shrinkage

Experimental tiles were submerged in water for five (5) days, and their masses determined as M_{sat}, the specimens were then oven dried at 110°C for 24 hours [16]. The mass of the specimens were then recorded after cooling to room temperature as M_f, the drying shrinkage S_d was then determined [14].

$$Sd = \frac{M_{sat} - M_f}{M_f} \times 100 \quad (3)$$

2.2.4 Modulus of rupture

The specimen was supported horizontally on two vertical supports of known span. The test specimens were centrally loaded by means of wire cord and a hanger suspended from the wire. The specimens were loaded with known masses until failure occurred [17]. The modulus of rupture for each tile specimen was then obtained using equation (4):

$$R = \frac{3PL}{2bd^2} \quad (4)$$

Where:
R = modulus of rupture,

P = applied load at failure,
L = length between the supports
b = width of the specimen; and
d = thickness of the specimen

2.2.5 Compressive strength

The test specimen was loaded into a universal testing machine M500-100AT. The load was slowly and carefully applied centrally on the tile specimens until the first sign of crack was observed and the load was then recorded. The compressive strength of each tile specimen was calculated using equation (5):

$$\sigma = \frac{P_C}{A_C} \quad (5)$$

Where:
Pc = maximum load on the specimen at failure,
Ac = calculated cross-sectional area of the specimen; and
σ = compressive strength of the test specimen.

2.2.6 Impact resistance

In order to ascertain the level of shock load the experimental tiles can withstand, impact resistance test was carried out by measuring the coefficient of restitution, of the colliding body, according to ENISO10545 [18]. A steel ball of mass 5 kg was dropped from a distance 1 m, time taken to make impact was recorded and time taking for rebound was also recorded. It was ensured that the mass hit the tiles at the centre of the dimension. The damages visible from 1 m were observed, and the coefficient of restitution was found as thus:

$$e = \frac{V_{fb}}{V_{ib}}$$

(6)
Where:
e = coefficient of restitution
V_{fb} = final velocity of steel ball
V_{ib} = initial velocity of steel ball

3. Results and Discussion

The physical properties of materials used as aggregate in the production of paving tiles are presented in Table 2.

Table 2: Physical properties of aggregates

Physical properties	Lateritic soil	Silica sand	Pulverized animal bones
Color	Reddish brown	Light brown	Whitish ash
Specific gravity	2.50	2.60	1.83
Moisture content as received (%)	24.67	32.72	38.10
Maximum mass retained on sieve (%)	23.20	27.20	92.80
Mesh size of occurence	1000μm	500μm	4750μm
Liquid limit (%)	39.30	-	-
Plastic limit (%)	25.00	-	-

IOP Conf. Series: Materials Science and Engineering **413** (2018) 012046 doi:10.1088/1757-899X/413/1/012046

Plasticity index (%)	14.30	-	-

Table 2 and Figure 1 shows that the maximum grain size distribution for lateritic soil, silica sand and pulverized animal bones occurring at 1000μm, 500μm and 4750μm sieve sizes respectively; and these were used in the production process in order to maximise the bulk of material resources.

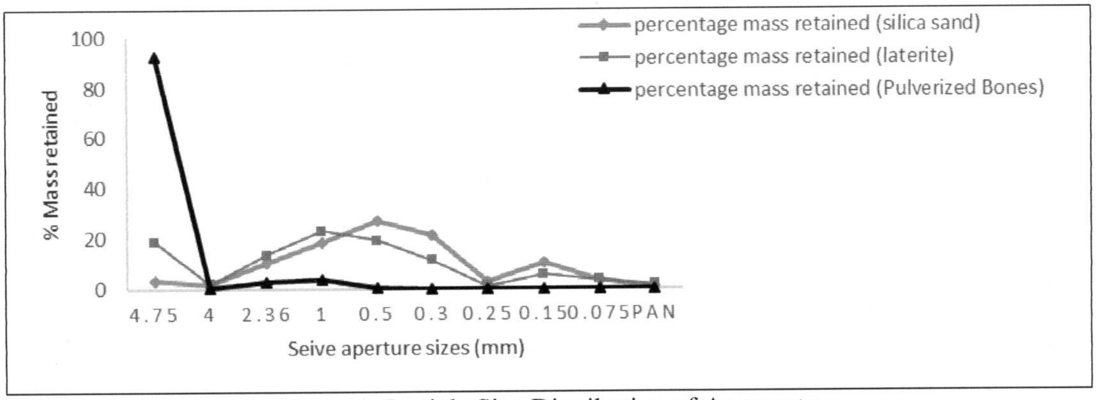

Figure 1: Particle Size Distribution of Aggregates

3.1 Bulk Density

Figure 2 is a graph of bulk density against varying % weight (wt) compositions of cement, laterite, silica sand and pulverized cow bone. The figure shows cement at 20%, 15% and 10% wt compositions at different concentrations of laterite: silica sand: pulverized cow bone. When cement was at 20%, the laterite: silica sand: pulverized cow bone ratios of 40:10:30 produced bulk density of 2.45 g/cm^3 and at 20:55:5 ratios the bulk density increased to 2.65 g/cm^3, an increase of about 8.16%. For situation when cement content was reduced to 15% wt, similar trend could be seen from the figure as laterite: silica sand: pulverized cow bone ratios of 35: 20: 30 had 2.42 g/cm^3 bulk density, and became 2.62 g/cm^3 at 30: 50: 5 ratios another increases of 7.85%. Though, there was a similar situation when the cement content was further reduced to 10%, but lower values of bulk density were observed for all the ratios used. For ratios of 30: 30: 30 (laterite: silica sand: pulverized cow bones), the bulk density is 2.38 g/cm^3 and increased to 2.58 g/cm^3 at ratios 40: 45: 5 giving an 8.4% increase. With the increase in the percentage composition of cement in the aggregate, there is a corresponding increase in the bulk density of the samples. The highest recorded value of bulk density was 2.65 g/cm^3 at 20% cement and 20: 55: 5 ratios of laterite: silica sand: pulverized cow bone respectively. It was observed that with increase in the % wt compositions of cement and silica sand in the aggregate, there is corresponding increase in the bulk density. Conversely, with increase in the pulverized cow bone, there is decrease in the bulk density as seen in Figure 2. This shows that an increase in addition of silica sand and cement led to the increase in the bulk density, while an increase in the pulverized cow bone produced a decrease in the bulk density.

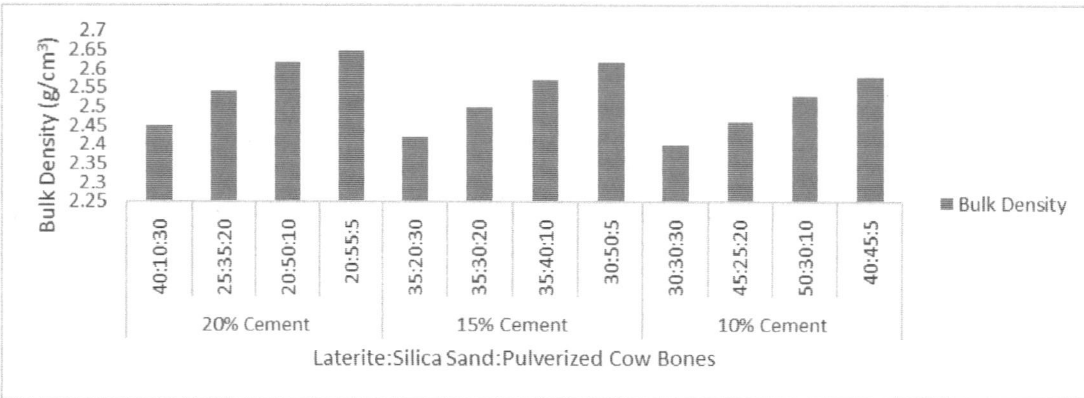

Figure 2: Bulk density of experimental tiles

3.2 Water Absorption

Figure 3 shows a graph of water absorption against varying % wt compositions of cement, laterite, silica sand and pulverized cow bone. With 20% cement weight, laterite: silica sand: pulverized cow bone ratios of 40:10:30 produced water absorption capacity of 3.01%, and at 20:55:5 ratios the water absorption capacity decreased to 2.5%, a decrease of about 16.9%. When cement content was reduced to 15% wt, laterite: silica sand: pulverized cow bone ratios of 35: 20: 30 had 3.93% water absorption capacity and became 3.34% at 30: 50: 5 ratios with a decrease of 15%. When the cement content was further reduced to 10%, for ratios of 50: 30: 10 (laterite: silica sand: pulverized cow bones), the water absorption capacity became 5.85% and decreased to 3.92% at ratios 30: 30: 30 giving a 33% decrease. The highest recorded value of water absorption was 5.85% at 10% cement for 50: 30: 10 ratios of laterite: silica sand: pulverized cow bone respectively. It was observed that with increase in the % wt compositions of laterite in the aggregate, there is corresponding increase in the water absorption. With increase in the silica sand, there is a decrease in the water absorption capacity as observed in Figure 3.

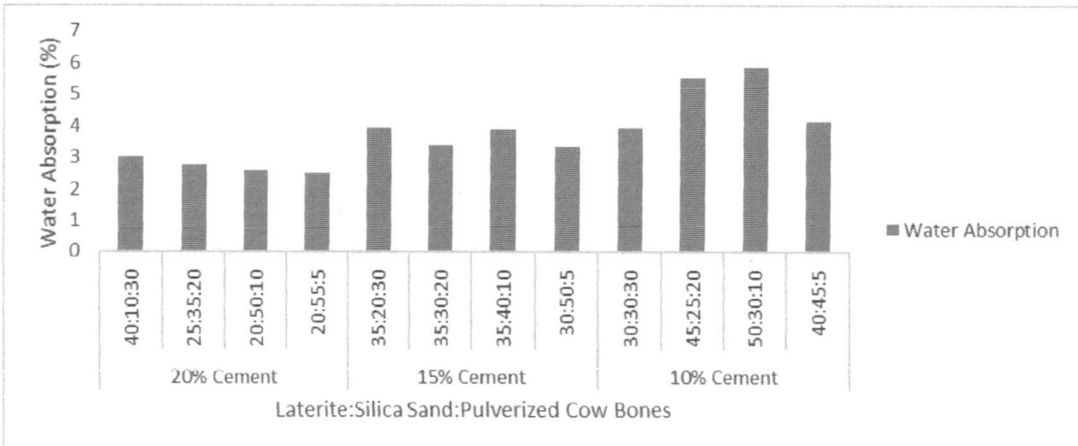

Figure 3: Water absorption capacity of experimental tiles

3.3 Drying Shrinkage

Figure 4 represents the graph of drying shrinkage against varying % wt compositions of cement, laterite, silica sand and pulverized cow bone. Determined different drying shrinkage levels, with cement at 20%, the laterite: silica sand: pulverized cow bone ratios of 40:10:30 had drying shrinkage value of 3.53% and at 20:55:5 ratios the drying shrinkage decreased to 2.78%, a decrease of 21.25%. For condition when cement content was reduced to 15% wt, similar trend could be seen from the figure as laterite: silica sand: pulverized cow bone ratios of 35: 20: 30 had 4.2% drying shrinkage and rose to 3.82% at 30: 50: 5 ratios another decrease of 9.05%. When the cement content was further reduced to 10%, for ratios of 50: 30: 10 (laterite: silica sand: pulverized cow bones), the drying shrinkage was 6.21% and decreased to 4.19% at ratios 30: 30: 30 giving a 32.53% decrease. As the percentage composition of laterite in the aggregate increased, there is an increase in the drying shrinkage capacity of the samples. The peak recorded value of water absorption was 6.21% at 10% cement and 50: 30: 10 ratios of laterite: silica sand: pulverized cow bone respectively. It was observed that with increase in the % wt compositions of laterite in the aggregate, there is corresponding increase in the drying shrinkage.

The drying shrinkage, was observed to follow same pattern as that of the water absorption tests, which was also observed by [14], however the percentage of shrinkage is slightly higher than that of the water absorbed, which is in agreement with [7].

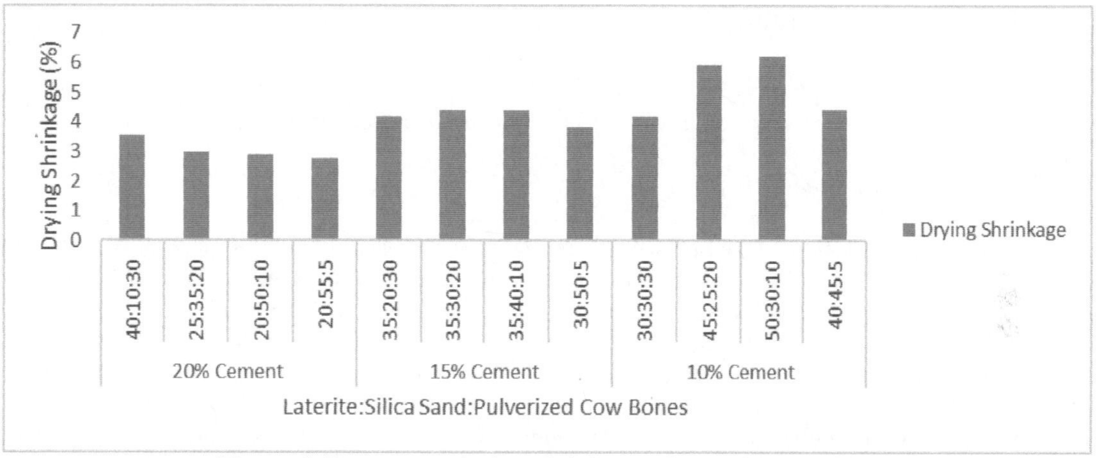

Figure 4: Percentage drying shrinkage of experimental tiles

3.4 Modulus of Rupture

Figure 5 represents a graph of cement at 20%, 15% and 10% wt compositions at different concentrations of laterite: silica sand: pulverized cow bone. When cement was at 20%, the laterite: silica sand: pulverized cow bone ratios of 40:10:30 produced modulus of rupture of 1.83 MPa, and at 20:55:5 ratios the modulus of rupture decreased to 1.7 MPa, a decrease of about 7.1%. But, with reduced cement content to 15% wt, laterite: silica sand: pulverized cow bone ratios of 35: 20: 30 had 1.21 MPa modulus of rupture, and dropped to 0.86 MPa at 30: 50: 5 ratios, another decrease of 28.93%. Similar situation was observed when the cement content was further reduced to 10%, but lower values of modulus of rupture were observed for all the ratios used. For ratios of 30: 30: 30 (laterite: silica sand: pulverized cow bones), the modulus of rupture is 0.26 MPa and decreased to 0.08 MPa at ratios 40: 45: 5 giving a 69.23% decrease. With the increase in the percentage composition of cement in the aggregate, there is increase in the modulus of rupture of the samples. The highest recorded value of modulus of rupture was 1.83 MPa at 20% cement and 40: 10: 30 ratios of laterite: silica sand: pulverized cow bone respectively. It could be observed that with increase in the % wt compositions of cement and pulverized cow bones in the aggregate, there is corresponding increase in the modulus of rupture. This shows that increase in addition of pulverized cow bones and cement led to the increase in the modulus of rupture. Cow bones been a good reinforcement material, and provides enhanced bonding strength with other additives.

IOP Conf. Series: Materials Science and Engineering **413** (2018) 012046 doi:10.1088/1757-899X/413/1/012046

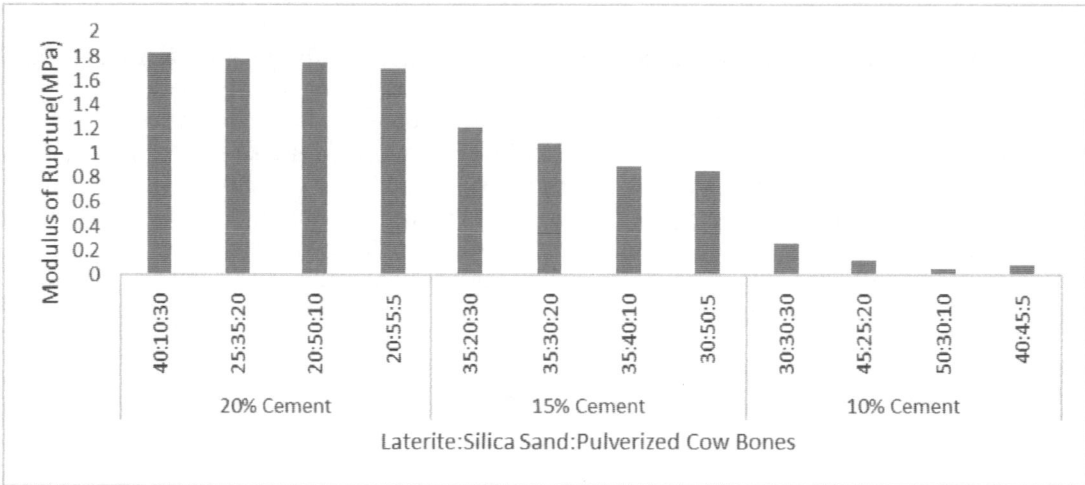

Figure 5: Modulus of Rupture of experimental tiles

3.5 Compressive Strength

Figure 6 is a graph of compressive strength against varying % wt compositions of cement, laterite, silica sand and pulverized cow bone. The figure shows cement at 20%, 15% and 10% wt compositions at different concentrations of laterite: silica sand: pulverized cow bone. When cement was at 20%, the laterite: silica sand: pulverized cow bone ratios of 40:10:30 produced compressive strength of 5.05 MPa, and at 20:55:5 ratios the compressive strength decreased to 2.42 MPa a decrease of about 52.08%. When cement content became 15% wt, similar trend could be seen from the figure as laterite: silica sand: pulverized cow bone ratios of 35: 20: 30 had 5.05 MPa of compressive strength and reduced to 2.4 MPa at 30: 50: 5 ratios another decrease of 52.48%. Though, there was a similar situation when the cement content was further reduced to 10%, but lower values of compressive strength was observed for all the ratios used. For ratios of 30: 30: 30 (laterite: silica sand: pulverized cow bones), the compressive strength is 3.08 MPa, and decreased to 2.38 MPa at ratios 40: 45: 5 giving a 22.73% reduction. With the increase in the percentage composition of cement in the aggregate, there is increase in the compressive strength of the samples. The highest recorded value of compressive strength was 5.05 MPa, at 20% cement and 40: 10: 30 ratios of laterite: silica sand: pulverized cow bone respectively, and that of 15% cement of ratios 35: 20: 30 of laterite: silica sand: pulverized cow bone of same compressive strength value. It was observed that with increase in the % wt compositions of cement and pulverized cow bone in the aggregate, there is a corresponding increase in the compressive strength.

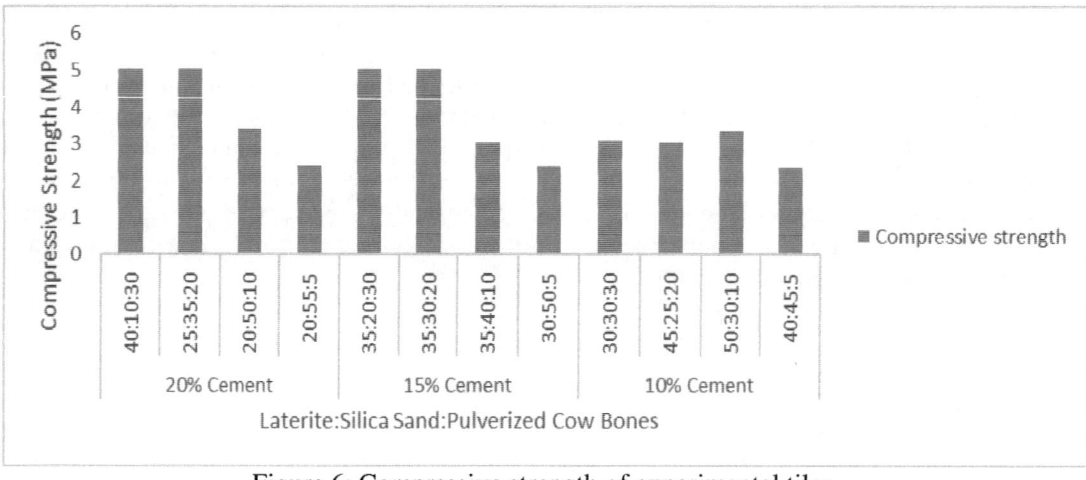

Figure 6: Compressive strength of experimental tiles

IOP Conf. Series: Materials Science and Engineering **413** (2018) 012046 doi:10.1088/1757-899X/413/1/012046

3.6 Impact Resistance

Impact resistance was analysed, the coefficient of restitution was calculated for each mix and was plotted against varying %wt composition of materials. From observation, tiles with 20% cement had better resistance to impact than those of 15% and 10% cement content.
Figure 7 shows the impact resistance of various mixes, the figure shows cement at 20%, 15% and 10% wt compositions at different concentrations of laterite: silica sand: pulverized cow bone. When cement was at 20%, the laterite: silica sand: pulverized cow bone ratios of 40:10:30, 0.55 maximum value of coefficient of restitution was observed and ascertained to be sufficient for low stress use following the European standard [18], and decreases to 0.22 at 20:55:5 ratios, a decrease of 60%. For situation when cement content was reduced to 15% wt, similar trend could be seen from the figure as laterite: silica sand: pulverized cow bone ratios of 35: 20: 30 had value of 0.4 and decreases to 0.21 at 30: 50: 5 ratios, another decrease of 47.5%. Though, there was a similar situation when the cement content was further reduced to 10%, but lower values of coefficient of restitution was observed for all the ratios used. For ratios of 30: 30: 30 (laterite: silica sand: pulverized cow bones), the coefficient of restitution is 0.28 and decreased to 0.08 at ratios 40: 45: 5 giving a 71.43% decrease. Cement was observed to play a major role in this, due to the good bonding strength it provides, as cement content reduces, the impact resistance reduces.

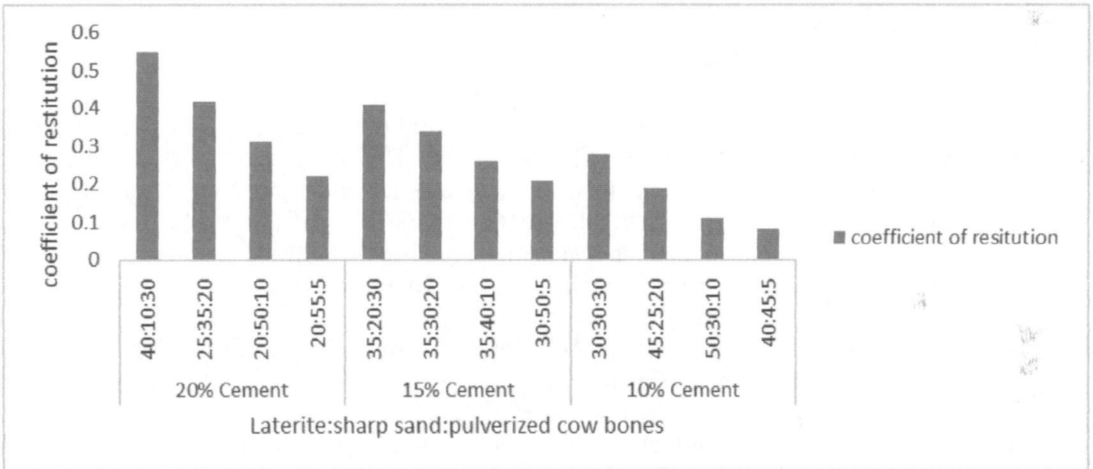

Figure 7: Impact resistance of experimental tiles

3.7 Quality of Paving Tiles

Enhanced mechanical and physical strength were observed with 30% pulverized bones and 20% cement additions. Tiles with ratio 4:3:1:2 of laterite: pulverized cow bones: silica sand: cement, were observed to have the highest mechanical and physical strength of 5.05 MPa of compressive strength and 1.83 MPa of modulus of rupture.

4. Conclusion

1. The capacity of lateritic based tiles to absorb water is same as its ability to loose water i.e. the increase in laterite content corresponds to an increase in water absorption capacity and drying shrinkage, in agreement with Ohijeagon *et al.* [7].
2. The bulk density of experimental tiles was greatly affected by the presence of silica sand, where higher percentage content leads to increase in bulk density. Pulverized cow bones did

not contribute significantly in the bulk density determination, due to the lower specific gravity it possesses.

3. The maximum compressive strength and modulus of rupture were found to be 5.05 MPa and 1.83 MPa respectively. Resulting in an optimum ratio of 4:1:3 of laterite:pulverized cow bones:silica sand mixed with a 20% cement content respectively, compared to Owolabi [13] who used palm kernel shells, an agricultural waste and obtained a compressive strength of 4.7 MPa in an optimum mixing ratio of 1:4 of kernel shells:laterite respectively and compressive strength of plain laterite of 4 MPa after 28 days of curing.
4. The impact strength of the used pulverized cow bones was observed as there was improved strength of experimental tiles in terms of compressibility, and modulus of rupture with increase in pulverized cow bones percentage in the mix. Cement composition of 20% was found as optimum for pulverized bones addition in the composition of paving tiles.

References

[1] Sridhar M K, Hammed T B. Turning Waste to Wealth in Nigeria: An Overview. J Hum Ecol. 2014, 46 (2): 195-203.

[2] Blench R. The Nigerian National Livestock Resources Survey: A Personal Account. Nigerian Livestock survey. 1999, 627-648.

[3] Staughton J. Science of the Skeleton, Science ABC. 2016 [online]. http://www.scienceabc.com/humans/skeleton-mystery-dont-bones-decay-decompose.html. [Accessed 6th June 2018].

[4] Lorprayoon L C. Phases of Cattle Bones at Elevated Temperatures. J. Sci. Soc. 1986, 12: 159-170.

[5] Kim W K, Donalson L M, Herrera P, Woodward C L, Kubena L F, Nisbet D J, Ricke S C. Effects of Bone Preparation Methods (Fresh, Dry, Fat-Free Dry) on Bone Parameters and the Correlation Between Bone Breaking Strength and other Bone Parameters. Poultry Science Association Report. 2004, 83: 1663-1666.

[6] Falade F, Ikponmwosa E and Fapohunda C. Potential of Pulverized Bones as a Pozzolanic Material. International Journal of Scientific and Engineering Research. 2012, 3 (7).

[7] Ohijeagon I O, Olusegun H D, Adekunle A S, Adewoye O S, Oladiji A O. Impact of Suitable Replacement of Granite-Particles on Interlocking Tiles. Journal of Engineering Science and Technology Reviews. 2012, 5 (2): 51-56.

[8] Ojuri O O. Interlocking Paving Stones Pavement as a Solution to Marshy Roads, Modern Applied Science. 2012, 6 (5): 91-99.

[9] Olusegun H D, Adekunle A S, Ogundele O S, Ohijeagbon I O. Composite Analysis of Laterite-Granite Concrete Tiles, Epistemics in Science, Engineering and Technology. 2011, 1 (1): 53-59.

[10] Okonkwo U N, Odiong I C, Akpabio E E. Effect of Eggshell Ash on Strength Properties of Cement Stabilized Lateritic. International Journal of Sustainable Construction Engineering and Technology. 2012, 18-25.

[11] Adebayo B, Olajide O M. Evaliuation of Suitability of Selected Dredged Sand and Slurry for Engineering Application. Frontiers in Geotechnical Engineering. 2013, 2 (1): 1-6.

[12] Raheem A A, Falola O O, Adeyeye K J. Production and Testing of Lateritic Interlocking Tiles, Journal of Construction in Developing Countries. 2012, 17 (1): 33-48.

[13] Owolabi A. Assesment of Palm Kernel Shell as Aggregate in Concrete and Lateritic Blocks. Journal of Engineering Studies and Research. 2012, 18 (2): 88-93.

[14] Ajao K S, Ohijeagbon I O. Recycling of Agro-Waste to Produce Sustainable Paving Tiles, International Conference on Commercializing Engineering for Diversification in Developing Economy: Bridging Town and Gown 2016 Proceedings,). Awka, Nigeria: Faculty of Engineering, Nnamdi Azikiwe Univeriesty. 2016, 485-495.

[15] Agunwa J J. Study of Compressive Strengths of Laterite-Cement Mixes as Building Materials. AU J T technical Report 2009, 13 (2): 114-120.

[16] ASTM C373. Standard Test Methods for Determination of Water Absorption and Associated Properties by Vacuum Method for pressed Ceramic Tiles and Boil Method for Extruded Ceramic Tiles and Non-Tile Fired Ceramic Whiteware Products. 2015.

[17] ASTM C293. Standard Test Method for Determination of Flexural Strength of Concrete Using Simple Beam with Center Point Loading. 2015.

[18] ENISO10545:5. Determination of Impact Resistance by Measurement of Coefficient of Restitution, Italy. 1997.

IOP Conf. Series: Materials Science and Engineering **413** (2018) 012047 doi:10.1088/1757-899X/413/1/012047

Performance Evaluation of Polydimethylsiloxane-Solvent Blends as Defoamer for Crude Oil Foam

Ohimor Evuensiri Onoghwarite[1], Erude Abraham Okeoghene[1], Onocha Ovonomo[1], Oreko Benjamin Ufuoma[2], Ononiwu Prosper Ikechukwu[3]

Department of Chemical Engineering, Federal University of Petroleum Resources, Effurun, Delta State, Nigeria[1]
Department of Mechanical Engineering, Federal University of Petroleum Resources, Effurun, Delta State, Nigeria[2]
Michael Okpara University of Agriculture, Umudike, Umuahia, Abia State, Nigeria[3]
E-mail address: ohimor.evuensiri@fupre.edu.ng

Abstract. The study is directed at the use of polydimethylsiloxane-based defoamer formulated with different solvent blends that are capable of breaking and preventing crude oil foam. The control or elimination of the foam that arises in oil and gas industrial processes is a crucial factor in the smooth and efficient operation of the plant. Foam can reduce available capacity of processing vessels, and interfere with the efficiency of the process. Foaming problem is common in the three-phase separators in crude oil flowstations; it can result in level control difficulties and carryover of valuable crude oil into the gas scrubbers and eventually to the gas flare, resulting in loss of production. The effectiveness of these solvent blended defoamers in controlling crude oil foam was determined using foam-rise or Bikerman test method. The collapse rates of the foam with the application of each defoamer sample at various treatment rates (10ppm, 15ppm and 20ppm), were observed as against a blank as control. It was observed that at 10ppm treatment, defoamer A had the highest foam collapse rate (0.0108mm/s). At 15ppm treatment, defoamer G had the highest foam collapse rate (0.0110mm/s), and at 20ppm treatment, defoamer E had the highest foam collapse rate (0.0108mm/s). This shows that defoamer A, which is the conventional formulation (i.e. 20 %vol. silicone + 80 %vol. kerosene), had the highest foam collapse rate at the lowest dosage of 10ppm. On the overall, the highest foam collapse rate (0.0110mm/s) was obtained with defoamer G at 15ppm treatment. Defoamer G is 20 %vol. Silicone + 80 %vol. of a (1:1 %v/v) blend of kerosene and crude palm kernel oil. This suggests that the presence of crude palm kernel oil in the solvent blend had a remarkable effect on the crude oil foam collapse rate. Conclusively, the performance of defoamers varies considerably with the kind of solvent or blend of solvents used for the dispersion of the silicone fluid. It is also worthy of note that the treatment dosage of a defoamer, together with the foam collapse rate are key factors that influence the selection of an effective defoamer.
Keywords: production chemical; flowstations; silicone products; crude oil emulsion

1. Introduction

Petroleum-based foams are amongst the most important and abundant types of non-aqueous foams. Furthermore, the complexity of the continuous phase in these foams is higher than in other non-aqueous foams. This is because the composition of crude oil varies greatly depending on the well location and age. Crude oil is a naturally occurring mineral oil consisting of a variety of organic compounds, mainly saturated and aromatic hydrocarbons, as well as complex compounds such as resins and asphaltenes [1].

Foams may occur in any stage of oil recovery and processing, and they can be desirable or a nuisance; since it is the bubbles that have sufficient buoyancy to rise to the surface that are described as foams [1]. For example, oil foams can be useful as drilling fluids or gas blocking agents during extraction from porous media, but undesirable during gas/oil separation, oil/water separation and during refining.

An additional complexity of petroleum foams, apart from oil composition, is that they usually contain water, particles (sand, clay, corrosion products, paraffin crystals, precipitated asphaltenes, etc.) or even additives introduced during the oil extraction phase (surfactants commonly used as bactericides, anti-corrosion, anti-oxidants, emulsion breakers, asphaltene dispersants, anti-scale, etc). These factors produce petroleum foams that may exhibit different types of behaviour, showing that foaming of liquids arises from the presence of impurities [2,3]. The amount of foam that a system can create, the stability of the foam and, even, the capacity to create gas/ liquid films depend on the characteristics of the gas dissolved in it. Gases that do not have an affinity for the crude oil tend to form unstable foams. However, for gases soluble in the oil, foam can be formed, and the extent of formation will depend on the pressure and temperature of the system.

The composition of the crude oil is equally important to the eventual foaming properties. Several constituents can promote the formation of foam and its stabilization. Mechanical problems that may generate foam are leaky seals on pumps, high-pressure flows, poor system design (tank, pump inlet, outlet and manifold design) and pressure release.

A defoamer is a chemical additive that reduces the foam in foaming liquids in industrial processes. The terms 'anti-foam agent' and 'defoamer' are often used interchangeably. In industrial processes, foams pose serious problem; they cause defects on surface coatings, they prevent the efficient filling of containers, reduction of pump efficiency (cavitation) and reduce capacity of storage tanks. Modern anti-foams are complex formulated specialty chemicals but generally consist of an insoluble oil phase containing hydrophobic solids dispersed within it. The anti-foam is typically provided in a dispersed form, with droplets of controlled particle sizes and distribution in the surfactant solution [4]. The significant property of a defoamer is that it is generally insoluble in the foaming medium and has surface-active properties. An essential feature of a defoamer product is its low viscosity and ability to spread rapidly on foamy surfaces. It has affinity for the air-liquid surface where it destabilizes the foam lamellas. This causes rupture of the air bubbles and breakdown of surface foam. Anti-foams are introduced in surfactant solutions to prevent the formation of excessive foam [5]. A typical anti-foam consists of an oil (polydimethylsiloxane or hydrocarbon), dispersed hydrophobic solid particles, or a mixture of both.

Among the five types of product-based defoamers, silicone-based defoamers are the widely used anti-foaming agent in non-aqueous foaming systems such as in the crude oil production and oil refining systems [1]. Silicone-based defoamers are polymers with silicon backbones. The silicone compound often contains hydrophobic silica dispersed in an organomodified silicone fluid. Polydimethylsiloxane is a particularly efficient defoamer ingredient because of its low surface tension, spreading capability, thermal stability, chemical inertness and water insolubility. Silicone based defoamers can be formulated as 20 %vol liquid products, as solutions and as aqueous emulsions. Organic modification of the polydimethylsiloxane with polyethers or other organic groups renders the products more compatible and allows the formulation of highly efficient defoamers with excellent compatibility characteristics for high performance applications. Emulsifiers or suitable solvents are added to aid handling and efficient dispersion within the foaming medium.

It is a well-known fact that the performance of foam-control agents varies considerably with the type of foaming surfactants [4].

Three steps can characterize the working mechanism of a defoamer: first, a defoamer droplet has to enter the foam lamella. Due to its highly surface-active properties, the defoamer droplet has a strong tendency to migrate to the air/water-surfactant interface. Secondly, the defoamer droplet spreads and pushes aside the stabilizing surfactants. Without the stabilizing surfactants, the foam lamella locally becomes much less elastic usually referred to as "reverse Marangoni effect". Thirdly, this kind of destabilization facilitates rupture of the foam lamella. To increase defoamer efficiency, it is a common approach to add finely divided hydrophobic particles, such as hydrophobic silica or waxes, to the formulation. The reason for this is that the hydrophobic solid particles are able to destabilize the foam lamella by a de-wetting process. In simple terms, the surfactant film, which stabilizes the foam lamella, is not able to wet the hydrophobic solids and collapses (i.e. shrinks back). An unstable area results where the lamella is thus destabilized [4].

The most noticeable form of foam is foam floating on the stock surface, thus, the case with crude oil formation. Surface foam may cause problems with liquid levels resulting in overflow and carryover. This might reduce the process efficiency and availability of process equipment. Hence, this paper seeks to formulate different blends of silicone-based defoamers using solvents and combination of solvents; afterwards, determine their effectiveness in controlling crude oil foam formation. This will proffer solution that will control foaming in the crude oil flow station using defoamers.

2. Materials and method

The materials used are; glass capillary tube, micropipette, air sparger, glass measuring cylinder (500ml), crude oil, silicone, solvents such as kerosene., diesel, refined palm kernel oil (RPKO) and crude palm kernel oil (CPKO); stopwatch, hydrometer and pH meter. The properties of the solvents and their blends, such as

IOP Conf. Series: Materials Science and Engineering **413** (2018) 012047 doi:10.1088/1757-899X/413/1/012047

turbidity, colour and phase separation, were determined through visual examinations while measuring instruments like hydrometer and pH meter were used to determine the specific gravity and pH respectively.

The method used for the defoamer test is the foam-rise or Bikerman test method [6].

A 100ml of fresh crude oil was poured into 500ml glass cylinder. The glass capillary tube was inserted in the glass cylinder with the air sparger tube connected to it, with the other end of the tube connected to the air sparger compressor. The air sparger system, which has capacity to deliver air at a constant flow rate of 94 ml/min, was operated for a period of 2 min before been switched off for observation. The following were determined: height of foam generated immediately after 2 min, height of foam that collapsed after 30 sec and the time taken for the entire foam to collapse. The first run (without defoamer) is the blank. The procedure above was repeated with the addition of 10ppm, 15ppm and 20ppm respectively, of the various defoamers.

Table 1 Defoamers and their composition

S/N	Defoamer products	Composition
1	Defoamer A	20 %vol. Silicone + 80 %vol. kerosene
2	Defoamer B	20 %vol. Silicone + 80 %vol. diesel
3	Defoamer C	20 %vol. Silicone + 80 %vol. refined palm kernel oil
4	Defoamer D	20 %vol. Silicone + 80 %vol. crude palm kernel oil
5	Defoamer E	20 %vol. Silicone + 80 %vol. of a (1:1 %v/v) blend of kerosene and diesel
6	Defoamer F	20 %vol. Silicone + 80 %vol. of a (1:1 %v/v) blend of kerosene and refined palm kernel oil
7	Defoamer G	20 %vol. Silicone + 80 %vol. of a (1:1 %v/v) blend of kerosene and crude palm kernel oil
8	Defoamer H	20 %vol. Silicone + 80 %vol. of a (1:1 %v/v) blend of diesel and refined palm kernel oil
9	Defoamer I	20 %vol. Silicone + 80 %vol. of a (1:1 %v/v) blend of diesel and crude palm kernel oil
10	Defoamer J	20 %vol. Silicone + 80 %vol. refined palm kernel oil

3. Results and discussion

Table 2 Physical properties of the Defoamers and their components

Components/ defoamers	Colour	pH	Turbidity (visual examination)	Specific Gravity (SG)	Phase Separation
Crude palm kernel oil	Light yellow	4.38	Slightly turbid	0.915	No Separation
Kerosine	Yellow	4.99	Slightly turbid	0.820	No Separation
Refined palm kernel oil	Yellow	5.24	Clear liquid	0.921	No Separation
Silicone	Colourless	9.11	Clear liquid	0.970	No Separation
Diesel	Medium brown	9.28	Slightly turbid	0.810	No Separation
Defoamer A	Colourless	4.60	Very turbid	0.9106	No Separation
Defoamer B	Brown	5.03	Slightly turbid	0.901	No Separation
Defoamer C	Yellow	4.84	Clear liquid	0.8486	No Separation
Defoamer D	Light brown	6.32	Very turbid	0.841	No Separation
Defoamer E	Orange	4.06	Clear liquid	0.924	No Separation
Defoamer F	Yellow	5.00	Clear liquid	0.840	No Separation
Defoamer G	Colourless	5.20	Clear liquid	0.842	No Separation
Defoamer H	Light brown	5.19	Clear liquid	0.843	No Separation
Defoamer I	Amber	5.27	Clear liquid	0.840	No Separation
Defoamer J	Yellow	5.85	Clear liquid	0.845	No Separation

The foam heights and foam collapse rates for the blanks were the same in each sets of experiment. There were variations in foam heights and foam collapse rates for the various defoamers at the different treatment rates of 10ppm, 15ppm and 20ppm. The collapse rates $R_c(mm/s)$ were obtained using the equation below.

$$R_c(mm/s) = \frac{\text{Height of the entire foam column}}{\text{Time taken for complete collapse}} \qquad \text{Eqn. 1.0}$$

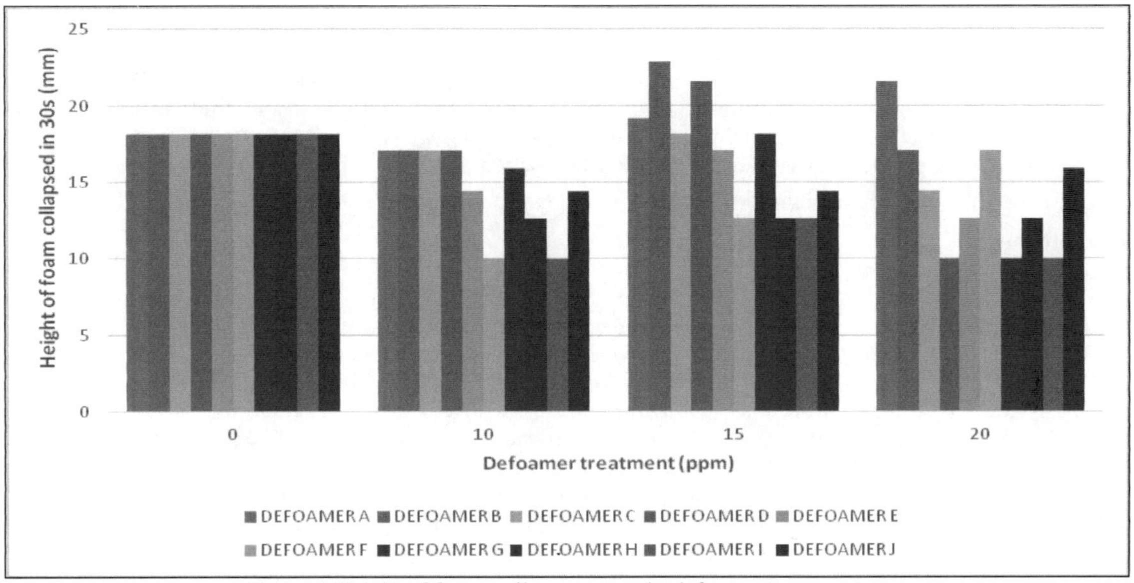

Figure 1 Height of foam-collapse versus in defoamer treatment

At various injection doses (0, 10, 15, 20ppm), the different defoamers applied were observed for the height of foam-collapse in 30sec. From Figure 1.0, defoamer A showed increasing order of foam collapse within 30 sec with respect to increasing defoamer dose. Defoamer B showed decreasing order of foam collapse within 30 sec with respect to increasing defoamer dose. Other trends can be, readily seen in the figure. However, defoamer B had the highest height of foam-collapsed in 30sec of 22.5mm at 15ppm dosage.

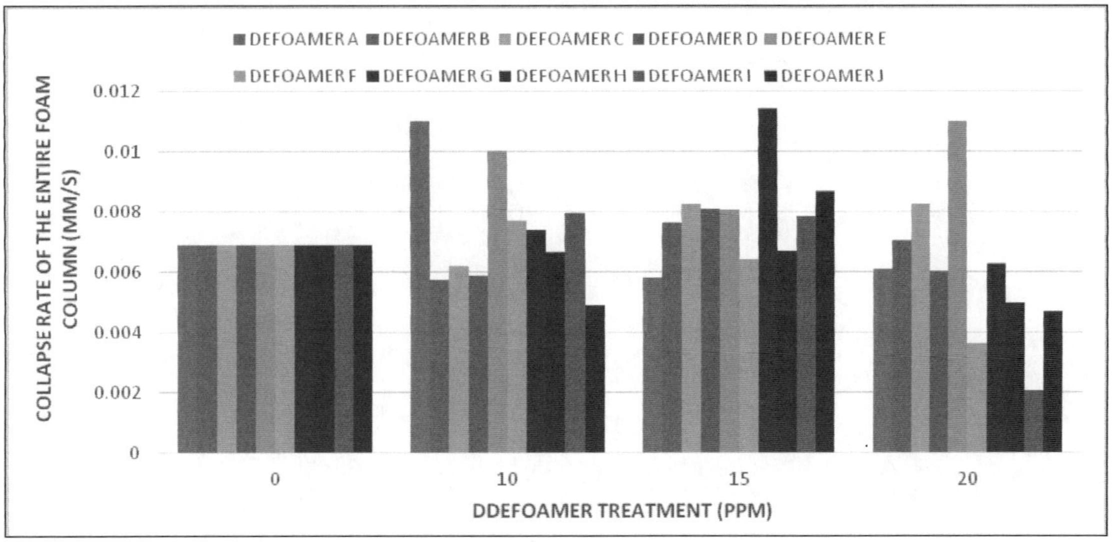

Figure 2 Collapse rate of the entire foam column versus defoamer treatment

At various injection doses (0, 10, 15, 20ppm), the different defoamers applied were observed for the collapse rate of the entire foam column. From Figure 2.0 It was observed that at 10ppm treatment, defoamer A had the highest foam collapse rate (0.0108mm/s). While at 15ppm treatment, defoamer G had the highest foam collapse rate (0.0110mm/s), and at 20ppm treatment, defoamer E had the highest foam collapse rate (0.0108mm/s). This shows that defoamer A, which is the conventional formulation (i.e. 20%vol. silicone + 80%vol. kerosene), had the highest foam collapse rate at the lowest dosage of 10ppm. On the overall, the highest foam collapse rate (0.0110mm/s) was obtained with defoamer G at15ppm treatment. Defoamer G is 20%vol. Silicone + 80%vol. of a (1:1 %v/v) blend of kerosene and crude palm kernel oil.

4. Conclusion

This research has enabled the study of silicone dispersed in readily available solvents such as diesel, kerosene, and palm kernel oil (crude and refined) and their blends as defoamers for foamy crude oils. Some of the defoamers were effective in resolving crude oil foam with high foam collapse rates. The performance of defoamer G can be further evaluated in a crude oil flowstation pilot plant. This is to further validate the research findings and for possible commercial application.

References

[1] Tarek, K. (2015). I - Anti-foaming agents. 1st ed. Available at: http://tarek.kakhia.org/books_eng/Defoamer.Tarek_kakhia.pdf [Accessed 16 Feb.2017].
[2] Shaban, H. I (1995). A Study of Foaming and Carry-Over Problems in Oil and Gas Separators. *Gas Separation and Purification. 9* (5) pp 81-86 https://doi.org/10.1016/0950-4214 (95)93944-F
[3] Lewis, V.E and Minyad, W.F. (1994) Antifoaming and Defoaming in Refineries. *Foams: Fundamentals and Application in the Petroleum Industry* (Advances in Chemistry, Vol. 242) Chapter 12, pp 461–483.doi: 10.1021/ba-1994-0242.ch012
[4] Jha, B., Christiano S. and Shah D. (2000). Silicone Anti-foam Performance: Correlation with Spreading and Surfactant Monolayer Packing. *Langmuir, 16* (24), pp.9948
[5] Garrett, P. (1993). *Defoaming: Theory and Industrial Applications,* Garrett, PR, Ed, Marcel Dekker, New York.
[6] Denkov N. (2004). Mechanisms of Foam Destruction by Oil-Based Anti-foams, *Langmuir 20* (22), pp.9467-9468.

IOP Conf. Series: Materials Science and Engineering **413** (2018) 012048 doi:10.1088/1757-899X/413/1/012048

Thermodynamic Analysis of Raw Mill in Cement Industry Using Aspen Plus Simulator

[1]Anthony I. Okoji, [1,2]Damilola Elizabeth Babatunde, [1]Ambrose N. Anozie, [1]James A. Omoleye

[1]Department of Chemical Engineering, Covenant University, Km 10, Idiroko Road, Canaan Land, PMB 1023, Ota, Ogun State, Nigeria

[2]Corresponding author: Damilola Elizabeth Babatunde;
damilola.babatunde@covenantuniversity.edu.ng

Abstract. This study investigates the appropriateness of exergy calculation using Aspen Plus Process Simulator which has a robust data library and powerful engineering calculation capabilities. The simulator was used for the thermodynamic performance of a raw mill (RM) and raw materials preparation unit in a cement plant in Nigeria using actual operating data. The raw mill has a capacity of 240,000 kilogram-material per hour. Also, both exergy and exergetic efficiency of raw mills from three literature sources were investigated and compared with the simulation results from Aspen Plus process model. The results were subjected to statistical analysis using ANOVA. The exergy efficiency for the raw mill studied using Aspen plus simulator modelling technique was found to be 21.4%. It was found that the difference in exergy efficiencies of the simulation results of the three-literature data vary within ±2.5% of the published results. The present method using the Aspen plus simulator is suggested as a useful tool in making informed decisions for developing energy policies and exergy utilization, providing energy conservation measures in improving the efficiency of the system.
Keywords: Cement Industry, Raw Mill, Exergy, Exergy Efficiency, Process Modelling and Simulation

1. Introduction

It has been asserted that the energy resources in the world are finite in nature and this consciousness has necessitated prudent utilization of energy sources [1]. Energy conversion and energy utilization processes are being closely considered to effectively utilise the limited energy resources. For economic, environmental and sustainability reasons, more attention is being paid to improvement of processes and efficient use of energy [2 -4]. The first law of thermodynamics has been applied extensively for performance evaluation of energy system. The first law of thermodynamics though helpful in certain ways, is not sufficient to handle the intricacies of energy transformation as it fails to consider entropy generations and irreversibilities associated with real processes [5].

All real processes are accompanied by a level of entropy generation leading to irreversibilities and as energy is being transformed, it gets degraded in quality even though the quantity of energy is conserved. Irreversibilities and degradation of energy reduces the quantity of energy that is available to be transformed to useful work. A measure of useful work extractable from energy sources is therefore of more importance than the quantity of the energy itself. With respect to this, exergy, a concept which stems out of the first and second laws of thermodynamics is an invaluable tool for evaluating thermodynamic performance of energy requiring and energy producing systems as it typifies the moiety of energy that can be transformed into the maximum useful work. Dissimilarly to energy, exergy is not conserved but depletes due to irreversibilties as transformations take place. In terms of useful work extracted from an energy source, performance evaluations based on exergy analysis give true efficiency of a system. In which case, the thermodynamic imperfections arising from irreversibilities are considered to be exergy destroyed and this represents losses in energy usefulness or quality. With exergy analysis, efficiencies that truly measure how nearly actual performances tend towards the ideal are obtained and the locations, types and causes of thermodynamic losses are more clearly identified [6]. The units or areas where there are significant differences between the actual performance and the ideal performance are the potential areas where engineers and scientist target for improvements.

Methods of exergy analysis have been developed over the years and exergy related calculations require values of some thermodynamic properties [7]. Despite the vast amount of thermodynamic data available, there are few pure substances for which we have thermodynamic data in the range of temperature of interest, therefore, it is usually necessary to calculate thermodynamic properties of pure component and mixtures from basic data, correlations and equations of state. In past works, most of the data estimations and exergy calculations have been semi-automated or carried out manually. More often than not, these processes of data generation via semi-automation or manual calculation and estimations are laborious and error prone. Error in data collection will inevitably lead to erroneous and misleading exergy values thereby defeating the essence of exergy analysis. Avoiding errors due to omission and commission in exergy analysis is therefore paramount.

The cement production being a complex process combines various endothermic, exothermic reactions with heat transfer in the solid, liquid, and vapour phases of different materials. Therefore, for sustainable and more efficient processes, the use of process engineering tools, such as process modelling software is inevitable as an alternative. In this study, the peculiarity of the material and exergy balance in a cement industry makes ASPEN Plus an essential tool for the analysis. The process equipment design, simulation or modelling and sensitivity analysis utilized ASPEN Plus as a process simulator. One of the key features of Aspen Plus software is the availability of large numbers of the physical, chemical and thermodynamic property data, which enables modelling of most of the complex industrial processes. This study investigates the appropriateness of using Aspen Plus process simulator in thermodynamic analysis of raw mill in cement industry. A local raw mill in Nigeria and three others from literature were analysed.

2. Theoretical Analysis of Mass, Energy and Exergy

The work and heat interactions, including rate of exergy decrease, irreversibility, energy and exergy efficiency are required at steady state.

2.1 Mass balance

The mass balance equation at steady state can be written in the rate form as

$$\Sigma \dot{m}_{in} = \Sigma \dot{m}_{out} \tag{1}$$

Where, $\dot{m}$ is the mass flow rate for both inlet and outlet.

2.2 Energy balance

The general energy balance can be expressed as

$$\Sigma E_{in} = \Sigma E_{out} \tag{2}$$

$$Q_{net\ in} + \Sigma \dot{m}_{in} h_{in} = W_{net\ in} + \Sigma \dot{m}_{out} h_{out} \tag{3}$$

Where, ΣE_{in} is the total sum of the energy transferred in, ΣE_{out} is the total sum of the energy transferred out. It is assumed that $Q_{net\ in} = 0$ and $W_{net\ in} = 0$. Then, Equation (3) reduces to:

$$\Sigma \dot{m}_{in} h_{in} = \Sigma \dot{m}_{out} h_{out} \tag{4}$$

2.3 Exergy balance

Assuming that other parameters are negligible, then total exergy of a system are evaluated using both physical exergy, ($\dot{E}x_{phy}$), and chemical exergy, $\dot{E}x_{ch}$

IOP Conf. Series: Materials Science and Engineering **413** (2018) 012048 doi:10.1088/1757-899X/413/1/012048

The physical exergy can be stated as:

$$\dot{E}x_{phy} = (h - h_0) - T_0(s - s_0) \quad (5)$$

While the chemical exergy of the ideal gas with liquid mixtures is calculated from:

$$\dot{E}x_{ch} = \Sigma x_i((\dot{E}x_{choi}) + RT_0(In\, x_i)) \quad (6)$$

Where x_i represent the species i molar ratio, and $\dot{E}x_{choi}$ also represents the standardized chemical exergy. Exergetic performance is expressed as:

$$\in 1 = \frac{\dot{E}x_{out}}{\dot{E}x_{in}} \quad (7)$$

Anergy, Φ, is expressed as:

$$\Phi = \frac{\dot{E}x_{losses}}{\dot{E}x_{input}} \quad (8)$$

The irreversibility of the system is expressed as:

$$I_{sys} = \dot{E}x_{input} - \dot{E}x_{output} = T_o S_{gen} \quad (9)$$

Where, I_{sys} is known as exergy destroyed or irreversibility and S_{gen} is the entropy generated.

3. Cement Production Process

Cement industry is usually located very close to the deposit of a naturally occurring rock material such as limestone or chalk or calcium carbonate ($CaCO_3$), which provides a major constituent when extracted from quarries. Iron ore (Fe_2O_3), bauxite (Al_2O_3) , shale, clay, slag or sand (SiO_2) are considered as minor materials which may be needed in very small quantity to provide the extra mineral ingredients for cement production. The mined material is conveyed with the aid of dump truck to primary and secondary crushers and reduced into less than 10 centimetre pieces.

The crushed raw materials are mixed and stored for homogenization purpose, then milled together using raw mill (ball or vertical) to produce 'raw meal' under a strict quality control of the material chemistry. Hot flue gases coming from the rotary kiln, which is in the opposite direction with the material flow, preheat the powdered raw meal at the preheater tower before it finally enters the kiln. In these preheater cyclones, raw meal is preheated with heat exchange taking place to improve the process efficiency and less fuel consumption. The raw material moisture content and heat recovery efficiency determines the investment type of a rotary kiln and stages of cyclone required. The additional fuel introduced at the calciner, is partly used to increase the raw meal temperature to calcining level and primarily to implement actual calcination process (decomposition of limestone into lime CaO and CO_2).

Precalcined meal enters the rotary kiln at temperatures of approximately 1050°C. The rotary kiln is being fired directly or indirectly to a temperature above 1800°C to ensure a well prepared material. The raw material flows down the kiln due to an inclined positioning, to the burning zone, the hottest region of the kiln. Both chemical and physical reactions occur as a result of an intense heat which partially melts the meal into clinker. From the kiln, the hot clinker is moved to the grate cooler where it is cooled to a temperature between 85°C to 120°C with cooling fans, part of which is used as combustion air.

The clinker formed is mixed with 4-5% gypsum, with small quantity of limestone or slag (filler). All these are milled into a final product called cement, or Portland Composite Cements depending on the type and quantity of fillers added.

3.1 Raw materials Preparation

The raw meal preparation flowsheet is shown in Figure 1. The most common and vast raw materials is limestone ($CaCO_3$) of varied quality, in addition with much smaller quantities of clay, shale and sand (as a source of silica, aluminum and iron). A common belt conveyor transfers the different raw material components from the weigh feeders directly to the mill inlet. The grinding rollers are forced downwards onto the materials.

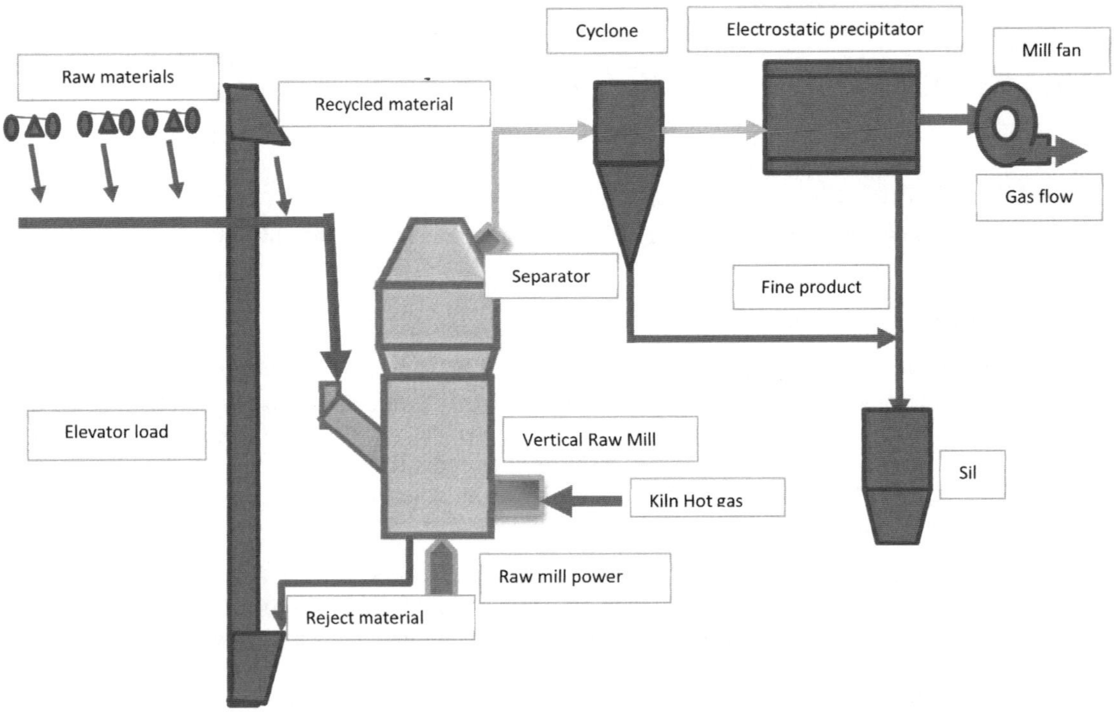

Figure 1. Raw Meal Preparation Flowsheet

As the material is ground by the rollers, it continues to move to the periphery of the grinding table. As the material spills over the dam ring, it is suspended in the air stream by the hot gases from the kiln or preheater. The hot gases enter the base of the mill and pass upwards through a louvered annular ring around the grinding table. The entrained fine material is carried by the air stream upwards into the classifier forming the upper part of the mill. The fine particles pass the separator and leave the mill as final product. The coarser particles will be rejected by the separator and recycled to the grinding table for further grinding.
The major part of the fine material leaving the mill separator with the outgoing gases is precipitated in cyclones. This material is conveyed to the blending storage silo for raw meal. The remaining fine dust from the mill not precipitated in the cyclones is carried to the electrostatic precipitator, where the fine cleaning of the gases takes place, before the gases (now cleaned) are sent into the atmosphere by the filter fan. The raw meal is a milled fine powder from the proportioned raw materials with correct chemical balance. To ensure consistency in product quality, homogenization of material is essential. The production rate of the raw mill is 240,000 kilogram per hour.

4. Process Simulation Model of Raw Mill

The raw mill as a unit operation in cement manufacturing process was simulated using Aspen Plus version 8.4. "SOLID Model" inbuilt template in Aspen plus was used as a basis for the simulation model. Selection of an appropriate property package which accurately reproduces the various physical properties for the system in question is a key requirement of process modelling. Redlich-Kwong equation of state method is being used to find the vapour phase properties, while Henry's Law handles the supercritical components present in the liquid phase using asymmetric convention. The equilibrium constants and enthalpy are solved using Kent-Eisenberg method. All simulations are performed in steady state.

Modelling of solids anywhere in a process sheet has been made easy using Aspen Plus. A wide range of unit operation models for solids handling equipment is available and this includes crushers (mills), screens (separator), cyclones and electrostatic precipitator. Material and energy streams are represented using blocks which are placed on a flowsheet. An extensively built-in databank of both physical and chemical properties was used for the simulation calculations. Properties and component flows of different types such as liquid, solid and vapour are reported in ‘Solids Template’. Mixed conventional solids with particle size distribution (MIXCIPSD) is present. The following assumptions were made in the development of the models:

a) The process operates in steady state conditions and ignores pressure losses, turbulent motions, and air leakages.
b) Both the atmospheric and pressure drop of the process is considered negligible.
c) No heat losses through the system
ASPEN Plus model could be developed through these processes:

• Property selection method and stream classes ae described.

• Comprehensive databank for system component description.

• Process flow sheet are defined base on the unit operation of blocks with the connecting material and energy stream. This also involves the thermodynamic conditions with both the physical and chemical reactions in the blocks.

• Feed stream description, which includes mass or volumetric flowrate, feed component, with particle size distribution.

The process model is based on a raw mill plant operation with capacity of 240,000 kg per hour, and includes the required physical property parameters to simulate this type of system. The raw mill process consists of three-unit operations, namely: drying, grinding and separation. The raw materials were crushed/ milled and the entrained fine material is carried by the air stream upwards into the classifier forming the upper part of the mill. The rich inlet stream is defined as “Solid-Liquid” and “Solid-Vapour”, meaning that it is treated as a two-phase stream by Aspen Plus. The stream leaving the mill is led into a screen (separator), where the fine dust is separated from the course and the coarse recycled back to the mill for regrinding. The mixed fine dust (product) and gas is transported to cyclones where the fine dust in gas stream is separated and stored in the silo as kiln feed. The gas stream laden with dust is led to the electrostatic precipitator to trap the fine dust while the cleaned gas leaves to the environment.

Crusher (mill)

Crusher (mill) model is a dry grinding continuous operation that assumes homogeneity of inlet feed. The raw material is crushed/milled and the entrained fine material is carried by the air stream upwards into the classifier forming the upper part of the mill. Both the feed stream and the outlet particle stream are of the same composition, and no chemical reaction takes place. The operation only involves size reduction to fine particles.

Screen (Separator)

The stream leaving the mill is led into a screen (separator), where the fine dust are separated from the coarse and the coarse recycled back to the mill for regrinding. The simulation of the screen (separator) sizes of entrained fine material will determine the quantity recycled or allow passing as final product and this depict the separation efficiency of the screen.

IOP Conf. Series: Materials Science and Engineering **413** (2018) 012048 doi:10.1088/1757-899X/413/1/012048

Cyclone

Further separation of gas laden with dust is simulated with cyclone separators. The fine dust particles is removed from a dust entrained gas stream with the aid of a centrifugal force of a gas vortex obtained from cyclone.

Electrostatic precipitator (ESP)

Final separation of the gas stream entrained with dust particles is done using electrostatic precipitator before the clean air is released to the atmosphere. ESP, as Electrostatic precipitator is used for the simulation. Positioning of the wires in parallel and in between the plates makes it easy for the electrostatic field of the collecting plate electrodes to remove the dust particles from the gas stream.

4.1 Simulation Data

The main purpose of the raw mill plant simulation is to make a simple but realistic model to evaluate the exergy efficiency of the overall raw mill plant. The raw mill with a capacity of 240,000 kilogram-material per hour has been simulated with Aspen Plus. The chosen property package for the simulations performed in this paper was the "SOLID", however, in order to cater for the thermodynamic requirements of the two phase streams, the components are distributed over the sub-stream of types MIXED and CISOLID as this is the most accurate property method for the simulated processes.
Specifications for the calculation in ASPEN PLUS Simulation are listed in Tables 1 to 4. The flowsheet of the Aspen Plus process model for raw meal preparation is presented in Figure 2.

Table 1: Specification for Raw mill simulation

	Unit	
Inlet material flow	kg/h	240,000
Inlet moisture flow	kg/h	48,000
Inlet hot gas flow	kg/h	477,086
Inlet dust flow	kg/h	19.353
Inlet hot gas temperature	°C	290
Inlet material temperature	°C	30
Operating Pressure	atm	1
Cyclone efficiency	%	96
Separator efficiency	%	86
Electro-static precipitator efficiency	%	84

Table 2: Typical composition of Raw Material

Component	Weight Percentage (%)
$CaCO_3$	75
SiO_2	22
Al_2O_3	2
Fe_2O_3	1

Table 3: Material Particle size distribution for raw mill inlet

Sieve size	Unit	Weight fraction (%)
20	*mm*	10
40	*mm*	25
60	*mm*	30
80	*mm*	30
100	*mm*	5
120	*mm*	0

Table 4: Sieve Analysis for raw mill outlet

Sieve size	Unit	Weight fraction (%)
20	Mm	80
40	mm	15
60	mm	5
80	mm	0

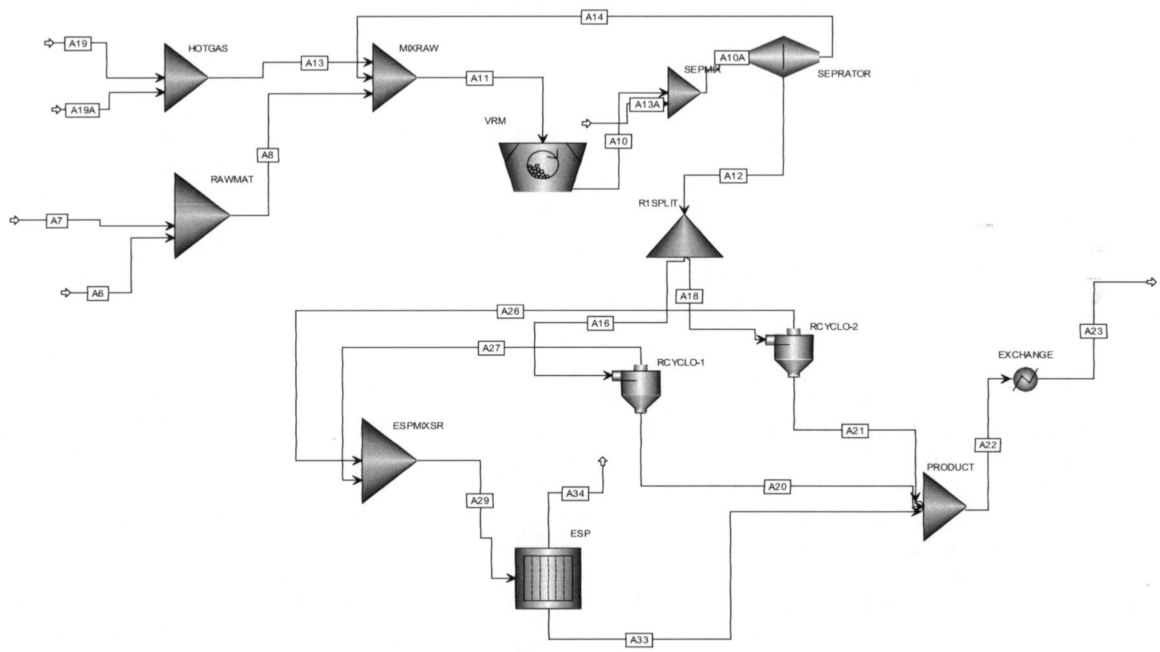

Figure 2: **Aspen Plus process model flowsheet for raw meal preparation**

4.2 Material and Exergy Balance Calculations

The raw mill mass balance is a representation of the law of conservation of mass of Eq. (1), which on application on the raw mill becomes:

$$\Sigma \dot{m}_{in} = \dot{m}_{raw\,feed} + \dot{m}_{hot\,gas} + \dot{m}_{moisture} + \dot{m}_{return\,separator} + \dot{m}_{dust\,in\,hot\,gas} \quad (10)$$

$$\Sigma \dot{m}_{out} = \dot{m}_{raw\,meal} + \dot{m}_{gas} + \dot{m}_{moisture} + \dot{m}_{steam} \quad (11)$$

The feed stream for each component is supplied with other necessary parameter for each block and the software in turn generates for the outlet mass balance. ASPEN Plus calculates for each of the stream components the exergy value. Exergy efficiency of the raw mill can be calculated using equation (7).

5. Results and Discussion

The results of the raw mill thermodynamics analysis with Aspen Plus process simulation model using available data from a local cement plant and three literature data are presented in this section.

5.1 Local raw mill simulation and exergy analysis

The Aspen Plus process simulation results of material balances of the local raw mill are presented in Table 5.

Table 5: Simulation results for the Mass balance of the local raw mill

Mass Balance of Raw mill							
Input				Output			
Material	Unit	Plant data	Simulation	Material	Unit	Plant data	Simulation
Raw feed	kg/h	240000	240000	Raw meal	kg/h	235300	231977
Moisture in raw feed	kg/h	48000	48000	Gas	kg/h	555260	558487
Hot gas from kiln	kg/h	477086	477086	moisture	kg/h	3266	3564
Dust in hot gas	kg/h	19353	19353	Steam	kg/h	65612	65412
Return from separator	kg/h	75000	75000				
Total		859439	859439	Total		859439	859439

It was observed that the input of the material simulation results is the same as that of the actual plant data and that the output of the material simulation results varied though not excessively with the actual plant data since at significance level set at 0.05, the calculated *F* value (8.14E-12) was less than *Fcritical* (5.987378). This is an indication that the software can simulate the material balance of the raw mill in cement industry within acceptable limits of accuracy. The results of the exergy balance for the local raw mill are shown in Table 6.

Table 6: Simulation result for the exergy balance of the raw mill

Exergy balance of raw mill					
Input			Output		
Material	Unit	Simulation	Material	Unit	Simulation
Raw feed	kJ/h	3194	Raw meal	kJ/h	2024720
Moisture in raw feed	kJ/h	-9053	Gas	kJ/h	7545475
Hot gas from kiln	kJ/h	47474430	moisture	kJ/h	48152

Dust in hot gas	kJ/h	1655660	Steam	kJ/h	926255
Return from separator	kJ/h	173780			
Total input Exergy		49298012	Total output Exergy		10544602
Exergy efficiency		21.4%			

The exergy efficiency of the local raw mill is 21.4%. It was found from Table 6 that at the operating conditions, exergy loss is (38753410 **kJ/h**) and this corresponds to 78% of the total exergy input to the raw mill. The exergy and mass flow diagram of the local raw mill is presented in Figure 3 and shows that most of the exergy input to the system was due to hot gas and dust (96.6%) followed by the returned material (3.3%) with the exergy of raw materials contributing less than 1%.

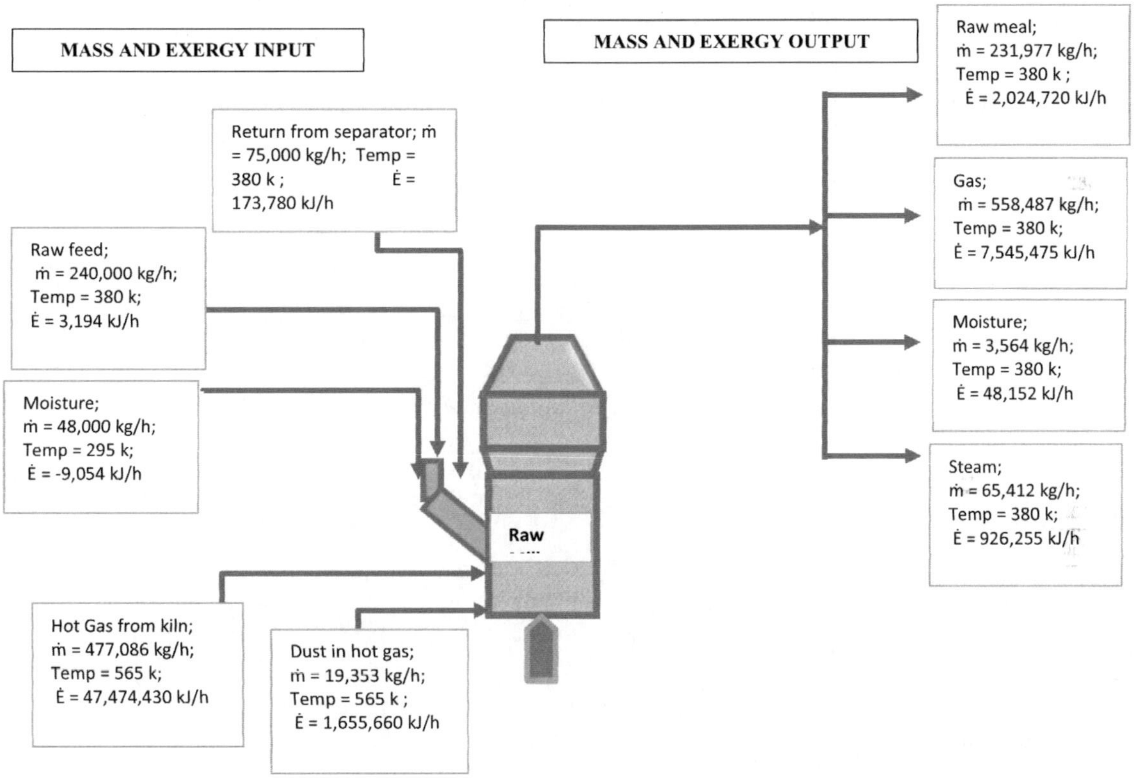

Figure 3. Exergy and mass flow diagram of the raw mill

5.2 Simulation and exergy analysis of raw mills from literature

The Aspen Plus process simulation results of material balances of three raw mills from literature, namely: Utlu et al. [8], Dyuthi [9] and Atmaca et al. [10] are presented in Table 7.

Table 7: Comparison of simulation results and literature values of the mass balance of raw mill

Mass balance of the raw mill from the literature

		Utlu et al. [8]		Dyuthi [9]		Atmaca et al. [10]	
Input	Unit	Literature data	Simulation	Literature data	Simulation	Literature data	Simulation
Raw feed	kg/h	82515	82515	117352	117352	135016	135016
Moisture in raw feed	kg/h	6065	6065	3696	3696	15038	15038
Hot gas from kiln	kg/h	49464	49464	98714	98714	66346	66346
Dust in hot gas	kg/h	3067	3067	3043	3043	2103	2103
Return from separator	kg/h	36000	35999	242000	242143	27104	27218
Total		177111	177110	464805	464948	245607	245721

		Utlu et al. [8]		Dyuthi [9]		Atmaca et al. [10]	
Output	Unit	Literature data	Simulation	Literature data	Simulation	Literature data	Simulation
Raw meal	kg/h	121582	120853	362395	362499	162223	162076
Gas	kg/h	49464	51559	98714	98161	68346	68592
Moisture	kg/h	715	475	605	513	6430	6435
Steam	kg/h	5350	4223	3091	3775	8608	8618
Total		177111	177110	464805	464948	245607	245721

It was observed that the input of the material simulation results are the same as that of the literature data except for the return from separator stream which varied very slightly with literature data; and that the output of the material simulation results varied though not excessively with literature data. The same trend was observed for the material simulation results for the local raw mill and the raw mills from literature data. The results of exergy balance of the raw mills from literature and simulation using Aspen Plus process simulator are shown in Table 8.

Table 8: Comparison of simulation results and literature values of the exergy balance of the raw mill

		Utlu et al. [8]		Dyuthi [9]		Atmaca et al. [10]	
Input	Unit	Literature data	Simulation	Literature data	Simulation	Literature data	Simulation

Raw feed	kJ/h	-1041	1140	3724	1758	4428	4
Moisture(feed)	kJ/h	-381	2929	641	10397	2664	11315
Hot gas from kiln	kJ/h	7405586	7897690	11912266	13099390	7950060	9614140
Dust in hot gas	kJ/h	310257	267223	338809	36031	180144	147835
Return from separator	kJ/h	150310	299892	831427	312559	160164	242785
Total input exergy		7864732	8468874	13086866	13460135	8297460	10016079
		Utlu et al. [8]		Dyuthi [9]		Atmaca et al. [10]	
Output	Unit	Literature data	Simulation	Literature data	Simulation	Literature data	Simulation
Raw meal	kJ/h	1087322	1006770	1744570	1954000	1733616	1445710
Gas	kJ/h	758158	810269	562588	438187	1151136	1840790
Moisture	kJ/h	29475	84671	14843	13983	350856	355597
Steam	kJ/h	106453	138715	35467	64414	47592	369549
Total output Exergy		1981408	2040426	2357467	2470584	3283200	4011646
Exergy efficiency		25.2%	24.1%	18.0%	18.4%	39.6%	40.1%

It was observed that the exergies of the streams in literature data and simulation varied. The differences in exergy of streams are due to the differences in the datum levels used in calculation of thermodynamic data in the literature data and Aspen Plus process simulator. The exergy efficiencies of the raw mills from literature were calculated with the electrical work included in the exergy input. The exergy of electrical work is excluded in this work so as to focus on the suitability of using Aspen Plus process simulator to evaluate the exergy efficiency of the raw mills based on exergy of material streams and make a comparison between exergy efficiencies from literature and this work. As shown in Table 8, average deviations of exergy efficiencies for the literature data and simulation vary within ±2.5% for the three individual literature data. The Aspen Plus simulation model can therefore be considered as a reliable and efficient tool to predict the exergy of material streams and exergetic efficiency of raw mill operation.

6. Conclusion

This study focused on exergy utilization, exergy balance and irreversibility for a raw mill in the cement industry using the real plant data and literature data. Exergy efficiency of the raw mill studied using Aspen plus simulator modelling technique was found to be 21.4%. It was found that the simulation results of the three literature data vary within ±2.5% of the published results. From the operating conditions, the raw mill exergy efficiency is very poor. The study is limited only to the material streams exergy of the system

IOP Conf. Series: Materials Science and Engineering **413** (2018) 012048 doi:10.1088/1757-899X/413/1/012048

investigated. The present method using the Aspen plus simulator is suggested as a useful tool in making informed decisions for developing energy policies.

References

[1] Bejan A. Advanced engineering thermodynamics. Wiley – Interscience Publication, John Wiley & Sons, New York, 1988

[2] Boroum J. G., Rismanchi B., Saidur R. A review on exergy analysis of industrial sector. Renewable and Sustainable Energy Reviews, 2013, 27: 198– 203.

[3] Duflou J. R, Sutherland J. W., Dornfeld D., Herrmann C., Jeswiet J., Kara S., Hauschild M., Kellens K. Towards energy and resource efficient manufacturing: A processes and systems approach. CIRP Annals - Manufacturing Technology, 2012, 61(2): 587–609.

[4] Sciubba E., Wall G. A brief commented history of exergy from the beginnings to 2004. Int. J. Thermodynamics, 2007, 10 (1): 1-26.

[5] Narayanan K. V. A text book of chemical engineering thermodynamics. PHI learning private limited, Delhi, 2013.

[6] Dincer, I., Ratlamwala, T.A.H. Importance of exergy for analysis, improvement, design, and assessment. WIREs Energy Environ, 2013, 2: 335–349 doi: 10.1002/wene.63

[7] Rosen M. A. Enhancing ecological and environmental understanding with exergy: concepts and methods. In: Proceedings of the 4th IASME/WSEAS Int. Conference on Water Resources. Hydraulics & Hydrology (WHH'09), 2009, Pp 94-103.

[8] Utlu Z., Sogut Z., Hepbasli A., Oktay Z. Energy and exergy analyses of a raw mill in a cement production. Appl Therm Eng, 2006, 26: 2479-2489.

[9] Dyuthi. Assessment of energy conservation in Indian cement industry and forecasting of CO2 emissions PhD thesis, Cochin University of Science and Technology, India, 2012.

[10] Atmaca A., Kanoglu, M. Gadalla M. Thermodynamic analysis of a pyroprocessing unit of a cement plant: a case study. International Journal of Exergy, 2012, 11(2)

IOP Conf. Series: Materials Science and Engineering **413** (2018) 012049 doi:10.1088/1757-899X/413/1/012049

Nigeria Vision 20:2020; Review of Current Status Using Re Development and Gross Domestic

Inegbenebor Anthony[1], Okolie Sociis TA[1], Udoye Nduka E[1], Enesi Salawu Y[1], Onyisi Ozuor[1]

Department of Mechanical Engineering, Covenant University[1], Km. 10 Idiroko Road, Canaan Land, Ota, Ogun State, Nigeria E-mail address: okolie.sociis@covenantuniversity.edu.ng,

Abstract. The Nigerian government has put several economic tools into place to improve on the economic productivity of the nation. To mention but a few, we have vision 2010 and the seven-point agenda. Many of these economic reforms failed as a result of inadequate plans to improve the technology. The current economic agenda tagged vision 20:2020 which Nigeria is becoming one of the top 20 world largest economy for year 2020. However, the area in which the GDP of the largest world economics are leading is in their continuous research and development of their engineering technology. The study therefore critically reviewed the current state of the Nigeria renewable energy development (RE) and its contribution towards gross domestic product in order for the country to achieve Vision 20:2020. The study also evaluated the current impact of GDP from 2012-2017 in order to know the area needed for improvement. The result of the assessment showed that Nigeria will need about US$900 billion of GDP and per capital income of about US$4000 per annum for successfully become one of the 20 largest economics in the world. This analysis can only be possible with keen interest in developing the technology.
Keywords: Production, Technology, Energy, Failure, Economy

1. Introduction

Nigeria vision 20:2020 is a conscious economic strategy that was planned to improve the sustainability of the developmental goals in order to build a robust economy and become one of the top 20 countries in the world. However in an attempt to achieve a good and sustainable economic development, the nation failed to understand that developing a reliable power and energy sector is the easiest way. A number of research works have been carried out on the need to enhance the energy and power sector. For instance, formulation of policies and implementation of research outcome can spur the energy industries of developing country like Nigeria towards sustainable development, thereby boosting the economic activities of the country as well as improvement in the general economic activity and the welfare of the citizens. For a developing nation like Nigeria blessed with renewable energy resources, vision 20:2020 can be a reality if we adopt a highly decentralized, efficient and dependable economy through harnessing and utilizing the available energy sources, thereby building a robust sustainable development [1, 2]. A comparative study of India and Nigeria had shown the trend of progression made by India in their sustainable development goal. This was actually achieved as a result of consistency in the economic and development plan. On the contrary, Nigeria has not made the targeted progress due to instability in her plans and decision [3].

Although, in our quest to achieve this sustainability in year 2020, the Nigerian government must realize the impact of the climate change caused by emission of Co_2 into the atmosphere. This is because a developing nation is expected to have more of these emissions and thus conscious effort must be made towards maintaining the strategy for sustainable development [4]. In fact study has shown that sustainable industrialization policies can actually boost the economic productivity of Nigeria as a nation through harnessing the technical and potential resources available in the country. Hence review of policies guiding electricity generation and distribution becomes very important towards achieving vision 20:2020 [5].
According to [6], to achieve efficient sustainable development in the energy sector, there is need to democratize the energy sector so that we can have increase in the competition. Electricity should be divided into different sectors such as generation, transmissions and distribution. In this case, the participation of

the consumers in decision making would be more, thereby having a decentralized system of running the sector and a better sustainable development as a result of competition. Study has shown that growth in wind energy, geothermal energy and nuclear energy can only be assessed when data are stored and analyzed. Proper assessment of these resources and good policies to improve them will give room for foreign investors to invest in Nigeria and this will greatly affect the country achieving sustainable development [7].

A focus on renewable energy resources available in the country will help Nigeria to combat the epileptic power supply, instability in price of petroleum resources as well as attracting industrialization. This can only be possible when government revisit the current and future outlook of the renewable energy in the country. The current problem, as well as strong solution towards achieving sustainability [8]. About 40% of the people are connected to national grid and even those connected are not enjoying the power as expected. This is as a result of the poor performance of the power sector. For the country to enjoy sustainable development there is need to review the performance of the infrastructures available to the people as well as the industries so that better economic productivity could be achieved [9].

Furthermore, the renewable energy development cannot be achieved without taking conscious effort to invest in the R&D. Research and development is the bedrock of industrialization. For Nigeria to achieve her vision 20:2020, there is need to encourage investors by ensuring a stable policy which will have direct impact on the renewable energy development such as introduction of tax relief thereby making it possible for investors to cut down cost [10]. According to [11, 12, 13, 14], Nigeria as a nation can achieve sustainable development by allowing the energy commission of her country to engage in better programs and strategic plans that can improve the energy generation from the renewable energy resources. The assessment of the future state of Nigeria's Vision 20:2020 cannot be completed without evaluating the renewable energy development. This work focused on the future state of the renewable energy and the total assessment of the Vision 20:2020 using the Gross Domestic Product.

2. Review on Gross Domestic Product Impact From 2012-2017

Nigeria has greatest economy in the West African sub locale. Given the nation's extensive asset and waterfront area there is potential for solid development. The nation directly positions among the 11 nations distinguished by Goldman Sachs to have the potential for accomplishing worldwide intensity in view of their financial and statistical settings and the establishment for changes effectively laid. In any case, Nigeria has acknowledged extremely little of these possibilities. The historical backdrop of monetary stagnation, declining welfare and social insecurity, has undermined improvement for the greater part of the previous 30 years [11].

However, development planning is essential for the country to achieve a total development. The different ways that our country Nigeria goes about the development plan has brought failure of most of their plans. Vision 20:2020 is one target plan to put Nigeria among the best 20 countries in the year 2020. The question of whether this vision is realizable becomes very important because of the failure experience in the past year [12]. According to [13], failure of the development plan can be traced to instability in the socio-political sector of the economy making it impossible for the investors to come for the harnessing and investment in our natural resource. However adequate implementation of the policies and experience from the development plans of the advanced countries such as the Asian tigers will help in making the country's sustainable development plan a success.

Study has shown that Nigeria vision 20:2020 can become a reality, if adequate knowledge of economic improvement is done. This can only be possible if the agricultural industry is increasing the economic growth in alignment with vision target, hence the domestic and infrastructural part of the economy can improve the economic growth while government policies will eventually make this growth sustainable

[14]. More so, to achieve the vision, GDP must grow geometrically from US$752 to $30,000 as well as transformation of the rural areas into urban areas. Education, health, agriculture and power must be better in 5 years compared to the past [15]. Actualizing the vision also calls for the development of the human resources as well as the physical amenities to curb the area of poverty [16].

3. Impact of GDP annual growth rate

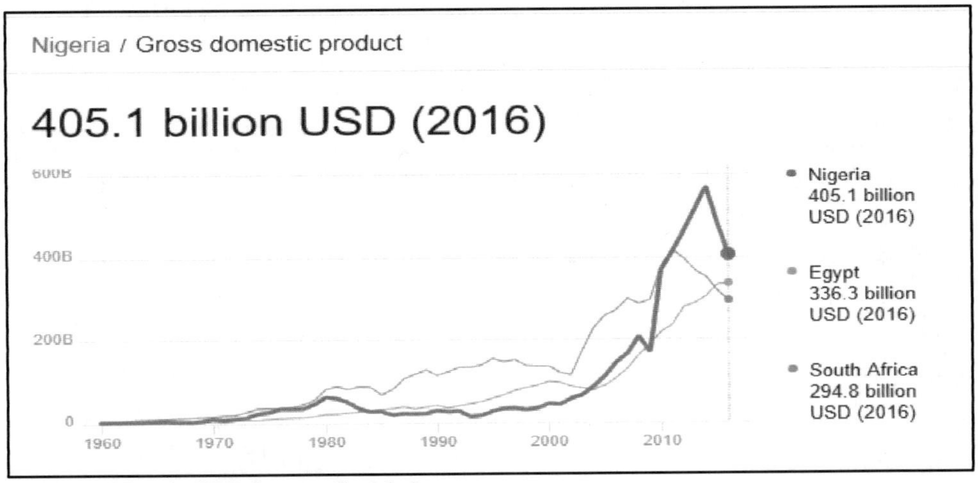

Figure 1: Nigeria GDP Source: Google images

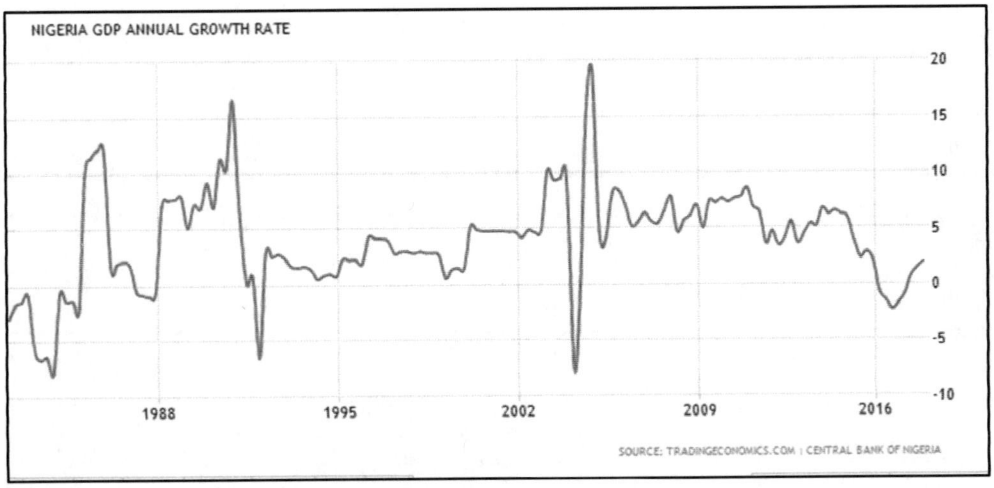

Figure 2: Nigeria GDP trend from 1982 till date Source: tradingeconomics.com CENTRAL BANK OF NIGERIA

Figure 3 Nigeria GDP over the past 10 years, Source: tradingeconomics.com CENTRAL BANK OF NIGERIA

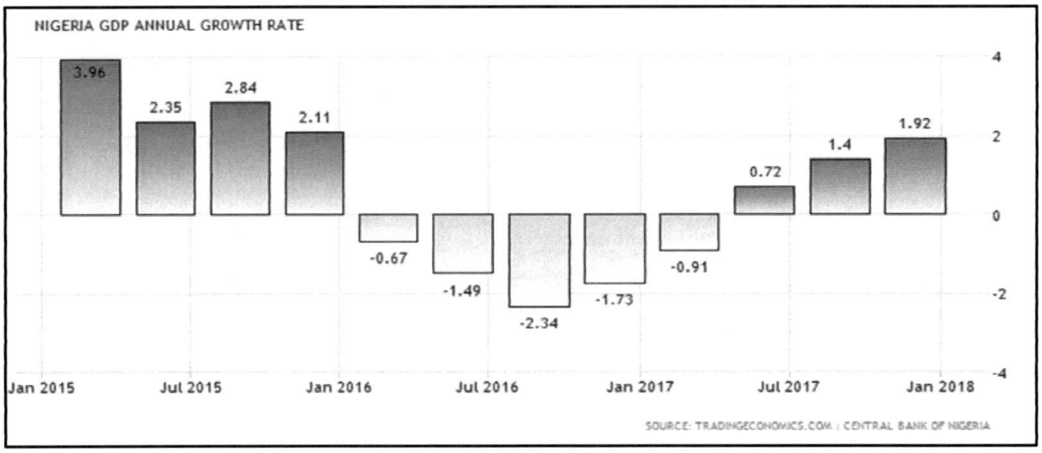

Figure 4 GDP growth rate from Jan 2015, Source: tradingeconomics.com CENTRAL BANK OF NIGERIA

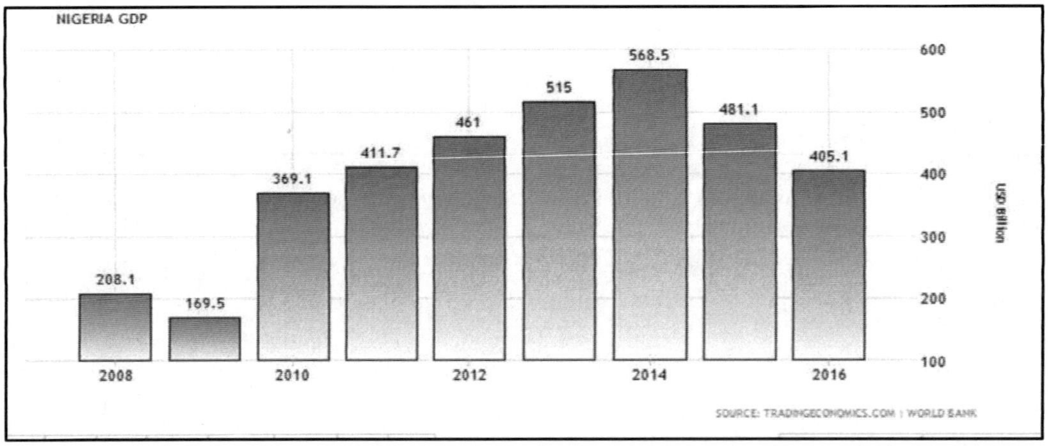

FIGURE 5: GDP RAW VALUES FROM 2008 IN USD BILLION, SOURCE: TRADINGECONOMICS.COM CENTRAL BANK OF NIGERIA

IOP Conf. Series: Materials Science and Engineering **413** (2018) 012049 doi:10.1088/1757-899X/413/1/012049

Table 1: Top economies in the world according to the IMF, World Bank and the United Nations

Per the International Monetary Fund (Estimates 2017)[1]			Per the World Bank (2016)[20]			Per the United Nations (2016)[21][22]		
Rank	Country	GDP (US$MM)	Rank	Country	GDP (US$MM)	Rank	Country	GDP (US$MM)
	World[19]	**75,278,049**		*World*	**75,543,543**		*World*[24]	**75,648,448**
1	United States	19,362,129	1	United States	18,624,475	1	United States	18,624,475
—	European Union[n 1][19]	17,112,922	—	European Union[n 1][23]	16,397,980	—	European Union[n 1][25]	16,332,631
2	China[n 2]	11,937,562	2	China[n 5]	11,199,145	2	China[n 5]	11,218,281
3	Japan	4,884,489	3	Japan	4,940,158	3	Japan	4,936,211
4	Germany	3,651,871	4	Germany	3,477,796	4	Germany	3,477,796
5	France	2,574,807	5	United Kingdom	2,647,898	5	United Kingdom	2,647,898
6	United Kingdom	2,565,051	6	France	2,465,453	6	France	2,465,453
7	India	2,439,008	7	India	2,263,792	7	India	2,259,642
8	Brazil	2,080,916	8	Italy	1,858,913	8	Italy	1,858,913
9	Italy	1,921,139	9	Brazil	1,796,186	9	Brazil	1,795,925
10	Canada	1,640,385	10	Canada	1,529,760	10	Canada	1,529,760
11	South Korea	1,529,743	11	South Korea	1,411,246	11	South Korea	1,411,245
12	Russia[n 3]	1,469,341	12	Russia[n 3]	1,283,162	12	Australia	1,304,463
13	Australia	1,390,150	13	Spain	1,237,255	13	Russia[n 3]	1,246,015
14	Spain	1,307,170	14	Australia	1,204,616	14	Spain	1,237,255
15	Mexico	1,142,453	15	Mexico	1,046,922	15	Mexico	1,076,914
16	Indonesia	1,010,937	16	Indonesia	932,259	16	Indonesia	932,259
17	Turkey	841,206	17	Turkey	863,711	17	Turkey	863,711
18	Netherlands	824,480	18	Netherlands	777,227	18	Netherlands	777,227
19	Switzerland	680,645	19	Switzerland	668,851	19	Switzerland	668,851
20	Saudi Arabia	678,541	20	Saudi Arabia	646,438	20	Saudi Arabia	639,617
21	Argentina	619,872	21	Argentina	545,866	21	Argentina	545,866
22	Taiwan	571,453	22	Sweden	511,000	22	Sweden	514,475
23	Sweden	541,889	23	Poland	469,509	23	Poland	471,402
24	Poland	509,955	24	Belgium	466,366	24	Belgium	467,955
25	Belgium	491,672	25	Iran	425,326	25	Iran	425,402
26	Thailand	437,807	26	Thailand	406,840	26	Thailand	407,026
27	Iran	427,666	27	Nigeria	405,083	27	Nigeria	404,649
28	Austria	409,316	28	Austria	386,428	28	Austria	390,799
29	Egypt	408,045/Na	29	Venezuela	371,337	29	Norway	371,068
30	Nigeria	394,818	30	Norway	370,557	30	United Arab Emirates	399,451
31	Norway	392,052	31	United Arab Emirates	348,743	31	Colombia	377,740
32	United Arab Emirates	378,656	32	Egypt	336,297	32	South Africa	349,819

Source: Wikipedia 2018

4. Discussion

From figure 1, it can be deduced that, the Gross Domestic Product (GDP) in Nigeria was about 405.10 billion US $ which represent 0.65% of the world economy in the year 2016. This tremendous rise in the GDP showed the performance of economic activities towards gearing sustainable development agenda.

Figure 2 also present the rise in GDP from 1982 till date. It can be seen that the GDP experienced geometric increase in growth between 1985-1990 and 2002 to about 2004. However compare to the GDP rise from 2008-2016, we experienced a fall in GDP which can be tagged to failure in

economic activities. Further to this, figures 3, 4 and 5 represent GDP growth rate in 10 years, GDP growth rate form January 2015-2018 and GDP raw values from 2008-2016.from figure 3, annual growth rate was witnessed between 2008-2011 and this also was repeated 2012-2015.
However, figure 5 showed that the only annual growth rate was experienced in January 2015 about 3.96. But figure 5 showed that the GDP growth rate took a geometric increase from 2010-2014 and dropped geometrically from 2015-2016. Table 1 above is a table showing the top countries in the world that Nigeria is aspiring to belong. These countries are called top in the world because of their advancement in technology and their economy which has led to a better sustainable development. They were rated by IMF, World Bank and United nation.

5. CONCLUSION

Looking at the above table, for Nigeria to achieve Vision 20:2020, the GDP must rise beyond 65% to compete with Saudi Arabia which is the least on the list. Even as that with the sudden fall in GDP of Nigeria to 8.67% from 2012 -2017, the probability of becoming one of the 20 world best in economy is in doubt. However, overcoming the challenges highlighted in the renewable energy development and gross domestic product can make the Nigeria vision 20:2020 a success.

Acknowledgments
The authors will like to acknowledge the support of Covenant University Ota for Conference paper

References

[1] Asaju, A. Vision 20:2020 Realities And Challenges. Research Gate [J] 275-281, 2012..
[2] Ajayi, O. O., Ajayi, O. O. Nigeria's energy policy: Inferences, analysis and legal ethics toward RE development. Energy Policy [J] 2013, 60, 61-67,
[3] Aliyu, A. S., Dada, J. O., Adam, I. K. Current status and future prospects of renewable energy in Nigeria. Renewable and sustainable energy reviews [J], 2015, 48, 336-346.
[4] Ayodele, O. E. Challenges facing the achievement of the Nigeria vision 20:2020 . Global Advanced Research Journals [J], 2013, 144-157.
[5] Ajogwu.. Nigeria Vision 2020 and a Challenging Economic Situation: Is Good Governance a Panacea? . Centre for Values in Leadership Lecture [J], 2014, 1-20
[6] Adhekpukoli, E. The democratization of electricity in Nigeria. The Electricity Journal [J], 2018, 31(2), 1-6.
[7] Brycz.. Nigeria Vision 20:2020 Can Dream Become Reality? Evidence from National Accounts. Research Gate [J], 2014. 162-170.
[8] Brimmo, A. T., Sodiq, A., Sofela, S., Kolo, I. Sustainable energy development in Nigeria: Wind, hydropower, geothermal and nuclear (Vol. 1). Renewable and Sustainable Energy Reviews [J], 2017, 74, 474-490.
[9] Eneh.. Nigeria's Vision 20:2020-Issues, Challenges and Implications for Development Management. Asian Journal of Rural Development [J], 2011, 21-40.
[10] Ejiogu, A. R. A nuclear Nigeria: How feasible is it. Energy Strategy Reviews [J], 2013, 1(4), 261-265.
[11] Elum, Z. A., Momodu, A. S. Climate change mitigation and renewable energy for sustainable development in Nigeria: A discourse approach. Renewable and Sustainable Energy Reviews [J], 2017. 76, 72-80.
[12] Falebita, O., Koul, S. From developing to sustainable economy: A comparative assessment of India and Nigeria. Environmental Development [J]. 2017
[13] Ikpe, E., Torriti, J. A means to an industrialisation end? Demand side management in Nigeria. Energy Policy [J] 2018

[14] Maji, I. K. Does clean energy contribute to economic growth? Evidence from Nigeria. Energy Reports [J], 2015, 1, 145-150.

[15] Oyedepo, S. O. On energy for sustainable development in Nigeria. Renewable and sustainable energy reviews [J], 2012, 16(5), 2583-2598.

[16] Oyedepo, S. O. Towards achieving energy for sustainable development in Nigeria. Renewable and sustainable energy reviews [J], (2014, 34, 255-272.

[17] Osunmuyiwa, O., Kalfagianni, A. The Oil Climax: Can Nigeria's fuel subsidy reforms propel energy transitions. Energy research & social science [J], 2017, 27, 96-105

[18] Oseni, M. O. An analysis of the power sector performance in Nigeria. Renewable and Sustainable Energy Reviews, 2011, 15(9), 4765-4774.

[19] Usman, Z. G., Abbasoglu, S., Ersoy, N. T., Fahrioglu, M. Transforming the Nigerian power sector for sustainable development. Energy Policy [J], 2015, 87, 429-437.

[20] Waziri, B.. Actualizing Nigeria's Vision 20:2020: Imperatives of the Construction Sector . Civil and Environmental Research www.iiste.org [J], 2014, 69-74.

IOP Conf. Series: Materials Science and Engineering **413** (2018) 012050 doi:10.1088/1757-899X/413/1/012050

Numerical Modeling of Cooling Rate and Hardness Prediction in Axi-Symmetric Flash Butt -Welded C45 Steel Circular Long Bars

Olabamiji T.S.[1] **and** **Adedayo S.M.** [2]

[1]Mechanical Engineering Department, Landmark University, Omu- Aran, Nigeria
[2]Mechanical Engineering Department, University of Ilorin, Ilorin, Nigeria

E – mail: olabamiji.taye@lmu.edu.ng

Abstract. FE investigation into the effect of welding variables on the cooling rate, $\Delta t_{8/5}$ (cooling time from 800 to 500 °C), around the weld and heat affected zone (HAZ) areas of C45 steel was carried out. Weldment material hardness property changes was examined using Brinell hardness test method, variation in hardness at known distances from the weld line in the HAZ arising from different cooling rates was the main focus. There exists in literature empirical correlations that relate cooling rates and hardness vis-a-vis welding variables, these were used for the computational evaluation of hardness. To establish the effects of welding variables on the cooling rate, a computer programme was written with Matlab. Generalized numerical prediction of temperature distribution, peak temperature, cooling rate and thermal cycles in solid bars joined by butt arc welding process was carried out and subsequently used to predict hardness of C45 steel. Increase in arc duration reduced the cooling rate, which consequently reduced the hardness by as much as 47%. The cooling rate decreased with increase in preheat, results in a significant decrease in hardness of the HAZ by as much as 49%. The results also indicated that the cooling rate increased slightly with an increase in coefficient of convection and caused an insignificant increase in hardness. It was discovered from the results that metal pre-heat temperature prior to welding has minimal effect on the cooling rate. Simulated weldment hardness obtained was in good agreement with the experimental value within 9 %.
Keywords: cooling rate; welding variable; arc duration; preheat temperature; hardness

1. Introduction

It is important to evaluate the cooling rate and hardness at particular locations of a weldment so that it can be controlled in a proper way for prevention of cracks. Cooling rate dictates the soundness of the weld metal in respect of porosity, inclusion and the weld bead geometry. Cooling rate also affects the mechanical properties such as hardening and embrittlement of the hard enable steel and the metallurgical structure [1,2]. In the heat-affected zone, the metallurgical structure can be controlled favorably in order to avoid such defects as embrittlement, cracks or excessive hardening of the metal. Typical materials such as stainless steels where the chromium carbide formation takes place in the heat affected zone under the slow cooling conditions within a sensitive temperature range lead to the loss of corrosion resistance of the stainless steel. Control is made in form of faster cooling through the sensitive temperature range where the chromium carbide precipitation takes place. The cooling rate in the heat affected zone affects the metallurgical structures that consequently influence the mechanical properties as well as the corrosion behavior.

Allied welding processes such as friction stir welding had witnessed extensive research works in the examination of temperature distribution [3,4]. In 2009, Martinson et al., stated that phase transformation due to different cooling rates is important when dealing with high strength or advanced high–strength steel, which are being used more frequently in the car industry [5]. In 2012, Dean obtained the transient temperature distributions, residual stress and deformation induced by electro-slag welding [6]. In 2014, Mato et al., examined numerical and experimental residual stresses and distortions induced in T-joint welding of two plates [7]. A simplified method of predicting stresses in surfaced steel rods was carried out by Jerzy Winczek [8]. In his research, the dependence of stresses and strains was assumed on the basis of tension curves considering temperature. In 2010, Hwa et al., examined the effect of temperature field on the sensitization of Alloy 690 welds fabricated using the gas tungsten arc welding method and the laser beam welding method respectively [9]. The welding thermal cycle of the two welding methods were based upon a moving heat source model. Finite element modeling of temperature cycles in axi-symmetric flash butt-welded thin steel rods and experimental validation was carried out by Olabamiji and Adedayo, 2016. Welding process variables examined include; effect of pre-heat temperature, flash temperature, flash duration, and material geometries on temperature profile at various points along the rod and length of the rod [10]. The present work modeled cooling rates in Arc-welded bars with experimental validation.

2. Heat Treatment Analysis

a. Geometrical Configuration

Figure 1 depicts the welding process of two relatively long circular bars whose lengthwise dimension is large compared with the diameter. (L>>d).

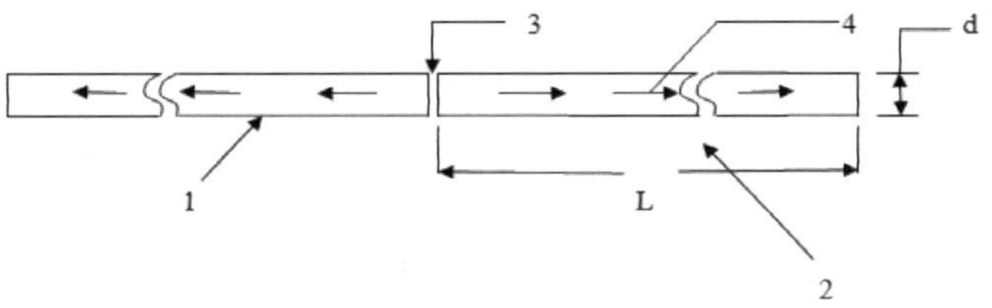

1,2Welded steel bars 3. Weld centre line 4. Heat flow direction

Figure 1: Flash butt - welded bar

With the application of flash as obtained in electrical resistance welding, sparks are produced which generate the necessary heat input for welding of materials. Conductive heat flow is principally along the rod length (L) while some are lost to convection and radiation at rod boundaries and circumferential surfaces.

b. Transient Temperature Distribution Modeling

The following assumptions are made in the modeling equations:

(i) Latent heat associated with phase transformation is not factored in the temperature modeling.
(ii) The heat transfer process is assumed to be symmetrical about the centre line of the welded joint.
(iii) Heat transfer by conduction is principally along the axial direction.
(iv) Effect of fixture and related clamping devices are neglected.
(v) Internal heat generation associated with metallurgical changes is assumed to be negligible.
(vi) Heat loss by convection is considered only at the circular end and circumferential curved surfaces of the rod [11].

Generalized governing differential equation (1) representing temperature distribution profile in one-dimension is depicted by the partial differential parabolic equation:

$$\frac{\partial}{\partial x}\left(kA\frac{\partial T}{\partial x}\right)+Ag=\rho cA\frac{\partial T}{\partial t}+P\beta(T_s-T_\infty)\cdot \qquad (1)$$

3. Finite Element Modeling of Temperature and Hardness Prediction

3.1 Finite Element Equations

Application of the Galerkin's Weighted Residual Method (GWRM) to the governing equation (1) neglecting internal heat generation gives the expression:

$$\sum_1^{n-1}\int_0^{h_e} N^T\left[w_e\frac{\partial T}{\partial t}+\frac{\partial}{\partial x}\left(a_e\frac{\partial T}{\partial x}\right)+(c_eT)\right]dx=0 \qquad (2)$$

Where:

NT - transpose of the shape function,

we -$\rho.c.A$

he - element length

The second order term in eq. (2) need to be reduced to the first order equivalent using integration by parts;

$$\sum_1^{n-1}\int_0^{h_e}\left[N^Tw_e\frac{\partial T}{\partial t}+a_e\frac{\partial N^T}{\partial x}\frac{\partial T}{\partial x}+c_eN^TT\right]dx+\left[a_eN^T\frac{\partial T}{\partial x}\right]_0^{h_e} \qquad (3)$$

Evaluation of the term $\left[a_eN^T\frac{\partial T}{\partial x}\right]_0^{h_e}$ results in Neumann's boundary condition. Its interior nodes vanish during element assembly except for the end nodes. Therefore, the term is dropped at this point because its contribution is incorporated in Neumann's boundary specification [12,13].

The discretized function, $T(x)=[N]\{T\}$ is substituted into eq. (3) to yield:

$$\sum_1^{K-1}\int_0^{h_e}\left[N^Tw_e\frac{\partial}{\partial t}([N]\{T\})+a_e\frac{\partial N^T}{\partial x}\frac{\partial}{\partial x}([N])\{T\}+c_e\,N^T([N])\{T\}\right]dx=0 \qquad (4)$$

The addition of the three component terms of eq. (4) and presentation in matrix form gives:

$$\frac{\rho cAh_e}{6}\begin{bmatrix}2&1\\1&2\end{bmatrix}\begin{Bmatrix}\dot{T}_1\\\dot{T}_2\end{Bmatrix}+\frac{kA}{h_e}\begin{bmatrix}1&-1\\-1&1\end{bmatrix}\begin{Bmatrix}T_1\\T_2\end{Bmatrix}+\frac{P\beta h_e}{6}\begin{bmatrix}2&1\\1&2\end{bmatrix}\begin{Bmatrix}T_1\\T_2\end{Bmatrix}=0 \qquad (5)$$

The assembled global matrix equation can assume the following general form:

$$[A]\{\dot{T}\}+[E]\{T\}=0 \qquad (6)$$

$[A]$= Because of the time assemblage of the first matrix

$[E]$= Addition of the assemblage of the second and third matrices.

In derivation of $\{\dot{T}\}$ in equation (6), there is a need to carry out some form of approximation to the derivatives to reduce the equation to a system of algebraic equivalent before solution can be advanced.

3.2. Evaluation of Time Derivatives

The concept of extending the finite element to include the time domain is found in literature. The approach is to regard the approximation function as being dependent on time as well as the domain as given in equation (7).

$$\frac{\partial T(x,t)}{\partial t} = \frac{\partial N_i(x,t)}{\partial t} T_i \quad (7)$$

Where,

$T(x,t)$ = dependent variable, *T,* as a function of space, *x* and time, *t*;

$N_i(x,t)$ = approximation function at node, *i,* as a function of space and time.

When the time derivatives are approximated with finite differences, either a central in time (Crank-Nicholson) or a backward/forward in time (implicit) scheme can be used [13].

$$\text{Forward difference, } \dot{T} = \frac{T^{n+1} - T^{n}}{\Delta t} \quad (8)$$

Where: n is time level.

Δt = time step,
Substitution of eq. (8) into eq. (6) yields:

$$[A]\left\{\frac{T^{K+1} - T^{K}}{\Delta t}\right\} + [E]\{T^K\} = 0 \quad (9)$$

If Θ is dimensionless time – weighting factor. A value of $\Theta = 0.55$ is chosen which minimizes distortion while conservatively ensuring theoretical stability [13].

The introduction of dimensionless time – weighting factor, Θ into eq. (9) gives:

$$[A + \Delta t\theta E]T^{K+1} = [A - \Delta t(1-\theta)E]\{T\}^K \quad (10)$$

3.3. Computational Procedure

After power activation, an interface plastic state temperature was assumed for all points located at this interface boundary. In the numerical simulation of the butt-arc welding it is assumed that the two bars are welded symmetrically. The arc point is the symmetric line and thus only half of the welded bar is modeled. With the assumption of symmetry, the weld line is modeled as an adiabatic boundary [6,9,10,14]**.** The weldment zones are divided into a number of elements.

In the element discretization of the bar length, the specimen was divided into two parts based on element sizes. The first portion close to weld line around the HAZ has fine and even mesh sizes while the other coarse part

IOP Conf. Series: Materials Science and Engineering **413** (2018) 012050 doi:10.1088/1757-899X/413/1/012050

contains dispersed elements.
The preferred mesh size for the modeling within the refined mesh region was taken to be 0.5 mm. This spacing provides a good compromise between accuracy and computation time. The mesh in the remaining part was 10 mm.
Parameters used for the trial program are:
Convection coefficient = 50 [W m-2 0C-1].
Initial temperature at weld line = 1,600 0C
Initial temperature at other points on rod length = 35 0C.
Diameter = 30 mm
Length = 240 mm
Number of elements = 39
Number of nodes = 40
Arc duration= 5 seconds
Material type = 0.45% C Steel (k = 54 M/m oC , c = 465 J/Kg oC , ρ = 7833 kg/m3).

A computer program written in Matlab_Windows_R2012a on a microcomputer (Processor Speed = 1.6 GHz, RAM = 1.00 GB, Hard Disk Capacity = 148 GB) was used for the computational analysis.

3.4. Prediction of Hardness and Experimental Validation.

A major area of application of simulated temperature and cooling rates is in the prediction of some mechanical properties of materials. Hardness can be obtained from the relationship between Jominy-distance and Rockwell C scale Hardness (HRC) value [15,16]. The Jominy-distance is obtained from standard Jominy-distance versus cooling-time curve. Characteristic cooling that is critical for microstructural phase transformation in ferrous materials is the time of cooling from 800 0C to 500 0C often designated as Δt8/5.. Tables 1 and 2 are cooling time (Δt8/5) versus Jominy- distance and Jominy distance versus hardness (HRC) for a typical (0.45 % C) carbon steel. Information from Tables 1 and 2 is used in obtaining the predicted hardness results.

Table 1 Cooling time versus Jominy distance (Source: Rose, 1958)

Cooling time from 800 °C to 500 °C (s)	**0**	**20**	**40**	**80**	**120**	**140**	**160**
Jominy distance (mm)	0	10	15	28	42	50	60

Table 2: Jominy Test Results of C45 Carbon Steel (HRC) (Source: Rose, 1958)

Jominy distance (mm)	**1.5**	**3**	**5**	**8**	**11**	**15**	**20**	**25**	**30**	**35**	**40**	**80**
Hardness (HRC)	56	55	55	53	52	45	38	35	32	31	30	22

A 240 mm length and 30mm diameter medium carbon steel bar of 0.45% carbon was used in the experiment. Current intensity and voltage were chosen as 80 A and 150 V. After welding, the specimen was air-cooled. Hardness test was carried out using a Brinell Hardness Tester model and Serial No (EEDB 0006/13). Hardness at 1 mm from the weld line was obtained. Applied force was 29.42KN and dwell time was 15 seconds.

4. Numerical Results and Discussions

4.1 Effect of distance from weld line on temperature variation

Figure 2 shows simulated thermal cycles at different distances from the weld line. Peak temperatures of 1,250.9, 965.4, 737.2 and 574.6 0C were attained at distances 0.5, 1.5, 2.5 and 3.5 mm respectively from weld line. Time to attain peak temperatures gradually increased with distances due to time taken at such points to experience the conductive heat wave propagation. The lower peak temperature at farther distances from weld line is due to continual heat losses between the interval surfaces. Also, at distance 0.5 mm from weld line, cooling was rapid immediately after the attainment of peak temperature and subsequently slow cooling obtains. Convection accounted largely for cooling in weldments and is dependent upon the temperature differential between the material and ambient. This differential is highest around peak temperature thus accounting for highest cooling rate around peak temperature.

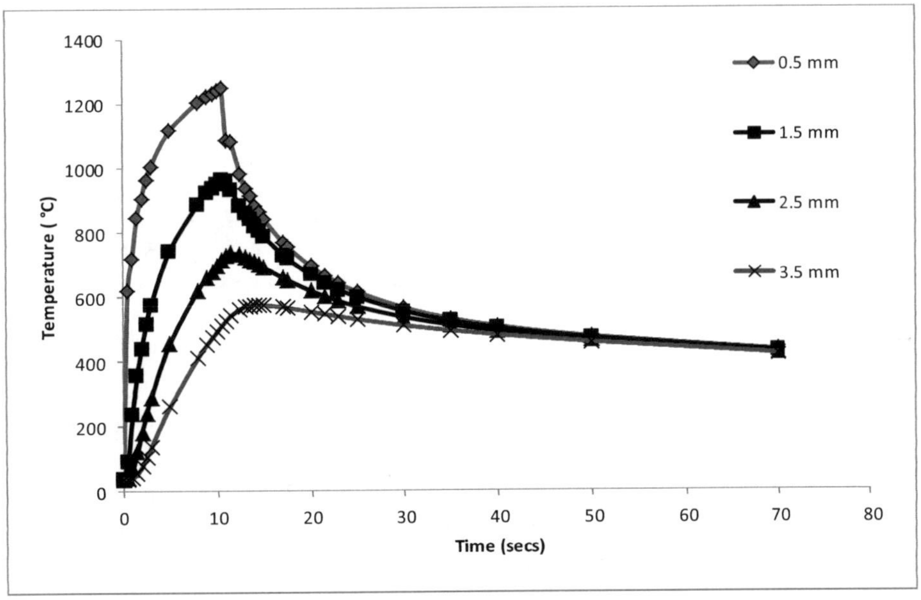

Fig. 2 Temperature variations with distances from weld line.

4.2 Effect of Arc Duration on Cooling Rate at point 1mm

Figure 3 shows the effect of weld duration on cooling rate at 1 mm from weld line. It is observed that cooling rate decreases with increase in weld duration as a result of increase in energy input and distribution across the weldment. This is in accordance with the findings of Martinson *et al.*, 2009 and Suresh, 2014 [5,16]. At 5 seconds weld duration and 800 °C, cooling rate was 76 °C/s, while at 15 second weld duration, cooling rate was from 33.4 °C/s. A similar trend was observed in cooling rate obtained at lower metal temperature values. With further increase in arc duration, the HAZ tends towards equilibrium temperature within its vicinity thereby reducing the

IOP Conf. Series: Materials Science and Engineering **413** (2018) 012050 doi:10.1088/1757-899X/413/1/012050

cooling rate.

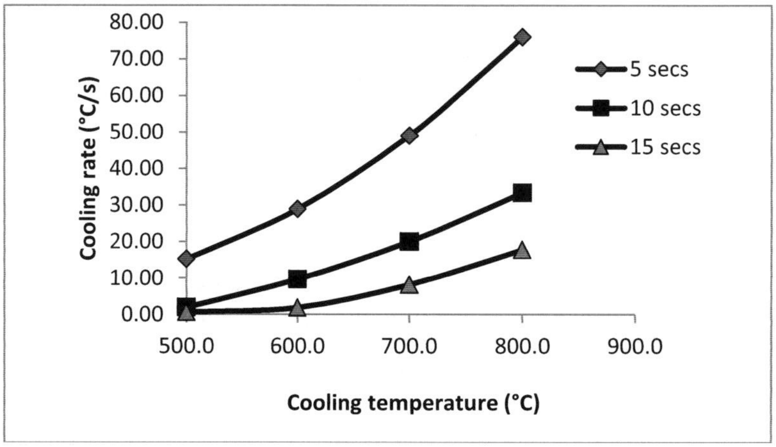

Fig 3 Effect of Flash Duration on Cooling Rate During Cooling Process.

4.3 Effect of Metal Peak Fusion Temperature on Cooling Rate at Distance 1mm

Figure 4 shows the effect of metal fusion temperature on cooling rate at 1 mm from weld line. It is observed that cooling rate decreases with increase in metal fusion temperature. At 1800, 1600 and 1400 °C, peak temperatures cooling rate at temperature 800 °C was 24.3 °C/s, 33.2 °C/s and 47.7 °C/s respectively. Cooling rate increased with decreasing peak fusion temperature. Similar trend of cooling rate were obtained at HAZ within 800 °C and 500 °C cooling temperature. Higher cooling rate at lower metal peak fusion temperature was as a result of higher rate of heat loss through conduction to the neighboring region near the HAZ as a result of high temperature gradient between them. With the increase in metal temperature, the HAZ attains equilibrium temperature within its vicinity thereby making the cooling rate slower.

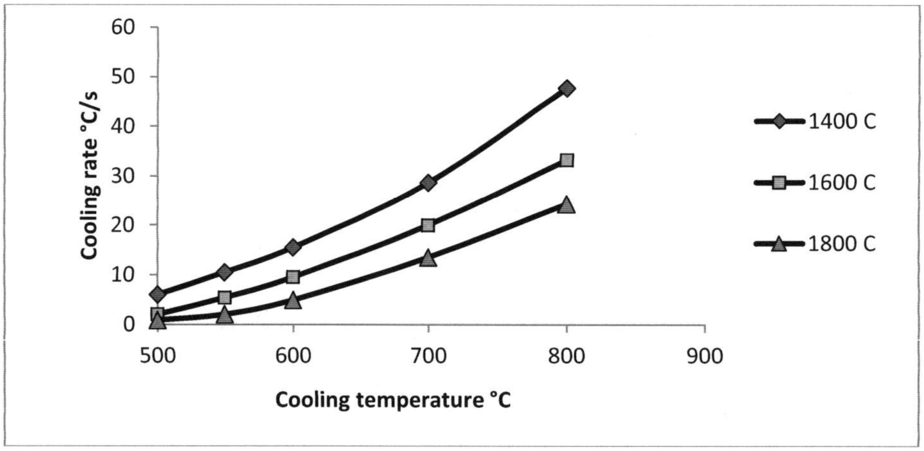

Fig. 4 Effect of Cooling Temperature on Cooling Rate at different fusion temperatures

IOP Conf. Series: Materials Science and Engineering **413** (2018) 012050 doi:10.1088/1757-899X/413/1/012050

4.4 Effect of Boundary Convection on Cooling Rate.

Figure. 5 shows the effect of boundary convection on cooling rate at distance 1mm from weld line. It is observed that cooling rate increases with the increase in convection coefficient. At 50 1,000 and 2,000 W/(m2 °C), cooling rate at temperature 800 °C was 33.4, 36.5 and 39.5 °C/s respectively. Cooling rate increased with increasing convection coefficient. Similar trend of cooling rate obtains at lower cooling temperatures. At higher convective coefficient faster heat removal from the HAZ ocurs.

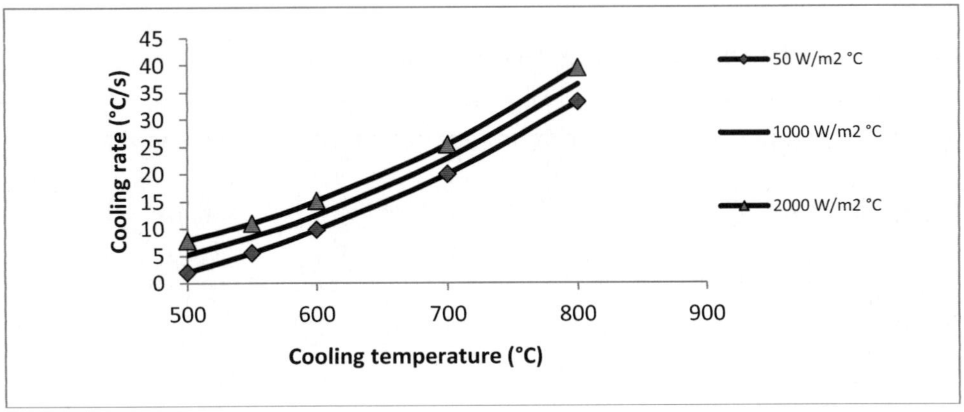

Fig. 5 Effect of Cooling Temperature on Cooling Rate at different Convective Boundary.

4.5 Effects of Preheat Temperature on cooling rate at 1mm from weld line.

Figure 6 shows the effects of preheat temperature on cooling rate at distance 1 mm from weld line. It is observed that cooling rate decreases with increase in preheat temperature. At 35, 200 and 300 °C pre-heat temperatures, peak temperature cooling rate at temperature 800 °C was, 44.5 and 29 °C/s respectively. Cooling rate decreased with increasing preheat temperature. Similar trend of cooling rate obtains at other lower cooling temperatures. This is in accordance with the findings of [17].

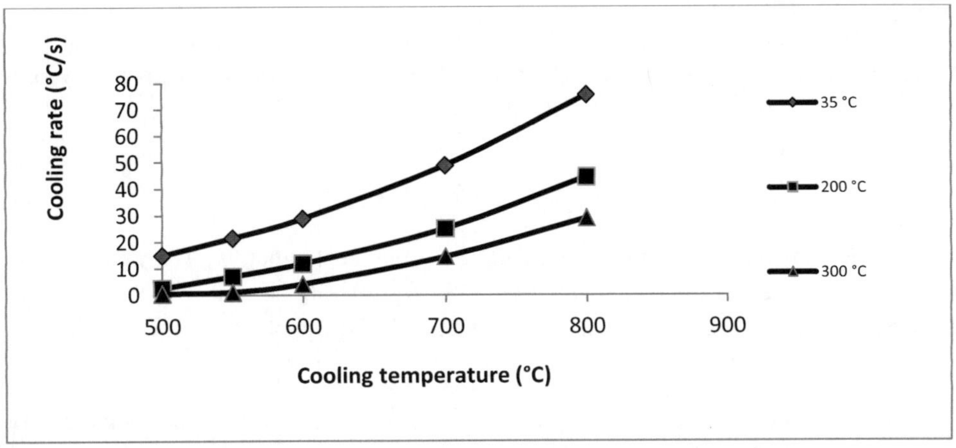

Fig. 6 Effect of Pre-Heat on Cooling Rate under Cooling Temperature Conditions.

IOP Conf. Series: Materials Science and Engineering **413** (2018) 012050 doi:10.1088/1757-899X/413/1/012050

5. Hardness Prediction

The validity of simulated hardness value was made by comparing with experimentally obtained result. At distance 1 mm from weld line, simulated hardness value of 29.1 (HRC) was obtained as shown in Table 3. Experimentally obtained hardness value of 32 (HRC)at the same distance was observed under the same welding condition as stated in the computational procedure with a discrepancy of 9%. This agrees with the findings [18.19]. The disparity is accounted for based on some assumptions made and heat losses not accounted for in the one-dimensional simulation.

Table 3: Predicted Hardness for Various Welding Parameters at Distance 1 mm.

S/n	Welding variable			
1	Arc duration (Sec)	5	10	15
	Hardness (HRC)	55	49.5	29.1
2	Metal temperature (°C)	1400	1600	1800
	Hardness (HRC)	55.3	55	53.5
3	Boundary Convection (W/ m2 .°C)	50	1000	2000
	Hardness (HRC)	55	55.2	55.5
4	Preheat Temperature (°C)	35	200	300
	Hardness (HRC)	55	50.5	28

6. Conclusion

The following conclusions can be drawn

- Cooling rate decreases with increase in arc duration, which leads to a significant decrease in hardness of the HAZ. There is a 47% reduction in hardness when the arc duration is increased from 5 to 15 seconds.
- Cooling rate decreases with increase in metal temperature, which leads to a slight decrease in hardness of the HAZ. A slight reduction inhardness of 3.3% was archived with the increase in metal temperature from 1,400 to 1,800°C.
- Cooling rate increases negligibly with increase in boundary convection, which leads to a slight increase in hardness of the HAZ.
- Cooling rate decreases appreciably with increase in preheat temperature, which leads to a considerable reduction in hardness of the HAZ. A 49% reduction in hardness was achieved when the preheat temperature was increased from 35 to 300 °C

References

[1] Olga, L. and Americo, S. Determination of the gross heat input in Arc welding, Journal of Material Processing Technology, 2015, Vol. 225, pp. 139-150.

[2] Aweda, E. O., Dauda, M., Dagwa, I. M and Dauda, E.T. Effects of Continuous Cooling on Impact and Micro Structural Properties of Low Carbon Steel Welded Plate, International Journal of Modern Engineering Research, 2015, Vol. 5, issue 1, pp. 1-10.

[3] Armin, R and Mehdi, I. Prediction of asymmetric transient temperature and longitudinal residual stress in friction stir welding of 304L stainless steel, *Journal of Material and Design*, 2014, *Vol. 55* pp. 812-820.

IOP Conf. Series: Materials Science and Engineering **413** (2018) 012050 doi:10.1088/1757-899X/413/1/012050

[4] Sirajuddin, E.K., Krishnan, K.N. and Mohd, A.W. Study of Transient Temperature Distribution in a Friction welding, 2012.
[5] Martinson P., S. Daneshpour, M. Kocak, S. Riekehr and P. Starson, Residual stress analysis of laser spot welding of steel sheets, Materials and Design, 2009, Vol. 30, pp. 3351-3359.
[6] Dean Deng and ShoichiKiyoshima. Numerical simulation of welding field, residual stress and deforma-tion induced by electro slag welding, Computational Material Science, 2012, Vol. 62 pp. 23-34.
[7] Mato P., Zdenko T., Alan R., Martin S., Ivica G., Ivanka B. and Srec ́k.Š. Numerical analysis and experimental investigation of welding residual stresses and distortions in a T-joint fillet weld, Journal of material and design, 2014, Vol. 53, pp. 1052-1063.
[8] Jerzy Winczek. A simplified method of predicting stresses in surfaced steel rods, Journal of Material Processing Technology, 2012, Vol. 212 pp. 1080-1088.
[9] Hwa Teng Lee, ChunTe Chen and Jia Lin Wu (2010). Numerical and experimental investigation into effect of temperature field on sensitization of Alloy 690 butt welds fabricated by gas tungsten arc welding and laser beam welding, Journal of Material Processing Technology, Vol. 210, pp. 1636- 1645.
[10] Olabamiji T.S. and Adedayo S.M. Finite Element Modeling of Temperature Cycles In Axi-Symmetric Flash Butt Welded Thin Steel Rods And Experimental Validation, Journal of Engineering Research, 2016, Vol. 21(1) pp. 31- 40.
[11] Adedayo S.M. and Irehovbude S.O. Numerical simulation of transient temperature in flash butt-welded axi-symmetric circular sections, Journal of Naval Architecture and Marine Engineering, 2013, Vol. 10, pp. 33-40.
[12] Reddy JN. An introduction to the finite element method, 3rd edition, Texas, McGraw Hill, 2006, pp. 1-575.
[13] Ify Nwaogazie L. Finite element modeling of engineering systems, 2nd edition, University of Port Harcourt Press, 2008, pp. 1-179.
[14] Li Chaowenand Wang Yong. Three-dimensional finite element analysis of temperature and stress distributions for in-service welding process, Material and Design, 2013, Vol. 52, pp. 1052-1057.
[15] Rose A. Atlas zurWarmebehandlung der Stahle I, VerlagStahleisen, Dussldorf. (Atlas for Heat Treatment of Steels), 1958, (German).
[16] Daniel H..Jominy testing- the Practical Side, Archive Industry Heating. Internet 2017, www.industrialheating.com.
[17] K.Suresh Kumara. Analytical Modeling of Temperature Distribution, Peak Temperature, Cooling Rate and Thermal Cycles in a Solid Work Piece Welded By Laser Welding Process, Procedia Materials Science, 2014, Vol. 6, pp. 821-834.
[18] Joseph O.O. and Alo F.I. An Assessment of the Microstructure and Mechanical Properties of 0.26% Low Carbon Steel under Different Cooling Media",Industrial Engineering Letters, 2014, Vol.4 (7).
[19] Innocent, U. Evaluation of the hardness and microstructure of a tungsten inert gas weld of a mild steel pipe joint, International Journal of Research in Engineering & Technology, 2014, Vol. 2(8), pp. 117-126.

IOP Conf. Series: Materials Science and Engineering **413** (2018) 012051 doi:10.1088/1757-899X/413/1/012051

Predicting the Effect of Well Trajectory and Production Rates on Concurrent Oil and Gas Recovery from Thin Oil Rims.

[1]Olabode Oluwasanmi; [2] Egeonu Gerald; [3] Ojo Temiloluwa; [4] Oguntade Tomiwa; [5] Bamigboye Oreofeoluwa

[1,2,3,4] Department of petroleum engineering covenant university, ota, Ogun state, [5] Department of management information systems, covenant university, ota, Ogun state
E-mail address: Oluwasanmi.olabode@covenantuniversity.edu.ng

Abstract Thin oil rim reservoirs are those with less than a 100 ft. of oil thickness underlain and overlain by an aquifer and gas cap respectively. Irrespective of the sizes of the aquifer or gas cap, oil rims productivity is usually low due to excessive coning of water and gas. Production optimizations are usually carried out before recovery schemes are initiated. One of such is a concurrent oil and gas production which is used in comparison with sequential production in this study in conjunction with production rates and well geometry to ascertain the extent of recovery and to improve the overall field development plan. Production of gas is slightly restricted to meet gas demands and also to prevent an adverse effect on oil production. A case study of an oil rim from the Niger delta was studied by running reservoir simulation using Eclipse software to check the effect of production schemes, well type and production rates on oil recovery. From the studies, concurrent production using a horizontal well and a production rate of 1000 stb/day is desired for an appreciable incremental oil recovery Further studies on limiting oil rim movements and thusly increasing oil recovery can be done by initiating secondary and enhanced oil recovery schemes.
Key words; oil recovery, horizontal well, vertical wells, concurrent production, sequential production

1. Introduction

Due to the reservoir, geological and operational uncertainties of oil rims, production challenges as described by the coning of water and gas are inevitable. In 2012, Masoudi described the pros and cons of the strategies of developing oil rims. A gas cap blow down will ultimately mean production of oil is ignored, sequential production neglects the early production of gas, swing development offers cyclic production of gas and a low oil recovery factor and concurrent oil and gas production which might suit both investor and host country[1]. The conventional way to exploit oil from a thin oil rim is to produce from the oil column while keeping the gas cap in place as a drive mechanism or pressure support. For reservoirs with gas cap expansion as the dominant drive mechanism, a concurrent oil and gas development is less attractive. It is however feasible and should still be considered. With increased focus on LNG in the world, we are faced with the challenge that some key gas reservoir that are crucial for the timely LNG supply have underlying oil rims. Concurrent oil and gas production involves the simultaneous production of oil and gas from the same reservoir from the start of production through either separate conduits (i.e. by separate wells or dual string completion) or through a single conduit (i.e. by commingling or with a smart completion).

In 2011, Ibukun described some of the major challenges facing the performance of simultaneously producing oil and gas especially for this case study includes: the size of the gas cap, the presence, activity , and size of the aquifer, production rates, well geometry, geometry of the reservoir, fluid properties, reservoir heterogeneities, phase behaviors and gas off takes just to mention a few[2]. It is also of importance to make mention or to understand the movement of the oil over the life of the field which is typically determined by two dominant drives:

- Water drive (the aquifer)
- Gas cap expansion drive (ie the size of gas cap)

Reservoir simulation has demonstrated the ability of concurrent wells to enable simultaneous oil and gas production with minimal impact on oil recovery. The proposed concept can significantly impact the portfolio of available gas reservoirs by delivering a cost effective technology solutions. As a result significant cost benefits can be realized (i.e. one concurrent smart well can potentially replace two conventional dedicated oil and gas wells).

The applications of horizontal well technology in developing hydrocarbon reservoirs have been widely used in recent years. One of the main objectives of using this technology is to improve hydrocarbon recovery from water and/or gas-cap drive reservoirs. The advantages of using a horizontal well over a conventional vertical well are their larger capacity to produce oil at the same drawdown and a longer breakthrough time at a given production rate. Reduced water and gas coning because of reduced drawdown in the reservoir for a given production rate, thereby reducing the remedial work required in the future and Increased production rate because of the greater wellbore length exposed to the pay zone.

Sensitivity analysis have majorly been carried out by changing parameters and evaluating their effects on estimated oil recovery. In 2005, Wanye developed a qualitative set of screening criteria for sequential and concurrent development of oil and gas resources in thin oil reservoirs by sensitivity analysis on gas cap size and oil rim thickness. Their suggestions are summarized as:

- M-factor > 2 and Ho >30 ft : concurrent oil and gas development is feasible
- M- factor < 2 and Ho > 30 ft : delay gas development till the end of oil development
- For all m- factor and Ho < 30 ft. : gas development only oil development is not feasible[3]

In 2012, Iyare and Marcelle estimated oil recovery by initiating sensitivity analysis by combining the effects of gas cap, aquifer sizes and optimal well locations for oil rims[4].

In 2008, Sascha and Marc proposed the concurrent production of oil and gas from oil rim reservoirs through a single well conduit smart well (intelligent well). Using a case study, the authors postulated that a smart well solution to concurrent production of oil and gas from oil rim reservoirs results in significant costs benefits (reduction in costs), as one concurrent smart well can potentially replace 2 conventional dedicated wells[5].

An oil rim reservoir in the Niger delta region of Nigeria is used to analyze the effects of production rates on vertical and horizontal wells on concurrent and sequential production strategies on oil recovery in oil rim reservoirs.

2. Methodology.

An oil rim reservoir from the Niger delta is used in this study to verify the effects of oil flow rates and well configurations on oil recovery. The subsurface map of the reservoir is seen in figure 1 while figure 2 is the discretized sub surface map showing the 5 producing wells. The eclipse software was used to build a black oil model by incorporating Pressure, Volume and Temperature properties as well as solution properties from the software manual for the active cells in the grid. Table one describes the reservoir properties of the oil rim.

2.1. Model Construction

The GridSim Module was used to build the reservoir grid model. The reservoir lies between 10000 to 10450 feet. The Cartesian grid is set into 75 x 50 x 20 grid-cells with 74000 cells set active by boundary definition and shaped according to the cross-sections drawn across the reservoir sand map in both X and

IOP Conf. Series: Materials Science and Engineering **413** (2018) 012051 doi:10.1088/1757-899X/413/1/012051

Y directions. There are 32420 active cells and 41580 inactive cells. GridSim was also used to assign the petro physical properties (arrays) such as horizontal permeability (PERMX), porosity (PORO), active grids (ACTNUM), net-to-gross (NTG) ratio, and the initial water saturation (SWATINIT) to the reservoir grid model. The model was assumed to be homogenous, so average values were used to populate the grid model.

There are five producing wells named: "04L/E2", "06L/E2, ", "09L/E2", "10L/E2, and " -13L/E2," located at I, j coordinates: (62, 34), (50, 22), (27, 13), (49, 38) and (40, 33) respectively on the grid model. All wells penetrate the grid blocks in Z-direction and were set as producers. Analytical (Fetkovich) aquifer system was connected at the eastern flank of the grid

Fig. 1 Subsurface map of E2 reservoir

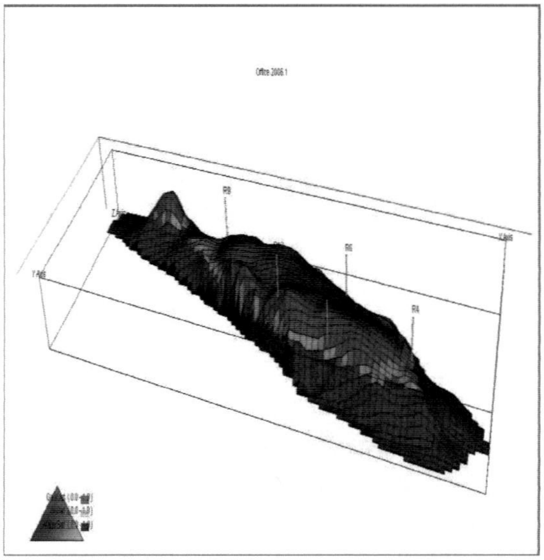

Fig. 2 Discretized sub surface structure of E2 showing fluid saturations

Table 1 Oil rim properties

Properties	Value	Properties	Value
Oil Sand Porosity (%)	0.26	Oil Rim Thickness (ft)	38
Gas sand Porosity (%)	0.24	Swc	0.14
Sh Oil	0.9	Oil Viscosity (cP)	0.43
Sh Gas	0.77	Oil Gravity (°API)	32
NTG	0.8	Boi (bbl/stb)	1.48
Permeability (mD)	400 – 1500	Rsi (scf/stb)	1035
Sand Thickness (ft)	71	Ti (°F)	179
Pi (psi)	4394	Np (MMstb)	10.73
Pb (psi)	4394	Free Gas In Place (bcf)	128.5
STOIIP (MMbbls)	35.9	ΔP/Pi @ Np/N	7%@26%
No. of Drainage Points	5		

IOP Conf. Series: Materials Science and Engineering **413** (2018) 012051 doi:10.1088/1757-899X/413/1/012051

2.2. WELL EVENTS SCHEDULE

The observed production history data for oil, gas and water from E5 "reservoir were made into a Schedule production data format and imported into Schedule Module. In addition, well events data such as perforation intervals, bottom-hole pressure measurements, well tests, stimulations, plugs and sidetracks, squeezes and re-working were specified on Schedule interface interactively. Well geometries, positions and connections with the grid were equally specified. Finally, the simulation options and tuning were configured and the interface submitted for the simulator to generate an output schedule file used for the history matching. This section is the core of simulation study. All the 5 wells in question were at some point shut in due to excessive GOR (gas oil ratio) and high water cuts leading to low ultimate oil recoveries.

2.3. History Matching

The history matching was done in two phases: first the wells were set to produce the correct reservoir fluid volume rate for each time step. This was done by setting the control mode to calculate the reservoir fluid volume rate from the observed phase flow rates in the keyword WCONHIST. The resulting pressures that are calculated by ECLIPSE* were compared to the observed BHP field measurements. The volume of reservoir fluids was tuned for the model to produce pressure results similar to those measured; this was not perfectly matched due to very sparse BHP pressure data available. In the second phase of the history matching, wells were set to produce the correct oil rates. Again, this was controlled to calculate oil rates from the observed oil rates in the keyword WCONHIST. Subsequently, vertical permeability was varied and gas relative permeability modified to tune the model until the produced water and gas matches the historical rates. The oil initially in place must first be matched with the simulated data before the cumulative oil production is matched. The simulated and field gas oil ratio and water cut are matched as shown in (figure 3). The table 2 shows the summary of the initial fluid in place data after history matching. The simulator gave an increase of 0.88% in original oil in place while the gas in place also increased by 2%. While table 3 shows total well productions before and after history matching.

Table 2 Simulated original oil and gas in place before production

	Oil (stb)	Water (stb)	Gas (Mscf)
Currently in place	36,216,568	797,816,825	131,098,739
Outflow through wells	0	0	0
Analytical aquifer	0	0	0
Well material balance	0	0	0
Originally in place	36,216,568	797,816,825	131,098,739

Table 3 Total well production

WELLS	OIL (STB) Before history match	After history match	GAS (MSCF) Before history match	After history match
Well 4L	164,367.23	4,692,793	922,498.81	9,494,806
Well 6L	237,827.52	2,628,576.3	613,888.4	4,903,431
Well 9L	325,913.44	3,416,676.5	911,229.5	9,172,350
Well 10L	37,369.77	344,142.5	361,014.22	358,795
Well 13L	8560.623	397,169.94	115,643.89	819,233.62
TOTAL	**634,559.25**	**11,479,358.**	**2,924,274.8**	**24,748,616**

IOP Conf. Series: Materials Science and Engineering **413** (2018) 012051 doi:10.1088/1757-899X/413/1/012051

Figures 3 shows that the production rate for oil is very low indicating a higher water and gas influx while figure 4 shows the history matched production rates. describes the historical production rate plots. A high gas production suggests an initial gas cap or has high quantity of initial solution gas. The field historical production rate also follows similar trend. Figure 3 shows the field production total with 11,522,593 barrels of oil and 25 MM scf of gas produced during 44 years of production. Water production was very high which indicates the existence of a large aquifer accompanied with a large gas cap. Figure 3 shows the history matched cumulative production which closely corresponds to the oil produced given in table 1.

The static pressure decline in the aquifer from 0 to 10000 days (approximately for up to 24 years of production) is about 22% of initial reservoir pressure. A finite acting (fetkovich) aquifer was modeled was modelled in accordance to the grid dimensions, pressure, compressibility and reservoir fluid contacts to match observed pressure level in the field. Adding an aquifer to the reservoir though helped to maintain or balance the reservoir pressure.

Figure 3 also shows the history matched profiles for water cut and gas oil ratios respectively. This is a major step in reservoir simulation before the commencement of prediction.

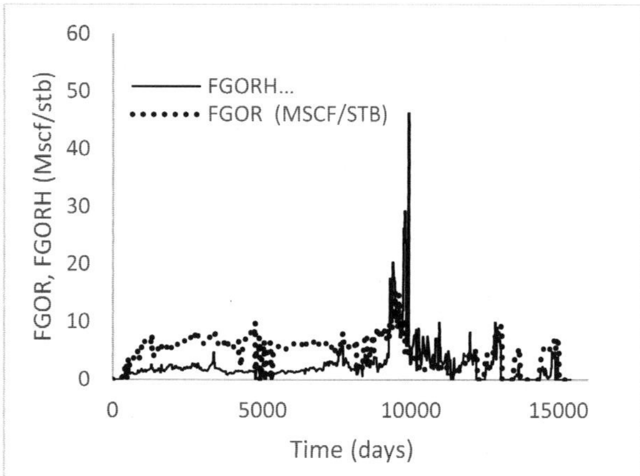

(a) Field gas oil ratio match

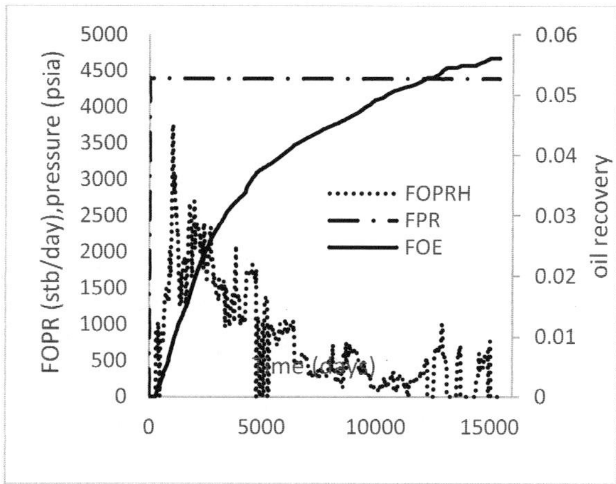

(b) Field oil recovery

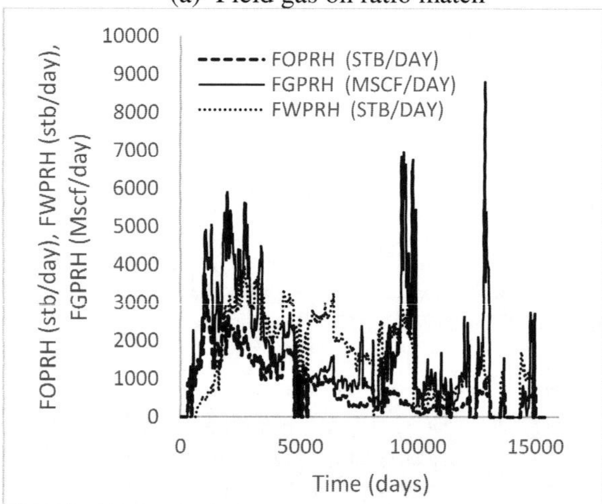

(c) Field production rate history

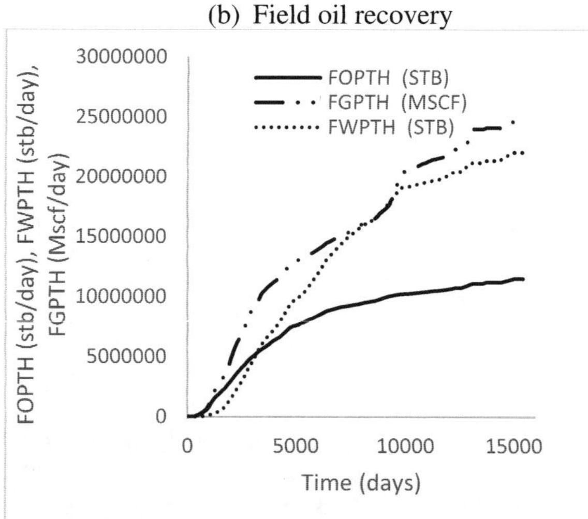

(d) Field production total history

IOP Conf. Series: Materials Science and Engineering **413** (2018) 012051 doi:10.1088/1757-899X/413/1/012051

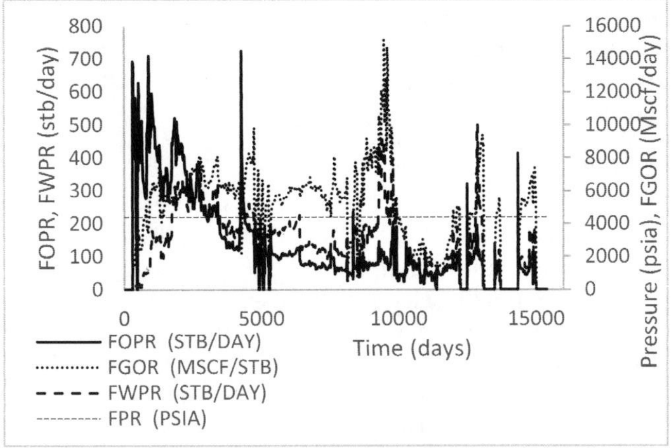

(e) Field production rates

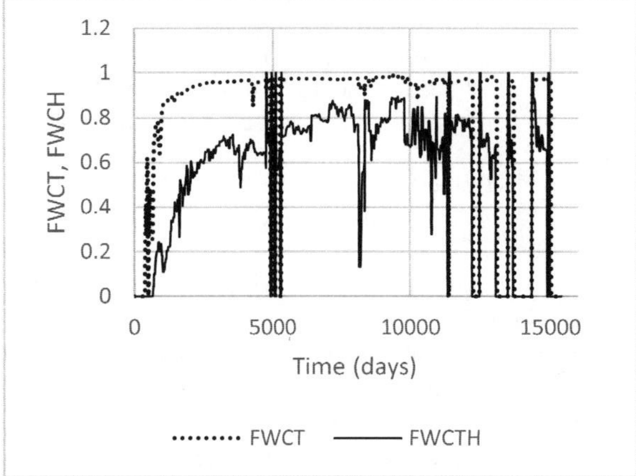

(f) Field water cut match

Fig. 3 E2 production history and profiles

2.4. Prediction.

Having history- matched the simulated model to the observed data and concluded on having a near perfect simulation run with an average of 2% increase, a number of prediction runs were made on the history matched model. 3 sensitivity runs were made for the following case scenarios that were investigated:

1. The type of completion (vertical or horizontal) we try to figure out which completion gives best oil recovery.
2. Expected rate recoveries from sequential production i.e. produce oil first then gas.
3. Effect of concurrent production on overall field performance (i.e. recovery of both oil and gas)

The base case scenario is the history matched model. The sequential production scenario was run as a benchmark for comparing the performance of concurrent production and also to estimate the expected performance of such scenarios.

The concurrent production cases were selected to investigate typical oil rim reservoir development scenario in which the reservoirs serve as an important source of gas supply for an available market. This will economically improve the overall field development from sales of gas. In concurrent production unlike sequential production the gas production is not differed till after recovery as the gas is needed to meet demands.

2.5. Description of the Cases

1. Base case: this describes the current situation of the reservoir after history matching (table 3).
2. Case 1: studies the effect of type of well i.e. (vertical or horizontal) used in the field right from production inception. The historical wells in the field were modelled to be 5 vertical wells. In this case however the field was put on 5 horizontal wells. The well location and trajectory were identified with respect to initial oil saturation distribution in the field before the start of production. This case was developed to investigate the draining field with horizontal wells from inception of production and if it would deliver a better result than vertical.
3. Case 2: oil recovery from concurrent oil and gas production. All existing 5 wells were converted to horizontal wells.

4. Case 3: oil recovery from sequential production (producing the oil rim first and then producing the gas subsequently after the oil wells reach abandonment). Apart from the first case scenario, all other subsequent cases were run with 3 multiple oil rate sensitivities. A total number of 19 prediction runs asides history match were made.

A total number of 19 prediction runs asides history match were made.

2.6. Analysis of Prediction

The predictions simulations were made on the history matched models after normal production for a period of 15 years thus making overall field life 57 years. (i.e. normal production commenced 1968 and ended 2010, prediction started 2010 and ended 2025). Some of the economic limit used for the field and these wells are maximum water cut of 100%, the minimum allowable oil rate for the wells was 100bopd while that for the field too was 500bopd multiplied by the number of wells in the prediction model. The simulator was programmed to shut in wells that violated stipulated well penalties and stop simulation run once any of the field penalties was violated. Oil rates of (2000, 1500, 1000 and 500 stb/day were used on different case scenario)

For the gas wells producing from the gas cap, the minimum gas rate was set to be 10 Mscf/d. The bottom hole target for all wells was assumed to be 2000 psia and put under a minimum field constraint of 1200 psia.

The figure 9 shows the location of new or predicted wells (pred1, pred2, pred3, pred4, and pred5) and location of the 5 history matched wells) with respect to oil saturation grid function.

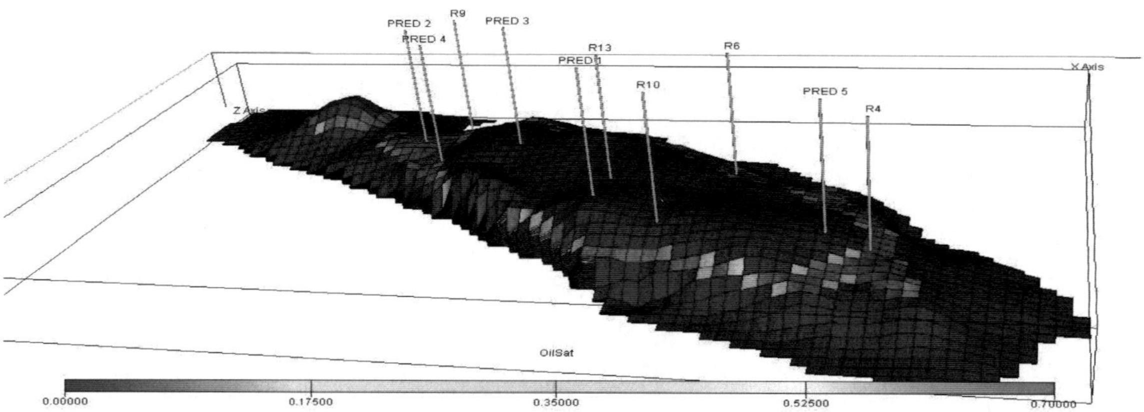

Fig. 4 Location of history match and prediction wells

3. Results

Here the analysis of the results according to their case specifications are highlighted and discussed below:

3.1. Base Case: this case is the do-nothing case. It models the case where the current historical wells for the reservoir in question are reopened for production without any stimulation, production enhancement or intervention or advanced recovery runs made. This case records a low 0.4% increase in oil recovery over the historical production. Most of the wells in this case scenario didn't produce much as he wells were not properly located within region of high reserves.

Firstly, a change of well production trajectory from vertical to horizontal or deviated showed a significant increment in oil and gas production and it was also noticed that this increments occurred for lower oil production rates. To also further prove these two vertical and horizontal wells were added and simulated at different occasions and the simulated result from the horizontal wells had higher deliverability. Horizontal wells also proved higher productivity for both sequential and concurrent production as compared to vertical wells with horizontal concurrent production having higher performance.

3.2. Case 1: Converting All Previous Vertical Wells to Horizontal Wells.

This case scenario considers converting all the initial history matched wells from vertical completions to horizontal completions. 2 rate sensitivities were still used in this case scenario just as in case scenario 1. The 1500 stb/day gave the highest oil recovery of about 33.04% an equivalent of an additional of 351,521 barrels of oil produced. Actual field total oil production was 11,861,144 barrels of oil was produced as compared to 11,779,557 barrels of oil for vertical wells. Comparison of the results from this case study suggests that vertical completion of wells in this reservoir were not or shouldn't have been the best completion to be used in the initial field development. Horizontal completions should have been initiated.

Also, a recovery factor of 79.51% was recorded for gas. The simulation ran for an average of 35 years for all the horizontally completed wells and stopped due to low predicted well pressure.

3.3. Case 2: Concurrent Production of Gas and Oil After History (5 Horizontal Wells).

This last case on concurrent production of oil and gas via considers 5 horizontal wells. The rate sensitivities used are same as in case 1 above namely (2000, 1500 and 1000 stb/day). The 1000 stb/day produced or performed best. A total of 1,343,576 barrels was produced over an average period of 43 years. Rate sensitivities of 2000 and 1500 stb/day produced 12,425,652 barrels and 12,464,984 barrels respectively. Oil recoveries for oil rates 2000 stb/day, 1500 stb/day and 1000 stb/day were 34.30%, 34.42% and 37.1% respectively. The gas recoveries for the same rates were 44.2%, 44% and 35.32% respectively.

This case clearly shows the dynamic effect of the displacement efficiency of the two dominant drive mechanisms in the reservoir (i.e. water influx and gas cap expansion). Analysis of field performance with respect to oil recovery shows that gas cap expansion is the predominant drive actively followed by water drive. However even though gas cap expansion is the predominant drive, the displacement of oil by gas is relatively less efficient as compared to the displacement of oil by water because of relatively less mobility ratio. Since the aquifer water is generally more viscous than reservoir gas, the mobility ratio for the water oil displacement process (i.e. water influx).

At 1000 stb/day per well oil rate water influx had relatively larger effect on oil displacement. at oil recovery of 37.1% a cumulative of 909 MMSTB of water was produced with a gas recovery factor of 35.32%. At a higher oil rate 2000 stb/day the gas recovery was 44.2% which indicates that the gas cap expanded more rapidly to occupy the voidage caused by increased underground withdrawal since it had more energy than the aquifer.

It is also worth knowing that with respect to oil recovery, this case performed better than the case of sequential production of oil and gas which shall be proved in a later case scenario. In sequential production, the gas cap is not produced until the oil has been produced to economic rates. This implies gas cap expansion been the dominant reservoir drive throughout field oil production life. This makes oil displacement process for sequential production case less as compared to concurrent production due to the fact that in the latter case the production is also supported by an aquifer support in conjunction with the gas cap expansion. Also the simultaneous production of gas cap with oil rim reduces the effect of gas cap energy and thereby resulting to a relative corresponding increase in effect of aquifer energy on reservoir drive.

An analogy of increasing the aquifer size (aquifer support) was also carried out on this concurrent scenario. After running the simulation, an average of 39% oil recovery and 46.6% gas recovery was derived from all the respective oil rates i.e. (500, 1000, 1500 and 2000 stb/day). This signifies that a larger aquifer support helps in additional oil recovery.

On a further note, it is worthy to note that at a lower rate i.e. 500 stb per day for this case scenario an oil recovery factor of 38% was recorded but at a lower gas recovery factor.

3.4. Case 3: Sequential Production (5 Horizontal Wells).

This case scenario considers sequentially producing the oil rim first and then the gas cap using 5 horizontal wells. This case was also developed to test the relative productivity of concurrent production of the gas and oil so that gas sales supply will meet demands without delay. Analysis of concurrent production case vis-a-vis the sequential production case would proffer to what extent concurrent production would adversely affect oil recovery. Two oil sensitivity rates were used for this case scenario (1000 stb/day and 2000 stb/day) were run for all the 5 horizontal wells in this case. An oil recovery of 34.3% was recorded for 1000 stb/day as compared to 38% of concurrent production while a recovery of 34% was recorded for 2000 stb/day. The simulation run for theses horizontal wells ran for an average of 15 years and stopped due to low field pressure. As earlier explained, sequential production is not as effective as concurrent production in terms of oil recovery for the E2 reservoir This shows that concurrently producing the field does not have adverse effect on oil recovery. Indeed results of the study shows that concurrent production is a desirable development option for field oil and gas reserves.

Below is tabulated summary of the result analysis of each case scenario.

Table 4: Result for case 2

OIL RATE (STB/DAY)	CUM OIL (STB)	CUM GAS (MSCF)	CUM WATER (BBL)	OIL RF(%)	GAS RF (%)	Incremental oil RF (%)	Incremental gas RF (%)	NFA
500	13,773,703	43,909,304	5.11E+08	38.364	65.82	6.30	0.64	2025
1000	13,435,767	41,792,712	8.87E+08	37.425	67.48	5.37	2.30	2025
1500	11,779,557	43,362,576	1.68E+08	34.721	66.25	2.66	1.07	2025
2000	11,781,926	43,165,032	1.77E+08	34.611	66.41	2.55	1.23	2025
(extended aquifer size)	14,073,079	38,626,416	2.11E+09	39.2	69.94	7.14	4.76	2025

Table 5: Result for case 3

OIL RATE (STB/DAY)	CUM OIL (STB)	CUM GAS (MSCF)	CUM WATER (BBL)	OIL RECOVERY (%)	GAS RECOVERY (%)	CHANGE IN OIL RF WRT HISTORY MATCH	CHANGE IN GAS RF WRT HISTORY MATCH	SIMULATION END YEAR
1000	12410337	34272896	1.42E+08	34.57	73.33	2.64	8.15	2025
1500	12357815	31643808	1.10E+08	34.42	75.37	2.36	10.19	2025

IOP Conf. Series: Materials Science and Engineering **413** (2018) 012051 doi:10.1088/1757-899X/413/1/012051

4 Conclusions

From the three cases considered, a production rate of 1000 stb/day is considered suitable in developing oil rim reservoirs under concurrent oil and gas production.

Horizontal completions generally performed better than the vertical completions for oil recovery for E2 reservoir both for cases that considered different scenarios at initial development of the field i.e. without history and for cases that considered re-development of the field after history.

This is consistent with what has been reported in literature, but the applicability of horizontal wells has to be considered for each oil rim reservoir. The economic importance and analysis of horizontal completions also have to be run to know the viability of such completions for the field and optimal number required for adequate field drainage. Other crucial factors that may affect feasibility of horizontal completions are the average net oil column thickness and formation anisotropy.

The results also show that the concurrent oil and gas production is the best strategy to be considered in developing oil rim. Sequential production is not necessarily a bad development option, it was not a better option for this particular field. It may be applicable where gas cap expansion is the only active source of reservoir energy, or has overriding effect on the other sources of reservoir energy.

Nomenclature

GOR: gas oil ration

Stb: stock tank barrel

Scf: standard cubic feet

References

[1] Masoudi, R. How to get the most out of your Oil Rim Reservoirs? Reservoir management and hydrocarbon recovery enhancement initiatives. *SPE.* (2012)

[2] Ibukun, S. *"Evaluation of Oil Rim Reservoirs Development – A Case Study".* ibadan: Master of Science Thesis of the University of Ibadan, Department of Petroleum Engineering. (2011)

[3] Wanye, M.. *Nun river mode: Gas Recovery Mechanism and oil rim development strategy.* Rijswijk: shell. (2005)

[4] Iyare, U., Marcelle-De silva, J., & UWI.. effect of gas cap and aquifer strenght on the opyimal well location for thin oil rim reservoirs. *SPETT 2012 Energy Conference and Exhibition* . port of spain, trinidad: SPE. (2012)

[5] Sascha, v., & marc, N.. concurrent oil and gas development wells: a smart well solution to thin oil. *international petroleum technology conference* . kuala lumpur: IPTC. (2008)

IOP Conf. Series: Materials Science and Engineering **413** (2018) 012052 doi:10.1088/1757-899X/413/1/012052

Investigation of Co and So2 from Acid Clay Treatment of used Lubricating Oil

Oladimeji Temitayo[1], Sonibare Jacob[2], Omoleye James[1]

[1]Department of Chemical Engineering, Covenant University, Ota, Ogun state, Nigeria
[2]Department of Chemical Engineering, Obafemi Awolowo University, Ile-Ife, Osun State, Nigeria
E-mail address: temitayo.fatoki@covenantuniversity.edu.ng

Abstract: Recycling and treatment processes is one of the acceptable means to reduce the effect of pollution from used oil. These recycling methods save some amount of time and money and also helps to conserve our natural resources. However, various methods have been used for used oil treatment which has resulted in toxic air emissions. This work monitored the emissions of Carbon Monoxide (CO) and Sulphur Dioxide (SO_2) from used lubricating oil using a combustion analyzer. CO emission was found to be the highest emitted pollutant with MEC of 167.07 mg/m^3 due to incomplete combustion. Also, CO concentration at dehydration exceeded the set limits which could cause adeverse effect to receptor locations while SO_2 was within the limit. Hence, quantifying the emissions at each stage of this process is needed to control emissions in the sector. The predicted results in this study can help decision makers to formulate policy for controlling the impact of air emissions from used lube oil treatment plants using acid-clay treatment method.
Keywords: Acid-Clay, Emissions, Used lubricating oil, combustion analyzer

1. Introduction

One of the major problems that developing countries in Africa face is pollution. Improper disposal of Used Lubricating Oil (ULO) is one of the contributing factors involved in environmental pollution. Toxic and carcinogenic materials found in the used lubricating oils corroborates the point on how improperly disposed lubricating oil can cause debilitating effects to man and the environment at large [1]. Such materials include poly-aromatic hydrocarbons, Polychlorinated Biphenyl (PCB), poly-cyclic benzenes, lead, zinc, arsenic [2]. As a result, recycling the used lube oil became one of the best methods of effective disposal. Recycling of used lube oil can help reduce the environmental threats that improper disposal poses as well as helping to preserve crude oil reserves [3].

Acid-clay method is one of the famous existing treatment methods of ULO. Several works have been done on used oil treatment using acid clay method. These include [4] , [5], [6], [7]. [8] states that regardless of the disadvantages in acid-clay, it also has several distinguishing features from other technologies which include low capital investment, non-sophisticated and simple process, requires no skilled operators, low operating cost. This treatment process generate air pollutant which include Carbon Monoxide (CO), Sulphur Dioxides (SO2). CO is produced from the partial oxidation of carbon-containing compounds; it forms when there is not enough oxygen to produce carbon dioxide (CO2), it is also formed from incomplete combustion of various other fuels [9]. CO is a deadly poison which binds to hemoglobin molecules in blood, reducing the amount of oxygen carried to body tissues and organs [10]. Fossil fuel combustion accounts for almost all anthropogenic (human-caused) sulphur emissions. [11] reports that exposure to sulphur dioxide in the ambient air has been associated with reduced lung function, increased incidence of respiratory sympt oms

and diseases, irritation of the eyes, nose, and throat, and premature mortality. CO and SO2 can be emitted from the dehydration stage of used oil treatment.
Detailed information concerning the air pollution emissions for used oil treatment is very important for used oil treatment evaluation of air emissions sources and pollutants quantification. In order to achieve this aim, this study investigate CO and SO2 emission from the treatment of ULO using acid-clay method.

2. Materials and Method

2.1. Materials

The used lubrication oil was collected from a mechanic workshop. Equipment used include Sieve, Weighing balance (RADWAG,WLC 0,6/B1), Filtering flask, Measuring cylinders (500 ml, 100ml), Pyrex Beakers (250 ml, 100ml), Pyrex Conical Flask (100ml, 500ml, 250 ml), Pyrex Round bottom Flask (500 ml), Thermometer (Uniscope), Stir Hot Plate (Maple Scientific Limited), Seperating Funnels (200 ml, 500 ml- BS 2021 Borosilicate 24/29), Retort stands, Gallenkamp Centrifuge, Buchner Funnel and Combustion analyzer (E-instrument E8500). The chemical, Acetic acid (98%) was from Sigma Aldrich.

2.2. Methods

The used lubricating oil sample was allowed to settle for 24 hrs before it was filtered to remove solid impurities such as metal chips, sand and dust. The filtered oil was measured in a beaker covered with a a metal which has a sampling port where the combustion analyzer was connected and heated on hot plate at 250°C for about 1 hour for dehydration (plate 1). Impurities in form of antifreeze, water and other solvent present in the oil were significantly removed. Emission concentration for the pollutant present in the process were displayed on the analyzer screen. The dehydrated oil was then treated with acetic acid (10:1 oil-acid ratio), heated at 50o C and left for a day in a seperating funnel. Two layers were formed- filterate and residue. The filterate was treated with Kaolin clay to remove the odour and dark colour and the mixture was heated at 240oC. Emission concentration for all the process was obtained using the analyzer. CO and SO2 emission concentration were recorded at different range of oil temperatures (Toil). The concentration was measured in mg/m3.

Plate 1: Emission Measurements Set-up at Dehydration stage

3 Result and Discussion

3.1 Emissions at Dehydration stage

The temperature of used oil was monitored at temperature range of 110 oC to 250oC. The combustion analyzer was connected to the dehydration apparatus setup as shown in plate 1 and the readings of pollutant

IOP Conf. Series: Materials Science and Engineering **413** (2018) 012052 doi:10.1088/1757-899X/413/1/012052

identified was displayed on the analyzer screen as presented in Table 1. The pressure was constant at 760 mmHg while Carbon monoxide (CO) and Sulphur dioxide (SO2) were the emission concentration measured in mg/m3. The mean concentration of emissions of each pollutant and averaging time was also calculated. A sample calculation of dehydration process is presented below.

Table 1: Dehydration Stage of Acid-clay treatment

S/N	Ta (°C)	T.oil (°C)	Time (mins)	CO (mg/m3)	SO2 (mg/m3)
1	28	110	2	5.443	0.000
2	28	120	2	9.154	6.340
3	28	130	2	9.550	6.550
4	28	140	3	12.324	7.530
5	28	150	3	19.675	7.530
6	29	160	3	26.854	9.555
7	29	170	4	70.143	9.578
8	30	180	4	95.546	9.578
9	30	190	4	99.879	11.255
10	30	200	5	105.564	13.543
11	30	210	5	115.789	14.677
12	30	220	5	459.223	17.540
13	31	230	6	495.675	17.955
14	31	240	6	535.775	20.934
15	31	250	6	445.458	21.967

Note: 0 means no pollutant was identified for that activity
Where T.oil = Temperature of oil in oC Ta = ambient temperature in oC measuring temperature of the room

$$\text{Mean emission concentration for CO} = \frac{\text{total emission rate of CO}}{no\ of\ emissions} = \frac{\epsilon fx}{\epsilon f} = \frac{2506.052}{15} = 167.07 \text{ mg/m3}$$

$$\text{Mean emission concentration for SO2} = \frac{\text{total emission rate of NO2}}{no\ of\ emissions} = \frac{\epsilon fx}{\epsilon f} = \frac{174.532}{14} = 12.467 \text{ mg/m3}$$

Averaging time = 1 hour

3.2 Emissions at Acid Treatment stage

Acid treatment is the stage of reaction of dehydrated oil with acid (acetic acid). The temperature was monitored at 50oC to ensure adequate mixture of the sample. The combustion analyzer was connected to the acid treatment apparatus setup and readings of pollutant identified was displayed on the analyzer screen as presented in Table 2. At constant pressure, 760 mmHg only CO was identified as the pollutant while no readings was displayed for SO2 .

Table 2: Acid Treatment Stage of Acid-clay treatment

S/N	Ta (°C)	T.oil (°C)	Time (mins)	CO (mg/m3)
1	28	30	6	23.457

2	28	35	8	20.389
3	29	40	10	22.788
4	29	45	16	23.734
5	29	50	20	24.134

Mean emission concentration for CO = $\frac{114.502}{5}$ = 22.9 mg/m3

Average time = 1 hour

3.3 Emissions at Clay Treatment stage

This stage removes dark colour and odour in the used oil. The temperature was monitored from 90 – 240oC. The combustion analyzer was connected to the clay treatment apparatus setup and readings of pollutant identified was displayed on the analyzer screen as presented in table 3. The pressure was constant at 760 mmHg. Also CO was identified has pollutant and SO2 was not emitted

Table 3: Clay treatment Stage of Acid-clay treatment

S/N	Ta (°C)	T.oil (°C)	Time (mins)	CO (mg/m3)
1	27	90	1	2.899
2	27	100	1	0.000
3	28	110	1	2.156
4	28	120	1	2.156
5	28	130	1	3.233
6	28	140	1	3.233
7	29	150	1	4.067
8	29	160	1	5.154
9	29	170	2	4.123
10	29	180	2	4.225
11	29	190	2	1.987
12	30	200	2	2.013
13	30	210	2	2.114
14	31	220	4	7.897
15	31	230	4	10.556
16	31	240	4	16.768

Note: 0 means no pollutant was identified for that activity

Mean emission concentration for CO = $\frac{72.581}{15}$ = 4.83 mg/m3

Average time = 30 mins

3.4 Air quality standards

The mean Emission concentration of the pollutant at each stage was compared with National Air Quality Standards (NAQS) in Nigeria (Table 4) to measure the level of compliance of Acid-clay treatment method of used lubricating oil

IOP Conf. Series: Materials Science and Engineering **413** (2018) 012052 doi:10.1088/1757-899X/413/1/012052

Table 4: National Air Quality Standards (NAQS) in Nigeria

Air Pollutants	Emission Limits	Averaging time
CO		
	9ppm (10.31 mg/m3)	8-hour
	35ppm (40.08 mg/m3)	1-hour
SO2	75 mg/m3	1-hour

[12]; [13]

Table 5: Mean Emission Concentration (MEC) of CO and SO2 compared with NAQS

S/N	Treatment stages	Air Pollutants	MEC (mg/m3)	NAQS (mg/m3)
1	Dehydration	CO	167.07	40.08
		SO2	12.467	75
2	Acid Treatment	CO	22.9	40.08
3	Clay Treatment	CO	4.83	40.08

Comparing the MEC of CO and SO2 with NAQS from Table 5, it was observed in dehydration stage that SO2 was within the set standard limit which means no health risk is present in the receptor locations while CO exceeded the set standard limits and it means there is health risk in the receptor locations. Also in acid and clay treatment, CO is within the set limit.

4. Conclusion

In this study, the emissions of pollutant in the treatment of used oil using acid-clay was investigated at varying oil temperature CO emission was found to be the highest emitted pollutant with MEC of 167.07 mg/m3 due to incomplete combustion. When compared with NAQS standards, CO concentration at dehydration exceeded the set limits which could cause adeverse effect to receptor locations while SO2 was within the limit. The findings indicated that there is a need to quantify the effects of the emissions from used lube oil treatment plants. However, the predicted results of this study will help decision makers to formulate policy for controlling the impact of emission from used lube oil treatment plants using acid-clay method.

Acknowledgement

The authors appreciates Covenant University for the partial sponsorship of the paper

References

[1] Yang, C.-Y. (. (2008). The Feasibility Studies onSonochemical Processes For Treating Used Oil: Toxin Reduction For Eliminating Recycle Interference. California, Usa: University Of Southern California - Faculty Of The Graduate School.
[2] Kanokkantapong, V., Kiatkittipong, W., Panyapinyopol, B., Wongsuchoto, P., Pavasant, P., (2009). Used lubricating oil management options based on life cycle thinking. Resour. Conserv. Recycl 53, 294e299.
[3] Wanga X. L, Zhang G. N, Zhanga J. Y., Yin Y. L, and Xu Y. (2013). Failure Analysis and Regeneration Performances Evaluation on Engine Lubricating Oil. Physics Procedia 50:473 – 479

[4] Udonne, J. D., & Bakare, O.A. (2013). Recycling of Used Lubricating Oil Using Three Samples of Acids and Clay as a Method of Treatment. Int. Arch. App. Sci. Technol., Vol 4 [2] 08- 14.

[5] Princewill Nemibo Josiah, Sunday Sunday Ikiensikimama. The Effect of Desludging and Adsorption Ratios on the Recovery of Low Pour Fuel Oil (LPFO) from Spent Engine Oi. Chemical Engineering Research Bulletin 2010;25-28.

[6] Falah Bani Hani and Hussien Al-Wedyan (2011). Regeneration of Base-Oil from Waste-Oil under Different Conditions and Variables. *African Journal of Biotechnology*, **10(7)**:1150-1153.

[7] Hamawand Ihsan, Talal Yusaf and Sardasht Rafat. Recycling of Waste Engine Oils Using a New Washing Agent. Energies 2013; pp. 1023-1049. Inc. 2010.

[8] Falah Bani Hani and Hussien Al-Wedyan (2011). Regeneration of Base-Oil from Waste-Oil under Different Conditions and Variables. African Journal of Biotechnology; 1150-1153.

[9] Haynes, William M. (2010). *Handbook of Chemistry and Physics* (91 ed.). Boca Raton, Florida: CRC Press. p. 9-33. ISBN 978-1439820773.

[10] Prockop LD, Chichkova RI (Nov 2007). Carbon monoxide intoxication: an updated review. Journal of the Neurological Sciences. **262** (1–2): 122–30. *doi:10.1016/j.jns.2007.06.037. PMID 17720201*

[11] WHO (2000). Air quality guidelines for Europe, 2nd ed.Co-penhagen, World Health Organization Regional Office for Europe,(WHO Regional Publications, European Series No. 91).

[12] IDEM (2012). Criteria Pollutants Air Quality Trends Analysis Report. Office of Air Quality. Indiana Department of Environmental Management.

[13] Federal Environmental Protection Agency (FEPA), National interim Guidelines and Standards for Industrial effluents, Gaseous emissions and Hazardous wastes. Environmental Pollution Control Handbook. Lagos, FEPA. 1991.

IOP Conf. Series: Materials Science and Engineering **413** (2018) 012053 doi:10.1088/1757-899X/413/1/012053

Electric Power Crisis in Nigeria: A Strategic Call for Change of Focus to Renewable Sources

[1]Olatunji Obafemi, [2]Akinlabi Stephen, [3]Oluseyi Ajayi, [3]Abioye Abiodun, [3]Ishola Felix, [2]Peter Mashinini and [1]Madushele Nkosinathi

[1]Mechanical Engineering Science, University of Johannesburg, South Africa
[2]Mechanical & Industrial Engineering Technology, University of Johannesburg, South Africa
[3]Mechanical Engineering, Covenant University, Ota, Nigeria
tunjifemi@gmail.com

Abstract. In all ramifications, the development and economic prosperity of a nation substantially depend on her power or energy sector. Reliable access to the electric power needed by several homes and businesses remains a great obstacle facing Nigeria. More than ever, the social-economic wellbeing of Nigeria is under severe threat due to formidable challenges in her Power sector. The manufacturing industries are folding up, while others are relocating their production plants to more friendly havens because of the high cost of doing business. This paper reviews various renewable energy options available to Nigeria and the National Renewable Energy and Energy Efficiency Policy (NREEEP). It also identified the major factors militating against renewable energy development in Nigeria, proffering possible way forward to assist renewable energy policy implementation.
Keywords: Electric power; GDP; development; NREEEP; Renewable energy

1. Introduction

Nigeria is the largest economy in Sub Saharan Africa, but the Nigeria economy is characterized by a continuously increasing rate of inflation, high unemployment rate, stagflation, receding stock market index and full-blown recession which is substantially due to limitations in the power sector[1]. Several years of negligence and improper policy implementation strategy has culminated into the present economic quagmire. The strong nexus between the energy and socio-economic wellbeing of a nation cannot be overemphasized. Notwithstanding her large oil and gas reserve, which compete favourably among the largest reserve in the world, Nigeria is battling with the severe energy crisis. The unpalatable effect of this situation is felt mostly by the masses that have to endure the excruciating effect of higher and arbitrary tariff including a prolonged period of power outage. Several nations who have heeded the warning of nature have diversified their power sector to eco-friendlier renewable sources and have since been reaping the dividend as revealed in their Gross Domestic Product. Electricity was first generated for public use in Nigeria in 1896 and over a century after, electricity demand in Nigeria is at present far more than the supply, thereby negatively affecting the country's socio-economic and technological developments [2, 3]. Based on the United Nation estimate as at 2017, Nigeria population is approximately 190 million [4] as against around 150million in the year 2012 [5], this is an indication of Nigeria ever-growing population. The larger part of this population lives in the rural area where there are no infrastructures for electricity supply with just around 36 % access to power [1]. In aggregate, only about 40 % of Nigerians are connected to the national grid, even those who are connected to the grid have access to power less than half of the time [6].

Nigeria's electricity consumption is one of the lowest in the world at 149 kWh per capita, about 7% of Brazil's and 3% of South Africa's [6]. Nigeria over-dependent on gas for the power plant is making

IOP Conf. Series: Materials Science and Engineering **413** (2018) 012053 doi:10.1088/1757-899X/413/1/012053

her so vulnerable, and this is perpetually leaving her at the mercy of the militant insurgents who vandalize the pipelines at will. For the second time in 2016, a case of zero generation was reported for some state in Nigeria[7]. The electricity generated in the first quarter of the year 2018 is a painful reminder of the persistently low electricity supply in Nigeria[8]. The cumulative power supply reduced to approximately 114GW as against 144GW recorded for the last quarter of the year 2017[8]. In the last 17years, several billions of dollars have been splashed on electricity power sector, yet the industry is in doldrums [7]. The reported peak electricity generation capacity of Nigeria as of February 2015 is 3342MW compared to an estimated peak demand of 12,800MW representing a power deficit of 74%[9] Despite the recent privatization of electric power generation and Distribution Company, there are no commensurate improvements in supply, rather it has gone worst. The Nigerian Electricity Regulatory Commission (NRC) website recalled that total energy at its peak in a particular that lead up to June 12 was 2,591MW, while off-peak power distributed was a paltry 1,547MW. This sort of power output cannot sustain the economy which was estimated to a Gross Domestic Product of $568.51 billion in 2014 by World Bank[10].
The manufacturing industries are folding up, while others are relocating their plants to more business-friendly countries because of the high cost of doing business. The MAN (Manufacturers Association of Nigeria) says it spends more than ₦1.8 billion (the US $ 11, 340 million) weekly in the running and maintenance of power generators[11, 12]. The use of these generators in the industries has resulted in high cost of production since energy consumption constitutes 40 % of the production cost. At present, the cost of production in Nigeria is nine times higher than that of China [12]. The unpalatable effect of this crisis is felt mostly by the masses that have to endure higher and arbitrary tariff coupled with the prolonged period of power outage and aching cost of living while they are left to pay for the power they never consumed. These and several other uncertainties surrounding the generation, distribution and use of electricity culminate into a phenomenon christened Electric power crisis in Nigeria. Against the backdrop of all these uncertainties, a possible change of approach and more focus on the renewable energy sources will bail the country out of this quagmire and help tame this age-long monster. Renewable Energy, RE can be defined as energy obtained from sources whose utilization does not result in the depletion of the earth's resources. The world is changing focus to renewable sources and energy efficient technologies which are more sustainable when compared to fossil fuel. Therefore, it is exigent that RE solutions be proffered to cater for Nigeria imminent industrialization. From the standpoint of sustainability, sources and challenges, several authors have discussed the renewable energy policy and options in Nigeria[13-24].Oyedepo [24] outlined the primary reasons why renewable energy is a vital and indispensable component of sustainable development. Presently, energy crisis is taking its toll on the available industries as many of them are running at high cost. Even the private homes are not excluded as so many homes do not have access to electric power. The biggest issues which have been identified as the major obstacle to the growth of power sector are ; Hostile macroeconomic forces, Weak and opaque regulatory mechanism, creditworthiness[1]. In Nigeria, the plan is underway to build a nuclear power plant, however as good as that may appear, there are hidden cost attached to coal and nuclear power; For instance, in South Africa, Eskom is currently building the Kusile coal-fired power plant, and it is estimated that the coal plant will cause damage of up to R60 billion for every year of operation. This will take the form of the costs of water pollution, human health impacts, the plant's huge water footprint, and climate change [25]. The Fukushima nuclear accident which occurred in the year 2011 is still very fresh in mind. A major 15meters tsunami disabled the power supply and abruptly ended the cooling of three Fukushima reactor, not less than 1000 people died in this incidence with several rendered homeless [26, 27].
Union of concerned scientist highlighted the merits of renewable energy as follows [10]:

- Little to no global warming emission,
- Improved public health and environment quality,
- A vast and inexhaustible energy supply,

IOP Conf. Series: Materials Science and Engineering **413** (2018) 012053 doi:10.1088/1757-899X/413/1/012053

- Job creation and other economic ripple effects,
- Reliable and resilient energy system and stable energy prices.

The renewable energy potential of Nigeria is well documented in the literature and its ability to revolutionize our power sector is not in doubt, so it is in this understanding that government has come up with different policy paper including National Renewable Energy and Energy Efficiency Policy, NREEEP, however, clear-cut implementation strategies are lacking.

2. Analytical Framework

In order to put the electricity issues of Nigeria in perspective and evaluate the recent trend in the development of renewable energy resources under the established implementation framework of NREEEP, the following objectives will be explored in this study as indicated in Figure 1;

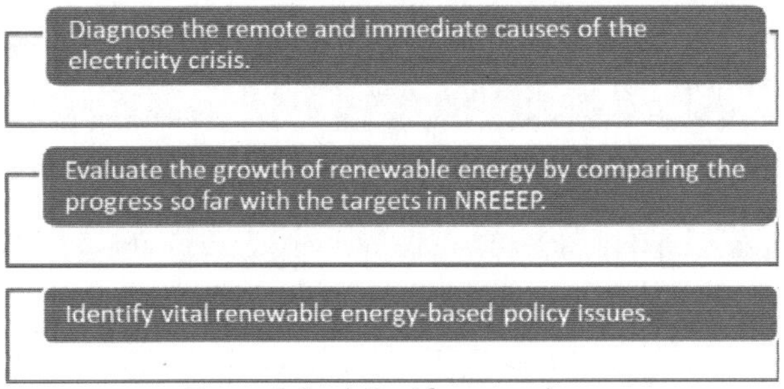

Figure 1:Analytical framework

3. Nigeria Electric Power Sector in Perspective

Electricity was first generated in Nigeria in the year 1896, although it was not until 1929 that the first utility company, the Nigerian Electricity Supply Company (NESCO) was established. In the 1950s and 1960s, the Nigerian government created the Electricity Corporation of Nigeria to control all existing diesel/coal fired isolated power plants across the country and the Niger Dams Authority to develop hydroelectric power in Nigeria. These two entities were amalgamated into the National Electric Power Authority (NEPA) in 1972. NEPA was characterized with a burden of subsidies, poor service quality and wretched collection of the tariff. The reform act of 2005 unbundled NEPA into 18 companies (under the flag of Power Holding Company of Nigeria): 6 generation companies, 1 transmission company, and 11 distribution companies.

The generating companies consist of three hydro and nine thermal (gas based) stations with their output shown in Figure 2 [28]. It can be observed that none of the present generating plants are made of renewable resources.

The federal government of Nigeria is seriously working towards generating 4000MW of electricity from nuclear energy, In June 2016; Nigeria signed a Memorandum of Understanding, MOU, with ROSATOM in order to partner in building a nuclear technology centre in the country. This MOU was supposed to signal the beginning of a new turn for the country to have clean energy, steady electricity supply and be able to do deep research in nuclear medicine[29]. However, these authors vehemently doubt the willingness of the government, given the inadequately trained manpower in this area combined with frightening insecurity and terrorism ravaging the country.

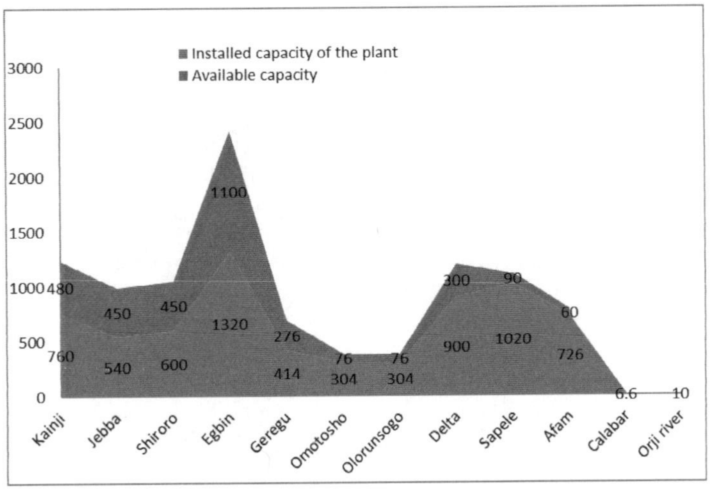

Figure 2: Nigeria power generation plant and their utilization capacity[28]

The PHCN (Power Holding Company of Nigeria) was structured to oversee the activities of the Managing Director/CEOs of the successor company for 5 years, which should be adequate to enable the companies privatized to be fully established [3]. The privatization exercise has suffered many drawbacks and its objectives have not been reasonably realized. Notwithstanding the existence of electricity in the country for over a century and in spite of all the reforms in this direction, its development has been at a slow pace. The minister of power who was part of the privatization committee in 2013 resigned from office, citing gross corruption and sabotage as his reasons for leaving the office [30].

According to Transmission Company of Nigeria (TCN), the power generation dropped to 2,662MW from initial 3,959MW as at January 2017. As usual, the reasons alluded to this include; difficulties in accessing the gas by generating company and low water level [10].

At the end of January 2016, the demand for electricity in Nigeria was estimated to be 12,800 MW [7]. Figure 3 indicates the wide margin between the energy generated and demanded over a period of three years. The historic gap between the demand for power in Nigeria and the electricity supply from the grid has led to widespread self-generation of power in the commercial, industrial and residential sectors. Many individuals and businesses own their generators to compensate for the lack of access to electricity supply [31]. Apart from the cost implication of diesel powered generator, environmental impact is also very significant [32].

A survey conducted by Adenikinju [33] showed that only 6.2 % of the firms in the Nigeria manufacturing sector exclusively rely on the grid electricity. Virtually all the firms have independent power generating set to support their activities. Figure 4 shows the contribution of power outages to capacity underutilization in the manufacturing sector. In this survey, 44 % of the firms attributed a fall of between 20 % and 49 % in their capacity underutilization to power outages. 24 % of small-scale firms lost over 50 % of their output to unstable electricity compared to 14 % and 17 % for the medium and large enterprises, respectively. Over 20million of Nigeria household are without power while around 36 % of the rural settler have access to electric power. This shows how devastating the effect of electricity unavailability has been on the country's economic survival. In order to improve current GDP growth and narrow the current electricity supply gap, market intervention, and fundamental power sector reform which is aimed at the implementation of renewable energy exploration strategies are needed.

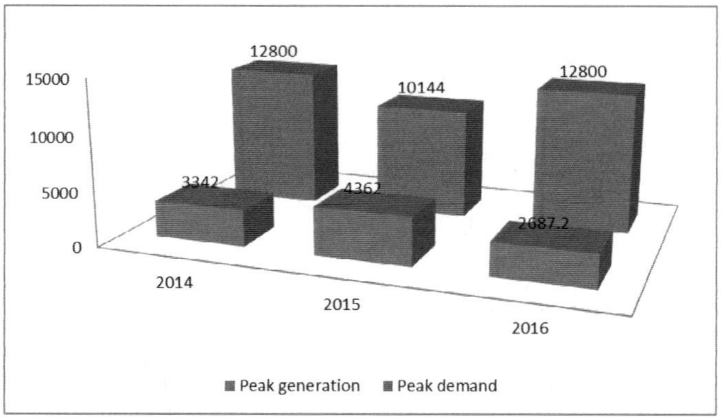

Fig 3: Effect of Electric power outages on capacity utilization[7]

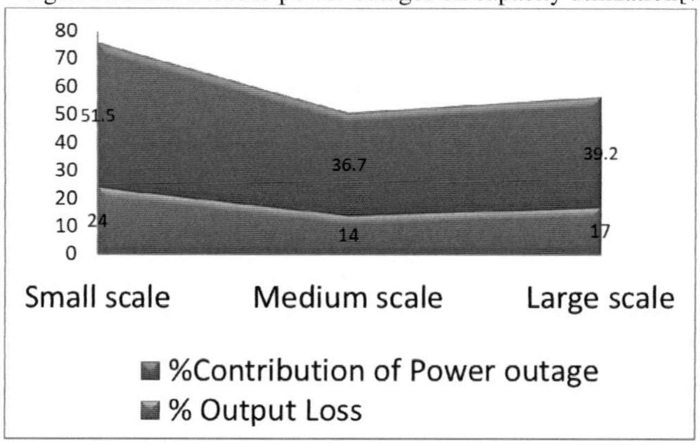

Figure 4: Effect of Electric power outages on capacity utilization. [33]

Idemudia and Nordstrom [34] succinctly highlighted some of the challenges facing the investors that acquired Nigeria electricity assets after privatization as follows;

- Insufficient gas supply due to poor gas infrastructure,
- Cost unreflective tariff regime,
- Obsolete transmission line assets and power sub-stations,
- Non-bankable gas supply agreements,
- Changes in government and uncertainties as to the future direction of government policies,
- The bureaucracy of government agencies,
- Lack of affordable and accessible long-term funding;
- Foreign exchange and currency issues;
- Vandalism of power plants, equipment, and transmission lines.

The following key players are also identified in Nigeria's reformed power sector: The Federal Ministry of Power; National Electricity Regulatory Commission (NERC).

Nigeria Electricity Bulk Trading Plc (NBET); The Gas Aggregation Company Nigeria Limited (Gas Aggregator); Nigerian Electricity Management Services Authority; National Power Training Institute of Nigeria; Bureau of Public Enterprises National Council on Privatization; Transmission Company of Nigeria; Nigeria Electricity Liability Management Company; Nigerian Gas Company; Operator of the Nigerian Electricity Market [35, 36].

4. Recent Trend in Renewable Energy

4.1. Biomass

There is a substantial opportunity for the exploitation of several types of biomass including municipal wastes in Nigeria. Biomass is a biological substance derived from living or recently living organisms. It is often referred to as plant-based substances, whereas biomass applies to both plant and animal-based materials. Not until recently, biomass has a proven capacity for renewable electricity when compared with wind and solar. [37]. In Nigeria, some of the biomass sources have been identified as animal waste, plant waste, municipal and industrial waste, aquatic biomass, and so on [14, 24, 37, 38]. The estimated biomass resources stand at 2.01 EJ [39]. Biomass derived from the plant can serve small-scale industries, it can also be processed to produce cheap fuel like biogas. Given the pace of the development of the technologies for harvesting biomass, it can provide an increasing amount of "biopower". The identified feedstock substrates for an economically feasible biogas production include water lettuce, water hyacinth, dung, cassava leaves and processing waste, urban refuse, solid (including industrial) waste, agricultural residues and sewage [38-40]. An estimate shows that Nigeria produces about 227, 700 tons of fresh animal waste. Apart from sawdust and wood waste derivable from lumber industries, Nigeria can potentially produce up to 6.8 million $m^3$of biogas from animal waste on a daily basis [39].

4.2. Solar Energy
The main source of energy which supports life on earth and all other surface phenomena come from the sun [23]. Solar energy is applied in households for activities such as drying of clothes, and hot water for domestic use. On the commercial scale, it is used in generating electricity and driving of turbine in thermal electricity plant. The efficiency of the photovoltaic cells is estimated to be about 40%, however, with the latest discovery in nanotechnology, it is expected that the efficiency would be tremendously increased. Most recently, solar energy has been converted to heat, which is then stored in the thermal tank. Another method of using solar energy to generate electricity also used in recent times is to convert the sun's energy into heat which is then stored in thermal tanks [23, 38]. The thermal tanks contain molten salts or paraffin wax which can absorb a great amount of heat energy at the same state. The heat energy derived from this source can be used to drive electric generators.

The adoption of solar renewable technology in the country is low compared to the large potentials widely reported. Nwofe [41] outlines the high initial capital cost of solar technologies, low levels of education and awareness about the social and economic benefits associated with solar technologies, and a lack of access to solar radiation data recording stations as some of the factors limiting the utilization of solar energy in Nigeria.

Ohunakin *et al.* [23] highlight the barriers to solar energy development and application in Nigeria based on Painuly [42] as follows: technical, social, cultural and behavioral; economic and financial; institutional; and political and market distortion.

4.3. Wind Energy
The energy sourced from the wind to generate electricity or another form of energy is called wind energy. The wind energy can be used to pump water and grinding grains. Previously electricity was generated from the windmill or ship sail. But this has evolved to wind turbine, which is presently used to convert the wind energy to electric power. In recent times numerous studies have been carried out to assess the wind speed characteristics and associated wind energy potentials in different locations in Nigeria. Promising attempts are being made in Sokoto Energy Research Centre (SERC) and Abubakar Tafawa Balewa University, Bauchi to develop the capability for the production of wind energy technologies [20].

4.4. Geothermal Energy
Geothermal energy can be defined as heat energy generated and stored in the earth; as a result of the radioactivity and progressive heat loss from the earth's crust. The thermal energy in the form of heat reaches the surface from the crust due to the temperature differential between the planet and the surface. The temperature of the crust can be as high as 4273^{0}C, hence heating the water and rock up to 370°C. This energy can be used for several purposes such as district and space heating, desalination, agricultural applications and most importantly to generate electricity. The world first large scale geothermal plant was built in the year 1911 [43]. Presently, United State is currently leading in geothermal energy exploration with 3450MW installed capacity and 2542MW generating capacity [44]. This form of energy is dependable with high sustainability due to enormous earth's internal heat content which is about 100 billion of the present world energy demand [43].

4.5. Waves and tidal energy

Waves are produced by winds blowing across the surface of the ocean. However, because waves travel across the ocean, their arrival time at the wave power facility may be more predictable than the wind. In contrast, tidal energy is driven by the gravitational pull of the moon and sun. Tidal energy is produced through the use of tidal energy generators. These large underwater turbines are placed in areas with high tidal movements and are designed to capture the kinetic motion of the ebbing and surging of ocean tides in order to produce electricity. Tidal power has great potential for future power and electricity generation because of the massive size of the ocean [45, 46]. Tidal power has a relative advantage in that it can generate power from the incoming and outgoing tides. Like most emerging energy technologies, wave, and tidal technologies are currently more expensive than traditional generating resources; but with further experience in the field, adequate R&D funding, and proactive public policy support then the costs of wave and tidal technologies are expected to follow the same rapid decrease in price that wind energy has experienced. Wave and tidal energy resources are among the various energy resources that are available in Nigeria. But this has not been explored, because of insufficient R&D in this area.

4.6. Hydro-Energy

This is energy generated from water conserved in a reservoir when a dam is constructed across flowing water like river or stream. Hydropower is generated by conversion of the potential energy of the high head water into electrical energy[14, 47]. The water which is conveyed by the penstock is made to fall on the blades of the turbine to run them by the conversion of the hydraulic energy to mechanical energy which in turn is converted into electrical energy by directly coupling the shaft of the turbine with a generator and the generator armature rotates to produce electricity [48]. Nigeria is significantly endowed with large rivers and some few natural falls. Small rivers and streams also exist within the present geographical spread of the country into eleven River Basin Authorities, some of which maintain minimum discharges all the year round. As it stands, hydropower accounts for about 29% of the total electricity supply [47]. In a study carried out by Aliyu and Elegba [19] in twelve (12) states and four (4) river basins, over 278 unexploited Small Hydropower (SHP) sites with total potentials of 734.3 MW were identified. However, Small Hydro Power potential sites exist in virtually all parts of Nigeria with an estimated total capacity of 3,500 MW. All the authors agreed that Nigeria possesses a potential renewable source of energy along with her numerous river systems[19, 24].

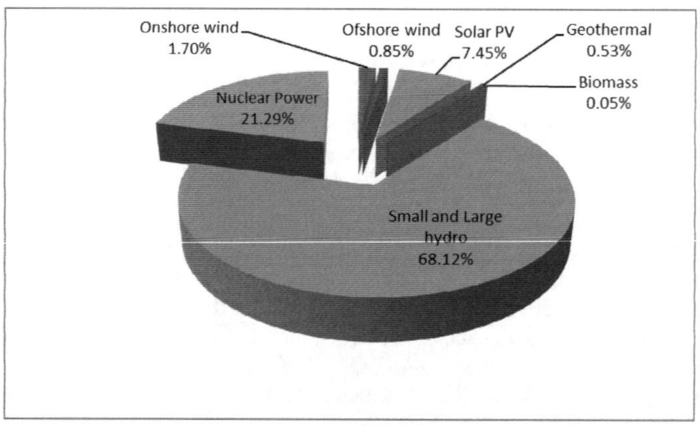

Figure 5: Renewable energy potentials of Nigeria [2, 49]

5. A brief review of RE in some selected Africa countries.

5.1. RE South Africa

In the entire Africa continent, South Africa has the most audacious renewable energy projection. Several authors have discussed the renewable energy potential and policy in South Africa [50-53]. The 1998 Energy Policy of South Africa states that the country will acquire 15% of its national supply from RE [54]. However, an analysis of the energy mix in South Africa by Pegels [53] shows that as at 2008, coal contributed 86%, nuclear contributed 5% of the country's energy mix. Other sources are hydro and gas which combined to contribute 9% of the energy mix. [53] observed that notwithstanding the vast potential for Renewable energy in South Africa, there has so far been little growth in the deployment. The author identified energy innovation system and the economics of renewable energy technologies as the main impediment on the path of Renewable energy development in South Africa. The government of South Africa has introduced several policies to support RE in the country. One of such policies is the feed-in tariff. The renewable energy feed-in tariff was launched in 2009. It requires the national electricity utility Eskom, to purchase renewable energy from qualifying generators at predetermined prices, this help to reduce the financial risk exposure of the generators and provide the market certainty [55-57].

5.2. RE Ghana

Ghana's Renewable Energy Development Program [57, 58] aimed to assess the availability of renewable energy resources and to examine the technical feasibility and cost-effectiveness of RE technologies in the country among other goals. The program highlighted and discussed the RE potential of Ghana grouping them into two areas; biomass and solar. Wereko-Brobby and Mintah [58, 59] identified major RE projects in the country and advised on how they could be made better. Ghana has been argued to achieve commendable access to modern energy services compared to her sub-Saharan peers [59]. Increase in industrialization and urbanization have resulted in high energy intensity in Ghana, therefore, Serwaa *et al.* [60] suggested that policies aimed at encouraging the production of less energy-intensive products and implementation of high energy efficient technologies in the manufacturing sector should be promoted. Ghana's renewable energy resources could be harnessed to play a role in supplying both rural and urban households. There are huge biomass resources in Ghana that have the potential for use as feedstock for biogas production to reduce the country's overdependence on fuelwood and fossil resources. He assessed the potential of biogas in Ghana and concluded that the country has the potential of constructing about 278 000 biogas plants and as of 2011, only about 100 biogas plants have so far been constructed. The development of RE has been hampered by the failure of the government to support biogas technology through well-articulated policy, poor design and construction, horrible maintenance culture on the part of the users [61].

5.3. RE Senegal

Senegal is far ahead of Nigerian in the promotion of RE in the rural areas. Since 2008 the new Energy Sector Development Policy Paper has been in place with a clear direction for RE [62]. The policy sets a penetration rate for renewable sources of energy and biofuels of at least 15% of internal energy consumption by 2020 [63]. The commitment to institutional reform and policy has positioned Senegal as a leader in RE. Promotion in the Economic Commission of Western Africa (ECOWAS) region leading to the country being tasked to develop solar energy projects in the sub-region by Heads of State and Government in the ECOWAS Summit held in July 2010, and subsequently chosen as one of the pilot countries to field-test the methodology being developed by International Renewable Energy Agency (IRENA) for the renewable readiness assessment [62].

6. Energy Mix in Nigeria

The past and future energy mix projection Figure 5 as indicated by Obadote [2] and [49] shows gas and hydro fuels will maintain their positions as the main drivers of the electricity sector in the short and

IOP Conf. Series: Materials Science and Engineering **413** (2018) 012053 doi:10.1088/1757-899X/413/1/012053

medium term. Renewable fuels like solar, Biomass and wind are expected to play roles in sustaining the Vision 20:2020; though the full potentials are not going to be tapped.

As of 2008, the energy consumption mix in Nigeria was dominated by fuelwood (50.45%), petroleum products (41.28%) and hydroelectricity (8%) [16]. Apart from hydro, Nigeria has not profitably explored other RE sources. This argument has been assertively situated by [2, 6, 15, 23, 24] About six years after, the empirical situation in the country is at variance with Ajao *et al.* [64] assertion that replacing fossil fuel with RE is the ultimate. These authors somewhat agree with the view expressed by Agbaitoro [43] that the notion which states that replacing fossil fuel with RE is the ultimate goal in Nigeria is somewhat hasty and lacking sufficient supporting empirical evidence. Agbaitoro [43] posited that governments at different levels and the private sector had given RE considerable attention recently. However, the current level of RE implementation does not suggest that stakeholders have prioritized it.

7. Nreeep 2015: Renewable Energy Policy in Perspective

There have been several policies on renewable energy in Nigeria which analyzed the renewable energy potential of the country and possible strategy for exploration. These include: National Energy Policy (NEP) [65-67]; Renewable Electricity Policy Guidelines (REPG) [68]; Renewable Electricity Action Programme (REAP)[69], National Economic Empowerment and Development Strategy (NEEDS)[70];Nigerian Biofuel Policy and Incentives (NBPI) [71]; Renewable Energy Master Plan (REMP) [47, 72]; and most recently approved National Renewable Energy and Energy Efficiency Policy (NREEEP), [73].

As stated in the policy document, the cardinal idea behind NREEEP is to diversify the present monotonous inefficient electric power source in Nigeria by improving the energy mix and increasing the share of renewable energy resources contribution to national utilization. The policy was tailored to align with the best renewable energy practices across the globe.

The purpose of NREEEP, [73] among other things can be summarized as follows;

i) Set out a framework for action to address Nigerians challenge of inclusive access to modern and clean energy resources, improved energy security and climate objectives. These will also incorporate provisions for renewable energy and energy efficiency generation activities into state policy statements and plans, and recognizes the importance of enabling a framework for private investment in renewable energy and energy efficiency;

ii) Declare that the proportion of Nigeria's electricity generated from renewable energy sources shall increase to a level that meets or exceeds the ECOWAS regional policy targets for renewable electricity generation and energy efficiency for 2020 and beyond and also set national targets for achievements in electricity from renewable energy and energy efficiency capacity addition by 2020 and beyond.

iii) Make it mandatory for the Ministry of power to facilitate the development of an integrated resources plan (IRP) and ensure the continuous monitoring and review of the implementation and effectiveness of the action plans prescribed under the national policy statement.

This policy, having identified the resources situation and the technology of the country, focuses on hydropower, biomass, solar, wind, geothermal, wave and tidal energy power plants and cogeneration plants for energy production, as well as the improvement of energy efficiency as an additional source of energy. For each element of renewable energy, the policies, objectives, and strategies were outlined as contained in the National Energy Policy (NEP) [65]. The short-term, medium-term and long-term objectives and target were well articulated.

The overall objective of NREEEP [73] is summarized as follows:

i. To ensure the development of the nation's energy resources, with diversified energy resources option, for the achievement of national energy security and an efficient energy delivery system with an optimal energy resource mix.

ii. To guarantee an adequate, reliable, affordable, equitable and sustainable supply of renewable energy at cost-reflective and appropriate costs and in an environmentally friendly manner, to the various sectors of the economy, for national development.
iii. To accelerate the process of acquisition and diffusion of technology, managerial expertise and indigenous participation in the renewable energy and energy efficiency sector industries, for stability and self-reliance.
iv. To guarantee efficient, location-specific and cost-effective consumption pattern of renewable energy resources and improved energy efficiency.
v. To promote increased investments and development of the renewable energy and energy efficiency sector, with substantial private sector participation.
vi. To ensure a comprehensive, integrated and well informed renewable energy and energy efficiency sector, with plans and programs for effective development.
vii. To foster international co-operation in trade and project development, in the ECOWAS, African Region and the World at large.
viii. To successfully use the nation's abundant energy resources to promote international cooperation.
ix. To bring abundant electricity access to almost half of the Nigerian population that is currently electricity abstinent, including more sustainable provisions for domestic use and cooking.
x. To develop the nation's renewable energy and energy efficiency resources through the establishment of an appropriate financing mechanism that supports private investment in the sub-sectors.
xi. To ensure effective coordination and collaboration

The document also summarized the implementation policies which have been applied successfully in other nations. The policy documents identify the regulation and fiscal instrument that have been applied. These can be summarized as follow;

i) Mandatory or voluntary Renewable Portfolio Standards (RPS), which define the percentage of energy generated that must come from renewables by a given target year;
ii) Generation Disclosure Requirement (GDR), which is applicable when consumers have a retail choice and have a preference for renewables;
iii) Power Production Tax Credit (PTC) to electricity generation companies, which is aimed at incentivizing the adoption of renewable energy;
iv) Feed-in tariffs (FIT), which typically incentivize electricity producers by offering more favorable pricing for electricity produced through renewables;
v) The adoption of a Public Benefits Fund (PBF), which requires that a certain percentage of the tariff is dedicated to supporting renewable energy generation projects on and off the grid;
vi) Bidding rounds through national renewable energy independent power producer procurement program;
vii) Provision of capital grants, tax holidays and exemptions, other incentives for renewable energy projects;
viii) Net metering framework.

It is observed that NREEEP policy document did not expatiate on the waves and tidal energy option despite its availability, the reason posited was that there were no sufficient Research and Development focuses on this subject. This is in line with the position of [22]. The NREEEP document projected the contribution of the RE based on 7% GDP growth over a short-term (2015), medium term (2020), and long-term (2030)[73]
These authors find the policy document very comprehensive and in line with the best practices around the globe. However, much has not been done in the area of implementation, and these authors are of the opinion that if serious attention is not focussed toward implementation, this report may go in the

way of other lofty policy papers which were not properly explored. The objectives should be to explore the relative renewable energy advantages for different state and local government [14, 20, 21]. With this, the country will be able to thoroughly analyze the area of strength of each state to tap their renewable resources. More so, this policy is capable of turning around the finances of so many states in Nigeria, which are currently battling with an insufficient allocation from the federal government.

8. Prospect of Renewable Energy to revolutionize Nigeria power sector: The obstacles and leverages.

It is very obvious from the current status of RE in Nigeria that the application of RE technologies for electric power generation has been slow and not well implemented despite an explicit policy paper. Some sort of stimulatory measures were put in place to promote RE service provision in Egypt, South Africa, and Malaysia, the same measures were outlined by Shaaban and Petinrin [21]. Therefore, for a successful penetration of renewable energy in Nigeria, which is capable of bailing the country out of her present energy crisis, it is apt to evaluate different international best practices and policies for RE integration. Therefore, these authors will discuss the obstacles that must be subdued and also lend their impetus to some policy implementation strategies, regulations and legislative frameworks that can speed up the development of RE for electric power generation in Nigeria.

8.1. Internal and international politics and policy (IIPP)

Internal politics of a country and its international goodwill could either serve as a stimulator or barrier to the implementation of renewable energy policy. As it is presently observed in Nigeria, the successive administrations are not always averse to continue with the policies of their predecessor, partly because of their quest to build a name for themselves. This is a counterproductive practice, which has left the country with a lot of abandoned projects. There is meant to be a continuum in government. To attain a favorable RE level, the enabling and systematic legal and regulatory framework is vital. Therefore, the progression of RE technologies demands a formidable political will via a well-articulated bureaucratic process that is void of polarity along political affiliation and personal interest. From an international perspective, economic-political risk factors play a role in determining the attitude of international associates to renewable energy implementation. Economic-political risk factors arise from the competition around scarce and valuable resources, tensions, conflicts and violence resulting from overly abundant, and the risk of the owner of a strategic resource using it as a tool a for achieving a political and economic advantage. On the contrary, liberal international relations theory presents a perspective and views interdependency as an important security-building factor [74, 75]. According to this theory, the more dependent countries are on each other, the more secure the world will be, which will also bring security to individual countries. This also includes energy security.

For instance, Lilliestam and Ellenbeck [76] analyzed the potential risks to the EU from the new interdependencies arising through large-scale imports of renewable electricity from North Africa. By comparing the relative importance of stable demand for energy to the supplier and stable supply of energy for the user, those authors showed that the EU can be vulnerable to coordinated political action from the supply countries, whereas a single country has more to lose than the EU from an action restricting the supply of solar electricity. From the illustration above it is obvious that the implementation is mostly based on the prejudicial view of the stakeholders. These authors align with the position of [75] and advise that renewable energy implementation should not be undermined based on prejudice. The larger picture and the common economic interest of the countrymen should engender the implementation.

8.2 Financial strategy

The initial start-up cost of renewable energy plant is always the major obstacle for the potential investors. Although the renewable energy resources have low operational and maintenance cost, the upfront capital investment is high compared to conventional energy alternatives. Another obstacle is the common perception about RE as been untested technologies. This notion has drastically reduced, thanks to the recent development in renewable energy applications. The investors are left to decide against the odds of economic risks and uncertainties. In a dwindling and recession infected economy like Nigeria, with very high demand on meager capital resources, the investor may face the high cost of the transaction and restricted access to capital. This inadequate access to financing, cum high-interest rates, culminate in the primary barriers hampering the RE market growth [13].

a. Fiscal issues

Fiscal incentives can be introduced to address some of the obstacles to RE development. This can be in the form of tax exception on RE equipment that is imported into the country, and a tax holiday on the RE incomes [36]. Also, the country can introduce environmental taxes, as it has proven to be very successful in Denmark. Although, this may be a gradual process in the case of Nigeria because of weak environmental policy and inadequate alternatives to fossil fuel. Denmark energy consumers have been charged a CO_2 tax and some of the revenue generated goes to generators of electricity from renewable energy [77]. In the Netherlands, industries received accelerated depreciation of investment in equipment when invested in renewable and energy efficiency projects [78]. Considering the environmental merits of RE, increased government support will cushion the current exorbitant capital investment requirement which is associated with RE technologies, hence encouraging the private sector to invest in RE. Fiscal and financial incentives should cut across business stages, from incorporation to production.

The fiscal support should be more in the form of production tax credit as against tax incentive on investment. This will motivate the owner to ensure the effective and efficient function of the power plant and reduce the possibility of abuse like it was previously observed in fuel importation subsidy regime and duty-free importation of goods which was severely abused for personal purposes.

b. Legislation and Relation matters

To achieve sufficient electric power supply where renewable play a vital role, relevant policy framework must be in place. This must include regulatory instruments which provide a conducive environment for both international and domestic investors. The legislation must be devoid of any ambiguity, hence the roles of all the stakeholders at every level of renewable energy development must be well clarified. [18].

In other countries, several legislative supports have been lent to RE development; this includes Feed-in Tariff in Germany and Brazil; Auction regime in Brazil; Quota obligation in the USA [22]. These legislative inputs have helped to position these countries in maximizing the benefit of RE.

With a proper legislation backed tendering procedure, the need for financial assistance is drastically reduced. In a situation where there is no well established tendering arrangement, subsidy arrangement presents itself as the main instrument. This has been successfully implemented as in the Swedish wind energy schemes [79]. However, such an approach must be done with a lot of caution because of Nigeria previous experience on petroleum products subsidy regime, which was thoroughly bastardized through corruption and rent taking.

c. Research and Innovation

The sustainability of any RE policy is contingent upon R&D Most of the previous R&D programs in Nigeria are typically in the form of grants which often do not have any expectations of the financial gain to the country. However, it is very important that funded R&D program develops patentable technologies. And the researcher must be held to their targets as it is presently obtainable in China. Also, the government of Australia has used their competitive grant scheme christened Renewable Energy Demonstration Programs (REDP) to encourage the development of large-scale RE and mini-grid project [80]. Nigeria can also key into such R&D strategy for accelerated RE development. Proper dissemination of information is also necessary to get the citizen more acquitted with the efforts of the government in the direction of RE.

9. Conclusion and recommendation

These articles review the renewable energy progress in Nigeria and also make a case for a proper implementation strategy. It is very clear from the current status of RE in Nigeria that its application in electric power generation has been very slow. The policy is well articulated, but innovative strategies are urgently needed for implementation. These measures which can give meaning to different policy paper will come in form of regulations, legislative framework, licensing arrangements for private-sector operators, Feed-in Tariffs and clarifying market rules for RE services and products.

The authors have discussed some policies, regulations and legislative frameworks that can speed up the development of RE for power generation in Nigeria. Moreso, RE could be the solution to the present problem of rural electrification. And it may be a major approach to get Nigeria out of the present economic situation.

Most of the Nigeria power generating facilities are operating at low efficiency. Therefore, the present Transmission and distribution network should be technically reviewed so as to accommodate renewable energy contribution to the national grid. Grid connected to the wind, solar, biomass power project can be developed on a fast track basis which can be commissioned within 1- 2 years. This could be a quick fix for the ongoing electricity crisis and contribute to improving the revenue and fiscal stability of the country. Since the plan is already underway for the implementation of the Nuclear power plant, the government of Nigeria should be careful to ensure a transparent feasibility study and analysis which is devoid of political interference. The safety of the nuclear power plant must be carefully addressed given the volatile security situation in the country.

The Implementation issues that have been identified should be holistically addressed. The implementation progress can be fully monitored by taking advantage of the Information Technology as have been successfully tried in other countries such as Malaysia. This will engender transparency and boost the confidence of the investors. Also, comprehensive studies should be conducted to implement renewable energy in rural and nongrid-connected areas of the country. This can be done through a participatory community-based business model. An energy policy which encourages efficiency should be devised. The manufacturing standard should be developed in such a way to eliminate energy-inefficient devices from the market.

A capacity building programs for all stakeholders at all levels of governance should be implemented. This will improve the performance of different governmental agencies that have roles in renewable energy implementation. The research funding for Renewable energy should be improved and monitored so as to identify the localized solution to disseminate the use of renewable energy.

REFERENCES

[1] USAID, "Nigeria Power Africa fact sheet," https://www.usaid.gov/powerafrica/nigeria *accessed on 13th June 2018,* 2018.

[2] D. Obadote, "Energy crisis in Nigeria: technical issues and solutions," in *Power sector prayer conference*, 2009, p. 4.

[3] A. S. Sambo, B. Garba, I. H. Zarma, and M. M. Gaji, "Electricity generation and the present challenges in the Nigerian power sector," *Journal of Energy and Power Engineering,* vol. 6, no. 7, p. 1050, 2012.
[4] D. o. E. a. S. A. United Nations, Population Division, " World Population Prospects: The 2017 Revision.," https://esa.un.org/unpd/wpp/Publications/Files/WPP2017_KeyFindings.pdf *(accessed on 25th, April, 2018),* 2017.
[5] O. Ismail SO, SJ, Olatunji OO, Ogunleye, IO. , " Design and Development of a Continuous Fluidized Bed Gari Dryer.," *International Journal of Engineering Innovations,* vol. 4, no. 3, pp. 54-65, 2012.
[6] N. Mellersh, "A scramble for power – the Nigerian energy crisis. African Law and Business. ," https://www.africanlawbusiness.com/news/5679 *accessed on 27/05/2016.,* 2015
[7] P. Newspaper, "Paralysis grips power sector," http://punchng.com/paralysis-grips-power-sector/ *accessed on 28/06/2016.,* 2016.
[8] V. Newspaper, "Electricity supply worsens in Q1' 2018," https://www.vanguardngr.com/2018/04/electricity-supply-worsens-q1-2018/ *accessed on 13th June 2018,* 2018.
[9] G. Ezirim, E. Onyemaechi, and O. Freedom, "The Political Economy of Nigeria's Power Sector Reforms: Challenges and Prospects, 2005-2015," *Mediterranean Journal of Social Sciences,* vol. 7, no. 4, p. 443, 2016.
[10] NAN, " Power generation drops to 2,662MW," *Punch newspaper, Sunday, 22 January, 2017. Available from:* www.punchng.com *Accessed on 22nd January,2017,* 2017.
[11] E. C. o. N. (ECN), ".FG to incur N177bn in electricity subsidy-NERC. ," www.energy.gov.ng *(accessed 4th.December.2011).*
[12] O. E., "Manufacturers need 2,000MW of electricity to Stay Afloat-MAN.," *Business Day,*www.Busnessdayonline.com*; 2009 (accessed 10th. December, 2011],* Tuesday, 21 July 2009.
[13] E. Efurumibe, "Barriers to the development of renewable energy in Nigeria," *Scholarly Journal of Biotechnology,* vol. 2, pp. 11-13, 2013.
[14] E. Okafor and C. Joe-Uzuegbu, "CHALLENGES TO DEVELOPMENT OF RENEWABLE ENERGY FOR ELECTRIC POWER SECTOR IN NIGERIA," *International journal of academic research,* vol. 2, no. 2, 2010.
[15] T. Olaoye, T. Ajilore, K. Akinluwade, F. Omole, and A. Adetunji, "Energy Crisis in Nigeria: Need for renewable energy mix," *American Journal of Electrical and Electronic Engineering,* vol. 4, no. 1, pp. 1-8, 2016.
[16] O. Omokaro, "Energy development in a fossil fuel economy: the Nigerian experience," *The report of a National Dialogue to Promote Renewable Energy and Energy Efficiency in Nigeria. 55p,* 2008.
[17] K. Bilen *et al.*, "Energy production, consumption, and environmental pollution for sustainable development: A case study in Turkey," *Renewable and Sustainable Energy Reviews,* vol. 12, no. 6, pp. 1529-1561, 2008.
[18] C. Chineke, R. Nwachukwu, O. Nwafor, E. Ugboma, and O. Ndukwu, "Much ado about little: Renewable energy and policy," *Journal of International Scientific Publications. Vol9,* 2015.
[19] U. Aliyu and S. Elegba, "Prospects for small hydropower development for rural applications in Nigeria," *Nigerian Journal of Renewable Energy,* vol. 1, no. 1, pp. 74-86, 1990.
[20] I. Vincent-Akpu, "Renewable Energy Potentials in Nigeria, Presented at the Energy Future the Role of Impact Assessment," in *32nd Annual Meeting of the International Association for Impact Assessment. Porto-Portugal: Centro de Congresso da Alfândega,* 2012.
[21] M. Shaaban and J. Petinrin, "Renewable energy potentials in Nigeria: meeting rural energy needs," *Renewable and Sustainable Energy Reviews,* vol. 29, pp. 72-84, 2014.

[22] Z. Abdmouleh, R. A. Alammari, and A. Gastli, "Review of policies encouraging renewable energy integration & best practices," *Renewable and Sustainable Energy Reviews,* vol. 45, pp. 249-262, 2015.
[23] O. S. Ohunakin, M. S. Adaramola, O. M. Oyewola, and R. O. Fagbenle, "Solar energy applications and development in Nigeria: drivers and barriers," *Renewable and Sustainable Energy Reviews,* vol. 32, pp. 294-301, 2014.
[24] S. O. Oyedepo, "Energy and sustainable development in Nigeria: the way forward," *Energy, Sustainability and Society,* vol. 2, no. 1, p. 15, 2012.
[25] Greenpeace, "Renewable energy myths," *http://www.greenpeace.org/africa/en/campaigns/Climatechange/renewable-energy-myths accessed on 21/08/2016.*
[26] P. Povinec, K. Hirose, and M. Aoyama, *Fukushima accident: radioactivity impact on the environment*. Newnes, 2013.
[27] B. B. Wittneben, "The impact of the Fukushima nuclear accident on European energy policy," *Environmental Science & Policy,* vol. 15, no. 1, pp. 1-3, 2012.
[28] J. Alimba, "Cost, demand and supply and price management in the power market," in *Workshop on Electricity Economics to PHCN Staff*, 2010, pp. 26-28.
[29] V. Newspaper, "Nigeria may bungle nuclear enrgy project," *http://www.vanguardngr.com/2016/06/nigeria-may-bungle-nuclear-energy-project/ accessed on 11/11/2016,* 2016.
[30] D. Post, "Defend Jonathan with facts, I resigned as Minister – Bart Nnaji blasts Omokri," *http://dailypost.ng/2017/08/18/defend-jonathan-facts-resigned-minister-bart-nnaji-blasts-omokri/ accessed on 19th August 2017.,* 2017.
[31] A. Castellano, A. Kendall, M. Nikomarov, and T. Swemmer, "Brighter Africa: The growth potential of the sub-Saharan electricity sector," *McKinsey Report. http://www. mckinsey. com/insights/energy_resources_materials/powering_africa,* 2015.
[32] H. Gitay, A. Suárez, R. T. Watson, and D. J. Dokken, "Climate change and biodiversity," 2002.
[33] A. F. Adenikinju, "Electric infrastructure failures in Nigeria: a survey-based analysis of the costs and adjustment responses," *Energy policy,* vol. 31, no. 14, pp. 1519-1530, 2003.
[34] I. Idemudia and D. Nordstrom, "Nigerian power sector: Opportunities and challenges for investment in 2016," *Latham & Watkins Africa Practice,* vol. 23, p. 1930, 2016.
[35] KPMG, "A Guide to the Nigerian Power Sector," *http://www.nigeriaelectricityhub.com/download/a-guide-to-the-nigerian-power-sector-kpmg-2013/. Accessed on 10/11/2016,* 2013.
[36] KPMG, "Taxes and Incentives for Renewable Energy.," *https://www.kpmg.com/Global/Documents/Taxes-Incentives-Renewable-Energy-2011.pdf 2011. accessed on 11th June, 2016,* 2011.
[37] E. I. A. (EIA), "Trends in Renewable Energy Consumption and Electricity,".*http://www.eia.gov/cneaf/solar.renewables. accessed on 30th August,2017,* 2008.
[38] O. S. Ohunakin, "Energy utilization and renewable energy sources in Nigeria," *Journal of Engineering and Applied Sciences,* vol. 5, no. 2, pp. 171-177, 2010.
[39] K. Simonyan and O. Fasina, "Biomass resources and bioenergy potentials in Nigeria," *African Journal of Agricultural Research,* vol. 8, no. 40, pp. 4975-4989, 2013.
[40] F. M. o. P. a. Steel, "Renewable Electricity Action Program (REAP).. ," *www.iceednigeria.org/backup/workspace/uploads/dec.-2006-2.pdfFederal Accessed on 6th August , 2015,* 2006.
[41] P. Nwofe, "Utilization of solar and biomass energy-A panacea to energy sustainability in a developing economy," *International Journal of Energy and Environmental Research,* vol. 2, no. 3, pp. 10-19, 2014.

[42] J. P. Painuly, "Barriers to renewable energy penetration; a framework for analysis," *Renewable energy,* vol. 24, no. 1, pp. 73-89, 2001.

[43] G. A. Agbaitoro, "Is having a robust energy mix a panacea for resolving the energy crisis in Nigeria?," *Renewable Energy Law & Policy Review,* vol. 7, no. 4, pp. 7-16, 2017.

[44] R. Bertani, "Geothermal power generation in the world 2010–2014 update report," *Geothermics,* vol. 60, pp. 31-43, 2016.

[45] R. Northwest, "Wave and Tidal Enegy technology," *http://www.rnp.org/node/wave-tidal-energy-technology accessed on 25/07/2015,* 2009.

[46] AENews, "Tidal Power," *http://www.alternative-energy-news.info/technology/hydro/tidal-power/ accessed on 25/07/2016,* 2018.

[47] E. C. o. N. E. a. U. N. D. P. (UNDP), "Renewable Energy master Plan (REMP) " *www.energy.gov.ng. Accessed on 6th August, 2015,* 2012.

[48] E. C. o. N. E. a. U. N. D. P. (UNDP), "Renewable Energy master Plan (REMP) (" *www.spidersolutionsnigeria.com/wp-content/. Accessed on August 6, 2015,* 2005).

[49] H. Gujba, Y. Mulugetta, and A. Azapagic, "Environmental and economic appraisal of power generation capacity expansion plan in Nigeria," *Energy policy,* vol. 38, no. 10, pp. 5636-5652, 2010.

[50] H. Winkler, "Energy policies for sustainable development in South Africa," *Energy for sustainable Development,* vol. 11, no. 1, pp. 26-34, 2007.

[51] H. Winkler, "Renewable energy policy in South Africa: policy options for renewable electricity," *Energy Policy,* vol. 33, no. 1, pp. 27-38, 2005.

[52] D. Banks and J. Schäffler, *The potential contribution of renewable energy in South Africa.* Sustainable Energy & Climate Change Project (SECCP), 2005.

[53] A. Pegels, "Renewable energy in South Africa: Potentials, barriers and options for support," *Energy policy,* vol. 38, no. 9, pp. 4945-4954, 2010.

[54] I. Bugaje, "Renewable energy for sustainable development in Africa: a review," *Renewable and Sustainable Energy Reviews,* vol. 10, no. 6, pp. 603-612, 2006.

[55] J. Krupa and S. Burch, "A new energy future for South Africa: The political ecology of South African renewable energy," *Energy Policy,* vol. 39, no. 10, pp. 6254-6261, 2011.

[56] B. Msimanga and A. Sebitosi, "South Africa's non-policy driven options for renewable energy development," *Renewable Energy,* vol. 69, pp. 420-427, 2014.

[57] D. R. Walwyn and A. C. Brent, "Renewable energy gathers steam in South Africa," *Renewable and Sustainable Energy Reviews,* vol. 41, pp. 390-401, 2015.

[58] C. Wereko-Brobby and I. Mintah, "Ghana's renewable energy development programme," in *Energy Conservation in Buildings*: Elsevier, 1991, pp. 172-176.

[59] G. S. Mensah, F. Kemausuor, and A. Brew-Hammond, "Energy access indicators and trends in Ghana," *Renewable and Sustainable Energy Reviews,* vol. 30, pp. 317-323, 2014.

[60] P. K. Adom and P. A. Kwakwa, "Effects of changing trade structure and technical characteristics of the manufacturing sector on energy intensity in Ghana," *Renewable and Sustainable Energy Reviews,* vol. 35, pp. 475-483, 2014.

[61] R. Arthur, M. F. Baidoo, and E. Antwi, "Biogas as a potential renewable energy source: A Ghanaian case study," *Renewable Energy,* vol. 36, no. 5, pp. 1510-1516, 2011.

[62] A. A. Z, "Senegal renewable readiness assessment," *The International Renewable Energy Agency (IRENA), Available from: (http://www.irena.org/Document Downloads/Publications/IRENA%20Senegal%20RRA.pdf); 2012.1– 76 accessed on 17th May,2017.,* 2012.

[63] D. Vilar, "Renewable energy in West Africa: status, experiences, and trends," *Casa África: ECREEE Canary Island Institute of Technology,* 2012.

IOP Conf. Series: Materials Science and Engineering **413** (2018) 012053 doi:10.1088/1757-899X/413/1/012053

[64] K. Ajao, H. Ajimotokan, O. Popoola, and H. Akande, "Electric energy supply in Nigeria, decentralized energy approach," *Cogeneration and Distributed Generation Journal,* vol. 24, no. 4, pp. 34-50, 2009.
[65] E. C. o. N. (ECN), " National Energy Policy (NEP) " *Federal Republic of Nigeria. www.energy.gov.ng. Accessed on 6th August, 2015.* , 2003.
[66] E. C. o. N. (ECN), "National Energy Policy (NEP)," *Energy Commission of Nigeria (ECN). Abuja: Federal Republic of Nigeria. Retrieved August 6, 2015, from www.energy.gov.ng,* (2006).
[67] E. C. o. N. (ECN), "National Energy Policy; Draft Revised Edition (NEP) " *Energy Commission of Nigeria (ECN), Abuja: Federal Republic of Nigeria. www.energy.gov.ng Accessed on 6th August, 2015, from,* 2013.
[68] F. M. o. P. a. Steel, "Renewable Electricity Policy Guidelines (REPG)," *www.iceednigeria.org/backup/workspace/uploads/dec.-2006.pdf. Accessed on 6th August, 2015,* (2006).
[69] F. M. o. P. a. Steel, "Renewable Electricity Action Program (REAP)," *International Center for Energy, Environment & Development, Abuja.* , 2006
[70] IMF, "National Economic Empowerment & Development Strategy (NEEDS) " *National Economic Empowerment and Development Strategy NEEDS (Vol. 2). International Monetary Fund.* , 2004.
[71] NNPC, " Draft Nigerian Bio-Fuel Policy and Incentives. ," *Nigerian National Petroleum Corporation, Abuja,* 2007.
[72] E. C. o. N. E. a. U. N. D. P. (UNDP), "Renewable Energy master Plan (REMP)," *www.spidersolutionsnigeria.com/wp-content/uploads/2014/08/Renewable-Energy-Master-Plan-2005.pdf Accessed on 6th August , 2015,* 2005.
[73] E. C. o. N. E. a. F. M. o. S. a. T. (FMST), "National Renewable Energy and Energy Efficiency Policy (NREEEP) " *www.energy.gov.ng.Retrieved Acessed on 6th August, 2015,* 2014.
[74] J. R. Oneal and B. Russett, "Assessing the liberal peace with alternative specifications: Trade still reduces conflict," *Journal of Peace Research,* vol. 36, no. 4, pp. 423-442, 1999.
[75] T. Nowotny, "Security and power through Interdependence: on the morality of globalisation," *Global Society,* vol. 21, no. 2, pp. 179-197, 2007.
[76] J. Lilliestam and S. Ellenbeck, "Energy security and renewable electricity trade—Will Desertec make Europe vulnerable to the "energy weapon"?," *Energy policy,* vol. 39, no. 6, pp. 3380-3391, 2011.
[77] D. Fouquet and T. Johansson, "Energy and environmental tax models from Europe and their link to other instruments for sustainability: policy evaluation and dynamics of regional integration," in *Presentation at the Eighth Senior Policy Advisory Committee Meeting, Beijing, China, November*, 2005, vol. 18.
[78] A. Ruijs and H. R. Vollebergh, "Lessons from 15 years of experience with the Dutch tax allowance for energy investments for firms," 2013.
[79] N. I. Meyer, "Learning from wind energy policy in the EU: lessons from Denmark, Sweden and Spain," *Environmental Policy and Governance,* vol. 17, no. 5, pp. 347-362, 2007.
[80] K. Hogg and R. O'Regan, "Renewable energy support mechanisms: an overview," *PricewaterhouseCoopers LLP, Globe Law and Business,* 2010.

Experimental investigation of thermal properties of Lignocellulosic biomass: A review

[1]Olatunji Obafemi, [2]Akinlabi Stephen, [3]Oluseyi Ajayi, [2]Peter Mashinini and [1]Madushele Nkosinathi

[1]Mechanical Engineering Science, University of Johannesburg, South Africa

[2]Mechanical & Industrial Engineering Technology, University of Johannesburg, South Africa

[3]Mechanical Engineering, Covenant University, Ota, Nigeria

Email address-tunjifemi@gmail.com

Abstract. Most recently, there is a renewed interest in the generation of energy from biomass given its abundance and its perceived ability as an alternative to a fossil fuel, which significantly contributes to greenhouse gas emission. Given this, lignocellulosic biomass has been acknowledged as the most abundant biomass stock for renewable energy generation across the globe. The socio-economic significance is also very overwhelming. This review article takes stock of different sources of lignocellulosic biomass and also different experimental procedure, which has been applied in the determination of the heating value and elemental composition. The review concluded that most of the research paper often echoed the enormous energy potential of biomass without thoroughly highlighting their contribution to greenhouse gas emission. It was also observed that it is only a few papers that have analyzed the effect of mixing on the heating value and elemental composition, while there are few direct comparisons of the biomass by geographical location. In overall the authors advocate for a more harmonized database for biomass.
Keywords: Lignocellulosic, proximate analysis, ultimate analysis, mixing, biomass.

1. Introduction

Among all the renewable energy options, biomass can be considered as the most abundant with the most significant potential, which can meet energy requirements of the globe and could provide energy supply in the future just as the world is earnestly seeking a way out of the consumption of fossil fuel. Also, sustainable development of a country decisively depends on the energy sustainability[1-3]. Biomass can be defined as an organic material which originates from plants and animals via a spontaneous or induced biological process. It is usually referred to certain types of wood, energy crops, marine algae, agricultural and silvicultural waste, and certain animal, industrial and human wastes[4, 5]. The phytomass and zoomass are often generally referred to as biomass.

Biomass exploration has the potential to employ thousands of people along the entire value chain. The growth of biomass power market has been supported by the gradual use of biomass in electricity generation. Based on political considerations, interest in bioenergy is driven by some major factors: increasing energy prices, energy security, rural development, in addition to the effect of climate change[6]. In the developing countries bioenergy may represent real time opportunities for domestic and industrial revolution towards economic growth given its abundance from various sources [7]. In a year, the production of biomass is around eight times the summation of other types of renewable energy sources [8] and it the only renewable resources which can be converted directly to liquid fuel [9].

Achieving sustainable fuel production means: developing fuels from raw materials that are locally sourced to avoid transportation over a long distance; that the production process does not encumber the land and take up so much water; and in overall, drastically reduce the Green House Gas (GHG) missions[10]. This is exactly where the biomass comes into play. Energy from biomass supplies 10% of global energy demands [7]. Biomass can be used in its natural state as fuel and can be refined to different kinds of biofuels, which are solid, liquid and gaseous fuels. These fuels can be applied in electricity production, transportation, heating and cooling across all the economic sectors of the society, considering criteria such as energy efficiency, applicability, environmental impact, and flexibility[11-13]. Therefore, bioenergy is now a global energy commodity.

As at 2011, European Commission (EU) had set a long-term goal to develop a competitive, resource efficient and low carbon economy by 2050 which is significantly built on bio-economy [13].From different scenarios that were analysed by EU, it was indicated that a cost effective economic pathway demands a 40% household decrease in Green House Gas(GHG) emissions for 2030 compared to 1990 levels, and 80% for 2050.The energy sector was identified as an important contributor to attaining cost effective economy[13]. Bioenergy production is expected to account for about 57 % of the renewable energy use in 2020, of which 45 % will consist in heat and electricity production from biomass, and 12 % will be provided by biofuels[14, 15]. Such was an audacious move intended to eradicate all non-bio-based material in European countries economy.

The aim of this present survey is to review the classification, availability and utilization of biomass, to review the properties, application and classification of lignocellulosic biomass; to describe the current trends in the experimental analysis of lignocellulosic biomass, to collect the most recent experimental results for the structural, elemental composition and heating value, and to identify the possible gaps in the experimental analysis of lignocellulosic biomass.

2. Biomass energy availability and utilization

The utilization of biomass has proven to be tremendously advantageous. The vast uncultivated land in most developing countries can be used to plant energy crops which can serve as a source of income and water retention of the soil under controlled conditions. The use of agricultural biomass residue can reduce the risk of forest burning and insect invasion while establishing a new market, but that has not been the case. For instance, billions of tonnes of straw and similar plant residues are annually available across the globe, but the utilization is very dismal. In each year, below 100 million tonnes are converted to energy while the remaining are usually burnt or left to decompose[7]. The consequent of this practice is the emission of greenhouse gases. This practice is the same, be it rice residue in Asia, sugar cane residue in Brazil, wheat stalk in Australia, soybean residue in Argentina, cotton in Egypt or palm oil in Indonesia and also the vast amount of higher moisture content biomass around the world, such as manures, abattoir wastes, and green leafy material[7].As shown in Figure 1, the global consumption of biofuel sourced from biomass has been on the increase, and this trend promises to continue provided all issues surrounding the exploration of bioenergy is frontally addressed.

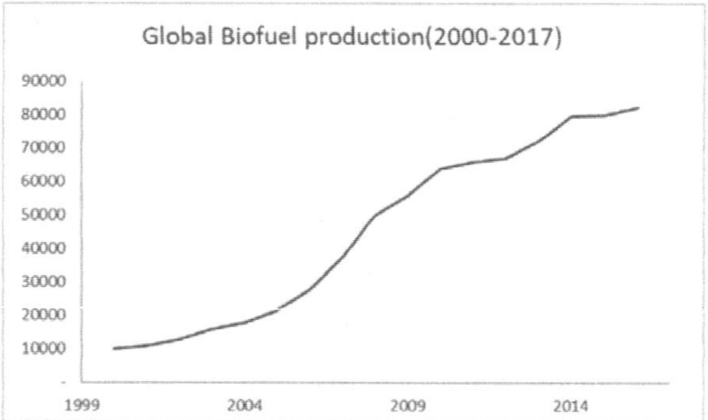

Figure 1:World biofuel production over sixteen years [16]

The growth of renewable energy applied across the globe in electricity generation since 1990 at the rate of 3.6 % per annum is slightly faster than total electricity generation growth rate, which stands at 2.9%. There has been a steady increase from 19.4 % in 1990 to 22.8 % in 2015.The contribution of the renewable energy used in the production of electricity increased from 1.3% in 1990 to 6.8% in 2015[16-18]. As shown in Figure 2, in 2015, renewable energy was the third largest contributor to global electricity production. They contributed 22.8% of world electricity generation, after coal (39.3%) and gas (22.9%) and ahead of nuclear (10.6%) and oil (4.1%)[16]. Apart from hydro which contributed 16% of renewable energy, biofuel and waste accounted for 1.9% of the total renewable energy contribution while others (solar, wind, geothermal, tidal) jointly accounted for 4.8%.

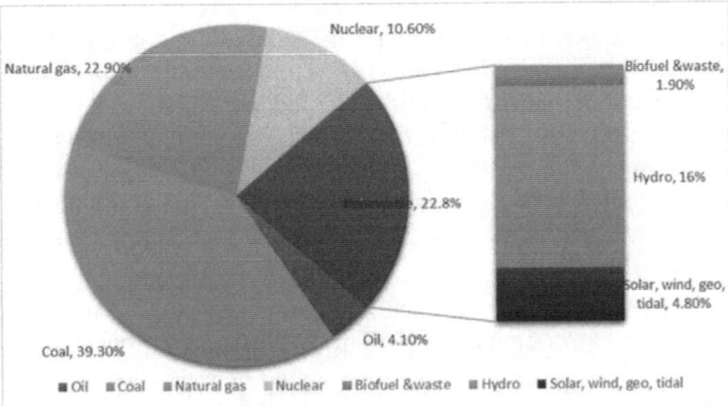

Figure 2:Fuel shares in global electricity production, 2015[16]

Globally, bioenergy which includes waste accounts for 14% of the world's energy consumption in 2012 with roughly 2.6 billion people dependent on traditional biomass to meet their energy needs[17]. The consumption and production pattern of bioenergy varies geographically. USA and Brazil lead the world in production and consumption of liquid biofuels for transportation (accounting for almost 80% of production). Figure 3, shows the biofuel production across the globe in the last sixteen years. The use of biomass for electricity which is predominantly sourced from forestry products and residues is common in Europe and North America [16]. The Europe and America contribute more than 70% of all consumption of biomass for electricity[19]. Although there has been some improvement in biomass to biofuel generation in Africa, it is interesting to note that Africa has not performed up to global expectation despite her enormous biomass resources. Africa has great potential to play a leading role regarding the biomass energy exploration given her large bio-resources, land mass and human capital. This implies that there is still so much to be done to improve the fortune of this continent in term of bioenergy.

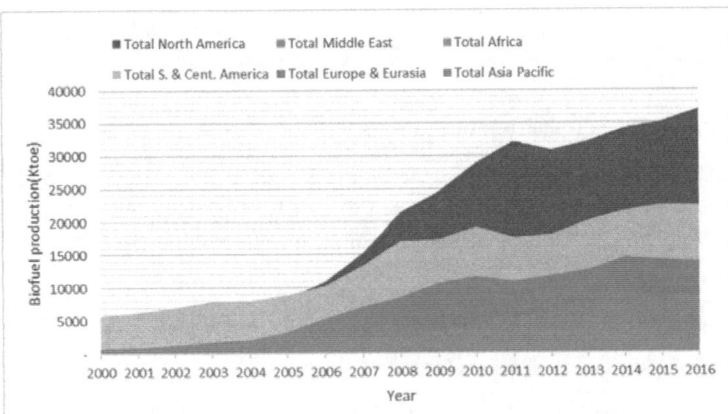

Figure 3:Biofuel Production in the last sixteen years for different regions in the world[18]

IOP Conf. Series: Materials Science and Engineering **413** (2018) 012054 doi:10.1088/1757-899X/413/1/012054

In the transport sector, the production of corn ethanol in USA and sugarcane ethanol in Brazil has increased significantly. The production of all biofuels in America increased from about 16 billion litres in 2000 to 79 billionlitres in 2012[16]. A significant sector for future use of biofuels is the aviation sector. Liquid biofuels are the only sustainable and viable option for replacing aviation fuel and efforts are underway where airlines, airports, finance institutions, and universities are coming together to explore sustainable aviation pathways. Commercial airlines using biofuels have already flown transatlantic routes. Other important sectors include heavy road and maritime where biofuels can play a big role. In 2013, 462 TWh of electricity was produced globally from biomass[19]. In the past few years, biomass uptake in developing countries such as Asia and Africa where significant population lacks access to electricity has increased. Biogas and decentralized bioenergy systems are becoming more cost competitive. Already co-generation plants using agricultural residues like Bagasse in India, Mauritius, Kenya, and Ethiopia are successful. Currently, the major use of biomass is in the form of heat in rural and developing countries[20].

The usability of these biomasses has been a subject of intense investigation. Several researchers have attempted to investigate different properties of biomass using various approaches. They include the elemental composition, the heating value, physical composition, chemical composition and little about the environmental impact of the exploration of these biomasses for energy production[21-38].

3. Classification of Biomass Resource

In view of the diversity of the sources of biomass, a classification which eases the process of prediction of the properties of biomass is very vital. Khan *et al.*[39]proposed two classifications of biomass; based on origin and properties. These have been discussed in[11] and it is outlined here as follows;

Table 1:Classification of biomass

Based on Origin	Based on properties
Primary waste, e.g. grain waste, wood waste	Wood and woody fuel (newly harvested soft and hardwood
Secondary waste, e.g. food processing residue and sawdust.	Agricultural biomass (Stray, stalk, grass, Stover)
Tertiary waste, e.g. wood used in construction	Aquatic biomass (Algae)
Energy crop, e.g. cassava tuber	Animal and human waste (bones, manure, etc.)
	Municipal biomass (solid waste, organic waste, sludges, refuse.
	Biomass mixtures

3.1 Factors affecting the composition of biomass.

The composition and thermal properties of the biomass is determined by an interplay of several factors. These factors affect the characteristics of biomass to a different extent. While some play very significant roles, the effect of others is negligible. These factors have been outlined as follows[28];

- Types of biomass
- Parts of biomass plant and the species
- Growth process and growing conditions
- Geographical location
- Climate and seasons
- Soil types

- Water and nutrients and soil pH
- Age of plant
- Presence of pollutants and contaminants and the proximity to the pollutants
- Harvesting time and technique
- Transport and storage condition
- Mixing of biomass from multiple sources

The wood species grown in different geographical location showed slight differences in their elemental composition. Also, contaminants are introduced during transportation and industrial processing of natural biomass. This contaminant maybe dust particles and various left over from construction material, plastic rubbers, metals, chemicals, char, detergent, paints, coloured paper and so on[27, 40]. With regards to the mixing of biomass and presence of contaminants, combination of biomass from diverse sources maybe an advantage when it concerns the generation of sufficient fuel which can be applied on an industrial scale. Since a single source and the cost of sorting of biomass from multiple sources may not be sustainable in the long run, the effect of mixing and contamination should be a subject of more scientific investigations.

From the above and other related factors Vassilev *et al.*[28]identified some issues concerning the investigation on biomass. These can be summarized as follows: there is incomplete information in regards of life cycle assessment of most biomass which has been reported in the literature; in most cases, investigations do not detail the environmental impact of the selected biomass. However, there has been deliberate attempt to standardize the testing procedure so as to ensure uniformity in testing procedure and reportage[25, 41, 42]. Also, there is insufficient information on the biomass processing technique and ambiguous definition of resources potential. For instance, the location of the biomass, the manner in which they are collected, the storage process, and the exact condition of the biomass before analysis not often clearly outlined. This information is very vital in the analysis and application of a particular class of biomass on the industrial scale.

4. Lignocellulosic Biomass

Lignocellulosic materials are abundantly available and usually not expensive. From the literature survey, global annual production of Lignocellulosic biomass stands at around 181.5 billion tonnes[43]. About 50 % of the global biomass resources are made of Lignocellulosic biomass which is estimated at 3×10^{14} kg[44]. Several studies have confirmed that Lignocellulosic biomass holds great potential for sustainable energy generation and it is globally available, be it in the developing countries or developed countries[45]. Also, cellulose which is a vital component of Lignocellulosic biomass has a real potential to displace petroleum based fuel products[46]. Lignocellulose is a generic term for describing the main constituents in most plants, namely cellulose, hemicellulose, and lignin. It is a complex matrix, comprising many different polysaccharides, phenolic polymers, and proteins. Cellulose, the major component of cell walls of land plants, is a glucan polysaccharide containing large reservoirs of energy that provide real potential for conversion into biofuels. Lignocellulosic biomass consists of a variety of materials with distinctive physical and chemical characteristics. It is the non-starch based fibrous part of plant material.

First-generation biofuels (produced primarily from food crops such as grains, sugar beet, and oilseeds) are limited in their ability to achieve targets for oil-product substitution, climate change mitigation, and economic growth. Their sustainable production is under the scanner since it has the possibility of creating undue competition for land and water used for food and fibre production.

These concerns have increased the interest in developing biofuels produced from non-food biomass feedstock from lignocellulosic materials which includes cereal straw, bagasse, forest residues, and purposely grown energy crops such as vegetative grasses and short rotation forests [28]. These second-

IOP Conf. Series: Materials Science and Engineering **413** (2018) 012054 doi:10.1088/1757-899X/413/1/012054

generation biofuels could avoid many of the concerns facing first-generation biofuels and potentially offer greater cost reduction potential in the longer term.

Lignocellulosic materials consist mainly of three polymers: cellulose, hemicellulose, and lignin. These polymers are associated with each other in a hetero-matrix to different degrees and varying relative composition depending on the type, species, and even source of the biomass. The relative abundance of cellulose, hemicellulose, and lignin are key factors in determining the optimum energy. Extensive discussion has been done on lignocellulosic biomass by several researchers.[34, 37, 43, 47, 48]. Therefore, a brief description of the classes is sufficient in this survey.

4.1 Brief description of the major components of lignocellulosic biomass

4.1.1 Cellulose

Cellulose is the most abundant member of the trio, followed by hemicellulose and lignin[43]. Its structure is made of tightly knighted hydrogen bond due to complex inter and intramolecular relationship. Since about half of the organic carbon in the biosphere is present in the form of cellulose, the conversion of cellulose into fuels and valuable chemicals is a vital process in energy generation from biomass[49].

4.1.2 Hemicellulose

Hemicellulose is the second most abundant polymer. Unlike cellulose, hemicellulose has a random and amorphous structure, which is composed of several heteropolymers including xylan, galactomannan, glucuronoxylan, arabinoxylan, glucomannan and xyloglucan. It is a heterogeneous group of branched polysaccharides which serve as a connecting medium between the lignin and cellulose[34]

4.1.3 Lignin

Lignin is a complex, large molecular structure with a cross-linked phenolic polymer which is linear with no regular repeating units. Lignin serves as a binder between hemicellulose and cellulose in the cell wall [37]. Softwoods have a greater lignin content than hardwoods[37].

In energy generation from biomass, especially biochemical conversion process, the relative amount of cellulose and lignin are major determining factors for their suitability[50].

Table 2 shows the structural composition of some Lignocellulosic biomass. For most of these biomasses, the biodegradation of cellulose is greater than lignin and Hemicellulose. It should be noted that the conversion of cellulose is higher than lignin due to its biodegradation[51].

Table 2: Lignocellulosic biomass and their structural composition[37, 52-54]

Lignocellulose biomass class	Source	Lignin	Hemicellulose	Cellulose
Wood biomass	Oak	35.4	21.9	43.2
	Eucalyptus	21.5	18.4	54.1
	Pine	26.8	24.0	45.6
	Spruce	27.5	21.2	50.8
	Poplar	15.5-16.3	26.2-28.7	58.8-53.3
	Douglas Fir	27.0	11.0	44.0
	Ailanthus wood	22.2	22.6	46.7

IOP Conf. Series: Materials Science and Engineering **413** (2018) 012054 doi:10.1088/1757-899X/413/1/012054

	Albizia	33.8	6.7	59.5
	Birchwood	15.7	40.0	25.7
	Beechwood	21.9	31.8	45.8
	Furniture sawdust	22.16	32.63	37.23
	Subabul wood	24.7	24.0	39.8
	Oak	35.4	21.9	43.2
	Pine	26.8	24.0	45.6
	Spruce	27.5	21.2	50.8
	Wood chips	19.0	31.8	31.8
	Wood bark	31.0	47	22
Agricultural residue	Corn cob	15-16.6	28.7-35	40.3-45
	Cornstalk	15.59	43.01	22.82
	Corn Stover	14.4	30.7	51.2
	Bagasse	18.3	22.6	41.3
	Cashew nutshell	40.1	18.6	41.3
	Banana waste	14.0	14.8	13.2
	Barley straw	27.7	29.7	48.6
	Tea waste	40.0	19.9	30.2
	Rice Husk	14.3	24.3	31.3
	Rice straw	13.6	22.7	37.0
	Millet husk	14.0	26.9	33.3
	Sorghum bagasse	10.0	24.0	41.0
	Hazelnut shell	51.5	15.7	22.9
	Hazelnut seed coat	53.0	15.7	29.6
	Groundnut shell	30.2	18.7	35.7
	Coconut shell	28.7	25.1	36.3
	Nuts shell	30-40	25-30	25-30
	Flax straw grape residue	28.9	34.40	36.70
		12.1	34.40	36.30
	Tobacco leaf	27.0	28.2	42.4
	Tobacco stalk			

	Bast fibre seed flax	23	25	47
	Bast fibre jute	21-26	18-21	45-53
	Bast fibre kenaf	15-19	22-23	31-39
	Coffee pulp	18.8	46.3	35
	Leaf fiber Abaca	8.8	17.3	60.8
	Leaf fiber Sisal	7-9	21-24	43-56
Energy grasses	Giant reed	19.0-17.6	30.0-29.5	37.0-36.1
	Switchgrass	19.1-17.8	27.2-27.8	38.2-36.5
	Pennisetum	18.5-16.5	22.53-21.93	41.8-40.9
	Silver grass	17.5-17.1	26.2-25.6	44.1-43.3
	Cat grass(orchard)	4.70	32	40
	Phalaris arundinacea	7.6	42.6	29.70
	Alfa grass	17-19	27-32	33-38
	Willow copies	20	49.3	14.1
	Orchard grass	4.7	40	32
	Water hyacinth	3.5-3.55	48.7-49.2	18.2-18.4
Energy crop	Bamboo	21-31	15-26	26-43
	Sugarcane	22.9	31.3	45.8
	Jerusalem artichoke (October)	5.70	25.99	4.50
	Jerusalem artichoke (September)	5.05	20.95	5.48
	Hemp	8.76	53.86	5.18
	Silage	9.02	39.27	25.96
Municipal Waste	Urban wood	-	-	-
	Urban greening	22.73	22.96	6.86
	Municipal biomass	-	-	-
	Sewage sludge	-	-	-
	Waste material	24.7	29.2	50.6

Newspaper	18-30	25-40	40-55
Kraft paper	20.8	9.9	57.3
General MSW	10-14	9-16	33-49
Food waste	11.4	7.2	55.4
Sorted refuse	20	20	60

4.2 Main Lignocellulosic biomass conversion route

Figure 4 shows the main conversion route for lignocellulose biomass. Three main processes through which energy can be generated from Lignocellulosic biomass were identified. Combustion is mostly used since it accounts for almost 97% of the bioenergy which is harnessed across the world[55]. It is particularly important in developing countries [56]. Combustion can be divided into three stages, which are drying, pyrolysis, and reduction. The last stage can generate up to 70% of the total heat of all the stages[37]. It is advised that only true end wastes should be applied in energy production in order to ensure that fixed CO_2 is not returned early into the atmosphere[57].

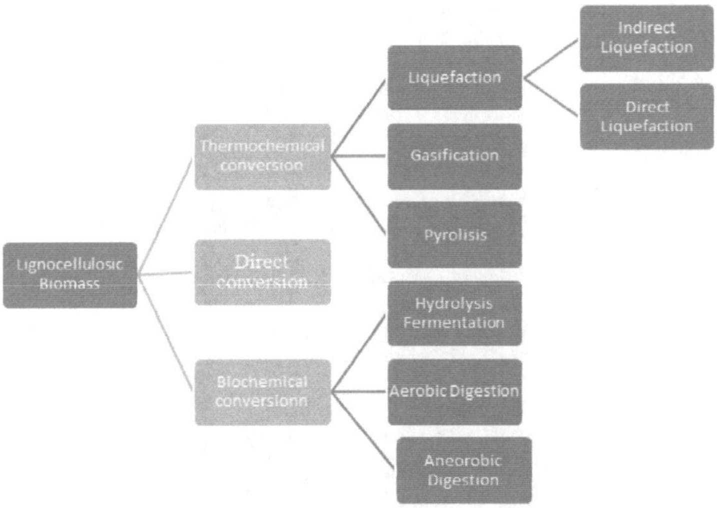

Figure 4:Main Lignocellulosic biomass conversion route [37, 57]

5. Environmental impact assessment of Lignocellulosic biomass

While researchers often focus on the thermal properties, elemental composition and chemical composition of Lignocellulosic biomass, little emphasis is laid on the environmental impacts about their contribution to greenhouse gas emission. Performance analysis of any biomass source must include a detailed life cycle assessment[42], which should include its environmental impact. It should be noted that greenhouse gas (GHG) is present in biomass even if it is in traces. The multiple effects of these maybe very significant when considering biomass for industrial energy generation.

One important factor which is often overlooked when considering the use of biomass to reduce global warming is the time lag between the instantaneous release of CO_2 from burning fossil fuels and its eventual uptake as biomass, which can take many years[25]. One of the dilemmas facing the developed world is the need to recognize this time delay and take appropriate action to mitigate against this lag period. An equal dilemma faces the developing world as it consumes its biomass resources for fuel but does not implement a programme to ensure replacement planting.

The advocates of biomass energy based their argument on the "carbon neutrality" of biomass. This means there is no net contribution of carbon dioxide to global warming [25, 58]. As much as this argument may be valid, biomass is not just a lump of carbon; it contains sulphur, nitrogen and several other vital nutrients in different ratios. An intensive cultivation of biomass and usage of such for energy generation has other implications apart from carbon capture. As at 1999, agricultural activities have been reported to generate more than 75 % of emitted reactive nitrogen compounds[59].The significance of this value will be appreciated more if it is understood that each molecule of nitrogen oxide has 300 times global warming potential when compared with carbon dioxide[60]. Therefore, the contribution of each potential GHG elemental component of biomass must be accounted for because of the cumulative effect for industrial-scale biomass energy plant.

The combustion of biomass produces significantly fewer nitrogen oxides and sulphur dioxide than the burning of fossil fuels. But, if a forest region is indiscriminately cleared for fuel, the CO_2level will increase because its released into the atmosphere is not recycled for new growth[61].On the basis of Table 3, an attempt was made to review the research articles on the experimental investigation of the heating value and elemental composition of lignocellulosic biomass and to identify the geographical locations from where the biomasses were sourced. Also, the review considers the extent to which emphasis was placed on the GHG emission potential from biomass utilization for energy generation.

Table 3: Review of the contribution on geographical location and GHG

Author	Lignocellulosic Biomass samples group and variety	Geographical Location of biomass samples	Contribution to GHG emission	Comment
	Wood and woody biomass			
[62]	Downy Birch bark	South Norway	Not highlighted	The information supplied on the location of biomass is insufficient.
[63]				
[23]	Olive husk	Turkey	Not highlighted	
[64][12]	Hazelnut husk Hazelnut seed coat Spruce wood Beechwood Ailanthus wood	Baghchung reserve forest, Assam, India		
[64]	Albizia	Baghchung reserve forest, Assam, India	Not highlighted	
[65]	Furniture sawdust	Not stated	Not highlighted	
[66, 67]	Oak	Not stated		
[17]	Pinewood	Saskatchewan forest, Canada	Not in the objective.	The study site was clearly described
	Pine sawdust	Huazhong University, China	Not in the objective.	
[68]	Eucalyptus bark	Undonthani, Thailand		
[26]	Eucalyptus wood Subabul wood			
	Agricultural residue Banana waste			
[48]	Paddy straw	Indonesia		

[37, 48, 69, 70]	Bagasse	Wakiso District, Uganda, Ombo forest, Kakamega frest, Aloso forest,		
[71, 72]	Sugarcane leaves Sugarcane tops Cassava stalk Cassava rhizome	North east, Thailand		
[37, 73]	Corncob	Rosenberg, Germany and Ado Ekiti, Nigeria		
	Tea waste	Turkey		
[23, 69]	Corn Stover Corn stalk Tobacco leaf Tobacco stalk	Turkey		
[23, 66]	Wheat straw	Turkey and Saskatchewan forest, Canada	Not in the objective.	The study site, Test procedure and the equipment were clearly discussed.
[66]	Barley straw	Saskatchewan forest, Canada	Not in the objective.	The study site was clearly described
[66]	Flax straw	Saskatchewan forest, Canada		
[69]	Coconut coir Coconut shell Millet husk Rice husk Rice straw Groundnut shell	Bombay, India	Not in the objective.	The study site was not clearly described
[28]	Oat straw			
[74]	Cotton waste Soybean waste	Maharashtra, India	Not in the objective	The information supplied on the location of biomass is insufficient.
[75]	Kernel olive Eucalyptus sawdust	Morocco	Not in the objective	The information supplied on the location of biomass is insufficient.
	Energy grasses			
[76]	Willow coppice Switchgrass Elephant grass Orchard grass Esparto grass Wild reed	Not state	Not in the objective	The study site was not clearly described
[77]	Timothy grass Hemp Straw Reed Reed canary grass	Latvian	Not in the objective	The information supplied on the location of biomass is insufficient.
[48]	Imperata cylindrical Eragrostisairoides	Assam and Manipur, North -East India	Not in the objective	The study site was clearly described.

	Typha angustifolia Arundinella khasiana Echinochiastagnina			
	Municipal waste Newspaper	Not stated	Not in the objective	The information supplied on the location of biomass is insufficient
[78]	Sewage sludge Physical chemical sludge Biochemical sludge	Nanjing, China	Not in the objective	
	Cardboard Plastic Synthetic rubber	Morocco	Not stated	The information supplied on the location of biomass is insufficient.
[77]	**Biomass Mixture** Hemp and peat (15%,30%,50%) Reed and peat (15%,20%,30%) Reed canary grass (15%,30%,50%)	Latvian	Not stated	The information supplied on the location of biomass is insufficient.
[79]	Wood-agricultural residue Wood-almond residue Wood and straw residue	Not known	Not stated	The study site was not clearly described
[75]	Biochar and kernel olive (25%/75%:50%/50%:75%/25%) Biochar and manure (25%/75%:50%/50%:75%/25%) Biochar and Eucalyptus sawdust (25%/75%:50%/50%:75%/25%) Biochar and sugar cane (25%/75%:50%/50%:75%/25%) Biochar and wood sawdust (25%/75%:50%/50%:75%/25%) Activated Biochar and Kernel oil (25%/75%:50%/50%:75%/25%) Activated biochar and cattle manure (25%/75%:50%/50%:75%/25%) Activated biochar and eucalyptus sawdust (25%/75%:50%/50%:75%/25%) Activated biochar and sugar cane (25%/75%:50%/50%:75%/25%) Activated biochar and wood sawdust (25%/75%:50%/50%:75%/25%) Activated biochar and alfa (25%/75%:50%/50%:75%/25%)	Morocco	Not stated	The information supplied on the location of biomass is insufficient.
[79-82]	**Contaminated biomass** Shredded currency Mixed waste paper Demolition wood Greenhouse-plastic waste Refuse derived fuel.	Not stated	Not stated	

IOP Conf. Series: Materials Science and Engineering **413** (2018) 012054 doi:10.1088/1757-899X/413/1/012054

5.1 Proximate and Ultimate analysis of Lignocellulosic biomass

The proximate analysis, ultimate analysis and High temperature composition of biomass have been detailed in [28, 37]. It is briefly reviewed in this article;

5.1.1 Proximate analysis

Proximate analysis can give an extensive array of information on the characteristics of biomass samples. It classifies the biomass in term of moisture, fixed carbon, volatile matter, and ash content. The moisture content (M) is an indication of the amount of water present in a biomass sample. It is frequently used in the proximate analysis to show that the water content of biomass is a mineralized aqueous solution containing cation, anion or non-charged species. Moisture content determines the conversion factor between the different bases of analysis of biomass data. This can be expressed as a percentage of the biomass weight which are as received, air dried, oven dried and dry ash free basis[34, 83]. As received basis is determined from the gross weight of the sample as it arrived at the laboratory and prior to any pre-treatment. Air dried basis is based on the equilibrium condition between the sample and atmospheric humidity with only consideration been the inherent biomass moisture. Dry basis does not consider all the moisture component of biomass, be it external or inherent. Dry ash free basis is on the condition that there is neither moistures nor ash. Figure 4, shows the transformation principle between the different bases of biomass moisture content analysis.

Ash (A) content is the leftover of a complete combustion of biomass with the primary components which includes; oxides of silica, aluminium, iron, calcium, titanium, sodium, potassium. Ash content is among the most investigated properties of biomass[28].However, the researchers have not gained full understanding due to its complex behaviour. Ash content can be used to determine the bulk inorganic matter, prevailing affinity of elements and compound of organic and inorganic matter, and the probability of contamination[28]. The understanding of the physicochemical characteristics of ashes is helpful in determining the propensity of a biomass source to form a deposit in the boiler[84].

Table 4: Transformation between different basis of biomass moisture content analysis [34]

Wanted / Given	As received (AR)	Air dried (AD)	Dry (DB)	Dry ash free (DAF)
As received	1	$\frac{1-m_{AD}}{1-m_{AR}}$	$\frac{1}{1-m_{AR}}$	$\frac{1}{1-m_{AR}-A_{AR}}$
Air dried	$\frac{1-m_{AR}}{1-m_{AD}}$	1	$\frac{1}{1-m_{AD}}$	$\frac{1}{1-m_{AD}-A_{AD}}$
Dry basis	$1-m_{AR}$	$1-m_{AD}$	1	$\frac{1}{1-A_{AD}}$
Dry ash free	$1-m_{AR}-A_{AR}$	$1-m_{AR}-A_{AR}$	$1-m_{DB}$	1

5.1.2 Ultimate analysis

The ultimate analysis is the determination of elemental composition of biomass. The result from the ultimate analysis is more detailed than proximate analysis. The result of this analysis can be used to determine the heating value of biomass [85].The aim of ultimate analysis of biomass is to determine the organic components which are C, H,N,S, Cl and O and inorganic constituents which are Si, Al, Ti, Fe, Ca, Mg, Na, K, S and P [37, 38, 48].The amount of N, S, Cl can give a clue on the environmental impact of biomass consumption. Different models of the elemental analyser can be used in the ultimate analysis of biomass [86]. However, this procedure is more expensive in comparison with proximate analysis.

The Table 5shows the heating value and elemental composition of various kinds of biomass. The mixing ratios were described in [75, 76]. However,[75]results show that mixing of several biomass materials has some positive effect on the energy content and the heating value but nothing was reported on the elemental composition and proximate values. Meanwhile [76], only reported the elemental composition of the biomass blends which was selected without making mention of the heating value.

Table 5: Literature review of Experimental analysis of biomass

Author	**Lignocellulosic Biomass group**	**Heating value(MJ/kg)**	**Proximate Analysis (%)**				**Elemental composition (%)**				
			M	**VM**	**FC**	**Ash**	**C**	**H**	**N**	**S**	**O**
	Wood and woody biomass										
[62] **[63]**	Downy Birch bark	18.50-18.72	11.13	78.7	20.9	0.3	48.4	5.6	0.2	-	-
[23]	Olive husk	19.2	9.2	70.3	26.1	3.6	50.0	6.2	1.6	-	**42.2**
[64][12]	Hazel nut husk		9.0	69.3	28,3	1.4	52.9	5.6	1.4		**42.7**
	Hazelnut seed coat	19.4	6.8	71.2	27.0	1.8	51.0	5.4	1.3	-	**42.3**
	Spruce wood	19.7	7.6	70.2	28.3	1.5	51.9	6.1	0.3		**40.9**
	Beech wood	17.4-18.9	7.4-9.7	74.0	24.6	0.4	49.5	6.2	0.4	-	**40.9**
	Ailanthus wood	19.0	8.1	73.5	24.8	1.7	49.5	6.2	0.3	-	**41.0**
[64]	Albizia	17.4	9.7	72.7	25.5	1.8	46.4	5.8	0.6	1.7	**45.5**
[65]	Furniture sawdust	15.79	7.1	79.43	12.69	0.78	47.42	5.67	0.2	-	**46.71**
[67]	Oak		0.6	89.7	9.5	0.2	-	-	-	-	-
[66]	Pinewood	19.6	5.8	82.4	10.3	1.5	49.0	6.4	0.14	0.01	**44.4**
	Pine sawdust	20.54	13.58	70.23	15.06	1.12	50.54	7.08	0.15	0.57	**41.11**
[68]	Eucalyptus bark	15.7	10.7	76.1	19.7	10.7	38.7	4.5	0.3	<1.8	**54.9**
[26]	Eucalyptus wood	18.6	7.7	82.6	16.4	1.0	48.7	6.2	0.3	-	**44.8**
	Subabul wood	19.78	-	85.6	-	0.9	48.2	5.9	0	-	**45.1**
	Agricultural residue										
	Banana waste										
[48]	Paddy straw	-	7.3	56.4	15.4	20.9	48.75	5.98	1.99	-	**43.28**
[37, 48, 69]	Bagasse	14.67-16.29	13.2	71.0-84.2	13.8	2.1-2.9	43.8-51.71	5.32-5.8	0.33-0.4		**42.64-47.1**
[71, 72]	Sugarcane leaves	18.4	6.7	79.0	8.6	5.7	48.9	6.5	0.2	-	**44.4**
	Sugarcane tops	18.3	6.6	74.9	12.5	6.0	49.0	6.6	0.6	-	**43.8**
	Cassava stalk	18.1	8.5	69.7	14.7	7.1	48.8	6.7	1.1	-	**43.4**
	Cassava rhizome	21.7	8.8	65.0	15.0	11.2	49.5	6.5	1.1	-	**42.9**
[37]	Corncob	16.6	12.1	86.5	12.5	1.0	49.0	5.4	0.3	-	**44.6**

	Tea waste	16.8	6.5	85.0	13.6	1.4	48.6	5.5	0.5	<0.1	**40.9**
[23, 69]	Corn Stover	17.6	10.6	78.7	17.6	3.7	42.6	5.06	0.83	0.09	**36.52**
	Corn stalk	16.54		80.1	-	6.8	41.9	5.3	0	-	**46.0**
	Tobacco leaf	16.3	8.4	72.6	11.2	17.2	-	-	-	-	**-**
							-	-	-	-	**-**
	Tobacco stalk	17.6	8.9	79.6	18.0	2.4	-	-	-	-	**-**
[23, 66]	Wheat straw	18.7-20.3	6.0-8.5	63.0-78.3	14.4-23.5	1.3-13.5	41.6-45.5	5.1-6.1	0.14-1.8	0.06	**34.1-52.115.7**
[66]	Barley straw	15.7	6.9	78.5	4.8	9.8	41.4	6.2	0.63	0.01	**51.7**
[66]	Flax straw	17.0	7.9	80.3	8.8	3.0	43.1	6.2	0.68	0.09	**49.9**
[69]	Coconut coir	14.67	-	82.8	-	0.9	47.6	5.7	0.2	-	**45.6**
	Coconut shell	20.50	-	80.2	-	0.7	50.2	5.7	0.2	-	**45.6**
	Millet husk	17.48	-	80.7	-	18.1	42.7	6.0	0.1	-	**33.0**
	Rice husk	15.29	-	81.6	-	23.5	38.9	5.1	0.6	-	**32.0**
	Rice straw	16.78	-	80.2	-	19.8	36.9	5.0	0.4	-	**37.9**
	Groundnut shell	18.65	-	83.0	-	5.9	48.3	5.7	0.8	-	**39.4**
[28]	Oat straw		8.2	73.9	12.5	5.4	48.8	6.0	0.08	0.5	**44.6**
[74]	Cotton waste	16.65	4.8	72.05	20.05	3.1	40.6	6.0	0.19	0.9	**53.2**
	Soybean waste	18.77	5.8	70.5	19.0	4.7	43.8	6.3	1.4	0.8	**48.5**
[75]	Kernel olive	20.54	-	-	-	-	47.63	6.41	0.97	0.08	**44.92**
	Eucalyptus sawdust	16.30	-	-	-	-	44.80	6.03	0.84	0.03	**48.30**
	Energy grasses										
	Willow coppice	20	2.8	87.6	17.3	6.3	49.9	6.5	0.2	-	**39.9**
[76]	Switch grass	16.3	9.84	69.14	12.93	8.09	42	5.24	0.60	0.17	**33.97**
	Elephant grass	15.61	10.04	65.0	14.66	6.0	44.50	5.4	1.4	-	**31.8**
	Orchard grass	-	-	-	-	-	56.1	5.9	0.4	0.03	**37.6**
	Esparto grass	19.1	5.2	80.5	16.8	2.2-8	46.94	6.44	0.86	0.0	**43.56**
	Wild reed	24.98	4.34	78.57	12.54	4.54	48.7	6.96	0.36	-	**42.96**
[77]	Timothy grass	16.7	5.0	77.9	16.0	1.1	42.4	6.0	1.03	0.15	**50.4**
	Hemp	15.54	8.75	-	-	-	2.97	-	-	-	**-**
	Straw	16.02	8.40	-	-	-	2.10	-	-	-	**-**
	Reed	15.94	9.16	-	-	-	2.76	-	-	-	**-**
	Reed canary grass	15.12	15.68	-	-	-	5.13	-	-	-	**-**
[48]	Imperata cylindrical	-	8.55	84.14	0.36	6.95	50.04	5.92	1.15	-	**42.9**
	Erasgostis airoides	-	8.275	86.84	1.23	3.6	41.02	6.72	1.14	-	**51.12**
	Typha angustifolia	-	13.95	80.06	2.18	3.81	52.90	5.84	1.22	-	**40.04**

	Arundinella khasiana	-	10.37	80.06	2.18	8.12	41.26	5.40	1.25	-	**52.10**
	Echinochia stagnina	-	10.27	83.19	0.41	6.13	44.98	5.66	1.86	-	**45/50**
	Municipal waste										
	Newspaper	-	-	-	-	-	-	-			
[78]	Sewage sludge	6.56	-	29.01	3.49	67.50	12.79	1.74	1.20	0.55	**16.22**
	Physical chemical sludge	19.10	-	69.1	0.5	30.4	32.78	5.81	0.04	0.91	**60.09**
	Biochemical sludge	17.60	-	65.53	6.6-1	27.86	31,75	5.76	1.54	1.88	**59.08**
	Cardboard	13.81	-	-		-	38.49	5.68	0.82	0.09	**54.92**
	Plastic	44.81	-	-	-	-	82.61	14.01	0.74	0.00	**2.64**
	Synthetic rubber	37.82	-	-	-	-	83.96	7.93	0.97	1.14	**6.00**
	Biomass Mixture										
[77]	Hemp and peat (15%,30%,50%)	15.52,15.79,16.25	9.08, 10.30,11.53	-	-	2.60,2.92,2.88		-	-	-	--
	Reed and peat (15%,20%,30%)	15.84,15.68,14.89	10.33,10.53, 14.10			3.20,2.83.2.50	-	--	-	-	-
	Reed canary grass (15%,30%,50%)	15.16	16.13	-	-	4.33	-	-	--	-	--
[79]	Wood-agricultural residue	-	30.3	54.7	12.7	2.3	56.7	6.6	2.7	0.87	**33.1**
	Wood-almond residue	-	22.7	59.7	12.3	5.3	50.9	5.9	0.6	0.08	**50.9**
	Wood and straw residue		7.3	69.6	15.5	7.6	51.7	6.3	0.4	0.13	**41.5**
[75]	Biochar and kernel olive (25%/75%:50%/50%:75%/25%)	23.0/25.0/26.0	-	-	--	-	-	-	-	-	-
	Biochar and manure (25%/75%:50%/50%:75%/25%)	16.5/19.0/23.5	-	-	--	-	-	-	-	-	-
	Biochar and Eucalyptus sawdust (25%/75%:50%/50%:75%/25%)	17.0/21.0/24.0	-	-	--	-	-	-	-	-	-
	Biochar and sugar cane (25%/75%:50%/50%:75%/25%)	17.5/21.0/23.5	-	-	--	-	-	-	-	-	-
	Biochar and wood sawdust (25%/75%:50%/50%:75%/25%)	21/22/26.5	-	-	--	-	-	-	-	-	-
	Activated Biochar and Kernel	25.0/28.0/31.5	-	-	--	-	-	-	-	-	-

	oil(25%/75%:50% /50%:75%/25%)										
	Activated biochar and cattle manure (25%/75%:50%/50%:75%/25%)	17.0/24.0 /28.0	-	-	--	-	-	-	-	-	-
	Activated biochar and eucalyptus sawdust (25%/75%:50%/50%:75%/25%)	21.0/26.0 /30.0	-	-	--	-	-	-	-	-	-
	Activated biochar and sugar cane (25%/75%:50%/50%:75%/25%)	21.0/25.0 /30.0	-	-	--	-	-	-	-	-	-
	Activated biochar and wood sawdust (25%/75%:50%/50%:75%/25%)	23.0/26.0 /32.0	-	-	--	-	-	-	-	-	-
	Activated biochar and alfa (25%/75%:50%/50%:75%/25%)	23.0/26.0 /31.0	-	-	--	-	-	-	-	-	-
[79-82]	**Contaminated biomass**										
	Shredded currency	-	4.7	79.0	11.1	5.2	45.4	6.3	1.9	0.32	**46.1**
	Mixed waste paper	-	8.8	76.8	6.8	7.6	52.3	7.2	0.2	0.08	**40.2**
	Demolition wood	-	16.3	63.4	14.5	5.8	51.7	6.4	1.1	0.09	**40.7**
	Greenhouse-plastic waste	-	2.5	61.0	5.5	31.0	70.9	11.2	1.5	0.01	**16.4**
	Refuse derived fuel.	**-**	**4.2**	**70.3**	**0.5**	**25.0**	**53.8**	**7.8**	**1.1**	**0.47**	**36.8**

5.2 Outlook for further research

Most of the research findings from78 articles that were reviewed did not give information on the season, time of storage and the geographical location of the biomass samples. Approximately 5 % of the article reviewed described the geographical location from where the biomass was sourced. This may pose a difficulty on the reproducibility of data and may hinder future research that may want to compare the effect of location on the biomass properties. Developing a robust software for prediction means that all the variables which could affect the heating value and elemental composition have been properly investigated experimentally. Therefore, it may be necessary to properly describe the season, time of storage and geographical location from where the biomass is harvested.

Estimation of greenhouse gas (GHG) emission potential of biomass sources should be included in the assessment of the energy potential of biomass sources, most researchers did not account for this valuable parameter in the same context of biomass properties. Although this may not be significant at the experimental level, the cumulative impact of this may be so large in the case of industrial energy generation.

It is impossible to sustain the intensive and continued generation of biomass per unit land area envisaged in the biomass-based energy production programmes on the basis of the native nitrogen stocks in the soil as they are inadequate to provide sufficient nutrients to sustain non-nitrogen-fixing crops. It is expected that more researchers should be conducted on the effect of mixing and contamination on the thermal properties and elemental composition of biomass.

Combination of biomass from diverse sources would be required to generate sufficient fuel which can be applied on an industrial scale. There are two implications of this;

1) The researchers may need to understand the extent to which the presence of contamination and biomass mixing would possibly affect the thermal properties, proximate value, and elemental composition. The understanding of this scope may eventually eliminate the need for sorting of some biomass while alleviating the fear of contamination. This will lead to the reduction in the sorting cost of biomass for energy generation.

2) Greenhouse gas emission as a result of this large-scale biomass mixing must be investigated. Under the normal condition biomass is not expected to contribute substantially to GHG, but in a scenario where the rate of harvesting of biomass resources does not correspond to the rate of cultivating, then there would be carbon deficit which will eventually culminate in greenhouse gas. Therefore, possible magnitude of this eventual deficit must be understood. This would provide information to the users and ensure sustainable use of biomass resources.

6. Conclusion

A drive towards sustainable and renewable energy generation triggered researchers to look into the potential of biomass as an alternative to fossil fuel. Lignocellulosic biomass has been identified due to its abundance across the globe. However, the environmental issues surrounding the exploration should be thoroughly investigated, knowing fully well that these may be magnified when the biomass is applied on a large scale. Also, the effect of biomass blending and contamination as related to the production of energy from Lignocellulosic biomass should be investigated for thorough understanding. In the long run, biomass would provide an efficient, cost-effective and sustainable alternative energy, which would assist the global community in reducing greenhouse gas emission. However, a complete understanding of different dynamics in biomass thermal and elemental composition must be well understood.

References

[1] M. Ozturk *et al.*, "Biomass and bioenergy: An overview of the development potential in Turkey and Malaysia," *Renewable and Sustainable Energy Reviews,* vol. 79, pp. 1285-1302, 2017.

[2] M. A. Rosen, "Energy sustainability: a pragmatic approach and illustrations," *Sustainability,* vol. 1, no. 1, pp. 55-80, 2009.

[3] E. Toklu, M. Güney, M. Işık, O. Comaklı, and K. Kaygusuz, "Energy production, consumption, policies and recent developments in Turkey," *Renewable and Sustainable Energy Reviews,* vol. 14, no. 4, pp. 1172-1186, 2010.

[4] R. Saidur, G. BoroumandJazi, S. Mekhilef, and H. A. Mohammed, "A review on exergy analysis of biomass based fuels," *Renewable and Sustainable Energy Reviews,* vol. 16, no. 2, pp. 1217-1222, 2012.

[5] A. Callejón-Ferre and J. López-Martínez, "Briquettes of plant remains from the greenhouses of Almería (Spain)," *Spanish journal of agricultural research,* vol. 7, no. 3, pp. 525-534, 2009.

[6] F. GBEP, "A review of the current state of bioenergy development in G8 þ5 Countries," *Rome: FAO,* 2007.

[7] World Energy Council (WEC), " Waste to Energy. https://www.worldenergy.org/wpcontent/uploads/2017/03/WEResources_Waste_to_ Energy_2016.pdf (Accessed on 16th January,2018)," 2016.

[8] S. Tewfik, "Biomass utilization facilities and biomass processing technologies," *Energy Education Science and Technology,* vol. 14, pp. 1-19, 2004.

[9] Y. Yanli, Z. Peidong, Z. Wenlong, T. Yongsheng, Z. Yonghong, and W. Lisheng, "Quantitative appraisal and potential analysis for primary biomass resources for energy utilization in China," *Renewable and Sustainable Energy Reviews,* vol. 14, no. 9, pp. 3050-3058, 2010.

[10] P. S. Nigam and A. Singh, "Production of liquid biofuels from renewable resources," *Progress in energy and combustion science,* vol. 37, no. 1, pp. 52-68, 2011.

[11] J. M. Vargas-Moreno, A. J. Callejón-Ferre, J. Pérez-Alonso, and B. Velázquez-Martí, "A review of the mathematical models for predicting the heating value of biomass materials," *Renewable and Sustainable Energy Reviews,* vol. 16, no. 5, pp. 3065-3083, 2012.

[12] C. Sheng and J. L. T. Azevedo, "Estimating the higher heating value of biomass fuels from basic analysis data," *Biomass and Bioenergy,* vol. 28, no. 5, pp. 499-507, 2005.

[13] N. Scarlat, J.-F. Dallemand, F. Monforti-Ferrario, and V. Nita, "The role of biomass and bioenergy in a future bioeconomy: Policies and facts," *Environmental Development,* vol. 15, pp. 3-34, 2015.

[14] C. Hamelinck, I. De Lovinfosse, M. Koper, C. Beestermoeller, C. Nabe, and M. Kimmel, "Renewable energy progress and biofuels sustainability. Report for the European Commission," *Ecofys, Utrecht, Netherlands, Tech. Rep. ENER/C1/463-2011-Lot2,* 2012.

[15] M. Banja, N. Scarlat, F. Monforti-Ferrario, and J.-F. Dallemand, "Renewable energy progress in EU 27 (2005–2020)," *Ispra, Italy: Joint Research Centre,* 2013.

[16] U. S. E. I. A. (USEIA), ", Renewable energy explained. https://www.eia.gov/energyexplained/?page=renewable_home (accessed on 27th March 2018)," 2017.

[17] I. D. w. o. r. e. e. W. (2013), "Global Bioenergy Statistics. (Accessed on 26th December,2017)," 2013.

[18] BP, "Statistical review of world energy 2016.https://www.bp.com/en/global/corporate/energy-economics/statistical-review-of-world- energy.html (Accessed on 26th March 2018)," 2016.

[19] T. Kar and S. Keles, "Environmental impacts of biomass combustion for heating and electricity generation," *Journal of Engineering Research and Applied Science,* vol. 5, no. 2, pp. 458-465, 2016.

[20] S. Karekezi, W. Kithyoma, and M. Kamoche, "9. EVALUATING BIOMASS ENERGY COGENERATION OPPORTUNITIES AND BARRIERS IN AFRICA: THE CASE OF BAGASSE COGENERATION IN THE SUGAR INDUSTRY," *BIO-CARBON OPPORTUNITIES IN EASTERN & SOUTHERN AFRICA,* p. 179, 2009.

[21] M. F. Askew and C. A. Holmes, "The potential for biomass and energy crops in agriculture in Europe, in land use, policy and rural economy terms," (in English), *Aspects of Applied Biology,* no. No.65, pp. 365-374, 2001.

[22] M. K. Bahng, C. Mukarakate, D. J. Robichaud, and M. R. Nimlos, "Current technologies for analysis of biomass thermochemical processing: a review," *Anal Chim Acta,* vol. 651, no. 2, pp. 117-38, Oct 5 2009.

[23] A. Demirbaş, "Calculation of higher heating values of biomass fuels," *Fuel,* vol. 76, no. 5, pp. 431-434, 1997.

[24] M. F. Demirbas, M. Balat, and H. Balat, "Potential contribution of biomass to the sustainable energy development," *Energy Conversion and Management,* vol. 50, no. 7, pp. 1746-1760, 2009.

[25] T. Abbasi and S. A. Abbasi, "Biomass energy and the environmental impacts associated with its production and utilization," *Renewable and Sustainable Energy Reviews,* vol. 14, no. 3, pp. 919-937, 2010.

[26] M. Guerrero, M. P. Ruiz, Á. Millera, M. U. Alzueta, and R. Bilbao, "Characterization of biomass chars formed under different devolatilization conditions: differences between rice husk and eucalyptus," *Energy & Fuels,* vol. 22, no. 2, pp. 1275-1284, 2008.

[27] S. Vassilev, C. Braekman-Danheux, P. Laurent, T. Thiemann, and A. Fontana, "Behaviour, capture and inertization of some trace elements during combustion of refuse-derived char from municipal solid waste," *Fuel,* vol. 78, no. 10, pp. 1131-1145, 1999.

[28] S. V. Vassilev, D. Baxter, L. K. Andersen, and C. G. Vassileva, "An overview of the chemical composition of biomass," *Fuel,* vol. 89, no. 5, pp. 913-933, 2010/05/01/ 2010.

[29] S. V. Vassilev and C. Braekman-Danheux, "Characterization of refuse-derived char from municipal solid waste: 2. Occurrence, abundance and source of trace elements," *Fuel processing technology,* vol. 59, no. 2-3, pp. 135-161, 1999.

[30] D. R. Nhuchhen, "Prediction of carbon, hydrogen, and oxygen compositions of raw and torrefied biomass using proximate analysis," *Fuel,* vol. 180, pp. 348-356, 2016.
[31] D. R. Nhuchhen and M. T. Afzal, "HHV Predicting Correlations for Torrefied Biomass Using Proximate and Ultimate Analyses," *Bioengineering (Basel),* vol. 4, no. 1, Jan 24 2017.
[32] E. Akkaya, "ANFIS based prediction model for biomass heating value using proximate analysis components," *Fuel,* vol. 180, pp. 687-693, 2016.
[33] G. E. Acquah, B. K. Via, O. O. Fasina, S. Adhikari, N. Billor, and L. G. Eckhardt, "Chemometric modeling of thermogravimetric data for the compositional analysis of forest biomass," *PLoS One,* vol. 12, no. 3, p. e0172999, 2017.
[34] J. Cai *et al.*, "Review of physicochemical properties and analytical characterization of lignocellulosic biomass," *Renewable and Sustainable Energy Reviews,* vol. 76, pp. 309-322, 2017.
[35] J. Cai *et al.*, "Processing thermogravimetric analysis data for isoconversional kinetic analysis of lignocellulosic biomass pyrolysis: Case study of corn stalk," *Renewable and Sustainable Energy Reviews,* vol. 82, pp. 2705-2715, 2018.
[36] C. Huang, L. Han, Z. Yang, and X. Liu, "Ultimate analysis and heating value prediction of straw by near infrared spectroscopy," *Waste Manag,* vol. 29, no. 6, pp. 1793-7, Jun 2009.
[37] V. Dhyani and T. Bhaskar, "A comprehensive review on the pyrolysis of lignocellulosic biomass," *Renewable Energy,* 2017.
[38] S. K. Sansaniwal, K. Pal, M. A. Rosen, and S. K. Tyagi, "Recent advances in the development of biomass gasification technology: A comprehensive review," *Renewable and Sustainable Energy Reviews,* vol. 72, pp. 363-384, 2017.
[39] A. Khan, W. De Jong, P. Jansens, and H. Spliethoff, "Biomass combustion in fluidized bed boilers: potential problems and remedies," *Fuel processing technology,* vol. 90, no. 1, pp. 21-50, 2009.
[40] J. Koppejan and S. Van Loo, *The handbook of biomass combustion and co-firing*. Routledge, 2012.
[41] E. Gidarakos, G. Havas, and P. Ntzamilis, "Municipal solid waste composition determination supporting the integrated solid waste management system in the island of Crete," *Waste Manag,* vol. 26, no. 6, pp. 668-79, 2006.
[42] A. L. Borrion, M. C. McManus, and G. P. Hammond, "Environmental life cycle assessment of lignocellulosic conversion to ethanol: a review," *Renewable and Sustainable Energy Reviews,* vol. 16, no. 7, pp. 4638-4650, 2012.
[43] S. Paul and A. Dutta, "Challenges and opportunities of lignocellulosic biomass for anaerobic digestion," *Resources, Conservation and Recycling,* vol. 130, pp. 164-174, 2018/03/01/ 2018.
[44] J. Lynch, "Utilization of lignocellulosic wastes," *Journal of Applied Microbiology,* vol. 63, no. s16, 1987.
[45] C. Somerville, H. Youngs, C. Taylor, S. C. Davis, and S. P. Long, "Feedstocks for lignocellulosic biofuels," *science,* vol. 329, no. 5993, pp. 790-792, 2010.
[46] Y. Ahn *et al.*, "Electrospinning of lignocellulosic biomass using ionic liquid," *Carbohydrate polymers,* vol. 88, no. 1, pp. 395-398, 2012.
[47] J. B. Sluiter, R. O. Ruiz, C. J. Scarlata, A. D. Sluiter, and D. W. Templeton, "Compositional analysis of lignocellulosic feedstocks. 1. Review and description of methods," *J Agric Food Chem,* vol. 58, no. 16, pp. 9043-53, Aug 25 2010.
[48] Y. D. Singh, P. Mahanta, and U. Bora, "Comprehensive characterization of lignocellulosic biomass through proximate, ultimate and compositional analysis for bioenergy production," *Renewable Energy,* vol. 103, pp. 490-500, 2017.
[49] C.-H. Zhou, X. Xia, C.-X. Lin, D.-S. Tong, and J. Beltramini, "Catalytic conversion of lignocellulosic biomass to fine chemicals and fuels," *Chemical Society Reviews,* vol. 40, no. 11, pp. 5588-5617, 2011.
[50] P. McKendry, "Energy production from biomass (part 1): overview of biomass," *Bioresource technology,* vol. 83, no. 1, pp. 37-46, 2002.
[51] S. Mahalaxmi and C. Williford, "Biochemical conversion of biomass to fuels," in *Handbook of Climate Change Mitigation*: Springer, 2012, pp. 965-999.

[52] M. Raud, M. Tutt, J. Olt, and T. Kikas, "Effect of lignin content of lignocellulosic material on hydrolysis efficiency," *Agronomy Research,* vol. 13, no. 2, pp. 405-412, 2015.
[53] H. Zabed, J. Sahu, A. Boyce, and G. Faruq, "Fuel ethanol production from lignocellulosic biomass: an overview on feedstocks and technological approaches," *Renewable and Sustainable Energy Reviews,* vol. 66, pp. 751-774, 2016.
[54] L. Kou, Y. Song, X. Zhang, and T. Tan, "Comparison of four types of energy grasses as lignocellulosic feedstock for the production of bio-ethanol," *Bioresource Technology,* vol. 241, pp. 424-429, 2017/10/01/ 2017.
[55] A. V. Bridgwater, "Renewable fuels and chemicals by thermal processing of biomass," *Chemical Engineering Journal,* vol. 91, no. 2-3, pp. 87-102, 2003.
[56] L. Zhang, C. C. Xu, and P. Champagne, "Overview of recent advances in thermo-chemical conversion of biomass," *Energy Conversion and Management,* vol. 51, no. 5, pp. 969-982, 2010.
[57] V. Menon and M. Rao, "Trends in bioconversion of lignocellulose: biofuels, platform chemicals & biorefinery concept," *Progress in Energy and Combustion Science,* vol. 38, no. 4, pp. 522-550, 2012.
[58] A. Demirbaş, "Biomass resource facilities and biomass conversion processing for fuels and chemicals," *Energy conversion and Management,* vol. 42, no. 11, pp. 1357-1378, 2001.
[59] V. Smil, "Nitrogen in crop production: An account of global flows," *Global biogeochemical cycles,* vol. 13, no. 2, pp. 647-662, 1999.
[60] H. Gitay, A. Suárez, R. T. Watson, and D. J. Dokken, "Climate change and biodiversity," 2002.
[61] Y. Sürmen, "The necessity of biomass energy for the Turkish economy," *Energy Sources,* vol. 25, no. 2, pp. 83-92, 2003.
[62] J. Dibdiakova, L. Wang, and H. Li, "Heating Value and Ash Content of Downy Birch Forest Biomass," *Energy Procedia,* vol. 105, pp. 1302-1308, 2017.
[63] R. Zanzi, K. Sjöström, and E. Björnbom, "Rapid high-temperature pyrolysis of biomass in a free-fall reactor," *Fuel,* vol. 75, no. 5, pp. 545-550, 1996/04/01/ 1996.
[64] R. Kataki and D. Konwer, "Fuelwood characteristics of some indigenous woody species of north-east India," *Biomass and Bioenergy,* vol. 20, no. 1, pp. 17-23, 2001.
[65] M. Jindal and M. Jha, "Catalytic hydrothermal liquefaction of waste furniture sawdust to bio-oil," *Indian Chemical Engineer,* vol. 58, no. 2, pp. 157-171, 2016.
[66] S. Naik, V. V. Goud, P. K. Rout, K. Jacobson, and A. K. Dalai, "Characterization of Canadian biomass for alternative renewable biofuel," *Renewable energy,* vol. 35, no. 8, pp. 1624-1631, 2010.
[67] J. Yu, N. Paterson, J. Blamey, and M. Millan, "Cellulose, xylan and lignin interactions during pyrolysis of lignocellulosic biomass," *Fuel,* vol. 191, pp. 140-149, 2017/03/01/ 2017.
[68] B. Pidtasang, P. Udomsap, S. Sukkasi, N. Chollacoop, and A. Pattiya, "Influence of alcohol addition on properties of bio-oil produced from fast pyrolysis of eucalyptus bark in a free-fall reactor," *Journal of Industrial and Engineering Chemistry,* vol. 19, no. 6, pp. 1851-1857, 2013.
[69] K. Raveendran, A. Ganesh, and K. C. Khilar, "Influence of mineral matter on biomass pyrolysis characteristics," *Fuel,* vol. 74, no. 12, pp. 1812-1822, 1995.
[70] G. O. Mosiori, C. O. Onindo, P. Mugabi, S. B. Tumwebaze, S. Bagabo, and R. B. Johnson, "Characteristics of potential gasifier fuels in selected regions of the Lake Victoria Basin," *South African Journal of Science,* vol. 111, no. 5-6, pp. 1-6, 2015.
[71] A. Pattiya, S. Sukkasi, and V. Goodwin, "Fast pyrolysis of sugarcane and cassava residues in a free-fall reactor," *Energy,* vol. 44, no. 1, pp. 1067-1077, 2012.
[72] S. Sirijanusorn, K. Sriprateep, and A. Pattiya, "Pyrolysis of cassava rhizome in a counter-rotating twin screw reactor unit," *Bioresource technology,* vol. 139, pp. 343-348, 2013.
[73] A. M. Azeez, D. Meier, J. r. Odermatt, and T. Willner, "Fast pyrolysis of African and European lignocellulosic biomasses using Py-GC/MS and fluidized bed reactor," *Energy & Fuels,* vol. 24, no. 3, pp. 2078-2085, 2010.
[74] K. A. Motghare, A. P. Rathod, K. L. Wasewar, and N. K. Labhsetwar, "Comparative study of different waste biomass for energy application," *Waste Manag,* vol. 47, no. Pt A, pp. 40-5, Jan 2016.

[75] I. Boumanchar *et al.*, "Effect of materials mixture on the higher heating value: Case of biomass, biochar and municipal solid waste," *Waste Manag,* vol. 61, pp. 78-86, Mar 2017.
[76] J. S. Tumuluru, C. T. Wright, R. D. Boardman, N. A. Yancey, and S. Sokhansanj, "A review on biomass classification and composition, co-firing issues and pretreatment methods," in *2011 Louisville, Kentucky, August 7-10, 2011*, 2011, p. 1: American Society of Agricultural and Biological Engineers.
[77] A. Kakitis, I. Nulle, M. Ozollapins, and J. Kjakste, "ASSESSMENT OF COMBUSTION PARAMETERS OF BIOMASS MIXTURES," in *14th International Scientific Conference Engineering for Rural Development, Jelgava, Latvia, 20-22 May, 2015*, 2015, pp. 133-139: Latvia University of Agriculture.
[78] Y. Li, L. W. Zhou, and R. Z. Wang, "Urban biomass and methods of estimating municipal biomass resources," *Renewable and Sustainable Energy Reviews,* vol. 80, pp. 1017-1030, 2017.
[79] T. R. Miles, T. Miles Jr, L. Baxter, R. Bryers, B. Jenkins, and L. Oden, "Alkali deposits found in biomass power plants: A preliminary investigation of their extent and nature. Volume 1," National Renewable Energy Lab., Golden, CO (United States); Miles (Thomas R.), Portland, OR (United States); Sandia National Labs., Livermore, CA (United States); Foster Wheeler Development Corp., Livingston, NJ (United States); California Univ., Davis, CA (United States); Bureau of Mines, Albany, OR (United States). Albany Research Center1995.
[80] A. T. Masiá, B. Buhre, R. Gupta, and T. Wall, "Characterising ash of biomass and waste," *Fuel Processing Technology,* vol. 88, no. 11-12, pp. 1071-1081, 2007.
[81] D. A. Tillman, "Biomass cofiring: the technology, the experience, the combustion consequences," *Biomass and Bioenergy,* vol. 19, no. 6, pp. 365-384, 2000.
[82] A. Pettersson, M. Zevenhoven, B.-M. Steenari, and L.-E. Åmand, "Application of chemical fractionation methods for characterisation of biofuels, waste derived fuels and CFB co-combustion fly ashes," *Fuel,* vol. 87, no. 15-16, pp. 3183-3193, 2008.
[83] E. Dahlquist, *Technologies for converting biomass to useful energy: combustion, gasification, pyrolysis, torrefaction and fermentation*. CRC Press, 2013.
[84] L. Nunes, J. Matias, and J. Catalão, "Biomass combustion systems: A review on the physical and chemical properties of the ashes," *Renewable and Sustainable Energy Reviews,* vol. 53, pp. 235-242, 2016.
[85] S. Channiwala and P. Parikh, "A unified correlation for estimating HHV of solid, liquid and gaseous fuels," *Fuel,* vol. 81, no. 8, pp. 1051-1063, 2002.
[86] W. Kirmse, *Organic elemental analysis: Ultramicro, micro, and trace methods*. Elsevier, 2012.

Implications of Lack of Maintenance of motorcycles on Ambient Air Quality

Olayemi A Odunlami[1], FrancisB. Elehinafe[1], Temitayo E. Oladimeji[1], Muyiwa A. Fajobi[2], Oyetunji B. Okedere[3], Bamidele S. Fakinle[4]

[1]Department of Chemical Engineering, Covenant University, Ota, Nigeria
[2]Department of Mechanical Engineering, Covenant University, Ota, Nigeria
[3]Department of Civil Engineering, College of Science and Engineering, Osogbo, Nigeria
[4]Department of Chemical Engineering, Landmark University, Omu-Aran, Nigeria[4] E-mail address: olayemi.odunlami@covenantuniveristy.edu.ng

Abstract. This paper reported the results of the investigation of motorcycle emissions in selected roads in Abeokuta, Nigeria. Five roads: Ijaye, Kuto, M.K.O. Abiola way, Olabisi Onabanjo, and Olorunsogo roads were considered with five different motorcycle types (Bajaj, Day Long, Frajend, Hajue Suzuki and Honda) on each road. Motorcycles were parked 8.0m away from the main road in downwind direction. The motorcycle exhaust emissions monitored were CO_2, O_2, CO, and HC emissions. The total mean concentration of measured CO_2emission from motorcycle exhaust on all the roads, ranged between 18800ppm for Frajend and 37400ppm for Bajaj, while O_2ranged between 139200ppm for Honda and 171800ppm for Frajend, also CO ranged between 14500ppm for Hajue Suzuki and 30700ppm for Bajaj, and HC ranged between 674ppm for Honda and 3289ppm for Day Long. It could be noted that the air pollutants are high for almost every type of the motorcycles, and significantly higher than the National standards. This implies that these motorcycles, which are mostly used for commercial purposes are poorly maintained. This study therefore concluded that motorcycle pollution in Abeokuta is significant with possible serious environmental and health consequences.
Keywords: motorcycle; emission; carbon II oxide, pollution

1. Introduction

Transportation is a significant part of development, which provides the economic, social and political interaction, (Button and Hensher, 2001) and serves as compartment for passengers and goods. Vehicle ownership is low, and the dependency on public transport is high in developing countries. Financial conditions and efficiency of the government-organized public transport is discouraging and declining (Kumar, 2011). Hence a resort to motorcycles either for personal mobility or for public transportation. However, transportation a major source of air pollution contributes to greenhouse gas emissions that cause climatic change (Button and Hensher, 2001).

Emissions by automobiles which includes motorcycles results in air pollution which causes a reduction in the air quality thereby resulting in eye irritation, cardiac, respiratory disorders smog, acid rains, and other diseases (Narayanan, 2007). An average motorcycle is known to pollute as much as 15 times per mile than passenger car or high truck (Carpenter, 2008). The environmental friendliness of motorcycles depends on the age, engine capacity and the maintenance (Owate*et al.*, 2005).

The environment is daily endangered by motor bikes which is majorly used for commercial purposes. Hence the importance of annual exhaust emission test and awareness (Peters, 2000) for environmental friendliness.

This has brought about the need to assess motorcycle emissions along Ijaye, Kuto, M.K.O. Abiola way, Olabisi Onabanjo, and Olorunsogo roads in Abeokuta by measuring the exhaust concentrations of pollutants from the exhaust pipes of different kinds of motorcycles plying the roads to determine the levels of emissions released into the ambient air.

2. Sampling of vehicular emission/pollutants

The exhaust emission tests were carried out at five different locations within Abeokuta. The Portable, Hand-Held, battery operated Kane automotive 4-gas analyzer with detector tube (Model Auto 4-1) was used to measure the vehicular emission. The machine is capable of measuring CO (a resolution of 0.01 % with an accuracy of ± 0.5% volume). HCs (a range of 1.0 ppm), CO_2 (a resolution of 0.1 % with an accuracy of ± 0.5% volume), and O_2 (a resolution of 0.1% with an accuracy of ± 0.5% volume). The Kane Automotive 4-Gas analyzer model Auto 4-1 has been designed to be used on petrol, LPG or CNG powered engines. All measured and calculated parameters can be printed on the optional infrared printer or saved to the analyzer's memory.

The analyzer was switched on and allowed to fully initialize to display zero setting, then the main menu was accessed and fuel type selected (diesel, petrol, CNG, and LPG). Having done these, the analyzer was returned to the zero settings with the main button, to check if the setting was within the allowable range (especially oxygen, O_2 which should be between 20.9 % and 21%). At an idle state, the exhaust was observed for a while to ensure that the smoke was steady, then the analyzer's probe was completely inserted into the exhaust pipe and clamped. During the analysis, CO was recorded at the maximum level and stored as soon as it started reducing, after which the probe was removed from the exhaust pipe.

Idle method was used for this work. Five motorcycle types were selected for this paper: Bajaj, Day Long, Frajend, Hajue Suzuki and Honda.Emissions from their exhaust were reported in concentration units (parts per million (ppm) or percent (%) where 1% = 10,000ppm). Also, raw exhaust samples were taken from each of the motorcycles.

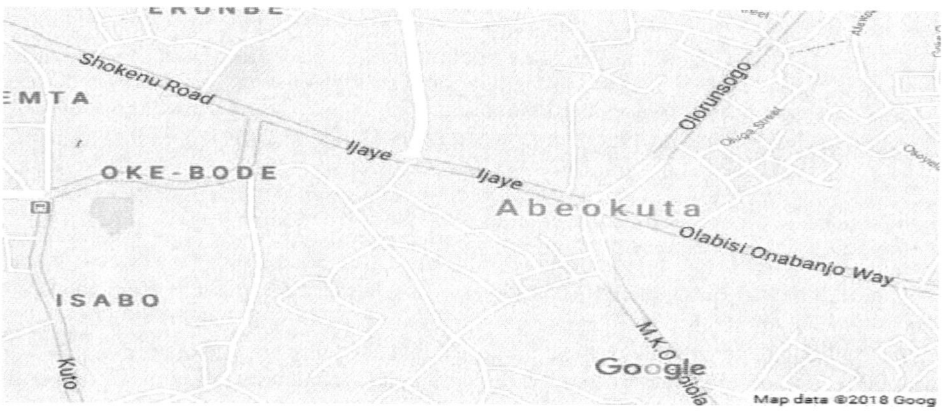

Fig.1The study locations in Abeokuta, Ogun State, Nigeria (Google map)

3. Results and Discussion

The mean concentrations of measured motorcycle exhaust emissions at sampling points on all the selected roads are summarized in Tables1 to 5. The chart representations of the measured emissions are shown in Figures 1 - 6. At the sampling point on Ijaye Road, the mean concentrations of CO_2ranged between 19000ppm for Honda and 35000ppm for Bajaj, while the mean concentrations of O_2ranged between 139300ppm for Hajue Suzuki and 178300ppm for Honda, also the mean concentrations of CO ranged between 15100ppm for Honda and 37600ppm for Hajue Suzuki, and the mean concentrations of HC ranged between 157ppm for Day Long and 1643ppm for Hajue Suzuki.

At the sampling point on Kuto Road, the mean concentrations of CO_2 ranged between 13000ppm for Frajend and 51000ppm for Bajaj, while the mean concentrations of O_2 ranged between 123600ppm for

Bajaj and 184600ppm for Frajend, also the mean concentrations of CO ranged between 7600ppm for Honda and 28200ppm for Bajaj, and the mean concentrations of HC ranged between 541ppm for Honda and 2622ppm for Day Long, while at the sampling point on M.K.O. Abiola way, the mean concentrations of CO_2 ranged between 10000ppm for Frajend and 33000ppm for Honda, while the mean concentrations of O_2ranged between 135600ppm for Bajaj and 181400ppm for Frajend, also the mean concentrations of CO ranged between 3000ppm for Hajue Suzuki and 38900ppm for Honda, and the mean concentrations of HC ranged between 590ppm for Honda and 3246ppm for Day Long.

At the sampling point on Olabisi Onabanjo road, the mean concentrations of CO_2 ranged between 11000ppm for Frajend and 66000ppm for Honda, while the mean concentrations of O_2ranged between 179700ppm for Frajend and 92100ppm for Honda, also the mean concentrations of CO ranged between 6000ppm for Hajue Suzuki and 38600ppm for Honda, and the mean concentrations of HC ranged between 630ppm for Honda and 5050ppm for Day Long, while at the sampling point on Olorunsogo Road, the mean concentrations of CO_2 ranged between 26000ppm for Honda and 49000ppm for Hajue Suzuki, while the mean concentrations of O_2ranged between 133300ppm for Day long and 150200ppm for Frajend, also the mean concentrations of CO ranged between 16600ppm for Hajue Suzuki and 38700ppm for Honda, and the mean concentrations of HC ranged between 1073ppm for Honda and 5370ppm for Day Long.

The total mean concentrations of measured vehicular exhaust emissions from all Roads, the total mean concentrations of CO_2ranged between 18800ppm for Frajend and 37400ppm for Bajaj, while the total mean concentrations of O_2ranged between 139200ppm for Honda and 171800ppm for Frajend, also the total mean concentrations of CO ranged between 14500ppm for Hajue Suzuki and 30700ppm for Bajaj, and the total mean concentrations of HC ranged between 674ppm for Honda and 3289ppm for Day Long.

All the ranges of each emission for every type of motorcycle are greatly high. The adverse implication of these emissions is so much. The results of tables 1-6 (in ppm) did not compare favorably with Table 7, as the values of tables 1-6 are higher compared to the National standards. This shows that most of the motorcycles on Nigerian roads pose a great risk to the environment.

Table 1: The Mean Concentrations of Measured Motorcycle Exhaust Emissions from Ijaye Road.

Motorcycle Type	**Mean Concentrations of Motorcycle Exhaust Emissions**			
	CO_2 (ppm)	**O_2 (ppm)**	**CO (ppm)**	**HC (ppm)**
Bajaj	35000	148100	22900	1002
Day Long	21000	169300	17600	157
Frajend	23000	163000	20600	197
Hajue Suzuki	25000	139300	37600	1643
Honda	19000	178300	15100	534

Table 2: The Mean Concentrations of Measured Motorcycle Exhaust Emissions from Kuto Road.

Motorcycle Type	Mean Concentrations of Motorcycle Exhaust Emissions			
	CO_2 (ppm)	O_2 (ppm)	CO (ppm)	HC (ppm)
Bajaj	51000	123600	28200	1015
Day Long	29000	167700	10700	2622
Frajend	13000	184600	12000	583
Hajue Suzuki	49000	145400	9300	1960
Honda	39000	143400	7600	541

Table 3: The Mean Concentrations of Measured Motorcycle Exhaust Emissions from M.K.O. Abiola way.

Motorcycle Type	Mean Concentrations of Motorcycle Exhaust Emissions			
	CO_2 (ppm)	O_2 (ppm)	CO (ppm)	HC (ppm)
Bajaj	32000	135600	37500	1135
Day Long	26000	160600	17900	3246
Frajend	10000	181400	20200	623
Hajue Suzuki	30000	177400	3000	2049
Honda	33000	138900	38900	590

Table 4: The Mean Concentrations of Measured Motorcycle Exhaust Emissions from Olabisi Onabanjo Road.

Motorcycle Type	Mean Concentrations of Motorcycle Exhaust Emissions			
	CO_2 (ppm)	O_2 (ppm)	CO (ppm)	HC (ppm)
Bajaj	36000	155800	32800	1348
Day Long	43000	135900	27500	5050
Frajend	11000	179700	18600	1290
Hajue Suzuki	29000	162600	6000	2098
Honda	26000	143100	38700	1073

Table 5: The Mean Concentrations of Measured Motorcycle Exhaust Emissions from Olorunsogo Road.

Motorcycle Type	Mean Concentrations of Motorcycle Exhaust Emissions			
	CO_2 (ppm)	O_2 (ppm)	CO (ppm)	HC (ppm)

Bajaj	33000	140100	32300	1402
Day Long	47000	133300	28600	5370
Frajend	37000	150200	16800	3172
Hajue Suzuki	49000	136700	16600	2424
Honda	26000	143100	38700	1073

Table 6: The Total Mean Concentrations of Measured Motorcycle Exhaust Emissions from all Roads

Motorcycle Type	**Mean Concentrations of Motorcycle Exhaust Emissions**			
	CO_2 (ppm)	**O_2 (ppm)**	**CO (ppm)**	**HC (ppm)**
Bajaj	37400	140600	30700	1180
Day Long	33200	153400	20500	3289
Frajend	18800	171800	17600	1173
Hajue Suzuki	36400	152300	14500	2035
Honda	36600	139200	27800	674

Table 7: National Standards for Exhaust Emissions

Emission Standard	**CO (ppm)**	**HC (ppm)**
New model	35000	600
Existing model	45000	800

Source: NAC, 2011

Table 8: Effect of Pollutants on environmental and health.

Pollutants	Environmental	Health
CO (carbon II oxide)	Greenhouse gas contributing to global warming.	At the blood flow, oxygen flow decreases but the likelihood increases to exercise that are related to heart pain for heart disease (coronary) people. Neurobehavioral function and impair concentration can occur when the doses are low.
CO_2 (carbon IV oxide)	The global warming are resulted from the gases of green house	Non

NO_X (nitrogen oxide)	Ozone (smog) are formed which causes corrosion to crop by destroying them. Also leads to global warming as acid rain are formed through greenhouse gases.	Causes asthma, cough, and respiratory disorder.
HC (unburned hydrocarbon)	Ozone level compound	Causes cough, irritates the eyes. Carcinogenic comes by weight compound of high molecule
SO_2 (sulfur oxides)	Causes mainly acid rain	Respiratory disorder and irritates the eyes

(Source: Ghio et al. 1999; Pooley et al. 1999).

4. Conclusion

This study has shown that the concentration of pollutants from motorcycles with poor maintenance is much higher than the National standards (see Table 7), this implies that such motorcycles are very harmful to the environment and climate (see Table 8).

5. Recommendation

This study suggests the following:

- Government should provide mass transit transportation such as good train system, and big busses.
- Strict adherence to standard regulations should be a must.
- Motorcycle owners should be made to understand why they should regularly go for checks and maintenance, so that exhaust emissions could be reduced.

If these measures are properly put in place, greenhouse gases and other harmful substances will be reduced and Nigeria will be making a move towards a green economy.

Acknowledgment

The authors acknowledge the financial support of Covenant University (CUCRID) for open access publication.

References

[1] Button K.J., Hensher. Handbook of Transport Systems and Traffic control. [T]. Pergamon, UK. 2001.
[2] Carpenter, S. Inconvenient truth about motorcycles and smog. [I] (htt-articles hy-throttle), 2008. http://www.physorg.com/news9197:html retrieved may 2010.
[3] Ghio, A.J., Samet J. Air pollution and health: metals and air pollution particles [J]. Academic Press, p.635-651, 1999.
[4] Kumar, A., World, B. Understanding the Emerging Role of Motorcycles in African Cities: A Political Economy Perspective [J]. SSATP Discussion Paper No. 13. Urban Transport Services.Int. Bank Reconstr. Dev., / the World Bank. 2011.

IOP Conf. Series: Materials Science and Engineering **413** (2018) 012055 doi:10.1088/1757-899X/413/1/012055

[5] Miller, G.T. Living in the environmental. [J]. Wards worth publishing company Belmont California, 1997

[6] Narayanan, P. In a textbook of environmental pollution principles, analysis and control. [T]. 1st ed. CBC, New Dehli, 2007.

[7] National Automotive Council. [J]. Federal Republic of Nigeria Official Gazette: National Environmental Regulations, vol.98, B615-635, 2011.

[8] Owate, I. O., Nte, F. U., Johnson, N. [T]. In a textbook of energy resources and environmental crisis Pearl Pub. Lagos, 2005.

[9] Peters, S. W. In textbook of Nigerian environmental Education and Management. [T]. University Press, Calabar, 2000.

[10] Pooley, F. Mille M. Air pollution and health: Composition of air pollution particles. [J]. Academic Press, p. 619-634, 1999.

Design of a Pick and Place Serial Manipulator

Oyetunji Oluremilekun Ropo[1], Oyawale Festus Adekunle[2], Odebode Adetola Oluwaseun[1] and Pele, Babatope Olabanji [3]

[1]Mechanical Engineering Department, Faculty of Engineering, Ladoke Akintola University of Technology, Ogbomoso, Oyo state, Nigeria. E-mail: oroyetunji@lautech.edu.ng
[2]Mechanical Engineering Department, Covenant University, Ota.
[3]Mechanical Engineering Department, University of Ibadan, Ibadan.

Abstract- Repetitive tasks and high accuracy have become the two contradictory needs of any industrial process. By introducing autonomous robotic applications, simple repetitive tasks can be accomplished keeping the demands of the accuracy and speed in mind which will result in optimum utilization of the machine and manpower. However, developing these applications for industries specific to countries like Nigeria, where cheap labour is available, becomes a major problem to be tackled in terms of cost. This project deals with the design, fabrication and control of a 3 degrees of freedom serial manipulator used to pick and place ferromagnetic metals from one place to another. The serial manipulator mechanism was designed using a direct drive mechanism. The links of the arm were designed using Blender CAD software and fabricated using 3D printing technology. Two servo motors control the motion of the arm while another one controls the rotation of the base. Robot motion is controlled using three knobs switch and a toggle switch for the electromagnetic end effector. The motor control is achieved using a microcontroller. The electromagnetic end effector can be moved in a workspace which is in a hemisphere of 40cm. The serial manipulator was able to carry a payload of 15 N with relative ease.
Keywords: Serial manipulator, Electromagnetic, Robotic and Automation

1. Introduction

In today's world, robots are all around us, they come in a variety of different shapes and sizes, designed to do an extraordinary range of tasks. When we think of robots we picture these big humanoid robots from science-fiction films like "Transformers", "I Robot" or the big robotic arms seen manufacturing cars on automated assembly lines. When we talk about robots, what we are really talking about is machines that will

do what human beings would normally be expected to do. These machines mimic the operation of the human being or at least certain parts of it.
RIA (1977) defined a robot as a re-programmable multi-functional manipulator designed to move material, parts, tools, or specialized devices, through variable programmed motions for the performance of a variety of tasks". The term 'robot' got prominence way back in the 1920s when a Czechoslovakian playwright, Karl Capek in his play R.U.R. (Rossum's Universal Robots) denoted the birth of a superior race that had intelligence similar to that of humans, then later a well-known Russian science fiction writer Isaac Asimov coined the word robotics. (William, 2016).

In general, robotics, which is the science of robots, can be divided into two areas, industrial and service robotics. IFR (1987) defines a service robot as a robot which operates semi- or fully autonomously to perform services useful to the well-being of humans and equipment, excluding manufacturing operations. In general, service and industrial robots are in widespread use, performing jobs more cheaply and with greater accuracy and reliability than humans. They are also employed for jobs that are too dirty, dangerous or boring to be suitable for humans. Robots are widely used in manufacturing, assembly and packing, earth

and space exploration, surgery, weaponry, laboratory research and mass production of consumer and industrial goods. As robotic technology develops and becomes cheaper, domestic robots for cleaning or mowing the lawn are available, along with robotic toys for children of all ages.
Craig (2016) stated that industrial robots is growing at a fast pace, in the last few years the demand for industrial robots is increasing. A major reason for the growth in the use of industrial robots is their declining cost. Through the decade of the 1990s, robot prices dropped while human labor costs increased. Also, robots are not just getting cheaper; they are becoming more effective, faster, more accurate and more flexible. If we factor these quality adjustments into the numbers, the cost of using robots is dropping even faster than their price tag is. As robots become more cost effective at their jobs, and as human labor continue to become more expensive, more and more industrial jobs become candidates for robotic automation. This is the single most important trend propelling growth of the industrial robot market.

The most common robot used in industry today is the 'serial manipulator' also known as robot arm. These arms are used to weld, package, paint, position and assemble a host of products that we use daily. Basically, a robot arm is a series of linkages that are connected in such a way that a motor can be used to control each joint.

2. Mechanism

Tsai (1999) defined a serial manipulator as a mechanism consisting of "several links connected in series by various types of joint, typically revolute and prismatic joints". Another common name for serial manipulator is 'robotic arm'. Serial manipulators are sometimes referred to as open-loop manipulators simply because one end (or the base) of the manipulator is fixed and the other end (known as gripper or end effector) is free to move in space.

Serial manipulators can be classified into direct drives and non-direct drives. Kazerooni (1989) stated that direct drive robots have two major advantages over the non-direct drive systems. The first advantage being that the direct-drive arms are free from mechanical backlash and friction due to elimination of transmission systems. A small mechanical backlash in the transmission system would cause the gear teeth to wear faster. The high rate of wear in the gear would develop an even larger backlash. About 25 percent of the torque in non-direct drive arms is used to overcome the friction. The second advantage is such that the structural stiffness of the direct-drive arms is greater than the non-direct drive systems. About 80 percent of the total mechanical compliance in most non-direct drive industrial robots is caused by transmission systems. The high structural stiffness allows for wide-bandwidth control, the low structural stiffness of non-direct drive arms due to the existence of many mechanical elements in the transmission system is a limiting factor on the achievement of a relatively wide bandwidth control system.

Since continuous rotation is not required, therefore servo motors will be used as the actuator in this project.

Components of the Serial Manipulator

This serial manipulator was designed using the direct drive method. To accomplish this goal, careful analyses have been made concerning the materials that will be suitable for this project. The components include the mechanical, electronic and computer component.

Below is a list of the materials used for the mechanical part of the serial manipulator.

- Actuators (Servo motor)
- 3mm thick Polylactic acid
- Fastening screws
- Couplers
- KK-P20/15 Electromagnet
- Spur gears

IOP Conf. Series: Materials Science and Engineering **413** (2018) 012056 doi:10.1088/1757-899X/413/1/012056

Actuators: A lightweight servo motor is used as the actuator for the arm because they are not heavy and can be easily controlled. The servo motor used in this project is the TowerPro MG995R servo motor. It produces a torque of 9.218 Nm at 4.8V and 10.787 Nm at 6.0V.

Link: The links were designed using the CAD software called Blender before been sliced using another software called Dicer, after which it was then printed using a 3D printer. After the printing of all the links to be used, the links were coupled together using screws and fasteners. Polylactic acid of thickness 3 mm was used as links due to their light weight.

End-Effector: The electro-magnetic end effector used in this project is the KK-P20/15 which is rated 2.5kg at 12V DC, consuming 3W. The electromagnet works at a lower voltage with reduced pulling power.
For the electronics and computer, the materials used are:

- ATMegaMicrocontroller (ATMega328P);
- Programmer board;
- SPDT switches;
- LEDs;
- Diodes;
- Resistors;
- Capacitors;
- Voltage regulators and
- Arduino Integrated Development Environment

Joint Torque Calculation

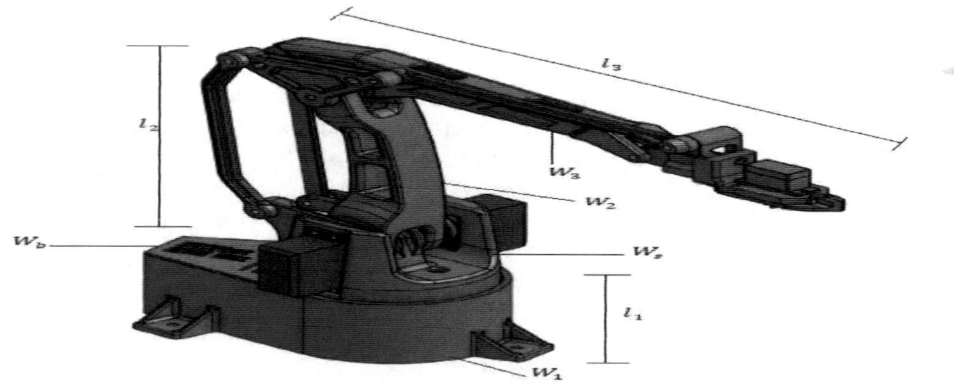

Figure 1: Simple Overview of the Manipulator

Where W_1, W_2 and W_3 are the weight of the links l_1, l_2 and l_3 respectively
W_s *and* W_b is the weight of the shoulder joint and base joint respectively,
W is the weight of the object to be carried.
With good engineering decision, we take:
$l_1 = 0.05\ m,\ l_2 = 0.20\ m$ and $l_3 = 0.29\ m$
Also,
$W_1 = 3.3034\ N,\ W_2 = 0.9565\ N,\ W_3 = 0.5334\ N,\ W_s = 1.0791\ N,\ W_b = 0.5395\ N,\ W = 14.7150\ N$

Torque about joint 3,

$$T_3 = (W \times L_3) + \left(W_3 \times \frac{L_3}{2}\right)$$

$$T_3 = (14.7150 \times 0.29) + \left(0.5334 \times \frac{0.29}{2}\right)$$

$$T_3 = 4.345\ Nm$$

Torque about joint 2,

$$T_2 = (W \times [L_2 + L_3]) + \left(W_3 \times \left[L_2 + \frac{L_3}{2}\right]\right) + (W_s \times L_2) + \left(W_2 \times \frac{L_2}{2}\right)$$

$$T_2 = (14.7150 \times [0.2 + 0.29]) + \left(0.5334 \times \left[0.2 + \frac{0.29}{2}\right]\right) + (1.0791 \times 0.2) + \left(0.9565 \times \frac{0.2}{2}\right)$$

$$T_2 = 7.210\ Nm$$

Torque about joint 1,

$$T_1 = (W \times [L_1 + L_2 + L_3]) + \left(W_3 \times \left[L_1 + L_2 + \frac{L_3}{2}\right]\right) + (W_s \times [L_1 + L_2]) + \left(W_2 \times \left[L_1 + \frac{L_2}{2}\right]\right) + (W_b \times L_1) + \left(W_1 \times \frac{L_1}{2}\right)$$

$$T_1 = (14.7150 \times [0.05 + 0.2 + 0.29]) + \left(0.5334 \times \left[0.05 + 0.2 + \frac{0.29}{2}\right]\right) + (1.0791 \times [0.05 + 0.2]) + \left(0.9565 \times \left[0.05 + \frac{0.2}{2}\right]\right) + (0.5395 \times 0.05) + \left(3.3034 \times \frac{0.05}{2}\right)$$

$$T_1 = 8.679\ Nm$$

Since T_1, T_2 and T_3 are lesser than the 10.787 Nm torque provided by the servomotors at each joint, the suggested length of the links is justified to be used.

3. Results

From the calculations, the torque needed at the elbow, shoulder and base joints are 4.345 Nm, 7.210 Nm and 8.679 Nm respectively for it to move optimally. Since the motors at each joint produces 10.787 Nm torque according to the producer specification, which is more than the needed torque, then the serial manipulator should move optimally.

The serial manipulator was able to carry a payload of 15 N which was made possible by the use of a direct drive mechanism. With a direct drive, the arm is made lighter, and torque is increased.

4. Conclusion

Robotic arm or serial manipulator, as demonstrated in this project, can be made to carry heavier loads without forfeiting the movement speed using direct drive mechanism. It can also be seen that the project is justified as it will help reduce human efforts and improve efficient material handling with precision and accuracy.

References

[1] Asada, H., and Kanade, T. (1983). Design of direct-drive mechanical arms. *ASME J. Vibrations, Acoustics, Stress, and Reliability in Design,* 105(3), 312-316. Atmel ATmega32 Datasheet. (2011). AVR Corporation.

[2] Brooks, R. A., Connell, J. H., and Ning, P. (1988). *Herbert: a second generation mobile robot.* Cambridge, CA: MIT Press.

[3] Craig, J. J. (2005). *Introduction to robotics: mechanics and control* (3rd ed.). Upper-Saddle River, NJ: Pearson Prentice-Hall. Electrical4u (2014). Theory and working principle of servomotor. (n.d.). http://www.electrical4u.com/servo-motor-servo-mechanism-theory-and-working-principle/. Accessed on 27 June, 2016.
[4] Ferretti, G., Magnani, G., and Rocco, P. (1999). Force oscillations in contact motion of industrial robots: an experimental investigation. *IEEE/ASME Trans. Mechatron.,* 4(1), 86-91.
[5] Hanselman, D., Hung, J., and Keshura, J. M. (1992). Torque ripple analysis in brushless permanent magnet motor drives. *Proc. Int. Conf. Electric Machines,* 823-827.
[6] Kazerooni, H. (1989). Statically balanced direct drive manipulator. *Robotica,*7, 143-149.
[7] Kircanski, N. M., and Goldenberg, A. A. (1997). An experimental study of non-linear stiffness hysteresis, and friction effects in robot joints with harmonic drives and torque sensors. *Int. J. Robot. Res,* 16(2), 214-239.
[8] Robotics Institute of America. (1977). *Proceedings of the NBS/RIA workshop robotics research.*
[9] Tsai, L. (1999). *Robot analysis: the mechanics of serial and parallel manipulators* (1st ed.). New York, NY: John Wiley and Sons.
[10] Vischer, D., and Khatib, O. (1995). Design and development of high performance torque-controlled joints. *IEEE Trans. on Robotics and Automation,* 11(4), 537-544.
[11] Williams, B (2016). Mechanics and control of robotic manipulators. *Bob Production*
[12] 3Dprinting (2015). What is 3D Printing. https://3dprinting.com/what-is-3d-printing/ . Accessed on 21 March, 2017.

IOP Conf. Series: Materials Science and Engineering **413** (2018) 012057 doi:10.1088/1757-899X/413/1/012057

Model Selection Process in Time Series Analysis of Production System with Random Output

Ongbali S. O. [1], Igboanugo A. C. [2], Afolalu S. A. [1], Udo M. O. [1], Okokpujie I. P. [1].

[1]College of Engineering, Mechanical Engineering Department, Covenant University, Ota, Ogun State, Nigeria.

[2]Production Engineering Dept. faculty of Engineering, University of Benin, Edo state, Nigeria.

E-mail: samson.ongbali@covenantuniversity.edu.ng

Abstract. In time series investigation of characteristics of production system, different competing models are generally obtained particularly in production settings with stochastic output attributable to bottleneck problems. Consequently, selecting the best model that describes a production system becomes challenging and critical because some models that fit observed data most accurately may not predict future values correctly on account to model complexities. This research desires to demonstrate the procedure for model selection in production system with random output via the use of Adjusted Coefficient of Determination $(\overline{R}^2)$, Akaike and Schwarz criteria tools. Production output measurements obtained serve as input data to Autocorrelation Function and Partial Autocorrelation Function to obtain the order of Autoregressive, Autoregressive Moving Average and Autoregressive Integrated Moving Average models. The model parameters were estimated and used for predictions and compared with original and transformed data to obtain Sum of Squared Error (SSE). Afterward, the models were subjected to adequacy evaluation and subsequently tested with Akaike and Schwarz criteria. Among the competing models, ARIMA (3, 1, 1) model explain 66% variance of the dataset and wielded the lowest Akaike and Schwarz values of 534.41m and 534.34m respectively and thus selected as the model that represents the production system under investigation. The approach establishes that Adjusted Coefficient of Determination in conjunction with Akaike and Schwarz criteria are adequate tools for model selection in time series investigation particularly in stochastic situation
Keywords: Model selection, Time series, Random production output

1. Introduction

The objective of model selection is to discover a model that optimizes a process because exact model that describes a system is unknown to man. The primary objective of model selection is to compare competing models and select the best that describes a system. Model selection is a critical stage in time series investigation because we are always confronted with competing models particularly in production system with stochastic output attributable to bottleneck problems [12]. Inappropriate process of model selection results in choice of poor model with consequence. Modeling is approximation of reality, thus, model selection is to reject a model far from reality and select that which is close to reality [5]. [2] asserted that primary reason of model selection is to evaluate the performance of different models and choose the best among the models for specific dataset. Hence, overlooking model selection procedures result in worthless conclusion in statistical reasoning [29].

Researchers in the topic area over the decades have proposed different methods for model selection, for example, [2] argued that model performance is a function of its prediction ability and its selection is exceptionally imperative because it guides the choice of the quality of a selected model. Improved model performance obtained via model selection provide dependable future forecast of a system [1]. [3] claimed that aside model adequacy testing, objective of model selection include finding a good predictor that describes a system and Akaike Information Criterion (AIC) is a principal model selection method [4] and [14]. Akaike Information Criterion (AIC), Bayesian information criterion (BIC) and Structural Risk Minimization (SRM) where methods proposed by [2] for model selection. [5] derived Regularization Information Criterion (RIC) from Kullback-Leibler information number which is an extension of Takeuchi's Information Criterion (TIC) and AIC and then used it for model selection. [7] emphasized that Expert System (ES) can guide model selection for forecasting.

Similarly, [8] presented procedure for predictive testing of predicted residuals and noted that only parameter consistency is required for model selection. There are arguments whether to use model testing or model selection procedures in deciding the model that best fits a specific dataset but Arbitrage Pricing Theory (APT) model resolve the argument [9] and [20] asserted. [11] proposed the use final prediction Error (FPE) for model selection which is the expected variance of prediction error when Autoregressive is fitted. [10] used different methods for model selection including hypothesis testing, diagnostic tests, goodness-of-fit methods, Bayesian approaches and forecast evaluation methods. [13] asserted that cross-validation and Bayesian approach can be used for model selection. [15] used SURE-Autometrtcs Algorithm for Model Selection and asserted that the method performed well. [16], [18] and [24] proposed hypothesis tests and Selection Criteria using final prediction error (FPE) for model selection.

In addition, [17] proposed mean squared error of prediction as criterion for selecting model and claimed that it is better than using residual sum of squares. [19] recommended the use of single statistic as the sum of squared residuals for model selection. [21] offered unbiased Akaike Information Criterion (AICu) for model selection and claimed that it outperforms biased correction Akaike Information Criterion (AICc) but [28] claim that AICc outperform other forms of model selection tools. [23] projected penalized profile likelihood method for model selection and claimed it outperform the adaptive methods.

Furthermore, [25] develop a Cp statistic with a known distribution for model selection. [31] emphasize advantage of AIC for model selection in bootstrapping. [27] proposed AIC and BIC, Predictive Least Squares (PLS) and Sequentially Normalized Least Squares (SNLS) for model selection. [29] proposed model selection method based on multiple hypothesis testing including AIC, BIC, FPE and Minimum Description Length (MDL). [30] asserted that there are four model selection methods namely, parametric vision, data generating process (DGP), evaluation based on fit, and ignoring model uncertainty on inference and subsequently recommended semiparametric method for model selection.

The forgoing presented different methods of model selection procedures but with application to specific situations different from area we are exploring. It is evidence also that there is no unified method for model selection but some of the procedures recommended in literature are complex, laborious, time consuming and more theoretical than empirical but our approach is simple and easy to apply in practice. Section one reviews literature in the topic area, section two presents methodology, and section three depicts results and discussions and finally section draw conclusion of the study.

2. Methodology

The key variables required for model selection using Adjusted Coefficient of Determination $(\overline{R}^2)$ Akaike and Schwarz are the number of model parameters and the Sum of Squared Errors (SSE). Production output measurement obtained from a manufacturing company serve as input data to Autocorrelation function and Partial autocorrelation function computations. The order AR(p), ARMA (p, q), ARI(p, d) and ARIMA(p, d, q) models were identified and their values were estimated using OLS technique in conjunction Matlab pseudocode. The parameter values were used for prediction and compared with the original and the transformed data to obtain Sum of Squared Error (SSE). The identified order of parameters (p, d, q) along with SSEs serve as input variables to the Adjusted Coefficient of Determination $(\overline{R}^2)$, Akaike and Schwarz criteria test to select the model that represents the system under investigation

A Brief Framework of Adjusted Coefficient of Determination, Akaike and Schwarz Criteria

Adequacy of a model is evaluated by using Adjusted Coefficient of Determination $\overline{R}^2$ which gives idea of how many data points fall within the regression line to examine the relationship within a dataset. Adjusted coefficient of determination $\overline{R}^2$ shows the percentage of variation explained by the independent variable that affects the dependent variable.

Equation for Adjusted Coefficient of Determination is given by,

$$\overline{R}^2 = 1 - \frac{\sum_{1}^{n}(y_t - \hat{y})^2/(n-k)}{\sum_{1}^{n}(y_t - \overline{y})^2/(n-1)} \qquad (1)$$

Were, k = number of parameters
n = number of dependent variables.

Equation for Akaike and Schwarz criteria are given respectively by,

Akaike (AIC) = $n\log(SSE) + 2k$ (2)

Schwarz $BIC = n\log(SSE) + k\log n$ (3)

Where, k = number of parameters that are fitted in the model
Log = natural logarithm
n = number of observations in the series
SSE = sum of squared errors, given as, $SSE = \sum_{t=1}^{n}(y_t - \hat{y}_t)^2$

3. Results

Four models – AR (4), ARMA (4, 3) ARI (3, 1) and ARIMA (3, 1, 1) were obtained from analysis and were subjected to model adequacy evaluation and subsequently model selection.

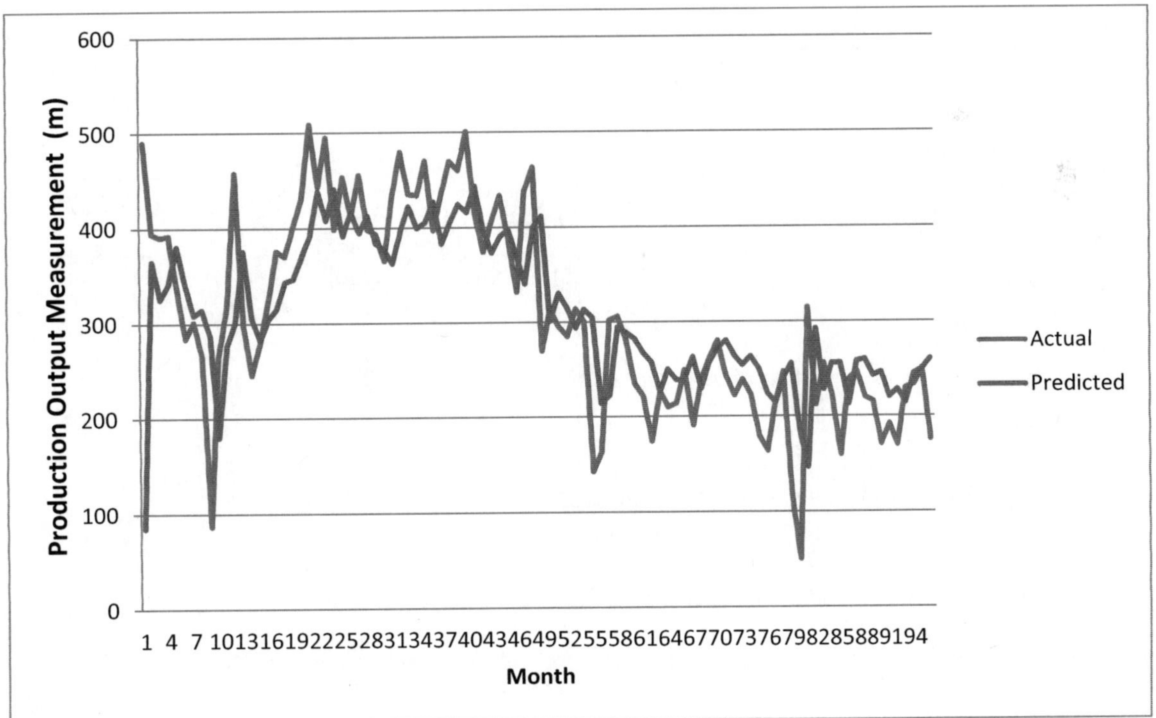

Figure 1. AR (4) model prediction superimpose on actual data with SSE = 551610.804m.

IOP Conf. Series: Materials Science and Engineering **413** (2018) 012057 doi:10.1088/1757-899X/413/1/012057

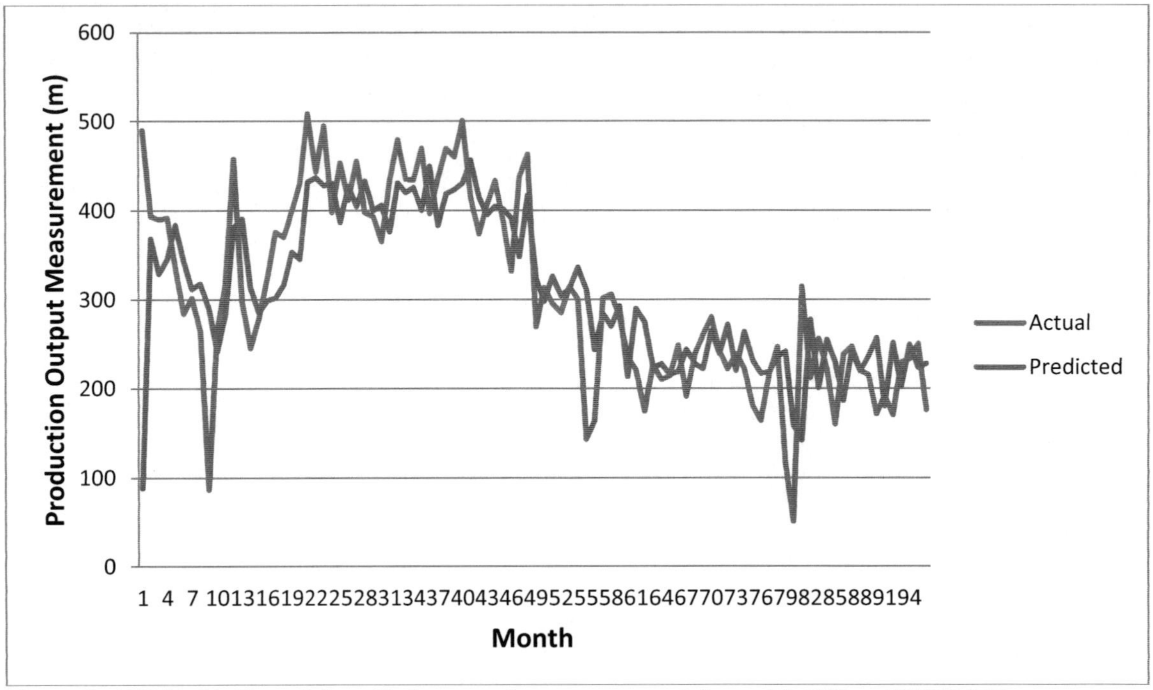

Figure 2. ARMA (4, 3) model prediction superimpose on actual data with SSE = 487623.122m

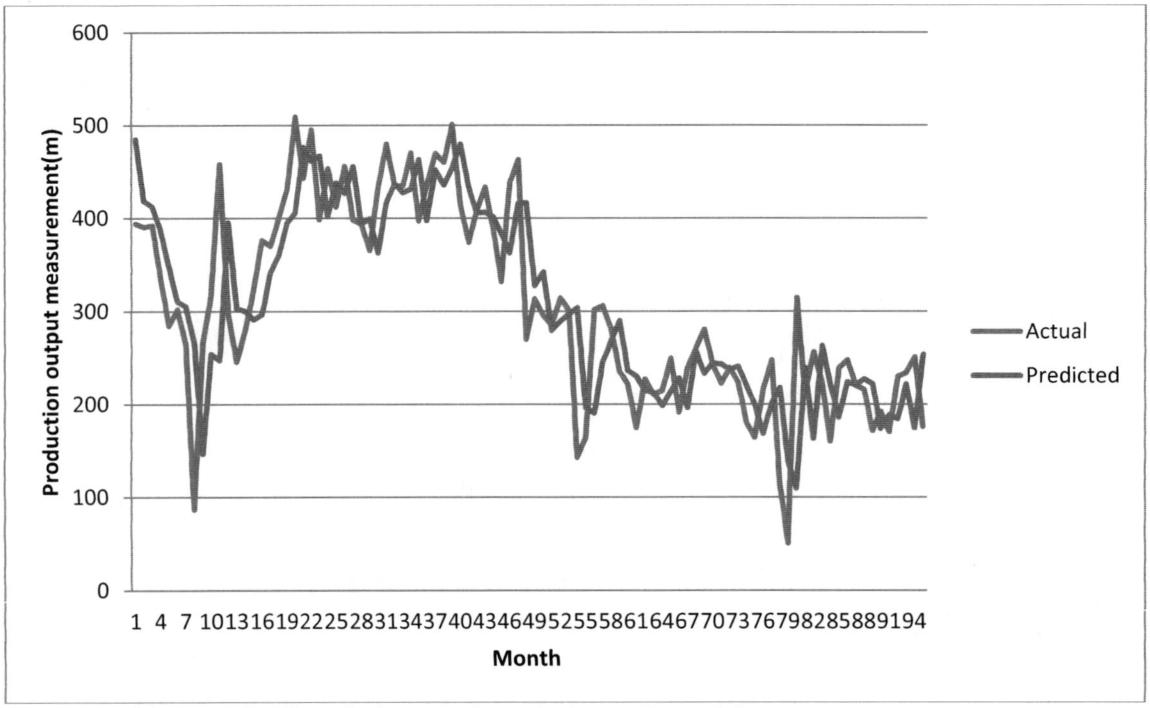

Figure 3. ARI (3, 1) Model prediction superimpose on transformed data

Predicted values in figure 3 tracked the transformed data closely at some points with SSE = 385822.75m

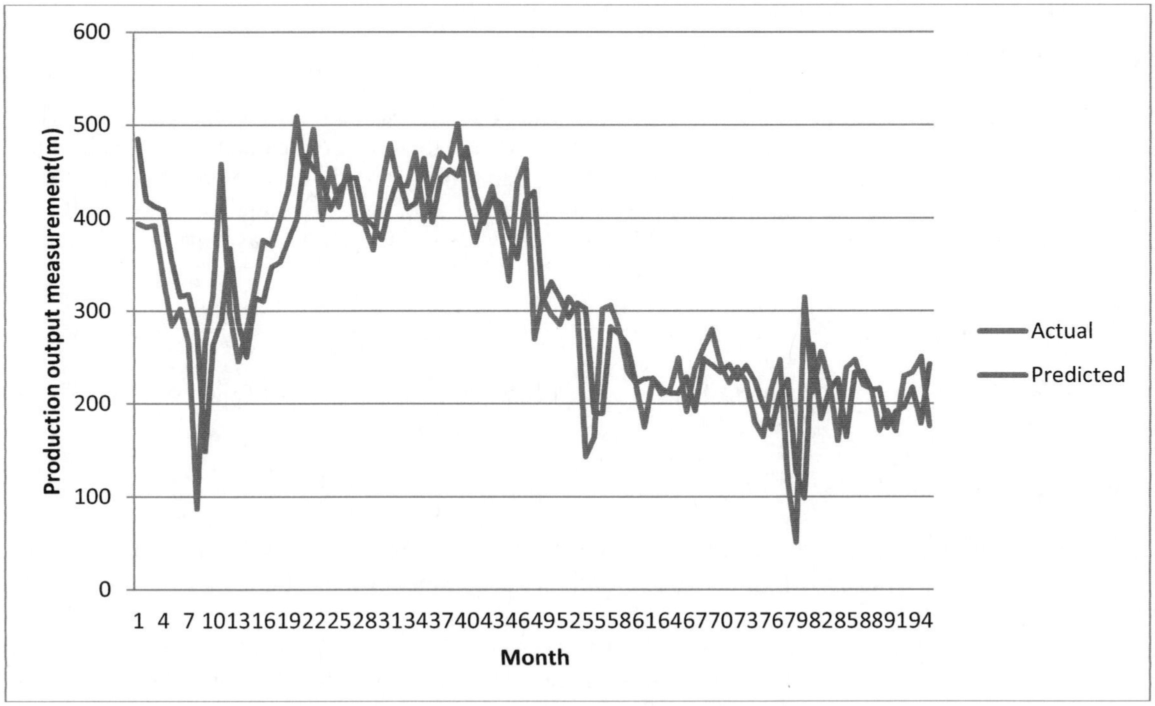

Figure 4 ARIMA (3, 1, 1) Model prediction superimpose on transform data

The predicted values in figure 4 tracked the transformed data closer than ARI (3, 1) model with SSE = 364910.024m

Table 1. Evaluation of Model Adequacy Using Adjusted Coefficient of Determination ($\overline{R}^2$)

Model	Adjusted Coefficient of Determination ($\overline{R}^2$)
AR (4)	52%
ARMA (4, 3)	54%
ARI (3, 1)	62%
ARIMA (3, 1, 1)	66%

ARMA(3, 1, 1) model in table 1 outperformed other competing models as indicated in table 1 by wielding ($\overline{R}^2$) value of 66% signifying that the model explains 66% of the variance of the independent variable that affects the dependent variables

Table 2. Akaike and Schwarz Criteria Values

Model	Akaike (AIC)	Schwarz (BIC)
AR (4)	559.20	559.13
ARMA (4, 3)	554.056	553.985
ARI (3, 1)	536.71	536.64
ARIMA (3, 1, 1)	534.41	534.34

The AIC and BIC values in table 2 show that ARIMA (3, 1, 1) model surpasses other models by wielding the lowest Akaike and Schwarz values of 534.41 and 534.34 respectively. The Adjusted Coefficient of Determination, Akaike and Schwarz values support ARIMA (3, 1, 1) model to be the best model that described the production system under investigation. Thus, ARIMA (3, 1, 1) is selected and fitted to the time series data analysed. Consequently, ARIMA (3, 1, 1) model can be fitted to the production system and use for out-of-sample prediction to approximate future production output.

4. Conclusion

Selecting the right model leads to accurate prediction and guarantee sustainable production system. The foregoing demonstrates that Adjusted Coefficient of Determination in conjunction with Akaike and Schwarz criteria are powerful tools for model selection particularly in production setting with random output. Most times, Adjusted Coefficient of Determination points out the appropriate model to be fitted to a data set and usually supported by Akaike and Schwarz criteria for final selection as demonstrated in this study.

Acknowledgements

We acknowledge the financial support offered by Covenant University for actualization of this research work for publication.

References

[1] Eija Ferreira. Model Selection in Time Series Machine Learning Applications [D]. Academic dissertation of Technology and Natural Sciences of the University of Oulu, Linnanmaa, 2015.
[2] Hastie T., Tibshirani R. and Friedman J. The Elements of Statistical Learning: Data Mining [J]. Inference and Prediction. Springer Series in Statistics. Springer-Verlag, New York, NY, 2nd edition, 2009.
[3] Hannan E.J., Krishanaiah P. R. and Rao M. M. Various Model Selection Techniques in Time Series Analysis[J]. Eds., Handbook of Statistics, 1985, Vol. 5,
[4] Clifford M. Hurvich and Chih-Ling Tsai. Regression and time series model selection in small samples [J]. Biometrika, 1989, 76, 2, pp. 297-307
[5] Shibata R. Statistical Aspects of Model Selection. Working Paper [C]. International Institute for Applied Systems Analysis, 1989, A-2361 Laxenburg, Austria
[6] Markus B., Wolfgang A., Peter F., Tim L., Silvia M., and Alexander R. Visual Analytics for Model Selection in Time Series Analysis [J]. IEEE Computer Society, 2013
[7] Delurgio Stephen A (1998). Forecasting Principles and Applications [T]. Irwin McGraw-Hill New York.
[8] Dufour J, Ghysels E. and Hall A. Generalized Predictive Tests and Structural Change Analysis in Econometrics [J]. International Economic Review, 1994, Vol. 35, No. 1, pp. 199-229
[9] Bharti M. I., Al-Shanfari H and Hossain M. Z. Econometric Analysis of Model Selection and Model Testing [J]. Journal of the American Statistical Association, 2002
[10] Dufour J-M. Model selection. [T] New Palgrave Dictionary of Economics, 2nd Edition, 2007.
[11] Akaike H. Fitting autoregressive models for prediction [J]. Annals of the Institute of Statistical Mathematics, 1969, 21, 243–247.
[12] Igboanugo A. C. and Ongbali S. O. A Factorial Study of Bottleneck Problems in Multi-Stage Production System: Results from Textile Industry [J]. University of Benin Journal of Science & Technology, 2012, Vol.1, No. 1
[13] Burnham K. P. and Anderson D. R. Model Selection and Multimodel Inference A Practical Information-Theoretic Approach [J]. Second Edition. (Springer-Verlag New York, Inc., 175 Fifth Avenue, New York, 2002, NY 10010, USA),
[14] Burnham K. P., Anderson D. R. and Huyvaert K. P. AIC model selection and multimodel inference in behavioral ecology: some background, observations, and comparisons [J]. Behav Ecol Sociobiol, 2011, 65:23–35
[15] Norhayati Y. SURE-Autometrtcs Algorithm for Model Selection in Multiple Equations [D]. PhD Thesis, Universiti, Utara, Malaysia, 2016.

IOP Conf. Series: Materials Science and Engineering **413** (2018) 012057 doi:10.1088/1757-899X/413/1/012057

[16] Hannes L. and Benedikt M. P. The Model Selection [J]. Yale University, 2000.

[17] **Allen D. M.** Mean Square Error of Prediction as a Criterion for Selecting Variables [J]. Technometrics, 2000, Volume 13, 1971 - Issue 3

[18] Allen D. M. The Relationship Between Variable Selection and Data Augmentation and a Method for Prediction [J]. Technometrics, 1974, Vol. 16, No. 1, pp. 125-127

[19] Amemiya T. Selection of Regressors [J]. International Economic Review, 1980, 21(2):331-54

[20] Bhatti M. I. and Al-Shanfari H. Economic Analysis of Model Selection and Model Testing [J]. 2006, Routledge

[21] McQuarrie A., Shumway R., and Tsai C-L. The model selection criterion AICu [J]. Statistics & Probability Letters, 1998, Vol. 34, Issue

[22] McQuarrie A. D. R. and Tsai C-L "Regression and Time Series Model Selection," [C] World Scientific, Singapore City, (2000).

[23] Chen Z., Tang M-L., Gao W and Shi N-Z. New Robust Variable Selection Methods for Linear Regression Models [J]. Scandinavian Journal of Statistics, 2014.

[24] Liebscher E A. Universal Selection Method in Linear Regression Model [J]. Open Journal of Statistics, 2012, Vol.2, No. 2

[25] Riani M. and Atkinson A. C. The Selection of Time Series Models Perhaps with Regressors [J]. London School of Economics, London WC2A 2AE, UK, 2010.

[26] Berkowitz J. and Kilian L. Recent Developments in Bootstrapping Time Series [J]. Econometrics Reviews, 2000, 19(1) 1 – 48

[27] Maatta J. Model Selection Methods for Linear Regression and Phylogenetic Reconstruction [D]. PhD Thesis, Series of Publications A, Report A-2016-3, 2016.

[28] Hurvich C. M and Tsai C-L. Regression and Time Series Model Selection in Small Samples [J] Biometrika, 1989, 76, 2, pp. 297-307

[29] Leeb h. and Pötscher B. M. Model Selection and Inference: Facts and Fiction [J]. Econometric Theory, 2005, 21, 21-59

[30] Hansen B. E. Challenges for Econometric Model Selection [J]. Econometric Theory, 21, 2005.

IOP Conf. Series: Materials Science and Engineering **413** (2018) 012058 doi:10.1088/1757-899X/413/1/012058

Thermal Optimization of Heat Sink for Inverter Applications

F. Onoroh[1], M. Ogbonnaya[1], J. L. Chukuneke[2], C. B. Echeta[2]

[1]Department of Mechanical Engineering, University of Lagos, Akoka, Yaba, Lagos state, Nigeria
[2]Department of Mechanical Engineering, Nnamdi Azikiwe University, Awka, Anambra state, Nigeria
E-mail: onofrankus@yahoo.com

Abstract. Power inverters enable efficient conversion of DC voltage and current to AC voltage and current and are intermediate device in renewable energy technology application especially solar, wind and thermoelectric energy. The key technologies behind power inverters switching operation are the semiconductor devices such as the IRF 3205 MOSFET chip. During operation the switches dissipate heat and if the dissipated heat is not properly managed, thermally induced failure may occur. The research work involves generation of a heat sink model and its optimization for a 24VDC/2.5K.V.A inverter. The thermal resistance, fval, of the optimized heat sink was found to be equal to 0.0345 °C/W which is less than the thermal resistance of the heat sink to ambient of the IRF 3205 MOSFET switch of 0.5063 °C/W. The thermal management of the power inverter package yields experimental and modeled package temperatures of 34.7°C and 36.5°C respectively. The optimized heat sink using the model created leads to reduction in size, weight and reliable operation of the power inverter.
Keywords. Power inverter; Renewable energy; Reliability; Heat Sink; Thermal resistance; MOSFET chip; Failure.

1.Introduction

Power inverter involves the conversion of electric power from direct voltage and current levels to alternating voltage and current levels. Power electronics is now a major part of power engineering; at least 25% of all power generated passes through some form of electronic system before being utilized [1].

In the inverter, DC power from the battery bank is inverted to AC power via a set of solid state switches – MOSFETs or IGBTs that essentially flip the DC power back and forth creating AC power [2][3]. In the course of the inversion, electrical power is dissipated as thermal energy, and reliability of the power inverter can be compromised if accurate thermal management is not put in place [4].

Problems associated with the thermal dissipation and heat transfer through power inverters is one of the most urgent issues that requires significant attention in order to produce efficient and reliable electrical energy conversion systems [5].

In a typical MOSFET base power inverter, power is dissipated as heat. To avoid unacceptable temperature rises to prevent thermal problems, improvements in cooling methods are obligatory [6]. The heat generated as a result of these losses must be conducted away from the power devices and into the environment using a heat sink. If an appropriate thermal system is not used, elevated temperatures can adversely affect electronic device operation; power handling capability, reliability and even risk of electrical fire outbreak. It has been showed that more than 55% of the electronic failures are caused by heat [7].

2.Thermal Dissipation in a Power Inverter

IOP Conf. Series: Materials Science and Engineering **413** (2018) 012058 doi:10.1088/1757-899X/413/1/012058

Heat energy can be transferred from one place to another by conduction, convection and radiation. These phenomena may take place in a given system singly or simultaneously. However, the majority of heat generated by the IRF 3205 MOSFET switches in a power inverter is transported via conduction to the bottom of the switches. Other mechanisms are seldom sufficient to cause any noticeable change [8][9].
As a result of heat dissipation during the switching operation of the IRF3205 MOSFET chips, thermal management system must be optimized to attain the highest performance of the inverter. However in other to optimize the overall heat sink, thermal performance is usually maximized and the weight satisfying constraints are minimized.

The main components that make the power module stack are the die, the solder and the base-plate. The IRF 3205 MOSFET chip is fastened to the heat sink with the aid of screw. In this configuration, the heat generated in the die is conducted down through the stack to the base-plate which is then cooled by the heat sink via air circulation. Extended surface is used specially to enhance the heat transfer rate between a solid and an adjoining fluid. The heat sink is a high thermal mass attached to the base-plate enabling it to help reduce the die temperature during operation [10][11].
Variations to the heat sink geometry and material allow the heat sink to be optimized. An understanding of the way in which the material of the heat sink affects its performance is required to inform about cost and performance as given by the coefficient of heat penetration, (CHP).

Heat sink with a higher CHP reduces the die temperature more as the material have a low thermal resistance and high thermal capacitance. The CHP describes how much thermal capacity is available within the early period of the heat flow and is obtained directly from the material properties as:

$$CHP = \sqrt{k\rho C_p} \qquad (1)$$

Where k is the thermal conductivity of the material, W/mK, ρ is the density of the material, kg/m^3, Cp is specific heat capacity of the material, J/kgK.

3.Generation of Heat Sink Model

Figure 1 shows the geometry of an extruded heat sink and the associated thermal resistance network. Using the definitions, of the heat sink geometry as shown in figure 1, the fin spacing ratio k_f is defined as [12]:

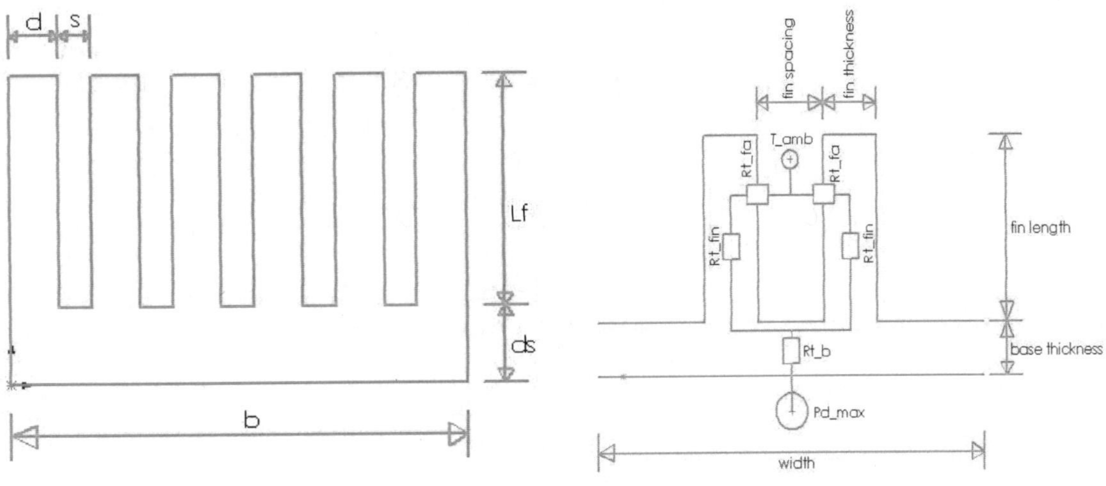

Fig. 1 Geometry of an Extruded Heat Sink

$$k_f = \frac{s(n-1)}{b} \tag{2}$$

Where s is the air passage width, m, b is the distance between the first and last fins, m, n is the number of fins. With values between zero and one, the fin spacing ratio characterizes the cross section of the heat sink available for air flow. The hydraulic diameter, d_h, is a commonly used term when handling flow in noncircular tubes and channels. It can be expressed as a function of the heat sink geometry as:

$$d_h = \frac{4A}{p} \tag{3}$$

Where A is the area of duct section perpendicular to the flow, it can be expressed as:

$$A = (n-1)s \cdot L_f \tag{4}$$

P is the wetted perimeter of the duct, it can be expressed as:

$$P = 2(n-1)(s+L_f) \tag{5}$$

In the light of equation (4) and equation (5), the hydraulic diameter of the fluid channel is obtained as [13]:

$$d_h = \frac{2sL_f}{s+L_f} \tag{6}$$

Where L_f is the fin length, m. For a given fin geometry, the total air flow through all channels creates a pressure drop along the heat sink length, L, into the paper which is given for laminar flow as [14]:

$$\Delta P_L = 1.5\frac{32\rho\upsilon L}{n(sL_f)d_h^2}\dot{V} \tag{7}$$

Where ρ is the density of air, υ is the kinematic viscosity of air, m^2/s, $\dot{V}$ is volume flow rate of air m^3/s. The 1.5 correction factor takes into account the non quadratic channel shape characterized by $s \ll L_f$. In the case of turbulent flow, the pressure drop is given as [14]:

$$\Delta P_T = \frac{L\frac{s+L_f}{2sL_f}\rho\frac{1}{2}\left(\frac{\dot{V}^2}{n(sL_f)}\right)^2}{\left(0.79\ln\left(\frac{2\dot{V}}{n(s+L_f)\upsilon}\right)-1.64\right)^2} \tag{8}$$

For laminar flow the Nusselt number, Nu, can be obtained as [14]:

$$Nu_L = \frac{3.657\left[\tanh\left(2.264X^{1/3}+1.7X^{2/3}\right)\right]^{-1}+\frac{0.0499}{X}\tanh X}{\tanh\left(2.432Pr^{1/6}X^{1/6}\right)} \tag{9}$$

With

$$X = \frac{L}{d_h RePr} \tag{10}$$

$$Pr = \frac{C_p\mu}{k} \tag{11}$$

μ is the dynamic viscosity, kg/ms.

While the Nusselt number, Nu, for Turbulent flow is [14]:

$$Nu_T = \frac{\left[8(0.79\ln Re-1.64)^2\right]^{-1}(Re-1000)Pr}{1+12.7\sqrt{[8(0.79\ln Re-1.64)^2]^{-1}}\,(Pr^{2/3}-1)\left(1+\left(\frac{d_h}{L}\right)^{2/3}\right)} \tag{12}$$

Therefore, the heat transfer coefficient is obtained from the Nusselt number as [15][16]:

$$h_c = \frac{Nuk}{d_h} \tag{13}$$

The Reynolds number, Re, of the fluid can be expressed as a function of the heat sink geometry is as follows:

$$Re = \frac{\rho u D_h}{\mu} \tag{14}$$

Using the concept of kinematic viscosity, the Reynolds number can be expressed as:

$$Re = \frac{u D_h}{\upsilon} \tag{15}$$

From the definition of the hydraulic diameter of equation (3), equation (15) can be expressed as:

$$Re = \frac{4\dot{V}}{\upsilon P} \tag{16}$$

Substituting equation (5) into equation (16), the Reynolds number as a function of the heat sink profile of figure 1 is then obtained as:

$$Re = \frac{2\dot{V}}{\upsilon(n-1)(s+L_f)} \tag{17}$$

If the Reynold's number, Re, is less than 2300, the flow is laminar, otherwise it is turbulent. All properties of air are evaluated at the mean film temperature, T_f, as [17][18]:

$$T_f = \frac{T_b+T_\infty}{2} \tag{18}$$

The significant thermal performance index for efficient heat dissipation is the thermal resistance, $R_{t,hs}$, of the heat sink, which comprises the thermal resistance of the heat sink base, $R_{t,b}$, thermal resistance between fin surface and the air channel, $R_{t,fa}$, the thermal resistance of the fin, $R_{t,fin}$, as shown in figure 1, for an air channel.

The resultant thermal resistance of the heat sink by application of series and parallel combination of resistance yields the thermal resistance of the heat sink as:

$$R_{t,hs} = \frac{1}{n}\left\{\frac{\left(\frac{1}{h_c A_{fa}}+\frac{\frac{1}{2}L_f}{\frac{1}{2}\delta L k_{al}}\right)\left(\frac{1}{h_c A_{fa}}+\frac{\frac{1}{2}L_f}{\frac{1}{2}\delta L k_{al}}\right)}{2\left[\left(\frac{1}{h_c A_{fa}}\right)+\left(\frac{\frac{1}{2}L_f}{\frac{1}{2}\delta L k_{al}}\right)\right]}\right\} + \left(\frac{d}{\frac{1}{n}A_b k_{al}}\right) \tag{19}$$

Under operating conditions, the thermal resistance between the IRF 3205 MOSFET chip/heat sink assembly, is defined as:

$$R_{t,sa} = \frac{T_j - T_\infty}{P_{dmax}} - \left(R_{t,jc} + R_{t,cs}\right) \quad (20)$$

For effective heat dissipation, the necessary condition to be satisfied is:

$$R_{t,hs} \leq R_{t,sa} \quad (21)$$

The convective heat transfer is obtained from the relation:

$$\dot{Q} = \frac{T_j - T_c}{R_{t,hs}} \quad (22)$$

Where d is the heat sink base thickness, m, A_{fa} is the effective convective surface area of air channel, m^2, k_{hs} is the thermal conductivity of heat sink material, W/mK, L is the heat sink length, m, L_f is the fin height, m, δ is the fin thickness, m, T_j is the junction temperature, °C, T_∞ is the ambient temperature, °C, T_c is the case temperature, °C, P_{dmax} is the maximum dissipated power, W, $R_{t,jc}$ is the junction to case thermal resistance, °C/W, $R_{t,cs}$ is the case to heat sink thermal resistance, °C/W.

4.Results and Discussion

The critical thermal performance parameters of the generated heat sink model are fin spacing, fin thickness, length of fin, length of heat sink and number of fins. The heat sink model is simulated in MATLAB, to show the influence of each parameter on the heat sink performance. The calculation procedure is as shown in the flow diagram of figure 2.

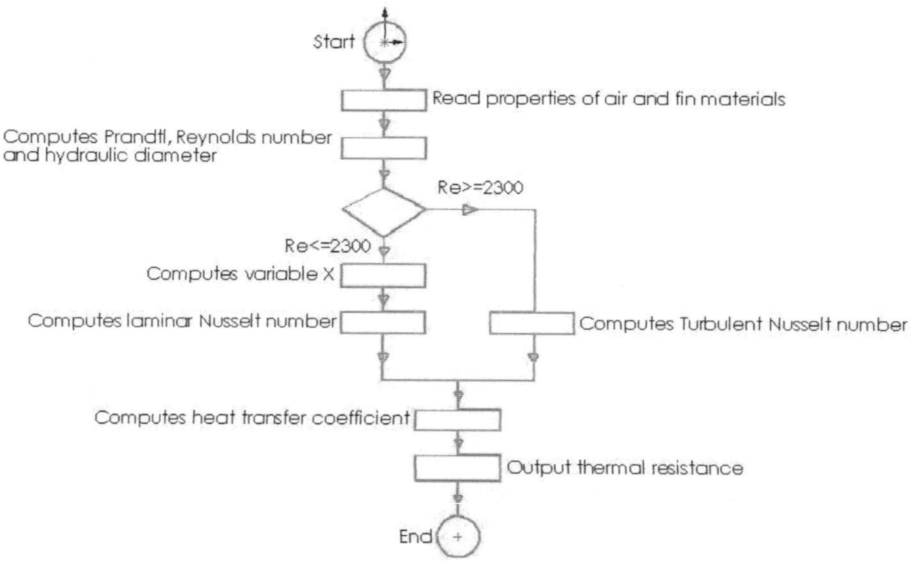

Fig. 2 Heat Sink Model Simulation Flow Diagram

4.1 Variation of Thermal Resistance with Fin Spacing and Number of Fins

Figure 3 shows the 3-D and 2-D plot of thermal resistance with fin spacing and number of fins. From the plots increasing the fin spacing and number of fins, results in a decrease in the thermal resistance of the heat sink, and thus an increase in heat dissipation capability of the heat sink. This is as expected as increasing fin spacing leads to an increase in effective heat transfer area for heat transfer to take place.

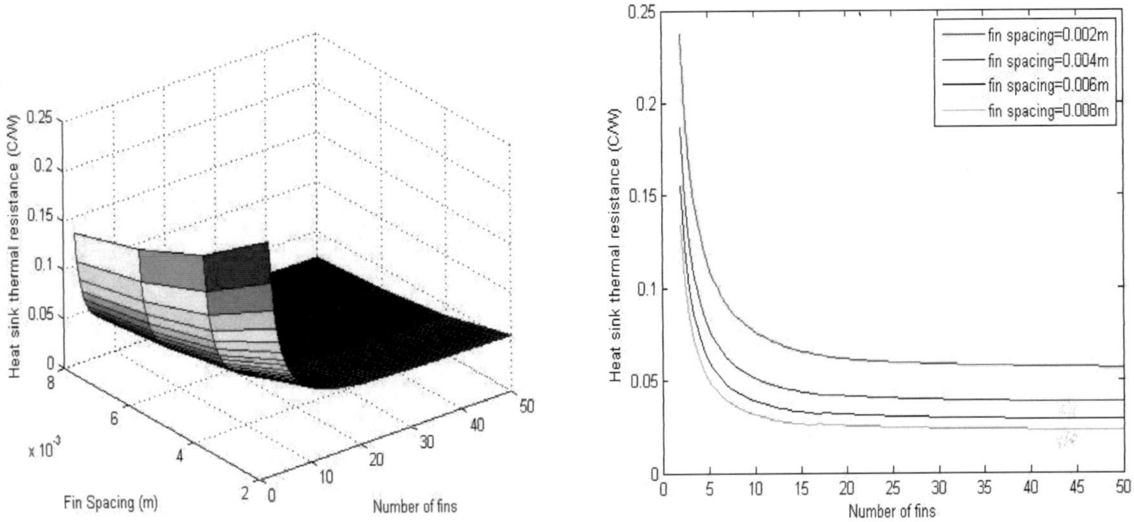

Fig. 3 Thermal Resistance Curves at various Fins spacing and number of Fins

However, for a constant heat sink width, increasing the number of fins means a corresponding decrease in the fin spacing/or thickness of the fin. Thus, the fin spacing ratio characterizes the cross section of the heat sink available for air flow.

4.2 Variation of Thermal Resistance with Length of Fin and number of Fins

Figure 4 shows the 3-D and 2-D plot of thermal resistance with length of fins and number of fins. Clearly, there is a progressive decrease in thermal resistance with an increase in length of fin and an increase in number of fins. At a large length of fin, there is an increase in hydraulic diameter of the channel which will leads to a decrease in convective heat transfer coefficient. However there is a corresponding increase in heat transfer surface area which leads to reduction in the heat sink thermal resistance.

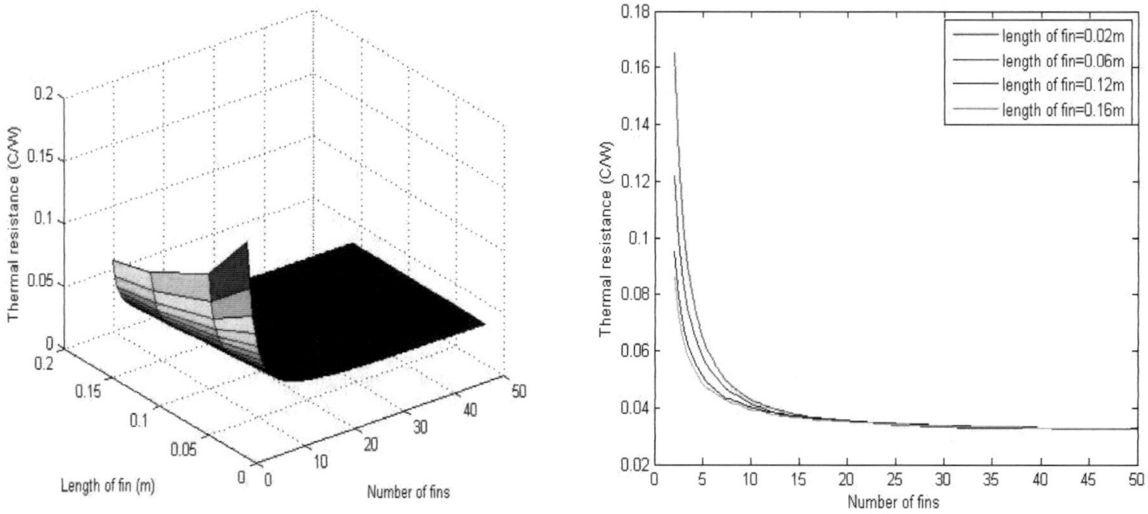

Fig. 4 Thermal Resistance Curves at various Length of Fin and number of Fins

Also at large numbers of fins, changes in length of fins have little influence on the thermal resistance of the heat sink. It must be pointed out that as the length of fin increases, and at large number of fins, there is a reduced heat exchange at the fin top, due to the temperature of the fin top approaching the ambient temperature.

4.3 Variation of Thermal Resistance with Length of Heat Sink and number of Fins

Figure 5 shows the 3-D and 2-D plot of thermal resistance with length of heat sink and number of fins. Clearly, low thermal resistance is achieved with a corresponding increase in length of heat sink and number of fins.

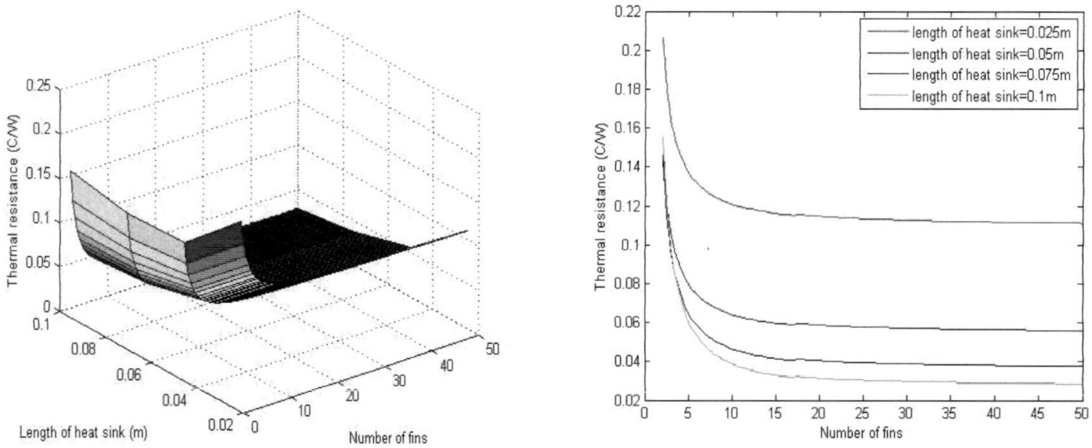

Fig. 5 Thermal Resistance Curves at various Length of heat sink and number of fins

The decrease in thermal resistance is due to the increase in the total fin surface area, and a corresponding increase in the average convective heat transfer coefficient. However, there would be an increase in pressure drop in the channel, which in turn leads to an increase in pumping power.

IOP Conf. Series: Materials Science and Engineering **413** (2018) 012058 doi:10.1088/1757-899X/413/1/012058

4.4 Variation of Thermal Resistance with Fin Thickness and number of Fins

Figure 6 shows the 3-D and 2-D plot of thermal resistance with fin thickness and number of fins. There is a gradual decrease in thermal resistance with an increase in fin thickness at constant number of fin.

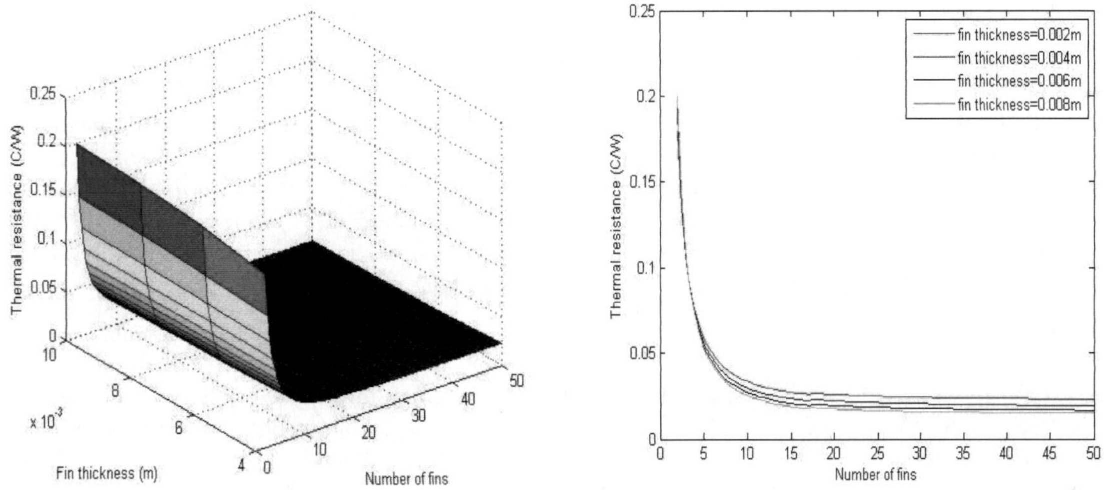

Fig. 6 Thermal Resistance Curves at various Fin Thickness and number of Fins

An increase in fin thickness allows heat flow through the whole length of the fin. A small fin thickness will not allow equal distribution of heat over the fin surface. This reduces the effective fin surface area, makes a part of the fin useless, and has associated manufacturing constraints. For a constant heat sink width, a compromise has to be made, as an increase in fin spacing will lead to corresponding decrease in fin thickness.

4.5 Variation of Thermal Resistance with Cross Sectional area and number of Fins

Figure 7 shows the 3-D and 2-D plot of thermal resistance with cross sectional area and number of fins. There is a gradual decrease in thermal resistance with an increase in cross sectional area at constant number of fin.

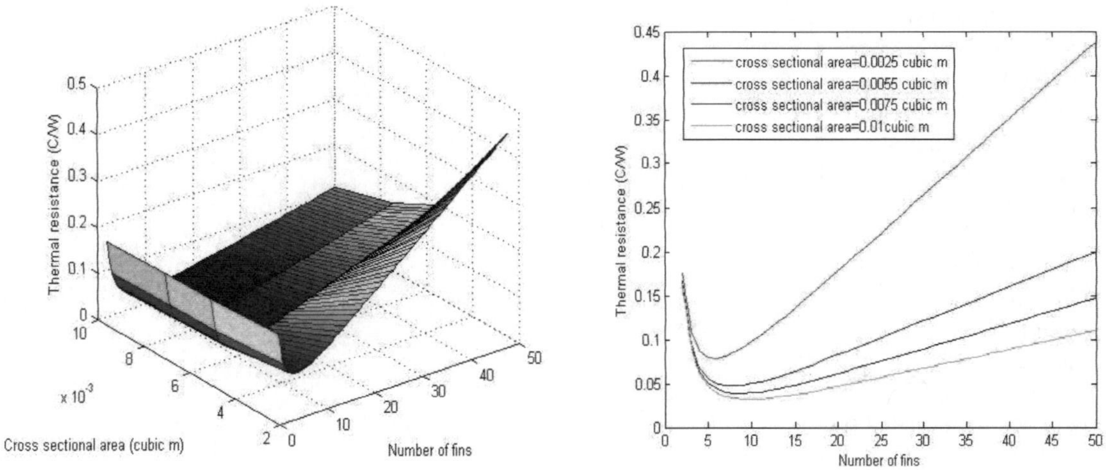

Fig. 7 Thermal Resistance with Cross Sectional Area and number of Fins

IOP Conf. Series: Materials Science and Engineering **413** (2018) 012058 doi:10.1088/1757-899X/413/1/012058

However, for each cross sectional area, there is an optimum number of fins beyond which the thermal resistance increases. Thus, small number of fins is desired as the cross sectional area increases.

The presented analysis shows that the performance of the heat sink can be maximized by selecting the correct profile. Due to the large number of variables, and the desire to minimize cost, weight and volume of heat sink, the least material approach was adopted.

In the optimization, the non dimensional parameters that are function of the independent variables are passed as global variables to the objective function. The resulting optimized heat sink variables are; fin spacing equal to 0.005m, fin length equal to 0.02m, number of fins equal to 10, fin thickness equal to 0.003m, heat sink length equal to 0.1m, heat sink base thickness equal to 0.004m, from which width of the heat sink equal to 0.075m, fin spacing ratio of 0.6667. The thermal resistance, fval, is equal to 0.0345 °C/W which is less than the thermal resistance of sink to ambient of the MOSFET Switch of 0.5063 °C/W, which clearly satisfy the heat sink model. The thermal management of the power inverter package yields experimental and modeled package temperatures of 34.7°C and 36.5°C respectively at air flow rate of 0.03065m^3/s.

5. Conclusion

Power inverters are essentially for supplying energy for today's society in a more efficient, sustainable and controllable manner. Thermally induced failures modes are the main causes of power inverter reliability issues. Thus the principal parameters that influence the performance of the heat sink are optimized for effective dissipation of the power losses in the inverter. In optimization of the heat sink, thermal performance is maximized and the weight satisfying constraint is minimized. Result of an actual test at 1500W corresponding to 75% of the inverter rating yields an experimental and modeled package temperatures of 34.7°C and 36.5°C respectively, recorded with the aid of MTM–380SD 3 Channels Temperature Monitor with SD Card Data logger, which shows that the optimized heat sink model works well and the inverter can safely and reliably be operated.

Nomenclature--

A	: Cross sectional area of heat sink, m^2
A_b	: Base area of heat sink, m^2
A_{fa}	: Convective surface area of air channel, m
B	: Distance between first and last fins
C_p	: Specific heat capacity, J/kgK
D	: Heat sink base thickness, m
D_h	: Hydraulic diameter, m
h_c	: Convective heat transfer coefficient, W/m^2K
k	: Thermal conductivity, W/mK
k_{al}	: Thermal conductivity of heat sink material, W/mK
L	: Heat sink length, m
L_f	: Length of fin, m
n	: Number of fins
Nu	: Nusselt number
$\dot{V}$	: Volume flow rate, m^3/s
v	:Kinematic viscosity, m^2/s
P	: Wetted perimeter, m
P_{dmax}	: Maximum dissipated power, W
Pr	: Prandtl's number
$\dot{Q}$	: Convective heat transfer, W
Re	: Reynold's number

IOP Conf. Series: Materials Science and Engineering **413** (2018) 012058 doi:10.1088/1757-899X/413/1/012058

$R_{t,hs}$: Thermal resistance of heat sink, °C/W
$R_{t,sa}$: Thermal resistance of sink to ambient, °C/W
$R_{t,jc}$:Thermal resistance of junction to case, °C/W
$R_{t,cs}$: Thermal resistance of case to heat sink, °C/W
S : Fin spacing, m
T_b : Base temperature, °C
T_c :Case temperature, °C
T_∞ :Ambient temperature, °C
T_f : Film temperature, °C
T_j : Junction temperature, °C
U : Velocity, m/s
μ : Dynamic viscosity, kg/ms
δ : Fin thickness, m
ρ : Density, kg/m^3

References

[1] Singh R., Baliga B. (1998). *Cryogenic Operation of Silicon Power Devices*. Kluwer Academic Publishers.
[2] Hannemann R., Fox L.R., Mahalingham M., (1991). Thermal Design for Microelectronic Components in Cooling Techniques for Components. *W. Aung, Hemisphere*, New York, pp. 245-276.
[3] Whitaker J. C. (1996). *The Electronics Hand Book* CRC Press pp. 441, 455, 484-486, 973-975.
[4] Ellison G.N., (1987). Thermal Computations for Electronic Equipment. *Van Nostrand Reinhold*, New York.
[5] Pang Y. E., Scott E. P., Chen J. E., Thole K. A. (2005). Thermal Design and Optimization Methodoly for Integrated Power Electronics Modules. *Journal of Electronic Packaging*, ASME, vol. 127, pp. 59-66.
[6] Mandel S. K., Sen D. (2016). A brief review on mixed convection heat transfer in channel flow with vortex generator for electronic chip cooling. *International Journal of Engineering Research and Application*, vol. 6, issue 6, pp. 74-82
[7] Ericsson (2000). Thermal Aspects on DC/DC Power Modules Design note: *Ericsson Microelectronics AB*, no. 004 pp. 1-8.
[8] Lostetter A., Barlow F., Elshabini A., (2000). An overview to Integrated Power Module design for high Power Electronics Packaging. *Microelectronics Reliability*, Vol.40, pp. 365-379.
[9] Xiao C., Ngo K., Lu G.-Q., (2009). Thermal Design of power Module to minimize peak Transient Temperature. In: *International Conference on Electronic Packaging Technology and High Density Packaging* (ICEPT-HDP), Xi'an, China pp. 248-254.
[10] Mokheimer E. M. A. (2003). Heat transfer from extended surfaces subject to variable heat transfer coefficient. *Heat and mass transfer* vol. 39, pp. 131-138.
[11] Shrirao P. N., Sherekar R. M., Bhalerao S. V. (2014). Evaluation of surface heat transfer coefficient for electronic packages. *International Journal of Research in Advent Technology*, vol. 2. Issue 1, pp. 59-66.
[12] Furkay S.S. (1984). Convective heat transfer in Electronic Equipment: An overview, In Thermal
[13] Management Concepts in Microelectronic Packaging. *ISHM Technical Monograph* Series 6984-003,
[14] Silver spring, MD., pp. 153-179.
[15] Kern D.Q., Kraus A.D. (1972). *Extended Surface Heat transfer*. McGraw-Hill, New York.
[16] Baehr H. D., Stephen K., (1998). *Heat and Mass Transfer*. ISBN 3-540-64458-X, 3rd Edition, Springer, pp. 220-221,259-262, 350-355.
[17] Buller M.L., Kilburn R.F., (1981). Evaluation of surface heat transfer coefficients for Electronic modular packages in Heat Transfer in Electronic Equipment. *HTD*, Vol. 20, pp. 25-28, ASME, New York.
[18] Tritt T.M. (2004). *Physics of Solids and Liquids Thermal Conductivity: Theory, properties and Applications*. Second Edition, New York, NY: Kluwer academic/plenum publishers Vol. 1.
[19] Rogers G.F.C., and Mayhew Y.R., (1966). *Engineering Thermodynamics Work and Heat Transfer*. 4th edition, Longman pp. 509-518.
[20] Eastop T.D., Mckonkey A. (1999). *Applied Thermodynamics for Engineering Technologies*. 5th edition, Pearson Education, pp. 577-583.

IOP Conf. Series: Materials Science and Engineering **413** (2018) 012059 doi:10.1088/1757-899X/413/1/012059

Development of a Counting System Method for Managing Crowd Using Image Sensing Device

Owodolu Akinbode A, Christian Bolu A, Samuel Ayuba and Babaremu Kunle O.

Mechanical Engineering Department, Covenant University, Ota, Ogun State, Nigeria

Email; kunle.babaremu@covenantuniversity.edu.ng

Abstract. Counting of people and collecting good data is a major problem in every organization. Counting of people may be manual or automatic. In order to create an efficient counting system, this study focused on automatic counting system to sense, count and collect data using image sensing device technology. The image sensing device was able to sense, and count people. Also, adaptive algorithm was developed to ensure accurate counting for both indoor and outdoor counting. Background Subtraction is also considered for varying shadows and lighting conditions. This was done using MATLAB. Finally the method of image processing, gives good result of quick sensing, and rapid counting.
Keywords: Counting, Detecting, Image sensing device, and Manual

1. Introduction

Counting of people and collecting data is a major problem in every organization. However, conventional method of people counting system is manual. Manual counting of people is time consuming, labour intensive and highly costly. Humans have limited attention span and reliability when large amount of data has to be collected and analyzed over a long period of time, especially in a condition where resources are limited [1]. The defects of labour waste and subjectivity could not meet the need for public safety [2] [3]. This problem can be solved by the utilization of fast growing computer vision technology and high speed computer as a processor for counting. Due to rapid development image sensing device is possible to count people with high degree of accuracy and efficiency. This image processing comprises of segmentation, background extraction and blobs detection and is done in MATLAB. Accurate counting of people and data collection will enable the administrators to have data of people in attendance for proper planning. Image sensing device captures images [4].In this paper, background Subtraction is considered for varying shadows and lighting conditions. Background Subtraction is based on colour and texture. Its application is to count people only and is placed at overhead position. The result of analysis of data will enable the knowledge of statistical information on the people flow at different periods of time.

Also, the counting system is important especially in areas like people flow analysis, video surveillance, behavior analysis access control or event management. The methodology is more flexible to count. Likewise, adaptive algorithms are developed to ensure accurate counting for both indoor and outdoor counting using image sensing device counting. In order to create an efficient system, this study focused on counting system that provides a quick method of sensing, counting and collecting data with an image sensing device

2. Previous Works

Crowd counting based on count estimates in a multi-view camera environment was done using k-NN classification from the feet maps. The proof of the foreground likelihood was obtained in each view through a robust bio-inspired mechanism which has self-organizing background subtraction against foreground detection challenges. It was able to detect both stationary and moving foreground objects [5]. Proposed multi-view Computational complexity of a people counting system is linear and depended on views only in the computational feet map. However, the denser and larger the crowd, the harder individual detection and tracking becomes. This makes crowd counting inaccurate.

An approach was proposed based on perspective weight model to segment the crowd from the background for dense crowd counting in public, the setting of region of interest and background subtraction are proposed first. Likewise, a perspective weight model was produced through feature extraction, [6]. The results indicate that the proposed approach is effective for crowd counting but suffer from poor effects of a dynamic background in outdoor crowd counting and real-time performance. In scene invariant multi camera, the algorithm of people counting designed operates across multiple view cameras [7]. Camera calibration approach was used to normalize features to compensate for regions of overlap between viewpoints. The evaluation of the combination of multi camera crowd counting and scene invariance was evaluated using system on footage obtained from the QUT camera network which was trained and tested on three major cameras from the 2009 database of PETS. Crowd counting observed indicates less than 10% mean relative error. A proposed automated counting of people in crowd with low-resolution visual and infrared cameras was carried out. The experiment conducted by mounting the low cost thermal imager and visual imaging device looking vertically shows that the results of 18 experiments of the system over a wide range of lighting conditions is accurate within 3% [8]. A label distribution learning for common complexity scenes was proposed for crowd counting in public based on video surveillance [9]. The existing crowd counting approaches based on regression models learn a mapping between class labels and low-level features. Experimental result confirmed the effectiveness of label distribution learning crowd counting method and the robustness to various data sets contains imbalanced and insufficient training data. Two main approaches were covered in recent survey on crowd density estimation and counting for visual surveillance. They are direct approach (object based target detection) and indirect approach (corner points, pixel, and texture-based analysis) [10]. The task is becoming complex and difficult when detecting people with the presence of occlusions due to highly denser crowd [11]. Therefore count estimation is performed only on blobs labeled as groups. Counting is done for tracking group as follows:

Estimate the count for the current frame of the group being tracked in intersected area.
Update the estimate of the group tracker for the blob tracked, if not group tracker is initialized for the group with initial estimate of the count.
Sum up the count estimates for all groups in the scene by compute the total number of people in the scene, along with the people present. A frame-by-frame basis is done in real-time. This method ensures a simple solution to the problem encounter in group counting of people, though it fails in some cases.
(1) Density determines a group estimated count. Since the number of people occupying the same area is less than the average. When significant gaps are created, the technique will overestimate the count.
(2) If changes occur in configuration with this method or there is dynamics of the explicit group, even though the number of people remains the same, the projected area it occupies may change, [12].
Fixed head plane at a constant height of 160 cm is used in this method. Although relatively stable results can be produced with conservative estimate in some cases still height of a group may be different significantly, which affects the count accuracy. In order to deal with these challenges, a more flexible methodology is proposed to generate count estimates with consideration given to the shape of the group as presented in detail in the consequent section.

Adaptive algorithms developed were to ensure accurate counting for both indoor and outdoor counting using video image counting. Background Subtraction considered is based on colour and texture and is robust algorithm for variation in shadows and conditions of light. Mostly many works recently by-pass the task of individual detection even with the present pattern recognition techniques and computer vision advancement which can save time spent on processing. In order to create an efficient system, this paper focused on automatic counting system to find a quick method of sensing, counting and collecting data using image sensing device.

3. Methodology

Crowd density estimation and counting basically comprises of direct counting method and indirect counting method. Indirect counting method as adopted comprises of pixel analysis, texture analysis and corner points analysis.

IOP Conf. Series: Materials Science and Engineering **413** (2018) 012059 doi:10.1088/1757-899X/413/1/012059

1 Lighting systems

The most critical and essential aspect in image processing is lighting system used to receive a quality image. This provides homogeneous illumination for quality lighting set-up to make the information from images sufficient for any machine vision application [13].

2 The images from video camera are taken for analysis. Background of the image is estimated for segmentation. Then information is passed through a decision making algorithm for tracking and counting the people to stop the counting process.

3 Processes of image segmentation involves the use of MATLAB Image processing

Video camera has been introduced here for automatic people counting. The images from video camera are processed on computer (RAM 4.00GB; Processor; Intel (R) Celeron CPU 2.16GHz) via USB connection for analysis. Details of the process have been discussed in the following sections.

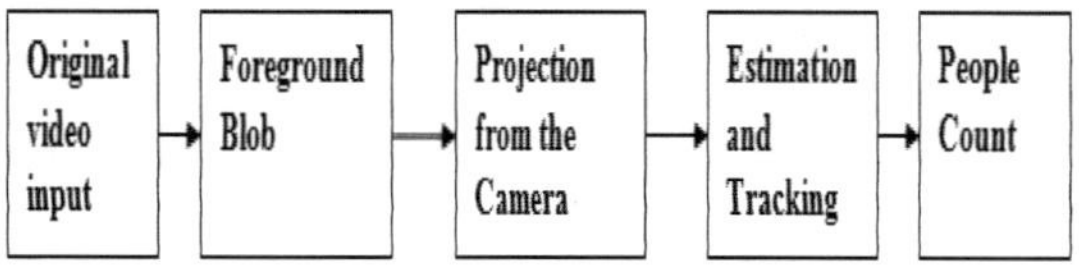

Fig 1: Flow Chart of People Counting System

The algorithm, include estimation received for every foreground 'blob' in image segment which make total estimate for the scene which also is the sum of the estimates for every blob. Ground truth annotation is done to train the system, by explicitly people labeling as represented by every blob formed in image. For this reason, every frame generates many instances of ground truth. Feature extraction is done subsequently to group segmentation which is used to achieve normalization across scenes by weighing the features extracted from pixel by pixel.

The categories of features used for counting are:

1. Size or magnitude of interested segments extracted from a relevant image.
2. Shape as related to the orientation and descriptors of the image segments.
3. Edges is measured using a binary edge detector. It is relatively change in pixel intensities across an image,
4. Key points are any other interesting points detected in an image such as corners

Thus, normalization of the features is important in order to reduce effects of perspective.

Foreground segment features. Moreover the foreground segment is obtained by the background subtraction .The image formation diagram is shown below.

a. Reading of the Image

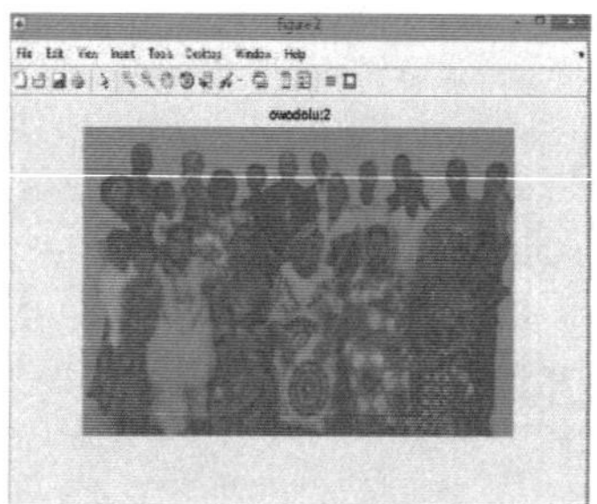

b. changing the color space

c. Equalizing brightness to get skin area

d. Image conversion rgb2gray

e. Converting Gray Scale to Binary Image

f. Complementing the Binary Image

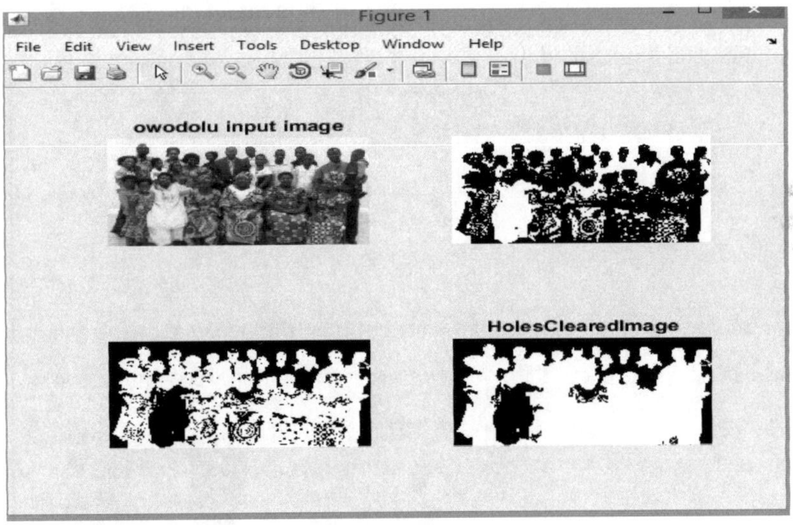

g. Image processing flow display

Fig 2(a-g): Image Processing

4. Discussion

The new methodology proposed for the system contains several steps:

1. Background extraction which uses improved algorithm to get a pure background.
2. Background subtraction which uses background subtraction algorithm, the foreground is separated from the background and forms a foreground mask.
3. The system detects blobs that having more than one in them.

4. Segmentation. These include: image reading, morphological opening to background, background image differencing, balancing of image contrast, image thresholding and real image formation.

The morphology opening is composed of erosion and dilatation. Erosion and a dilatation can be made using the same structure of element based on identity. Image is required at the starting of this people counting system and for this reason image has been taken from video image sensing device (camera) and transformed into gray-scale image for analyzing. Frame differencing is the first steps of people counting algorithm [14]. Frame differencing is the process of making order of pixel absolute differencing between the two consecutive or simultaneous frames which in turns results in a new image which depicts all observed differences between this two consecutive frames. Motion can be detected by this image difference. There must be modifications between the consecutive frames if it is not an empty image formed. Therefore in the field of the video camera there is motion. Background estimation is another important process of people counting system. A reference image for estimating background part of the scene is created by background estimation algorithm. Detecting and tracking moving objects in the background image is necessary and this is used to separate the foreground and the background [15]. During the days intensity of light of the sun varies and for this reason some objects can be added or removed in the scene. For this reason estimation of the background should be updated time to time otherwise people counting algorithm will fail to detect the variations. Basically, the background image is to detect objects in motion and also track them and this is used to separate the foreground and the background.

The differences between these two consecutive frames from the operation results in a new image. The background image must be updated because the background estimation is dynamic. Also, the morphological operation can remove small objects created by noise likewise the binary image. For example, if the people counting system is placed in the entrance of one building, the good algorithm for people counting will bring about good result. In fact, all the objects on the scene is to be identified in order to track objects, which is the first step and all their features are then calculated. This process is known as a blob analysis. The Tracking and Counting is then performed.

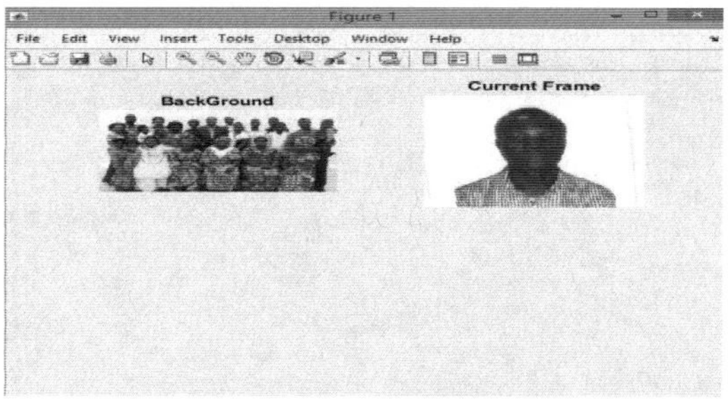

Fig 3: Frame differencing

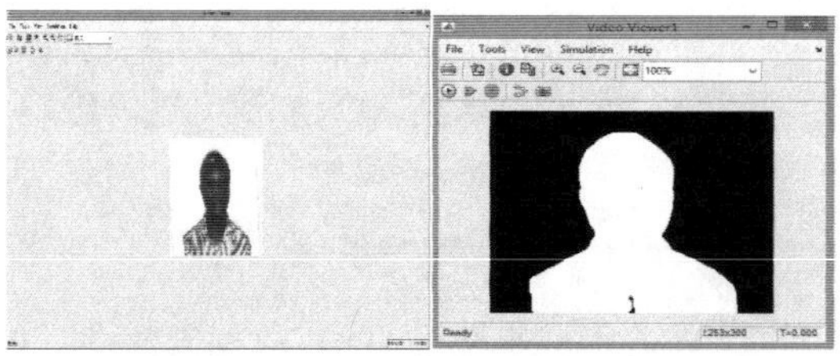

Fig 4: Segmentation and Blob formation

The reference image is updated when there is no motion. In other words, if there is existence of motion in the video camera frame, the background image will not be updated but will be in estimation of the background. In the estimation of the background, the algorithm is applied for separating the foreground and the background .At this point, large variation between the images of the video camera and the present background is represented by the foreground. Algorithm for frame differencing is quite similar with this process but have some differences. As we have frame differencing algorithm faces some difficulties to separate the background and the foreground of the image especially for current and previous frames due to the tendency to select the edges of objects in the foreground which makes the image analysis more difficult subsequently. Standing objects can be detected by differencing algorithm which is another limitation of this algorithm. Therefore, objects are detected by background subtraction process and motion is detected by differencing frame in people counting system. Erosion is the next morphological operation which is used for eroding away the boundaries of the several regions of the foreground. As a result, objects of the foreground region will become smaller (some of them will entirely be disappeared) and holes in objects will be bigger. Combination of two basic operations Erosion and Dilatation are called opening operation and the primary objective of this operation is to remove noise as well as separation of blobs that are linked with small layer. The important steps of the image processing algorithm are the Blobs detection for efficient tracking of the objects in scene. Typically, area (number of pixels which compose the blob), perimeter, and location and blob shape are the blobs features usually calculated. Segmentation quality maintains the performance of the blob analysis algorithm. After that is counting of people.

5. Conclusion

The image sensing device technology or people counting system involves image input, frame differencing background estimation, segmentation, tracking and counting stages. The paper studied the people counting methodology in a specific location or place. MATLAB is the programming tool or interface on computer that is adaptable for counting implementation in real time processing. This method is an alternative to old or manual counting system .This counting system can be used in various places and also for further research.

References

[1] M. Ashkanani, Ali Roza, Ali Sobhy M Naghavipour, and Hadi ''A Design Approach of Automatic Visitor Counting System Using Video Camera,'' 2015, Vol 10, pp. 62-67

[2] K. Kopaczewski, M. Szczodrak and A. Czyzewski ''A method for counting people attending large public events,''2015, 74: 4289

[3] J. Han ''Employing an rgb-d sensor for real-time tracking of humans across multiple re-entries in a smart environment, IEEE Trans.,'' 2012, Consum. Electron. 58 255–263.

[4] S. Ikemura, H. Fujiyoshi ''Real-time human detection using relational depth similarity features, in: Asian Conference on Computer Vision,'' 2010, pp. 25–38.

[5] Lucia Maddalena, Alfredo Petrosino, and Francesco Russo ''People counting by learning their appearance in a multi-view camera environment,'' 2014, Vol. 36, pp.125-134.

[6] S. Bahadori, L. Iocchi, G. Leone, D. Nardi ''Real-time people localization and tracking through fixed stereo vision,'' Appl.Intel.,2007,Vol.26,pp83–97.

[7] Xuemin Hu, Hong Zheng, Yuzhang Chen, Long Chen ''Dense crowd counting based on perspective weight model using a fisheye camera,'' 2015, Vol.126, pp.123-130

[8] D. Ryan, S. Denman, S. Sridharan, C. Fookes, ''an evaluation of crowd counting methods, features and regression models,'' Comput. Vis. Image Und. 2014, Vol. 130 pp.1–17.

[9] I.J. Amin, A.J. Taylor, F. Junejo, A. Al-Habaibeh, R.M. Parkin) ''Automated people-counting by using low-resolution infrared and visual cameras,''2008, Vol.41, pp.589-599

[10] Zhaoxiang Zhang, Mo Wang, Xin Geng Crowd counting in public video surveillance by label distribution learning, 2015, Vol.166, pp.151-163

[11] Sami Abdulla ,Mohsen Saleh, Shahrel Azmin Suandi Haidi Ibrahim Recent survey on crowd density estimation and counting for visual surveillance, 2015, Vol.41,pp.103-114

[12] A. Bevilacqua, L.D. Stefano and P. Azzari, ''People tracking using a time-of-flight depth sensor,'' in 2006 IEEE International Conference on Video and Signal Based Surveillance,pp.89–93.

[13] M. Enzweiler, A. Eigenstetter, B. Schiele, and D. Gavrila. ''Multi-cue pedestrian classi- fication with partial occlusion handling'', in 2010, IEEE Conference on Computer Vision and Pattern Recognition, pp. 990–997.

[14] F. Fleuret, J. Berclaz, R. Lengagne, and P Fua., Multicamera people tracking with a probabilistic occupancy map, 2008, IEEE Trans. Pattern Anal. Mach. Intell., 30 267–282.

[15] W. Choi, C. Pantofaru, and S. Savarese ''A general framework for tracking multiple people from a moving camera,'' 2013, IEEE Trans. Pattern Anal. Mach. Intell. 35, 1577–1591.

IOP Conf. Series: Materials Science and Engineering **413** (2018) 012060 doi:10.1088/1757-899X/413/1/012060

The Role of Research in Economic Development

Okokpujie I. P. [1*], Fayomi O.S.I. [1, 2], Leramo R. O. [1]

[1]Department of Mechanical Engineering, College of Engineering, Covenant University Ota, Lagos state, Nigeria
[2]Department of Chemical, Metallurgical and Materials Engineering, Faculty of Engineering and the Built Environment, Tshwane University of Technology, Pretoria, South Africa.
*Imhade.okokpujie@covenantuniversity.edu.ng, imhadeprincess@gmail.com ojo.fayomi@covenantuniversity.edu.ng

Abstract- This study channels via the association between analysis utilizations, and financial advancement at the national level in creating country. Innovative works are basic driving advancement, national improvement and financial development, yet there likewise should be objective will to succeed. There must be a solid collaboration amongst the scholarly world and the industries. At the same time, hypothetical, as well as experimental writing has radically demanded that the thing of interests in examine is an irreplaceable part of economic development. The affiliation between inquire about uses and monetary execution has pulled in a lot of consideration in the scholastic field for a long while. Notwithstanding that, the subject has been for some time stressed by governments and private undertakings. The vast majority of the modern financial experts have ascribed the sustained development in developingnations to their escalated inquiresin research works. The dialog of the connection between economic development and research expenses has an uncommon direness, for instance Turkey's aspiring financial targets. The emphasis on human capital as a driver of economic development for developing nations has prompted undue consideration on school accomplishment. Developing nations have made impressive advance in shutting the hole with regarding school accomplishment, yet ongoing exploration has underscored the significance of subjective aptitudes for economic development. This outcome shifts regard for issues of school qualityaround the developing nations has been significantly less fruitful in shutting the holes. Without enhancing school quality, developing nations will think that it is hard to enhance their long run financial performance. This paper exposes the impact of research in economic development of developing nations where it is taken in high-regard, shows where research has taken them to, influencing their GNP, NDP, new inventions, established standards.

Keywords: Research, Human Capital, Economic Development

1. Introduction

Research University in low-and centre wage nations have critical parts to play in creating separated and successful scholastic frameworks, and in making it workable for their nations to join the worldwide learning society and contend in complex information economies [1]
An in-depth study of something with a focus and undivided attention to get intense knowledge about it for addition to already existing stock of knowledge is known as Research.

It is apparent that for accomplishing long haul financial aspects development the measure of Research is profoundly critical. The immediate impact can be estimated with trouble due to the slack and the genuine relationship. So, following the genuine impacts of the fruitful item advancements, improvements will be increasingly essential yet additionally troublesome in light of the fasten mechanical changes also, the expanding number of new customers and markets. In any case, even of them, still numerous of the change tries occurred without the perception of its ease of use or the absence of sources [2].

Research and development are key long determinant of profitability and buyer welfare. Be that as it may, the amount Research and Development is sufficient? A great part of the hypothetical writing has concentrated on reasons, for example, imposing business model evaluating and people in general amiable attitude of information—why there might be too minimal private Research and Development. Notwithstanding, there are similarly convincing reasons, identified with the contorted impetuses from patent races and the exchange of rents through imaginative obliteration, to feel that there might be excessively Research and Development. In the end, the inquiry with regards to the sign and extent of the deviation of Research and Development venture from its socially ideal level is an experimental one that can be drawn closer from various headings. [3]
Jones and Williams [4] demonstrate to translate existing econometric assessments of profits to Research and Development with regards to endogenous development models and locate that ideal Research and Development venture is no less than four times more noteworthy than genuine spending. This approach yields a reasonable answer with regards to the level of over-or underinvestment in Research, however it doesn't reveal which factors decide the general outcome, the learning of which may have critical arrangement suggestions.

2. Literature Review on the Purpose of Research and the Significance of Research

Falk [5] inspected the impact of Analysis and Development uses on monetary development for OECD nations with board information investigation strategy by utilizing the information identified with the period 1970-2004. As per the Findings, both the proportion of organization's Research and Development uses to GDP, and the offer of Analysis and Development interests in cutting edge innovation segments, has solid beneficial outcomes on both per capita pay and normal hourly profit per capital in the long haul.

Nadir and Kim [6] examined the spread of global Research and Development, business and proficiency for G-7 nations. In this unique circumstance, he dissected the impact of universal Analysis and Development overflow on the expansion of aggregate factor profitability, impact of overflow on the generation structure, impact of innovation exchanges on fare and import models, and computing the personal come back rates of Research and Development speculations and physical capital also social come back rates of Research and Development for seven noteworthy industrialised nations. Because of the examination, it was presumed that there are huge contrasts between a nation's own Research and Development endeavours and the level of advantage picked up from the Research and Development ventures of different nations. It was additionally presumed that the course of worldwide overflow consequences for factor requests and yield is steady between the nations, yet their extent fluctuates as per distinctive nations and periods.

Nunes et al. [7] directed an examination to decide if there is a comparable similarity as far as Research force and development amongst little and average size endeavours with high innovation and the individuals who need high innovation. As per the discoveries, Research force confines the development of ventures with high innovation at bring down levels of power, and urges them to develop at higher Research Expenditures and Economic Growth Relationship in Turkey 191 levels.

In any case, Research and Development power confines the development of endeavours without high innovation paying little heed to the level of Research and Development level. In their investigations on global Research and Development overflow and associations, Coe et al. [8] came to the summary that hierarchical contrasts are huge factors for add up to determinant profitability and Research and Development overflow influences its level.

Horowitz [9] has examined the relations amongst Research and Development increment and provincial monetary development for different states in United States utilizing the information for the period 1920-1964 and inferred that development condition, connection estimation coefficients are reliable with development, and locales are happy with the way that Research and Development exercises are very predictable with development rates.

Kim [10] hypothesized the impact of Analysis and Development exercises on monetary development in Korea by utilizing Analysis and Development based Cobb-Douglas creation work and the information for the period 1976-2009. As indicated by observational discoveries got because of the examination, components of creation (work and capital) make a commitment to monetary development by around 65%. Additionally, commitment rate of Research and Development stocks on monetary development is around 35%. After investigation, it is seen that estimated commitment of private and open Research and Development stocks on monetary development are separately 16% and 19%.

Funke and Niebuhr [11] dissected the connection amongst Research and Development overflow and financial development for West Germany utilizing the information for the period 1976-1996. As indicated by the information acquired, dispersal of data is past local limits. Likewise, in geological terms, overflow between close areas were for the most part viewed as essential.
Sadraoui et al. [12] broke down the causality amongst Research and Development coordinated effort and financial development by dissected the connection amongst Research and Development consumptions and monetary development for China and achieved the summary that GDP will expanded by roughly 0.92% if there is 1% expansion of Research and Development uses [13]. Ulku [14] examined the connection amongst Research and Development, advancement and financial development with board information technique for 20 OECD nations and 10 nations, which are not individuals from OECD by utilizing Patent and Research and Development information for the time of 1981-1997. The information got propose that impact of Research and Development stocks on advancement is huge just in OECD nations with expansive markets, and there is a positive connection between GDP per capita and advancement in both OECD nations and nations, which are not individuals from OECD.

Altın and Kaya [15] broke down the connection amongst Research and Development uses and monetary development by utilizing the information for Turkey for the period between 1990-2005 and found that there isn't a causality connection amongst Research and Development uses and financial development in the long haul, yet there is a causality similarity from Research and Development uses to financial development in the long haul. Korkmaz [16] dissected the connection amongst Research and Development speculations and monetary development with Co-joining strategy by utilizing the information for the period 1990-2008 and arrived at the decision that there is a co-reconciliation between the two variations and the two variations influence each other in the long haul. Also, ŞimşekveBehdioğlu [17] led an investigation on the significance of financial development by breaking down the Research and Development exercises in Turkey and OECD nations with bunch examination for the period from 1999-2002 and arrived at the conclusion that Turkey falls behind OECD nations concerning Research and Development markers.

IOP Conf. Series: Materials Science and Engineering **413** (2018) 012060 doi:10.1088/1757-899X/413/1/012060

Taban and Şengür [18] clarified the connection amongst Research and Development and monetary development by utilizing the information for the period between 1990-2012 in Turkey and coincorporation models, and went to a position that Research and Development consumptions emphatically influence financial development over years. According to Gülmez and Yardımcıoğlu [19] broke down the connection amongst Research and Development uses and monetary development in OECD nations by utilizing the information for the period 1990-2010 and reached the decision that there is a huge intuitive connection amongst Research and Development consumptions and financial development variations in the long haul. One strand of models incorporates those by Barro [20], Barro and Lee [21] and Baumol [22], which contend that human capital assumes a critical part as an encouraging variable on the universal exchange of innovation from advancing nations to 'replicating' ones, helping them to 'make up for lost time' with the created nations. Jin [23] develops a growth mode, assuming that the creation of new thoughts/plans is an immediate capacity of the human capital (which has the type of logical information). Hence, interest in human capital, by enhancing innovative work, involves a development in physical capital speculation, which thusly brings about higher genuine development rates.

In all actuality, from a theoretical viewpoint, though specific change may grow the enthusiasm for human capital and along these lines the wage premium of expert over unskilled specialists, a bigger measure of direction should provoke a higher supply of talented work and a short time later to a fall in the wage differential.

2.1 Purpose of Research

Research is a logical and deliberate scan for appropriate data on particular theme. It is a workmanship and in addition art of examination. The propelled Learners' Dictionary of momentum English characterized as "a watchful examination or request extraordinarily through scan for new actualities in any branch of information." Redman and Mary characterize look into as a "systematized push to increase new learning". A few people consider examine as a movement, a development from the known to the obscure. Exploration is a scholarly procedure whereby an issue is seen, separated into its constituent components, and examined in the light of certain fundamental presumptions (J.H.Shera).

2.2 Significance of Research

As indicated by Hudson Maxim Significance as, "All advances is conceived of request. Uncertainty is frequently superior to anything carelessness, for it prompts request and request promptsexamination"Research instils logical and inductive reasoning and it advances the improvement of coherent propensities for considering and association. The part of research in a few fields of connected financial matters, regardless of whether identified with business or to the economy in general, has significantly expanded in present day times. Research gives the fundamental to about all administration strategies in our economic framework.

Research gives the premise to almost all administration approaches in our economic framework. Research has its uncommon centrality in taking care of different operational and arranging issues of business and industry. In a few ways, activities investigate, statistical surveying and motivational research are essential and their outcomes help with taking business choices. Research is similarly critical for social researchers in concentrate social connections and in looking for answers to different social issues. It gives scholarly fulfilment of knowing things for information. It likewise has the down to earth utility for the social researcher to pick up learning in order to have the capacity to improve or in a more proficient way.

IOP Conf. Series: Materials Science and Engineering **413** (2018) 012060 doi:10.1088/1757-899X/413/1/012060

2.3 Impact of Research in Economic Development in respect to Establishment of Research, New Inventions, Increase Gross National Profit and Impact on Country's Human Capital

2.3.1 Establishment of Research

Kwack and Yang [24] broke down Korean development encounter from neoclassical development point of view. Yearly information for the period 1971-2002 was utilized for this reason. Speculation rate, R&D, Education and size of government developed as real determinants of long-run financial development in Korea.

The Study gave confirmation to endogenous development hypothesis. The examination demonstrated that open and family unit's consumption on instruction and Research venture are significant supporters of development and enhancing nature of work. Higher reliance proportion of young and more established individuals antagonistically influenced Korean development rate. Change of political, social and social establishments was prescribed to accomplish long-run managed monetary development rate.

2.3.2 New Inventions

The Research and development uses might be considered as an interest in new innovations, and information base, which can be then, changed into more effective creation techniques for accessible assets. Should the more elevated amount of Research and development consumptions succeed, the larger amount of development rates may be normal. In light of the overflow impacts, the active advantages of new thoughts may not precisely pay back to the general population who really improve them, and this circumstance focuses to the way that private segment will be socially more averse to lead Research exercises on ideal level, giving that no political intercession is accessible. With a specific end goal to empower the Research exercises of private part, this condition may genuine some pubic mediations, for example, backhanded measures like expense motivating forces and insurance of protected innovation rights, and in addition giving direct help, for example, supply and fund accommodations (OECD, [25])

Grossman [26] built up a model to discover the commitment of Research endowments and openly gave science instruction to financial development. The examination presumed that Research sponsorships may not contribute monetary development and open welfare and the inter-temporal learning overflows are the externalities of Firms use on Research. The sponsorships to Research increment salary disparity.

As state funded instruction adds to monetary development more proficiently than Research sponsorships, in this manner, it was proposed to create Research through advance government funded training of researchers and other talented people.

2.3.3 Increase Gross National Profit

Khan and Khattak [27] took a shot at the part of human asset administration in financial aggressiveness in South Asian economies with exceptional reference to Pakistan. It was watched that Pakistan profited not the brilliant open doors made by globalized world. Powerless aptitude base is one of the main considerations fending off Pakistan of taking preferences from worldwide markets. The investigation proposed change in ability base, extension of instruction and preparing, and advancement of Research for better monetary outcomes. Fundamental instruction is critical to for human asset improvement in Pakistan. It was additionally recommended to quicken the labour fare to diminish neediness and enhance macroeconomic pointers.

Falk [5] built up a dynamic observational model to know criticalness of Research interest in long run monetary development of OECD nations utilizing board informational index. The investigation gave another confirmation to Research-monetary development relationship. The outcomes were determined through GMM (summed up strategy for moments). The think about explored whether higher Research venture push financial development keeping speculation proportion, mechanical Research force and human capital steady. Five yearly and ten yearly midpoints were utilized. Higher Research speculation was discovered decidedly identified with GDP development in working age populaces. The outcomes were vigorous in both 5-year and 10-year cases.
Goel, et al [28] planned the patterns in different segments of and its commitment to economic development in USA by utilizing broken-down information of 50 years. Shockingly, USA experienced decrease in safeguard Research costs and government Research consumption. The nongovernment Research financing extended pointedly amid the period. Strikingly, the estimation demonstrated solid relationship of monetary development to government Research use instead of non-elected Research expenses. The economic development portrayed solid relationship to guard research rather than non-government research. The investigation proposed significant push up in protection Research and non-government Research in USA for economic development.

2.3.4 Impact on Country's Human Capital

Kuoet al [29] analysed the impacts of information capital and innovation overflow on territorial financial development in China. The outcomes demonstrated that Research, capital and innovation import contribute altogether too monetary development in China. Flexibility of Research to economic development was as substantial as of innovation demonstrating same commitment to financial development of China. An investigation proposes the presence of Research overflow and global learning overflows. Assessment impetuses, money related help and Research gifts can be useful instruments to energize look into exercises and advancement in economy. Likewise, arrangements which energize interest in instruction and employment preparing were suggested for China. Human capital collection and Research Development are unquestionably two of the most imperative motors of development and, for some odd reason, over the most recent couple of years just couple of endeavours have been made keeping in mind the end goal to coordinate them inside a bound together and homogeneous system. In any case, in these papers abilities amassing occurs through learningbydoing and at work preparing in the generation movement instead of a different development (or training) part. Grossman and Helpman [30] additionally indigenise both human capital and specialized change.

3 Methods and Methodology

3.1 Research Goal

In the present extreme aggressive air of economic process, the countries are dynamically turning their face to in the present intense aggressive atmosphere of globalization, the nations are progressively turning their face to in the present extreme focused atmosphere of globalization, the nations are progressively turning their face to innovative work exercises keeping in mind the end goal to save and fortify their aggressive edges all through the globe. Since, there is most likely that any expansion in innovative work exercises causes an outstanding ascent in efficiency.

IOP Conf. Series: Materials Science and Engineering **413** (2018) 012060 doi:10.1088/1757-899X/413/1/012060

Table 1: Data Collection (Source: [31])

Dependent Variables		Variables		
Log (GDP)		Log(employee)	Log (patient)	Log (RD)
Coefficients		0.5070	-0.2134	0.0148
		(2.7531)	-2.3134	2.1175
R^2	0.85	Adjusted R^2	0.78	
F-statistic	51.59			
DW-Stat.	1.77			

Table 1 Presents the Panel Regression comes to fruition. In perspective of backslide comes to fruition the illustrative vitality of the model is extremely basic (R2=0.85) and there is no autocorrelation (DW=1.77). As demonstrated by this model the coefficient of delegate is sure and essential at the 1% level. Of course, the coefficient of Research and improvement theory is sure and critical at the 5% level. Additionally, the coefficient of the amount of patent is negative and gigantic at the 5% level. The outcome demonstrated that there is a strong association between the number of R&D agents and GDP. Be that as it may, there is a frail association between's R&D endeavours and GDP. When we looked coefficient of number of licenses there is a negative association among GDP and number of licenses. Every research has its own method, all the method is tending towards increase in production in all manufacturing industry [32-34]

3.2 Analysis and Results

These results from the data cumulated over the years for a developing country, Turkey. In this exploration in which the connection between R&D uses and financial development is investigated, as a matter of first importance the stationary of arrangement incorporated into the examinations is tried. From the above Table 1, under Significance at 1% level ** Significance at 5% level

4 Conclusion

Research is, subsequently, a unique commitment to the current load of learning making for its progression. It is the quest for truth with the assistance of study, perception, examination and test. To put it plainly, the look for learning through goal and orderly strategy for discovering answer for an issue is inquire about. The deliberate approach concerning speculation and the definition of a hypothesis is likewise investigated.

The surveys of hypothetical and exact examinations important to the part of Research in financial development of nations around the globe concur on the huge part of various frame research in efficiency or monetary development. Along these lines, it can be presumed that the creating nations should focus on research to accomplish the supported monetary development. Research centres in developed nations are at the highest point of the scholarly pecking order and are fundamental to the accomplishment of any advanced learning-based economy. Every single developed nation requires these establishments to take an interest in the globalized condition of advanced education. In this manner, understanding the attributes of the exploration college and building the foundations and the scholarly condition required for fruitful research colleges is a best need."

References

[1] Frantzen D., Philips C. (2000). R&D. Human Capital and International Technology Spillovers: A CrossCountry Analysis. The Scandinavian Journal of Economics, 102 (1), 57-75. DOI: 10.1111/14679442.00184.

[2] Czarl J, Evenson, R.E. (2007). Economic Growth, International Technological Spillovers and Public Policy: Theory and Empirical Evidence from Asia. Yale University, Economic Growth Center, Discussion Paper No. 777.

[3] Charles G, Engelbrecht, H.J. (2000). International R &D spillovers, human capital and productivity in OECD economies: An empirical investigation, European Economic Review, 41,1479-1488.

[4] Jones C.I., Williams, J.C. (1999). Too Much of a Good Thing? The Economics of Investment in R&D.

[5] Falk, M. (2007). R&D spending in the high-tech sector and economic growth. Research in Economics, Vol. 61, pp. 140–147. http://dx.doi.org/10.1016/j.rie.2007.05.002

[6] Nadiri, M.I., Kim, S. (1996). International R&D Spillovers, Trade and Productivity in Major OECD Countries.

[7] Nunes, P.M., Serrasqueiro, Z., Leitao, J. (2012). Is there a linear relationship between R&D intensity and growth? Empirical evidence of non-high-tech vs high-tech SMEs. Research Policy 41, 36-53

[8] Coe F, Chou, Y. K. (2002). The Australian Growth Experience (1960-2000), R & D based, Human Capital-Based or Just Steady State Growth? Research Paper No. 855, ISSN 0819-2642, Department of Economics, University of Melbourne. http://econpapers.repec.org/RePEc:mlb:wpaper:85

[9] Horowitz, I. (1967). The Relationship Between Interstate Variations in the Growth of R&D and Economic Activity. IEEE Transactions of Engineering Management, EM-14, No.3.

[10] Kim, L.W. (2011). The Economic Growth Effect of R&D Activity in Korea. Korea and the World Economy, 12(1), 25-44.

[11] Funke, M., Niebuhr, A. (2000). Spatial R&D Spillovers and Economic Growth-Evidence from West Germany. Hamburgiches Welt-Wirtschafts-Archiv Discussion Paper 98.

[12] Sadraoui, T., Ali, T.B., Deguachi, B. (2014). Economic Growth and International R&D Cooperation:

[13] Ulku, H. (2004). P.S. (2000). R&D, Innovation, and Economic Growth: An Empirical Analysis. IMF Working Paper, WP/04/185.

[14] Altın, O., Kaya, A. (2009). Türkiye'deAr-Ge HarcamalarıveEkonomikBüyümeArasındakiNedenselİlişkininAnalizi. EgeAkademikBakış 9(1), 251259.

[15] Korkmaz, S. (2010). Türkiye'deAr-Ge YatırımlarıveEkonomikBüyümeArasındakiİlişkininVar

[16] Şimşek, M., Behdioğlu, S. (2006). Araştırma-geliştirme (AR-GE) faaliyetlerininTürkiyeOECD ülkelerindekümelemeanaliziileincelenmesiveekonomikbüyümedekiönemi. İktisatİşletmeveFinans 21(245), 123-137.

[17] Taban, S., Şengür, M. (2013). Türkiye'deAr-Ge veEkonomikBüyüme. AİBÜ SosyalBilimlerEnstitüsüDergisi, 14(1), 355-376. International Journal of Economics and Financial Issues, Vol. 5, No. 1, 2015, pp.188-198

[18] Gulmez, A., Yardimcioglu, F. (2012). OECD ÜlkelerindeAr-Ge HarcamalarıveEkonomikBüyümeİlişkisi: Panel Eşbütünleşmeve Panel NedensellikAnalizi (19902010). MaliyeDergisi, 163, 335-353.

[19] Borensztein E, Gregorio, J. D., & Lee, J-W. 1998. How Does Foreign Direct Investment Affect Economic Growth. Journal of International Economics, 45, 115–135, DOI. 10.3386/w5057

[20] Borro, J.Y., Chuang Y.C., Lai, W.W., Yang, M.C. (2012). OECD A dynamic general equilibrium model for public R&D investment in Taiwan. Economic Modelling 27, 171-183.

[21] Jin, J. C., 2009. Economic research and economic growth: Evidence from East Asian economies, Journal of AsianJournal of Policy Modeling, 30, 237–250. http://dx.doi.org/10.1016/j.jpolmod.2007.04.008

[22] Samimi, A.J., Alerasoul, S.M. (2009). R&D and Economic Growth: New Evidence from Some Developing Countries. Australian Journal of Basic and Applied Sciences 3(4), 3464-3469.

[23] Kwack, S. U & Lee. Y. S . 2006. Analyzing the Korea's Growth Experience: The application of R&D and Human Capital Based Growth Models with Demography, National Public Expenditures on R&D. http://ec.europa.eu/invest- inresearch/pdf/download_en/spa4_final_report_final.pdf, (11.03.2014).

[24] Grossmann, V. 2007. How to promote R&D-based growth? Public education expenditure on scientists and engineers versus R&D subsidies. Journal of Macroeconomics, 29, 891–911. http://dx.doi.org/10.1016/j.jmacro.2006.01.001.

[25] Gerybadze, A. (2010). R&D Innovation and Growth: Performance of the World's Leading Technology Corporations. Innovation and International Corporate Growth. 7, 11-30.

[26] Khan, J., and Khattak, N. U. R. 2013. The Significance of Research and Development for Economic Growth: The Case of Pakistan, City University Research Journal, pp.175-86.

[27] Genç, M.C, Atasoy, Y. (2010). Ar-Ge HarcamalarıveEkonomikBüyümeİlişkisi: Panel VeriAnalizi

[28] Goel, R. K., Payne J. E., & Ram.. 2008. R&D expenditures and U.S. economic growth: A disaggregated approach.

[29] Kuo,C.C& Yang. C.H. 2008. Knowledge capital and spillover on regional economic growth: Evidence from China, China Economic Review, 19,594–604. http://dx.doi.org/10.1016/j.chieco.2008.06.004

[30] Grossman, G.M., &Helpman.E. 1989. Quality Ladders in the Theory of Growth. NBER Working Paper No. 3099 . http://www.nber.org/papers/w3099

[31] Awokuse, T.O., (2003). Is the export-led growth hypothesis valid for Canada? Canadian Journal ofEconomics, 36, 126–136.

[32] Salawu, E. Y., Okokpujie, I. P., Afolalu, S. A., Ajayi, O. O., & Azeta, J. (2018). Investigation of production output for improvement. *International Journal of Mechanical and Production Engineering Research and Development*, *8*(1), 915-922.

[33] Azeta, J., Okokpujie, K. O., Okokpujie, I. P., Osemwegie, O., & Chibuzor, A. (2016). A Plan for Igniting Nigeria's Industrial Revolution. *International Journal of Scientific & Engineering Research*, *7*(11), 489.

[34] Omoruyi, O. N., Omoruyi, M. G., Okokpujie, K. O., & Okokpujie, I. P. (2018). Electronic Fare Collection Systems in Public Transits: Issues, Challenges and Way-Forward. *COVENANT JOURNAL OF ENGINEERING TECHNOLOGY*, *2*(1).

IOP Conf. Series: Materials Science and Engineering **413** (2018) 012061 doi:10.1088/1757-899X/413/1/012061

Lifecycle Impact Assessment of an Engineering Project Management Process – a SLCA Approach

Israel Dunmade[1], Mfon Udo[2], Tunde Akintayo[3], Sunday Oyedepo[2], Imhade P. Okokpujie[2]

[1]Faculty of Science & Technology Mount Royal University, Calgary, AB T3K 0C3
idunmade@mtroyal.ca
[2]Mechanical Engineering Department Covenant University, Ota, Nigeria
[3]Mechanical Engineering Department Federal University of Technology, Akure, Nigeria

Abstract- Engineering projects are generally designed to solve societal problems, develop a community and/or to improve the quality of life of the beneficiaries. But sometimes there are hidden unintended consequences of projects on human life and the ecosystems. The society also affects implementation of projects. These impacts could be positive or negative. Similarly, the approach to managing engineering projects affects the extent to which project development process impact the workers, the community, the society and the ecosystems. This paper presents a framework for evaluating the potential lifecycle impacts of engineering project management process on workers, clients/consumers, the community and the society at large. The model is based on UNEP/SETAC guidelines on social lifecycle impact assessments. The analytical model is expected to be found helpful by project managers in identifying potential impacts of their projects on various categories of people, communities and the society. The findings will enable them to reinforce positive impacts and to eliminate or minimize negative impacts. Incorporation of a social lifecycle impact assessment in project management is expected to improve the overall value of the project, especially from the social sustainability point of view.
Keywords— social sustainability, lifecycle assessment, community impact assessment, Social impacts assessment, engineering project management, sustainable development, impact assessment

1. Introduction

There are various approaches to product development. One of them is project management approach. Project Management Institute defines project management as "the application of knowledge, skills, tools and techniques to a broad range of activities in order to meet the requirements of a particular project." A project lifecycle, according to [1], refers to "a series of phases and activities that a project passes through from its initiation to its closure and can differ from project to project depending on its area of application. It provides the basic framework for managing the project, regardless of the specific work involved, by defining how the project team envisages to start, plan, organize, carry out and close the project." Project management approach is often used for large size products such as aircraft, bridges, buildings, spacecraft, ship building, and road construction. The product realization process under product management approach consists of five groups of processes. The five process groups are initiating, planning, executing, monitoring & controlling, and closing. Figure 1 is an illustration of a typical project management process.

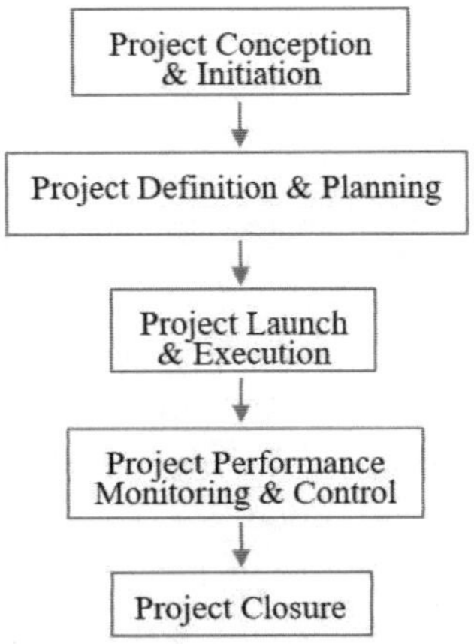

Fig. 1 An illustration of a typical project management process
Source: [2 - 4]

Project management process, just like any process, activity or product, has both positive and negative impacts on various stakeholders. The kind and severity of impact vary from one category of stakeholder to another. The focus of this study is on social impacts of project management process. According to [5], "society is taking increasing interest in assessing social impacts of various human activities". However, there are varieties of methods and approaches that are used in evaluating social impacts of our activities. The choice of social impact assessment approach depends on the object of interest. For example, one can utilize social impact assessment or health impact assessment if the focus is on a product, project or facility. According to [6], "social impacts are caused by changes." For instance, constructing a production facility, a flyover bridge or an airport where it has never been in existence is a change. Among the possible effects of such changes include increased employment opportunities, increased trade, increased noise level, and more traffic. These effects in turn have consequences such as improved standard of living from employment and death or injuries resulting from increased traffic. There is therefore a need to evaluate the overarching social impacts of a product and/or facility with the aim of providing information and shedding lights on areas where negative impacts needs to be addressed and where positive impacts should be reinforced. One of the tools for evaluating social impacts is social lifecycle assessment (SLCA). As the name implies, it is concerned with the social and socio-economic impacts of products, process, activities and facilities. It is a tool used to analyze the way products, projects and other business activities affects human well-being. According to [7], "Social Life Cycle Assessment (SLCA) is emerging as a powerful and necessary tool in sustainability science." However, the methodological aspects and applications of SLCA are not yet completely developed. They are currently at the evolutionary stage [8, 9]. SLCA is implemented in four steps, just like the counterpart lifecycle assessment tools. The four steps are: goal and scope definition, lifecycle inventory, lifecycle impact assessment, and lifecycle interpretation in Figures 2 and 3.

2. Goal and Scope Definition

A. The goal of the study

According to [10], the goal of a lifecycle assessment should specify the intended application, objectives of the study, and intended audience. The goal of this study therefore is to provide awareness of the potential social consequences of project management process so that policy makers, project managers and other stakeholders can make informed decisions regarding various stages of a project lifecycle. The work is also aimed at identifying hotspots in social sustainability aspect which will be useful in developing project plans and strategies to support the development of sustainable project management process.

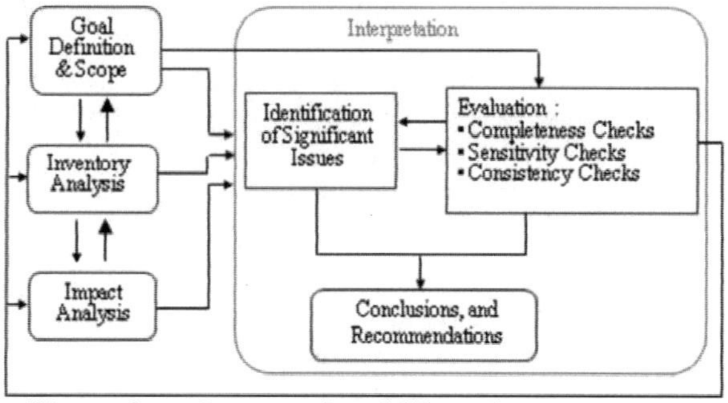

Fig. 2. An illustration of conventional lifecycle assessment process steps
Source: [7]

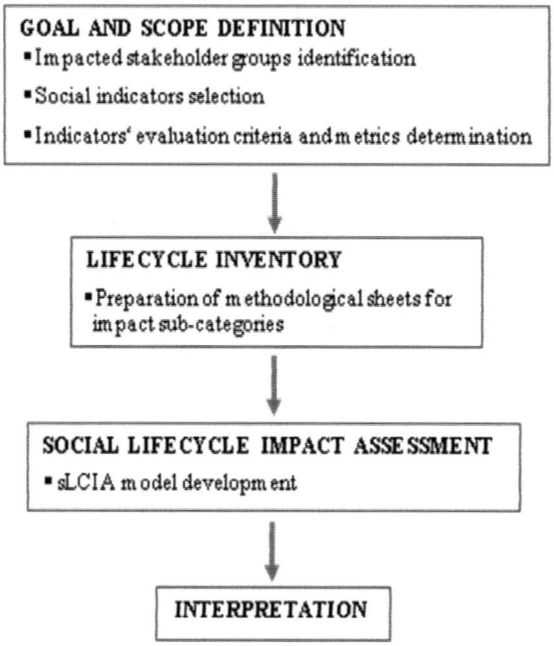

Fig. 3. Additional activities undertaken for sLCA at each LCA stage

To meet the goal of this research, the following questions will be answered:

- What are the appropriate social criteria that should be used to assess the social sustainability of a project management process?
- How should stakeholders assess the attainment of those criteria based on their experience in a specific case?
- What are the generic social sustainability hotspots within project management process lifecycle
- How can project stakeholders identify/foresee possible social sustainability hotspots at the conceptual/planning stages of a particular project?

B. Scope of the study

This sLCA study involves the development of a generic sLCA model for project management process and demonstrating it with a case study on an infant food production plant development project. At this stage of the study, we determined the function of the system, its functional unit, the system boundaries, data averaging, limitations and exclusions [11, 12]. We also identified affected stakeholder groups, impact categories, subcategories and indicator to be included in the analysis based on the goal of the study. Furthermore, we articulate the criteria for scoring the performance of the project on each indicator and determine the indicator scoring metrics in preparation for the lifecycle inventory.

B.1 A typical project management process

Figure 4 below is an illustration of the project management process of the pilot infant food production plant development project used as a case study. The project started with the recognition of the need by top management for locally produced nutritious baby food made from local ingredients. An infant food project development team/committee was then set up by the top management of the organization. Membership of the committee consists of relevant experts in the area of food science, microbiology, biochemistry, chemical engineering, techno-marketing, agricultural engineering, mechanical engineering, and planning. Project team are then assigned various responsibilities for the realization of the project objectives. The committee develop detailed plan for the project development. Quotations were invited from components, equipment and materials suppliers. After the funding release, selected materials and component suppliers were asked to supply their products. Scientists in the project team developed various formula and eventually came up with agreeable formula. Engineers designed and fabricated machinery for the demonstration infant food production pilot plant. Several experimental trials and modifications were carried out. At a point when the quality of infant food product and the pilot plant operations were considered adequate, the technology and the infant food product were taken to local and international trade fares. The infant food product was also sold to the public. In addition, seminars were also organized for possible transfer of the technology to potential entrepreneurs while further improvements were still progressively being carried out.

B.2 Function and functional unit

The function of a project management process is defined in this study as realization of the project purpose or providing the intended utility (deliverables) at the scheduled time and specified location(s). The functional unit proposed for measuring social impact of any project management process is the "net change in human wellness per person" (NCIHWPP) resulting from such project. This metric was chosen because according to [13], "everyone around the world, regardless of geography, age, culture, religion or political environment, aspires to live well." It is believed that every project is aimed at improving the human wellbeing. Thus, there is a need to measure how a project affect human wellbeing.

B.2.1 Human Wellness

According to [14], "Human well-being is a broad concept, one that includes many aspects of our everyday lives. It encompasses material well-being, relationships with family and friends, and emotional and physical health. It includes work and recreation, how one feels about one's community, and personal safety." [13] in sharing similar view stated that "Wellbeing is not necessarily bound by income, rather, it is an individual's thoughts and feelings about how well they are doing in life, contentment with material possessions and having relationships that enable them to achieve their goals." Quoting [15], [14] stated that "quality of life, welfare, well-living, living standards, utility, life satisfaction, prosperity, needs fulfillment, development, empowerment, capability expansion, human development, poverty, human poverty, land and, more recently, happiness" are other terms that are often used interchangeably with human well-being. [14] defined "human well-being as "the degree to which an individual, family, or larger social grouping (e.g. firm, community) can be characterized as being healthy (sound and functional), happy, and prosperous". [16] described wellness as "a full integration of physical, mental and spiritual well-being that leads to quality of life." [16] also described the seven dimensions of wellness, stating that "neglecting anyone dimension for a length of time has adverse effects on overall health." This study adopted the principles of the seven dimensions of wellness described by [16] for the evaluation of the social lifecycle impacts of project management process on human wellbeing.

B.3 System boundaries
The scope of the study covers an evaluation of how every stage in a project management process lifecycle affect human well-being. The focus of this study is on individual wellbeing, although according to [14], "the determinants of an individual's well-being can include characteristics that include characteristics of family, community, nation, and so forth". This evaluation is limited to impacts of a project on the wellbeing of project team and on the wellbeing of the immediate community where the project is implemented.

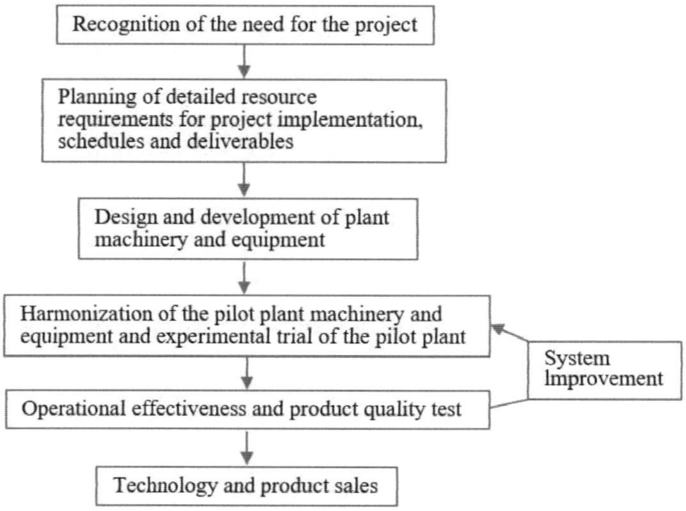

Fig. 4. The case studied Infant food pilot production plant project management process lifecycle

3. Lifecycle Inventory (LCI)

According to [17], lifecycle inventory "involves articulation and quantification of data to be used for impact analysis. Two major approaches have emerged in SLCA community regarding lifecycle inventory data to use. The first approach depends on company/site specific data. But they are often difficult to access. The second approach depends on generic data such as those obtainable from national censuses or public surveys which are easier to access [18]. According to [19], the development of social indicators that can be integrated into LCA depends on the sector that is monitored and the national context." Data used for this analysis is a site specific data relating to a research and development organization in Lagos State, Nigeria.

3A. Data Collection
There are three forms of social LCA data: quantitative, semi-quantitative (yes/no or rating scale responses) and qualitative (descriptive text) [20-26]. This is the stage where interview questions to obtain requisite data is prepared. The data collection is based on the UNEP/SETAC methodological sheets and focused on how project management process affect human wellbeing and ignite industries [43-45]. Stakeholder category of interest would have to be interviewed. The lifecycle inventory data reported would have to be an average of the data collected in each stakeholder category. Tables 1 and 2 are illustrative examples of lifecycle inventory analysis results for the case studied infant food production project management process. They show some of the project management process' social sustainability impact assessment criteria as they affect project team/workers and the local community where the project was implemented.

4. Social Lifecycle Impact Assessment (SLCIA)

According to [17], "Social Impact refers to the consequence of positive and negative pressure on social end points, i.e. the well-being of stakeholders. There are six steps in conventional (environmental) lifecycle impact assessment. The six steps are: definition of impact categories, classification, characterization, normalization, grouping and weighting. The same is transposed unto social lifecycle impact assessment." [17] explained that SLCIA stage of the lifecycle is still a work in progress and that this is in view of the "evolving understanding of possible social consequences of products, products and facilities." He further explained that there are several factors that bring variations in effects caused by products and facilities as experienced by individuals and groups of people. The author opinion is that the list of such impacts cannot be fixed but examples can be found in social lifecycle assessment literature.

A. Selection of impact categories and classification

In 2009 UNEP/SETAC published a set of SLCA guidelines [27]. The guidelines identified five stakeholders, out of which two of the stakeholder's categories are the focus of this study. The two stakeholder categories under consideration as earlier mentioned are project team/workers and local community where the project is implemented. The SLCA classification step in this study involved mapping impacts of project management process on the wellbeing of the project team/workers and on the local community. The selection of subcategories is based on the principles of UNEP/SETAC methodological sheets, currently relevant human wellbeing indicators project management practices in Nigeria and on the scope of the study. The indicators of the selected subcategories are defined by a set of semi-quantitative data.

4A. Characterization and normalization

The implementation of this step is as described in [17]. It involves converting the social information into interpretable indicators of a list of impacts. Quantitative approach was adopted in this study because it provides a platform for comparison with results. This will enable us to compare the results obtained from this study with future studies thereby enabling us to identify the improvements that had occurred over the period. The quantitative approach used in this study is the SIMSAW model developed by [17] and implemented in Microsoft excel program [7, 28-42].

4B. The SLCIA Calculation

The compilation of the potential social lifecycle impacts of a project management process performance for each stakeholder group was assessed using the SIMSAW model developed by [17]. The multi-layer model compiles normalized indicators scores at sub-category level and then sum up the weighted totals of all relevant sub-categories for each stakeholder group to obtain the facility's score with regard to the stakeholder group. The overall social impact (score) of a project management process is finally calculated by adding all relevant stakeholders' scores together. According to [17, "the normalization at subcategories level becomes necessary to avoid certain subcategories dominating the final result."

Using the SIMSaW model [17], the overall social impact (score) of the project management process under study is obtained from the expression,

$$S = T_1 + T_2 + \ldots + T_T \quad \ldots\ldots (1)$$

where T_k is the project management process performance score for stakeholder category t and obtained from

$$T_k = \frac{\sum_{j=1}^{J} w_j J_j}{\sum_{j=1}^{J} w_j} \quad \ldots\ldots (2)$$

IOP Conf. Series: Materials Science and Engineering **413** (2018) 012061 doi:10.1088/1757-899X/413/1/012061

Table 1: Sample LCI _ Stakeholder category: Project team/workers

Subcategories	Inventory Indicator	Normalized scoring metric	Averaged score	Maximum possible score
Physical Wellbeing	1. Working on this project provide me opportunity to exercise my body	Yes = 1, No = -1, N/A or Unverified = 0	1	1
	2. The level of exposure to injuries, harm and contagious disease while working on this project	High = -1, low = -0.5, Not at all = 0	-0.5	0
	3. Working on this project made me develop healthy habits (i.e. adequate rest, stop smoking, use safety equipment, etc.) .	Yes = 1, No = -1, N/A = 0	-1	1
Emotional wellbeing	1. Working on this project encouraged me to cultivate optimistic attitude	Yes = 1, No = -1, N/A or Unverified = 0	1	1
	2. Working on this project allowed me to freely express my feeling and share my views	Yes = 1, No = -1, N/A or Unverified = 0	0	1
	3. Working on this project enabled me to learn time management skills	Yes = 1, No = -1, N/A or Unverified = 0	1	1
Intellectual Wellbeing	1. Working on this project encouraged creativity and stimulated my mental activities	Yes = 1, No = -1, N/A or Unverified = 0	1	1
	2. Working on this project kept me up-to-date on current events and facilitated my participation in activities that arouse my mind	Yes = 1, No = -1, N/A or Unverified = 0	1	1
Social wellbeing	1. Working on this project helped me learn good communication skills (of my thoughts, feelings and ideas)	Yes = 1, No = -1, N/A or Unverified = 0	1	1
	2. Working on this project propelled me to get involved, share my talents and skills, and contribute to my community	Yes = 1, No = -1, N/A or Unverified = 0	1	1
Spiritual wellbeing	1. Working on this project encouraged me to meditate regularly and foster my commitment to my beliefs	Yes = 1, No = -1, N/A or Unverified = 0	0	0
	2. Working on this project gave me and those around me the freedom to be who we are	Yes = 1, No = -1, N/A or Unverified = 0	1	1

	3. Working on this project made me see opportunities for growth in the challenges that life brings	Yes = 1, No = -1, N/A or Unverified = 0	0	1
Environmental wellbeing	1. Working on this project increased my awareness regarding effects of our daily habits on the physical environment	Yes = 1, No = -1, N/A or Unverified = 0	0	1
	2. Working on this project encouraged me to conserve material, energy and water resources (i.e. reduce, reuse, recycle) thereby minimizing harm to the environment	Yes = 1, No = -1, N/A or Unverified = 0	1	1
Occupational/Vocational well being	1. Working on this project enhanced my vision for the future	Yes = 1, No = -1, N/A or Unverified = 0	0	1
	2. Working on this project facilitated my openness to change and learn new skills	Yes = 1, No = -1, N/A or Unverified = 0	1	1
	3. While working on this project I experienced satisfaction/pleasure in my employment and gave me a positive attitude to work	Yes = 1, No = -1, N/A or Unverified = 0	1	1

and J_k is the score of the project management process for subcategory j (under category t with regard to stakeholder group k)

Using similar procedure, maximum obtainable score for each stakeholder category is determined. The project management process score is then compared to the maximum obtainable score in assessing the social impact of the project.

Table 2 Sample LCI _Stakeholder category: Local community where the project is implemented

Subcategories	Inventory Indicator	Normalized scoring metric	Averaged score	Maximum possible score
Physical Wellbeing	1. Working on this project provide recreation opportunity for the community	Yes = 1, No = -1, N/A or Unverified = 0	1	1
	2. The level of exposure of the local community to injuries, harm and contagious disease as a result of project	High = -1, low = -0.5, Not at all = 0	-0.5	0
Emotional wellbeing	1. The project encouraged members of the local community to cultivate optimistic attitude towards the municipality	Yes = 1, No = -1, N/A or Unverified = 0	1	1

	2. The project enhanced freedom of expression by community members	Yes = 1, No = -1, N/A or Unverified = 0	0	0
Intellectual Wellbeing	1. The project encouraged creativity and stimulated mental activities of the members of the local community	Yes = 1, No = -1, N/A or Unverified = 0	1	1
	2. The project made the local community to be up-to-date by such facility being in their domain. It also foster the participation of community members in activities that arouse their curiosity	Yes = 1, No = -1, N/A or Unverified = 0	1	1
Social wellbeing	1. The project fostered good communication and rapport among the members and leaders of the local community	Yes = 1, No = -1, N/A or Unverified = 0	0	0
	2. The project propelled community members to get involved, share their talents and skills, and contribute to the local community	Yes = 1, No = -1, N/A or Unverified = 0	0	0
Spiritual wellbeing	1. The project encouraged members of the local community to be committed to their beliefs	Yes = 1, No = -1, N/A or Unverified = 0	0	0
	2. The project provided the local community increased opportunities to grow and to overcome its challenges	Yes = 1, No = -1, N/A or Unverified = 0	1	1
	3. The project enhanced community members' freedom to be who they are, thereby preserving their culture and tradition	Yes = 1, No = -1, N/A or Unverified = 0	1	1
Environmental wellbeing	1. The project increased the local community's awareness regarding effects of its activities on the physical environment	Yes = 1, No = -1, N/A or Unverified = 0	0	0
	2. The project facilitated communal effort to conserve material, energy and water resources (i.e. reduce, reuse, recycle) thereby minimizing harm to the environment	Yes = 1, No = -1, N/A or Unverified = 0	1	1
Sense of community well being	1. The project facilitated community's openness to change and to learning	Yes = 1, No = -1, N/A or Unverified = 0	1	1

IOP Conf. Series: Materials Science and Engineering **413** (2018) 012061 doi:10.1088/1757-899X/413/1/012061

	3. The project increased the community members' satisfaction /pleasure with the changes taking place in their local community	Yes = 1, No = -1, N/A or Unverified = 0	1	1

4C. Sample calculation

Taking the physical wellbeing sub-category under project team/workers stakeholder category as an example to demonstrate the calculation procedure:

The project's normalized score on physical wellbeing (pw) sub-category

$J_{pw} = {}^{1}/_{3}[1+ (-5) + (-1)] = -0.17$ .... (3)

The maximum obtainable normalized score by the project on physical wellbeing under project team/workers stakeholder category

$\text{Max } J_{pw} = {}^{1}/_{3}[1+ 0 + 1] = 0.67$ .... (4)

5. Results and Discussion

Tables 1 and 2 are the lifecycle inventory data resulting from the data collection and site observation while Figure 5 is an illustration of the lifecycle impact assessment results obtained from the analysis of the data from Tables 1 and 2 based on the SIMSaW model. The lifecycle impact assessment results were calculated by using equations 1 – 4 with the first step exemplified in the section above. The calculation showed that the social impact score for workers (wkr) = 4 while the score for local community = 3.5. The overall social impact score for the infant food production plant development process = 7.5. Similarly, the maximum obtainable overall social impact score by the project management process = 6.3 + 4.5 = 10.8. The overall social impact of the infant food production plant development process in relation to the possible achievable score, $S_R = 7.5/10.8 = 0.694 = 69.4\%$.

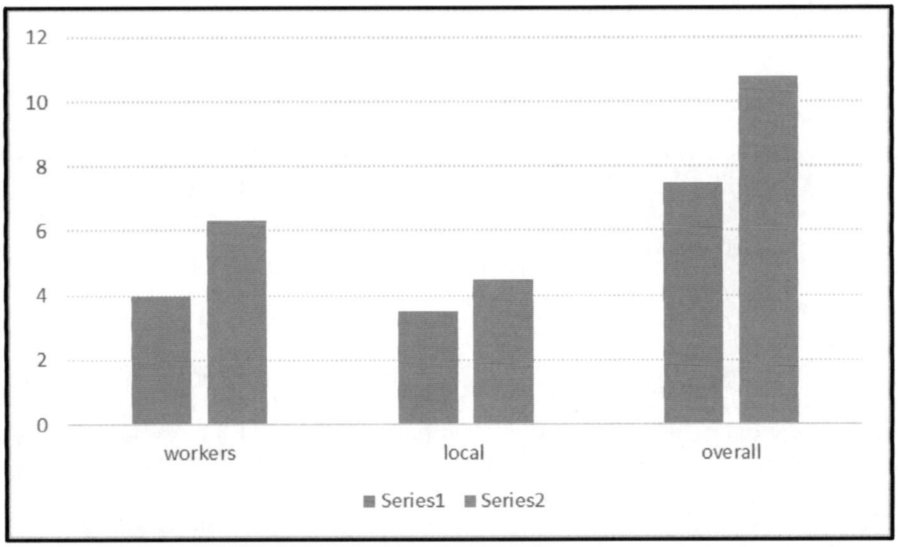

Fig. 5. Infant food production plant development process' social performance score in comparison with achievable score.

The SLCA Interpretation
According to ISO 1440/44, we are to examine the results obtained from lifecycle inventory and lifecycle impact assessment for interpretation of the results. An examination of the lifecycle inventory results in Tables 1 – 2 and lifecycle impact assessment results revealed that increase in workers' and local community's exposure to injuries, harm and disease is the significant issue. Further evaluation of the results shows the social performance of the facility under worker's stakeholder category and local community stakeholder category (in comparison with possible performance score) to be 63.16% and 77.78% respectively. The infant food production plant project management process' overall social impact score = 69.44%. Using the interpretation criteria table (Table 3), the social lifecycle impact assessment of the project management process showed that its social performance is very good both for the project team/workers and the local community. It however showed better performance towards the local community than to the project team.

Table 3 Interpretation criteria

Percentage score	**Remark**
≥ 80	Excellent
61 - 79	Very good
50 - 60	Good
21 - 49	Poor
0 - 20	Very poor

Social impact on workers
The impact of infant food production plant development on the project team was evaluated to promote the understanding of potential impacts of project management process on the project team/workers. The results showed that increased exposure to injuries, harm and diseases is the only outstanding issue that would need to be addressed.

Impact on the local community
Evaluation of the infant food production plant development impacts on the local community also showed increased exposure of the local community to harm, injuries and diseases.

Evaluation
According to [17], "the process of implementing the SLCA study needs to be evaluated to ensure conformity with the ISO 14040/44, UNEP/SETAC guidelines on SLCA and methodology sheets provided UNEP/SETAC on SLCA." "In addition, the scope of the study, data collected, and impact analysis steps undertaken were also examined in terms of their adequacy in meeting the goals of the study. Best practices were adopted where no definite guide was available."

6. Conclusion

A procedure for evaluating potential social lifecycle impacts of project management process on human wellbeing was developed. The requisite criteria for the evaluation was also articulated. The procedure was implemented with a case study on an infant food production plant process. On the overall, the analysis showed that the project management process has a fairly good social performance. The only area requiring attention is the need to take adequate steps to reduce the exposure of the project team and the local community to harm. The case study analyzed showed how stakeholders can assess the attainment of those criteria based on their experience in a specific case. Furthermore, it helped to pinpoint social sustainability hotspots within the project lifecycle that needs further improvement.

IOP Conf. Series: Materials Science and Engineering **413** (2018) 012061 doi:10.1088/1757-899X/413/1/012061

Moreover, the study is a good contribution to the progress in social sustainability studies that is at an infant stage in this country. The case study would also serve as an example and a good reference for further sustainability studies

References

[1] PMTA (Project Management Training Africa) (2014). The 47 PMBOK project management processes made simple. Accessed online at http://www.pmta.co.za/free-resources/the-47-pmbokr-project-management-processes-made-simple

[2] Duncan, W. R. (1993). The process of project management. *Project Management Journal, 24*(3), 5–10

[3] Metafuse and Core Performance concept (2018). Project Insight. Accessed online at https://www.projectinsight.net/projectmanagement-basics/basic-project-management-phases

[4] Bisk Education (2018). Five Phases of the Project Management Lifecycle. Accessed online at https://www.villanovau.com/resources/project-management/5-phases-project-management-lifecycle/#.WxZNZ0iFOM8

[5] C. Macombe, P. Leskinen, P. Feschet and R. Antikainen, "Social life cycle assessment of biodiesel production at three levels: a literature review and development needs," Journal of Cleaner Production, Vol. xxx, pp. 1-12, 2013. http://dx.doi.org/10.1016/j.jclepro.2013.03.026.

[6] F. Vanclay, "International principles for social impact assessment," Impact Assessment and Project Appraisal, Vol. 21, No. 1, pp. 5–11, 2003.

[7] Y. Fan, R. Wu, J. Chen and D. Apul, "A review of social life cycle assessment methodologies," In Muthu, S.S. (Ed) Social Life Cycle Assessment: An Insight. Springer Science+Business Media Singapore 2015. ISBN 978-981-287-296-8: Chapter 1.

[8] Y.H. Dong, and S.T. Ng, "A social lifecycle assessment model for building construction in Hong Kong," Int. J. Life Cycle Assess., Vol. 20, pp.1166–1180, 2015, DOI 10.1007/s11367-015-0908-5.

[9] J. Fava, R. Denison, B. Jones , Curran M, Vigon B, Selke S, Barnum J, editors. "A Technical Framework for Life-Cycle Assessment," Pensacola (FL): SETAC Press, 1991.

[10] I.S. Dunmade, "Environmental Profile Assessment of a Plastic Framed Tambourine Musical Instrument – A Lifecycle Approach," Resources and Environment, Vol. 3, No. 5, pp. 129-134, 2013. doi: 10.5923/j.re.20130305.03.

[11] I.S. Dunmade, "Lifecycle Assessment of a Stapling Machine," International Journal of Engineering & Technology, Vol. 4, No.1, pp. 12-19, 2015.

[12] L.C. Dreyer, "Inclusion of social aspects in life cycle assessment of products: development of a methodology for social life cycle assessment," PhD Thesis, DTU Management Engineering, Denmark, 2009.

[13] Ashton, K. and Jones, C. (2013). Geographies of human wellbeing. Accessed online at http://www.globaleducation.edu.au/verve/_resources/Global_Wellbeing_booklet.pdf

[14] PSI (Puget Sound Institute) (2018). The Nature of Human Well-being. Accessed online at https://www.eopugetsound.org/science-review/section-3-nature-human-well-being

[15] McGillivray, M., and M. Clarke. 2008. Understanding Human Well-being. Bookwell Publications, New Delhi, India.

[16] GRCC (2018). Seven dimensions of wellness. Accessed online at https://www.grcc.edu/humanresources/wellness/sevendimensionsofwellness

[17] Dunmade, I.S.; Onawumi, S; Loto, C. and Oyawale, F (2016). Social Lifecycle Assessment of Sachet Water: A Case Study on a Nigerian Facility's Impacts on Workers and the Local Community. European International Journal of Science and Technology, Vol. 5 No 6 p. 31-45

[18] A. Jørgensen, "Developing the social life cycle assessment: addressing issues of validity and usability," PhD Thesis, DTU Management Engineering, Denmark, 2008.

[19] S.A. Husseinijou, S. Mansour and M.A. Shirazi, "Social life cycle assessment for material selection: a case study of building materials," Int. J. Life Cycle Assess., Vol. 19, pp. 620–645, 2015, DOI 10.1007/s11367-013-0658-1.

[20] C. Benoît, D. Aulisio and G.A. Norris, "Identifying social impacts in product supply chains: overview and application of the social hotspot database," Sustainability, vol. 4, pp.1946–1965, 2012.
[21] M. Blom and C. Solmar, "How to socially assess bio-fuels: a case study of the UNEP/SETAC Code of Practice for social- economical LCA," Master thesis, division of quality and environmental management. Luleå University of Technology, Stockholm, 2009.
[22] L.C. Dreyer, M.Z. Hauschild and J. Schierbeck, "A framework for social life cycle impact assessment," Int. J. Life Cycle Assess. Vol.11, No. 2, pp. 88–97, 2006.
[23] L.C. Dreyer, M.Z. Hauschild and J. Schierbeck, "Characterisation of social impacts in LCA, Part 1: development of indicators for labour rights," Int. J. Life Cycle Assess., vol. 15, No. 3, pp. 247–259, 2010.
[24] L.C. Dreyer, M.Z. Hauschild and J. Schierbeck, "Characterisation of social impacts in LCA, Part 2: implementation in six company case studies," Int. J. Life Cycle Assess., vol. 15, No. 4, pp. 385–402, 2010.
[25] UNEP/SETAC, "Methodological sheets for 31 sub-Categories of impact for a social LCA of products," 2010. www.estis.net/sites/lcinit/default.asp?site=lcinit&page_id=A8992620-AAAD-4B81-9BACA72AEA281CB9. Accessed March 2016
[26] E. Ekener-Petersen and G. Finnveden, "Potential hotspots identified by social LCA—part 1: a case study of a laptop computer," Int. J. Life Cycle Assess., Vol. 18, No. 1, pp. 127–143, 2013.
[27] J. Franze and A. Ciroth, "A comparison of cut roses from Ecuador and the Netherlands," Int. J. Life Cycle Assess., Vol. 16, pp. 366–379, 2011.
[28] M.Z. Hauschild, C. Dreyer and A. Jørgensen, "Assessing social impacts in a life cycle perspective—lessons learned," CIRP Ann. Manuf. Technol., Vol. 57, pp. 21–24, 2008.
[29] A. Jørgensen, A. Le Bocq, L. Nazarkina and M.Z. Hauschild, "Methodologies for social life cycle assessment," Int. J. Life Cycle Assess., Vol. 13, No. 2, pp. 96–103, 2008.
[30] A. Jørgensen, L.C. Lai and M. Hauschild, "Assessing the validity of impact pathways for child labour and well-being in social life cycle assessment," Int. J. Life Cycle Assess., Vol. 15, No. 1, pp. 5–16, 2009.
[31] A.K. Kruse, A. Flysjo and N. Kasperczyk, "Socioeconomic indicators as a complement to life cycle assessment—an application to salmon production systems," Int. J. Life Cycle Assess., Vol. 14, pp.8–18, 2009.
[32] A. Manhart, "Key social impacts of electronics production and WEEE-recycling in China," Öko-Institut e.V, Freiburg, 2007.
[33] A. Manhart and R. Grießhammer, "Social impacts of the production of notebook PCs. Contribution to the development of a Product Sustainability Assessment (PROSA)," Öko-Institut e.V, Freiburg, 2006.
[34] Y. Manik, J. Leahy, and A. Halog, "Social life cycle assessment of palm oil biodiesel: a case study in Jambi Province of Indonesia," Int. J. Life Cycle Assess., Vol. 18, pp. 1386–1392, 2013.
[35] C. Benoît and B. Mazijn, "Guidelines for social life cycle assessment of products," UNEP/SETAC Life Cycle Initiative, Druk in de weer, Belgium, 2009.
[36] C. Benoit, and G.V. Niederman, "Social sustainability assessment literature review," The Sustainability Consortium, Arizona State University and University of Arkansas, 2010.
[37] I.S. Dunmade, " A Multi-Criteria Model for Sustainability Assessment of an Agri-Industrial Technology Meant for a Developing Economy," International Journal of Engineering Research and Applications, Vol. 3, No. 1, pp. 445-456, 2013.
[38] I.S. Dunmade, "The Use of Lifecycle Management Principles in Biosystems Engineering: A Pragmatic Approach to Solving Agri-Industrial Sustainability Problems," Journal of Agricultural Science and Technology A, Vol. 2, No. 3, pp. 357-362, 2012.
[39] I.S. Dunmade, "Development of System Models for Industrial Processes Selection with regard to Product Lifecycle Extension (PLETS Models)," Logos Verlag Berlin, 2001, ISBN 978-3-89722-744-6.

IOP Conf. Series: Materials Science and Engineering **413** (2018) 012061 doi:10.1088/1757-899X/413/1/012061

[40] G.A. Norris, "Social impacts in product life cycles: towards life cycle attribute assessment," Int. J. Life Cycle Assess., Vol. 11, No. 1, pp. 97– 104, 2006.
[41] M. O'Brien, A. Doig and R. Clift, "Social and environmental life cycle assessment (SELCA): approach and methodological development," Int. J. Life Cycle Assess., Vol. 11, No. 2, pp. 87–97, 1996.
[42] C. Reitinger, M. Dumke, M. Barosevcic and R. Hillerbrand, "A conceptual framework for impact assessment within SLCA," Int. J. Life Cycle Assess., Vol. 16, No. 4, pp. 380–388, 2011.
[43] O.N. Nwoke, I.P. Okokpujie, S.C. Ekenyem. Investigation of Creep Responses of Selected Engineering Materials. Journal of Science, Engineering Development, Environmen and Technology (JOSEDET). 2017;7(1):1-5.
[44] J. Azeta, K.O. Okokpujie, I.P. Okokpujie, O. Osemwegie, A.A. Chibuzor. Plan for Igniting Nigeria's Industrial Revolution. International Journal of Scientific & Engineering Research. 2016;7(11):489.
[45] E.Y. Salawu, I.P. Okokpujie, O.O. Ajayi, M.C. Agarana. Analytical Technique for the Determination of Hoop Stress and Radial Stress on the Tooth Spur Gear under Vertical Loading in a Food Packaging Machine. InProceedings of the International MultiConference of Engineers and Computer Scientists 2018 (Vol. 2).

IOP Conf. Series: Materials Science and Engineering **413** (2018) 012062 doi:10.1088/1757-899X/413/1/012062

Alternate Maintainability Evaluation Technique for Steering System of Used Automobiles

Ayodele Onawumi[1], Mfon Udo[2], Elijah Awoyemi[3], Imhade P. Okokpujie[2]

[1,3]Mechanical Engineering Department, Ladoke Akintola University of Technology, Ogbomoso, Oyo State. Nigeria.
[*2]Mechanical Engineering Department Covenant University, Ota, Nigeria
[3]Works and Maintenance Department, Osun State University, Osogbo, Nigeria.
*Corresponding author: mfon.udo@covenantuniversity.edu.ng

ABSTRACT: Automobile is an essential means of transporting people, goods and services which has impacted significantly on economic development of any nation. Maintainability and reliability in vehicles are design considerations that has the capacity to influence user's choice of owned vehicle for both private and commercial uses. Beyond the rigorous methods of evaluation maintainability of technological systems popular experience based opinion of operators of different makes of automobiles have proved to provide acceptable qualitative approach on critical decision making safe and moistenable vehicle. To demonstrate the application of participatory approach in maintainability structured questionnaires were developed and administered to operators (drivers and auto-mechanics) of selected vehicles (1997 models of Nissan Primera, Toyota Camry and Honda Accord) commonly used for private purposes in the study area. Prevailing system faults and the time to repair were determined and analysis were carried out to ascertain the significance of the result obtained which were also used to simulate some maintainability criteria of the steering system. From the result, Nissan Primera 1997 has the least Mean Time to Repair (MTTR) and Maximum Maintenance Time (M_{max}) of 30.4 and 52.6 minutes respectively, followed by Toyota Camry 1997 with MTTR and M_{max} of 39.9 and 128.9 minutes while Honda Accord 1997 has the highest MTTR and M_{max} of 93.2 and 143.4 minutes respectively. This work has demonstrated the evaluation of maintainability based on popular opinion of operators the system from the consideration of maintainability of steering system.
Key Words: Analysis of Variance, Maintainability, Repairable system, Participatory Approach.

1. Introduction

The automobile is an important means of transporting materials, machine, men and commodities from one place to another (Lucas and Onawumi, 2013). In Nigeria, road transportation is majorly used compared to other modes of transportation (Fapetu and Akinola, 2005; Azate et al 2016; Omoruyi et al 2018). As essential as transportation is to economic life only few Nigerian have financial strength to buy new vehicles to serve these purposes (Oke, 2012). The continuous use of motor vehicles reduces the life span as a results of general wear, tears and breakdowns (Akpakpavi, 2015). Maintainability of this complex system plays a vital role on the life cycle and its performance. Maintainability and durability of these vehicles have been of great concern to the owners owing to the fact that the cars have significant impacts on their economic life (Nwoke et al 2017; Ezugwu et al 2016). The US Military publication ML-STD-721 defines maintainability as the measure of the ability of an item to be retained or restored to specified conditions when maintenance is performed by a personnel having specified skill level using prescribed procedures and resources at each prescribed level of maintenance and repair. Ding (2009) and Okokpujie et al (2017) also defines maintainability as an important character which is given by product design, it makes easy to be repaired for the mechanical system. Abdullah *et al* (2006) reported that maintainability has unique effect on maintenance cost of mechanical system. Chen and Cai (2003) identified maintainability has a significant factor in the economic success of engineering systems and products while Gironimo *et al* (2004) viewed maintainability as synthesis of several characteristics. Coulibaly *et al* (2008) believed that maintainability criteria comprise of Intrinsic and Contextual, Intrinsic criteria includes repair-ability (ability to be repaired after failure or damage), accessibility, assemble-ability, disassemble-ability, standardization, interchangeability and survivability while contextual criteria include redundancy, competencies, tooling's, logistics, environment, detectability, testability and maneuverability.

Many researchers have worked on evaluation of maintainability, Chen and Cai (2003) used vector projection method (VPM) to evaluate maintainability of mechanical system in design review, VPM is a multi-objective assessment and is developed based on fuzzy set theory. Gong (1999) evaluated automobile maintainability base on the calculation repair time and maintenance cost. Blanchard *et al* (1995) identified Mean Time to Repair and limit for maximum repair time as the key feature of merit for maintainability. Vujosevic and Raskar (1995) developed procedures for identifying sequence of disassembly, animation of maintenance personnel while carrying out the process while Wanni and Gandhi (2002) considered tribology aspect. Ajide and Adegbola (2011) investigated maintainability of imported second hand (*Tokunbo*) gear drives from different sources and the outcome of the study shown that the same gear drive manufactured from different sources have different maintainability. Akinyemi *et al* (2015) worked on optimal repair time of municipal transit vehicle's clutches and come up with optimum repair time for clutches of municipal vehicle. Since the vehicle cannot remain new forever, the need to carry out necessary maintenance becomes necessary. Akinola and Ogedengbe (2005) defines Maintenance as repair activity carried out on vehicles and other equipment to keep them in good working condition, and if faulty, to restore it to a functioning state. In most engineering systems, repair is a common operation since the cost of repair or replacing constituent part is less than replacing the entire system and the system can be made available for a long period (Sukhwinder and Wadhwa, 2004). An automotive system is a group of related parts and assemblies that perform a specific function and deviation from its function leads to dissatisfaction, loss of man-hour, and can cause danger.
Automobile steering system has the sole aim of giving the driver a means of controlling the direction of motion of the vehicle. It is an input to suspension system (Reza, 2015) and works with suspension system to provide directional control with a comfortable amount of steering efforts. An interaction between the driver and of the steering system is a closed loop system (Marcos, 2007). Power steering was introduced to meet demands for low steering wheel torque and comfort. A traditional power-assisted steering system uses hydraulic power to assist the driver, referred to as Hydraulic Power Assisted Steering (HPAS). This system is a pure hydro-mechanical system and lacks the possibility to be activated other than by driver input (Alessandro, 2013). The rack-and-pinion system is mostly found in lighter vehicles and passenger cars while the worm gear steering box is found in heavy vehicles. The rack-and-pinion steering system is lighter, more compact, has higher efficiency and gives the driver a more accurate feeling of the tyre-ground interaction, Giancarlo and Lorenzo (2009). Some of the advantages of rack and pinion over other steering mechanism include its simplicity, having a play free and robust gear contact between rack and pinnion (Jazar, 2008).
Steering system of an automobile can be categorized as repairable system since repair action can be performed on the system. Lindqvist (2008) defined repairable system as system that can be repaired to original working condition when repair action is performed. For repairable systems, generally there are two main repair assumptions, either "as good as new" or "as bad as old" but Doyen (2005) reported that the system lies somewhere in between these two conditions, "better than old, but worse than new". Therefore, this study evaluated maintainability of steering system of selected vehicles in the study area with a view to establishing their repair ability.

2. Methodology

In this study, Pilot survey was conducted to identify three most common second hand cars (1997 model of Nissan Primera, Toyota Camry and Honda Accord). After the identification of these common cars, two

IOP Conf. Series: Materials Science and Engineering **413** (2018) 012062 doi:10.1088/1757-899X/413/1/012062

types of questionnaire were designed for users/owners of the identified cars. The first questionnaire was designed to capture information on automobile system with common fault, its severity and occurrence, road impact on their vehicle, the action taken on the faulty system were distributed to owners of three common second hand cars, frequency distribution were used to analyze the result of pilot survey.
The second questionnaire was designed for experienced mechanics to capture information on maintainability of the steering system. The steering faults were categorized into three: Noise, Harshness and Vibration, Performance and Fluid Issue. The second questionnaires were mechanics that have similar experience. (Ding, 2009) highlighted four approaches in evaluating maintainability, this study adopted maintainability evaluation sing quantitative approaches. (Dhillon 2008) identified Mean Time to Repair (MTTR), mean active preventive maintenance time, mean active corrective maintenance time, maximum corrective maintenance time, and mean maintenance downtime as measures used in maintainability analysis.
Analysis of Variance (ANOVA) was used to ascertain the significance of the data obtained on maintainability of selected vehicles at $p = 0.05$. The time to repair the faults of the steering system of the common vehicle was processed using SPSS Software 16.0 version. The time taken to repair were subjected to further analysis using the under-listed maintainability mathematical indices (Dhillon, 2008)

$$M_{ct} = \sum_{i=1}^{N} M_{cti} \qquad 1$$

$$MTTR = \frac{\sum_{i=1}^{N} M_{cti}}{N} = \frac{M_{cti}}{N} \qquad 2$$

For a log normal distribution

$$MTTR_G = anti\log\left[\frac{\sum_{1}^{N} \log M_{cti}}{N}\right] \qquad 3$$

$$M_{\max} = anti\log(\log MTTR_G + 1.645\Delta \log M_{cti} \qquad 4$$

Where

$$\Delta \log M_{cti} = \sqrt{\frac{\sum_{1}^{N} (\log M_{cti})^2 - \sum_{1}^{N} (\log M_{cti})^2 / N}{N-1}} \qquad 5$$

$$M = 1 - \exp(^{-t}/MTTR) \qquad 6$$

NOTATIONS
M_{ct} : Corrective Maintenance Time
MTTR : Mean Time to Repair

IOP Conf. Series: Materials Science and Engineering **413** (2018) 012062 doi:10.1088/1757-899X/413/1/012062

$MTTR_G$: Geometric Mean Time to Repair
M_{max} : Maximum Maintenance Time
M(t) : Maintainability factor

3. Results and Discussion

The time to repair (TTR) was subjected to Analysis of Variance (ANOVA) indicated significant difference at p =0.0224 and TTR was processed using maintainability mathematical indices. The results are shown in the Tables 1 and 2. The repair process of steering system starts with identifying the faulty component followed by removal of constituent parts, mending or replacing the faulty parts and re-assembling. It seen from the Table 1 shows that time to perform the repair operations varied from one vehicle to the other. The total time required in performing Disassembly, Accessibility and Manipulability analysis based on the type of suspension used on the repairability of steering system as shown in Table 2 Nissan under study uses multi links type for its independent suspension, Toyota that uses MacPherson strut for its suspension system while Honda uses Double Wish borne type also contributed variance in repair action as maintainability of any equipment has a direct link on design. The outcome of maintainability mathematical indices was presented in Table 2.0 Nissan Primera 1997 has the least maximum corrective time (52.6minutes) for repair of steering system, fault followed by Toyota Camry 1997 (128.9minutes) while Honda Accord has the highest maximum corrective time (143.4minutes) for repair. The time include time to locate the parts/components requiring repair, time to perform the needed repair and time to verify that the repair has been performed successfully

Table 1: Average time taken to repair faults in steering system of three selected vehicles in Minutes

Fault number	Nature of fault	Toyota Camry 1997	Nissan Primera 1997	Honda Accord 1997
1	Objectionable whistle while turning steering wheel when stationary or moving slowly	121	93	235
2	Rattle or Excessive clunk	45	32	99
3	Popping noise	42	30	125
4	Chirp or Squeal (power steering pump)	11	16	48
5	Whine, Growl, or Groan (Power Steering Pump)	5	4	6
6	Steeringwheel or Column has freeplay/lash/loosenes	94	72	219
7	Steering wheel has fore and aft looseness	15	14	67
8	Steering catches, surges or sticks or difficult to turn	16	10	50
9	Steering wheel does not return to centre	15	16	25
10	Fluid issues (leak, aerated, contamination	25	17	58

Table 2: Summary for Maintainability Parameters

Parameter	Toyota Camry 1997	Nissan Primera 1997	Honda Accord 1997
M_{CT}	399	304	932
MTTR	39.9	30.4	93.2
$MTTR_G$	25.27	20.87	62.36
M_{max}	128.9	52.6	143.4
M(t)	0.9605	0.8227	0.7852

IOP Conf. Series: Materials Science and Engineering **413** (2018) 012062 doi:10.1088/1757-899X/413/1/012062

4. Conclusion

In this study, quantitative approaches have been used to evaluate maintainability of steering system of common cars, the results have shown that Nissan Primera 1997 has the least MTTR, $MTTR_G$ and M_{max}, while Honda Accord 1997 has the highest MTTR, $MTTR_G$ and M_{max} . It was deduced from the study that despite using the same steering system maintainability of selected vehicles differ as a result of types of suspension system used. . The study also revealed that the steering system faults of all the selected vehicles can be repaired effectively but at different duration (repair time) while type of suspension used contributed to variance in time to repair. This study will be beneficial to intending car owners on life cycle while repair limit should be evaluated before embarking on repair of steering system.

References

[1] Abdullah. B, Yusoff. M, and Ripin. Z (2006) Integration of Design for Modularity and Design for Assembly to Enhance Product Maintainability, *Proceedings of 1st International Conference 7th AUN/SEED-Net Field-wise Seminar on Manufacturing and Materials Processing* ,263-269.

[2] Ajide, O.O. and Adegbola, A. A (2011) Equipment Maintainability Using Simulation, *Global Journal of Researches in Engineering.* 11(6): 8-16.

[3] Akinola, B. and Ogedengbe, T. (2005), Basic Automobile Technology, Olajuyin Printers, Akure,Nigeria

[4] Akinyemi,O. O., Ajayeoba, A.O., Adebiyi. K.A., Akintan. A. L and Jolaoso J.T, (2015), Optimal Repair Time for Municipal Transit Vehicle's Clutches, *International Journal of Science and Engineering and Technology*

[5] Akpakpavi, M. (2015), Modern automobile vehicle repair practice in Micro, Small and Medium Scale Garage in Ghana, *International Journal of Science, Technology and Society*, 2(6): 216-222

[6] Alessandro D. (2013). Pressure Control in Hydraulic Power Steering Systems, M.Sc Thesis, Linkoping University, Sweden

[7] Azeta, J., Okokpujie, K. O., Okokpujie, I. P., Osemwegie, O., & Chibuzor, A. (2016). A Plan for Igniting Nigeria's Industrial Revolution. *International Journal of Scientific & Engineering Research*, *7*(11), 489.

[8] Blanchard, B.S., Dinesh V.and Elmer L.P. (1995) Maintainability: A key to Effective Management, Prentice-Hall, Eaglewood, Cliffs, New Jersey.

[9] Chen L. and Cai J.(2003) Using Vector Projector Method to evaluate maintainability of mechanical system in design review, *Journal of Reliability Engineering and System Safety:* 147-154 retrieved on 24th June, 2012 available at: www.sciencedirect.com

[10] Coulibaly. A, Houssin. R, and Mutel, B (2008). Maintainability and Safety Indicators at Design Stage for Mechanical Products. Computers in Industry 59(5):438- 449.

[11] Dhillon, B.S. (2008) Engineering Maintainability Elsevier

[12] Ding Y, (2009) Product Maintainability Design Method and Support Tool Based on Feature Model, *Journal of Software Engineering and Applications.* 2:165-172

[13] Doyen, L., (2005). Repair Efficiency Estimation in the ARII imperfect repair model. Mod. Statist. Meth,. Reliability 10:15

[14] Ezugwu, C. A., Okonkwo, U. C., Sinebe, J. E., & Okokpujie, I. P. (2016). Stability Analysis of Model Regenerative Chatter of Milling Process Using First Order Least Square Full Discretization Method. *International Journal of Mechanics and Applications*, *6*(3), 49-62.

[15] Fapetu,O.P. and Akinola, A.O. (2008), Optimizing Auto -Repair Practice: Akure Metropolis as a case study, A U J.T 11(4): 232-238

IOP Conf. Series: Materials Science and Engineering **413** (2018) 012062 doi:10.1088/1757-899X/413/1/012062

[16] Giancarlo G. and Lorenzo M. (2009), The Automotive Chassis, Volume 1, Component Design, Springer. London.
[17] Gironimo G, Monacellia G and Palato S (2004), A Design Methodology for Maintainability of Automotive Components in Virtual Environment, *International Design Conference*
[18] Gong L. (1999). Study on the Evaluation of Automobile Maintainability. Liaoning Transportation Technology, 8 :46- 80
[19] Jazar, R.N. (2008), Vehicle Dynamics: Theory and Application, Springer.
[20] Lindqvist, B.H (2008), Maintenance of Repairable Systems, Complex System Maintenance Handbook, Springer, London: 657.
[21] Lucas, E. B and Onawumi, A.S. (2013). Ergonomic Evaluation of In-Vehicle Interface Design of Taxicabs in Nigeria. *International Journal of Engineering Research and Applications,* 3(4):566-572.
[22] Marco R. (2007), Hydraulic Power Steering in Road Vehicles: Analysis, Testing and Enhanced Functionality, PhD Thesis, Linkoping University, Sweden.
[23] Military Standardization (1966), Definition of terms for Reliability and Maintainability, US Department of Defense.
[24] Nwoke, O. N., Okokpujie, I. P., & Ekenyem, S. C. (2017). Investigation of Creep Responses of Selected Engineering Materials. *Journal of Science, Engineering Development, Environmen and Technology (JOSEDET)*, *7*(1), 1-15.
[25] Okokpujie, K. O., Abayomi-Alli, A., Abayomi-Alli, O., Odusami, M., Okokpujie, I. P., & Akinola, O. A. (2017). An automated energy meter reading system using GSM technology.
[26] Omoruyi, O. N., Omoruyi, M. G., Okokpujie, K. O., & Okokpujie, I. P. (2018). Electronic Fare Collection Systems in Public Transits: Issues, Challenges and Way-Forward. *COVENANT JOURNAL OF ENGINEERING TECHNOLOGY*, *2*(1).
[27] Reza, A. (2015), Topology Based Optimization of Suspension and Steering Mechanism of Automobiles, M.Sc Thesis, University of Waterloo.
[28] Sukhwinder, S.J and Wadhwa,S.S (2004),Reliability, Availability and Maintainability Study of High Precision Special Purpose Manufacturing Machines. *Journal of Scientific and Industrial Research* 63:512-517
[29] Vujosevic, R and Raskar,R.(1995), Simulation, Animation and Analysis of Design Disassembly for Maintainability Analysis. Prod Res; 33(11): 2999-3022.
[30] Wanni, M.F and Gandhi O.P (2002), Maintainability Design and Evaluation Mechanical System based on Tribology . *Reliability Engineering System Safety*; 77:181-188

Graphical Representations of Experimental and ANN Predicted Data for Mechanical and Electrical Properties of AlSiC Composite Prepared by Stir Casting Method

Philip O. Babalola[1*], Christian A. Bolu[1], Anthony O. Inegbenebor[1], Oluseun Kilanko[1]

Department of Mechanical Engineering, College of Engineering, Covenant University, Ota, Nigeria
E-mail address: phillip.babalola@covenantuniversity.edu.ng

Abstract. Artificial Neural network is a field of man-made intelligence that is able to undertake design prediction, mechanical property forecast, and process selection. In this paper, Aluminium Silicon Carbide composite was developed by reinforcing aluminium metal with silicon carbide powder using stir casting method. The produced aluminium matrix composites (AMC)were subjected to tensile, hardness and electrical tests to obtain tensile extension (mm), load (N), modulus (N/mm^2), yield strength (MPa), hardness (HV), ultimate tensile strength (MPa), tenacity at fracture (gf/tex), time at fracture (s), hardness (HV), conductivity(MΩ/m), and tensile stress (MPa) data. Artificial Neural Network (ANN) was then used to train, test, and validate the obtained experimental data and then predict new set of data. The experimental and ANN predicted data were represented using graphical illustrations. The results showed that ANN could be used to replace rigorous, costly and time consuming experimental exercise with minimal loss in accuracy.
Keywords: AlSiC composite, artificial neural network (ANN), mechanical properties, electrical properties, stir casting

1. Introduction

Aluminium metal is the most severely used non-ferrous metal in our world today and it is only second to steel, when it comes to automobile body frame [1], with 5xxx and 6xxx series on the lead. Light weight, high strength, and high resistance to corrosion in an aggressive environment make aluminium a suitable material for use in conventional and non-conventional applications. Aluminium also responds readily to strengthening mechanisms. However, aluminium does not perform very well in high temperature applications due to its low melting temperature. It also has very little resistance to abrasive wear because of small hardness value [2]. Improved mechanical properties is obtained by reinforcing aluminium alloy matrix with ceramic materials [3-6]. Some ceramic materials like alumina, B4C, SiC, Si3N4, AlN, TiC, TiB2, TiO2 and hard metals such as tungsten and titanium are used for this purpose [7]. The resultant aluminium matrix composites now have numerous use in reciprocating internal combustion engine members (cylinder liners, pistons, pushrods, cylinder blocks), rotors brake for high speed locomotives, transmission parts, turbocharger vanes, forks for gear shift, clutch plate, golf clubs, bicycles, electronic boards, and heat fins. Bringing aluminium and reinforcing material together may be done using processing methods such as powder metallurgy, cryomilling, vacuum infiltration, vacuum hot pressing, thixoforming, co-spray deposition process, compocasting, squeeze casting, centrifugal casting, laser alloying and stir casting methods [6, 8]. Stir casting method is used in this work due to its cost-effectiveness and ease of varying and monitoring of processing parameters. Many studies have been conducted on properties and characterization of Al/SiCp composites [9–13]. Evaluated properties of the composites depend on the particularly critical one for the material application. Some of these properties are tensile, hardness, thermal conductivity, electrical conductivity, fatigue, creep, wear rate and so on.

However, the same properties are now being predicted using Artificial Neural Network (ANN) method. Kavimani and Prakash used ANN and Taguchi method to predict wear rate properties of magnesium composite [14], while Rashed and Mahmoud in their work used ANN to predict the same property but with aluminium composite [15]. Apart from material properties, ANN is also used in process parameters in surface engineering [16], machining [17] and composite.

2. Materials and Methods

In this work, stir casting method was adopted to synthesize samples of AMCs using 1170Al reinforced with Silicon Carbide (SiC) particulates of 3 μm, 9 μm, 29 μm, and 45 μm sizes respectively. The chemical composition of Aluminium and Silicon Carbide are presented in Table 1 and Table 2 below.

Table 1 The Compositions in Percentage of Aluminium Ingot Used in the Experiment

Fe	Mg	Sn	Cu	Zn	Ti	Si	Pb	Mn	Al
0.232	0.0027	0.007	0.0006	0.0016	0.006	0.078	0.0012	0.000	99.66

Table 2 The Chemical Composition in Percentage of Silicon Carbide

Si	Al	Fe	C	SiO2	Magnetic Iron	SiC
0.80	0.30	0.20	0.50	0.60	0.04	97.6

The liquid metallurgy method (stir casting technique) was used to formulate AMC. An oil-fired tilting furnace was used to melt measured mass of 1170Al to 750 °C inside a graphite crucible. A K-type thermocouple was used to monitor the temperature of the melt to avoid overheating and energy wastage. A mould preheated to 450 °C was used to receive the melt where it was mixed with the help of an impeller to form a fine vortex. SiC particles heated to temperature of 1100 °C was then introduced into the melt simultaneously with mechanical stirring at 500rpm for about 5mins. Mixing occurs when the slurry is at semisolid form.
The melt behaves as a solid when no stress is involved but flows like a liquid when pressure is applied- this is thixotropic property. Uniform dispersion is produced by introducing particles when cooling of the melt is combined with rigorous agitation. The agitation helped in breaking the liquid-solid mixture by break down the dendritic structure. The AMCs having different particle sizes (3 μm, 9 μm, 29 μm and 45 μm) and each size with different weight percentage (2.5, 5.0, 7.5 and10 wt %) of SiC were fabricated by the same procedure.

2.1 Tensile test

All specimens produced through stir casting method had circular cross-section with dimensions of 110 mm ϴ and 30 mm length. Five tensile specimen with dimensions of 5 mm × 10 mm with a gauge length of 25 mm were machined out and tested in Universal Testing Machine (Instron:Model 3369) of 30kN load using ASTM International E8/E8M-09 standard. Five measurements (modulus) were recorded for each sample and the average was calculated.

2.2 Microhardness test

Microhardness measurements were carried according to ASTM Standard E 384. Test machine was LECO 700AT with a load of 492.3 mN and a dwell time of 10 s. Surface preparation was done with emery papers down to 1000 mesh. Six tests were conducted for each sample and the average recorded.

2.3 Electrical conductivity

Specimen of each cast were cut out and milled in the machine shop for electrical conductivity testing.. The working voltage of 20 mV was selected in 4-point probe machine on samples with dimensions 10 mm L × 10 mm B x 100 mm H. Keithley Instruments Model 2400 was used to generate current, voltage, conductivity and resistivity for each sample.

2.4 ANN predicted data

ANN was used for modelling and forecasting tensile, electrical conductivity and micro-hardness properties. ANN architecture as shown in Figure 1 was employed to train, test and validate the measured data and then generate new forecasted data. Figure 1 consists of number of joints, arranged in layers that are identified as output layer, hidden layer, and input layer. Iterative computations are done viz-a-viz input layers through network structure depicted by the hidden layers until it arrives at the output layers. The input is a 3x17 matrix, representing 17 samples of 3 elements (see Table 3). Three input elements are percentage weight of aluminium, percentage weight of SiC and size of SiC particle. The output (target) is a 11x17 matrix, representing 17 samples of 11 elements (see Table 4). The eleven output elements are microhardness, yield strength, tensile extension, modulus, ultimate tensile strength, tensile stress, time at fracture (break), load at maximum extension, tenacity, electrical resistivity and conductivity. Training and simulation steps used exactly 70% of the data, validation steps used 15% of the data, while remaining 15% was used for testing the network. The architecture of the network can be represented as (3, HL, 11), where HL is the hidden layer, hence the network topology (3,15,11), used the trained data to understand the weights, and records the Mean Square Error (MSE) values.

A dataset of experimental results was grouped into three categories: testing, training, and validation of the artificial neural network. The weights of all the connecting nodes is adjusted during training sessions until error level no longer improved. Coefficient of determination B (also called $R2$ coefficient) used to measure the effectiveness of ANN is denoted by:

$$B = 1 - \frac{\sum_{i=1}^{M}(O(p^{(i)}) - O^{(i)})^2}{\sum_{i=1}^{M}(O^{(i)} - O)^2} \qquad (1)$$

where O(p(i)) is the ith forecasted property characteristic, O(i) is the ith experimental value, O is the mean value of O(i), and M is the number of test data. Good output approximation competencies of ANN is measured with higher B coefficients. Hence, best quality could be deduced when B is equal or greater than 0.9. The relationship between outputs and targets is measured by the regression R values. A close relationship is known by an R value of 1, while 0 denotes random relationship and from Figure 2, training has R value of 1.0, validation has R value of 0.96298, test has R value of 0.90984 and all has R value of 0.96361. It can be inferred that there is close relationship between the outputs and the targets. **Training** data were used to modify the network during training using its error, **validation** samples were used to measure network generalization and to halt training when generation no longer improve, while **testing** data have no direct effect on training and so provided an autonomous measurement of network evaluation during and after training (Figure 2 and Figure 3). The summary of ANN results, a pattern recognizing tool is indicated in Figure 4.

Table 3 ANN Input Data

S/N	Al %wt.	SiC %wt.		SiC Size (μm)
1	100	0		0
2	97.5	2.5		3 x 10-6
3	95	5	3 x 10-6	
4	92.5	7.5	3 x 10-6	
5	90	10	3 x 10-6	
6	97.5	2.5		9 x 10-6
7	95	5	9 x 10-6	
8	92.5	7.5	9 x 10-6	
9	90	10	9 x 10-6	
10	97.5	2.5		2.9 x 10-5
11	95	5	2.9 x 10-5	
12	92.5	7.5	2.9 x 10-5	
13	90	10	2.9 x 10-5	
14	97.5	2.5		4.5 x 10-5

IOP Conf. Series: Materials Science and Engineering **413** (2018) 012063 doi:10.1088/1757-899X/413/1/012063

15	95	5	4.5 x 10-5
16	92.5	7.5	4.5 x 10-5
17	90	10	4.5 x 10-5

Table 4 ANN Output (Target) Data

S/N	Extension at Maximum Tensile Extension (mm)	Load at Maximum Tensile Extension (N)	Modulus (N/mm^2	Yield Strength (MPa)	Ultimate Tensile Strength (MPa)	Tenacity at Fracture (gf/tex)	Time at Fracture (Standard) (sec)	Hardness (HV)	Conductivity, (Mç/m)	Resistivity, (µΩ-m)	Tensile Stress at Maximum Tensile Extension (MPa)
1	20.783118	720.0062	402.413324	40.8	61.3	845.6629	40.8032	19.6	70.25378	**0.014234**	15.175522
2	10.685998	353.2760581	1293.428876	29.6	37.2	381.4674	21.3124	**20.05**	68.82136	**0.01453**	8.160014
3	10.84656	419.33323	1028.563265	35	53	438.4278	21.65	23.6	67.70123	**0.014771**	9.35811
4	7.945935	314.2605334	1517.59211	24.25	31.625	320.9439	15.883	24.75	64.14723	**0.015589**	6.76695
5	8.6769125	420.8433258	878.9286575	22.25	28.25	434.3051	17.3	25.9	48.74027	**0.020517**	5.649625
6	8.14303	688.626658	1290.11912	30.4	40	711.551	16.22	22.95	67.83901	**0.014741**	13.039944
7	7.156814	374.4512296	888.772108	16.8	24	383.5274	14.3	24.65	62.96254	**0.015882**	6.403696
8	7.01875	498.46985	1092.8752	21.625	26.5	519.2085	13.975	26.05	59.82686	**0.016715**	10.044365
9	6.4023433	291.1302844	760.3567167	11.667	13.867	301.7488	12.73333	26.2	48.96254	**0.020424**	4.4393267
10	11.061328	698.093155	1233.86522	30.75	41.625	721.5803	22.075	23.55	68.63504	**0.01457**	13.083903
11	9.46796	688.299396	969.405182	36	47.6	746.2332	18.7772	25.2	66.96015	**0.014934**	14.489662
12	6.3033333	62.09007	1326.21316	31.667	39.667	63.55679	12.57733	33.65	65.26178	**0.015323**	1.14747
13	13.491654	615.047286	990.415216	28	42.1	756.6384	46.5948	34.25	56.63504	**0.017657**	12.173872
14	12.404812	503.244	580.916218	22.8	33.46	610.1167	14.2568	23.85	67.80000	**0.014749**	8.624638
15	12.6141	453.2824975	793.2229175	34.5	44.75	497.3207	15.1	33.45	61.58468	**0.016238**	9.3490325
16	6.746216	762.192954	935.028496	19.8	27.62	780.5672	40.42	33.65	63.45271	**0.01576**	13.021274
17	10.785468	382.0129175	645.46291	29.125	39.625	398.2237	21.52	35.2	50.27102	**0.019892**	6.3503725

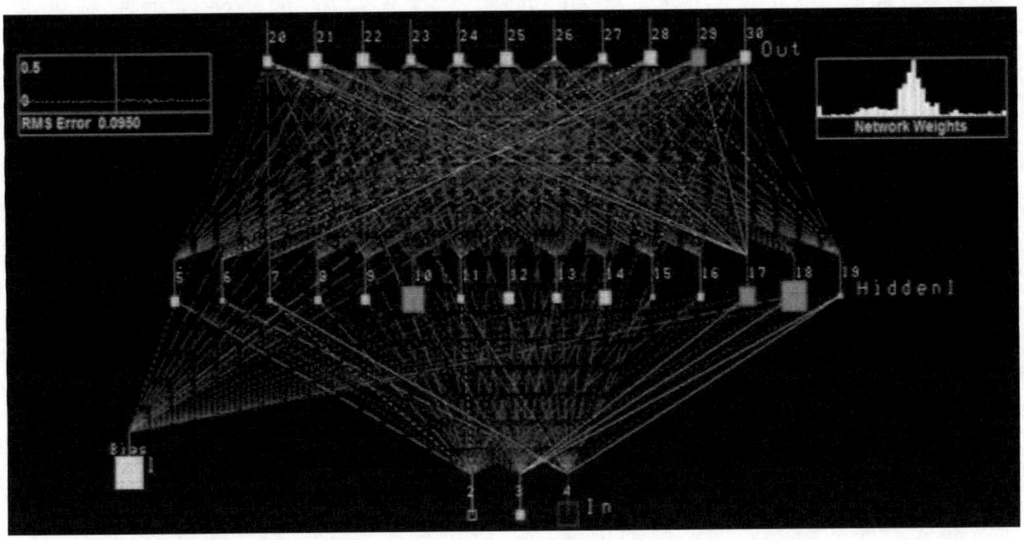

Fig. 1 ANN Architecture

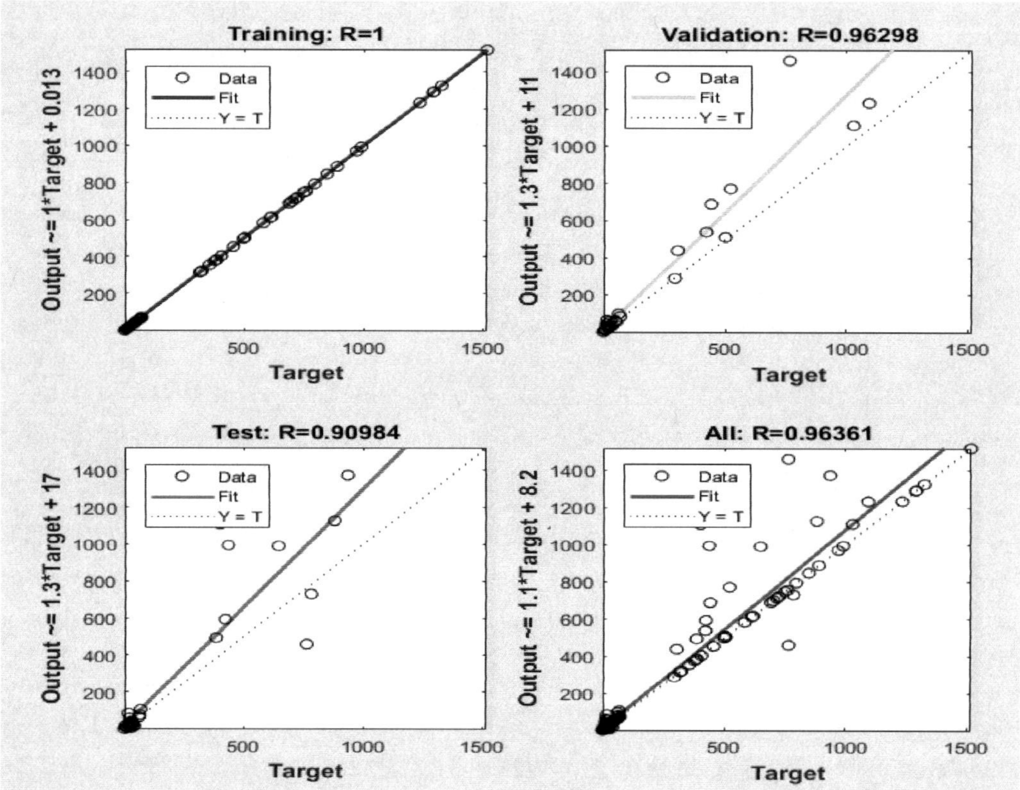

Fig. 2 ANN Output Graph for the Dataset

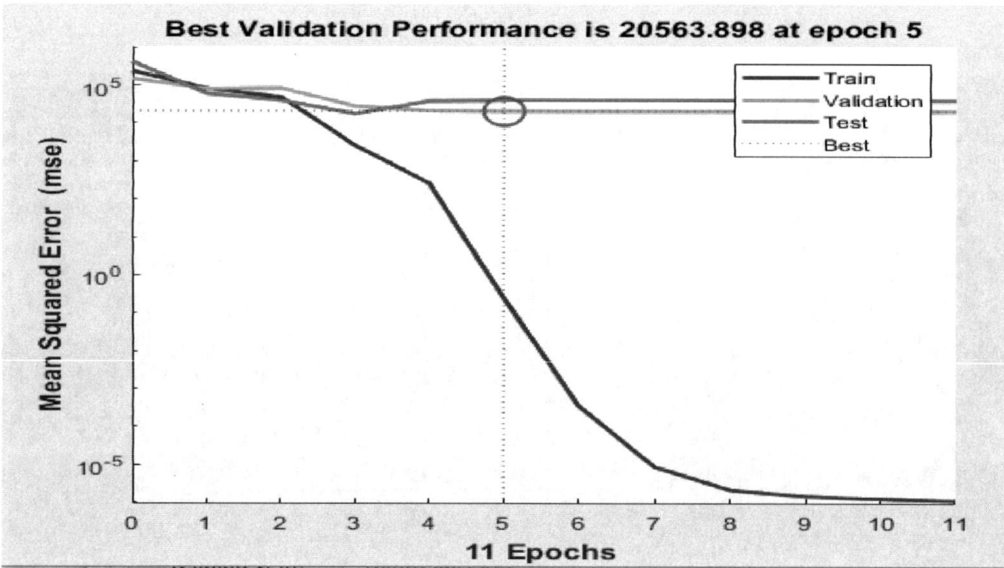

. Fig. 3 ANN Training Graph for the Data Used

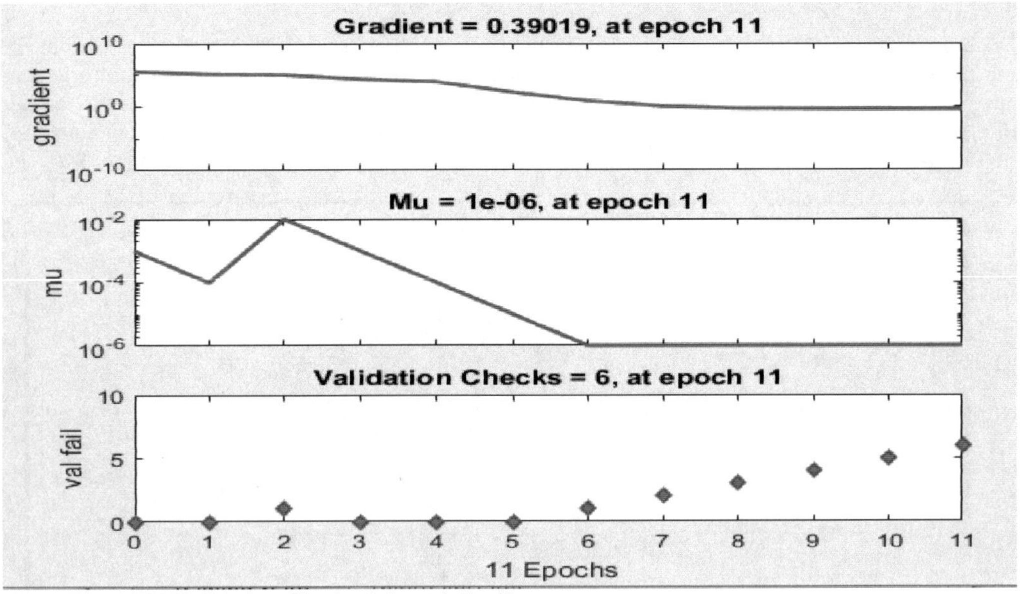

Fig. 4 The ANN Validation Checks, Mu and Gradient

3. Results and discussion

There were seventeen (17) specimens as indicated in Table 3. The first specimen contains no ceramic silicon carbide, whereas the remaining specimens were mixture of aluminium (97.5, 95.0, 92.5 and 90 wt %) and silicon carbide (2.5, 5.0, 7.5 and10 wt %) as indicated in the first two columns. The third column showed variation of particle sizes of silicon carbide (3 µm, 9 µm, 29 µm and 45 µm). The seventeen specimens were characterized to obtain the eleven material properties for Al/SiCp composites listed in Table 4. These measured data were presented to ANN for training, validation and testing. ANN subsequently generated new set of predicted properties for all the seventeen specimens.

Figures 5-15 showed graphically the measured and ANN predicted properties for tensile extension (mm), hardness and electrical tests to obtain tensile extension (mm), load (N), modulus (N/mm^2), yield strength (MPa), ultimate tensile strength (MPa), tenacity at fracture (gf/tex), time at fracture (s), hardness (HV), electrical conductivity(MΩ/m), electrical resistivity (µΩ-m) and tensile stress (MPa) respectively.

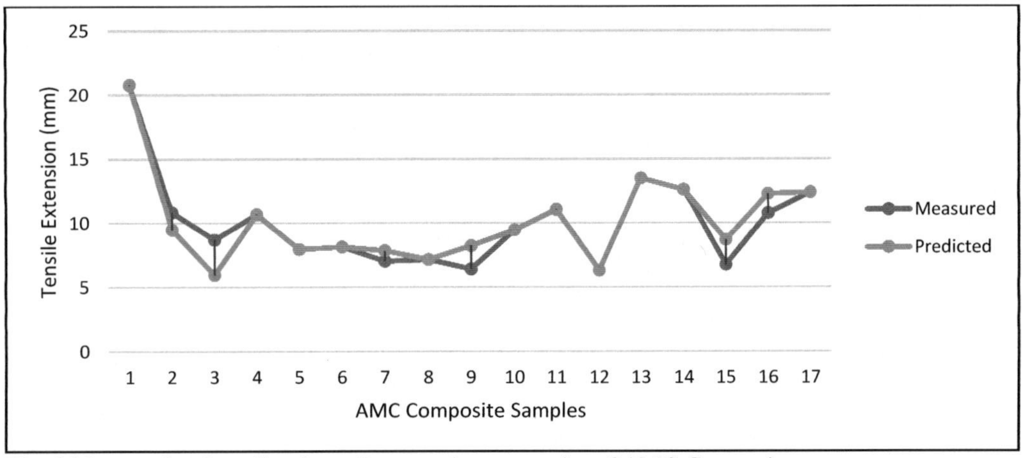

Fig. 5 Measured and predicted tensile extension properties of AMC Composites

It is observed that ANN predicted data is quite successful with perfect matches seen in ten (10) out of eleven (11) material properties. Aside resistivity, perfect predictions were seen in tensile extension (mm), load (N), modulus (N/mm^2), yield strength (MPa), hardness (HV), ultimate tensile strength (MPa), tenacity at fracture (gf/tex), time at fracture (s), electrical conductivity(MΩ/m), and tensile stress (MPa) respectively.

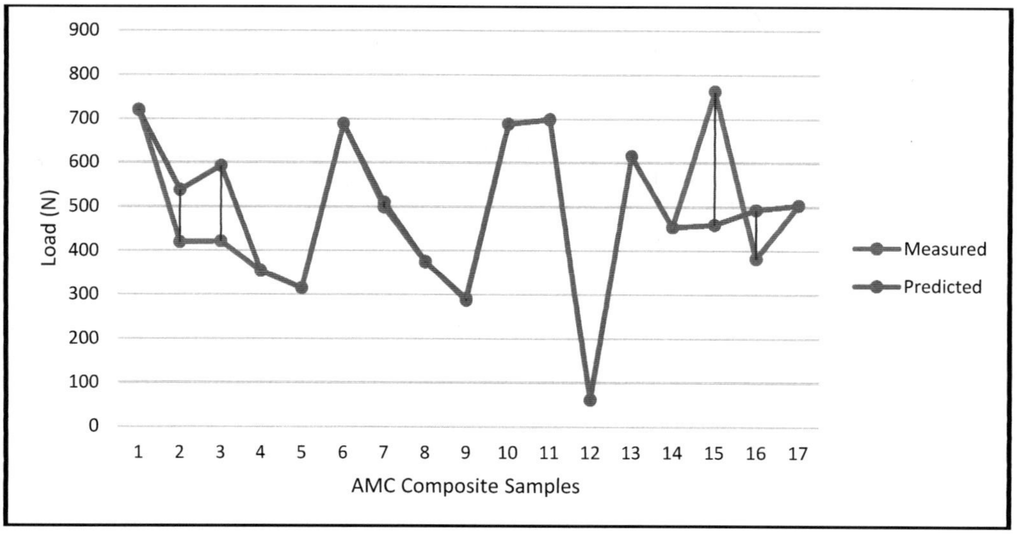

Fig. 6 Measured and predicted load of AMC Composites

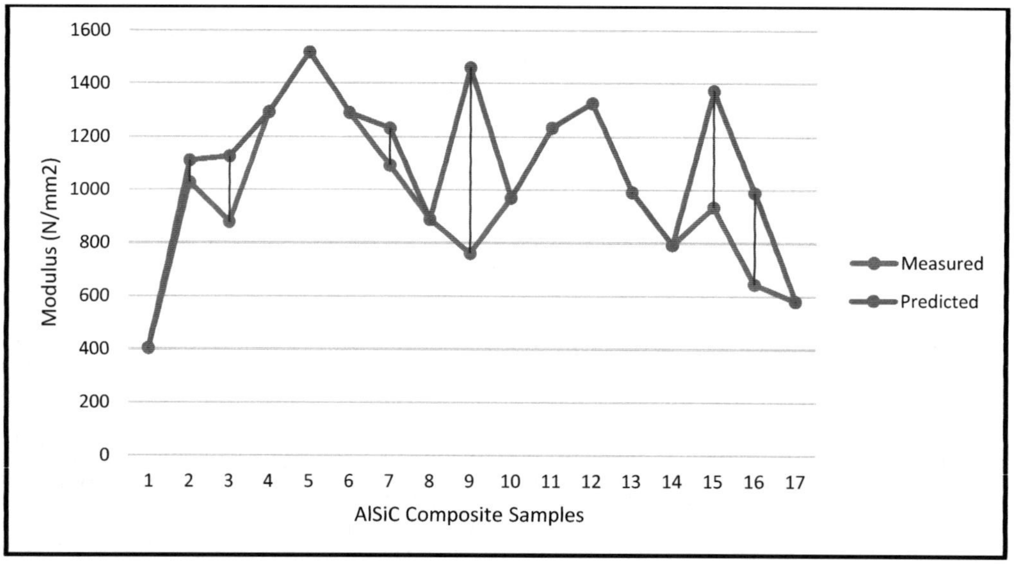

Fig. 7 Measured and predicted modulus of AMC Composites

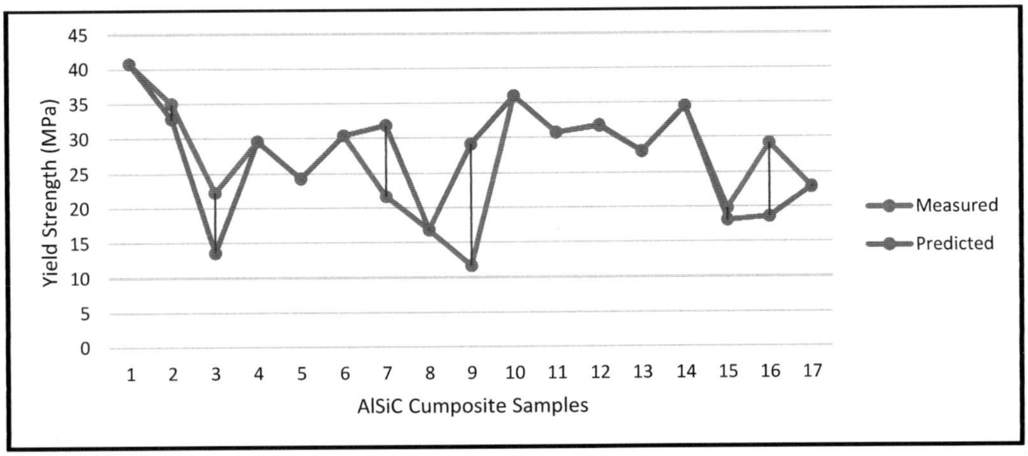

Fig. 8 Measured and predicted yield strength of AMC Composites

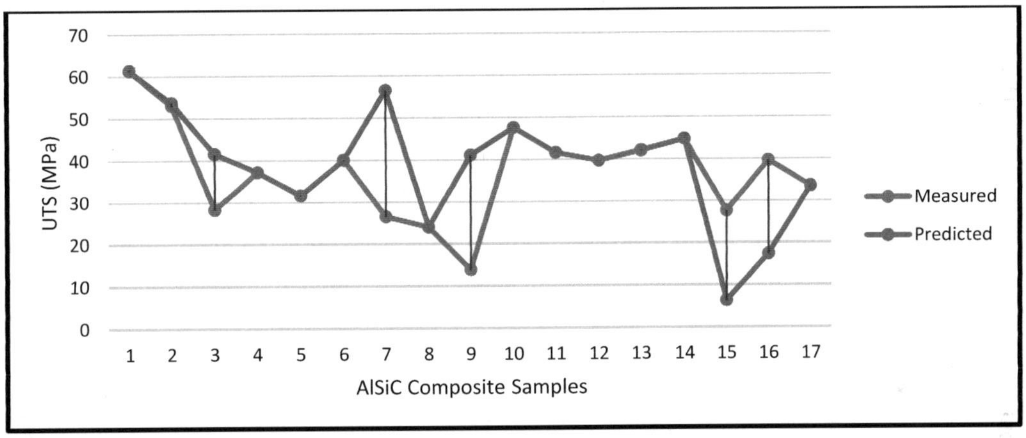

Fig. 9 Measured and predicted ultimate tensile strength of AMC Composites

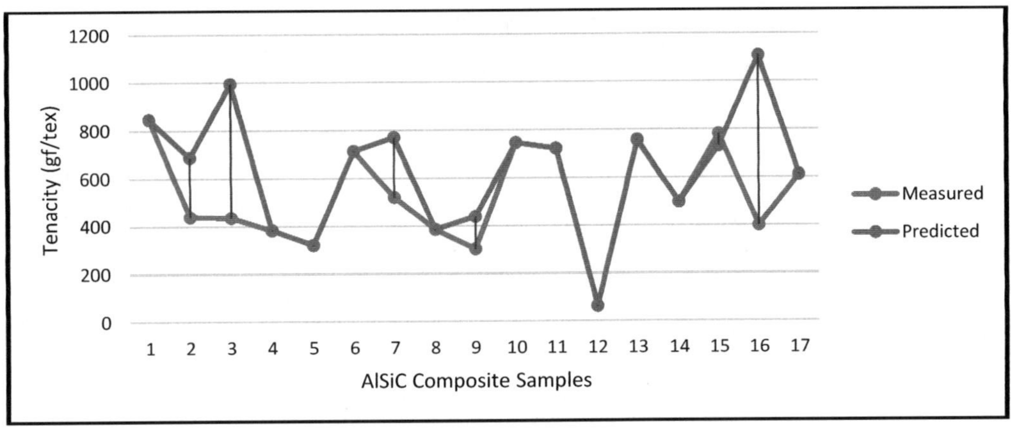

Fig. 10 Measured and predicted tenacity of AMC Composites

IOP Conf. Series: Materials Science and Engineering **413** (2018) 012063 doi:10.1088/1757-899X/413/1/012063

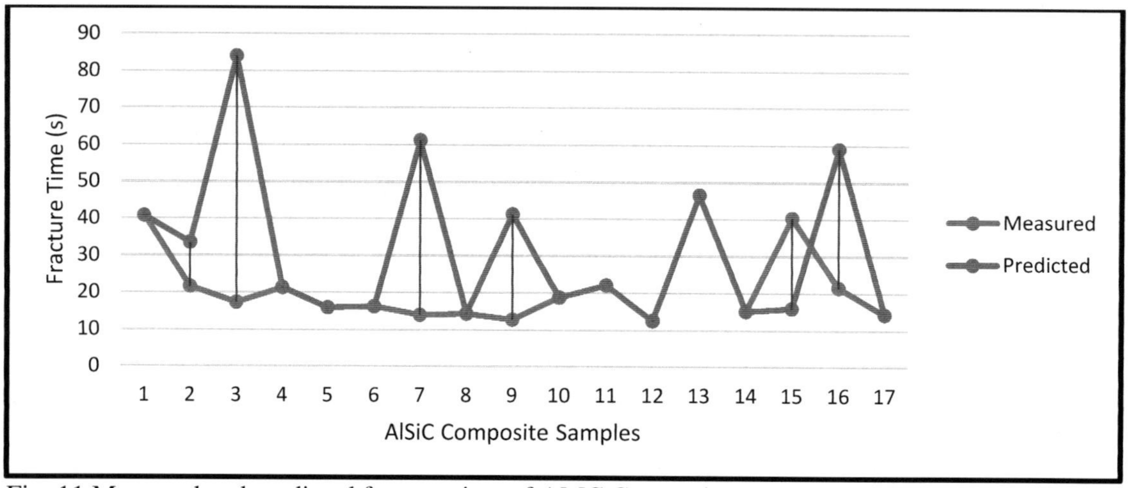

Fig. 11 Measured and predicted fracture time of AMC Composites

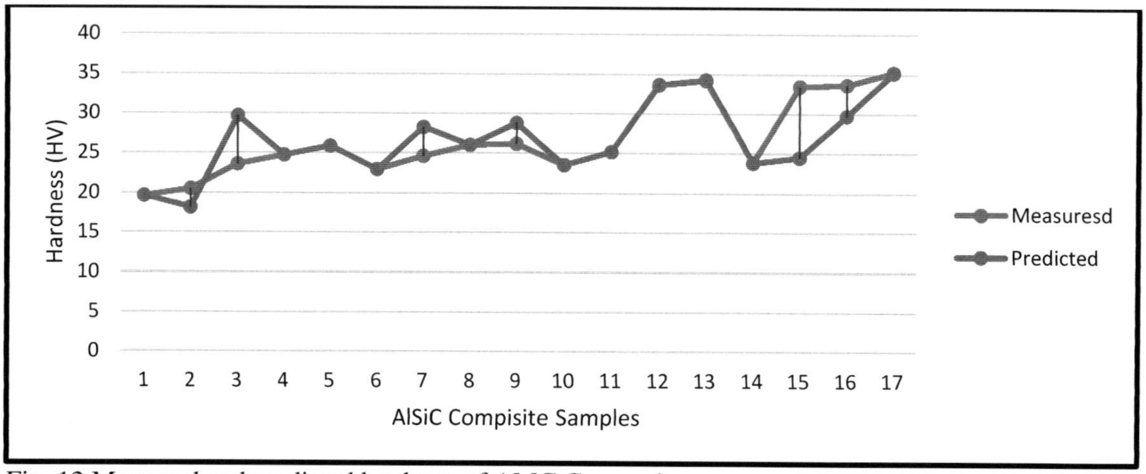

Fig. 12 Measured and predicted hardness of AMC Composites

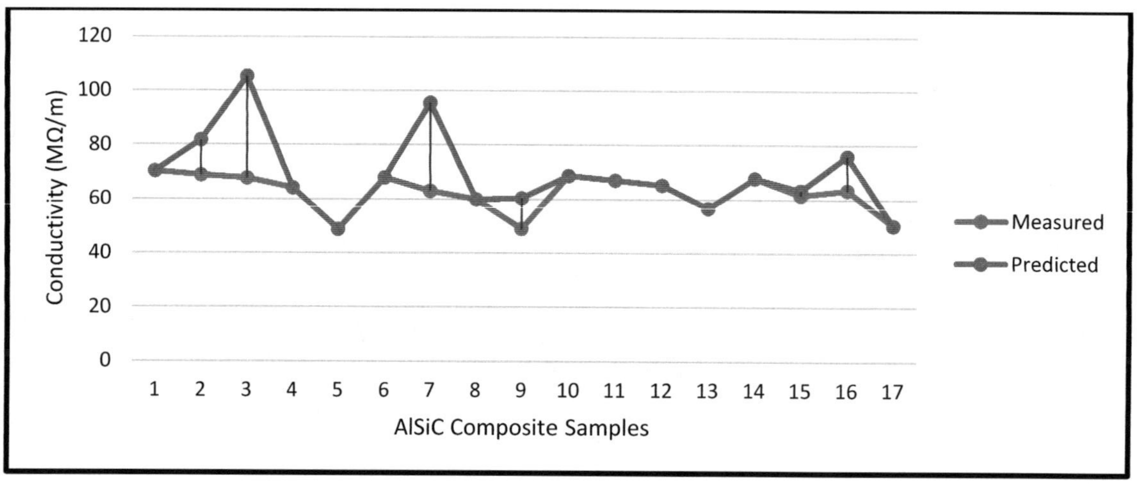

Fig. 13 Measured and predicted conductivity of AMC Composites

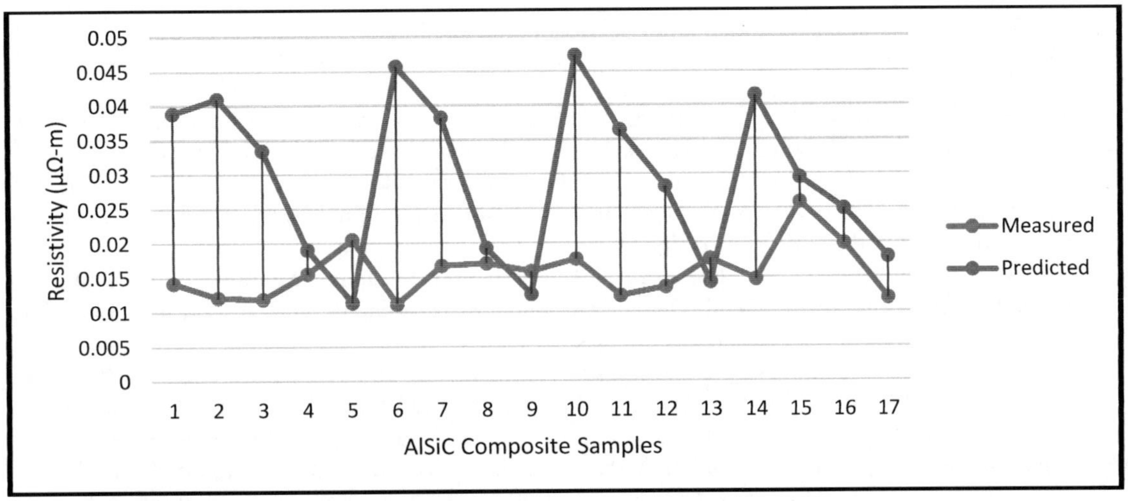

Fig. 14 Measured and predicted resistivity of AMC Composites

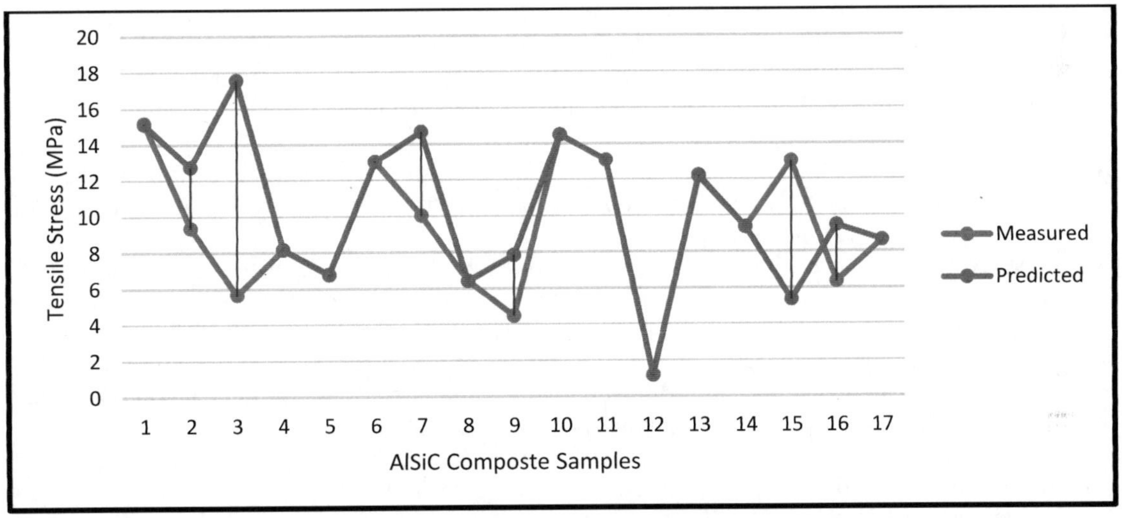

Fig. 15 Measured and predicted tensile stress of AMC Composites

4. Conclusions

In the present study, prediction of Al/SiCp composites with varied aluminium content, SiC content and silicon particle size was done. The following results were obtained:

i. Artificial Neural Network (ANN) is a versatile and effective tool in forecasting composite properties. In an established processing route and constituent materials, the resultant composite material properties could be predicted by designers and process engineers, thereby saving cost in the process.
ii. Forecasted tensile, electrical conductivity and hardness from ANN model were consistent and showed good agreement with measured results from the specimens.

Funding: The authors acknowledge Covenant University (CU) and Covenant University Centre for Research, Innovation and Discovery (CUCRID) Ota, Nigeria for the sponsorship and provision of research facilities for this work.

References

[1] Davies G. Materials for Automobile bodies. Butterworth-Heinemann, Oxford, MA 01803, 2003, pp. 90.
[2] Kok M, and Ozdin K. Wear resistance of aluminium alloy and its composites reinforced by Al2O3 particles. Journal of Materials Processing Technology, 2007, 183: 301–309.
[3] Prabu S.B, Karunamoorthy L., Kathiresan S., and Mohan B. Influence of stirring speed and stirring time on distribution of particles in cast metal matrix composite. Journal of Materials Processing Technology, 2006,171: 268–273.
[4] Inegbenebor A.O., Bolu C.A., Babalola P.O., Inegbenebor A.I. and Fayomi O.S.I. Aluminum Silicon Carbide Particulate Metal Matrix Composite Development Via Stir Casting Processing. Silicon, 2016, DOI 10.1007/s12633-016-9451-7.
[5] Inegbenebor A.O., Bolu C.A., Babalola P.O., Inegbenebor A.I. and Fayomi O.S.I. Influence of the Silicon Carbide Gritsize Particles on the Mechanical and Electrical Properties of Stir Casting Aluminum Matrix Composite Material. Silicon, 2015, DOI 10.1007/s12633-015-9305-8.
[6] Babalola P.O, Bolu C.A, Inegbenebor A.O and Odunfa K.M. Development of Aluminium Matrix Composites: A review. Online International Journal of Engineering and Technology Research, ISSN 2346-7452; 2014, 2: 1-11.
[7] Babalola P.O., Inegbenebor, A.O., Bolu, C.A., Inegbenebor, A.I. The Development of Molecular-Based Materials for Electrical and Electronic Applications. JOM, 2015, 67(4): 830-833.
[8] Luan B.F., Hansen N., Godfrey A., Wu G.H., and Liu Q. High strength Al–Al2O3p composites: Optimization of extrusion parameters. Materials and Design, 2011, 32:3810–3817.
[9] Singla M., Dwivedi D.D., Singh L., and Chawla V. Development of aluminium based silicon carbide particulate metal matrix composite. Journal of Minerals & Materials Characterization & Engineering, 2009, 8(6): 455-467.
[10] Naher S., Brabazon D., and Looney L. Examination of the semi-solid stir casting method for producing Al-SiC metal matrix composites. Materials Processing Research Centre, Dublin City University, Dublin 9, Ireland, 2012.
[11] Mohanavel V., Rajan K., Kumar S. S., Udishkumar S., Jayasekar C. Effect of silicon carbide reinforcement on mechanical and physical properties of aluminium matrix composites. Materials Today: Proceedings, 2018, 5:2938–2944.
[12] Alaneme K. K., Bodunrin M.O., Awe A. A. Microstructure, mechanical and fracture properties of groundnut shell ash and silicon carbide dispersion strengthened aluminium matrix composites. Journal of King Saud University – Engineering Sciences, 2018, 30: 96-103.
[13] Sulaiman S., Marjom Z., Ismail M.I.S, Ariffin M.K.A. and Ashrafi N. Effect of Modifier on Mechanical Properties of Aluminium Silicon Carbide (Al-SiC) Composites. Procedia Engineering, 2017, 184: 773 – 777.
[14] Kavimani V., Prakash K.S. Tribological behaviour predictions of r-GO reinforced Mg composite using ANN coupled Taguchi approach. Journal of Physics and Chemistry of Solids, 2017, 110: 409–419.
[15] Rashed F.S. and Mahmoud T.S. Prediction of wear behaviour of A356/SiCp MMCs using neural networks. Tribology International, 2009, 42: 642–648.
[16] Ali S. Hammood, Haider Mahdi. Development Artificial Neural Network Model to Study the Influence of Oxidation Process and Zinc-Electroplating on Fatigue Life of Gray Cast Iron. International Journal of Mechanical & Mechatronics Engineering, IJMME-IJENS, 2012, 12(5):74-78.
[17] Shandilya Pragya, Jain P.K., Jain N.K. RSM and ANN Modeling Approaches for Predicting Average Cutting Speed During WEDM of SiCp/6061 Al MMC. Procedia Engineering, 2013, 64: 767 – 774.

IOP Conf. Series: Materials Science and Engineering **413** (2018) 012064 doi:10.1088/1757-899X/413/1/012064

Production Of 100 Kg/Day of Zeolite a as a Builder for Powdered Detergent from Nigerian Ahoko Kaolin Using Locally Fabricated Mini Zeolite Plant.

Shehu Ahmed[1], Kovo Abdulsalami Sanni.[1], Oluwatosin S. Azeez[1], Abdulkareem Ambali Saka[1], Salako Oluwafemi.[1]

[1]School Of Infrastructural and Processing Engineering Technology, Federal University Of Technology, Gidan Kwano, Minna, Nigeria.
E-mail address: she_ahmed@futminna.edu.ng

Abstract. This project was aimed at assembling a plant already designed and fabricated, mainly furnace and crystallizer to produce a 100 kg/day of zeolite A. The produced Zeolite A was used to formulates a powdered detergent. The detergent grade Zeolite A was produced via hydrothermal process and used as substitutes for phosphates in the formulation of a powdered detergent. The formulation of the powder detergent was done by mixing and homogenising Sodium sulphate, linear alkyl benzene sulphonate acid (LABSA), Zeolite A, sodium tripolyphosphate (STTP), sodium silicate, and soda ash in a mixing unit. A 100 g of the detergent contains 55 % sodium sulphate (Na_2SO_3), 15 % LABSA, 8 % sodium silicate, 7 % soda ash (Na_2CO_3) and the remaining 15 % is formed by either (100 %) Zeolite A or a Zeolite A:STPP in a 50:50 ratio. In this formulation, aesthetics value additives such as optical brightener, perfumes, preservatives, active enzymes and stabilizer were not used. The Zeolite A produced was characterized using x-ray diffraction and scanning electron microscope. The characterisation shows that the peaks conform with standard Zeolite A, with peaks at 5, 10 ,13, 16, 20, 22, 24, 28, 30 and 34 ^{O}(2theta) respectively, while the SEM image reveals a particle size of 2.47 μm and a cubical surface morphology. The powdered detergent formulated was characterised and it shows a pH of between 9.8 - 10.2, while 9.5 for commercial detergent, a foam height of 85 - 90 cm, where as commercial detergent has 99 cm, percentage moisture content of 5.1 to 7.2 %, while 8.0 % was for commercial detergent. Therefore, this comparison with a commercial phosphate based detergents and standard organization of Nigeria (SON) standards shows that Zeolite A is a good substitute for phosphates in detergent formulations.

1. Introduction.

As the application of Zeolites which ranges from petroleum refining, to waste treatment increases, the need to increase production at low cost increases (Okoro, 2015). Detergents manufacturers are currently using Zeolites, (Zeolite A) as the builder material which has proved to be a substitute for phosphorus containing additives (Lijalem *et al*, 2016). Despite the remarkable potentials of Zeolite A, being more Eco friendly unlike the phosphates which contribute to eutrophication of aquatic life, its high cost has limited its effective use in detergent industries (Lijalem *et al,* 2016). Since one of the main hindrances in the application of Zeolite A is, its high cost of production, this has necessitated the diversification of more economic or a very low cost ways of producing Zeolite A on a large scale (Okoro, 2015).

Kaolin clay offers a unique and sustainable route toward the manufacturing of Zeolite A. This has ultimately been a factor in an effort to reduce the cost of zeolite A production. Since kaolin clay is a naturally occurring mineral, and it has been found in commercial deposit in many places in Nigeria including Ahoko, kogi state, Nigeria (Tatabu, 2015), this is considered as the most promising local sources of Zeolites in Nigeria (Okoro, 2015). The total reserve of kaolin in Nigeria as a whole is estimated to be about three billion metric tons in at least 45 known deposits distributed in some state of Nigeria (Alabi and Omojala, 2013).

Although Zeolite A has a wider range of application, the focus of this present work is to produce a detergent grade Zeolite A from Ahoko kaolin clay using the locally fabricated furnace and crystallizer as well as other equipment in order to use cheap route for production and its application in detergent formulation.

On the threshold of the new millennium, the attention of the detergent industry is focused on delivering against four challenges Economics Safety and Environment Technology Consumer Requirements The pressure of increasing costs and competition is forcing manufacturers to seek favourably priced raw materials and production processes. Laws and regulations with respect to environment and safety, as well as voluntary agreements, have already been affecting various detergent ingredients for years (example, phosphates and surfactants) it has also impacted upon packaging. New measures include a monitored self-regulation by detergent manufacturers (AISE: Code of Good Environmental Practice), and the introduction of an Eco - label for detergents. Both measures are having a direct impact on the composition and technology of detergents. In particular, efforts to save energy and to reduce detergent consumption have had an impact on the technological development of both detergents and washing machines. The crucial factor, however, remains the consumer. Although washing habits differ from country to country, there is a trend towards easier handling (compact powders, tablets). First and foremost, however, the consumer continues to demand a clean wash combined with maximum protection of the items laundered. The resulting requirement to continuously improve the performance of detergents is leading to the use of new or optimized raw materials. (Zeodet, 2000). To a large degree, the aforementioned challenges dominated the development of detergents in the 1980s and 90s. Zeolite, originally designed as a phosphate substitute for purely ecological reasons, increasingly had to meet the demands imposed by modified detergent composition and production technologies. In particular, the trend towards compact detergents increased the demand for builder systems with a high adsorption capacity for liquid components, especially for surfactants. Zeolite A, introduced approximately 20 years ago, proved to be a good carrier for surfactants and in the 90s advanced to become the builder leading to compact and super compact detergents. Nevertheless, the market demanded further improvements (Zeodet, 2000). The manufacturers of detergent Zeolites responded to the demand for higher standards of performance and processing by developing new grades of Zeolite. These include the Zeolites of types P, X and AX, which have all recently been introduced into the market. (Zeodet, 2000).

2. Material and methods

Purification (Pre-treatment) of Raw Kaolin

Beneficiation Process.

Beneficiation process is the first stage in the experiment; it involves beneficiation of the raw material (kaolin), which is the removal of debris, pebbles, clay particles and some inorganic matter (Kovo and Edoga, 2005). To start up this process the raw kaolin sample was collected from Ahoko village in Kogi state, Nigeria, mechanical methods such as separation and sedimentation was carried out to remove mainly quartz and other suspensions (Lijalem, *et al,* 2016). Two Hundred kilograms (200 kg) of raw kaolin clay was crushed and was soaked in deionized water (in batches) of 25 kg per batch, for four days during which it was vigorously mixed using stirrer (Kovo and Edoga, 2005).The Kaolin suspension was then transferred into sedimentation apparatus (10,000 ml) measuring cylinders and was allowed to settle out, on settling, the coarse component remained at the bottom, while the supernatant (Kaolin) settled on top (Kovo, 2010).The supernatant after ward was decanted into measuring cylinders and was allowed to settle for 24 hr, the recovered clay was then dried in an electric oven at 45 °C for 48 hr.

Calcination Process

One hundred kilogram (100 kg) of the beneficiated clay was calcined in batches of 10 kg per batch in the first unit, the locally fabricated pilot plant (Dehydroxylated furnace) at a temperature of 650 ^{O}C for 10 minutes. The metakaolin obtained was kept aside for further processing into Zeolite A. The calcination process was done by putting weighted quantity of the beneficiated clay in the rotating drum of the furnace, thereby subjecting it to a high temperature of 650 °C for the time range of 10 minutes. This process is meant to prepare the beneficiated clay to a more reactive form, before Zeolite A formation process.

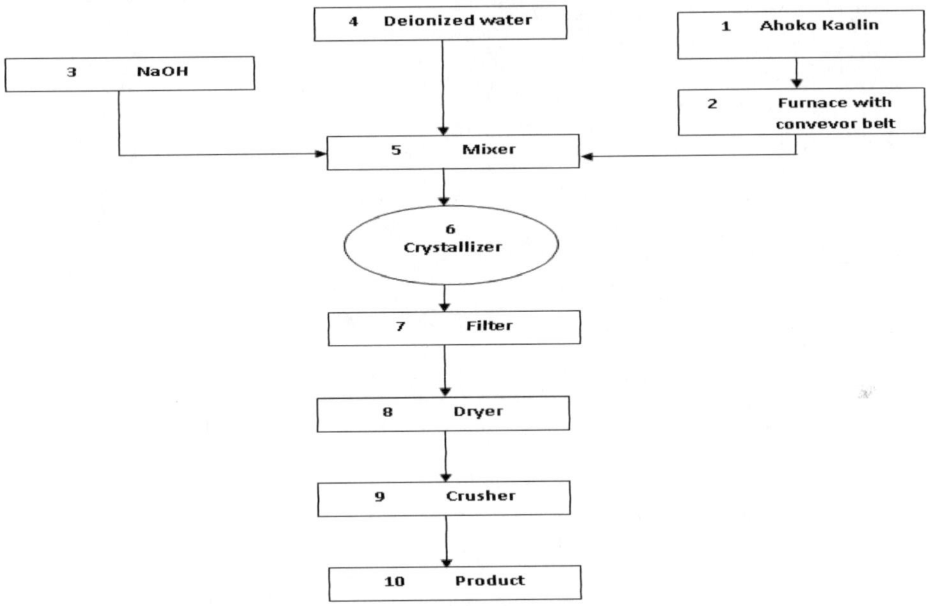

Figure 1: Step wise procedure for Zeolite A production

Zeolite A production procedure

Zeolite A production was carried out using Kaolin conversion process. The Kaolin conversion process makes use of the Meta Kaolin produced earlier, sodium hydroxide (pellets) and deionized water, as starting raw materials (Tatabu, 2015). Conventional hydrothermal procedure reported by (Kovo, 2011) was chosen due to its uniqueness in rapid achievement of the zeolite-A product. In the conventional hydrothermal process, the activation of chemically inert kaolin was performed by taking 10 kg each of purified kaolin separately and calcining it at 650 ^{O}C for 10 minutes in the locally fabricated furnace. Then the alkaline treatment was done in a mixer using NaOH (0.832 g:12 g:0.693 g) of (NaOH:deionized water:Mata kaolin respectively), with stirring (500 rpm) at 50 ^{O}C for 1 h for the complete homogenization of the reaction mixture. The gel formed undergoes ageing at room temperature under static conditions. (Lijalem *et al,* 2016). The molar gel composition was:

$$Na_2O:Al_2O_3:2SiO_2:37H_2O \quad (1)$$

Crystallization was carried out using hydrothermal method in the second unit of the locally fabricated pilot plant (autoclave crystallizer) for 3 hr at 100 ^{O}C. Once the time was attained, the reaction mixtures were filtered and washed with deionized water to remove excess alkali until the pH of the filtrate is below 10. Then, the samples were oven dried overnight at 80 ^{O}C.

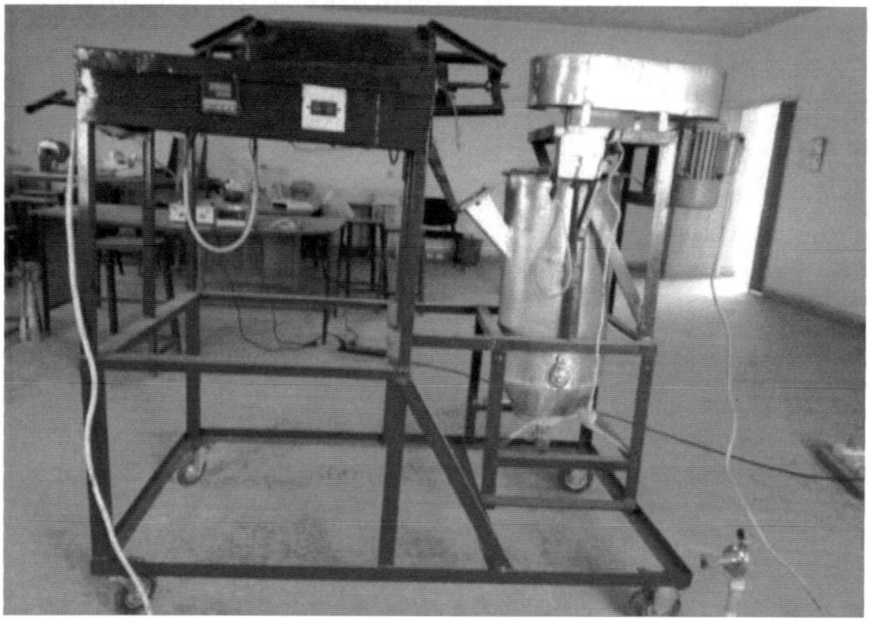

Plate 1: Locally Fabricated mini Zeolite Plant

Formulation of powdered detergent

Formulation of the powder detergent was done by mixing and homogenizing Sodium sulphate, linear alkyl benzene sulphonate acid (LABSA), zeolite A, sodium tripolyphosphate (STTP), sodium silicate, and soda ash in a mixing unit. The formulation of 100 g of detergent contains 55 % sodium sulphate (Na_2SO_3), 15% LABSA, 8% sodium silicate, 7% soda ash (Na_2CO_3) and the remaining 15% is formed by either 100 % Zeolite or a zeolite A:STPP in a 50:50 ratio. For this particular formulation (research work), aesthetic value additives such as optical brightener, perfumes, preservatives, active enzymes and stabilizer were not used. In addition, a commercial powder detergent having sodium tripolyphosphate as a builder was characterised for comparison. The Zeolite A sample synthesized from Ahoko kaolin were compared with commercial Zeolite A. In this study we have tried to substitute STTP from 50 % to 100 % with Zeolite A in the detergent formulation and composition. (Lijalem, *et al*, 2016)

3. Result and discussions

IOP Conf. Series: Materials Science and Engineering **413** (2018) 012064 doi:10.1088/1757-899X/413/1/012064

Characterization of Zeolite A samples

The characterization of detergent grade Zeolite A samples was carried out using X-ray diffraction (XRD) and scanning electron microscope (SEM). Figure 2 below shows the XRD of the synthesized Zeolite A, while figure 3 shows the SEM image. The XRD image shows that the peaks conform with standard, with peaks at 5, 10, 13, 16, 20, 22, 24, 28, 30 and 34 ^{O}(2theta) respectively, while the SEM image reveals a particle size of 2.47 μm and a cubical surface morphology. The XRD peaks were indexed with a reference to commercial Zeolite A sample obtained from Treacy and Higgins, 2001.

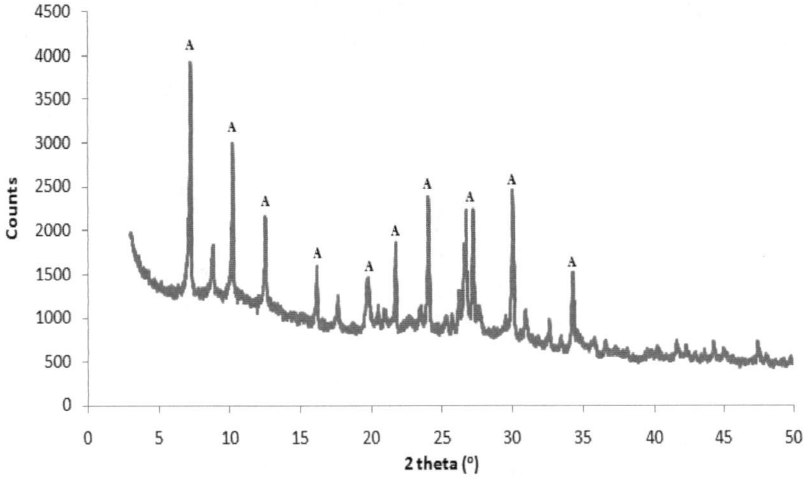

Figure 2: XRD pattern of Zeolite A synthesized from Ahoko kaolin

Figure 3 below shows the SEM image of the Zeolite A produced in this present work showing cubical morphology at 3h of crystallization which typical of Zeolite A.

IOP Conf. Series: Materials Science and Engineering **413** (2018) 012064 doi:10.1088/1757-899X/413/1/012064

Figure 3: SEM image of the ZeoliteA produced

Characterization of the formulated detergent powder

The formulated detergent sample was then tested using various physiochemical parameters such as the moisture content, foam height, pH value, alcohol and water insolubility, which are adapted from the Nigerian Standards for Soaps and Detergent. The same tests were also conducted on the known commercial powder detergent brand and the results of their comparative evaluation are presented. Foam height measurement was done by preparing 1 % detergent solution. 1 g of the detergent is dissolved in 99 ml of real hard water having moderate hardness (120 mg $CaCO_3$ per l). Then 10 ml of the prepared solution was taken in a 100 ml measuring cylinder with a glass stopper and shaken very well until the detergent is completely dissolved and forms foam. The foam height was measured after 10 minutes of stability of the foam. The pH of the 1 % detergent solution was measured using a pH meter at 25 ^{O}C following the calibration of the instrument. For the moisture and volatile content, 5 g of detergent sample was weighed, and dried to constant mass in an oven at 105 ± 2 ^{O}C. This was done until constant mass is attained when successive heating for one-hour period shows less than 5 mg difference. The percentage (%) moisture content was calculated using the following equation:

Error! Reference source not found. (2)

In this scenario insoluble matter in alcohol refers to inorganic salts, such as phosphates, sulphates, silicates and carbonates, which are usually present in non-soapy detergent preparations. This was done by weighing 5 g of the detergent into a beaker and digested with 50 ml of ethanol, followed by heating on a steam bath for about 2 minutes. Any hard lump was broken down with a glass rod flattened at one end. The solid matter was then allowed to settle and decanted through a sintered glass filter funnel. The alcoholic digestion was repeated in a similar manner with five further consecutive 30 ml portions of boiling ethanol. Each extract was filtered in turn through the same sintered glass funnel and the residue was washed several times with hot ethanol to remove all the alcohol soluble matter. Finally, the sintered glass funnel was dried with the residue at a temperature of 105 ± 2 ^{O}C until a constant weight is obtained. The insoluble matter in alcohol is determined using an equation:

Error! Reference source not found. (3)

For water insoluble matter analysis, 5 g of detergent sample was weighed and digested with 200 ml of freshly boiled water until the sample is completely dissolved. Then, the solution was filtered through the filter paper, dried at 105 ± 1 ^{O}C and mass recorded. The residue was washed several times with hot water. The filtrate and the residue were dried for 3 h at 105 ± 2 ^{O}C.

Finally, the water insoluble matter was calculated using the equation below;

Error! Reference source not found. (4)

The same characterisations were also carried out on a popular brand commercial powder detergent and results of their comparative evaluation are presented. Two (2) compositions were characterised: 50 % substitution of STTP by Zeolite A then followed by 100 % substitution of Zeolite A in the formulation. The results obtained are shown in tables 1 to 5.

Table 1 PH value

Detergent type	PH value
Detergent (50 % Zeolite: 50 % STTP)	9.8
Detergent (100 % Zeolite: 0 % STTP)	10.2
Commercial detergent (100 % STTP)	9.5
SON standard	9 - 11 maximum

* SON Standard, the pH of requirement for 1 % aqueous solution of industrial detergent determined at 25 ^{O}C shall not be less than 9 and more than 11.
Therefore, the value measured for the powder detergent formulated with the synthetic Zeolite A is in good standing order with the specification range and also with the commercial detergent.

Table 2 Foam height analysis (Lather formation)

Detergent type	Foam height (cm)
Detergent (50 % Zeolite: 50 % STTP)	90
Detergent (100 % Zeolite: 0 % STTP)	85
Commercial detergent (100 % STTP)	99
SON standard	80 minimum

From table 2 above is the result for foam height, it shows that the 50 % Zeolite substitution exhibited almost the same foam height with 100 % substitution by Zeolite A. The foam height result is comparable with the foam height of commercial powder detergent which is a very good development. The standard is 80 cm minimum.

Table 3 Percentage (%) moisture content analysis

Detergent type	Moisture content (%)
Detergent (50 % Zeolite: 50 % STTP)	7.2
Detergent (100 % Zeolite: 0 % STTP)	5.1
Commercial detergent (100 % STTP)	8.0
SON standard	13.0 maximum

The table 3 above, shows % moisture content test results on the detergents shows 7.2 %, 5.1%, 8.0 % percentage moisture on 50 % zeolite, 100 % zeolite, and commercial (100 % STTP) detergents respectively. The maximum moisture and volatile component content requirement of any commercial synthetic detergent is 13 % by mass.

Table 4 Percentage (%) matter insoluble in water

Detergent type	Matter insoluble in water (%)
Detergent (50 % Zeolite: 50 % STTP)	7.0
Detergent (100 % Zeolite: 0 % STTP)	5.0
Commercial detergent (100 % STTP)	6.0
SON standard	5.0 - 10.0 maximum

The above results (table 4) are in good agreement with commercial detergent and indicate the possibility of zeolite A to be used for detergent formulation. The appropriate solubility of detergent in water is one of the requirements in washing. The maximum insoluble matter in water should not exceed more than 5 – 10% by mass.

Table 5 Percentage (%) matter insoluble in alcohol

Detergent type	Matter insoluble in alcohol (%)
Detergent (50 % Zeolite: 50 % STTP)	71.0
Detergent (100 % Zeolite: 0 % STTP)	75.0
Commercial detergent (100 % STTP)	69.0
SON standard	80.0 maximum

Although 50% and 100% detergents show a higher values than the commercial detergent, all are still in the range of the standard, which is 80% maximum.
The summaries of tables 1 to 5 shown on table 6

Table 6 Summary of characterisations on 50 % Zeolite, 100 % Zeolite, and commercial powder 100 % (STTP) detergents

Detergent type	B	C	D	E	F
50 % Zeolite Detergent.	7.2	90	9.8	7.0	71.0
100 % Zeolite Detergent.	5.1	85	10.2	5.0	75.0
Commercial Detergent.	8.0	99	9.5	6.0	69.0
SON standard	13 %	80 min.	9-11 max.	5-10 max.	80.0 % max.

Key:
B- Moisture content (%)
C- Foam height (cm^3)
D- pH Value
E- Matter insoluble in water (%)
F- Matter insoluble in alcohol (%)
min - minimum
max - maximum

The detergency of the formulated powdered detergent

The formulated detergents both 50 % and 100 % has a high detergency or cleansing power as shown by the formation of foam in water as shown in table 2 which is actually the ability of the detergents to form lather when dissolve in water, and this makes the detergents very efficient in washing / cleaning.

4. Conclusions

The following conclusions were drawn;

1. A mini Zeolite plant was assembled from a locally fabricated units mainly furnace and auto clave crystallizer, Zeolite A was produced from Ahoko kaolin via the procedure reported by Kovo (2011) for 10 minutes calcination time, at a temperature of 650 ^{O}C using the assembled mini plant.

2. The X-ray diffraction (XRD) image of the Zeolite produced shows conformity with standard Zeolite A, with peaks at 5, 10, 13, 16, 20, 22, 24, 28, 30, and 34 ^{O}(2theta) respectively, while the scanning Electron Microscope (SEM) image reveals a particle size of 2.47 µm and a cubical surface morphology. Also the result of the XRD of the meta kaolin using the fabricated furnace acted as a guide to develop a new route for the production of meta kaolin with the main aim of reducing the time of exposure taken to obtain the amorphous material.

3. The Zeolite A produced was efficiently used as a builder in detergent formulation by substituting the environmentally unfriendly phosphate based builder sodium tripolyphosphate (STTP), This was done by formulating powder detergent using the Zeolite-A as substitute to STTP with 50 % and 100 % zeolite A. The result shows that the zeolite-A based detergent performs almost similar to a very popular brand of commercial powder detergent composed of sodium tripolyphosphate (STTP) detergent builder.

4. The powder detergent formulated was characterized and compared with commercially available phosphate based detergents and standards and the results shows that Zeolite-A is a better substitute for phosphates in detergent formulations.

References

[1] Alabi, F. M., and Omojola, M. O. (2013). Potentials of Nigerian Calcined kaolin as paint pigment. *African Journal of Pure and Applied Chemistry*,7(12), 410-417.

[2] Bekkum, V. H., Flanigen, E.M., Jacobs, P.A., & Jansen, J.C. (1991). *Introduction to Zeolite Science and Practice*, 2nd. Revised Edition., Amsterdam, Elsevie.

[3] Breck, D. W., (1974). Zeolite Molecular Sieves: *Structure, Chemistry and Use*. JohnWiley, New York, USA.

[4] Byrappa, k.and Masahiro,Y. (2001).*Handbook of Hydrothermal Technology*. Parkridge, New Jersey and William Andrew Publishing, LLC. Norwich, New-York. USA.

[5] Chandrasekhar, S. (1995). *Influence of Metakaolinization temperature on the formation of Zeolite A from kaolin clay minerals,* 31, 253-261.

[6] Coombs, T.A., Compbell, A.M., Ganney, I., Denis, S., Twardowski, T., and Dawson, B. (1998). *Journal of Materials Science and Engineering*, 53(1), 226-228.

[7] Flanigen, E. M., Beckum, H., Van, J. P.A., and Jenson, J.C. (2001). Zeolite and molecular sieves.An historical perspective, *in introduction to zeolite sciences and practice*.137, 11-35 Elsevier, Amsterdam.

[8] Foraminifera Venture (2010). Investment opportunity in Nigeria: Kaolin deposits in Nigeria, The opportunities. Retrieved from Farriconsultingng.blogspot.com is a subsidiary of Foraminifera Venture {www.foraminifera.com}, a business development firm and owners of nigeriabusinessplace.com

[9] Farag,I. H., and Zhang, F. (2012). Simulation of Synthetic Zeolites– 4A and 5A Manufacturing for Green Processing; *"IRACST – Engineering Science and Technology: An International Journal (ESTIJ),* 2 (2), 508-509.

[10] Georgiev, D., Bogdanov, B., Angelova, K., Markovska, I., and Hristov, Y. (2009), Synthetic Zeolites - structure, classification, current trends in zeolite synthesis review. *Economics and Society development on the Base of Knowledge,* 7 , 1-5. International Science conference 4th - 5th, Stara Zagora, Bulgaria.

[11] Haag, W. O., Lago, R. M., and Weisz, P. B., (1984). The active site of acidic aluminosilicate catalysts, Nature, *Mobile Research and Development Component* 309, 589-591

[12] Hirano, S. and Morikuni, M. (2003), *Method for packaging and sealing a Zeolite adsorbent with a dehydrating agent.* United States Patent No US 6,523,329 B2.

[13] Ilić, B. R., Mitrović, A. A., and Milićić, L. R. (2010). Thermal Treatment of Kaolin Clay to Obtain Meta kaolin. *Hem.ind .64*(4), 351-356.

[14] Kovo, A. S. and Edoga, M. O. (2005), Production and Characterization of Zeolite from Ahoko Clay in Kogi State, Nigeria. *Leonardo Electronic Journal of Practices and Technologies,* (7), 31-40.

[15] Kovo, A. S. and Holmes, (2010). Effect of ageing on the synthesis of kaolin-based Zeolite Y from Ahoko Nigerian,using a novel metakaolinization technique. J.Dispersion Sci. Technol, (31) 442-448.

[16] Kovo, A. S. (2011), *Development of Zeolites and Zeolite membranes from Ahoko Nigeria kaolin.* PhD. Thesis, University of Manchester, UK.

[17] Lijalem Ayele, Joaquim Perez-Pariente, Yonas Chebude and Isabel Diaze (2016). synthesis of Zeolite A using kaolin from Ethiopia and its application in detergents, NJC, Royal Society of Chemistry (40), 3440-3446.

[18] Mumpton, F.A. (1984). 'Zeolite exploration: the early days', in D. Olson and A. Bisio, eds, Proceedings of the 6th *International Zeolite Conference, Butterworth, 1984, 68-86.*New York.

[19] Murray, H.H., (2007). Bentonite Applications. In Applied clay mineralogy: Occurences, processing and Application of Kaolins, Bentonites, Palygoiskite-sepiolite and common clays, Murray, H.H. (Ed) Volume 2, chapter 6, Elsevier, Amsterdam, Netherland. ISBN-13:978-0444517012, page: 111-130.

[20] Okoro Enest Okwudiri (2015), Production of Zeolite Y from Nigerian Ahoko Clay using hydrothermal method. PGD Thesis, Federal University of Technology minna Nigeria.

[21] Pang, B., Zhou, X., Yu, H., Dong, M., Taghizadeh, K. and Wishnok, J.S. (2008). Lipid per oxidation dominates the chemistry of DNA adduct formation in a mouse model of inflammation. *Carcinogenesis,* 28 (8), 1807–1813.

[22] Parnham, E.R. and Morris, R.E. (2007). Iono thermal Synthesis of Zeolite, Metal–Organic Frame works, and Inorganic–Organic Hybrids.*Accounts of Chemical Research*, 40(10), 1005–1013. DOI:10.1021/ar700025k

[23] Perry, R.F and Green, D.W (1998); *Perry's Chemical Engineering Hand book* 7th edition, Mcgraw-Hill book company, New York, USA.

[24] Rabo, J.A., Pickert, P.E., Stamires, D.N. and Boyle, J.E. (1960). Molecular sieve catalysts in hydrocarbon reactions, in *Proceedings of the Second International Congress on Catalysis*, Ed. Tech, Paris, 2055.

[25] Santi, K. (2010). *Zeolites in Industrial Separation and Catalysis.* Edited by Santi Kulprathipanja.Verlag GmbH and Co. KGaA, Weinheim.

[26] Shivaji Sircar, S and Alan L. M. (2003). *Gas Separation by Zeolites,* M.sc thesis Lehigh University, Bethlehem, Pennsylvania, U.S.A.

[27] Sinnott, R. K. (2003). *Coulson and Richardson's Chemical Engineering Design* (3rd edition,Volume. 6). Butterworth-Heinemann.

[28] Siti A.B.I (2007). *Synthesis and characterization of zeolite* from sodium Alumino silicate solution.Masters Thesis.

[29] Tanabe, K. (1970), *Solid Acids and Bases*, Kodansha, Tokyo and Academic press,New York, chapter 2, 5-10.

[30] Tatabu, Muhammad (2015), Design and Fabrication of Crystallizer to Produce 100 kg/Day of Zeolite A From Ahoko Kaolin, M.ENG thesis, Federal University Of Technology Minna, Nigeria.

[31] Treacy, M.M.J. and Higgins, J.B. (2001), Collection of Simulated XRD Powder Patterns for Zeolites, Fourth Edition. Elsevier, New York.

[32] Towler, G. and Sinnott, R. (2008), Chemical Engineering Design, *Economic of Plant and Process Design Principles,*54 (11), 3034-3036.

[33] Ugal, J. R., Hassan, K. H. & Ali, I.H. (2010), *Preparation of type 4A Zeolite from Iraqi Kaolin characterization and properties measurement.* Assoc Arab University. Basic Applied Science, 9, 2-5.

[34] Van der Gaag F.J. (1987), ZSM-5 type Zeolites, *Synthesis and use in gas phase reactions with ammoni,* verlag, berlin.

[35] Walter, P. (1989). Natural Zeolite, *The effect Natural Zeolite on Excretion and Distribution of Radiocesium in Rats,* 34 (8), 67-74.

[36] Xu, R.W., Pang, J., Yu, Q.H and chen, J., (2007) *Chemistry of Zeolites and Related Porous Materials*: Synthesis and Structure, Clementi Loop Singapore John Wiley 679.

[37] Zeodet – Association of Detergents Zeolites Producers (2000), *Zeolites for Detergents as Nature Intended,*Brussels. http://www.zeodet.org/downloads/Zeolites.pdf, Europe Chemical Industry Council. CEFIC. (Zeodet, 2000).

IOP Conf. Series: Materials Science and Engineering **413** (2018) 012065 doi:10.1088/1757-899X/413/1/012065

14 Molar Concentrations of Alkali-Activated Geopolymer Concrete

Solomon Oyebisi[1], Joseph Akinmusuru[1], Anthony Ede[1], Olatokunbo Ofuyatan[1], Grace Mark[1], John Oluwafemi[1]

[1] Civil Engineering Department, Covenant University, Ota, Nigeria
E-mail address: solomon.oyebisi@covenantuniversity.edu.ng

Abstract. The increase in construction activities due to economic and population growth has led to the higher demand and utilization of cement. But cement production leads to the pollution of the environment. Consequently, this study examines the utilization of both ground granulated blast furnace slag (GGBFS) and corncob ash (CCA) as source materials in the production of geopolymer concrete (GPC). GGBFS was replaced by CCA in 0%, 20%, 40%, 60%, 80% and 100% respectively using Grade 30 (M30) mix design proportion. Alkaline liquids were prepared to obtain 14 molar concentrations and used to activate the source materials. Slump, density and compressive strength of GPC were determined and compared with Portland cement concrete (PCC). The research findings indicate that GPC has higher compressive strength than PCC. Based on the relationship between the compressive strength and the density, a model equation is established. And the equation is used to predict the compressive strength of GPC with respect to the density. Reprocessing of CCA and GBFS as emerging low-carbon footprints appears to be a feasible solution to the problem of environmental pollution.
Keywords: geopolymer concrete; corncob ash, ground granulated blast furnace slag; regression model; alkaline liquids; compressive strength

1. Introduction

Portland limestone cement has been the most widely used conventional binder in the construction industry particularly, in developing countries. Presently, while the utilization of cement is reducing in the major consuming nations such as China, India, Brazil, Turkey and Russia [1], the utilization of cement in Nigeria is increasing with 31.25 million metric tons per annum (MMTPA) and a total equipped production quantity of 45 MMTPA [2] compared with the annual utilization of 19.5 million metric tons in 2009[3]. Consequently, there has been an increase in ecological imbalance as a result of Portland limestone cement production. The production of Portland limestone cement emanates carbon dioxide (CO2) and other greenhouse gases. In 2002, *Malhotra* reported that one ton of Portland cement production emits of a ton of CO2, and also, contributes 7% of the total greenhouse gas emissions to the earth's atmosphere [4].

In 2017, the *United Nations Statistics Division Sustainable Development Goals* reported that in 2014, 9 in 10 residing in urban areas breathed air which did not conform to World Health Organizations air quality guidelines as a result of greenhouse gases and CO2 to the atmosphere [5]. These greenhouse gases lead to increase in global warming; increase in environmental damage; increase in respiratory diseases such as asthma and pulmonary ossification, lung cancer; kidney failure and other serious diseases to humans [6]. Therefore, safety, inclusiveness, resilience, and sustainability of cities and human settlements depend on the alternative utilization of Portland limestone cement in the production of conventional concrete in the construction industry [5] [7]. One of these alternatives is geopolymer concrete which does not use any Portland limestone cement in its production process and the materials are pozzolanic which are supplementary cementitious materials and they include ggbfs, cca, metakaolin, and fly ash [8-12]. Comparing with Portland cement production, *Davidovits* stated that there is 70-80% less CO2 emission and 43-59% less energy required in the production of geopolymer cement (slag byproduct) [13]. Geopolymer concrete is an eco-friendly, emerging and innovative binder produced by a polymeric reaction of silica and alumina present in the source materials inducting or activating by alkaline liquids to dissolve and form a geopolymeric gel or paste.

In 2017, *the United States Department of Agriculture* reported that Nigeria alone produced 7.20 million metric tons of corn in 2016/2017 calendar year [14]. Similarly, crude steel production in Nigeria was 0.3 million tons in 2016 while the consumption is above 20 million per annum [15]. It is obviously known that these industrial and agricultural products are treated as waste products. The global output of blast furnace slag in 2014 based on the production level of crude iron and steel was estimated to be 300 million metric tons – 350 million metric tons [16]. GGBFS is a by-product of steel and iron production and during the process, slag is formed, dried and ground.
Therefore, this study is significant because it removes high elevated temperature curing regime of fresh geopolymer concrete which is not economical and practicable in the construction field. The utilization of both GGBFS and CCA as source materials allows the fresh GPC to be cured at ambient conditions. Sodium silicate gel and sodium hydroxide solution were utilized as alkaline liquids. And the ratio of sodium silicate-to-sodium hydroxide solutions was chosen as 2.5 based on the relevant studies [9] [17]. The 14 molar concentrations were adopted because higher molar concentrations result in higher mechanical strength performance [17] [18]. In addition, percentage replacement levels were selected based on the applicable studies that replacement of cement with CCA at 20-30% is feasible with compressive strengths for structural applications [19-22]. Furthermore, weaknesses in the design of GPC mix design were removed because density and aggregates are normally assumed without considering the water absorption capacity, specific gravities, and the moisture contents of materials used.

2. Materials and Methods
2.1. Materials
Dangote 3X Portland limestone cement Grade 42.5R and aggregates were used and sourced from dealers in Ota, Nigeria. Russian made sodium hydroxide (NaOH) pellets with 99% purity, sodium silicate (Na2SiO3) gel, and the naphthalene-based superplasticizer (Conplast- SP 430) were also used and obtained from chemical dealers in Lagos, Nigeria. Water from laboratory source was used. Corncobs were got in Agbonle, Oyo State, Nigeria. And burnt by an open burning method in order to replicate the local production. The ash was then dried, ground, and sieved with BS sieve 90 μm. GGBFS was obtained from Dolphin Steel (Nigeria) Limited, Papalanto, Nigeria. It was dried, ground, and sieved with BS sieve 90 μm. The chemical compositions of both GGBFS and CCA were analyzed using X-Ray Fluorescence (XRF) at Sagamu plant of Lafarge Holcim Plc., Nigeria. The results of chemical compositions are presented in Table 1.

Table 1 Chemical compositions of GGBFS, CCA and cement

Composition	CaO	SiO2	Al2O3	Fe2O3	SO3	MgO	Na2O	M.C	LOI
GGBFS (%)	36.52	35.77	14.11	0.92	1.08	9.45	0.30	0.52	0.32
CCA (%)	12.62	60.50	11.78	9.13	1.25	1.23	0.65	1.25	0.49
Cement (%)	62.85	21.24	4.95	3.95	1.63	1.97	0.42	-	2.72

Notes: M.C (Moisture Content); LOI (Loss of Ignition)

2.2. Experimental methods
2.2.1 Design of concrete mix proportion
Both the Portland cement concrete and the geopolymer concrete mix proportions were designed in accordance with the *American Concrete Institutes* [23-24], taking into considerations, moisture contents, specific gravities, and water absorption capacities of the constituents used. Moreover, both fine aggregate (FA) and coarse aggregate (CA) were used at the saturated surface dry (SSD) conditions. The superplasticizer (Conplast SP-430) was added by 1.5% of the cementitious materials by mass. This was in accordance with the findings that 1.5% of the naphthalene-based superplasticizer by mass of the cementitious materials achieved the highest compressive strength of GPC [25-26]. The results of the volumetric computations for both PCC and GPC are presented in Table 2 and Table 3 respectively. Furthermore, PCC, 100% GGBFS + 0% CCA, 75% GGBFS + 25% CCA, 50% GGBFS + 50% CCA, 25% GGBFS + 75% CCA, and 0% GGBFS + 100% CCA were denoted as control experiment, GPC 1, GPC 2, GPC 3, GPC 4, GPC 5, and GPC 6 respectively.

Table 2 Volumetric computation of M30 PCC

S/N	Constituent	Weight (Kg/m3)	Specific Gravity	Absolute Volume (M3)	Adjusted Volume (M3)	Ratio
1	Cement	390	3.15	0.124	0.124	1.00
2	FA (SSD)	675	2.60	0.260	0.259	2.09
3	CA (SSD)	1031	2.64	0.390	0.388	3.13
4	Water	204.15	1.00	0.204	0.204	1.65
5	Air content	2.00	-	0.020	0.020	-
6	SP-430	5.85	1.20	0.005	0.005	0.04
	Total	2306		1.004	1.000	7.91

The mix ratio for PCC was 1: 2.09: 3.13

Table 3 Volumetric computation of M30 GPC

S/N	Constituent	Weight (Kg/m3)	Specific Gravity	Absolute Volume (M3)	Adjusted Volume (M3)	Ratio
1	GGBFS/CCA	390	2.90	0.134	0.142	1.00
2	FA (SSD)	675	2.60	0.260	0.276	1.94
3	CA (SSD)	1031	2.64	0.390	0.414	2.92
4	NaOH solution	60	1.49	0.040	0.042	0.30
5	Na2SiO3 solution	150	1.60	0.094	0.100	0.70
6	Conplast SP-430	5.85	1.20	0.005	0.005	0.04
7	Air content	2.00	-	0.020	0.021	-
	Total	2306		0.943	1.000	6.90

The mix ratio for GPC was 1: 1.94: 2.92

2.2.2 Preparation of alkaline liquids

The sodium hydroxide (NaOH) solid pellet (400g) was measured and dissolved in 600g of clean water based on the 14 molar concentration. This resulted in 1000g of sodium hydroxide solution [27]. The water-to-geopolymer ratio was 0.25. This solution was prepared 24 hours prior to casting in order to cool down the exothermic reaction of the solution. Thereafter, NaOH solution was added to Na2SiO3 solution two hours prior to casting of concrete in order to enhance its performance for a better result.

2.2.3 Mixing and casting

The dry constituents were mixed for about four minutes. The liquid components of the mixture were then added to the dry constituents and thoroughly mixed for another four minutes. The fresh concrete was cast manually, filled in the moulds, and compacted accordingly. The fresh mix and slump test of GPC are shown in Figure 1. The slump test was carried out in accordance with the *British Standards* [28]. After 24 hours of casting, PCC specimens were demoulded and immersed in water curing tank until testing day while GPC specimens were kept in rest period for 72 hours before being demoulded to allow for proper polymerization. The curing was done at the temperature (22 ± 30C) and the relative humidity (60 ± 5%).

(a) Freshly mixed GPC (b) Slump test

Fig. 1 Fresh mix and test on the fresh concrete

2.2.4 Tests on the hardened concretes

Both the compressive strength and the density tests were carried out on the hardened concretes using a digital 2000 KN capacity compressive machine and a digital weighing balance, and the summary is presented in Table 4 while the results are shown in Figure s respectively.

Table 4 Summary of specimens' tests and curing days

Test	Specification	No of tested Samples/mix	Sample size (mm)	Curing (day)
Compressive strength	*British Standards*[29]	3 cubes	150 × 150 × 150	7 and 28
Dry density	*British Standards*[30]	3 cubes	150 × 150 × 150	7 and 28

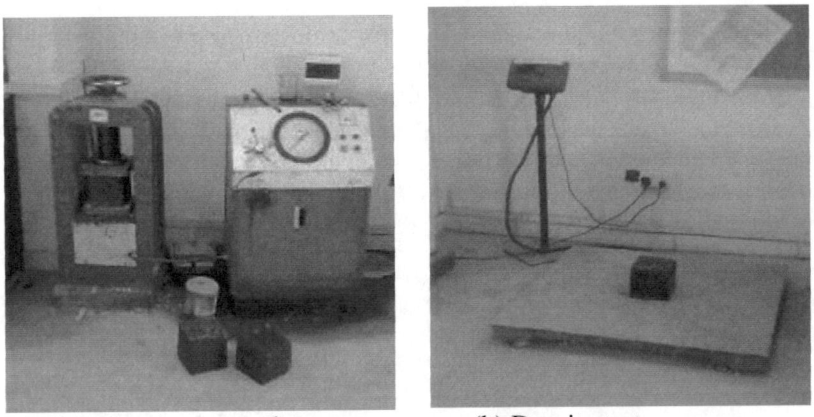

(a) Compressive strength test (b) Density test

Fig. 2 Tests on the hardened concrete

3. Results and Discussions

3.1. Chemical compositions of GGBFS, CCA and cement

The chemical compositions of the GGBFS used as presented in Table 1 met the requirements of *British Standards* [31] which specified that CaO+MgO+SiO2 ≥ 66.7; (CaO+MgO)/SiO2 ≥ 1.0; and CaO/SiO2 ≤ 1.4. The SO3, MgO, and LOI also met the maximum requirements of 2.5%, 14.0%, and 3.0% respectively. Moreover, chemical compositions of CCA used met the requirements of *American Society for Testing and Materials C* [32] which states that SiO2 + Al2O3 + Fe2O3 > 70%. Similarly, SO3, MgO, and LOI also satisfied the maximum requirements of 4%, 4%, and 10% respectively. In addition, chemical compositions of cement used satisfied the requirements of *British Standards* [33] which ranged the contents of SiO2, Al2O3, Fe2O3, and CaO as 18.0-24.0%, 2.6-8.0%, 1.5-7.0%, and 61.0-69.0% respectively.

3.2. Slump

Figure 3 shows the slump values for each mix proportion. The results indicate that slump value increases as the percentage replacement level of CCA increases. This is in agreement with the various findings that pozzolanic materials improve the workability of fresh concrete [34-36]. In addition, GPC extracts water from the reaction because it takes no part in the polymerization process. And during the curing process, the water is expelled from the geopolymer paste through nano-pores in the matrix [37].

IOP Conf. Series: Materials Science and Engineering **413** (2018) 012065 doi:10.1088/1757-899X/413/1/012065

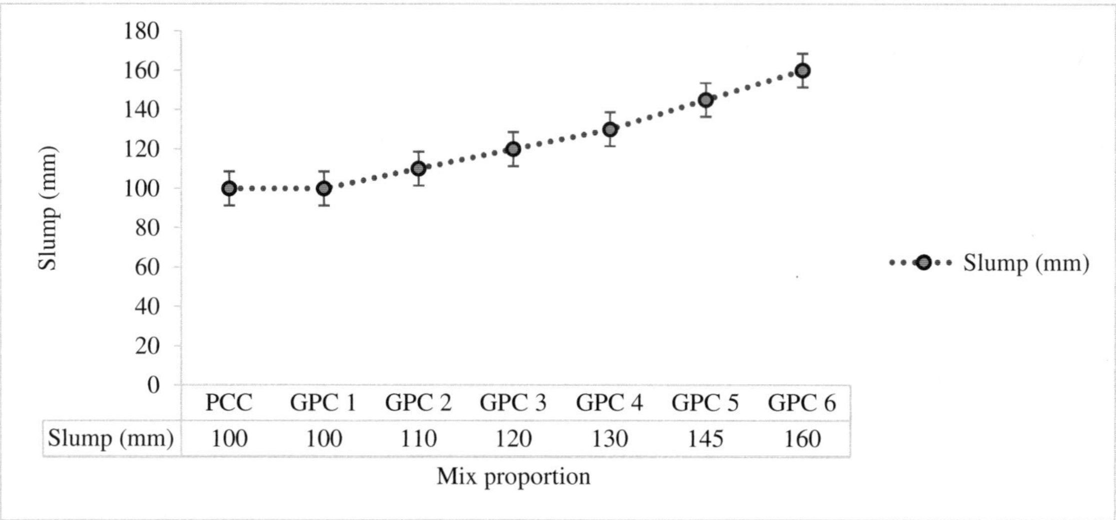

Fig. 3: Slump values (mm) for each mix proportion

3.3. Density

The density of PCC from Figure 4 ranges from 2401 kg/m3 in 7 days to 2371 kg/m3 in 28 days. The density slightly decreases with ages as a result of the hydration that takes place in the concrete. A similar pattern was noticed for GPC with a density ranging from 2410 kgm3 in 7 days to 2100 kgm3 in 28 days. It was observed that the addition of more CCA decreases the density in the GPC mix. This was as a result of the specific gravity of CCA which was less than that of GGBFS [38].

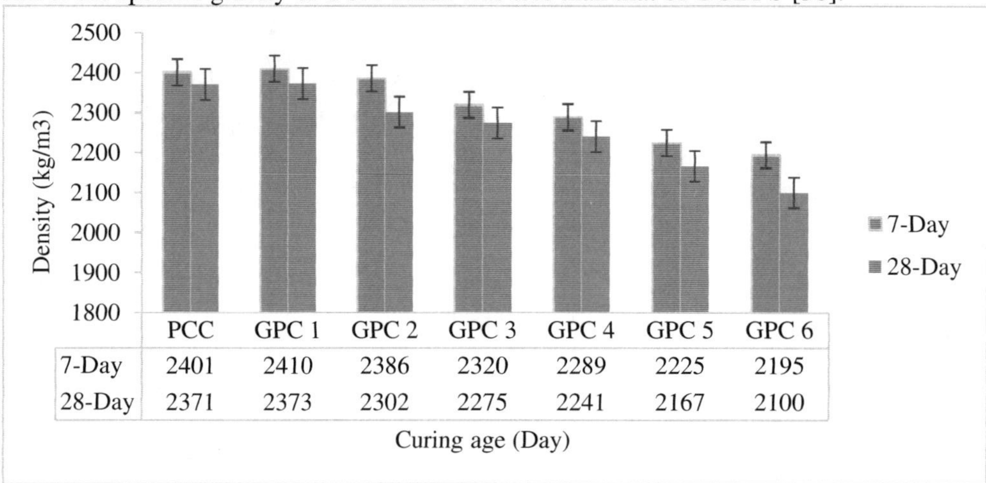

Fig. 4: Density (kg/m3) of PCC and GPC

3.4. Compressive strength

The comparison of the GPC with PCC at both 7 and 28 days curing from Figure 5 indicates that the compressive strength of GPC increases with CCA up to 40% replacement level and then at 60% CCA substitution, the strength decreases. Thus, 40% CCA replacement seems to be the optimal limit with compressive strength of 38.40 MPa for GPC when compared with 33.12 MPa for PCC at 28 days curing. Similarly, there is an increase in compressive strength by 22.92%, 17.96%, and 8.5% with GPC 1, GPC 2, and GPC 3 respectively when compared with the control concrete (PCC). The reactive presence of calcium-silicate-aluminate-hydrate (C-S-A-H) in the geopolymer paste and the continued-longer period of polymerization process contribute to the increase in both early and later strengths of GPC compared with PCC [39-42]. However, there is a decrease in compressive strength by 4.68%, 29.55%, and 48.31% with GPC 4, GPC 5, and GPC 6 respectively when compared with the control concrete (PCC). The decrease in compressive strength may be attributed to the reduction in the GGBFS's glassy content which reduces the CCA dissolution and retards the reaction product formation in ambient curing [37] [43-45].

IOP Conf. Series: Materials Science and Engineering **413** (2018) 012065 doi:10.1088/1757-899X/413/1/012065

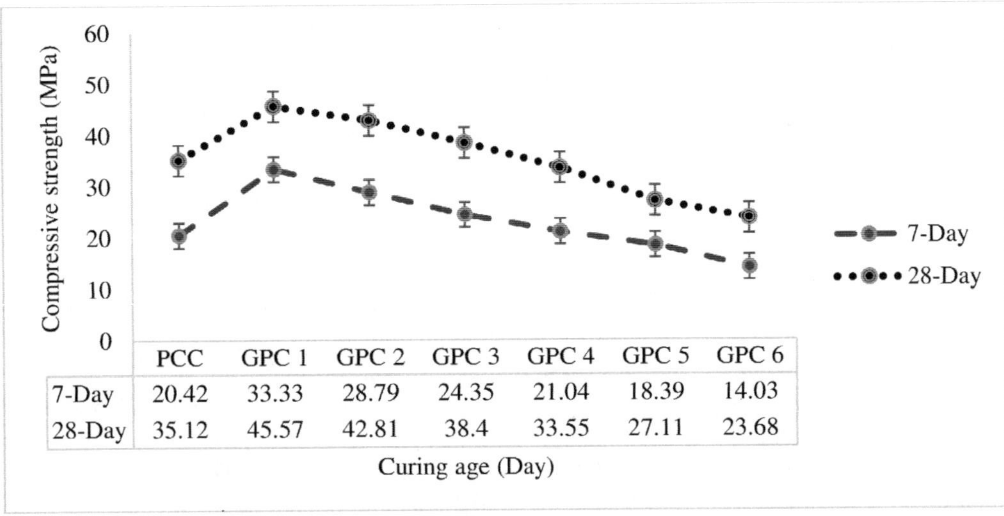

Fig. 5 Compressive strength (MPa) of PCC and GPC

3.5. Relationship between the compressive strength and the density of the GPC

Minitab 17 was used to examine the relationship that occurs between the compressive strength and the density of the GPC. Quadratic regression model of fitted line plot was engaged. And the results of regression equations at the end of 7 and 28 days curing are presented in Figure 6 and Figure 7 respectively. The coefficients of determination (R2) based on the compressive strengths are 97.6% and 96.7% for density at 7 and 28 days respectively. These indicate that the models are 97.6% and 96.7% significantly fit to predict the relationship. And the model equations can be 95% confident that the actual values of the compressive strength will fall within the predicted values of density at 95% confidence levels. Moreover, the sequence analysis of variance tables for these models indicate that F is 0.57 and P is 0.507 while F is 0.01 and P is 0.914 for both 7 and 28 days relationships respectively. Based on the level of significance (α) of 0.05, it is revealed that P-values > α. Thus, it is deduced that density has no zero coefficient. The ratio of variation in the response variable explained by the equation to variation left unexplained is referred to as F while P-value denotes a probability that the samples of the property data would have been obtained if the regression coefficients were all equal to zero.

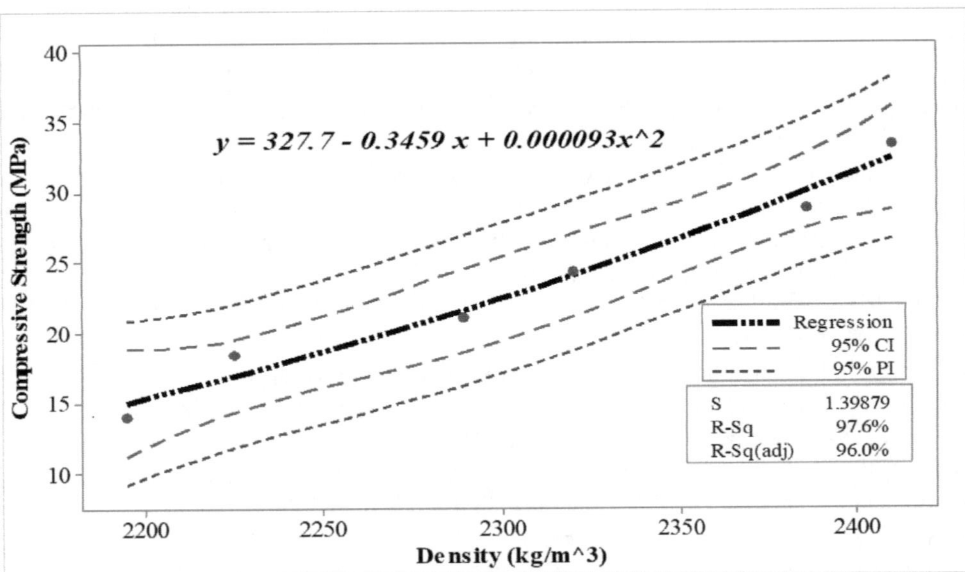

Fig.6: Relationship between the compressive strength and the density at 7 days

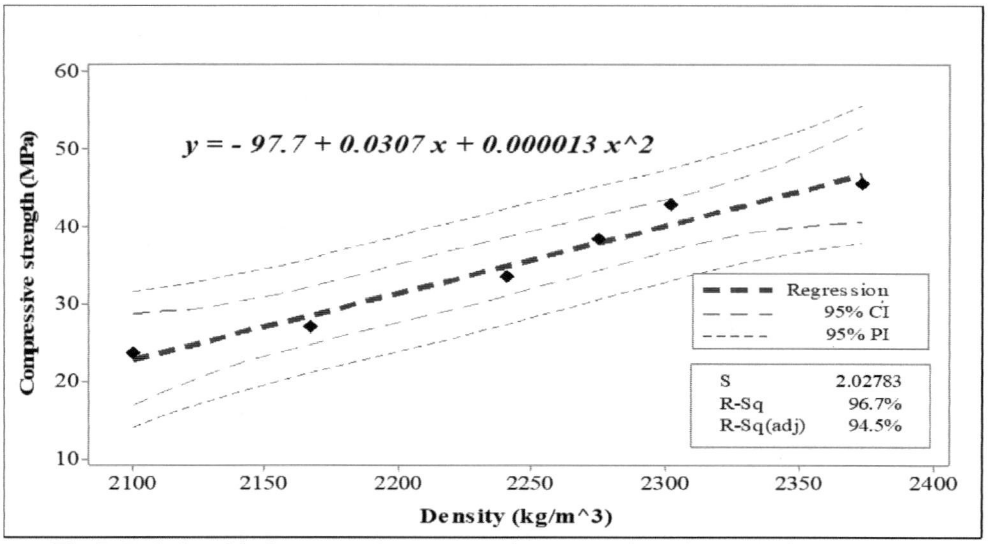

Fig.7: Relationship between the compressive strength and the density at 28 days

4. Conclusion

Geopolymer concrete shows a higher compressive strength when compared with the Portland cement concrete. Comparing with the PCC, the optimal replacement level of both GGBFS and CCA for optimum strength is obtained at 60% and 40% respectively. GGBFS and CCA-based GPC shows an emerging sustainability in place of PCC which can be utilized in general construction as a structural and non-load bearing concretes. Therefore, it is of great realistic importance to state that this study contributed to the engineering and emerging innovation for a sustainability world. It utilized the chemistry of materials for sustainable buildings, cities, and communities. Also, it established model equations to predict the compressive strength with respect to the density of the concrete.

Acknowledgements

The authors appreciate the Covenant University management for the financial support in course of the study and the provisions of laboratory facilities.

References

[1] International Energy Agency- Energy Technology Systems Anal Programme (IEA- ETSAP). Cement production. IEA ETSAP - Technology brief 103: Energy Technology Systems Anal Programme, June 2010. Retrieved 25 March 2018 from www.etsap.org.
[2] Cement Manufacturer Association of Nigeria. Gas shortage cut cement manufacturers output capacity by 36%, 2017.
[3] Franklin A. Nigeria: Minister's forecasts 2013 self-sufficiency in cement production. Vanguard Newspaper, Nigeria, 2009, 26th November: 23.
[4] Malhotra V M. Introduction: Sustainable Development and Concrete technology, ACI Concrete Journal, 2002: 1147-1165.
[5] The United Nations Statistics Division Sustainable Development Goals. The sustainable development goals report. United Nations publication issued by the Department of Economic and Social Affairs (DESA), New York, NY, 2017:10017.
[6] United Nations Environment Programme. The emissions gap report 2016. United Nations Environment Programme (UNEP), Nairobi, 2016. ISBN: 978-92-807-3617-5.
[7] Ede A N, Adebayo S O, Ugwu E I, Emenike C P. Life cycle assessment of environmental impacts of using concrete or timber to construct a duplex residential building. IOSR Journal of Mechanical and Civil Engineering (IOSR-JMCE), 2014, 11(2): 67-72.

[8] Akinwumi I I, Aidomojie O I. Effect of corncob ash on the geotechnical properties of lateritic soil stabilized with Portland cement. International Journal of Geomatics and Geosciences, 2015, 5(3): 375-392.
[9] Hardjito D, Rangan B V. Development and properties of low-calcium fly ash-based geopolymer concrete Research Report GC 1, Faculty of Engineering, Curtin University of Technology, Perth, Australia, 2005.
[10] Oyebisi S O, Olutoge F A, Ofuyatan M O, Abioye A A. Effect of corncob ash blended cement on the properties of lateritic interlocking blocks. Progress in Industrial Ecology-An International Journal, 2017, 11(4): 373-387.
[11] Oyebisi S O, Ede A N, Ofuyatan M O, Oluwafemi J O, Akinwumi I I. Comparative study of corncob ash –based lateritic interlocking and sandcrete hollow blocks. International Journal of Geomate, 2018, 15(51): 209-216.
[12] Raheem A A, Oyebisi S O, Akintayo S O, Oyeniran M O. Effects of admixtures on the properties of corncob ash cement concrete. Leonardo Electronic Journal of Practices and Technologies, 2010, 16: 13-20. ISSN 1583-1078.
[13] Davidovits J. Geopolymer cement: A review. [Online], 2013. Available at http://www.geopolymer.org/library/technical-papers/21-geopolymer-cement-review2013
[14] United States Department of Agriculture. World agricultural production: Circular series WAP 08-17, office of global analysis. International Production Assessment Division. South Building: Washington DC, 2017.
[15] Iron and Steel Senior Staff Association of Nigeria. Nigeria's steel sector gets $100m Chinese investment. The Vanguard, 9 November 2017 from https://www.vanguardngr.com/2016/12/nigerias-steel-sector-gets-100m-Chinese
[16] Hendrik G V O. Mineral Resource of the Month: Iron and Steel Slag. EARTH Magazine, 3 February 2018. Retrieved 5 March 2018 from https://www.earthmagazine.org/article/mineral-resource-month-iron-and-steel-slag
[17] Zarina Y, Mohd Mustafa A A, Kamarudin H, Khairul N I, Rafiza A, Andrei V S. Effect of solids-to-liquids, Na2SiO3-To-NaOH and curing temperature on the palm oil boiler ash (Si + Ca) geopolymerisation system. Materials Journal, 2015, 8: 2227-2242.
[18] Rangan B V. Low-calcium fly ash-based geopolymer concrete. Concrete Construction Engineering Handbook (Chapter 26), Second Edition, New York: CRC Press, 2008.
[19] Abdul-Manan D. Exploring the potential of alternative pozzolana cement for the northern savannah ecological zone in Ghana. American Journal of Civil Engineering, 2016,4(30): 74-79.
[20] Kamaul J, Ahmed A, Hirst P, Kangwa J. Suitability of corncob ash as a supplementary cementitious material. International Journal of Materials Science and Engineering, 2016, 4(4): 215-228.
[21] Oluborode K D, Olofintuyi I O. Strength evaluation of corn cob ash in a blended Portland cement. International Journal of Engineering and Innovative Technology (IJEIT), 2015b, 4(12): 14-17.
[22] Price A, Yeargin R, Fini E, Abu-Lebdeh T. Investigating effects of the introduction of corncob ash into Portland cements concrete: Mechanical and Thermal Properties. American Journal of Engineering and Applied Science, 2014, 7: 133-144.
[23] American Concrete Institute 211. 1. Standard practices for selecting proportions for normal, heavyweight and mass concrete. American Concrete Institute Committee, Detroit, Michigan, 2002.
[24] American concrete Institute 214R. Guide to evaluation of strength test results of concrete. American concrete Institute Committee, Farmington Hills, Michigan, 2011.
[25] Malathy R. Fresh and hardened properties of geopolymer concrete and mortar. Kongu Engg. College, Perundurai. India, 2009. Retrieved November 5, 2017, from www.schleibinger.com/cmsimple/?download=r2009_malathy_geopolymer.pdf
[26] Triwulan P W, Januarti J E. Addition of superplasticizer on geopolymer concrete. Asian Research Publishing Network Journal of Engineering and Applied Sciences, 2016, 11(24): 14456-14462.

[27] Rajamane N P, Jeyalakshmi R. Quantities of sodium hydroxide solids and water to prepare sodium hydroxide solution of given molarity for geopolymer concrete mixes. Indian Concrete Institute Technical Paper, SRM University, India, 2014.
[28] British Standard EN 12350-2. Testing fresh concrete: Method for determination of slump. British Standard Institution, 2 Park Street, London, 2009.
[29] British Standard EN 12390 -3. Testing hardened concrete: Compressive strength of test specimens. British Standard Institution, 2 Park Street, London, 2000.
[30] British Standard EN 12390 -7. Testing hardened concrete: Density of hardened concrete. British Standard Institution, 2 Park Street, London, 2000.
[31] British Standard 6699. Specification for ground granulated blast furnace slag for use with Portland cement. British Standard Institution, 2 Park Street, London, 1992.
[32] American Society for Testing and Materials C 618. Standard specification for coal fly ash and raw or calcined natural pozzolan for use as a mineral admixture in concrete. Annual book of ASTM standards: Philadelphia, PA, 2005.
[33] BS EN 196 – 2. Methods of testing cement – Part 2: Chemical analysis of cement. British Standard Institution (BSI), London, 1995.
[34] Deb P S, Nath P, Sarker P K. The effects of ground granulated blast-furnace slag blending with fly ash and activator content on the workability and strength properties of geopolymer concrete cured at ambient temperature. Materials and Design, 2014, 62: 32-39.
[35] Nath P, Sarker P K. Effect of GGBS on setting, workability and early strength properties of fly ash geopolymer concrete cured in ambient condition. Construction Building Materials, 2014, 66: 163-171.
[36] Provis J, Deventer JV. Alkali-activated materials: State-of-the-art report, RILEM TC 224-AAM Springe Netherlands, 2014.
[37] Davidovits J. Chemistry of geopolymeric systems, terminology. Proceedings of the 2nd International Conference on Geopolymer, Saint Quentin, 1999: 9-39.
[38] Rajini B, Narasimha Rao A V. Mechanical properties of geopolymer concrete with fly ash and ggbs as source materials. International Journal of Innovative Research in Science, Engineering and Technology, 2014, 3(9): 15944-15953.
[39] Diaz E I, Allouche E N, Eklund S. Factors affecting the suitability of fly ash as source material for geopolymers. Fuel, 2010, 89 (5): 992–996.
[40] Hou D, Li Z, Zhao T. Reactive force field simulation on polymerization and hydrolytic reactions in calcium aluminate silicate hydrate (C-A-S-H) gel: structure, dynamics and mechanical properties. RSC Adv. 2015, 5 (1): 448–461.
[41] Oh J E, Monteiro P J M, Jun S S, Choi S, Clark S M. The evolution of strength and crystalline phases of alkali-activated ground blast furnace slag and fly ash-based geopolymers. Cement Concrete Resources, 2010, 40 (2): 189-196.
[42] Temuujin J, Williams R, Van Riessen A. Effect of mechanical activation of fly ash on the properties of geopolymer cured at ambient temperature. Journal of Materials Processing Technology, 2009, 209 (12): 82-88.
[43] Palomo A, Grutzeck M W, Blanco M T. Alkali-activated fly ashes: A Cement for the future. Cement and Concrete Research, 1999. 29 (8): 1323-1329.
[44] Puligilla S, Mondal P. Role of slag in microstructural development and hardening of fly ash-slag geopolymer. Cement and Concrete Research, 2013, 43: 70–80.
[45] Xu H, Deventer J V. The geopolymerization of alumino-silicate minerals. International Journal of Mineral Processing, 2000, 59(3): 247-266.

Effects of 12 Molar Concentration of Sodium Hydroxide on the Compressive Strength of Geopolymer Concrete

Solomon Oyebisi[1], Anthony Ede[1], Olatokunbo Ofuyatan[1], Tolulope Alayande[1], Grace Mark[1], Joshua Jolayemi[1] Stephen Ayegbo[1],

[1] Civil Engineering Department, Covenant University, Ota, Nigeria
E-mail address: solomon.oyebisi@covenantuniversity.edu.ng

Abstract. Geopolymer concrete is an emerging and innovative alkali-activated concrete that has been growingly studied because of its superior mechanical strengths and durability properties. This study, therefore, investigates the utilization of both corncob ash (CCA) and ground granulated blast furnace slag (GGBFS) as source materials activating with both sodium hydroxide (NaOH) and sodium silicate (Na_2SiO_3) solutions in the production of geopolymer concrete (GPC). Sodium hydroxide was prepared in 12 molar concentration using Grade 30 MPa mix design ratio. GGBFS was replaced by CCA in varying percentages 20, 40, 60, 80, and 100% and cured in ambient conditions. Slump, density, and compressive strength of GPC were determined and compared with Portland Cement Concrete (PCC) of the same grade. The research findings indicate an optimal strength of 100% GGBFS with a compressive strength of 43.17MPa at 28 days curing for GPC compared with 35.12MPa for PCC. The result reveals that GPC has better strength than PCC and, CCA and GGBFS can be utilized as aluminosilicate materials to replace cement in the production of GPC.
Keywords: geopolymer concrete; corncob ash, ground granulated blast furnace slag; regression model; sodium silicate; sodium hydroxide; compressive strength

1. Introduction

Geopolymer concrete is an inorganic binder formed by the chemical reactions of aluminosilicate source materials of geological origin such as fly ash, slag and metakaolin activating with alkaline liquid to produce a geopolymeric gel [1]. To date, geopolymer concrete has been identified as a possible alternative binder to Portland limestone cement concrete due to its superior engineering properties and environmental benefits [1]. Portland limestone cement is the most usable binding agent in the production of conventional concrete in the construction industries, but it is generally known that cement production is a significant contributor to the emissions of carbon dioxide (CO_2) and greenhouse gasses to the atmosphere. In 2011, *Stewart, Wang, and Nguyen* established that the increase in carbon dioxide levels would globally raise the carbonation-induced corrosion in reinforced Portland cement concrete structures and that higher temperatures would possibly increase the deteriorating rates [2]. Similarly, *Saha and Eckelman* further forecast in 2014 that in construction projects, carbonation and chlorination penetrations in the existing Portland cement concrete structures would surpass current code-stipulated concrete cover thickness within 65 and 40 years respectively due to carbon dioxide emissions to the atmosphere [3]. In 2002, *Malhotra* approximately reported that Portland cement (PC) production contributes 7% of the total greenhouse gas emissions to the earth's atmosphere [4]. In comparison, there are about 70-80% less carbon dioxide emissions and 43-59% less energy required in the production of geopolymer cement, slag by-product [5]. Hence, the application of geopolymer concrete can significantly reduce the emissions of carbon dioxide into the atmosphere.

In 2017, *the United Nations Statistics Division Sustainable Development Goals (UNSDSDGs)* established that in 2014, 9 in 10 residing in urban areas breathed air which did not conform to World Health Organizations air quality guidelines as a result of greenhouse gasses and CO_2 into the atmosphere [6]. Moreover, rapid urbanization has brought great challenges such as increased air pollution, inadequate housing, and lack of opportunity for sustainable services and infrastructure most especially in developing countries like Nigeria. Therefore, safety, inclusiveness, resilience, and sustainability of cities and human settlements depend on the considerable alternative to the utilization of Portland limestone cement in the production of concrete in Nigerian construction industries [7], and one of these alternatives is geopolymer or green concrete which does not use any Portland limestone cement in its production process and the source materials are pozzolanic which are supplementary cementitious materials such as ggbfs, cca, metakaolin, and fly ash [8-12].

Geopolymer concrete has been reported to attain excellent strength and durability when it is cured at a higher temperature normally 60-85^{0}C because the ambient temperature will be too low to activate the aluminosilicate compound of the source materials and alkaline liquids [9] [13]. This type of elevated curing condition is not suitable and applicable for in-situ cast concrete and thus, it is important to develop a promising binder without curing at a higher temperature in order to establish a practical applicability of concrete in real fieldwork. Also, the energy requirement and cost connected with the elevated curing process will be minimally reduced.

The use of pozzolans such as CCA and low calcium fly ash as the only source material will retard the setting time, early and later age strengths. But in an attempt to achieve a geopolymer concrete cures at ambient conditions, *Nath and Sarker* in 2012, and *Parthiban, Saravanarajamohan, Shobana, and Bhaskar* in 2013 reported that an addition of ground granulated blast furnace slag significantly reduces the setting time and increases the early and later age strengths. It was discovered that the significant factor is the percentage replacement of slag with pozzolan in addition to the types of alkaline activators and molarities [14-15]. In 2013, *Pugilla and Mondal* observed that the addition of slag speeds up pozzolan dissolution and enhances the formation of reaction products in ambient curing condition [16]. Furthermore, higher compressive strength is achieved when a higher concentration of sodium hydroxide is used because more aluminosilicates will be dissolved thereby, forming stronger bonds [9] [17]. Thus, it is important to carry out a study focusing on the effects of sodium hydroxide concentration on the mechanical strength of GGBFS and CCA based-GPC.

Therefore, this study provides an understanding of slump, density and compressive strength of geopolymer concrete cured in ambient conditions at 7 and 28 days. CCA and GGBFS were used as binder materials while sodium hydroxide and sodium silicate solutions were used as alkaline activators and the investigation was done for grade 30 MPa concrete. Density and compressive strength were further analyzed by regression model to develop model equations and predict their relationships suitable for geopolymer concrete. The optimum scope of the mixture proportions was selected based on the relevant studies of *Rajini and Rao,* and *Fang, Ho, Tu, and Zhang* [18-19]. Finally, all the experimental works in this study were carried out at the Civil Engineering Department, College of Engineering, Covenant University, Ota, Nigeria.

2. Materials and Methods

2.1. Materials

Dangote 3X Portland limestone cement Grade 42.5R was used and obtained from a cement dealer in Ota, Ogun State, Nigeria. Both fine and coarse aggregates were sourced from tipper garage, Chelsea, Ota, Nigeria. The combined coarse aggregates used in this study were 12.5mm and 19mm sizes. Fine and coarse aggregates were used in saturated surface dry (SSD) condition in consonant with the *American Society for Testing and Materials* [20-21]. Russian made sodium hydroxide (NaOH) pellet with 99% purity and sodium silicate (Na_2SiO_3) solution were both used and sourced from Obi-Dan & Sons Chemicals Division, Lagos, Nigeria. The ratio of sodium silicate solution-to-sodium hydroxide solution was 2.5. Furthermore, the naphthalene-based superplasticizer (Conplast- SP 430) or high-range water-reducing admixture was sourced from Fine Coat Paint Industry, Lagos, Nigeria and administered at 1.0% of the binder materials in accordance with *Okoye, Durgaprasad, and Singh* [22]. Corncobs were obtained from the heaps of waste cobs which exist in large quantity in Agbonle (8° 53' 0" North, 3° 31' 0" East), Oyo State, Nigeria. Open air burning was adopted. The corncob ash was then sieved with a 90μm sieve and then analyzed for its oxides compositions using X-Ray Fluorescence (see Figure 1) while the granulated blast furnace slags was sourced from Dolphin Steel (Nigeria) Limited, Papalanto, Nigeria. It was further dried, ground (see Figure 1) and then sieved with a 90μm sieve, analyzed for its oxides composition using X-Ray Fluorescence. Finally, water was sourced from the laboratory tap and used for the production process.

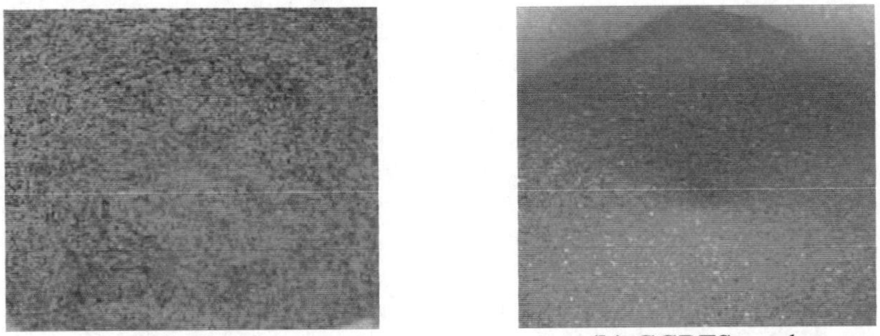

(a) CCA used **(b)** GGBFS used

Fig. 1 The source material used

2.2 Design of Concrete Mix Proportion

Both the Portland limestone cement concrete and geopolymer concrete mix proportions were designed in accordance with the *British Standards* [23-24], taking into considerations the specific gravities, water absorptions and moisture contents of the constituents in the mix. The mix proportions and mix number is presented in Table 1 while the results of volumetric computations for both PCC and GPC are presented in Table 2 and Table 3 respectively.

Table 1 The mix proportions for the concrete

S/N	Ingredient Proportions	Mix No.	Remarks
1	PCC	PCC	Control sample
2	100% GGBFS + 0% CCA	GPC 1	
3	80% GGBFS + 20% CCA	GPC 2	
4	60% GGBFS + 40% CCA	GPC 3	
5	40% GGBFS + 60% CCA	GPC 4	
6	20% GGBFS + 80% CCA	GPC 5	
7	0% GGBFS + 100% CCA	GPC 6	

Note: PCC (Portland Cement Concrete); GGBFS (Ground Granulated Blast Furnace Slag); CCA (Corncob Ash); GPC (Geopolymer Concrete)

Table 2 The volumetric computation of M30 PCC

S/N	Constituent	Weight (Kg/m^3)	Specific Gravity	Absolute Volume (M^3)	Adjusted Volume (M^3)	Ratio
1	Cement	390	3.15	0.124	0.124	1.00
2	FA (SSD)	675	2.60	0.260	0.259	2.09
3	CA (SSD)	1031	2.64	0.390	0.388	3.13
4	Water	204.15	1.00	0.204	0.204	1.65
5	Air content	2.00	-	0.020	0.020	-
6	SP	3.90	1.20	0.005	0.005	0.04
	Total	2306		1.004	1.000	7.91

Note: FA (Fine Aggregate); CA (Coarse Aggregate); SSD (Saturated Surface Dry); SP (Superplasticizer);M30 (Grade 30 Concrete); PCC (Portland Cement Concrete)

IOP Conf. Series: Materials Science and Engineering **413** (2018) 012066 doi:10.1088/1757-899X/413/1/012066

Table 3 The volumetric computation of M30 GPC

S/N	Constituent	Weight (Kg/m^3)	Specific Gravity	Absolute Volume (M^3)	Adjusted Volume (M^3)	Ratio
1	GGBFS/CCA	390	2.90/2.44	0.134	0.142	1.00
2	FA (SSD)	675	2.60	0.260	0.276	1.94
3	CA (SSD)	1031	2.64	0.390	0.414	2.92
4	NaOH solution	60	1.49	0.040	0.042	0.30
5	Na_2SiO_3 solution	150	1.60	0.094	0.100	0.70
6	Conplast SP-430	3.90	1.20	0.005	0.005	0.04
7	Air content	2.00	-	0.020	0.021	-
	Total	2306		0.943	1.000	6.90

Note: GGBFS (Ground Granulated Blast Furnace Slag); CCA (Corncob Ash) FA (Fine Aggregate); CA (Coarse Aggregate); SSD (Saturated Surface Dry); SP (Superplasticizer); M30 (Grade 30 Concrete)

2.3 Preparation of Alkaline Activators

The sodium hydroxide (NaOH) solid pellet (354g) was measured and dissolved in 646g of clean water based on the 12 molar concentration [25]. This correctly resulted in 1000g of sodium hydroxide solution. The sodium hydroxide solution was prepared 24 hours prior to cool down the solution up to ambient condition. Thereafter, NaOH solution was added to Na_2SiO_3 (water glass) two hours prior to casting of concrete to enhance its performance for the best result [17].

2.4 Mixing and Casting

The cementitious materials and aggregates were thoroughly mixed for about three minutes until a homogenous mixture was obtained. The liquid and dry components were added and the mixing continued for further 5 minutes. The fresh mix was manually cast, and then filled in the moulds and compacted accordingly. Workability of fresh concrete was measured by slump cone apparatus after making the homogeneous mix. PCC samples were removed from the moulds 24 hours after casting and immersed in water curing tank until testing day while GPC specimens were kept in rest period for 72 hours before being demoulded to allow for proper polymerization. All samples were cured at room temperature in ambient condition (23 ± 5 °C; 60% ± 5% RH). For each mixture, three samples were prepared for each testing age.

2.5 Experimental Test Methods

Slump, density and compressive strength were carried out in accordance with the procedures set out in *British Standards* [26-28] respectively. Slump test on freshly mixed concrete was carried out on 7 different samples. Compressive strength and dry density tests were conducted on the hardened concrete sample. Each test was examined on three samples with size 150 mm × 150 mm × 150 mm curing at 7 and 28 days.

3. Results and Discussions

3.1 CCA Oxides compositions

The results of the oxides compositions are presented in Table 4. It showed a silicon dioxide (SiO_2) content of 59.50% which is greater than the minimum requirement of 25.0% by mass recommended by the *American Society for Testing and Materials* [29]. Moreover, a total of ($SiO_2 + Al_2O_3 + Fe_2O_3$) content of 77.41% met the minimum specification of 70.0%. The magnesium oxide (MgO) and the sulphur oxide (SO_3) contents of 1.23% and 1.25% respectively are below the maximum requirement of 4.0%. The loss of ignition and the moisture contents of 0.49% and 1.25% are below the maximum specifications of 10% and 3% respectively. From the above analysis, it is inferred that the Corn

Cob Ash (CCA) used is a suitable material for use as a Pozzolan as it satisfied the required specifications [29].

Table 4: The oxides compositions of the CCA used

Composition	SO_3	SiO_2	Al_2O_3	Fe_2O_3	CaO	MgO	Na_2O	M.C	LOI
Properties (%)	1.25	59.50	8.78	9.13	18.23	1.23	0.65	1.25	0.49
ASTM C 618 Requirements	$\leq$ 4%	SiO_2+Al_2O_3+ Fe_2O_3 > 70%			-	$\leq$4%	> 0.70	$\leq$ 3%	$\leq$10%

3.2 GGBFS Oxides compositions

The oxides compositions of the GGBFS used is presented in Table 5 and the result indicates that GGBFS is suitable for use because its properties met the requirements of the *American Society for Testing and Materials* [30].

Table 5: The oxides compositions of the GGBFS used

Composition	CaO	SiO_2	Al_2O_3	Fe_2O_3	SO_3	MgO	Na_2O	M.C	LOI
Properties (%)	36.52	35.77	14.11	0.92	1.08	9.45	0.30	0.52	0.32
ACI 233R Requirements	32-45	32-42	7-16	0.1-1.5	0.7-2.2	5-15	-	-	-

3.3 Slump

Figure 2 shows the slump values of the geopolymer concrete (GPC) at various ground granulated blast furnace slag replacements. The results revealed that slump values were influenced by the replacement levels of ground granulated blast furnace slag (GGBFS) in the geopolymer concrete because the control sample (PCC) exhibited the similar slump to GPC (100% GGBFS). In the mixtures, the slump values increased with the decrease of ground granulated blast furnace slag content, and it is in agreement with the studies by [14] [18] [31] who reported an increase in a slump as the slag replacement levels decrease. The result can be attributed to the induced and stimulated reaction of calcium and the angular shape of ground granulated blast furnace slag when compared with the spherical shape of CCA [31]. Hence, the workability of geopolymer concrete is categorized as a highly workable concrete because the slump values exceed 90mm based on the condition of compaction [32].

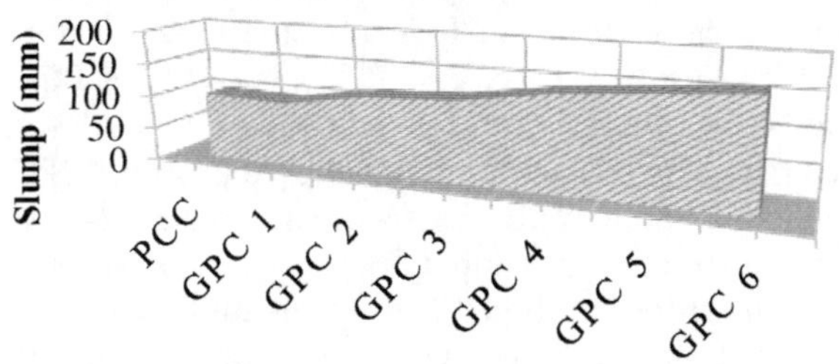

	PCC	GPC 1	GPC 2	GPC 3	GPC 4	GPC 5	GPC 6
Slump (mm)	100	95	115	120	140	150	160

Mix Proportions (kg/m3)

Fig. 2 Graph of a slump test

3.4 Density

The density of Portland cement concrete (PCC) from Figure 3 ranges from 2420kg/m^3 in 7 days to 2400kg/m^3 in 28 days. The density slightly decreases with ages as a result of the degree of hydration in the concrete. A similar pattern was noticed for geopolymer concrete (GPC 1 to GPC 3) with a density ranging from 2425kg/m^3 in 7 days to 2386kg/m^3 in 28 days. From GPC 4 to GPC 6, there is a slight decrease in density from 2373kg/m^3 in 7 days to 2250 kg/m^3 in 28 days. This infers that the addition of more corncob ash (CCA) marginally decreases the density in the GPC mix. This may be attributed to the specific gravity of CCA which is less than that of GGBFS.

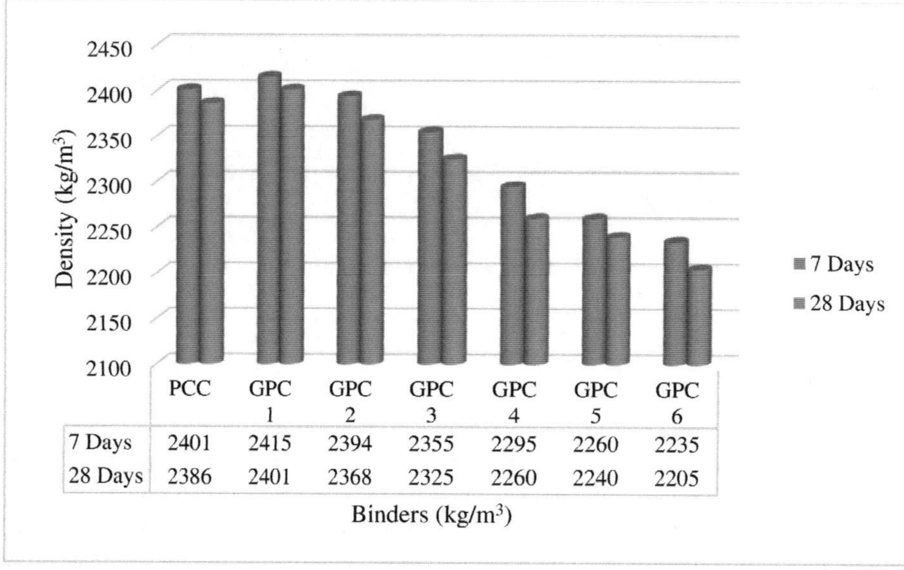

Fig. 3 Graph of density against the curing days

3.5 Compressive Strength

Comparing the compressive strengths of geopolymer concrete (GPC) to the compressive strengths of control mix (PCC) at 7 and 28 days curing in Figure 4, the results show that there is an increase in compressive strength of GPC up to 40% replacement levels over the PCC. Thus, 40% substitution of GGBFS with CCA seems to be the optimal limit for the Grade 30 mix design. The increase in both early and later strengths of GPC 1 to GPC 3 compare with the control mix (PCC) may be attributed to the reactive presence and formation of calcium-aluminate-silicate-hydrate (C-A-H-S) gels in the geopolymer paste, which compress the microstructure of geopolymer matrix, and reduce the void, and the continued-longer period of polymerization process of the GPC [33-34]. In addition, it is revealed that the compressive strength of the geopolymer concrete increased with the increased amount of ground granulated blast furnace slag in the mix. Furthermore, GPC 4, GPC 5, and GPC 6 exhibit relatively lower early age and later age strengths than the control mix (PCC) which may be imputed to the lower amount of reactive silica or aluminate in the matrix of the concrete to influence the formation of particle interaction that would yield better aluminosilicate bonds [35].

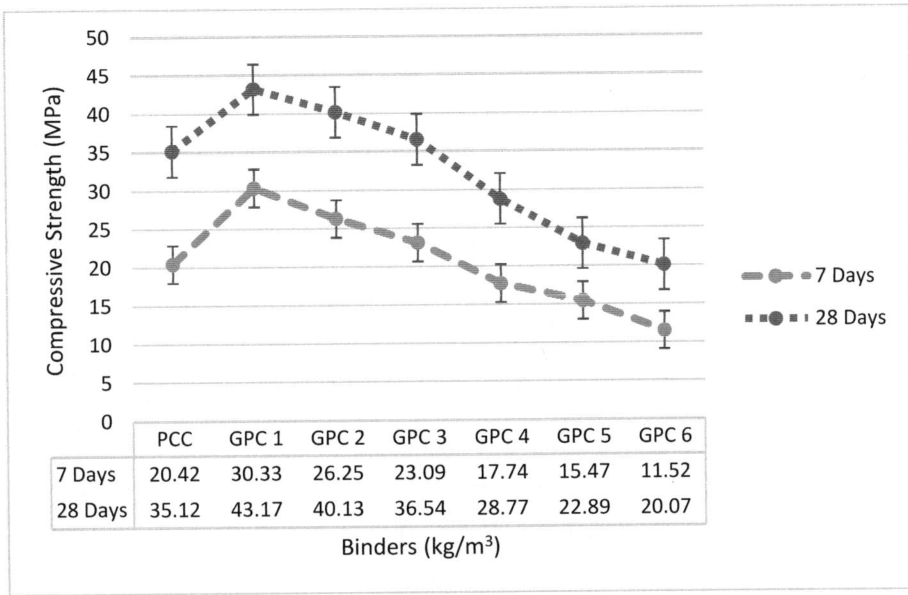

	PCC	GPC 1	GPC 2	GPC 3	GPC 4	GPC 5	GPC 6
7 Days	20.42	30.33	26.25	23.09	17.74	15.47	11.52
28 Days	35.12	43.17	40.13	36.54	28.77	22.89	20.07

Fig. 4 Graph of compressive strength against the curing days

3.6 The relationship between the compressive strength and the density of the GPC

A standard statistical software tool (Matlab 2017a) was employed to determine the relationship that exists between the compressive strength and the density of the geopolymer concrete at both 7 and 28 days curing. The polynomial regression model at degree 2 was used. The regression equations are presented in Figure 5 and Figure 6 for 7 and 28 days respectively. With respect to compressive strength, the coefficients of determination (R^2) are 82.28% and 92.19% for density at 7 and 28 days respectively. These infer that the models are 82% and 92% significantly fit to predict the 7 and 28 relationships respectively and also, compressive strength largely depends on the density at 95% confidence bounds.

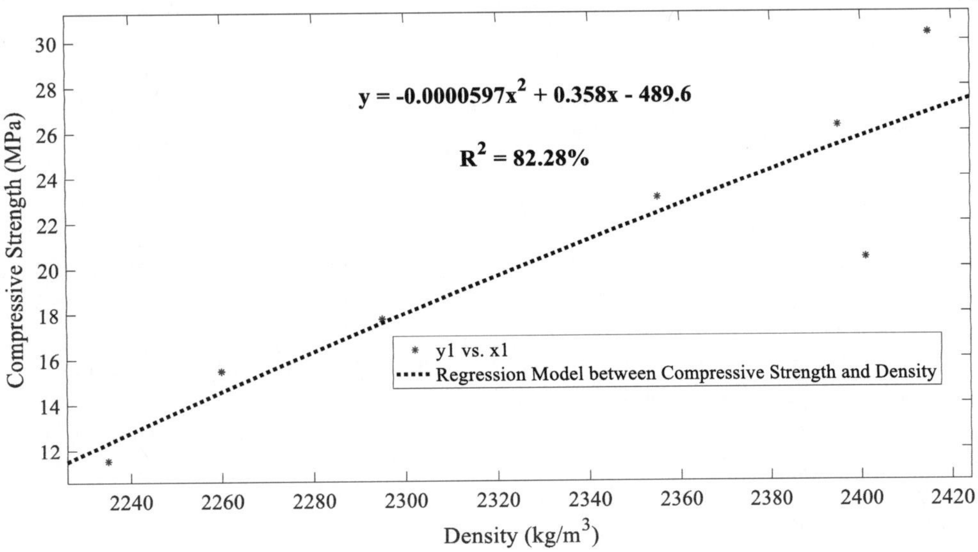

Fig.5: Relationship between the compressive strength and the density at 7 days

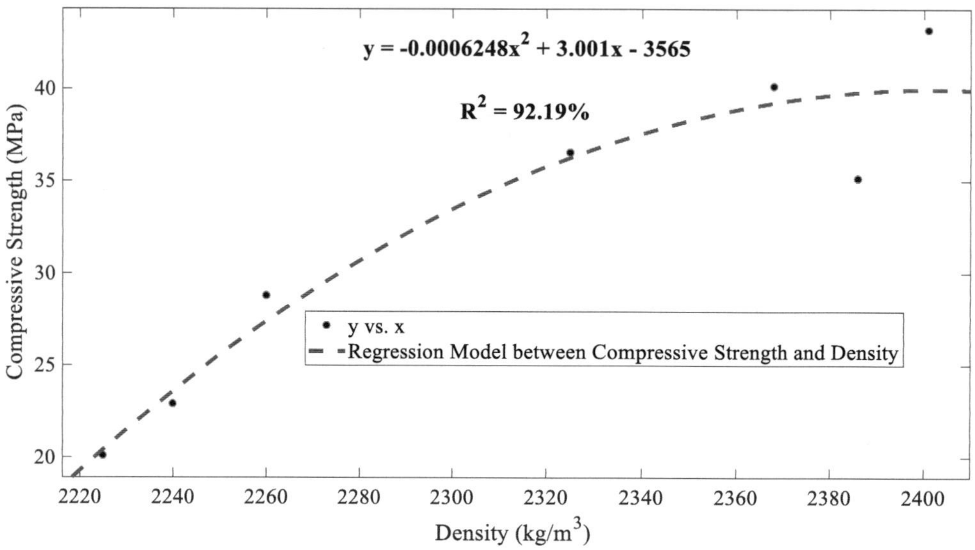

Fig.6 Relationship between the compressive strength and the density at 28 days

4. Conclusions

Geopolymer concrete shows a higher compressive strength when compared with the Portland cement concrete. Comparing with the PCC, the optimal replacement level of both GGBFS and CCA for optimum strength is obtained at 60% and 40% respectively. GGBFS and CCA-based GPC shows an emerging sustainability in place of PCC which can be utilized in general construction as a structural and non-load bearing concretes. Therefore, it is of great realistic importance to state that this study contributed to the engineering and emerging innovation for a sustainability world. It utilized the chemistry of materials for sustainable buildings, cities, and communities. Also, it established model equations to predict the compressive strength with respect to the density of the concrete.

Acknowledgements

The authors appreciate the Covenant University management for the financial support in course of the study and the provisions of laboratory facilities.

References

[1] Davidovits J. High-alkali cements for 21st century concretes. In concrete technology, past, present and future. Proceedings of V. Mohan Malhotra Symposium (pp.383-397). Editor: P. Kumar Metha, ACI SP- 144, 1994.
[2] Stewart M, Wang X, Nguyen M. Climate change impact and risks of concrete infrastructure deterioration. Engineering Structures, 2011, 33(4): 1326-1337.
[3] Saha M, Eckelman M. Urban scale mapping of concrete degradation from projected climate change. Urban Climate, 2014, 9: 101-114.
[4] Malhotra V M. Introduction: Sustainable Development and Concrete technology, ACI Concrete Journal, 2002: 1147-1165.
[5] Davidovits J. Geopolymer cement: A review. [Online], 2013. Available at http://www.geopolymer.org/library/technical-papers/21-geopolymer-cement-review 2013.
[6] The United Nations Statistics Division Sustainable Development Goals. The sustainable development goals report. United Nations publication issued by the Department of Economic and Social Affairs (DESA), New York, NY, 2017: 10017.

[7] Ede A N, Adebayo S O, Ugwu E I, Emenike C P. Life cycle assessment of environmental impacts of using concrete or timber to construct a duplex residential building. IOSR Journal of Mechanical and Civil Engineering (IOSR-JMCE), 2014, 11(2): 67-72.
[8] Akinwumi I I, Aidomojie O I. Effect of corncob ash on the geotechnical properties of lateritic soil stabilized with Portland cement. International Journal of Geomatics and Geosciences, 2015, 5(3): 375-392.
[9] Hardjito D, Rangan B V. Development and properties of low-calcium fly ash-based geopolymer concrete Research Report GC 1, Faculty of Engineering, Curtin University of Technology, Perth, Australia, 2005.
[10] Oyebisi S O, Olutoge F A, Ofuyatan M O, Abioye A A. Effect of corncob ash blended cement on the properties of lateritic interlocking blocks. Progress in Industrial Ecology-An International Journal, 2017, 11(4): 373-387.
[11] Oyebisi S O, Ede A N, Ofuyatan M O, Oluwafemi J O, Akinwumi I I. Comparative study of corncob ash –based lateritic interlocking and sandcrete hollow blocks. International Journal of Geomate, 2018, 15(51): 209-216.
[12] Raheem A A, Oyebisi S O, Akintayo S O, Oyeniran M O. Effects of admixtures on the properties of corncob ash cement concrete. Leonardo Electronic Journal of Practices and Technologies, 2010, 16: 13-20. ISSN 1583-1078.
[13] Fernández-Jiménez A, Palomo A., Sobrados I, Sanz J. The role played by the reactive alumina content in the alkaline activation of fly ashes. Microporous and Mesoporous Materials, 2006, 91(1-3): 111-119.
[14] Nath P, Sarker P K. Effect of GGBS on setting, workability and early strength properties of fly ash geopolymer concrete cured in ambient condition. Construction Building Materials, 2014, 66: 163-171.
[15] Parthiban K, Saravanarajamohan K, Shobana S, Bhaskar A A. Effect of replacement of slag on the mechanical properties of fly ash based geopolymer concrete. International Journal of Engineering Technology (IJET), 2013, 5 (3): 2555–2559.
[16] Puligilla S, Mondal P. Role of slag in microstructural development and hardening of fly ash-slag geopolymer. Cement and Concrete Research, 2013, 43: 70–80.
[17] Lloyd N, Rangan V. Geopolymer concrete-sustainable cement less concrete. ACI Special Publication SP-261, 10th ACI International Conference on Recent Advances in Concrete Technology and Sustainability Issues. American Concrete Institute, Farmington Hills, MI, 2009.
[18] Rajini B, Narasimha Rao A V. Mechanical properties of geopolymer concrete with fly ash and ggbs as source materials. International Journal of Innovative Research in Science, Engineering and Technology, 2014, 3(9): 15944-15953.
[19] Fang G, Ho W K, Tu W, Zhang M. (2018). Workability and mechanical properties of alkali-activated fly ash-slag concrete cured at ambient temperature. Construction and Building Materials, 2018, 172: 476–487.
[20] American Society for Testing and Materials C 128. Standard test method for relative density (specific gravity) and absorption of fine aggregate, ASTM International, West Conshohocken, PA, 2015.
[21] American Society for Testing and Materials C 127. Standard test method for relative density (specific gravity) and absorption of coarse aggregate, ASTM International, West Conshohocken, PA, 2015.
[22] Okoye F N, Durgaprasad J, Singh N B. Mechanical properties of alkali activated flyash/Kaolin based geopolymer concrete. Construction and Building Materials, 2015, 98: 685–691.
[23] British Standard EN 206- 1. Concrete specifications, performance, production and conformity. BSI, 389 Chiswick High Road, London, 2001.
[24] British Standard 8500- 1. Concrete: Complementary British Standard to BS EN 206-1- Part-1. Method of specifying and guidance for the specifier. BSI, 389 Chiswick High Road London, 2002.

[25] Rajamane N P, Jeyalakshmi R. Quantities of sodium hydroxide solids and water to prepare sodium hydroxide solution of given molarity for geopolymer concrete mixes. Indian Concrete Institute Technical Paper, SRM University, India, 2014.
[26] British Standard EN 12350-2. Testing fresh concrete: Method for determination of slump. British Standard Institution, 2 Park Street, London, 2009.
[27] British Standard EN 12390 -7. Testing hardened concrete: Density of hardened concrete. British Standard Institution, 2 Park Street, London, 2000.
[28] British Standard EN 12390 -3. Testing hardened concrete: Compressive strength of test specimens. British Standard Institution, 2 Park Street, London, 2000.
[29] American Society for Testing and Materials C 618. Standard specification for coal fly ash and raw or calcined natural pozzolan for use as a mineral admixture in concrete. Annual book of ASTM standards: Philadelphia, PA, 2005
[30] American concrete Institute 233R. Ground granulated blast furnace slag. American concrete Institute Committee, Farmington Hills, Michigan, 2002.
[31] Deb P S, Nath P, Sarker P K. The effects of ground granulated blast-furnace slag blending with fly ash and activator content on the workability and strength properties of geopolymer concrete cured at ambient temperature. Materials and Design, 2014, 62: 32-39.
[32] Junaid M T, Kayali O, Khennane A, Black J. A mix design procedure for low calcium alkali activated fly ash-based concretes. Construction and Building Materials. 2015, 79: 301–310.
[33] Oh J E, Monteiro P J M, Jun S S, Choi S, Clark S M. The evolution of strength and crystalline phases of alkali-activated ground blast furnace slag and fly ash-based geopolymers. Cement Concrete Resources, 2010, 40 (2): 189-196.
[34] Wan X, Hou D, Zhao T, Wang L. Insights on molecular structure and microproperties of alkali-activated slag materials: A Reactive Molecular Dynamics Study. Construction and Building Materials, 2017, 139: 430–437.
[35] Yip C K, Lukey G C, van Deventer J S J. The coexistence of geopolymeric gel and calcium silicate hydrate at the early stage of alkaline activation. Cement and Concrete Research, 2005, 35 (9): 1688–1697.

IOP Conf. Series: Materials Science and Engineering **413** (2018) 012067 doi:10.1088/1757-899X/413/1/012067

Characterisation of laser metal deposited titanium and molybdenum composite

Stephen A Akinlabi[1a], Madindwa P Mashinini[2a], Oluseyi O Ajayi[3c], Abiodun A Abioye[4c], Samuel O Fatoba[5b] and Esther T Akinlabi[6b]

[a]Mechanical & Industrial Engineering Technology, University of Johannesburg, South Africa
[b]Mechanical Engineering Science, University of Johannesburg, South Africa
[c]Department of Mechanical Engineering, Covenant University, Ota, Nigeria
E-mail: [1]stephenakinlabi@gmail.com; [2]mmashinini@uj.ac.za; [3]Oluseyi.ajayi@covenantuniversity.edu.ng; [4]Abiodun.abioye1@covenantuniversity.edu.ng; [5]drfatobasameni@gmail.com; [6]etakinlabi@gmail.com

Abstract. Laser Metal Deposition (LMD) is a unique way of building components by adding the material(s) layer by layer until the components are built. This process is based on the geometry path defined by the CAD model data form the component development. This paper reports on the characterization of Laser Metal Deposited Titanium alloy and Molybdenum Powder on Titanium alloy grade 5 substrate. The depositions are the process of delivering of the Titanium and Molybdenum powder from a powder feeder of two different hoppers containing each powder. The deposited samples were achieved through the combination of different ratios of Titanium and Molybdenum, i.e. 95%, 90% and 85% of Titanium and 5%, 10%, 15% Mo respectively. Three sets of experiments were conducted with multiple tracks at a constant laser power of 1.2kW, Scan speed of 0.5m/min and gas flow rate at 2 l/min. While the powder flow rate for both Ti6Al4V and Mo were varied. The deposits were sectioned through the thickness, metallographically prepared and characterized through microstructural evaluation and microhardness. The results from the microstructure revealed that columnar grains describe the layers, which gives an excellent interfacial bonding between the layers. The Micro Vickers hardness values were observed to decrease as the percentage of Mo increases. Through the EDS, the un-melted powders and pores were identified. Also, the grain sizes were found to reduce as the percentage of Molybdenum increases.

1. Introduction and Background

Laser Metal Deposition (LMD) is one of the Additive Manufacturing (AM) processes whereby new components can be produced and old components refurbished. LMD process remains very unique as a process that can achieve both the manufacture of a new component and the repair of old component part. It's an alternative manufacturing technique for special materials such as Titanium and its alloys that has found relevance in aerospace, medical, automotive, chemical and oil & gas applications. LMD employs a layer-by-layer process of manufacturing lightweight components part, with unique properties from the 3D CAD data. A typical schematic of an LMD process is illustrated with Figure 1. The manufacturing process employs the heat source from the laser beam and deposited powder through a feeder nozzle onto a substrate where the component part is built layer by layer.

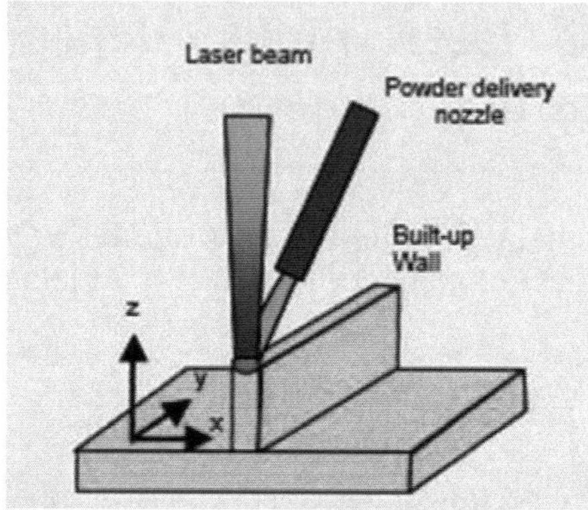

Fig. 1 Schematic of Laser Metal Deposition Process [1]

The LMD process has proven to be a unique alternative manufacturing process when compared to conventional manufacturing processes in which materials are always removed from the substrate which consequently lead to material wastages, high machine hour, high energy consumption and possible pollution to the environment. The use of conventional manufacturing process in the production of aerospace components parts has been associated with material wastages, making the process not sustainable because only about 10% of the input material eventually comes out as the final manufactured component. This is commonly referred to the as high buy-to-fly ratio in the aerospace industry, which has been attributed as one of the reasons for the high cost of aircraft [2]-[5]. On the other hand, LMD has the capability reducing the setup time, which consequently reduces the production cost, material wastages, energy consumption and man hour time.

Functionally Graded Material (FGM) is a special material developed with unique properties tailored for particular applications such as in aerospace, medical, automotive, energy and military etc. [6]-[7]. FGM has been developed to eliminate the point of stress initiation and concentration in a joining process of the manufactured component, which often acts as a point of reference and initiation for failure. These points are sharp interfaces observed in the joined material, which FGM has successfully eliminated through the developed material gradient in an LMD process [8]-[9]. Historically, FGM has been produced from different production processes such as Powder Metallurgy Approach, Centrifugal Casting Technique, Plasma Spraying Method, Self-propagating High-Temperature Synthesis, Physical or Chemical Vapour Deposition (PVD/CVD) and recently the Laser Metal Deposition Process [10]-[11].

Several studies into LMD process has been investigated, and some of the successes are now applied to the refurbishment and repairs of cracked components such as turbine blades [12]-[16]. More particularly, research into FGM of Titanium and its alloy has also gained good ground [7], [17]-[21]. Zhang et al. contributed to the development of FGM using LMD process and further characterized the graded material [22]. They deposited functionally graded Ti-TiC on a Ti6Al4V substrate. Through the preliminary study, the processing parameters for various volumes fractions of Ti-TiC composites were established. The results were then used to deposit thin wall of FGM by optimizing the process parameters during the process. The study revealed that the wear resistance of the Ti6Al4V substrate was improved by the addition of the TiC and show the produced FGM without sharp interfaces. Wang et al. [23] conducted another similar study of functionally graded Ti6Al4V-TiC deposit. In this study, Ti6Al4V wire and TiC powder were used, with the two materials fed simultaneously. An excellent compositional graded material was achieved with the wire feed rate of Ti6Al4V constant and varying the feed rate for TiC. Liu and DuPont [24] employed the controller in LENS to monitor the melt pool area of the deposited FGM of Ti-TiC composite using LMD. The controller in LENS was also used to regulate the laser power to achieve a constant melt pool area during the deposition. The control of the melt pool area is intended to control the dimensional accuracy in the deposited part while other properties of the deposit are not affected. The study of functionally graded Inconel 718 Nickel alloy and Ti6Al4V using laser metal deposition was conducted by Shah [25]. He investigated the effect of pulse parameters and powder flow rate on residual stress. The results revealed that the thickness of the layer has a contributory role in the crack formation in the deposit. It was also noted that increase in the powder flow rate increases the melt pool size. The solidification behaviour and microstructure of the functionally graded material of stainless steel-SS316L-super alloy-Rene88DT were studied through the laser metal deposition. The results show the development of Epitaxial growth and columnar dendrites microstructure [26].

The focus of this research is to incorporate varying percentages of Molybdenum on Titanium alloy through LMD, for biomedical application such as in knee and hip implant. Functionally graded Ti6Al4V- Molybdenum was developed through the laser metal deposition process. The process parameters employed for the deposition layer by layer as from the initial preliminary study, which was optimized to establish the process window for the deposition of porous free and dense deposit of Ti6Al4V-Mo composites. Molybdenum has been having been chosen because of its biocompatibility while Titanium alloy grade 5 is the most valuable among the grades of Titanium alloys.

The parameters considered in this study are Laser Power, Scan Speed, Powder Flow Rate, Gas Flow Rate and Percentage of Molybdenum. The produced deposited samples were characterized by mechanical, Microhardness and microstructural analysis using SEM microscopy.

2. Experimental Methods

The substrate employed in the laser metal deposition of the functionally graded Ti6Al4V-Mo composite is 5 mm Ti6Al4V, with a dimension of 72x72mm2. The Ti6Al4V samples were sandblasted to remove all oxide layers and dirt. It's cleaned with acetone before the deposition commences. In addition, the sandblasting is aimed to make the surface rough to aid laser power-material interaction and eliminate reflection of shining surfaces that may reflect the laser beam. The Ti6Al4V and Mo powder were used in the deposition of the functionally graded composites. The particle size of the Ti6Al4V powder is 150-300 μm and has 99.6% purity. While the particle size of the Mo powder is 45-120 μm and with a purity of 99.6%. The morphology of the parent materials – substrate, Ti6Al4V and Mo are shown in Figure 2 (a), (b) and (c) respectively.

(a) (b) (c)

Fig. 2 Characterization of parent materials showing the morphology of (a) Ti6Al4V substrate (b) Ti6Al4V powder (c) Mo Powder

IOP Conf. Series: Materials Science and Engineering **413** (2018) 012067 doi:10.1088/1757-899X/413/1/012067

During the LMD process, a 4.4kW ND: YAG laser system was employed, attached to a KUKA robot. A co-axial nozzle was equipped with the robot for the effective delivery of the powder to the melted pool. Both the Ti6Al4V and Mo powders are transported with the aid of Argon gas through the nozzle and also to protect the deposit from contamination from the environment. The beam spot size was kept at 2mm at a focal length of 195 mm above the Ti6Al4V substrate. The experimental setup and schematic of the LMD process are shown in Figure 3 (a) and (b) respectively.

(a)

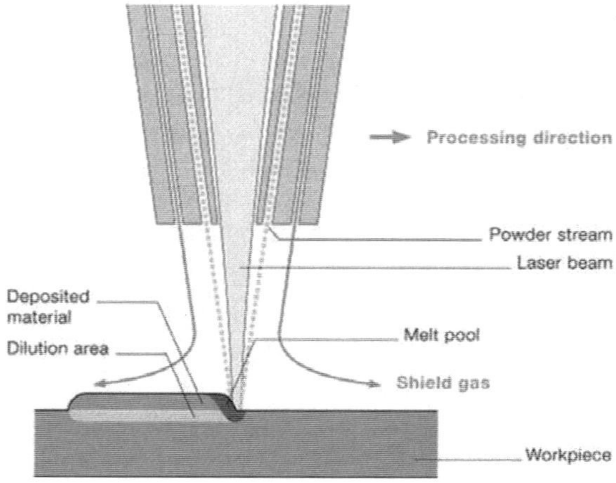

(b)

Fig. 3 (a) Experimental setup and (b) Schematic of the Laser Metal Deposition

The LMD was achieved through the developed melted tool on the surface of the Ti6Al4V substrate using the laser beam. The powders were emptied into the melted pool through the powder feeders. The Ti6Al4V substrate was positioned stationary while the robot was engaged in the deposition as depicted in the schematic shown in Figure 3 (b). The functionally graded material was produced from different percentages of both Ti6Al4V and Mo, the process parameters presented in Table 1.

Table 1 Process Parameters.

Sample #	**% Ti6Al4V**	**% Mo**	**P (KW)**	**V (m/min)**	**Powder flow rate (rpm)**		**Gas flow Rate (l/min)**
					Ti6A14V	**Mo**	
1	95	5	1.2	0.5	3.8	0.2	2
2	90	10	1.2	0.5	3.6	0.4	2
3	85	15	1.2	0.5	3.4	0.6	2

The deposited samples were after that characterized to determine the integrity of the deposited samples. The deposited samples selected for metallurgical examinations were sectioned to reveal the cross-sections. The cut samples were mounted in hot resin - polyfast and the mounted samples were ground and polished based on the metallurgical preparation standard for Titanium and its alloys. The polished samples were etched using Kroll's reagent. The microstructures were observed under an optical microscope and also through the Scanning Electron Microscope (SEM) with a capacity to conduct Energy Dispersion Spectrometry (EDS). The hardness of the deposited samples was also evaluated through microhardness profiling of the cross-section of the polished samples using the Vickers Microhardness indenter. A load of 500 g and dwell time of 15 sec was applied.

3. Result and Discussion

The micrographs of the Ti6Al4V substrate and the powders are shown in Figure 2 (a), (b) & (c). White and grey colourations, which are the alpha grains and Beta grains, were observed to be distributed across the whole cross-section of the micrograph. The Ti6Al4V powder is spherical; this is typical of a gas-atomized powder. The spherically shaped powder is always encouraged for laser metal deposition process because these maximize the absorption of the laser beam. The shapes of the Molybdenum powder, on the other hand, are irregular and rough; these shape characteristics are often typical of balled milled powder.

A sample photo of the Functionally Graded Ti6AL4V & Mo is shown in Figure 3 with multiple track depositions.

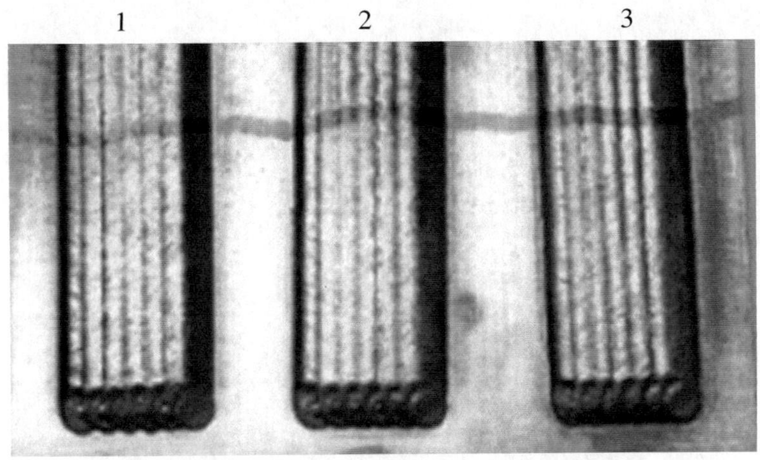

Fig. 4 Physical photo of Laser Melted Deposition at 5%, 10% and 15% of & Mo on the Ti6Al4V substrate

The microhardness test was conducted on the three deposited samples, and the average microhardness was presented in Table 2. The trend observed in Table 2 revealed that as the percentages of Molybdenum increases, the average microhardness decreases. This phenomenon is attributed to the Vickers hardness for Ti6Al4V and Molybdenum are 349 and 230 HV respectively. The minimum percentage of Mo has little influence on the deposited composite, having Ti6Al4V dominate and similarly, the increased portion of Mo had a more significant impact on the formed composite, which consequently reduced the microhardness.

Table 2 Average grain sizes of deposited samples at percentages of Mo.

% Mo	Average Grain Size (µm)
5	451.6632
10	279.7058
15	139.0304

The micrographs of the depositions of Molybdenum at 5%, 10% and 15% and 95%, 90% and 85% Ti6Al4V powder on Ti6Al4V substrate is shown in Figure 5. All the micrographs distinctively show the deposit the HAZ, HAZ and substrate. Also observed across the three deposits are porosity and un-melted Molybdenum powders, which may be attributed to the constant laser power employed in the deposition

IOP Conf. Series: Materials Science and Engineering **413** (2018) 012067 doi:10.1088/1757-899X/413/1/012067

process. The grains for the deposit with 10% molybdenum are elongated, which is different from the other deposits. This may be attributed to the cooling rate during the deposition process. The deposit produced with 15% of Mo reveal that the laser beam deployed to the process is low because this has resulted in un-melted Molybdenum. It would be expected that the laser power is increased for the effective melting of both the powder and formation of the melted pool.

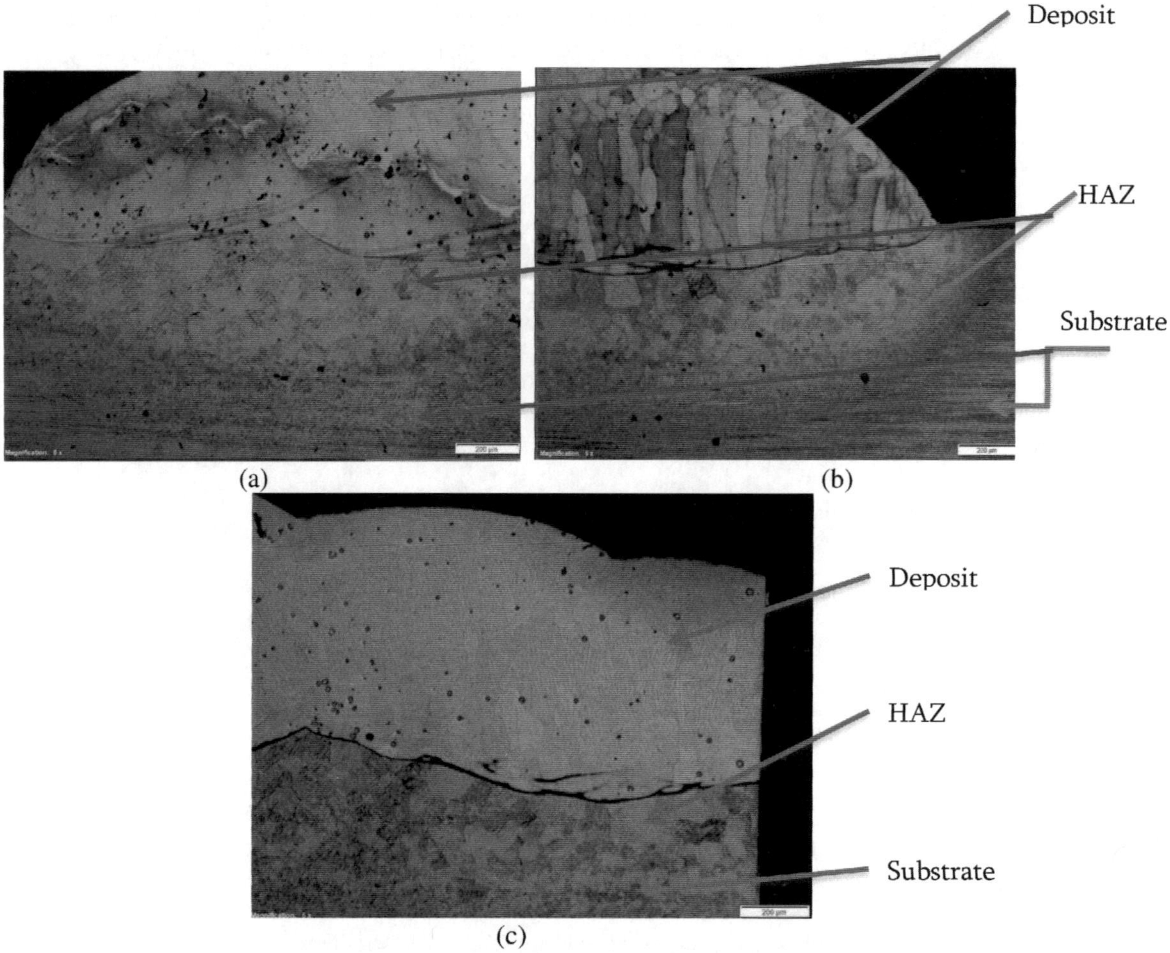

Fig. 5 Deposition at (a) 5% Mo, (b) 10% Mo and (c) 15% Mo

A typical SEM micrograph of the deposited sample with un-melted Molybdenum powder is shown in Figure 6 (a). The un-melted Molybdenum particles are confirmed through the EDS analysis of the selected point in (a).

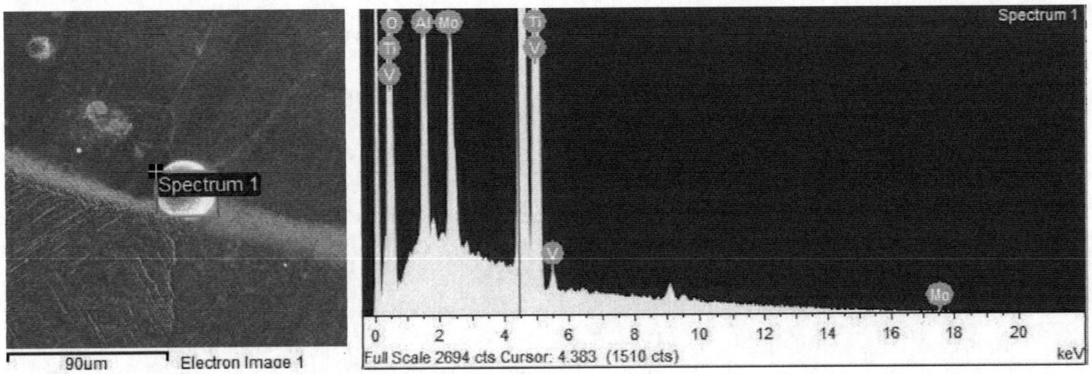

Fig. 6 (a) SEM Microstructure of a deposited sample (b) The EDS analysis of the labeled point in (a)

4. Conclusion

Through this study, Ti6Al4V-Mo Composites were successfully produced with the three sets of powder compositions at varying powder flow rates and constant laser power, scan speed and gas flow rates. The Vickers Microhardness values were found to be a function of the percentage composition of Molybdenum powder. The deposit produced at 10% Molybdenum show good bonding with the substrate and deformed columnar grains, which have been attributed to the adequate bonding between the deposit and the substrate. In addition, the columnar grains and the pores also provide good ose-integration of the body tissue when inserted into simulated body fluid.

Acknowledgements

The Rental Pool Programme of National Laser Centre, Council of Scientific and Industrial Research (CSIR), Pretoria, South Africa has supported this work.

References

[1] L. Xue, M. U. Islam, and A. Theriault, "Laser Consolidation Process for the Manufacturing of Structural Components for Advanced Robotic Mechatronic System – A State of Art Review," in 6th International Symposium on Artificial Intelligence and Robotics & Automation in Space, 2001.

[2] E. Brandl, V. Michailov, B. Viehweger, C. Leyens, Deposition of Ti–6Al–4V using laser and wire, part I: Microstructural properties of single beads, Surface & Coatings Technology, vol. 206, 2011, pp. 1120–1129.

[3] P. Kobryn and S. Semiatin, Laser Forming of Ti–6Al–4V: Research Overview, in: D. Bourell, J. Beaman, R. Crawford, J. Marcus, J. Barlow (eds.), Solid Freeform Fabrication Proceedings, University of Texas, Austin, TX, 2000, pp. 58–65.

[4] E. Brandl, A. Schoberth and C. Leyens, Morphology, microstructure, and hardness of titanium (Ti-6Al-4V) blocks deposited by wire-feed additive layer manufacturing (ALM) Materials Science and Engineering A, vol. 532, 2012, pp. 295– 307

[5] P.A. Kodryn and S.L Semiatin, Microstructure and texture evolution during solidification process of Ti6Al4V, Journal of Materials Processing Technology, vol. 135, 2003, pp. 330-339.

[6] P. Shanmugavel, G. B. Bhaskar, M. Chandrasekaran, P. S. Srinivasan. An overview of fracture analysis in functionally graded materials, Eur. J. Sci. Res. 68 (3) (2012) 412-439

[7] R. M. Mahamood, E. T. Laser Metal Deposition of functionally graded Ti6Al4V/TiC. Materials and Design 84 (2015) 402-410.

IOP Conf. Series: Materials Science and Engineering **413** (2018) 012067 doi:10.1088/1757-899X/413/1/012067

[8] S. S. Wang. Mechanics for delamination problem in composite materials. J. Compos. Mater. 17 (3) (1983) 210-233.
[9] M. Niino, T. Hirai. R. Watanabe. The functionally gradient materials. J. Jpn. Soc. Compos. Mater. 13 (1986) 257-264.
[10] R. Knoppers, J. W. Gunnink, J. Van den Hout, W. Van Vliet. The Reality of Functionally Graded Material Products, TNO Science and Industry. The Netherlands, 2004. Pg 38-43.
[11] M. Ivosevic, R. Knight, S. R. Kalidindi, G. R. Palmese, J. K. Sutter, Solid Particle erosion resistance of thermally sprayed functionally graded coatings for polymer matrix composites. Surf. Coat Technol. 200 (1617) (2006) 5145-5151.
[12] F. Weng, C. Chen, H. Yu. Research status of Laser cladding on Titanium and its alloys: A review. Mater. Des. 58 (2014) 412-425.
[13] K. Zhang, S. Wang, W. Liu, X. Shang. Characterization of Stainless Steel parts by Laser Metal Deposition Shaping. Mater. Des. 55 (2014) 104-119.
[14] T. Durejko, M. Zietala, W. Polkowski, T. Czujko. Thin Wall tubes with Fe3Al/SS316L graded structure obtained by using laser engineered net shaping technology. Mater. Des. 63 (2014) 766-774.
[15] S. H. Riza, S. H. Masood, C. Wen, D. Ruan, S. Xu. Dynamic behaviour of high strength steel parts developed through laser assisted direct metal deposition. Mater. Des. 64 (2014) 650-659
[16] G. Bi, G. L. Ng, K. M. The, A. E. W. Jarfors. Feasibility study on the Laser Aided Additive Manufacturing of die inserts for liquid forging. Mater. Des. 31 (2010) S112-S116.
[17] C. Meacock, R. Vilar. Laser powder micro deposition of CP2 Titanium. Mater. Des. 29 (2) (2008) 353-361.
[18] H. P. Qu, P. Li, S. Q. Zhang, A. Li, H. M. Wang. Microstructure and Mechanical property of laser melting deposition (LMD) Ti/TiAl structural gradient material. Mater. Des. 31 (1) (2010) 574-582.
[19] W. Liu, J. N. DuPont. Fabrication of functionally graded TiC/Ti composite by laser engineered net shaping. Scr. Mater. 48 (9) (2003) 1337-1342.
[20] K. Shah. Laser Direct Metal Deposition of Dissimilar and Functionally Graded Alloys PhD Thesis The University of Manchester, 2011.
[21] R. M. Mahamood, E. T. Akinlabi, M. Shukla, S. Pityana. Characterization of laser deposited Ti6Al4V/TiC Composite. Lasers Eng. 29 (3-4) (2014) 197-213.
[22] Y. Zhang, Z. Wei, L. Shi, M. Xi. Characterization of laser powder deposited Ti-TiC composites and functional gradient materials. J. Mater. Process. Technol. 206 (2008) 438-444.

Enhancing the Performance of Vapour Compression Refrigeration System using Nano Refrigerants: A review

T. O. Babarinde[1, 3*], S. A. Akinlabi[2], D. M. Madyira[1]

[1]Department of Mechanical Engineering Science, University of Johannesburg, South Africa
[2]Department of Mechanical & Industrial Engineering Technology, University of Johannesburg, South Africa
[3]Mechanical Engineering Department, Covenant University Ota, Nigeria
E-mail address: taiwo.babarinde79@gmail.com, stephenakinlabi@gmail.com, dmadyira@uj.ac.za

Abstract. Application of Nano particles as additives in refrigerants has been identified as a way of enhancing the performance of the vapour compression refrigeration system (VCRS) without modifying the system components. When Nano particles are dispersed in a refrigerant or lubricant, they are regarded as Nano refrigerants or lubricants. The improvement in heat transfer and tribology are responsible for the enhancement of VCRS performance. In this review paper, the effect of Nano particles on the coefficient of performance, power consumption, thermal conductivity, viscosity and heat transfer of refrigerants are discussed. It is shown that the application of Nano particles as additives in refrigerant and lubricant in VCRS is favourable and promising. Therefore, Nano refrigerants are expected to be the future refrigerants to be adopted by VCRS manufacturers.
Keywords: Nanofluids, COP, Power consumption, Thermal conductivity

1. Introduction

Today the world is confronted with the challenge of energy security with the governments of many nations implementing energy policies on the use of energy to decrease energy consumption levels for long term sustainability [1]. Government leaders and researchers have deliberated the problem of growth in energy consumption because the current generation of energy accompanies the burning of fossil fuels, which contributes to global warming. It is estimated that about 50% of energy consumption in commercial buildings goes to heating, ventilation and air-conditioning (HVAC) systems. Therefore, researchers are now focused on energy consumed by refrigeration and air-conditioning system [2]. Refrigeration and air-conditioning are needed for domestic and industrial comfort and preservation purposes. Refrigeration, air conditioning and heat pump appliances require significant amounts of energy for their operation. Due to this energy challenge, researchers are finding various means to improve performance and energy efficiency of refrigeration and air conditioning system. Although, some researchers have tried to improve the performance of refrigeration system [3]–[12], this has focused on the modification of the design components of the system. Choi et al. found that application of Nano particles to the refrigerant is capable of enhancing the thermal properties [13]. When Nano particles are dispersed in a refrigerant or lubricant base fluid, they are regarded as Nano refrigerant or Nano lubricant. The advantages of using Nano particles as additives in refrigeration and air conditioning system include enhancement of thermal conductivity and viscosity leading to the reduction in energy consumption of the system. The minimum widely available size of Nano particles ranges from 10–100 nm. However, the performance of Nano particles in refrigeration and air conditioning systems largely depends on the type and concentration of Nano particles in the base refrigerant or lubricant, Nano particles size and shape of the Nano Particles. The most common method of dispersing Nano particles in refrigeration and air conditioning applications is by using different technical methods such as homogenizing, ultrasonic agitation, high-shear mixing and magnetic force agitation [14]

2. Advancement of Nano Particles as Additives in Refrigeration and Air-Conditioning Systems

Thermal conductivity is a significant property by which Nano particles are being selected for application in Nano refrigerant applications. The heat transfer coefficient and heat transfer in the VCRS system are dependent on thermal conductivity and properties such as Nano concentration, particle size and flow rate. In addition, the viscosity of Nano particles imbedded fluid affects the pressure drop and hence the heat transfer of Nano fluid. Addition of Nano particles reduces the specific heat and increases density. Also, an increase in dynamic viscosity brings about decrease in convective heat transfer coefficient and frictional pressure drop therefore compressor pumping power increases (Bhattad et al, 2018). As a result of frictional drops in the compressor, the compressor work input reduces, which in turns increases the COP of the system. In this section, past and present works on the effect of Nano refrigerant, Nano lubricant and other Nano fluids on refrigeration and air-conditioning system are presented.

2.1 Effect of Nano particles additives on the thermal conductivity and viscosity of refrigerants in refrigeration system.

Thermal conductivity and viscosity are important properties of Nano refrigerants compared to the convectional working fluids in refrigeration and air-conditioning system application. Also, Nano particles additives have been proven by various researchers to improve the tribological performance of the compressor in refrigeration and air-conditioning systems. Some researchers have recently carried out investigations on Nano refrigerants as a drop-in replacement for conventional refrigerants in refrigeration systems. Selvam et al. [15] carried out an investigation using silver Nano fluids in ethylene glycol and water mixture. Thermal properties analyzer (KD2 Pro) was used to measure the thermo-physical properties such as, viscosity, thermal conductivity, density and specific heat. The thermal conductivity improved with an increase in Nano particle concentration and temperature. The enhancement of thermal conductivity observed was within the range 0.15%vol to 12 vol% at 50°C. The viscosity and density also increased with increase in Nano particle concentration and decreased with an increase in temperature. The specific heat decreased with increase in Nano particle concentration and increased with temperature increase. This decrease in specific heat is credited to the lower specific heat of added Nano particles. Akilu et al. [16] examined TiO_2-CuO Nano composite in ethylene glycol. The viscosity and thermal conductivity were measured regarding nanoparticles volume concentration and temperature. The result observed showed that thermal conductivity and viscosity largely depend on the Nano particles volume concentration and temperature. An enhancement of 16.7% and 80% was gained at 2.0% volume concentration compared to base ethylene glycol at 40.4° C. Akel et al. [17] evaluated a Single Wall Carbon Nano tube(SWCNT) water Nano fluid as a secondary fluid in a refrigeration system. The SWCNT-water Nano fluid was tested as a secondary fluid with the range of 4 - 9 kW cooling power. An R22 VCRS made up of electronic expansion valve and air-cooled condenser type was used as the test rig. Different volumetric concentrations of Nano fluid were used for the experiment. The resulting output showed that the thermal conductivity and refrigerating capacity of Nano fluid as secondary refrigerant were improved. Akhavan-behabadi et al. [18] carried out research on pure R600a refrigerant using CuO nanoparticles of different mass fraction (0.5, 1.0 and 1.5%) and compared it with baseline refrigerant (R600a). The result obtained revealed that the heat transfer of the system increased by 83% compared to the pure refrigerant. Ghorbani et al. [19] studied the performance of R600a/POE/CuO Nano refrigerant by comparing it to the base fluid (R600a/POE) in a refrigeration system. Separate fluids with pure R600a, R600a-oil with POE mass percentage of 1%, three different R600a-oil-Nano particle mixtures with mass percentages of 0.5, 1, and 1.5% were considered experimentally. The outcome revealed that addition of Nano particles resulted in 4.1%, 8.11%, and 13.7% average rise in condensing heat transfer coefficient when compared with the R600a-oil mixture. Also, the maximum amount of increase in heat transfer coefficient was recorded for 1.5% mass percentage. Nabil et al. [20] investigated heat transfer performance of TiO_2 Nano particles in water-ethylene glycol (EG) mixture. The study was carried out at different operating temperature conditions. The viscosity and thermal conductivity were considered under the temperature range of 30°C-80°C. The result showed an increment in thermal conductivity of 15.4% at 1.5vol % Nano concentration at 60°C. The Nusselt number also showed an improvement of 22.8% and 28.95% at 50°C and 70°c respectively. Maheshwary et al. [21] examined the effect of the shape of Nano particles on heat transfer properties and thermo physical properties of ZnO/R-134a Nano refrigerant in VCRS. A cubic and spherical shape of ZnO nanoparticles were considered for the test. The results revealed that cubic ZnO Nano particles in R134a had 42.5% increment thermal conductivity compared to pure R134a refrigerant. The author recommended the use of ZnO nanoparticles of cubic shape in R134a in VCRS. Zawawi et al. [22] investigated three different Nano composite (Al_2O_3-SiO_2, $Al2O_3$-SiO_2, TiO_2-SiO_2) in a polyakylene glycol (PAG) lubricant. The thermal conductivity was measured using KD2Pro. The results obtained reveal an enhancement of 20.5% increase in viscosity for Al_2O_3-TiO_2/PAG Nano lubricant in 1% Nano volume concentration at 34°C. Thermal conductivity increases of 2.4% was recorded for Al_2O_3-SiO_2/PAG for 0.1% Nano lubricant at 30°C. Sanukrishna et al. [23] studied the thermal and rheological behavior of TiO_2-PAG oil in VCRS. TiO_2-PAG Nano lubricant of 0.7% to 0.8% volume fractions were used in a temperature range of 20°C to`90°C. The result showed that thermal conductivity and viscosity were 10 and 1.38 times higher than pure lubricant for 0.6% and 0.8% volume fraction respectively with the optimal volume of 0.4%. Zawawi et al. [24] conducted an experiment on Al_2O_3 Nano lubricant in automotive air-conditioning system. The thermal conductivity and viscosity were measure using KD2Pro, thermal analyzer and LVDV-III Rheometer. The findings showed that Nano lubricant has higher thermal conductivity and viscosity of 1.45% and 7.8% respectively compared to pure PAG lubricant for 1.0% and 0.4%wt Nano concentration. Wang et al. [25] experimented to improve the coefficient of performance of air-conditioning system using a mixture of fullerenes (C70) and $NiFe_2O_4$ dispersed in compressor lubricant oil after solid grinding. The result showed that the coefficient of friction reduced from 0.13 to 0.06 which translate to a reduction in wear. The coefficient of performance of the air-conditioning system also increased by1.23 %. In another research, Wang et al. [26] investigated tribological properties of $NiFe_2O_4$ and onion alike fullerenes as a Nano composite in reciprocating compressor to enhance the coefficient of performance and reliability of a refrigeration system. $NiFe_2O$ and Onion like fullerenes (OLFs) were modified to form a Nano composite using KFR22 solid grinding (SG) and dispersed in compressor lubricant oil. The morphologies of $NiFe_2O_4$ were also characterized using transmission electron microscopy (TEM). Furthermore, the tribological properties of the refrigerator Nano lubricant were considered using a multi-functional reciprocating frictional wear tester. The results revealed that the frictional coefficient of the compressor was reduced from 0.15 to 0.04. Melnyk [27] presented experimental data for the solubility, viscosity and density of working fluid (R600a) in a refrigeration system with a constant capillary tube length. Al_2O_3 and TiO_2 nanoparticles were tested in R600a refrigerant using a mineral oil as the compressor lubricant and compared with pure R600a refrigerant under a varied range of concentrations and temperature. The result showed that the R600a/TiO_2 and R600a/$Al2O_3$ Nano refrigerant increased in viscosity with a reduction in surface tension. Sun et al. [28] examined heat transfer of multi-walled CNT of MWCNT-COOH/R141b and MWCNT-OH/R141b. The volume concentration of the Nano refrigerants was 0.059, 0.117 and 0.1765% and a vapor quality rang of 0.2-0.7 with a mass flow rate of 100-350 kgm^2s^1. The findings showed that the heat transfer rate of MWCNT-COOH/R141b Nano refrigerant performed better than that of MWCNT-OH/R141b Nano refrigerant.

2.2 Effect of nanoparticles on the COP, cooling capacity and power consumption of refrigeration system

Power consumption has been one of the major criteria always considered for purchasing refrigeration and air-conditioning units due to the high cost of energy consumption attached to it. Refrigerators and air-conditioners manufactures also consider the cooling capacity and coefficient of performance of a refrigeration unit. Some recent works were done on Nano refrigerants to substitute convectional refrigerants in VCRS. Balaji et al. [29] conducted research in order to reduce the air conditioning compressor load. A shell-and-coil heat exchanger was introduced as an intercooler using water and ethylene glycol of 70:30 mixture ratio as the base fluid. Al_2O_3 Nano particles of different Nano concentration by volume were used for the experiment. The result revealed that the coefficient of performance (COP) of the system increased with decrease in temperature as a result of decrease in the compressor work input. An increase in COP of 49.32% was achieved for the 0.75% of Nano fluid compared with the base fluid without intercooler. A significant drop in power consumption of 12.24% was also recorded. Nabil et al. [30] conducted a comparative study on thermo-physical properties of Al_2O_3 and SiO_2 Nano particles dispersed in polyalkyene glycol (PAG) lubricant oil to improve the performance of an automobile air conditioning system. The Nano particles were dispersed in the lubricant oil of the automobile air conditioning compressor to enhance the thermal conductivity and viscosity of the system. The SiO_2 Nano particles were dispersed in PAG lubricants oil of 0.2–1.5% volume concentrations at 30–80°C working temperature conditions and compared with performance of Al_2O_3 Nano lubricant. The result showed that thermal conductivity and viscosity increased with increase in Nano concentration of the lubricant oil but decreased with temperature. However, SiO_2 Nano lubricants with concentration of less than 1.0% by volume were the most suitable in the system. Ohunakin et al. [31] worked on performance of SiO_2, TiO_2, and Al_2O_3 Nano lubricants in household refrigerator system using liquefied Petroleum Gas (LPG) as working fluid. TiO_2, SiO_2 and Al_2O_3 nanoparticles were dispersed in a mineral oil in the LPG as the working fluid. The performances such as power consumption, viscosity and thermal conductivity at the compressor discharge and suction, the pull-down time was also considered. The result revealed that Nano lubricant-based LPG had a lower temperature with a pull-down time of 180 min. Moreover, SiO_2 and TiO_2 led to 12% and 13% reduction in power consumption respectively in comparison with the base LPG refrigerant. Also, there were improvement in the cooling COP, the COP of SiO_2 and TiO_2 Nano particles in LPG are higher than of pure LPG with 2.06% and 2.97% respectively, while a drop in COP of 2.91% was experienced with $A12O_3$ Nano particles in LPG at steady state. In the experiment, Al_2O_3 Nano lubricant was not suitable for LPG due to an increase in power consumption compared to the base LPG refrigerant. In a related study by Adelekan et al. [32], the author investigated TiO_2 Nano particles in 40, 50, 60 and 70 g of liquefied petroleum gas with varied Nano concentration of 0.2, 0.4 and 0.6g/L in domestic refrigerator designed to work with R134a refrigerant. The system pull down time, compressor power input, cooling capacity, coefficient of performance (COP) and power consumption were analyzed. The outcome of results obtained indicated equivalent evaporator air temperatures with decline in power consumption for 40, 50, 60g Nano lubricant concentration but an increased in 70g charge of LPG in all Nano-lubricant concentration. Bhattacharyya et al. [33], [34] studied an energy saving of a geothermal Nano technology in an air-conditioning system. A Nano fluid of Al_2O_3/R134a was used as the working fluid. The findings showed that a geothermal air-conditioning system using Nano fluid of Al_2O_3/R134a had better energy savings and COP compared with conventional air conditioning refrigerant using R134a refrigerant. Ajayi et al. [35] investigated the fluid flow of Cu Nano particles in R134a and R600a in adiabatic capillary tube of a VCRS. The result showed that pure R134a and R600a have more isothermal regions which mean that the heat transfer is not fast enough to increase the temperature of the connecting region compared to the Cu/R134a and CU/R600a Nano refrigerants. Also, the base refrigerant had higher work input than Cu/R134a and CU/R600a, which implies a decrease in coefficient of performance of the system. Hady et al. [36] performed an experiment to replace water in chilled water air-conditioning unit with Al_2O_3-water Nano fluid using an Al_2O_3 Nano-concentration of 0.1, 0.2, 0.3 and 1% wt. The result showed that Al_2O_3-water Nano fluid has higher COP of 5% and 7% for 0.1%wt and 1%wt respectively in the system. Also, there was an increase in cooling capacity and decrease in power consumption of Al_2O_3-water in the chilled water air-conditioning unit. Kumar and Singh [37] worked on ZnO Nano particles effect on a mixture of hydrocarbon made up of 50% R290 and 50% R600a in a vapour compression system. The nanoparticles were introduced via the lubricant oil in the compressor using different percentage by weight of ZnO/R290/R600a Nano refrigerant. The outcome showed that energy consumption of the system was reduced by 7.8% using ZnO/R290/R600a Nano refrigerant of 0.2 -1.0 wt% range of Nano concentration. The COP of the system also increased by 46% compared to base refrigerant (R290/R600a).

3. Present state of Nano refrigerant in refrigeration and air conditioning system.

The application of Nano particles in refrigeration systems has shown significant improvement in terms of COP, power input, cooling capacity, power consumption, thermal conductivity and viscosity Nano refrigerants in VCRS compared to the conventional refrigerants.

Table 1 Summary of the previous work on enhancing the performance of VCRS using Nano refrigerants

Author	Type of experiment	Nano refrigerant/ Nano lubricant	Performance evaluation
Maheshwarry et al., 2018	Effect of shape of nanoparticles on heat transfer performance	ZnO/R134a	The experiment revealed that cubic shape ZnO/R134a Nano refrigerant has higher transfer of 42. % in the system.
N.N Zawawi et al., 2018	Thermal conductivity performance of Nano hybrid in PAG lubricant oil.	Al2O3-SiO2, Al2O3-TiO2, TiO2-SiO2, PAG	The result showed an increase in thermal conductivity of 20.5% and 2.4% for Al2O3-TiO2 and Al2O3-SiO2 respectively.
Sanukrishna et al., 2018	Experimental study of thermal conductivity of Nano lubricant in VCRS	TiO_2- PAG	The study showed 1.3% increase in thermal conductivity
Nabil et al., 2017	Comparative study of Al_2O_3 and SiO_2 in PAG in automobile air-conditioning system	Al_2O_3 and SiO_2 in PAG	The experimental analysis showed that Nano lubricant has higher thermal conduction and viscosity than pure PAG with SiO2/PAG having highest
Ohunakin et al., 2017	Performance of Nano lubricant in VCRS	TiO2 Al_2O_3 and SiO2 in LPG	The study showed 12% and 13% reduction in power consumption for SiO2/LPG and TiO2/LPG respectively
Adelakn et al., 2017	Investigation of Nano lubricant in VCRS working with LPG	LPG/TiO_2	The investigation revealed that 40. 50. 60g mass charge of LPG/TiO_2 perform better in terms of COP and power consumption
Bhattacharya et al., 2017	Energy saving using Nano technology in geothermal air-conditioning system	Al_2O_3/R134a	The research showed a better energy savings with Al_2O_3/R134a compare to pure R134a
Oluseyi et al., 2017	Investigation of nanoparticles in VCRS	Cu/R600a, Cu/R134a	The outcome of the result showed Cu/R600a and Cu/R134 have higher than pure R134a
Hady et al., 2017	Replacement of chilled water-air conditioning unit	Al_2O_3-water	The result revealed a higher COP of 5% and 7% for Al2O3-water0.1% and 1% wt Nano concentration respectively.
Akilu et al., 2017	Experimental investigation of Nano composites in ethylene glycol	TiO2-CuO/ ethylene glycol	Performance evaluation of the Nano refrigerant showed that an increase in thermal conductivity and viscosity of 16.7% and 80% respectively.
Akel at al., 2017	Investigation of a single wall carbon nanotube in VCRS	SWCNT/R22	The experimental outcome revealed an improvement in thermal conductivity of the VCRS.
Ghorbani etal., 2017	Analysis of Nano refrigerant in vapour compression system.	R600a/POE/CuO	The experimental analysis showed an increase in 4.1%, 8.11% and 13.7% for 0.5, i% and 1.5% mass percentage of Nano refrigerant.
Nabil et al., 2017	Performance of Nano particle in water-ethyl glycol	TiO_2/water-ethyl glycol	The study showed an increase in thermal conductivity by 15.4% and increase in nusselt number by 22.8% and 28.95% at 50°C and 70°C respectively.
Akilu et al., 2017	Experimental investigation of Nano composites in ethylene glycol	TiO_2-CuO/ ethylene glycol	Performance evaluation of the Nano refrigerant showed that an increase in thermal conductivity and viscosity of 16.7% and 80% respectively.

Akel at al., 2017	Investigation of a single wall carbon nanotube in VCRS	SWCNT/R22	The experimental outcome revealed an improvement in thermal conductivity of the VCRS.
Ghorbani etal., 2017	Analysis of Nano refrigerant in vapour compression system.	R600a/POE/CuO	The experimental analysis showed an increase in 4.1%, 8.11% and 13.7% for 0.5, i% and 1.5% mass percentage of Nano refrigerant.
Nabil et al., 2017	Performance of Nano particle in water-ethyl glycol	TiO_2/water-ethyl glycol	The study showed an increase in thermal conductivity by 15.4% and increase in nusselt number by 22.8% and 28.95% at 50°C and 70°C respectively.
Zawawi et al., 2017	Performance of nano particle in PAG lubricant oil.	TiO_2/PAG	The investigation showed higher thermal conductivity and viscosity of 1.45% and 7.8% respectively
Selvan et al., 2016	Enhancement of thermal conductivity and viscosity of ethylene –water.	Ag/ethylene-water	The finding showed an enhancement in thermal conductivity within the range of 12% to 0.15% for 50°C.
Balaja et al., 2015	Experimental research on air-conditioning compressor load.	Al_2O_3/ water ethylene glycol	The COP of the system and thermal conductivity increased by 49.32% and 12.24 % respectively

But currently, apart from R600a refrigerant, only few and scanty researches have been done on eco-friendly Nano refrigerants such as LPG and other hydrocarbon mixture refrigerants. Moreover, more researches need to be carried out on thermodynamics and heat transfer properties of hybrid nanoparticles in hydrocarbon mixtures and R152a refrigerants in vapor compression refrigeration systems.

4. Conclusion

Application of Nano particles as additives in refrigerant, lubricant and other fluids for refrigeration and air-conditioning systems has been proved to be successful according to the literature review without modification to the physical design components of the VCRS as a result of the significant improvement, especially in energy enhancement. It is believed that VCRS manufacturers will adopt Nano refrigerants as working fluids in refrigeration and air-conditioning system.

The following conclusions can be drawn from this current study:

- A significant reduction in energy consumption is achieved
- Work input of Nano refrigerant reduces compared to the base refrigerant thereby promotes increase in COP of the system.
- A shorter pull-down time in refrigeration system is possible
- The thermal conductivity increases with temperature increase up to the optimal concentration
- The viscosity increases with increase in temperature up to the optimal concentration
- Nano refrigerants specific heat increases with increase in temperature and concentration
- A reduction in compressor friction and wear can be achieved.

Reference

[1] A. M. K. P. T. Ã, "Science review of internal combustion engines $," vol. 36, no. 2008, pp. 4657–4667, 2010.

[2] W. H. Azmi, M. Z. Sharif, T. M. Yusof, R. Mamat, and A. A. M. Redhwan, "Potential of nanorefrigerant and nanolubricant on energy saving in refrigeration system – A review," *Renew. Sustain. Energy Rev.*, vol. 69, no. December 2015, pp. 415–428, 2017.

[3] J. Gill and J. Singh, "Energy analysis of vapor compression refrigeration system using mixture of R134a and LPG as refrigerant Analyse énergétique d ' un système frigorifique à compression de

vapeur utilisant un mélange de R134a et de GPL comme frigorigène," *Int. J. Refrig.*, vol. 84, pp. 287–299, 2017.

[4] T. O. Babarinde, O. S. Ohunakin, D. S. Adelekan, S. A. Aasa, and S. O. Oyedepo, "Experimental study of LPG and R134a refrigerants in vapor compression refrigeration," *Int. J. Energy a Clean Environ.*, vol. 16, no. 1–4, 2015.

[5] S. O. Oyedepo, R. O. Fagbenle, T. Babarinde, and A. Tunde, "Effect of Capillary Tube Length and Refrigerant Charge on the Performance of Domestic Refrigerator with R12 and R600a," *Int. J. Adv. Thermofluid Res.*, vol. 2, no. 1, pp. 2–14, 2016.

[6] B. O. Bolaji, "Experimental study of R152a and R32 to replace R134a in a domestic refrigerator," *Energy*, vol. 35, no. 9, pp. 3793–3798, 2010.

[7] S. J. Sekhar and D. M. Lal, "HFC134a/HC600a/HC290 mixture a retrofit for CFC12 systems," *Int. J. Refrig.*, vol. 28, no. 5, pp. 735–743, 2005.

[8] R. Saidur, S. N. Kazi, M. S. Hossain, M. M. Rahman, and H. A. Mohammed, "A review on the performance of nanoparticles suspended with refrigerants and lubricating oils in refrigeration systems," *Renew. Sustain. Energy Rev.*, vol. 15, no. 1, pp. 310–323, 2011.

[9] M. Fatouh and M. El Kafafy, "Experimental evaluation of a domestic refrigerator working with LPG," *Appl. Therm. Eng.*, vol. 26, no. 14–15, pp. 1593–1603, 2006.

[10] M. M. Joybari, M. S. Hatamipour, A. Rahimi, and F. G. Modarres, "Exergy analysis and optimization of R600a as a replacement of R134a in a domestic refrigerator system," *Int. J. Refrig.*, vol. 36, no. 4, pp. 1233–1242, 2013.

[11] M. Rasti, S. Aghamiri, and M. S. Hatamipour, "Energy efficiency enhancement of a domestic refrigerator using R436A and R600a as alternative refrigerants to R134a," *Int. J. Therm. Sci.*, vol. 74, pp. 86–94, 2013.

[12] A. S. Dalkilic and S. Wongwises, "A performance comparison of vapour-compression refrigeration system using various alternative refrigerants," *Int. Commun. Heat Mass Transf.*, vol. 37, no. 9, pp. 1340–1349, 2010.

[13] S. U. S. Choi, "Jam 1 1 19s5," 1995.

[14] A. Bhattad, J. Sarkar, and P. Ghosh, "Improving the performance of refrigeration systems by using nanofluids: A comprehensive review," *Renew. Sustain. Energy Rev.*, vol. 82, no. August 2016, pp. 3656–3669, 2018.

[15] C. Selvam, D. M. Lal, and S. Harish, "Thermophysical properties of ethylene glycol-water mixture containing silver," vol. c, no. 3, pp. 1271–1279, 2016.

[16] S. Akilu, A. T. Baheta, and K. V. Sharma, "Experimental measurements of thermal conductivity and viscosity of ethylene glycol-based hybrid nanofluid with TiO2-CuO/C inclusions," *J. Mol. Liq.*, vol. 246, pp. 396–405, 2017.

[17] A. Akel *et al.*, "Experimental evaluation of SWCNT-water nanofluid as a secondary fluid in a refrigeration system," *Appl. Therm. Eng.*, vol. 111, pp. 1487–1492, 2017.

[18] M. A. Akhavan-behabadi, M. K. Sadoughi, M. Darzi, and M. Fakoor-pakdaman, "Experimental study on heat transfer characteristics of R600a / POE / CuO nano-refrigerant flow condensation," *Exp. Therm. FLUID Sci.*, vol. 66, pp. 46–52, 2015.

[19] B. Ghorbani, M. A. Akhavan-behabadi, S. Ebrahimi, and K. Vijayaraghavan, "Experimental investigation of condensation heat transfer of R600a / POE / CuO nano-refrigerant in fl attened tubes," *Int. Commun. Heat Mass Transf.*, vol. 88, no. October, pp. 236–244, 2017.

[20] M. F. Nabil, W. H. Azmi, K. Abdul Hamid, R. Mamat, and F. Y. Hagos, "An experimental study on the thermal conductivity and dynamic viscosity of TiO2-SiO2nanofluids in water: Ethylene glycol mixture," *Int. Commun. Heat Mass Transf.*, vol. 86, pp. 181–189, 2017.

[21] P. B. Maheshwary, C. C. Handa, and K. R. Nemade, "Effect of Shape on Thermophysical and Heat Transfer Properties of ZnO/R-134a Nanorefrigerant," *Mater. Today Proc.*, vol. 5, no. 1, pp. 1635–1639, 2018.

[22] N. N. M. Zawawi, W. H. Azmi, A. A. M. Redhwan, M. Z. Sharif, and M. Samykano, "PT US CR," *Int. J. Refrig.*, 2018.

[23] S. S. Sanukrishna, M. Murukan, and P. M. Jose, "PT US CR," *Int. J. Refrig.*, 2018.

[24] N. N. M. Zawawi, W. H. Azmi, A. A. M. Redhwan, M. Z. Sharif, and K. V. Sharma, "Propriétés thermo-physiques du nanolubrifiant composite Al2O3-SiO2/PAG pour les systèmes frigorifiques," *Int. J. Refrig.*, vol. 80, pp. 1–10, 2017.

[25] R. Wang, Y. Zhang, and Y. Liao, "Performance of rolling piston type rotary compressor using fullerenes (C70) and NiFe 2 O 4 nanocomposites as lubricants additives," 2017.

[26] R. Wang, Y. Zhang, and H. Wu, "Tribological properties of onion like fullerenes and NiFe 2 O 4 nanocomposites in reciprocating motion," 2017.

[27] A. V Melnyk, "A complex investigation of the nanofluids R600 a -mineral oil-AL 2 O 3 and R600 a -mineral oil-TiO 2 . Thermophysical properties Une étude complexe des propriétés thermophysiques des nanofluides R600a-huile minérale-Al 2 O 3 et R600a-huile," *Int. J. Refrig.*, vol. 74, pp. 488–504, 2017.

[28] B. Sun, H. Wang, and D. Yang, "E ff ects of surface functionalization on the fl ow boiling heat transfer characteristics of MWCNT / R141b nanorefrigerants in smooth tube," *Exp. Therm. Fluid Sci.*, vol. 92, no. November 2017, pp. 162–173, 2018.

[29] N. Balaji, P. S. Mohan, and K. R. Velraj, "Experimental Investigations on the Improvement of an Air Conditioning System with a Nanofluid-Based Intercooler," pp. 1681–1693, 2015.

[30] M. F. Nabil, W. H. Azmi, K. A. Hamid, N. N. M. Zawawi, G. Priyandoko, and R. Mamat, "Thermo-physical properties of hybrid nanofluids and hybrid nanolubricants: A comprehensive review on performance," *Int. Commun. Heat Mass Transf.*, vol. 83, pp. 30–39, 2017.

[31] O. S. Ohunakin, D. S. Adelekan, T. O. Babarinde, R. O. Leramo, F. I. Abam, and C. D. Diarra, "Experimental investigation of TiO2, SiO2 and Al2O-lubricants for a domestic refrigerator system using LPG as working fluid," *Appl. Therm. Eng.*, vol. 127, 2017.

[32] D. S. Adelekan *et al.*, "Experimental performance of LPG refrigerant charges with varied concentration of TiO2 nano-lubricants in a domestic refrigerator," *Case Stud. Therm. Eng.*, vol. 9, no. October 2016, pp. 55–61, 2017.

[33] S. Bhattacharyya, P. Das, A. Haldar, Á. G. Á. Experimental, Á. Nano, and Á. Air, "Performance Analysis of a Geothermal Air Conditioner Using Nano fl uid," pp. 89–95.

[34] A. Bhattad, J. Sarkar, and P. Ghosh, "Exergetic analysis of plate evaporator using hybrid nanofluids as secondary refrigerant for low-temperature applications," *Int. J. Exergy*, vol. 24, no. 1, pp. 1–20, 2017.

[35] O. O. Ajayi, P. O. Babalola, O. Kilanko, D. O. Oyegbile, and D. Lawson-jack, "A Comparative experimental study on performance of domestic refrigerator using r600a and lpg with varying refrigerant cha ... Authors," p. 1615.

[36] M. R. A. Hady, M. Salem Ahmed, and G. Abdallah, "Experimental investigation on the performance of chilled - water air conditioning unit using alumina nanofluids," *Therm. Sci. Eng. Prog.*, 2017.

[37] R. Kumar and J. Singh, "Effect of ZnO nanoparticles in R290 / R600a (50 / 50) based vapour compression refrigeration system added via lubricant oil on compressor suction and discharge characteristics," *Heat Mass Transf.*, vol. 53, no. 5, pp. 1579–1587, 2017.

Exergetic Performance of Household Refrigerator using R600a and LPG

T. O. Babarinde[1, 3*], S. A. Akinlabi[2], D. M. Madyira[1]

[1]Department of Mechanical Engineering Science, University of Johannesburg, South Africa
[2]Department of Mechanical & Industrial Engineering Technology, University of Johannesburg, South Africa
[3]Mechanical Engineering Department, Covenant University Ota, Nigeria
E-mail address: taiwo.babarinde79@gmail.com, stephenakinlabi@gmail.com, dmadyira@uj.ac.za

Abstract. This paper presents a case study of coefficient of performance (COP), exergetic performance and irreversibility of four major components (compressor, condenser, capillary tube and evaporator) of a household refrigerator using R600a and LPG. The results showed that the COP and exergetic efficiency of LPG were 9.5% and 7.1% higher than that of R600a respectively. Also, the total irreversibility of LPG was lower than that of R600a with 22.4%. In general, LPG performed better than R600a in terms of COP, power consumption, exergetic efficiency and irreversibility in the system.
Keywords: Exergetic efficiency, COP, R600a, LPG

1. Introduction

Majority of the household refrigerators used today are working on vapour compression principle and halogenated refrigerants such as Chlorofluorocarbons (CFCs) and Hydrofluorocarbon (HFCs) are mostly used as the working fluids in household refrigerator due to their characteristics such non-flammability, non-toxicity and stability with materials vapour compression refrigerator are made up of. However, the use of CFC's in household refrigerator has been banned worldwide and also the use of CFC's refrigerant has been banned in Europe due to their high ozone depleting potential (ODP) and global warming potential (GWP). In line with the Montreal and Kyoto protocols, HFC's refrigerants such as R134a will be phase out fully in the future [1-2]. As these refrigerants are being prohibited and phased out in Europe, they are gradually gaining access into Sub Sahara region as fairly used products. Recently the use of Hydrocarbon refrigerants such as R600a has been embraced as working fluid in household refrigerator in Europe and other hydrocarbons mixture are rapidly gaining attention due to their good characteristic and better environmental features and such as miscibility with mineral oils, compatibility with materials used in vapour compression refrigeration system. Hydrocarbon refrigerant is less expensive and available in abundance in major oil producing nation of the world. Hydrocarbons have good thermodynamic properties; therefore, they are energy efficient [3-4]. The disadvantage of hydrocarbons is it high flammability. However, a maximum mass charge of 150 g has been recommended in household refrigerator [5].

Some researchers have carried out researches on replacement of R134a with an eco-friendly refrigerant in a domestic refrigerator. Mahmood *et al* [6] investigated the use of R600a as a drop-in replacement for R134a in vapour compression system the result revealed that R134a has higher exergy destruction in the system than R600a with the destruction more pronounced more in the compressor, condenser, evaporators and expansion valve decreasingly. In a similar way, Rasti *et al* [7] also performed experimental analysis on energy efficiency improvement of a household refrigerator using R436A and R600a as a substitute for R134a refrigerant in a domestic refrigerator originally designed to work with R134a as the working fluid. The results obtained showed that with R134a compressor the energy consumption of R600a and R436A were reduced. Babarinde et al [8] conducted an experimental study on the use of liquefied petroleum gas (LPG) as a drop in replacement for R134a in a domestic refrigerator system. The results obtained indicated that the coefficient of performance (COP) of the system was better and the power consumption reduced when compared with R134a. Bolaji [9] carried out a research on the exergetic and performance of a domestic vapour compression refrigerator using R134a and R152a to replace R12, the performance was compared with the performance of R12 in the refrigerator. The results indicated that the COP of R152a was almost the same in comparison with that of R12 and the highest exergetic efficiency was also obtained for R152a refrigerant in the system. The total efficiency defect of R152a in the refrigerator was the lowest compared to that of R12 and R134a. Babarinde at [10] investigated the coefficient of performance (COP) exergetic destruction and exergetic performance of a household refrigerator R134a and R600a the result outcome showed that the coefficient of performance and exergetic efficiency is higher than that of R134a and the total exergetic defect of R600a is lower than that R134a in the system.

The goals of this research are to investigate the more economical refrigerant between the two hydrocarbons (R600a and LPG) in terms of evaporator temperature energy efficiency in a household refrigerator.

2. Experimental procedure

The experiments were carried out using refrigerator designed to work with R134a. The capillary tube was extended by 30% in order to enhance the performance of the hydrocarbon refrigerants. The experiment was tested using 60 g charge of R600a and 40g charge of LPG; this is because the latent heat of hydrocarbon is much higher than that of R134a and also, the density of LPG is lower than R600a. Type K thermocouples were connected to measure the inlet and outlet temperature of the four components (compressor, evaporator, condenser and capillary tube) of the refrigerator. Two pressure gauges were also connected to the suction and discharge of the compressor, see Figure 1. The measurements of uncertainties of the measuring instrument used are in Table 2. Table 3 shows the experimental condition and ranges for continuous and cycling experiment. The experiment was repeated four times. The thermodynamic properties of the results obtained were analyzed using Refprop, version 9 [11]. The experiment was conducted at an average an environmental temperature of 27°C at no load conditions.

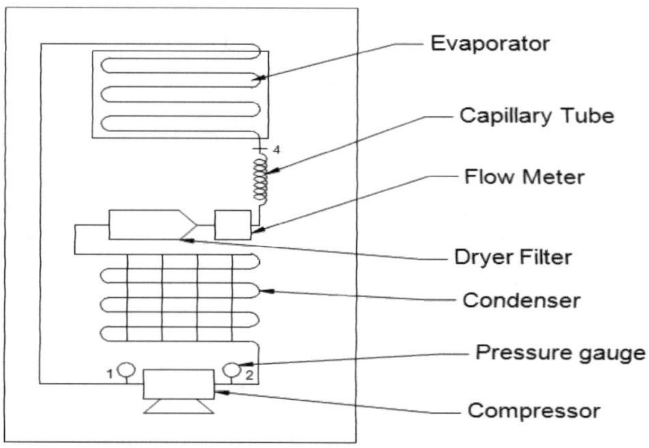

Fig. 1 Schematic diagram of experimental setup

Table 1: characteristic properties of hydrocarbon refrigerants

Refrigerants	NPB(°C)	T(°C)	GWP	Safety group	Molar mass (kg/kmol)
R134a	-26	101	1300	A1	102.03
R600a	-12	135	20	A3	58.12
R600	-0.49	151.98	20	A3	58.12
R290	-42	97	20	A3	44.09
LPG	-25.4	101	20	A3	48.8

Table 2: Characteristics of the measuring instruments

S/N	Measured Data	Manufacturers Specification	Range	Uncertainty
1	Temperature	Digital Thermocouple K	-50°C - 750°C	±3°C
2	Pressure	Digital pressure gauge	5 - 5000 Pa	±1%
3	Power consumption	Digital Watt/Watt-h-meter	1-3000W (0.0001-999.9 kWh)	±1

Table 3: Range and conditions of experiment

S/N	Parameters	Range of Experiment
1	R600a refrigerant charge	60g
2	LPG Refrigerant	40g
3	Compressor lubricant	Mineral oil
6	Test environment temperature	27°C
7	Capillary tube length	1.5m
8	Condenser type	Air cooled
9	Evaporator size	72litres

The exergy efficiency of the cycle is calculated as follows:

Exergy is a degree of how a system deviates from it given and reference state. The total Exergy that is obtainable at the system output (Xout) is less than total Exergy at the system inlet (Xin). This amount of Exergy used up in the system is as a result of irreversibility. The summation of irreversibility in each component of the system is called total irreversibly of the system. The following assumptions below were made to calculate the performance of the refrigerator system.

1. Steady-state operation is assumed in all the system components.
2. Pressure losses along each component are neglected.
3. There is no heat gain and heat loss from or to the system.
4. Potential and Kinetic energy are neglected.
5. There is no pressure drop along the condenser and evaporator of the system.

The exergy efficiency of the cycle is evaluated as follows:

$$I = X_{in} - X_{out} \tag{1}$$

When equation (1) is applied to each (evaporator, compressor, condenser and expansion valve) of the system

Exergy of the evaporator

Where ṁ is the refrigerant mass flow rate (in, kg/s), h_4 represents the specific enthalpy (in kJ/kg) of the refrigerant at the evaporator inlet and, h_1 is the specific enthalpy (in kJ/kg) of the refrigerant at the evaporator outlet, S_1 is the specific entropy (kJ/kg.K) of the refrigerant at the outlet of the evaporator, S_4 is the specific entropy (kJ/kg.K) of the refrigerant at the evaporator inlet, Q_e (kW) is heat transfer rate through the boundary at T_{evap}. Thus Exergy efficiency of the evaporator can be written as

$$I_{evap} = \dot{m}(h_4 - T_O S_4) + Q_{evap}\left(1 - \frac{T_0}{T_r}\right) - \dot{m}(h_1 - T_O S_1) \tag{2}$$

Exergy of the compressor

$$I_{comp} = \dot{m}(h_1 - T_O S_1) + W_{comp} - \dot{m}(h_2 - T_O S_2) \quad (3)$$

Where S_2 is the specific entropy (kJ/kg.K) of the refrigerant at the compressor outlet W_{comp} is compressor power input (kW).

Exergy of the condenser

Exergy at the condenser are calculated from equation (9) and (10)

$$I_{cond} = \dot{m}(h_2 - T_O S_2) - \dot{m}(h_3 - T_O S_3) - Q_{cond}\left(1 - \frac{T_0}{T_{cond}}\right) \quad (4)$$

Where S_3 is the specific entropy (kJ/kg.K) of the refrigerant at the condenser outlet Q_{cond} is heat transfer rate at the condenser (kW).

Exergy of the expansion valve

Where S_4 is the specific entropy (kJ/kg.K) of the refrigerant at the expansion valve outlet.

Total irreversibility of the refrigerator

The total exergy of the refrigerator system is the summation of the components.

$$I_{total} = I_{evap} + I_{comp} + I_{cond} + I_{exp} \quad (5)$$

Exergetic efficiency

The second law efficiency of vapour compression system is written as follows:

The total exergy efficiency (n_X) is the ratio of exergy output (X_{out}) to exergy input

$$n_X = \left(\frac{X_{out}}{X_{in}}\right) X\ 100\% \quad (6)$$

$$X_{out} = X_{in} - X_{total} \quad (7)$$

$$n_X = \left(1 + \frac{X_{total}}{W_{comp}}\right) \quad (8)$$

$$COP = \frac{m(h_1 - h_4)}{m(h_2 - h_1)} \quad (9)$$

3. Results and discussion

The exergetic performance analysis of household refrigerator system is discussed below:

According to ISO (1991), the pull-down can be defined as the required time for the evaporator air chamber to attain the steady state condition from ambient temperature.
Figures 2 show the pull-down time of R600a and LPG in the refrigerator. The operating temperatures of –12 and –14°C were obtained in the refrigerator at a pull-down time of 240 and 210 min for R600a and LPG respectively.

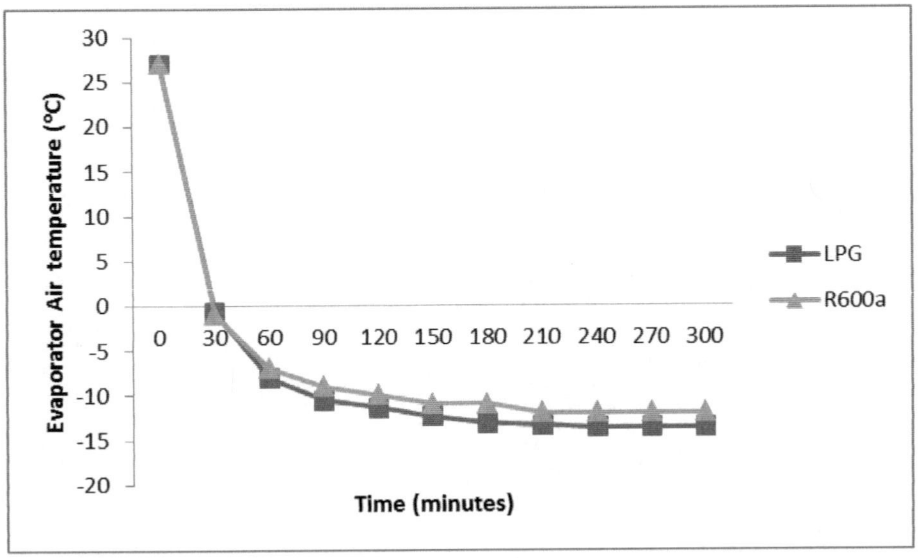

Fig. 2 pull down time of R600a and LPG

Figure 2 shows the cooling capacity and power consumption by the compressor of the system. The power 236 and 111 W were recorded for cooling capacity and power consumption power respectively for R600a while the power 205 and 90 W were recorded for cooling capacity and power consumption respectively for LPG. The cooling capacity and power consumption of LPG is 13.4% and 18.9% lower than R600a refrigerant respectively in the system.

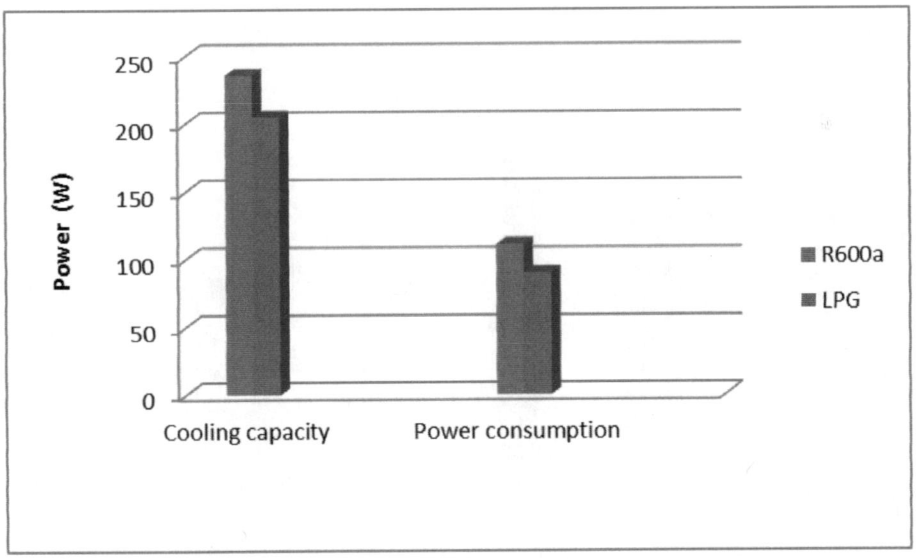

Fig. 3 Comparison of cooling capacity and power consumption of the system

Figure 4 shows the comparison of coefficient of performance (COP). The results showed that the COP of R600a and LPG are 2.1 and 2.3 respectively. The COP for LPG is 9.5% higher when compared to R600a. This was due to decrease in compressor work of LPG. With lower COP of R600a it implies it requires more electric power to produce the same refrigerating effect, which is not cost effective.

IOP Conf. Series: Materials Science and Engineering **413** (2018) 012069 doi:10.1088/1757-899X/413/1/012069

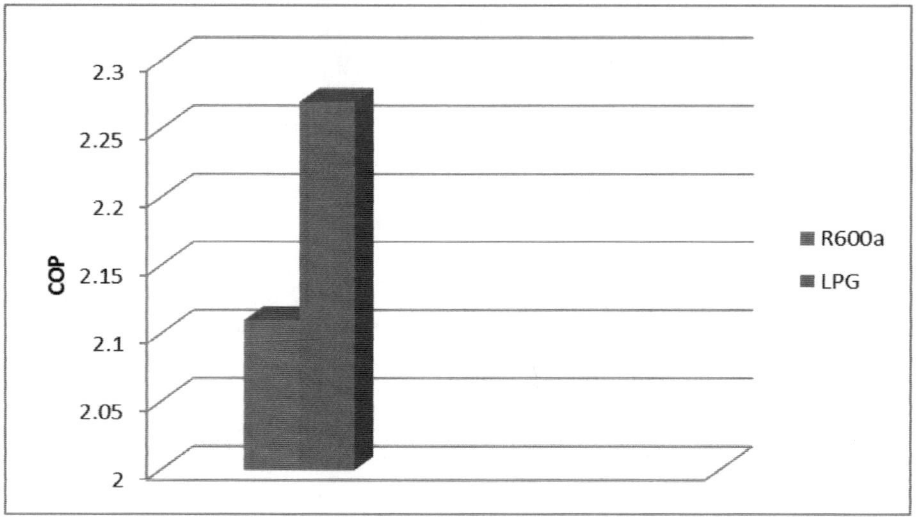

Fig. 4 Comparison of COP of R600a and LPG refrigerant

Figure 5 and 6 shows variation of irreversibility in compressor, condenser, capillary tube and evaporator and total irreversibility of R600a and LPG refrigerants. The figure shows that irreversibility in the compressor, condenser, capillary and evaporator are 0.265, 0.258, 0.09 and 0.031W respectively for R600a refrigerant, while irreversibility in the compressor, condenser, capillary and evaporator are 0.224, 0.054, 0.071 and 0.149W respectively for LPG refrigerant. The irreversibility reduces for LPG compared to R600a. The irreversibility in evaporator for LPG is higher than R600a refrigerant. However the total irreversibility of LPG is 21.9% lower than that R600a in the system.

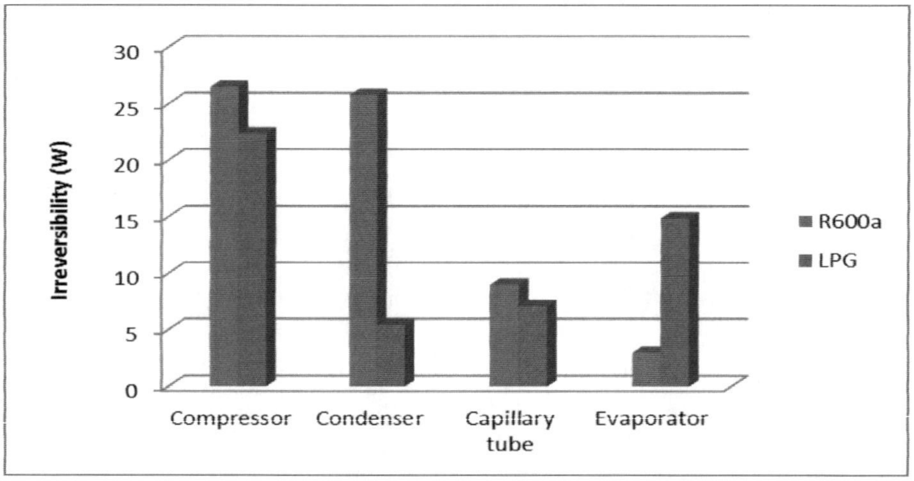

Fig. 5 comparison of irreversibility of the refrigerator components

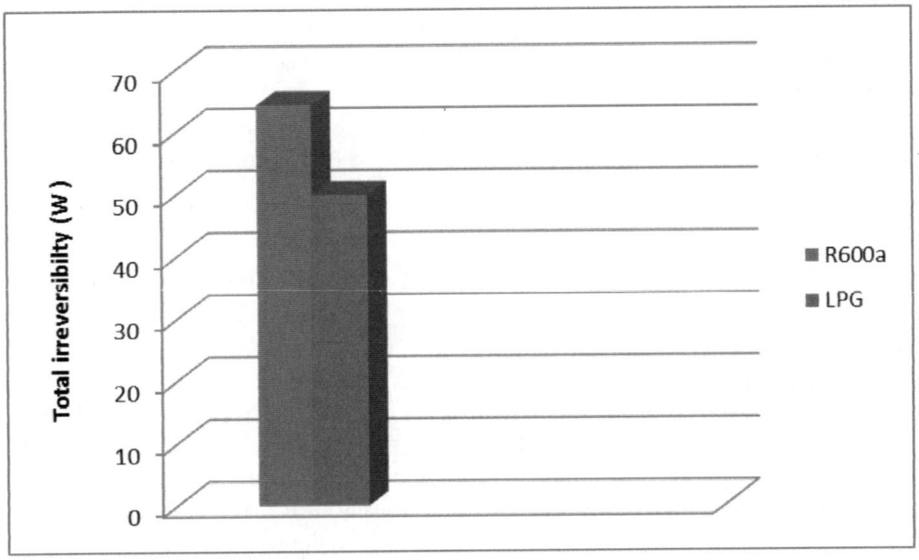

Fig. 6 comparison of total irreversibility of the refrigerator components

Figure 3 shows exergetic efficiency for R600a compared with LPG. The average exergetic efficiency for LPG is 7.1% higher compared to that of R600a. Exergetic efficiency of 45 and 42% were attained at evaporator temperature of -14 ˚C and -12° C for LPG and R600a, respectively. This is as a result of the irreversibility of the components and evaporator temperature increase.

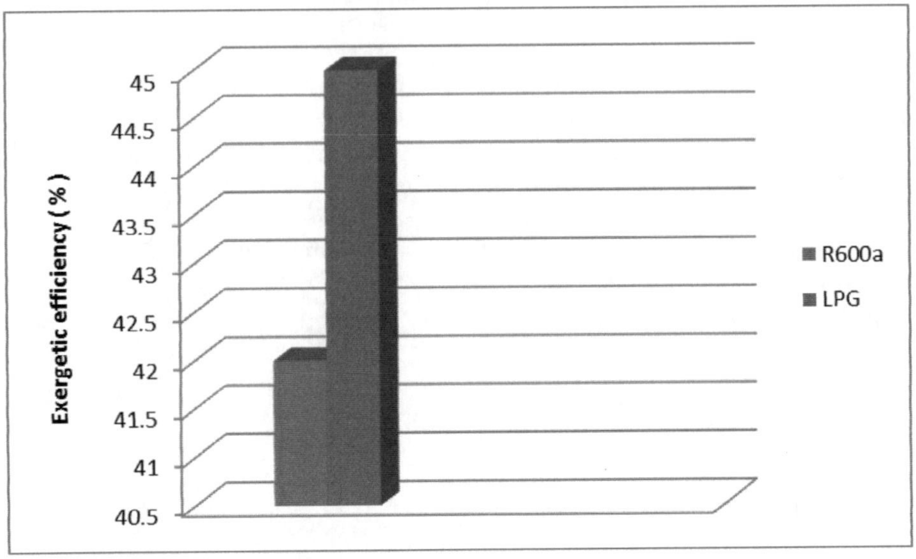

Fig. 7 comparison of exergetic efficiency of the refrigerator components

4. Conclusion

The performance of refrigeration system is studied using R600a and LPG, environment-friendly refrigerant after the analysis on the exergetic performance of R600a and LPG in the system, the following conclusions are made:

- The COP of the household refrigeration system using R600a as a refrigerant was considered as baseline. The COP obtained using LPG higher than that of R600a.
- Lower power consumption was obtained for LPG in the refrigerator.

- The highest exergetic efficiency was obtained using LPG in the system.
- The total irreversibility of LPG is lower than R600a, which indicates improving energy management.
- Generally, the refrigeration system performance improved with LPG than R600a in the system. Hence household refrigerator is more economical to run with LPG working fluid than R600a.

References

[1] Montreal Protocol, United Nations (UN), New York, NY, USA (1987 with subsequent amendments), 1987.
[2] Kyoto Protocol to the United Nations Framework Convention on Climate Change, United Nations (UN), New York, NY, USA, 1997.
[3] Hammad M.A, Alsaad M.A. The use of hydrocarbon mixture as refrigerants in domestic refrigerator. Applied Thermal Engineering 1999; 19:1181-9.
[4] Jung D, Park B, Lee H. Evaluation of supplementary/retrofit refrigerants for automobile air-conditioners charged with CFC12. International Journal of Refrigeration 1999; 22:558-68.
[5] Hewitt NJ, McMullan JT, Mongey B. Some aspects of using hydrocarbons as
.
[6] Mahmood Mastani Joybari , Mohammad Sadegh Hatamipour , Amir Rahimi, Fatemeh Ghadiri Modarres, Exergy analysis and optimization of R600a as a replacement of R134a in a domestic refrigerator system, *International Journal of Refrigeration* 36 (2013) 1233-1242
[7] Mehdi Rasti, SeyedFoad Aghamiri, Mohammad-Sadegh Hatamipour. Energy efficiency enhancement of a domestic refrigerator using R436A and R600a as alternative refrigerants to R134a, *International Journal of Thermal Sciences* 74 (2013) 86-94
[8] T. O. Babarinde, O. S. Ohunakin, D. S. Adelekan, S. A. Aasa, O. Oyedepo. Experimental Study of LPG and R134a Refrigerants in Vapor Compression Refrigeration System. *International Journal of Energy for a Clean Environment* 15(2), 1–10 (2014)
[9] B.O Bolaji. Exergetic Performance of a Domestic Refrigerator using R12 and its Alternative Refrigerants. *Journal of Engineering Science and Technology* Vol. 5, No. 4 (2010) 435 - 446.
[10] T.O. Babarinde, S.A. Akinlabi, D.M. Madyira, O.S. Ohunakin, D.S. Adelekan, & S.O. Oyedepo. Comparative analysis of the exergetic performance of a household refrigerator using R134a and R600a. *International Journal of Energy for a Clean Environment* 19(1–2):37–48 (2018)
[11] NIST Thermo-physical properties of hydrocarbon mixture database (Version 9), National Institute of Standards and Technology, Gaithersburg, Maryland, USA.

Optimizing Laser Parameters of Functional Graded Grade 5 Titanium Alloy (TI6AL4V) and Titanium Carbide (TIC)

P. Ratilal[1], D. M. Madyira[1*], E. T. Akinlabi[1], T.O Babarinde[1,2]

[1]Department of Mechanical Engineering Science, University of Johannesburg, South Africa
[2]Mechanical Engineering Department, Covenant University Ota, Nigeria
E-mail address: dmadyira@uj.ac.za etakinlabi@uj.ac.za, taiwo.babarinde79@gmail.com,

Abstract. Functionally graded materials (FGMs) are materials in which the spatial chemical composition and microstructure gradually change resulting in a corresponding spatial change in the material properties. For this study, grade 5 titanium alloy (Ti6Al4V) and titanium carbide (TiC) powders were co-sintered using laser metal deposition (LMD) on Ti6Al4V substrate. The challenge of this process was determining the process parameters that minimizes or prevent macro cracking due to high thermal gradients and hence the residual stresses. The aim was to determine the optimum parameters. The laser energy density (LED) was used to characterize the processing parameters applied. It was found that when the energy density is in the range of 15 – 25 MJ/m^2, the FGM cracked. When the energy density is in the range of 12 – 16 MJ/m^2, the FGM did not crack.
Key words: Laser metal deposition, FGMs, Laser energy density, Ti6Al4V, TiC

1. Introduction

Functionally graded materials (FGMs) are materials in which the spatial composition and structure gradually change resulting in a corresponding spatial change in the material properties. FGMs are not only characterized by the appearance of compositional or other gradients but also by their sophisticated behaviour in comparison with conventional materials [1]. FGMs find use in applications such as aerospace components, medical implants, automobile components, sports equipment, sensors, optoelectronic etc. FGMs applied to titanium and titanium alloys have the potential to reduce the need of thermal protection in extreme environments like those experienced in aerospace vehicles. Functionally graded titanium alloys are also used to reduce wear in hip prosthesis implants which are exposed to three body abrasive wear (titanium, bone cement and poly ethylene cup) [2, 3]. The need to bring in FGM's is due to the low usage of titanium and its alloys. Although titanium and its alloy have beneficial properties (high strength – to – weight ratio, corrosion resistance, etc.), their use and application is considered low when compared to other commercial materials such as steel. This is due to titanium and its alloys having a high production cost and being difficult to machine. The high cost of titanium alloy production is due to its high reactivity even with tool materials, high strength that is maintained at elevated temperatures thus leading to high tool wear and hence high cost of titanium components. Most titanium alloys are poor thermal conductors, thus heat generated during cutting does not dissipate through the part and machine structure but is concentrated in the cutting area [7]. Therefore, the high temperatures generated during the cutting process may cause a work hardening phenomenon that affects the surface integrity of component. This could lead to inaccuracies in the part and severe reduction in its fatigue strength. FGM's can solve this problem by eliminating the issue of titanium being difficult to machine. Also the processes used to form FGM's will produce more accurate parts with lower material wastage. FGMs can be produced using different fabrication processes. The process used depends on whether the FGM is thin or thick. Thin sections are classified as surface coatings whereas thick sections are those that require a build-up. For thin sections or surface coatings, Physical or Chemical Vapour Deposition (PVD/CVD), Plasma Spraying, Self-propagating High Temperature Synthesis (SHS) and others are used [4, 5]. Thick sections are commonly produced using such processes as Powder Metallurgy, Centrifugal Casting and Solid Freeform Technology [4, 5]. For the purpose of this investigation solid freeform technology, meaning laser metal deposition (LMD), was used to form the FGM. Various investigations have been conducted using LMD for FGMs [6, 7]. One of the major issues related to forming FGMs with LMD is cracking of the deposit. The main reason for cracking is the differential thermal straining induced during the high temperature process. In this work the cracking of functionally graded Ti6Al4V and TiC was studied to determine processing parameters that optimize the FGMs and hence reduce or remove crackTo form functionally graded grade 5 titanium alloy (Ti6Al4V) and titanium carbide (TiC), appropriate. Laser parameters have to be selected. Laser parameters for FGMs are not standard as most FGMs are still being developed and processing parameters depend on the material being produced. LMD processing parameters are grouped into three categories: laser, process and material related parameters [8]. It is important to know and understand the processing parameters as they influence the resulting properties (hardness, microstructure, density, strength, etc.) and surface integrity of the LMD manufactured part.

Laser spot size is the diameter of the laser beam. This has an effect on surface finish, accuracy of the completed part and delivered heat input. The smaller the beam diameter, the higher the laser energy density available to melt the powder or wire material. The available laser energy is then concentrated on a smaller area. This has been reported to produce better surface finish. Laser scanning speed is the rate at which the laser interacts with the material. The scanning speed can be achieved in two ways: by moving the laser relative to the material or (2) the laser remains stationary while the material moves meaning the base plate on which the components is placed moves [8] the laser scanning speed is inversely proportional to LED. This implies that the slower the scanning speed the larger the LED and vice versa. The scanning speed can have significant effect on the material usage efficiency, microstructure and properties of the fabricated part. Laser power is a function of the laser source machine. LED is directly proportional to laser power, meaning the greater the power the higher the LED. *LED* is the most important factor among laser parameters as it determines the amount of energy that is available per unit area to fabricate the material. The interaction time, which is the rate at which the heat is delivered, also determines the amount of energy delivered per unit area. From equation (1), LED is dependent on laser power, scanning speed and laser spot diameter. The energy calculate determines whether the powder will be fully or partially melted or not melted at all. Process parameters are related to powder and shield gas flow rate. Shield gases are used to protect the powder from environmental attack (molten titanium is highly reactive and if the melt pool is not shielded by an inert gas severe oxidation will take place). The powder flow rate also has a major influence on the properties of the part being fabricated. If the powder flow is too high, material is delivered which cannot be fully melted as the energy might not be sufficient to melt all the material. This then leads to poor coagulation and hence poor mechanical properties and surface finish. And if the powder flow rate is too low, excessive melting of the substrate may occur which may cause dilution of the deposited material. This results in higher dissolution and hence unintended properties [8]. These factors need to be carefully determined prior to deposition to achieve desired mechanical performance and hence reduce scrap. The shape and size of powder particles affects the rate at which the laser energy will be absorbed or reflected. The smaller the powder particles, the better the density of the final part produced [8]. The substrate surface condition also has a considerable influence on the properties of the resulting part. Too smooth surface increases the reflectivity of the material surface causing significant energy loses [8]. Therefore, smooth substrate surfaces should be avoided during laser additive manufacturing process. With regards to titanium, the substrate condition is of relatively little importance as titanium has a relatively high absorption. Moreover, after the laser additive manufacturing processes, some parts may require secondary or finishing operations depending on the service requirement of the part. The finishing can either be completed by traditional or non-traditional manufacturing processes.

The aim of this work was to determine the optimum parameters for the LMD of Ti6Al4V with TiC. The parameters which must be controlled include the laser power, laser scanning speed, beam diameter and the powder flow rate. Grade 5 titanium alloy (Ti6Al4V) plate was used as substrate for this investigation. The plate dimensions were $102 \times 102 \times 7$ mm. Ti6Al4V and titanium carbide (TiC) powder were then used to produce the FGM. Ti6Al4V powder was supplied by TLS Technik Gmbh & Co. and the particle sizes ranged between 45 – 100 μm. The shape of these particles appears to be spherical as it is gas atomized. TiC powder was supplied by Weartech (Pty) Ltd and had a particle size of -100/+40 microns. TiC appears to be irregular as it is ball milled powder.

IOP Conf. Series: Materials Science and Engineering **413** (2018) 012070 doi:10.1088/1757-899X/413/1/012070

2. Material and method

LMD process was conducted at the National Laser Centre (NLC) at Council for Scientific & Industrial Research (CSIR) using a Kuka robot with a Nd: YAG laser. The laser had maximum power of 3 kW. The angle of incidence of the laser used was 12°. The type of nozzle used was co-axial nozzle. The carrier gas and shield gas employed was Argon. Fig. 1 illustrates the equipment used and Figure 2 illustrates the laser deposition process setup.

(a) Kuka Robot (KR 30) with laser head

(b) Kuka SmartPad (controller) (c) GTV powder feeder (Hopper)

Fig. 1 Kuka Robot (KR 30) with laser head, Kuka SmartPad (controller) and GTV powder feeder (Hopper)

2.1 Experimental procedure

Table 1 presents the initial laser parameters that were used for the specimen preparation. These parameters were selected on the basis of the percentage volume of TiC that was deposited. The TiC composition was varied from 5% to 20% in increments of 5 percentage points. Four tracks were prepared with the laser power remaining fixed at 1.2 kW, the gas flow rate at 2 l/min and the beam diameter at 4 mm. An overlap of 50% was applied and minor variations allowed in scanning speed and powder flow rate.

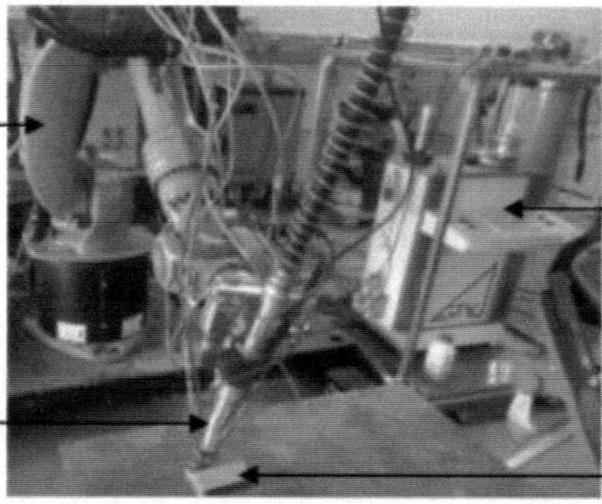

Fig.2 LMD process setup

Table 1 Initial laser parameters

Vol. % TiC	Laser Power (kW)	Scanning Speed (m/min)	Powder Flow Rate (X) (g/min)	Gas Flow Rate (l/min)	Beam diameter (mm)	Overlap percentage (%)	Number of tracks
5	1.2	0.552	4.6	2.00	4	50	4
10	1.2	0.45	3.34	2.00	4	50	4
15	1.2	0.45	3.34	2.00	4	50	4
20	1.2	0.432	3.34	2.00	4	50	4

Parameters from Table 1 were used to determine powder flow rate for both Ti6Al4V and TiC. To determine the powder flow for Ti6AlAV the following equation was used:

$$\text{Powder flow rate Ti6Al4V (Y)} = X \times \text{vol.\% of Ti6Al4V} \quad (1)$$

To determine the powder flow rate of TiC the flowing equation was used:

$$\text{Powder flow rate TiC} = X - Y \quad (2)$$

at machine independent values)

Table 2 and table 3 present the powder flow rate for 4.6 g/min and 3.34 g/min which were calculated using equation 2 and 3 (RPM units are machine specific and proprietary. Calibration work is currently in progress or arrives at machine independent values)

Table 2: Powder flow rate for Ti6Al4V and TiC at 4.6 g/min

Constant parameters	Ti6Al4V (%)	Powder flow rate for Ti6Al4V (Y = Ti6Al4V /100 * 4.6 g/min)	TiC (%)	Powder flow rate for TiC (4.6 g/min - Y)
Laser Power: 1.2 kW	100	-	-	-
Scanning Speed: 0.552 m/min	95	4.37	5	0.23
Powder flow rate: 4.6 g/min	90	4.14	10	0.46
Spot size: 4 mm	85	3.91	15	0.69

Percentage overlap: 50%	80	3.68	20	0.92
Number of tracks: 4	75	3.45	25	1.15
Track length: 90 mm	70	3.22	30	1.38
Gas flow rate: 1.5 RPM	65	2.99	35	1.61
Shielding and carrier gas: Argon	60	2.76	40	1.84

Table 3: Powder flow rate for Ti6Al4V and TiC at 3.34g/min

Constant parameters	Ti6Al4V (%)	Powder flow rate for Ti6Al4V (Y= Ti6Al4V /100 * 3.34 g/min)	TiC (%)	Powder flow rate for TiC (3.34 g/min - Y)
Laser Power: 1.2 kW	100	-	-	-
Scanning Speed: 0.45 m/min	95	3.17	5	0.17
Powder flow rate: 3.34 g/min	90	3.00	10	0.33
Spot size: 4 mm	85	2.83	15	0.50
Percentage overlap: 50%	80	2.68	20	0.67
Number of tracks: 4	75	2.51	25	0.84
Track length: 90 mm	70	2.34	30	1.00
Gas flow rate: 1.5 RPM	65	2.17	35	1.17
Shielding and carrier gas: Argon	60	2.00	40	1.34

When powder was deposited at a scanning speed of 0.552 m/min with a laser power of 1.2 kW, it was found that cracks were forming when the 3rd layer (90% Ti6Al4V and 10% TiC) was deposited. Thereafter, when powder was deposited at a scanning speed of 0.45 m/min at a power of 1.2 kW it was found that cracks were forming when the 4th layer (85% Ti6Al4V and 15% TiC) was deposited. The deposition was repeated three more times by keeping the scanning speed constant at 0.45 m/min and the power was reduced to 1 kW and further to 800 W and 600 W using the same parameters. Table 4 presents the results of all the trial parameters used for the deposition.

Table 4: Parameterization of functionally graded Ti6Al4V with TiC

Trial	1	2	3	4	5	6	7	8	9	10	11	12	13
Laser power (kW)	1.2	1.2	1	0.8	0.6	1.5	1.5	1.2	1.5	1.5	1.2	1.5	2.5
Scanning speed (m/min)	0.55	0.45	0.45	0.45	0.45	3	3	1.5	1.5	3	1.5	2.5	3
Laser spot size (mm)	4	4	4	4	4	4	4	4	4	2	2	2	2
Overlap percentage (%)	50	50	50	50	50	50	50	50	50	50	50	50	50
Number of tracks	4	4	4	4	4	4	4	4	4	4	4	4	8
Gas flow rate (l/min)	1.5	1.5	1.5	1.5	1.5	1.5	1.5	1.5	1.5	1.5	1.5	1.5	1.5
Powder flow rate (g/min)	3.34	3.34	3.34	3.34	3.34	3.34	6.68	6.68	6.68	6.68	6.68	6.68	6.68
Cracking	Yes	Yes	Yes	Yes	Yes	No	No	Yes	No	No	No	No	No

From the trials it was found that as the laser power and scanning speed were increased, cracking was minimal. Cracking occurred when the laser power and scanning speed were low due to re-melting of the powder as it is deposited on the substrate.

3. Results and discussion

The energy density was calculated as per equation 1. Table 5 shows the energy densities of the 13 trials conducted.

Table 5 Energy densities of the 13 trials

Trial	1	2	3	4	5	6	7	8	9	10	11	12	13
Energy density (MJ/m^2)	20.76	25.5	21.22	16.98	12.7	4.77	4.77	7.64	9.55	9.55	15.28	11.46	15.92
Laser power (P) (kW)	1.2	1.2	1	0.8	0.6	1.5	1.5	1.2	1.5	1.5	1.2	1.5	2.5
Scanning speed (V) (m/min)	0.552	0.45	0.45	0.45	0.45	3	3	1.5	1.5	3	1.5	2.5	3
P/V	2.18	2.67	2.22	1.78	1.33	0.5	0.5	0.8	1	0.5	0.8	0.6	0.83
Cracking	Yes	Yes	Yes	Yes	Yes	No	No	Yes	No	No	No	No	No

From Table 5 the ratio of laser power and scanning speed was calculated. It can be seen that when P/V > 1 cracking does occur but when P/V ≤ 1 then cracking does not occur. Fig (a) and (b) show plots of the energy densities for the trials where cracking was visible and where cracking was not visible respectively. From Figure 3a it can be seen that the maximum energy density is 25.46 MJ/m^2 and minimum is 7.64 MJ/m^2. A higher energy density is obtained when the power is relatively high (1.2 kW) and when the scanning speed is relatively low (0.45 m/min) as the power is inversely proportional to scanning speed. From Figure 3b it can be seen that the maximum energy density is 15.92 MJ/m^2 and minimum is 4.77 MJ/m^2. It can be concluded that if the energy density lies in the range of 9 – 16 MJ/m^2 cracking will not occur provided that the laser spot size is 2 mm when the energy density is greater than 12 MJ/m^2.

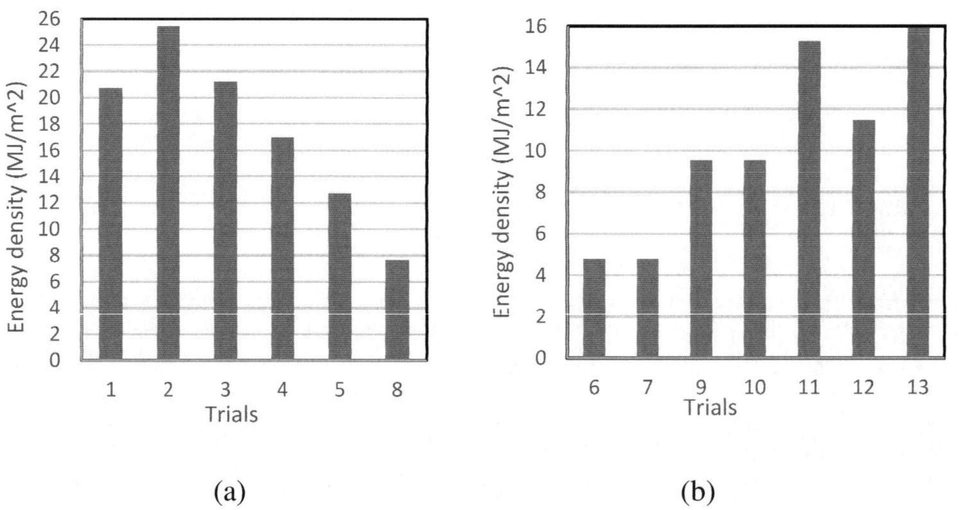

Fig. 3 (a) Energy densities of the cracked samples and (b) Energy densities of the samples that did not crack

IOP Conf. Series: Materials Science and Engineering **413** (2018) 012070 doi:10.1088/1757-899X/413/1/012070

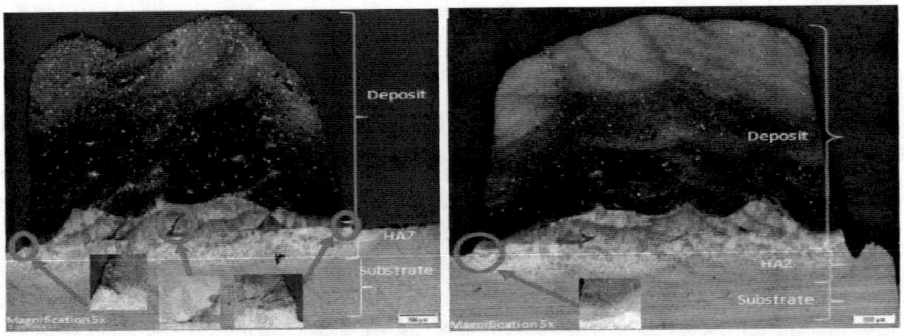

(a) **(b)**

In Fig. 4(a) this case, cracking occurs in the edge and bottom centre of the deposit indicated by red circle. In figure 4(b) cracking was visible at the base of the deposit similar to Fig. 4(a) mainly at the top and middle sections**.**

To visualize the microstructure of the specimen an Olympus DP25 optical microscope was used. Microstructural analysis was conducted on trials 1, 2, 11, 12 and 13. The micrographs of the respective trials are illustrated in Figure 4. The micrographs have been fully described in the figure to get a better understanding on the behaviour of Ti6Al4V and TiC when deposited to form an FGM.

For this deposit, no cracks were visible or no initiation of cracks was seen. This is due to lack of fusion. It can also be seen that when a single layer is deposited the deposit appears to have discontinuities and lumps

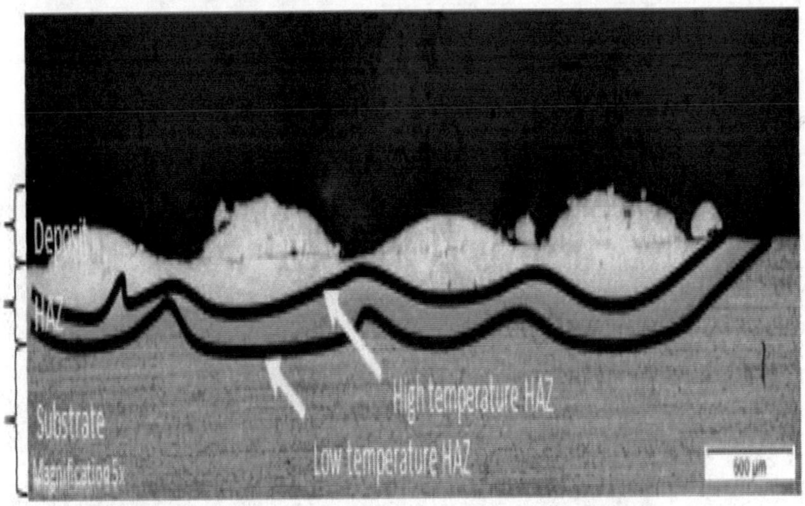

(c)

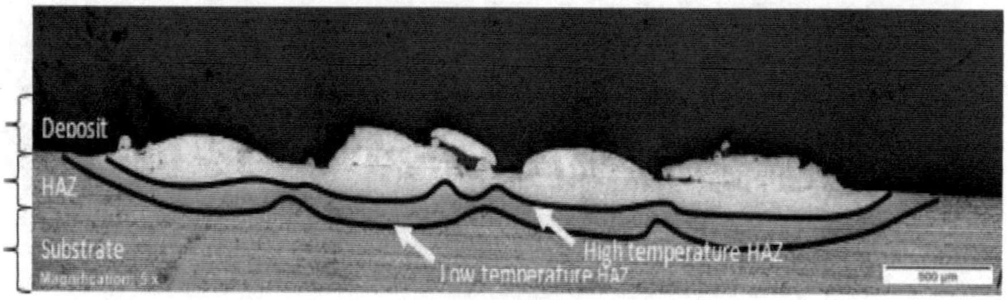

(d)

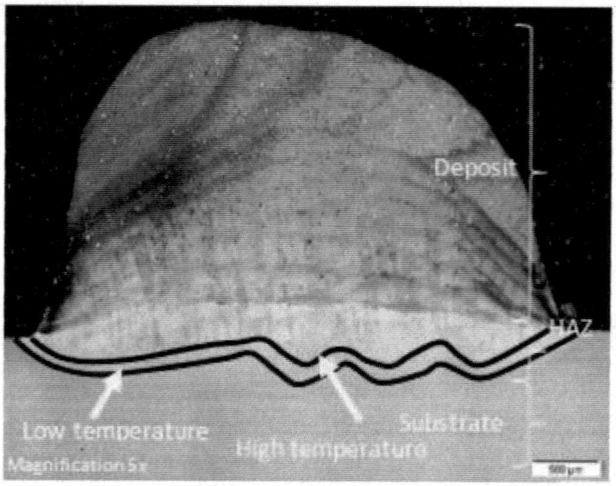

(e)

No cracks were observed for this trial. Once again, the deposit appears to be discontinuous like trial 11.

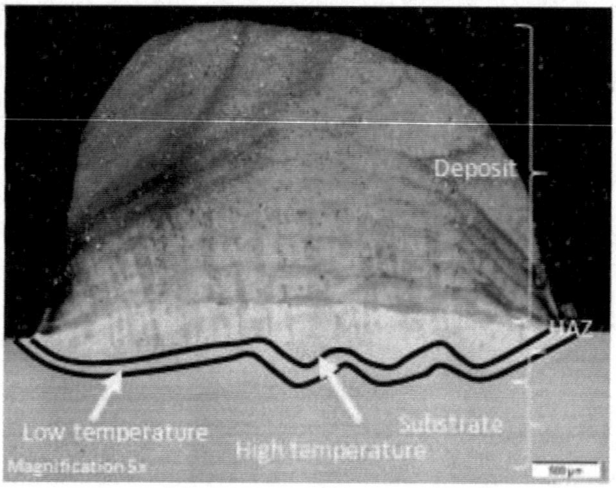

(e)

No cracks were visible on the base of the deposit or on the entire deposit
Fig. 4 Micrographs of various trials: a) Trial 1, b) Trial 2, c) Trial 11, d) Trial 12 and e) Trial 13

The microstructures of all the trials had a similar characterization. The heat affected zone appeared to have a 2 phases structure with large grains. The grains appeared to be needle like structures. These structures were visible throughout the deposit. However, when the TiC was fused with Ti6Al4V, dendritic structures were observed. On the top half of the deposit where a greater vol. % of TiC was deposited, re-solidified TiC was observed as a result of TiC not melting.

4. Conclusion

In this work, FGM was produced by additive deposition of grade 5 titanium alloy power with titanium carbide powder on a grade 5 titanium alloy substrate. Titanium carbide fraction was varied from 5 to 20 vol .% in increments of 5 percentage points. In addition other parameters such as laser power, laser spot size and power flow rate were also varied. The processing parameters were reduced to laser density in characterising the process. Analysis of the resulting materials using microstructural examination led to the following conclusions:

1. If the laser density is in the range 15 – 25 MJ/m^2 produced specimens cracked for laser spot size of 4 mm
2. If the laser energy density is in the range 9 – 16 MJ/m^2, no cracking occurs as long as the laser spot size is reduced to 2 mm for LED greater than 12 MJ/m^2
3. If the laser power to scanning speed (P/V) ratio is less than or equal to 1 then cracking will not occur
4. Cracking is mainly concentrated in the interface region between the substrate and the FGM which suggests a discontinuity in thermal behavior of the FGM and substrate

However, the process parameters reported as optimum may be influenced by other external factors such as ambient conditions which were not monitored in this work. It is suggested that future work be conducted with full characterisation of environmental conditions together with more detailed analysis of changes in mechanical and thermal properties of the materials.

Acknowledgements

The authors would like to thank Prof. Sisa Pityana for his assistance with preparing the samples at the National Laser Centre, CSIR.

Reference

[1] Y. Bilotsky en M. Gasik, Local fields in functionally graded materials, *Functionally Graded Materials 1996,* vol. 1, pp. 21-28, 1996.

[2] A. E. Adoba, H. B. Mcshanes en R. D. Rawlings,The structure and properties of titanium matrix composites reinforced with bioactive particles, *Seventh European Conference on Composite Materials: realising their commercial pontential,* vol. 7, pp. 461-465, 1996.

[3] Y. Miyamoto, M. Niino en M. Koizumi,FGM reserch programs in Japan from the structual to fuctional uses, *Functionally Graded Materials 1996,* vol. 1, pp. 1-8, 1996.

[4] R. Mahamood, E. T. Akinlabi, M. Shukla en S. Pityana,Functionally Graded Material: An overview, in *International Conference in Mechanical Engineering, (WCE2012)*, London, UK, 3-5 July 2012.

[5] C. Ferguson,Historical Introduction to the Development of Material Science and Engineering as a teaching discipline, UK center of Materials education, London, 2012.

[6] Y. Li, P. Bai, Y. Wang, J. Hu en Z. Guo,Effect of TiC content on Ni/TiC composites by direct laser fabrication, *Materials and Design,* vol. 30, pp. 1409-1412, 2009.

[7] F. Wang, J. Mei en X. Wu,Compositionally graded Ti 6Al 4V + TiC made by directed laser fabrication using powder and wire, *Materials and Design,* vol. 28, pp. 2040-2046, 2006.

[8] R. Mahamood, E. Akinlab, M. Shukla en S. Pityana, *Laser Additive Manufacturing,* Lecture notes: University of Johannesburg, 2013.

Thermal Investigation of an Indirectly Heated Rotary Kiln

T. F. Mokoena[1], M.D Madyira[1], T.O Babarinde[1,3], S. A Akinlabi[1]

[1]Department of Mechanical Engineering Science, University of Johannesburg, South Africa
[2]Department of Mechanical & Industrial Engineering Technology, University of Johannesburg, South Africa
[3]Mechanical Engineering Department, Covenant University Ota, Nigeria
E-mail address: taiwo.babarinde79@gmail.com, stephenakinlabi@gmail.com, dmadyira@uj.ac.

Abstract. Rotary kilns are very important to the construction industry and many others. These machines are very inefficient, and this is a big concern since they rely on fossil fuels to operate. The aim of this research is to use the power of computational fluid dynamics to save costs and time to optimize the indirectly heated rotary kiln. The research will mainly focus on the thermal analysis of the rotary kiln. The significance of this research mainly lies in the ever-growing need to preserve natural resources and in this case, it is by optimizing highly inefficient machinery such rotary kilns. The numerical model study was conducted using steady state conditions and was also validated. It was found that the numerical model results deviated from experimental results by 10.63%. The optimization process was conducted where the mass flow rate of the air was varied and while the fuel flow rate was kept constant. The reverse was done where the mass flow rate of fuel was varied while the flow rate of air was kept constant. It was concluded that there exists an inverse relationship between the increase of the flow rate of air and the temperature distribution and there exists a directly proportional relationship between the increase of fuel flow rate and the temperature distribution.
Keywords: Rotary Kiln, Optimization, CFD

1. Introduction

Fossil fuels play a very significant role in modern society and continue to do so because of their versatile nature. These precious commodities prove to be the most convenient way of transforming one form of energy to another and more specifically, chemical energy to thermal energy. The need for thermal energy has grown significantly due to power generation and thermal processing to name a few. Thermal processing is defined as the application of heat to bring about a certain chemical and physical change to material [1]. There are numerous ways in which thermal processing can be executed and a rotary kiln is one of them. Rotary kilns are long rotating drums that are used for thermal processing of certain materials over a specified amount of time [2]. These machines use fossil fuels such as natural gas as their source of fuel for thermal processing. They do however come with big challenges such as dust emissions and high thermal inefficiency [3]. It is therefore, important that the process in rotary kilns is carefully studied and from any results obtained, recommendations can be made to future designs to make them more efficient and ultimately use natural resources such as fossils fuels sparingly. Most conventional rotary kilns use fossil fuels as a source of energy to carry out their processes. However, due to their high poor thermal efficiency and their reliance on the price of fossil fuels, there is a need to make them more efficient so that more work can be done with less energy supplied to the kiln.

A rotary kiln is a device that is used in pyro processing to raise temperatures of material in a continuous process [4]. Due to the development and evolution of rotary kilns, they have had synonyms such as cement kilns and lime kilns [4]. Rotary kilns are devices that are used to process materials at high temperatures to create a chemical change or physical change to the heated material [5]. These days, rotary kilns have acquired various applications such as reduction of oxide ore, reclamation of hydrated lime, calcining of petroleum coke, hazardous waste reclamation, and others. Rotary kilns can operate at temperatures above 2000K with the aid of natural gas as a fuel source [1]. Rotary kilns are slightly tilted and at an angle and rotate at a controlled rate by a motor. They are tilted at an angle so that they material can be extracted with

IOP Conf. Series: Materials Science and Engineering **413** (2018) 012071 doi:10.1088/1757-899X/413/1/012071

the aid of gravity and the kiln is rotated so that the material which is being processed can be consistent in terms uniformity. An industrial rotary kiln usually has a diameter of 2-2.5m and a length of 50-80m [3]. These machines, however, come with a number of problems. They have poor thermal performance; they create a lot of dust and have a certain amount of inconsistency in the uniformity of the products that they produce [4]. The issue of inconsistency in the uniformity of the products that are processed has influenced the design of the interior of the kilns and the speed at which the kiln is rotating which will be further discussed in the bed phenomenon of the kiln. There are many types of kilns in the rotary kiln practice namely: long dry kilns, short dry kilns and indirectly fired rotary kilns. The long dry kiln shares a common trait with the wet kiln and that is the fact that the preheating, drying and calcination all occurs in one vessel. They normally operate well when the feed particles are large. Applications include lime kilns and lightweight aggregate kilns where the mined stones are crushed to about 1.3–5cm before feeding them into the kiln [4]. Short dry kilns are usually accompanied by an external preheater or pre-calciner in which the feed is dried, preheated, or even partially calcined prior to entering the main reactor. Short dry kilns normally have feed material which is almost calcined.

The applications are mostly in cement and lime processing [7]. Indirectly heated rotary kilns take a different approach to the way in which they are heated compared to the directly heated kilns. These types of kilns are designed this way for the case where direct contact between the material of the gas or heat source and the material which is being processed is undesirable. These kilns are relatively small due to their low efficiency and are used for applications of calcinating specialty materials. Indirectly fired rotary kilns have multiple and compartmentalized temperature-controlled zones which can be electrically or gas heated separately [4]. Sass [8] developed a computer a modelling using empirical relationship for heat transfer radiation in a rotary kiln. Ghoshdastidar et al. [9] also carried out simulation results on an improved heat transfer model for a rotary kiln that is being used for preheating and drying of wet iron ore. Schmidt and Nikrityuk [10] developed a 2D numerical simulation model of a transient temperature distribution by granular mixing in a horizontal rotary kiln. Cundy [11] investigated an analytical model between the solid in a rotary kiln and heated wall of a rotary cylinder with initial moisture content of 2-7% of mass of the dry the solids. . Ghoshdastidar and Unni [12] also developed a simulation model based on steady state heat transfer of drying and pre-of moisture solids in the non-reacting zone of a cement rotary kiln. In another study, Ghoshdastidar and Agarwal [13] carried out a numerical simulation study of heat transfer in a rotary kiln that was being used for drying and preheating of wood chips with superheated steam. Due to growing need for solid natural resources there is need for energy management and energy efficiency system. This research presents a steady state thermal analysis of a rotary kiln and develops a numerical model for optimization of the thermal performance of the rotary kiln system.

The aim of the project is to develop a numerical model that will characterize the thermal performance of an indirectly heated rotary kiln with acceptable accuracy using computational fluid dynamics and use the results from the numerical model to optimize the rotary kiln.

2. Results and discussion

The experimental results obtained from the data logger are from the various locations of the thermocouples. The results are summarized in graphical form in Figure 1.

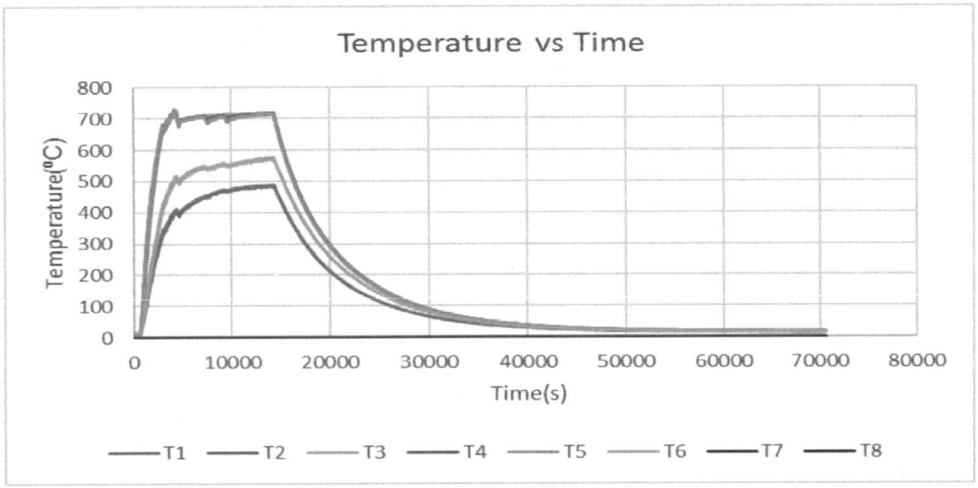

Fig.1 Experimental result of variation of temperature of the kiln with time

The graph is plotted with temperature as a function of time in seconds. The experiment was run over a course of 1 day. Since the temperatures of the different thermocouples are a function of time, it can be concluded that the results are transient. Initially, the kiln was at a temperature of 11^{0}C due to the surrounding environment. When the burner is switched on, the temperature measured by the different thermocouples rises sharply due to high temperature difference. As the time progresses to approximately 3000 seconds, the temperature at each location starts to stabilize. This is due to the fact that the burner has reached the maximum amount of power it can give out and therefore, it can be assumed that the kiln was at steady state at this point. After 15000 seconds, the power supply from the burners was cut off and the kiln was allowed to cool down. The CFD model was run under steady state conditions and therefore the most crucial results were at the time where the temperatures do not vary significantly with time. These temperatures, along with their locations on the kiln are important and were used to validate the model. The highest temperature observed at each location was be used. These results are summarized in Table 1.

Table 1: Temperatures at different locations

Location (mm)	355	850	1255	1760	2700
Temperature(^{0}C)	478.8	705.9	711.4	714.7	560

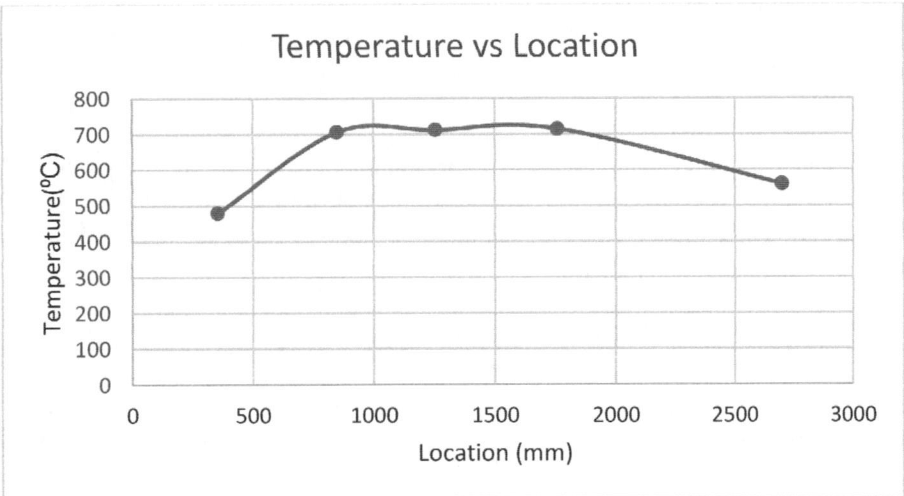

Fig 2 Experimental results at steady state conditions based on location

Computational fluid dynamics is a very complex tool which uses numerical analysis to produce results. This tool does this by creating a mesh which is applying cells to the part which is to be studied. The CFD software analyses the interaction between the cells based on the study which is being performed. The size of the cells can be easily changed to any desirable size. This in return changes the number of iterations done by the software based on the size of the mesh. This means that the smaller the cell sizes, the more computational effort required, and more iterations done by the computer. The inverse of this happens as the mesh size is increased. This however, has a very large impact on the accuracy of the results from the computations. The mesh sensitivity for the model was conducted by measuring the temperature at the exit point of the drum without changing any other conditions but the mesh size.

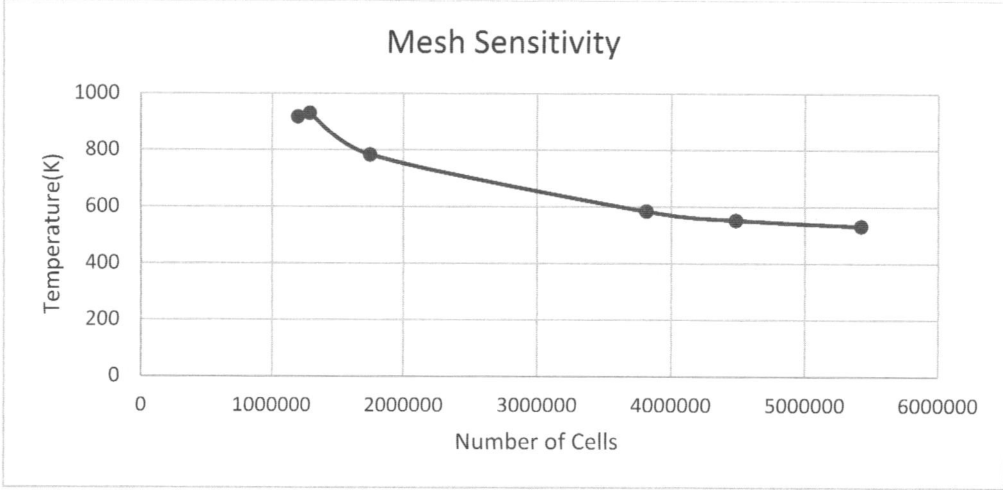

Fig. 3 Mesh sensitivity study

Table 2: The summary of the model boundary

Parameter	Value	Unit
Outside Walls Temperature	11	^0C
Ambient air convection coefficient	10	W/m^2. K
Mass flow rate of air	0.01	kg/s
Mass flow rate of fuel	4.11×10^{-4}	kg/s
Pressure	82.5	kPa
Initial air and fuel temperature	25	^0C

2.1 Model Validation

After the mesh sensitivity study was conducted, the boundary conditions were applied to the model and to see whether the model results correlate with the experimental results. The model results at the locations of interest are summarized in

Table 3. The results were further plotted in Fig and compared to those of the experimental results.

Table 3 Numerical model results and experimental results comparison

Location (mm)	355	850	1255	1760	2700
Temperature(^0C) (Experimental)	478.8	705.9	711.4	714.7	560
Temperature(^0C) Model	391.94	672	705	698	409
Error (%)	18.1	4.8	0.899	2.4	27
Average error (%)			10.63		

IOP Conf. Series: Materials Science and Engineering **413** (2018) 012071 doi:10.1088/1757-899X/413/1/012071

From the comparison at the different locations, it can be concluded that the average difference between the model results and the experimental results is 10.63%. The validation is illustrated in a graph.

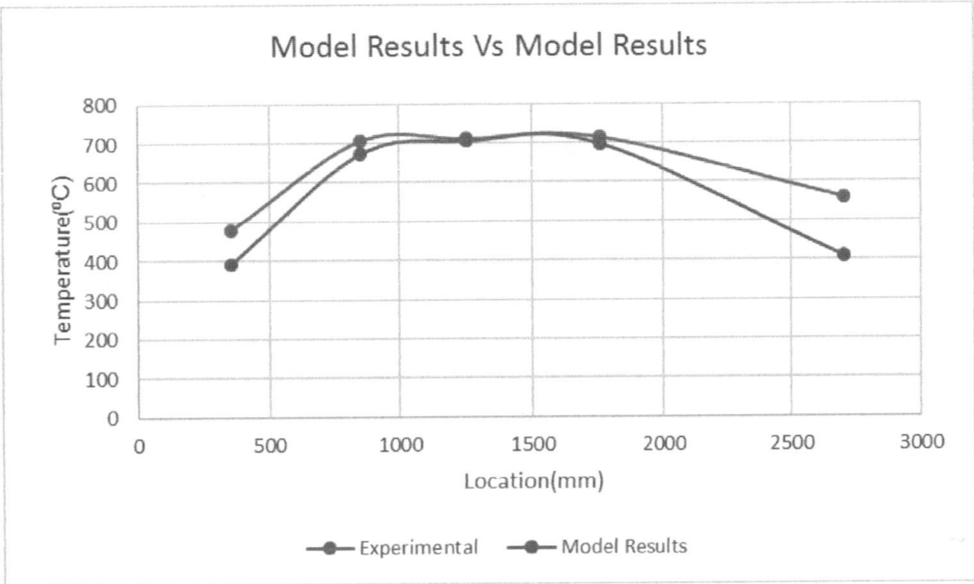

Fig. 4 Comparison of experimental result with model result

The flow field given by Figure 5 illustrates the flame and temperature distribution of the kiln. it can be observed that some of the thermal energy produced by the combustion of proprane was used to heat up the insulation of the kiln. Though most of the energy was used to heat up the drum and the material in the drum.

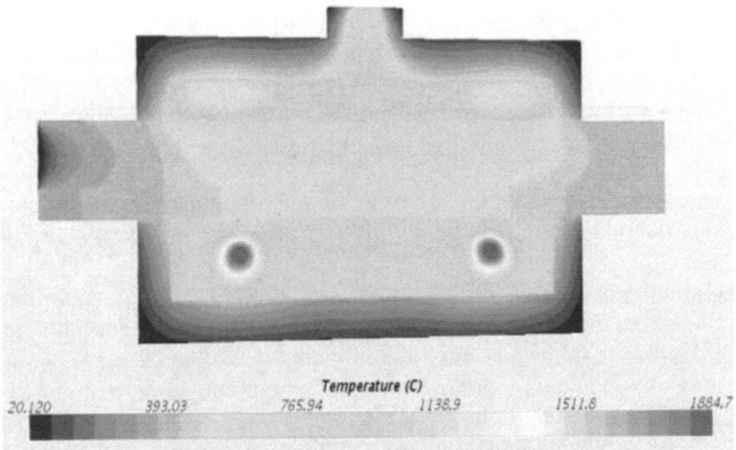

Fig. 5 Temperature distribution in the drum

IOP Conf. Series: Materials Science and Engineering **413** (2018) 012071 doi:10.1088/1757-899X/413/1/012071

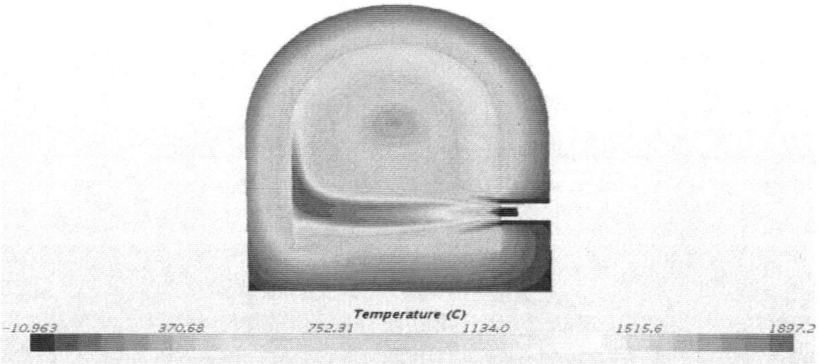

Fig. 6 Flame distribution from experimental boundary conditions

2.2 Model Optimization

The main aim of the experimental program was to analyze the changes that take place when the mass flowrate of the fuel is varied while the mass flowrate of the air is kept constant. The second part of the experimental program was to analyze the changes that take place when the mass flow rate of the air is varied while the mass flowrate of the fuel is kept constant. This was done on the model since it is very close proximity to the experimental results and thus the same would be observed if it was to be conducted in an experiment.

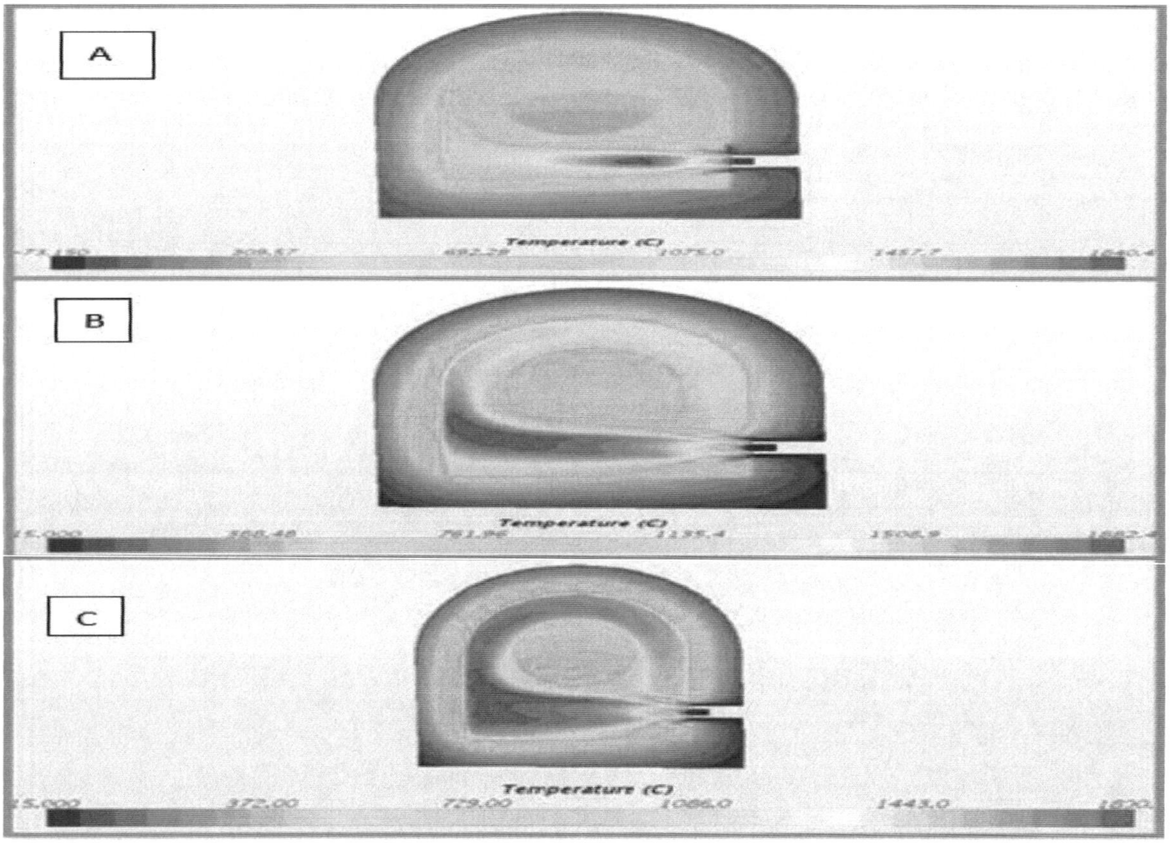

Fig. 7 Temperature distributions for fuel mass flow rates of (A) 3x10^-4kg/s, (B) 5x10^-4kg/s, (C)6.5x10^-6kg/s

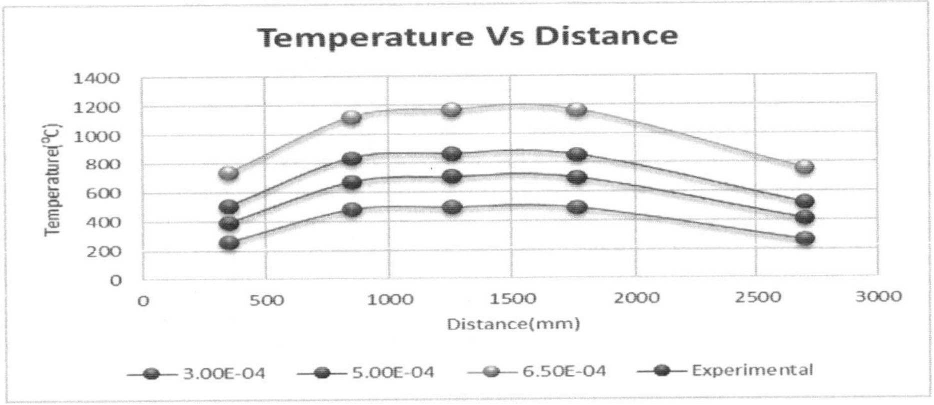

Fig. 8 Temperature distribution graphs along the kiln for fuel optimization

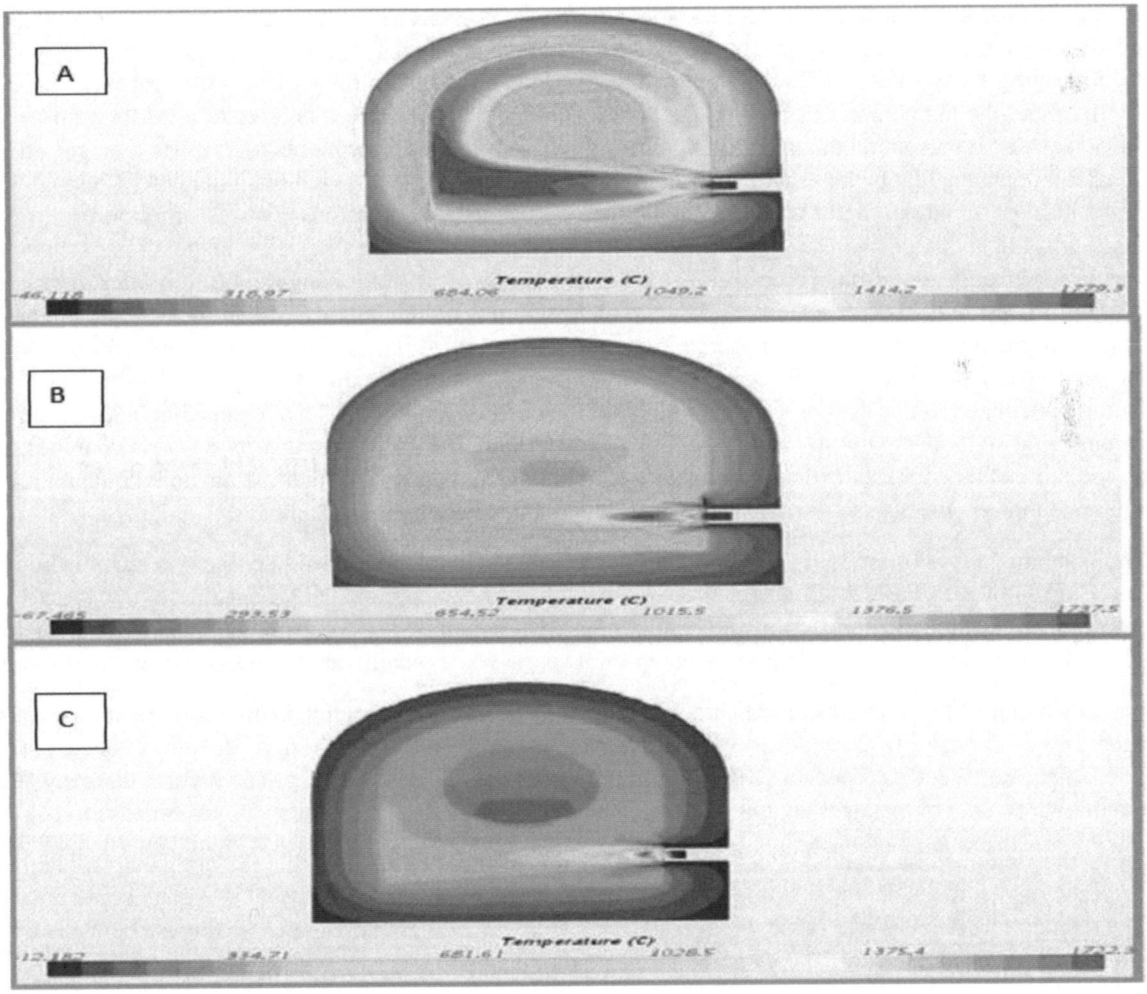

Fig. 9 Temperature distribution for air mass flow rate (A) 0.0075kg/s, (B) 0.03kg/s, (C) 0.05kg/s

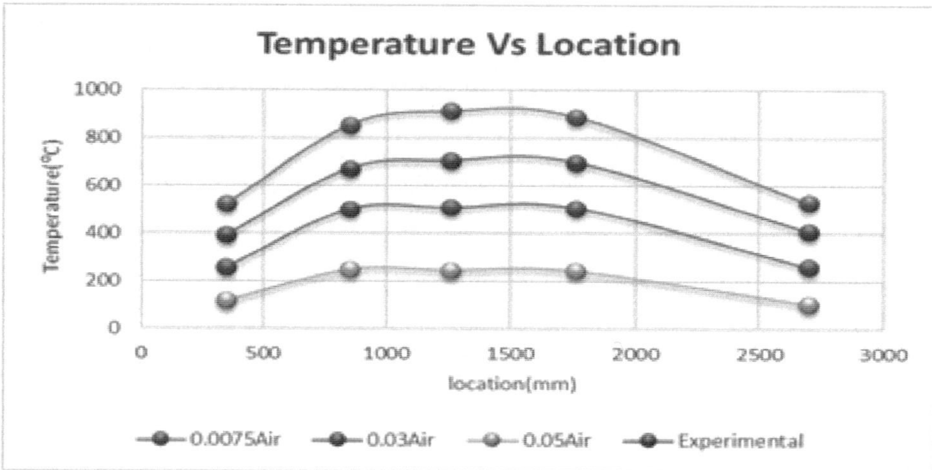

Fig. 10 Temperature distribution along the kiln length for air optimization

After the modelling process was complete, the validation of the model was conducted to see how close of an approximation the model is to the experimental results. The model was assumed to have been done at steady state conditions. This means that the validation of the model had to be done on using the steady state points from the experimental results. The average of 300 points from the experimental data was used to approximate the temperature at each of the five locations. These temperatures were used to validate the model results. It was found that the model results deviate from the experimental results by a magnitude of 10.63%. There are few factors that constituted to the error that was observed from the model results. The first and most important factor was the assumption of steady state conditions. This assumption means that the results are not extensively analysed as the kiln is in the process of heating up. The next factor that took part in the error of the results is the simplification of the model by not having the drum rotates during the heating process. This simplification meant that there will not be full mixing of the air during the heating process in the drum. The last factor that constituted to the error of the model was the choice of modelling the sand inside the drum as a solid component, even though it had the material properties of the sand. This factor contributed to the error at a larger scale than the other assumptions. This means that in the locations where the sand model is not in the section of the drum where the drum is in the combustion chamber, the temperature of those places will only rise due to the conductive nature of both the drum and the sand model. This was a problem because the material was not modelled as a flowing material.

The insulating layer of the kiln shell was also modelled. This was important because this part of the kiln has a big impact in the overall performance of the kiln. This is because any heat that is not lost to the environment due to both convection and conduction, will be supplied to the drum and thus making the kiln more efficient. However, the kiln insulation was not studied extensively during the experimental program.

From the optimization results, it can be seen that the overall temperature in the combustion chamber decreases as the mass flow rate of the air is increased while the mass flow rate of the air is kept constant. The temperature decreases because the overall flame from the burner decreases. The reverse happens when the mass flow rate of the air is kept at a constant while the mass flow rate of the fuel is increases.

From the optimization results, it can be seen that the overall temperature in the combustion chamber decreases as the mass flow rate of the air is increased while the mass flow rate of the air is kept constant. The temperature decreases because the overall flame from the burner decreases. The reverse happens when the mass flow rate of the air is kept at a constant while the mass flow rate of the fuel is increases.

3. Conclusion

The numerical model was found to have deviated from the experimental results by a magnitude of 10.63%. The reason for the error was discussed in the discussion section of the report. The main reason for developing a numerical model was to save costs by running multiple tests on the numerical model without the cost of using fuel and raw materials. Since this numerical is within the acceptable range of similarity to the actual rotary kiln, this means that it can be used to vary certain parameters and to see what will happen and those occurrences will be observed on the actual rotary kiln. This process is known as the optimization. When the optimization was conducted, the mass flow rate of air was varied while the flow rate of the fuel was kept constant. When the mass flow rate of air was 0.0075kg/s, the highest temperature observed in the drum was 850^{0}C. When the mass flow rate of air was 0.03kg/s, the highest temperature observed in the drum was 450^{0}C. Lastly, when the mass flow air is 0.05kg/s the highest temperature observed in the drum is 220^{0}C. It can be concluded that there is an inverse relationship between the air flow rate and the temperature distribution. When the mass flow rate of fuel is increased, there is a direct relationship with the temperature distribution.

References

[1] Karl-Heinz, M. Roeb, P. Schwarzboezl and H. Warnecke, "Journal of Themal Sience and Thermal Applications," *Aluminium Remelting using solar heated rotary kilns,* vol. 123, no. 1, p. 8, 2001.

[2] G. Handler, "FEECO International," FEECO International, 2017. [Online]. Available: http://feeco.com/rotary-kilns/. [Accessed 25 February 2017].

[3] Z. Zhang and T. Wang, "Journal of Fluids Engineering," *Simulation Of Combustion and Thermal flow inside a petroleum coke rotary kiln,* vol. 2, no. 1, p. 12, 2010.

[4] A. A. Boateng, Roraty kilns-Transport Phenomena and, New York: Butterworth-Heinemann, 2008.

[5] FEECO International, "The rotary kiln hand book," 2015. [Online]. Available: http://feeco.com/rotary-kilns/. [Accessed 25 February 2017].

[6] "SEW Gear Manufacturer," 2017. [Online]. Available: http://www.sewgears.net/rotary-kiln-shell-manufacturer-in-india-new-delhi. [Accessed 25 02 2017].

[7] P. S. Ghoshdastidar, "Journal of Fluids Engineering," *Simulation and optimization of drying of wood and chip with superheated steam in a rotary kiln,* vol. 1, no. 1, p. 6, 2009.

[8] A. Saas, Simulation of the heat transfer phenomenon in a rotary kiln, Ind. Eng.

[9] P.S. Ghoshdastidar, G. Bhargava, R.P. Chhabra, Computer simulation of heat transfer during drying and preheating of wet iron ore in a rotary kiln, Drying Technol. 20 (1) (2002) 19–35

[10] R. Schmidt, P.A. Nikrityuk, Numerical simulation of the transient temperature distribution inside moving particles, Can. J. Chem. Eng. 90 (2012) 246–262.

[11] C.A. Cook, V.A. Cundy, Heat transfer between a rotating and a moist granular bed, Int. J. Heat Mass Transfer 38 (3) (1995) 419–432.

[12] P.S. Ghoshdastidar, V.K. Anandan Unni, Heat transfer in the non-reacting zone of a cement rotary kiln, J. Eng. Ind., Trans. ASME 118 (1996) 169–172.

[13] P.S. Ghoshdastidar, A. Agarwal, Simulation and optimization of drying of wood chips with superheated steam in a rotary kiln, J. Therm. Sci. Eng. Appl., Trans. ASME 1 (2009) 024501

IOP Conf. Series: Materials Science and Engineering **413** (2018) 012072 doi:10.1088/1757-899X/413/1/012072

Thermal Performance and Energy Consumption Analyses of R290 and R22 Refrigerants in Air-conditioning System

Teddy I. Okolo, *Oluseyi O. Ajayi

Department of Mechanical Engineering, Covenant University, P.M.B 1023, Ota, Ogun State, Nigeria
*Correspondence: oluseyi.ajayi@covenantuniversity.edu.ng; +234-8036208899

Abstract. The study investigated the comparative performance of R290 and R22 refrigerants in a domestic air-conditioning system. It used a 2 hp window split air-conditioning system with Capella D oil as lubricant. Performance analysis which include coefficient of performance, cooling capacity, energy consumption and thermal conductivity and viscosity at vapour and liquid phases were determined. The results were employed to carry out the comparative analyses of the efficiency of R290 as a replacement alternative to R22 refrigerant in a domestic air-conditioning unit. The findings show that although R22 performed better in terms of its coefficient of performance and cooling capacity, R290 however showed better results in terms of its energy consumption, thermal conductivity and viscosity at the liquid and vapour phases. Further to this, there is no marked difference in the pull-down time results. The consequence of the results is that R290 can be a suitable alternative air-conditioning refrigerant to R22, the performance notwithstanding. However, the performance can be further improved with a reconfiguration of the thermo-physical properties of the thermodynamic system such as the lubricant. This can be achieved by taking advantage of the ability of metallic and non-metallic nanoparticles to improve the thermal and physical properties of the lubricant and by extension the thermodynamic system. Thus, with improvement in the thermal conductivity and viscosity of the lubricant, there can be further improvement in the performance of R290 as air-conditioning refrigerant. Going by the findings and coupled with the fact that R22 is environmentally malignant, R290 can become a suitable alternative in the long run.
Key **words**: Energy system, Nano-refrigerant, Nano-lubricants, Power consumption, Heat transfer, Nano-technology, Thermal system

1. Introduction

Globally, the issues of sustainable development and environmental protection have brought about the need to rethink and redesign many of the engineering systems that have shaped the world over. This is done in a bid to reduce energy consumption, thereby encouraging the use of more energy efficient appliances and also reduce the incidents of reliance on fossil fuels for power generation. Based on this, several debates and United Nations Conventions have favoured the adoption of renewable energy and environment friendly sources for power generation (Ajayi *et al.*, 2014; Fagbenle *et al.*, 2011; Ohijeagbon & Ajayi, 2014; 2015). They have also favoured the use of energy efficient appliances, redesign of high energy consumption appliances to make them more energy efficient and the creation of opportunity to promote access to modern energy. These suggestions have placed major focus on energy systems and thermal appliances. Thus, while efforts are geared towards redirecting the resources for power generation to embrace renewable resources, those of thermal appliances/systems have been faced with the need to redesign the systems' energy use pattern and also ensure they operate with environment friendly thermodynamic systems. For instance, refrigeration and air-conditioning systems are two thermal systems that place a very high demand on power. It is reported that air-conditioning and refrigeration appliances consumed the largest amount of electricity domestically (Kubota *et* al., 2011). Also, in major office spaces, the combined wattage demand of air-

IOP Conf. Series: Materials Science and Engineering **413** (2018) 012072 doi:10.1088/1757-899X/413/1/012072

condioners and refrigerators exceed those of other appliances. Moreover, cooling systems are a necessity, most especially in tropical climatic countries, and during summer periods. Hence, systems design that favours energy efficiency and employs environmentally benign thermodynamic systems will be a welcome improvement. Consequently, some researches have been undertaken to improve the systems' efficiency and also to reconfigure the thermodynamic system for reduced power consumption. Some of such include the improvement in the efficiency of air-conditioners and heat pumps to above 50%. However, other research results were drawn to focus on the outcome of the Kyoto Protocol where refrigerants that have global warming consequences were targeted to be phased out. Such refrigerants include those of hydrofluorocarbons (HFCs). It is reported that HFCs emissions are increasing by 7% per annum and could reach 8.8 gigatons of CO_2 by the year 2050 if it is not controlled (News, 2018). Similar to the HFCs are the hydrochlorofluorocarbons (HCFCs). For instance, the United States of America has set targets for partial to complete phase out of refrigerants with configurations of HFCs and HCFCs between the years 2020 and 2030. Also, the European Union has since ban the use of R134a (fluoromethane), a refrigerant mostly used in mobile air-conditioners (Bhatkar *et al.*, 2013).

Going by the aforementioned, some research efforts geared towards seeking alternatives to HFCs and HCFCs are ongoing. Based on this, various alternative refrigerants such as hydrocarbons, carbon dioxide, ammonia, artificially created refrigerants and refrigerant mixtures have been proposed (Bhatkar *et al.*, 2013; Choudhury *et al.*, 2013; Pearson, 2005). The drawbacks of some of these refrigerants are their safety issues and potential flammability. The proponent of hydrocarbons more so, have brought about the suggestions that hydrocarbon refrigerants can be used as alternatives to the environment malignant refrigerants. This is because hydrocarbons are majorly environment friendly substances. However, the shortcomings of employing this class of refrigerants have always been due to their low thermodynamic characteristics and sometimes their flammability when compared to the HCFC and HFC counterpart. Moreover, in selecting alternative refrigerants for a system, there is no general rule of thumb. This is due to the fact that, refrigerants have different thermo-physical properties and therefore exhibits different thermal responses or performance characteristics. One basic and accepted method of alternative refrigerant selection is through comparison, to observe and measure the relative performance of the refrigerants.

This study is focused on this. It aimed at comparing the thermodynamic performance of R290 (propane) as an alternative air-conditioning refrigerant to R22 (dichlorodifluoromethane). R22 is an HCFC compound with ozone layer depletion and minimal global warming potentials. Its ozone depletion and global warming potentials are 0.55 and 1700 respectively, and it is non-flammable. Presently, it is being phased out and new compound cannot be produced, especially in the European Union countries. R290 on the other hand, is purely hydrocarbon, it is natural, has zero ozone depletion and negligible global warming potentials, bearable toxicity but flammable with ignition limits between 1.7 and 10.9% vol. in air. A similar research to this study has been carried out by Choudhari and Sapali (2017). The study however considered employing R290 for refrigeration systems and focused majorly on using already developed models and simulation procedures as basis for comparisons. This study moreover pushes this further by experimentally investigating the thermodynamic performance of R290 as compared to R22 in air-conditioning system.

2. Materials and Method

A domestic split air-conditioning unit was used as the test rig. The compressor suction and discharge pressures were evaluated by means of pressure gauges. The refrigerant temperatures were measured by means of thermocouples fixed at the inlet and outlet of the major components of the air-conditioner. Service ports were used at the inlet of the compressor for charging and recovering the refrigerant. The experiment was repeated three different times and the average employed as the expected value of results. The base oil employed was Capella D, while the refrigerants were R22 and R290 respectively. Fig. 1 shows the experimental setup while Table 1 displays the split air-conditioning unit specifications and the material component of the working fluids incorporated. Table 2 displays the properties of the lubricant oil (i.e. Capella D).

Figure 1: Experimental setup (Test-rig)

Table 1: Component specification of the refrigerator system and the working fluids

Material	Specification
Unit	2hp Panasonic Air conditioning unit
Conventional Refrigerant/Lubricant	R22 & R290/Capella D
Compressor	Rotary compressor
Evaporator	Finned heat exchanger
Expansion valve	Capillary tube
Charged mass	300 g

Table 2: Physical properties of the Capella D oil

Property	**Value**
ISO Viscosity Grade	46
Density at 15°C, kg/L	0.9
Flash Point, COC, °C	238
Freon Floc Point, °C	-50
Freon Haze Point, °C	-43
Pour Point, °C	-45
Total Acid Number	0.02
Viscosity at 40°C	42.54
Viscosity at 100°C	6.45
Viscosity Index	100
Breakdown Voltage, kV min	40

Determination of the Coefficient of Performance (COP)

This is a measure of performance of the system. It is the ratio of the heat absorbed from the refrigerated space.

IOP Conf. Series: Materials Science and Engineering **413** (2018) 012072 doi:10.1088/1757-899X/413/1/012072

$$COP = \frac{Desired\ output}{Required\ input} = \frac{Q_L}{W_{net.in}}$$

It can also be represented as:

$$COP = \frac{\dot{m}(h_1 - h_4)}{\dot{m}(h_2 - h_1)} = \frac{(h_1 - h_4)}{(h_2 - h_1)}$$

where: h_1 and h_2 are enthalpies of compressor inlet and outlet, h_4 = enthalpy of evaporator inlet

Determination of the Refrigeration Effect (RE) or Cooling Capacity

Refrigerating effect is the rate of heat removal from the refrigerated space (space to be cooled). In the case of air conditioners, this may be an entire room, office or auditorium. Mathematically, it is the product of the coefficient of performance and compressor work.

$$RE = COP \times W_c$$

where: W_c = compressor work given by the product of mass flow rate and change in the inlet and outlet enthalpies of compressor ($\dot{m}(h_2 - h_1)$).

Determination of the Thermal Conductivity and Viscosity at the Vapour and Liquid Phases

These were evaluated from the combination of experimental values of temperature of refrigerated space, suction and discharge pressures with REFPROP and Genetron Refrigerants Modelling Software

3. Results and Discussion

Fig. 2 displays the comparative variation of the values of the coefficient of performance between R22 and R290 refrigerants in the air-conditioning system. It shows that the system with R22 consistently performed better than that with R290. The degree of performance improvement of R22 over R290 was estimated to be about 44.5%. Further to this, the Figure shows that while the thermal performance of R290 refrigerant was fairly uniform over the period of experiment, that of R22 consistently increased across the period. This invariably suggest that R22 is a better air-conditioning refrigerant than R290.

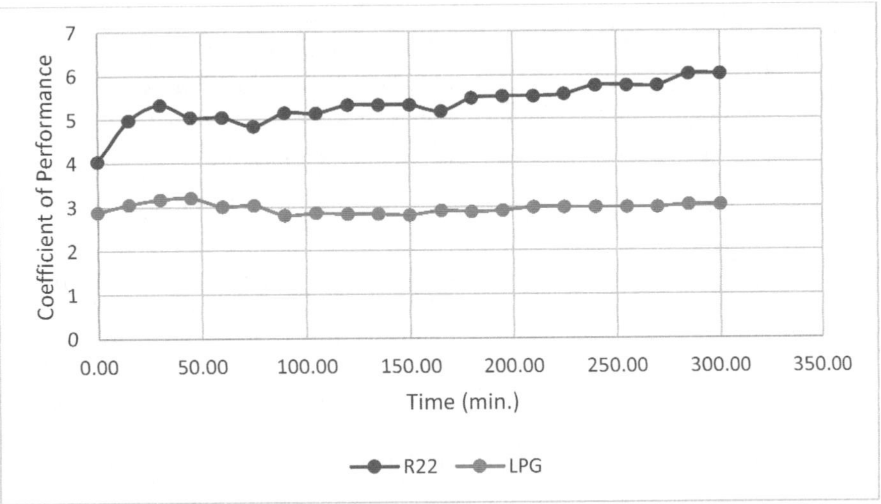

Fig. 2: Variation of the coefficient of performance of the refrigerant variants with time

Similarly, the cooling capacity which depicts the refrigeration effect of the refrigerant on the refrigerated space is displayed in Fig. 3. The result corresponds to the measure of the system's ability to remove heat. It shows that R22 has a higher heat removal capacity than R290. Hence, Fig. 3 reveal that, like the results of COP, R22 demonstrates a better refrigeration effect than R290 across the entire period of experiment. The cooling capacity increased periodically for R22 while it was fairly uniform for R290. The difference in their cooling capacities demonstrates that R290 was 48% lower than R22.

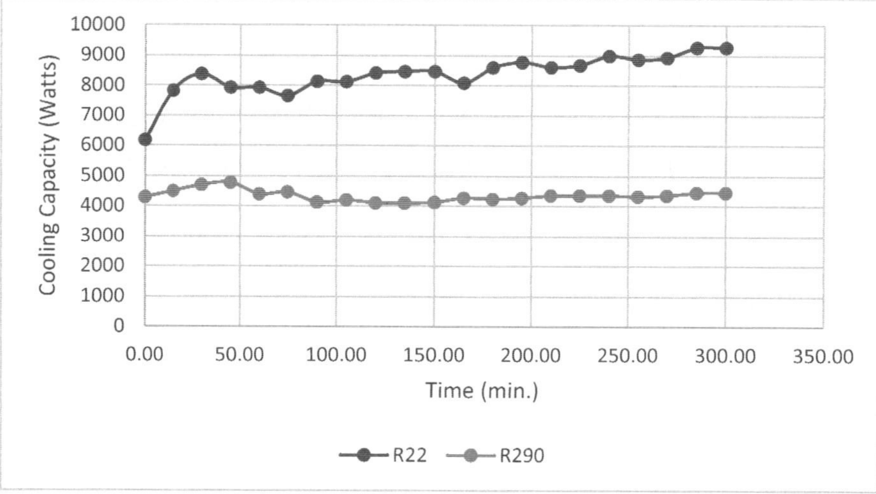

Fig. 3: Variation in the cooling capacities of the refrigerants

Moreover, analysis of the power consumption shows that, despite the performance of R22, it consumed higher amount of energy for every unit drop in the temperature of the refrigerated space. Fig. 4 displays the comparative analysis of the energy consumption of the refrigerants. It reveals that R22 consumed between 1.53 and 1.59 kW power throughout the period compared to that of R290 which was found to lie between 1.45 and 1.49 kW. The energy saving capacity of R290 is about 6.3% better.

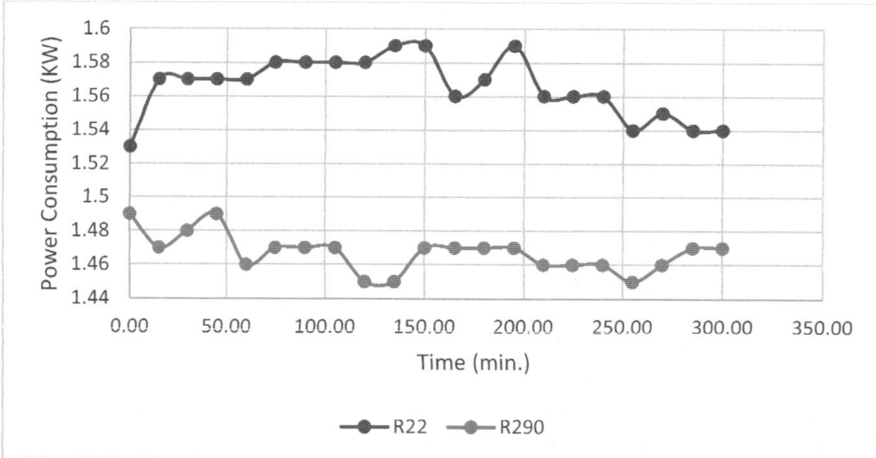

Fig. 4: Comparison of energy consumption between the refrigerants

The implication of the results of energy consumption is indicative of the fact that if there is a means of improving the thermal performance of R290, such that its refrigerating efficiency is enhanced, the refrigerant may be a better option on the long run. Such enhancement may be achieved with the use of metallic/non-metallic nanoparticles (Ajayi *et al.*, 2017) in a way that reconfigures the working fluid and makes it thermally more efficient with higher coefficient of performance. If this can be achieved, it may bring about a further reduction in the value of energy requirement per degree change in temperature of the system. Furthermore, the time required to achieve the temperature of the refrigerated space for a given period is referred to as the pull-down time. Fig. 5 display the results and show that there is no marked difference in the pull-down time of the two refrigerants. The difference in the temperature of the refrigerated space is not more than 2°C per period of the experiment. Hence, suggesting that, although R22 is thermally more efficient than R290, it however consumes more energy and produces minimal effect of not lower than 2°C per period.

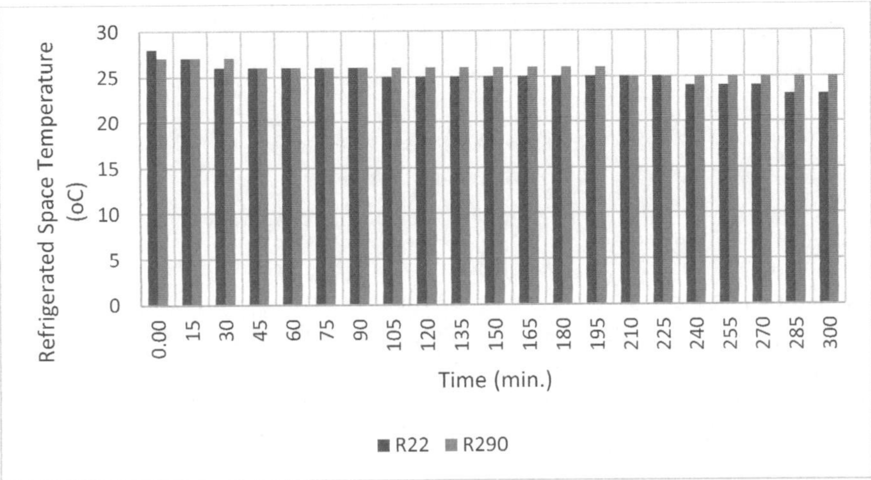

Fig. 5: Variation of pull-down time across the period of experiment

Considering the aforementioned, in order to ascertain the reason behind the difference in performance of the refrigerants, it was worthwhile to determine the variation in the thermal conductivities and viscosities of the refrigerants at the vapour and liquid phases. Figs. 6 and 7 display the results of thermal conductivity analyses for both refrigerants at the liquid and vapour phases.

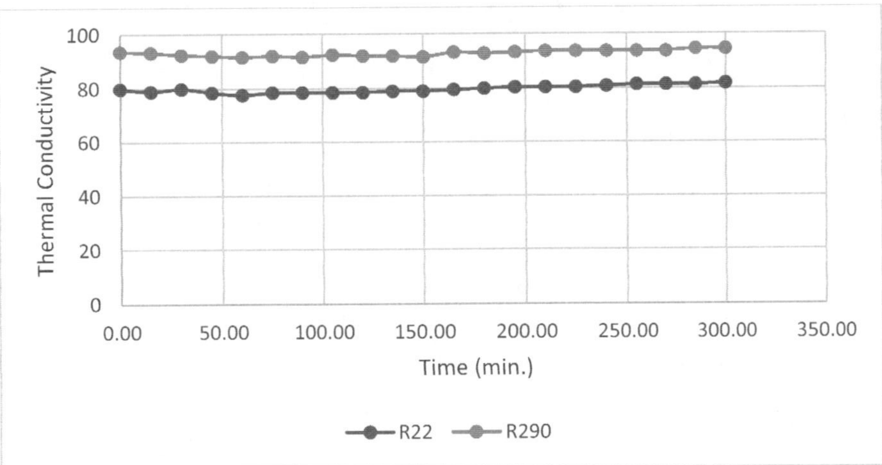

Fig. 6: Variation of the thermal conductivities of the refrigerant at the liquid phase

IOP Conf. Series: Materials Science and Engineering **413** (2018) 012072 doi:10.1088/1757-899X/413/1/012072

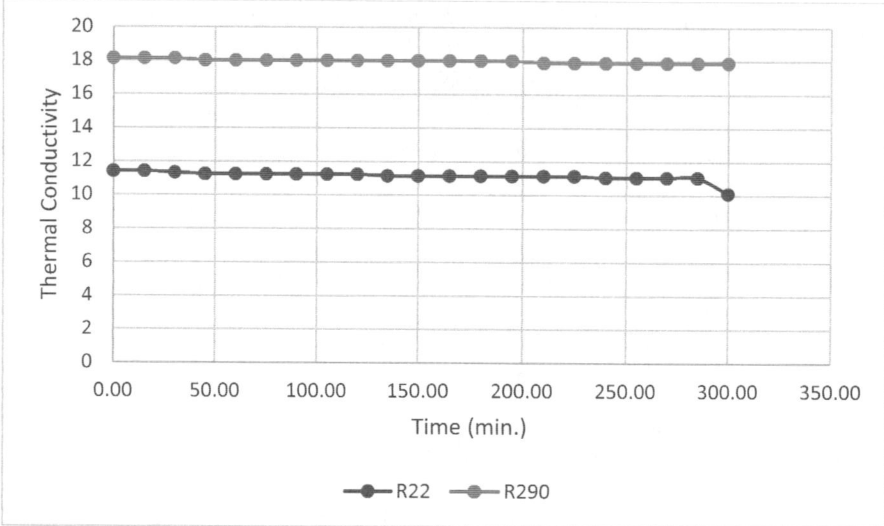

Fig. 7: Variation of the thermal conductivities of the refrigerant at the vapour phase

The Figures demonstrate that the thermal conductivity of R290 at both liquid and vapour phases is higher than that of R22. The results depict that, despite the thermal performance of R22, R290 has a better capacity to convey a higher magnitude of heat per unit time. Hence, with little improvement in the thermal efficiency of R290, its coefficient of performance may exceed that of R22. This improvement as suggested earlier, may be achieved by reconfiguring the thermodynamic system. In addition to this, the values of viscosity analyses (Figs. 8 and 9) for both refrigerants at the liquid and vapour phases depicts that R290 has a better visco-energetic performance than R22.

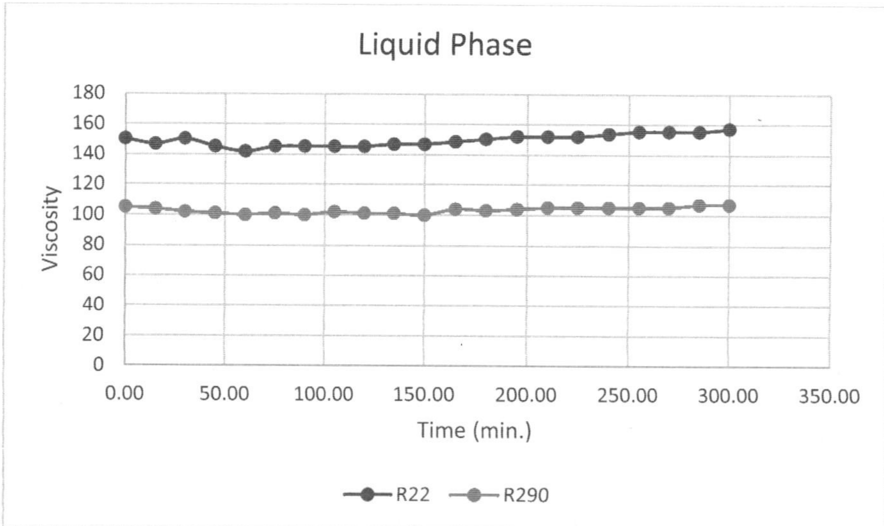

Fig. 8: Variation of the viscosity index of the refrigerant at the liquid phase

IOP Conf. Series: Materials Science and Engineering **413** (2018) 012072 doi:10.1088/1757-899X/413/1/012072

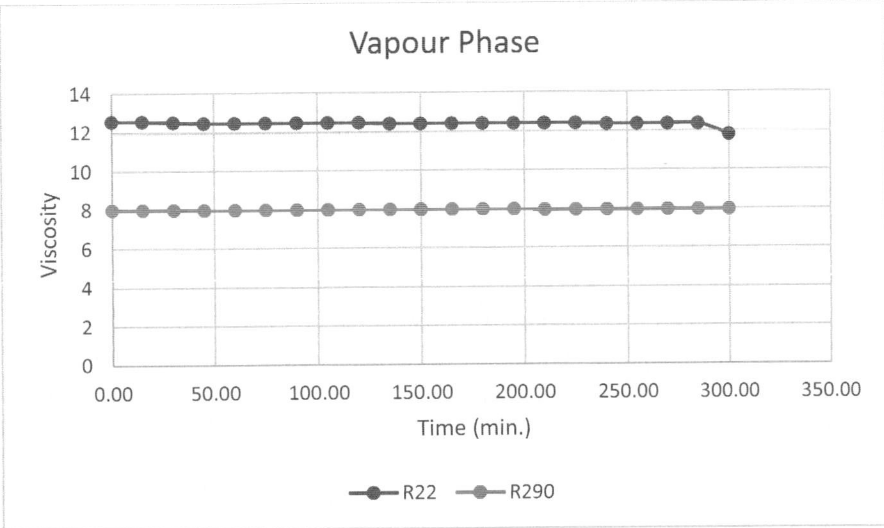

Fig. 9: Variation of the viscosity index of the refrigerant at the vapour phase

The results show that the viscosity of R290 at both liquid and vapour phases is lower than that of R22. Worth noting is the fact that, the value of viscosity is proportional to the power requirement of the compressor for fluid transport. Hence, the higher the viscosity, the higher the power required to transport the fluid through the compressor passage. More so, the ease with which the fluid passes through the compressor depicts the ease of heat expulsion from the compressor. Thus, the lower values of the viscosity of R290 at both the liquid and vapour phases show the stronger ability of the refrigerant to expel heat from the system and also reduce the compressor work. The results corroborate those of Fig. 4. However, comparing the results of the Figs. 6 to 9 with those of Figs. 3 to 4 clearly demonstrate that, although R290 has a better thermal conductivity and viscosity than R22, the COP and Cooling Capacity is lower. The reason can be adduced to the effect of the thermodynamic system on the performance of R290. The Capella D lubricant variant of mineral oil may not be very compatible with the R290 as refrigeration fluid system. Hence, an improvement in the thermo-physical properties of the lubricant may enhance the performance of R290.

4. Conclusion

The study investigated the performance of R290 as a replacement alternative to the environment malignant R22 refrigerant. It used a 2 hp window split air-conditioning system with Capella D oil as lubricant. Comparative analyses of the thermodynamic performance of the refrigerants were carried out and the effects of the lubricant on the refrigerant performance was highlighted. The findings show that despite that R22 outperformed R290 in terms of its coefficient of performance and cooling capacity by 44.8% and 48%, R290 showed better results in terms of its energy consumption, thermal conductivity and viscosity at the liquid and vapour phases. Further to this, there is no marked difference in the pull-down time results. Going by the findings, when consideration is given to the fact R22 adds to the burden of the environment and also the fact that the manufacture of new products of it have been banned around the globe, employing R290 with improved lubricant property can be a good decision in the long run.

IOP Conf. Series: Materials Science and Engineering **413** (2018) 012072 doi:10.1088/1757-899X/413/1/012072

References

[1] Ajayi, O.O., Fagbenle, R.O. Katende, J., Ndambuki, J.M., Omole, D.O., Badejo, A.A. (2014). Wind energy study and energy cost of wind electricity generation in Nigeria: past and recent results and a case study for South West Nigeria, *Energies*, 7 (12), 8508 – 8534.

[2] Ajayi, O.O., Ibia, D.E., Ogbonnaya, M., Attabo, A & Agarana, M.A. (2017). CFD analysis of nanorefrigerant throght adiabatic capillary tube of vapour compression refrigeration system, *Procedia Manufacturing*, 7, 688-695.

[3] Bhatkar, V. W., Kriplani, V. M., Awari G. K. (2013). Alternative refrigerants in vapour compression refrigeration cycle for sustainable environment: a review of recent research, *International Journal of Environmental Science and Technology*, 10(4), 871-880.

[4] Choudhari, C.S. Sapali, S.N. (2017). Performance Investigation of Natural Refrigerant R290 as a Substitute to R22 in Refrigeration Systems, *Energy Procedia* 109, 346 – 352.

[5] Choudhury, B., Saha, B. B., Chatterjee, P. K., & Sarkar, J. P. (2013). An overview of developments in adsorption refrigeration systems towards a sustainable way of cooling. *Applied Energy*, 104, 554-567.

[6] Fagbenle, R.O., Katende, J., Ajayi, O.O., Okeniyi, J.O., (2011). Assessment of wind energy potential of two sites in North - East, Nigeria, *Renewable Energy*, 36, 1277 – 1283.

[7] Kubota, T., Jeong, S., Toe, D.H.C., Ossen, D.R. (2011). Energy consumption and air-conditioning usage in residential buildings in Malaysia, *Journal of International Development and Cooperation*, 17(3), 61-69.

[8] Ohijeagbon O.D. Ajayi, O.O. (2014). Potential and Economic viability of stand-alone hybrid systems for a rural community of Sokoto, North-West Nigeria, *Frontiers in Energy*, 8 (2), 145 – 159.

[9] Ohijeagbon, O.D., Ajayi, O.O. (2015). Solar regime and LVOE of PV embedded generation systems in Nigeria, *Renewable Energy*, 78, 226 – 235.

[10] Pearson, A. (2005). Carbon dioxide - new uses for an old refrigerant. *International Journal of Refrigeration*, 28(8), 1140-1148.

[11] The News (2018). Montreal Protocol sets global HFC Phasedown. Available online [https://www.achrnews.com/articles/131056-montreal-protocol-sets-global-hfc-phasedown, accessed June 27, 2018]

IOP Conf. Series: Materials Science and Engineering **413** (2018) 012073 doi:10.1088/1757-899X/413/1/012073

FTIR Investigation of the Effect of Storage on Ogogoro-Gasoline Blend's Stability

Uguru-Okorie D. C.[1], Ikpotokin I.[1], Ajiboye M. O.[1], Ojediran M. E.[1]

The Department of Mechanical Engineering, Landmark University, Omu-Aran, Kwara State, Nigeria 1.

E-mail address: uguru-okorie.daniel@lmu.edu.ng

Abstract. Ethanol has wide application as spark ignition (SI) engine fuel and its usage comes as blend with gasoline or as sole fuel. Industrially produced anhydrous ethanol has been the type of ethanol used for this purpose. There has been conscious drive by governments and researchers over the years for a shift from environmentally detrimental energy sources to more environmentally friendly and renewable energy sources. Fourier-Transform Infrared (FTIR) spectrums were obtained from ogogoro in its pure state and when mixed with 90% by volume of gasoline before and after a year of storage. The spectrums obtained before and after storage were compared and the results showed that there was no significant change in their chemical compositions. The FTIR results and physical observation showed that the ogogoro and its blends were chemically and physically stable, confirming its suitability as fuel for spark ignition engines.

1. Introduction

Ethanol has been found to be a promising fuel, suitable for use by internal combustion engines. Internal combustion engines will remain major players in the automobile and related industries in the years to come [1, 2], due to increased demand to develop fuels to power them. Over the years it has been observed that the fuel from non-renewable sources are fast depleting and this has necessitated a shift from non-renewable energy sources to renewable sources [3,4]. It is worthy of note that with the growing world population, the demand for this limited resource will continue increase. As a result, other sources of fuel for these engines are required.

In recent years, there has been a conscious drive by the government and researchers to shift from toxic and environmentally detrimental energy sources to more sustainable and renewable energy sources [5-7]. World agreements such as the Kyoto Protocol [8], Paris agreement [9], etc. have all been reached and signed to ensure this goal is realised.

To reduce the contribution of greenhouse gases made by spark ignition engines, the use of ethanol or ethanol-gasoline blends seems to be the answer. Ethanol is obtained from renewable sources, it has better antiknock quality compared to gasoline and it gives more environmentally friendly by-products [10, 11]. The following qualities have made ethanol the future fuel for SI engine.

While anhydrous ethanol is commercially used, there is limited literature on hydrated ethanol and the effect of its use as blend with gasoline, on the performance as fuel in SI engines. An experimental study by Uguru-Okorie et al. [12] has shown that a hydrated ethanol obtained from fermented palm tree sap, known as Ogogoro, could be used as fuel for SI engines when blended with gasoline.

Crudely produced ethanol (also known as ogogoro) blended with gasoline in an SI engine has been experimentally studied with results published, showing an improved engine performance in its 1 to 5% by volume blend with gasoline. The stability of this blend in storage overtime, has not been investigated and this finding led to the research on the FTIR investigation of the stability of Ogogoro-gasoline blend over time.

2. Production of Ethanol Locally

Ethanol is popularly produced in Nigeria from the sap of raffia palm and oil palm tree [13]. The sap obtained is further processed to a substance with higher ethanol content known as Ogogoro. The processing involves harvesting the sap of a raffia palm and oil palm tree and thereafter, it is left to ferment and this is done by leaving it in an enclosure for seven (7) days. The solution is subsequently heated up to steam and the steam is condensed to liquid whose distillate is consumed as locally produced alcohol known as Ogogoro [14, 15].

This locally produced ethanol (ogogoro) is known to have between 30-60% alcohol content [16-20] and it's consumption as drink is said to be harmful because of the unhygienic mode of production [14][21, 22] Ethanol is also produced from other biological components which are referred to as feedstock [23,24]. These feedstocks are classified into three major groups: simple sugar, starches and lignocellulosic biomass. The sources of raw materials in each of the groups are as follows: simple sugar: sugarcane, sugar beet and sorghum, starches: corn, wheat, potatoes [25] and cassava and lignocellulosic biomass: mix of lignocellulosic.

The extraction of ethanol from simple sugar raw materials starts with a milling process for sugar extraction which is later fermented. Subsequently, fermentation, distilling, rectification and finally dehydration processes take place [26,27]. For the starches group, which comprise the grains, it starts with the milling of the grain, which could be done in wet or dry condition. If the milling is done in wet condition, the starch in the grain is extracted and then fermented and if the milling is done in dry condition, a mash is formed by adding water to the product and it is then allowed to ferment. For the lignocellulosic group, a more complex technology compared to simple sugar and starches is employed for the production of ethanol. The process starts with pre-treatment, hydrolysis, fermentation and finally product separation.

2.1 Use of Ethanol-Gasoline Blends in SI engines

In recent years, ethanol-gasoline blends have been used as fuels in spark ignition engines with the "E" describing the percentage of ethanol by volume in the mixture. Examples are E5 denotes that the fuel contains 5% anhydrous ethanol and 95% gasoline by volume while E85 denotes that the fuel contains 85% anhydrous ethanol and 15% gasoline [3,4].Research has shown that more than 10% content of ethanol in ethanol-gasoline mixture as fuel in SI engines requires modification of such engine [28].

2.2 Performance of Ogogoro-Gasoline Blend

Anhydrous ethanol blended with gasoline has been widely used in spark ignition engines while the hydrous ethanol used as sole fuel is reported to have about 96% content of ethanol and about 4% of water by volume [28-31]. While ethanol used as blends with gasoline are industrially produced ethanol with high purity, there is no records of studies done with ogogoro (which is reported to contain between 30 to

IOP Conf. Series: Materials Science and Engineering **413** (2018) 012073 doi:10.1088/1757-899X/413/1/012073

60% ethanol by volume) mixed with gasoline, as fuel for spark ignition engines apart from the one carried out by Uguru-Okorie et. al. [12].

The investigation on the potential of the blends of ogogoro-gasoline as spark ignition engines fuel, was carried out in a four-stroke, single cylinder, SI engine with between 1 to 5% by volume ogogoro content in gasoline fuel. Pure regular gasoline was the base fuel and was assigned E0 while E1, E3 and E5 were gasoline fuels with 1, 3 and 5% content of ogogoro by volume. The study showed improved performance of the blends over pure gasoline while the best engine performance was with gasoline with 3% by volume content of ogogoro. Results obtained are shown in figures (1) and (2).

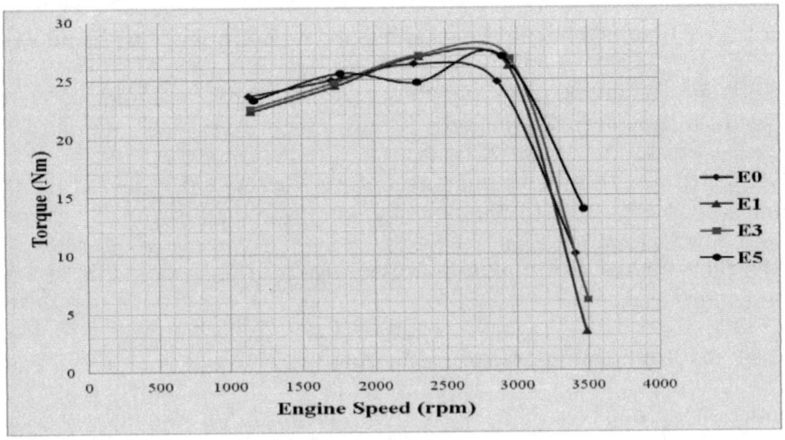

Fig. 1 Engine Torque at Varying Speeds [12]

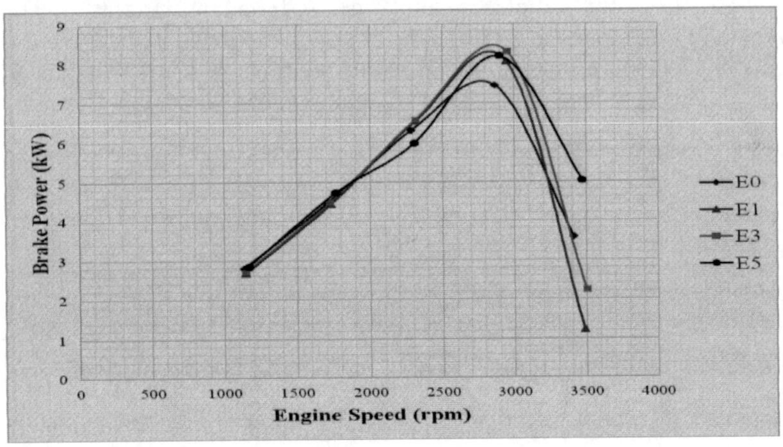

Fig. 2 Brake Power at varying Speeds [12]

3. Methodology

This research is based on Fourier-Transform Infrared (FTIR) Spectroscopy test for the stability of ogogoro-gasoline blend as fuel for spark ignition engine, over a period of fourteen (14) months. The stability of the fuels was investigated by the study of functional groups present in the pure fuels (gasoline and ogogoro), as well as their blends before and after storage. The following steps were taken in the investigation: -

a. FTIR spectrum from ogogoro (obtained from palm wine source) was obtained before storage.

b. FTIR spectrum from unleaded gasoline was obtained.

c. FTIR spectrum of ogogoro and gasoline blend in the proportion of 10 and 90% by volume respectively was obtained.

d. FTIR spectrum from ogogoro, in storage, for fourteen (14) months, in a thermo-set container, was obtained.

e. FTIR spectrum of ogogoro and gasoline blend in the proportion of 10 and 90% by volume respectively stored for fourteen months, in a thermo-set container, was obtained.

f. FTIR spectrums obtained from pure gasoline and ogogoro-gasoline blend were compared.

g. FTIR spectrums obtained from ogogoro-gasoline blend and old ogogoro-gasoline blend in storage for 14 months were compared.

h. The FTIR spectrums from pure ogogoro before storage and after fourteen (14) months storage were compared to determine if there were variations in the functional groups.

i. FTIR spectrums obtained from pure gasoline, ogogoro-gasoline blend and old ogogoro-gasoline blend were compared.

4. Analysis of FTIR Spectrum of Gasoline Fuel

The following functional groups were observed in the spectrum for pure unleaded gasoline seen in figure (3): terminal (vinyl) C-H stretch at 3026.31cm-1 at 97.73%, methyl C-H asymmetric or symmetric stretch at 2956.87cm-1 at 79.14% transmittance, methylene C-H asymmetric or symmetric stretch at 2924.01cm-1 at 80.40% transmittance, methylene (CH2)n rocking at 731.02cm-1 at 85.18% transmittance, aromatic ring stretch at 1458.18cm-1 at 86.54% transmittance, mono-substitution (phenyl) at 696.30cm-1 at 88.87% transmittance, meta-di-sub benzene at 881.47cm-1 at 97.58% transmittance [32-34].

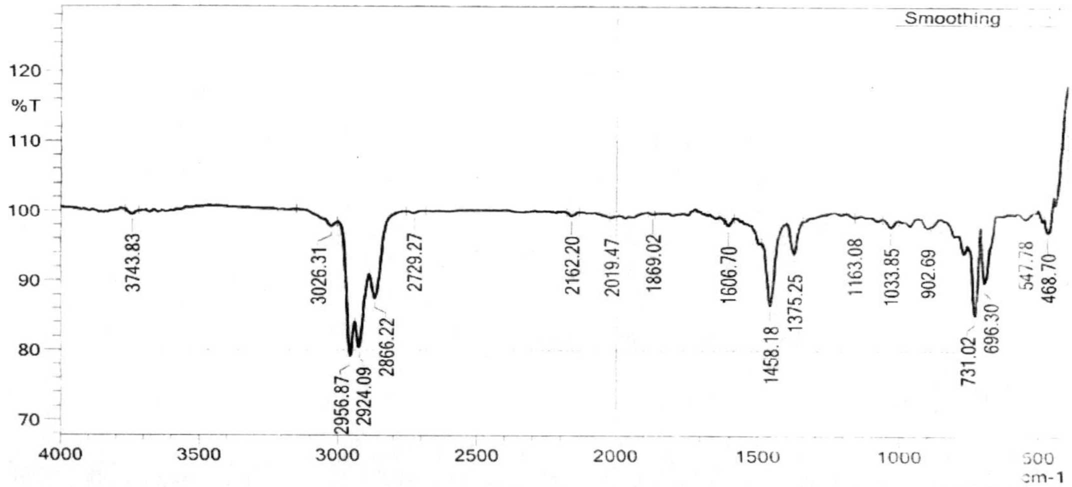

Fig. 3 FTIR Spectrum for Pure Gasoline

4.1 FTIR Comparison of Pure Gasoline and Ogogoro-Gasoline blend

The FTIR spectrums for gasoline and ogogoro-gasoline blend were compared as shown in figures (4) and (5). The FTIR data obtained from pure gasoline was similar to that obtained from the FTIR data of ogogoro-gasoline blend, when compared. With 10% by volume of ogogoro in gasoline, the alcohol content in the ogogoro-gasoline blend was visible on the FTIR spectrum as shown in figure (9).

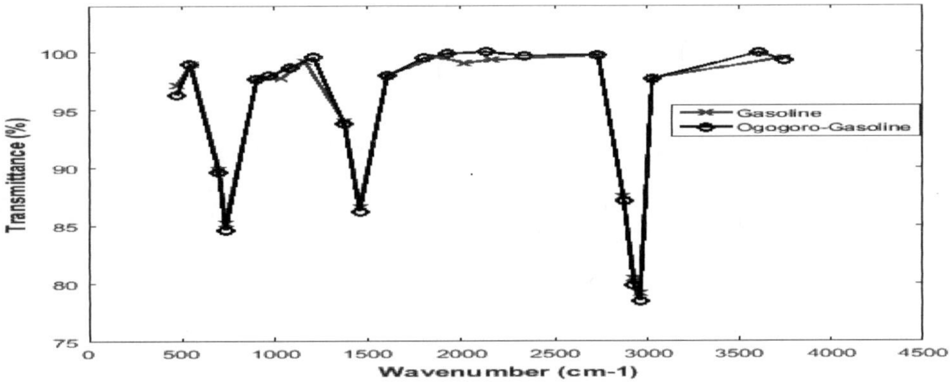

Fig. 4 Comparing FTIR Spectrums of Pure Gasoline and New Ogogoro-Gasoline Blend

4.2 Storage Effect on Ogogoro and Ogogoro-Gasoline Blend

The storage effects on pure ogogoro and a blend of ogogoro with gasoline were investigated with the FTIR spectrometer to determine if long storage gave rise to degradation.

4.2.1 Storage Effect on the FTIR Spectrum of Ogogoro

The strong intensity of alcohol in both the fresh and old ogogoro in storage were maintained at transmittance between 57.96 to 60.38%; between the wavenumbers of 3200-3400 cm-1 and also other intensities at other wavenumber in the FTIR spectrums for the new and old ethanol in storage remained the same as shown in figures (5) and (6). The results from the comparison of the FTIR data obtained from the fresh and old ogogoro in storage and physical observation show that there was no noticeable change in both chemical and physical composition of ogogoro after over fourteen (14) months in storage. The peaks from the spectrums were compared with no significant difference as shown in figure (7).

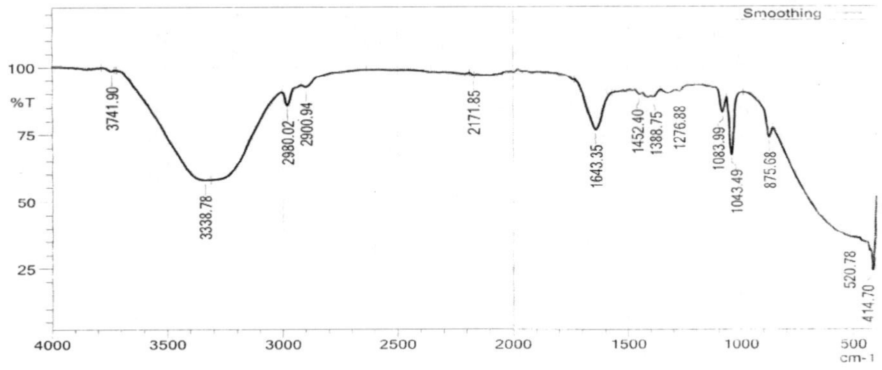

Fig. 5 FTIR Spectrum for Ogogoro before storage

IOP Conf. Series: Materials Science and Engineering **413** (2018) 012073 doi:10.1088/1757-899X/413/1/012073

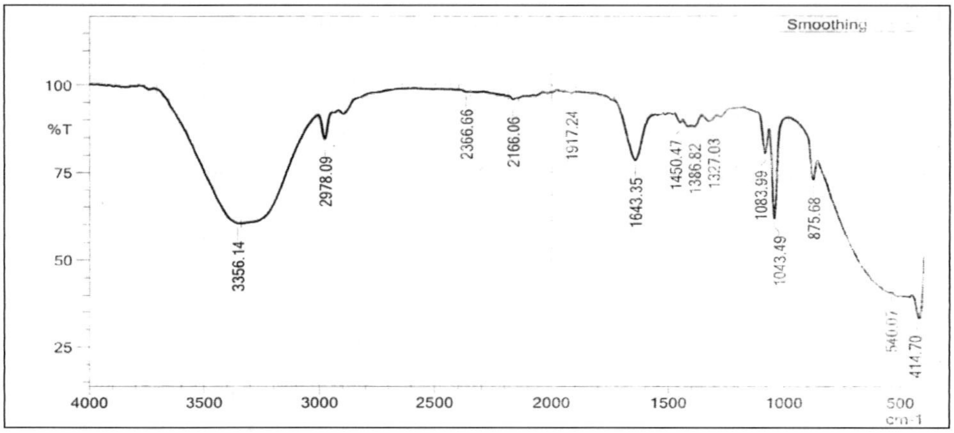

Fig. 6 FTIR Spectrum for Ogogoro after Storage (14 months)

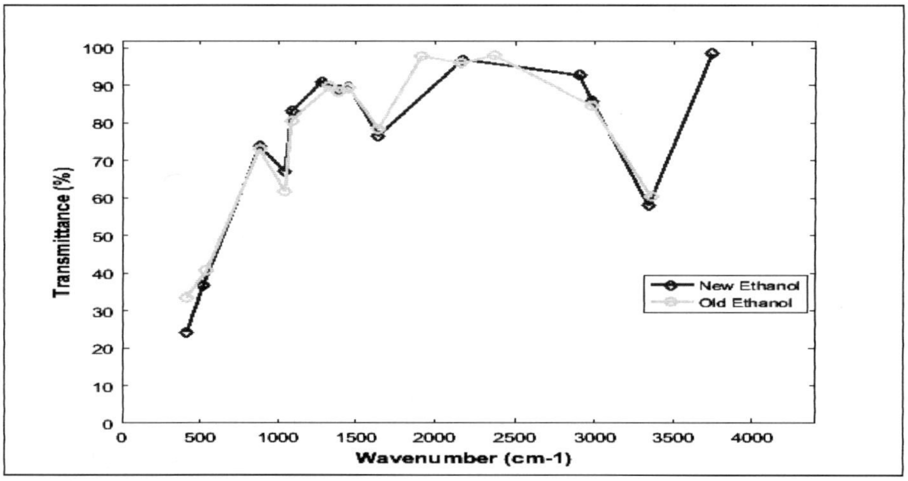

Fig. 7 FTIR Spectrum Peaks of Old and New Ogogoro

4.2.2 Storage Effect on Ogogoro-Gasoline Blend

The FTIR spectrums for the fresh ogogoro-gasoline blend and the old ogogoro-gasoline blend in storage for fourteen (14) months, when compared, showed very similar structures with slight variations in the intensities of the functional groups present, as seen in figures (8) to (10). The graph for fresh ogorogo-gasoline is represented with "New Etha-Gas" while ogogoro-gasoline in storage is represented with "Old Etha-Gas".

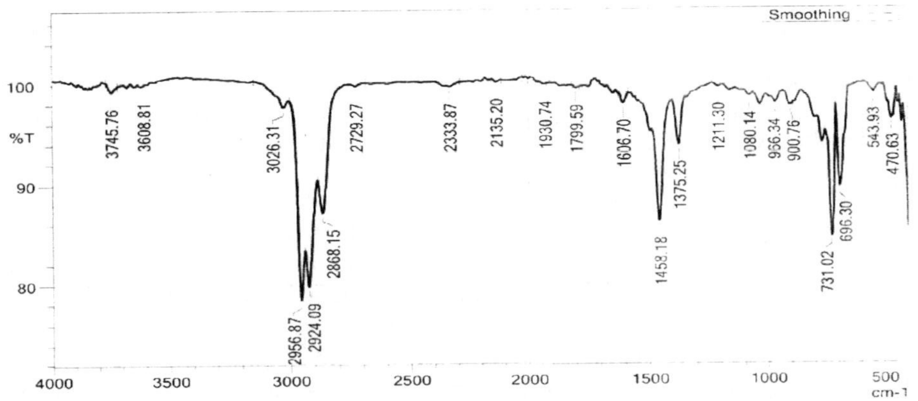

Fig. 8 FTIR Spectrum for Ogogoro–Gasoline Blend

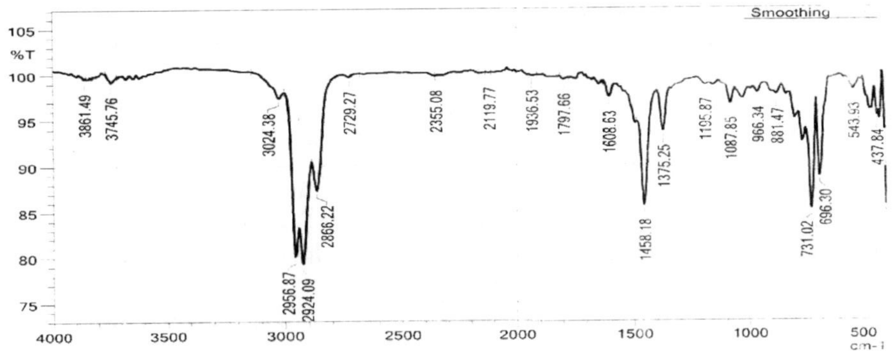

Fig. 9 FTIR Spectrum for Ogogoro–Gasoline Blend in Storage (14 months)

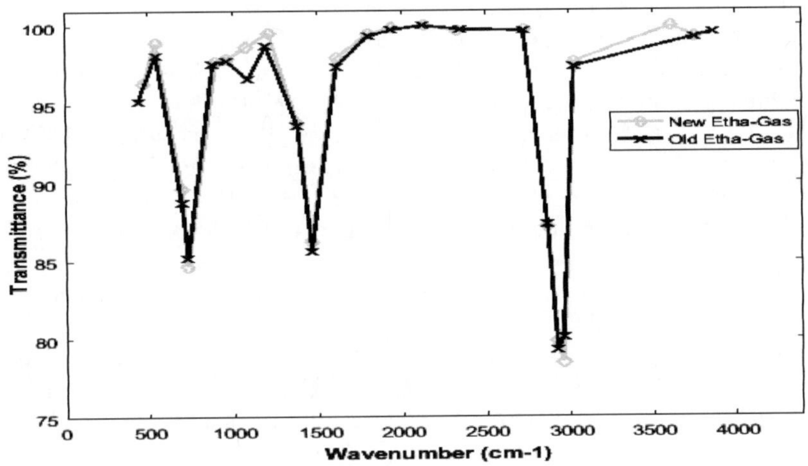

Fig. 10 FTIR Spectrum Peaks of Old and New Ogogoro-Gasoline Blends

IOP Conf. Series: Materials Science and Engineering **413** (2018) 012073 doi:10.1088/1757-899X/413/1/012073

5. Conclusions

The aim of the research work was to investigate the effect of prolonged storage of ogogoro-gasoline blend on its chemical stability, which was investigated using an FTIR spectrometer. The FTIR spectrums of the fuels were obtained and analyzed, before and after storage in its pure and blended states. At the end of the investigation, the following findings were made:

(i) FTIR spectrum obtained, showed negligible trace of alcohol in 10% ogogoro-90% gasoline blend, when compared with pure gasoline.
(ii) The FTIR spectra obtained from the fresh ogogoro and ogogoro in storage for 14 months and physical observation showed no change in both chemical and physical composition. The result proves that ogogoro is chemically and physically stable.
(iii) FTIR spectrum obtained from new ogogoro-gasoline blend when compared with the spectrum obtained from ogogoro-gasoline in storage (14 months), showed no significant change in the structure of the fuels.
(iv) The findings show that ogogoro-gasoline is a stable fuel and is resistant to degradation in storage.
(v) This observation rules out the need for preservatives to maintain the stability of ogogoro-gasoline blend in storage and this makes the use of this fuel promising for SI engines.

References

[1] Uguru-Okorie D. C., Dare A. A. and Burluka A. A. Effect of Supercharging on Cycle-to-Cycle Variation in a Two-Stroke Spark Ignition Engine [C]. SAE Technical Paper, 2016, 2016-01-0688.
[2] Ormsby, Matthew P. Turbulent flame development in a high-pressure combustion vessel [B]. PhD Thesis, University of Leeds, 2005.
[3] Iliev Simeon A Comparison of Ethanol and Methanol Blending with Gasoline Using a 1- D Engine Model [J]. Procedia Engineering, 2015, 100: 1013 – 1022.
[4] Ozsezen A.N., Canakci M. Performance and combustion characteristics of alcohol-gasoline blends at wide-open throttle [J]. Energy, 2011, 36: 2747-2752. doi:10.1016/j.energy.2011.02.014.
[5] Hussin, A. New and Renewable Energy: Renewable Fuels in Internal Combustion Engines [B]. PhD Thesis, University of Leeds, 2012.
[6] Bardin M. E., Hussin A. M. T. A. El-Dein, Gushchin P. A., Vinokurov V. A. and Burluka A. A. Technical aspect of Ethyl Tert-Butyl Ether (ETBE) for Large-Scale Use as Gasoline Improver [J]. Energy Technology (Wiley Online Library) 2014, 2: 194 -204.
[7] Zhiyou, W., John, I. and Jactone, A. Fuel Ethanol [B]. Virginia Polytechnic Institute and State University, Communications and Marketing. Petersburg: College of Agriculture and Life Sciences, 2009. https://vtechworks.lib.vt.edu/bitstream/handle/10919/50132/442-884.pdf?sequence=1
[8] Anderson J.W. The Kyoto Protocol on Climate Change [B]. Resources for the Future: Washington, DC, 1997.
[9] Jayaraman T. The Paris Agreement on Climate Change: Background, Analysis and Implications [J]. Review of Agrarian Studies, 2015, 5(2): 42-59.
[10] Thakur A. K., Kaviti A. K., Mehra R. et al. Progress in Performance Analysis of Ethanol-Gasoline Blends on SI Engine [J]. Renewable and Sustainable Energy Reviews, 2017, 69: 324 – 340.
[11] Jeuland N., Montagne X. and Gautrot X. Potentiality of Ethanol, as a Fuel for Dedicated Engine [J]. Oil & Gas Science and Technology – Rev. IFP, 2004, 59(6): 559 -570.
[12] Uguru-Okorie D. C., Ikpotokin I., Efemwenkiekie U. K and Idiku U. D. Performance Characteristics of Blends of Crudely Produced Ethanol with Gasoline in a Four-Stroke SI Engine [J]. International Journal of Trend in Research and Development, 2017, 4(4): 78-81.

[13] Ohimain E. I., Tuwon P. E. and Ayibaebi E. A. Traditional Fermentation and Distillation of Raffia Palm Sap for the Production of Bioethanol in Bayelsa State, Nigeria [J]. Journal of Technology Innovations in Renewable Energy, 2012, 1:131-141.

[14] Iwuoha C. I. and Eke S.O. Nigerian Indigenous Fermented Food: Their Traditional Process Operation, Inherent Problems, Improvement and Current Status [J]. Food Research International, 1996, 29: 527 -540.

[15] Akinshilo A. How Ogogoro is Produced in Nigeria [W]. Accessed December 6, 2016, https://www.naij.com.

[16] Iwegbue Chukwujindu M. A., Ojelum Anwuli L. and Bassey Francisca I. A Survey of Metal Profiles in some Traditional Alcoholic Beverages in Nigeria [J]. Food Science & Nutrition, 2014, 2(6): 724–733.

[17] Uzogara S. G. and Goma A. J. Tyramine and Ethanol Content of some Alcoholic Beverages Produced in Nigeria: A Research Note [J]. Nigerian Journal of Biochemistry, 1991, 6: 1-13.

[18] Amaziah W. O., Ukeh F. S. and Nkani C. M. Compressive Strength Improvement and Admixture Potential of Palm Tree Liquor Juice Extract (Ogogoro) in Cement Mortar and Concrete [J]. American Journal of Civil Engineering and Architecture, 2017, 5 (6): 245-252. DOI: 10.12691/ajcea-5-6-4.

[19] Zakpaa, H. D., Mak-Mensah, E. E., and Avio, O. A. Effect of storage conditions on the shelf life of locally distilled liquor (Akpeteshie) [J]. African Journal of Biotechnology, 2010, 9(10): 1499-1509. DOI: 10.5897/AJB09.1001.

[20] Obiora S. E., Bruna B., Ju¨rgen R. and Dirk W. L. Composition of Surrogate Alcohol from South-Eastern Nigeria [J]. African Journal of Drug & Alcohol Studies, 2007,6(2):65-74.

[21] Péléi T., Elolo O., Marie-Laure F. and Courdjo L. Aromatic Composition of "Sodabi", a Traditional Liquor of Fermented Oil Palm Wine [J]. Advance Journal of Food Science and Technology, 2018, 14(1): 15-22. DOI:10.19026/ajfst.14.5421

[22] Idonije O.B., Festus O.O. Asika E.C.A., Ilegbusi M.I. and Okhiai O. A Comparative Biochemical Analysis of Local Gin (Ogogoro) from Different Parts of Nigeria and Imported Gin (Dry Gin) - Toxicogenic, Carcinogenic and Sociopolitical Implications [J]. Science Journal of Medicine and Clinical Trials, 2012, 2012:1-4.

[23] Akpan U.G., Kovo A.S., Abdullahi M., and Ijah J. J. The Production of Ethanol from Maize Cobs and Groundnut Shells [J]. AU J.T, 2005, 9(2): 106-110.

[24] Jitendra K, Dhananjay T ,Prakash M, et al. Performance Study of Ethanol Blended Gasoline Fuel in Spark Ignition Engine [J]. IOSR Journal of Mechanical and Civil Engineering (IOSR-JMCE), 2013, 7: 71 - 78.

[25] Masoud Vazirzadeh, Reza Robati. Investigation of Bio-ethanol Production from Waste Potatoes [J]. Annals of Biological Research, 2013, 4(1): 104-106.

[26] Basso Luiz Carlos, Basso Thiago Olitta and Rocha Saul N. Ethanol Production in Brazil: The Industrial Process and Its Impact on Yeast Fermentation [B]. Biofuel production-recent ,2011, 85 - 100. www.intechopen.com. DOI: 10.5772/17047

[27] Mino Alexander Kiyoshi. Ethanol Production from Sugarcane in India: Viability, Constraints and Implications [B]. Master of Science thesis in the Department of Natural Resources and Environmental Sciences in the Graduate College of the University of Illinois at Urbana – Champaign. 2010, 1 – 142.

[28] Regina C.O.B. D., Antonio S. A., Valter J. F. Properties of Brazilian gasoline mixed with hydrated ethanol for flex-fuel technology [J]. Fuel Processing Technology, 2007, 88: 365–368.

[29] HE Blends. Abstract Anhydrous Ethanol Blends [W]. Accessed June 3, 2018, http://www.heblends.com/index.php?option=com_content&task=view&id=12&Itemid=27.

[30] Nguyen H. H. Modelling and Simulation of a Pervaporation process using Tubular Module for Production of Anhydrous Ethanol [C]. AIP Conference Proceedings, 2017, 1878(1):020029.

IOP Conf. Series: Materials Science and Engineering **413** (2018) 012073 doi:10.1088/1757-899X/413/1/012073

[31] Augoye A.K. and Aleiferis P.G. Characterization of Flame Development with Hydrous and Anhydrous Ethanol Fuels in a Spark-Ignition Engine with Direct Injection and Port Injection Systems [C]. SAE Technical Paper, 2014, 2014-01-2623.
[32] John Coates Interpretation of Infrared Spectra, A Practical Approach [B]. Encyclopedia of Analytical Chemistry, R.A. Meyers (Ed.), John Wiley & Sons Ltd, Chichester, 2000, pp. 10815–10837.
[33] Mikhail A. V., Dilyara I. A., Aliya Z. G., Boris N. S. FTIR study of H-bonds cooperativity in complexes of 1,2-dihydroxybenzene with proton acceptors in aprotic solvents: Influence of the intramolecular hydrogen bond [J]. Spectrochimica Acta Part A, 2010, 77: 965–972.
[34] Larkin P. Infrared and Raman Spectroscopy Principles and Spectral Interpretation [B]. Elsevier Inc., 2011.

IOP Conf. Series: Materials Science and Engineering **413** (2018) 012074 doi:10.1088/1757-899X/413/1/012074

Sensitivity analysis on a variant machine guard compliance productivity model (VMGPEM) for a bottling plant using the OFAT technique

Uzor Chukwunedum [1], Oke Sunday [2], Okolie Sociis T. A [1]

[1]Covenant University, Ota, Nigeria
[2] University of Lagos, Yaba, Nigeria E-mail address: chukwunedumuzor@gmail.com

Abstract. Certainly, the operating errors of a given model can be assuaged by proper selection of its working parameters and distinctly identifying the most critical for optimization. To achieve this objective, sensitivity analysis is introduced to the system to show relevant strengths and avoidable lapses. For this study, the one-factor-at-a- time (OFAT) sensitivity analysis technique is employed to the model to disclose the input parameters in form of guard categories that is most significant in optimizing the machine guard compliance productivity for a particular period. The result of the experiment reflected fixed guards as the most significant guard category by which machine guard compliance productivity for the bottling plant can be optimized.
Keywords: machine guard; critical parameter

1. Introduction

Machine guards are protections against injuries, attached on workshop equipment as safety mechanisms. They shield machine operators' body from interaction with material particles, noise and harms, which is a huge progress towards achieving the no-accident goal in any work station. As the need to measure operators' compliance arose in the bottling plant, a variant machine guard productivity evaluation model was designed. This model was to measure the productivity of the operator with respect to machine guard compliance having the start period of the experiment as a root period. The model is designed to have input and output sections. The numbers of functional guards for the categories engaged are carefully observed for their use during production processes and recorded during the investigation span. The model shows how the output to inputs ratio can be used as a determining value. The farther the value for machine guard compliance productivity is away from unity is indicative of a poor performance and the closer it gets the better the performance by the operator in complying to guard usage.

This model helps the operations manager to evaluate the operators on their compliance to guard usages and to quickly spot damaged guard for replacement in other to reduce injury tendencies and keep proper inventory of guard parts. It was observed that machine guard compliance productivity over the selected periods were within the range of 0.60-0.70. This appears not to be completely satisfactory as it is interpreted as 60%-70% guard compliance productivity. This prompted the employment of sensitivity analysis using the one-factor-at-a-time technique to investigate this situation. The one-factor-at-a-time sensitivity analysis also known as the local sensitivity analysis investigates the importance of parameters by altering one parameter after another, while doing this it keeps all other parameters fixed. This increases the comparability of the results. This regularly chosen method is employed by many scientists as it gives room to immediately spot out input parameters that initiated poor performances in a system and by what magnitude and vice-versa. The sixth period was selected for the application of the OFAT technique as it showed the most productive period for machine compliance with a productivity of 0.6794, in other words 67.94% productivity percentage. The OFAT technique was systematically applied on the parameters for the sixth period to identify the most crucial parameter, vary it by certain percentages and note its optimal compliance productivity value. For this study, the OFAT technique worked adequately for optimization of machine guard compliance productivity.

2. Brief literature review

This section will be presenting a brief overview of few works found in literature relating to machine guard usage. *Eckhardt* in 2001 investigated on the diverse enhancements achieved in the employment of guards in machinery [1]. To reveal the implication of working in systems without guards, the aim of securing the employment of machine guards as well as the benefits, with particular instance drawn from the huge system used to guard a 4-bolt shaft was emphasized. In 2012, *Soranno* emphasized the importance of machine guarding in production systems [2]. The exposition centered on the need for machine guarding, the concept of design as well as machinery installation. In 2013, *Rubinger and Hamilton* centered discussion on matters associated with the guarding of machines and aid to assist in the achievement of a safe working relationship with woodworking machines [3]. In 2006, *Cordier* emphasized on promoting machine guarding usage with report given on the injuries sustained in production centers in relation to rotating parts of machines [4]. Furthermore, *Berke* in 2008 discussed four general methods in machine guarding: guarding devices, warnings, barrier guards and guarding by location [5]. Barrier guard was emphasized as a necessary requirement in most industrial settings and that its effective usage ensures a separation of the hazards from the operator of the machines. With the introduction of the variant machine guard productivity evaluation model (VMGPEM) this gap was breeched with room for improvements because of unsatisfactory results thus a need to investigate the situation using the OFAT sensitivity analysis technique. A brief overview of sensitivity work in literature follows.

In 2013, *Tian* said that methods used for sensitivity analysis can be categorized into local and global methods [6]. Local sensitivity analysis is associated with one-factor-at-a-time (OFAT) methods by *Navid et al.* in 2018[7]. OFAT technique understudy's system results with the changes in input parameter values. With respect to this technique, one parametric input value is varied, while the other parametric values remain fixed main values. In 2016, *Paleari and Confalonieiri* affirmed that this method can reoccur for other parameters [8]. OFAT sensitivity analysis techniques tends to establish dependency on parametric values, and relevant in understudying challenges with a few compromising factors. Furthermore, *Navid et al.* in 2018 conferred that the gross weakness of local sensitivity analysis techniques is that they are inaccurate for nonlinear models [7]. For this investigation OFAT sensitivity technique was proposed to identify crucial parameters.

3. Methods

The current analysis was carried out to investigate the effect on the output for productivity from a Variant Machine Guard Productivity Evaluation Model VMGPEM analysis when its main factors are altered. The VMGPEM analysis in this work was done using the first month of the data collection interval as the base or root period and the results obtained for productivity of machine guard compliance were all less than 70% of the original status value. From table 1 below, the highest compliance percentage was obtained in the sixth month with 67.94% compliance productivity and on the other hand the lowest compliance productivity was observed from the first, eighth and eleventh month having compliance productivity of 65.46% apiece. The sensitivity analysis carried out in this investigation is aimed at studying the guard categories considered for the model makeup and altering their respective percentage compliances by -25%, -20%, -15%, -10%, -5%, 5%, 10%, 15%, 20% and 25% in the VMGPEM that by so doing the category that gives optimal compliance productivity will be identified. With respect to the objective of the study, the VMGPEM analysis for the sixth period was considered separately for sensitivity since it had the highest compliance productivity. Therefore, altering the guard categories for this period will not only display the behavioral response of the machine guard compliance productivity output to changes but will also indicate the category which will yield optimal productivity when increased to a certain acceptable level. In the sensitivity analysis 25% is selected as the peak altering value as a percentage compliance above 100% will occur if it is exceeded with respect to the percentage arithmetic progression of 5% used in this work. Furthermore, in the course of the analysis the percentage compliance of categories in the base period were increased simultaneously by 5%, 10%, 15%, 20% and 25% respectively. The same was also done for the percentage compliance for all the compared periods. The results from this analysis will be graphically represented.

The data used for this work was obtained from a bottling process plant situated in the western region of Nigeria. The data was retrieved from the company's safety monitoring unit and spanned for a period of 12 months as it was considered to be quite sufficient to getting hold of relevant test results. The data gotten was arranged in sequential months and coded according production process lines. The data was summed up, averaged and tabulated. This information was further transferred to a Variant Machine Guard Productivity Evaluation Model (VMGPEM) appropriately and results of productivity was shown and this was done by retaining the first month as a root period in comparing other months. The basic results were gotten for all compared periods without any alteration. However, for the sake of this test alterations were employed to identify the guard category that will give maximum productivity when improved by a certain percentage.

Based on the data in

Table 1: Considering the impact of external forces on oil spills, this paper analyzes the assessment of the effective rescue levels of each rescue point relative to the oil spill batches, and we can calculate the effective rescue degree matrix P of six bases relative to each oil spill batch.

		Period 1			Period 6		Biased Variation Ratio		
Current Status	αi	Θi	αiΘi	αi	Θi	αiΘi	α2Θ1/α1Θ1	α2Θ2/α2Θ1	α2Θ2/α1Θ1
F	487	0.71	345.77	487	0.73	355.51	1	1.028169	1.028169
I	354	0.58	205.32	354	0.62	219.48	1	1.068966	1.068966
E	279	0.79	220.41	279	0.79	220.41	1	1	1
L	103	0.64	65.92	103	0.73	75.19	1	1.140625	1.140625
T	208	0.71	147.68	208	0.73	151.84	1	1.028169	1.028169
								Δ =	1.037895
TOTAL			985.1			1022.43			
Original Status									
F	490	1	490	490	1	490	1	1	1
I	420	1	420	420	1	420	1	1	1
E	280	1	280	280	1	280	1	1	1
L	105	1	105	105	1	105	1	1	1
T	210	1	210	210	1	210	1	1	1
TOTAL			1505			1505			
Productivity for period 1 =		0.654551		productivity for period 6 =			0.679355		
Change In Guard Compliance productivity =				1.037895					

IOP Conf. Series: Materials Science and Engineering **413** (2018) 012074 doi:10.1088/1757-899X/413/1/012074

4. Results

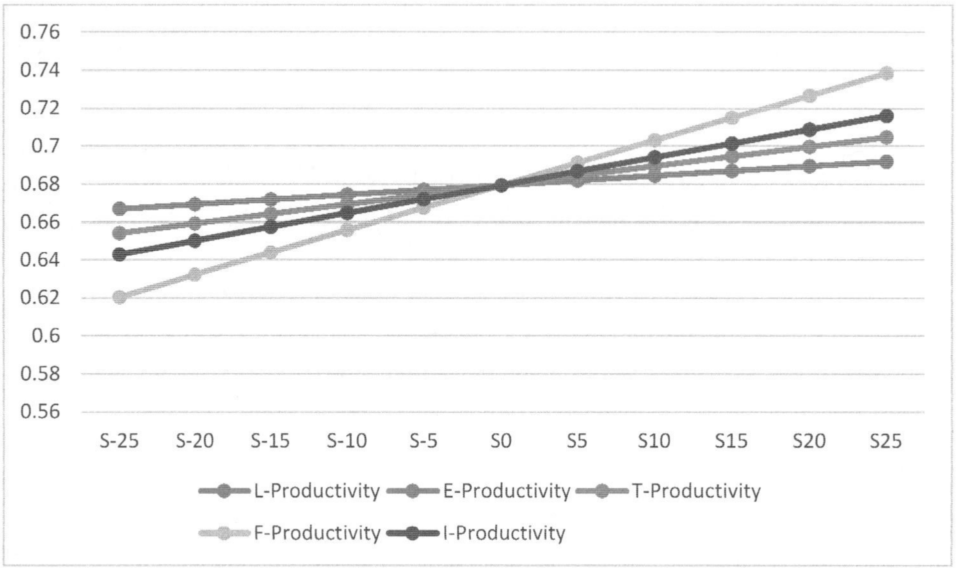

Figure 1: The effect of reductions and increments of guard categories on the machine guard compliance productivity

Results from the VMGPEM analysis shows that period six is the period with maximal guard compliance productivity. The sensitivity analysis was therefore carried out on the individual guard categories in period-6 to determine the guard category with the potential of further increasing the period's compliance productivity and to what optimal percentage increase the guard category can be adjusted to. The guard categories considered here are; Fixed guards (F), E-Stop guards (E), Lagging guards (L), Trip devices (T) and Interlocking guards (I). Figure 1 shows the effect of reductions and increments of guard categories on the machine guard compliance productivity. The maximum productivity of 73.84% was obtained when fixed guards where upped by 25%. While the guard that had less impact on the productivity was lagging guard, having 69.18% after 25% increment. This implies that a 25% increase in the percentage compliance of fixed guards will optimize the machine guard compliance during the sixth period to 73.84%.

5. Conclusion

The sensitivity analysis carried out on the variant machine guard productivity evaluation model is imminent to figure out lapses to be adequately tackled for a profitable work experience in the bottling plant. This study being a maiden work on optimizing machine guard compliance productivity breached the gap successfully for the bottling plant used for the experiment. The OFAT technique employed for this work provides a clear indication of the guard category that needs to be given attention to by machine operators with respect to compliance to its engagement during processing periods. Advanced work will be advised to be carried out by the safety and maintenance department using other adoptable methods to improve results from the VMGPEM by understudying how improvement on the number of functional machine guards across the guard categories will affect the machine guard compliance productivity values.

IOP Conf. Series: Materials Science and Engineering **413** (2018) 012074 doi:10.1088/1757-899X/413/1/012074

Acknowledgments
Research is supported by the Covenant University Center for Research, Innovation and Discovery (CUCRID).

References

[1] ECKHARDT B. Machine guard improvements. Concrete Products, 104(9): 29.
[2] SORANNO C. Five serious machine guarding problems. Professional Safety, 2012, 57(2): 64.
[3] RUBINGER J. AND HAMILTON J. Machine guard guidelines. Wood and Wood Products, 2013, 118(2): 18.
[4] CORDIER P J. Machine guarding remains a safety concern. Pulp and Paper, 2006, 80(3): 27.
[5] BERKE L. Safe but slow: How long' till operators remove the guard? Machine Design, 2008, 80(20): 40.
[6] TIAN W. A review of sensitivity analysis methods in building energy analysis, Renew. Sustain. Energy Rev, 2013, 20(30): 411-419.
[7] NAVID D, BEHRANG S, KHADIJEH A, SAEED D. Sensitivity analysis of building energy performance: A simulation-based approach using OFAT and variance-based sensitivity analysis methods. Journal of Building Engineering, 2018, 15(17): 181-193.
[8] PALEARI L, CONFALONIERI R. Sensitivity analysis of a sensitivity analysis: We are likely overlooking the impact of distributional assumptions. Ecological Modelling, 2016, 340:57–63.

IOP Conf. Series: Materials Science and Engineering **413** (2018) 012075 doi:10.1088/1757-899X/413/1/012075

An alternate productivity evaluation practice to depict machine guarding performance in a process plant

Uzor Chukwunedum [1], Oke Sunday [2], Okolie Sociis[1],

Covenant University, Ota, Nigeria 2 University of Lagos, Yaba, Nigeria E-mail address: chukwunedumuzor@gmail.com

Abstract. Varying from the notion that the significant expenditure channeled on machine guarding has little or no benefit, this paper through an innovated model, argues that the machine guarding productivity could be evaluated for progress in a bottling production company. This paper contributes to the literature on machine guarding by developing a novel variant machine guarding productivity evaluation model which was validated with dataset from a process plant in Nigeria. Well descriptive results from the model shows that when period 5 is compared with period 1, the model showed an increase in % compliance for trip guards, fixed guards and e-stop guards by 2.82%, 1.41% and 2.53% respectively, while the number of functional guards remained unchanged across all guard categories. As a result of the increment in % compliance for the afore mentioned guard categories, there was an increment in machine guard compliance output for the affected guard types, thus a rise in total machine guard compliance output from 985.1 to 993.08. The machine guard productivity for periods one and five showed a 34.54% and 34.01% decrease for the respective periods from the original status. It also reveals a 0.81% change in guard compliance productivity from period 1 to period 5. The model showed the efficiency of the framework in evaluating operator's productivity performance as a useful source of information for operations managers to trace and adjust quantifiable productivity targets of the machine compliance scheme.
Keywords: machine guard; compliance

1. Introduction

Scholars have increasingly directed interest to a kind of safety process called the machine (equipment) safety [1-3]. Their investigations have revealed that machine safety is largely associated with other safety categories in industry, such as the human safety, occupational safety [4], environmental safety, and the degree of attention paid to these safety categories will determine the level of safety in the industrial enterprise [5]. While the safety literature has made substantial advancement in bringing out interesting ideas concerning the dynamics of safety and the intricate nature of machine safety, numerous principal questions about machine safety stands unanswered [6-7]. In particular, despite the fact that previous research has strongly emphasized the significance of high productivity in the sustenance of machine safety guards, researchers have, in general restricted their interest to the design aspect: an area of research that is believed to lessen the hazards provided by the machine through continuous weight reductions in new designs, enhancement in the overall cost of the equipment and enhancement in equipment noise production and other machine safety guard quality enhancement issues.

There are a number of probable problems with this method of machine safety guarding. The collective emphasis on design has robbed the safety community of the immense benefits of evaluating both the operator and the workplace. The operations manager is unclear about how to achieve optimum results from the guarding influences and compliance of the operator as there is absolute confusion on what to get done in terms of productivity since no concrete measurements are being done concerning inputs and the responses of the machine safety to guarding system. Secondly, a huge challenge exists in properly evaluating the value of investment on machine guards and reports are difficult to create and this in turn leads to long period of waiting for requests for machine guards related issues by operator due to tactical delay by management not to expend on the activity. The suggestion is that a radical pursuit of productivity measurement programs for the machine safety operator directed at the employee's compliance to the use of machine safety guarding from the productivity evaluation perspective may cause a turnaround of the unattractive situation. In this research, these principal challenges are resolved by drawing inspiration from the theory of productivity to develop a machine guarding compliance model based on the ability and willingness of the operator to comply with the judicious utilization of materials via surrogate measures of incorporating the functionality of guards and this will clearly help create laid out procedure for evaluating the operators productivity and the operator's productivity is made and the operator is aware of his/her periodic performance. Little dependence on the operator is made for certain information on machine guards and request by management or third parties could easily be satisfied by the operations/safety manager instead of putting burden on the operator for information.

The model is principally segregated into inputs and outputs. The numbers of functional guards of the diverse categories are carefully observed for their activations in use by the operator during any production activity through a physical count of the "on" and "off" states of machine guarding usage by the operator over the study period. The responses are then taken as the amount of guards in the right state of usage, indicating the output of the system. The model reveals how the ratio of output to inputs could be used as a parameter for the judicious resource usage in comparison with unity. A value less than unity is not satisfactory indicating poor performance while unity indicates acceptable performance and a value greater than unity shows outstanding performance that is desirous to be sustained and improved upon.

2. Brief review of literature

In 2009, *Parker et al.* published the report of an investigation that centered in interventions for safety and the primary attention of the researchers were directed to how the fabrication industries could be assisted with respect to safety programs [8]. The current investigators noted with interest that this report adds knowledge to the domain of interest in this investigation from the perspective of inputs for the manufacturing system, to enhance the machine guarding performance appraisal. In 2012, *Peabody* in their contribution that broadly centered on the movable as well as fixed guards (i.e. Categories as well as B), considered the protection of operators from any kind of hazards both in the whole span of life of the machine as well as for the daily and short-term interactions with the machines [9]. The interest of *Rubinger and Hamilton* in 2013 was a machine guarding in wood product manufacturing systems. A safety-conscious mindset was perceived as the gateway to avoiding accidents [10]. The review of this investigative document is relevant to the current researchers' study focus on evaluating the performance (productivity) of machine guard operators in a manufacturing system. *Foo and Wong* in 2015 appraised the risk in a manufacturing organization devoted to cutting tool usage [4]. In 2017, *Kyslinger* produced a novel roof bolting equipment for underground mining wherein guarding was apprehended to restrict the operator's reach area to those outside the possible accidental contact with a drill [11]. This research work is relevant to the current study of productivity evaluations of the machine guarding operator wherein the proper consideration of the guarding would lessen the risk of accident by the operator. A study was carried out in 2016 by *Wong and Lee* studying the intention of employees to comply with safety regulations among manifold ethnic classes of people [12]. In 2016 *Hu and Bertuleit* further focused on the integration of the theory of technological acceptance in the workplace [13]. Drawing insights from the prior literature concerning machine guarding, it is ascertained that a huge volume of studies has been contributed on machine guarding fundamentals and the use of basic compliance or non-compliance verifications of the operators at the workstation and not on established performance metrics of productivity indices. The bridging of this huge gap is an effort made in this research as the study focuses on the quantification of productivity of the machine guarding operator in terms of utilization of the scarce resources provided to the operator from the perspective of output to input vis-à-vis the machine guarding function.

3 Methods

The main objective of this innovative investigation is to ascertain the machine guard efficiency in the bottling plant used as a case study; to identify lags in the company's safety status as well as guard maintenance routines for adequate attention, to see what periods had the optimal and minimal functional number of machine guards and also the periods with greater operator's compliance to machine guard usage. To get this accurately done the Variant Machine Guard Productivity Evaluation Model (VMGPEM) was employed to achieve the desired purpose. This is gotten by the evaluation of different types of machine guards used on equipment along all production process lines in the plant and further the efficiency values are obtained. Literatures were analyzed to point out gaps and inefficiencies in the measurement of machine guard compliancy. The case outfit is a bottling plant located in western region of Nigeria. The firm comprises of four production process lines with each having eight processing machines for the entire production. Data for the machine guards were obtained from the maintenance inspection unit for a period of 12 months. From the data it was observed that equipment in each process line comprised of five different types of guards; fixed guards, Interlock guards, trip guards, lagging guards and E-stop guards. These guards and its compliance by machine operators are major determinants in measuring machine guard compliance productivity using the innovated VMGPEM. The relevant data after taken were properly collated. The VMGPEM follows a systematic tabulated procedure to give desired output.

An overview of relevant terms and model construct is as follows:

Table 1 Variant machine guard productivity evaluation model (VMGPEM)

	Period 1			Period 2			Biased Variation Ratio		
Current Status	αi	Θi	$\alpha i\Theta i$	αi	Θi	$\alpha i\Theta i$	$\alpha 2\Theta 1/\alpha 1\Theta 1$	$\alpha 2\Theta 2/\alpha 2\Theta 1$	$\alpha 2\Theta 2/\alpha 1\Theta 1$
F									
I									
E									
L									
T									
								$\Delta =$	
TOTAL									
Original Status									
F									
I									
E									
L									
T									
TOTAL									

In VMGPEM, a base period is taken and used to compare outputs for other periods. Table 1 reflects a part of the model where:

- Period 1: is the selected base period used to compare the other periods. It is also the first month from the data.
- Period 2: indicates the next period being compared with period 1. F: symbolizes fixed guards I: signifies interlock guards
- E: represents e-stop guards L: stand for lagging guards
- T: is for trip guards
- αi: denotes number of functioning guards in a period
- Θi: is the % compliance in the usage of machine guards
- $\alpha i\Theta i$: represents guard compliance output
- $\alpha 2\Theta 1/\alpha 1\Theta 1$: is the weighted guard compliance output ratio having % usage compliance constant to reflect actual changes in number of functional guards
- $\alpha 2\Theta 2/\alpha 2\Theta 1$: signifies weighted guard compliance output ratio with the number of functional guards constant to reveal the actual variation in % compliance

IOP Conf. Series: Materials Science and Engineering **413** (2018) 012075 doi:10.1088/1757-899X/413/1/012075

- $\alpha2\Theta2/\alpha1\Theta1$: is for guard compliance ratio Δ: change in compliance to machine guard Current Status: indicates the present number of functional machine guards for each period. Original Status: this shows the actual number of guards that are meant to be functional for each.

4. Result

Table 2 VMGPEM analysis of period 1 and period 5

	Period 1			Period 5			Biased Variation Ratio		
Current Status	αi	Θi	αiΘi	αi	Θi	αiΘi	α2Θ1/α1Θ1	α2Θ2/α2Θ1	α2Θ2/α1Θ1
F	487	0.71	345.77	487	0.72	350.64	1	1.014084507	1.014084507
I	354	0.58	205.32	354	0.57	201.78	1	0.982758621	0.982758621
E	279	0.79	220.41	279	0.81	225.99	1	1.025316456	1.025316456
L	103	0.64	65.92	103	0.61	62.83	1	0.953125	0.953125
T	208	0.71	147.68	208	0.73	151.84	1	1.028169014	1.028169014
								Δ =	1.0081007
TOTAL			985.1			993.08			
Original Status									
F	490	1	490	490	1	490	1	1	1
I	420	1	420	420	1	420	1	1	1
E	280	1	280	280	1	280	1	1	1
L	105	1	105	105	1	105	1	1	1
T	210	1	210	210	1	210	1	1	1
TOTAL			1505			1505			
Productivity for period 1 =		0.654551		productivity for period 5 =			0.65985382		
Change In Guard Compliance productivity =				1.0081007					

Table 2 above shows a layout of the model comparing period-5 with period-1. The model showed that there was an increase in % compliance for fixed guards, e-stop guards and trip guards by 1.41%, 2.53% and 2.82% respectively, while the number of functional guards remained unchanged across all guard categories. Due to the increase in % compliance for certain guards there was an increment in machine guard compliance output for the affected guard types, hence an increase in total machine guard compliance output from 985.1 to 993.08. Productivity being the ratio of the current status value for total machine guard compliance of periods to the total the original status value of total machine guard compliance, we have that the machine guard productivity for period one and two is 0.6546 and 0.6599 respectively. This can be interpreted as a 34.54% and 34.01% decrease in productivity for the respective periods from the original status. It also reveals a 0.81% change in guard compliance productivity from period 1 to period 5. This obviously speaks to the point that an increased percentage compliance to guard usages by operators will also increase the machine guard productivity.

5 Conclusion

The existent literature on machine guarding so far has considered the design of machine guards to attain the required safety limit standards by supervisory bodies, but issues concerning the performance of operators monitoring the machine workplaces is almost absent. In this contribution, this particular challenge of productivity quantification for the machine guarding operator has been overcome through the employment of the competent productivity quotients. Visualizing parameters as inputs and other as outputs, while the quotients are carefully analyzed. Data from a soft bottling processing organization was considered to establish the competence of the proposed methodology. An advanced approached in this research is largely appropriate to solve and attain productivity indices for process industries such as those in the drinks manufacturing system. The results demonstrate that the productivity appraisal approach is competent in diagnosing the productivity performance in terms of being poor or good and causal factors may be further investigated in subsequent studies as well as other human factors such as fatigue could be analyzed to further understand the dissimilarities in the productivity indices.

Acknowledgments

Research is supported by the Covenant University Center for Research, Innovation and Discovery (CUCRID).

References

[1] GOOMAS D T, YEOW P H P. IT-assisted equipment safety checks system to improve compliance: A case study at a distribution center. Safety Science, 2013, 60: 77-86.
[2] PALAZZI E, CURRÒ F, FABIANO B. A critical approach to safety equipment and emergency time evaluation based on actual information from the Bhopal gas tragedy. Process Safety and Environmental Protection, 2015, 97: 37-48
[3] VAHDATIKHAKI F, LANGARI S M, TAHER A, EL-AMMARI K, HAMMAD A. Enhancing coordination and safety of earthwork equipment operations using multi-agent system. Automation in Construction, 2017, 81: 267-285.
[4] FOO S H. WONG K Y. Occupational safety and health improvement in a cutting tools manufacturing company. Proceedings of the 26th International Business Information Management Association Conference: Innovation Management and Sustainable Economic Competitive Advantage: From Regional Development to Global, 2015: 446-462.
[5] QUARTEY S H. Examining employees' safety behaviors: an industry-level investigation from Ghana. Personnel Review, 2017, 46(8): 1915-1930.
[6] MOHAMED S, CHINDA T. System dynamics modelling of construction safety culture. Engineering, Construction and Architectural Management, 2011, 18(3): 266-281.
[7] RIPAMONTI S C, SCARATTI G. Safety learning, organizational contradictions and the dynamics of safety practice. Journal of Workplace Learning, 2015, 27(7): 530-560.
[8] PARKER D L, YAMIN S C, BROSSEAU L M, XI M, GORDON R, MOS I G, STANLEY R. National machine guarding program: Part 1, Machine safe-guarding practices in small metal fabrication businesses. American Journal of Industrial Medicine, 2015, 58: 1174-1183.
[9] PEABODY J. The dos and don'ts of fixed and moveable machine guards. Today, 2012, 5(1): 39.
[10] RUBINGER J, HAMILTON J. Machine guard guidelines. Wood Products, 2013, 118(2):18.
[11] KYSLINGER B. Development of a six drill head roof bolting machine," SME Annual Conference and Expo 2017: Creating Value in a Cyclical Environment, 2017, 196-200.
[12] WONG D B, & LEE S G. Modeling the predictors of intention in workplace safety compliance of a multi-ethnic workforce. Safety Science, 2016, 88: 155-165.
[13] HU X, GRIFFIN M A, BERTULEIT M. Modeling antecedents of safety compliance: Incorporating theory from the technological acceptance model. Safety Science, 2016, 87: 292-298.

IOP Conf. Series: Materials Science and Engineering **413** (2018) 012076 doi:10.1088/1757-899X/413/1/012076

Investigating the Production Quality of Electrical Porcelain Insulators from Local Materials

V.C. Nwachukwu and S.A. Lawal*

Department of Mechanical Engineering, School of Infrastructure, Process and Engineering Technology, Federal University of Technology, Minna Nigeria
Corresponding author email: lawalsunday@futminna.edu.ng

Abstract- Electrical porcelain insulators were manufactured from Agai and Kutigi Clay and other locally available raw material i.e. Feldspar (Kagara), Quartz (Kadna) and Talc (Kagara). Compositions were formulated using a universal triaxial porcelain composition (kaolin, 28%; ball clay, 10%; feldspar, 35% and quartz, 25%) to produce test pieces. The pulverized and mixed compositions were molded and fired at a temperature of 1250°C. The physical and electrical properties were investigated. The results show that the samples have electrical resistance between 3.52 x 10^8 Ω and 2.05 x 10^9 Ω, and dielectric constant between 8.7 and 11.4, with respect to the frequency considered. The results of the physical tests i.e. linear shrinkage, bulk density, water absorption and failing load were also in agreement with the standard for porcelain insulators.
Keywords: Porcelain; Shrinkage; Bulk density; Water absorption.

1. Introduction

Locally developed technology is considered an essential index for exploring and promoting underutilized resources, technological and economical potential of a nation's industrialization processes. It has been established that abundant raw materials are available for the manufacture of electrical porcelain insulators as well as heated ceramic wares in Nigeria. Insulators are extensively used for high and low voltage applications in generation, transmission and distribution of electrical power. The need for insulation is essentially to prevent passage of electricity to some other device or area, so that the electricity does not cause harm or cause death to those who touch areas or devices connected to the electrical insulators. Despite the enormous wide range of application and availability of raw materials, most used insulators are still being imported to Nigeria. It is hereby imperative to facilitate manufacturing of locally produced electrical insulators to meet increasing demand. Hence, the development of techniques to produce high quality porcelain units that are commercially viable and meet required standards cannot be overemphasized. Porcelain is a ceramic material made by heating raw materials, generally including clay in the form of kaolin, in a kiln to temperatures between 1,200 °C (2,192 °F) and 1,400 °C (2,552 °F). Porcelain is primarily composed of clay, feldspar and filler material, usually quartz or alumina. The clay [$(Al_2Si_2O_5(OH)_4$] gives plasticity to the ceramic mixture, flint or quartz [SiO_2] maintains the shape of the formed article during firing and feldspar [$KxNai\text{-}x(AlSi_3)\ O_8$] serves as flux. These three constituents place electrical porcelain in the phase system in terms of oxide constituents, hence the term triaxial porcelain. Most existing literature on body compositions and processing conditions for porcelains of all kinds such as Norton [1] applies mainly to foreign raw materials, which can be quite different from the local ones in terms of chemical, mineralogical and physical characteristics. Therefore, more efforts should be channeled to establish data and procedures on the development of electrical porcelain with local raw materials. Attempts was made by Onaji and Usman [2] to develop a slip casting technique for the electrical porcelain body using ball clay materials and the result was fruitful. According to Biffi [3], the compression pressure to produce porcelain tiles must be between 350 and 450 kg/cm^2 because the density of particles must allow oxidation of organic substances and removal of gases that are generated during firing. Therefore, this study aimed at producing and characterizing porcelain insulators utilizing kaolin, plastic clay, quartz and feldspar.

IOP Conf. Series: Materials Science and Engineering **413** (2018) 012076 doi:10.1088/1757-899X/413/1/012076

2. Materials and Methods

2.1 Materials

The raw materials used in this research include Plastic Clay, Quartz. Feldspar, Kaolin and Talc as shown in Fig. 1. Other equipments used are Glost kiln (for firing the product), Electric dryer (for drying), Universal weighing machine, Crusher, Mesh (300µm), Mixer, Mould ball mill and Blunger.

(a) Kaolin	(b) Feldspar	(c) Quartz	(d) Plastic Clay

Fig. 1. Production materials

2.2 Methods

2.2.1 Processing of kaolin, quartz, feldspar, talc and plastic clay

The raw materials were sourced from the respective location and then purified by washing and drying to remove deleterious material. This was followed by crushing operation to reduce the materials into a smaller lump. According to the percentage composition presented in Table 1, batch weighing of each materials was carried out, some quantity of water was then added to the paste and then charged into the ball mill for wet milling. Each of the ten batches were milled for 5hrs, 6hrs, 8hrs, 9hrs, 10hrs, 11hrs, 12hrs, 13hrs, 14hrs and 15hrs respectively. Thereafter, the resulting slip was made to undergo blunging for two hours to produce a liquid slurry having the tendency to flow. Blunging operation promotes balance plastic charge and better integration. The slip from the blunger was emptied into a sieve of about 300µm size to determine the particle and then passed through a classifier to ascertain appropriate particle size. After classifying the slip, it which was satisfied okay was then magnetized by passing it through a magnetic sieve to remove the iron content. The resulting material was then kept for 24hrs for efficient ageing and degassing. Other material percentage composition was varied except for talc that was kept constant throughout the whole process.

2.2.2 Casting process

Solid state casting was the method employed in this process. A mould made of Plaster of Paris (POP), was used in each of the batch casting. The suspension was poured into porous plaster moulds where capillary forces sucked the water into the moulds - from the slip leading to a steady dispersion of clay particles in dense face to face packing on the inside surface of the moulds in a process called hollow casting process. It is called a hollow casting process because draining of water from the slip is required. After draining, the resulting mixture could stay in the mould for 3hrs before pulling it out from the plaster mould. The resulting casted samples were arranged in iron trays and allowed to air dry for three days and then - sent into a ceramic electric dryer to further drying for six hours. The mould used in the casting process is shown in Fig 2.

IOP Conf. Series: Materials Science and Engineering **413** (2018) 012076 doi:10.1088/1757-899X/413/1/012076

Fig. 2: Casting mould made of POP

2.2.3 Fettling of porcelain insulators

Fettling is a post casting operation carried out to remove rough edges and contours from the casted ceramic (Fig. 3a) products to get them ready for drying and firing. In this experiment, the surfaces of the resulting casted products were then fettled with the help of a grit paper and then a little foam, socked with water was used to remove the deposits before glazing. Figure 3b is a sample of the fettled product.

(a) Casted specimen; unfired, unglazed and unfelted.

(b) Fettled specimen; unfired and unglazed

Fig. 3: Produced samples

2.2.4 Glazing

The fettled products were then subjected to glazing prior to loading into Glost kiln furnace for firing. Zinc oxide was mixed with powdered calcium carbonate and milled. The resulting mixture was then sprayed on the produced specimens.

2.2.5 Drying and firing

The glazing operation was followed by a drying process. The products from the casting operation, were placed on a wooden board for 24 hours for air drying. This was followed by oven drying at 105^0C for 6 hours to eliminate vapour. Fig. 4 shows the fired samples.

IOP Conf. Series: Materials Science and Engineering **413** (2018) 012076 doi:10.1088/1757-899X/413/1/012076

Fig. 4: Fired sample

2.3 Determination of insulator properties

The linear shrinkage was determined using the three equations below [1; 4; 5] while the value of the water absorption was obtained using Equ. 4.

$$\text{Drying shrinkage} = \frac{L_W - L_d}{L_d} X\ 100 \quad 1$$

$$\text{Fired shrinkage} = \frac{L_W - L_f}{L_d} X\ 100\% \quad 2$$

$$\text{Total shrinkage} = \frac{L_W - L_f}{L_W} X\ 100\% \quad 3$$

Where L_W = Wet length, L_F = Fired length and L_D = Dry length.

$$\text{Water Absorption} = \frac{W_s - W_d}{W_d} X\ 100\% \quad 4$$

Where W_S = Soaked weight after boiling for 100^0C for 2 hours and W_d = Dry weight.

Also, the strength of the porcelain insulators was investigated by determining their failing load according to [4]. The failed load was determined using Equ. 5, while loss in ignition was determined using Equ. 6. The percentage residue was determined using Equ. 7:

$$\sigma_T = \frac{\rho}{A} \text{ but } A = b * t \text{ hence } \rho = \sigma_T(b * t). \quad 5$$

And the young modulus can be calculated using the formula below: $E = \sigma_t/e$, where σ_t = tensile strength of the composite material, ρ = Maximum load, b = width and t = thickness.

$$L.O.I = \left(\frac{m_2 - m_3}{m_2 - m_1}\right) x\ 100 \quad 6$$

Where m is the mass of the porcelain.

$$\%Residue = \frac{Dry\ Weigth\ x\ 100}{A} \quad 7$$

Where A, is the mass of the sample obtained using; A = (Litre)(Weight) – $1000dm^3$ x 1.625.

2.4 Insulation resistance, dielectric strength and x-ray fluorescence spectrometer (XRF)

A 5kV Megger (model No:AM-5KV) insulation tester was used to measure the insulation resistance of each sample at the Enugu Electricity Distribution Company (EEDC) after firing at 1250°. Copper plates with terminal heads were placed on both side of the porcelain material and voltage of 5000V was passed through the porcelain material and the resistance of the insulator was then measured from device in MΩ and GΩ. The dielectric strength test was measured at Enugu Electricity Distribution Company (EEDC) Achalla Layout office in Enugu, Enugu State using Hipotronics (model number 880/PL-10mA-B). An x-ray fluorescence spectrometer was used in determining the chemical compositions of the raw materials. This test was conducted on the five samples using an X-Ray fluorescence spectrometer method available at National Geoscience Research Laboratory (NGRL) Kaduna and the result of the experiment is tabulated in the appendices.

IOP Conf. Series: Materials Science and Engineering **413** (2018) 012076 doi:10.1088/1757-899X/413/1/012076

3. Results and Discussions

3.1 Constituent charge materials and chemical analysis of constituent materials

The constituent percentage composition of each material used in the production of each of the ten specimen is presented in Table 1, while the ore composition and the result are presented in Table 2.

Table 1: Constituent percentage compositions of porcelain insulators

Specimen	Feldspar	Quartz	Clay		Talc
			Kaolin	Plastic clay	
	%	%	%	%	%
1	0	0	72.21	25.79	2
2	5	5	64.84	23.16	2
3	10	10	57.47	20.53	2
4	15	15	50.10	17.90	2
5	20	20	42.74	15.26	2
6	25	25	35.37	12.63	2
7	30	30	28.00	10.00	2
8	35	35	20.63	7.37	2
9	40	40	13.26	4.73	2
10	35	25	28.00	10.00	2

Table 2: Chemical characterization of the materials (kaolin, plastic clay, feldspar, quartz and talc).

Oxide Composition %	Feldspar	Quartz	Kaolin	Plastic Clay	Talc
SiO_2	68.4	94.5	66.7	58.75	52.9
Al_2O_3	17.9	0.06	23.71	27.6	17.38
SO_3	ND	3.07	ND	1.2	ND
K_2O	9.33	0.39	0.31	0.38	0.31
CaO	0.19	0.6	0.8	0.16	2.1
Fe_2O_3	0.73	0.422	0.478	0.637	8.4
MgO	0.16	ND	0.08	ND	3.73
SrO	0.032	ND	0.024	0.014	0.088
PbO	0.017	0.015	ND	0.019	ND
BaO	0.065	0.03	ND	0.08	ND
MnO	0.08	0.025	0.005	0.018	0.37
TiO_2	0.12	0.11	0.133	0.11	0.559
V_2O_5	0.01	ND	ND	0.035	0.02
Cr_2O_3	ND	0.01	ND	0.032	0.772
CuO	0.02	0.029	0.016	0.024	0.077
Cl	ND	ND	ND	0.604	ND
ZnO	ND	0.055	0.002	0.004	0.01
Na_2O	2.13	ND	0.23	ND	0.02
ZrO_2	0.022	0.054	0.02	0.171	ND
Ta_2O_5	ND	ND	ND	ND	0.07
NiO	ND	ND	ND	ND	1.28
L O I	0.73	0.63	7.49	10.16	11.91

ND = NOT DETECTED

It was observed from Table 2, that the percentage of silica present in the quartz is about 94.5% which follows the requirement. The clay is also an alumino-slilicate clay judging from the silica and alumina content of the clay which was 58.75 and 27.6% respectively. Ta_2O_5 and NiO were detected in Talc but absent in the other constituents. About 0.604% of chlorine was found in plastic clay and then absent in other constituent materials.

3.2 Physical tests of the produced specimens

Several physical tests, which includes bulk density determination, shrinkage test (total shrinkage, fired shrinkage and dry shrinkage), failing load and water absorption were carried out on the produced specimen and the result is presented in Table 3, it should be noted that physical test was not carried out on Specimen 1 and 2, because they did not come out well after the casting operation (they stuck on the mold after casting).

Table 3: Physical properties of produced porcelain insulators

	Specimen							
Properties	3	4	5	6	7	8	9	10
Total Shrinkage (%)	8.70	4.19	3.20	3.16	3.01	2.97	5.76	2.60
Fired Shrinkage (%)	0.6	0.98	1.06	0.8	0.37	0.32	1.84	0.60
Dry Shrinkage (%)	8.1	3.21	2.14	2.36	2.64	2.65	3.92	2.00
Water Absorption (%)	3.72	2.89	2.82	2.63	2.60	2.59	3.46	2.52
Bulk Density (g/cm3)	2.14	2.16	2.29	2.37	2.42	2.49	2.74	2.52
Failing Load (kN)	1.97	2.55	2.57	2.67	2.69	2.73	1.92	3.01

From Table 3, it was observed that average total shrinkage for each of the samples was within the recommended value for porcelain production [6]. Higher shrinkage values result in warping and cracking of the porcelain wares resulting in reduction in its strength. Specimen 9 with a shrinkage percentage of 5.76% warped after the firing operation while Specimen 3 with a shrinkage percentage of 8.70% came out well but with cracks.

These observed anomalies are also as a result of the high shrinkage values, high feldspar content (in the case of specimen 9), and low amount of feldspar (in the case of specimen 3) and quartz since quartz serves as the main refractory backbone. But generally, the shrinkage values are still within the accepted ranges. Also, from Table 3, it was noted that water absorption decreases with increase in kaolin content and decrease in feldspar content. Feldspar as a flux which permits better flow of the mixture and allows for reaction with impurities to form slag which are burnt off at elevated temperatures during firing.

It was noted that too low amount of feldspar result in the formation of immature wares characterized with the formation of cracks and warpage. It could also result in high water absorption and consequently, high shrinkage. It is therefore advisable to use moderate amount of feldspar in porcelain insulator preparation. Similarly, the bulk density experienced a general slight reduction with increasing amount of kaolin as feldspar was reducing as shown in Table 3. This was because of higher value of loss on ignition of kaolin (7.49) as against feldspar (0.73) given by the chemical analysis shown in Table 4. Bulk density is an important property in porcelain wares. Bulk densities of the mixed samples lie within the range of 2.14 – 2.74g/cm^3 which fall in standard requirements for porcelain body [7]. It was observed from Table 3 that the failing loads increases with increase in percentage of kaolin, this goes to show that porcelain strength are enhanced as kaolin content is increased.

3.3 Electrical properties of the porcelain insulators
3.4

The results of the test for electrical resistivity and di-electric strength carried out for the porcelain specimens are shown in Table 4.

Table 4: Dielectric properties of the produced insulator samples

Sample	Frequency (Hz)	Dielectric Constant	Loss Angle	Dissipation Factor
3	50	11.4	8.06	0.133
4	100	11.1	7.21	0.127
5	1000	10.9	7.01	0.119
6	10000	10.3	6.93	0.112
7	20000	9.7	5.47	0.079
8	30000	9.4	4.22	0.061
9	40000	11.1	3.8	0.056
10	50000	8.7	3.2	0.052

It can be seen from Table 4 that the values of the di-electric test ranges between 8.7 and 11.4. Though increases were observed in the value of dielectric constant for samples 3, 4 and 11 dues to the physical defects associated with the porcelain insulators. The ranges of the values of dielectric constant observed for the samples suggestive of a good insulating material. Good insulating materials are materials with dielectric constants below 12 [8-10]. The significance of the reduced dielectric constant is that the charge storage capacity of the insulator is low which also result in low possibility of electrical shock. The values of the electrical resistance of the tested insulators is presented in Table 5.

Table 5: Measure electrical resistance and the corresponding voltage applied and leaked voltage

Specimen	Voltage (Volts)	Resistance (Ω)	Leakage Current
3	5000	3.52×10^8 Ω	1.42×10^{-5} A
4	5000	4.06×10^8 Ω	1.23×10^{-5} A
5	5000	5.25×10^8 Ω	9.52×10^{-6} A
6	5000	5.27×10^8 Ω	9.48×10^{-6} A
7	5000	9.13×10^8 Ω	5.48×10^{-6} A
8	5000	1.15×10^9 Ω	4.35×10^{-6} A
9	5000	3.62×10^8 Ω	1.38×10^{-5} A
10	5000	2.05×10^9 Ω	2.44×10^{-6} A

It can be seen from Table 5 that the values of the electrical resistance of the produced specimen ranges between 3.52×10^8 Ω and 2.05×10^9 Ω. This shows that the electrical resistance range of the produced materials is more than the recommended range of 1.0×10^6 Ω [11-12]

4. Conclusions

This research focused on local development of electrical porcelain using kaolin, ball clay, feldspar, quartz and talc, sourced from Niger State, Nigeria. Mechanical, physical and electrical tests were carried out on the produced material. The outcomes of the research are hereby summarized as follows;

1. Good porcelain bodies were produced from kaolin clay, ball clay, feldspar, talc and quartz. Sample 8 with composition 20.63% kaolin, 7.37% plastic clay, 35% feldspar and 35% quartz and 2% talc, was found to possess the highest failing load of 8.0kN, water absorption of 2.5%,

IOP Conf. Series: Materials Science and Engineering **413** (2018) 012076 doi:10.1088/1757-899X/413/1/012076

bulk density of 2.49 g/cm^3, with appreciable insulation resistance of 1.15 Giga ohms at injection of 5,000 volts.

2. That using the locally available raw materials, electrical porcelain with good dielectric properties can be produced since it has dielectric constant below 12.
3. The insulators may be employed for usage at high frequencies where the dissipation factor is correspondingly lower.

References

[1] Norton, F.H (1980), Ceramics for the Artist Potter, Addison-Wesley Publishing Company Inc., Massachusetts, USA.

[2] Onaji, P. B. and Usman M, (1988), Development of Slip Cast Electrical Porcelain Body: Bomo Plastic Clay, Nigeria Journal of Engineering (NJE), Vol. 5, No.2, pp.89-96.

[3] Biffi GO (2002) Grit Porcelain: Manual of Manufacture and Techniques of Employment. Ed. Faenza Editrice do Brasil.

[4] ASTM C356-17, Standard Test Method for Linear Shrinkage of Preformed High-Temperature Thermal Insulation Subjected to Soaking Heat, ASTM International, West Conshohocken, PA, 2017, www.astm.org

[5] Clay Shrinkage Testing (2010), [online], Accessed 13th August 2017, Available from World Wide Web: http://www.potters.org/subject01835.htm

[6] Chester J.H (1973), Refractories, Production and Properties. The Iron and Steel Institute, London, Pp 4-13, 295-315.

[7] Omotoyinbo.J.A and Oluwole O.O (2008) "Working Properties of Some Selected Refractory Clay Deposits in South Western Nigeria" Journal of Minerals and Materials Characterization and Engineering, Michigan Technological University, USA. 7(3),233-245

[8] Anih, L.U (1988). Characterization of Kaolin-Quartz-Feldspar Triaxial Porcelain for Insulation Application (Msc). Obafemi Awolowo University, Ile-Ife.

[9] Salawu, E. Y., Okokpujie, I. P., Afolalu, S. A., Ajayi, O. O., & Azeta, J. (2018). INVESTIGATION OF PRODUCTION OUTPUT FOR IMPROVEMENT. International Journal of Mechanical and Production Engineering Research and Development, 8(1), 915-922.

[10] Azeta, J., Okokpujie, K. O., Okokpujie, I. P., Osemwegie, O., & Chibuzor, A. (2016). A Plan for Igniting Nigeria's Industrial Revolution. International Journal of Scientific & Engineering Research, 7(11), 489.

[11] Anih, L. U. (1997) "Characterization of kaolin- feldspar quartz triaxial porcelain for insulator applications" Proceedings of Electric Power Engineering Conference University of Nigeria, Nsukka

[12] Anih L.U. (2005) "Indigenous manufacturer and Characterization of Electrical Porcelain Insulator" Nigerian Journal of Technology, 24 (1).

IOP Conf. Series: Materials Science and Engineering **413** (2018) 012077 doi:10.1088/1757-899X/413/1/012077

Direct Air Capture: A Review of Carbon Dioxide Capture from the Air

A. A. Okesola, A. A. Oyedeji, A. F. Abdulhamid, J. Olowo, B. E. Ayodele, T. W. Alabi

National Agency for Science and Engineering Infrastructure (NASENI), Idu Industrial Area. PMB 391 Garki, Abuja
Corresponding Author: yomiokesola@yahoo.com
Mobile No: 08188229756

Abstract- The level of carbon dioxide as well as other noxious gases in the atmosphere has been on the increase and this has caused a lot of issues relating to the effect of these gases to human and animal health. A process for capturing carbon dioxide directly from the atmosphere will go a long way in making the world a safer and healthier place for the entire inhabitant. Different methods and processes have been postulated and promoted by various researchers to mitigate climate change by reducing the amount of carbon dioxide in the atmosphere. This paper aims to consider a process to capture the carbon dioxide directly (Direct Air Capture) from the atmosphere at a cheap, sustainable and affordable way. The process aims to capture the carbon dioxide, concentrate it and pressurize it for storage or further usage. This is a deviation from the process of carbon capture from point source such as power plants, industries etc. The process focusses on capturing carbon dioxide that have found their way into the atmosphere through other smaller sources such as automobile exhaust, domestic power generators, fossil fuel burning equipment and many other sources not reported or captured in the international data base. A summation of the above-named sources will amount to a huge impact on the environment and a method to remove the contaminant at a cheap, affordable and sustainable way will be of immense benefit worldwide most especially developing countries in tackling climate change.
Keywords: carbon dioxide, carbon capture, mitigate, climate change, DAC, sustainable

1. Introduction

The growth in world economies attributed to the emergence of new technologies has brought about environmental problems along with it. With problems such as global warming, acid rain formation, smog formation, etc mentioned in areas where they were never mentioned before. Among all these problems, global warming, resulting from the emission of greenhouse gases mainly carbon dioxide, is considered as the most serious one.
The Carbon dioxide (CO2) which is the major constituent of the greenhouse gases can be produced from various sources, such as fossil fuel combustions (in cars, trucks, power generators), biomass energy facilities, large power plants in industrial processes, etc. And this is due to the increasing consumption of fossil fuels by humans as needed for industrialization. Carbon dioxide concentration in the atmosphere has increased from 270 ppm in pre-industrial period to nearly 400 ppm at present [1]. This increase in carbon dioxide concentration has led to numerous health concerns and increased cost of removing the pollutant from the atmosphere. To remove the carbon dioxide generated by heavy fossil fuel burning industries, the released carbon dioxide is capture and stored, but this has not taken care of the carbon dioxide released from smaller sources scattered all over to the atmosphere, this has led to a new approach of capturing (or removing) the carbon dioxide directly from the atmosphere.

The process of capturing waste carbon dioxide from large point sources such as fossil fuel power plants, transporting, storing and depositing it where it will not be allowed to enter back into the atmosphere like an underground geological formation is known as carbon capture and storage (CCS) [2]. The major reason for CCS is to prevent the release or emission of large quantities of carbon dioxide into the atmosphere as a result of burning fossil fuel to generate power and in other industrial uses.

On the other hand, carbon dioxide removal (CDR) from the air is different from reducing emissions. Capturing waste carbon dioxide from large point source decreases the inlet of carbon dioxide to the atmosphere while carbon dioxide removal produces an outlet of carbon dioxide from the earth's atmosphere.CDR is a different approach to removing carbon dioxide from the stack emissions of large fossil fuel point sources, such as power stations.

Capturing waste carbon dioxide can only reduce the level of emission to the atmosphere but will not have any control over the amount of carbon dioxide already present in the atmosphere. Carbon dioxide removal (CDR) creates negative emission by removing carbon dioxide from the atmosphere, cancelling out the emissions from small and dispersed point sources such as domestic heating systems, domestic power generation, airplanes and vehicle exhausts.

Both processes (carbon dioxide removal and carbon dioxide capture) works towards achieving the same net effect in the long run but to achieve carbon dioxide concentration levels below the present level, carbon dioxide removal using the direct air capture process is very important. Carbon dioxide capture from the air is becoming more popular as it shows signs of providing the only possibility to fill the gap between the needed reductions to meet expected mitigation targets and global emission trends in higher concentration levels.

The objective of the carbon dioxide removal (CDR) process is to remove carbon dioxide from the atmosphere on large scale bases [3]. There are several technologies captured under the CDR process, some of which are; bio-energy with carbon capture and storage, biochar, ocean fertilization, enhanced weathering and direct air capture when combined with storage [2].

2. Direct Air Capture (DAC)

It is possible to remove carbon dioxide from the air through chemical processes. The captured carbon dioxide can then be sequestered, stored and possibly re-used. The rate of increase of atmospheric carbon dioxide concentration can be slowed down by the traditional modes of carbon capture such as pre-combustion and post-combustion CO2 capture from large point sources. However, only the direct removal of carbon dioxide from the air, or direct air capture (DAC) can actually reduce the global atmospheric carbon dioxide concentration if combined with long term storage or usage. Work in this area of research is still not well developed. [4]

Some of the main technologies proposed for direct air capture are; causticization with alkali-earth hydroxides [5], carbonation [6] and organic-inorganic hybrid sorbents consisting of amines supported in porous adsorbents.

Several scientists, engineers and researchers have proposed different methods for direct air capture. One of these methods is the use of artificial trees [7][8]. This method aims to create artificial trees around the world which can suck up more carbon dioxide from the air than the natural trees. The carbon dioxide will be captured in a filter, removed and stored.

Another method proposed can be carried out at temperature of only 40 o C, this method uses a polymer based ion exchange resin, which takes advantage of changes in humidity to prompt the release of the carbon dioxide captured from the air instead of using a kiln for the release process, thereby reducing the energy requirement for the process.[9].

Sodium hydroxide-based chemistry is used in different variants to achieve the direct air capture of carbon dioxide in all the literatures reviewed above.

This review introduces the use of Temperature Swing Adsorption (TSA) system of capturing carbon dioxide from the atmosphere directly using the most suitable adsorbent for optimum performance. The choice of the process is due to the objective of achieving the direct air capture of carbon dioxide at low temperature and at affordable cost.

3. Methodology

Temperature swing adsorption uses temperature to regenerate the adsorbent. At low temperatures, adsorbents can retain significant amounts of water while at temperatures above 200°C, adsorbents hold almost no water. By swinging the temperature from low to high, it is possible to adsorb large

quantities of moisture at low temperature, such as 40°C, and release it at high temperature. The adsorption and desorption curve will be analyzed to determine the rate at which the system functions and used for further optimization.
The software, Labview can be used to optimize the process to determine the optimum condition of the adsorbent to be used and at what conditions/parameters the best results will be obtained before deploying the equipment to capture the carbon dioxide from the atmosphere. In the search for the ideal adsorbent, the selection will be based around materials with high carbon dioxide adsorption capacity, high adsorption selectivity towards carbon dioxide (as opposed to other gases in the atmosphere), low energy requirements for regeneration, stability to prolonged adsorption-desorption cycling and tolerance to the presence of moisture.
A very large volume of data will be generated as a result of carrying out the research work in direct air capture that will ultimately lead to creating awareness within the populace within Nigeria and Diaspora. This paves way for a better mitigation method to climate change rather than the adaptive method currently popular in Africa.

4. Conclusion

The project is very relevant to current events in the mitigation against climate change and environmental degradation. The data and results from the work will not only help in understanding the process of removing carbon dioxide from the atmosphere better but will increase the knowledge in the general area of removing dangerous gases from the atmosphere.
The research work is of paramount importance to Nigeria as a country at this particular time. Claims have been made by researchers estimating the average number of homes with generating sets in Nigeria to 60 million10. We can imagine the enormous amount of carbon dioxide that will be pumped into the atmosphere with such a huge number of fossil fuel burning generators.
With the projected increase in the expected amount of carbon dioxide released into the atmosphere by 2060, the world needs a better means of removing the gas out of the atmosphere as fast as possible.

References

[1] Eisenberger, P.M., et al., Global Warming and Carbon-Negative Technology: Prospects for a Lower-Cost Route to a Lower-Risk Atmosphere. Energy & Environment, 2009. 20(6): p. 973-984.
[2] Fanchi, John R; Fanchi, Christopher J (2016). Energy in the 21st Century. World Scientific Publishing Co Inc. p. 350. ISBN 9789813144804. Retrieved 29th December,2018
[3] Geoengineering the climate: science, governance and uncertainty". The Royal Society. 2009. 29th December, 2018.
[4] Sanz-Pérez, E. S.; Murdock, C. R.; Didas, S. A.; Jones, C. W. (2016). "Direct Capture of CO2 from Ambient Air". Chem. Rev. 116 (19): 11840–11876.
[5] Lackner, K. S.; Ziock, H.; Grimes, P. (1999). Carbon Dioxide Extraction from Air: Is It an Option?. Proceedings of the 24th Annual Technical Conference on Coal Utilization & Fuel Systems. pp. 885–896.
[6] Nikulshina, V.; Ayesa, N.; Gálvez, M. E.; Stainfeld, A. (2016). "Feasibility of Na–Based Thermochemical Cycles for the Capture of CO2 from air. Thermodynamic and Thermogravimetric Analyses". Chem. Eng. J. 140 (1–3): 62–70.
[7] "New Device Vacuums Away Carbon Dioxide". LiveScience. 2007-05-01. Retrieved 23rd January 2018
[8] Adam, David (2008-05-31). "Could US scientist's 'CO2 catcher' help to slow warming? | Environment". London: The Guardian. Retrieved 23rd January 2018
[9] Lenfest Center for Sustainable Energy". Energy.columbia.edu. Retrieved 1st February 2018
[10] '60 Million Nigerians now own power generators' Energy Comission of Nigeria. www.energy.gov.ng retrieved on the 1st of April 2018

IOP Conf. Series: Materials Science and Engineering **413** (2018) 012078 doi:10.1088/1757-899X/413/1/012078

Challenges of Research in Contemporary Africa World

Fayomi O.S.I.[1, 2], **Okokpujie I. P.** [1], **Kilanko O.**[1]

[1]Department of Mechanical Engineering, College of Engineering, Covenant University Ota, Lagos state, Nigeria
[2]Department of Chemical, Metallurgical and Materials Engineering, Faculty of Engineering and the Built Environment, Tshwane University of Technology, Pretoria, South Africa.
Corresponding author: ojosundayfayomi3@gmail.com, ojo.fayomi@covenantuniversity.edu.ng, fayomio@tut.ac.za Tel: (+2348036886783) (+27835828119)

Abstract- Research is the vital source of technological progress over the decades and the advance of scientific knowledge has helped to develop countries Research is a structured inquiry that utilizes acceptable scientific methodology to solve problem, sand creates new knowledge that is generally applicable. Colleges and research institutes over the world are in charge of research, information creation, scholarship and advancement as they serve as channels for the exchange, adjustment, and dispersal of learning produced over the world. Universities alone never again stay sole fortresses of research exercises, as private or government organizations are progressively associated with information creation and spread. The globalization of advanced learning, combined with developing understudy versatility and expanded rivalry for financing, has as of late occasioned endeavors to rank colleges as far as their scholarly quality and profitability at national, provincial and worldwide levels. New technical knowledge presently offer helps for African colleges to re-build and re-position themselves to meet these consistently expanding societal requests successfully. A good research is the concretized solution to the challenges of research in contemporary Africa. Educational research is the vital force towards solving enormous challenges faced by our research institution in developing African Nations. On this note, this paper will examine the difficulties of doing research in Africa and its main focal point is on colleges in sub-Saharan Africa, barring North Africa. This is because their education is already generally affected by hones in Europe and already has a genuinely all-around created mechanical and modern economy which makes them in front of other African nations.
KEYWORDS: Research, Research Projects, Research Team, Research Supervision,

1. Introduction

Research is essential in the improvement and effective development of any country. The literacy level of any country is measured by the economic development and progressive level that cut across social wellbeing of the people, human, capital development, environmental factors, health issues, security and entrepreneurship success as deduced by the economics (O'Sullivan and Sheffrin, 2003). Research simply means search for knowledge through objective and systematic

IOP Conf. Series: Materials Science and Engineering **413** (2018) 012078 doi:10.1088/1757-899X/413/1/012078

method of finding solution to a problem. It is also a scientific and systematic search for pertinent information on a specific topic. Research is an original contribution to existing stock of knowledge making for its advancement. Research is a process where you can discover something new by adding information to wealth of knowledge. It is the pursuit of truth with the help of study, observation, comparison and experiment (Kothari, 2004).

Research is the systematic way of organizing an essay for proper investigation required by the students for the award of degree with the study of scientific articles needed for teaching staff in higher institutions. Research is the vital source of technological progress over the decades and the advance of scientific knowledge has helped developed countries (Goldin and Katz, 1999). Research is a patient study and investigation or establish facts (more facts) or principles (more principles). It is a discovery (rediscovery) from known facts to the unknown facts. AccordingtoSchultz, P. T. (2001),'Researchisastructuredinquirythat utilizes acceptablescientificmethodologytosolveproblemsandcreatesnewknowledgethatisgenerallyapplicable'.

2. Theoretical Frame Work

2.1 The Purpose of Research

- Research is a necessary tool for advancing knowledge
- It is a tool for promoting personal and societal progress
- Research is an important tool used to accomplish one's purpose
- Research helps the individual to live effectively well in his or her society
- Research is a tool used for resolving man's conflict
- Research conducts greatly help to monitor the progress of educational enterprises. (journal of teacher education)

Through research we can advance in knowledge to boost the effort of Africa World toward competing with the world in terms of technology and economic development (Meagher and Lyall, 2013). Research will help through promoting personal and societal progress in driving Africa development agenda. Institutions will encourage his students to achieve their purpose by doing research that will relate to boosting the image of Africa. Research is a crucial tool that will help an individual to live a comfortable life in their region (Pahl-Wostl et al, 2013). It can also assist to resolve conflict among Africa nations, States and people living in a particular community. Research conducts will help to develop Africa through monitoring of student's project and to encourage them to do more research that will bring positive progress to educational enterprises. Ralph Dahrendorf's harsh words stated that the existence of tertiary institutions does not meet the high expectation of their role in the modern world. Universities have continuously been an instrument of change where research can be supported by government and sponsored to achieve development agenda of Africa (Spaargaren, 2011) and (Schäfer, 2010).

3. Challenges in the Supervision of Research Projects

Postgraduate and undergraduate students usually face challenges in conducting research. There are three categories of challenges face by students in carrying out their research project.

1. Supervisor related challenges
2. Student related challenges
3. Institution related challenges. (Chiome and Chabaya, 2009)

3.1 Supervisor Related Challenges

These are a few of the supervisor-related challenges within the supervision of students' investigateprojects.

- When the supervisor is not having regular contact with the students.
- Supervisor lackadaisical attitude toward students
- If he is not interested in the topic
- If the supervisor did not give enough practical assistance
- Inadequate direction to the students
- Delay and slow in reading students work
- Lack of research experience and regular absent from duty
- When the supervisor lack research skill, experience and knowledge
- The method of forcing research topics on the students militates against student performance in research projects. (Thomas and Nelson, 2001)

3.2 Student-Related Challenges in the Supervision of Research Projects

The following are the problems faced in the supervision of student research projects.

- When students do not have time for their projects.
- Lack of capital
- Inability to source required information from the internet
- If adequate information is not obtained from the library facilities
- Students lack of competent and research motivation
- Lack of theoretical background in the area of research
- Challenges in the family and commitments (Mapolisa, 2007) and (Anderson et al., 2006)

3.3 Institution-Related Challenges

In addition to the students and supervisor-related challenges, students' capacities to conduct investigate research are compelled by institution-related challenges (Nyawararnda, 2005).

- Non-exposure of supervisors to research skills
- If the institution does not have internet services
- Inability to use computers and programs
- If they do not have correct research material. (Thondhlana et al., 2011)

In this study, we are able to deduce that institution-related challenges, students and supervisor-related challenges were impacting negatively on the studentresearch output in African world.
Other challenges faced by students are as follows:

- Challenge of choosing the correcttopic
- Choosing the proper Methodology
- Assembling a Research team
- Finding Study Participants
- Getting Institutions to Participate
- Staying propelled and Working Your Plan
- Dealing together with your information (www.waldenu.edu)

IOP Conf. Series: Materials Science and Engineering **413** (2018) 012078 doi:10.1088/1757-899X/413/1/012078

4. Proffered Solution Towards Curbing the Challenges or Research in Contemporary African World

4.1 Choosing the Correct Topic

The step towards doing a good research work starts with the foundation of choosing the right topic.Start with your resources at hand which include time, money and personnel in choosing the topic you can handle. Study and read literature review, articles, journal papers and write up that is related to the topic and area of research work (Muchiri, 2011). Try to find the theory that will support your topic and area of specialization. Ensure that the topic suit your area of interest. To make a difference, then look for a niche to offer something new to the field of study. Try to add to the body of knowledge by coming up with more than one dissertation. Fine-tune your theme based on input from others since the topic you started with is not the topic you will end up with (Belfiore, 2015).

4.2 Choosing the Proper Methodology

After choosing the right topic the next is to find procedure to carry out the research work (Christie, 2010). The methodology to utilize comes from the investigative question and not from the individual preferences for one plan or another. Decide what kind of plan and strategy can best answer your research questions. Determine between quantitative or qualitative research Methodology before going into detailed study. Know within yourself if you have the capacity and knowledge to carry out the research work (Brandt et al, 2013)

4.3 Assembling a Research Team

You will require a research group and think about member to back your work once you know your Point and Techniques. Make companions with individuals who are willing to assist you towards your project view point from a distinctive angle or perspective (Kuada, 2010). Be clear about your needs since individual are exceptionally willing to assist once you are well organized.

4.4 Finding Study Participants, Getting Institutions to Participate, Staying Motivated and Working with Your Plan

Don't be afraid to reach out and leverage the power of a network by meeting established researchers in your field for direction and organizing. Also target a specific group, locating advocates within that social network, who can link other group members (UNESCO, 2012).

Recruiting study participants will require going through institutions which is a serious challenge especially if your research is controversial or sensitive. Don't stop at the first rejection but Pursue alternative routes. Persevere and build relationships with the people who will assist you gain access into the institution (Ynalvez and Shrum, 2011).

Maintaining the inspiration to keep going in spite of impediment is a motivating factor in your research. The pressure of work and individual commitments is the greatest challenge during the research project (Harle, 2010). Follow your passion and your purpose. The first part of inspiration is working on something you're energetic about and accept to do in the event that you're passionate almost what you're researching about, the research will increase your energy to complete your research work. Monitor your attitude and share your most prominent dreams and objectives with individuals who are going to be positive and supportive (Harle, 2011).

4.4.1 Dealing Together with Your Information

When you've completed your research work, the ultimate challenge is to know how to form sense of the information you've collected. Ground yourself in the research by connecting your research to the existing research (Sargent, 2004). When you are grounded in a colossal overview of the literature, it is simpler to have the parameters to organize your research. Your data ought to be displayed in a way that illustrates how your research adds to the body of knowledge. Stay focused against being distracted by irrelevant data as you do your analysis. Always keep a really close eye on your research questions and your speculation since some of the data you collect will take you away from your objectives (NCST, 2012).

The duties of research institution assigned to them are the manufacturing of high skilled intellectual and research giant to meet perceived developmental need of Africa World. Universities and their constituents have not played key role to see that education system have being resuscitated from the shackle of cankerworm ravaging the educational system but it has experienced slight changes in its curriculum. The economy of Africa can be improved through introduction of skilled and entrepreneurial courses into the education system. It is good to facilitate new cultural values, and in training students to achieve success in different field of endeavor. Research grant should be given to costly project and sponsored to the final result. Projects performed by the student that will move Africa forward should be encouraged and supported to encourage more students to work harder (Harris, 2013).

The job of the university in the area of research will contribute valuable knowledge and will be transferred to the future generation. The impact of the university can contribute to social change where the next generation will experience dramatic turn around to compete with the rest of the world (Kashorda and Waema, 2014). To produce potential people at all level of discipline, we need to help gifted children by training them through experienced trained personality that will act as a pillar of support in nation building. Institutions will assist to modernize courses which have lost its true value to impact worthwhile knowledge to the society. Academics should enable the community to experience free and open discussion on ideas that will bring value to Africa. (TFHE, 2000) A 2002 World Bank report has identified four essential functions of higher education in supporting knowledge driven economic growth:

- Training and equipping experience labor force at all level of work force through the high-level researchers, experts, professionals, instructors, auxiliary instructors as well as future government;
- To produce good research work for human and capital development.
- The importance of sourcing for knowledge and information from the world that will be of benefit to African Nations.

- The solid foundation for good governance and democratic elected system is the only way the country can experience changes through change in terms of values, attitudes and ethics as the foundation of the social capital necessary to construct healthy civil societies and cohesive cultures (Harrison et al, cited in World Bank, 2002).

To consider how University system has been transformed in the series of event in the society

- Improving quality and standards in the curriculum;
- Diversify;
- Introducing open policies, student information and experiences;
- Academic reaction to change.
- The need to study different cases to know the one that will benefit Africa (Brennan et al, 2004).

Academic will assist in the area of diversification in research. There is need to improve quality and standard in the curriculum. To introduce open policies, student information and experiences and academic response to change, there should be proper orientation on our students. It is through study and research on the way global entities carry out their research work and adapt the same to Africa can we strive to meet up with the world. There is urgent need to train a qualified and adaptable labor force – including high level scientists, professionals, technicians, teachers for basic and secondary education, as well as future government. Public and private organizations should establish private research institution where major research work can be further developed and give patent to the original researcher (Lee and Bozeman, 2005).

This way academic can contribute to meet up with the demand of competing with the rest of the world to drive Africa's Development Agenda. The role of academic in research change appears to be both complex and contradictory (Adams et al,2010). The proper demand is to remove the old way of doing the research work and introduce new techniques in respect of the latter. Academic should introduce symposium and workshop to sensitize the student to choose research project that will boost and make Africa to be giant of Africa. Indigent student and those living in river line areas should be encourage to study and enroll in research organization where they can contribute information on what they learn from their environment to support the building of giant of Africa (Adams et al,2014).

5. Conclusion

Research is changing dramatically becoming more multifaceted, farther collaborative, more global, and more reliant on the abilities offered through cutting edge systems and huge information stockpiling; there are new opportunities and challenges for information professionals within higher education. Fortifying associations with different researchers outside our institution and with those inside our own institution is a critical way to keep up relevance and contributing to the research struggle of our institution. Funding of research work will encourage many in African nation to engage in research finding to better our society. This is one of the most important aspects that will help in making the door of research finding open both in private and public sectors of the economy. Through proper funding of research work, university scholars will be encouraged to help make research a way forward to numerous challenges of research in contemporary African world. Researchers ought to be urged to self-archive pre-prints as well as post-prints of their papers in open access files or institutional vaults to help address the scarcity of research materials they face.

IOP Conf. Series: Materials Science and Engineering **413** (2018) 012078 doi:10.1088/1757-899X/413/1/012078

References

[1] Adams, J., King, C., &Hook, D. (2010) Global Research Report: Africa. Thomson Reuters, Leeds, UK.

[2] Adams, J., Gurney, K., Hook, D., &Leydesdorff, L. (2014). International collaboration clustersin Africa. Scientometrics 98 (1), 547–556

[3] Anderson, C. Day, K. & MacLaughlin, P. (2006). Mastering the Dissertation: Lecturers' Representations of the Purposes and Processes at Masters Level Dissertation Supervision'. Studies in Higher Education. Vol.31, No.2, 20.

[4] Belfiore, E., (2015). 'Impact', 'value' and 'bad economics': Making sense of the problem Ofvalue in the arts and humanities. Arts Hum. Higher Educ. 14 (1), 95–110.

[5] Brandt, P., Ernst, A., Gralla, F., Luederitz, C., Lang, D.J., Newig, J., Reinert, F., Abson,D.J., &Von Wehrden, H., (2013). A review of transdisciplinary research in sustainabilityscience. Ecol. Econ. 92, 1–15.

[6] Brennan J., King R. & Lebeau Y. (2004). The role of Universities in the transformation of societies. An international research project synthetic report. Centre for higher education research and innovation, 56.

[7] Chabaya, O., Chiome, C. &Chabaya, R.A. (2009).A Case Study from Masvingo Regional Centre at the Zimbabwe Open University. Open Learning, Vol.24, No.3, 211-221 Nov 2009 (EJ856759) ephost@epnet.com

[8] Christie, P (2010). Landscapes of leadership in South African schools: Mapping the changes. Educational Management Administration and Leadership 38(6): 696–711

[9] Goldin C. & Katz L. F. (1999) The shaping of higher education: The formative years in the

[10] United States, 1890-1940. Journal of Economic Perspectives Vol. 13, Number 1, 37–62

[11] Guilloches Z. Issues in Assessing the Contribution of Research and Development to

[12] Productivity Growth. Bell J Econ [Internet] 1979 Apr 1;10(1):92–116. [cited 2018 June 04]; Available from: http://www.jstor.org/stable/3003321.

[13] Harle, J., (2010). Growing Knowledge: Access to Research in East and Southern AfricanUniversities. British Academy and Association of Commonwealth Universities, London.

[14] Harle, J., (2011). Foundations for the Future: Supporting the Early Careers of AfricanResearchers. British Academy and Association of Commonwealth Universities, London.

[15] Harris, R., (2013). The impact of research on development policy and practice: An Introductionto a review of the literature. Available from: http://www.researchtoaction. Org/2013/07/the-impact-of-research-on-development-policy-andpractice/[3rd June, 2014].

[16] https://www.waldenu.edu/-media/Walden/general-media/newsroom/alumni- magazine.2010.

[17] Kashorda, M.,& Waema, T., (2014). E Readiness Survey of Kenyan Universities 2013 Report.KENET, Nairobi

[18] Kothari, C., R. (2004). Research methodology methods and techniques, New age international Publishers, 4835/24, Ansari Road, Daryaganj, New Delhi – 110002, ISBN (13) : 978-81-224-2488-**1,** 18.

[19] Kuada, J. (2010). Culture and leadership in Africa: A conceptual model and research agenda. African Journal of Economic and Management Studies 1(1): 9–24

[20] O'Sullivan, A. &Sheffrin, S. M. (2003). Economics: Principles in action. New Jersey Pearson Prentice Hall, Upper Saddle River.

[21] Lee, S., & Bozeman, B., (2005). The impact of research collaboration on scientific productivity. Social Studies. Sci. 35 (5), 673–702.

[22] Mapolisa, T.& Mafa, O. (2007). challenges being experienced by undergraduate students in conducting research in open and distance learning international journal of Asian social science 2(10):1672-1684

IOP Conf. Series: Materials Science and Engineering **413** (2018) 012078 doi:10.1088/1757-899X/413/1/012078

[23] Meagher, L.&, Lyall, C., (2013). The invisible made visible Using impact evaluations to Illuminateand inform the role of knowledge intermediaries. Evidence Policy 9 (3), 409–418

[24] Muchiri, M. K (2011). Leadership in context: A review and research agenda for sub-Sahara Africa. Journal of Occupational and Organizational Psychology 84(3): 440–452

[25] NCST, (2012). National Council for Science and Technology Annual Report, Government Printer, Nairobi

[26] Nyawaranda, V.A. (2005). A Paper Delivered at the Zimbabwe Open University (ZOU) Workshop, Mashonaland Central Region, Bindura on 'Supervising Research Projects/Dissertations'

[27] Pahl-Wostl, C., Guiponni, C., Richards, K., Binder, C., de Sherbinin, A., Sprinz, D., Toonen,T.,& van Bers, C., (2013). Transition towards a new global change science: requirements for methodologies, methods, data and knowledge. Environ. Sci. Policy 28, 36–47

[28] Sargent, L.D. & Waters, L.E., (2004). Careers and academic research collaborations: Inductiveprocess framework for understanding successful collaborations. J. Vocational. Behaviour. 64(2), 308–319.

[29] Schäfer, M., Ohlhorst, D., Schön, S.& Kruse, S., (2010). Science for the future: challenges

[30] Andmethods for trans disciplinary sustainability research. Afr. J. Sci. Technol. Innovation.Dev.(AJSTID) 2 (1), 114–137

[31] Schultz, P. T. (2001). Higher education in Africa: Monitoring efficiency and improving equity. Paper for Yale Center for International and Area Studies, African Studies and Economic Growth Center Conference on International Higher Education and African Development, New Haven, CT

[32] Spaargaren, G., (2011). Theories of practices: agency, technology, and culture: exploringthe Relevance of practice theories for the governance of sustainable consumption practices in the new world-order. Global Environ. Change 21 (3), 813–822

[33] Thondhlana, S., Mawere, D. & Weda, Z. (2011). Factors Affecting Completion of Research Projects by Students: A Study of Three Zimbabwe Open University Regions. In the Zimbabwe International Journal of Open and Distance Learning Vol.1, No.1,

[34] Thomas, J.. R. & Nelson, J.K. (2001). Research Methods in Physical Activity (4th Edition).Albany: Human Kinetics.

[35] Task Force on Higher Education and Society (TFHES) (2000) Peril and Promise: Higher Education in Developing Countries, Washington DC, The World Bank.

[36] UNESCO, (2012). Global Investments in R & D, 2012, Factsheet No.22. UNESCO Institutefor Statistics

[37] Ynalvez, M., & Shrum, W., (2011). Professional networks, scientific collaboration, and Publication productivity in resource-constrained research institutions in a developing country. Res. Policy 40 (2), 204–216.

IOP Conf. Series: Materials Science and Engineering **413** (2018) 012079 doi:10.1088/1757-899X/413/1/012079

Design of a high Temperature 'Anaerobic Gas-Furnace' suitable for Pyrolysis

Ishola Felix A[1], Oyawale F. A[1], Inegbenebor A. O[1] and Boyo Henry[1]

[1]Covenant University, Ota, Ogun State, Nigeria
Correspondence to: felix.ishola@covenantuniversity.edu.ng

Abstract. The need for a high temperature anaerobic furnace for pyrolysis is fast becoming indispensable for technological growth in Nigeria. In this project, a high temperature furnace was designed and built using the principle of heat transfer. The furnace was made of casing of low carbon steel sheets in four rectangular shapes. The refractories made of bricks filling the cavities of the furnace. This help to preserve heat loss. The pyrolysis capsule (chamber) was made of high carbon steel tube with flanges which can be opened for loading the products and evacuate. The mechanism to regulate the temperature and the gas input were installed. The furnace was gas-fired due to its high combustibility and providing a solution to the limited supply of electricity in the country. The prototype furnace produced is easy to use and multi-versatile with the innovation of the detachable pyrolysis capsule which can be modified by intent of pyrolysis products and by-products. The pyrolysis capsule is surrounded by the flame jacket (of the furnace main chamber) created to heat up the capsule at a regulated rate to any desired temperature up to 1,200 °C with the automated control system for the gas supply unit. The designed high temperature anaerobic gas furnace is deployable for a wide range of pyrolysis processes and can be used for different precursors to produce an achievable end-product within the temperature limit.
Keywords: Pyrolysis, Furnace, design

1. Introduction

Nigeria, at this point in time, is focusing on technological development which is her only chance at catching up with the back logs in her race towards Industrialization and economic advancement. Undisputedly, a lot of efforts will have to go into researches in the field of Metallurgy and Material development using local contents. This reality brought about this project; design and development of an efficient high temperature furnace using basic principles of Furnace development using locally available materials. In as much as the pilot scale furnace can be used for numerous conventional furnace purposes it can be modified for anaerobic heating of materials, a process known as pyrolysis. Practically, the efficiency of a furnace is directly related to the ability to minimize heat loss. Heat loss is the heat from the surface of a pyrolysis furnace to its surroundings by natural convection and radiation [6]

This project focused on building an anaerobic furnace which could be used for pyrolysis purposes. When a solid material heats up, it eventually reaches a temperature threshold where it begins to chemically break down. This process is called pyrolysis; the more the absence of air the more the process flourishes. Pyrolysis is the simultaneous change of chemical composition of a material (eg, long hydrocarbon chains to shorter chains) the process is irreversible and it has been found to be the most promising thermochemical process for the waste treatment [3].

The particular interest for this project is in the production of activated carbon and graphite from different locally available precursors such as bamboo, coconut husk, palm kennel shell and some inorganic waste materials. Nigeria had heavily relied on importation of these essential Engineering materials (activated

IOP Conf. Series: Materials Science and Engineering **413** (2018) 012079 doi:10.1088/1757-899X/413/1/012079

carbon and graphite) thereby neglecting the exploration of the abundant opportunities available in our local resources. Adopting this model of furnace can help in preservation of foreign reserve by producing such furnaces for the local industries, without importing them. Also, this project, aimed at providing furnace suitable for experimental purposes on a primary objective and on a secondary objective for Small and medium Scale foundry industries in the country.

On a preliminary basis; the authors identified several types and categories of furnaces according to the fuel they use (natural gas, electricity, oil), and their technology (single-stage, dual stage or multi-stage) [2].
Gas furnaces: Gas furnace utilizes fossil fuel to transfer heat and are commonly considered the most efficient and economical method of heating. Typical efficiencies of coal, oil and gas furnaces have been investigated and the system with natural gas had been found to be most efficient [8]. Essentially, the furnace converts gas into heat by either using burners with natural air mixture, when necessary the air supply may be aided with blowers or using burners with oxygen-gas mixture.

Oil furnaces: An oil furnace is a heating device which burns heating oils, diesel fuel or other similar fuels to produce heat sufficient to carry out industrial activities. New oil furnaces can be fairly efficient and relatively clean and safe. There are now oil furnaces mimicking close-sealed combustion high efficient gas furnaces, able to provide high levels of efficiency.

Electrical Furnaces: a heating chamber with electricity as the heat source for achieving very high temperatures to melt metal alloy. The electricity has no electrochemical effect on the metal but simply heats it. The cleanliness of electric furnaces and the precise control that they afford is of distinct advantages - Even when the cost of electricity exceeds the cost of other forms of heating [1].

2. Design Philosophy

Design philosophy is simply defined as an individual's vision on a particular concept i.e. what an individual is trying to accomplish with the design, and, more importantly the principal aim and objectives of such venture. As it's well recognized during the course of many design activities, the use of past experiences, and previously acquired knowledge, either from the designer's own experiences or from available resources is an important part of the design process [4]. In Design philosophy, we try to relate with the assumptions, foundations, and implications of design. For simplicity sake the design is aimed at being a single stage, as well as the furnace structural dimensions being cubic.

Design Initial Specifications

A	=	Area, m^2	L	=	Length, m	B	=	Breath, m
V	=	Volume, m^3	H	=	Height, m			

<u>Area of Sheet metal</u>

Length of Sheet metal for Casing = 0.8m

Breath of Sheet metal for Casing = 0.8m

Area of sheet Metal = L x B = 0.8 x 0.8 = 0.64 m^2

<u>Casing Volume</u>

This involve joining 4 sheet metals of the given area to form a box like structure.

Height of Casing = 0.8m

$V = L \times B \times H = 0.8 \times 0.8 \times 0.8 = 0.512\ m^3$

<u>Internal Frame</u>

Length of wood for framing = 0. 8m

IOP Conf. Series: Materials Science and Engineering **413** (2018) 012079 doi:10.1088/1757-899X/413/1/012079

Breath of Wood f or framing = 0.5m

Area of framing = 0.8 x 0.5 = 0.40m^2

Taking into consideration the fact that the Furnace is expected to reach a high temperature needed for the pyrolysis process special attention to fuel requirements of the furnace is important. Making use of some foundation data for thermal efficiency calculation for different types of furnaces. Summers [8] investigated the thermodynamic efficiencies for types of fuel used in an energy conversion device. The schematic representation of the summary of results gotten is as depicted below.

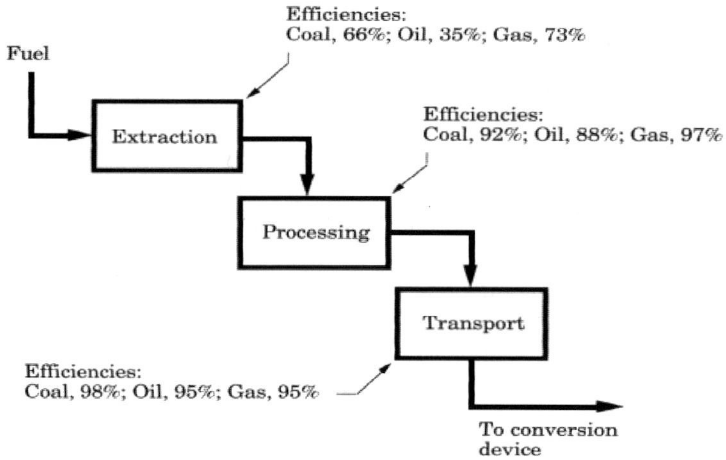

Fig. 2.1: A typical thermodynamic efficiencies of Fossil Fuels (Summers, 1999)

The design takes a turn towards a gas fired furnace because gas is a cleaner, affordable, safer and efficient energy choice. Apart from the fact that gas costs less than electricity, electrical power is poorly supplied in the country.

Working Drawing

The working drawing of the design of the built furnace are shown below as assembly drawing, exploded drawing and detail drawing as shown in the figures below using SolidWorks® Software:

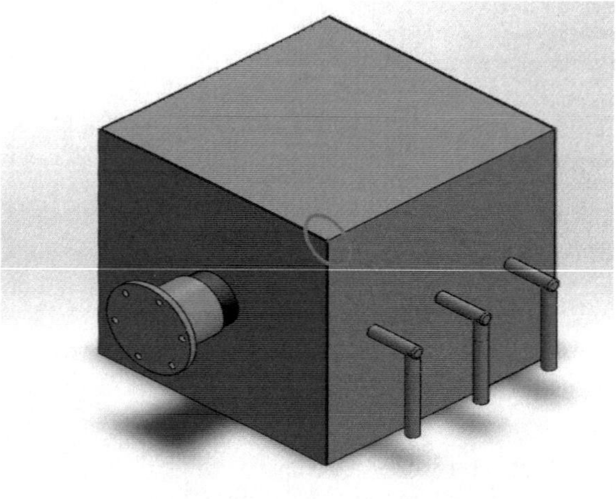

Figure 2.2: Assembly drawing of furnace

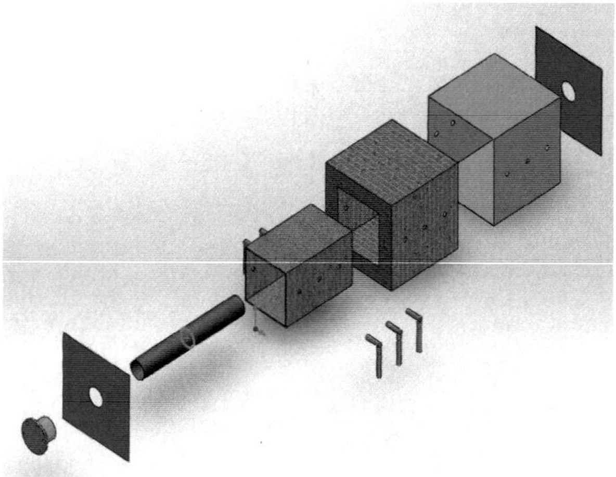

Figure 2.3: Exploded view of furnace

Design Procedures

This stage involved the design of the anaerobic furnace from scratch to completion. The materials used in the design was carefully selected with utmost consideration of various design factors such as wear, cost, reliability, weight, life, recycling, lubrication, friction, size, maintenance, safety, liability and availability;

- The sheet metal of 2mm which would be used for the furnace framing was cut into the desire shaping and size in the workshop.
- The sheet metal was then joined to give a box like shape. Welding technique was used. Steel sheets was used for the fabrication of the furnace casing,
- Next, the mould was constructed which would aid the casting of the refractories into lay-able bricks. The inner part of any furnace is generally expected to be a dense-phase zone with refractory materials to conserve and thus concentrate the heat generated on the enclosure.
- Next the internal framing which would serve as a support for the refractories was constructed. This was done with a wood of 3.5 inches. The wooden part is expected to be removed or burnt off at the first fire introduction into the furnace.
- After that, the area of the spacing in which the refractory bricks would come in was calculated to determine the structural dimensions of the inner chamber where combustion is expected to be taking place.
- The refractories bricks are made of Kaolin made to cure naturally. Kaolin is a locally available ceramic known to have a very good refractory property.
- The cargo chamber, cylindrical in shape in which the raw material would be placed was gotten and a flange and blind as well.
- The flange was welded to both ends of the cargo chamber.

3. Principle of Operation

The principle of operation for this furnace is simple. The furnace is gas powered and thus produces heat by burning gas and heat transfer occurs.

Given the three modes of heat transfer; Conduction, Convection and Radiation.

- Conduction is the mode of heat transfer in which energy exchange takes place from the region of high temperature to that of low temperature. This mode of heat transfer is prominent vertically along the length of the pyrolysis chamber (Cylindrical pipe), there is an energy transfer from the high-temperature region (the points of direct heating) to the low-temperature region along the pipe length. [7]
- Convection occurs when fluid flows over a solid body or inside a channel while temperatures of the fluid and the solid surface are different, heat transfer between the fluid and the solid surface takes place as a consequence of the motion of fluid relative to the surface [5].
- Radiation heat transfer involves the transfer of heat by electromagnetic radiation that arises due to the temperature of the body. Radiation does not need matter. All bodies continuously emit energy because of their temperature, and the energy thus emitted is called thermal radiation. After the gas nozzles are turned on, the flame produced is emitted by radiation to the cargo chamber where the material which would be processed stays and conduction helps to transmit the heat all through the chamber [9].

4. Results and Discussion

The result of the test running the anaerobic furnace, the performance involved in running the furnace, are here presented.

Time VS. Temperature

When the design and fabrication of the project was done, the testing of the furnace was carried out. The furnace was heated up and the time taken for the furnace to reach 1200°C was considered. The readings were read from the digitally programmed control mechanism.

Furnace incumbent Temperature

After running the furnace, it was noticed that the furnace's temperature is a function of time. Below is a representation of the furnace temperature as it varies with time.

Table 4.1: Time against Temperature

TIME (min)	TEMPERATURE (^{0}C)
5	80
10	160
15	200
20	350
25	415
30	520
35	610
40	700
45	775
50	850
55	985
60	1200

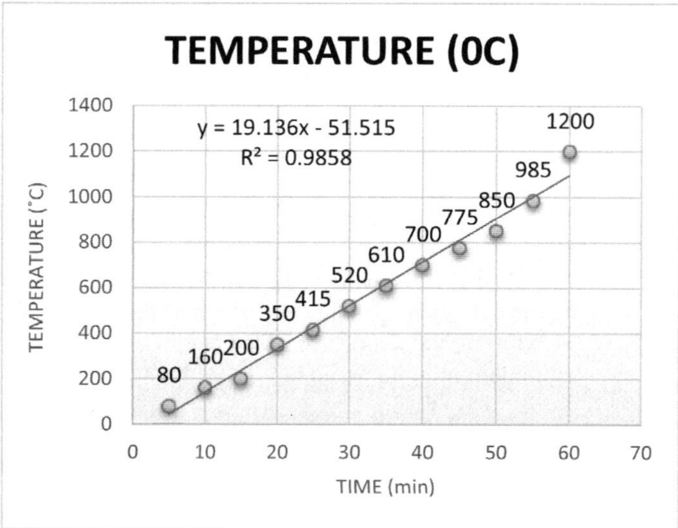

Figure 4.1: Showing the furnace temperature against time

After starting the furnace, we waited 5 minutes before taking the initial temperature reading of the furnace. And we recorded 80°c after just 5 minutes. We kept on taking the readings after every 5 minutes for the next 1 hour. From the graph we are the temperature increases drastically as time goes on. From the data given above we see that in an hour the temperature gets to 1200°c.

The furnace was able to reach 1200˚C in an hour using 12.5g methane gas. The furnace consumed 12.5g of methane gas from the gas cylinder in an hour.

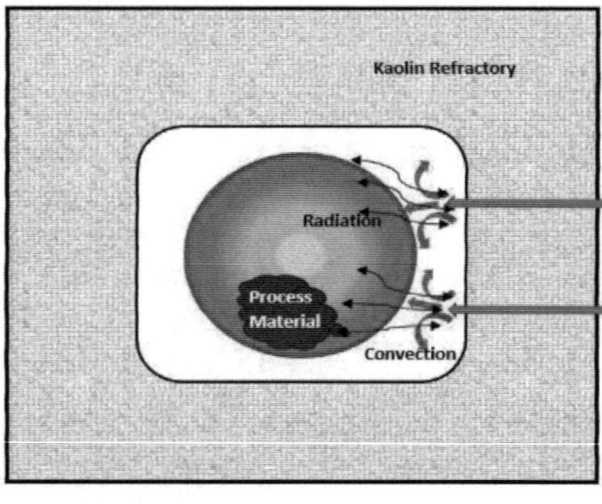

Fig 4.2: Sectional View to show Basic Features of the Furnace with the Pyrolysis

Fig. 4.3: Test- Running the model Anaerobic furnace

5. Conclusion and Recommendation

The project produced a prototype anaerobic furnace, which can "pyroduce" 6 – 10 kg of activated carbon from locally accessible precursors. The furnace built can adopt changeable anaerobic chamber and can

IOP Conf. Series: Materials Science and Engineering **413** (2018) 012079 doi:10.1088/1757-899X/413/1/012079

deliver high thermal energy. Hence, higher temperature within limited time. It is useful when rapidly high temperatures are required moreover it is an anaerobic chamber furnace which is not common for its size. The heat loss is minimal in the furnace and thus this makes it efficient in the sense that in little time and in less consumption of fuel it can attain a very high temperature.
For further works, a chimney exit would be manufactured and incorporated with the furnace. This would discharge the exhaust gas to the desired location.

References

[1] Bashforth, G. R. (2005). *THE FUTURE OF ELECTRIC FURNACES IN THE METAL INDUSTRIES OF INDIA*. Retrieved from eprints.nmlindia.org/3219/1/263-267.PDF
[2] Carina, K. (1985). *Metallurgical furnaces. J. Wiley & Sons* (Vol. 15). https://doi.org/10.1016/0042-207X(65)90017-5
[3] Chhabra, V., Shastri, Y., & Bhattacharya, S. (2016). Kinetics of Pyrolysis of Mixed Municipal Solid Waste-A Review. *Procedia Environmental Sciences*, *35*, 513–527. https://doi.org/10.1016/j.proenv.2016.07.036
[4] Crowder, R., Bracewell, R., Hughes, G., Kerr, M., Knott, D., Moss, M., … Waterson, P. (2003). A Future Vision For The Engineering Design Environment: A Future Sociotechnical Scenario. *Scenario*, 1–10. Retrieved from http://eprints.ecs.soton.ac.uk/8188/
[5] D. Q. Kern. (1983). *Process Heat Transfer* (Internatio). McGraw-Hill International Book Company.
[6] Kodera, Y., & Kaiho, M. (2016). Model calculation of heat balance of wood pyrolysis. *Nihon Enerugi Gakkaishi/Journal of the Japan Institute of Energy*, *95*(10), 881–889. https://doi.org/10.3775/jie.95.881
[7] Serth, R. W. (2007). *Process Heat Transfer Principles and Applications*. Elsevier Ltd.
[8] Summers, C. M. (1999). The Conversion of Energy. In *Scientific American* (Vol. 225, pp. 148–160). https://doi.org/10.1038/scientificamerican0971-148
[9] Vondál, J., & Hájek, J. (2015). Wall heat transfer in gas-fired furnaces: Effect of radiation modelling. *Applied and Computational Mechanics*, *9*, 67–78.

IOP Conf. Series: Materials Science and Engineering **413** (2018) 012080 doi:10.1088/1757-899X/413/1/012080

An Experimental-assessment of Human Exposure-levels to Aircraft Noise-hazards in the Neighbouring-environments of four Nigerian Airports

Ibhadode Osagie[1], **Oyedepo O. S.**[2], **Ogunro A. S.**[1], **Azeta Joseph**[2], **Solomon O. Banjo**[2], **Umanah, I. I.**[1], **Apeh, E.S.**[1], **Ayoola A. R.**[1]

[1] Nigerian Building and Road Research Institute (NBRRI), Km 10 Idiroko Road, P. M. B. 1055, Ota; Ogun state, Nigeria.
[2] Covenant University (CU), Km 10 Idiroko Road, Ota; Ogun state, Nigeria.
E-mail address: osagie.ibhadode@gmail.com

Abstract. This paper reports a scientific-assessment of the exposure-levels of aircraft noise-hazards suffered by persons living/working within the neighbourhoods of four Airports [Ibadan, Benin-City, Warri and Owerri] in Nigeria. Physical measurements of selected aircraft and environmental noise parameters [Ambient Noise Level (ANL), Sound Pressure Level (SPL), Aircraft Take-off Noise-level ($ATNL$) and Aircraft Landing Noise-level ($ALNL$)]; were carried-out using the integrated CR811C Noise meter, during one hundred and twenty (120) periodic noise sampling-surveys; performed [from January to December 2017] at thirty (30) randomly-selected study-locations, within the vicinity of each of these four(4) airports in accordance with the Method/Standard-procedures specified by the International Standards Organisation's (ISO) relevant standards—ISO 3891, ISO 1996-1 and ISO 1996-2. The results showed that: while SPL ranged from 103-115 $dB(A)$, ANL ranged from 52.3 – 64.1 $dB(A)$, the $ATNL$ ranged from 69.6 – 87.7 $dB(A)$, and $ALNL$ ranged from 66.2 – 82.7 $dB(A)$. Actually, these results are alarming, since they significantly exceed the WHO Standard Recommended Maximum Noise-levels of: 35 $dB(A)$ [Indoor], 55 $dB(A)$ [Outdoor] to prevent Speech-intelligibility, Noise-annoyance & Sleep-disturbance; and 90 $dB(A)$ [being the Permissible Noise level/limit for 8hour daytime safe human exposure]. These and the results of the Statistical analysis (Wilcoxon Sign Rank Test) carried-out, clearly prove the existence of a generally ignored, but yet dangerous problem of continuous human-exposure to excessively high-levels of Aircraft noise-hazards to which residents of Airports' neighborhoods are subjected.
Keywords: Aircraft Noise-hazard, Airport, Environment, Exposure-level, Noise-level

1. Introduction

The problem of the marine environment is gaining widespread social attention. Bohai Sea is the The history of our modern world is unarguably incomplete without the repetitive mention of the 'generic word' "Aircrafts"—which refers to aeroplanes, helicopters, drones and the likes. Suffice it to say that, in almost every metropolitan city of the world today, [including Nigeria's big cities], the regular sights of flying aircrafts traversing our earth's atmosphere with their accompanying noise emissions, seem to have become a part of the daily lives of the citizenry.

Noise pollution—be it aircraft-generated or not, has been repeatedly proven to be a global health hazard, which may have motivated governmental leaders all over the world to champion the noble cause for 'Quieter Cities'—a worthwhile move, which would not only enhance human health-conditions, but would also foster positive country-to-country relations, while ensuring that tourism-related economic gains are reaped etc. [1]. Research evidence has proven that an alarming minority percentage of people living in big cities of developing countries like Nigeria, are gradually becoming partially aware of some of the dangers/health risks associated with hazardous noise exposure [2].

In 2008, Kaltenbach, M. et al showed that there is a dose-response relationship between arterial hypertension occurrence and exposure to aircraft noise; this was based on the findings of their empirical study on the 'Health Consequences of Aircraft Noise', from which they found out that: 'Increased risk of hypertension [for their human subjects] was linked to outdoor aircraft noise-induced equivalent Sound Pressure Levels of $45dB(A)$ in the night and $60dB(A)$ during the day[3]. There is a higher probability for persons living in areas with medium and high background environmental noise to suffer the same level of aircraft noise hazard, than their counterparts residing in low background environmental noise areas [4].

In 2004, Fransen, E. A. M. et al asserted that, there is a link between aircraft noise exposure level and certain health indicators [such as the use of medicinal drugs for high blood pressure or cardiovascular ill-health, general wellbeing/health condition, and the use of sedatives or sleep medicine(s), during the daytime; while the use of self-prescribed sedatives and sleep medicine(s) was prevalent among the experimental subjects when exposed to aircraft noise hazard late in the evenings [5].

It has now been established that, particularly at high levels of exposure to aircraft noise, there exists statistically significant relationship between the human cardiovascular health/wellbeing and exposure-level to aircraft noise-hazard [6]. This may have necessitated the world's apex health body—the World Health Organization (WHO) to adopt $40dB(A)$ as the desirable threshold value for people/persons subjected to chronic exposure to aircraft noise hazard [7]. At the moment, there is enough of scientific evidence to assert that, there is a strong positive relationship between several health challenges/issues [and particularly high blood pressure] and exposure to aircraft noise hazard—which according to several schools of thought, is generally considered a "nuisance and an environmental stressor" [8].

Whether in the civil aviation sector or in the military aviation sector of a nation's economy, aircrafts have visibly shaped the course of human history and more often than not, positively impacted virtually all sectors of the global economy including transportation, defence, Healthcare, Information Technology, Trade & Commerce and Power etc. However, it is sad to mention that, the merits of the aircraft notwithstanding, they also come with the accompanying demerits of aircraft-related environmental noise pollution—a situation which at the moment, cannot be completely eliminated, but can only be controlled or mitigated.

The Wikipedia online free encyclopaedia defined Aircraft Noise as that pollution generated by an aircraft or its parts, which occurs at different stages of an air flight i.e. on the ground when parked such as auxiliary power units, when taxiing, on run-up from the propeller and the jet-exhaust, while taking-off, under and beside the arrival and over-flying while enroute [9]. According to Mauskar J. M. (2008), Aircrafts are one of the major sources of noise, especially when they take-off and land. Thus, it is ideally in the health interest of the nearby residential human populace that, airports should be located at reasonable distances far away from people, to prevent them from being negatively affected by traffic noise [10].

The World Health Organisation (WHO) is seriously concerned about the many negative consequences of the aviation industry operation on the health conditions/general wellbeing of man. It has expressed genuine fears that a prolonged exposed to $65dB(A)-75dB(A)$ of aircraft noise for periods ranging from 5 – 30 years, could resulting hypertension and increased blood pressure [11]. So many research works have been carried-out, which reported the negative effects of aircraft noise exposure on human health. Some researchers have established the existence of a strong link/relationship between aircraft noise exposure and sleep disturbance, hypertension, headaches, stress & fatigue, increased risk of heart attack and noise-annoyance etc. [5], [12], [13], [14], [15].

According to Gualandi, N. and Manteccchini, L. (2008), aircraft operations actually constitute a major source of environmental externalities, particularly at the level of the local residents; also, noise pollution is now a serious concern to persons who reside near airports. It has been shown that, Noise— [including that emitted by an aircraft] can significantly affect the development of the cognitive capacity of children [16], [17].

Thus, from the above discourse, it is clear that aircraft noise negatively affects human health. Although, several studies have been conducted to investigate the noise pollution levels around some international airports in Nigeria [e.g. Muritala Mohammed International Airport in Lagos, Nnamdi Azikiwe

International Airport in Abuja, Port-Harcourt International Airport in Port-Harcourt, and Margaret Ekpo International Airport in Calabar etc.]; little or no work has been done in relation to domestic airports in Nigeria.

Our primary research hypothesis was that: "Persons residing/working within the proximate neighbouring-environments of domestic airports in Nigeria are continuously subjected to hazardous levels of aircraft noise pollution, which on prolonged daily exposure, will most certainly result in a number of physiological and psychological ill-health conditions—as is scientifically proven by a multiplicity of scholarly research literatures, globally".

Consequently, the aim of this study is to carry-out a scientific assessment of the aircraft noise hazard-levels [in terms of quantitative empirical estimates] that people/persons working/residing within the proximate neighbouring-environments of four domestic airports in Nigeria—Ibadan Airport in Ibadan city of Oyo state, South-Western Nigeria; Benin Airport in Benin-City of Edo state, South-Southern Nigeria; Owerri Airport [also called "Sam Mbakwe Airport"] in Owerri city of Imo state, South-Eastern Nigeria and Warri Airport [also called "Osubi Airport"] in Warri city of Delta state, South-Southern Nigeria.

2. Materials and Method

(i) About the Study-areas

The four (4) Study-areas were the Neighbourhoods [i.e. Neighbouring-environments] of Ibadan Airport, Benin Airport, Owerri Airport and Warri Airports—all in Nigeria.

Ibadan Airport [IATA: 'IBA', ICAO: 'DNIB'] is located at coordinates (7°21'35''N, 3°58'33''E) has an average Mean Sea Level elevation of 221m (i.e. 725ft), and 2,400m (i.e. 7,874ft) cumulatively long asphalt run-ways. It is owned and operated by the Federal Airports Authority of Nigeria (FAAN), and mainly serves the civil aviation needs of residents of Ibadan—the economically viable metropolitan capital city of Oyo state in South-Western Nigeria. Ibadan is Africa's second largest city [only next to 'Cairo' in Egypt], West-Africa's largest city by land area (6,800Km^2 i.e. 2,600sq mi), and is Nigeria's third most populous city with an estimated population of 3,565,108 as at June 2018 [18], [19].

Benin Airport [IATA: 'BNI', ICAO: 'DNBE'] is located at coordinates (6°19'00''N, 5°36'00''E) has an average Mean Sea Level elevation of 79m (i.e. 258ft), and 2,400m (i.e. 7,874ft) cumulatively long asphalt run-ways. It is owned and operated by the Federal Airports Authority of Nigeria (FAAN), and mainly serves the civil aviation needs of residents of Benin-City—the economically viable metropolitan capital city of Edo state in South-Southern Nigeria. Benin-City is Nigeria's sixth most populous city with an estimated population of 1,125.058 as at June 2018 [19], [20].

Owerri Airport also known as Sam Mbakwe Airport or Imo state Airport [IATA: 'QOW', ICAO: 'DNIM'] is located at coordinates (5°25'35''N, 7°12'20''E) has an average Mean Sea Level elevation of 114m (i.e. 373ft), and 2,700m (i.e. 8,858ft) cumulatively long asphalt run-ways. It is owned and operated by the Federal Airports Authority of Nigeria (FAAN), and mainly serves the civil aviation needs of residents of Owerri—the economically viable metropolitan capital city of Imo state in South-Eastern Nigeria. Owerri is Nigeria's forty-third most populous city with an estimated population of 215,038 as at June 2018 [19], [21].

Warri Airport also known as Osubi Airport [IATA: 'QRW', ICAO: 'DNSU'] is located at coordinates (5°35'50''N, 5°49'10''E) has an average Mean Sea Level elevation of 8.2m (i.e. 27ft), and 1,800m (i.e. 5.906ft) cumulatively long asphalt run-ways. It is owned and operated by the Shell Petroleum Development Company (SPDC) Nigeria, and mainly serves the civil aviation needs of residents of Warri—the economically viable, metropolitan city and oil-rich business hub in Delta state in South-Southern Nigeria. Warri is Nigeria's eighteenth most populous city with an estimated population of 536,023 as at June 2018 [19] [22].

IOP Conf. Series: Materials Science and Engineering **413** (2018) 012080 doi:10.1088/1757-899X/413/1/012080

Fig 1: Map of Nigeria showing the Ibadan, Benin, Owerri and Warri Airports [23]

(ii) <u>Experimental Programmme</u>

Real-time physical measurements of two environmental noise parameters [Ambient Noise Level (ANL) and Sound Pressure Level (SPL)] and two aircraft environmental noise parameters [Aircraft Take-off Noise Level ($ATNL$) and Aircraft Landing Noise Level ($ALNL$)] around each of four(4) domestic airports in Nigeria—Ibadan Airport (Lat. 7.3584°N, Long. 3.9751°E) in Oyo state, South-west Nigeria; Benin Airport (Lat. 6.3172°N, Long. 5.6037°E) in Edo state, South-South Nigeria; Owerri Airport (Lat. 5.4274°N, Long. 7.2029°E) in Imo state, South-East Nigeria; and Warri Airport (Lat. 5.5945°N, Long. 5.8193°E) in Delta state, South-South Nigeria; were repeatedly measured nine (9) times on a weekly basis during the one (1) year study-period, using the integrated $CR811C$ Noise meter in accordance with and ISO 3891:1978 standards [24], ISO 1996-1:2016 [25], and ISO 1996-2:2016 [26].

The integrated $CR811C$ Noise meter is a type 1 integrated sound level meter which is specially designed and equipped with the capacity to obtain and record precise field measurements of time-integrated sound level values. Having a measurement accuracy of $\pm 1dB(A)$, it measures sound levels from a minimum of $21dB(A)$ to a minimum of $140dB(A)$, with a measurement frequency which ranges from $25Hz$ to $16Hz$. Furthermore, this precision sound level meter measures sound levels using the ' A ' weighting scale sound frequency, which is preferable to sound frequency weighting scales ' B ' and ' C ', since it is most suitably similar to the usual behaivoural response (reaction) of the human ear to most sound frequencies.

Three hourly measurements [morning (7am – 11am), afternoon (12noon -3pm) and evening (4pm – 8pm)] were carried-out at thirty (30) selected locations in the neighbouring environments of each of these four (4) airports, thrice every week, on any weekday [from Monday through Friday], for twelve (12) months [from January to December 2017]. All measurements were taken and recorded in '$dB(A)$', where 'A' represents the A-weighting frequency for sound measurements.

Aircraft Take-off Noise Level ($ATNL$) and Aircraft Landing Noise Level ($ALNL$) were measured when the aircraft were taking-off and landing respectively. The Ambient Noise Level (ANL)—being the background noise level, was measured when the airports and the particular environments (neighbourhoods) were observed to be least busy, with no major activity taking place. The Sound Pressure Level (SPL) was measured when the normal and/or major activities/operations were carried-out in the airports.

For the purpose of these experimental measurements, this measuring device—the $CR811C$ Noise level meter, which was equipped with the inherent operational capacity to take multiple precision measurements of sound levels for the duration of a preset time, and then compute and record the average value; was first calibrated, and specifically set at the following configurations: (a) the 'A-weighting frequency network', (b) one (1) hour measurement time operation, and (c) the 'fast' response range—[corresponding to a time constant of $0.125s$]; after which, it was tripod-mounted at a height of $1.5m$ from the ground, with its microphone always pointing in the direction of the airport. When all necessary measurements of the four desired parameters had been performed, the daily/weekly averages were used to compute the monthly averages, and then these monthly averages were used to compute the yearly averages for each of both airports. Finally, the obtained yearly overall airport average values were analyzed by comparing them the three reference values stated by the World Health Organization (WHO), i.e.: WHO Standard Recommended Maximum Indoor Noise-level [$RINL_{WHO} = 35dB(A)$], WHO Standard Recommended Maximum Outdoor Noise-level [$RONL_{WHO} = 55dB(A)$], and the WHO Standard Recommended Noise-level/Sound Pressure Level for eight (8) hour daytime safe human exposure [$SPL_{WHO} = 90dB(A)$]. This was done to experimentally determine whether or not they conformed to these three standard values, and in the event that they did not conform, a rigorous statistical analysis was carried-out to evaluate the levels of deviations from the global standard values.

Also, the experimentally obtained data of the four acoustic parameters ('ANL', 'SPL', '$ATNL$' and '$ALNL$') were collated and tabulated in tables 1-4. As was earlier mentioned, the collated and tabulated data was then subjected to statistical analysis, using the 'SPSS-23 (Software Package for Statistics & Simulation-23)'. This was carried-out to determine the levels (extents) by which the ANL as a parameter deviates (differs/varies) from the 'World Health Organisation (WHO) Standard Recommended Maximum Outdoor Noise Level ($RONL_{WHO}$)' of $55dB(A)$, and also to ascertain the different levels (extents) by which each of the three other parameters [i.e. 'SPL', '$ATNL$' and '$ALNL$'] deviate (differ/vary) from the 'World Health Organisation (WHO) Standard recommended Sound Pressure Level for 8 hour daytime safe human exposure (SPL_{WHO})' of $90dB(A)$

Thus, the levels (extents) of deviations (variations) of these four acoustic parameters 'ANL'and ('SPL', '$ATNL$' and '$ALNL$'), from the WHO Standard Recommended Maximum Noise-levels [i.e. '($RONL_{WHO}$)' and '(SPL_{WHO})' respectively], were obtained, by carrying-out a non-parametric statistical test called "The Wilcoxon Sign Rank Test" on the values of these four acoustic parameters ('ANL', 'SPL', '$ATNL$' and '$ALNL$'). Thus, with respect to the 'World Health Organisation (WHO) Standard recommended Outdoor Noise Level ($RONL_{WHO}$)' of $55dB(A)$, one(1) 'Null-Hypothesis (H_0)' and one(1) 'Alternative Hypothesis (H_a)' were all stated to a significance-level (α), as follows:

For parameter 'ANL', $\rightarrow H_0 : ANL = 55dB(A)$, $H_a : ANL > 55dB(A)$, $\alpha = 0.05$ [95% confidence level]

But with respect to the "World Health Organisation (WHO) Standard recommended Sound Pressure Level for 8 hour daytime safe human exposure (SPL_{WHO})' of $90dB(A)$, three (3) 'Null-Hypotheses (H_0)' and three (3) 'Alternative Hypotheses (H_a)' were all stated to a significance-level (α), as follows:

For parameter 'SPL', $\rightarrow H_0 : SPL = 90dB(A)$, $H_a : SPL > 90dB(A)$, $\alpha = 0.05$ [95% confidence level]

For parameter '$ATNL$', $\rightarrow H_0 : ATNL = 90dB(A)$, $H_a : ATNL > 90dB(A)$, $\alpha = 0.05$ [95% confidence level]

For parameter '$ALNL$', $\rightarrow H_0 : ALNL = 90dB(A)$, $H_A : ALNL > 90dB(A)$, $\alpha = 0.05$ [95% confidence level]

3. Results and Discussion

Below are nine (9) tables and eight (8) figures in which are contained and illustrated the results obtained during this twelve (12) months study. Specifically, Table 1 and Figure 2 display the results for Ibadan domestic Airport in Oyo state, South-Western Nigeria; Table 2 and Figure 3 display the results for Benin domestic Airport in Edo state, South-Southern Nigeria; Table 3 and Figure 4 display the results for Owerri domestic Airport in Imo state, South-Eastern Nigeria and Table 4 and Figure 5 display the results for Warri domestic Airport in Delta state, South-Southern Nigeria.

Table 1: Average Values of Noise Parameters Measured at Thirty (30) Study-locations within the Neighbourhoods of Ibadan Domestic Airport, in Oyo state, South-Western Nigeria.

Airport Neighbourhood Study-location S/No.	Ambient Noise Level (ANL) [dB(A)]	Sound Pressure Level (SPL) [dB(A)]	Aircraft Take-off Noise Level (ATNL) [dB(A)]	Aircraft Landing Noise Level (ALNL) [dB(A)]	SPL-RONL$_{WHO}$ [dB(A)]	% by which SPL exceeds RONL$_{WHO}$ (%)
1	63.5	107.0	77.0	85.7	17.0	94.5
2	64.0	116.0	88.4	89.4	26.0	110.9
3	62.0	113.0	81.7	78.0	23.0	105.5
4	57.2	116.0	84.6	92.8	26.0	110.9
5	65.2	108.0	84.2	70.6	18.0	96.4
6	61.3	115.0	90.7	85.4	25.0	109.1
7	64.5	113.0	78.6	77.3	23.0	105.5
8	54.5	112.0	84.4	84.9	22.0	103.6
9	74.2	114.0	85.0	75.4	24.0	107.3
10	54.9	110.0	77.0	84.1	20.0	100.0
11	59.8	115.0	73.9	80.0	25.0	109.1
12	65.9	111.0	80.6	81.7	21.0	101.8
13	61.1	114.0	80.8	74.2	24.0	107.3
14	68.0	112.0	77.5	92.3	22.0	103.6
15	47.5	113.0	87.8	75.3	23.0	105.5
16	72.1	115.0	71.4	91.4	25.0	109.1
17	56.0	108.0	86.5	82.7	18.0	96.4
18	71.4	109.0	77.8	84.5	19.0	98.2
19	54.8	113.0	77.3	83.0	23.0	105.5
20	69.1	114.0	77.0	80.8	24.0	107.3
21	55.6	110.0	84.2	79.5	20.0	100.0
22	66.4	117.0	89.5	83.7	27.0	112.7
23	59.8	115.0	78.3	84.0	25.0	109.1
24	73.3	115.0	83.5	84.8	25.0	109.1
25	61.2	113.0	77.0	79.8	23.0	105.5
26	58.6	104.0	86.7	77.7	14.0	89.1
27	67.5	98.0	84.1	88.9	8.0	78.2
28	56.3	112.0	79.3	87.8	22.0	103.6
29	72.1	106.0	85.8	75.3	16.0	92.7
30	73.8	112.0	89.6	81.1	22.0	103.6
Average	**63.1**	**112.7**	**82.0**	**82.4**	**21.7**	**103.0**

Ibadan Domestic Airport:

As could be seen from Table 1, the mean Ambient Noise level (ANL) within thirty (30) neighbourhoods of Ibadan domestic Airport ranged from a minimum value of $47.5dB(A)$ to a maximum value of $74.2dB(A)$, with an average value of $63.1dB(A)$. Also, the mean Sound Pressure Level (SPL) within these same thirty (30) airport neighbourhoods ranged from a minimum value of $107.0dB(A)$ to a maximum value of $117.0dB(A)$, with an average value of $112.4dB(A)$. Thus, it is obvious that, all six (6) values of both environmental noise parameters were far greater than the Standard Maximum Values of $35dB(A)$ and $55dB(A)$ recommended by the World Health Organisation (WHO) for indoor and outdoor activities respectively.

The mean Aircraft Take-off Noise level ($ATNL$) within thirty (30) neighbourhoods of Ibadan domestic Airport ranged from a minimum value of $71.4dB(A)$ to a maximum value of $90.7dB(A)$, with an average value of $82.0dB(A)$. Also, the mean Aircraft Landing Noise level ($ALNL$) within these same thirty (30)

Table 2: Average Values of Noise Parameters Measured at Thirty (30) Study-locations within the Neighbourhoods of Benin Domestic Airport, in Edo state, South-Southern Nigeria.

Airport Neighbourhood Study-location S/No.	Ambient Noise Level (ANL) [dB(A)]	Sound Pressure Level (SPL) [dB(A)]	Aircraft Take-off Noise Level (ATNL) [dB(A)]	Aircraft Landing Noise Level (ALNL) [dB(A)]	SPL-RONLWHO [dB(A)]	% by which SPL exceeds RONLWHO (%)
31	69.4	114.0	78.5	81.5	24.0	107.3
32	54.6	114.0	78.1	87.5	24.0	107.3
33	69.0	105.0	83.2	76.2	15.0	90.9
34	53.1	117.0	72.9	94.5	27.0	112.7
35	63.8	114.0	87.8	81.9	24.0	107.3
36	58.9	116.0	73.4	72.9	26.0	110.9
37	67.7	113.0	83.9	73.6	23.0	105.5
38	57.9	110.0	74.9	74.0	20.0	100.0
39	63.3	109.0	70.8	80.0	19.0	98.2
40	61.7	109.0	85.6	76.3	19.0	98.2
41	60.8	114.0	81.4	84.1	24.0	107.3
42	71.4	106.0	76.6	69.6	16.0	92.7
43	61.8	104.0	86.3	85.9	14.0	89.1
44	57.3	110.0	78.2	79.6	20.0	100.0
45	63.4	117.0	86.5	70.4	27.0	112.7
46	64.7	109.0	87.8	72.7	19.0	98.2
47	61.6	115.0	82.9	80.8	25.0	109.1
48	64.2	104.0	82.1	79.6	14.0	89.1
49	58.8	111.0	75.8	81.6	21.0	101.8
50	59.9	110.0	79.5	85.6	20.0	100.0
51	63.5	112.0	76.8	78.4	22.0	103.6
52	58.6	114.0	73.6	89.7	24.0	107.3
53	67.0	107.0	91.6	70.9	17.0	94.5
54	59.4	113.0	78.4	92.3	23.0	105.5
55	69.3	116.0	83.0	77.7	26.0	110.9
56	63.7	114.0	90.8	93.8	24.0	107.3
57	70.3	109.0	83.6	85.7	19.0	98.2
58	68.8	114.0	89.7	92.1	24.0	107.3
59	65.9	112.0	91.4	93.8	22.0	103.6
60	67.2	110.0	73.5	92.4	20.0	100.0
Average	**63.2**	**111.4**	**81.3**	**81.8**	**21.4**	**102.5**

airport neighbourhoods ranged from a minimum value of $70.6dB(A)$ to a maximum value of $92.8dB(A)$, with an average value of $82.4dB(A)$. Thus, it is obvious that, all six (6) values of both aircraft noise parameters and environmental noise parameters were far greater than the Standard Maximum Values of $35dB(A)$ and $55dB(A)$ recommended by the WHO for indoor and outdoor activities respectively; but were lesser than the Standard Maximum Noise level of $90dB(A)$ recommended by the WHO for eight (8) hour daily safe human exposure.

As is pictorially represented in Figure 2 and shown in Table 1, twenty-one (21) out of thirty (30) neighbouring-environments of Ibadan domestic airport had mean SPL values that deviated upward from the $RONL_{WHO}$ [i.e. WHO Recommended Outdoor Noise Level of $55dB(A)$] by over 100%, while seven (7) out of same thirty (30) neighbouring-environments had SPL values that deviated downward from the $RONL_{WHO}$ by less than 100%, and the remaining two (2) neighbouring-environments had SPL values that deviated upward from the $RONL_{WHO}$ by exactly 100%.

Table 3: Average Values of Noise Parameters Measured at Thirty (30) Study-locations within the Neighbourhoods of Owerri Domestic Airport, in Imo state, South-Eastern Nigeria.

Airport Neighbourhood Study-location S/No.	Ambient Noise Level (ANL) [dB(A)]	Sound Pressure Level (SPL) [dB(A)]	Aircraft Take-off Noise Level (ATNL) [dB(A)]	Aircraft Landing Noise Level (ALNL) [dB(A)]	SPL-RONLWHO [dB(A)]	% by which SPL exceeds RONLWHO (%)
61	53.8	107.0	78.3	76.6.	17.0	94.5
62	64.3	114.0	88.9	86.3	24.0	107.3
63	71.4	116.0	78.8	73.2	26.0	110.9
64	59.2	106.0	81.3	75.4	16.0	92.7
65	64.2	114.0	73.2	69.1	24.0	107.3
66	55.8	107.0	72.5	82.7	17.0	94.5
67	70.0	108.0	88.5	85.7	18.0	96.4
68	55.5	111.0	73.4	71.5	21.0	101.8
69	63.5	112.0	86.7	87.9	22.0	103.6
70	51.7	116.0	89.5	79.1	26.0	110.9
71	69.5	106.0	74.3	82.4	16.0	92.7
72	67.7	112.0	82.0	81.6	22.0	103.6
73	51.0	110.0	91.6	71.7	20.0	100.0
74	66.2	113.0	77.6	81.8	23.0	105.5
75	53.7	111.0	90.6	68.5	21.0	101.8
76	74.6	114.0	89.5	86.7	24.0	107.3
77	59.7	107.0	76.2	92.3	17.0	94.5
78	65.5	115.0	83.4	87.3	25.0	109.1
79	68.0	109.0	85.3	78.6	19.0	98.2
80	58.9	111.0	79.9	87.7	21.0	101.8
81	53.6	104.0	90.5	72.3	14.0	89.1
82	61.1	112.0	81.4	83.5	22.0	103.6
83	66.8	115.0	82.0	79.3	25.0	109.1
84	53.1	108.0	82.3	82.6	18.0	96.4
85	72.2	110.0	92.6	80.9	20.0	100.0
86	73.7	106.0	87.4	69.2	16.0	92.7
87	68.5	102.0	91.8	89.6	12.0	85.5
88	57.2	99.0	88.5	85.7	9.0	80.0
89	71.7	110.0	90.7	86.4	20.0	100.0
90	69.9	105.0	86.4	91.1	15.0	90.9
Average	**63.1**	**109.7**	**83.8**	**81.0**	**20.5**	**101.3**

Benin Domestic Airport:
The experimental results shown in Table 2, revealed that, for all thirty (30) neighbourhoods of Benin domestic Airport, mean Ambient Noise Level (ANL) peaked at $71.4dB(A)$, averaged at $63.2dB(A)$, and recorded its five lowest values of $53.1dB(A)$, $54.6dB(A)$, $57.3dB(A)$, $57.9dB(A)$ and $58.6dB(A)$. It was noted that, the peak ANL, average ANL and three(3) of the five(5) lowest ANL values, [not to mention the other twenty-five (25) ANL values] were higher than the Maximum Value of $55dB(A)$ recommended by the World Health Organisation (WHO) for outdoor activities ($RONL_{WHO}$). In the same vein, while mean Sound Pressure Level (SPL) peaked at $117.0dB(A)$, averaged at $111.4dB(A)$, and recorded its five lowest values of $104dB(A)$ [twice], $105.0dB(A)$, $106.0dB(A)$ $107.0dB(A)$ and $109.0dB(A)$ [four times]. It was noted that, the peak SPL, average SPL and all five(5) lowest SPL values, [excluding the other twenty-five (25) SPL values] were higher than the Standard Maximum Noise level of $90dB(A)$ recommended by the WHO for eight (8) hour daily safe human exposure.

Table 4: Average Values of Noise Parameters Measured at Thirty (30) Study-locations within the Neighbourhoods of Warri Domestic Airport, in Delta state, South-Southern Nigeria.

Airport Neighbourhood Study-location S/No.	Ambient Noise Level (ANL) [dB(A)]	Sound Pressure Level (SPL) [dB(A)]	Aircraft Take-off Noise Level (ATNL) [dB(A)]	Aircraft Landing Noise Level (ALNL) [dB(A)]	SPL-RONLWHO [dB(A)]	% by which SPL exceeds RONLWHO (%)
91	57.1	112.0	69.5	73.3	22.0	103.6
92	69.8	105.0	81.7	82.7	15.0	90.9
93	56.3	112.0	82.9	85.4	22.0	103.6
94	75.7	109.0	73.7	78.7	19.0	98.2
95	64.8	103.0	78.6	88.9	13.0	87.3
96	71.5	113.0	80.7	80.2	23.0	105.5
97	53.1	114.0	82.9	81.7	24.0	107.3
98	62.1	112.0	87.8	72.4	22.0	103.6
99	57.9	111.0	82.8	85.2	21.0	101.8
100	72.4	116.0	80.9	78.8	26.0	110.9
101	55.1	116.0	80.5	78.5	26.0	110.9
102	75.2	105.0	82.7	89.0	15.0	90.9
103	64.9	117.0	88.9	92.3	27.0	112.7
104	69.0	108.0	79.8	80.1	18.0	96.4
105	62.3	114.0	76.5	87.8	24.0	107.3
106	59.6	108.0	72.2	79.5	18.0	96.4
107	63.5	108.0	87.3	74.6	18.0	96.4
108	62.5	109.0	73.4	83.1	19.0	98.2
109	65.9	114.0	87.6	80.0	24.0	107.3
110	73.2	110.0	76.2	77.2	20.0	100.0
111	64.2	113.0	83.8	82.8	23.0	105.5
112	66.9	110.0	72.4	74.6	20.0	100.0
113	67.6	106.0	89.6	92.3	16.0	92.7
114	75.6	103.0	84.2	87.8	13.0	87.3
115	62.8	102.0	77.0	79.5	12.0	85.5
116	73.6	107.0	85.5	90	17.0	94.5
117	74.9	109.0	78.2	91.7	19.0	98.2
118	75.4	110.0	80.3	78.6	20.0	100.0
119	72.8	108.0	87.6	83.7	18.0	96.4
120	74.6	109.0	88.1	85.2	19.0	98.2
Average	**66.7**	**109.8**	**81.1**	**82.5**	**19.8**	**100.0**

Similarly, in this same Table 2, is presented the mean Aircraft Take-off Noise level ($ATNL$) and mean Aircraft Landing Noise level ($ALNL$) results, from which it was observed that, the mean $ATNL$ peaked at $91.6dB(A)$, averaged at $81.3dB(A)$, and recorded its five lowest values of $70.8dB(A)$, $72.9dB(A)$, $73.4dB(A)$, $73.5dB(A)$ and $73.6dB(A)$. It was noted that, the peak $ATNL$, average $ATNL$ and all five(5) lowest $ATNL$ values, [not to mention the other twenty-five (25) $ATNL$ values] were higher than the Maximum Value of $55dB(A)$ recommended by the World Health Organisation (WHO) for outdoor activities $(RONL_{WHO})$. Likewise, the mean $ALNL$ peaked at $94.5dB(A)$, averaged at $81.8dB(A)$, and recorded its five(5) lowest values of $69.9dB(A)$, $70.4dB(A)$, $70.9dB(A)$, $72.7dB(A)$ and $72.9dB(A)$. It was noted that, although, all these values of $ALNL$ [with exception to its peak value] were lower than the Standard Maximum Noise level of $90dB(A)$ recommended by the WHO for eight (8) hour daily safe human exposure (SPL_{WHO}); yet, they all significantly exceeded the Standard Maximum Values of $35dB(A)$ and $55dB(A)$ recommended by the World Health Organisation (WHO) for indoor and outdoor activities respectively.

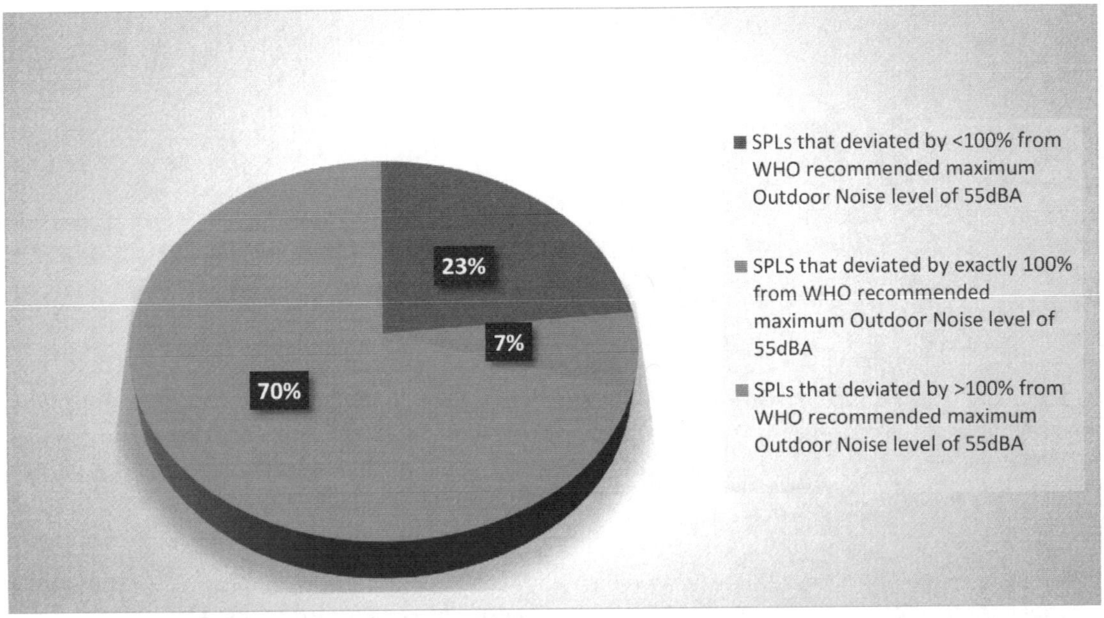

Fig. 2: Percentages by which the SPL-values in thirty (30) Neighbouring-environments of Ibadan Airport exceeded the WHO Recommended Maximum Outdoor Noise-level of 55dB(A)

Also, depicted in Figure 3 and shown in Table 2, is the fact that: seventeen (17) out of thirty (30) neighbouring-environments of Benin domestic airport had mean *SPL* values that deviated upward from the $RONL_{WHO}$ [i.e. WHO Recommended Outdoor Noise Level of $55dB(A)$] by over 100%, while nine (9) out of same thirty (30) neighbouring-environments had *SPL* values that deviated downward from the $RONL_{WHO}$ by less than 100%, and the remaining four (4) neighbouring-environments had *SPL* values that deviated upward from the $RONL_{WHO}$ by exactly 100%.

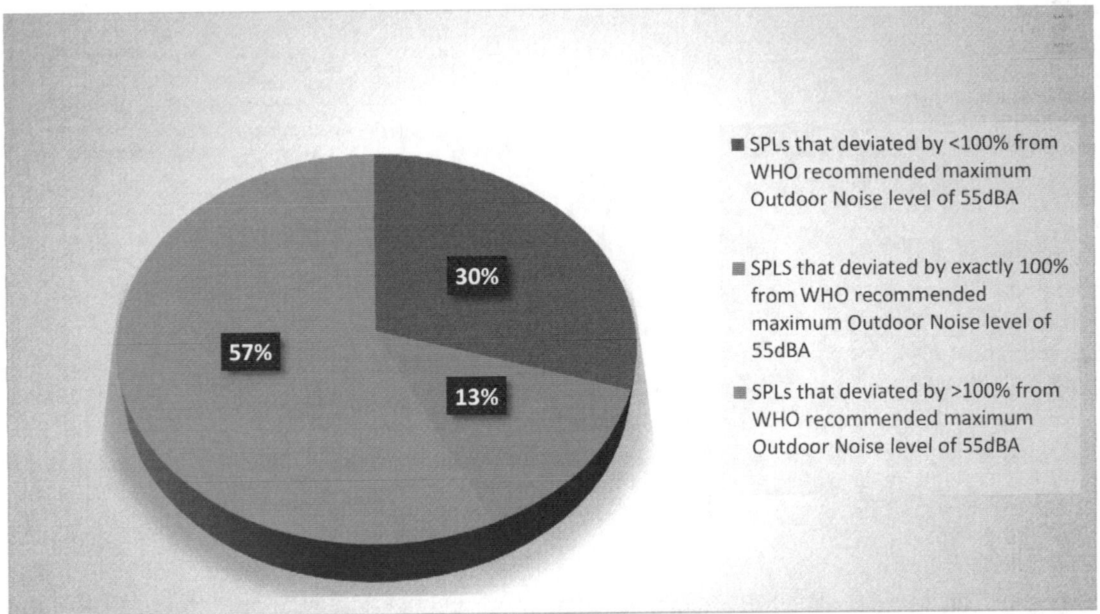

Fig. 3: Percentages by which the SPL-values in thirty (30) Neighbouring-environments of Benin Airport exceeded the WHO Recommended Maximum Outdoor Noise-level of 55dB(A)

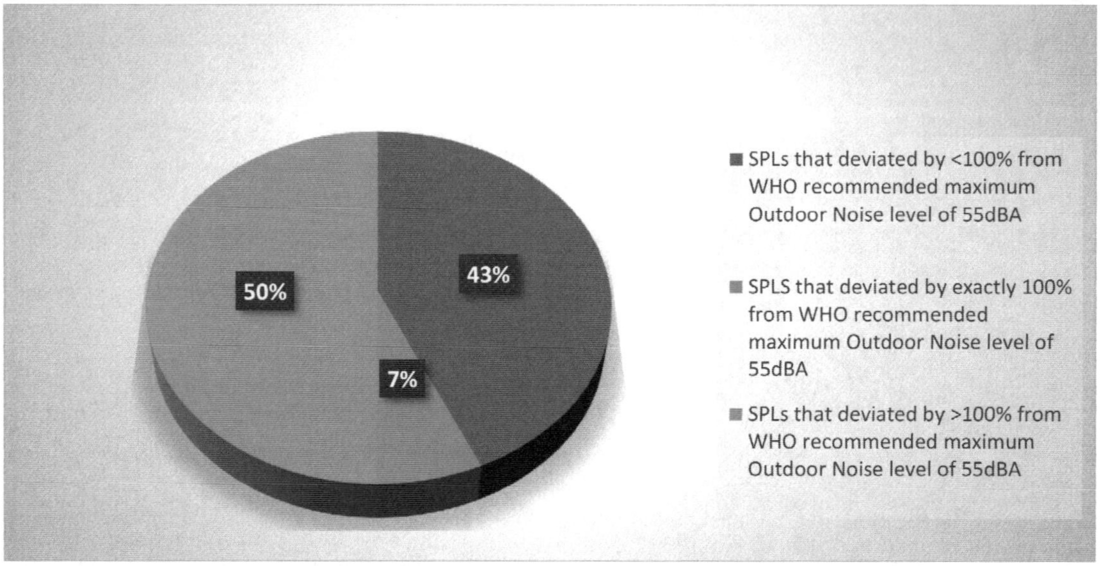

Fig. 4: Percentages by which the SPL-values in thirty (30) Neighbouring-environments of Owerri Airport exceeded the WHO Recommended Maximum Outdoor Noise-level of 55dB(A)

Owerri (Sam Mbakwe) Domestic Airport:

As could be seen from Table 3, the mean Ambient Noise level (ANL) within thirty (30) neighbourhoods of Owerri domestic Airport ranged from a minimum value of $51.0dB(A)$ to a maximum value of $74.6dB(A)$, with an average value of $63.1dB(A)$. Also, the mean Sound Pressure Level (SPL) within these same thirty (30) airport neighbourhoods ranged from a minimum value of $104.0dB(A)$ to a maximum value of $116.0dB(A)$, with an average value of $109.7dB(A)$. Obviously, all six (6) values of ANL and SPL were larger than the Standard Maximum Values of $35dB(A)$ and $55dB(A)$ recommended by the World Health Organisation (WHO) for indoor and outdoor activities respectively.

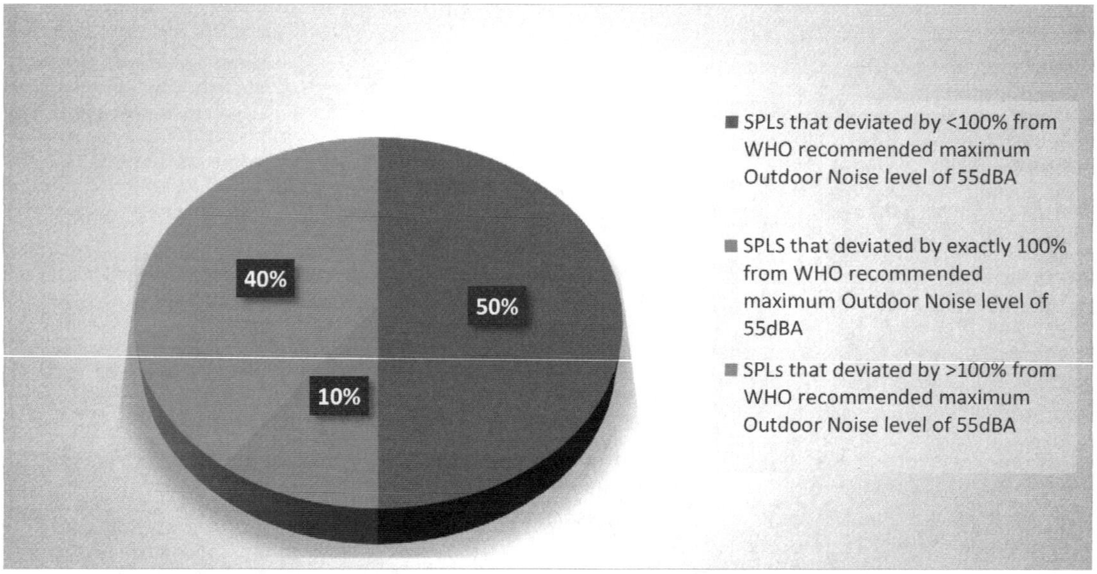

Fig. 5: Percentages by which the SPL-values in thirty (30) Neighbouring-environments of Warri Airport exceeded the WHO Recommended Maximum Outdoor Noise-level of 55dB(A) The mean Aircraft Take-off Noise level ($ATNL$) within thirty (30) neighbourhoods of Owerri domestic Airport ranged from a minimum value of $72.5dB(A)$ to a maximum value of $92.6dB(A)$, with an average value of $83.8dB(A)$. Also, the mean Aircraft Landing Noise level ($ALNL$) within these same thirty (30) airport neighbourhoods ranged

from a minimum value of $68.5dB(A)$ to a maximum value of $92.3dB(A)$, with an average value of $81.0dB(A)$. Again, it is obvious that, all six (6) values of both aircraft noise parameters and environmental noise parameters were greater than the Standard Maximum Values of $35dB(A)$ and $55dB(A)$ recommended by the WHO for indoor and outdoor activities respectively; but were lesser than the Standard Maximum Noise level of $90dB(A)$ recommended by the WHO for eight (8) hour daily safe human exposure.

Moreover, based on the experimental results illustrated in Figure 4 and presented in Table 3; fifteen (15) out of thirty (30) neighbouring-environments of Owerri domestic airport gave SPL values that deviated upward from the $RONL_{WHO}$ [i.e. WHO Recommended Outdoor Noise Level of $55dB(A)$] by over 100%, while thirteen (13) out of same thirty (30) neighbouring-environments had SPL values that deviated downward from the $RONL_{WHO}$ by less than 100%, and the remaining two (2) neighbouring-environments had SPL values that deviated upward from the $RONL_{WHO}$ by exactly 100%.

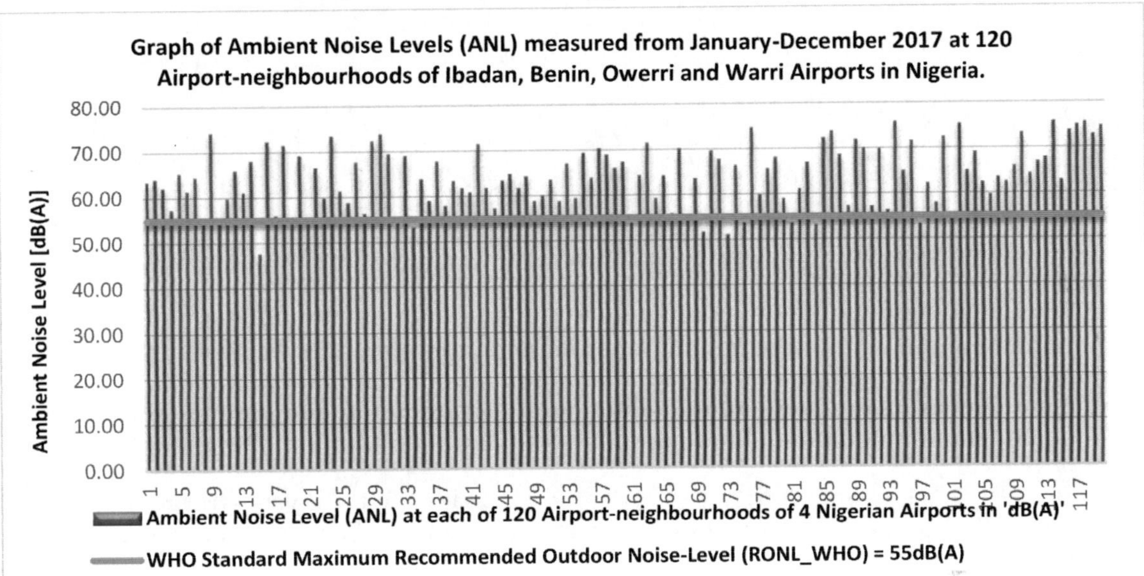

Figure 6: Graph of Ambient Noise Levels (ANL) measured from January-December 2017 at 120 Airport-neighbourhoods of Ibadan, Benin, Owerri and Warri Airports in Nigeria.

Warri (Osubi) Domestic Airport:

The experimental results shown in Table 4, revealed that, for all thirty (30) neighbourhoods of Warri domestic Airport, mean Ambient Noise Level (ANL) peaked at $75.7dB(A)$, averaged at $66.7dB(A)$, and recorded its five lowest values of $53.1dB(A)$, $55.1dB(A)$, $56.3dB(A)$, $57.1dB(A)$ and $57.9dB(A)$. It is noteworthy that, the peak ANL, average ANL and four(4) of the five(5) lowest ANL values, [not to mention the other twenty-five (25) ANL values] were higher than the Maximum Value of $55dB(A)$ recommended by the World Health Organisation (WHO) for outdoor activities $(RONL_{WHO})$. Furthermore, while the mean Sound Pressure Level (SPL) peaked at $117.0dB(A)$, averaged at $109.8dB(A)$, and recorded its five lowest values of $102dB(A)$, $103dB(A)$ [twice], $105.0dB(A)$ [twice], $106.0dB(A)$ and $107.0dB(A)$. It was noted that, the peak SPL, average SPL and all five lowest SPL values, [excluding the other twenty-three (23) SPL values] were higher than the Standard Maximum Noise level of $90dB(A)$ recommended by the WHO for eight (8) hour daily safe human exposure.

Again, in this same Table 4, is presented the mean Aircraft Take-off Noise level ($ATNL$) and mean Aircraft Landing Noise level ($ALNL$) results, from which it was observed that, the mean $ATNL$ peaked at $89.6dB(A)$, averaged at $81.1dB(A)$, and recorded its five lowest values of $69.5dB(A)$, $72.2dB(A)$, $72.4dB(A)$, $73.4dB(A)$ and $73.7dB(A)$. Thus, It was noticed that, the peak $ATNL$, average $ATNL$ and all five lowest $ATNL$ values, [not to mention the other twenty-five (25) $ATNL$ values] were higher than the Maximum Value of $55dB(A)$ recommended by the World Health Organisation (WHO) for outdoor activities $(RONL_{WHO})$. Similarly, the mean $ALNL$ peaked at $92.3dB(A)$, averaged at $82.5dB(A)$, and recorded its five lowest values of $72.4dB(A)$,

$73.3dB(A)$, $74.6dB(A)$ [twice], $77.2dB(A)$ and $78.5dB(A)$. Thus, it is worthy of note that, despite the fact that all these values of $ALNL$ [with exception to its peak value] were below the Standard Maximum Noise level of $90dB(A)$ recommended by the WHO for eight (8) hour daily safe human exposure (SPL_{WHO}); yet, they all were considerably higher than the Standard Maximum Values of $35dB(A)$ and $55dB(A)$ recommended by the World Health Organisation (WHO) for indoor and outdoor activities respectively.

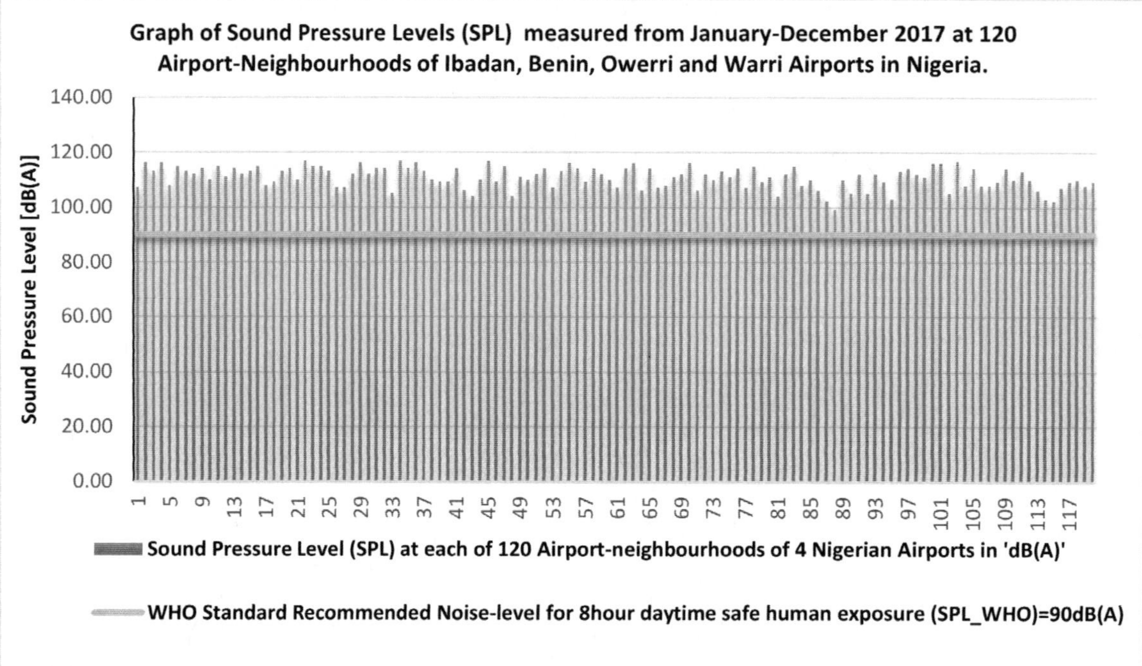

Figure 7: Graph of Sound Pressure Levels (SPL) measured from January-December 2017 at 120 Airport-neighbourhoods of Ibadan, Benin, Owerri and Warri Airports in Nigeria.

Furthermore, from the pictorial depiction in Figure 5 and the experimental results contained in Table 4, it could be inferred that: twelve (12) out of thirty (30) neighbouring-environments of Warri domestic airport gave SPL values that deviated upward from the $RONL_{WHO}$ [i.e. WHO Recommended Outdoor Noise Level of $55dB(A)$] by over 100%, while fifteen (15) out of same thirty (30) neighbouring-environments had SPL values that deviated downward from the $RONL_{WHO}$ by less than 100%, and the remaining three (3) neighbouring-environments had SPL values that deviated upward from the $RONL_{WHO}$ by exactly 100%.
Summarily, a careful look at Figures 6 - 9, shows that, for all one hundred and twenty (120) neighbourhoods of four (4) domestic airports in Nigeria, this periodic repetitive experimental acoustic assessment clearly revealed amongst other things that: 89.2% of the investigated Airports' neighbourhoods [i.e. 107 of 120 neighbourhoods] had ANL values which were higher than the WHO-

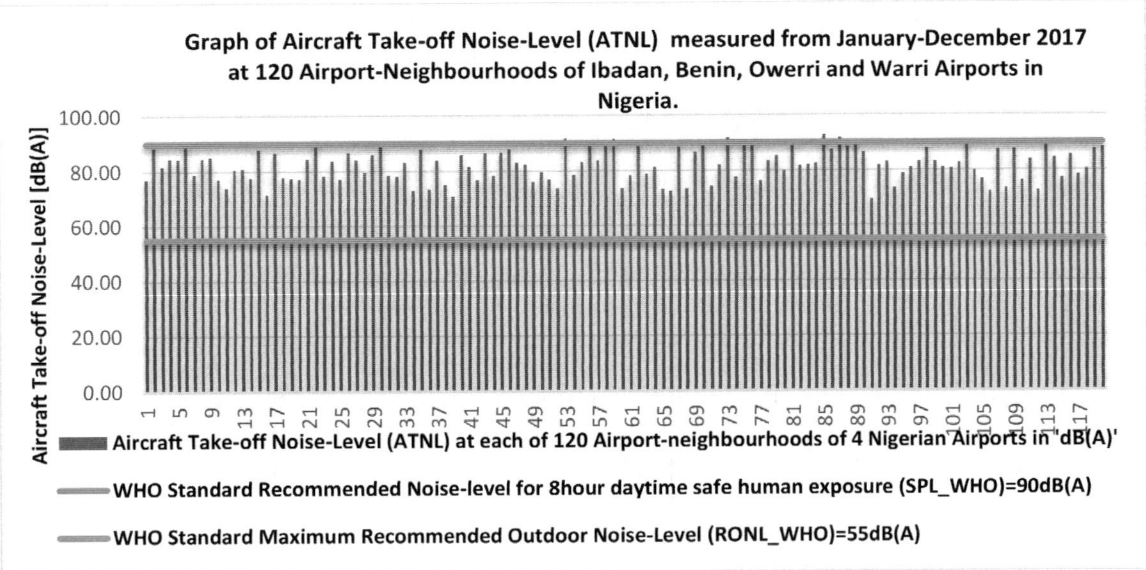

Figure 8: Graph of aircraft Take-off Noise-Level (ATNL) measured from January-December 2017 at 120 Airport-neighbourhoods of Ibadan, Benin, Owerri and Warri Airports in Nigeria.

recommended value ($RONL_{WHO}$) of $55dB(A)$; 100% of the investigated Airports' neighbourhoods [i.e. 120 of 120 neighbourhoods] had SPL values which were higher than the WHO-recommended value (SPL_{WHO}) of $90dB(A)$; 100% of the investigated airports' neighbourhoods [i.e. 120 of 120 neighbourhoods] had $ATNL$ values which were higher than the WHO-recommended value (SPL_{WHO}) of $90dB(A)$; and 87.5% of the investigated airports' Neighbourhoods [i.e. 105 of 120 neighbourhoods] had $ALNL$ values which were higher than the WHO-recommended value (SPL_{WHO}) of $90dB(A)$. Very importantly also, from a comparative stand-point, it was keenly observed from Tables 1-4, that: the highest level of exposure to Airport environmental noise hazard was recorded at Ibadan domestic Airport [with an SPL average value of $112.7dB(A)$], followed by Benin domestic Airport [$111.3dB(A)$], and Owerri domestic Airport [$110.7dB(A)$] and then, Warri domestic Airport [$110.0dB(A)$].

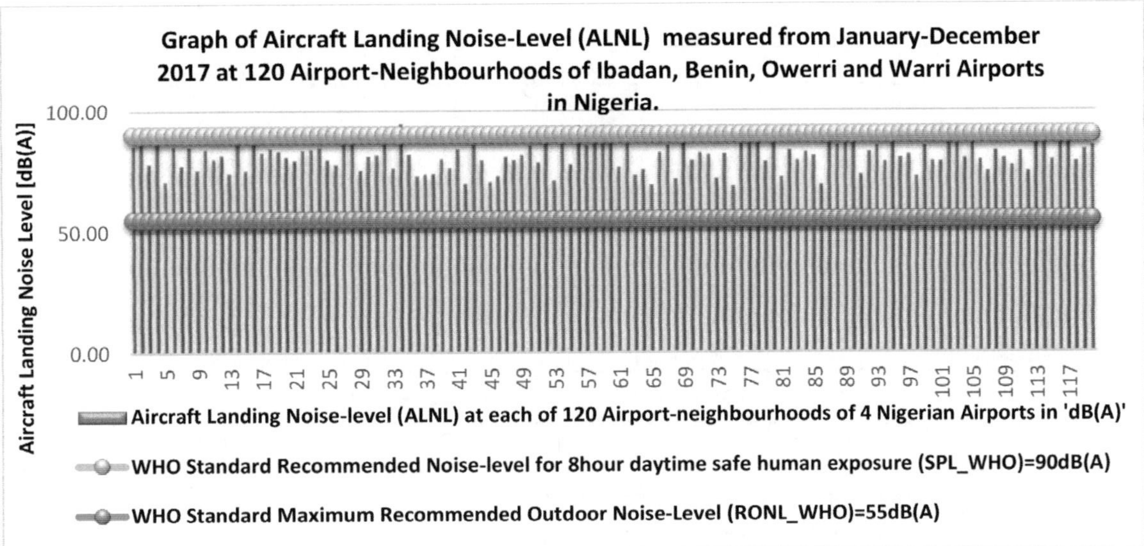

Figure 9: Graph of aircraft Landing Noise-Level (ALNL) measured from January-December 2017 at 120 Airport-neighbourhoods of Ibadan, Benin, Owerri and Warri Airports in Nigeria.

IOP Conf. Series: Materials Science and Engineering **413** (2018) 012080 doi:10.1088/1757-899X/413/1/012080

Combined Statistical Analyses of Experimental Values of all Four (4) Acoustic Parameters ['ANL', 'SPL', 'ATNL' and 'ALNL'] Measured at 120 Distinct Airport Neighbourhoods [of Ibadan, Benin, Owerri and Warri Airports] in Nigeria:

Table 5: Basic Descriptive Statistics of all Ambient Noise Levels (*ANL*), Sound Pressure Levels (*SPL*), Aircraft Take-off Noise Levels (*ATNL*) and Aircraft Landing Noise Levels (*ALNL*) measured at 120 distinct Airport Neighbourhoods of Ibadan, Benin, Owerri and Warri Airports in Nigeria.

		ANL [dB(A)]	SPL [dB(A)]	ATNL [dB(A)]	ALNL [dB(A)]
N	Valid	120	120	120	120
	Missing	0	0	0	0
Mean		64.0075	110.8083	82.0600	81.9125
Median		64.1000	111.0000	82.2000	81.7500
Mode		63.50	114.00	77.00	92.30
Std. Deviation		6.63767	3.87904	5.76042	6.47985
Variance		44.059	15.047	33.182	41.988
Range		28.20	18.00	23.10	26.00
Minimum		47.50	99.00	69.50	68.50
Maximum		75.70	117.00	92.60	94.50
Sum		7680.90	13297.00	9847.20	9829.50

Shown in Table 5 above, are nine (9) common descriptive statistics of the combined data of these four acoustic parameters at all 120 investigated airport neighbourhoods in Ibadan, Benin, Owerri and Warri Airports. Although, some of these parameters may have been separately mentioned earlier, however, they will be re-discussed here briefly. In summary, while ' ANL ' ranges from $47.5-75.5\ dB(A)$, with a mean of $64.0dB(A)$, a mode of $63.5dB(A)$, and a median of $64.1dB(A)$; ' SPL ' ranges from $99.0-117.0\ dB(A)$, with a mean of $110.8dB(A)$, a mode of $114.0dB(A)$, and a median of $111.0dB(A)$; Also, while ' $ATNL$ ' ranges from $69.5-92.6dB(A)$, with a mean of $82.1dB(A)$, a mode of $77.0dB(A)$ and a median of $82.2dB(A)$; and ' $ALNL$ ' ranges from $68.5-94.5dB(A)$, with a mean of $81.9dB(A)$, a mode of $92.3dB(A)$ and a median of $81.8dB(A)$.

Table 6: Wilcoxon Signed Ranks Tests of measured *ANL* at 120 distinct Airport Neighbourhoods [of Ibadan, Benin, Owerri and Warri Airports] in Nigeria with respect to $RONL_{WHO}$

Ranks

		N	Mean Rank	Sum of Ranks
ANL [dB(A)] - RONL_WHO [dB(A)]	Negative Ranks	13[a]	15.50	201.50
	Positive Ranks	107[b]	65.97	7058.50
	Ties	0[c]		
	Total	120		

a. ANL [dB(A)] < RONL_WHO [dB(A)]
b. ANL [dB(A)] > RONL_WHO [dB(A)]
c. ANL [dB(A)] = RONL_WHO [dB(A)]

Test Statistics[b]

	ANL [dB(A)] - RONL_WHO [dB(A)]
Z	-8.979[a]
Asymp. Sig. (2-tailed)	.000

a. Based on negative ranks.
b. Wilcoxon Signed Ranks Test

Table 7: Wilcoxon Signed Ranks Tests of measured *SPL* at 120 distinct Airport Neighbourhoods [of Ibadan, Benin, Owerri and Warri Airports] in Nigeria with respect to SPL_{WHO}

Ranks

		N	Mean Rank	Sum of Ranks
SPL [dB(A)] - SPL_WHO [dB(A)]	Negative Ranks	0[a]	.00	.00
	Positive Ranks	120[b]	60.50	7260.00
	Ties	0[c]		
	Total	120		

a. SPL [dB(A)] < SPL_WHO [dB(A)]
b. SPL [dB(A)] > SPL_WHO [dB(A)]
c. SPL [dB(A)] = SPL_WHO [dB(A)]

Test Statistics[b]

	SPL [dB(A)] - SPL_WHO [dB(A)]
Z	-9.515[a]
Asymp. Sig. (2-tailed)	.000

a. Based on negative ranks.
b. Wilcoxon Signed Ranks Test

Following the 'SPSS-23' generated results of the 'Wilcoxon Signed Ranks Tests' analysis of the ' *ANL* data', shown in Table 6 above and Figures 10-12, with respect to the WHO Standard Recommended Maximum Outdoor Noise-level ($RONL_{WHO}$) of $55dB(A)$: for the 120 airport neighbouring-environments investigated around Ibadan, Benin, Owerri and Warri Airports, n is 119, *tie* is 0, Positive-Ranks i.e. $ANL[dB(A)] > RONL_{WHO}[dB(A)]$ is 107, and the Negative-Ranks i.e. $ANL[dB(A)] < RONL_{WHO}[dB(A)]$ is 13. Thus, the alternative hypothesis [$H_a : ANL > 55dB(A)$] is statistically significant and should be accepted at $Z = -8.979,\ n = 119,\ p = 0.000 < 0.05;$ while the null hypothesis [$H_0 : ANL = 55dB(A)$] should be rejected.

Table 8: Wilcoxon Signed Ranks Tests of measured *ATNL* at 120 distinct Airport Neighbourhoods [of Ibadan, Benin, Owerri and Warri Airports] in Nigeria with respect to SPL_{WHO}

Ranks

		N	Mean Rank	Sum of Ranks
ATNL[dB(A)] – SPL [dB(A)]	Negative Ranks	120[a]	60.50	7260.00
	Positive Ranks	0[b]	.00	.00
	Ties	0[c]		
	Total	120		

a. ATNL [dB(A)] < SPL [dB(A)]
b. ATNL [dB(A)] > SPL [dB(A)]
c. ATNL [dB(A)] = SPL [dB(A)]

Test Statistics[b]

	ATNL [dB(A)] – SPL [dB(A)]
Z	-9.507[a]
Asymp. Sig. (2-tailed)	.000

a. Based on negative ranks.
b. Wilcoxon Signed Ranks Test

From the 'SPSS-23' generated results of the 'Wilcoxon Signed Ranks Tests' analysis of the ' *SPL* data', shown in Table 7 above and Figures 10-12, with respect to the WHO Standard Recommended Noise-level/Sound Pressure Level for eight (8) hour daytime safe human exposure (SPL_{WHO}) of $90dB(A)$]: for

IOP Conf. Series: Materials Science and Engineering **413** (2018) 012080 doi:10.1088/1757-899X/413/1/012080

the 120 airport neighbouring-environments investigated around Ibadan, Benin, Owerri and Warri Airports, *n* is 119, *tie* is 0, Positive-Ranks i.e. $SPL[dB(A)] > SPL_{WHO}[dB(A)]$ is 120, and the Negative-Ranks i.e. $SPL[dB(A)] < SPL_{WHO}[dB(A)]$ is 0. Thus, the alternative hypothesis [$H_a : SPL > 90dB(A)$] is statistically significant and should be accepted at $Z = -9.515,\ n = 119,\ p = 0.000 < 0.05$; while the null hypothesis [$H_0 : SPL = 90dB(A)$] should be rejected.

Table 9: Wilcoxon Signed Ranks Tests of measured *ALNL* at 120 distinct Airport Neighbourhoods [of Ibadan, Benin, Owerri and Warri Airports] in Nigeria with respect to SPL_{WHO}

Ranks

		N	Mean Rank	Sum of Ranks
ALNL[dB(A)] - SPL_WHO[dB(A)]	Negative Ranks	105[a]	65.53	6881.00
	Positive Ranks	14[b]	18.50	259.00
	Ties	1[c]		
	Total	120		

a. ALNL [dB(A)] < SPL_WHO [dB(A)]
b. ALNL [dB(A)] > SPL_WHO [dB(A)]
c. ALNL [dB(A)] = SPL_WHO [dB(A)]

Test Statistics[b]

	ALNL [dB(A)] - SPL_WHO [dB(A)]
Z	-8.780[a]
Asymp. Sig. (2-tailed)	.000

a. Based on positive ranks.
b. Wilcoxon Signed Ranks Test

Likewise, based on the From the 'SPSS-23' generated results of the 'Wilcoxon Signed Ranks Tests' analysis of the ' *ATNL* data', shown in Table 8 above and Figures 10-12, with respect to the WHO Standard Recommended Noise-level/Sound Pressure Level for eight (8) hour daytime safe human exposure (SPL_{WHO}) of $90dB(A)$]: for the 120 airport neighbouring-environments investigated around Ibadan, Benin, Owerri and Warri Airports, *n* is 119, *tie* is 0, Positive-Ranks i.e. $ATNL[dB(A)] > SPL_{WHO}[dB(A)]$ is 120, and the Negative-Ranks i.e. $ATNL[dB(A)] < SPL_{WHO}[dB(A)]$ is 0. Thus, the alternative hypothesis [$H_a : ATNL > 90dB(A)$] is statistically significant and should be accepted at $Z = -9.507,\ n = 119,\ p = 0.000 < 0.05$; while the null hypothesis [$H_0 : ATNL = 90dB(A)$] should be rejected.
Also, the 'SPSS-23' generated results of the 'Wilcoxon Signed Ranks Tests' analysis of the ' *ALNL* data', shown in Table 9 above and Figures 10-12, with respect to the WHO Standard Recommended Noise-level/Sound Pressure Level for eight (8) hour daytime safe human exposure (SPL_{WHO}) of $90dB(A)$]: for the 120 airport neighbouring-environments investigated around Ibadan, Benin, Owerri and Warri Airports, *n* is 119, *tie* is 1, Positive-Ranks i.e. $ALNL[dB(A)] > SPL_{WHO}[dB(A)]$ is 105, and the Negative-Ranks i.e. $ALNL[dB(A)] < SPL_{WHO}[dB(A)]$ is 14. Thus, the alternative hypothesis [$H_a : ALNL > 90dB(A)$] is statistically significant and should be accepted at $Z = -8.780,\ n = 119,\ p = 0.000 < 0.05$; while the null hypothesis [$H_0 : ALNL = 90dB(A)$] should be rejected.

4. Conclusion

The average values of two (2) environmental noise parameters [Ambient Noise Level and Sound Pressure Level] and two (2) Aircraft Noise parameters [Aircraft Take-off Noise Level and Aircraft Landing Noise Level] have been scientifically assessed and studied for twelve (12) months, at one hundred and twenty (120) purposively selected neighbourhoods/neighbouring-environments of four (4) domestic airports [Ibadan, Benin, Owerri and Warri] in Nigeria. The results obtained showed that: at all one hundred and twenty (120) domestic airport neighbourhoods investigated, majority of the values of both 'the mean values of all Ambient Noise Levels (ANL)' [i.e. 89.0%] and 'the Aircraft Landing Noise Levels ($ALNL$)' [i.e. 87.9%], measured were far higher than 'the WHO Standard Recommended Maximum Noise-levels for both indoor and outdoor activities [i.e. ($RINL_{WHO}$) and ($RONL_{WHO}$)]' and 'the WHO recommended SPL value for 8hour daytime safe human exposure [(SPL_{WHO})]' respectively. Also, all of both 'the mean values of all Sound Pressure Levels (SPL)' [i.e. 100%] and 'the mean values of Aircraft Take-off Noise Levels ($ATNL$)' [i.e. 100%] measured, were higher than the WHO recommended SPL value for 8hour daytime safe human exposure (SPL_{WHO}). Interestingly, the authenticity of the fact that, majority of some of the ANL and $ALNL$ values exceed the $RONL_{WHO}$ and SPL_{WHO} , while all the SPL and $ATNL$ values exceed the SPL_{WHO} ; is further proven/validated by above published results of the descriptive Statistics and Non-parametric statistical test analyses. Also, the highest environmental noise hazardous exposures were experienced at the neighbouring environments of: Ibadan airport, followed by Benin airport, and Owerri airport, with the lowest at Warri airport. Consequently, people/persons living and/or working in the neighbourhoods of Ibadan, Benin, Owerri and Warri domestic airports in Nigeria, are most likely to experience frequent cases of Speech-unintelligibility, Noise-annoyance, Headache and Sleep-disturbance; after prolonged exposure to the environmental aircraft noise etc.

5. Recommendations

In order to mitigate the human exposure levels to aircraft noise-hazard in the neighbouring-environments of these four airports and others in Nigeria, we hereby make the following recommendations:

- Functional geo-physical planning of cities/towns hosting airports by professional experts such as Urban & regional planners etc.
- Adoption of Aero-polis instead of a simple airport, as applicable is most developed countries.
- Effective Land-use planning & management
- Adaptive architectural designs & orientation of buildings, with respect to local prevalent wind directions
- Use of anechoic and sound-proof wedges as wall-claddings
- Pre-design considerations for low noise emissions of aircraft
- Further research in certain related technical areas/fields such as Nano-technology, Reverse-engineering and Sound-spectra etc.

Acknowledgments

The authors are immensely grateful to GOD almighty who by HIS infinite grace, has made what seemed to be a herculean task to become today's success story. Furthermore, time will fail us to fully express our heartfelt gratitude to the Management and Staff of Hafalix Nigeria Limited, and a host of others, for their never wavering support during the course of this historic research adventure.

References

[1] Ibhadode, O.; Tenebe, I. T.; Emenike, P. C.; Adesina, O. S.; Okougha, A. F. Aitanke, F. O. (2018) Assessment of noise-levels of generator-sets in seven cities of South-Southern Nigeria. African Journal of Science, Technology, Innovation and Development, 10:2, 125-135, DOI: 10.1080/20421338.2017.1400711

[2] Ibhadode, O., P. Ibhadode, A. F. Okougha, I. I. Umanah, F. O., Aitanke, and S. A. B. Fiyebo. (2016) Hazards Assessment Analyses of Fossil-Fuel Generators: Holistic-Study of Human

Experiences and Perceptions in South-Southern Nigeria. Journal of Sustainable Development Studies 9 (2): 153–242. http://www.infinitypress.info.

[3] Kaltenbach, M.; Christian. M.; Kilinke R.., (2005) Health Consequences of Aircraft Noise. *Deutsches Arzteblatt International* 105(31-32), 548-556

[4] Issarayangyun, T.; Black. J.; Black D.., Samuels, S., (2005) Aircraft Noise and Methods for the Study of Community Health well-being. *Journal of the Eastern Asia Society for Transportation Studies* 6(1), 3293-3308

[5] Franssen, E. A. M.; Van-Wiechen, C. M. A. G.; Nagelkerke, N. J. D.; Lebret, E. (2004) Aircraft Noise around a large International Airport and its Impact on General Health and Medication use. *Occup Environ Med* 61(1), 405-415, doi: 10.1136/oem.2002.005488

[6] Correla, A. W.; Peters, J. L.; Levy, J. I.; Melly, S.; Dominici, F. (2013) Residential exposure to aircraft noise and hospital admissions for cardiovascular diseases: multi-airport retrospective study. *BMJ* 347:15561, 1-11

[7] Charakida, M. and Deanfield, J. E. (2013) Nighttime Aircraft Noise Exposure: Flying Towards Arterial Disease. *European Heart Journal* 34, 3472-3474

[8] Rizk, S. A. M.; Sharaf, N. E.; Mahdy-Abdallah, H.; Abd-ElGell, K. S. (2014) Some Health effects of Aircraft noise with special reference to shift work. *Technology and Industrial Health* 1-7. sagepub.co.uk/journalsPermissions.nav DOI: 10.1177/0748233713518602

[9] Wikipedia free Encyclopaedia (2015). Aircraft Noise, Retrieved from https://en.wikipedia.org/wiki/Aircraft_noise on 1st November 2017, 05:00am.

[10] Mauskar, J. M. (2008). Ambient Noise Level due to Aircrafts: Requirement and Procedure for Monitoring (Annexure-III). Indian Central Pollution Control Board-Ministry of Environment and forest, 22, 19-23. http://www.cpcb.nic.in

[11] European Parliamentary Research Service (2013). Airport noise pollution. Retrieved from http://www.epithinkthank.eu/2013/01/airport-noise-pollution on 1st November 2017, 05:30am.

[12] Eriksson, C., Rosenlund. M., Pershagen. G., Hilding, A., Ostenson, C. G., Bluhin, G. (2007) Aircraft noise and incidence of hypertension. *Epidemiology* 18(1), 716–721.

[13] Jarup, L., Babisch, W., Houthuijs, D., Pershagen, G., Katsouyannai, K., Cadun, E., Dudley, M. L., Svigny, P., Seiffert, I., Swart, W., Bregelmans, W., Blum, G., Selander, J., Haralabidis, A., Dimakopoulou, K., Sourtai, P., Velonakis, M., Vigna-Tagllanti, F. (2008) Hypertension and exposure to noise near airport: *The NYENA Study. Environ. Health Perspect.* 116 (1), 329–333.

[14] Huss, A., Spoerri, A., Egger, M., Roosli, M. (2010) Aircraft noise air pollution, and mortality from myocardial infarction. *Epidemiology* 21(1), 829–836.

[15] Black, D. A., Black, J. A., Issarayangyun, T., Samuels, S. E. (2007) Aircraft noise exposure and residents stress and hypertension: A public health perspective for airport environmental management. J. *Air Transp. Manage.* 13(1), 264–276.

[16] Gualandi, N., & Mantecchini, L. (2008) Aircraft Noise Pollution: A Model of Interaction between Airports and Local Communities. *International Journal of Mechanical Systems Science and Engineering* 2(2), 137-141.

[17] Osueke, C. O., & Ofondu, I. O., (2011) Study of The Adverse Effect of Aircraft Noise in Nigeria & Discovered Solutions. *International Journal of Scientific & Engineering Research* 2(9), 1-4.

[18] Wikipedia online encyclopaedia, *Facts about Ibadan airport.* Retrieved July 1, 2018, from https://en.wikipedia.org/wiki/Ibadan_Airport

[19] Google Internet search engine, *Population data of Nigerian Cities.* Retrieved July 1, 2018, from http://worldpopulationreview.com/countries/nigeria-population/cities

[20] Wikipedia online encyclopaedia, *Facts about Benin airport.* Retrieved July 1, 2018, from https://en.wikipedia.org/wiki/Benin_Airport

[21] Wikipedia online encyclopaedia, *Facts about Owerri airport.* Retrieved July 1, 2018, from https://en.wikipedia.org/wiki/Owerri_Airport

[22] Wikipedia online encyclopaedia, *Facts about Warri airport.* Retrieved July 1, 2018, from https://en.wikipedia.org/wiki/Warri_Airport

[23] Google Internet search engine, *Map of Nigeria showing Airports.* Retrieved July 1, 2018, from www.mapsoftheworld.com

[24] ISO 3891 Standard: Acoustics—Procedure for Describing Aircraft Noise heard on the ground (1978).
[25] ISO 1996-1 Standard: Acoustics—Description, measurement of environmental noise— Basic quantities and assessment procedures (2016).
[26] ISO 1996-2 Standard: Acoustics—Description, measurement and assessment of environmental noise—Determination of environmental noise levels (2016).

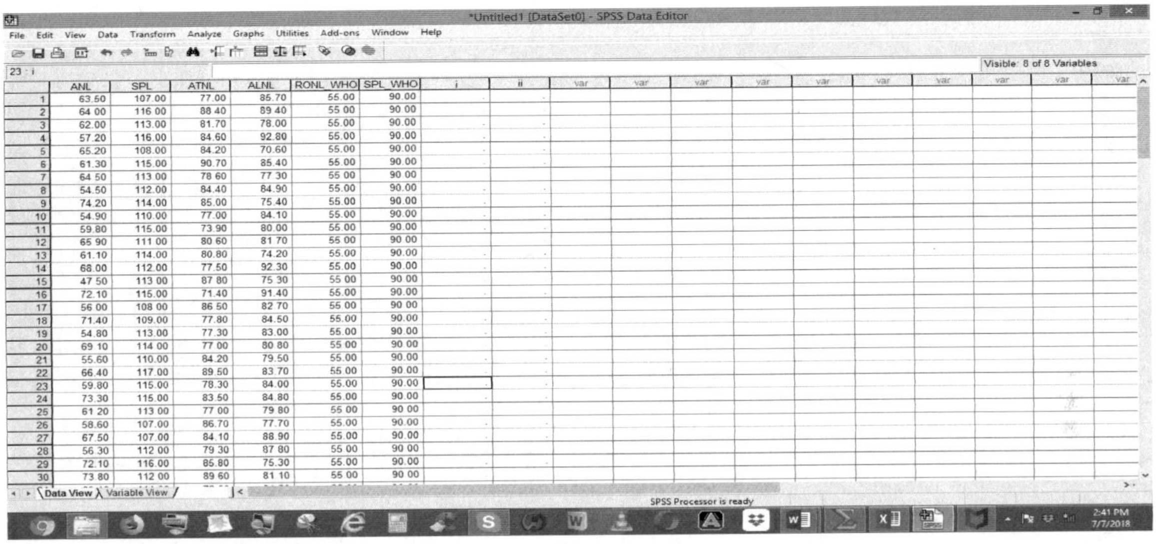

	ANL	SPL	ATNL	ALNL	RONL_WHO	SPL_WHO
1	63.50	107.00	77.00	85.70	55.00	90.00
2	64.00	116.00	88.40	89.40	55.00	90.00
3	62.00	113.00	81.70	78.00	55.00	90.00
4	57.20	116.00	84.60	92.80	55.00	90.00
5	65.20	108.00	84.20	70.60	55.00	90.00
6	61.30	115.00	90.70	85.40	55.00	90.00
7	64.50	113.00	78.60	77.30	55.00	90.00
8	54.50	112.00	84.40	84.90	55.00	90.00
9	74.20	114.00	85.00	75.40	55.00	90.00
10	54.90	110.00	77.00	84.10	55.00	90.00
11	59.80	115.00	73.90	80.00	55.00	90.00
12	65.90	111.00	80.60	81.70	55.00	90.00
13	61.10	114.00	80.80	74.20	55.00	90.00
14	68.00	112.00	77.50	92.30	55.00	90.00
15	47.50	113.00	87.80	75.30	55.00	90.00
16	72.10	115.00	71.40	91.40	55.00	90.00
17	56.00	108.00	86.50	82.70	55.00	90.00
18	71.40	109.00	77.80	84.50	55.00	90.00
19	54.80	113.00	77.30	83.00	55.00	90.00
20	69.10	114.00	77.00	80.80	55.00	90.00
21	55.60	110.00	84.20	79.50	55.00	90.00
22	66.40	117.00	89.50	83.70	55.00	90.00
23	59.80	115.00	78.30	84.00	55.00	90.00
24	73.30	115.00	83.50	84.80	55.00	90.00
25	61.20	113.00	77.00	79.80	55.00	90.00
26	58.60	107.00	86.70	77.70	55.00	90.00
27	67.50	107.00	84.10	88.90	55.00	90.00
28	56.30	112.00	79.30	87.80	55.00	90.00
29	72.10	116.00	85.80	75.30	55.00	90.00
30	73.80	112.00	89.60	81.10	55.00	90.00

Fig. 10 Screen-shot of Acoustic data inputted into the SPSS sheet, prior to performing the Descriptive-Statistics' and non-parametric 'Wilcoxon Signed Ranks test' Analyses.

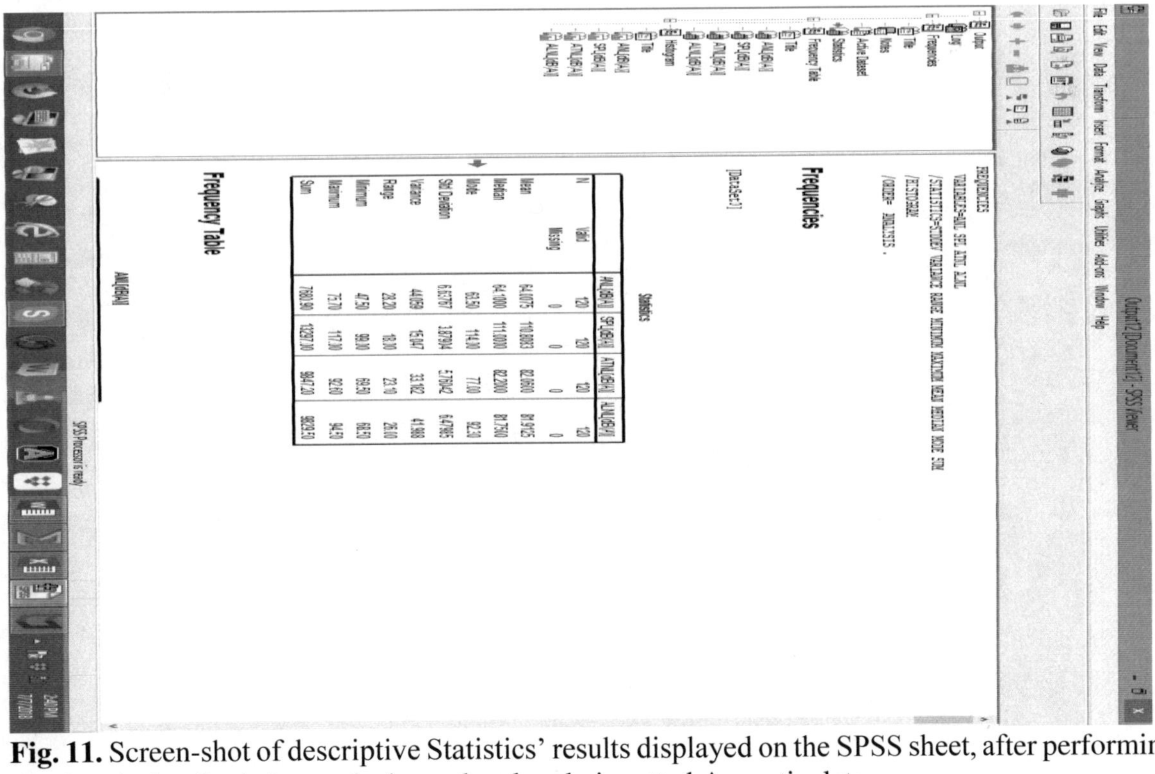

Statistics

		ANL[dB(A)]	SPL[dB(A)]	ATNL[dB(A)]	ALNL[dB(A)]
N	Valid	120	120	120	120
	Missing	0	0	0	0
Mean		64.0075	110.8083	82.0600	81.9125
Median		64.1000	111.0000	82.2000	81.7500
Mode		63.50	114.00	77.00	92.30
Std. Deviation		6.63767	3.87904	5.76042	6.47985
Variance		44.059	15.047	33.182	41.988
Range		28.20	18.00	23.10	26.00
Minimum		47.50	99.00	69.50	68.50
Maximum		75.70	117.00	92.60	94.50
Sum		7680.90	13297.00	9847.20	9829.50

Fig. 11. Screen-shot of descriptive Statistics' results displayed on the SPSS sheet, after performing the descriptive Statistics analysis on the already inputted Acoustic data.

IOP Conf. Series: Materials Science and Engineering **413** (2018) 012080 doi:10.1088/1757-899X/413/1/012080

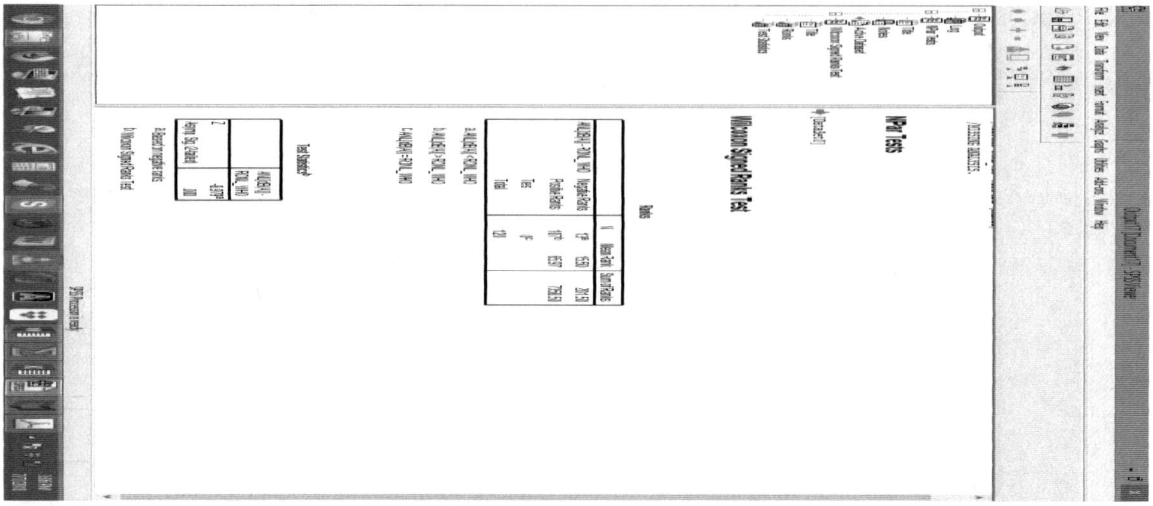

Fig. 12. Screen-shot of 'Wilcoxon Signed Ranks test' results displayed on the SPSS sheet, after performing the non-parametric two-related sample test analysis on the already inputted Acoustic data

Numerical Study of the Effects of Nanoparticles on Fuel Diffusion Combustion

M. A. Waheed

Department of Mechanical Engineering, College of Engineering Federal University of Agriculture, P.M.B 2240, Abeokuta, Ogun State, Nigeria
E-mail: waheedma@funaab.edu.ng

Abstract. The work investigates diffusion combustion of nanofuel made from the mixture of the base fuel and Al_2O_3 nanoparticles in a two-dimensional combustor with stationary horizontal walls. The numerical study was carried out by solving the problem governing equations including the continuity, momentum, energy and species concentration equations formulated for two-dimensional unsteady-state laminar convection using the finite volume method. The fully implicit scheme was used for the discretization in time; while upwind differencing and central differencing were used to discretize the convective and diffusive terms respectively. The results show that the values of the temperature and fuel concentration throughout the domain were higher during combustion of nanofuel than that of the base fuel only. Specifically, for flow streams at Re = 100, the addition of nanoparticles of volume fraction, ϕ = 0.05, enhanced the outlet average temperature of the combustor by 13% over that of ordinary fuel. Also, the wall temperature at steady state for nanofuel combustion is 7.4742 compared with 6.5897 for the base fuel. The high concentration of nanofuel is an indication of incomplete combustion which implies that more heat and higher temperature can be generated by increasing the flow stream of air.
Keywords: Combustion, diffusion, nanofuel, numerical modelling

1. Introduction

There is renewed interest in the development of alternative fuels due to the depletion of fossil fuels, increasing fuels prices, environmental considerations and the need to improve the efficiencies of energy systems [1]. Nanofuels were conceived as a class of nanofluids which serve as alternatives to conventional fuels. They are made by adding metallic nanoparticles additives to base fuel resulting in fuels with higher calorific value, high energy density with high reactive interfacial area which allows increased fuel-oxidiser reaction to produce potential output and thus help in the reduction of soot formation, shortened ignition delay and combustion time [2 – 4]. The high heat of formation of nanofuel is released rapidly thereby saving on specific fuel consumption and significant reduction in green-house gases [5].

Sonawane et al. [6] performed experiments to investigate the thermo-physical properties and the heat transfer rate of Al_2O_3/aviation turbine nanofluids. The thermal conductivity, dynamic viscosity and heat transfer rate were enhanced due to the addition of nanoparticles. Tyagi et al. [7] studied the ignition properties of diesel fuel by varying the size and quantity of Al and Al_2O_3 nanoparticles in a diesel fuel. Increased ignition probability and shortened ignition delay were observed. Sajeevan and Sajith [8] conducted an experiment to investigate the catalytic activity of nanoparticles in an IC engine combustion. It was observed that fuel properties such as viscosity, flash and fire point increase with the addition of nanoparticles. A 30% in NO_x reduction was observed.

The diffusion combustion of Al_2O_3-based nanofuel in a square combustor is studied in this work by solving the problem governing equations numerically. The diffusion combustion of fuel involves convection, diffusion and reaction with respect to the characteristic time scales in which the fuel and oxidizer are not mixed before they enter the combustion chamber and are separated by a thin boundary layer [9 - 10]. Mixing between the two substances occurs at intermolecular level. The entropy of the molecular interactions and mixing increases with the ambient temperature of the chamber, leading to

the rise in internal energy. This continues to rise until the activation energy for chemical reaction is reached. Fuel combustion is exothermic in which the unlocked energy content of fuel is released.
There are a couple of parameters that indicate the progress of combustion. These include the temperature, fuel and oxygen concentration distribution within the flow domain, the inlet velocity of the fuel and air streams, ambient temperature, inlet temperature, heat of formation and boundary conditions of the combustor. High energy is released when the temperature within the flow domain is high. This indicates near or completes combustion and high combustion efficiency. The presence of high fuel concentration however means combustion is not progressing as it should, and combustion efficiency is low. When this happens, soot formation occurs. This work therefore studies the concentration and temperature fields in the diffusion combustion of fuel and nanofuel made from the mixture of the base fuel and Al_2O_3 nanoparticles of 0.05 volume fraction in a two-dimensional combustor with stationary horizontal walls. The fields will show insight into the combustion process of the fuels.

2. Problem Description and Theoretical Model

The schematic representation of the physical configuration of the problem is presented in Figure 1.

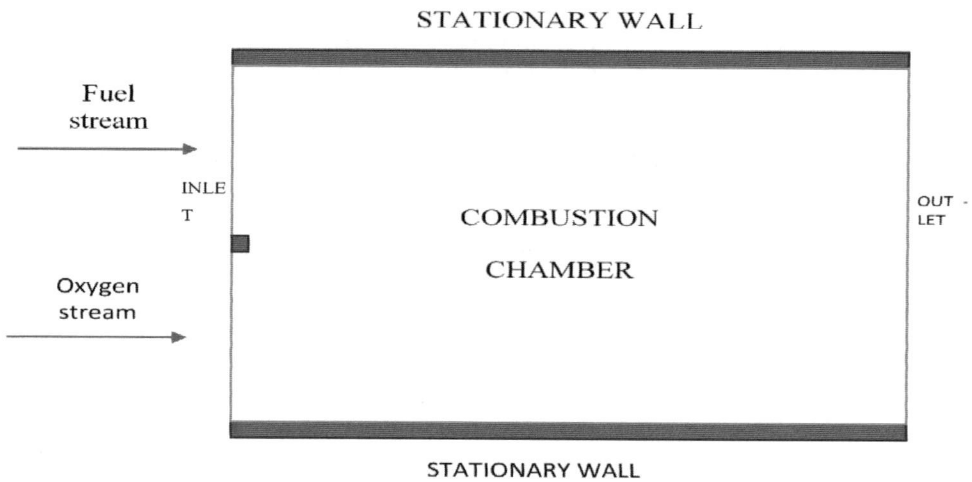

Figure 1: The reaction domain

At the inlet, there are two ports of equal length. One port is for the inflow of fuel while the other is for the oxygen stream. The ambient temperature is high enough for activation energy to be attained. Combustion takes place inside the combustion chamber or flow passage and the products of combustion flow out at the outlet. The walls of the combustion chamber are stationary, so no-slip condition applies at the inner surfaces. The walls are also insulated so there is no heat loss.
The governing equations for species concentration, temperature and fluid flow are given as [11].
Fuel concentration transport equation:

$$\frac{\partial C_f}{\partial t} + u\frac{\partial C_f}{\partial x} + v\frac{\partial C_f}{\partial y} = \frac{1}{Pe_f}\left(\frac{\partial^2 C_f}{\partial x^2} + \frac{\partial^2 C_f}{\partial y^2}\right) + C_f^+ \tag{1}$$

Oxygen concentration transport equation:

$$\frac{\partial C_a}{\partial t} + u\frac{\partial C_a}{\partial x} + v\frac{\partial C_f}{\partial y} = \frac{1}{Pe_a}\left(\frac{\partial^2 C_a}{\partial x^2} + \frac{\partial^2 C_a}{\partial y^2}\right) + C_a^+ \tag{2}$$

Energy transport equation:

$$\frac{\partial T}{\partial t} + u\frac{\partial T}{\partial x} + v\frac{\partial T}{\partial y} = \frac{1}{Pe}\left(\frac{\partial^2 T}{\partial x^2} + \frac{\partial^2 T}{\partial y^2}\right) + T^+ \tag{3}$$

Kinematics equation:

$$\frac{\partial^2 \psi}{\partial x^2} + \frac{\partial^2 \psi}{\partial y^2} = \omega \tag{4}$$

Vorticity transport equation:

$$\frac{\partial \omega}{\partial t} + u\frac{\partial \omega}{\partial x} + v\frac{\partial \omega}{\partial y} = \frac{1}{Re}\left(\frac{\partial^2 \omega}{\partial x^2} + \frac{\partial^2 \omega}{\partial y^2}\right) \tag{5}$$

The variables C_f and C_a represent the concentration of air and oxygen respectively while T represents temperature. The equations for fluid flow are given by the vorticity-stream function formulation for continuity equation in Eq. (4) and momentum equation in Eq. (5). The reacted fuel and oxygen are denoted by the source terms, C_f^+ and C_a^+. The relationship between the two parameters are given in Eq. (6)

$$C_f^+ = C_a^+ = -\frac{L}{U}kC_fC_a \tag{6}$$

where k is the reaction rate constant. The expression for k is given in Eq. (7)

$$k = k_0 e^{-E/RT} \tag{7}$$

where E is the activation energy, R is the gas constant and T is the temperature. The product RT is the kinetic energy of the reaction.

If we make L and U the reference scales for length and velocity respectively, we can write $k_0L/U = K$ and $E/R = T_E$. Based on these relations, Eq. (6) can be expressed as

$$C_f^+ = C_a^+ = -kC_fC_a e^{-T_E/T} \tag{8}$$

The heat source, T^+ in Eq. (3) is the generation of heat due to the combustion reaction. It can be expressed as

$$T^+ = -H_tC_fC_a e^{-T_E/T} \tag{9}$$

where H_t is the rate of heat generation.

Reference Scales

Following Kotake and Hijikata [11] the non-dimensional variables used to normalize the problem governing equations are:

1. Length: The width of flow inlet is L. It is constant all through the flow passage.
2. Velocity: The average velocity of flow at inlet is U. The inlet velocity for fuel and oxygen is denoted by U_f and U_a respectively. The width of the oxygen port can be expressed as $y_0 = Y_0/L$. Hence, the relationship between U, U_f and U_a is
$$U = y_0U_{a0} + (1 - y_0)U_{f0} \tag{10}$$
3. Concentration: The concentration of fuel and oxygen at inlet is denoted by C_{f0} and C_{a0} respectively.
4. Temperature: The temperature of fuel and oxygen at inlet is denoted by T_f and T_a respectively.

Nondimensional Parameters

The nondimensional variables which occur in the problem include the Reynolds number, which is expressed as

$$Re = UL/v \tag{11}$$

The Peclet number of the flow is given as

$$Pe = RePr = {}^{C_p\mu}/_{k} \tag{12}$$

For fuel: $Pe = {}^{C_{pf}\mu_f}/_{k_f}$

For oxygen: $Pe = {}^{C_{pa}\mu_a}/_{k_a}$

The concentration ratio at inlet can be expressed as

$$R_{fa} = C_{f0}/C_{a0} \tag{13}$$

The velocity ratio can be expressed as

IOP Conf. Series: Materials Science and Engineering **413** (2018) 012081 doi:10.1088/1757-899X/413/1/012081

$$U_{fa} = U_{f0}/U_{a0} \tag{14}$$

Boundary Conditions

The domain boundaries can be divided into five sections following Kotake and Hijikata [11]. The boundary conditions for these sections are given in Eqs. (15) – (19).

Inlet (oxygen stream):

$$\left.\begin{array}{c} C_a = C_{a0} \\ C_f = 0 \\ u = u_a(y) = 6U^o{}_{a0}(y^o - y^{o2});\, U^o{}_{a0} \equiv \frac{1}{y_0 + (1-y_0)U_{fa}};\; y^o \equiv \frac{y_0 - y}{y_0} \\ T = T_a \\ \psi = y_0 U^o{}_{a0}[1 - (3y^{o2} - 2y^{o3})] \\ \omega = \frac{6U^o{}_{a0}}{y_0}(1 - 2y^o) \end{array}\right\} \begin{array}{c} x = 0; \\ 0 < y < y_0 \end{array} \tag{15}$$

Inlet (fuel stream):

$$\left.\begin{array}{c} C_f = C_{f0} \\ C_a = 0 \\ u = u_f(y) = 6U^o{}_{f0}(y^o - y^{o2});\, U^o{}_{f0} \equiv \frac{1}{y_0 + (1-y_0)U_{fa}} U_{fa};\; y^o \equiv \frac{y - y_0}{(1-y_0)} \\ T = T_f \\ \psi = (1 - y_0)U^o{}_{f0}(3y^{o2} - 2y^{o3}) + y_0 U^o{}_{a0} \\ \omega = \frac{6U^o{}_{f0}}{(1-y_0)}(1 - 2y^o) \end{array}\right\} \begin{array}{c} x = 0; \\ y_0 < y < 1 \end{array} \tag{16}$$

Outlet:

$$\left.\frac{\partial C_f}{\partial x} = 0;\; \frac{\partial C_a}{\partial x} = 0;\; \frac{\partial u}{\partial x} = 0;\; \frac{\partial T}{\partial x} = 0;\; \frac{\partial \psi}{\partial x} = 0;\; \frac{\partial \omega}{\partial x} = 0\right\} \begin{array}{c} x = 1; \\ 0 < y < 1 \end{array} \tag{17}$$

Top wall:

- Stationary wall

$$\left.\frac{\partial C_f}{\partial y} = 0;\; C_a = 0;\; u = 0; v = 0;\; \frac{\partial T}{\partial y} = 0;\; \psi = 1;\; \omega = \frac{\partial^2 \psi}{\partial y^2}\right\} \begin{array}{c} 0 < x < 1; \\ y = 1 \end{array} \tag{18}$$

Bottom wall:

- Stationary wall

$$\left.C_f = 0;\; \frac{\partial C_a}{\partial y} = 0;\; u = 0;\; v = 0;\; \frac{\partial T}{\partial y} = 0;\; \psi = 0;\; \omega = \frac{\partial^2 \psi}{\partial y^2}\right\} \begin{array}{c} 0 < x < 1; \\ y = 0 \end{array} \tag{19}$$

Nanofuel Properties

For nanofluids in general, the thermophysical properties can be estimated using empirical expressions in the absence of experiments. The most important properties are viscosity, thermal conductivity, specific heat and density. It is assumed that the nanofluid (in this case, fuel with nanoparticles) is homogenous in nature. The nanoparticles are evenly dispersed and free of coagulants. The properties are constant all through the fluid volume and the effect of radiation is negligible.

Viscosity

Brinkman [12] presented an empirical model for estimating the dynamic viscosity of a nanofluid. The model is presented in Eq. (20).

$$\mu_{nf} = \frac{\mu_{bf}}{(1-\emptyset)^{2.5}} \tag{20}$$

Thermal conductivity

The thermal conductivity of a nanofluid mixture can be estimated with the model of Mintsa et al. [13].

$$k_{nf} = k_{bf} + 3\emptyset\left(\frac{k_p - k_{bf}}{2k_{bf} + k_p - \emptyset(k_p - k_{bf})}\right)k_{bf} \quad (21)$$

Specific heat

A commonly used nanofluid model for estimating specific heat capacity is that of Zhou and Ni (2008).

$$C_{pnf} = \emptyset C_p + (1 - \emptyset)C_{bf} \quad (22)$$

Density

The density of nanofluid was approximate using the empirical expression of Pak and Cho (1998) in Eq. (23).

$$\rho_{nf} = \emptyset_p \rho_p + (1 - \emptyset)\rho_{bf} \quad (23)$$

In Eqs. (20) to (23), the subscripts *nf* and *bf* represent nanofluid and basefluid respectively while $\emptyset$ represents the volume fraction of nanoparticles in the nanofluid.

Effect on Activation Energy

Activation energy is the least amount of energy required to start combustion. The activation energy can be determined from reaction rate at constants different temperatures based on the Arrhenius equation [14].

$$\ln\left(\frac{H_{t_{nf}}}{H_{t_f}}\right) = \frac{E_a}{R}\left(\frac{1}{T_f} - \frac{1}{T_{nf}}\right) \quad (24)$$

Where the subscripts *f* and *nf* denote fuel and nanofuel respectively. The variables H_t, E_a, R and T respectively represent reaction rate, activation energy, gas constant and reference temperature.

3. Numerical Solution

The finite volume method (FVM), which is a suitable and capable numerical technique for solving various types of simulation problems involving conservation laws [15], was used to solve the species concentration, temperature and flow equations.

The domain is divided into control volumes over each of which the conservation equation is integrated. A linear algebraic set of equations is obtained and solved to get the parameter distribution over the domain. For any generic conservation equation with diffusive and convective terms, the equation can be represented as

$$\frac{\partial \Gamma}{\partial t} + u\frac{\partial \Gamma}{\partial x} + v\frac{\partial \Gamma}{\partial y} = \left(\frac{\partial^2 \Gamma}{\partial x^2} + \frac{\partial^2 \Gamma}{\partial y^2}\right) + f(x, y) \quad (25)$$

The convective terms are discretized using Upwinding Differencing Scheme (UDS) while the diffusive terms are discretized using Central Differencing Scheme (CDS). For the discretization in time, the fully implicit scheme is used. The fully implicit scheme is unconditionally stable, so we can employ any time step size [16].

Eq. (25) becomes:

$$(\alpha\Gamma)_{i,j} - (\alpha\Gamma)^o_{i,j} = F_w\Gamma_{i-1,j} - F_e\Gamma_{i,j} + F_s\Gamma_{i,j-1} - F_n\Gamma_{i,j} + D_e(\Gamma_{i+1,j} - \Gamma_{i,j}) - D_w(\Gamma_{i,j} - \Gamma_{i-1,j}) + D_n(\Gamma_{i,j+1} - \Gamma_{i,j}) - D_s(\Gamma_{i,j} - \Gamma_{i,j-1}) + S_u\Delta x\Delta y \quad (26)$$

where $\alpha = {}^{\Delta x \Delta y}/_{\Delta t}$; $F_w = u_{i-1,j}$; $F_e = u_{i+1,j}$; $F_n = u_{i,j+1}$; $F_s = u_{i,j-1}$; $S_u = f(x,y)$

also, $D_e = D_w = {}^1/_{\Delta x}$ and $D_n = D_s = {}^1/_{\Delta y}$

Δx, Δy and Δt are the spatial step in x-direction, spatial step in y-direction and time step respectively

By substituting for the appropriate boundary conditions, the link equation in Eq. (27) is arrived at.

$$A_{i,j} = A_{i-1,j} + A_{i+1,j} + A_{i,j+1} + A_{i,j-1} + A^o_{i,j} + (F_e - F_w) + (F_n - F_s) - S_{i,j} \quad (27)$$

where $A_{i-1,j} = F_w + D_w$; $A_{i+1,j} = D_e$; $A_{i,j+1} = D_n$; $A_{i,j-1} = F_s + D_s$; $A^o_{i,j} = \alpha$;

$S_{i,j} = -S_u/\Gamma_{i,j}$

For each control volume, the algebraic equation in Eq. (27) is solved

$$\Gamma_{i,j} = \frac{A_{i-1,j}\Gamma_{i-1,j} + A_{i+1,j}\Gamma_{i+1,j} + A_{i,j+1}\Gamma_{i,j+1} + A_{i,j-1}\Gamma_{i,j-1} + A^o_{i,j}\Gamma^o_{i,j} + S_u}{A_{i,j}} \quad (28)$$

4. Results

The results of the numerical experiments are presented in the forms of the temperature distribution, fuel and oxygen concentrations within the flow passage and the domain boundaries, together with the combustion products concentration profiles at the outlet.

IOP Conf. Series: Materials Science and Engineering **413** (2018) 012081 doi:10.1088/1757-899X/413/1/012081

Starting with the thermo-physical properties of the nanofuel, the volume fraction of nanoparticles added to the fuel, $\phi = 0.05$. The sample nanoparticles chosen for this study is Al_2O_3. A generic fuel is chosen with all properties with the value of unity. The properties of both these fuel and nanoparticles are presented in Table 1.

Table 1: Properties of nanoparticle material and fuel

Properties	Fuel	Al_2O_3	Nanofuel	Oxygen
Density (kg/m^3)	1	3950	198.45	1.33
Thermal Conductivity (W/K)	1	18	1.13	0.0337
Specific heat (W/kg K)	1	880	44.95	0.91
Viscosity (kg/m^2)	1		1.14	0.000029

The estimated properties of the nanofuel are also presented. The density of the fuel increased by about 198%, and the thermal conductivity was enhanced by 13%. There are also increase in values of specific heat capacity and viscosity. The large percentage in the values of the properties can be explained by large differences in the property values of both fuel and nanoparticles.

The inlet temperature profile at inlet is shown in Figure 2. The lower section shows the profile for oxygen stream while the upper section profile is for the fuel stream. The concentration profiles for fuel and oxygen are also shown in Figure 3a and Figure 3b.

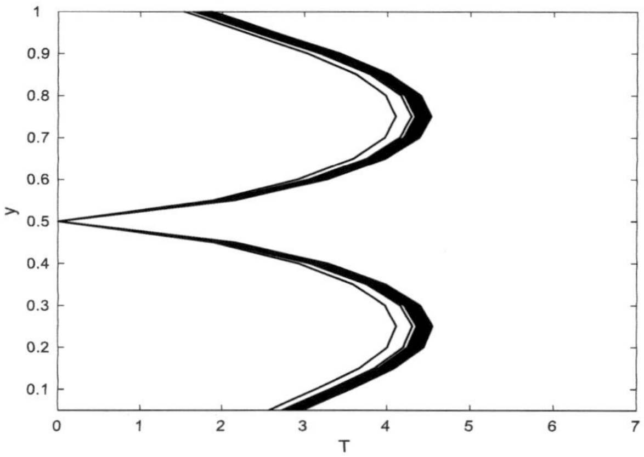

Figure 2: Inlet temperature profile up to 5secs with Δt = 0.2

The parameter values are:

Fuel: $T_E = 4;\ H_t = 10;\ U_{f0} = 1;\ U_{a0} = 1;\ T_f = 1;\ T_a = 1;\ T_0 = 2; nx = ny = 20$

Nanofuel: $T_E = 3.53;\ H_t = 10;\ U_{f0} = 1;\ U_{a0} = 1;\ T_f = 1;\ T_a = 1;\ T_0 = 2; nx = ny = 20$

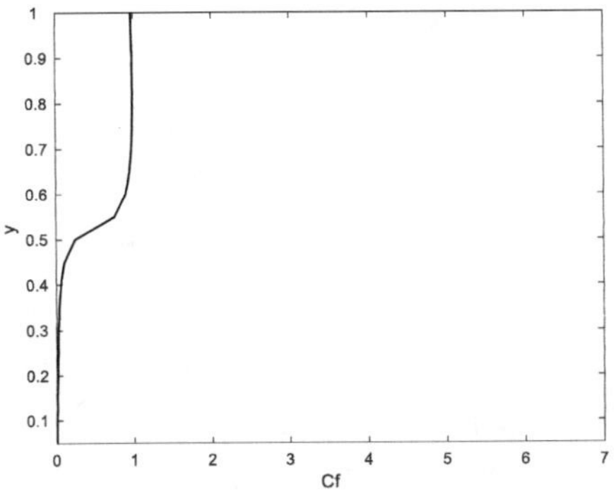

Figure 3a: Fuel concentration profile at inlet

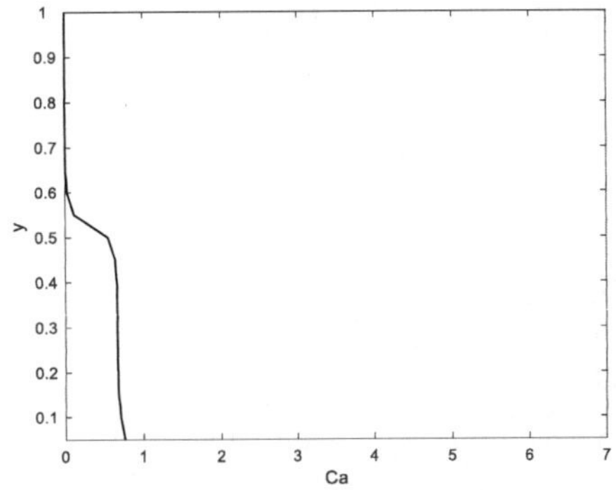

Figure 3b: Oxygen concentration profile at inlet

The profiles show that the variables T, C_f and C_a are functions of y along the inlet width. Fuel and oxygen flow along positive x-direction and diffuse into one another before combustion takes place. Combustion started when the temperature of the mixture is high enough that the heat generated is greater or equal to the activation energy. At the outlet of the domain, which is far upstream, combustion reaction is complete, and the amount of oxygen present is almost non-existent. With the addition of nanoparticles, the activation energy is reduced by approximately 11%. The nanoparticles serve as catalysts by increasing the reaction rate, increasing thermal capacity and thermal conductivity. Figure 4a and 4b shows the concentration profiles of the products at the outlet of the domain for a stationary wall. It can be clearly seen that the steady state concentration for nanofuel products is higher than that of fuel products after combustion. This can be attributed to higher heat generation during combustion which results in a more complete combustion and higher amount of product generation from nanofuel.

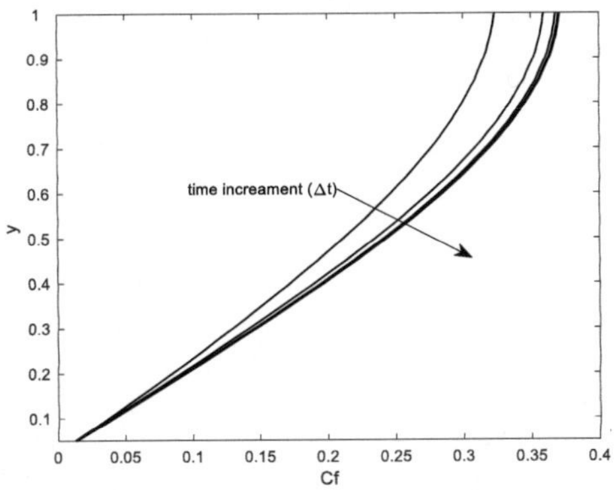

Figure 4a: Fuel concentration profiles at outlet, $\Delta t = 0.2$

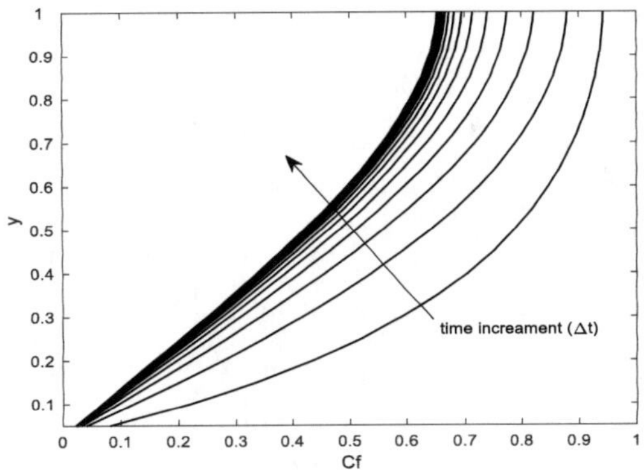

Figure 4b: Nanofuel concentration profiles at outlet, $\Delta t = 0.2$

The distribution of fuel concentration is shown in Figure 5a and while that of nanofuel is shown in Figure 5b at t = 5 secs. The concentration of nanofuel is higher in any part of the domain than the corresponding concentration of fuel. Nanofuel achieved better distribution and circulation within the domain. While we can conclude on better fuel efficiency, the chance of incomplete combustion is however higher in the presence of higher nanofuel distribution because oxygen is rapidly consumed within the domain.

IOP Conf. Series: Materials Science and Engineering **413** (2018) 012081 doi:10.1088/1757-899X/413/1/012081

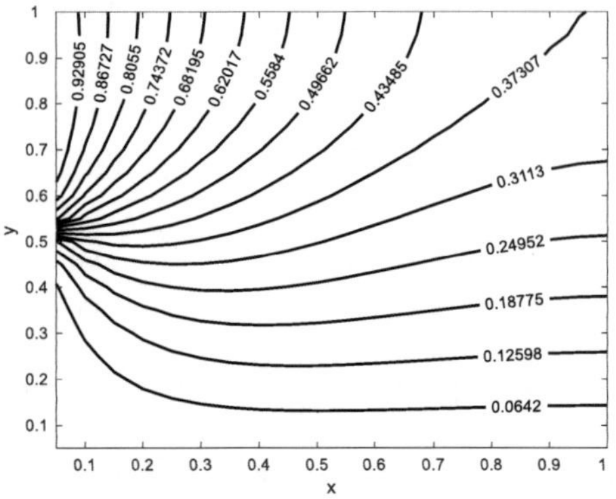

Figure 5a: Fuel concentration within the domain at t = 5 secs

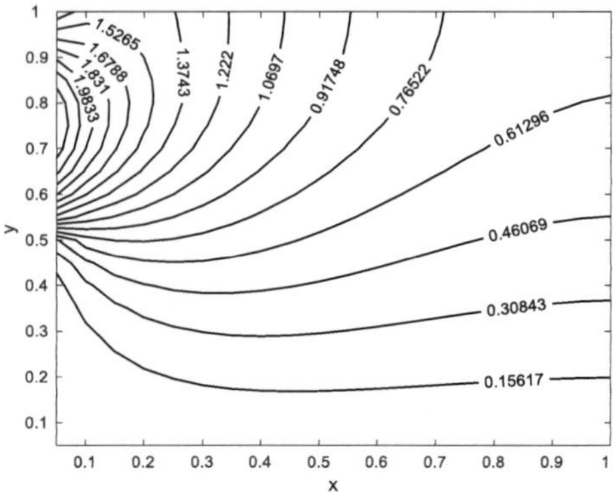

Figure 5b: Nanofuel concentration within the domain at t = 5 secs

Figure 6 shows the oxygen distribution with the domain for stationary wall combustion. The difference in the distribution of oxygen for fuel and nanofuel combustion cases is negligible. Oxygen is immediately consumed as it enters the domain at about 20% of the length of the passage. The same observation is made for both fuel and nanofuel. Reducing the fuel-oxygen concentration ratio by increasing the oxygen concentration by a large value had little effect on the distribution.

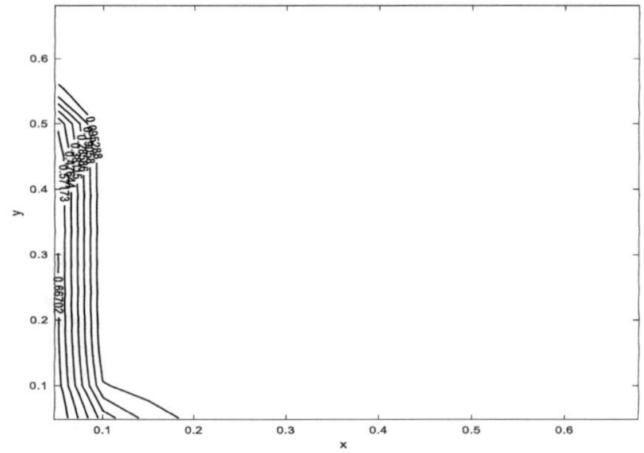

Figure 6: Oxygen distribution within the domain

The rate of heat generation is because of lower activation energy and faster release of the locked energy content of the fuel. A higher temperature distribution highlights better fuel efficiency which can lead to improved system performance. At the start of combustion, temperature increases rapidly with time and approached an asymptotic steady state value. Figure 7 presents the variation in the average temperature values with time for fuel and nanofuel combustion at the outlet of the domain. At t = 5 secs, the outlet temperature is at a steady state value. The value for nanofuel is about 13% higher than that of fuel combustion. This is a clear indication of the higher release of energy in nanofuel combustion. The metallic particles contained within the bulk fuel has very high specific heat capacity and their energy density is high as well. It is expected that the temperature along the walls follow the same trend.

IOP Conf. Series: Materials Science and Engineering **413** (2018) 012081 doi:10.1088/1757-899X/413/1/012081

In Figure 8, the upper wall temperature is presented for fuel and nanofuel combustion at different time levels with $\Delta t = 2$. With increasing time step, the difference between the inlet end (at x = 0) and the outlet end (at x = 1) keeps increasing until an asymptotic value is attained. The maximum temperature for nanofuel combustion is 7.4742, while that of fuel combustion is 6.5897. This is another affirmation of greater heat release in nanofuel.

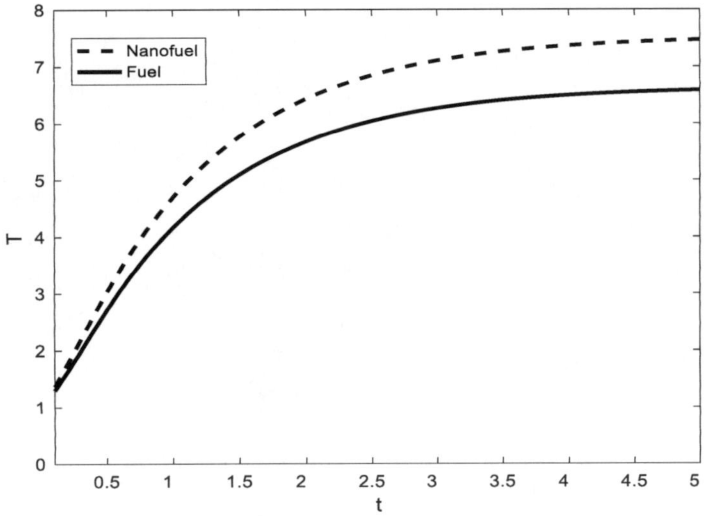

Figure 7: The temperature profile of the domain

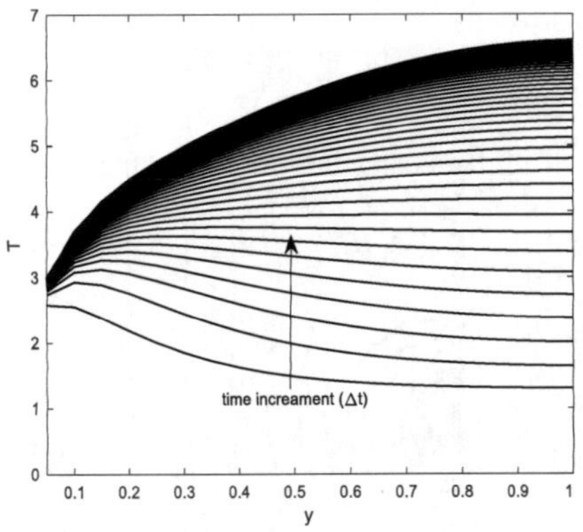

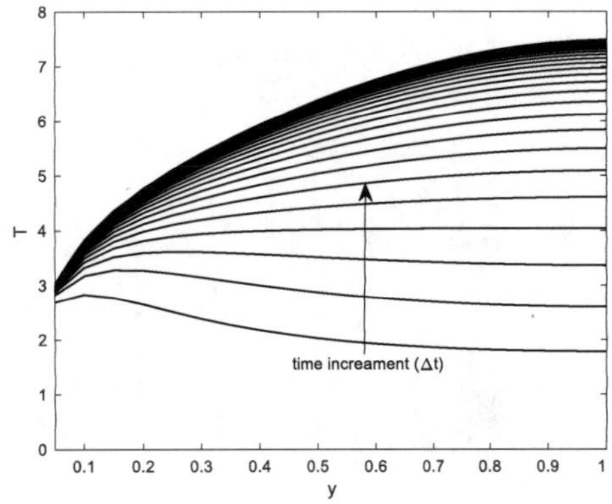

Figure 8: Temperature profile along the upper wall for fuel combustion (left) and nanofuel combustion (right)

Figure 9 shows the temperature distribution within the combustion chamber for different time frames for stationary wall boundary condition. At t = 1 sec, the distribution within the chamber looks chaotic. Combustion is just starting because the temperature at inlet is higher than the outlet and inner temperatures.

Fuel **Nanofuel**

t = 1 sec **t = 1 sec**

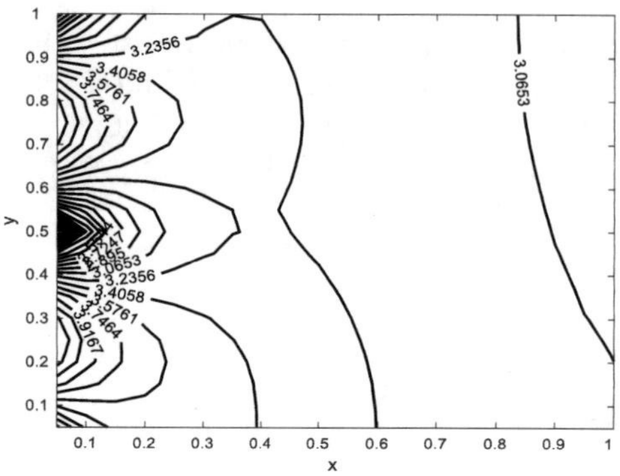

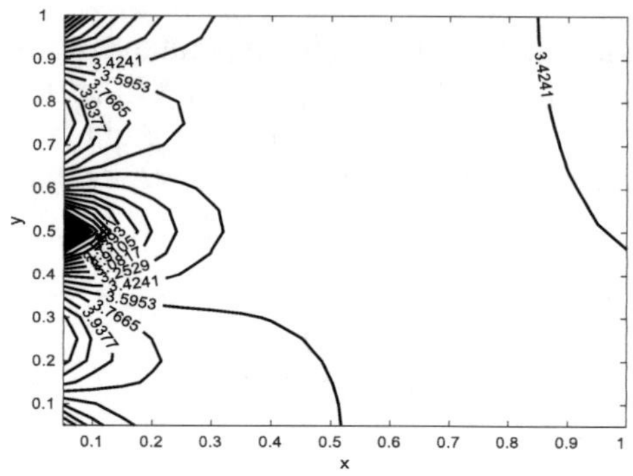

t = 3 secs **t = 3 secs**

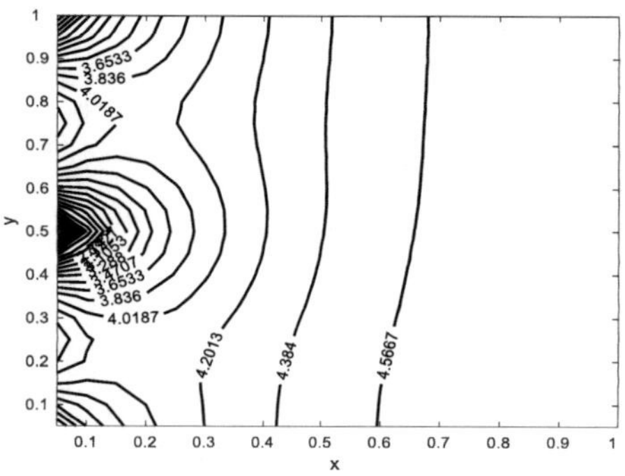

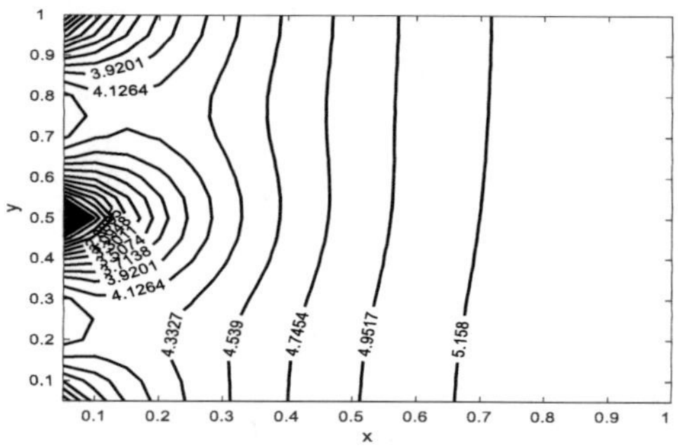

t = 5 secs **t = 5 secs**

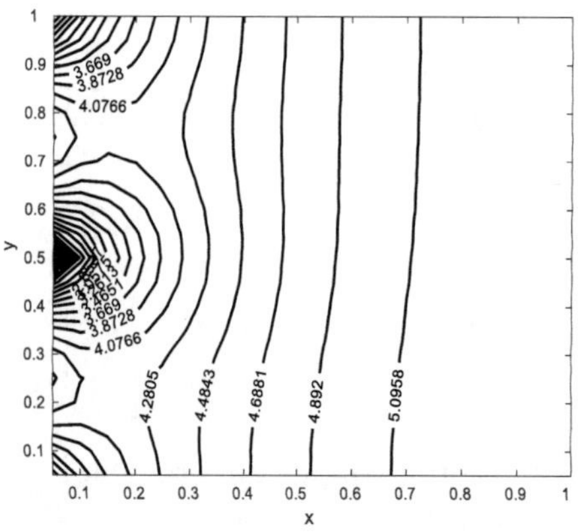

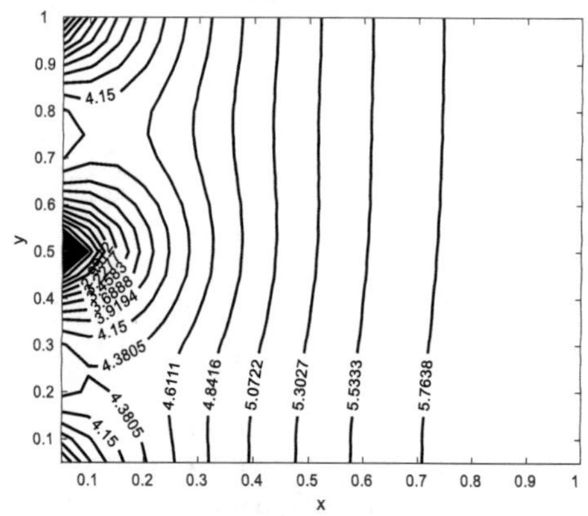

Figure 9: Temperature distribution for fuel combustion (left) and nanofuel combustion (right) within the combustion chamber at different time

The temperatures at any cross section are also largely uneven. As combustion reaction commences, the temperature distribution pattern changes at $t = 3$ secs onwards. The temperature rises as we move along the length of the domain from left to right. The increase in temperature gets more obvious with increase in time step. Generally, a higher temperature is observed in Nanofuel combustion.

5. Conclusion

The simulation of fuel and nanofuel combustion in a diffusion flame setup was performed, examined and compared. Adding nanoparticles to fuel increases the thermal conductivity, heat capacity and viscosity of the nanofuel mixture and causes a reduction in the activation energy. The nanoparticles

IOP Conf. Series: Materials Science and Engineering **413** (2018) 012081 doi:10.1088/1757-899X/413/1/012081

volume fraction used is 0.05. Temperature is observed to increase rapidly as combustion commences and slows gradually until a steady state temperature is achieved. Fuel concentration distribution within the domain is well dispersed. The distribution of oxygen however shows near total consumption as the presence of oxygen is only observed in like 20% of the domain length. Comparing nanofuel combustion to fuel combustion, the steady state concentration of products in the former case is higher. Temperature is also higher with about 13%. Fuel concentration distribution is higher, but oxygen concentration distribution differences are negligible. The results reflect higher heat generation in nanofuel combustion. The results show that the use of nanofuel can lead to better fuel efficiency, faster and complete combustion, and less formation of pollutants.

References

[1] Gan, Y. and Qiao, L. Combustion characteristics of fuel droplets with addition of nano and micron-sized aluminium particles. Combustion and Flame, vol. 158, pp. 354–368, 2011.

[2] Mehta, R. N., Chakraborty, M. and Parikh, P. A. Nanofuels: Combustion, engine performance and emissions, Fuel, vol. 120, pp. 91–97, 2014.

[3] Yetter, R.A., Risha, G.A. and Son, S.F. Metal particle combustion and nanotechnology. Proceedings of the Combust Inst, vol. 32, pp. 1819-38, 2009.

[4] De Luca, L. T., Galfetti, L., Severini, F., Meda, L., Marra, G., Vorozhtsov, A. B., Sedoi, V. S. and Babuk, V. A. Combustion of composite solid propellants with nanosized aluminium. Combustion, Explosion & Shock Waves, vol. 41, pp. 680–692, 2005.

[5] Basu, S. and Miglani, A. Combustion and heat transfer characteristics of nanofluid fuel droplets: A short review, International Journal of Heat and Mass Transfer, vol. 96, pp. 482–503, 2016.

[6] Sonawane, S, Patankar, K, Fogla, A, Puranik, B, Bhandarkar, U. and Kumar, S. S., An Experimental investigation of thermo-physical properties and heat transfer performance of Al2O3-Aviation turbine fuel nanofluids. Applied Thermal Engineering, vol. 31. No. 201, pp. 2841–849, 2011.

[7] Tyagi, H., Phelan, P. E. and Prasher, R. Increased Hot -plate Ignition Probability for Nanoparticle –Laden Diesel Fuel. Nano Letters, vol. 08, pp. 1410-1416, 2009.

[8] Sajeevan, A. C., Sajith,V. Diesel Engine Emission Reduction Using Catalytic Nanoparitcles: An Experimental Investigation, Journal of Engineering, Article ID 589382, 2013.

[9] Peters, N. Laminar diffusion flamelet models in non-premixed turbulent combustion. Prog. Energy Combust. Sci., vol. 10, pp. 319-339, 1984.

[10] Poinsot, T. and Veynante, D. Theoretical and Numerical Combustion, 2_{nd} Ed., R.T. Edwards. Inc., Philadelphia, USA, 2005.

[11] Kotake, S. And Hijikata, K. Numerical simulations of heat transfer and fluid flow on a personal computer, Elsevier Science Publishers B. V., Amsterdam, The Netherlands, 1993.

[12] Brinkman, H. C. The viscosity of concentrated suspensions and solutions, Journal of Chemical Physics, vol. 20, No. 4, pp. 571-581, 1952.

[13] Mintsa, H. A., Roy, G., Nguyen, C. T. and Doucet, D. New temperature dependent thermal conductivity data for water-based nanofluids. International Journal of Thermal Sciences, 48(2), pp. 363-371, 2009.

[14] Petrou, A. L., Roulia, M and Tampouris, K. The use of the Arrhenius equation in the study of deterioration and of cooking of foods and some scientific and pedagogic aspects, Chemistry Education: Research and Practice in Europe, vol. 3, No. 1, pp. 87-97, 2002.

[15] Patankar, S.V. Numerical heat transfer and fluid flow. Series in Computational Methods in Mechanics and Thermal Sciences, Hemisphere Publishing Corporation, USA, 1980.

[16] Versteeg, H. K. and Malalasekera, W. An introduction to computational fluid dynamics: The finite volume method, Pearson Education Ltd, Harlow, England, 2007.